# Human – Computer Interaction:

## Theory and Practice (Part I)

### Volume 1

## Human Factors and Ergonomics
### *Gavriel Salvendy, Series Editor*

**Bullinger, H.-J., and Ziegler, J.** (Eds.) : Human–Computer Interaction: Ergonomics and User Interfaces, *Volume 1 of the Proceedings of the 8th International Conference on Human–Computer Interaction*

**Bullinger, H.-J., and Ziegler, J.** (Eds.) : Human–Computer Interaction: Communication, Cooperation, and Application Design, *Volume 2 of the Proceedings of the 8th International Conference on Human–Computer Interaction*

**Hollnagel, E.** (Ed.) : *Handbook of Cognitive Task Design*

**Meister, D.** (Ed.) : *Conceptual Foundations of Human Factors Measurement*

**Meister, D., and Enderwick, T.** (Eds.) : *Human Factors in System Design, Development, and Testing*

**Smith, M. J., Salvendy, F., Harris, D., and Koubeck, R. J.** (Eds.) : *Usability Evaluation and Interface Design: Cognitive engineering, Intelligent Agents and Virtual Reality*

**Smith, M. J., and Salvendy, G.** (Eds.) : *Systems, Social and Internationalization Design of Human–Computer Interaction*

**Stephanidis, C.** (Ed.) : *Universal Access in HCI: Towards an Information Society for All*

**Stephanidis, C.** (Ed.) : *User Interfaces for All: Concepts, Methods, and Tools*

**Ye, N.** (Ed.) : *The Handbook of Data Mining*


**Jacko, J., and Stephanidis, C.** (Eds.) : *Human-Computer Interaction: Theory and Practice (Part I)*

**Stephanidis, C., and Jacko, J.** (Eds.) : *Human-Computer Interaction: Theory and Practice (Part II)*

**Harris, D., Duffy, V., Smith, M., and Stephanidis, C.** (Eds.) : *Human-Centred Computing: Cognitive, Social and Ergonomic Aspects*

**Stephanidis, C.** (Ed.) : *Universal Access in HCI: Inclusive Design in the Information Society*

# Human – Computer Interaction:

## Theory and Practice (Part I)

*Volume 1 of the Proceedings of HCI International 2003*
*10th International Conference on Human - Computer Interaction*
*Symposium on Human Interface (Japan) 2003*
*5th International Conference on Engineering Psychology*
*and Cognitive Ergonomics*
*2nd International Conference on Universal Access*
*in Human - Computer Interaction*
*22 – 27 June 2003, Crete, Greece*

Edited by

**Julie Jacko**
Georgia Institute of Technology

**Constantine Stephanidis**
ICS-FORTH and University of Crete

2003
LAWRENCE ERLBAUM ASSOCIATES, PUBLISHERS
Mahwah, New Jersey          London

Lawrence Erlbaum Associates, Inc., Publishers
10 Industrial Avenue
Mahwah, New Jersey  07430

Human-Computer  Interaction : Theory and Practice (Part I) / edited by Julie Jacko and Constantine Stephanidis.

p.  cm.

Includes bibliographical references and index.
ISBN 0-8058-4930-0 (cloth : alk. paper) (Volume 1)

ISBN 0-8058-4931-9 (Volume 2)
ISBN 0-8058-4932-7 (Volume 3)
ISBN 0-8058-4933-5 (Volume 4)
ISBN 0-8058-4934-3 (Set)

2003

Books published by Lawrence Erlbaum Associates are printed on acid-free paper, and their bindings are chosen for strength and durability.

10  9  8  7  6  5  4  3  2  1

# Preface

The 10th International Conference on Human-Computer Interaction, HCI International 2003, is held in Crete, Greece, 22-27 June 2003, jointly with the Symposium on Human Interface (Japan) 2003, the 5th International Conference on Engineering Psychology and Cognitive Ergonomics, and the 2nd International Conference on Universal Access in Human-Computer Interaction. A total of 2986 individuals from industry, academia, research institutes, and governmental agencies from 59 countries submitted their work for presentation, and only those submittals that were judged to be of high scientific quality were included in the program. These papers address the latest research and development efforts and highlight the human aspects of design and use of computing systems. The papers accepted for presentation thoroughly cover the entire field of human-computer interaction, including the cognitive, social, ergonomic, and health aspects of work with computers. These papers also address major advances in knowledge and effective use of computers in a variety of diversified application areas, including offices, financial institutions, manufacturing, electronic publishing, construction, health care, disabled and elderly people, etc.

We are most grateful to the following cooperating organizations:

- Chinese Academy of Sciences
- Japan Management Association
- Japan Ergonomics Society
- Human Interface Society (Japan)
- Swedish Interdisciplinary Interest Group for Human Computer Interaction - STIMDI

- Asociación Interacción Persona Ordenador - AIPO (Spain)
- Gesellschaft für Informatik e.V - GI (Germany)
- European Research Consortium for Information and Mathematics - ERCIM

The 258 papers contributing to this volume (Vol. 1) cover the following areas:

- Design Approaches, Methods and Tools
- HCI Methodologies and Perspectives
- Usability
- Design and Evaluation Studies

- Web Design and Usability
- Learning and Edutainment
- Virtual, Mixed and Augmented Environments

The selected papers on other HCI topics are presented in the accompanying three volumes: Volume 2 by C. Stephanidis and J. Jacko, Volume 3 by D. Harris, V. Duffy, M. J. Smith and C. Stephanidis, and Volume 4 by C. Stephanidis.

We wish to thank the Board members, listed below, who so diligently contributed to the overall success of the conference and to the selection of papers constituting the content of the four volumes.

Eduardo Salas, *USA*
Dirk Schaefer, *France*
Neville A. Stanton, *UK*

**Universal Access in Human-Computer Interaction**
Julio Abascal, *Spain*
Demosthenes Akoumianakis, *Greece*
Elizabeth Andre, *Germany*
David Benyon, *UK*
Noelle Carbonell, *France*
Pier Luigi Emiliani, *Italy*
Michael C. Fairhurst, *UK*
Gerhard Fischer, *USA*
Ephraim Glinert, *USA*
Ion Gunderson, *USA*
Ilias Iakovidis, *EU*

Arthur I. Karshmer, *USA*
Alfred Kobsa, *USA*
Mark Maybury, *USA*
Michael Pieper, *Germany*
Angel R. Puerta, *USA*
Anthony Savidis, *Greece*
Christian Stary, *Austria*
Hirotada Ueda, *Japan*
Jean Vanderdonckt, *Belgium*
Gregg C. Vanderheiden, *USA*
Annika Waern, *Sweden*
Gerhard Weber, *Germany*
Harald Weber, *Germany*
Michael D. Wilson, *UK*
Toshiki Yamaoka, *Japan*

We also wish to thank the following external reviewers:

Chrisoula Alexandraki, *Greece*
Margherita Antona, *Greece*
Ioannis Basdekis, *Greece*
Boris De Ruyter, *Netherlands*
Babak Farschian, *Norway*
Panagiotis Karampelas, *Greece*
Leta Karefilaki, *Greece*

Elizabeth Longmate, *UK*
Fabrizia Mantovani, *Italy*
Panos Markopoulos, *Netherlands*
Yannis Pachoulakis, *Greece*
Zacharias Protogeros, *Greece*
Vassilios Zarikas, *Greece*

This conference could not have been held without the diligent work and outstanding efforts of Stella Vourou, the Registration Chair, Maria Pitsoulaki, the Program Administrator, Maria Papadopoulou, the Conference Administrator, and George Papatzanis, the Student Volunteer Chair. Also, special thanks to Manolis Verigakis, Zacharoula Petoussi, Antonis Natsis, Erasmia Piperaki, Peggy Karaviti and Sifis Klironomos for their help towards the organization of the Conference. Finally recognition and acknowledgement is due to all members of the HCI Laboratory of ICS-FORTH.

Constantine Stephanidis
ICS-FORTH and University of Crete, GREECE

Julie A. Jacko
Georgia Institute of Technology, USA

Don Harris
Cranfield University, UK

Vincent G. Duffy
Mississippi State University, USA

Michael J. Smith
University of Wisconsin-Madison, USA

June 2003

## HCI International 2005

The 11th International Conference on Human-Computer Interaction, HCI International 2005, will take place jointly with:

Symposium on Human Interface (Japan) 2005
6th International Conference on Engineering Psychology and Cognitive Ergonomics
3rd International Conference on Universal Access in Human-Computer Interaction
1st International Conference on Virtual Reality
1st International Conference on Usability and Internationalization

The conference will be held in Las Vegas, Nevada, 22-27 July 2005. The conference will cover a broad spectrum of HCI-related themes, including theoretical issues, methods, tools and processes for HCI design, new interface techniques and applications. The conference will offer a pre-conference program with tutorials and workshops, parallel paper sessions, panels, posters and exhibitions. For more information please visit the URL address: http://hcii2005.engr.wisc.edu

General Chair:

Gavriel Salvendy
Purdue University
School of Industrial Engineering
West Lafayette, IN 47907-2023 USA
Telephone: +1 (765)494-5426  Fax: +1 (765) 494-0874
Email: salvendy@ecn.purdue.edu
http://gilbreth.ecn.purdue.edu/~salvendy
        and
Department of Industrial Engineering
Tsinghua University, P.R. China

The proceedings will be published by Lawrence Erlbaum and Associates.

# Table of Contents

## Section 2. HCI Methodologies and Perspectives

**Section 4. Design and Evaluation Studies**

## Section 5. Web Design and Usability

## Section 7. Virtual, Mixed and Augmented Environments

# Section 1

Design Aproaches, Methods and Tools

# Use Case Maps: A Visual Notation for Scenario-Based User Requirements

*A. Alsumait, A. Seffah, and T. Radhakrishnan*
Human-Centered Software Engineering Group
Computer Science Department, Concordia University,
Maisonneuve Blvd. W. Quebec, Canada
{alsumait, seffah, khrishnan}@cs.concordia.ca

## Abstract

Among their popular and potential applications, scenarios and use cases provide an integrative picture of the user tasks with its context (users and work). Our goal is to build a complete and consistent user interface and usability requirement model that is simple, intuitive, unambiguous and verifiable by extending the prospective standard the Use Case Maps (UCMs). In particular, we provide step-by-step guidance for developing UCMs from scenarios.

## 1    Introduction and Background

During the last two decades, several HCI research efforts tried to develop appropriate representations to support user interface and usability requirements. Among the various representations, scenarios and use cases were proposed as a working design representation of user experiences with, and reactions to, system functionality in the context of pursuing a task. Other efforts have tried to combine use cases and task analysis as vehicle for bridging the current gaps between user interface/usability and functional requirements. For example, Forbrig (1999) introduced a framework for combining task models, user models, and object models. These multiple dimensions require tools enabling the manipulation of all these models along the different phases of software development (Sutcliffe, 1998; Weidenhaupt, 1998). Related to this work, Constantine and Lockwood (1999) proposed a use case model description that does not contain any implicit user interface decisions; essential use cases. Later on, user interface designers can use these essential use cases as input to create the user interface without being bound by any implicit decisions. Our research aims to complement the scenario-based requirement approach using an extended use case maps (UCMs) notation. We describe the process of building two UCM models that provide an easy way to understand and validate the user interface and usability requirements.

## 2    The User Interface and Usability Requirement (UIUR) Model

The proposed UIUR model makes use of the prospective standard Use Case Maps (UCMs) to bridge the gap the between users' statements of requirements in an informal natural languages and the target of formal specifications. UCMs are intended to be a useful tool for functional requirement of scenario-based software development. However, their support for designing user interfaces is still acknowledged to be insufficient (Alsumait et all., 2002). Our strong belief that UCMs are powerful for user interface and usability requirements is based on the simplicity of UCMs

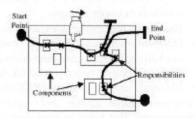

**Figure 1**: example of a Use Case Map (Buhr, 1998)

notation (Buhr, 1998). The basic UCM notation is very simple (See Figure 1), and consists of *start-points* (filled circles, representing pre-conditions), *responsibilities* (crosses, representing tasks to be performed), *end- points* (bars, representing post-conditions). The responsibilities can be bound to *components*, which are the entities or objects composing the system. The wiggly lines are *paths* that connect start points, responsibilities, and end points. UCMs notation is easily understandable by both the user and the user interface designer. In fact, this helps the designers to handle different users' understanding and expectation of the interface, and bridge the gap by refining the requirements earlier. Moreover, several studies are focusing on how to integrate UCMs to UML and how to formalize UCMs in XML (Buhr, 1998). Finally, we will show that by incorporating the extended UCMs to create our UIUR, we can express user interface and usability requirements much better.

## 2.1   UIUR Process and Notation

The proposed UIUR model divides the process of transforming requirements stated described in an informal language to formal specifications through a number of iterative phases. Each phase aims at increasing the formality of representation by a small step. The use of UIUR model includes four typical phases: (1) scenario analysis, (2) conceptual use case maps, (3) physical use case maps, and (4) formal specification.

## 2.2   Scenario Analysis

At this phase, only very limited information is available. A process model of scenario analysis consists of the following iterative and interactive activities: identification of scenarios, elicitation of information and description of scenarios as well as building use cases.

## 2.3   Build Use Case Maps

By extending the UCM notation, we find that it can be used in the following four dimensions:
- Task Dimension: *represents tasks that are relevant for interactions. Thus, UCMs are used as a simple and expressive visual notation that allows describing task scenarios at an abstract level in terms of sequences of responsibilities and tasks over a set of components.*
- Dialog Dimension: *explains the style of human-computer interaction and also describes the sequence of dialogs that can take place between the user and the system. New notations are introduced to provide UCMs with more expressive power for interaction design (Alsumait et. al, 2002).*
- Structural Dimension: *identifies the objects comprising the user interface, their grouping and specifies their layout, e.g. by indicating approximate placement or by indicating topological relations between groups.*
- Usability Dimension: *Measure the usability of the use case maps by using usability metrics. Different metrics such as task simplicity and task performance can be used to measure the usability of the use case map early in the requirement phase.*

*In the second phase, two types of UCMs are created: the* conceptual use case map *(CUCM) and the* physical use case map *(PUCM). The CUCM and PUCM together integrate the four dimensions stated above and capture a complete picture of user interface and usability requirements. This helps in analyzing the consistency between requirements of different use cases, and discovers if any conflicts between different types of users, different purposes of use and different operating conditions.*

4

## Conceptual Use Case Map (CUCM)

The CUCM helps provide details on what information will be needed from the user and the system to accomplish a task. Steps to create CUCM are:
- Partition the use cases: *Consider only the use cases with human actors.*
- Create Task Model: *Decompose use cases into tasks and subtasks.*
- Define components of the use case: *Tasks can be bounded to components. This will help in determining the entities and objects that compose the system*
- Create Dialog Model: *Tasks represent decision points in the use case. At any decision point, a dialog will take place between the system and the user.*
- Analysis of the consistency: *Two properties of consistency are examined. The consistency among a set of use cases where the requirements captured by a set of use cases are not in conflict with each other. The consistency of a use case with respect to a requirements model where the information contained in the use case does not conflict with the requirements model and the information contained in the use case is a subset of the information contained in the model. Both properties can be validated in our proposed use case model. If the use cases are inconsistent, it will backtrack to the information elicitation and scenario description (phase 1) to resolve the conflicts*
- Analysis of completeness. *Completeness is the property where all the information contained in a requirement model is covered by at least one use case and does not contain any information that cannot be inferred from the information contained in the use cases. If incompleteness is discovered, it will backtrack to the scenario identification phase to identify additional scenarios that will cover the missing situation.*

## Physical Use Case Map (PUCM)

The extended UCMs not only describes the sequence of tasks and dialogs that can take place between the user and the system but also help to understand and reason about the requirements of the user interface, including usability aspects. The PUCM represents the space within the user interface of a system where the user interacts with all the functions, containers, and information needed for carrying out some particular task or set of interrelated tasks. Moreover, successive display of different screens and interactive objects are presented. The PUCM can greatly benefit from the graphical representation of use cases. Steps to create PUCM:
- *Identify the objects that make of the user interface.*
- *Convert the use case maps into an active prototype using a graphical user interface development tool.*
- *Measure the usability of the screens*

Further information on validating CUCM and PUCM can be found in (Seffah and J.A Poll, 2003). Once the CUCM and PUCM are constructed and validated, further requirement analysis and specification activities can start to produce formal requirement specifications according to the requirement models. More research and investigations is required in this area.

# 3   A Case Study- Movie Recommender Software (MRS)

This section illustrates the description and analysis of the UIUR model using a case study that simulates Movie Recommender Software (MRS) for a PDA. The software should provide recommendations for movies based on preferences of other users.

The movie Recommender software allows the user to carry out the following scenarios: (1) View a list of the latest movies by selecting (Any Genre, Action, Comedy, Drama), (2) Search for movies of (any genre, For Action, For comedy, For drama), (3) View or add a movie to the  (To See) list; a list of movies to be seen in the future, (4) Rate a movie. On choosing the options 1 or 2, the user can get detailed information of a movie generated by the Recommender system

As shown in Figure 2, the MRS system can be decomposed into five main tasks, (1) view latest movies, (2) search for movie recommendations, (3) create user to see list, (4) rate a movie, and (5) view movie detailed information. Different scenarios can be drawn, and details are delayed to sub-UCMs. Figure 3 depicts the second level of the requirement model when a user rates a movie. A *dialog* notation illustrates that a conversation between the user and the system is taking place. The CUCM should be validated against the scenarios by analyzing whether the use cases are consistent with each other. Examples of such validation are shown in (Seffah and J.A Pool, 2003).

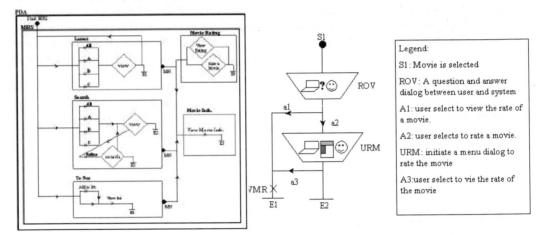

Figure 2: **CUCM for the movie Recommender**          Figure 3: **UCM for movie rating**

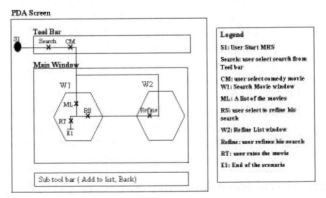

Figure 4: **PUCM for movie Recommender software**

The PUCM in Figure 4 describes the scenario where a user searches for a certain comedy movie and rates the movie. The user interface consists of a tool bar, Main Menu Window, and a sub-

6

toolbar. Next, the PUCM can be converted into a prototype. Both the PUCM and this prototype can be used by usability expert to predict the usability of the interface.

# 4    Conclusions

Our research effort leads to the development and analysis of a complete model for user interface and usability requirements (UIUR). This model ensure that: (1) a consistent and complete requirement specification can be captured using scenarios, (2) the specification is a valid reflection of user requirements, (3) the derivation of early design artifacts such as low fidelity prototypes is possible. The new model extends the UCMs to accommodate four dimensions of requirements including the task, the dialog, the structure, and the usability of the user interface. New notations are introduced to provide UCMs with more expressive power for interaction design. As a consequence, the extended UCM helps not only describe the sequence of tasks and dialogs that can take place between the user and the system but also assists to understand and reason about the requirements of the user interface, including usability aspects. Future work includes more investigation on formalizing the requirements according to the UIUR model.

# References

Alsumait, A.; Seffah, A.; Radhakrishnan, T. (2002), Use Case Maps: A Roadmap for Usability and Software Integrated Specification, In: *Proceedings of IFIP World Computer Congress 2002*, August 25-30, Montreal, Canada.

Buhr, R. (1998), Use case maps as architectural entities for complex systems, *IEEE Transactions on Software Engineering*, 24(12), 1131-1155

Constantine, L.; Lockwood, A. (1999), Software for use: a practical guide to the models and methods of usage-centered design. Reading, Massachusetts: Addison-Wesley.

Forbrig, P. (1999), Task and object-oriented development of interactive systems - how many models are necessary? In: *Proceedings of Design Specification and Verification of Interactive Systems Workshop-DSVIS'99*, Braga, Portugal, 225-237.

Jarke, M. (1999), Scenarios for modeling. *Communications of the ACM*, 42(1), 47-48.

Seffah A., J.A, Pool Combining UCMs and Formal Methods for Representing and Checking the Validity of Scenarios as User Requirements, to appear in interact 2003. Human-Centered Software Engineering Technical Paper 01-0203, Concordia University

Sutcliffe A.G., Maiden N.A., Minocha S. & Manual D. (1998), Supporting scenario-based requirements engineering. *IEEE Transactions on Software Engineering*, 24(12), 1072–1088.

Weidenhaupt K., Pohl K., Jarke M. & Haumer P. (1998), Scenario usage in system development: a report on current practice, *in Proceedings of the International Requirements Engineering Conference (ICRE '98)*.

# A Socio-centric Model of User Interactions

*David Ambaye*

Interaction Design Group (IDC),
Middlesex University, London, UK
d.ambaye@mdx.ac.uk

## Abstract

The investigation described in this paper is motivated by the paucity of socio-centric models of user behaviour in relation to complex interactive systems such as collaborative technologies. It argues that existing approaches to modelling users, such as those underlying task based techniques, do not take sufficient account of the psychological and social issues surrounding the usability of such systems. This fundamental weakness means that user-interface design and evaluation are currently less holistic and effective than they ought to be. Two key premises are thus presented. The first is that such holistic models form a proper basis for creating more effective evaluation tools. The second is that the construction of such models needs to be rooted in empirically derived understanding of interactions which occur between users and collaborative systems. The paper reports on results from an empirical study of users interacting with collaborative technologies known as groupware, where social and psychological issues are often as important as other factors. It describes the research framework underlying a multi-method ethnographic case study and summarises the observations.

## 1    Introduction

Few will dispute that collaborative software implementations such as groupware commonly fail to fulfill both end-user and organisational expectations. Where failures have occurred, it is not unusual to find that commentators and users alike point to root causes that have multifarious socio-technical and psychological dimensions. This is not surprising in light of the fact that collaboration or teamwork is a social phenomenon, and that collaborative technologies are aimed at enriching activities that occur within such a context. This paper argues that such difficulties are largely the result of a paucity of understanding of how users are really likely to behave when interacting with such systems. Poor understanding makes it difficult to create effective collaborative systems, in terms of users' and organisational needs. The result is that organisations resist such technologies because they feel them to be constraining, instead of augmenting and supporting their 'natural' communication and co-ordination processes. Moreover, systems which appear to have few usability problems during design and evaluation, can prove awkward to use and learn when placed in real work contexts (Pinelle & Gutwin, 2000). The importance of getting to grips with this issue cannot be underestimated. The drivers within the business world for the development and utilisation of collaborative technologies are unlikely to lessen with time. Organisations are likely to continue to restructure using a variety of team constructs, and globalisation is unlikely to decrease. Consequently, organisations will come to rely on such technologies as critical technological levers for augmenting the creative and productive potentials of teams and individuals. Such technologies will also be useful in the transitional role, enabling organisations to cope with the downsides of rapid changes, which are increasingly becoming the norm. Sudden change in structure or environment can have a chaotic effect on existing intra-team co-ordination and communication strategies. Inspite of early hopes, the simpler types of groupware such as e-mail, video-conferencing and shared diaries appear to have gained significant foothold within the workplace today. In sharp contrast, technologies which have a more rigourous dependence on principles of team collaboration, such as work-flow systems, group decision support systems, shared document editing and application sharing systems appear to be much less successful.

# 2   Evaluation Implications

There are two main conclusions to be drawn from the above discussion. The first is the pressing need to develop new approaches and techniques which are better suited to assessing the usability of technologies oriented towards teams. The second is that such assessment approaches and techniques must be firmly rooted on improved understanding and identification of the critical social psychological factors and mechanisms that can contribute to the success or failure of such implementations. Success in this respect will mean that designers and usability experts can feel confident that they are designing *team-aware* technologies which fulfill functional, psychological and social needs of teams and individuals in organisations (Baker, 2002). The next section describes results from an investigation which attempts to assist the aforementioned objectives. In particular, it describes the 3 Phase model which characterises observed interactions between team-base users and a variety of collaborative technologies.

## 2.1   The research framework

Users in three organisations were observed over a period of four years. An ethnographically derived analytical framework known as 3D was utilised to structure the observations. In brief, 3D attempts to relate system usability to individual and team effectiveness (Ambaye 2002). The framework postulates the existence of three usability components: Individual-empowerment, Productive-empowerment and Cultural-empowerment. Taken together, these represent psychological, productivity and social impacts of technology intervention on the effectiveness of team-based users. This made it possible to categorise observed usability issues into one or more of these components. The insight gained from applying the 3D framework has led to the identification of three key emergent properties about how users may interact with collaborative technologies. That is, the bulk of interactions observed in this investigation exhibited three key characteristics:

- Team attributes were transformed in different ways through the intervention.
- The interaction between groupware and users occurred in temporally differentiated phases which were directional, inter-dependent and cyclic.
- Perceived net-benefit was the main motivation that sustained and encouraged user commitment towards a system.

### 2.1.1   Property 1: Team Attributes are Transformed

When collaborative technology is introduced into a team, team members and the team as a collective may begin to undergo a process that changes many existing attributes, as well as causing the development of new ones. For instance, it may necessitate team members learning new skills, re-negotiating existing working relationships, adopting new norms, communicating differently and generally experiencing changes to the nature of their work activities.

### 2.1.2   Property 2: The Existence of Temporally Differentiated Phases

The quality of interactions between the technology and users appears to be dependent on, not only the impact a system has on users, but also when it has this impact. For instance, the response time of a video-conference system under extreme conditions (such as large number of users, or intensive use) is unlikely to be of paramount importance to first-time users. Rather, high in their minds will be concern about issues such as: learn-ability, ease of use, video and audio quality, as well as the consequences the system will have on the way they communicate with each other. Performance under high load conditions, as influential as it can be, will more likely become important at a later time when:

a) users are in such a position as to exploit and appreciate the full potential of its capabilities
b) there is sufficient load on the system.

However, this is not meant to imply that such factors cannot be important in more than one 'phase' of an interaction. Rather, it is simply a matter of priorities. Issues such as reliability, underlying task-model or flexibility are important throughout the lifetime of a system, but may have their most profound impact at particular times and less so in others. This suggests that such interactions may progress through various inter-dependent, but temporally differentiated, phases. More specifically, it is suggested that there may be an initial phase, where users learn and explore the system's characteristics. In subsequent phases, users will evaluate the extent to which they can rely on the system for augmenting or performing mission-critical tasks and will begin to exploit it according to their conclusions.

### 2.1.3   Property 3:  Perceived Net-Benefit is a Key Driver of Entrainment

The notion of **entrainment** is introduced here to describe the process of inclusion of and reliance on, a system for accomplishing routine work. The above examples suggest that the net benefit a system provides has a great influence on the extent of its entrainment. That is, it greatly influences users' level of desire to use a system and whether they will continue to rely upon it for performing routine work. So in the case of  a shared diary system, initial usage difficulties (perhaps because of poor training, policy deficiencies etc.) might prevent a majority of users from exploring and using the system. In turn, this may lead to a situation where the system cannot gain the needed critical mass of users (Greenberg, 1991) and by implication critical mass of information. Not surprisingly, the remaining users find that a shared diary system that is used by only a small proportion of users (and infrequently), is unlikely to contain information of value to them. Eventually, they too become non-users.

## 3    Three Phases of Entrainment (3P)

Taken together the three emergent properties described above form the basis of the 3P interaction model proposed in this paper. 3P conceptualises user interactions as progressing through three key phases of development (property 2 above) that are temporally differentiated, directional and have cyclic characteristics. These are described below.

From system developers' perspective, and in an ideal case, the level of entrainment of a system and the resulting user experience in Phase 3, will generally be greater than for the other two. Conversely, the Orientation phase will reflect a stage when systems have the lowest level of entrainment. Each of these phases is described below. It should be noted that the term 'user' is utilised here to mean primarily team-based users.

## 3.1    An Orientation Phase

Johansen (1988) observed long ago that a major obstacle to the adoption of groupware is the lack of understanding of its purpose, or how it can be exploited.  The Orientation phase in many ways conforms to this observation.  This is because it is an impressions phase, very much characterised by whatever initial concerns, questions or opinions team members may have regarding a system.

In this phase, it is not uncommon to find that team member's differing roles, backgrounds and experiences, lead users to react differently to a given system.  As a result the issues of concern may also differ between users.  If a system is being introduced for the first time, then the Orientation phase reflects first impressions. This may have been gained by observing the way that others have used the system before, information they have read or heard about the system (such as product reviews in trade journals) or through an organisational orientation program (i.e. training). These issues may have even come to mind by simple analogy to previous similar systems users may have been exposed to. If the system has been in use for some time, this phase largely reflects impressions based on experiential evidence that users may have gained from previous contact with the system, in the same or similar roles.

## 3.2    An Exploration-Evaluation Phase

### 3.2.1    Phase 2.1: Exploration

One element of Phase 2, Exploration, is intrinsically an experiential stage.  It is a stage, where team members actively investigate in greater detail whether the general impressions and concerns which were gained in the Orientation phase are well grounded.  If users do not perceive any benefit they are unlikely to enter the Exploration phase.  Through this exploration, users may attempt to determine whether there are any benefits or consequences which have been overlooked. Many of the issues that users will be eager to investigate further, will be those that were identified as being important during Orientation.  New issues may also emerge that were not considered previously.  The means for attempting to gain this understanding involve, exploratory use, a training program, or even careful observation of how others are currently implementing the system.  In this phase, users also consider how they could adapt a system's attributes to suit their own work process.  They question what would be involved in terms of maintenance and ease of use of the system.

### 3.2.2    Phase 2.2: Evaluation

A process of evaluation is also important in the second phase.  Here, team members may make decisions about what level of entrainment is warranted.  This will be mediated in relation to the experiential and other evidence collected about the system.  In this sense, users weigh up the costs and benefits of entraining the system (if at all), after having appraised how the variety of system attributes will impact their work.  For instance, two important considerations could be: the degree of changes required to the way that existing tasks are performed and the way that users might have to re-organise their work life. Based on the costs and benefits identified users make a decision as to how deeply they will entrain the system into their work.  If they are already using the system, they decide whether they should increase or decrease the level of entrainment. Generally this exploration and evaluation of a system occurs simultaneously.

## 3.3    An Exploitation Phase

Here team members begin the process of entraining a system into their work processes according to the 'contract' details of the preceding phase. Users will begin to exploit whatever attributes of the system they have decided to entrain. This phase can be characterised as follows:

• Users interact with the groupware as per the 'contract' they concluded in Phase 2.
• System is used here as a part of routine work processes.
• Users input just enough effort to maintain the contract type agreed (relative to Phase 2).

To summarise, each of the three phases described above are depicted together in figure 1. When the interaction progresses in time from orientation towards exploitation, the directionality of the phases is such that user understanding and experience of the system increases in each phase. Moreover, knowledge, experience and attitudes acquired in one phase are likely to influence others which follow it. Entrainment is highest near the outer circle and lowest at the inner circle. The outer and inner circles can also be viewed as representing high and low levels of commitment, respectively. The dimension of time is depicted by the clockwise arrow. Users will cycle through each phase (clockwise) on a continuous basis, so long as users continue to entrain the system, even at a minimum level. This cyclic characteristic also implies that the usability of a system can vary substantively within the life cycle of a system. That is, each time users interact with a system, ever changing circumstances will lead to it having a changed level of usability. Results have demonstrated that representation to discuss and represent conditions under which this cyclic trend can grow, decrease or terminate.

## 4    Conclusions

The 3P framework consolidates much of what has been learned from observations of users, in real contexts. Its potential value as a conceptual and practical analysis tool for gaining insight into the possible factors and mechanisms surrounding usability problems is currently the focus of a number of researchers. At a base level, the framework attempts to provide a plausible description of the essential elements of how users interact were observed to behave when interacting with groupware. It models these interactions as interactive processes that progresses through several temporally differentiated phases, that are directional and cyclic. Future work will concentrate on exploring the generalisability of the key assumptions underlying the framework. In particular, more work needs to be done in understanding whether the three emergent properties that form the basis of the 3P framework are plausible. Work also will focus on how to derive practical design and evaluation tools from the framework.

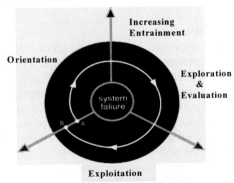

Figure 1: Three Phases of Interaction

# Tools for task-based interaction and collaboration analysis

*Nikolaos Avouris[1], Georgios Fiotakis[1], Nikolaos Tselios[1], Vassilis Komis[2]*

[1]Electrical & Computer Eng. Dept, HCI Group,  [2]Early Childhood Educ. Dept.,
University of Patras, 26500 Rio Patras, Greece
{ N.Avouris, Fiotakis, NiTse }@ ee.upatras.gr, komis@upatras.gr

## Abstract

An innovative framework of interaction and collaboration analysis is proposed, jointly with tools to support the process. The proposed framework is based on a *collaboration analysis* first and *individual task analysis* subsequently, approach. The objective of this framework is to facilitate understanding of the group's and individual user's tasks and goals and associate the artifacts used with usability problems. An innovative aspect of the framework is the association of tasks to artifacts (tools) engaged by the users during the activity. The typical use of this framework is in interactive systems evaluation and design. The framework and the tools functionality are described in the paper. The framework is inspired by the *Activity Theory* perspective, which recognizes the importance of artifacts, actors and the context in which an activity takes place.

## 1    Introduction

Tools and techniques to support interaction and collaboration analysis have been proposed in the frame of usability evaluation studies for many years now (Dix et al., 1998, Nielsen 1994). These techniques involve analysis of data that are collected from field studies in various forms. Stream media like audio and video, logfiles, as well as notes and comments of observers are used in these studies. Discrete data items, like files containing solutions to problems, drawings, etc. may also be used. All these data need to be correlated and processed in order the researchers to extract useful patterns of behaviour of the actors involved, identify usability and conceptual flaws in the design of the artifacts used and evaluate the approaches that have been pursued. This analysis process has become tedious, since the high volume of data has made it more time-consuming and complex. The need for adequate tools to support the analysis has therefore increased recently. A framework of interactive systems analysis is proposed in this paper, which involves two complementary views over an activity involving multiple actors, supported with relevant tools:

(a) The first one concerns analysis of collaboration, which involves collection of field data, annotation of these observations and building of an abstract description of observed collaborative task execution. A tool has been built for annotation of data and building of abstract multi-level annotated views. This tool (Collaboration Analysis Tool, ColAT) bears interesting characteristics among which the support for various annotation schemes, the capability to commend and annotate at various levels of abstraction a sequence of events, interrelation of multiple media (video, audio, log files, snapshots) to the multilevel annotation and to the cognitive structures built through the second view.

(b) The second view is a cognitive one, which involves building of typical task structures, as anticipated by the designers of the artifact (e.g. the computer tools that are used in the activity), subsequently relating the observed task execution to the anticipated model by individuals who participate in the activity. A tool has been built to support this task-based approach, the Cognitive Modelling Tool (CMTool), also described in (Tselios & Avouris 2003). Analysis characteristics of

13

task models are stored in the CMTool database, together with qualifying information. The possibility of analysing further the observed system usage according to a number of dimensions permits evaluation of the artifacts both in terms of usability and effectiveness.

The proposed framework and analysis tools have been applied in non-routine task domains (Tselios & Avouris 2003). The findings reported indicate the effectiveness of this structured technique in identifying usability and interaction design problems. In the reported studies, we focused on analysis and evaluation of collaborative tasks, involving a number of individuals and artifacts engaged in problem solving either at a distance or in the same place. The ethnographic methodological approach used in the evaluation studies of this nature proved to be compatible to the proposed theories underlying the framework and the tools.

## 2 Analysis of observed collaborative activity

The first phase of analysis involves collaboration analysis study. A new integrated environment of analysis, the *Collaboration Analysis Toolkit* (ColAT), which integrates multiple sources of behavioural data of multiple logging and monitoring devices is used in this phase. The main emphasis of the ColAT environment is on the analysis of situations involving more than one actors. Special attention has been put on scenarios of synchronous computer-supported collaborative problem solving, in which the actors are spatially dislocated, a factor which imposes additional complexity in the analysis task.

The most important phase of analysis relates to the interpretation and annotation of the collected data, as well as generation of aggregate data of interpretative nature. An innovative feature of the ColAT approach is the support for creation of a multi-level structure that describes and interprets the logfile events. In figure 1 the concept of the multi-level logfile is shown.

The original sequence of events contained in the master logfile is level 1 (*events level*) of this multilevel structure. The keystrokes or raw observations are included in this level. An example is the event "User X selects option Y from the menu" or "User Z said ...."in case of a dialogue event. A number of such events can be associated to an entry at *level* 2 by the analyst. Such an entry can have the following structure:

$$< ID, User, entry\_type, comment >$$

where *ID* is a unique identity of the entry, *type* is a classification of the entry according to a typology that has been defined by the researcher, followed by a textual comment or attributes that are relevant to this type of task entry. Examples of entries of this level are:" User X inserts a link in the model", or "User Y contests the statement of User Z". In a similar manner the entries of the higher levels are also created, which describe the activity at the strategy level as a sequence of interrelated goals of the actors involved.

An implication of this approach is that the associated stream media are related to this multi-level view of the activity and therefore the user of ColAT can decide to view the activity from any level of abstraction he/she wishes. This approach results in a powerful analytical tool, since the possibility of viewing a process from various levels of abstraction supports its deeper understanding and interpretation.

A *"ColAT project"* is stored in a database to facilitate processing and navigation of the source data and annotations. The integrated logfile can be exported in XML form to other applications and data processing tools for further analysis. The main activity of this phase involves creation of the higher-level logfile entries. In these higher levels of the logfile the typology of the Object-Oriented Collaboration Analysis Framework (OCAF), see (Avouris et al. 2003a), has been used. This framework is particularly suitable for analysis of collaborative activity, which involves interleaving of actions and dialogue. OCAF puts emphasis on the objects of the jointly developed solution. Every object is assigned its own history of events (actions and messages) related to its

14

existence, as a sequence of events according to the following functional types: I = Insertion of the item in the shared space, P= Proposal of an item or proposal of a state of an item, C= Contestation of a proposal, R= Rejection / refutation of a proposal, X= Acknowledgement/ acceptance of a proposal, T= Test/Verify using tools or other means of an object or a construct.

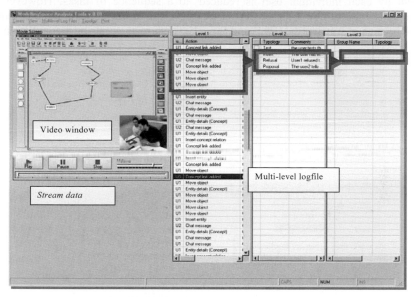

**Figure 1:** Overview of the ColAT data navigation environment

As an example of an OCAF event, the introduction of a new object in the shared space, is indicated as *Object (X)= $I_{UI}$*, i.e. User 1 inserted the object (X) in the shared space.

The ColAT environment that supports navigation of the constructed multilevel logfiles is shown in figure 1. A video window permits viewing of streaming data in association to selected events in any level of the logfiles. There are different modes of use of this environment: In the first mode, *navigation is controlled through the video*. When a position of the video file is selected, the corresponding event of the log hierarchy that the video is related to, is highlighted. In the second mode, *navigation is controlled from the logfiles*. In this case the user can select any event in the first level of the log file and the video starts from that event onwards.

The user can hide the levels of abstraction he/she wishes to ignore, thus defining the desired view over the field data. The ColAT navigation tool has been proven particularly useful in analysing the data of reported experiments (Avouris etal. 2003b), following the OCAF framework.

## 3. The task analysis of individual actors

During the second phase, each individual actor is studied in relation to the tasks she undertook and the goals she attempted to accomplish. The task analysis method adopted is the Hierarchical Task Analysis (HTA), proposed by Shepherd (1989), accordingly modified, as discussed here. The intention is to build a conceptual framework reflecting the way the user views the system and the tasks undertaken in order to accomplish set goals. So the main objective of our analysis is to reflect on the observable behaviour of users and identify bottlenecks in the human-system dialogues rather than explicitly understand the internal cognitive processes as one performs a given task. So we include in our analysis, even incorrect problem solving approaches, caused either by limitations in the provided tools design or by conceptual misunderstandings related to domain knowledge and use of available tools. These errors may lead to not satisfactory solutions to a given problem, but often do contribute to better and deeper understanding of concepts.

Through the proposed task analysis technique, the typical or expected use of the software environment by each individual user is reduced, to a sequence of tasks. However, this task analysis can be achieved if the right level of abstraction is used, as fed from the previous phase of the study. Kirwan and Ainsworth (1992) and Boy (1999) show methods of accomplishing this.

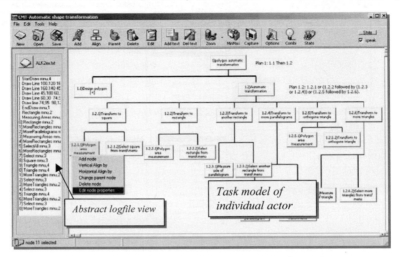

**Figure 2:** The Cognitive Modelling Tool (CMTool) Environment

One important aspect of our analysis framework is the classification of observed unexpected or incorrect user behaviour, see Tselios et al. (2002) and Tselios & Avouris (2003). Through the application and analysis of this technique, we have identified and classified five main categories of errors: *Severe syntactic error, Minor syntactic error, Conceptual error, Inconsistency, Design principle violation*. These errors can be identified during observation and analysis of problem-solving activity during this phase. Also in *plans* (Shepherd, 1989), simple expressions are included such as (!) and (x!) to denote subtasks containing errors, and { } to indicate occurrence of unforeseen tasks according to the original designer model.

Display of an annotated log file at selected level of abstraction, shown on the left of fig. 2, next to the task model structure, is supported. So both interaction details and cognitive goal hierarchies are displayed simultaneously to the user of CMTool. Task models are stored in a relational database, grouping the various tasks analysed, with additional identification information (designer's model or revised designer's task model (DTM) or user's task models (UTM)). In addition, quantitative analysis tools are supported to extract useful metrics related to the analysed tasks. Examples of these metrics are the number of keystrokes required to achieve a specific goal or sub-goal, the mean time required and the interaction complexity of specific user model, compared to the original designer's expectations or to the revised and adapted model.

In CMTool, the evaluator can select parts of a task structure representing a specific problem solving strategy, which can be stored for future reference or comparison with other users' strategies. In addition, the possibility to analyse system's usage in five dimensions is a contribution of the tool to the evaluation process. These are: (i) High level tasks, (ii) users, (iii) specific strategies, (iv) tools used, (v) usability problems detected. This process is carried out through a visual information query environment. Field experiments could be analysed across any possible element of the five different dimensions discussed (e.g. "Show all encountered problems related to tool X", etc.). Complex analysis can be carried out according to any of these dimensions, supporting study and analysis of encountered problems.

# 4 Conclusions

This paper describes the main functionality of tools to support multilevel analysis of field data collected during evaluation studies of group collaboration. Both *ColAT* and *CMTool* have been recently applied in evaluation studies of group problem solving activities. The theoretical underpinning of this approach is Activity Theory (Kaptelinin etal. 1999). This is a conceptual approach, that considers as the unit of analysis the activity, consisting of a subject (an individual or group), an object or motive, artefacts, and sociocultural rules. Understanding thus human activity requires a commitment to a complex unit of analysis. The multi-view approach proposed covers this aspect. In the proposed framework we move from the group level analysis to the individual human-computer interaction study in a smooth way. The multi-level annotation scheme described here permits change of point of view and relates the observational data to the annotations. These annotations can comply with a typology imposed by a methodological framework, like the OCAF scheme used in our example. The framework permits shifting from bottom up annotation of group activity data to top-down task level description of the observed human-computer interaction

The concepts and tools discussed here are relevant to researchers who are involved in analysis and evaluation of complex collaboration-support activities, in design and evaluation of new tools and in support of users' meta-cognitive activities.

# Acknowledgement

Partial funding of project IST-2000-25385 ModellingSpace, by the EC is acknowledged.

# References

Avouris N.M., Dimitracopoulou A., Komis V., (2003a). On analysis of collaborative problem solving: An object-oriented approach, Computers in Human Behavior, 19 (2), March, pp. 147-167.

Avouris N., Komis V., Margaritis M., Fiotakis G., (2003b). Tools for Interaction and Collaboration Analysis of learning, Proc. Conf. CBLIS 2003, July, Nicosia, Cyprus.

Boy, G. A. (1998). Cognitive Function Analysis. Ablex Publ., Greenwood, Westport, CT, USA.

Dix A., Finlay J., Abowd G, Beale R., (1998), Human-Computer Interaction, Prentice Hall

Kaptelinin, V., Nardi B., Macaulay C., (1999). The Activity Checklist: A Tool for Representing the "Space" of Context, Interactions, July, 27-39.

Kirwan, B. and Ainsworth, L.K. (1992). A Guide to Task Analysis. Taylor & Francis, London.

Nielsen, J., (1994). Usability inspection methods. In J.Nielsen, R.L. Mark (Ed.), Usability Inspection Methods, John Willey, New York.

Paterno', F. (2000) Model-based design and evaluation of interactive applications, Springer Series in Applied Computing, Springer-Verlag, London.

Shepherd, A. (1989). Analysis and training in information technology task. In Diaper, D.(Ed.) Task Analysis for human computer interaction. Ellis Horwood Limited, 15-55.

Tselios N., Avouris N., (2003), Cognitive Task Modeling for system design and evaluation of nonroutine task domains, E. Hollnagen (ed.) Handbook of Cognitive Task Design, LEA, 307-332.

Tselios N., Avouris N., Kordaki M., (2002), Student Task Modeling in design and evaluation of open problem-solving environments, Education and Information Technologies, 7:1, 19-42.

# The Semiotic Engineering Use of Models for Supporting Reflection-in-Action

*Simone D. J. Barbosa, Clarisse Sieckenius de Souza, Maíra Greco de Paula*

Departamento de Informática, PUC-Rio
Rua Marquês de São Vicente, 225
Gávea, Rio de Janeiro, RJ
Brasil, 22453-900
{simone, clarisse, mgreco}@inf.puc-rio.br

## Abstract

Diverse models have been proposed to cope with the complexity of HCI design and increase software usability, most of them based on cognitive theories. In this paper, we argue for the need of complementary theories of HCI and present an alternative set of HCI design representations and models based on Semiotic Engineering. These comprise extended scenarios and task models, and a novel semiotically-based interaction model. In the latter, signs and communication breakdowns handle cohesion and coherence issues for prospect conversations between users and the *designer's deputy* – an entity capable of participating in all and only the designed conversations. Our goal is to empower designers, supporting reflection about their interactive solutions.

## 1   Introduction

Diverse models have been proposed to represent HCI design, taking into account users' characteristics, preferences and needs. Most of them are based on cognitive theories (Norman & Draper, 1986). They focus mainly on the individual interacting with an application, considering his/her physical (motor and perceptive) and mental processes. Their main goal is to understand how these processes take place during interaction, in order to predict and avoid possible cognitive problems. But software applications are intellectual artifacts, resulting from designers' understanding, reasoning, and decision-making processes. Thus, we should support not only the designers' intellectual processes (Schön, 1983), but also the expression and usage of such processes' results through the user interface. This double enterprise is the object of study of Semiotic Engineering (de Souza, 1993; de Souza, in preparation). In this paper, we describe how a specific set of representations and models can support HCI design. We are concerned with building a coherent and cohesive overall interface message, attempting to maximize the chances that intellectual artifacts such as software will be interpreted as meant. As is the case with all interaction about what goes on in one's mind, HCI should support *conversation*, even if non-textual codes are extensively used to convey the designers' ideas and the products' affordances.

Semiotic Engineering views the interface as a designer(s)-to-users message, representing the designers' response to what they believe are the users' problems, needs, preferences and opportunities. In this message, they are telling users, directly or indirectly, what they have in mind. In fact, user–system interaction is viewed as a *conversation* between the user and the "designer's deputy" – a *programmed* representative of the designer. Since designers are no longer there at the time such conversations are carried out, they must anticipate every possible conversation. Thus, the designer's deputy is capable of participating in all and only the designed conversations.

The semiotic engineering of human-computer interaction is focused on building the designer's deputy, which will be implicitly or explicitly represented in the application's interface. And this imposes certain additional requirements on design representations and models. For example, we need to represent the nature of signs that conversational parties may use, the remedial actions for interaction breakdowns, as well as the potential threads of conversation that can be followed during interaction. We propose and use a classification of *signs* to describe every piece of information relative to some concept in the domain or in the application itself. Signs are extracted from scenario representations and provide a thread that binds all design models and representations with the actual user interface. Also, because we take *communication breakdowns* as an inherent part of HCI, we aim at helping designers to tell users not only how to perform their tasks under normal conditions, but also how to ask for help or take remedial action in mistaken or unsuccessful situations. To this end we propose a classification of communication breakdowns, to be used in extended task models.

## 2   Extending scenarios and task modelling

Scenarios can be used throughout the development process, starting from the analysis of the domain and the users' tasks and characteristics (Carroll, 1995). By means of scenarios, designers not only learn about the users' vocabulary and thought processes, but they have a chance to make this knowledge explicit in a language that is accessible to users. When writing scenarios, the designer should have in mind what the purpose of each scenario is, allowing for user involvement in: investigating certain aspects of a domain or socio-cultural environment; proposing a task structure or an interaction solution; contrasting alternative task structures or interaction solutions; and so on.

Our extension to scenarios is one of adding a semiotic dimension to their use. Designers should carefully extract from scenarios the domain and application *signs* that are meaningful to users. While these signs are usually treated as data, in Semiotic Engineering they acquire a new status. They are analyzed and classified according to the degree of familiarity users are expected to have with them. *Domain* signs are those directly transported from the users' world, such as "full name". *Application* signs are those that have been introduced by the application and had no previous meaning to the user, such as "your login account". Finally, *transformed* signs are those derived from an existing sign in the world that has undergone some kind of transformation (e.g. based on an analogy or metaphor) when transported to the application, such as "backup files".

Different kinds of signs may require different kinds of sense-making support. In general, a domain sign would only require explanation if specific constraints are imposed by the application. For example, the concept of "full name" is clear to users, but a restriction about the number of characters allowed in the corresponding piece of information might not be, and thus need some clarification from the designer's deputy. A transformed sign would require an explanation about the boundaries of the analogy. For example, a folder in the desktop metaphor might require an explanation about its "never" getting full, except when the hard disk which contains it lacks space. Finally, an application sign may require a complete explanation about its purpose, utility and the operations that can manipulate it. An example might be a sign for "zooming" in a graphics application. There are of course some signs that can be classified in either group. In these cases, it is the designer's responsibility to choose, based on the analyzed characteristics of users and their tasks, the amount of support to provide in order to have users fully understand each sign.

From such semiotically-enriched scenarios, users' goals are identified and may be structured in a hierarchical goal diagram. Each goal may be further decomposed into tasks that are necessary to achieve it. We made simple adaptations to the structure chart notation in order to represent the

19

hierarchical decomposition of tasks, some task properties and relations between tasks, and discrimination among sequential, alternative, and iterative tasks. We can thus represent tasks that do not follow a predefined sequence, tasks that may be performed anywhere within the goal, preconditions for performing a certain goal or task, optional tasks, and stereotypes as a reuse mechanism.

Our approach is somewhat similar to Hierarchical Task Analysis (Annett & Duncan, 1967), but a major difference is that our adaptations are specifically suited to represent signs and potential breakdowns involved in each task. We allow designers to classify and represent the kind of support they intend to provide in each problematic situation: passive prevention (PP: problems that are prevented by documentation or online instructions); active prevention (AP: errors that will be actively prevented by the system); supported prevention (SP: situations which the system detects as being potential errors, but whose decision is left to the user); error capture (EC: errors that are identified by the application and that should be notified to users, but for which there is no possible remedial action within the application); and supported error handling (EH: errors that should be corrected by the user, with application support).

## 3  A Semiotically-based Interaction Model

Task models describe the structure and organization of the tasks that will be supported by the system, but more is needed to promote reflection about how design sense-making and decisions will be conveyed to users through the interface, for example if explicitly or not, in detail or not.

The interaction model described here was conceived to be applied between the initial analysis and the interface specification phases. It has shown to be useful in helping designers grasp a global view of the conversations that comprise the application (Paula, 2003), and thus design a coherent designer-to-users message, keeping in mind the communicative role of the designer's deputy. It has a diagrammatic view coupled with a textual specification. The diagrammatic view gives designers a global perspective on the interactive discourse, promoting reflection by representing all of the potential user–system conversations. The textual specification delves into the details of each user–system dialogue, and sets the stage for formal user interface specification.

The interaction model comprises scenes, system processes, and transitions between them. A scene represents a user–deputy conversation about a certain matter or topic, in which it is the user's "turn" to make a decision about where the conversation is going. This conversation may comprise one or more dialogues, and each dialogue is composed of one or more user/deputy utterances. In a GUI, for instance, it may be mapped onto a structured interface component, such as window or dialog box, or a page in HTML. Figure 1 illustrates the representation of a scene called Search documents, which contains two dialogues: [inform search criteria] and [choose presentation format].

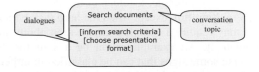

**Figure 1:** Diagrammatic representation of a scene called "Search documents".

The signs comprised in dialogues and scenes are included in a textual specification. When a sign is uttered by the deputy, i.e., presented to the user, it is represented by the sign name followed by an exclamation mark (e.g. date!). A sign about which the user must say something, i.e., which must be manipulated by the user, is represented by a name followed by an interrogation mark (login?, password?). Apart from signs, the remedial action for each predicted breakdown situation is also represented. Figure 2 illustrates the textual specification of a "Search documents" scene[1]:

```
Search document {
        [inform search criteria:
                title?, author?, date? ([PP] dd/mm/yyyy)]

        [choose presentation format:
                format?(table, report)]
}
```

**Figure 2:** Textual representation of a scene, including its corresponding signs.

Going back to the diagrammatic view, in it a system process is represented by a black rectangle. By using "black-boxes" for representing something users do not perceive directly, we encourage the designer to carefully represent the deputy's utterances about the results of system processing, as the only means to inform users about what goes on during interaction.

Transitions are changes in topic, and are represented by labeled arrows. An outgoing transition from a scene represents a user's utterance that causes the transition, whereas an outgoing transition from a process represents the result of the processing as it will be "told" by the designer's deputy. In case there are pre- or post-conditions, they should be represented within the transition label, following the keywords pre: and post:, respectively.

Some scenes may be accessed from any point in the application, i.e., from any other scene. The access to these scenes, named ubiquitous access, is represented by a transition from a grayed scene containing an identifier in the form U<number>. Moreover, there are portions of the interaction that may be reused in various diagrams. Figure 3 presents a diagrammatic representation of part of an interaction model corresponding to a goal of searching documents.

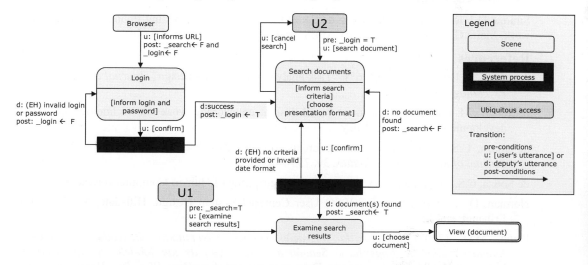

**Figure 3:** Sample interaction diagram.

---

[1] The complete definition of the textual specification may be found in (Paula, 2003)

# 4 Concluding remarks

The semiotically-enriched scenarios and the proposed interaction model are epistemic design tools. They support Schön's *reflection in action* epistemology (Schön, 1983) in that they explicitly represent dimensions of semiotic classification for communication-centered design. The centrality of communication and the importance of conversational affordances are justified by the intellectual nature of software, be it in the traditional format of desktop applications or in contemporary applications for mobile devices. In both, *signs* (visual, textual, gestural, aural, etc.) must be used to convey ideas, create conventions and facilitate usage. The proposed epistemic tools are so formatted that the construction of storyboards and prototypes should be as straightforward as in the case of other popular notations.

Semiotically-enriched scenarios support communication with users, allowing for the identification, exploration and verification of the application's purposes as far as designers and users are concerned. Signs bring attention to what is known and unknown, what remains the same and what is changed, what new language(s) or conventions must be learned, and so on. The extended task model incorporates elements that are crucially important for sense-making, namely the kinds of actions that handle breakdown situations. And the interaction model expands a blueprint of the potential conversations that may be carried out between the user and the designer's deputy during interaction. They help designers gain a sense of what are *all* and *the only* conversations (or communicative exchanges, to incorporate non-textual interaction) that the designer's deputy is equipped to entertain.

The design models under investigation have been used in the design of some Web and Windows applications, and in undergraduate class assignments. We have gathered some informal indications that they succeed in promoting the designer's reflection and as a result help produce better interactive applications. However, further studies are necessary to provide stronger evidence.

## Acknowledgments

Simone D.J. Barbosa and Clarisse S. de Souza thank CNPq for providing financial support to this work. Maíra Greco de Paula thanks CAPES for the scholarship granted for her M.Sc. program.

## References

Annett, J., & Duncan, K. D. (1967). "Task analysis and training design". *Journal of Occupational Psychology*, 41, 211-221.

Carroll, J. M. (ed) (1995). Scenario-based design: envisioning work and technology in system development, New York, Wiley.

de Souza, C.S. (1993). The Semiotic Engineering of User Interface Languages. *International Journal of Man-Machine Studies*, 39, 753-773.

de Souza, C.S. (in preparation). The Semiotic Engineering of Human-Computer Interaction.

Norman, D. e Draper, S. (eds., 1986) User Centered System Design. Hillsdale, NJ. Lawrence Erlbaum.

Paula, M.G. (2003) *Projeto da Interação Humano-Computador Baseado em Modelos Fundamentados na Engenharia Semiótica: Construção de um Modelo de Interação* (in Portuguese). Master's dissertation. Departamento de Informática, PUC-Rio, Brazil. Available online from http://www.serg.inf.puc-rio.br/

Schön, D. (1983). The Reflective Practitioner: How Professionals Think in Action. New York: Basic Books.

# Levels of Guidance

*Andrew Basden*

Information Systems Institute, University of Salford,
Salford, M5 4WT, U.K.
A.Basden@salford.ac.uk

## Abstract

Many sets of guidelines have been offered for, for example, web site design. But how complete are they? Often their categories overlap or lack an ordering principle. This paper shows how philosophy (that of Hart and Dooyeweerd) can be used to identify guidelines and their categories for web site design.

## 1    Introduction: Categories of Guidelines

Sets of guidelines for designing user interfaces, web sites, etc. abound. But, after using them for some time, we find ourselves wondering whether they cover everything important, we find the categories within some sets overlap or seem haphazard. This paper suggests a framework that can help us critique and refine categories of guidelines that is based on philosophy. We focus on the UI found in web sites, but at the end suggest how the framework may be extended to guidelines for other types of software.

The Evidence-Based Guidelines on Web Design and Usability Issues by the National Cancer Institute (NCI, 2003), contains 60 guidelines in the categories: Design process, Design considerations, Titles and headings, Page length, Layout, Font/text size, Reading and scanning, Links, Graphics, Searching, Navigation, Software and hardware, Accessibility. But these categories exhibit problems by virtue of having been assembled from empirical findings and the preferences of individuals found in other sources. Considerable overlap is evident, e.g. between Links, Navigation and Search, and some guidelines, e.g. "Establish a high-to-low level of importance for each category ..." occurs several times. The categories are not homogeneous (properties of site mix with user and design activity) and their order appears haphazard. We need categories in which the ordering principle is clear, not only to minimize overlap, but to allow us to judge, as guidelines proliferate, whether new categories are needed or categories may be merged or split.

The Web Content Accessibility Guidelines 2.0 (W3C, 2003) contains 21 guidelines under the categories: Perceivable, Operable, Navigable, Understandable, Robust. Though there is no overlap, their content is unbalanced in scope and depth. Navigable, for example, has six guidelines ('checkpoints') of general applicability while Operable has only three, all relevant only to specialised conditions. We also wonder whether other categories should be included, covering such issues as aesthetics, quality of content and cultural acceptability. We need categories that take on their intuitive meaning, are balanced in scope and depth, and cover all important issues of web site design.

# 2   Philosophy

Categorization involves ontological commitment. Hart (1984) has argued that, only philosophy, which concerns itself with the diversity and coherence of our experience, can provide us with both wide coverage and an ordering principle, and explores the philosophy of the late Dutch thinker, Herman Dooyeweerd (1955). We cannot rehearse his arguments here, but will examine how this philosophy can provide categories of guidelines. Dooyeweerd's ideas are interesting because he radically questioned the presuppositions underlying Western thinking and constructed a philosophical approach based on Meaning-oriented presuppositions that, while still in need of some refinement, is breathtaking in its scope and diversity.

## 2.1   Aspects

Perhaps the best known portion of Dooyeweerd's thought is his notion of irreducible aspects. Each aspect has a set of laws that enable meaningful functioning. Based on long reflection on both day to day living and scholarly writing, Dooyeweerd offered the following suite of aspects (though he made it clear his suite was to be criticised and refined):

- Quantitative aspect, of amount
- Spatial aspect, of continuous extension
- Kinematic aspect, of flowing movement
- Physical aspect, of energy and mass
- Organic aspect, of life functions and maintenance of organisms
- Sensory/psychic aspect, of sense, feeling and emotion
- Analytical aspect, of distinction, abstraction
- Formative aspect, of history, culture, creativity, achievement and technology
- Lingual aspect, of symbolic meaning and communication
- Social aspect, of social interaction, relationships and institutions
- Economic aspect, of frugality, skilled use of limited resources
- Aesthetic aspect, of harmony, surprise and fun
- Juridical aspect, of 'what is due', rights, responsibilities
- Ethical aspect, of self-giving love, generosity, care
- Pistic aspect, of faith, commitment and vision.

While the earlier (at least first four) aspects have determinative laws, the later aspects are normative, enabling freedom in our functioning but acting as guides for it. Guidelines may be seen as verbal expressions of selected norms in the aspects. To Dooyeweerd, human activity and life involves, in general, every aspect, in rich coherence. This includes using a web site. He claimed that human activity works well when we function well in all aspects. Thus the web site works well only if the user functions well in every aspect when using it. No aspect, nor its laws, may be reduced to any other. This gives multiple levels of design freedom. But it also means we cannot assume that good functioning in one aspect will automatically engender good functioning in any other. So designers must consider each and every aspect in which the user will function and, since much of our aspectual functioning is tacit (Polanyi, 1967), care must be taken not to overlook the less obvious aspects.

The first proposal in this paper is that guidelines for web site design can be based on the norms of the aspects of the user's use of the web site, which also provide useful categories for grouping them. Being irreducible, each aspect provides a distinct level of analysis.

# 3    Identifying the Aspects of Web Site Use

But how do we identify which these aspects are? Normally, we use a web site for some purpose in life, such as finding legal advice so that we might challenge genetically modified crops. We must separate the aspects that are contingent on the purpose (which in this case include the juridical, biotic and economic) from those of web browsing itself, those of searching, scanning, reading, understanding, etc., which pertain across all purposes. (The set of contingent aspects, as we will call them, might overlap with the set of aspects of web browsing.) To identify the aspects of web browsing itself, we make use of three further elements of Dooyeweerd's thought.

First, in many human activities, especially specialised activities carried out for wider purposes as web browsing is, one aspect is of primary importance; Dooyeweerd called this the qualifying aspect. It is the lingual aspect that qualifies web browsing (and indeed most texts, communications, etc.) because, whatever its purpose, its role is to convey meaning via symbols.

Second, Dooyeweerd claimed that proper functioning in any aspect depends on and involves proper functioning in earlier aspects. The lingual aspect, therefore, depends on the formative, analytical, sensory/psychic aspects. (We start at the sensory because it is where medium differentiates into visual, aural and haptic, but a more detailed treatment would continue back at least to the physical aspect.) These aspects correspond approximately with the levels found in linguistics: semantics, syntax, lexics and sensory.

Third, each aspect contains 'echoes' of all the others that anticipate later aspects and retrocipate earlier ones. Often, an aspect anticipates most strongly the one that follows it, paving the way for it. Thus the lingual aspect strongly anticipates the social; so cultural context is important in interpreting text etc.

Our second proposal is that in the case of web sites the relevant aspects for which we must provide guidelines, the 'levels' of guidance, are the sensory/psychic, analytical, formative, lingual and social, and the contingent aspects taken as a single category.

# 4    Aspects of a Web Site: Levels of Guidance

The sensory/psychic aspect is about sensing (seeing, hearing) and feeling (emotion). This offers us guidelines about matching UI devices used with the sensory capabilities of people. Thus both NCI and W3C sets contain guidelines about colour, colour blindness, sound, flicker, etc. Hardware considerations can be of this aspect, since it is phenomena that emerge from hardware that we sense. Long download times, while sometimes having economic repercussions, is mainly an annoyance to users, so guidelines about this are usually expressions of the norm of feeling. Page layout has a sensory aspect, but is an example of a multi-aspectual guideline that speaks of how one aspect of the web site facilitates another; see below.

The main norm of the analytical aspect is that we should be able to make clear distinctions that are meaningful. Thus the W3C guideline, "Ensure that foreground content is easily differentiable from background for both auditory and visual presentations" expresses this norm, as do the NCI guidelines, "... Headings provide strong cues ..." and "Always use underlines or some other visual indicator ... to indicate that words are links." Guidelines about font size, graphics, screen resolution, etc. where the thrust is about helping the user distinguish things express this norm. Guidelines about helping the user to identify structures in meaning anticipate the next aspect.

25

The formative aspect is about formative power. Part of this focuses on structure. To understand the site, users of a web site form their own structured mental model of its content, and this is facilitated if the syntactic structure of the site supports its content. The NCI guideline "Put as much important content as close to the top of the hierarchy as possible" is an expression of this. So are those about splitting into paragraphs, sections and subsections (for which the headings are analytical-aspect cues), about how pages are linked, etc. Part of this aspect is to do with achievement. A major area of achievement in web site use is finding relevant knowledge. Therefore guidelines about helping the user find the knowledge they want are expressions of the norms of this aspect.

Dooyeweerd maintained that though the aspects are distinct, they 'resist' being separated from one another, because they were meant (by their Creator) to form a coherent spectrum of Meaning. So it is often difficult to clearly identify under which aspect a guideline belongs because it seems to have several aspects. Such guidelines concern how earlier aspects provide the means by which later aspects may be implemented, and how later aspects provide the purpose for implementing it in the earlier aspects. As mentioned above, page layout has several aspects. Sometimes it is aesthetically important, but most guidelines about it focus on how layout (itself of the sensory aspect) can help the user distinguish what is important to them (analytical aspect) and thus reflect the structure of the content (formative aspect). Recognising the various aspects in even a single phenomenon like this helps us in design of that phenomenon.

The lingual aspect concerns meaning conveyed by symbols, and is the primary aspect of web sites. As one guideline (NCI) says, "Content is the most critical element of a Web site." Some general guidelines exist, such as NCI's "Use only graphics that enhance content or that lead to a better understanding of the information being presented" and W3C's "Write as clearly and simply as is appropriate / possible for the purpose of the content." But the norm of conveying meaning goes further. Content should be accurate, consistent, interesting, relevant, polite, coherent, etc.; some of these are Grice's (1967) maxims. Such quality issues are seldom expressed as guidelines, but perhaps they should be.

The social aspect concerns social interaction and institutions. While web sites might help towards these, the aspect's importance in web site design lies in it being strongly anticipated by the lingual aspect. Cultural assumptions, expectations, etc. of the reader determine how the reader interprets content, and are often of a tacit nature, so misunderstandings or offence can occur. The norm for site designers is "Consider the person who will read this; do not be satisfied with just conveying information." This aspect urges us to be careful and creative in use of humour (thus anticipating the aesthetic aspect).

NCI offers guidelines about site goal, such as "Provide useful and usable content that supports the Web site goal on each page." But does this refer to a communicative goal of what information is conveyed or the goal of the user in accessing the site? Though usually not made clear, it is a crucial difference. A communicative goal is covered by the aspects above, but the user's goal is covered by the contingent aspects. Guidelines need to be made for each contingent aspect. When the designer knows the purpose(s) for which the site will be used, then the contingent aspects can be identified, and guidelines devised that will express their norms. For example, for a site that will be used to give legal advice to campaigners, juridically-oriented guidelines might be useful, such as "Because it will be used in courts of law, be particularly careful about the accuracy and completeness of the content." But for many sites the uses may be diverse and not clearly known in advance. So contingent-aspect guidelines will be more general, such as, for the ethical aspect, "As far as possible, steer the user towards ethical rather than anti-ethical use of the information."

## 5    Discussion

We have shown how Dooyeweerd's (1955) theory of aspects can help identify categories of guidelines for web site design, and also how, by considering each aspect in turn, we can be stimulated into devising specific guidelines for aspects we might have overlooked. Our two-part proposal is that guidelines are expressions of the norms of relevant aspects, and that we can identify these as: the qualifying aspect, some of its preceding ones, its immediate successor, and the set of contingent aspects considered as one.

By doing this, we avoid the problems identified earlier. The clear principle behind the categories is to include relevant aspects that form the framework of Meaning for human activity, and the order among the categories is that of the aspects. Minimizing overlap between categories is ensured by the irreducible distinctness of the aspects. Some guidelines are multi-aspectual, so may be included in several aspects, but there is now a rationale for such duplication. Categories based on aspects are all of wide scope, none being applicable only in very specialised situations, but the aspects offer us a framework for considering specialised situations. And, because Dooyeweerd claimed that the kernel meanings of the aspects may be grasped intuitively, the categories take on their intuitive meaning. Finally, by reference to the aspects, we have a scheme that can 'tap us on the shoulder' when we have overlooked something important.

A similar approach might be used to provide guidelines for other types of software: devise guidelines that express norms of its qualifying aspect, and those nearby. For example, computer games are qualified by the aesthetic aspect, of surprise, play and fun, which is preceded by the economic aspect of limited resources (as seen in the importance to many games of limiting players' resources of ammunition, stamina, time, etc.). The approach outlined here shows considerable promise and is worthy of further research.

## References

Dooyeweerd, H. (1955). A New Critique of Theoretical Thought. Ontario: Paideia Press.

Grice, H.P. (1967). Logic and conversation. In Cole, P., Morgan, J. (Eds) *Syntax and Semantics* Vol 3. New York: Academic Press.

Hart H, (1984), Understanding Our World: An Integral Ontology, University Press of America.

NCI, (2003). Evidence-Based Guidelines on Web Design and Usability Issues. National Cancer Institute, Retrieved January 31, 2003 from http://usability.gov/guidelines/.

Polanyi, M. (1967). The Tacit Dimension. London U.K.: Routledge and Kegan Paul.

W3C. (2003). Web Content Accessibility Guidelines 2.0. Retrieved January 31, 2003 from http://www.w3.org/TR/WCAG20/.

# Guidelines and Freedom in Proximal User Interfaces

*Andrew Basden*

Information Systems Institute, University of Salford,
Salford, M5 4WT, U.K.
A.Basden@salford.ac.uk

## Abstract

There is a tension between guidelines and freedom. It arises from certain philosophical assumptions that we do well to question. An alternative view of freedom as 'ability to respond appropriately to the diversity of the situation' is considered. It is shown how this can work out within the guideline principles of Proximal User Interface for three classes of software.

## 1    Introduction: Guidelines and Freedom

Guidelines for the design of user interfaces (UIs) - such as style guidelines - can provide a coherent look and feel to software across a platform, so that novices quickly become familiar with new software. But they can be restrictive, constraining the design in unnecessary and inappropriate ways. This is especially so when the user progresses from being a novice to an everyday user. Somehow, we need to relate freedom to guidelines in such a way that both have appropriate power and both are meaningful.

There are two main ways to approach the issue of freedom: 'freedom from' and 'freedom to'. The 'freedom from' view is epitomised by the British thinker Hobbes' stress on personal freedom from interference and the German thinker Habermas' norm of emancipation from unwarranted constraints, as adopted by the critical system community (Lyytinen & Klein, 1985). In this view, guidelines can be seen as interference and constraint, and thus inherently opposed to freedom. But it often results in antinomy; as seen in Hobbes' monarchial Leviathan and in the paradox of enforced emancipation (cf. Wilson, 1997).

We adopt the alternative notion, of 'freedom to'. This is found, for example in Milton's idea of freedom as not being beholden (to lords) and thus being able to be what we should be. More recently the notion has been worked out more precisely by the Dutch thinker, Dooyeweerd (1955), who developed an idea of freedom as the ability to respond meaningfully to the diversity around us. In this view, guidelines can enable and encourage such freedom rather than constrain - but only high quality guidelines do so.

What are high quality guidelines? To Dooyeweerd, the ability to respond is made possible by the 'law side' of reality, which is a framework of Meaning in which everything exists and occurs. He accounts for diversity by suggesting that this framework of Meaning comprises a number of irreducibly distinct aspects, each with a different kernel meaning. Each aspect has a distinct set of laws that enable a distinct type of meaningful functioning to which we respond in life (e.g. lingual laws of syntax, semantics, etc. enable meaningful communication). The laws of earlier aspects (quantitative, spatial, kinematic and physical) are determinative, while in the later aspects (biotic, sensory, analytic, formative, lingual, social, economic, aesthetic, juridical, ethical and credal) they are normative, allowing freedom. Human activity - including engaging with software via a user interface - involves all aspects.

Guidelines for designing UIs can now be seen as an expression of this diversity of aspectual Meaning. To be of high quality and facilitate freedom, they should be rooted in the aspects and take all into account. This need not mean devising completely new guidelines or methodologies. Basden & Wood-Harper (2002) have shown how Dooyeweerd's philosophy can underpin and enrich soft systems methodology. Here we examine how it might underpin and enrich principles of proximal user interface, discussed by Basden, Brown, Tetlow & Hibberd (1996).

## 2      Principle of Proximal User Interfaces

"The real problem with the interface," said Norman (1990), "is that it is an interface. Interfaces get in the way. I don't want to focus my energies on an interface. I want to focus on the job." A proximal user interface (PUI) is one that is so 'natural' that it does not "get in the way". It embodies the norm that the software should be able to become, as it were, part of the user or, as Polanyi (1967) stated, 'proximal'. By contrast, conventional UIs are 'distal', relating to the user via a 'dialog' of commands, messages, clicks, menus, etc. To achieve proximality, the UI should not consume the user's thinking and attention nor interrupt the flow of thinking within the task (Basden & Hibberd, 1996).

The principles of PUI recognise that the real task for which the tool is being employed might be more fluid, flexible, dynamic and sophisticated than its formal specification (given in design of system, or in training of the user and work definition) might suggest. Of the nine principles, two concern the overall relationship between user, task and tool, and bring together the notions of direct manipulation and affordance, three concern what the user experiences via the UI, and the remaining four with how the user can act. We focus on visual interfaces, but the principles apply equally to aural and haptic devices. We illustrate how the principles apply to three different types of software: word processors (WPs) (where Dooyeweerd's lingual aspect is important), computer games of the Doom type, where the user moves about, finding objects, being ambushed and fighting enemies (spatial and aesthetic aspects), and mind-mapping toolkits such as Istar (Basden & Brown, 1996) that allow the user to build a knowledge base as a box and arrows diagram (the analytical aspect).

### 2.1     Principle of Direct Semantics

The principle of direct semantics states that there should be a direct mapping from lexical phenomena in the interface to the semantics of the user's task. For example, in computer games the enemy is a recognisable shape on screen and sound from speakers, and pressing the joystick button is directly linked to fighting it. In Istar, each item is a box and each relationship, an arrow, and pressing the mouse button draws a new box (creates a new item). This principle is not unlike later versions of direct manipulation (Shneiderman, 1992), but it concerns the user's view (e.g. shapes seen, sounds heard) as well as user action (e.g. mouse gestures). In conventional, distal, UIs, the mapping is indirect, via syntax of a 'dialog' between user and machine. Interpreting and managing syntactic paraphernalia like toolbars, menus, dialog panels, commands, etc., incurs a heavy penalty in cognitive load. Though excellent for novice users, they "get in the way" during everyday use, so they are to be avoided.

29

## 2.2   Principle of Appropriateness

The principle of appropriateness states that mapping between lexical phenomena and semantics of the application should be appropriate, or 'natural'. This implies three things. What is 'simple' in the application domain should be expressed by simple lexical gestures and shapes. Each lexical phenomenon should 'afford' the semantic meaning it expresses (Greeno, 1994). And the diversity encountered in the situation of use should be respected. If diversity is vested in distinct aspects (Dooyeweerd, 1955) the UI must reflect this by a  different lexical-semantic mapping for each. For example, a line expresses a relationship in Istar, underlining in WPs and, say, the amount of damage sustained in the game. Dragging the mouse draws a new relationship in Istar or selects a region of text in WPs.

## 2.3   Principle of Large Easels

The 'easel' is what the user sees (or hears) at the user interface. In WPs, it is textual, not graphical, in form. This art-related term 'easel' is chosen to reflect two notions: holistic access and active engagement.

Just as the artist is aware of the whole even when focusing on a tiny part, so should the user be. The UI should encourage, not hinder, such holistic awareness, and let the user access any part with minimum cognitive effort. This implies a large easel within immediate reach. In Istar, this is achieved by a large, smooth-scrolling screen. In Doom-like games the main screen shows a limited view, but this easel is 'enlarged' to a more holistic awareness by provision of a map and by sounds. In WPs holistic access is via markers in the text. But windowing systems can tend to parcel up the detail separate from the whole, which reduces proximality.

Just as the artist not only views and examines the easel, but is continually making modifications, so the easel of PUI is not just a representation but a stimulation of thinking, and not just a stimulation but an invitation to make changes at the very instant of the stimulated thought. The user is 'situated' in the easel. So, in the design of the PUI, there must be no barrier between viewing and action. The page of a WP is a good example of this: as you read you type, without having to think about it.

## 2.4   Principle of Visual Cues

Visual cues (or auditory or tactile) are those indicators of relevant meaning that occur on the easel. Formal cues (e.g. colour codes) are mediated by the software. But informal cues are often more important - such as the pattern of lines around a box in Istar, the size of a paragraph and the shape of its ragged right hand edge in text. As the user draws, types or plays, such cues are left both on the easel and in the user's memory, forming a direct link lexical phenomena with the semantics in the user's mind. Visual cues should be encouraged and respected but desire for tidiness often works against them. Automatic layout and tidying facilities, common in visual programming software, destroy them. Icons all the same size in neat array, as on Windows, are less easily located than those with gross visual differences (Byrne, 1993). However, in computer  games, disguise is important so cues should be such as to mislead.

## 2.5   The Clutter Principle

Clutter is lexical phenomena that distract from, rather than support, the meaning in the application. The clutter principle recognises that clutter will always occur when the user is engaged with ill-structured knowledge of non-trivial size, and thus demands proximal mechanisms to prevent and manage it. In box and arrows diagrams, congestion is one type of clutter and in Istar it is first minimized by a visually simple genre and then managed by making it very easy for the user to move things around and make space (see Basden, et. al. 1996). In WPs, clutter is of other sorts, such as unformatted paragraphs, spurious characters and bits of <html> that litter our emails, and semantic clutter, where the content of what is written is not well structured. These are managed by automatic formatting and spelling, grammar and style checkers. In computer games, clutter might be employed to mislead.

## 2.6   Principle of the Wide Input Channel

When we consider user action, we must choose lexical gestures for each semantic operation, for example, to create a new item (Istar), move along a corridor (game), start a new paragraph (WPs). In each type of application, many different operations will be needed because of the diversity of meaning in real life. But a two-button mouse allows only three operations - a very narrow lexical input channel. This principle says that we must give careful consideration to how we widen the lexical input channel. In street-fighter games complex gestures are used. In WPs, a multi-key keyboard is used. In Istar the free (non-mouse) hand presses keys to qualify the operation and different parts of the visual object give different operation.

## 2.7   Principle of Graded Effort

But remembering complex gestures, qualifying keys or the significance of different parts can result in extra cognitive effort or interrupt flow of thought when compared with the simple mouse operations of move, click or drag. The principle of graded effort recognises this, and says that we must carefully design each operation so that those operations that 'deserve' less effort will have it. Usually the most frequent operations should incur least effort, and thus mapped to the simplest click or drag. (Note that, since in each type of application different types of operation will be more common, this principle goes against established norms of implementing a standard 'look and feel' across all types of application.)

## 2.8   The Danger Principle

The danger principle modifies the previous principle by saying that operations that are more dangerous 'deserve' greater cognitive effort, and so should be designed to be more distal. But danger is of different forms. In Istar deletion is dangerous, so is made distal with an 'Are you sure?' dialog box (see Basden & Brown, 1996). But a game that kept asking 'Are you sure?' before an enemy was killed would fall into disuse! In games, danger has a different meaning.

## 2.9   The Principle of Tentative Action

Tentative action is user action that might not have a significant semantic result but is an important part of the user's thinking processes. Examples include 'doodling' while the user thinks something through, or the setting down of fleeting thoughts before they are forgotten. The principle of tentative action says that good, proximal support should be provided for such actions in the software. In Istar, fleeting thoughts are recorded simply by creating a new box anywhere near. In WPs, they can be written down. Doodling was observed in Istar (see Basden, et. al., 1996): the user picked up a box with the mouse and wiggled it about. Doodling is not usually supported in WPs, but maybe should be. Both types are seldom necessary in games.

# 3    Guidelines for Freedom in PUI

We can see that the principles of proximal user interface may be applied to very different types of software, taking different forms in each. We have noted how some of them run counter to attempts to standardize look and feel, as they support a diversity of lexical-semantic mappings. But, as discussed in Basden (2000), this approach might, in the end, lead to a richer, more natural type of standard. The main thread running through this has been seeing freedom as the ability to respond appropriately to diversity, a notion derived from the philosophy of Dooyeweerd (1955). Reviewing the discussion above, we can see that much of the flexibility with which the principles are applied is influenced by Dooyeweerd's notion of aspects. Most pieces of software are aimed at aiding human tasks in which certain aspects predominate. It is these aspects that should guide the design of the UI, if we wish it to be proximal. We suggest that high quality guidelines, that will enable rather than hinder freedom, may be constructed by a knowledge of the relevant aspects: the analytical for mind-mappers, the lingual for word processors, and the spatial for games.

# References

Basden, A., (2000). Guidelines for a 'Proximal' User Interface. In J. Vanderdonckt, C. Faraenc (Eds.) *Tools for Working with Guidelines* (p.339-56). Springer-Verlag.

Basden, A., Brown, A.J., (1996). Istar - a tool for creative design of knowledge bases. *Expert Systems*, 13, (4), 259-276.

Basden, A., Brown, A.J., Tetlow, S.D.A., Hibberd, P.R., (1996). Design of a user interface for a knowledge refinement tool. *Int. J. Human Computer Studies*, 45, 157-83.

Basden, A., Hibberd, P.R., (1996). User interface issues raised by knowledge refinement. *Int. J. Human Computer Studies*, 45, 135-55.

Basden, A., Wood-Harper A.T., (2002). A philosophical enrichment of CATWOE. *ANZSYS'2002*, Australia and New Zealand Systems Conference, 10-12 Dec 2002, University of Sunshine Coast.

Byrne, M.D., (1993). Using icons to find documents: simplicity is critical. *Proc. InterCHI '93*, April 1993, 446-453, ACM, USA.

Dooyeweerd, H., (1955). A New Critique of Theoretical Thought, Vol. I-IV., Jordan Station, Ontario: Paideia Press.

Greeno, J., (1994). Gibson's Affordances. *Psychological Review*, 101, 336-42.

Lyytinen, K.J., Klein, H.J., (1985). The critical theory of Jurgen Habermas as a basis for a theory of information systems. In E. Mumford, R.A. Hirschheim, G. Fitzgerald, A.T. Wood-Harper (Ed.), *Research methods in information systems* (p.219-31), North Holland.

Polanyi, M., (1967). The Tacit Dimension. London: Routledge and Kegan Paul.

Shneiderman, B., (1992). Designing the User Interface: Strategies for Effective Human-Computer Interaction. Addison Wesley.

Wilson, F.A., (1997). The truth is out there: the search for emancipatory principles in information systems design. *Information Technology and People*, 10, (3), 187-204.

# It's all in a days work of a software engineer

*Inger Boivie[1], Jan Gulliksen[1] and Bengt Göransson[1,2]*

[1]Dept. for IT/HCI, Uppsala University, PO Box 337, SE-751 05 Uppsala, Sweden
[2]Enea Redina AB, Smedsgränd 9, SE-753 20, Uppsala, Sweden
Jan.Gulliksen@hci.uu.se

## Abstract

This paper reports on two studies that have been performed in order to explore if and how software engineers make design decisions that affect the usability of the software. Our conclusion is that the main goal for software engineers is to produce program code. They consider usability being the responsibility of others, for instance, the users or usability experts. They do not feel that their code contributes to the overall usability of the software. Nevertheless, software developers make numerous decisions that affect usability as part of their work with analysis, design and programming. The conclusion one must draw from this is that there is a need for a strategic change of attitude towards usability among software engineers.

## 1    Introduction

Despite the attention usability[1] has received in the last number of years, poor usability is still a major problem for users. According to Clegg, et al. (Clegg, Axtell, Damodaran, Farbey, Hull, Lloyd-Jones, Nicholls, Sell and Tomlinson, 1997) usability issues are on the agenda in most IT development projects. The respondents in their study estimate that as many as 60-70 % of the projects addressed usability matters "successfully". They go on to say, however, that there is criticism that the area is still not sufficiently understood, and that the view of usability is rather mechanistic. This indicates that we need to know more about how usability is addressed in software development projects. Not least, we need to know how software engineers think about and work with usability.

We have conducted a couple of studies of how software engineers work, with particular focus on if and how they address usability issues. If we can understand more about how the software engineers work, we may be able to identify new tools for addressing usability in software development or improve existing ones.

## 2    Method

In order to understand what type of support software engineers need to develop usable systems we must study and analyse their work, i.e. their work practices, and how they think about their work. It is difficult to analyse work practices since most workers, including software engineers, are not fully aware of what they do. Their work practices are partly based on tacit knowledge and cannot

---

[1] We use the ISO 9241-11 (1998) definition where usability is defined as follows:
"The extent to which a product can be used by specified users to achieve specified goals with effectiveness, efficiency and satisfaction in a specified context of use."

be described in all detail. Work studies must therefore be based on observations of the workers, combined with interviews.

The first study consisted of a series of *observation interviews* (Åborg, Sandblad, Gulliksen & Lif, 2003) with 7 software developers (4 men and 3 women). The software developers worked in an in-house development organisation with designing and developing user interfaces. The respondents were software engineers, programmers, project managers and process owners at the IT department of a large government organisation. Most of them worked in Stockholm but a few worked in sub-departments in other towns around Sweden.

This interview study was part of a project developing an on-line interactive corporate style guide (Olsson and Gulliksen, 1999). The main purpose of the study was to capture user requirements and the software developers were interviewed as potential users of the style guide. The project worked in accordance with the human-centred approach described in ISO 13407 (ISO, 1999). ISO 13407 prescribes activities for acquiring a good understanding of the users and their context of use, in this case the software developers and the software development process.

The second study consisted of a series of open-ended interviews with eight people (six women and two men) involved in software development projects in two in-house development organisations (one of which was the same as in the first study). The respondents had backgrounds in and worked with usability, user interface (UI) design and/or software development. The purpose of this study was to investigate how design decisions, affecting usability, are made. Who makes them, when and on what grounds?

# 3    Results

The two studies describe the software development process on two different levels. In the first study, we identified how software engineers work on a day-to-day basis, solving particular problems. Somebody, for instance, the project manager, typically hands down the design problem to them. Each design problem can usually be solved within two days. The result of the task is a piece of software that is delivered for integration with the rest of the system.

The observation interviews indicate that the developers could not fully understand the result of their programming task. The quality of their work was determined by the robustness and stability of the program code, rather than usability. The developers did not feel that they designed user interaction. The design simply emerged as if by magic, as a result of the integration of the various pieces of code produced by the developers. Nobody admitted to having designed a particular part of the user interaction. The below figure illustrates the workflow for the software developers.

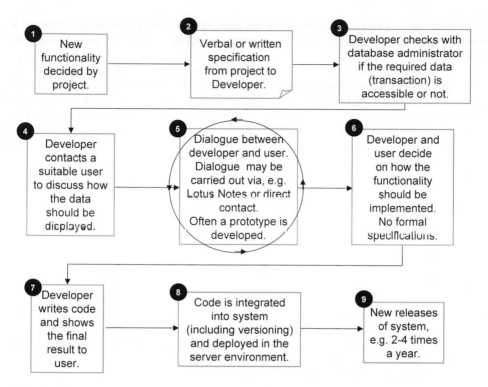

Figure 1: Model of how developers work. The model is a general model, created after studying several developers in one in-house development organisation.

In the second study, the respondents described the software development process. The typical project starts with some kind of requirement specification or description from the client. Requirements are captured by means of use case modelling and expressed mainly by use case models Users and/or domain experts[2] participate in the requirements capture process. The user interface modelling is done in parallel with the use case modelling, with the same group of users and domain experts. The use case models are then refined to analysis models and design models. The user interface developers/use case specifiers develop a solution, which is reviewed by the users and domain experts, and modified accordingly. The design solutions are often evaluated or reviewed by users a few times, before they are implemented. The number of iterations/reviews is determined by the date the implementation must start.

The focus of the interviews in the second study was design decisions. The respondents were asked what kind of decisions they are allowed to make, and whether design decisions are made by them or by other people.

When asked who makes the design decisions, none of the respondents said "I do". Instead, they pointed to the user representatives and the client representatives. According to several of the respondents the usability of the resulting system depends to a large extent on the user representatives. Quite a few of the design decisions are made in workshops (modeling sessions)

---

[2] Domain experts are representatives from the user organisations with expertise in the domain, e.g. laws and regulations affecting the new system. These people will not necessarily be active users of the system.

together with the user representatives. There are also matters over which the development teams do not have control, despite the fact that these matters are likely to affect the perceived usability of the system.

The use cases are the most important information carrier in the projects. They are the basis of the interaction design and thus limit the design "space". They specify, for instance, the workflow, what information should be displayed when, how the information should be grouped, etc. The respondents in both organisations reported difficulties with the use case modelling. One respondent said that the people in her project had misunderstood the concept, resulting in far too many, far too detailed system operations instead of real use cases. *"I still don't know what a use case is."*

# 4    Discussion

The two studies gave us a rich picture of the problems involved in addressing usability issues in the software development process

*The development process often fails to provide the necessary support for addressing usability*
The two descriptions of how software developers work differ on some point, mainly regarding the extent of formalisation. Figure 1 shows a rather informal process, where contacts with "a suitable" user are made by the individual software developer whenever he/she feels the need. In the process described in the second study, users participated on a more formal basis, in modelling sessions and by reviewing specifications, etc. The differences may, at least to some extent, ascribed to the different data collection methods used in the studies. The first study is based on observations whereas the other is based on interviews. When interviewed, the developers recount what they think they do at work, which is probably fairly close to the official version of how they should work, i.e. the software development model they are supposed to comply with. In an observation, you can see what they really do at work. These two views of work correspond to an "explicit" view and a "tacit" view as described by Kuhn (1996).

Software development work is often thought of as engineering-oriented, rational and described by means of workflow models consisting of well-defined steps. But our results indicate that certain parts of it are based on tacit knowledge, which is not easily captured and expressed in models. Development work also seems less formal, less ordered and less rational than we would expect from the "explicit" view of it.

The two pictures of software development do not have to be contradictory. Some of the respondents in the second study also contacted users on an informal basis even though they also described formal meetings and contacts. Thus, there is a need for quick, informal contacts with users in software development. This indicates that the models used to capture the requirements and other inputs are insufficient. They do not provide enough information for the software developers to get on with their work. One interesting question is, what information is missing.

*Decisions affecting usability just "happen" as part of the process instead of being "made"*
The software developers, and basically everybody else in the project, make usability decisions without being aware of it. These decisions encompass all levels from what is to be included in a certain release (i.e. what the user will be allowed to do in the system, in terms of tasks or functionality) to the use case modelling, to detail decisions about input/output controls.

The respondents claimed that the user representatives and domain experts in the projects were responsible for most of the interaction design decisions. Nevertheless, we would like to argue that the developers make a lot of the decisions. This phenomenon may be explained by the fact that in many cases design decisions just "happen" as part of a process, where the design slowly grows and develops. Design decisions are explicitly made in those cases where a design issue has been discussed, for instance, when not complying with the style guide or when there has been disagreements.

*Frustration with user involvement*

The respondents in both studies expressed some frustration with how users were involved, regardless of whether it was on an informal or formal basis. There was a great reluctance towards making contacts with the users among most of the developers. They did not consider it their responsibility to involve users. Nevertheless, they did contact users when they felt they had to, as described above. The developers also expressed frustration with the disproportionate influence some users get (for instance, the formally appointed user representatives).

In order to facilitate a focus on usability and the users' real needs in software development, we need to:

- Shift the attitudes towards a more user-centred design and development process (UCSD), and educate or train all software developers in UCSD.
- Make usability activities become Standard Operating Procedures within the organisation.
- Make management aware of their obligation to put usability on the agenda in development work
- Provide success stories that can serve as a examples for successful deployment of UCSD
- Make commercial systems development processes support and prescribe usability activities.
- Resolve the conflict between usability and deadlines.

# References

Clegg, C. Axtell, C. Damodaran, L. Farbey, B. Hull, R. Lloyd-Jones, R. Nicholls, J. Sell, R. Tomlinson, C. (1997). Information technology: a study of performance and the role of human and organizational factors. *Ergonomics*, Vol. 40. No 9. pp. 851-87

Kuhn, S. (1996). Design for people at work. In Winograd, T. *Bringing design to software.* ACM press. Addison Wesley. New York. pp. 273-289.

ISO 9241-11 (1998), Ergonomic Requirements for Office Work with Visual Display Terminals (VDTs). Part 11: Guidance on Usability. International Organization for Standardization, Geneva.

ISO 13407 (1999), Human Centered Design Processes for Interactive Systems. International Organization for Standardization, Geneva.

Olsson, E. & Gulliksen, J. (1999) A corporate style guide that includes domain knowledge. In G. Salvendy, M.J. Smith, & M. Oshima (eds.) International Journal of Human-Computer Interaction. Vol. 11, No. 4, Ablex Publishing Corporation, Norwood, New Jersey.

Åborg, C., Sandblad, B., Gulliksen, J. & Lif, M. (in press) Integrating work environment considerations into usability evaluation methods – the ADA approach. Accepted for publication in Interacting with Computers.

# Semiotic Conference: Work Signs and Participatory Design

*Rodrigo Bonacin*

Institute of Computing, State University
of Campinas, Caixa Postal 6176
13083 970 Campinas, SP - Brazil
+55 19 32958178
ra000470@ic.unicamp.br

*M. Cecilia C. Baranauskas*

Institute of Computing, State University
of Campinas, Caixa Postal 6176
13083 970 Campinas, SP - Brazil
+55 19 37885870
cecilia@ic.unicamp.br

## Abstract

In this paper we propose a participatory design technique named Semiotic Conference as a way to explore process-oriented issues related to the social context of the workplace and its connections with aspects of the product being designed. This technique was proposed to enable a design process in which workers (final users) and designers construct a new organizational context embedding the system in it. The software is a product of their social interaction and a new social context appears as a consequence of the software design and use. Results of applying the proposed technique to a real context are discussed.

## 1  Introduction

Literature in Software Engineering and Information Systems (Floyd, 1987; Muller et al 1998; Mahemoff & Johnston, 1998) has shown two basic perspectives to software design: the product-oriented, which focuses on the software artifact, and the process-oriented one, which focuses on the human work processes that the computer artifact is intended to support. In agreement with Floyd (1987), in this paper we argue that these approaches could be complementary to each other when considered in the proper balance. According to Floyd, the process-oriented perspective views the software in connection with human learning, work and communication, taking place in an evolving world of changing needs.

The close contact between designers and users, and the direct participation of the users during the system design are pointed out as way to promote the process-oriented design. In this paper we propose a Participatory Design (PD) technique named *Semiotic Conference* as a way to explore process issues from the social context of the human workplace and their connections with aspects of the product being designed. The *Semiotic Conference* was proposed in order to promote the participation and signification aspects (Baranauskas et al, 2002) during the design process.

This new PD technique enables a design process in which workers (final users) and designers construct a new organizational context embedding the system in it. The software is a product of their social interaction and a new social context appears as a consequence of the software design and use. This technique was motivated by the concept of "mutual learning" in design defined by Kyng (1991), where the application domain practitioners learn about the new possibilities brought about by the technology while designers learn about the application domain. The mutual learning is promoted through the participation of the users during the design using a semiotic model that addresses at the same time the application being designed and the intended social context that the application is supposed to support.

The paper is organized in the following way: Section 2 presents the theoretical background based on the Participatory Design approach and methods from Organizational Semiotics, Section 3 describe the proposed technique, Section 4 shows results of using the technique in the design of a CSCW system interface, and discuss the drawbacks and direct benefits of using it, and Section 5 concludes.

## 2 Participatory Design and Organizational Semiotics

The *Semiotic Conference* is based on two main theoretical backgrounds: the Participatory Design, which promotes the participation of the user during the design and the Organizational Semiotics that enables to model the signs system at the social level. The Participatory Design approach was developed initially in the Scandinavia and it employs a variety of techniques to carry out design with the user, rather than for the user (Muller et al 1997). The Participatory Design approach stresses the importance of democracy in the workplace to improve the work methods, the efficiency in the design processes (with the users backgrounds and feedback), to improve the systems quality, and "to carry on" formative activities.

In order to develop a common view of both the technology and the organization, that is necessary for users and designers to explore new organizational structures, we combined the use of PD with Organizational Semiotics (OS) methods in a new PD technique. OS is defined as "the study of organization using the concepts and methods of Semiotics" ("OSW," 1995). The study of the organization is based on the observation that people affect all organized behaviors through the communication of signs, individually and in groups.

A set of methods proposed by Stamper (1973) enables to study the use of signs in organizations and their social effects. The two main methods of the OS used in the *Semiotic Conference* are the Semantic Analysis and the Norm Analysis. Semantic Analysis (SA) focuses on the agents and their pattern of behaviors to describe the organization. Some basic concepts of SA adopted in this paper based in Liu (2000) are: (a)"The world" is socially constructed by the actions of agents, on the basis of what is offered by the physical world itself, (b)"Affordance", the concept introduced by Gibson (1979) can be used to express the behavior of an organism made available by some combined structure of the organism and its environment, (c)"Agent" can be defined as something that has responsible behaviour, an agent can be an individual person, a cultural group, a language community, a society, etc., and (d)"An ontological dependency" is formed when an affordance is possible only if certain other affordances are available.

A norm analysis is usually carried out on the basis of the result of the semantic analysis (Liu 2000). At the pragmatic level the norm analysis describe the relationships between an intentional use of signs and the resulting behaviour of responsible agents in a social context. At the social level it describes beliefs, expectations, commitments, contract, law, culture, as well as business.

## 3 The *Semiotic Conference*

In this paper we describe the proposed technique in terms of the attributes proposed by Muller (1998) for Participatory Design techniques: the object model, the process model, the participation model, expected results and position of the technique in the whole product life cycle.

***Object Model:*** Copies of Organizational Semiotic Models: ontology charts (semantic analysis) and norm analysis (pragmatic and social level analysis), prototype screen shots, and pens are the material used for running the participatory practice. The ontology charts are the graphic representation of relationships among concepts of the semantic analysis. The ontology charts represent agents (circles), affordances (rectangles), ontological dependencies (lines from left to right), role-name (parentheses) and whole-part (dot). The norm analysis are represented by Deontic logic expressions with the following basic form: <Norm>::= If <Condition> then <D> <Agent> <Action> Where <D> is a deontic operator that specifies that the action is obligatory, permitted or prohibited. The norms are not necessarily obeyed by all agents in all circumstances; they are social convention (laws, roles and informal conventions) that should be obeyed. A more detailed description of ontology charts and norm modelling can be found in Liu (2000).

***Process Model:*** In the format of a Conference the designer shows concepts of the workplace modeled in ontology charts that include the affordances made available or/and that would made available by the system. This model contains the concepts compiled by the designers during the initial definition (the diagram must contain only the terms used by the users) of the system obtained from application of other PD techniques or from the last model of the former prototype. Copies of the models are distributed to all practitioners and if necessary the designer clarifies the notation and concepts described by the diagram. For each concept quoted in the model that any person of the group judge important, the practitioners discuss the semantic dependencies with other concepts and the social norms associated. Members of the group, as a result of discussion, propose changes in the semiotic models. Alternative solutions are raised on prototype screen shots to reflect the conceptual changes in the model. The changes in the prototype can be directly drawn in the prototype screen shots or, if necessary, other PD techniques can be used to define them.

***Participation Model:*** Members from different levels and functions in the organization together with the design team participate in the Conference. Facilitators mediate the interaction among the group.

***Results:*** Semantic and Norm models constructed during the conference with people from the organizational context are produced. The establishment of social norms that will be applied at the workplace considered with the prospective application is also produced as a result of the technique. The user interface is also produced through the "mutual learning" promoted by the technique.

***Position in the product life cycle:*** The technique applies iteratively to several phases of the design life cycle: Problem Identification & Clarification, Requirements & Analysis, High-Level Design, Evaluation and Re-Design (after the creation of the first prototype).

# 4    Results of Using the Technique

The proposed technique was explored during the design of Pokayoke: a system prototype constructed in the context of a partnership between our Institution and an automotive industry, for supporting corrective and preventive actions related to problem solving in the factory (Bonacin & Baranauskas, 2003). This technique was applied to five iterations of the prototype cycles during the Pokayoke design: at the starting of each iteration, in order to review the concepts of the former prototype and to redirect the design of the next prototype, and at other moments when necessary for reviewing concepts. Figure 1 shows an example of using the *Semiotic Conference* during the Pokayoke design. It refers to concepts in one of the steps of problem solving process (Step IV). Figure 1a (left) presents part of the ontology chart for the second iteration of the prototype. It shows the drawing made by a practitioner suggesting changes in the semantic model (He proposed to include Brainstorming in this step). Figure 1b (right) shows the ontology chart of the third prototype including the changes proposed by the practitioner.

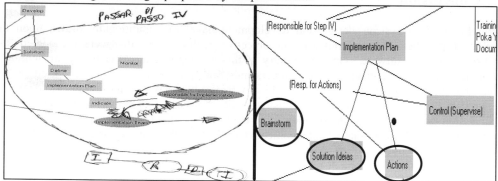

**Figure 1(a and b):** Changes in Ontology Charts

Figure 2 shows changes in the user interface caused by the alterations in the ontology charts presented in Figure 1. The affordances marked with circles in Figure 1 are associated to the interface constructions in Figure 2. The ontological dependencies of the objects are represented by the conditions of enabling or disabling functions in the software system; for example, it is necessary to exist a Multifunctional Team to enable the Brainstorming functionality. A norm interpretation package was developed to interpret formal norms (rules of the organization) that specify the responsibility, duties and permissions in the problem solving method. These norms specify the permission to access some functions of Pokayoke, for example: any user is allowed to include an idea in a brainstorming session but s/he is not allowed to specify actions to correct the problem. These norms also specify who could be responsible for which task at different phases of the problem solving processes. Changes in the norms specifications proposed during the *Semiotic Conference* are reflected in attributes configured by the norm interpretation module.

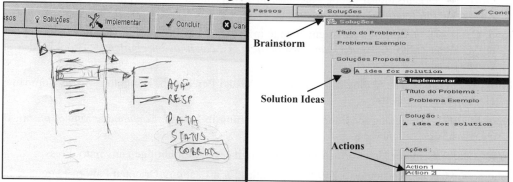

**Figure 2(a and b):** Changes in the system interface

The fifth prototype of Pokayoke has substituted the paper-based form of the "five steps" process. A group of seven workers of different levels and areas at the organization have participated in the *Semiotic Conferences*. These workers were in charge of training the other users of the system showing evidence of the mutual learning occurred. The mutual learning allied with the feeling of authorship was fundamental to the system acceptance. The workers have expressed this feeling many times during the system design and use, such as: *"... we have defined this in this way to avoid..."*. The main drawback of the proposed technique noticed during the Pokayoke design was the necessity of explaining the semantic and norm analysis concepts at the first applications.

# 5  Conclusion

Methodologies for systems design and development have been traditionally drawn upon the objectivist paradigm, which considers an objective reality to be discovered, modeled and represented in the software. On the contrary to this assumption, the Organizational Semiotics adopt the subjectivist paradigm, which understands reality as a social construct based on the behavior of agents participating on it. The subjectivist paradigm has shown usefulness to deal with the process-oriented issues during the Pokayoke design. People in different positions at the organization have different opinions about how the software should support the work practices. Therefore the approach considered not an objective truth about the best way of supporting problem solving, but this truth was a social construct based on negotiations occurred during the participatory practice. The formalism of the semiotic models and the use of the prototype screen-shots in the Semiotic Conference have been useful to explore the connections between the Process-Oriented and the Product-Oriented issues in the software design.

## Acknowledge
This work was partially supported by grants from Brazilian Research Council (Capes 2214/02-4, Cnpq 301656/84-3, Fapesp 2000/05460-0). The authors also thank Delphi Automotive Systems in Jaguariúna, Brazil, and Nied-Unicamp for collaboration and partnership in the Pokayoke project.

## References
Baranauskas, M. C. C., Bonacin, R. & Liu K., 2002. Participation and Signification: Towards Cooperative System Design. in *Proceedings of V Symposium on Human Factors in Computer Systems*, 3-14.

Bonacin, R. & Baranauskas, M. C. C. (2003) Designing Towards Supporting and Improving Co-Operative Organisational Work Practices, in *Proceedings 5th International Conference on Enterprise Information Systems*, to appear.

Floyd, C. (1987). Outline of a paradigm change in software engineering. In G. Bjerkners, P. Ehn, & M. Kyng (Eds), *Computers and democracy: A Scandinavian challenge*. Brookfield VT: Gower.

Gibson,J. J., (1968). The Ecological Approach to Visual Perception. Houghton Miffin Company, Boston, Massachusetts.

Kyng, M.(1991). Designing for Cooperatin: Cooperating in Design. *Communications of the ACM* 34, 12, 65-73.

Liu K., (2000). Semiotics in information systems engineering, Cambridge University Press.

Mahemoff, M. J. & Johnston, L. J. (1998). Principles for a Usability-Oriented Pattern Language In Calder, P. & Thomas, B. (Eds.), *OZCHI '98 Proceedings*, (pp.132-139). Los Alamitos, CA.

Muller, M. J., Haslwanter, J. H., & Dayton, T. (1997) Participatory Pratices in the Software Lifecycle. In M. Helander, T. K. Landauer, P. Prabhu (eds.), *Handbook of Human-Computer Interaction*, (pp. 255-297), Elsevier Science, 2 ed.

Muller, J. M., Matheson, L., Page, C. & Gallup, R. (1998) Participatory Heuristic Evaluation. *Interactions*. september + october 1998. V. 5, Issue 5, ACM Press, 13-18.

OSW, (1995). *The circulation document in the Organizational Semiotic Workshop*, Enschede.

Stamper, R. K., (1973). Information in Business and Administrative Systems, John Wiley & Sons.

# Creative Design of Interactive Products and Use of Usability Guidelines – a Contradiction?

*Michael Burmester*

University of Applied Research –
Hochschule der Medien
Wolframstr. 32, D-70191 Stuttgart
burmester@hdm-stuttgart.de

*Joachim Machate*

User Interface Design GmbH
Teinacher Str. 38,
D-71634 Ludwigsburg
joachim.machate@uidesign.de

## Abstract

Often the use of guidelines for the design of interactive products is criticized for several reasons. A major point of criticism is that guidelines are not flexible enough, so that the design can be adapted to the context of use. Furthermore, there is an anxiety that guidelines hinder creative design solutions and innovations. In the following paper, the strategic use and definition of guidelines associated with a user centred design process (UCDP) is proposed.

## 1    Guidelines

Guidelines have been used since the beginning of human-computer interaction (HCI). The main goal of guidelines is to increase the usability of interactive products when designing them. Guidelines are a way to put theories, empirical findings or good practice in to a form which is intended to be useful for and adapted to the process of designing interactive products.

Two main categories of guidelines in HCI can be defined: design guidelines and process guidelines (Stewart & Travis, 2002). Design guidelines are providing support for design decisions and process guidelines to plan and structure design processes.

Design guidelines are differentiated by Stewart and Travis (2002) between guidelines, standards and style guides. According to Stewart and Travis, guidelines can be seen as recommendations of good practice relying on the credibility of their authors. Standards are "formal documents published by standard making bodies that are developed through some form of consensus and formal voting process" (Stewart & Travis, 2002, p. 992). Finally, style guides are "sets of recommendations from software providers or agreed within development organizations to increase consistency of design and to promote good practice within a design process of some kind" (Stewart & Travis, 2002, p. 992). This categorisation is based on the question, who has generated these guidelines.

Another possibility to categorize guidelines is along the degree of precision in guiding the design, and the degree of freedom in interpretation for taking design decisions. We have identified five categories: principles, rules, interaction patterns, standard interactions and user interface building blocks.

(1) Principles or heuristics (Nielsen, 1993) describe basic usability aspects which are important when designing user interfaces. In order to apply principles to design, the designer has to understand the rationale behind the principles. For all design decisions, the meaning of a principle for a specific design situation is a matter of interpretation. Examples for principles are the dialogue principles for visual display terminals described in the ISO 9241 part 10, the ten usability heuristics of Jakob Nielsen (1994).Ben Shneiderman's (1998, pp. 74-75) eight "golden rules" of dialogue design, or Alan Cooper's principles for polite software (Cooper, 1999, p. 162). All of these are principles in the sense of our definition.

(2) Rules focus on specific design decisions, e.g. if upper and lower case should be treated equivalent in search dialogues or not (Smith & Mosier, 1986, cit. Steward & Travis, 2002, p. 995). It is up to the designer to decide whether a particular rule is relevant and must be applied for the design decision under discussion. Interpretation is still necessary.

(3) Interaction patterns (Tidwell, 1999; van Welie, van der Veer & Eliëns, 2000) describe a complete context of an interaction and do not focus on a single design decision. Usually, interaction patterns are specified in a specific language. van Welie, van der Veer & Eliëns (2000) are describing for each pattern the name of the pattern, the usability principle behind it, the context in which the pattern should be applied, the force influencing the user, the design solutions within the pattern, examples where the pattern has been applied already, the usability impact, rationale behind the pattern and known uses (e.g. products using the described interaction pattern). Example interaction patterns are editor dialogues, browsers, warning message dialogues, wizard dialogues etc. (for more see Tidwell, 1999 or van Welie, 2002). The designer has still freedom in interpretation and in finding own solutions, but support is provided for a complete interaction task.

(4) Standard interactions are fully defined with regard to the way which steps need to be taken in order to accomplish a particular input task, e.g., single selection from a list There is no possibility left for interpretation of the dynamic input and output structure. These structures are formally described formalisms (e.g. in flow chart).Although the way of interaction is fixed, a user interface designer still has the freedom to define the visual presentation and style. An example for a description of standard interactions are the FACE guidelines for the design of electronic home devices (Burmester, 1997).

(5) User interface building blocks are representing interactions which are completely defined with regard to their interaction and sensual, e.g. visual and acoustic, presentation (Görner, Burmester & Kaja, 1997). No freedom of interpretation is left to the designer. Building blocks are sometimes also referred to as idioms (e.g. Sun Microsystems, 2001, pp. 65). Only selecting the right building block for a specific user task and setting the right options of the building is required. Building blocks are described in style guides (e.g. Microsoft Windows XP, 2002).

## 2    Problems with Guidelines

It is quite common that problems are reported in properly applying guidelines for a specific design situation(e.g. Görner, Burmester & Kaja, 1997). One reason for that is that there is an inflation of guidelines, which makes a designers' live quite hard: Smith & Mosier (1986, 944 HCI guidelines)[1], Brown (1988, 302 HCI guidelines)[1], Mayhew (1992, 288 HCI guidelines)[1], Burmester (1997, ~400 CE guidelines), ISO 9241 12 − 17 (1996-1999 ~500 HCI guidelines),

---

[1] cit. Nielsen (1993)

Nielsen Norman Gr. (2002, ~600 Web guidelines), etc. Furthermore, guidelines are difficult to understand, difficult to interpret, often too simplistic or too abstract, can be difficult to select, can be conflicting, often have authority issues concerning their validity (see Spool, 2002) or are made for a specific contexts (e.g. specific input output technology, tasks, user groups) which are not obvious (van Welie, van der Veer & Eliëns, 2000).

# 3 Guidelines as part of the user centred design process

## 3.1 Guidelines as input for user centred design

ISO 9241-11 formally defines usability as "the extent to which a product can be used by specified users to achieve specified goals with effectiveness, efficiency and satisfaction in a specified context of use". In other words, usability describes the fit between products, users, tasks and environment. In order to achieve a good quality of use, or usability, there is a quite common understanding in the international community of usability experts how this should be achieved. It is the user centred design process (UCDP). Several more or less detailed process models are available (e.g. ISO 13407, 1999; Mayhew 1999; IBM, 2002; Rosson & Carroll, 2002). Across the different process models, four central phases of a UCDP can be identified: a) analysis of context of use, b) designing, c) prototyping, and d) evaluation of the user interface.

All process models focus on the three main aspects of user interface design defined by Gould and Lewis (1985): early focus on users and tasks, empirical measurement and iterative design. The iterative process runs until a defined criteria derived from user requirements is achieved (for further details see ISO 13407, 1999).

Guidelines of the different levels described above are an important input for a user centred design process. Derived from project experience and some research literature (e.g. Rosenzweig, 1996) we see the involvement of guidelines in the following UCDP phases.

(1) Context of use analysis: When having finished the context of use analysis the relevant principles for the quality of use for a product can be selected and prioritized. Often it turns out that not all available principles have the same importance. Basic principles like "suitability for the task" (ISO 9241-10) are of central importance in most cases, but e.g. "suitability for individualisation" (ISO 9241-10) has not always top priority. (2) Design phase: The high prioritised principles should guide the process of finding design solutions. The first step in the design phase is to design the main views of dialogues according to the actions of the important and frequently performed tasks of the identified user roles (e.g. Rosson & Carroll, 2002). Here rules and interaction patterns are important input for designing information views. The second step in the design phase is to design single interactions. Here rules, interaction pattern, standard interactions and user interface building blocks can be used as input for design solutions and supporting design ideas. (3) Prototyping phase: Some prototyping tools provide user interface building blocks, e.g. as libraries, which can be used for developing a prototype of the user interface. Style guides help to ensure consistency with the underlying platform, and idioms provide building blocks which are already available for a particular look and feel. (4) Evaluation phase: In UCDP, usability testing is a crucial method to detect usability problems (Nielsen, 1995). Before running a usability test, quite often expert based evaluation methods, e.g. heuristic evaluation (Nielsen, 1993) are applied. For heuristic evaluation the selected and prioritized principles set in the context of use analysis phase can be used to guide the evaluation. The outcome of the evaluation phase is also be used to decide whether a further design iteration is

necessary. For this decision metrics are needed. The selection of such metrics (see ISO 9241-11) can be guided by the top prioritized principles, e.g. if learnability is important, it might be useful to measure the time and repetitions until a specific task has been carried out correctly.

## 3.2 Creativity, the user centred design process and guidelines

An important criticism concerning the use of guidelines is that innovations and creative ideas are hindered by guidelines. On the one hand it is clear that guidelines have the goal to decrease the number of possible design solutions in order to sort out solutions which are not usable or inconsistent. On the other hand, new design solutions which are more appropriate for the context of use might not be found. The idea behind that is that guidelines are treated as laws. In UCDP it is more appropriate to take guidelines as an input for the design process in order to help the selection of usable design alternatives. Design decisions can be seen as design hypotheses which have to proved during the evaluation phase. This holds also for guidelines. If certain guidelines, e.g. interaction patterns or even user interface building blocks have been used to construct a dialogue, it is also the guidelines which must be tested in the evaluation phase. In many cases it will turn out that the solution is good. Then it can be kept. In other cases. usability problems will be detected although the design was based on sound guidelines. In these cases the design and also the related guidelines have to be revised. Product design can also be seen as theory development (Carroll, Singley & Rosson, 1992). In this view the product is a theory on how the user can be optimally supported in a certain context of use. By continuous testing this theory is refined step by step until it covers the main requirements of the context of use.

## 3.3 Guidelines as output of user centred design

If a UCDP brings about design solutions which contribute to an improved quality of use, the design solutions can be formulated as new guidelines. Now, the underlying principles should be described and interaction patterns, standard interactions or user interface building blocks should be defined. These guidelines have proved their compatibility with the context of use in question. Now, guidelines can ensure that the design solutions developed in a UCDP are consistently applied during implementation. Style guides ensure consistent dialogues within a product or a product family or platform. Rosenzweig (1996) pointed out that it is important to explain the rationale of the guidelines to the development team and give them the chance to provide feedback from their point of view. Our experience is that development teams must have the possibility to discuss with the guideline authors upcoming questions and remarks (see also Gale, 1996).

## 4    Conclusion

We showed that treating guideline documents as living documents in the sense of Rosenzweig and providing possibilities for continuously testing the guidelines and making revisions and improvements will lead to a sustained increase of the acceptance of guidelines. By using and generating guidelines in the framework of user centred design creative design of interactive products and use of usability guidelines is NOT a contradiction.

## References

Burmester, M. (Ed.). (1997). Guidelines and Rules for Design of User Interfaces for Electronic Home Devices. Stuttgart: IRB.

Carroll, J.M., Singley, M.K. & Rosson, M.B. (1992). Integrating theory development with design evaluation. Behaviour & Information Technology, 11[5], 247-255.

Cooper, A. (1999). The Inmates Are running the Asylum. Indianapolis, Indiana: SAMS Publ.

Gale, S. (1996). A Collaborative Approach to Developing Style Guides. In Proc. of the ACM Conference on Human Factors in Computing Systems (pp. 361-366). New York: ACM Press.

Görner, C., Burmester, M. & Kaja, M. (1997). Dialogbausteine: Ein Konzept zur Verbesserung der Konformität von Benutzungsschnittstelle mit internationalen Standards. In Tagungsband Software-Ergonomie 97 (pp. 157-166). Stuttgart: Teubner.

Gould, J.D. & Lewis, C.H. (1985). Designing for usability: key principles and what designers think. Communications of the ACM, 28[3], 300-311.

IBM (2002). IBM Ease of Use: What is User-Centered Design? Retrieved Feb., 5 2003, from www-3.ibm.com/ibm/easy/eou_ext.nsf/EasyPrint/2.

ISO 9241 (1996-1999). Ergonomic requirements for office work with visual display terminals (VDTs) Part 10, 12-17. Berlin: Beuth-Verlag.

ISO 13407 (1999). Human-centred design processes for interactive systems. International Organization for Standardization. Berlin: Beuth-Verlag.

Mayhew, D.L. (1999). The usability engineering lifecycle. A practitioner's handbook for user interface design. San Francisco, CA: Morgan Kaufmann.

Microsoft (2001). Windows XP - Guidelines for Applications. Retrieved February 5, 2003, from www.microsoft.com/hwdev/windowsxp/downloads/

Nielsen, J. (1993). Usability Engineering. Boston, San Diego: Academic Press.

Nielsen, J. (1994). Enhancing the explanatory power of usability heuristics. Proc. of the ACM. CHI'94 Conference (pp. 152-158). New York: ACM Press.

Nielsen, J. (1995). Getting usability used. In Proc. of Human Computer Interaction- INTERACT '95 (pp. 3-13). London: Chapman & Hall.

Nielsen – Norman – Group (2003). Retrieved Feb., 5 2003, from www.nngroup.com/reports

Rosenzweig, E. (1996). A Common Look and Feel or a Fantasy? Interactions, 5, 21-26.

Rosson, M.B. & Carroll, J.M. (2002). Usability Engineering – Scenario-based development of human-computer interaction. San Francisco: Morgan Kaufmann.

Shneiderman, B. (1998). Designing the User Interface: Strategies for Effective Human-Computer Interaction (3rd edn.). Reading Mass: Addison-Wesley.

Spool, J.M. (2002). Evolution Trumps Usability Guidelines. Retrieved Nov., 23 2002, from www.uie.com/Articles/evolution_trumps_usability.htm

Steward, T. & Travis, D. (2002). Guidelines, Standards and Style Guides. In J.A. Jacko & A. Sears (Eds.), The Human-Computer Interaction Handbook (pp. 991-1005). Mahwah: L.E.A.

Sun Microsystems (2001). Java Look and Feel Design Guidelines: Advanced Topics. Boston, Mass.: Addison Wesley.

Tidwell, J. (1999). Common Ground - A Pattern Language for Human-Computer Interface Design. Retrieved Feb., 5 2003, from www.mit.edu/~jtidwell/common_ground_onefile.html.

van Welie (2002). Interaction Design Patterns. Retrieved Feb., 5 2003, from www.welie.com/patterns/index.html

van Welie, M., van der Veer, G.C. & Eliëns, A. (2000). Patterns as Tools for User Interface Design: In: Int. Workshop on Tools for Working with Guidelines (pp. 313-324). France. Retrieved Feb., 5 2003, from www.cs.vu.nl/~martijn/gta/docs/TWG2000.pdf

# Study Of Spatial Biological Systems Using a Graphical User Interface

*Nigel J. Burroughs, George D. Tsibidis*
Mathematics Institute,
University of Warwick, Coventry,
CV47AL, UK
njb,tsibidis@maths.warwick.ac.uk

*William Gaze, Liz Wellington*
Department of Biology,
University of Warwick, Coventry,
CV47AL, UK
w.gaze,e.m.h.wellington@warwick.ac.uk

## Abstract

In this paper, we describe a Graphical User Interface (GUI) designed to manage large quantities of image data of a biological system. After setting the design requirements for the system, we developed an ecology quantification GUI that assists biologists in analysing data. We focus on the main features of the interface and we present the results and an evaluation of the system. Finally, we provide some directions for some future work.

## 1. Introduction

Biologists experience difficulties in managing and analysing large amounts of data. To tackle this problem, we designed a GUI, which incorporates tools for a systematic analysis of image data. The objective of the interface is to enable biologists to manage and analyse a large number of images by aiding classification of, and viewing of object groups that are present in image set. Its design is based on a consideration of system requirements and an assessment was conducted by allowing experts to evaluate the system. This paper focuses on three main aspects of the development of the GUI:

- The requirements of the system design.
- The design and development of the interface.
- An evaluation of the system focusing on the quality of the results.

## 2. System requirements

The design of the system should be based on the following requirements: The interface should

- offer a rapid data overview
- require minimum intervention by the user for object classification,
- allow fast error identification,
- be dynamic and adaptive to changes,
- be reliable and robust.

## 3. Application and User Interface Design

A biology experiment was carried out in the Department of Biology at the University of Warwick whose objective was the study of the ecology of a biological system (*Acanthamoeba polyphaga*, a ubiquitous unicellar amoeba, and *Salmonella typhimurium*, an enteric bacterial pathogen on non nutrient on agar). A large number of images were obtained (a series of at least 3,000 spatial images, each image about 2MB, 1022 x1024) and within each image a number of biological objects were present (see Figure 1). The aim of the analysis is to extract and classify all objects in the data set, distinguish dead from live objects, group objects into classification categories and finally carry out a detailed quantitative and qualitative spatial analysis of the results.

Basic Image analysis techniques (Weeks, 1996) were used and modified to locate and count all objects. The refractive halo was used to locate all objects in each image by intensity thresholding, giving morphological information about the objects such as perimeter, centre of mass and shape details. To distinguish dead from live objects, a delayed image (4secs) comparison (image subtraction) was performed.

**Figure 1:** Typical image containing amoeba and cysts, (perimeter and centre of mass shown) and bacteria, e.g. filaments.

However, four problems are encountered with the automated procedure:

- Classification error,
- Incorrect perimeter,
- Missing objects,
- Multiple objects (refractive halo objects comprising more that one biological object)

To overcome these problems, we built a GUI (see Figure 2) using Matlab 6, which is one of a pair of graphical interfaces we have developed, an amoeba classification GUI (this paper) and a bacterial quantification GUI (not shown). The amoeba classification GUI allows the user to intervene and correct interactively classification and morphological errors.

There exist two types of classification for the objects of the experiment, one is based on the type of the object (amoeba, cyst, etc) while the second is based on the state of the object (live or dead). The GUI can help the user to avoid a time consuming manual classification for every individual object by running an automated classification procedure. The basis of these algorithms is illustrated in the two plots in Figure 3; cysts and dead amoebae are clustered in the lower section of the respective graphs and can thereby be classified based on approximate clustering. As a result, the interface offers the facility to create clustering patterns that allow a faster classification and decision making process. This is very important in minimising the effort of the user to achieve object classification throughout the entire data set.

The GUI consists of viewing facilities and modification and analysis tools. It operates as a management tool allowing the user to view summary information for individual objects or whole categories of objects, and also conduct a detailed analysis. The modification tools enable the user to either redraw the perimeter of an object if it is incorrect, introduce new objects if they were missed by the automated image analysis and to modify an object's classification at any time. All these manual changes are recorded and an overall analysis is adapted to the changes making the GUI a dynamic interface.

The analysis tools are used to facilitate a fast quantitative and qualitative analysis for objects belonging to a particular category by enabling the user to view graphical representations of the results in terms of plots (see Figure 3).

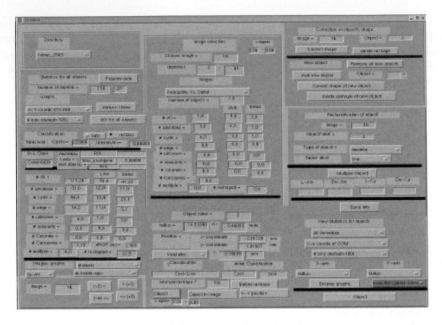

**Figure 2:** The Graphical User Interface

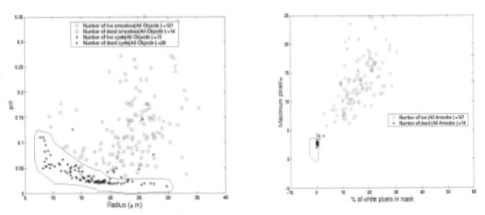

**Figure 3:** Amoeba-cyst and Dead-live classification. *Left panel*: shape analysis, using relative, standard deviation of object by radius. *Right panel*: maximum pixel intensity, in object (relative to an estimated intensity standard deviation $\sigma$) by fraction of live, pixels in object (live being significantly above noise, 95%).

The viewing facilities aids the user to scan quickly a library of objects belonging to each category (see Figure 4), compare them and decide whether a object classification change is required. The decision of the user can be assisted by the fact he/she is able to compare all objects that are in the same category and determine an optimal object classification criterion. If the classification of an object has been modified, it is indicated to inform the user of the change. In the dead or live object category, a library of images resulting from the delayed image comparison for all objects is produced and it provides a clear and fast view of the dead-live classification.

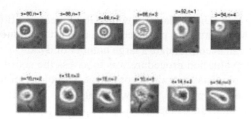

**Figure 4:** Library of images (upper row: cysts, lower row: amoeba)

The user can also employ the viewing tools to obtain data, measurements, confidence statistics and images for individual objects. In Figure 5, the first row shows the image of an extracted object (left box) and the same figure with the perimeter of the object drawn (right box) while the second row provides a summary of measurement and results (left box) and the delayed image subtraction for live or dead analysis (right box).

Another useful feature of the viewing facility is that it allows to view the full image sequence displaying all existing objects and their perimeters and centres of mass. The advantage of supplying a tool that allows to go through the whole set of images and see this type of characteristics provides a fast process of determining which objects will be used in the analysis and which objects must be included and redrawn manually.

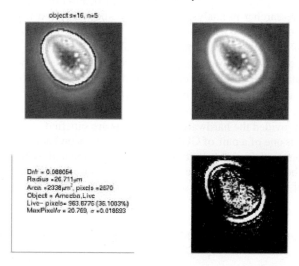

**Figure 5**: Images and measurements for an individual object

Both analysis and viewing facilities are very helpful tools because they simplify data management and reduce the number of objects requiring manual reclassification. This is a very important issue in terms of speeding up the analysis of a huge data set.

Finally, the efficiency of the interface can be tested by comparing the results derived from the automated and manual classifications of the objects and measuring the level of accuracy and confidence level of the automated procedure.

## 4. Evaluation

The interface was run on a Sun (Unix) platform with 1Gb of memory. It is mobile in the sense that it can run on Windows or Mac platforms provided there is substantial memory and the computer is equipped with a drive that can accommodate a large number of data.

The main objective of the evaluation procedure was to assess the set of requirements against the decision of experts. Two biologists from the Department of Biology at the University of Warwick helped the evaluation of the system. The focus was centred on the efficiency of the GUI and the simplicity it provided in the analysis of the data:

- The accuracy in the object classification was very high; More than 85% of the objects in the dataset did not require manual classification.
- The number of the objects that required their shape to be redrawn or had to be introduced manually was very small ($\approx$5% of the total number of objects).
- The pattern recognition for the objects through the ROI clustering led to a faster object classification.
- The GUI is user friendly.

These demonstrate that the GUI leads to an efficient fast classification, minimises the need for the user intervention and simplifies the analysis.

## 5. Conclusions-Future Work

This GUI helped to analyse data collected in the Department of Biology at the University of Warwick. The aim of this project was primarily the simplification of object analysis by providing the means to assist a researcher in analysing a biology experiment in a fast and efficient way. The fundamental findings of our inquiry are that the GUI:

- operates as an information management tool,
- helps to group data and produce graphical representations that simplify the analysis,
- enables pattern recognition for object identification,
- minimises user intervention in object classification,
- is dynamic and adaptive to changes,
- is mobile provided the hardware requirements are satisfied

The present interface is one of a pair of GUIs (the other focuses on bacteria extraction and bacteria coverage in a huge set of images) that have been implemented with excellent results. An improvement would be to build an on-line version of the interface. A further study is ongoing that analyses sequences of images (videos) by studying trajectories of biological objects.

## 6. References

Weeks A.R. Jr., (1996). Fundamentals of Electronic Image Processing. IEEE press.

# The use of Metaphors for Interaction between Children and Children's sites

*Alessandra Carusi*

PUC – Catholic University
of Rio de Janeiro
Rua Henrique Stamile Coutinho,
470/
201 – Rio de Janeiro – RJ – Brasil
alecarusi@alternex.com.br

*Vera Nojima*

PUC – Catholic University
of Rio de Janeiro
Rua Marquês de São Vicente, 225
Rio de Janeiro – RJ   Brasil
nojima@dsg.puc-rio.br

## Abstract

This project puts in evidence the importance of observing principles and presupposes related to communicational issues, mainly in what concerns the production of meaning in which web design projects are involved. Hence, defining message is the starting point, until its minimal unit – the sign. Thus, this study focus on the construction of meaning, which is accomplished through the use of metaphors in graphic interfaces.

## 1   Introduction

The Internet is gradually becoming more and more present in the routine of those children who have a computer at hand in their environment. It can be considered that the construction of meaning of graphic representations, which comprises the interfaces of these sites, is the issue which is responsible for the achievement of these communicative goals. These children take decisions once they realize the design elements, emphasizing the designer's need of knowing their tendencies of perception while producing web sites for them.

This assembly of children is growing up surrounded by digital media, spending their upbringing years in a context and environment quite different from those of their parents. They learn, communicate and play with video games, CD-ROMs and computers connected to the Internet. Novelties are plenty. Knowledge isnot only acquired through reading and doing written exercises or by the presentation of the information by their teachers. It is believed that the computer intervenes in these children's routine and intelligence development.

The Internet operates as a means for children to find the latest news and have fun in children'sweb sites, which are presented by their parents, friends or television. TV networks see the Internet as a vehiclefor presenting, advertising and interacting with its public. Therefore, children's sites linked to television channels are increasing in number and working as a complement to TV programs and as a publicityentreaty. However, the rise of available sites for children on the Web does not mean that these sitesare effectively stimulating their visits or interactions with the messages displayed on their interfaces. This fact will only take place under the condition of the graphic interfaces presenting a language, which is understood by its public. The designer does not always know the expectations and repertoire of his or her children's public.

## 2      The Use of Metaphors in Interfaces

Graphic interfaces facilitate the way human beings interact with computers, making the communicative process feasible between them. The graphic interface was the feature, which contributed to make the usage of computers something simpler, acting like a translator of a signals and symbols language, proper to its functioning. In people's homes and in daily life there was a mounting of the presence of textual and pictorial interface computers, which helped user and machine interact in a semantic relationship, characterized by meaning and expression.

According to Peirce (1977), all thinking has subjective and emotional qualities, which are attributed to its sign, in an absolute or relative way, or through a convention. The sign comes up to the human mind through words, concepts, images, sounds and associations. Johnson (2001) declared that a man's visual memory of pictorial contents is far more lasting than verbal one. Therefore it is easier to forget a name than a face or remember the place where a specific text segment was located on a site, even if what was written there was forgotten. When it comes to graphic interfaces the value of images is considerable. The user associates the images on the screen to objects and actions of his or her daily life, and thus, considers pleasant the environment of the site.

The representations of graphic interfaces frequently take the form of metaphors. According to Fiorin & Savioli (1998), the metaphor is characterized by changing the meaning of a word by attaching a second meaning which shares a relation of similarity between thefirst and the second one. That is, when theyshare semantic characteristics. A command line constituted of 'zeros'and 'ones' is substituted by a metaphorof a virtual folder in a virtual desktop. This manner of representing the message expanded the capacity of using a computer amongst people who do not have the knowledge ofthe antecedent language expressed by 'command lines'.Johnson (2001) comments that visual metaphors, presented for the first time in the sixties, probably had more to do with the digital revolution than any other advance ever recorded within the software field.

## 3    The Relevance of the Representation Forms of Information

The metaphors on the interfaces are elements which, whether associated with others or not, lead to a meaning which is constituted by the associations of other meanings. Its interpretations are influenced by their public'sculture and context. A child, as a receptor of information, while accessing a site, not only decides if she orhe should click an icon, but should also decide which one to click. Understanding text combinations and images in their different colors and shapes, motioned or not, and transform them in meaning is a process which demands the understanding of each of these elements as well as the relations amongst them. To interact in a site, this child needs to feel encouraged. Nevertheless,motivation will only turns out if there is enough comprehension of the content of the commu nication object that is being presented.

Fiorin & Savioli (1998) gave the definition of a communication object: it is any product which has the aim of communicating something to someone, once it must establish a relation between its parts. What will determine a unique sense to the product will be the many relations between these parts and not the mere adding of them. When analyzing a text it is possible to be noticed that it is a context in which smaller units, as sentences, are part of and function as context to other words. Each word has a particular meaning, denotation, or a second meaning, connotation, which can vary according to the context in which it isinserted.

> 'A hat in a peasant's head is just a sun protecting outfit; over a lady's head in a ceremony, it is an adornment; in front of a clergymanit is a symbol of power; in a beggar's supplicating hand it means a demand for help. To sum up: meaning is defined by relations'. (Fiorin & Savioli, 1998)

Therefore, context influences the reading of any product, pictorial or verbal. It is not possible to base oneself on isolated fragments, because the meaning of the parts is determined by the whole to which they belong. For this reason, graphic elements such as texts, images, colors, and contrasts of a communication object are fundamental to constitute a exclusive element with an intelligible voice. This will be expressed by language.

Andrade & Medeiros (1997) state that the language used for interpersonal communication is also helpful for man to structure his inner world. Consequently, language is used to think and communicate thoughts and emotions. Communication needs to use a common language between the transmitters and receptorsof the message, being it a language or a dialect, spoken or written, gestures, taps, colors, light or sound signals, etc. Communicating is not only transmitting information. An effective communication process occurs when the receptor, after having understood, makes a move or changes his or her behavior. The importance of the message isequal as its power of persuasion, which is the act of leading another person to accept what isbeing stated. Consequently, communicating is not only the knowledge of how to say something but, more than that, make others believe it and do what the transmitter has in mind. The acceptance depends on a series of factors such as emotions, feelings, values, ideology, vision of the world, political convictions etc. These values should be organized in a language known by the speakers of the message. There is no such thing as neutral communication. Somehow, all communication aims to convince the receptor of something.

Andrade & Medeiros (1977) claim that language, knowledge and communication are mixed, being on that account any knowledge is considered incomplete if it is not capable of being communicated through language.

A message may be constituted by verbal and non-verbal language, or by one of them separately. Fiorin & Savioli (1998) remark that there are many forms of language. Men, besides articulated verbal language, also display of other language systems, some of each are non-verbal ones. According to these authors, language is structured in two levels: content, which comprehends the meanings conveyed, and expression, which is constituted by the vehicle of these meanings. In what concerns expression, two voices are implicated: one which affirms and another which refutes former information. This is how contrast is fully and clearly expressed. This opposition of the senses provides unityto the elements of the message, because it establishes a relation among each element within the context. To understand the organization of these relations is what will determine the apprehension of concrete meaning within an specific context. An isolated figure does not have only one meaning. It can imply many ideas.

The smallest unit in a language system is the sign. According to Nöth (1995), the sign does not have the role of a class of objects but the function of an object in the communication process. The interpretation of a signis a dynamic process that happens within the receptor's mind. The apprehension process of a sign is called semiosis. Eco (1973) used to comment that the sign is used to transmit an information, to indicate to someone, something known by another person, who by his or her turn, wants to share this information with other people.Fiorin & Savioli (1998) believe that this relation is constituted by three elements: the objector referent, the sign or representant, and the interpreter. The relation, which takes place between signand referent, determines the existence of three types of signs: whether the relation is arbitrary, there is a symbol (word). If there is contiguity, proximity between sign and object (smoke – fire; footsteps – people; yell – pain; wet ground – rain), there is an index. If the relation is based on similarity, there is an icon (map, photograph, drawings, comic strip, caricature, metaphor).

Being the sign the unit of a message, it can be concluded that its significant, its information and its meaning are based on the relations amid the signs which are part ofthis message. The organization of the relations between these signs is formed by coherent nets, which are comprised of the theme of the message or product. Fiorin & Savioli (1998) allege that coherence is a tight union amongst many parts. Furthermore, it is the relation between ideas which are in harmony, causing an absence of contradiction. Coherence distinguishesa product amidst an assemble of things, creating a unit of meanings. In a graphic interface, for example,the elements enchain themselves in a special way to express a particular theme. Thus, they must be compatible ones with others, on the contrary the user will not distinguish which theme is being conveyed. A very same theme may be treated in many different ways, but its coherence is fundamental for it to be plausible. A graphic interface for kids may lose its verisimilitude if it contains a design constituted of a gray scale of colorstogether withfew iconic images. If this occurs, the power of persuasion becomes weaker because the user will not believe in the message, consequently, he or she will not make the expected moves.

## 4  Semiotic applied to the Language of Sites

Santaella (2002) makes a commentary that the classification of signs is the division of Peirce's most important semiotictheory 'When it isintendedto analyze manifest languages in a semiotic way, it can be observedthat semiotics furnishes us with the definitions and general classifications of all types of codes, languages, signals andsigns of any species and the main aspects which embrace them, for instance: signification, representation, setting objectives and interpretation'. Therefore, semiotic offersbacking for the relations established in a semiosis, allowing the foreshadowing of meaning and the applicability of what was elaborated.

Semiotic provides information and knowledge, to put in other words, semiotic theoretical base allows the elaboration of a webdesign project based on the solutions of its communicational issues and meaning, in the relation between product – user. Any webdesign project contains elements which are signs and which convey meaning to the public to whom this project is somehow addressed to. Through the study of signs, which interact in the project, the process of production of meaning can be amended and evaluated if it will meet its public'needs.

## 5  Conclusion

A webdesign project organization should not be based on formal resources as color sets, shapes and words, with the mere intention of adorning it. The most important circumstance is to ascertain the best manner of displaying signs because doing it exclusively this way, it will be possible to generate coherent and cohesive messages.

While creating a communication object composed of many metaphors, due to the fact that it occurs in many sites for children, the designer and his or her target public must share multiple information.In the intersection ofthis knowledge, the metaphor constitutes and provides, for the children's public, the meaning idealized by its creators. This metaphoric content within graphic interfaces will be likely to unchain feelingsandemotions which are necessary to arouse the child to take many moves while accessing a site addressed to children.

An analysis of how children identify structured images as language could reveal the importance of doing a specific study about howinformation will be presented in projects. One should be cautions when publicity for children is being conveyed and it should avoided, for instance, pages with too much information, incorrect use of typography or colors,metaphors which are out of context or are repetitive etc. The way of presenting graphic elements is fundamental so they do not become mere points of exploitation. Consequently,it is possible to conclude that , facing all new possibilities offered by the latest technologies of communication, the most adequate solution would be the observation of children's behavior and their interaction with sites which already exist. It would also be useful to observe children's interaction with projects which are being developed. Once this observation takes place, it would be possible to make propositions for further projects.

Adults often forget that children are children. The designer can manage the elements present on graphic interfaces to facilitate a child's perceptive process, terminating with obstacles that make reading and understanding difficult. If the designer, during the project, is aware of the mechanisms that construct the meaning of these elements, the child will understand their content in a better manner.

Thereafter, the designer's message to the computer user by the intermediacy of the graphic interface of a site becomes an effective communication process, where the targets outlined during the design project are achieved.

## References

Andrade, Maria Margarida de & Medeiros, João Bosco (1997). *Curso de lingua portuguesa: para a área de humanas: enfoque no uso da linguagem jornalística, literária, publicitária*. São Paulo: Atlas.

Eco, Umberto (1973). *O Signo*. Lisboa: Editorial Presença.

Fiorin, José Luiz & Savioli, Francisco Platão (1998). *Lições de Texto: Leitura e Redação*. São Paulo: Ática.

Johnson, Steven (2001). *Cultura da interface: como o computador transforma nossa maneira de criar e comunicar*. Rio de Janeiro: Jorge Zahar.

Noth, Winfried (1995). *Panorama da Semiótica*. São Paulo: Annablume.

Peirce, Charles S. (1997). Tradução : COELHO NETO, José Teixeira. *Semiótica*. São Paulo: Perpectiva.

Sataella, Lúcia (2002). *Semiótica Aplicada*. São Paulo: Pioneira Thomson Learning.

Tiski-Franckowiak, Irene T. (1997). *Homem, comunicação e cor*. São Paulo: Ícone, 1997.

# Can novice designers apply usability criteria and recommendations to make web sites easier to use?

*Aline Chevalier*

Research Center in Psychology of
Cognition, Language and Emotion
University of Provence
29, avenue Robert Schuman
13621 Aix en Provence, France
alinech@up.univ-aix.fr

*Melody Y. Ivory*

The Information School
University of Washington
Seattle, WA 98195-2840 USA
myivory@u.washington.edu

## Abstract

We present two studies that we conducted to determine the effect of a short training program in applying usability criteria and recommendations on the evaluation and improvement of a web site, which contained usability problems. This training program presented web site usability criteria and recommendations. We conducted the two studies with novice web site designers, who had or had not participated in the short training program, to compare their performances. Study results showed that the trained novices identified 30.7% more usability problems than the untrained novices. Moreover, the trained novices managed to make several usability-related modifications to the web site to make it easier to use.

These first results have to be verified by other studies, but they are encouraging. They suggest that web site designers, who have no university training in ergonomics, can use usability criteria. Furthermore, results suggest that usability criteria should be integrated into web site design training.

## 1    Introduction

In an experimental study, Scapin and Bastien (1997) asked 17 cognitive ergonomics and human factors students to evaluate a music application that had usability problems. Students conducted evaluations in one of three conditions:

> Condition 1: Six students used a list of usability criteria for the design of hypermedia interfaces (defined by Bastien, Scapin, & Leulier, 1997).
>
> Condition 2: Five students used the ISO/DIS 9241-10 standard (International Organization for Standardization, 1994).
>
> Condition 3: Six students used no guidelines (control group).

Results showed that students identified 51% more usability problems in the condition with the usability criteria than in the conditions without guidelines or with the ISO/DIS 9241-10 standard. Nevertheless, we must underline that in this study, participants were familiar with ergonomic and usability aspects (they were in a cognitive ergonomics university program). We argue that it remains to be determined whether people, who do not have a background in ergonomics or human factors, can similarly apply usability criteria.

# 2 General presentation of the two studies

## 2.1 Objective

Bastien *et al.* (1998) adapted the usability criteria for the design of hypermedia interfaces (used in the study described above) to the design of web sites. We wanted to determine whether web site designers could apply the adapted usability criteria to improve web sites; thus, we conducted two studies with novice designers. We chose this domain, because in most cases, web site designers have not had any training in human factors or in usability, yet these aspects are very important in their design activities. Indeed, designers must address the needs of the client (person who requests the site) and of the users (Chevalier, 2002; Chevalier & Ivory, 2003) while designing web sites.

These two studies aim at determining the effect of a short training session on a designer's ability to use usability criteria to evaluate and improve a web site that had a set of usability problems.

## 2.2 Web site designers

Numerous HTML authoring tools (e.g. Netscape Composer®, Macromedia Dreamweaver®) are available and their use, after a short training time, is not very complex. Therefore, many short training programs are available for people who want learn how to create web sites. So, more and more persons can be considered as web site designers. On this basis, the two studies were led with novice designers, who had just followed a short training program in using the Macromedia Dreamweaver® authoring tool. These designers had not yet dealt with real web site clients, and they had created only one site during their training program.

We chose to focus on novice designers for two main reasons:
- A request from the University of Mediterranean (Marseille -France) for training students in how to consider and apply usability criteria during web site design was suggested to us.
- Novice designers, who had never dealt with clients, would not have developed any procedure or skill linked with clients' needs (see Chevalier & Ivory, 2003).

## 2.3 Web site to evaluate

For the two studies, we created a web site in HTML for a music store, which commercialised music products (CD, concert tickets, etc.). The web pages were hyperlinked such that we could navigate on the site. The site comprised 18 pages and had a total of 337 usability problems (34 of them were unique); example problems included: the absence of page titles, the lack of color changes to indicate visited links, the use of too much text, and so on. We chose the nature and the number of usability problems based on a previous study, wherein we analysed similar HTML web sites produced by web site designers (see Chevalier & Ivory, 2003).

## 2.4 Hypothesis

One specificity of web site design is that designers are also web users, which is not the case for most design situations (e.g., the design of aerospace products). Nevertheless, designers can become, with experience, expert web users. This statute will allow them to navigate easily on the Web. So, they may not be able to recognize usability problems, which can be considered as such by non-expert web site users. On the contrary, the training program in applying usability criteria and recommendations will support novice designers to activate user knowledge as opposed to client knowledge (by guiding their evaluation activity). Therefore, they should identify more usability problems in the web site than untrained novices. On this basis, we hypothesized that trained designers will manage to apply user knowledge to improve certain usability problems that exist in the study web site. To test this hypothesis, we led two studies with novice designers.

## 3  First study: Evaluation without receiving training in the application of usability criteria

### 3.1  Procedure and objective

Seven novice designers evaluated the web site without receiving training in the application of usability criteria. They had completed the same training in using the Macromedia Dreamweaver® authoring tool. All the designers had to evaluate the study web site while using the same Personal Computer (PC). They had to inform the experimenter of the positive or negative aspects of the web site (i.e., "to think aloud;" see Ericsson & Simon, 1993); There was no time constraint for this task. We videotaped their verbal protocols and their use of the site. More precisely, this study aimed at determining:

- The number of web pages that designers did not visit in order to identify their navigational behaviour.
- The number of usability problems that designers could identify in the web site

### 3.2  Results

We recorded designers' verbal protocols and then transcribed them for analysis by two different persons. Study results showed that novice designers did not visit in mean 8.2 of the 18 web pages (see Table 1). Moreover, they identified very few usability problems introduced in the web site to evaluate (see Table 1). One natural question that arises from these study results is whether or not designers, who completed a training program in the application of usability criteria, would manage to identify and to improve more usability problems than the untrained novice designers did.

**Table 1:** Number of identified usability problems and unvisited web pages for untrained novice designers (in mean and percentage)

|  | Identified usability problems | Unvisited web pages |
|---|---|---|
| Mean | 16.2 of 337 | 8.2 of 18 |
| Percentage | 4.8% | 45.6% |

## 4  Second study: Evaluation with receiving training in the application of usability criteria

### 4.1  Procedure and objective

Six novice designers evaluated and modified the web site. These designers had recently completed the training program on developing web sites with the Macromedia Dreamweaver® authoring tool.

In our first study, seven novice designers evaluated the web site without receiving training in the application of usability criteria. In this second study, the novice designers completed a short training session in the application of usability criteria. During the six-hour training session, the instructor used several existing web sites to demonstrate how to identify usability problems as well as how to modify sites to mitigate problems (i.e., how to respect usability criteria). After the training session, the designers evaluated the web site. To further determine if designers could apply the usability criteria, we asked them to rectify usability problems on one web page in the site. We instructed designers to indicate positive or negative aspects of the site and to make the home page easier to use; Thus, the corrected page had to have fewer usability problems than the original page.

## 4.2 Results

Verbal protocols were analyzed by two different persons. Study results showed that the designers visited all the web pages. They identified, in mean, 35.7% of the usability problems introduced in the site. They also managed to address usability problems of the home page by implementing, in mean, 6.3 modifications. The home page had thirteen usability problems, and the designers managed to correct 6.3 (48.6%) of them. (see Table 2). For example, if we compare the original home page (Figure 1) with the one modified by a novice designer (Figure 2), we can see that the designer modified certain usability problems like reducing the page's length or modifying the fonts used However, other problems still exist in the modified design (e.g., the page's background color or the presence of animations may impede user's navigation). Therefore, we can conclude that the designers, who had no background in ergonomics and human factors, were able to successfully apply the usability criteria, because the modified home pages had fewer usability problems than the original pages.

**Table 2:** Number of identified usability problems, usability problems in the modified home page, and unvisited web pages for trained novice designers (in mean and percentage)

|  | Identified usability problems | Modified usability problems | Unvisited web pages |
|---|---|---|---|
| Mean | 119.5 of 337 | 6.3 of 13 | 0 of 18 |
| Percentage | 35.7% | 48.5% | 0% |

Figure 1: **Original home page**

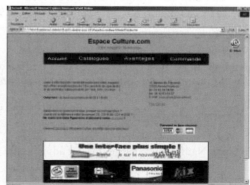

Figure 2: **Example of a modified home page**

# 5 Discussion and future work

Although these preliminary results need to be verified by additional studies, they are encouraging. Designers who completed a short training in applying usability criteria and recommendations were able to identify more usability problems in the study site than untrained designers did (119.5% *vs* 16.2%; $F(1,11)=95.399$; $p<.0001$). Moreover, trained designers visited all pages, whereas untrained designers did not. Untrained designers were hindered by the usability problems on some pages, which kept them from visiting all pages. For instance, they did not see a hyperlink to visit another page, because this hyperlink was at the bottom of a web page. Moreover, the trained designers managed to modify the home page, but we do not know whether these modifications actually improved usability. Future work entails asking users to evaluate the modified home pages.

These results suggest that novice web site designers, who have no background in human factors, can apply usability criteria with some training. Therefore, usability criteria need to be integrated into web site development training. In addition, usability criteria need to be adapted such that designers, who do not have a background in human factors, can apply them.

Toward this end, several questions could be explored. First, these two studies have to be replicated with professional web designers to determine whether they produce the same results. According to the obtained results, a new study could be led. It would aim at determining the influence of usability criteria, not in evaluation or modification situations, but in a design situation with both professional and novice web designers.

## References

Bastien, J. M. C., Leulier, C., Scapin, D. L. (1998). *L'ergonomie des sites web. Créer et maintenir un service web.* Cours INRIA, ADBS, Paris, 111-173.

Chevalier, A. (2002). *Le rôle du contexte et du niveau d'expertise des concepteurs de sites web sur la prise en compte de contraintes.* Ph. D. thesis, University of Provence (France), Cognitive Psychology Department.

Chevalier, A., & Ivory, M. Y. (2003). Web Site Designs: Influence of Designer's Experience and Design Constraints. *International Journal of Human-Computer Studies, 58,* 57-87.

Ericsson, K. A., & Simon, H. A. (1993). *Protocol analysis: Verbal reports as data* (Revised edition). Cambridge (MA): MIT Press.

International Organization for Standardization (1994). *ISO 9241-10. Ergonomic requirements for office work with visual display terminals (VDTs) -- Part 10: Dialogue principles.* Draft International Standard.

Scapin, D. L., & Bastien, J. M. C. (1997). Ergonomic criteria for evaluating the ergonomic quality of interactive systems. *Behavior & Information Technology, 16,* 220-231.

## Acknowledgements

We thank Nathalie Bonnardel for her advices and the web site designers for their participation.

# Learning from Museum Visits: Shaping Design Sensitivities

*Luigina Ciolfi  & Liam J. Bannon*

Interaction Design Centre
Department of Computer Science & Information Systems
University of Limerick, Ireland
Luigina.ciolfi@ul.ie, liam.bannon@ul.ie

## Abstract

This paper describes an observational study of visitors interacting with artefacts in a museum, and attempts to draw from these studies a number of design considerations. Gaining a thorough understanding of the context, and the way visitors move through the exhibitions and interact around the objects on display, is crucial in designing effective museum installations. In our research, we are designing novel installations that will engage visitors in a natural and unobtrusive way. Our designs – which are ongoing at the moment – are informed by our field study work at the museum site. Our purpose here is to show how our observations can indeed be made relevant to design concerns, a topic that is fundamental to the development of successful pervasive technology.

## 1    Introduction

Despite the increasing prevalence of computer-based museum installations, visitor experiences with computer technology remain mixed, at best. Many of the most insipid and uninspiring multimedia kiosk-type installations that proliferated in the mid-nineties have thankfully been removed. These kiosk installations tended to separate the person from the actual artefacts, and called attention to the computer interface itself as the object of interest, rather than the actual artefacts. Far from augmenting the artefacts, they served to distance visitors from the items in the collection. People are not happy about moving away from the exhibitions to consult a kiosk, they would prefer to focus on the displayed objects. These kiosk technologies also provide essentially an individual experience, whereas visitors usually go to museums in small groups. Nowadays, there is a greater realization within the computing community that the technology needs to be in the service of the exhibition, rather than stand out from it. The increasing interest in such areas as ubiquitous or pervasive computing, where computational power is taken "out of the box", distributed and attached to material objects, provides interesting new possibilities for how one might augment museum exhibits. "Traditional" desktop computers are not an effective form of technology for supporting visitor behaviour within such a rich, interesting and complex environment as a museum. Novel mobile, "ultralight" appliances and "Disappearing Computer" technologies could be more effectively employed in order to design better technological installations in the museum, both for supporting the visit and for creating interesting visitor experiences or educational activities.

This paper provides a number of insights derived from observing visitors, and interviewing them, at a specific site - the Hunt Museum in Limerick, Ireland. The purpose of these studies is to provide some background material that will inform and inspire our creativity in designing new augmented museum exhibits. However, within the design community, while there is interest in ethnographic and other forms of field studies, there is increasing concern as how one can take the results of such studies and make them relevant for design. Often, social scientists focus on issues that do not seem of relevance to design, and they do not have the training or expertise to be able to move from observation to the specification of requirements. Our own expertise crosses the human and social sciences, but also extends to concept and software design, a somewhat unusual combination. Thus we are especially interested in ensuring our observational studies are of relevance to the design process. In an effort to show how such observations can indeed be shown to be relevant for design, we, along with colleagues, have been involved in exercises that attempt to shape the design process so as to be sensitive to various issues, as determined from the observational studies. In a word, we are looking for "design sensitivities", rather than requirements *per se*.

Our work has been conducted within the EU-Disappearing Computer SHAPE[1] project (Situating Hybrid Assemblies in Public Environments). The SHAPE project is not strictly concerned with introducing technology within museums and galleries, rather the project focus is on creating hybrid public environments that allow visitors to actively interact with features of the physical, and of the digital, space. However, museums and exploratoria are the chosen context to inspire and support the development of such installations. The SHAPE team at the IDC, University of Limerick, jointly with the Hunt Museum, are currently developing a full design scenario informed and inspired by the analysis of a corpus of field studies aimed at understanding human behaviour within the museum environment, and, specifically, the way visitors approach and make sense of particular exhibits and specific objects. In this paper, we attempt to match specific vignetttes culled from our video studies, and pair them to design sensitivities, to show how particular user behaviours around existing objects and exhibits might sensitize us to certain issues, which we may wish to explore further in our computer-augmented installation.

## 2    Cabinets of Curiosities

The Hunt Museum has three "cabinets of curiosities" exhibits: they are wooden closets where objects are arranged both on the upper shelves and in the drawers that the visitors can open if they wish. The objects are protected by glass on the top of each drawer. The nature of the objects contained in the cabinets is varied: ivory pieces, tapestry, drawings, coins, etc (see Figure 1).

We conducted numerous observations on the behaviour of people in the "Study Collection" room, where one of the most important of the cabinets is located. We were interested in understanding the way visitors relate to the exhibit and their interaction with the drawers. These cabinets are particularly appreciated by the visitors. An evident reason is the potential of drawers, chests and boxes to stimulate curiosity and exploration. Containers suggest the presence of secrets (Elsner & Cardinal, 1994), or of objects that have some kind of symbolic relevance for being sheltered from the eyes of the public (Bachelard, 1958). "Curiosity is a major factor in determining whether environments are appealing, and indeed curiosity triggers interaction towards its object." (Falk & Dierking, 2000, p. 115). The interactions around the drawers reveal interesting visitor activities, often involving collaborative understanding and appreciation of the objects.

---

[1] Members of the SHAPE Consortium are: the Royal Institute of Technology-KTH (Sweden; Coordinating Partner), King's College London (UK), the University of Nottingham (UK) and the University of Limerick (Ireland).

Fig.1. A cabinet of curiosities in the Hunt Museum

# 3    From Field Studies to Design Sensitivities

In what follows, we list in more detail a few of the interesting observations and our interpretation of their relevance to a potential design process.

**Observation 1.** People tend to interact with the drawers in one of two ways: they open either a few drawers or all of them in sequence. Those who cannot resist opening all the drawers usually spend a longer time on each, and tend to comment more on the objects with their companions.

**Design concern 1.** *To keep the user's interest and engagement high, we must envision ways to support different 'layers of activity': this could provide participants with the ability to engage in a progressive sequence of actions (both alone and with others )in order to provide successive surprises and discoveries. For example, more, and varied information on assemblies of objects and their mutual relationships is provided to those who want to explore all the parts of the exhibit.*

**Observation 2.** In the case of couples visiting the exhibit, each person usually takes control of one side of the cabinet, resulting in two drawers being open at the same time. This pattern of interaction facilitates the exchange of opinions and comments among the two visitors, usually comparing the contents of the respective drawers. Gestures accompany verbal comments on objects' similarities, possible relations, visual features, etc.

**Design concern 2.** *The augmented cabinet of curiosities should provide the possibility of parallel discovery and of making comparisons in order to support collaborative understanding of objects: the interactivity we support should not be limited to the one between individual and exhibit, but we should consider the different degrees and combinations of verbal gestural interaction amongst individuals. The installation should also provide some kind of added value associated with collaborative interaction around the drawers.*

**Observation 3.** The glass surface on top of each drawer is often used for sketching or taking notes: children especially spend time in taking notes and drawing sketches of the objects contained in the drawer, using the drawer itself as support. In many cases, we observed children visiting the Study Collection with their parents, and literally taking the lead in discovering the content of the drawers and showing to and commenting on it with their parents (see Figure 2).

Figure 2. A child is leading his parents through the exploration of the drawers

**Design concern 3.** *Children should be allowed to take part in the activity and be able to take notes or sketches around the augmented cabinet of curiosities. The installation should also give children the possibility to lead the process of discovery and to show it to their companions.*

**Observation 4.** Some people think that it is not possible to touch or open the drawers and they might open one only after seeing somebody else doing it, and in this way being reassured they are actually allowed to interact with the exhibit. Even if one of the drawers is left slightly open to suggest its real use, these visitors did not seem to investigate further, assuming the open drawer was the result of some work being done by museum personnel. When visitors proceed to open the drawers and discover what they contain, all of them express their surprise.

**Design Concern 4.** *The technologically augmented cabinet should provide clues, triggers and adequate affordances to make visible which actions the visitors are allowed to perform on each component of the installation. We must consider possible ways of encouraging interaction with, and around, the exhibition, and specifically collaborative interaction.*

**Observation 5.** Unlike couples, members of larger groups of visitors cannot all simultaneously open the drawers: usually, two people act as "openers" on the two sides of the cabinet, but all the people in the group comment on the objects and tend to draw each other's attention to what has been discovered.

**Design Concern 5.** *The augmented cabinet should support the group visit experience with appropriate feedback that all the members of the group can appreciate. The possibility for the visitors to talk to each other must also be insured, as discussing the objects together is an essential part of the group experience around the cabinet. This means that devices as head-mounted displays or headphones are not appropriate for such an installation.*

## 4    Developing an augmented cabinet of curiosities

Following the data analysis sessions and an initial phase of reflection on our collected material, we are developing an augmented reality Study Collection room. The existing cabinet has some important features that make it a successful and inspiring installation: it is robust, the visibility is good, it allows for collaborative interactions and supports small groups. It is also easily accessible by children. The cabinet adds value to the experience of museum visit as it includes an element of active discovery and plays on a natural curiosity towards drawers and cabinets. Each object is interesting by itself, but multiple objects are connected by similarities or unexpected connections. What can pervasive technologies do to achieve such features in the technologically enhanced exhibits? At a minimum, we need to seamlessly integrate aspects of the exhibit with a visually pleasing structure, react to collaborative behaviour, and provide ambient feedback. We need to create a space that integrates with the rest of the museum. This is our design brief. Currently several researchers at the Interaction Design Centre, are engaged in concept design, inspired by the material presented above. We have found that the video material has helped us in creating outline scenarios, and storyboards. In the process of defining what kind of added value a technology-enhanced cabinet of curiosities could provide to people, we noticed that the only element currently missing in the process of interaction with the cabinet and its objects is the "physical" one: what if the visitors could handle the objects and explore them directly? Experiencing the exhibit has a physical aspect as well as a reflective one. As museum researchers have noted: "Exhibits … allow people to see, touch, taste, feel, and hear real things from the real world. … Visitors devote most of their time to looking, touching, smelling, and listening, not to reading. Visitors tend to be very attentive to objects, and only occasionally attentive to labels." (Falk & Dierking, 1995, p. 77). As emerged in our series of observations, interaction among visitors, collaborative content discussion and discovery all happen *around* the objects. The current cabinets in the Museum do not provide a way to experience features such as the look and feel of the surface of an object, the light and shadow it presents, the shape, texture,  weight, etc. Our scenarios and prototypes are exploring narratives and progressive unfolding of events, attempting to focus attention on aspects of the artefacts themselves and their context of discovery and use. Space does not permit us to detail our designs here, but we will show some of our prototypes during the Conference presentation.

# 5 Conclusion

The field studies conducted at the Hunt Museum have provided a useful corpus of material concerning the nature of exhibits, and visitor interaction around these exhibits, that is still under analysis. The elaboration of the "cabinet of curiosities" represents the first prototype we are developing, where we wish to augment these physical cabinets in a variety of ways with new technologies. Outline scenarios include cabinets where people can actually handle replicas of the real objects and experience some of the characteristics of these real objects. Our interest in this scenario has been driven by analyses of visitor engagement with exhibits and their interactions during specific "handling sessions" supervised by museum personnel that we have analysed (see Ciolfi & Bannon, 2002.)

This scenario, in which the objects themselves are envisioned as interfaces through which visitors can make sense of the object, of their history and their multiple relationships and features, poses a challenge for designers. Several projects are underway regarding the design of graspable interfaces and physical icons instead of GUIs (e.g. Ishii & Ullmer, 1997), but usually the object itself is not the locus of information, nor the focus of attention. Rather, objects are essentially tools for interacting with a computer system, and they are intended to act as a physical representation of surface interface elements such as icons and pointers. Thus our approach would be distinct, as we are interested in objects as both material and symbolic devices in their own right, with a history, context of use, etc, both mediating, and being the object of, interaction. Work in the coming months will involve further exploration of a variety of technical devices and platforms that may assist us in achieving some of our objectives for the prototypes, such as RFID tags, use of accelerometers and potentiometers for sensors, projection surfaces, webcam tracking etc., and we will show the results of our design and development work in the Conference presentation.

# References

Bachelard, G. (1958). The Poetics of Space, Boston: Beacon Press.

Ciolfi, L. & L.J. Bannon (2002). Designing Interactive Museum Exhibits : Enhancing visitor curiosity through augmented artefacts. Proceedings of ECCE11, European Conference on Cognitive Ergonomics, Catania (Italy) September 2002, pp. 311-317.

Elsner, J. & Cardinal, R. (eds.) (1994). The Cultures of Collecting, Cambridge, MA: Harvard University Press.

Falk, J.H. and Dierking, L.D. (1995). The Museum Experience, Washington, D.C.: Whalesback.

Ishii, H. & Ullmer, B. (1997). "Tangible Bits: Towards seamless Interfaces between People, Bits and Atoms", Proceedings of CHI'97, New York: ACM Press.

# Scenarios in the model-based process for design and evolution of cooperative applications

*Bertrand David, René Chalon, Olivier Delotte, Gérald Vaisman*

Laboratoire ICTT, Ecole Centrale de Lyon
36, avenue Guy de Collongue, 69134 ECULLY, France
{Bertrand.David, Rene.Chalon, Olivier.Delotte, Gerald.Vaisman}@ec-lyon.fr

## Abstract

In this paper we describe the role of the scenario-based approach which we integrate in a more global perspective of a well-organized process based on a collection of models which are used in design and evolution to transform scenarios in an operational cooperative application.

## 1    Introduction

CSCW (computer supported cooperative work) [Ellis-94, Andriessen-03] research proposes a new type of software, called groupware, which is an interactive multi-participant application allowing participants to carry out a "joint" task working from their own workstations. It is now a question of managing not only the man-machine interface but also the man-man interface mediated by the machine. The relationship between the participants can be considered from various points of view. Ellis et al. [Ellis-94] proposed a matrix which classifies the nature of cooperation in regard to time - synchronous or asynchronous, and to distance - local or remote aspects of cooperation. This classification was extended later, introducing awareness of cooperation, foreseeability or unpredictability of collaboration and location. The possibility of bringing together geographically distant people is an important contribution of groupware. The first aim of groupware is thus to propose a support for the abolition of space and time distances. Moreover, knowledge and management of the interventions of the multiple participants appear necessary. In fact, the participants constitute a work group that has to be organized with respect to working conditions, time and location. The organization can lead to the definition of different roles, sub-groups and phases of project work. The success of cooperative work can be measured by the way in which the groupware is able to create and support good group dynamics, which contributes to the disappearance of the virtuality of participants' presence. The project must be able to proceed as naturally as in collocation and without IT support. It must even take advantage from an organization of more effective work based on the new possibilities offered by information technologies (IT). The technological devices used should not interfere with the work or the group dynamics needed for project accomplishment. When designing cooperative systems, it is thus necessary to be aware that the usability aspect, the aim of which is to validate the environment suggested, is at least as significant as the engineering aspect. The evolution of users' practices during the project life-cycle must be taken into account in order to provide an effective and adaptable environment.

In-depth analysis of cooperation reveals several dimensions which must be examined, as initially proposed by Ellis [Ellis-94] with the Clover model, i.e. a support of production, conversation / communication and coordination between participants.

# 2    Scenario-based Design and CAB Model

Carroll's view of Scenarios-Based Design [Carroll-00] is an interesting starting point for collaborative system design and evolution. However, it seems important to go further. We propose to approach this scenario-based approach in a more organized way. In particular, in respect with the Clover model, it is important to locate each scenario in production, coordination or conversation space or in their intersections. More globally, it seems important to synthesize these scenarios in a model integrating collaborative application behaviors. We call this model a CAB model (Collaborative Application Behavior Model).

"Scenarios are stories", as Carroll wrote [Carroll-00], which, to be useful, can be expressed more or less freely or formally. We consider that it is important to integrate them as soon as possible in CAB perspective i.e. to ask the scenario writers to express explicitly the position of the scenario in relation with the CAB Model. The main goal of the CAB model is to describe explicitly the structure of actors, artifacts, contexts and tasks that characterize the behavior of the cooperative application in three Clover spaces (co-production, coordination, conversation). Each scenario expressing a task, might indicate its position in relation with these actors, processes, artifacts and contexts. In this way it is possible to elaborate progressively the CAB model for a given application. The CAB model for a specific collaborative application, contains concrete actors, artifacts, tasks and contexts which the cooperative application will take into account. Tasks expressed in different scenarios are studied in order to organize them. The goal is to eliminate redundancies and to elaborate a task tree and a task process. The task tree can be expressed in ConcurTaskTree formalism proposed by Paterno [Paterno-00]. The process view is a workflow view with temporal and logical dependencies between tasks. The context view is an expression of different contexts (logical or physical) related to environment and devices constraints, if any. Once we have collected several scenarios, we can start to elaborate the CAB model. This model will be used in the elaboration stage (development or prototyping).

# 3    Software infrastructure

With respect with software engineering considerations, the cooperative application cannot be carried out from scratch. It is necessary to identify different levels of development which are more or less dependent on the application. Usually, three functional layers are recognized.

The top layer corresponds to the collaborative application level. It contains all the cooperative software employed by the users. This level is definitely user-oriented, which means that it manages interaction control and proposes interfaces for notification and access controls. This model is called CUO-M (Collaboration User-Oriented Model). It uses multi-user services provided by a second layer called the groupware infrastructure, called CSA-M (Collaborative System Architecture Model). It is a generic layer between applications and the distributed system. This layer contains the common elements of group activities and acts as an operating system dedicated to groups. It supports collaborative work by managing sessions, users, it groups and provides generic cooperative tools (e.g. telepointer) and is responsible for concurrency control. It also implements notification protocols and provides access control mechanisms. The last layer is essentially in charge of message multicast and consistency control. We call it DSI-M (Distributed System Infrastructure Model). Usually, it is a computer-oriented layer which provides transparent mechanisms for communication and synchronization of distributed components which misfit with CSCW aims but which are very useful.

69

The degree of generality (and genericity) is not the same for these tree layers and models. The lowest layer (DSI-M) is for the most part independent from the collaborative applications. The middle level (CSA-M) can be dependent on a category of applications, but is stable for each category and each application during its life-cycle. The highest level (CUO-M) is, by construction, dependent on the application, because it is constructed (or specialized) with respect to the CAB-M.

# 4    AMF-C as CUO Model

An appropriate CUO model should fulfill three main objectives. Firstly, it organizes the software structure to improve implementation, portability and maintenance. Secondly, it helps identify the functional components, which is essential during the analysis and design process. Its third role is to facilitate the understanding of a complex system, not only for designers, but also for end-users.

AMF-C (the French acronym for Collaborative Multi-Faceted Agent) [Tarpin-Bernard-97] is our proposal for the CUO model for collaborative software which fulfills all these objectives. AMF-C is a generic and flexible model that can be used with design and implementation tools. It includes a graphical formalism that expresses the structures of software, and a run-time model that allows dynamic control of interactions.

The current trend in software engineering is to identify design patterns [Gamma-95], which help developers to share architectural knowledge, help people to reuse architectural style, and help new developers to avoid traps and pitfalls traditionally learned only as a result of costly experience.

AMF proposes a multi-faceted approach, in which each facet can be a pattern. Each new identified behavior which seems to be reusable, can be formalized as a new facet, i.e. a new pattern. AMF also proposes a very powerful graphical formalism which helps understand complex systems. This formalism is used as a design tool by editors and builders. It represents agents and facets with overlapped boxes, communication ports with rectangles which contain the associated services, and control administrators with symbols which express their behavior.

# 5    Evolution during the life-cycle

During application life-cycle, users progressively change their perception of the system and their use. The cooperative application could, during its use, take into account requirements concerning evolution expressed explicitly or implicitly by its users. This is particularly the case in the context of Capillary CSCW (cooperative work using handheld devices) [David-03, Brown-02] in which behavior evolution is more important related to context condition (connected or disconnected work) and the device used. To take into account this evolution in an explicit way, new scenarios can be presented which upgrade or extend initial the COB model. The Cooperative Application Behavior model evolves smoothly and it is important to be able to take this evolution into account. This evolution can vary in importance and its impact can be taken into account in different ways:
- If the evolution is within the scope of the system, i.e. these adjustments have been imagined and they are at the disposal of the user in the "configuration panel" (use of different interaction modalities, modification of awareness, choice of different WYSIWIS relaxation, plasticity of user interface, etc.)
- If the evolution is more important and leads to modification of the CAB model, two different solutions are possible:

- Change of interaction pattern with the same algorithmic behavior: this evolution is relatively easy and can be performed by the end-user (i.e. by visual programming using AMF-C graphic formalism),
- Change is more important also with algorithmic behavior evolution: in this case a developer intervention seems inevitable.

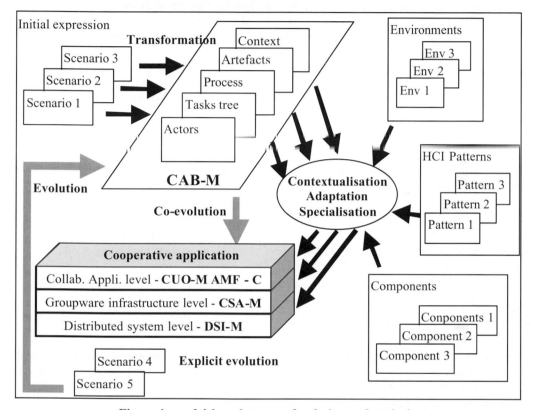

**Figure 1: model-based process for design and evolution.**

To take into account this re-configurability, adaptability and flexibility, we need new scenarios, which can also replace or modify existing ones. Their processing leads to modification of the CAB model and has an impact on the CUO model. The evolution can either be implemented on the existing IT support or this support can evolve too. We expect that the latter evolution, which concerns the CSA and DSI models, will seem to be out of scope of the evolutions which can be taken into account dynamically. To be able to modify dynamically CUO, we need to have at our disposal a meta-model of the CUO model to create dynamically new AMF-C agents in the relation with the re-configuration of the COB model. This co-evolution can be implemented either by the end-user himself or at least expressed by him, on his directives or under his control, by the developer.

# 6 Conclusions

We propose a development process which is based on a scenario-based approach and a collection of well-organized models. The synthesis of collected scenarios leads to the definition of a CAB model. These scenarios contribute to the elaboration of task and process models which are the

71

components of the CAB model. It synthesizes different scenarios and constructs a comprehensive model of the cooperative application with its users, processes, tasks, data and contexts.

Initial design is in this way properly assisted and the evolution during the application life-cycle needs new concepts. In actual fact, the CAB model seems to be appropriate for the evolutions. Lower level software models (CSA and DSI) are generic enough to be able to support these evolutions. The problem is mainly between CAB and CUO models. Dynamic evolution of the CUO model in relation to changes in the CAB model requires access to the meta-model of CUO, to be able to construct the new CUO model. We are working in this direction so as to take into account the evolution in a comprehensive way.

The second direction of our research relates to the automation, or at least assistance, of the design and evolution processes. We aim at assisting the initial transformation between a set of scenarios and the CAB model, between the CAB model and the CUO model in relation with design patterns which can be used, and the projection of the CUO on CSA and DSI. We would also be able to assist evolutions expressed by new or alternative scenarios and their impact on other models. Consideration of this joint evolution (known as co-evolution) is a significant challenge in the field of CSCW research.

## References

Andriessen J.H.E., Working with Groupware: Understanding and Evaluating Collaboration Technology, Springer, CSCW Series 2003

Carroll J. M., Making Use : Scenario-Based Design of Human-Computer Interactions, The MIT Press 2000

David B., Chalon R., Vaisman G., Delotte O., Capillary CSCW, to be published at HCI International 2003

Dewan P., An Integrated Approach to Designing and Evaluating Collaborative Applications and Infrastructures, Journal os Computer-Supported Cooperative Work, 2001, Vol; 10, N°1, p. 75-111, Kluwer Academic Publishers

Ellis C.A., Wainer J., A Conceptual Model of Groupware, ACM CSCW'94 Conference, p. 79-88, ACM Press

Gamma E., Helm R., Johnson R., Vlissides J. (1995), Design Patterns : Elements of Reusable Object-Oriented Software, Addison Wesley, Reading, MA.

Paterno F., Model-Based Design and Evaluation of Interactive Applications, Applied Computing Series, Springer, 2000

Seffah, A.and Forbrig, P. Multiple User Interfaces: Towards a Task-Driven and Patterns-Oriented Design Model, Proceedings of DSVIS 2002 (Rostock, Germany, June 2002) p. 160-174, to appear in Lecture Notes in Computer Science.

Tarpin-Bernard F., David B.T., AMF a new design pattern for complex interactive software ?. International HCI'97, Elsevier, ISBN: 0 444 82183 X. San Francisco. Vol. 21B. p. 351-354. August 1997.

Tarpin-Bernard F., David B.T., Primet P., Frameworks and patterns for synchronous groupware : AMF-C approach. IFIP Working Conference on Engineering for HCI : EHCI'98. Kluwer Academic Publishers. Greece. p. 225-241. September 1998

# Configuring the Design Process –
# Applying the JIET Design Process Framework

*Helmut Degen*

Vodafone Holding GmbH
Global Products & Services
Berger Allee 25
D-40213 Duesseldorf,
Germany
Emails:
helmut.degen@vodafone.com
h.degen@acm.org

*Sonja Pedell*

The University of Melbourne
Department of Information
Systems
111, Barry Street
3010 Victoria, Australia
Emails:
spedell@studentmail.dis.unim
elb.edu.au, pedell@acm.org

*Stefan Schoen*

Siemens AG
Corporate Technology
Otto-Hahn-Ring 6
D-81730 Munich, Germany

Email:
stefan.schoen@siemens.com

## Abstract

To be able to offer appropriate and efficient design processes, the JIET design process framework developed for web based e-business applications is presented. By means of a configuration model, the appropriate design process can be tailored respective to the conditions of industrial Tippfehler projects. The configuration model consist of four parameters to identify a suggest process: availability of users (or not), existing user interfaces (or not), available resources, and intended quality standard. The JIET design process framework and one part of the configuration model is outlined.

## 1 Introduction

Several publications deal with design processes, also known as usability engineering processes (e.g. 2002; Constantine & Lockwood, 1999; Beyer & Holtzblatt, 1998). These design processes seem to be carried out from one starting point in a very similar way. However, in some industrial projects user interfaces already exist, while in others they do not. Sometimes users can be involved, sometimes not. Also the goal, i.e. the intended quality of user interfaces, may differ from project to project. One of the key questions in industrial projects is: Which parts of the design process should be carried out, if you have only 70 %, 50 %, or even 30 % of necessary resources available?

The published design processes do not consider different starting conditions explicitly and directly applicable for tailoring user interface projects in industry with a short time to market. Beyer & Holtzblatt offer an heuristic approach to tailor the process, but this is focussed on the contextual inquiry method only (Wood, 2002). Constantine (2000, 2002) offers some methods around his model based approach, but it is only implicitly described and therefore not directly applicable.

Therefore a user interface design process that takes these different conditions into account, and that can be customized to the needs of individual projects, is needed (Vredenburg, Isensee & Righi 2002). How does a design process look like when different steps have to be left out because of restricted time or resources, or where steps have already been carried out? The idea is to develop a design process framework together with a configuration model. With this configuration model, an appropriate design process can be tailored according to existing project conditions.

This paper is structured in four further sections: In the following section related work is sketched. Then the development of the JIET Design process is described. The third section describes the

JIET design process framework, and why it is a usable design process. The last section summarizes and shows future work.

For better understanding, we would like to introduce some basic terms. There is a *provider* of an e-business solution, e.g. Siemens. The *client* is the sponsor of the user interface designers and often responsible for developing the application, e.g. Boston Consulting Group. The *users* are the persons who use the e-business application. User interface designers are people with two core competencies: The *interaction designer* is responsible for the interaction architecture as well as usability work. The *visual designer* is responsible for finding and designing the visual language of the user interface.

The presented design process framework was elaborated during industrial practice in an inductive way over the last 2.5 years. We followed the following steps: 1. New e-business methods modules were developed or existing modules were adapted to the specific requirements of e-business projects. We optimized them during many e-business projects. 2. The structure of the new design process framework was reorganized. In order to find a structured framework, a literature research was conducted. The most relevant result of the literature research was the process model of ISO 13407. We used this model as a starting point and enhanced it according to the requirements of e-business design processes. 3. This design process framework was applied in every new e-business project followed by a systematic evaluation and improvement. 4. Within a 2-day process workshop, eight usability experts from the Competence Center "User Interface Design" (Siemens AG, Munich) reviewed and improved the process framework.

## 2    JIET Design Process Framework

### 2.1    Overview

We call this new design process framework for e-business applications "JIET Design Process Framework" (herein called JIET = Just in ETime). Goal of the framework is to offer a set of method modules in order to be able to deliver user interfaces with the highest possible quality according to project conditions and goals. The JIET framework consists of nine phases that are shortly described (see Table 1).

Project setup (phase 0) is the start of the user interface design project. The user interface designers obtain an understanding of the project, the client, the audience, the application and the expected kind of support regarding the intention of involving user interface experts. The starting conditions of the project will be identified, and the decision process will be tailored. After obtaining the signed contract, the audience identification (phase 1) is the next phase. The purpose of the audience identification phase (1) is to understand the business of the provider and to identify the user roles. Within the use context analysis (phase 2), the use context of the target audience will be analyzed. For each user role we identify job-related goals, tasks, core tasks, scenarios, processes, tools, and content objects. The purposes of the requirements gathering phase (3) is to gather use scenarios and requirements for each scenario and user role. The purpose of the user interface design phase (4) is to create the user interface architecture. Within the evaluation phase (5), the designed user interfaces are evaluated. The results of the evaluation phase are considered to improve the user interface architecture within the redesign phase (6). All existing results of the design phase (4) will be improved. Within the design specification phase (7) the design specification is compiled based on the results of the last design phase. In the last phase (8 Process Improvement), the applied design process is analyzed, and improvement potentials of applied method modules are identified, based on the results of the usability evaluation (phase 6).

**Table 1: JIET Design Process Framework**

| Phase | Available method modules |
|---|---|
| 0 Project Setup* | 0.1 Project Characteristics (Client)**<br>0.2 Project Planning**<br>0.3 Proposal** |
| 1 Audience Identification* | 1.1 Enterprise Profile (Provider)**<br>1.2 Business Process Profile**<br>1.3 Business Customer Profile (Provider)<br>1.4 Private Customer Profile (Provider)<br>1.5 Supplier Profile .(Provider)<br>1.6 Competitor Profile (Provider) |
| 2 Use Context Analysis* | 2.1 Enterprise Profile (User)**<br>2.2 Business Process Profile (refined)**<br>2.3 Job Profile (User)**<br>2.4 Activity Profile (User)**<br>2.5 User Profile (User)<br>2.6 Competitor Profile (User)<br>2.7 Process Profile (User)<br>2.8 Content Profile (User)<br>2.9 Tool Profile (User)<br>2.10 User Role Profile (User)*<br>2.11 Usability Goals |
| 3 Requirements Gathering* | 3.1 NOGAP**<br>3.2 Card Sorting |
| 4 User Interface Design* | 4.1 Interaction Architecture**<br>4.2 Content Architecture<br>4.3 User Interface Architecture**<br>4.4 Prototype |
| 5 Evaluation | 5.1 Usability Check<br>5.2 Optimization Profile**<br>5.3 Usability Inspection<br>5.4 Usability Test (summative)<br>5.5 Usability Test (formative) |
| 6 Redesign | 6.1 Optimized Interaction Architecture**<br>6.2 Optimized Content Architecture<br>6.3 Optimized User Interface Architecture**<br>6.4 Optimized Prototype |
| 7 Design Specification | 7.1 Design Specification** |
| 8 Process Improvement | 8.1 Process Control** |

**Legend**

\*  This is a duty phase in each user interface project and should be applied in any case

\** This method module is a duty module within its phase and should be applied in any case

## 2.2 Configuration Model

Each project has several starting conditions. In order to be able to offer the most efficient design process for certain project conditions an adapted design process is necessary. For this purpose, a configuration model is required. One particular task in creating a configuration model means to identify all of these starting conditions that help to select an appropriate design process. These relevant starting conditions are herein called *starting parameters*. All in all, four parameters could be identified that enable us to select the appropriate design process. Two of the four parameters are basis parameters. One parameter stands for the status of user interfaces (do they already exist or not), the other for the availability of users during the process (they are involved or not). With these two starting parameters, we have four different configuration variants, shown in Table 2.

**Table 2: Configuration Variants, based on two basis parameters**

|  | User Interfaces do not exist | User Interfaces exist |
|---|---|---|
| **Users are not available** | Configuration Variant 1 | Configuration Variant 3 |
| **Users are available** | Configuration Variant 2 | Configuration Variant 4 |

**Table 3: Configuration Variant 3**

| Phases | Basis Process 1: Usability Assessment Process | Small | Medium |
|---|---|---|---|
| **Project Setup** | 0.1 Project Characteristics (Client) | ID,C | ID,C |
|  | 5.1 Usability Check | ID | ID,VD |
|  | 0.2 Project Planning | ID | ID,VD |
|  | 0.3 Proposal | ID | ID,VD |
| **Audience Identification (Provider)** | 1.1 Enterprise Profile (Provider) | ID,P | ID,P |
|  | 1.2 Business Process Profile | ID,P | ID,P |
|  | 1.3 Business Customer Profile (Provider) | ID,P* | ID,P* |
|  | 1.4 Private Customer Profile (Provider) | ID,P* | ID,P* |
|  | 1.5 Supplier Profile (Provider) | ID,P* | ID,P* |
| **Use Context Analysis (User)** | 2.1 Enterprise Profile (User) | ID,P | ID,VD,P |
|  | 2.2 Business Process Profile (refined) | ID,P | ID,VD,P |
|  | 2.3 Job Profile (User) | ID,P | ID,VD,P |
|  | 2.4 Activity Profile (User) | ID,P | ID,VD,P |
|  | 2.10 User Role Profile (User) | ID,P | ID,VD,P |
| **Evaluation** | 5.2 Optimization Profile | ID,P | ID,VD,P |
|  | 5.3 Usability Inspection |  | D,VD,P** |
|  | **Basis Process 2: User-Centered Design Process** |  |  |
| **Audience Identification (Provider)** |  | ID,P**** | ID,P**** |
| **Use Context Analysis (User)** |  | ID,P**** | ID,VD,P**** |
| **Requirements Gathering** |  | ID,P**** | ID,VD,P**** |
| **User Interface Design** | 4.1 Interaction Architecture | ID | ID,VD |
|  | 4.2 Content Architecture |  | ID,VD |
|  | 4.3 User Interface Architecture | ID | ID,VD |
| **Evaluation** | 5.3 Usability Inspection |  | ID,VD,P |
| **Redesign** | 6.1 Optimized Interaction Architecture |  | ID,VD |
|  | 6.2 Optimized Content Architecture |  | ID,VD |
|  | 6.3 Optimized User Interface Architecture |  | ID,VD |
| **Design Specification** | 7.1 Design Specification |  | ID,VD |
| **Required Resources** |  | Low | Medium |
| **Achievable Quality** |  | Medium - | Medium o |

**Legend**

\*     Will be carried out, if this profile is relevant for the project.

\*\*    Will be carried out, if the usability check (phase 0) established that the user interface quality is sufficient for a usability test.

\*\*\*  Will be conducted, if a separate order is requested for the second basis process (user centered design);

\*\*\*\* Will be carried out, if additional information is required.

ID: Interaction designer involved; VD: Visual designer involved; C: Client involved; P: Provider involved; U: User involved

For each configuration variant, the configuration model offers different alternatives. Each of the further alternatives is identified by the third and fourth parameter: Resources and Quality. To describe how to apply this configuration model, we selected the configuration variant 3 (user interfaces already exist, but users are not available). A usability test is required to achieve a high quality standard. Since users are not available we cannot conduct such tests and therefore the achievable quality can be only "medium" for variant 3. Therefore, we offer two alternatives (see Table 3), a "Small" and a "Medium" alternative.

The "small" version begins with the project setup (phase 0). To ensure that the existing user interfaces have a necessary quality standard, a short usability check has to be carried out. If the quality is too bad we cancel the rest of the usability assessment and suggest to design new user interfaces. If the usability quality achieves a certain quality standard, we carry out the usability assessment. After setting up the project the target audience (phase 1) will be identified. We carry out a use context analysis (phase 2). These results are used for the evaluation (phase 5). Within the "small" variation the evaluation is an open interview (5.2 Optimization Profile). In the "medium" variation a usability inspection is carried out (5.3 Usability Inspection). In both cases the interview or inspection partners are representatives of the provider.

The second basis process uses the results of the usability assessment. The purpose is to improve the user interfaces. The first four phases (Project Setup, Audience Identification, Use Context Analysis, Requirements Gathering), or parts of it, are only repeated if necessary information is missing. Within the "small" variation, the user interface design consists of the interaction and the user interface architecture. Within the "medium" variation, the content architecture will be carried out in addition. Since we do not evaluate the user interfaces by means of a usability test, we do not need a (interactive) prototype. Only the "medium" version consists of another evaluation (phase 5) and redesign (phase 6) phase. If required a design specification (phase 7) will be compiled.

Apart from these different method modules, competence of the people is a relevant distinction between the "small" and "medium" version . In the "small" version, only an interaction designer is involved, because of the intended quality and available resources, whereas in the "medium" version, a visual designer is involved in addition.

# 3 Conclusions and Future Work

Compared to ISO 13407 user-centered design process, two important phases are added: The "Audience Identification" (Phase 1) because in e-business projects the target audience is not obvious. Another new phase is the "Process Improvement" (Phase 8), which is applied to evaluate and improve the applied design process.

During industrial practice, the JIET design process framework should be improved. We assume the framework itself can be applied to other domains outside e-business.

# 4 References

Beyer, H. & Holtzblatt, K. (1998): Contextual Design. San Francisco, USA: Morgan Kaufmann.

Constantine, L. L. & Lockwood, L. D. ( 1999): Software for Use: Reading, USA: Addison-Wesley, ACM Press.

Constantine, L. (2000): Cutting Corners: Shortcuts in Model-Driven Web Design. Retrieved February 10, 2003, from http://foruse.com/articles/shortcuts.htm.

Constantine, L. (2002): Process Agility and Software Usability: Toward Lightweight Usage-Centered Design. Retrieved February 10, 2003, from http://foruse.com/articles/agiledesign.htm.

Vredenburg, K.; Isensee, S. & Righi, C. (2002): User-Centred Design. Upper Saddle River, USA: Prentice Hall.

Wood, S. (2002): Targeted Contextual Design — Scope Your Project. Document sent by email at 12th of November 2002.

# Finding Decisions Through Artefacts

*Alan Dix, Devina Ramduny, Paul Rayson, Victor Onditi,*
*Ian Sommerville and Adrian Mackenzie*

Lancaster University, Computing Department
Lancaster, LA1 4YR, UK

alan@hcibook.com, devina@comp.lancs.ac.uk, paul@comp.lancs.ac.uk,
v.onditi@lancaster.ac.uk, is@comp.lancs.ac.uk, a.mackenzie@lancaster.ac.uk

`http://www.hcibook.com/alan/papers/HCII2003-artefacts/`

## Abstract

This paper addresses the use of artefacts as a resource for analysis. Artefacts are particularly useful in situations where direct observation is ineffective, in particular for infrequent activities. We discuss two classes of techniques: focusing on the 'artefact as designed' as a means of recovering designers' explicit and implicit knowledge and 'artefacts as used' as a means of uncovering the trail left by currently inactive processes. These techniques have been applied using a meeting capture system and meeting minutes as the artefact resources. This is part of a wider study to understand the nature of decisions and so reduce rework.

## 1    Introduction

The ethnographic literature is full of the importance of artefacts as the means with which individuals represent, mediate and negotiate work in collaborative settings (Hughes, O'Brien, Rouncefield, Sommerville & Rodden 1995). This is also recognised in approaches such as distributed cognition (Hutchins, 1990) and situated action (Suchman, 1987) as well as some more traditional cognitive models (Howes & Payne, 1990). In our previous work, we have studied the way in which artefacts in their setting act as *triggers* for action (Dix, Ramduny & Wilkinson, 1998) (Dix, Ramduny & Wilkinson, 2003) and *placeholders* for formal and informal processes (Dix, Wilkinson & Ramduny, 1998). In related work, we emphasised the centrality of artefacts as the focus of work and as the locus of communication *through the artefact* or feedthrough (Dix, 1994).

Because of this we have proposed various forms of artefact-centred analysis to run alongside more direct methods of observation (Dix, Ramduny, Rayson & Sommerville, 2001). We consider both:

- the artefact *as designed* – looking at the ways in which the explicit and implicit knowledge of the designer are exposed in artefacts
- the artefact *as used* – looking at the way on which people have appropriated, annotated and located artefacts in their work environment

Artefact centred sources are particularly useful where an activity occurs or is only active infrequently so that direct observation may fail to record any instance or part of the activity at all. In the Tracker[1] project we are seeking to understand the nature of decisions in teams and

---

[1] http://www.comp.lancs.ac.uk/computing/research/cseg/projects/tracker/

organisations (Rayson et al., 2003); in particular the way past decisions are acted on, referred to, forgotten about and otherwise function as part of long term organisational activity. In one of the ethnographic studies we carried out, we found that 'real' decisions were made between meetings or implicitly assumed, but rarely explicitly voiced during official meetings. Furthermore, interactions between decisions take place over periods of month or years. In other words, they have exactly the properties that make artefact-centred approaches attractive.

## 2    The artefacts as a resource

Like the fossil left where the soft parts of the body have decomposed, artefacts act as a residual record of work done and work in progress. In and of themselves they form a resource for analysis. Furthermore, just like the palaeontologist looking at fossils, there are a variety of circumstances in work domains where the 'soft tissue' of lived work, the ephemeral actions and words, are difficult or impossible to collect and so the matrix of artefacts that remains needs to be interpreted.

This may be because the actions have already taken place and so the physical remains are our only resource. In the Tracker project we have access to a corpus of meeting minutes. The meetings have long past; we cannot go back and observe what happened; at best we can interview some of the participants; but the formal minutes remain – fossils of the moment. We will return to these formal minutes later.

Perhaps more fundamentally there are some classes of human activity that direct observation cannot, or cannot easily, capture. Where a class of activity is frequent and short lived we can expect that periods of direct observation, such as ethnographic studies, will completely capture some instances of the activity from end to end. Where activities are longer lived, direct observation can at best hope to capture aspects of the activity at different points and so piece together the complete story from parts. But where a class of activity happens infrequently or is only active infrequently, direct observation can only record the activity if it happens to overlap with one of the infrequent episodes.

However, these activities, even when inactive must in some way still have a representation within the organisational ecology: in people's memories *and* in physical or electronic artefacts. The 'and' in the previous sentence is not just in the sense that both will be present, but in the more holistic recognition that the interpretation of artefacts is itself invested within the human understanding of the context. Artefacts tell a story to the extent that they invoke stories. To some extent as analysts we may understand the contexts well enough to 'read' artefacts, in others the artefacts can form the prompts to evoke memories during formal and informal interviews.

## 3    The artefact as designed

Long lasting artefacts: tools, procedures, documentation, buildings, organisational structures, have all by explicit action been 'designed'. As we know these designs can often fail and so are not paradigmatic. However, they are a powerful resource embodying the knowledge, skills and assumptions of the original designer. We call this *archaeologically-inspired artefact analysis*. An archaeologist will look at the artefacts produced by long-dead civilisations and by considering the design infer the patterns of use, work and social activity that surrounds those artefacts. This process is problematic as we may draw tenuous conclusions from meagre evidence, but is in fact more robust as a contemporary technique as we are in a better position to understand the target context and may also be in a position to use this as a resource in participative critique.

In the early stages of the Tracker project, we reviewed a number of meeting support systems. We analysed in greatest detail TeamSpace[2] (Richter et al., 2001), which is related to the very successful Classroom2000 (eClass) system (Abowd, 1999). In the version of TeamSpace we tested, we found various classes of context assumptions. Some are explicitly embedded in the software; for example, TeamSpace requires meetings to be scheduled. Some are explicit in the documentation but not enforced; for example, the suggestion that a facilitator is necessary. Some are implicit in the software; for example, if you stop and then restart a meeting, the audio recording for part of the meeting is lost, implicitly assuming meetings do not break and reconvene.

## 4    The artefact as used

In previous work we have focused especially on the fact that artefacts encode the state and trigger action not just by their explicit content or significance, but also by their disposition in the environment. A piece of paper at a particular location on the desk may mean "file me"; in another location, perhaps in a straight pile means "in progress"; and on the same pile, but higgledy-piggledy at an odd angle means "to be read". By taking an office at the end or beginning of a day we can use these artefacts to tell the story of the activities that are, in a temporal sense, passing through the office at that moment. Most significantly this includes activities which are not currently captured in the 'official' systems or whose state is indeterminate or intermediate between 'official' stages. We call this *transect analysis* as it is similar to the field biologist's use of a transect through an ecosystem such as a shoreline.

Unfortunately meetings are an extreme case of 'clean desk policy'. The documents and artefacts are removed from the room with the participants – the only remnant of the meeting is the explicit records and the changed memories and attitudes of the participants.

The one obvious artefact that is left behind by a meeting is the formal minutes. These are problematic as they are not a record of *what happened* at the meeting, but rather a *sanitised account* prepared for a purpose, by an individual. Although problematic the minutes are significant as they are the foci by which the participants agree (or are forced to agree) to a fiction that in some way legitimises future actions. In the extreme, in certain legal situations, minutes of meetings are created which never occurred – quite literally legitimising the desired end state by an agreed legal fiction of the process.

To some extent the artificial nature of the formal minutes reflects the artificial nature (in the sense of artifice) of collaborative activity. Ethnomethodology makes a strong focus on the accountability of individuals – that they can make stories (accounts) about their actions that legitimise them socially. We have to read formal minutes carefully, more like an *historical document*, written by someone, for a purpose, but nonetheless exposing aspects of the real process.

As noted our focus is on decisions and this has proved even more problematic. In ethnography of actual meetings one of the marked results was the fact that decisions did not 'happen' in the meeting. This is not to say that formal minutes would not record decisions (or their consequences), but that there are not clear points of decision-making. Instead decisions have either clearly been made prior to the meeting and are merely brought into the meeting to validate them, or alternatively decisions are 'made' implicitly by simply discussing an issue that the minute taker reads later as a particular outcome.

---

[2] http://www.research.ibm.com/teamspace/

This problematic nature is also evident in the minutes themselves. Formal minutes do not explicitly record 'decisions' but instead either note agreed statements or 'actions', usually relating to formally numbered items in the meeting. Whereas formal actions are explicitly marked there is no such explicit marking for decisions (or related topics such as options, issues etc.). Instead an extensive manual analysis was required to identify salient features.

When the analysis started we had some discussion about the level of structure required in the analysis. The minutes we studied themselves had a fairly consistent formal structure: date, participant list, numbered items, comments and listed actions against each item. Also there are a number of ontologies of decision making from the design rationale and decisions support literature (e.g. IBIS (Conklin & Begeman, 1988), QOC (MacLean, Young, Bellotti & Moran, 1991), DRL (Lee & Kai, 1991). Based on these a database structure was created to record decisions, actions, issues and relations between them. So, for example, a decision would have associated actions, actions would have responsible persons and optionally a deadline.

As the analysis proceeded, it became increasingly clear that the reality of the 'formal' minutes was, perhaps not surprisingly, far less structured and far more ad hoc than our predefined structure. Even the explicit 'actions' sometimes turn out to be more comments or statements of intent and some actions are not marked as such. Decisions are far more complicated as they are sometimes explicit in the text and sometimes inferred from context (e.g. an action presupposes a decision to take action).

In the end the rigid structure has been dropped, except for the record of the explicit structure of the minutes themselves, and the analysis uses a simple recording (in a database to make it amenable to search and analysis) of 'things' and relations between them. With this more flexible structure, the analyst is no longer restricted to a fixed repertoire of concepts. While the analyst is reading the minutes, if she feels that any issue ought to be recorded she can now easily do so by adding a 'thing' with as many named attributes as desired. Only a short title/description, link to the raw transcript and 'type' field are required. The last of these is to enable the recording of terms such as 'decision', 'action' and the like, but not constrained to a predetermined vocabulary. The aim is to see an ecologically valid ontology *emerge* from the ongoing analysis.

# 5   Conclusion

We have seen using the example of TeamSpace, how it is possible to use the artefact as designed as the analytic resource. Similarly, our analysis of minutes is an example of using artefacts as used. In both cases, the permanent record embodied in artefacts has given us an understanding of the nature of decisions which otherwise prove elusive to direct observation.

# 6   Acknowledgement

The Tracker project is supported under the EPSRC Systems Integration programme in the UK, project number GR/R12183/01.

# 7   References

Abowd, G. (1999). Classroom 2000: An Experiment with the Instrumentation of a Living Educational Environment. *IBM Systems Journal, Special issue on Pervasive Computing*, 38(4) 508-530.

Conklin, J., & Begeman, M. L. (1988). gIBIS: A hypertext tool for exploratory policy discussion. In *Proceedings of the Second Conferences on Computer-Supported Co-operative Work*, New York, USA. ACM Press, 140-152.

Dix, A. (1994). Computer-supported cooperative work - a framework. In D. Rosenburg and C. Hutchison (Eds.) *Design Issues in CSCW* (pp 23-37). Springer Verlag.

Dix, A., Ramduny D., & Wilkinson, J. (1998). Interaction in the Large. *Interacting with Computers*, 11(1), 9-32.

Dix, A., Wilkinson, J., & Ramduny, D. (1998). Redefining Organisational Memory – artefacts, and the distribution and coordination of work. In *Understanding work and designing artefacts* (York, 21st Sept., 1998). http://www.hiraeth.com/alan/papers/artefacts98/

Dix A., Ramduny, D., Rayson, P., & Sommerville, I. (2001). Artefact-centred analysis - transect and archaeological approaches. In *Team-Ethno Online, Issue 1 - Field(work) of Dreams*, Lancaster University, UK. http://www.teamethno-online.org/Issue1/Dix.html

Dix, A., Ramduny-Ellis, & D., Wilkinson, J. (2003). Trigger Analysis - understanding broken tasks. In D. Diaper & N. Stanton (Eds.) *The Handbook of Task Analysis for Human-Computer Interaction*. Lawrence Erlbaum Associates.

Hughes, J., O'Brien, J., Rouncefield, M., Sommerville, I., & Rodden, T. (1995). Presenting ethnography in the requirements process. In *Proceedings of IEEE Conf. on Requirements Engineering, RE'95*. IEEE Press, 27–34.

Howes, A., & Payne, S. (1990). Display-based competence: towards user models for menu-driven interfaces. *International Journal of Man-Machine Studies*, 33, 637-655.

Hutchins, E. (1990). The Technology of team navigation. In Gallagher, J., Kraut, R. and Egido, C., (Eds.) *Intellectual teamwork: social and technical bases of collaborative work*. Lawrence Erlbaum.

Lee, J., & Lai, K.-Y. (1991). What's in Design Rationale? *Human-Computer Interaction*, 6(3&4), 251-280.

MacLean, A., Young, R., Bellotti, V., & Moran, T. (1991). Questions, options and criteria: Elements of design space analysis. *Human-Computer Interaction*, 6(3 & 4), 201-250.

Suchman, L. (1987). Plans and Situated Actions: The problem of human–machine communication. Cambridge University Press.

Rayson, P., Sharp, B., Alderson, A., Cartmell, J., Chibelushi, C., Clarke, R., Dix, A., Onditi, V., Quek, A., Ramduny, D., Salter, A., Shah, H., Sommerville, I., & Windridge, P. (2003). Tracker: a framework to support reducing rework through decision management. To appear in *5th International Conference on Enterprise Information Systems ICEIS2003*, Angers, France, April 23-26, 2003.

Richter, H., Abowd, G., Geyer, W., Fuchs, L., Daijavad, & S., Poltrock, S. (2001) "Integrating Meeting Capture within a Collaborative Team Environment". In *Proceedings of the International Conference on Ubiquitous Computing, Ubicomp 2001*, Atlanta, GA, September, Springer, 123-138.

# The Constrained Ink Metaphor

*Björn Eiderbäck*
CID/KTH
SE-100 44 Stockholm
Sweden
bjorne@nada.kth.se

*Sinna Lindquist*
CID/KTH
SE-100 44 Stockholm
Sweden
sinna@nada.kth.se

*Bosse Westerlund*
CID/KTH
SE-100 44 Stockholm
Sweden
bosse@nada.kth.se

## Abstract

In this paper we describe a novel metaphor for developing interactive computer applications, *the constrained ink metaphor*. Crucial to the development of the constrained ink was an aim to find simple and natural means for defining and implementing interaction among persons. We will describe how we was lead to considering this metaphor, some basic inks following the metaphor, and finally some typical applications and their impact on the development of the metaphor.

## 1 Background, Goals and Influences

### 1.1 The interLiving Project

The interLiving project aims to study and develop, together with families, technologies that facilitate generations of family members living together with the objectives: to understand the needs of diverse families; to develop innovative artefacts that support the needs of co-located and distributed families; to understand the impact such technologies can have on families (Beaudouin-Lafon, 2002, Hutchinson, 2003).

### 1.2 Goals and Research Questions

A key objective of the interLiving project is to experiment with different design methodologies. We would like to develop better ways of letting the family members directly influence and shape the design of communication technologies we develop together with them.

The *premiere goal* with the particular work described in this paper is: to develop an infrastructure and metaphor that will enable us to build applications were we leave as much us possible open to the co-development with families, even late in the development process. *Secondary goals* are: i) that it should be easy and natural to develop all our intended applications by means of this infrastructure and metaphor; ii) that the metaphor should encourage development of applications that are fun to use (and develop!)

*Research Questions* are:
- Is it possible to create an infrastructure and metaphor of the type we strive for in the goals?
- For which types of applications is the metaphor well suited and for which types is it not naturally applicable?

### 1.3 Technology Probes

As inspiration and triggering techniques we have used technology probes. A 'technology probe' combines the social science goal of collecting data about the use of the technology in a real-world setting, the engineering goal of field-testing the technology and the design goal of inspiring users (and designers) to think of new kinds of technology (Beaudouin-Lafon, 2002 Chapter 2). The probe that influenced the development of the applications we currently are working on most is *The Message Probe*, a simple application that enables members of a distributed family to communicate with digital notes using a pen and tablet interface. Already at early stages of the development of applications inspired by these probes we realized that we required something that both was fun to use and easily adoptable to various and changing requirements. This in turn led us to the development of *The Constrained Ink Metaphor*.

## 1.4 Influencing Approaches

There are of course a lot of achievements in the history that has inspired, or at least influenced, our development. For instance Ivan Sutherlands pioneering work on Sketchpad (Sutherland, 1963), the NLS system in the SRI project (Engelbart, 1975), the very direct manipulated A Reality Toolkit (ARK) (Smith, 1987), editors for drawing and animation like Macro Mind Director, Calendaring facilities (Beaudouin-Lafon, 2002), and the more recent KidPad (Benford, 2000). We have also been inspired by work done in CSCW and design patterns (Eiderbäck 2001).

# 2 Ink of Various Kind and their Usage

## 2.1 The Metaphor: What, Why, and How

*The constrained ink metaphor* is a novel metaphor for developing interactive computer applications. The idea of it sprung from an attempt to develop a common base for a message central and a distributed shared drawing editor, intended for communication between family members possible living in different households. In the former case we focus on the same place different time aspects were we want to provide for submitting shared notes visible within certain time frames. In the latter case we focus on same time different place aspects were we for instance want to provide for co-operative drawing, communication and address awareness aspects. Our intention is to enabling communication of both important facts and more informal chatting in a way youngsters, adults, and elder members of the family, computer literate or not, could find useful and "fun"! We discussed the concept together with the families and agreed that it seemed to be promising, useful and fun.

## 2.2 Ink

Central to the Constrained Ink Metaphor, as its name suggests, is the Ink!

### 2.2.1 Natural Inks

There are a lot of different types of ink that could be considered natural in the sense that they more or less have their counterparts in the real world. For instance, we have the invisible ink that even a small children most likely have experiences from using a special purposes pen with ink that only appears after one heat the paper it is written on. Another natural ink is the aging ink; actually this is the way all inks work, where the ink slowly disappears from the material it is written on. However in our computerized versions we have speeded up and made the aging more controllable.

The Coloured Ink

As a basis we use ordinary coloured ink, i.e. all inks have a defined colour or texture. On top of this basic ink all the other inks was developed, by applying various constraining schemas that made them behave and response to external events.

## The Invisible Ink

*The Invisible Ink* is the most natural of all the constrained inks.

### Context

The user wants to write a note that should be presented at a specified time in the future. Thereafter the note should stay until someone actively removes it.

### Problem

How could we provide model providing a means to construct entities that should appear at a specific time in the future? How could we develop a model that fits into and is suitable for all the various applications we are developing within the project?

### Forces

The model should be natural to use. The usage of the model should not constrain the process or the interaction. The model should be natural for handling constrained entities of various kinds as graphical one, e.g. lines and ovals, and non graphical ones, e.g. email and speech. It must be feasible to implement the model in software.

### Solution

Make a computerized version of an invisible ink. For convenience for programmers incorporate the ink model into the system's ordinary model of drawing with various colours and textures, i.e. it should be possible to use the ink for colouring objects even in "non-ink aware" applications. Therefore separate parts for handling the interaction with the ink from ones handling its behaviour and ones handling its visible appearance. In this way one could easily change or adopt new behaviour to ink and at a very fined grained level control its constraints.

## The Aging Ink

The Aging Ink is ink that disappears after a pre-defined time. It works as ordinary ink, but we have speeded up the decaying process and also made it more abrupt.

### Context

The user wants to write a note that is valid from the time the note is written until a certain time in the future.

### Problem and Forces

The problem forces are the same as for the Invisible Ink but now the entities should disappear after a while instead.

### Solution

The solution follows the same lines as the one for the Invisible Ink. With the separation of controlling and behaviour from appearance we only has to replace the constraint controller for one that makes the ink disappear after a certain time, instead of appear as for the former ink.

## 2.2.2 Generally Constrained Inks

After discussing applications, and reflecting on our earlier prototypes among ourselves but also with our families we considered the ink metaphor in more dept. We realised that the natural inks would not solve all the problems that we intended. We require to entities responding to general events, as someone pushing a button or joining a family's network. Therefore we decided to expand the metaphor further to see if it could be useful even in ways that not have their direct counterparts in ink from the natural world.

## 2.2.3 Asymmetric Inks

We also want to be able to show things differently, or at different times, at diverse platforms. Sometimes everything should be visible to all users in the same way at other times some parts are not visible to all users or just presented differently to some of them. Entities could even be handled on dissimilar platforms and by different media by various users, i.e. on use speech at a

PDA whereas another user has a graphical platform with a text interface. Therefore we try to investigate the impacts these situations has on the ink and try to develop ink that also are suitable for them.

### 2.2.4   Inks Intended for Sharing

In some senses we could use the previously described inks for sharing. We have inks visible at all platforms, inks that appear differently for diverse users, etc. However, only relying on these inks makes sharing of artefacts required in a more general sense very clumsy. To address this we have played with inks that could define certain (filled) areas where all other inks painted on the area should be visible by a shared and connected community. Thereby we could easily, within the limits of the constrained ink metaphor, even provide for shared desktops and other means of co-operative work. Therefore we also investigate how this type of ink is usable and fits into the metaphor.

## 3   Some Typical (Ink Based) Applications

### 3.1   Type of Application

The applications we currently are working on affect the type of ink required in different ways. In this section we very briefly exemplify of the various types of applications we consider. These considerations are a basis for our further development and exploration of the constrained ink metaphor. Some typical kinds of applications are:

- *Synchronous vs. Non-Synchronous Applications.* There is an obvious difference between synchronous and non-synchronous applications. In the former case communication takes effect momentarily whereas the latter case is more indirect, probably taken its way via some server, storage medium, or alike.
- *Shared vs. Non-Shared Applications.* Another situation we must consider is if the application should be shared, i.e. everyone manipulates a shared set of entities, or non-shared where different users could manipulate their own restrictive set of the entities.
- *(Just) Graphical vs. Multimedia.* Typical shared applications of today also provide for other media than graphics. Examples are telephony over IP, and videoconferences.
- *Sinking Ships.* An archetypical application where different users at certain times sees different parts of the entities or even presented in different ways is the famous game Sinking Ships.

### 3.2   Applications

Currently we are focusing on two different applications. The InkPad and the Door. We also explore some types of interaction, not central in the other two, in a Pie Diagram framework. In the sense of exploring the constrained ink metaphor the InkPad is the most central and new kinds of ink and constraints are first tested within this application.

### 3.2.1   A Shared (Drawing) Editor, InkPad

The InkPad is a tool with the main aim to enabling free and non-formal communication among family members of all ages. To support free communication we try to make InkPad an as relaxed environment as possible. The focus on this prototype is on enabling communication of both important facts and more informal chatting in a way both youngsters, adults, and elder members of the family, computer literate or not, could find useful and "fun".

The user could choose ink from any of the previously described types of ink. In this way the user could achieve effects as writing messages and notes that will appear or disappear at specific times. We have also considered other media, such as audio, video, and speech.

### 3.2.2  Message Central, the Door

We also develop a message central nick named *The Door*, from the first intended placement in the household. The Door prototype is an effort to improve the communication and scheduling of activities among family members. At the start we concentrate on communication between members living in the same household. In this case we use the ink metaphor for controlling and delivering messages.

### 3.2.3  Pie Diagrams

Pie Diagrams are just like ordinary pop up menus but circular. In particular we investigate how invisible ink could be used to supporting expert users that now the relative location of certain submenus and the items they want to chose. The ink is constrained to only paint a certain sub-pie if the user "fires a certain event", by for instance stopping the movement more than a pre-defined time limit. In such a case the ink reacts by switching from transparent colour to non-transparent ones and thereby makes the pie visible.

## 4    Conclusions and Future Work

In this paper we have described the *Constrained Ink Metaphor* by describing various forms of (constrained) inks and their usage. We demonstrated that the metaphor is both natural and useful for developing a various set of interactive and distributed applications.

From now on we will investigate the metaphor further by using it to full extent while continuing the development of a various set of applications within the interLiving project.

## 5    References

Beaudouin-Lafon M., Bederson B, Conversy S., Druin A., Eiderbäck B., Evans H., Hansen H., Harvard Å., Hutchinson H., Lindquist S., Mackay W., Plaisant C., Roussel N., Sundblad Y., Westerlund B (2002). *Co-design and new technologies with family users*, Deliverable 1.2 & 2.2, 2002-09-23.

Benford, S., Bederson, B., Akesson., K., Bayon, V., Druin, D., Hansson, P., Hourcade, J., Ingram, R., Neale, H., O'Malley, C., Simsarian, K., Stanton, D., Sundblad, Y., and Taxen, G (2000). *Designing Storytelling Technologies to Encourage Collaboration Between Young Children*. Proceedings of CHI, pp. 556-563.

Eiderbäck B. (2001) *Object Oriented Frameworks with Design Patterns for Building Distributed Information Sharing*. PhD thesis, ISBN 91-7265-240-3, http://www.nada.kth.se/~bjorne/th/.

Engelbart, D.C. (1975), *NLS tele-conferencing features: the journal and shared-screen telephoning*, in Proceeding Fall COMPCON, Sept. 1975.

Hutchinson, Mackay, Westerlund, Bederson, Druin, Plaisant, Beaudouin-Lafon, Conversy, Evans, Hansen, Roussel, Eiderbäck, Lindquist, Sundblad. (2003). *Technology Probes: Inspiring Design for and with Families. Proceedings of ACM CHI 2003*.

Smith, R (1987). *Experiences with the alternate reality kit: an example of the tension between literalism and magic*, Proceedings of CHI + GI.

Sutherland, I (1963). *SketchPad: A man-machine graphical communication system*, AFIPS Spring Joint Computer Conference, No 23.

# Meta—Design: Beyond User-Centered and Participatory Design

*Gerhard Fischer*

University of Colorado, Center for LifeLong Learning and Design (L3D)
Department of Computer Science, 430 UCB
Boulder, CO 80309-0430 – USA
gerhard@cs.colorado.edu

## Abstract

Meta-design characterizes objectives, techniques, and processes for creating new media and environments that allow the owners of problems to act as *designers*. A fundamental objective of meta-design is to create socio-technical environments that empower users to engage in informed participation rather than being restricted to the use of existing systems. The seeding, evolutionary growth, reseeding model is a process model that supports meta-design. We have explored and assessed meta-design approaches in the development of innovative computational environments and in our teaching and learning activities.

## 1 Introduction

Our research interest is in designing the social and technical infrastructures in which new forms of collaborative design can take place. For most of the design domains that we have studied over many years (ranging from urban design to graphics and software design) [Arias et al., 2000], the knowledge to understand, frame, and solve problems is not given, but is constructed and evolved during the problem-solving process.

## 2 Design Time and Use Time

In all design processes, two basic stages can be differentiated: design time and use time. At *design time*, system developers (with or without user involvement) create environments and tools. In conventional design approaches they create complete systems. At *use time*, users or "stakeholders" use the system but their needs, objectives, and situational contexts can only be anticipated at design time, thus, creating a system that often requires modification to fit the user's needs. In order to accommodate unexpected issues at use time, systems need to be underdesigned at design time. *Underdesign* in this context does not mean less work and demands for the design team, but is fundamentally different from creating complete systems. The primary challenge of underdesign is in developing environments and not the solutions allowing the "owners of problems" at use time to create the solutions themselves. This can be done by providing a context and an interpretive background against which situated cases coming up later can be interpreted. Underdesign is a defining activity for meta-design by creating design spaces for others.

# 3   User-Centered Design and Participatory Design

*User-centered design* approaches [Norman & Draper, 1986] (whether done *for* users, *by* users, or *with* users) have focused primarily on activities and processes taking place at design time in the systems' original development and have given little emphasis and provided few mechanisms to support systems as *living* entities which can be evolved by their users. In user-centered design, designers generate solutions placing users mainly in a reactive role.

*Participatory design* approaches [Schuler & Namioka, 1993] seek to involve users more deeply in the process as co-designers by empowering them to propose and generate design alternatives themselves. Participatory design supports diverse ways of thinking, planning, and acting making work, technologies, and social institutions more responsive to human needs. It requires the social inclusion and active participation of the users. Participatory design has focused on system development at design time by bringing developers and users together to envision the contexts of use. But despite the best efforts at design time, systems need to be evolvable to fit new needs, account for changing tasks, and incorporate new technologies.

# 4   Meta-Design

*Meta-design* [Fischer & Scharff, 2000] extends the traditional notion of system design beyond the original development of a system to include an ongoing process in which stakeholders become *co-designers*—not only at design time, but throughout the whole existence of the system. A necessary, although not sufficient condition for meta-design is that software systems include advanced features permitting users in creating complex customizations and extensions. Rather than presenting users with closed systems, meta-design provides them with opportunities, tools, and social reward structures to extend the system to fit their needs. Meta-design shares some important objectives with user-centered and participatory design, but it *transcends* these objectives in several important dimensions and it has changed the processes by which systems and content is designed. Meta-design has shifted the control from designers to users and empowered users to create and contribute their own visions and objectives. Meta-design is a useful perspective for projects where 'designing the design process' is a first-class activity, meaning that creating the technical and social conditions for broad participation in design activities is as important as creating the artifact itself [Wright et al., 2002].

**The Seeding, Evolutionary Growth, and Reseeding (SER) Process Model.** The major conceptual framework which we have developed to support meta-design is the *seeding, evolutionary growth, and reseeding (SER) process model* [Fischer & Ostwald, 2002]. The SER model is a descriptive and prescriptive model for large evolving systems and information repositories postulating that systems that evolve over a sustained time span must continually alternate between periods of activity and unplanned evolutions and periods of deliberate (re)structuring and enhancement. The SER model encourages designers to conceptualize their activity as meta-design, thereby supporting users as designers in their own right, rather than restricting them to being passive consumers. Figure 1 provides a graphical illustration of the SER model.

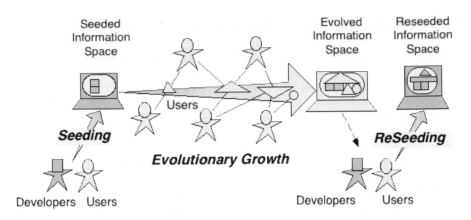

**Figure 1:** The Seeding, Evolutionary Growth, and Reseeding Process Model

**Informed Participation and Unselfconscious Cultures of Design.** *Informed participation* [Brown & Duguid, 2000] is a form of collaborative design in which participants from all walks of life—not just skilled computer professionals—transcend beyond the information given to incrementally acquire ownership in problems and to contribute actively to their solutions. It addresses the challenges associated with open-ended and multidisciplinary design problems. These problems involving a combination of social and technological issues, *do not have right answers*, and the knowledge to understand and resolve them changes rapidly. To successfully cope with informed participation requires social changes as well as new interactive systems that provide the opportunity and resources for social debate and discussion rather than merely delivering predigested information to users.

Being ill-defined, design problems cannot be delegated (e.g., from users to professionals), because they are not understood well enough that they can be described in sufficient detail. Partial solutions need to "talk back" [Schön, 1983] to the owners of the problems who have the necessary knowledge to incrementally refine them. Alexander [Alexander, 1964] has introduced the distinction between an unselfconscious culture of design and a selfconscious culture of design. In an *unselfconscious* culture of design, the failure or inadequacy of the form leads directly to an action to change or improve it. This closeness of contact between designer and product allows constant rearrangement of unsatisfactory details. By putting owners of problems in charge, the positive elements of an unselfconscious culture of design can be exploited in meta-design approaches by creating media that support people in working on their tasks, rather than requiring them to focus their intellectual resources on the medium itself.

## 5   Environments Supporting Meta-Design

The goal of making systems modifiable by users does not imply transferring the responsibility of good system design to the user. In general, "normal" users do not build tools of the quality a professional designer would since users are not concerned with the tool per se, but in doing their work. *Domain-oriented design environments* support meta-design by advancing human-computer interaction to *human problem-domain interaction*. Because systems are modelled at a conceptual level with which users are familiar, the interaction mechanisms take advantage of existing user knowledge and make the functionality of the system transparent and accessible so that the computational drudgery required of users can be substantially reduced. The *Envisionment and*

*Discovery Collaboratory* [Arias et al., 2000] is a second generation design environment focused on the support of *collaborative design* by integrating physical and computational components to encourage and facilitate informed participation by all stakeholders in the design process.

## 6 Application of Meta-Design Approaches

**Social Creativity.** Complex design problems require more knowledge than any single person can possess, and the knowledge relevant to a problem is often distributed among all stakeholders who have different perspectives and background knowledge, thus providing the foundation for *social creativity* [Arias et al., 2000]. Bringing together different points of view and trying to create a shared understanding among all stakeholders can lead to new insights, new ideas, and new artifacts. Social creativity can be supported by innovative computer systems that allow all stakeholders to contribute to framing and solving these problems collaboratively.

**Open Source.** *Open source development* [Raymond & Young, 2001] is an activity in which a community of software developers collaboratively constructs systems to help solve problems of shared interest and for mutual benefit. The ability to change source code is an enabling condition for collaborative construction of software by changing software from a fixed entity that is produced and controlled by a closed group of designers to an open effort that allows a community to design collaboratively based on personal desires following the framework provided by the seeding, evolutionary growth, and reseeding process model. Open source invites passive consumers to become active contributors [Fischer, 2002].

**Learning Communities.** *Courses-as-seeds* [dePaula et al., 2001] is an educational model that explores meta-design in the context of university courses by creating a culture of informed participation. Courses are conceptualized as seeds, rather than as finished products, and students are viewed as informed participants who play an active role in defining the problems they investigate. The output of each course contributes to an evolving information space that is collaboratively designed by all course participants, past and present.

**Interactive Art.** *Interactive art*, conceptualized as meta-design, focuses on collaboration and co-creation. The original design (representing a seed in our framework) establishes a context in which users can create content. Interactive art puts the tools rather than the object of design in the hands of users. It creates interactive systems that do not define content and processes, but the conditions for the process of interaction. Interactive art puts the emphasis on different objectives compared to traditional design approaches, including shifts from (1) guidelines and rules to exceptions and negotiations; (2) from content to context; (3) from objects to process, and (4) from certainty to contingency (these "cultural shifts" have been developed jointly with Elisa Giaccardi who has explored the concept of meta-design in interactive arts [Giaccardi, 2003]).

## 7 Conclusions

We have evaluated our approaches in different settings, with different task domains, and with different stakeholders. While meta-design is a promising approach to overcome the limitations of closed systems and to support social creativity, it creates many fundamental challenges: in the technical domain as well as in the social domain including the need for social capital, the willingness of users to engage in additional learning to become designers, and the additional efforts to integrate the work into the shared environment. Meta-design addresses one of the fundamental challenges of a knowledge society [Florida, 2002]: to invent and design a culture in which all participants in a collaborative design process can express themselves and engage in personally meaningful activities.

## 8 Acknowledgements

The author thanks the members of the Center for LifeLong Learning & Design at the University of Colorado, who have made major contributions to the conceptual framework described in this paper. The research was supported by (1) the National Science Foundation, Grants (a) REC-0106976 "Social Creativity and Meta-Design in Lifelong Learning Communities", and (b) CCR-0204277 "A Social-Technical Approach to the Evolutionary Construction of Reusable Software Component Repositories"; (2) SRA Key Technology Laboratory, Inc., Tokyo, Japan; and (3) the Coleman Initiative, San Jose, CA.

## 9 References

Alexander, C. (1964) *The Synthesis of Form,* Harvard University Press, Cambridge, MA.

Arias, E. G., Eden, H., Fischer, G., Gorman, A., & Scharff, E. (2000) "Transcending the Individual Human Mind—Creating Shared Understanding through Collaborative Design," *ACM Transactions on Computer Human-Interaction,* 7(1), pp. 84-113.

Brown, J. S., & Duguid, P. (2000) *The Social Life of Information,* Harvard Business School Press, Boston, MA.

dePaula, R., Fischer, G., & Ostwald, J. (2001) "Courses as Seeds: Expectations and Realities," *Proceedings of the Second European Conference on Computer-Supported Collaborative Learning (Euro-CSCL' 2001)*, Maastricht, Netherlands, pp. 494-501.

Fischer, G. (2002) *Beyond 'Couch Potatoes': From Consumers to Designers and Active Contributors, in FirstMonday (Peer-Reviewed Journal on the Internet),* Available at http://firstmonday.org/issues/issue7_12/fischer/.

Fischer, G., & Ostwald, J. (2002) "Seeding, Evolutionary Growth, and Reseeding: Enriching Participatory Design with Informed Participation," *Proceedings of the Participatory Design Conference (PDC'02)*, Malmö University, Sweden, pp. 135-143.

Fischer, G., & Scharff, E. (2000) "Meta-Design—Design for Designers," *3rd International Conference on Designing Interactive Systems (DIS 2000)*, New York, pp. 396-405.

Florida, R. (2002) *The Rise of the Creative Class and How It's Transforming Work, Leisure, Community and Everyday Life,* Basic Books, New York, NY.

Giaccardi, E. (2003) *Meta-Design,* Available at http://x.i-dat.org/~eg/research.htm.

Norman, D. A., & Draper, S. W. (Eds.) (1986) *User-Centered System Design, New Perspectives on Human-Computer Interaction,* Lawrence Erlbaum Associates, Inc., Hillsdale, NJ.

Raymond, E. S., & Young, B. (2001) *The Cathedral and the Bazaar: Musings on Linux and Open Source by an Accidental Revolutionary,* O'Reilly & Associates, Sebastopol, CA.

Schön, D. A. (1983) *The Reflective Practitioner: How Professionals Think in Action,* Basic Books, New York.

Schuler, D., & Namioka, A. (Eds.) (1993) *Participatory Design: Principles and Practices,* Lawrence Erlbaum Associates, Hillsdale, NJ.

Wright, M., Marlino, M., & Sumner, T. (2002) *Meta-Design of a Community Digital Library, D-Lib Magazine, 8 (5),* Available at http://www.dlib.org/dlib/may02/wright/05wright.html.

# Usability Patterns in Software Architecture

*Eelke Folmer and Jan Bosch*

Department of Mathematics and Computing Science
University of Groningen, PO Box 800, 9700 AV the Netherlands
mail@eelke.com, Jan.Bosch@cs.rug.nl

## Abstract

Over the years the software engineering community has increasingly realized the important role software architecture plays in fulfilling the quality requirements of a system. Practice shows that for current software systems, most usability issues are still only detected during testing and deployment. To improve the usability of a software system, usability patterns can be applied. However, too often software systems prove to be inflexible towards such modifications which lead to potentially prohibitively high costs for implementing them afterwards. The reason for this shortcoming is that the software architecture of a system restricts certain usability patterns from being implemented after implementation. Several of these usability patterns are "architecture sensitive", such modifications are costly to implement due through their structural impact on the system. Our research has identified several usability patterns that require architectural support. We argue the importance of the relation between usability and software architecture. Software engineers and usability engineers should be aware of the importance of this relation. The framework which illustrates this relation can be used as a source to inform architecture design for usability.

## 1    Introduction

In the last decades it has become clear that the most challenging task of software development is not just to provide the required functionality, but rather to fulfil specific properties of software such as performance, security or maintainability, which contribute to the *quality* of software (Folmer & Bosch, 2002). Usability is an essential part of software quality; issues such as whether a product is easy to learn to use, whether it is responsive to the user and whether the user can efficiently complete tasks using it may greatly affect a product's acceptance and success in the marketplace. Modern software systems are continually increasing in size and complexity. An explicit defined architecture can be used as a tool to manage this size and complexity. The quality attributes of a software system however, are to a large extent determined by a system's software architecture. Quality attributes such as performance or maintainability require explicit attention during development in order to achieve the required levels (Bosch & Bengtsson 2002). It is our conjecture that this statement also holds for usability. Some changes that affect usability, for example changes to the appearance of a system's user interface, may easily be made late in the development process without incurring great costs. These are changes that are localised to a small section of the source code. Changes that relate to the interactions that take place between the system and the user are likely to require a much greater degree of modification. Restructuring the system at a late stage will be extremely and possibly prohibitively, expensive. To improve on this situation, it would be beneficial for knowledge pertaining to usability to be captured in a form that can be used to inform architectural design, so that engineering for usability is possible early in the design process. The usability engineering community has collected and developed various design solutions such as usability patterns that can be applied to a system to improve usability. Where these prescribe sequences or styles of interaction between the system and the user, they are likely to have architectural implications. For example, consider the case where the software allows a user to perform a particularly complex task, where a lot of users make mistakes. To address this usability issue a wizard pattern can be employed. This pattern guides the users through the complex task by decomposing the task into a set of manageable steps. However implementing such a pattern as the result of a design decision made late on proves to be very costly. There needs to be provision in the architecture for a wizard component, which can be connected to other relevant components, the one triggering the operation and the one receiving the data gathered by the wizard. The problem with this late detection of usability issues is that sometimes it is very difficult to apply certain usability patterns after the majority of a system has been implemented because these patterns are 'architecture sensitive'. The contribution of this paper is to make software engineers aware that certain 'design solutions' that may improve usability are extremely difficult to retro-fit into applications because these patterns require architectural support. Therefore being able to design architectures with support for usability is very important. The framework that we present in the next section can be used as an informative source during design.

## 2    Usability Framework

Through participation in the STATUS[1] project we have investigated the relationship between usability and software architecture. Before the relationship between usability and software architecture was investigated an accurate definition of usability was tried to obtain by surveying existing literature and practice. Initially a survey was undertaken to try and find a commonly accepted definition for usability in terms of a decomposition into usability attributes. It was soon discovered that the term usability attribute is quite ambiguous. People from industry and academia have quite different perceptions of what they consider to be a useful usability attribute. The number of "usability attributes" obtained in this way grew quite large therefore we needed a way to organise and group the different attributes. Next to the need for organising these different interpretations of usability attributes, a relation between usability and software architecture was tried to discover. The only 'obvious' relation between usability and architecture is that there are some usability patterns that have a positive effect on usability and that are architecture sensitive. However, it was soon discovered that it was extremely difficult to draw a direct relationship between usability attributes and software architecture. An attempt was made to decompose the set of usability attributes into more detailed elements such as: "the number of errors made during a specific task", which is an indication of reliability, or "time to learn a specific task" which is an indication of learnability, but this decomposition still did not lead to a convincing connecting relationship with architecture. The reason is that traditionally usability requirements have been specified such that these can be verified for an implemented system. However, such requirements are largely useless in a forward engineering process. For example, it could be stated that a goal for the system could be that it should be easy to learn, or that new users should require no more than 30 minutes instruction, however, a requirement at this level does not help guide the design process. Usability requirements need to take a more concrete form expressed in terms of the solution domain to influence architectural design. To address to these problems discussed a framework has been developed. Figure 1 shows the framework developed so far. It shows a collection of attributes, properties and patterns and shows how these are linked to give the relationship between usability and software architecture. This relation is illustrated by means of an example. Figure 1 shows the wizard pattern linked to the "guidance" usability property which in

[1] STATUS is an ESPRIT project (IST-2001-32298) financed by the European Commission in its Information Society Technologies Program. The partners are Information Highway Group (IHG), Universidad Politecnica de Madrid (UPM), University of Groningen (RUG), Imperial College of Science, Technology and Medicine (ICSTM), LOGICDIS S.A.

turn is linked to the "learnability" usability attribute. The wizard pattern guides the user through a complex task by decomposing the task into a set of manageable subtasks. To implement a wizard a provision is needed in the architecture for a wizard component, which can be connected to other relevant components: the one triggering the operation and the one receiving the data gathered by the wizard. The wizard is related to usability because it uses the primitive of guidance to guide the user through the task. Guidance on its turn has two "obvious" relations with usability. Guidance has a positive effect on learnability and a negative effect on efficiency. The concept of "guidance" is defined as a usability property; a usability property is a more concrete form of a usability requirement specified in terms of the solution domain. Patterns relate to one or more of these usability properties. Properties on their turn relate to one or more usability attributes. This relation is not necessarily a one to one mapping. The relationship can be positive as well as negative. To avoid the table becoming too cluttered, and the risk of possibly producing a fully connected graph, only the links thought to be strongest and positive

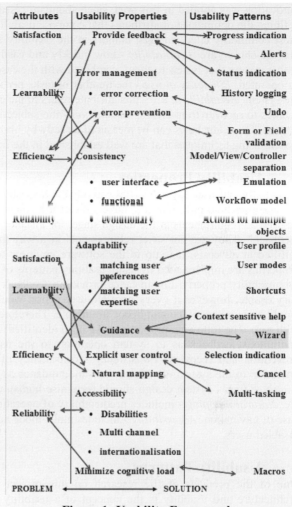

**Figure 1: Usability Framework**

are indicated in the table. The framework relates the problem to the solution domain; a usability attribute can be measured on a completed solution, whereas a usability property exists in the problem domain, and can be used as a requirement for system design. A usability pattern bridges the gap between problem and solution domains, providing us with a mechanism to fulfil a requirement, providing us with a solution for which the corresponding usability attribute can be measured. The next sections enumerate the concepts of usability attributes, properties and patterns which comprise our framework.

## 3    Usability Attributes

A comprehensive survey of the literature (Folmer & Bosch, 2002) revealed that different researchers have different definitions for the term usability attribute, but the generally accepted meaning is that a usability attribute is a precise and measurable component of the abstract concept that is usability. After an extensive search of the work of various authors, the following set of usability attributes is identified for which software systems in our work are assessed. No innovation was applied in this area, since abundant research has already focussed on finding and defining the optimal set of attributes that compose usability. Therefore, merely the set of attributes most commonly cited amongst authors in the usability field has been taken. The four attributes that are chosen are: *Learnability* - how quickly and easily users can begin to do productive work with a system that is new to them, combined with the ease of remembering the way a system must be operated. *Efficiency of use* - the number of tasks per unit time that the user can perform using the system. *Reliability in use* - this attribute refers to the error rate in using the system and the time it takes to recover from errors. *Satisfaction* - the subjective opinions that users form in using the system. These attributes can be measured directly by observing and interviewing users of the final system using techniques that are well established in the field of usability engineering.

## 4    Usability Properties

Essentially, our usability properties embody the heuristics and design principles that researchers in the usability field have found to have a direct influence on system usability. These properties can be used as requirements at the design stage, for instance by specifying: "the system must provide feedback". They are not strict requirements in a way that they are requirements that should be fulfilled at all costs. It is up to the software engineer to decide how and at which levels these properties are implemented by using usability patterns of which it is known they have an effect on this usability property. For instance providing feedback when printing in an application can be very usable, however if every possible user action would result in feedback from the system it would be quite annoying and hence not usable. Therefore these properties should be implemented with care. The following properties have been identified: *Providing feedback* - the system provides continuous feedback as to system operation to the user. *Error management* - includes error prevention and recovery. *Consistency* - consistency of both the user interface and functional operation of the system. *Guidance* - on-line guidance as to the operation of the system. *Minimise cognitive load* - system design should recognise human cognitive limitations, short-term memory etc. *Natural mapping* - includes predictability of operation, semiotic significance of symbols and ease of navigation. *Accessibility* - includes multi-mode access, internationalisation and support for disabled users.

## 5    Usability Patterns

One of the products of the research on this project into the relationship between software architecture and usability is the concept of a usability pattern. The term "usability pattern" is chosen to refer to a technique or mechanism that can be applied to the design of the architecture of a software system in order to address a need identified by a usability property at the requirements stage. Various usability pattern collections have been defined (Welie & Trætteberg 2000), (Tidwell 1998). Our collection is different from those because we only consider patterns which should be applied during the design of a system's software architecture, rather than during the detailed design stage. (Bass et al, 2001) have investigated the relationship between the usability and software architecture through the definition of a set of 26 scenarios. These scenarios are in some way equivalent to our properties and usability patterns. However there are some differences. They have used a bottom up approach from the scenarios whereas we have taken a top down approach from the definition of usability. Our approach has in our opion resulted in a more clearly documented and illustrated relationship between those usability issues addressed by the design principles and the software architecture design decisions required to fullfill usability requirements. Another difference is that our patterns have been obtained from an inductive process from different practical cases (e-commerce software developed by the industrial partners in this project) whereas their scenarios result from personal experience and literature surveys. Their work has been useful to support our statement that usability and software are related through usability patterns. A full catalogue of patterns identified is presented on http://www.designforquality.com. There is not a one-to-one mapping between usability patterns and the usability properties that they affect. A pattern may be related to any number of properties, and each property may be improved (or impaired) by a number of different patterns. The choice of which pattern to apply may be made on the basis of cost and the trade off between different usability properties or between usability and other quality attributes such as security or performance. This list of patterns presented in Figure 1 is not intended to be exhaustive, and it is envisaged that future work on this project will lead to the expansion and reworking of the set of patterns presented here, including work to fill out the description of each pattern to include more of the sections which traditionally make up a pattern description, for instance what the pros and cons of using each pattern may be.

## 6 Summary and conclusions

Our research has argued the importance of the relation between usability and software architecture. A framework has been developed which illustrates this relation. The list of usability patterns and properties identified/defined in our framework is substantial but incomplete, new usability patterns or properties that are developed or discovered can be fitted in the existing framework. Future research should focus on verifying the architectural sensitiveness of the usability patterns that have been identified. For validation only e-commerce software provided by our industrial partners in this project has been considered. To achieve more accurate results our view should be expanded to other application domains. The usability properties can be used as requirements for design, it is up to the software architect to select patterns related to specific properties that need to be improved for a system. It is not claimed that a particular usability pattern will improve usability for any system because many other factors may be involved that determine the usability of a system.. The relationships in the framework indicate potential relationships. Further work is required to substantiate these relationships and to provide models and assessment procedures for the precise way that the relationships operate. This framework provides the basis for developing techniques for assessing software architectures for their support of usability. This technique allows for iteratively designing for usability on the architectural level.

## References

Bass, Lenn; Kates, Jessie & John, Bonnie. E.(2002) Achieving Usability through software architecture, http://www.sei.cmu.edu/publications/documents/01.reports/01tr005.html

Bosch, J. (2000). Design and Use of Software Architectures: Adopting and Evolving a Product Line Approach, Pearson Education (Addison-Wesley and ACM Press).

Bosch, J. & Bengtsson, P. O. (2002), Assessing optimal software architecture maintainability. In proceedings of Fifth European Conference on Software Maintenance and Reengineering (CSMR'01), IEEE Computer Society Press, Los Alamitos, CA pp. 168-175

Folmer, E. & Bosch, J. (2003). Architecting for usability; a survey. Submitted for the Journal of systems and software. Accepted for the Journal of systems and software.

Tidwell, J. (1998). Interaction Design Patterns. Conference on Pattern Languages of Programming.

Welie, M. & Trætteberg, H., (2000). Interaction Patterns in User Interfaces. 7th Conference on Pattern Languages of Programming (PloP).

# Bridging the Gap between Scenarios and Formal Models

*Peter Forbrig and Anke Dittmar*

University of Rostock, Department of Computer Science
Albert-Einstein-Str. 21, D-18051 Rostock, Germany
[pforbrig | ad]@informatik.uni-rostock.de

## Abstract

A software design process has to consider both the user- and system-oriented aspects. On the one hand, scenarios are often used to illustrate the interface between a system and its application context. On the other hand, abstract design models with a preferably constructive character are requested for implementing software systems. This paper shows a combination of the scenario- and model-based approaches to fulfil the demands mentioned above.

## 1    Introduction

Scenario-based design (SBD) as well as model-based design (MBD) aim to support a user-centred design process. Both approaches emphasize that software developers have to pay attention to the application context to get usable systems. It is necessary to analyse the current working situation to know about the activities users have to perform, the objects they manipulate, tools they apply, and the way they cooperate. The analysis results in a set of goals showing which current practices have to be changed and which things are worth to maintain. Goals describe an intended state, called envisioned working situation. Thus, it is not enough to design a software system fulfilling these requirements only but the whole context of use has to be designed.

By taking this point of view one has to accept software design as a process which involves many stakeholders. Although MBD approaches catch different perspectives on the system design in different kinds of models, they don't really support a participatory design. The process is rather specification driven. That is, the main focus is on the construction of models with a more or less formalized structure and it is not guaranteed that all stakeholders are able to participate. SBD approaches claim to give every stakeholder an active role by "telling stories" (scenarios) which he can understand and which evoke his interest and empathy. Nevertheless, the system-centred side of the design process cannot be ignored. A precise constructive specification of the software system under development which can serve as a direct input to the implementation is necessary. Hence, SBD approaches have to be seen as reasonable supplement to a specification driven design [Car02]. The problem of linking scenarios and formal models effectively is still an active field of research. In this paper, it is argued that an effective combination of ideas coming from SBD and MBD is possible. It is shown how task models can mediate between scenarios and software specifications.

## 2    Scenarios

*Prof. Thiel goes through the results of the examination the 213 students of the 1.term had to write to pass the course about programming techniques. The students had to work on 5 exercises. He sighs. 51 students got less than 20 points. That means they have to repeat the exam to be allowed to continue their studies. Prof. Thiel asks his assistant Peter Meyer who compiled the list of the total numbers of points to check the list once more. While Peter is checking the list, Anne Holz who had to mark the second exercise sent him, he discovers that 6 students got no points. He wonders if Anne forgot to record some results. Furthermore, he decides to ask Prof. Thiel how to treat the 12 students who got 18 or 19 points. Perhaps it would be reasonable to review their exam papers.*

A "scenario is a story about people and their activities" [RC02]. The above example gives an impression. It can be seen as a *problem scenario* describing aspects of a current practice. Problem scenarios are developed during a requirements analysis. Together with claims which give an evaluation of the current setting they serve as basis for the description of requirements on the system under design. It is further distinguished between activity, information, and interaction scenarios on design level.

Rosson and Carroll emphasize that scenarios are rather constructions based on general descriptions about actors, tasks, artifacts etc. than stories based on pure observation or imagination. Perhaps this is surprising. Surely, SBD aims to evoke the interest and participation of all stakeholders by "concrete" stories – but not without purpose. This participation should lead to a better system design. Hence, the stories have to be told in a way which people let think more consciously about the impact a system design would have on working practices. The stories should force people to think about alternative design ideas. Scenarios also comprise planning and evaluating activities of the actors.

A slightly different view on scenarios is used in the object-oriented software development. Here, a scenario describes a path through a use case. Thus, the focus is on the description of the interaction between the actor and the system. There is nothing told about mental activities of the users and the description of the application context is more restricted. Whereas Rosson and Carroll say "much of the richness of a scenario is in the things that are not said" and allow partial task descriptions and the involvement of more than one task in a scenario, scenarios as instances of use cases are more rigid. They are suitable for test cases in later phases of the system development.

# 3     Models in MBD

In MBD, one can distinguish between two main groups of sub-models. One group forms the specification of the software system itself. There exists at least an application model to describe the application core and a dialogue model to specify the UI. The other group constitutes the task model. Here, the same concepts are considered as in SBD but in an abstract and formalized way to make a transformation of task knowledge to formal system specifications (as finally requested in the software development) easier. Consequently, task models are suited to mediate between scenarios and system specifications. In this section, the sub-models of a task model are explained shortly. [FD03] gives a more detailed description of the interdependencies between task models and appropriate system specifications.

Task models consist of sub-models specifying goals, actions, business objects, and actors. A goal is represented by a network of sub-goals each describing an intended sub-state of the domain. The refinement of a goal has to result in an action structure. Actions are hierarchically decomposed into sub-actions until the level of basic actions. The objects of a business object model character-ized by a name and a set of attribute-value-pairs serve to specify states of the domain. In contrast to most object-oriented modelling approaches, it is not distinguished between objects and classes but objects can be instances of other ones as will be explained in the next section.

There are inner and outer relationships between elements of the sub-models. For example, two sub-goals can be related by operators describing that both/at least one/at most one of them have to be achieved (and-/or-/xor-operator) etc.. Temporal constraints between the sub-actions of a super-action are specified by using temporal operators in a temporal equation with the super-action on the left hand side and the sub-actions on the right hand side. Attributes of objects reflect relation-ships between different objects.

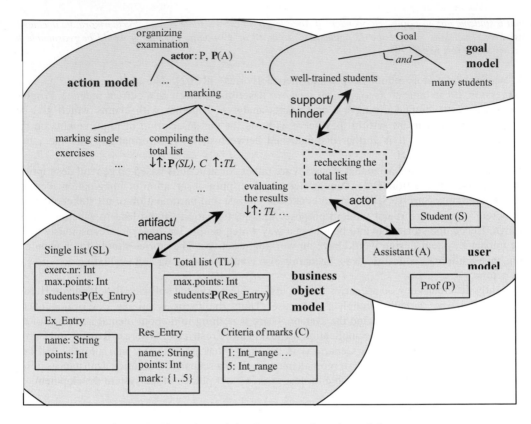

**Figure 1:** The sub-models of an example task model

A goal cannot be fulfilled without performing some appropriate actions changing the state of the domain. In order to emphasize the central role of actions only the relationships between the action model and the other ones are sketched out in the example of Fig.1 which partly explores the task of educating students. Every sub-action supports or hinders some sub-goals. Thus, the action hierarchy reflects the compromises made in planning the achievement of the goal.

Executing an action means to create, destroy, use, or change some business objects. In other words, an action needs a set of objects in a certain state (precondition) and supplies a set of objects in a certain state (post condition or effect). Actions at a higher level of the hierarchy and, finally, the goals represent approximations of or views on affected business objects and their states. For example, the effect of *compiling the total list* in Fig.1 is the object *Total list* (abbrev. *TL*, $\downarrow / \uparrow$ mark pre-/post conditions). With respect to an action a business object can play the role of an artefact or a means of work. Means are applied to change the state of the artefact which is essential for the achievement of the goal. They are further divided into tools used and resources consumed in actions. The artefact of *compiling the total list* is *Total list,* a set of *Single lists* and the object *Criteria of marks* are used as tools.

A task can demand the participation of a whole group of people. In this paper, the division of labour is simply modelled by assigning actors to the nodes of an action hierarchy. Consequently, it is reasonable not to stop the action decomposition until the basic actions (the leaf nodes) are executed by single persons. Otherwise, questions concerning the division of labour are still open. In Fig.1, *Prof.* and (set of) *Assistant(s)* are the actors in *organizing examination*. Task models are described e.g. in [Dit02] more formal and detailed.

# 4  Task Models as Mediator between Scenarios and Formal System Specifications

## 4.1  Scenario Elements are Instances of Model Elements

An action model describes a set of sequences of basic actions which all lead to task completion. Each sequence is a kind of instance of the action model. An object $O_1$ is an instance of an object $O_2$ in the business object model (denoted by $O_1::O_2$) if for all attributes $a_i:v_i$ of $O_2$ there is an attribute $a_i:v'_i$ in $O_1$ and $v'_i$ is an instance of $v_i$. Further, instance relations between actors are assumed.

Let $T$ be a task model with the action model $A$, the business object model $O$ and the user model $R$. A *complete scenario* is a sequence $\langle a_1(r_1,\{o^1_1,...,o^1_{n1}\}),...,a_m(r_m,\{o^m_1,...,o^m_{nm}\})\rangle$ where $a_i$ are basic actions in $A$ and $\langle a_1(R_1,\{U^1_1,...,U^1_{n1}\}),...,a_m(R_m,\{O^m_1,...,O^m_{nm}\})\rangle$ is a possible sequence in $A$ with

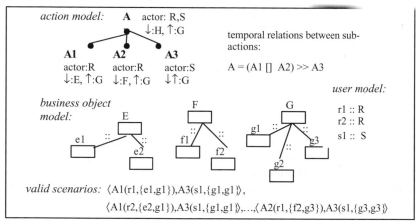

**Figure 2:** Complete scenarios derived from a task model

actors $R_i$ from $R$ and objects $O_i$ from $O$ assigned to $a_i$ (for brevity pre-and post conditions, artefacts and means are not distinguished). Further, $r_i$ resp. $o^j_i$ have to be instances of $R_i$ resp. $O^j_i$ $(i=1,..,m, j=1,2,...)$. The notion $\mathbf{P}(O)$ (e.g. $\mathbf{P}(SL)$ in Fig.1) describes a set of instances of $O$.

A task model describes a (mostly infinite) set of complete scenarios. Fig.2 visualizes this idea by a small abstract example. The used temporal operators are [] for alternative and >> for sequential sub-actions.

Comparing the scenario at the beginning of Sect.2 with the task model in Fig.1 it should be clear now that *Prof. Thiel* is an instance of *Prof.*, *Peter Meyer* and *Anne Holz* are instances of *Assistant*, *Prof. Thiel* is *evaluating the results* of an instance of *Total list* and so on.

## 4.2  Scenarios as Stories vs. Scenarios as Test Cases

This paper proposes a distinction between two kinds of scenarios.

*Scenarios as stories* have an incomplete character in that sense that they are constructed from different parts of the appropriate task model. This kind reflects rather the approach taken by Rosson and Carroll. Scenarios can involve different actors and actions, and can even refer to goals or pieces of action models to describe mental activities of actors. Scenarios as stories are mainly used to improve and illustrate (initial) task models. For example, the scenario in Sect.2 was constructed on the basis of the task model in Fig.1 and revealed the sub-action *rechecking the total list* which was added afterwards to the task model (indicated with dotted lines in the figure). "In SBD new activities are always grounded in current activities" [RC02]. Scenarios as stories support the transition from current to envisioned models. Taken the scenario in Sect.2, for example, one can easily imagine that in the envisioned practice the *compiling of the total list* out of the single lists for each of the 5 exercises can be done automatically by the software system under development. It is also of advantage that the system can label the "borderline cases" (the entries of the students who miss 1 or 2 points to pass the examination). An envisioned task model has to reflect the application of the software system by the actors.

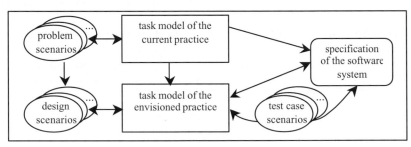

**Figure 3:** A combination of MBD and SBD

*Scenarios as test cases* are complete scenarios in the sense of Sect. 4.1. They can be generated from sub-trees of the action hierarchy as should be underlined by Fig.2. They are useful for validating a system specification with respect to a task model and vice versa because they describe the interaction between a user and the system precisely. Scenarios of this kind are rather comparable with scenarios as paths through use cases.

Fig.3 illustrates possible applications of scenarios to supplement a model-based design process.

## 5    Summary

A combination of SBD and MBD to improve the software design process was shown. Task models play a mediating role. They supply a general description which is useful for constructing scenarios on the one side but also for deriving formal system specifications on the other side. A distinction between scenarios as stories and scenarios as test cases was proposed and their different fields of application were explored. There are a lot of open questions. The problem of choosing a representative set of test case scenarios out of a possibly infinite set is only one of them. The interested reader is referred to [FD03] where tool support is described.

## References

[Car02] J.M.Carroll: Scenarios and Design Cognition, Proc. of the Int. Conf. On Requirements Engineering RE2002, Essen, Germany.

[RC02]  M.B.Rosson, J.M.Carroll: Usability Engineering – Scenario-Based Development of Human-Computer Interaction, Morgan Kaufmann Publishers, 2002.

[FD03]  P.Forbrig, A.Dittmar: Interfacing Business Object and User Models with Action Models in this Proceedings of HCI 2003.

[Dit02]  A.Dittmar: Ein formales Metamodell für den aufgabenbasierten Entwurf interaktiver Systeme, PhD-Thesis, Universität Rostock, 2002.

# A Layered Approach for Designing Multiple User Interfaces from Task and Domain Models

*Elizabeth Furtado, João José Vasco Furtado, Quentin Limbourg\*, Jean Vanderdonckt\*, Wilker Bezerra Silva, Daniel William Tavares Rodrigues, and Leandro da Silva Taddeo*

Universidade de Fortaleza, NATI - Célula EAD - Fortaleza, BR-60455770 Brazil
{elizabet, vasco, wilker, danielw, taddeo}@unifor.br
\*Université catholique de Louvain, BCHI – B-1348 Louvain-la-Neuve (Belgium)
{limbourg,vanderdonckt}@isys.ucl.ac.be

## Abstract

More frequently, design of user interfaces covers design issues related to multiple contexts of use where multiple stereotypes of users may carry out multiple tasks, possibly on multiple domains of interest. Existing development methods do not necessarily support developing such user interfaces as they do not factor out common parts between similar cases while putting aside uncommon parts that are specific to some user stereotypes. To address this need, a new development method is presented based on three layers: (i) a conceptual layer where a domain expert defines an ontology of concepts, relationships, and attributes of the domain of discourse, including user modeling; (ii) a logical layer where a designer specifies multiple models based on the previously defined ontology and its allowed rules; and (iii) a physical layer where a developer develops multiple user interfaces from the previously specified models with design alternatives depending on characteristics maintained in the user models.

## 1    Introduction

Universal design (Savidis, Akoumianakis & Stephanidis, 2001) adheres to a vision where user interfaces (UIs) of interactive applications are developed for the widest population of users in different contexts of use by taking into account differences such as preferences, cognitive style, language, culture, habits, conventions, and system experience. Universal design of UIs poses some difficulties due to the consideration of these multiple parameters depending on the supported differences. In particular, the multiplicity of parameters dramatically increases the complexity of the design phase by adding many design options among which to decide. In addition, methods for developing UIs do not mesh well with this variety of parameters as they are not necessarily identified and manipulated in a structured way nor truly considered in the design process. The method structures the UI design in three levels of abstraction as represented in fig. 1 so as to delegate specific conditions and decisions the latest possible in the whole development life cycle:

1.  The *conceptual level* enables a domain expert to define ontology of concepts, relationships, and attributes involved in the production of multiple UIs.
2.  The *logical level* allows designers to capture requirements of a specific UI design case by instantiating concepts, relationships, and attributes with a graphical editor. Each set of instantiations results in a set of models for each considered design case (*n* designs in fig. 1).
3.  The *physical level* helps developers in deriving multiple UIs from each set of models thanks to a model-based UI generator: in fig. 1, *m* possible UIs are obtained for UI design #1, *p* for UI design #2,..., *r* for UI design #*n*. The generation is then exported to imported in a traditional development environment for any manual edition (here, MS Visual Basic).

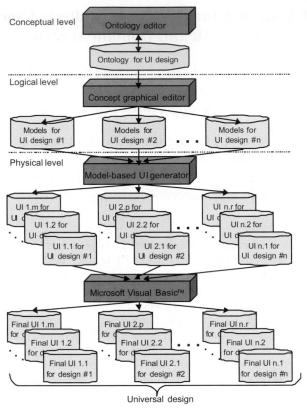

**Figure 1:** The different levels of the proposed method for universal design of user interfaces

## 2    Ontology Editor

An ontology (Guarino, 1995) explicitly defines any set of concepts, relationships, and attributes that need to be manipulated in a particular situation, including universal design. The ontology notion comes from the Artificial Intelligence context where it is identified as the set of formal terms with one represents knowledge, since the representation completely determines what "exists" for the system. We define a *context of use* as the global environment in which a user population, perhaps with different profiles, skills, and preferences, are carrying out a series of interactive tasks on one or multiple semantic domains. In universal design, it is expected to benefit from the advantage of considering any type of the above information to produce multiple UIs depending on the varying conditions. These pieces of information of a context of use can be captured in different models (Puerta, 1997). In order to input model properly, an ontology of these models is needed so as to preserve not only their syntactical aspects, but also their semantics. For this purpose, an ontology defines a specification language with which any model can be specified. The concepts and relations of interest at this layer are here introduced as meta-concepts and meta-relations belonging to the meta-modeling stage. An ontology defines a kind of reference meta-model. Of course, several meta-models can be defined, but only one reference is used here. Moreover, the ontology may allow the transformation of concepts expressed according to one meta-model to another one. Fig. 2 exemplifies how these fundamental concepts can be defined in an ontology editor that will serve ultimately to constrain any model that will be further defined and instantiated. The core entity of fig. 2 is the concept, characterized by one or many attributes, each having here a data type (e.g., string, real, integer, Boolean, or symbol). Concepts can be related to each other thanks to relations. Some particular, yet useful, relations include inheritance (i.e., "is"), aggregation (i.e., "composed of"); and characterization (i.e., "has"). At the meta-modeling stage, we do know that concepts, relations, and attributes will be manipulated, but we do not know yet of what type. Any UI model, can be expressed in terms of the basic entities as specified in fig. 2.

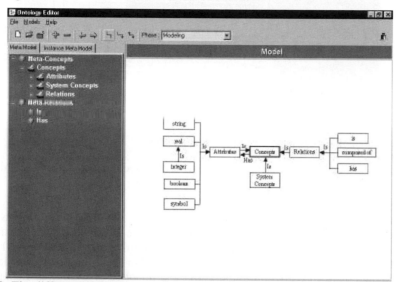

**Figure 2:** The different basic elements (concepts, relations, attributes) in the ontology editor

# 3    Concept Graphical Editor

Once the basic elements have been defined, any instance of these elements can be used at the subsequent level to define the models used to capture the specifications of a single UI or for several ones. In particular, it is important to separate, when appropriate, aspects that are independent of the context of use (e.g., a task or a sub-task that needs to be carried out) from aspects that are dependent of this context of use. Therefore, it is likely that each model (i.e., task, domain, user, presentation, dialogue) (Puerta, 1997) will be subject to identifying and separating dependent parts from independent parts. The concept graphical editor is now used to instantiate the context of use, the relationships and attributes of models for the *Medical Attendance* domain involved in patient admission. Fig. 3 graphically depicts the *Urgency Admission* context of use and the attributes of models of task, user and domain. There are two tasks instantiated: *to admit patient* and *to show patient data*. The first one is activated by a *secretary* and uses *patient information* during its execution. For the user model of the secretary, the following parameters are considered: her/his experience layer, input preference, information density with the enumerated values *low* and *high*. The information items describing a patient are the following: date of the day, first name, last name, birth date, address, phone number, gender, and civil status. Information items regarding insurance affiliation and medical regimen can be described similarly. The parameters of an information item of a domain model depend on the UI design process. For instance, parameters and values of an information item used to generate UIs are (Vanderdonckt, 2000): data type (date, Boolean, graphic, integer, real, or alphanumeric), length ($n>1$), domain definition (know, unknown, or mixed), interaction way (input, output, or input/output), orientation (horizontal, vertical, circular, or undefined), number of possible values ($n>1$), number of values to choose ($n>1$), and precision (low or high).

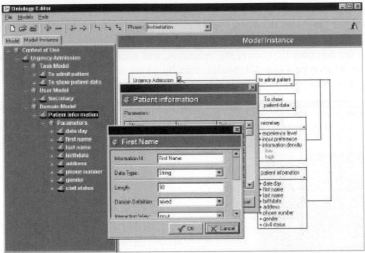

**Figure 3:** Instantiating the basic elements for a particular model in the concept graphical editor

# 4    Model-based generator

Once a particular model is obtained, each model can initiate a UI generation process. Here, the SEGUIA (Vanderdonckt and Bodart, 1993) tool is used (fig. 4): it consists of a model-based interface development that is capable of automatically generating code for a running UI from a file containing the specifications defined in the previous step. For this purpose, any model stored in the graphical editor can be exported from this editor and imported in SEGUIA, as a simple file or a DSL file (Dynamic Specification Language) (fig. 4a). Of course, any other tool which is compliant with the model format and/or which can import the specification file may be intended to produce running UI for other design situations, contexts of use, user models, or computing platforms. SEGUIA is able to automatically generate several UI presentations to obtain multiple UIs. These different presentations are obtained

- In an automated manner, where the developer only launches the generation process by selecting which layout algorithm to rely on (e.g. two -column format or right/bottom strategy).
- In a computer-aided manner, where the developer can see at each step what are the results of the generation, can cooperate in a mixed-initiative manner, and govern the process before reaching a final status.

Once a file is opened, the designer may ask the tool to generate a presentation model, if not done before by relying on the concepts of presentation units and logical windows (fig. 4b). A presentation unit is a set of logical windows all related to a same interactive task, but not all these logical windows should be presented simultaneously. For instance, a first input logical window may be presented and a second one afterwards, depending on the status of information acquired in the first one. Logical windows (e.g., a window, a dialog box, a tabbed dialog box) can be determined according to various strategies: maximal, minimal, input/output, functional, and free. Once a presentation model is finalised, the generation process is performed into 4 steps (fig. 4c):

1. Selection of Abstract Interaction Objects: the tools selects interaction objects to input/output information items contained in the model. The objects are tried to be independent from any context of use.
2. Transformation of Abstract Interaction Objects into Concrete Interaction Objects: once a context of use is known, the abstract part can be mapped onto a concrete part that is context -dependent. For instance, once a computing platform is targeted, it univocally determines the possible objects, turned into widgets, upon availability on this platform.

3.  Placement of Concrete Interaction Objects: once the context is fixed, the physical constraints imposed by this context can restrict the design space for producing the running UI. Widgets can be laid out in containers, containers can be arranged together so as to create a final UI. Several algorithms exist that lay these widgets out.

4.  Manual edition of Concrete Interaction Objects: after being laid out by the algorithm, the layout can be manually retouched using any traditional graphical editor, such as the one we can find in the Microsoft Visual Basic Development environment. This step is sometimes referred to as *beautification*, as it attempts to improve the final layout. Fig. 5 shows a sample of UIs generated for the example above.

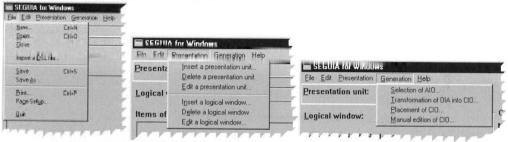

**Figure 4:** Menus of the model-based generator to generate a running UI

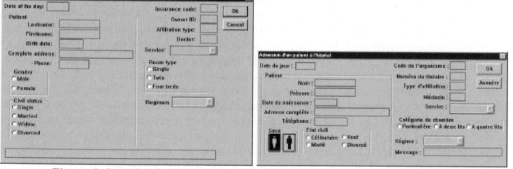

**Figure 5:** Sample of user interfaces generated for the task "patient admission"

# References

Guarino, N. (1995). Formal Ontology, Conceptual Analysis and Knowledge Representation: The Role of Formal Ontology in the Information Technology. *International Journal of Human-Computer Studies*, 43(5-6), 625-640.

Savidis, A., Akoumianakis, D., and Stephanidis, C. (2001). The Unified User Interface Design Method. Chapter 21. In C. Stephanidis (Ed.), *User Interfaces for All: Concepts, Methods, and Tools* (pp. 417-440). Mahwah: Lawrence Erlbaum Associates Pub.

Puerta, A.R. (1997). A Model-Based Interface Development Environment. *IEEE Software*, 14( 4), 41-47.

Top, J. and Akkermans, H. (1994). Tasks and Ontologies in Engineering Modelling. *International Journal of Human-Computer Studies*, 41( 4), 585-617.

Vanderdonckt, J. and Bodart, F. (1993). Encapsulating Knowledge for Intelligent Automatic Interaction Objects Selection. in S. Ashlund, K. Mullet, A. Henderson, E. Hollnagel, and T. White (Eds.), *Proceedings of the ACM Conf. on Human Factors in Computing Systems INTERCHI'93 (Amsterdam, 24-29 April 1993)* (pp. 424-429). New York: ACM Press.

Vanderdonckt, J. (2000). A Small Knowledge-Based System for Selecting Interaction Styles. In *Proceedings of International Workshop on Tools for Working with Guidelines TFWWG'2000 (Biarritz, 7-8 October 2000)* (pp. 247-262). London: Springer-Verlag.

# A Pattern Framework for Eliciting and Delivering UCD Knowledge and Practices

*A. Gaffar, H. Javahery and A. Seffah*

Human-Centered Software Engineering Group, Concordia University, Montreal

{gaffar, h_javahe, seffah}@cs.concordia.ca

**Abstract**: In the software and usability engineering community, there exist various tools for gathering and disseminating design knowledge. These tools aim to capture the best practices about the design of usable systems and to disseminate, or distribute, this knowledge. Examples of such tools include guidelines, patterns, various databases, and repositories of information. Guidelines and patterns are by far the most popular tools. This paper proposes a methodology supported by a tool for capturing and disseminating UCD knowledge and practices.

## Introduction

There are many approaches to knowledge management and patterns is just one of those. Patterns do more than capture good UCD practices; they are also useful for documenting process and organizational strategies. They are an ideal vehicle for transferring, by means of software development tools, the design expertise of human-factor experts and UI designers to software engineers, who are usually unfamiliar with UI design and usability principles. However, the lack of common fulcrum and central repository for patterns make it hard to achieve this goal. Several pattern writers have introduced their own terminology and ontology of patterns. Our goal is to develop and validate a methodology, which will highlight a path through the heterogeneous pattern world, and result in a comprehensive framework for disseminating and sharing HCI patterns. In addition to the above, a further challenge is the lack of tool support, which makes it difficult to capture, disseminate and apply patterns effectively and efficiently [7]. Pattern writers, who are most often usability engineers with a background in psychology and cognitive science, must try to convey complex information to describe the user problem and design solutions in a comprehensible and comfortable way. Pattern users, who are most often software developers unfamiliar with usability engineering techniques, need also tools to understand when a pattern can be applied (context), how it works (solution), why it works (rationale), and how it should be implemented. To address this problem, our team is developing MOUDIL (Montreal Online Usability Patterns Digital Library) which is a method + a tool for capturing and disseminating patterns.

## UCD Patterns Variety

Like the whole software engineering community [Gamma et al., 1995], the user interface design community has been a forum for a vigorous discussion on pattern languages for user interface design and usability engineering. It has been also reported that patterns are useful for a variety of reasons [10, 7].

Within our approach, we have been using three different categories of patterns:

- Interactive system design patterns. The aim of these patterns is to discuss and show a number of object-oriented design techniques and architectures for building flexible, portable, modifiable and extensible systems. A classical pattern of this category is the Observer that decouples between the user interface and the model [4}. For example, you might have a spreadsheet that has an underlying data model. Whenever the data model changes, the spreadsheet will need to update the spreadsheet screen and an embedded graph. In this example, the subject is the data model and the observers are the screen and graph. When the observers receive notification that the model has changes, they can update themselves.

- Human computer interaction design patterns. These are proven user experiences and solutions to common usability problems. Different pattern languages have been developed and are being used as an alternative/complementary design tool to guidelines [3, 5, 9, 10].

- Process patterns [8] that describe user-centered design best practices. Such patterns have the potential to support UCD practices such as iterative design, low-fidelity prototyping as well as conducting a user satisfaction-oriented test using a questionnaire.

- Organizational patterns that depicted an effective organizational strategy for establishing the usability function in an organization. Basic patterns include training a champion, motivating developers, building a usability group.

- Software process improvements that describe a maturity scale for the assessment of an organization's progress towards human-centredness in system development and management. The scale is based on a number of models including Flanaghan's Usability Leadership Maturity Model, Sherwood-Jones' Total Systems Maturity Model and ISO 13407 human-centered Design Processes for Interactive Systems.

- UCD to software engineers' communication patterns which describes proven, successful approaches for organizing and managing the multidisciplinary team generals needed. Examples of this category of patterns include engage user patterns, build and use a persona, etc.

- Pedagogical patterns that implement effective approach for training developers in UCD skills. For example, the DIRR (Design-Implement-Redesign-Reimplement) pattern described in Table 6 attempts to explain new concepts and methods based on legacy concepts (for instance, learning object-oriented fundamental concepts using structural software design methods).

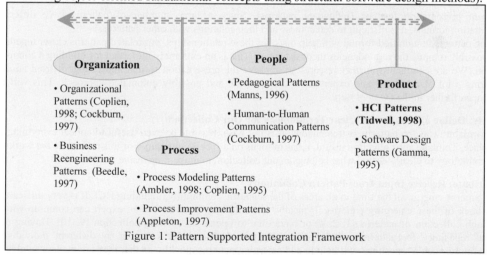

Figure 1: Pattern Supported Integration Framework

We use the term UCD pattern to refer to any of these categories of patterns. We also define a pattern supported integration framework as a collection of patterns and the relationships between them. A UCD pattern supported integration framework is a proven solution for integrating UCD practices and knowledge at the organizational, process, product and people levels (Figure 1). Each pattern is a three-part rule, which expresses a relation between a certain context, a certain system of forces that occurs repeatedly in that context, and a solution that allows these forces to resolve themselves.

# The Seven C'S Methodology

However, each of these pattern collections introduces its own terminology, classification system and notation/format [8]. Our methodology called "The seven C's" aims of centralizing and organizing patterns into one repository, as well as disseminating pattern knowledge to the HCI community. Our methodology distinguishes seven steps:

- Collect: Place Different Research Work on Patterns in One Central Data Repository
- Cleanup: Change from Different Formats/Presentations into One Style
- Certify: Define a Domain and Clear Terminology for Our Collection
- Contribute: Receive Input from Pattern Community
- Categorize: Define Clear Categories and relationships for our Collection
- Connect: The Second Level of Complexity – Establishing Semantic Relationships between Patterns in a Relationship Model
- Control – Machine Readable Format for Future Tools

### Collect: Place Different Research Work on Patterns in One Central Data Repository
Numerous works on patterns have been developed in the HCI community, however they are scattered in different places. A central repository of patterns will allow the user to concentrate on knowledge retrieval, rather than wasting time on searching for patterns. For this reason, we are collecting known references on patterns into one corpus. Currently our corpus includes more than 300 patters.

### Cleanup: Change from Different Formats/Presentations into One Style
Ideally, different works on patterns deal with different problems. However, as we went through step 1, we were able to identify that some patterns are dealing with different sides of the same problem (correlated patterns), some patterns are offering different solutions to the same problem (peer patterns/competitors) and some are even presenting the same solution to the same problem (similar), only in different collections with different presentation formats (redundant patterns). Since a large number of patterns have differing presentation formats, it is difficult to detect these and their relations with other patterns.

Putting patterns in a unified format will help discover these relationships, put related patterns closer together, and possibly remove the redundancies/inconsistencies. This is an extremely important for building a common ground. We are conducting further research to improve the presentation format and change it from an *art*, requiring a lot of creativity and expertise, to a systematic and possibly automated approach. This will be developed further in the *Control* step.

### Certify: Define a Domain and Clear Terminology for Our Collection
The available work on patterns is tremendous; it is not wise, feasible or even useful to collect everything in one place. To make any collection useful, it has to focus on a specific domain. For this reason we are working on terminology to clearly define what belongs in our collection to make it inclusive and concise.

### Contribute: Receive Input from Pattern Community
New patterns emerge all the time in all areas of the scientific community, including HCI. It is very difficult to keep track of these emerging patterns. Typically, it would take years before an expert can come up with a thorough collection of patterns [1, 2, 4] or have time to update an existing collection [9, 10]. Having one central repository for patterns will help to unify pattern knowledge captured by different individuals. Furthermore, such a repository will help to add emerging patterns quickly, so that they are made available to the community. We will therefore have a continuously evolving collection of patterns.

### Categorize: Define Clear Categories for our Collection
Within our collection, we need to be able to create a hierarchy of categories to make them manageable. The first goal of categorization is to reduce the complexity of searching for, or understanding, the relationship between patterns. The second, and more important goal, is to build a model for our categories. We are inspired by the evolution process in other domains like C++ (hierarchies in I/O classes, STL hierarchies, etc) and Java (evolving hierarchies in event handling models, etc).

**Connect: The Second Level of Complexity – Establishing Semantic Relationships between Patterns in a Relationship Model**

A significant part of knowledge associated with patterns lies in the relationships between them. Finding and documenting these relationships will allow developers to easily use patterns as one integrated part to develop an application, instead of relying on their common sense and instinct to pick up some patterns that seem to be suitable. A proven model for the pattern collection will help to define an ontology for the pattern research area with all proper relationships such as inference, equivalence and subsumption between patterns.

**Control – Machine Readable Format for Future Tools**

Once a model is established, it will enhance the process of automating the UI design. The ultimate goal of many applications is to interact with the machine as a viable partner that can read, understand our work, and then contribute to it in an intelligent way. In short, having a machine-readable format can help automate the process of UI design using patterns.

## Tools Support

In order to accelerate the implementation of our seven C's methodology, tool support is necessary. Each step requires a certain tool. Our research team is currently developing an Integrated Pattern Environment (IPE), called MOUDIL, which will unite all necessary functionality. MOUDIL was originally designed with two major objectives: Firstly, as a service to UI designers and software engineers for UI development. Secondly, as a research forum for understanding how patterns are really discovered, validated, used and perceived. One last objective for MOUDIL, in addition to the above two, is to use it as a prototypical implementation of an IPE. It will be able to provide functionality that supports every step of the seven C's methodology.

In particular, MOUDIL as medium for delivering patterns provides the following key features:

- At the heart of MOUDIL is a database designed specifically to serve as a central repository for patterns. These patterns can be backed up with examples and supported by other content areas – guidelines, case studies, checklists, reusable assets and components, templates, and resources.
- MOUDIL has been designed to accept proposed or potential patterns in many different formats or notations. Therefore patterns in versatile formats can be submitted for reviewing.
- Reviewing tools that can allow an international editorial board to evaluate patterns. Before publishing, collected and contributed patterns must be accessed and acknowledged by the editorial committee.
- Pattern Ontology editor captures our understanding of pattern concepts and puts them into relation with each other (Taxonomy).
- The MOUDIL Pattern Editor allows us to attach semantic information to the patterns. Based on this information and our ontology, patterns will be placed in relationships, grouped, categorized and displayed.
- The pattern navigator provides different ways to navigate through patterns or to locate a specific pattern. The pattern catalogue can be browsed by pattern groups or searched by keyword. Moreover, a pattern wizard will find particular patterns by questioning the user.
- The pattern viewer provides different views of the pattern, adjusted to the preferences of the specific user.

MOUDIL should include the following features:

- Quality assurance checklists
- A glossary of pattern terminology
- Booklist defined by category and includes ratings and links to an internal company book site or external book site, such as amazon.com
- Resources for developing patterns-oriented designs. Examples of resources available to developers and users include graphics, tables, and templates
- Case studies that add contextual meaning. These patterns can be backed up with examples and supported by MOUDIL other content areas such as case studies, checklists, and resources.

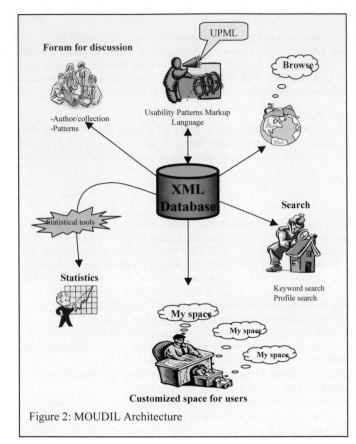

Figure 2: MOUDIL Architecture

## Conclusion

Patterns are useful in gathering and documenting experiences for future developers. A great deal of work has been done on HCI patterns by many different individuals. Patterns do not only provide help in applying UCD practices, but also provide support for understanding and mastering UCD concepts and applying related methods. The lack of knowledge centralization, however, requires users to hunt for suitable patterns, and extract them for their own use. Gathering relevant patterns in one repository will help overcome this difficulty. In our future research, we want to use the gathered information from the collected and analyzed patterns to come up with a formal pattern notation. Such a notation will help capture and disseminate pattern knowledge effectively.

## References

1. Alexander, C. *The Timeless Way of Building.* New York: Oxford University Press, 1979.
2. Alexander C., Ishikawa S., Silverstein M., Jacobson M., Fiksdahl-King I., and Angel S., *A Pattern Language: Towns, Buildings, Construction.* New York: Oxford University Press, 1977.
3. Coram, T., Lee, J. *Experiences - A Pattern Language for User Interface Design.* Available at http://www.maplefish.com/todd/papers/experiences/Experiences.html.
4. Gamma, E., Helm, R., Johnson, R. and Vlissides, J. *Design Patterns: Elements of Reusable Object-Oriented Software.* Addison-Wesley, 1995.
5. Griffiths, R. *Brighton Usability Pattern Collection.* Available at http://www.it.bton.ac.uk/cil/usability/patterns/.
6. HCI Department, Concordia University, Montreal. *MOUDIL: Montreal Online Usability Digital Library.* Available at http://hci.cs.concordia.ca/moudil/homepage.php.
7. Seffah, A., Javahery, H. *A Model for Usability Pattern-Oriented Design.* In Proceedings of TAMODIA 2002, (Bucharest, Romania, July 2002).
8. Seffah, A., Javahery, H. *On the Usability of Usability Patterns.* Workshop entitled *Patterns in Practice*, CHI 2002, (Minneapolis, Mi, April 2002).
9. Tidwell, J. *COMMON GROUND: A Pattern Language for Human-Computer Interface Design.* Available at http://www.mit.edu/~jtidwell/interaction_patterns.html
10. Welie, M. *Interaction Design Patterns.* Available at http://www.welie.com/patterns/.

# Towards Virtual Intuitive Tools for Computer Aided Design

*Yvon Gardan*

CMCAO Team / IFTS
Pôle de Haute
Technologie
08000 Charleville-
Mézières, France
gardan@infonie.fr

*Erwan Malik*

CMCAO Team / IFTS
Pôle de Haute
Technologie
08000 Charleville-
Mézières, France
erwan.malik@univ-
reims.fr

*Estelle Perrin*

CMCAO Team /
University of Metz
Île du Saulcy
57045 METZ Cedex 01,
France
perrin@sciences.univ-
metz.fr

## Abstract

The main objective of the DIJA project is to provide a CAD system through the internet and to help designers from the early phases of the design with an intuitive interface. Distributing a CAD system over the World Wide Web implies that this system has not any knowledge about its user (his computer and CAD skills, his trade, etc.). The particularity of our system is that it is based on a new design method, called the "synthetic" approach. In this paper, we focus our attention on the man-machine interface part of the DIJA project: our goal is to provide the user with the ability to express his intent. We propose a set of virtual tools used to control the design and we consider the importance of the media in the user-system dialog.

## 1   Introduction

Current CAD systems are monolithic and their computer-human interface is not easily accessible for a non-expert user. It is common that the latter has to learn about a CAD system before use. The reason is that CAD systems do not integrate properly functions from requirements. In fact, they completely forget and hide functional intent, design intent and trade know-how. They just deal with features, which is inadequate to really take into account design intent. Features are very close to classical geometric models (such as Boundary representation, Constructive Solid Geometry, ...), so they are enclosed in a way of modeling that imposes the user a method to build models, essentially based on basic objects and operations. At the same time, CAD systems have a rustic interface that is based on standard devices (such as mouse) and number of menus soars with increase of services.

We propose, in this paper, an interface to provide the user with the ability to express his intent, based on a modeling method, called the "synthetic method". Both interface and modeling method are parts of the DIJA project (Danesi, Denis, Gardan & Perrin, 2002). The main objective of the DIJA project is to provide a CAD system through Internet and to help designers from the early phases of the design of an object with an intuitive interface. The basic idea of the synthetic method is that the user, even by giving a function or by choosing a form, obtains a very approximate shape. This shape, from a certain point of view, only represents the topology of the object. The user can then deform this shape using some dedicated tools. The initial form can be very close to or far away from the goal. It does not have really influence, the method being always the same, requiring more or less steps. Obviously, an expert user will choose a form necessitating only a little number of steps.

The synthetic method respects a top down methodology which seems very interesting for many applications. It considers basic deformable objects depending on domain of the application. Each domain has its proper rules of design and its own vocabulary. The words are different and even the same words can lead to different signification according to the trade. Then the language which is at the disposal of the user is domain dependent. A word (specifically a verb) of the language can lead to a general behaviour (that means that this word has a shared meaning, which is the case of those we describe hereafter) or to a domain dependent behaviour. Three kinds of behaviour or tool are defined:

- deformation: the object is deformed following some rules described later;
- dividing: an object is divided in several parts of the same type;
- transmutation: the type of the object changes (for instance a cylinder becomes a box).

In this paper, we only focus on the deformation tool of the DIJA project. The user deforms the object, by stretching one of the shape surfaces (surface deformation), or by deforming the shape globally (shape deformation). Both deformations perform the same interactions which are fully described in the next section: selection of a shape, choice of a tool and applying the tool. In the last interaction, the user has to describe how the tool will affect the shape in terms of direction and strength. Such notions are not easily received by a standard device. In this context, we present (section 3) an interface based on a camera that we think to be more intuitive.

## 2 Virtual Intuitive Tools of Dija Project

Traditionally, with CAD systems, the user has to express deformation by using a geometric or mathematic description (references). On the contrary, by using tools, the user can focus on his design rather than on geometric aspects of the object he wants to achieve. Then he can consider his design more like a virtual object and thus, he can use deformations close to those from real world. In DIJA, to deform an object, the user has to define the shape of the tool he will use – like bending a bar using the sharp edge of a table corner or the smooth shape of a large cylinder. Through this, we want to increase the relation between the user and his working environment by very simple meanings.

We define a tool by associating a shape with some vocabulary terms. For example, the tool shape represented in figure 1a is associated with the term "to swell" in a knowledge based model. This knowledge based model associates a shape to a term in function of a trade context. For example, the maximum or minimum of swelling depends on mechanical considerations. Considering two different trades, the system would not propose the same shape for the same term. This is realised by using fuzzy parameters (Zadeh, 1997). Nonetheless, other terms fully specify the tool shape like "sharp" or "narrow". Obviously, an internal geometric representation is also used by the system (by example, "to swell" is represented by conics such as a circle or ellipse) but the user is not aware of it. He only manipulates a trade vocabulary that protects him from geometric reflections.

The method we use to get the tool shape follows the same synthetic approach than the rest of our project. The user starts the process by choosing a trade and a term which qualifies the deformation. Our system deals with this term to propose a standard shape as we mentioned above. Finally, the shape is refined by using adverbs and/or adjectives belonging to the tool's vocabulary through an iterative process. Once again, the system reactions will depend on the previous user interactions. If the user asks for "more sharp" for the fourth time, the result will not be the same as the first time. As for trade-oriented tool definition, this is realised by fuzzy parameters.

In the next paragraph we present an example illustrating a current implementation of tools design in the DIJA system.

First, the user chooses to design the "to swell" tool which will be used to deform the virtual object he has designed. The system shows him the standard shape for the bend deformation that is a circle with a set of terms that can modify the prior shape. This set is composed by the adverbs "more" and "less" applied to adjectives:

- "swell" (which represents the intensity of the deformation and it's global direction),
- "sharp" (which represents the quality of the shape's edge),
- "narrow" (which represents the width of the shape),
- "inclined" (which modifies the direction of the deformation).

Figure 1b shows a possible evolution between the initial tool shape and the final one. Each time the user selects a term the system is able to adapt the proposed shape: the circle turns into a parabola if the user selects "less slender", then the axis of the parabola changes when the user selects "more inclined".

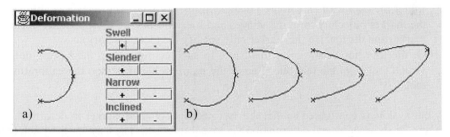

**Figure 1**: a) the interface for designing tools, b) a possible evolution of the tool in function of user's interactions.

## 3   Visual Perceptive Interface

When the user has defined a tool, he can use it in order to deform a virtual object. We have first implemented a traditional mouse/keyboard interaction type because the DIJA system is dedicated to any hardware configuration. But we have found that it was not intuitive as it might seems. This is due to device manipulation by the user. So we investigate the field of perceptive interface to find interactions not based on device manipulation, which is led to visual device and image tracking. We have chosen to use camera (see figure 2a) because it tends to be a more and more standard device (principally due to the democratisation of web-cams).

Visual perceptive interface is a domain of growing interest since the last decade (Porta, 2002). One of the main goals behind this research is the replacement of standard devices – i.e. keyboard, mouse – by camera (Zhang, Wu, Shan & Shafer, 2001). In fact, the use of a mouse as an input device induces an indirect manipulation of any virtual element (Kjeldsen & Hartman 2001). In this kind of configuration, the user makes a gesture to move the physical device which in turn moves the virtual pointer with which it is associated. Standard computer user, do not fell any difficulties with this kind of gestures: they became familiar. Unfortunately, it is not the case for neophyte computer users who may be disappointed with them. In opposition, by using visual interface we can propose interactions based on direct manipulation. Gestures made by the user are directly translated into the virtual environment (figure 2b). We believe that such interactions may help a non-skilled user in their task.

The specifications of the DIJA project impose strong constraints in term of real-time interaction. The delay between gestures and their consequences must be as small as possible. (Bérard, 1999) speaks of a maximal threshold in milliseconds. After it, the interactions might become annoying for most of the users. This constraint guides us for choosing an algorithm to analyse the images. Basically we can identify two classes: those based on learning and on statistical model (PDM, etc.) (O'Hagan, Zelinsky, Rougeaux, 2002) and those based on low-level image processing (Bérard, 1999) – like thresholding, colour analysis or region growing for example. The former approach is not proper to our needs: our system must be usable instantly by any user. Finally we have chosen the colour threshold algorithm because it offers simplicity and suits well our needs: our efforts are not on the development of image analysis but on interaction process.

In fact, we seek to extract the trajectory of a gesture and use it in association with a tool. Both of theses parameters are used to deform the virtual object. We work on two different ways to deform an object with our interface:

- Either the user wears a coloured marker in his/her hand. The system is able to extract the marker trajectory form the images and it establishes a correspondence between this marker and the tool (so the marker only specify the tool trajectory).
- Or the user has a real tool in his hand. In such configuration, the system defines the virtual tool from the real one. One of the main problems is then the calibration of the tool.

Currently our system is configured so that the user needs a coloured marker to deform the virtual object. Figure 2c shows a typical deformation via visual interface. The user starts by selecting the element of the virtual object he/she wants to deform. Then we see the shape of the marker which is used as a tool.

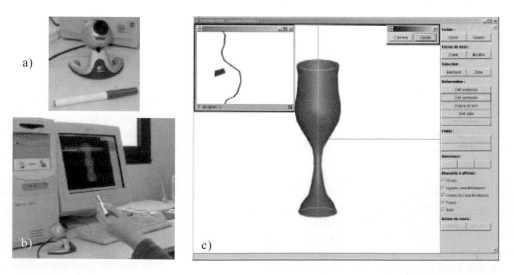

**Figure 2**: a) a camera and a coloured marker, b) the user in front of the DIJA system, c) a snapshot of the DIJA application.

# 4    Conclusion and Future Works

In this paper, we present our research on the man – machine interface of the DIJA project. We introduce the notion of virtual tools and how they are constructed. Our primary goal is to provide the user with the ability to design virtual tools via a language based interface. By using words, the tool's shape is modified in an intuitive manner. Then we focus on the meanings to control such tools. We think that visual perceptive interface may offer an intuitive way to do it by letting the user directly manipulates the virtual object. A prototype has been implemented that validates our concepts.

We intent to pursue our investigation on dealing with words based interface in a more accurate way. This can be achieved by extending the vocabulary associated with a type of deformation. Besides we also want to guide the user during the stage of the tools design. We plan to do this by combining the knowledge of the user's previous actions and heuristic methods. Thus, we will be able to provide the user with a panel of choices classified by probability. The different proposals made by the system are intended to enhance the user's design view (Liu, 2000)(Lawson & Loke, 1997).

In the field of visual perceptive interface we need to extend our image treatment algorithm to include specific gesture – like extruding or twisting gesture. Moreover we are interested in the couple visual – vocal interface. Thus the interest of the words based interface would take its full meaning.

# 5    References

Bérard, F. (1999). Vision par ordinateur ordinateur pour l'interaction homme-machine fortement couplée. *Ph.D. Thesis.*

Danesi, F., Denis, L., Gardan, Y., Perrin, E. (2002). Basic components of the Dija. *In Proceedings of the seventh ACM symposium on Solid Modeling and Applications.* ACM Press.

Kjeldsen, R., Hartman, J. (2001). Design Issues for Vision-based Computer Interaction Systems. *In Workshop on Perceptive User Interfaces.* ACM Digital Library.

Lawson, B. (1997). Computers, words and pictures. *Design Studies,* 18 (2), 171-183.

Liu, Y.-T. (2000). Creativity or novelty?. *Design Studies,* 21 (3), 261-276.

O'Hagan, R.G., Zelinsky, A., Rougeaux, S. (2002). Visual gesture interfaces for virtual environments. *Interacting with Computers,* 14 (3), 231-250.

Porta, M. (2002). Vision-based user interfaces: methods and applications. *International Journal of Human-Computer Studies,* 57 (1), 27-73.

Zadeh, L.A. (1997). Toward a theory of fuzzy information granulation and its centrality in human reasonning and fuzzy logic. *Fuzzy Sets and Systems,* 90 (2), 111-127.

Zhang, Z., Wu, Y., Shan, Y., Shafer, S. (2001). Visual Panel: Virtual Mouse, Keyboard and 3D Controler with an Ordinary Piece of Paper. *In Workshop on Perceptive User Interfaces.* ACM Digital Library.

# Engineering the HCI profession or softening development processes

*Jan Gulliksen[1], Stefan Blomkvist[1] and Bengt Göransson[1,2]*

[1]Dept. for IT/HCI, Uppsala University, PO Box 337, SE-751 05 Uppsala, Sweden
[2]Enea Redina AB, Smedsgränd 9, SE-753 20, Uppsala, Sweden
Jan.Gulliksen@hci.uu.se

## Abstract

Despite of all knowledge on usability and user centered design this knowledge is poorly integrated into the software development process. The purpose of this session is to bring together authors with a long tradition of working to bridge the gap between software engineering and HCI. The session will include papers on methods, tools, processes and principles for integrating user centered systems design in the software development process as well as present some illustrative case studies. The purpose of this introductory chapter is to discuss the engineering nature of usability work and the consequences this might have, both for the credibility of the work within the software engineering community and for the success in development projects.

## 1   Introduction

How can we make HCI and usability have an actual impact on the software development process. What we see today is an increasing amount of applications, products and systems that suffer from severe usability deficits. In addition to this we see an increasing amount of users that are forced to use technology that largely does not support the fulfilment of the user's tasks. Even to the extent that they get hurt through using the system. Well, obviously there is still a lot to be achieved to be able to systematically work with a focus on usability.

A common view within the software engineering community is that a field of research is not mature enough until it becomes engineering. So what does it mean that a field becomes an engineering field? According to the Webster Dictionary Engineering means the application of scientific and mathematical principles to practical ends such as the design, manufacture, and operation of efficient and economical structures, machines, processes, and systems.

Formal commercial software development processes, such as Rational Unified Process – RUP (Kruchten, 1998), gives no or little support for addressing usability matters. RUP is the most widely used software development process in Sweden today and it is explicitly architecture-centred. To be able to do usability work today is very difficult given that it is not explicitly promoted in the development processes.

## 2   The engineering nature of usability work

Usability as a concept has a commonly agreed upon definition (ISO 9241-11, 1998). However, the various approaches focusing on usability are a lot more vague in their definition and general interpretation.

The way usability is defined in ISO 9241-11, it prescribes an engineering approach to be able to measure and achieve the desired level of usability. To be able to achieve this one needs to specify usability metrics – that is quantitatively measurable goals of effectiveness, efficiency and satisfaction (see for instance Whiteside, Bennet & Holtzblatt,1988). Our opinion is that it is of utmost importance to try to specify usability goals, since otherwise the focus on usability might get lost, but even so that it is very difficult to do so. Usability metrics contributes to the engineering nature of usability work.

Usability engineering (Nielsen, 1993; Mayhew, 1998, Faulkner, 2000, Carroll, 2002) concerns itself with providing methods that can be used with the purpose of increasing the system usability, mainly evaluation methods. This approach has, however, had little influence on the commercial development processes as such. It gives detailed guidance on how to perform a specific activity, but it gives a rather limited advice on how the activity contributes to the whole

Finally there is the problem with the representations. Current software development practices require multiple sets of representations to serve as a useful tool both for the software developers and the users. Users require concrete notations, such as scenarios, prototypes or role play to be able to understand the impact that the new system might have on them. On the other hand, software developers require abstract notations such as UML diagrams or use cases that give them a clear picture of what to implement. Running two different parallel representations means that you take on a gigantic translation job maintaining consistency between the different representations.

# 3    The soft part of usability

There are several phases of the development process that clearly would suffer by being forced into a formal development process. Unfortunately there is not very much room for this in existing approaches. Following we will describe some examples of activities that does not fit very well with a formal process:

## 3.1    The tradition of participatory design and active user participation

There seem to be a common agreement that users are an important source in the design process. However, the extent to which users will be allowed to make decisions or be an active participant in the development process is still an issue of much debate. Our experiences and observations are in line with the participatory design tradition that requires the users to take an active part in not only the analysis and the evaluation parts of the development projects but also in the making of design decisions. Given the unpredictable nature of participation and accommodation to individual preferences and abilities the formal process gets more in the way than supporting these activities.

## 3.2    The non engineering nature of design

Design is "a creative activity that involves bringing into being something new and useful that has not existed previously" [Jones, 1981]. There is an on-going discussion about to what extent the procedures by which designs are created could be specified, described and thereby communicated and made repeatable for others. There is not much evidence that design can be made independent of talent and reliability of all aspects that are involved in craft work.

Efforts have been made to try and specify design as an engineering task. For example, "Engineering design is the use of scientific principles, technical information and imagination in the definition of a mechanical structure, machine or system to perform pre-specified functions with maximum economy and efficiency. Design refers to both the process of developing a product, artefact or system and to the various representations (simulations or models) of the product that are produced during the design process" [Preece, Rogers, Sharp, Benyon, Holland, & Carey, 1994].

For a software engineering process to support design activities it must allow some flexibility to be able to handle design considerations with the degrees of freedom it takes to maintain it as the creative and unbounded task it needs to be. Providing the necessary place in the development process for design is a challenge.

# 4    Agile approaches to software development

Recently, Agile approaches to software development has emerged as an alternative to the formal commercial processes. Agile does not deny the value of engineering or model-based methods in software engineering, but would like to shift the overall focus to a more agile or "lightweight" perspective. Agile developers embrace individual skill, communication, fast adaptation to change and delivering of working, useful software. They are trying to find a balance between a pragmatic, soft approach and a non-rigorously, sufficiently use of engineering methods. One interesting subject for the HCI community is how well usability matters are handled by Agile software development.

## 4.1    The Agile view on Engineering

"The trouble with using engineering as a reference is that we, as a community, do not know what that means" (Cockburn, 2002). Engineering is about applying science and math and addressing problems in a repeatable and consistent manner. But this does not explain what doing engineering is about. There is confusion about the act of doing engineering and the result of engineering work. The typical outcome of doing engineering is the factory – measured quality and control of products. Another perspective of "doing engineering" is looking up previous solutions in code books, based on known solutions.

The engineering perspective does not fit the current state of software development especially well:
- Software development is more about design solutions and tradeoffs between solutions, not measured quality and control of products in a factory way. The whole solutions are seldom repeatable.
- Technology and usage changing so fast that few code books exist.

According to Cockburn (2002), to think of software development as model building leads to inappropriate project decisions, as the interesting part of what we want to express doesn't get captured in those models. Models are useful in many ways, but not as an overall metaphor. That's because software development is not model building in essence. A model is a medium of communIcation, and is sufficient as soon as it permits the next person to move on with his/her work.

The advocates of Agile development, argues that the answers of successful software development has much more to do with craft, community, pride and learning. The essence of Agile is the pragmatic use of light but sufficient rules of project behaviour and the use of human and communication oriented rules.

Although the different methods collected under the Agile concept differ from each other[1], they more or less share a common set of values, described in the Agile Manifesto (Agile Alliance, 2002):

- Individuals and interactions over processes and tools. It is people, not processes and tools, who develop software. Therefore, individual's skill and communication between people is important. The Agile movement argues that face-to-face communication is the most effective (Cockburn, 2002). Pair programming is a consequence of these arguments.
- Working software over comprehensive documentation. Requirements, documents, models and other intermediate products, are only means of communication during development. They can be very useful but they should only be worked on as long as they serve a purpose in delivering working and useful software (Cockburn, 2002).
- Customer collaboration over contracted negotiation. This value describes the relations between the people who wants the system  and the people who builds it
- Responding to change over following a plan. Agile methods include planning activities. Although, plans should not be followed strictly in every situation. The important thing is to plan for and adapt to changes. This is necessary, as the prerequisites for most systems will change during the development. As communication between people always is more or less incomplete, the initial requirements will also change.

## 4.2  Agile and usability

The Agile movement hasn't paid much attention to usability and user-centred design. On the other hand, they have an open and pragmatic view of collaboration with different stakeholders, which should include real end-users. So Agile is a promising platform for integrating user-centred system design with a soft perspective together with engineering methods.

The Agile values emphasizes people, skill and communication, which also applies to user-centred principles, such as:

- Direct participation of real users in the development process is the most effective way of communicating the needs of the users.
- To involve skilled usability experts and talented designers, benefits the system development.

In both Agile and user-centred design, responding to change is critical. Iterative and evolutionary design, as well as prototyping, are ways to handle this demand that fit well with the Agile values.

In user-centred design, the process is of great importance, which doesn't seem to agree with the Agile point of view. But in Agile, process still have an important part to play, but it must be flexible and actively support the work of skilled individuals and communication on all levels. This is a risk with to formal, detailed and heavyweight processes.

## 5  Discussion

The point we do want to raise in this paper is a point of trying to discuss to what extent we can move towards an engineering of our profession. Usability professionals today have no natural role

---

[1] The best-known Agile methods are Extreme programming (XP) and Dynamic Systems Development Method (DSDM).

in the commercial software development processes. Their efforts and activities are mainly regarded as add-on activities that easily can be left out if they risk delaying or disturbing the project, as all activities that are not properly integrated in the process. The agile software development approaches shows a promising attempt at relieving the software from the formal burdens of processes such as RUP. However, at this point they do not provide sufficient support for usability work.

Judging from the application of these approaches in practice it is quite obvious that the software engineering community needs to be able to view the field of HCI as an engineering discipline. Otherwise they would have problems understanding how the knowledge could be incorporated. On the other hand judging from the end-user perspective the tools and procedures used in the engineering traditions give no or little space for the users to express their needs and requirements. Finally, design, as a craft discipline, needs a larger space to manoeuvre than what is provided in the engineering tradition. From the application point of view we think that to be able to shift the attitude in the software development process we need to educate all participants in the development process in usability and user centred design, we need to relate the essential activities to the development process and we need better and improved tools to facilitate communication in the process.

# References

Agile Alliance (2002) Manifesto for Agile Software Development, http://www.agilealliance.org [referred on 13 February 2003].

Carroll, J. (2002) Usability Engineering.

Cockburn, A. (2002) Agile Software Development, Boston, MA, Addison-Wesley.

Constantine, L.L. & Lockwood, L.A.D. (2002) User-Centered Engineering for Web Applications. *IEEE Software, Vol. 19, No. 2,* pp. 42-50.

Faulkner, X. (2000), Usability Engineering, Palgrave, New York, N.Y.

ISO 9241-11 (1998), Ergonomic Requirements for Office Work with Visual Display Terminals (VDTs). Part 11: Guidance on Usability. International Organization for Standardization, Geneve.

Jones, J.C. (1981). Design Methods: Seeds of Human Futures, 2nd Edition, London: Wiley.

Kruchten, P. (1998). *The Rational Unified Process—An Introduction,* Addison Wesley Longman Inc., Reading, Mass., USA.

Mayhew D.J. (1999), *The Usability Engineering Lifecycle, A Practitioner's Handbook for User Interface Design,* Morgan Kaufmann Publishers Inc., San Francisco, CA.

Muller, M.J., Haslwanter, J.H. & Dayton, T. (1997). Participatory practices in the software lifecycle. In M. Helander, T.K., Landauer & P. Prabhu (Eds), Handbook of Human-Computer Interaction (s.255-297). Elsevier Science B.V.

Nielsen, J. (1993), Usability Engineering, Academic Press Inc., San Diego.

Whiteside J., Bennet J. & Holtzblatt K., (1988), Usability engineering: our experience and evolution, in Helander (ed.), Handbook of Human-Computer Interaction, North-Holland.

Preece, J., Rogers, Y, Sharp, H., Benyon, D., Holland, S., & Carey, T. (1994). Human-Computer Interaction. Addison-Wesley Publishing Company, Wokingham, England.

# Experiences with User Centered Development (UCD) for the Front End of the Virtual Medical Campus Graz

*Andreas Holzinger*

Institute of Medical Informatics, Statistics and Documentation (IMI)
Graz University, Engelgasse 13, A-8010 Graz Austria
andreas.holzinger@uni-graz.at

## Abstract

This paper offers some experiences with User Centered Development (UCD) during the subproject User-Interface Development for the Virtual-Medical Campus (VMC) Graz This is a modular multi-medial Information System to make a new curriculum human medicine digitally accessible and supports students and teachers in their workflows. The UCD in combination with well accepted usability methods including thinking aloud, cognitive walkthrough and video analysis proved again as a well-suited design approach.

## 1    The Virtual Medical Campus Graz

The general objective of the Virtual Medical Campus Graz (vmc.uni-graz.at) is the realization of a modular multi-medial Information System to make the totally new curriculum of human medicine digitally accessible. The contents of the various disciplines are presented in a logical, clinically oriented context, resulting in an integrated curriculum. Since the target population includes a large number of both students and teachers (approx. 4500 people) it was essential to design the front-end best suited to their specific needs.

We decided to use the User Centered Design (UCD) method and followed the general research recommendations of (Stephanidis et al., 1999). In agreement with (Marcus, 2002) we referred to this process as User Centered Development, analogous to Software Development, since it covers all aspects – from the requirements analysis to the working prototype. Experience with UCD was already available (Holzinger, 2002b). A main point is securing sustainability – thus, such a project can only be successful when it is supported by the entire faculty and when students and teachers are integrated in the design activities by using a User-Centered approach. It is not just a project but a strategy which raises awareness of the possibilities of e-Learning. We consider multimedia as being one of many possible elements of an integrated solution (Holzinger, 2002a).

We developed the System on the basis of Microsoft Windows 2000 server, MS SQL-server 7.0 and MS visual studio .net including MS active server pages .net (asp.net) and MS visual basic .net (vb.net). This decision was driven by the fact that this software was available at the university.

## 2    The subproject "User Interface Development"

The project start was on 4th April 2002 with a total running time of two years, whereby the first part was devoted to the technical development of a multi-media repository in the back-end, the related middleware which includes the VMC logic and the front-end. The other parts of the project concentrate on content including the incorporation of existing study material and the production of new study material (e.g. on the basis of reusable learning objects). The new curriculum became effective on 1st October 2002. Therefore, the first goal was to implement the basic functionalities not later than 1st October 2002. It was a primary concern to insure that the subsequent development of the curriculum coincided with the development of the VMC. Nobody uses a system that lags behind. In only 25 weeks a fully functional version was to be implemented with a full commitment to UCD. Due to this extreme time pressure, our guiding principle was first to create a fully functioning core which contained all the necessities and could be expanded at a later date. This goal resulted from the fact that we realized during our field studies that our system was the only access point for the students to gather information about the contents of the new curriculum. Also, we were aware of the fact that the acceptance by the teachers would only be possible if we could provide our system in advance of their respective lectures.

## 3    User Centered Development in Practice

It is a generally known fact that many software developers rarely use (any) usability-engineering methods on software development projects in real life. This includes even such basic usability engineering techniques as early focus on the user, empirical measurement and iterative development of prototypes (Nielsen, 1993). It is rare that projects adopt a fully integrated User Centered Design approach in one strategic shift (Dray & Siegel, 1998). We committed ourselves in this project to a full UCD (Vredenburg, Isensee, & Righi, 2002), (Torres, 2002), (Norman & Draper, 1986).

### 3.1    Requirements Analysis

In field studies (observation of users at work, interview users and stakeholders, survey users), together with a group of UCD-participants (N=49) having different roles in the VMC, we gathered information about the context of its use including the characteristics of the intended users. These were e.g. knowledge, skill, experience, education, training, physical attributes, as well as habits and capabilities. Also it was necessary to define different types of users, for example with different levels of experience or performing different roles (students, teachers, module administrators, subject coordinators etc.). We carefully examined the specific tasks that the users have to perform including typical scenarios, frequency and duration of performance etc. (Carroll, 2000), (Diaper, 2002). As deliverables, we got the requirements document including Use Case Scenarios (with details of tasks, user profiles, workflows etc.) and the respective usability goals.

### 3.2    Conceptual Design

At this stage we created a general understanding of how the new system will operate and called it top-level view. We defined the main functions of the system and began to consider certain interface designs. As deliverables, we produced paper mock-ups and conducted walk-through and thinking aloud sessions with our UCD-participants.

## 3.3    Detailed Design

At this stage as a deliverable we produced a functional online prototype (hi-fi prototype) based on the results of the previous UCD-sessions in combination with existing GUI standards and principles of good screen design.

## 3.4    Implementation

The development stage included the programming of the application following the feedback from results of the former UCD-sessions.

## 3.5    Usability Testing and Re-Engineering

With the fully functional version, further usability inspection was carried out. During this stage we were able to concentrate on more specific details including the fonts, font sizes, colors, size of entry-fields, design of buttons, shapes and so on. After a thorough re-engineering the final version was put into operation.

# 4    Usability Methods Applied

During the prototyping process, we adapted one room of our Institute as a usability lab and used well accepted research methods including cognitive walkthrough, thinking aloud and video analysis. We used the same methods to carry out UCD-sessions with the users in their respective environments (e.g. teachers in their rooms, students in public computer labs etc.) We also observed how our online-prototype worked at students' homes, due to the fact that the interaction should function immediately even when using a slow dial-up modem. Thus simplicity was one of our main concerns.

## 4.1    Cognitive Walkthrough

A task-oriented cognitive walkthrough focuses explicitly on the learnability of the system, i.e. it is based on a first-time use without formal training (de Mul, S., & van Oostendorp, H. (1996). The big difference to thinking-aloud (section 4.2) is that it is non intrusive. This method is based on the cognitive model of human exploratory learning. Exploratory Learning implies that users often prefer to learn a new system by "trial and error" rather than read manuals or attend courses (Carroll, 1987), (John & Packer, 1995). We used this method from the beginning with the mock-up and later with the working prototype analogous to a structured walkthrough in software engineering (Satzinger, Jackson, & Burd, 2002). We specified correct action sequences for each task, according to our task analysis. For each action we constructed possible success or failure scenarios (e.g. "will the user know that the correct action is available and if the correct action is taken, will the user see that things are going right") and tested how we could solve the given task. With this method we detected at first the rough problems and we carried on with the thinking aloud method.

## 4.2    Thinking aloud

The UCD-participants were asked to verbalize their thoughts ("think aloud") while performing specific tasks. Thus we defined representative tasks, for example " ... please store the learning object X to the repository ..." This provided an abundance of relevant data  already with a relatively small number (approx. 3 to 6) of participants (Nielsen, 1994a). This method is based on asking the users to talk freely about what they are momentarily trying to do, questions that arise in their mind, things they find confusing and the decisions they make. It is necessary to prepare the user by demonstrating thinking aloud during the briefing – this can be achieved best by showing a short video clip of a previous thinking aloud test session. It is important to encourage the user to keep up the flow of information with neutral, unbiased prompts, e.g. "Please tell me what you are doing now." That way, we not only located usability problems  but also were able to find out why they occur.

## 4.3    Video analysis

According to  (Brun-Cottan & Wall, 1995)  video is uniquely suited to support the UCD process. With the consent of the UCD-participants we recorded their interaction, first with the paper-mock up and then with the working prototype. It was vital that the programmers watched and discussed these recordings together with the analysts. It was astonishing that things what were absolutely clear to the programmers caused problems to the user.

## 5    Conclusion

Amongst the abundance of information we gathered were some which we considered to be vital for our design. Some of them include the following:
The current user-mode must be immediately recognizable to the user. They should never find themselves wondering if they are in student-mode or edit-mode! This can only be achieved through clear and consistent information regarding the current user-mode. We recognized that consistency is one of the most important design goals. We therefore tried to arrange all information always in the same style and the tasks carried out by the users always in the same order. We noticed that every action a user takes should respond in a feedback from the interface. We attained that goal by partly giving messages such as "Thank you, the learning object was successfully created" or by visually changing buttons.  In order to achieve a well usable interface the method of UCD proved once more to be the right solution, although it was difficult and hard at the beginning to gain the awareness amongst our programmers and directors due to the fact that they had never before used this method and found at the beginning that "it is not necessary to make such efforts". It also was not easy to establish a good UCD-participants pool which was accessible throughout the subproject. However, the fact that the users were involved from the very beginning until the release of the working system proved to be a well-suited approach. Finally (Nielsen, 1994b) argued that many software developers do not apply usability engineering techniques, as they are considered expensive. We made the experience that using UCD is not expensive, rather that the pay-off is higher due to less corrections in the final stage of development. Above all, the user satisfaction is higher and exactly this can be regarded as a further step in reaching the goals of an information society for all (Stephanidis & Savidis, 2001).

# 6    References

Brun-Cottan, F., & Wall, P. (1995). Using Video to Re -Present the User. *Communications of the ACM, 38*(5), 61-71.

Carroll, J. M. (1987). *Interfacing Thought: Cognitive Aspects of Human-Computer Interaction.* Boston (MA): MIT.

Carroll, J. M. (2000). *Making use: Scenario-based Design of Human–Computer Interactions.* Cambridge (MA): MIT Press.

Diaper, D. (2002). Task scenarios and thought. *Interacting with Computers, 14*(5), 629-638.

Dray, S. M., & Siegel, D. A. (1998). User-centered design and the "vision thing". *interactions, 5*(2), 16-20.

Holzinger, A. (2002a) *Multimedia Basics, Volume 2: Learning. Cognitive Fundamentals of multimedial Information Systems.* New Delhi: Laxmi. (Available also in German by Vogel Publishing, Wuerzburg, Germany)

Holzinger, A. (2002b). User-Centered Interface Design for disabled and elderly people: First experiences with designing a patient communication system (PACOSY). In K. Miesenberger, J. Klaus & W. Zagler (Eds.), *Springer Lecture Notes on Computer Science Vol. 2398* (pp. 34-41). Berlin et.al.

John, B. E., & Packer, H. (1995). *Learning and Using the Cognitive Walkthrough Method: A Case Study Approach.* Paper presented at the CHI '95, Denver (CO).

Marcus, A. (2002). Dare we define user-interface design? *interactions, 9*(5), 19-24.

Nielsen, J. (1993). *Usability Engineering.* San Francisco: Morgan Kaufmann.

Nielsen, J. (1994a). Estimating the number of subjects needed for a thinking aloud test. *International Journal of Human-Computer Studies, 41*(3), 385-397.

Nielsen, J. (1994b). Using discount usability engineering to penetrate the intimidation barrier. In R. G. Bias & M. D. J. (Eds.), *Cost Justifiing Usability.* Boston (MA): Academic Press.

Norman, D. A., & Draper, S. (Eds.). (1986). *User Centered System Design.* Hillsdale (NY): Erlbaum.

Satzinger, J. W., Jackson, R. B., & Burd, S. D. (2002). *Systems Analysis and Design in a Changing Word, Second Edition.* Boston (MA): Thomson.

Stephanidis, C., Salvendy, G., Akoumianakis, D., Arnold, A., Bevan, N., Dardailler, D., Emiliani, P. L., Iakovidis, I., Jenkins, P., Karshmer, A., Korn, P., Marcus, A., Murphy, H., Oppermann, C., Stary, C., Tamura, H., Tscheligi, M., Ueda, H., Weber, G., & Ziegler, J. (1999). Toward an Information Society for All: HCI challenges and R&D recommendations. *International Journal of Human-Computer Interaction, 11*(1), 1-28.

Stephanidis, C., & Savidis, A. (2001). Universal Access in the Information Society: Methods, Tools and Interaction Technologies. *Universal Access in the Information Society, 1*(1), 40-55.

Torres, R. J. (2002). *User Interface Design and Development.* Upper Saddle River (NJ): Prentice Hall.

Vredenburg, K., Isensee, S., & Righi, C. (2002). *User Centered Design: an integrated approach.* Upper Saddle River (NJ): Prentice Hall.

# Interface Metaphors for Automated Mobile Phone Services

*Mark Howell, Steve Love*
Department of Information Systems
and Computing,
Brunel University,
Uxbridge UB8 3PH.
mark.howell@brunel.ac.uk
steve.love@brunel.ac.uk

*Mark Turner, Darren Van Laar*
Department of Psychology,
University of Portsmouth,
King Henry Building,
Portsmouth PO1 2DY.
mark.turner@port.ac.uk
darren.van.laar@port.ac.uk

## Abstract

This paper describes a study designed to elicit categories of interface metaphors that can be applied to voice-based automated mobile phone services. The effects of gender and previous computing experience on users' preferences for these metaphor categories were also investigated. The motivation for this work is the potential improvement in usability and performance that interface metaphors may have for voice user interfaces (VUIs), especially if designed to accommodate for the individual differences of users. Through a user-centred design process metaphors were generated, selected, developed and utilised by participants to explain how to perform tasks with 2 different automated telephone services. Five different categories of metaphor were apparent, and were named 'Hierarchical', 'Shopping venue', 'Transport system', 'Information provider', and 'Natural circular'. In addition, fixed line telephone competence and gender were significant factors affecting preference for the 'Transport system' and 'Shopping venue' metaphor categories respectively.

## 1   Introduction

The graphical user interface (GUI) can consist of many different simultaneous components, including text, images and video, which allow the user to visually scan for information. GUIs are commonly designed according to an underlying metaphor, as a way of helping the user to understand the system in terms of objects and actions they are familiar with from their real world experience (Carroll & Mack, 1985). In contrast, the voice user interface (VUI) relies on speech and simple sounds as output, and has not yet been commercially designed using an interface metaphor, and therefore lacks these potentially important contributors to usability. Speech represents a natural and widespread mode of human communication. However, speech is also sequential and transient, which for the user means listening to the whole dialogue rather than skimming for keywords, and having to remember the content, which leads to slow interaction and a burden on short-term memory (Sawhney & Schmandt, 1998).

This paper focuses on a specific type of VUI, the automated mobile phone service. A study conducted by Dutton, Foster and Jack (1999) on the design of an automated telephone shopping service found that participants were significantly more positive towards versions of the service implemented using interface metaphors. It may therefore be possible that interface metaphors can be usefully applied to VUI design. The aims of the experiment reported in this paper were firstly to evaluate which metaphors are most relevant to 2 different automated telephone services (a telephone internet service (TIS), and a telephone city guide service (TCGS)). Secondly, to investigate whether gender, previous telephone experience, and previous computing experience affect participants' metaphor preferences.

### 1.1 Interface metaphors

From the framework of linguistic metaphor categories proposed by Lakoff and Johnson (1980), structural metaphors are of most relevance to this paper, and are characterised by the representation of the structure of one concept by an implicit comparison with a real world object. An effective interface metaphor will lead the user to develop a mental model of the system that is closely related to the system image. This process involves 2 types of mapping, the conceptual mapping of ideas, and the independent perceptual mapping of the representations of these ideas (Gaver, 1989). In terms of GUI metaphors, a visual interface designed according to a real world object allows for spatial navigation, and leads to a subsequent conceptual mapping between the functionality of the system and that of the real world object. However for a VUI metaphor, for example 'the telephone internet service is a department store', it is not apparent what the similarities are between a non-visual voice-based telephone service and a real world department store. There exists no visual representation of the system for a VUI, making spatial navigation difficult. For this study it was therefore necessary to shift the focus from the users' perception of the system to the users' conception of the system by allowing them to produce a visual representation of the system, thus helping them to make conceptual mappings between the structure and potential spatial metaphors.

But what is the difference between a linguistic metaphor and a VUI metaphor? One of the major differences is that linguistic metaphors highlight similarities between existing objects, whilst VUI metaphors create new similarities that must be discovered through active exploration. A linguistic metaphor such as 'Argument is War' allows the actions we perform in arguing to be talked about and understood in terms of another well understood activity, namely 'War'. For a VUI to work it is necessary to add 2 additional factors, visualisation and interactivity. The user must be able to visualise the system structure in order to spatially navigate to their goal, and must interact with the system to understand its limitations.

## 1.2 Generating, selecting and developing interface metaphors

Palmquist (2001) carried out a study to analyse users' choice and description of web metaphors, and found that users preferred not to generate their own metaphors, and would rather choose from a pre -defined list. It was therefore decided that the experimenter and 3 other HCI experts would use a brainstorming methodology (Alty, Knott, Anderson & Smyth, 2000) to generate metaphors from which participants could choose. To facilitate users' selection of metaphors they were required to graphically represent their conception of the structures of the TIS and the TCGS by using a card sorting, and a sketching methodology respectively (Howell, Love, Turner & Van Laar, 2003). These methodologies have been used within HCI for simple prototyping, and were considered appropriate ways for users to represent both the semantic and perceptual structures of the services. To develop the features of the chosen metaphor it was decided to use a model called POPITS (Cates, 2002) originally proposed for GUI metaphor development, and in the process test its applicability to VUI development. The POPITS model proposes systematically identifying the related Properties, Operations, Phrases, Images, Types, and Sounds of the metaphor that are relevant to the system, thus allowing users to select features that resonate with their real world experience.

## 1.3 Individual differences and metaphor preference

Maglio and Matlock (1998) investigated people's metaphorical conceptions of the WWW and discovered that novice web users tended to use mixed metaphors more often than experienced users, suggesting that previous computing experience may impact metaphor selection and usage. In another study, Palmquist (2001) explored users' metaphor preferences for the WWW, and found a significant effect of gender on metaphor choice. With reference to automated telephone services, Dutton et al. (1999) found that women had a significantly more positive attitude towards a telephone service implemented using a magazine metaphor, suggesting that gender could be a contributing factor to metaphor preference.

## 2 Experimental design

The experiment involved 18 participants, including 10 females and 8 males with ages ranging from 18 to 39. A 2-condition within subjects design was used with the service type (TIS or TCGS) as the independent variable. Each condition involved a series of stages to aid the identification and evaluation of metaphors for each telephone service. All participants had previous experience of using automated telephone services, and completed both conditions working in pairs. The order of presentation of the two conditions was counterbalanced.

At the beginning of the study participants completed a technographic questionnaire, in order to collect data about gender, and previous telephone and computing experience. For the first experimental condition, participants were given a set of cards, on each of which was printed a menu option from an existing TIS. The top-level menu options from the TIS were: People, Shop, and Web Channels. The task was to arrange the cards on a large Velcro board into a structure which they believed would be most suitable for the TIS. Having constructed a prototype system, participants were presented with a list of 60 potential metaphors and were asked to rank 5 of the metaphors that they considered to share a similar structure to their TIS. Alternatively, participants could self-generate their own metaphors if they felt these would be more appropriate to the structure. Participants were then asked to develop their highest ranked metaphor by writing relevant words and phrases within each of the 6 POPITS categories. Finally, participants were asked to verbally explain how to perform 2 tasks using their prototype service to a naïve user, using language derived from their top ranked metaphor. The second experimental condition was similar to the first, but required participants to construct a prototype system by sketching rather than card sorting (Howell et al., 2003), and using menu options from a TCGS rather than a TIS. The top-level menu options from the TCGS were: Arts, Eating out, and Nightlife. All sessions were video-recorded, and digital photos were taken of each participant's card based prototype. This allowed an analysis of the prototypes to be conducted, and for the metaphor relevant language produced by users to be identified.

## 3 Results

### 3.1 Metaphor categories

The ranking data from each telephone service was scored (Rank 1 = 5 points…Rank 5 = 1 point), and the metaphors with the 10 highest mean scores were extracted. Table 1 lists the top 10 metaphors, the number of times each metaphor was selected, and the mean rank score.

**Table 1:** Top 10 metaphors for each telephone service

| Top 10 position | Telephone internet service | | | | Telephone city guide service | | | |
|---|---|---|---|---|---|---|---|---|
| | Metaphor | No of ranks | Mean score | SD | Metaphor | No of ranks | Mean score | SD |
| 1 | Filing cabinet | 9 | 1.72 | 2.08 | Talking pages | 6 | 1.44 | 2.12 |
| 2 | Department store | 8 | 1.44 | 1.89 | Computer | 6 | 1.11 | 1.78 |
| 3 | Talking pages | 6 | 1.28 | 1.90 | Department store | 5 | 1.06 | 1.86 |
| 4 | Computer | 6 | 1.28 | 2.02 | The brain | 5 | 1.06 | 1.95 |
| 5 | Road system | 6 | 1.17 | 1.92 | Filing cabinet | 6 | 1.00 | 1.71 |
| 6 | Shopping | 6 | 1.11 | 1.78 | Road system | 5 | 0.89 | 1.71 |
| 7 | Supermarket | 7 | 0.94 | 1.39 | Transport | 4 | 0.78 | 1.66 |
| 8 | TV | 3 | 0.50 | 1.29 | Tree | 3 | 0.67 | 1.57 |
| 9 | Shopping centre | 2 | 0.50 | 1.47 | Supermarket | 6 | 0.61 | 1.20 |
| 10 | Transport | 4 | 0.44 | 0.92 | Circulatory system | 3 | 0.50 | 1.20 |

Overall, 7 out of the top 10 metaphors were common to both services. The 'Filing cabinet' metaphor was ranked most highly for the TIS, and the 'Talking pages' metaphor for the TCGS. It was evident from examining the 10 metaphors that some of them were strongly related, sharing similar structures and features. To investigate similarities between different metaphor types, a series of affinity diagrams (Beyer & Holtzblatt, 1998) were constructed. These suggested 5 different metaphor categories to exist, which also included single metaphors not ranked amongst the 10 highest. The 5 metaphor categories, example metaphors from each category, and the category definitions can be seen in Table 2.

**Table 2:** The 5 metaphor categories and their descriptions

| Metaphor category | Examples | Description |
|---|---|---|
| Hierarchical | Filing cabinet<br>Talking pages<br>Computer | The overall structure is a pyramid consisting of different levels of information, with each level being linked to the levels above and below it |
| Shopping venue | Department store<br>Shopping centre<br>Supermarket | A venue consisting of different sections and levels, with different means of travelling between them (walking, elevator etc), characterised by the division and sub division of goods into categories, and categories into shops |
| Transport system | Road system<br>Bus route<br>Transport | A complex network of routes connecting towns and cities. These routes are of different sizes and levels of importance, with different rules governing their use. The routes become more concentrated around urban hubs |
| Information provider | Television<br>Website<br>Internet | An information space with different channels and categories, with which it is possible to engage at different degrees of interactivity, and by using different input modes |
| Natural circular | The brain<br>Circulatory system | A biological system separated into different functional areas, but connected by the process of circular flow |

## 3.2 Effective use of metaphor

Participants' metaphor based explanations of how to complete a task were analysed and referenced against their digital photos and sketches. Three criteria were used to establish whether the metaphor had been used effectively (1) consistent use of language derived from the metaphor (2) whether the language used correlated with the prototype the participant had designed (3) whether the metaphorical description was comprehensible to the other participant. For the TIS, 15 participants used metaphor effectively and 3 did not. For the TCGS, 10 used metaphor effectively and 8 did not. A McNemar test showed no significant difference between service types in the number of effective metaphorical explanations given ($p = 0.063$). It was therefore concluded that the majority of participants were capable of producing effective metaphorical explanations.

## 3.3 Individual differences

The technographic data were treated as 3 separate variables: mean of the mobile phone scores, mean of the fixed line telephone scores, and mean of the computer scores. The mean ranks of the metaphors within each metaphor category were also calculated. Correlations were performed between the technographic variables, and the average rank for each metaphor category. For the TIS, the only significant correlation was between the 'Transport system' metaphor category and fixed line telephone experience ($r = -0.48$; $n = 18$; $p = 0.04$). This result suggests that participants with high fixed line telephone competence ranked metaphors from within the 'Transport system' metaphor category significantly more highly for the TIS. There were no significant correlations for the TCGS. In order to investigate the effect of gender on metaphor preference, an independent samples t-test was performed on the 5 metaphor category variables for each service. The only significant difference between males (mean=0.16) and females (mean=0.80) was found for the 'Shopping venue' metaphor category for the TCGS (t(16) = -2.47; p = 0.03). This indicates that female participants ranked metaphors from within the 'Shopping venue' metaphor category significantly more highly than male participants.

# 4 Conclusion

The results from this study provide an important first step towards investigating the potential of interface metaphors for VUIs. The present data suggest five categories of metaphor, from which metaphors based around a 'Filing cabinet'; 'Department store' or 'Road system' may be most applicable to the mobile phone services examined. However, further consideration needs to be given to the specific content and user-based assessment of such services before metaphor-based systems can become a practical reality. Differences in metaphor preference were observed a s a function of participants' gender and level of previous phone use, but these differences were not consistent across both telephone services. This may highlight the importance of selecting and customising a metaphor to accommodate the individual characteristics of the users, and suggests that the service domain to which the metaphor is applied may influence users' attitudes towards the interface metaphor.

# 5 References

Alty, J. L., Knott, R. P., Anderson, B., & Smyth, M. (2000). A framework for engineering metaphor at the user interface. *Interacting With Computers*, 13, 301-322.

Beyer, H., & Holtzblatt, K. (1998). Contextual design: Defining customer-centred systems . San Francisco, California: Morgan Kaufmann.

Carroll, J. M., & Mack, R. L. (1985). Metaphor, computing systems, and active learning. *International Journal of Man-Machine Studies*, 22 (1), 39-57.

Cates, W. M. (2002). Systematic selection and implementation of graphical user interface metaphors. *Computers and Education*, 38, 385-397.

Dutton, R. T., Foster, J. C., & Jack, M.A. (1999) Please mind the doors – do interface metaphors improve the usability of voice response services? *BT Technology Journal*, 17 (1), 172-177.

Gaver, W. W. (1995). Oh what a tangled web we weave: Metaphor and mapping in graphical interfaces. *Proceedings of ACM CHI '95*, Denver, CO, USA.

Howell, M. D., Love, S., Turner, M., & Van Laar, D. L. (2003). Generating interface metaphors: a comparison of 2 methodologies. To appear in: *Contemporary Ergonomics 2003*. Taylor & Francis, London.

Lakoff, G., & Johnson, M. (1980). Metaphors we live by. University of Chicago Press.

Maglio, P. P., & Matlock, T. (1998). Metaphors we surf the web by. In *Workshop on Personal and Social Navigation in Information Space*, Stockholm, Sweden, 138-149.

Palmquist, R. A., (2001). Cognitive style and users metaphors for the web: an exploratory study. *The Journal of Academic Librarianship*, 27 (1), 24-32.

Sawhney, N., & Schmandt, C. (1998). Speaking and listening on the run: design for wearable audio computing. In *Proceedings of ISWC '98*, Pittsburgh, PA.

# Ecological Interface Design in Practice: A Design for Petrochemical Processing Operations

*Greg A. Jamieson*

*Wayne H. Ho*

*Dal Vernon C. Reising*

University of Toronto
Toronto, Canada
jamieson@mie.utoronto.ca

IBM Canada Ltd.
Markham, Canada
who@ca.ibm.com

Honeywell Ltd.
Minneapolis, USA
dalvernon.reising@honeywell.com

## Abstract

This paper describes the design of an ecological interface for an existing petrochemical process. We describe the iterative design process by example and identify the challenges of applying the Ecological Interface Design framework in industry using a user centred design approach.

## 1   Introduction

Ecological Interface Design (EID) is a framework for designing graphical user interfaces (that is, ecological interfaces) for complex systems (Vicente & Rasmussen, 1992). Over the past decade, researchers have reported substantial progress in applying EID to a variety of work domains of increasing complexity (see Vicente, 2002). However, the literature lacks depth in two key areas. First, it offers few applications of the framework to real work domains. Second, it tends to focus on the design product rather than on the design process (cf. Reising & Sanderson, 2002). These two characteristics limit the usefulness of the EID literature for industry practitioners who might consider designing ecological interfaces for applied problems. The two objectives of the work discussed in this paper were: a) to extend the literature by applying EID to an existing industrial process, and b) to identify the challenges of applying EID in a user-centred design process in industry.

We designed a novel graphical user interface for an ethylene manufacturing process. EID formed the basis for an iterative design approach that incorporated several types of user feedback. Although EID does not contradict other user-centred design approaches, its emphasis on using work domain analysis to identify information requirements may reduce the attention that designers devote to acquiring user feedback in the design process. This has been compounded by the fact that there have been no expert users of the laboratory microworld simulations for which most ecological interfaces have been developed. Given that we were working in an industry setting with a population of domain experts as users, stakeholder acceptance of the designs was critical.

The target domain is a sub-process of an ethylene refining plant. The reactor converts acetylene in a hydrocarbon stream into ethylene by reacting acetylene and hydrogen in the presence of a catalyst. However, other hydrocarbons in the process stream will react with hydrogen as well, giving off heat and destabilizing the reactor. The operator must balance hydrogen consumption, reducing the acetylene concentration in the product stream to below 5 ppm and minimizing excess hydrogen. This is done by regulating the flow rate of the mixed hydrogen stream.

# 2 Iterative Design Approach

We employed an iterative approach when developing the functional interface. At several stages, target users and other domain experts provided feedback that factored into subsequent revisions. In the following sections, we describe three iterations of the display and the user feedback solicited.

The results of a work domain analysis (Miller & Vicente, 1998) and two task-based analyses (Miller and Vicente, 1999; Jamieson, Reising, & Hajdukiewicz, 2001) formed the input to the design process. Information requirements derived from those analyses were compared, filtered and assimilated into a comprehensive list that drove the design process (see Jamieson et al, 2001). In reviewing this list, the critical role of hydrogen was apparent. Hydrogen serves as the facilitating and throttling agent in the chemical reaction. Providing hydrogen to the reaction is a necessary function, and understanding its disposition in the reaction is the key to maintaining effective control. Therefore, hydrogen became the central focus of the display.

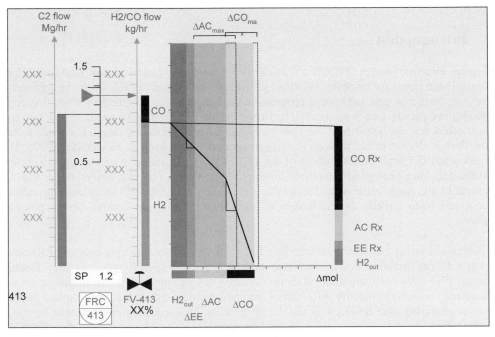

**Figure 1:** Balance Display, version 1.

Version 1 of the display focused on the stages of hydrogen functionality (see Figure 1). The hydrogen flow into the reactor must be regulated and accounted for in the reaction. A ratio controller (FRC) establishes a setpoint for hydrogen flow based on the hydrocarbon (C2) flow. Thus, we showed process flow (first column at left) and used a scale (0.5-1.5) to show how the setting of the FRC would result in a setpoint for the hydrogen flow (H2/CO flow). The scale was intended to move up and down with the C2 flow column. The setting of the FRC is shown on the scale with an arrow that points to the desired setpoint on the hydrogen flow column. We next drafted a graphic to show how the hydrogen was being used in the reactor. A mol-mol scale shows hydrogen coming into the reactor on the vertical scale and hydrogen "sinks" on the horizontal scale (that is, outlet hydrogen, acetylene conversion ($\Delta$AC), ethylene conversion ($\Delta$EE, and carbon monoxide conversion ($\Delta$CO)). The thickness of three grey blocks corresponds to the number of moles of hydrogen given up to each sink. A sloped line is drawn from the hydrogen inlet point across the blocks. It descends at a slope of $-1$ for the first three reactions because they each use up one hydrogen mole for each mole converted. When the consumption line reaches the carbon monoxide (CO) sink block, it descends at a slope of $-3$ because 3 moles of hydrogen are required for each mole of CO. In other words, the consumption line reflects the stoichiometrics of the various reactions.

The widths of the three rectangles are repeated as a vertical column on the right hand side of the consumption graphic. A line drawn across the graphic from the hydrogen inflow to the stacked column represents the comparison of hydrogen inflow and known outflows. A positive (excess hydrogen) or negative (missing hydrogen) deviation of this line from the horizontal represents a discrepancy in the hydrogen accounting.

We produced paper prototypes of our design concept and conducted design walkthroughs with two process engineers and two senior operators. The engineers found that the hydrogen balance design showed graphically what they had been trying to teach operators about the reaction process. However, they pointed out that the CO reaction represented in the graphic does not take place as called out in the work domain analysis. The correction to this error is shown in version 2. The operators commented that the balance graphic gave them greater insight into a process that they knew from experience was central to the effective operation of the reactor. We were warned that small deviations in the balance were common and important. They were also interested to see the behaviour of the FRC graphic and the balance graphic with actual process data. This would turn out to be important because a casual observer later noted that the FRC design was flawed. Notably, neither the designers nor the domain experts detected this error until several weeks later.

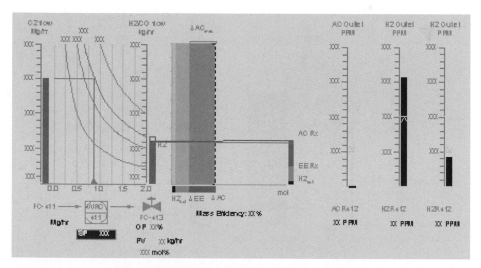

**Figure 2:** Hydrogen Balance Display, version 2.

In version 2 (Figure 2), we introduced a new graphic for the representation of the FRC. The original FRC concept showed an additive relationship between C2 flow and FRC setting when a multiplicative relationship was required. The replacement graphic shows the setpoint for the FRC in the form of a triangle along the horizontal axis of a two-dimensional plot. The vertical axis represents the C2 flow. A series of constant value curves is drawn in this surface, such that the

135

product of the values of the horizontal and vertical axes meets on that curve. The current value of that product should be the setpoint for the H2/CO flow. This is represented by the green curve, which intersects the H2/CO scale at the intended setpoint. The graphic thereby allows the user to assess whether the control task of regulating the hydrogen to process flow ratio is being met.

Note also in version 2 that the CO reaction components no longer appear in the balance graphic and the consumption slope has been removed. This line was only marginally useful given that all of the reactions taking place in the reactor have the same stochastic relationship, thus resulting in a monotonic consumption curve.

Based on the feedback from our initial design review, we expanded the scope of the display to address not only the hydrogen reactions, but also the performance of the reactor as a whole. Three analyzer columns were added on the right side of the display, thereby enabling the design to display the inlet, reaction, and outlet process flow through the reactor.

Once the revisions were complete, a rapid prototyping tool was used to create a dynamic Flash-based prototype driven with recorded process data. Twenty-two professional operators conducted a design walkthrough with the prototype display. Many of the operators requested a single graphic to monitor all of the functions related to the reactor performance. This included the settings of upstream and downstream control valves that had been represented in other displays.

In version 3, in response to operator requests, we moved information from other views into the reactor view to support the control task. We produced a design specification for the view and hired a software contractor to implement it (see Figure 3).

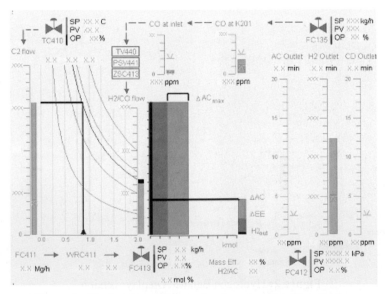

**Figure 3:** The implemented version (version 3) of the display.

# 3    Results

We were able to successfully apply EID to the design of a graphical user interface for an industrial process. A subsequent simulator evaluation of the displays confirmed that the designs led to improved operator performance on representative control tasks (Jamieson, 2002). We did not encounter any substantial difficulties in scaling up the framework and we have shown that EID can make valuable contributions to industry design problems.

One of the many challenges that we faced was prototyping and implementing the atypical designs. Although we knew that paper prototypes are useful tools for evaluating more traditional designs (Nielsen & Mack, 1994), we were unsure if they would work well with the novel integrated graphics that characterize ecological interfaces. In fact, they were almost as effective as the Flash prototype. Moreover, implementing the Flash prototypes proved difficult and time-consuming, and they were only of marginal added value compared to the paper. The implementation challenge also carried through to the final user interface controls. We had to identify an implementer who was able to develop new and unusual widgets from scratch, as well as establish the data structure to connect the controls to an instrumentation and control system. The short supply of implementers with this skill set should be considered in attempting to practice EID in an industry setting.

Despite these challenges, applying EID within a user-centred design approach led to improved display designs. The task analysis (which is not a traditional component of EID; see Jamieson et al, 2001) and design walkthroughs provided valuable feedback from domain experts. We believe that this represents the first case in which representative users were consulted in the design of an ecological interface. Previous applications of the framework have typically been completed for work domains for which expert users did not yet exist. Thus, the present example serves as proof that the user-centred design approach compliments the EID approach.

# 4    References

Jamieson, G. A. (2002). Empirical evaluation of an industrial application of ecological interface design. In *Proceedings of the 46th Annual Meeting of the Human Factors and Ergonomics Society* (pp. 536-540). Santa Monica: HFES.

Jamieson, G. A., Reising, D. V. C., Hajdukiewicz, J. R. (2001). EID design rationale project: case study report (CEL-01-03). University of Toronto: Cognitive Engineering Lab.

Miller, C. A., & Vicente, K. J. (1998). Abstraction decomposition space analysis for NOVA's E1 acetylene hydrogenation reactor (CEL-98-09). University of Toronto: Cognitive Engineering Lab.

Miller, C. A., & Vicente, K. J. (1999). Comparative analysis of display requirements generated via task-based and work domain-based analyses in a real world domain: NOVA's acetylene hydrogenation reactor (CEL-99-04). University of Toronto: Cognitive Engineering Lab.

Nielsen, J., Mack, R. (1994). Usability Inspection Methods. New York: John Wiley & Sons.

Reising, D. C., & Sanderson, P. (2002). Ecological interface design for Pasteurizer II: A process description of semantic mapping. *Human Factors*, 44, 222-247.

Vicente, K. J. (2002). Ecological interface design: Progress and challenges. *Human Factors*, 44, 62-78.

Vicente, K. J., & Rasmussen, J. (1992). Ecological interface design: Theoretical foundations. *IEEE Transactions on Systems, Man, and Cybernetics, SMC-22*, 1-18.

# Lightweight Contextual Design –
# a Case Study in Process Control Environment

*Kirsi Kontio, Juhani Rauhamaa*
ABB Oy
Kirsi.Kontio@iki.fi
Juhani.Rauhamaa@fi.abb.com

*Marko Nieminen, Toni Koskinen*
Helsinki University of Technology,
Sofware Business and Engineering Institute
Marko.Nieminen@hut.fi
Toni.Koskinen@hut.fi

## Abstract

Various monitoring and control systems used in process industry are operated by several people with different motives, backgrounds and motivations. Understanding the context of use is therefore important for developing usable systems. This paper describes the process that was used to evaluate the context of use of ABB's paper web and metal surface inspection systems. The project's aim was to discover differences and similarities in the users tasks, backgrounds, and work environment. Holtzblatt & Beyer's Contextual Design (1998), was chosen as the method for the field visits and analysis of the information. However, the method was seen too extensive for this specific study, and therefore it was applied as a lightweight version. Four paper mills and three metal factories in Finland were visited and totally 50 persons were observed or interviewed. The objective was also to compare the differences in the context of use in these industries. A multidisciplinary group was not available for the project because of the small size of the whole product team. The field visits were made in a team of one or two, but the analysis of the collected information was conducted in a varying group with participants from the whole product team. Different types of sharing sessions, were experimented with. As a result of the inquiries and analysis of the data the user groups in paper and metal industry are identified and the experiences using the method are being discussed. Lessons learned from conducting the inquiries and analysing the data are reported. The different sharing sessions are compared and their effectiveness and suitability for a relatively small product team is evaluated.

## 1    Introduction

In industrial processes constant monitoring and control of product quality is required. Various systems utilised for these purposes are tools for people with different backgrounds and motives. ABB's Industrial IT Web Imaging (WIS) and Industrial IT Surface Imaging (SIS) are systems for detection of defects in paper web or on surface of metal strip. Understanding the context of use is essential in efforts to enhance the usability of these systems. Contextual Design by Beyer and Holtzblatt (Holtzblatt & Beyer, 1998) is considered an effective but heavy method for gathering and analysing context and user data (Kujala, 2002). In the real settings, a product development team is not always in a position to name a multidisciplinary group that can adopt a new method and concentrate on customer data for a few weeks as suggested by Beyer and Holtzblatt. Thus in some occasions the method needs to be adapted into a more lightweight and incremental process.

## 2 Objective

This paper assesses whether it is feasible to apply Contextual Design when the research team has limited resources available. The feasibility is evaluated on the basis of a case study, whose aim was to describe the context of use of ABB's WIS and SIS.

## 3 Implementation of the study

The adaptation of Contextual Design into a more lightweight process was carried out as a case study taking place in autumn 2002 (Kontio, 2003). Four paper mills and three metal factories were visited, and totally 50 persons using the imaging system were observed or interviewed. The research team had one person concentrating full-time on the project and the other members took part only in the phases described in the following chapter. The most active team members represented research and development department and marketing. They had only little previous knowledge of user centred design and its methods.

The mills were chosen to represent different geographical areas in Finland and major paper companies. Imaging technology has only recently become advanced enough for the needs of metal factories. The mills chosen for the study had at least some experience in using a surface inspection system. Operators were, prior the visits, recognised as the main users and therefore they were selected as the starting point for the contextual inquiries. The other user groups were identified during the first visits and more attention was paid to them in the later visits

The extent into which the method was implemented was restricted to include only the first four phases of Contextual Design. Contextual Inquiries (Holtzblatt & Jones, 1993) in the field and the analysis of the data using work modelling and consolidation received the main emphasis. The implementation of results was started as a workshop with the target to generate ideas and to redesign the product. User environment design and prototyping were left out from this project.

## 4 Description of the case

### 4.1 Preparing for the site visits

The lightweight contextual design was conducted mainly by one person, the facilitator, but teamwork was used as a supplement. The contextual inquiries were the facilitator's responsibility, but also the members of the research team were encouraged to participate in mill visits. The contacts of the marketing and sales people were utilised in finding suitable candidates for visit sites. The solution proved to be correct because most of the contacted mills and factories also participated in the project. The focus and objective of the field visits were carefully planned and discussed by the team before the first visit. The questions were prepared according to the guidelines in the literature and confirmed by the team members.

### 4.2 Site visits

Each visit lasted approximately one day, during which time several inquiries addressed to different persons were made. One inquiry lasted approximately one and a half-hour. The inquiries themselves followed the structure presented in the method, except for slight changes required by the process control environment. Some supervising persons were observed and interviewed in the middle of the operator's interview as they visited the control room. During major process disturbances the inquiry was interrupted and intervening inquiry was made at a different location. Audiotaping was used in some of the inquiries. A member of the research team participated in several inquiries. The facilitator did the other visits alone.

## 4.3　Analysing the results

Teamwork was effectively utilised in sharing the information from the field visits. The focus of the inquiries was re-examined in a group after each visit and it was analysed whether it was necessary to shift the emphasis. Due to restricted resources and lack of a fixed group, different types of sharing sessions were experimented with. Raw data from the field was typed out within 24 hours from a visit. The objective in the first sessions was, in addition to sharing the data, to familiarise the participants with the method and techniques used.

### 4.3.1　Sharing sessions for distributing and analysing the data

Prior some sharing sessions, the analysis of the data was outlined by the facilitator and presented with Contextual Design's graphical models. The results were still further processed and distributed in a varying group with participants from the whole product responsible team. Another type of session followed the guidelines presented by the method. During the session, the participants created flow and cultural models and made an affinity diagram from the remarks and ideas arising from the inquiry data and discussion during the session. As a rule, affinity diagramming was used individually for each site and consolidated in the end much the same way as the work models. It was necessary first to make a diagram separately from each site, because the workload would otherwise have become too heavy for one person. The groupings were partly reorganised when the diagrams were consolidated. The facilitator was responsible for the consolidation but the models and the affinity diagram were verified by the whole team in the last sharing session. The Third type of sharing session was targeted for a larger audience than just the people participating in the project. The results from the visits at paper mills were presented with an "open house" approach (Cleary, 1999) to the whole product team. The participants could add their comments to the models and affinity diagram with post-it notes.

### 4.3.2　Implemented work models

Each of the five work models
- flow model
- physical model
- artefact model
- cultural model
- sequence model

was used in the study. Flow models proved to be the most informative in this case, because the focus was in communication and responsibilities related to work. Sequence models were extremely informative in describing the task sequences related to the system. Their usefulness as aids for design was, though, limited by the small amount of sequences related to the system in operators' work. Artefact models supplemented flow and sequence models in addressing the work customs and problems. Physical models described the physical use environment. Clearly the artefact models were a useful addition to the flow and sequence models. Cultural models were

140

used to describe the overall power rules and attitudes affecting the system usage. An example of the usage of the models can be seen in figure1.

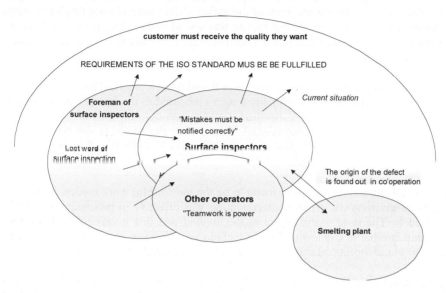

**Figure 1**. Example of a culture model created in the study

## 4.4　Implementing the results

The work redesign phase of the method was implemented as a workshop with group assignments for the whole product team. The structures of the groups were planned to break the existing organisational boundaries. In the beginning of the workshop, the Results of the inquiries were gone through by the groups. Each group collected user data concerning their assignment. A pre-determined goal and method　helped them in idea generation. One group was utilising brainstorming, another group card sorting, and two groups low level prototyping. In the end of the workshop the results were discussed with the whole product team.

## 5　Conclusions

Contextual Design, even in a lighter form, proved its effectiveness in assessing user requirements. The facilitator joined the research group with the assignment to make this study. She had no previous knowledge about the products concerned. This helped to guarantee less biased attitude to the role and usage of the product during the mill visits and inquiries, than which might have been the case if the visits would have been made only by those members of the product development team who had gained expertise about process industry during many years. The facilitator was provided with basic product information, but taking into account the limited amount of　time available for this phase, we believe that indoctrination to traditional way of thinking about the product's users could be avoided. Commission engineers participated in some sharing sessions and the workshop. They are representatives of the product group and have the closest relationships with the control room personnel operating the product. The results of visits were mostly agreed by them. However, it is unlikely that the same results could have been able to be gained by interviewing the commission engineers only. The conclusion of these observations is that Contextual Inquiries provided rich and accurate data about the context of use of the operators and other users of web or surface imaging solution. With the help of the work models and affinity diagramming even developers who did not participate in the visits or sharing sessions could get a clear picture of the users and their work. Parts of the gathered data were already familiar to some developers, but during the process the bigger picture was spread to the whole team. Analysis of the data also helped to emphasise which product attributes are important in a wider perspective from the user point of view.

To have only one person conducting the inquiries was found to be a practical solution. Time spent in the inquiry phase was obviously considerably longer compared to the whole team divided into pairs. The advantage was that the other members could concentrate on their work and massive training for interviewing skills was not needed. Though developers were not conducting the inquiries themselves, in this approach they still saw the users in their actual work context and experienced their problems.

The flow models and cultural models were found to be the most useful work models in describing the process control environment. Affinity diagramming was effective in processing the data left out from the models. The most useful aspect of the method was that it was easily digestible by users that were not familiar to Contextual Design techniques. Diagrams were also used to identify the emerged problematic aspects related to the product but not directly the context of use.

The participants considered the adaptation of the sharing session described in the Contextual Design method as the most effective way of transmitting user data. Data processed in the highly interactive sharing session were reported to be adopted and remembered considerably longer time than less interactive sessions. The interactive sharing session was, however, considered too heavy to be used after every inquiry. The session was considered to take too much time and a more cost-effective method was preferred. Thus alternation of sessions, where the data to be discussed was, in beforehand, highly analysed and only slightly analysed, was favoured. Still, the visits to the field were considered to be the most effective way to understand the actual way of use. The incremental process, where the focus and targeted user groups were continuously refined and the method was gradually introduced to the team, was found to be a functioning and suitable way to tackle the challenge of restricted resources

# References

[Cleary, Theresa. 1999. *Communicating Customer Information at Cabletron Systems Inc..* Interactions jan+feb 1999. p.45-49.
Available online: <http:www.csc.ncsu.edu/faculty/anton/csc/rep/cleary.pdf>]

[Holtzblatt Karen, Beyer Hugh. 1998. *Contextual Design: Defining Customer-Centered Systems.*Academic Press. p.469. ISBN 1-55860-411-1]

[Holtzblatt, K. & Jones, S. 1993. *Contextual Inquiry: Principles and Practice. Participatory Design: Principles and Practice.* Lawrence Earlbaum, New York NY.]

[Kontio Kirsi. 2003. *Defining the context of use for a web imaging solution using Contextual Design.* p.79. Master's thesis. Helsinki University of Technology]

[Kujala Sari. 2002. *User Studies: A Practical Approach to User Involvement for Gathering User Needs and Requirements.* Helsinki University of Technology. ISBN 951-22-5900-1]

# Process Snapshots Supporting Operators' Expertise Management

*Toni Koskinen, Marko Nieminen*

Software Business and
Engineering Institute (SoberIT)
Helsinki University of Technology
P.O.Box 9600, FIN-02015, FINLAND
Firstname.Lastname@hut.fi

*Hannu Paunonen, Jaakko Oksanen*

Metso Automation Inc.
P.O.Box 237, FIN-33101, FINLAND
Firstname.Lastname@metso.com

## Abstract

Computer systems supporting discussion and expertise mediation in organizations are becoming more common in process control. However, the earlier studies have shown that creating lengthy explanations by typing is not always a feasible solution in control rooms. The expressive power of these explanations can be increased with the help of multimodal interaction and multimedia documentation.

## 1    Introduction

When people communicate with each other they provide and process huge quantities of information in forms of spoken language and gestures. For instance, in process control operators can use gestures to communicate when they are fixing problems in noisy production facilities. However, the human to human interaction is restricted by spatial and temporal constraints – face-to-face communication requires that the communicating persons share the same space and time. Nowadays, with the help of modern information technology people are able to communicate and share their experiences in spite of the traditional human to human interaction constraints. The problem with these current systems supporting *asynchronous collaboration* in process control is that a significant amount of information is lost when compared to face-to-face communication. Most current systems are based on static text. They are not capable to lean on and reconstruct the situation and mediate the process control views or videos from the original situation. With the help of multimedia and multimodal interaction the expressive power of collaborative systems could be increased.

### 1.1    The Importance of Expertise

Process control as a work environment is complex. Decision-making requires expertise, communication, and co-operation between employees. Part of the expertise is gained through formal events and sources like training, documentation and guidance. However, a significant part of expertise is gained "on-site" through carrying out the work tasks. Auramäki & Kovalainen (1998) have recognized several classes of knowledge used in process control work. These classes are former incidents and actions, present situation, future tasks and incidents, process and equipment, normal situations, good control methods, disturbances, and general knowledge about automation and information systems.

The importance of knowledge and expertise in process control is evident since the process is not fully predictable; a great part of the decisions needs to be done in a real-time. The distances between control rooms are long and the uninterrupted production is carried out in shifts. The expertise sharing among employees supports organisational learning. The expertise does not need to be based on individuals' experiences. It can be gained through hearing others' experiences as well (B? rentsen, 1991).

## 1.2 Documenting Expertise alongside Work

The process control should be seen as a holistic system where organisation, automation system tools and production process are developing concurrently (Paunonen & Oksanen, 1999). The same factors also enable continuous development of the operations and expertise in the organisation. Process information provided by the automation system forms the basis for expertise acquisition. The automation system has the ability to play a significant role in storing and mediating the experiences of employees. Basically, the automation system can record operators' control activities and context information (process state), but it can also act as a means for discussion support: record the discussion, and offer it back to be used as support material for problem solving in later situations (Auramäki, Paunonen, & Kovalainen, 1995). Unlike spoken words, the electronic discussion is stored can be accessed later on (Koskinen, Nieminen, Paunonen, & Oksanen, 2002).

The electronic web-based diary supports discussion and expertise mediation in organizations (Figure 1). Using the diary the operators, production management and maintenance people can study what has happened in the previous shifts. They can also use the diary as a support for solving the problem by searching similar cases from the past. Because the operators occasionally work in hasty situations it is important that entering information to the diary is as easy as possible.

**Figure 1:** Sample picture of the DNAdiary (Metso Automation Inc.)

144

## 1.3  Towards Multimedia and Multimodal Computer Mediated Expertise Management

The contents of the early implementations of electronic diaries have consisted of written text. In a previous study, we found out that some operators find typing with a keyboard too laborious (Koskinen, 2002). In addition to that, some incidents are difficult and laborious to explain understandably using static text. More advanced means to share and store experiences are needed. In the near future, the electronic diaries are introducing multimodal features such as speech input. Therefore, they introduce capabilities to store and offer multimedia documentation.

The aim with multimodal interaction is to give users a more versatile interface for accessing and manipulating information and knowledge. The expressive power of information can be increased through multimedia output capabilities (Oviatt & Cohen, 2000). Multimedia documentation in electronic diaries means structured presentation of knowledge and experiences in forms of text, images, graphics, voice, other audio, and video.

## 2  Aim of the Study

In our earlier studies we have found out that creating lengthy explanations by typing is not always a feasible solution in control rooms. The incident descriptions and diary entries should be possible to create easily and quickly as a by-product of the work. However, the contents of these entries should be very informative and so reliable that the employees can trust them. Multimodal interaction and multimedia content aims to increase the expressive power of these entries.

The aim of this study is to study the role of process snapshot views (pictures) as a part of the expertise gathering, documenting and sharing. The main issues of the study are to address: 1) do operators find the storing of process views (snapshots) useful and informative 2) how should this functionality be integrated in a process control system?

## 3  Implementation and Method

The study consisted of two evaluations with two snapshot recording prototypes integrated in the operator interface of metsoDNA process control system. The first prototype was DNAdiary (Figure 2). The DNAdiary is basically a web-based discussion forum. The second prototype, the picture library, was especially designed for handling process snapshot views. It enabled quick navigation between different views with built-in tree-structure. It was also more closely integrated to the MetsoDNA operator desktop, which is the user interface of the process control system. Both prototypes allowed creating entries combining pictures and text.

The DNAdiary was evaluated with seven control room employees and the picture library with five employees. The evaluations consisted of following stages: the background interview, the presentation of the use scenario, the co-operative evaluation, and the post evaluation interview.

The background interview consisted of 26 questions related to users' education, skills, typical tasks, and user environment. The evaluation started with the presentation of the use scenario. The use scenario presented a situation where the users had to report the incident to the prototype utilizing the snapshot functionality. The conformance of the use scenario to reality was confirmed in advance with a pilot study. The co-operative evaluation method (Monk, Wright, Haber, & Davenport, 1993) is a technique for identifying usability problems in prototype products and processes. The basic idea of the co-operative evaluation is that users work with a prototype carrying out the tasks received from the evaluator. During this procedure users talk aloud and explain what they are doing. The evaluator records interesting events and user's comments regarding the prototype. The aim of the post-evaluation interview was to summarise the users' opinions and thoughts related to the snapshot functionality and its usability.

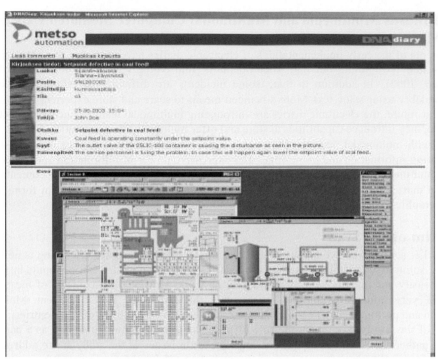

**Figure 2:** the sample picture of the DNAdiary prototype entry.

# 4 Results

The results of the first evaluation showed that the possibility to attach process snapshot views to entries was considered useful: all seven employees answered so. Most frequently, employees would store the views of process trends, overall process situation (Figure 2), and digital photos from the actual process. According to the employees the snapshot views would make the diary entries more explanatory. Furthermore, the employees stated that it was easier to formulate the entry text, because one had possibility to refer to the attached picture. Although all the employees were able to complete the evaluation, the background interview and the evaluation confirmed that most of the employees were not very familiar with using computers.

The second evaluation consolidated further the results of the first evaluation. The employees would use the snapshot functionality for the same purposes as presented above. Currently, they print the desktop views to the massive paper folders and they stated that it is virtually impossible to find the information later. A main difference in the results of the two evaluations was that unlike the employees evaluating the web-based DNAdiary prototype, the employees evaluating the picture library prototype stated that they could browse the entries for learning purposes like comparisons of different situations when they are not in a hurry.

# 5 Conclusions and Discussion

The study confirmed that the utilization of process snapshot views in expertise gathering and sharing was considered useful by the control room employees. One operator intensified: *"I'd*

*rather browse the pictures on the screen than take out the cardboard case full of dusty paper folders containing the printed pictures."*
Since the employees had novice computer skills and the entries are usually created under some time-pressure, the design of the snapshot feature should be kept tiny, simple, and easily accessible. Building on the last finding, the results gained in evaluating the two different prototypes indicate that rather than being a separate application the snapshot feature could be even more tightly and transparently integrated to the present functions of the operators' user interface. If the feature is always at hand, it is more likely to be used.

The results confirm further that multimedia content and multimodal interaction improve the expressive and explanatory power of computer mediated expertise management.

## Acknowledgements

The study was carried out in the research project PROKM funded by the Finnish Work Environment Fund (http://www.tsr.fi).

## References

Auramäki, E., & Kovalainen, M. (1998). In Search of Organizational Memory in Process Control. In Waern, Y., (ed.), *Co-operative Process Management: Cognition and Information Technology.* London: Taylor&Francis, 187-202.

Auramäki, E., Paunonen, H., & Kovalainen, M., (1995). Tools for Cooperative Work in Paper Mills. In Norros, L., (ed.), *Proceedings of 5$^{th}$ European Conference on Cognitive Science Approaches to Process Control.* Helsinki: VTT Symposium 158, 280-289.

B? rentsen, K. L. (1991). Knowledge and shared experience. 3rd European conference on cognitive science approaches to process control, Cardiff, UK, September 2-6, 1991, 217-232.

Koskinen, T. (2002). Understanding the Control Room Context: Usability Criteria for Knowledge Support Systems in Immediate Process Control. In Proceedings of 34th NES Congress Humans in a complex environment - Innovate, integrate, implement. Norrköping, Sweden. October 1-3, 2002.

Koskinen, T., Nieminen, M., Paunonen, H., & Oksanen, J., (2002). Talking to the Computer: A Feasibility Study about Making Speech Entries into the Electronic Diary in Process Control. In Proceedings of the Fourth International Conference on Machine Automation (ICMA'02). Tampere, Finland, 11.-13.9.2002.

Monk, A., Wright, P., Haber, J., & Davenport, L., (1993). Improving your human-computer interface: A practical technique. Prentice Hall International (UK) Ltd.

Oviatt, S., & Cohen, P. (2000). Multimodal Interfaces that Process What Comes Naturally, Communications of the ACM March 2000/Vol. 43, No. 3., 45-53.

Paunonen H. & Oksanen J., (1999). Usability criteria for knowledge support in process control (in Finnish). Prosessinohjausjärjestelmän tietämystuen käytettävyyskriteerit. Automaatiopäivät 99, 14.-16.9.1999. Helsinki, Finland.

# Aspect Model-Based Methods for Scenarios and Prototype Development

*Youn-kyung Lim*

Illinois Institute of Technology
350 N. LaSalle St.
Chicago, IL 60610 USA
yklim@id.iit.edu

*Keiichi Sato*

Illinois Institute of Technology
350 N. LaSalle St.
Chicago, IL 60610 USA
sato@id.iit.edu

## Abstract

Prototyping is a critical method for exploring, communicating and testing new concepts in an interactive system development process. However, the criteria used in prototype model evaluation often do not address the original project objectives and issues. To improve this situation, a prototyping specification mechanism with aspect models and scenarios created through the aspect models are introduced, using Design Information Framework (DIF) as a foundation of organizing information necessary for prototyping. Scenarios created through the aspect models are introduced for prototype testing context definition. This paper presents the prototyping specification mechanism with its application.

## 1    Introduction

In an interactive system development process, prototyping is a critical method for exploring, communicating and testing new concepts. However, there are no clear mechanisms for using prototyping to evaluate design concepts in relation to the issues and problems identified in the earlier phases of the process (Alavi, 1984). Not only the evaluation of the test outputs, the testing itself tends to be executed without a structured way.

This research developed a mechanism for specifying, constructing and evaluating prototype models with consistent and systematic views. This prototyping specification method selects aspects to be prototyped and transforms the represented aspects into real prototype models. If a specific aspect of a product is clearly defined and represented, more effective evaluation of the solution would be possible (Ribeiro & Bunker, 1988) (Sato 1991). A scenario addressing original user experience specifies the context and tasks for testing the prototype models. This paper explains the prototyping specification mechanism based on these concepts through a design case of a digital video camera.

## 2    Aspect Models and Scenarios with the Design Information Framework (DIF)

The Design Information Framework (DIF) has been introduced as a unified design information platform for supporting activities in the design process. The DIF provides a structure for organizing design parameters relating to those design activities. The prototyping specification mechanism proposed in this research uses the DIF to specify the variables of aspect models which are embodied in prototype models and to translate the aspect models into scenarios which identify context of a test and necessary tasks users should proceed in the test. The aspects for the prototype specification are derived from requirements for the system development that reflect the originally addressed issues and problems.

The DIF has two different levels for representing information. The lower level of representation consists of Design Information Primitives (DIPs) that cannot be further decomposed into smaller conceptual units. The higher level of representation consists of Design Information Elements (DILs) composed of those DIPs (Lim & Sato, 2001) (Figure 1). An aspect of the system can be represented by several design parameters which correspond to DILs or DIPs in the DIF. The DIF also provides an interpretive mechanism for integration among different aspect models of a product. It helps designers to create solutions with a holistic view by identifying clear relationships among different aspects. A set of the aspect models of a use situation can also be used to create scenarios. The DIF enables the translation of the aspect models into a narrative format for the scenarios through its contextual compatible structure.

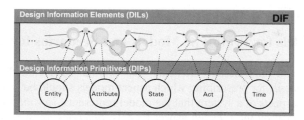

**Figure 1:** Structure of the Design Information Framework (DIF)

# 3 Prototyping Specification Mechanism with Scenarios and Aspect Models

Improving the usability of the digital video camera was the primary focus for this case. The process of this study is as follows: 1) set requirements for design, 2) identify problems through user observation, 3) generate a scenario for the representation of a use situation imposing the problems, 4) propose solutions to the identified problems, 5) develop prototype models to evaluate the solutions by using the prototyping specification mechanism, 6) set up the context and user tasks for testing with the created scenario, and 7) evaluate the solutions with test results.

## 3.1 Scenarios for Representing the Use Situation

For this case study, two requirements were given: 1) make relevant interface elements easier to find, and 2) help users to understand which mode they are in while accomplishing tasks. Aspect models to evaluate these requirements through prototype models were an operation sequence model, an interface layout model, and a state-transition model.

In the observation, the following problems were identified for the situation of taking still images with the digital video camera: 1) the user could not find the button for taking still images, identified from the operation sequence aspect, 2) it took time to find the relevant button to display the selected image, identified from the interface layout aspect, and 3) the user could not recognize whether the image was taken or not, identified from the state-transition aspect.

These requirements and problems were used as a foundation of creating solutions, and the requirements provide the rationalization of the solutions (McKerlie & MacLean, 1994).

The aspect models that revealed these problems were used to create a part of a scenario. The operation sequence model provided user's actions for the story of an event, the interface layout model provided the information of the system, and the state-transition model provided the reactions of the system to the user's actions.

These models formed the main story of the scenario. For the background information such as scene setting and actor description, the real situation observed through the user research was referred. Box 1 shows a part of the scenario created for this case. This scenario was used for preparing the test set up, which will be explained later.

**Box 1:** A part of the scenario created

> David, who is a college student, needs to take a digital picture of a classroom for his project. He is in the classroom now, and he has a digital video camera to take it. The room is bright enough to take a picture without a night shot function. He first finds the scene to take by zooming in and out with the zoom lever. After finding the right scene, he tries to take the photo, but he can not locate the photo button among many other buttons. He presses several buttons near the "photo" label, and finally figures out the button that seems like taking a picture. However, what he thought he took is not stored in the memory. (...)

## 3.2 Aspect Models for Prototype Models and Scenarios for Test Set-up

We identified important information elements that should be included in the prototyping specification: *requirements* for prescribing which aspect models should be selected, *aspect models* that best illustrate the aspects of a system that the model represents, *variables of a prototype model* for evaluation determined by the selected aspect models, *functions for simulation* that define the external elements on the prototype models such as interface elements, *functions for performing prototyping testing* for capturing data through the monitoring and recording of user inputs, *media of a prototype model* determined according to the nature of variables and functions, *benchmark tasks* that represent the tasks which can be identified by a scenario, and *assumed context* that should be considered in a test which can also be identified by the scenario.

In this case study for constructing prototype models, the following aspect models were concerned: an operational sequence model mapped onto the interface layout and a state-transition model within the task sequence were used. The first aspect generated three alternative solutions toward how to re-design the interface layout to make the photo button easy to find. Regarding the second aspect, the main concern was how to provide feedbacks/outputs for users to distinguish the focusing mode from the image taking mode. Two ideas were generated from this aspect.

The functions that should be embedded to the prototype model 1 (Figure 2(a)) were zooming in and out, and still photo taking. These functions determined the input elements for the directly related user tasks such as a zoom lever, a photo button, some other input elements that would affect the time required for users to find the photo button such as a "focus set" button, and a "night shot" switch, and the output elements such as a beeping sound for monitoring user's actions in the test. Table 1 shows a partial specification for prototype model 1 by adapting the DIF structure.

For prototype model 2 (Figure 2(b)), the important factors were the relationships between inputs and outputs of the product, as well as between the outputs and states of the product. To effectively capture time information, automatic time stamping and product state recording functions were implemented for testing purposes.

For the test set-up, the created scenario was used to define the benchmark tasks that the subjects need to proceed in the test. The following tasks are extracted from the scenario as the benchmark tasks: zooming in and out, focusing, and taking a still image. Besides the benchmark tasks, the scenario guided to define an assumed context for the test. The background information such as scene setting of the scenario was adapted to define it. In this case, the subjects assumed to be in a classroom to take a still picture, and the room was assumed to be bright enough to take the picture. In order to make consistency of subjects' task procedure, every subject followed the same operation sequence provided to them. After the test, several questions were asked for debriefing. During the test, subjects were asked to think aloud.

**Table 1:** A partial specification of prototype model 1 for concept A

| Entities for a prototype model | Attributes for the entities | Functions for simulation | Inputs | Outputs | Functions for performing the test | Inputs | Outputs |
|---|---|---|---|---|---|---|---|
| Photo button | - Shape: circle<br>- Size: R5.5(mm)<br>- Location: (x, y) =(45(mm), 10(mm))<br>- Pushable | | | | Notify if the photo button is pushed | Push the photo button | Make a beep sound |
| Zoom lever | - Shape: rounded rectangle<br>- Size: 23.3(mm) x 13(mm), R1(mm)<br>- Location: (x, y) =(65(mm), 10(mm))<br>- Slidable | Simulate zooming operation | Slide a zoom lever | Not needed | | | |
| Body | - Shape: rectangular parallelepiped<br>- Front plane size: 80(mm)x110(mm) | N/A | | | | | |

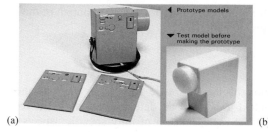
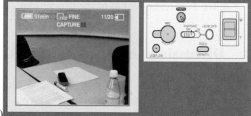

(a)  (b)

**Figure 2:** (a) Prototype model 1 and (b) prototype model 2

## 3.3  Evaluation of Test Results

For the evaluation, the test results can be represented through the selected aspect models that were related to the requirements. They were an operation sequence model, an interface layout model, and a state-transition model in this case. The interface layout model corresponded to the first requirement, "make relevant interface elements easier to find," and the state-transition model corresponded to the second requirement, "help users to understand which mode they are in while accomplishing tasks." The benchmark tasks provided to the subjects formed the operation sequence model, and the operation aspect was mapped onto the other aspects imposed by the prototype models to produce output models for the evaluation (Figure 3).

The outputs for prototype model 1 illustrates that concept B is best in terms of reducing the time taking to find the relevant interface elements (Figure 3(a)). The outputs for prototype model 2 showed that, for both concepts, repetition of inputting on one state happened (Figure 3(b)). However, with concept A, the subjects could not go to the auto-focus (AF) state easily unlike concept B with which the subjects could go to the AF state. The several trials of focusing operations was caused by the unrealistic simulation by the prototype model. The closure look to the real output data with debriefing revealed that the subjects could understand that a still picture was not taken on the AF state which the original product system failed to make users recognize.

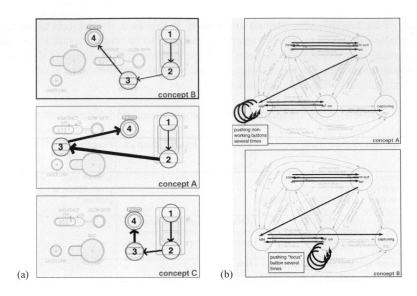

**Figure 3:** (a) The output of the aspect models for prototype model 1 and (b) the output of the aspect models for prototype model 2

## 4    Discussion and Future Studies

Aspect models supported the effective evaluation of the prototype models as well as their construction, and scenarios provided relevant context to address the selected aspects in the test. However, some problems might occur because of the representation of only specific aspects in prototype models. This case suggests that the prototyping specification mechanism can be improved by examining how related aspects could be implemented in the prototyping mechanism to enhance the evaluation. The structural relationships among the information elements defined in DIF could effectively support this mechanism. This issue will be investigated with more case studies of developing prototype models with this specification mechanism.

## 5    References

Alavi, Maryam. (1984). An assessment of the prototyping approach to information systems development. *Communications of the ACM*, 27 (6), 556-563.

Lim, Youn-kyung and Keiichi Sato. (2001). Development of Design Information Framework for interactive systems design. In *Proceedings of the 5th Asian Design Conference*. Seoul, Korea: International Symposium on Design Science.

McKerlie, Diane and Allan MacLean. (1994). Reasoning with design rationale: practical experience with design space analysis. *Design Studies*, 15 (2), 214-226.

Ribeiro, Rita A. and Ralph E. Bunker. (1988). Prototyping analysis, structured analysis, prolog and prototypes. In *Proceedings of the ACM SIGCPR Conference on Management of Information Systems Personnel*. College park, Maryland: ACM.

Sato, Keiichi. (1991). Temporal aspects of user interface design. In *Proceedings of the '91 International Symposium on Next Generation Human Interface*. Tokyo: Institute for Personalized Information Environment.

# DESK-H: Building Meaningful Histories in an Editor of Dynamic Web Pages

*José A. Macías and Pablo Castells*

E.P.S., Universidad Autónoma de Madrid
Ctra. de Colmenar Viejo Km. 15
Madrid, 28049 Spain
{j.macias, pablo.castells}@uam.es

## Abstract

DESK is a WYSIWYG editor for dynamic web pages that are generated from an explicit ontology-based domain model. DESK provides a significant scope of customization functionalities for dynamic pages with maximum ease of use, relieving users from having to edit the internal page generation code. In this paper we describe DESK-H, the monitoring module of DESK, that tracks all user actions in the DESK HTML editor and outputs a high-level semantic history of editing actions. The purpose of DESK-H goes beyond its current use in DESK: it provides an enriched model of user actions that can be exploited by other tools to carry through a Programming By Example (PBE) approach.

## 1 Introduction

Programming by Example (Cypher, 1993) seems to be a promising proposal since users only need to have knowledge about the domain of the problem rather than being programming experts. The PBE paradigm is based on the fact the user only has to provide examples of what s/he wants the system to do, and then the system creates automatically a program to automate procedural mechanisms from such examples. Anyway, there are some underlying difficulties that might turn out to be an obstacle for the success of the PBE paradigm (Lieberman, 2001, p. 61). One of this problems is a result of inferring user's intents from the comparison of the examples supplied to the system. Since comparing only a few examples (such as in initial and final example approach (McDaniel & Myers, 1999)) the inference process could result error-prone. One solution to overcome such problem consists in getting more information from user or asking him/her for ensuring the changes after performing them, but this can become somewhat cumbersome to end-user at interaction. The other main problem is the inherent difficulty to access application data in order to carry through a PBE strategy, due to most of the applications are close-source and they do not provide APIs for accessing internal data representation. Data characterization (Macias & Castells, 2003) is a main topic in PBE research, where the PBE system try to associate internal application data in order to infer more accurately procedural information. Data models provide an understanding of the knowledge behind the visual representation that the user manipulates in a PBE system. This knowledge can be extremely useful to make sense of the user's actions on visual objects.

DESK (Dynamic web documents by Example using Semantic Knowledge) (Macias & Castells, 2002) aims at filling these gaps by proving a WYSIWYG authoring tool on a demonstrational environment where the user is monitored for inferring accurately information at each user step, to get rid of ambiguities caused by comparing too few examples and improving wrong guesses from inference process. Our approach features PBE techniques such as user monitoring and data characterization using an explicit knowledge-based domain model. Such data model is used by our run-time system PEGASUS (Castells & Macias, 2001) which generates dynamic web pages by rendering domain objects codified in a presentation template. DESK acts as a client-side of PEGASUS system, and takes advantage of a clear separation between contents and presentation regardless of the domain being applied.

## 2    A WYSIWYG layer over domain model

DESK is an intelligent authoring tool that allows users to make changes to dynamic-generated web pages by means of a WYSIWYG user interface. The main goal of DESK is to relieve the user from having to edit the internal page generation code coming from web pages which are generated dynamically by web applications. With DESK, a non-expert user only has to provide the tool with changes to HTML documents as s/he usually achieves them in ordinary web edition tools (such as Internet Explorer, Netscape Composer and so on), the modifications being automatically inferred by DESK by firstly creating and historic-structured model of such changes, mainly composed by semantic and syntactic information meant to be added to the change's historic.

One of the main strength of DESK is to infer user's intents by analyzing changes previously achieved by him/her. To this end, DESK uses an mechanism, namely DESK-H conceived for data characterizations using PBE techniques (Macias & Castells, 2003), such as monitoring the user interaction and recognizing high-level user's tasks in HTML data manipulation, resulting in building an enriched model which relates user's actions and explicit semantics from underlying data model.

DESK-H works under the following assumptions:

- The web pages DESK-H operates on are dynamically generated by a complementary generation module as the user navigates over an information space.
- The information space is modelled as a semantic network (a graph) of domain objects with attributes and relations according to an explicit ontology.
- Hyperspace navigation is understood as a traversal of the semantic network, so that each time a node is attained, a web page is generated for that node, selecting pieces of the node contents, and possibly of surrounding nodes, to make up the page.

In our current implementation, the ontology-based data model must be provided in RDF, a widely accepted standard for ontology definition (Lassila & Swick, 1999), but it would be straightforward to support other ontology definition languages as well.

DESK-H lets the user edit a specific page that has been generated under these assumptions, recording all user actions in the WYSIWYG editing environment, and trying to find more meaning for the actions, in terms of both HTML display structures (paragraphs, lists, tables, widgets), and parts of the domain model (objects, relations, attributes). This enriched model is used by an inference module of DESK that is able to modify the page generation procedures in the underlying system (PEGASUS) to accommodate the changes made by the user to a specific page. In this paper we focus on the monitoring side of DESK, whose structured output can be useful beyond the scope of DESK/PEGASUS.

154

# 3 DESK-H architecture

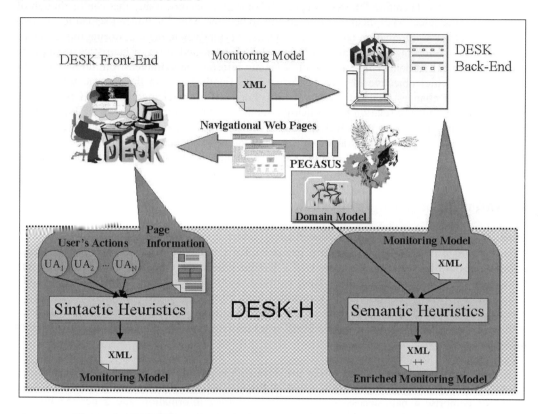

**Figure 1:** DESK-H engine embedded in the DESK client-server architecture

Figure 1 depicts DESK-H engine which consists, in the most basic sense, of two specialised tools, one located at client front-end (HTML editor) and another one located at server back-end. (inference engine) The client-side can be thought of as a WYSIWYG authoring tool which provides the user with an easy-to-use interface for accomplishing changes to web pages. This way, DESK monitors user interaction by capturing low level events generated by user's intents on a demonstrational environment. Rather than using only initial and final examples, DESK takes into account all actions performed by the user during the interaction, increasing accuracy and improving inference at each user step. As a result of that, DESK adds on (at server-side) the structured model of user interaction with enriched semantics from underlying constructor data model from PEGASUS. To this end, both tools involve different subsets of heuristic modules for obtaining:

- Syntactic information coming from user's actions. From each user's action occurred at front-end, DESK-H detects syntactic (HTML) blocks involved in such action. A block can be considered as a syntactic HTML fragment, in witch the context location being inferred by DESK (i.e. start and end positions as well as location before and/or after the context). User's actions are encapsulated into suitable constructor primitives for building up a monitoring model, that is a structured model containing filtered user's

155

actions as well as context information about such actions. Those heuristics also detect and manage widgets either being created by the user or just being created in previous sessions. Therefore DESK-H deploys a homogeneous processing that can be though of as the selection of meaningful fragments to be later identified under a semantic context.

- <u>Semantic information coming from syntactic primitives in the monitoring model</u>. Once the monitoring model has been sent to DESK back-end, a set of heuristics attempts to enrich the monitoring model with semantic information extracted from domain model. The chief objective of such heuristics is to finally build on the historic information about changes related to user's actions. For each user's change, DESK-H finds out a main object, a relation path witch the modified object comes from, as well as internal information about the affected object such as involved attributes and the class of such object. This process is achieved by applying several search and object-identification heuristics that search the ontology-based domain model for semantic knowledge and suitable relationships between related domain objects.

## 4  Inferring meaningful histories

DESK-H's internal heuristics are also intended to address inherent process difficulties such as ambiguities as a result of finding several semantics for a same syntactic context. To face those ambiguities, DESK-H arranges a subset of selection criteria to carry out the disambiguation process. Therefore DESK-H discards values not related to main object and chooses closest values to main object first, giving priority to values that have been involved in previous actions. User's interaction is required as long as there might not be any chance for automated disambiguation, prior user answers to the same questions being also taken into account for resolving similar future ambiguous situations.

As a result of the process, DESK-H finally provides with a historic model in order for any change accomplished by user to be codified using minimum amount of suitable semantics. Those semantics comprise a path access to domain and ontology information, the generated historic information is codified using XML language, thus changes can be processed by any web presentation system regardless of the presentation language used by such a system.

## 5  Related work and discussion

Tracking user's actions during the interaction is of common practice in plenty of PBE systems. Such strategy is mainly focused on inferring user's intents from observing her/him while s/he interacts with the system. Based on such philosophy, early works like Eager (Cypher, 2993) detect consecutive occurrences of a repetitive task in the sequence of a user's actions. Similarly, Mondirian (Cypher, 1993) creates program containing variables, iterative loops, or conditional branches from observing user's actions. SMARTedit (Lieberman, 2001, p. 209) represents a text-editing program as a series of functions that alter the state of the text editor. To do this, SMARTedit automates repetitive text-editing task by learning program by observing a user performing her or his task by means of applying techniques drawn from machine learning. TELS (Cypher, 1993) records high-level actions similar to the actions used in SMARTedit and implements a set of expert rules for generalising the arguments to each of the actions. Some other Agent's approaches like APE (Lieberman, 2001, p. 271) assist the user to detect loops and to suggest repetitive tasks automatically. Additionally, the use of a data model was already present in earliest PBE systems like Peridot (Myers, 1988), where the user can create a list of samples data to construct list of user interface widgets. In Gold (Myers et al., 1994) and Sagebrush (Roth et al.,

1994) the user can build custom charts and graphics by relating visual elements and properties to sets of data records. The data model in Peridot consists of list of primitive data types. Gold and Sagebrush assume a relational data model. Later works, as in HandsOn (Castells & Szekely, 1999), the interface designer can manipulate explicit examples of application data at design-time to build custom dynamic displays that depend on application data at run-time. In DESK we are interesting in lifting these restrictions and supporting richer information structures, supplying a knowledge-based domain model to represent contents, monitoring the user at each user step and obtaining an enriched structured model of semantic-network's references related to such user's actions. Therefore, it is able for DESK to create a more detailed historic of further changes.

## Acknowledgements

The work reported in this paper is being partially supported by the Spanish Interdepartmental Commission of Science and Technology (CICYT), projects number TIC2002-01948.

## References

Castells, P., & Szekely, P. (1999). HandsOn: Dynamic Interface Presentation by Example. *Proceedings of the HCI International'99*. Munich (pp. 188-1292).

Castells, P., & Macías, J. A. (2001). An Adaptive Hypermedia Presentation Modeling System for Custom Knowledge Representations. *Proceedings of the WebNet 01 Conference*. Orlando, Florida. Published by AACE, (pp. 148-153).

*Communications of the ACM* (2000). The Intuitive Beauty of Computer Human Interaction. Special issue on Programming by Demonstration, 43 (3).

Cypher A. (1993). Watch What I Do: Programming by Demonstration. The MIT Press.

Lassila, O., & Swick, R. (1999). Resource Description Framework (RDF) Model and Syntax Specification". W3C Recommendation from Http://www.w3.org/TR/1999/REC-rdf-syntax-19990222.

Lieberman, H (2001). Your Wish is my Command: programming by example. Morgan Kaufmann Publishers: Academic Press.

Macías, J. A. & Castells, P. (2002). Tailoring Dynamic Ontology-Driven Web Documents by Demonstration. *Proceedings of the IV 2002 Conference*. London (England), Springer Verlag.

Macias, J. A. & Castells, P. (2003). Using Domain Models for Data Characterization in PBE. *To appear in Proceedings of the End User Development Workshop at HCI 2003 Conference*. Ft. Lauderdale, Florida, USA.

McDaniel, R.G., & Myers, B. A. (1999). Getting More Out Of Programming-By-Demonstration. *Proceedings of the CHI 99 Conference*. Pittsburg, PA, USA.

Myers, B. A. (1998). Creating User Interfaces by Demonstration. Academic Press, San Diego.

Myers, B. A., Goldstein, J., & Goldberg, M. (1994). Creating Charts by Demonstration. *Proceedings of the CHI'94 Conference*. ACM Press, Boston.

Roth, S. F. et al. (1994). Interactive Graphic Design Using Automatic Presentation Knowledge. *Proceedings of the CHI'94 Conference*. Boston (pp. 112-117).

# MetroWeb: a Tool to Support Guideline-Based Web Evaluation

*Céline Mariage & Jean Vanderdonckt*

Université catholique de Louvain
Place des Doyens, 1 – B-1348 Louvain-la-Neuve
{mariage, vanderdonckt}@isys.ucl.ac.be

## Abstract

In this paper, we present the first version of the MetroWeb tool designed to manage usability knowledge and to exploit this knowledge throughout evaluation process phases. The tool supports usability knowledge manipulation in general and specifically supports heuristic inspection. The usability evaluation process supported by the tool is decomposed into 4 main phases: planning, assessment, finalisation and follow-up.

## 1    Introduction

A large variety of usability evaluation methods and tools exist today that are specific to the web or not (Ivory, 2001; Chi, 2002). They can be used with or without users, with a real or represented system, at different levels of granularity (Whitefield, 1991; Balbo, 1995). However, they are not all commonly used in the development life cycle of a web site because, up to now, the evaluation has mainly been considered as an external activity of the life cycle (Nielsen, 1993). To promote and assist the use of usability evaluation methods and tools, the MetroWeb tool is designed. In fact, this generic tool can manage usability knowledge so that it can be expanded at any time, disseminated at any time and explicitly used during design and evaluation in a continuous way. Although the tool can manage usability knowledge about any potential type of interface and a large spectrum of evaluation methods, this paper focuses on UIs for the Web with heuristic inspection (Nielsen, 1990) based on usability guidelines. The tool consists of a web-distributed system to manage usability knowledge in a flexible and autonomous manner that can be run on multiple computing platforms.

## 2    Conceptual Schema

Usability knowledge supported by the tool is articulated around the *guideline* (Fig.1), i.e. a design and/or evaluation principle to be observed to get and/or guarantee a usable user interface (Farenc et al., 1995). A guideline can be characterized by other concepts like *ergonomic criteria*, *linguistic level* (Nielsen, 1993) or *model*. The organization of fundamental concepts around the guideline enables browsing the knowledge base from multiple entry points toward any target information in a reversible way. In a guideline-based evaluation process, other interface information has to be specified: *interface and object types*, *context of use*, *development phase*. The evaluation methods and tools must also be specified. Positive and negative examples and bibliographic references reinforce the evaluation when present.

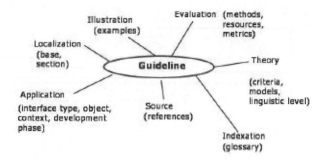

**Figure 1:** Concepts supported by the MetroWeb Tool

# 3 The Evaluation Process

The evaluation process particularized to heuristic evaluation (Fig.2) and supported by the tool is decomposed into four phases: (1) planning, (2) assessment, (3) finalization and (4) follow-up.

## 3.1 Evaluation Planning

### 3.1.1 Evaluation objectives specification

Evaluation objectives are specified, in consultation with people implied in the evaluation (e.g., user, designer or content responsible). It is necessary to accord to the *evaluation performance* (Denley & Long, 1997), function of two factors: *evaluation quality* and *implied costs of the activity*. The desired quality includes information like information type required by evaluation clients or data type required. The acceptable costs in resources implied in the evaluation include information related, e.g., to time, users or remuneration of actors implied in the evaluation. The evaluation objective depends on the desired use of the evaluation results by the site developer, i.e. proposing design changes or a diagnosis on difficulties encountered by user. The values influence the type of evaluation that can be taken in charge. Results of the evaluation can be qualitative or quantitative, objective or subjective, predictive or observed, explicative or corrective (Balbo, 1995).

### 3.1.2 Context specification

Types of user, task, environment and domain activity are specified.

### 3.1.3 Evaluation method selection

The selection is made regarding information collected in 3.1.2.

### 3.1.4 Evaluation protocol specification

This step specifies the variables and techniques to collect, the physical environment of evaluation, the users population, the hardware/software as basis of the evaluation and of the evaluation scenario. This point can result directly form the precedent phase.

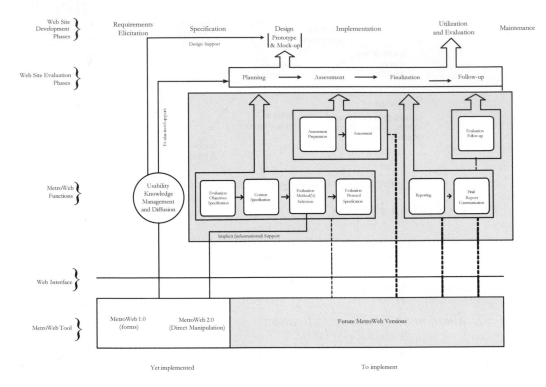

**Figure 2** Implementation of the evaluation process

## 3.2 Assessment

### 3.2.1 Preparation of assessment

The material necessary to evaluate the interface is prepared, as determined by the protocol.

### 3.2.2 Assessment

The assessment is guided by the specified protocol. Evaluator is guided by the evaluation scenario, trying to identify usability problems. This information is related to the evaluation objectives. Priority rate (Nielsen, 1993) must be dedicated to each detected problem. Information related to identify problems are collected and recorded to the end of the scenario or when an event occurs, representing other condition of evaluation suspension.

## 3.3 Finalisation

### 3.3.1 Reporting

Information recorded are reviewed, annotated, structured and consigned to present an evaluation report. Detected problems are documented and explained. Ideally, interface modification advices have to be formulated. Reporting is made in parallel of problems identification.

### 3.3.2 Report Communication

Report is communicated to people implied in the evaluation. User report profile determines the report content, presentation and diffusion.

## 3.4 Follow-up

A new interface version can be realised in regard of detected usability problems. To facilitate the comprehension of usability problems causes, it's important to link problems to established usability principles, using a formal classification of problems. A follow-up with client evaluation entity must be established, after evaluation result analysis. Interface redesign can be iterative; different solutions to usability problems can be proposed and evaluated.

## 4 The MetroWeb Tool

The MetroWeb tool is designed to assist significantly the web site usability evaluation. The tool supports generically and partially the evaluation for existing tools and methods and specifically and totally heuristic inspection, in particular design and diffusion of problems report. The MetroWeb tool (Fig.3 & Fig.4) manages fundamental usability evaluation knowledge.

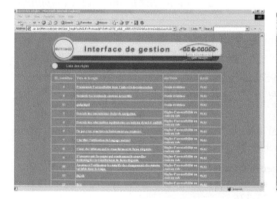

**Figure 3:** List of guidelines              **Figure 4:** Context management

Context of use of web pages to evaluate will determine a selection of methods and tools to be used by the evaluator. For example, a web page, belonging to such web site type and composed of such interactive objects, provides directly to the evaluator relevant information for the selection of a particular method or tool. The tool will provide report permitting:

- To record usability problems with information related (e.g. interface object, type of web site, development phase...). Information required will be configured by the evaluator.
- To use a usability problems categorization. The presentation of the problems could be made by problem types, chronological order or severity rate. Problems patterns can be formulated, for each activity domain.
- To have a traceability of the problems. Temporality is an important factor in the evaluation (Metzker, 2002).
- To facilitate communication between people implied in evaluation.

# 5 Conclusion and Future Work

The MetroWeb tool permits the management of usability knowledge necessary to usability problems reporting in the evaluation process. The production of usability problems report seems well-assisted, very flexible and among the most elaborated. Our future work will consist in designing and developing evaluation reporting module to support the evaluator in the usability problems reporting task. Different report types will be specified, regarding to the report user profile. Different presentation modes of usability problems and evaluation reports can be produced since all the needed usability knowledge is embodied in the tool and can be used on demand to create several levels of details. An evaluation report can be sorted by guideline, by screen, by widget, by usability problem and by ergonomic criteria. Reporting usability problems needs a well-established problem structure and categorisation. For each problem a title, one or many cause(s), effect(s) and remediation will be given. The problem will also be linked to a concrete interface object of the evaluated interface. A categorisation has to be established.

# References

Balbo, S. (1995). Software Tools for Evaluating the Usability of User Interfaces. *Proc. of 6th Int. Conf. on Human-Computer Interaction HCI International'95*, Elsevier, Amsterdam, 337-342.

Chi, E.H. (2002) Improving Web Usability Through Visualization. *IEEE Internet Computing*, April 2002, 64-70.

Denley, I. & Long, J. (1997), A Planning Aid for Human Factors Evaluation Practice Usability Evaluation Methods, *Behaviour and Information Technology*, 16(4/5), 203-219.

Farenc, C., Palanque, P. & Vanderdonckt, J. (1995), User Interface Evaluation: is it Ever Usable? *Proc. of 6th Int. Conf. on Human-Computer Interaction HCI International'95*, Elsevier, Amsterdam, 329-334.

Ivory, M.Y. & Hearst, M.A. (2001). The State of the Art in Automating Usability Evaluation of User Interfaces, *ACM Computing Surveys*, 33(4), 470-516.

Metzker, E. & Reiterer, H. (2002) Evidence-Based Usability Engineering, *Proc. of 3rd Int. Conf. on Computer-Aided Design of User Interfaces CADUI'99* (Louvain-la-Neuve, 21-23 October 1999), Kluwer Academic Publishers, Dordrecht, 323-336.

Nielsen, J. (1990). Heuristic Evaluation of User Interfaces Methodology. *Proceedings of ACM CHI'90 Conference on Human Factors in Computing Systems*, 249-256.

Nielsen, J. (1993). *Usability Engineering*, Academic Press, Boston.

Whitefield, J., Wilson, F. & Dowell, J. (1991). A Framework for Human Factors Evaluation, *Behaviour and Information Technology*, 10(1), 65-79.

# Deriving Manuals from Formal Specifications[1]

M. Massink

D. Latella

C.N.R.-ISTI
Via Moruzzi 1, Pisa, Italy
m.massink@cnuce.cnr.it

C.N.R.-ISTI
Via Moruzzi 1, Pisa, Italy
d.latella@cnuce.cnr.it

## Abstract

In this paper we propose the use (or re-use) of formal specifications for the development of user manuals and user instructions. We base our work on two observations. The first is that a considerable part of user manuals consist of series of user instructions that guide a user in the use of a device or a software application. The second is that the user interface in terms of user operations and feedback can often be described in the form of a finite state machine. In this paper we describe how model checking techniques can be used to derive automatically series of user instructions from the interface specification. The approach is illustrated by means of the specification of a well-known device such as a telephone which is then used to derive proper instruction sequences addressing relevant `how-to' and `what-if' questions about the operation of the system from a user point of view.

## 1  Introduction

The production of good user manuals and documentation of devices and software applications is not an easy task and leads in many cases to difficulties in the use of those artifacts by the end-users. This has been very well illustrated in several books (see e.g. (Norman, 1988)). The use of models of the interface, in particular models that specify the operations that a user can perform, the effect on the system and the corresponding feedback to the user have improved the analysis of the usability of an interface (see e.g. (Rushby, 2002)).

In this paper we propose to use and, where possible, to re-use formal specifications of the user interface for the derivation of series of user instructions for user manuals. Specifications of the user interface can often be presented in some form of finite state machine, transition matrix or labeled transition system that relates user operations to changes in system state and feedback to the user. In that form they can also serve as an input model for model checkers (Clarke, Grumberg & Peled, 1999). Model checkers are software tools that can perform behavioural analysis of formal models by checking the behaviour described by the model against temporal logic formulas (LTL). The result of this analysis is a yes/no answer indicating whether the model satisfies the logic formula, but also an example trace that shows how it was possible that a formula was not satisfied. This last feature of model checkers may be used in a somewhat indirect way to obtain series of user instructions from interface models by means of model checking. To that aim the relevant `how to' and `what if' questions of the user need to be expressed as proper temporal logic formulas in order to obtain traces that correspond to series of user instructions.

---

[1] This research has been partially financed by the C.N.R. coordinated project DimmiBene, CNRC00F31A.

## 2 POTS Phone Manual Case Study

In this paper we illustrate our approach to the derivation of user instructions by means of a simplified case study concerning the well-known plain old telephone system (POTS). This system has a hook and a sensor that signals when the phone is on-hook or off-hook. The hook also serves for voice communication in two directions and gives the user audible signals to indicate particular situations that may occur when the phone is used, such as a dial-tone that indicates that the user may start dialling a number, a busy tone etc. Furthermore, the phone has a number of buttons to dial phone numbers. The user operations and product responses are shown in Table 1. The behaviour is given by the transition system of Figure 1. The transitions are labelled by the user operations and device responses. Those marked by an exclamation mark (!) are generated by the phone, those with a question mark are operations that the phone may accept on initiative of the user.

**Table 1.** User operations and product responses

| User operations | | Product responses | |
|---|---|---|---|
| gof | Go off-hook | DTN | Present dial tone |
| gon | Go on-hook | VCE | Carry voice conversation |
| d | Dial number | BSY | Present busy tone |
| | | ROH | Present receiver off-hook tone |
| | | NRS | No response |
| | | NCT | Present no connection tone |
| | | RING | Present ringing tone |

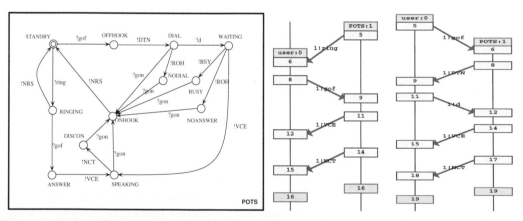

**Figure 1.** (a) Specification of the phone interface, (b) answering and (c) making a phone call.

## 3 Formal Model of the User Interface

The transition system specification of the previous section can be obtained by abstraction from more detailed formal specifications of the phone system that are intended to describe the internal operation of the system. The obtained model can be translated in a straightforward way in the input language of a model checker. We chose to use the model checker SPIN (Holzmann, 1997) and its related modelling language Promela for this paper because it is one of the widely used open source model checkers available for different platforms and with a graphical user interface. However, our approach is largely independent of the particular model checker chosen.

The Promela specification consists of two processes, one modelling the POTS phone behaviour and one modelling the generic user. The POTS process reflects the structure of the labelled transition system of Figure 1. The interaction between the user and the POTS phone is modelled by synchronous (rendez-vous) communication. The two processes are run in parallel, generating all possible states that would result from any interleaving execution. In the next section we formalize the higher level goals of the user in the form of `how to' and `what if' questions formulated as particular forms of temporal logic formulas. The Promela specification and the logic formulas together form the input that is given to the model checker such that it can produce the user instructions automatically. Details of the Promela specification and the definition of LTL formulas in SPIN can be found in (Massink & Latella, 2003).

## 4  `How to' and `What if' questions

A considerable part of user manuals consist of descriptions of how certain functionality of a device or a system can be activated. For example in a manual of a video recorder we find series of instructions on how to set the timer, tune the TV channels, various ways to program the recording, how to play back the tape and so on and so forth. Another part consists of the description of the basic interface points of the device (buttons, jacks, displays) and a list of problem shooting items for when unexpected results are encountered.  Most users will consult a manual as a reference document rather than reading it from the beginning to the end. Often they will consult the manual with one of the following kinds of questions: `How to' questions and `What if' questions. The first refer to the use of certain functionality that users expect the device to provide, and they  would like to know how this functionality can be activated. The second kind of questions may occur in situations where the device behaves differently than expected, and the user needs to know how to continue from such a situation to reach a more satisfactory one. In this section we shall address both kinds of questions, refine them into more detailed and concrete questions, and show how they may be formulated into appropriate LTL formulas. The aim is to arrive at certain recurring patterns of such formulas  that may be useful in a more general way.

### 4.1  How to make a phone call?

The first question we address is ``How to make a phone call". This corresponds to asking how, starting from the initial state, a state can be reached in which the user is speaking to someone else by phone. In terms of the Promela model we would like to obtain a sequence of instructions that guide the user from the initial state (the phone is silent and on hook) to a state in which a voice conversation is going on. This last state is characterised by process POTS being in state ``SPEAKING". SPIN provides automatically example traces when a universal property does **not** hold. So in order to obtain a trace, we should not verify the formula `eventually p holds' ($\Diamond p$), where p stands  for the phone being in the state SPEAKING, but rather its  negation: $!\Diamond p$. In this way we `provoke' the model checker to verify that from the initial state we will never eventually be able to reach a state in which the user is speaking by phone. The model checker will immediately provide us with a counter example showing that the SPEAKING state **can** be reached and this is what we were aiming at.  Note that the formula $!\Diamond p$ is equivalent to `always not p' ($[]!p$), this can be shown by a simple formal proof.  In order to get a usually better result, we may instruct SPIN to look for the **shortest** trace leading to the state SPEAKING. This is useful for obtaining the shortest set of instructions for the user. From the trace that Spin generates in plain text format (or as message sequence charts, see Figure 1 (b)) the series of symbolic user instructions, in terms of the actions used in the specification, can be obtained and gives us in this case the trace: RING, gof, VCE corresponding to the instructions saying that when the phone rings, take the hook off and start voice communication.

As can be noted, this trace is somewhat surprising at first. Although the outcome is correct, and is indeed the shortest trace, the user had probably in mind to **initiate** the phone call and not to wait for someone to call him/her. This means that a more accurate formulation of the 'how to' question is necessary in order to obtain the expected answer. Actually, the user would like to know how to initiate a phone call without waiting for someone else to call him/her. This can be expressed by means of the logical *until* operator (U). Let p be defined as before and define q as process POTS being in state RINGING, then we would like to obtain a series of instructions that lead to p without q becoming true before p. So, formulated as an LTL formula: (!q) U p. Of course, in order to obtain a trace from SPIN, we need to negate this formula when using the verifier, so we obtain !((!q) U p). The resulting trace is gof, DTS, d, VCE, (Figure 1 (c)) or, in words, take the hook off the phone , hear the dial tone, dial the number and start the voice communication.

Although the real trace generated by SPIN is containing more information than only the actions, the necessary information about the sequence of operations is easy to distil in an automatic way. A next step to produce a result that is resembling better the kind of text that can be found in user manuals is to relate each symbolic operation to a piece of text that informs the user in natural language what needs to be done. In principle the traces may also be used as input to an animator that illustrates the user in a graphically animated way how to proceed. In the following we give further examples of questions, their formalisations and resulting message sequence charts of the shortest series of instructions that can be given to instruct the user.

## 4.2  How to answer a ringing phone?

This question can be refined by requiring that the phone must be ringing, and in that case we are looking for the shortest series of instructions that lead to a situation that the user is speaking. Let p and q be defined as in the previous example, this 'how-to' question can then be formalised as $\Diamond$(q && $\Diamond$p). The negation of this formula gives [] ! (q && $\Diamond$ p). The resulting trace is the same as that in the first example.

## 4.3  What to do when the line is busy?

Using a very similar pattern of LTL formulas we can answer questions of the type 'What if', like ``What to do when the line is busy?''. In this case the user is probably intending that (s)he would like to proceed with other activities, like making another phone call, when (s)he discovers that the person (s)he was looking for is currently speaking to someone else. In this case, we can simply replace the definition of q by process POTS is in state BUSY and re-use exactly the same formula as in Section 4.2 i.e. []!(q && $\Diamond$p). The trace produced is gof, DTN, d, BSY, gon, NRS, RING, gof, VCE. This trace illustrates clearly that the user needs to put the hook on the phone after (s)he hears the busy tone before (s)he can receive another phone call. If we want to obtain the instructions when the user intends to **initiate** a phone call, we need to define r as process POTS is in state RINGING and use the LTL formula []!(q && ((!r) U p)). This gives the trace gof, DTN, d, BSY, gon, NRS, gof, DTN, d, VCE.

We could present many other examples, but the pattern of the LTL formulas would be essentially similar to the ones that we have presented. This means that the formulas do not need to be reinvented for every kind of question, thus facilitating their use.

# 5   Conclusions

In this paper we have shown how the technique of model checking can be used to obtain automatically series of user instructions from formal specifications of a user interface. Such series of instructions form an important part of user manuals where the user is informed about the operation of a device or system. The automatic derivation of user instructions has several important advantages. First of all, it guarantees a close relation between the specification of the interface and the user manual and therefore it helps to avoid the introduction of errors in the documentation for the end-user. When a formal specification of the interface already exists for the purpose of correctness analysis, its use for the generation of instruction sets for a manual makes the specification even more useful. A third advantage is that the developer of the manual does not need to rely only on natural language descriptions of the system, which may be rather imprecise. The use of a model checker could be integrated into a special tool for the preparation of user manuals and save the developer a lot of time. Finally, if a formal model of the interface is used also for the design of the system or device, changes in the model can be easier translated into corresponding changes in the user manual. We have outlined the basic idea for the use of model checking for the generation of user instructions. Much can be done to improve the generated output using for example techniques developed in the field of natural language generation. A more exciting extension would be to explore whether the model checker could be used interactively in for example on-line helpers. This would require an automatic translation of the on-line natural language question of the user into a proper logical formula and the immediate reply of instructions to the user. Our model checking based approach could be particularly helpful for the development of manuals for interactively complex systems (Thimbleby & Ladkin, 1997).

# References

Clarke, E. M., Jr., Grumberg, O., and Peled, D. (1999) *Model Checking*. The MIT Press, Cambridge, Massachusetts, London, U.K.

Holzmann, G. (1997) The model checker SPIN. *IEEE Transactions on Software Engineering* 23(5): pp. 279-295.

Massink, M., and Latella, D. (2003) *Deriving Manuals from Formal Specifications – Extended version*. ISTI-2003-TR-01 C.N.R.-ISTI Technical report.

Norman, D. (1988) *The Psychology of Everyday Things*. Basic Books, Inc, Publishers, New York.

Rushby, J., (2002). Using model checking to help discover mode confusions and other automation surprises. In *Reliability Engineering and System Safety*. Vol. 75, No. 2, pp. 167-177.

Thimbleby, H., and Ladkin, P. (1997). From logic to manuals. In *Proceedings of the IEEE Conference on Software Engineering*.

# Towards a Systematic Empirical Validation of HCI Knowledge Captured as Patterns

*Eduard Metzker[1], Ahmed Seffah[2] and A. Gaffar[2]*

[1]DaimlerChrysler Research and Technology
Dept. for Software Process Improvement
eduard.metzker@daimlerchrysler.com

[2] Human-Centered Software Engineering
Group, Concordia University
{seffah, gaffar}@cs.concordia.ca

## Abstract

There is an urgent demand for more tightly integrating HCI knowledge into the software development lifecycle. A valid HCI body of knowledge is a prerequisite to achieve this objective. This body of knowledge can provide a basis for addressing more objectively the concerns related to the integration of HCI knowledge including best practices and design rational into the mainstream software development processes. This paper reviews existing approaches to capture and disseminate HCI knowledge. The requirements for a pattern-oriented framework that facilitates empirical validation of HCI knowledge are set out.

## 1    Introduction

No matter which software engineering methods are selected for a particular project, in almost all cases, there is little hard evidence backing up these decisions, and their costs and benefits are rarely understood (Perry, Porter, & Votta, 2000). In fact, the fundamental factors and mechanisms that drive the costs and benefits of software tools and methods are largely unknown. Without this empirical evidence, choosing a particular technology or methodology for a project at hand essentially becomes a random act (Basili, Shull, & Lanubile, 1999).

This critique also applies to usability engineering as a discipline. The large variety of frameworks, techniques, patterns and guidelines that have been proposed over the years make it hard to select the appropriate method. Yet the kind of evaluation dedicated to all these proposals can often be only classified as "at best anecdotal". One side effect of this policy is, that research efforts proposed engineering methodologies that, in essence, had already been built – they varied primarily in the names chosen for constructs (Smith & March, 1995). This uncertainty about the validity of the constructs proposed – mainly caused by a lack of empirical evaluation – leads to problems for both researchers and practitioners.

For researchers, valid questions on how to optimally integrate HCI knowledge into the software development lifecycle are in danger to degenerate to "religious wars" (Paterno, 2002). A current prominent example is the dispute between the scenario-based and task-based camps (Benyon & Macaulay, 2002; Carey, 2002; Carrol, 2002; Paterno, 2002). Such disputes are largely caused by a lack of empirical evidence for supporting decisions on how, when, and in which cases to apply a method.

The situation is even worse for practitioners. Based on which data should a project manager, usability engineer or user interface software developer select an approach, methodology, technique, best practice, pattern or guideline? If there is no empirical data to justify the selection of a technique then why should anyone care about them at all?

In this paper we outline potential building blocks of the body of evidence called for. Furthermore we propose some quality criteria for adding a new piece of knowledge to an existing body.

Finally, we outline high-level requirements for a framework for assessing the validity of HCI knowledge captured in the format of patterns.

## 2    Existing Approaches for Capturing HCI knowledge

HCI has a long tradition in capturing knowledge for driving the design and evaluation of interactive systems. One of the first approaches developed were design guidelines. Their purpose is to capture design knowledge into small rules, which can then be used when constructing or evaluating new user interfaces. Vanderdonckt (Vanderdonckt, 1999) defines a guideline as a design and/or evaluation principle to be observed in order to get and/or guarantee the usability of a UI for a given interactive task to be carried out by a given user population in a given context.

A detailed analysis of the validation, completeness and consistency of existing guideline collections has shown that there are a number of problems (Vanderdonckt, 1999). Guidelines are often too simplistic/trivial or too abstract, they can be difficult to interpret and select, they can be conflicting and often have authority issues concerning their validity. One of the reasons for these problems is that most guidelines suggest a context-independent validity framework but in fact, their applicability depends on a specific context.

These limitations with guidelines motivated some researchers to introduce the concept of interaction patterns, also called HCI patterns. An HCI pattern is described in terms of a problem, context and solution. The solution is assumed to be a proven solution to the stated and well-known problem. Patterns have been introduced as a medium to capture and disseminate user' experiences and design practices/knowledge. They are a vehicle for transferring, by means of software development tools, the design expertise of human-factor experts and UI designers to software engineers, who are usually unfamiliar with UI design and usability principles. Many groups have devoted themselves to the development of pattern languages for user interface design and usability. Among the heterogeneous collections of patterns, *Common Ground*, *Experiences*, *Brighton* and *Amsterdam* play a major role in this field and wield significant influence.

Another difference is that patterns make both the context and problem explicit and the solution is provided along with a design rationale. Compared to guidelines, patterns contain more complex design knowledge and often several guidelines are integrated in one pattern. Patterns focus on "do this" only and therefore are constructive and less abstract. Guidelines are usually expressed in a positive and negative form; do or don't do this. Therefore guidelines have their strength for evaluation purposes. They can easily be transformed in questions for evaluating a UI.

Another interesting approach to capture HCI design knowledge are claims (Sutcliffe, 2001). Claims are psychologically motivated design rationales that express the advantages and disadvantages of a design as a usability issue, thereby encouraging designers to reason about trade-offs rather than accepting a single guideline or principle. Claims provide situated advice because they come bundled with scenarios of use and artifacts that illustrate applications of the claim. The validity of claims has a strong grounding in theory, or on the evolution of an artifact that demonstrated its usability via evaluation. This is also a weakness of a claim, because it is very situated to a specific context provided by the artifact and usage scenario. This limits the scope of any one claim to similar artifacts.

Most existing approaches for capturing and exploiting HCI knowledge have in common that their usage is largely focused on a small part of the overall development process for interactive systems, namely interface design and evaluation. They are still not integrated in important parts of the software development lifecycle such as requirements analysis. To bridge this gap, USEPACKs (Usability Engineering Experience Packages) have been proposed (Metzker & Reiterer, 2002a; Metzker & Reiterer, 2002b). USEPACKs are based on the idea of process patterns (Ambler, 1998). USEPACKs extend the idea of process patterns by providing a semi-formal notation and an explicit framework for the deployment, exploitation, evolution and validation of HCI patterns within the software development lifecycle (Metzker & Offergeld, 2001; Metzker & Reiterer, 2002a).

# 3    Quality Criteria for HCI Patterns

A number of frameworks have been proposed for identifying and classifying research approaches in the fields of software engineering and information systems. March and Smith distinguish two different areas of research activities in information technology (Smith & March, 1995): research which aims at understanding reality is referred to as natural science. The basic activities are discovery of theories and justification of discovered theories. This descriptive science generates knowledge. Natural science develops sets of concepts, or specialized vocabulary, with which to characterize phenomena. These concepts are used in higher order constructs such as laws, models, theories that make claims about the nature of reality. Explanations of phenomena (theories) are the most important achievements of natural science theory. The products of this research stream are evaluated against 'norms of truth, or explanatory power' (Smith & March, 1995). Progress is achieved by new theories that provide more comprehensive, and more accurate explanations of phenomena.

On the other hand design science – as defined by Simon (Simon, 1981) - uses knowledge to produce artifacts that serve human needs. The basic activities of this research stream are the construction of an artifact for a defined purpose and to measure the capability of the artifact. The capability of the artifact depends on the usage environment and the context of use. The criteria for the evaluation of an artifact in a certain context have to be defined.

Design science generates concepts, models, methods and implementations. Constructs and concepts form the vocabulary for characterizing problems and solutions in a certain domain. Design science produces advancement when existing technologies or methodologies are replaced by better ones. The properties of design science and natural science are set out in table 1.

**Table 1: Properties of Design and Natural Science**

| Property | Natural Science | Design Science |
|----------|-----------------|----------------|
| Concern of models | ▪ Truth | ▪ Utility |
| Quality | ▪ Consistency with observed facts <br> ▪ Ability to predict future observations | ▪ Evaluated against criteria of value or utility (difficult: depends on context of use) |
| Progress | ▪ New theories that provide more comprehensive, and more accurate explanations | ▪ Existing technologies, methods, tools are replaced by more effective ones |

Clearly, the disciplines of HCI and usability engineering, if classified via March's and Smith's framework, must be assigned to the area of design science and their models are thus subject to the quality criteria of design science. If we accept that the central quality criteria applicable to our models are utility and effectiveness, we can derive a set of interrelated quality factors for HCI process and product patterns. They are summarised under Q1-Q3:

- Q1: A patterns' level of empirical evaluation
  The level of empirical evaluation applied is the central quality criteria of any research artifact, whether it is a method, model, tool or HCI pattern. With respect to HCI patterns, empirical

evaluation means to examine whether a pattern or pattern language is useful, usable and effective. Such studies must exceed the level of "I tried it and I liked it". Empirical evaluation can include review and feedback from HCI experts who have applied the pattern. An invaluable source of feedback on utility, usability and effectiveness are project teams of industrial software development projects. Therefore, ideally, patterns are validated empirically through software development experience of independent of pattern users.

- Q2: The impact of a pattern on software engineering practice
  A solid contribution comprises substantial and valuable technical or methodical content. It provides substantial value to new software practitioners or experts not practicing the solution described. In the context of a pattern catalog or language, a pattern should contribute to successful software practice. HCI product and process patterns should deliver tangible value to the stakeholders in form of design rationale for products (e.g.: user metaphors, UI designs, or interactive systems) or in form of reusable best practices for usability engineering process improvement (e.g.: best practices, methods, templates, questionnaires, checklists).

- Q3: The usability and accessibility of a pattern language
  It is desirable for all HCI patterns to be published and publicly available. The popular acceptance and awareness of a pattern can be measured by objective criteria, such as publication circulation and usage by practitioners. A formal pattern representation should define the solution from multiple perspectives such as a solution's usability properties or effort required for implementation. A multi-perspective representation makes the pattern accessible and usable for a range of potential users such as usability engineers, user interface designers or user interface software developers.

# 4    First Steps Towards the Empirical Validation of HCI Patterns

The quality factors described above have to be kept in mind, when designing approaches and tools for supporting pattern oriented approaches. First steps towards meeting this objectives have been made. Evidence-Based Usability Engineering is a comprehensive approach for capturing, disseminating, exploiting, and evaluating HCI process patterns in industrial software development organizations (Metzker & Reiterer, 2002b). The concepts of Evidence-based Usability Engineering are embodied in a web-based support environment, ProUSE, which combines functions for patterns capturing, metrics-based selection of patterns and process guidance based on patterns (Metzker & Reiterer, 2002a).

A new approach for HCI product patterns is MOUDIL (Gaffar, Seffah, Javahery, & Sinning, 2003). MOUDIL aims at overcoming several shortcomings of existing HCI product pattern approaches such as sub-optimal pattern structure, inconsistent terminology, isolated pattern collections, lack of focus and redundant or contradicting solutions for the same problem. Finding a HCI pattern structure that is appealing to software engineering project teams is a vital goal, as they are the actual users of patterns. A user-optimized pattern structure will help to effectively communicate HCI patterns to project teams and thus increase the impact of HCI patterns on software engineering practice.

A future objective is to unify HCI process and product pattern repositories in a meaningful way. One open question is how the meta-data of pattern languages that involve both aspects should be structured. Are, for example, HCI process and product patterns influenced by the same forces?

# 5    Conclusion

HCI process and product patterns are useful in gathering and disseminating experiences for future developers. A great deal of work has been done on HCI patterns by many different individuals. The lack of knowledge validation, however, requires users to hunt for suitable patterns, and extract them for their own use. Gathering relevant patterns in one repository will help overcome this difficulty. In our future research, we want to use the gathered information from the collected and analyzed patterns to come up with a formal model for validating patterns. Such an approach will help capture and disseminate pattern knowledge effectively.

## References

Ambler, S. W. (1998). *Process Patterns - Building Large-Scale Systems Using Object Technology*: Cambridge University Press.

Basili, V. R., Shull, F., & Lanubile, F. (1999). Building Knowledge Through Families of Experiments. *IEEE Transactions on Software Engineering, 25*(4), 456-473.

Benyon, D., & Macaulay, C. (2002). Scenarios and the HCI-SE design problem. *Interacting with Computers, 14*(4), 397-405.

Carey, T. (2002). Commentary on "scenarios and task analysis" by Dan Diaper. *Interacting with Computers, 14*(4), 411-412.

Carrol, J. M. (2002). Making use is more than a matter of task analysis. *Interacting with Computers, 14*(5), 619-627.

Gaffar, A., Seffah, A., Javahery, H., & Sinning, D. (2003). *MOUDIL: A Comprehensive Framework for Disseminating and Sharing HCI Patterns.* ACM-CHI Workshop on Patterns.

Metzker, E., & Offergeld, M. (2001). *REUSE: Computer-Aided Improvement of Human-Centered Design Processes.* Paper presented at the Mensch und Computer, 1st Interdisciplinary Conference of the German Chapter of the ACM, Bad Honnef, Germany.

Metzker, E., & Reiterer, H. (2002a). *Evidence-based Usability Engineering.* Paper presented at the Computer-aided Design of User Interfaces (CADUI2002), Valenciennes, France.

Metzker, E., & Reiterer, H. (2002b). Use and Reuse of HCI Knowledge in the Software Development Lifecycle: Existing Approaches and What Developers Think. In J. Hammond, T. Gross, & J. Wesson (Eds.), *Usability - Gaining a Competitive Edge* (pp. 39-55). Norwell, Massachusetts: Kluwer Academic Publishers.

Paterno, F. (2002). Commentary on 'scenarios and task analysis' by Dan Diaper. *Interacting with Computers, 14*(4), 407-409.

Perry, D. E., Porter, A. A., & Votta, L. G. (2000). *Empirical Studies of Software Engineering: A Roadmap.* Paper presented at the International Conference on Software Engineering (ICSE2000), Limerick, Ireland.

Simon, H. A. (1981). *The Sciences of the Artifical.* (2. ed.). Cambridge, MA: MIT Press.

Smith, G. F., & March, S. T. (1995). Design and Natural Science Research on Information Technology. *Decision Support Systems, 15*(4), 251-266.

Sutcliffe, A. (2001). On the Effective Use and Reuse of HCI Knowledge. *ACM Transactions on Computer-Human Interaction, 7*(2), 197-221.

Vanderdonckt, J. (1999). Development Milestones Towards a Tool for Working with Guidelines. *Interacting with Computers, 12*(2), 81-118.

# User-centered Design in the Software Engineering Lifecycle: Organizational, Cultural and Educational Obstacles to a Successful Integration

*E. Metzker[1] and A. Seffah[2]*

[1]DaimlerChrysler Research and Technology,
Dept. for Software Process Improvement
eduard.metzker@daimlerchrysler.com

[2] Human-Centered Software Engineering
Group, Concordia University
seffah@cs.concordia.ca

## Abstract

Over the last 15 years, the user interface community has proposed a large variety of user-centered design (UCD) techniques and tools. However, these methods are still underused and difficult to understand by software development teams and organizations. This is because, software developers have their own techniques and tools for managing the software development lifecycle, including UI analysis and prototyping. They also do not see where exactly should the UCD techniques be integrated into the existing software development lifecycle to maximize benefits gained from both? This paper discusses the prevalent claims, beliefs, and obstacles promoted and supported by both the software and the usability engineering communities. Avoiding such obstacles is the first step towards a successful integration of user-centered design techniques into the entire software engineering lifecycle.

## 1    Motivation

With the considerable growth of interactive internet and mobile applications, software usability is no longer a luxury goal. It is the ultimate objective for developers and users productivity and satisfaction. However, knowledge how to integrate most efficiently and smoothly human-centered design (HCD) methods into established software development lifecycles is still missing (Mayhew, 1999). While some approaches such as the usability maturity model (UMM) (ISO/TC 159 Ergonomics, ) or the DATech-Model for usability process assessment (DATech Frankfurt/Main, 2001) provide means to assess an organization's capability to perform HCD processes, it remains unclear to common developers why certain UCD tools and methods are better suited in a certain development context than others (Welie, 1999) (Henninger, 2000). In fact, we know very little of the actual  value of the UCD methods. The way of evaluating the effects of software engineering methods on process and product quality so far is rated by statisticians as 'at best anecdotal' (American Statistics Assosication: Committee on Applied and Theroretical Statistics, 1996). The damaged merchandise of UCD methods that results of this policy is described in the Gray and Salzman seminal paper (Gray & Salzman, 1998).

The user-centered design has been marketed, mainly in the HCI community and by psychologists, as the right approach for designing usable interactive systems (Norman, 1998). Unfortunately, even if software development teams largely recognize its appropriateness and powerfulness, UCD techniques remain the province of visionaries ,few enlightened software practitioners (Constantine & Lockwood, 1999) and large organizations, rather than the everyday practice of computer scientists and software engineers. One barrier to the wider practice of user-centered design is that its structure and techniques are still relatively unknown, underused, difficult to master, and essentially inaccessible to small and medium-sized software development teams and common developers.

Indeed, while software developers may have high-level familiarity with basic UCD concepts and principles such as task analysis and design guidelines, few understand the complete UCD process at a level, which allows them to incorporate UCD techniques into the entire software development and deployment lifecycle (Constantine & Lockwood, 1999; Mayhew, 1999). Most often, a software team member has experience in UCD. However, he or she may see similarities between the UCD and traditional software engineering methods – for example use cases and task analysis - Sometimes, he try to substitute one by the other forgetting that UCD and software engineering techniques each have their own strengths and weaknesses. A few people are able to understand the overlap that exists between UCD and software engineering.

This paper discusses the prevalent beliefs, myths and obstacles on UCD techniques. Having a good understanding of such obstacles and beliefs can help software developers and HCI educators to improve their methodologies, and to learn techniques for communicating with usability "gurus" and supporters. Their understanding is also important for any person interested in promoting user-centered design in the software engineering community. It can help us to realize that user-centered design should be a core part of every software development activity yet, despite its well-documented paybacks, it has yet to be widely adopted.

## 2 Building a bridge between existing UCD methods and the overall software development lifecycle

Often, UCD techniques are regarded as being somehow decoupled from the software development process practiced by software engineers. The term "user interface" is perhaps one of the underlying obstacles in our quest for more usable interactive systems. The usability of "user interface" gives the impression that the UCD approach is only for developing a thin component sitting on top of the software or the "real" system. This Cartesian dichotomy that decouples the interface from the remaining system and that put, de facto, engineers and psychologists in different departments explains the `peanut butter theory of usability' (Lewis & Rieman, 1994). Lewis explains that usability is often seen as

"A spread that can be smeared over any software model, however dreadful, with good results if the spread is thick enough. If the underlying functionality is confusing then spread a graphical user interface on it. If the user interface still has some problems, smear some manuals over it. If the manuals are still deficient, smear on some training which you force users to take."

If usability is an engineering discipline, it has to share with software engineering some common values. Its specialists have to think and work like engineers (Mayhew, 1999). This dichotomy also explains why often it appears to software project managers that they have to control two separate processes: the overall system development process and the UCD process for designing the interactive components. As it remains unclear how to manage/synchronize these two processes, the UCD activities still are regarded as dispensable and have been skipped in case of tight schedules (Mayhew, 1999).

As a major consequence of this dichotomy, most UCD techniques and processes do not include a strategy for how to promote/apply/asses these methods for different software development organizations that can be at different level of maturity. For example, software engineering teams have a long tradition in using focus groups, rapid prototyping techniques, and scenarios during requirement elicitation. It might be a good strategy to try, as a first step, to enhance such methods before deploying the whole UCD toolbox.

Another consequence is the belief that experienced human factor specialists are easily accessible throughout the development team and therefore HCD methods can be performed ad hoc. Recent research investigations show that most often highly interactive systems are often developed without the help of in-house human factors specialists or external usability consultants (Metzker & Offergeld, 2001).

## 3   Establishing usability in a software development organization is a continuous organizational learning and improvement task

For small-size projects, software development teams can mostly avoid the direct involvement of usability experts, due in particular to the availability of design guidelines, usability patterns, and heuristics for evaluation or tasks flowcharts to supplement the functional requirements analysis. However, for large-scale projects, it is necessary, almost impossible not, to involve explicitly usability specialists, at least during the requirement analysis and usability testing steps.

However, the integration of external expertise is not enough for establishing the usability function. A learning strategy should supplement the continuous promotion, assessment and improvement of UCD methods in the mainstream software development lifecycle. In particular, whatever the approach chosen for involving usability engineers, the difficulties of communication between the software development team and the usability specialists has to be addressed. Among the reasons of such miscommunication, one can mention the educational gap, the use of different notations, languages and tools, as well as the perception of the role and importance of the software artifacts that have an impact on usability. In our opinion, such difficulties are the most critical obstacles that have to be avoided. In the transition from technology-centered practices to human-centered practices, organizations have to learn which knowledge/skills are required while gradually adopting/switching to the fundamental UCD best-practices.

Most often, software developers entrusted with interface design and evaluation lack fundamental UCD skills. Yet, as the need for usability was recognized by software development organizations, they tend to develop their own in-house guidelines, heuristics and sometimes define or reinvent the whole usability engineering toolbox or lifecycle. Recent research supports the claim that such behavior, in fact, compiled and used by software development organizations. Spencer (Spencer, 2000), for example, presents a streamlined cognitive walkthrough method that has been developed to facilitate efficient performance of cognitive walkthroughs under the social constraints of a large software development organization. Software engineering empirical studies (Basili, Caldiera, & Rombach, 1994) demonstrated that, in most cases, best practices like those suggested by Spencer's are unfortunately not published in either development organizations or scientific conferences. They are bound to the people of a certain project or, even worse, to one expert member "the guru" of this group, which decrease the availability of the UCD body of knowledge. Similar projects in other departments of the organization usually cannot take benefits from these experiences. In the worst case, the experiences may leave the organization if the expert quit his jobs. Therefore, an efficient integration of UCD techniques should not only support existing human factors methods but also allow the organizations to compile, develop and evolve their own practices.

Furthermore, tools are needed to support developers in acquiring and sharing and evaluating knowledge on how to perform effectively UCD activities. They should also help to refine and evolve basic methods to make them fit into their particular project context. Tools are needed for analyzing and visualizing the voluminous amount of observational data that we generally collect. UCD will be considered more seriously at large if and only if a computer-assisted usability engineering (CAUsE) platform is available. CAUsE should support the whole process for gathering and filtering the observational data, extracting relevant measures, and interpreting these measures. The central goal of such a tool is to make the behavioral aspects of UCD less time-consuming and tedious. The lack of such platform is one of the major technical obstacles that the HCI and software engineering communities together can easily avoid.

# 4    Educating software and usability professionals to communicate and work together

The UCD specialists, who have been trained in psychology departments, are sometimes regarded as mere nuisances who get in the way of "real/tradition" software/computer science education. Even, if the user interfaces code is often more than half of the whole code for a project and takes a comparable amount of development effort, courses on HCI are still very independent courses in software engineering programs. User interfaces design skills are often haphazard and regarded as unimportant by many software developers and managers. This explains why HCD methods often cannot be fully used. The necessary knowledge and training is not available within the development teams (Earthy, 1999; Mayhew, 1999).

Helping developers to understand and master UCD techniques in their own languages and cultural contexts is a successful pedagogy for them and to a certain extent for integrating/selling usability to software organizations. A very powerful case study is on usability and learnability of CASE tools (Seffah & Rilling, 2001.). This case study is an good starting point for helping software engineers to realize the importance and impact of usability in their own context. Such framework will probably improve also the usability expert capability to communicate with software engineers over the diversity of notations and tools they generally use.

# 5    Cross-pollinating disciplines

UCD techniques are still relatively unknown, underused, difficult to master, and essentially inaccessible to small and medium-sized software development companies, and to most software developers. The reasons for this lack are multiples including technical, organizational, economical, educational and social reasons. Basically however, the main reason is that UCD techniques and tools are not cost-effectively integrated into software engineering methods, processes and tools. As discussed in this paper, many prevalent myths and beliefs have to be avoided as a first step towards an effective integration of UCD techniques. A framework for promoting and improving UCD techniques is required. Adequate training for both developers and usability engineering on how to work together and understand each other over the diversity of tools and techniques is fundamental.

More than enthusiastic champions, supporters, disciples and believers, we need to investigate avenues for cross-pollinating disciplines. What should software and usability engineering learn from each other to facilitate and encourage their convergence and integration? A forum for sharing ideas about potential and innovative ways to cross-pollinate the two disciplines, successful and unsuccessful experiences related to integrating usability into the software engineering lifecycle and small and medium-sized organizations, as well as a future research agenda for building a tighter fit between UCD experiences and software engineering practices is needed.

# 6    Conclusions

Our findings indicate that we hardly need any new UCD methodology. Instead a meta-model is needed that supports the integration of UCD methods into mainstream software development processes. Furthermore evaluation of the "context of use" and the utility of the various existing

UCD methods is now most important. Regarding the utility of UCD methods, the acceptance of those methods by development teams is an important factor. The envisioned meta-model should consider the following central findings:

1. Integrating UCD methods into software development processes is an organizational learning process

2. Integrating UCD methods into software development processes is a process improvement task

3. The selection of UCD methods should be driven by the feedback of project teams regarding the utility of those methods.

## References

American Statistics Assosication: Committee on Applied and Theroretical Statistics. (1996). *Statistical Software Engineering*, [website]. National Academy of Sciences Available: http://www.nap.edu/readingroom/books/statsoft/index.html [2001]

Basili, V. R., Caldiera, G., & Rombach, H. D. (1994). Experience Factory. In J. J. Marciniak (Ed.), *Encyclopedia of Software Engineering* (Vol. 1, pp. 528-532). New York: John Wiley & Sons.

Constantine, L. L., & Lockwood, L. A. D. (1999). *Software for Use: A Practical Guide to the Models and Methods of Usage-Centered Design*: Addison-Wesley.

DATech Frankfurt/Main. (2001). *DATech Prüfbaustein: Qualität des Usability Engineering Prozesses*. Deutsche Akkreditierungsstelle Technik e.V. Available: http://www.datech.de/download/deutsch/13-2001_DATech_Baustein_UE-Prozess_1.1.pdf [2001]

Earthy, J. (1999). Human Centred Processes, their Maturity and their Improvement. *IFIP TC.13 International Conference on Human-Computer Interaction (INTERACT), 2*, 117-118.

Gray, W. D., & Salzman, M. C. (1998). Damaged Merchandise? A Review of Experiments that Compare Usability Evaluation Methods. *Human-Computer Interaction, 13*(3), 203-261.

Henninger, S. (2000). A Methodology and Tools for Applying Context-Specific Usability Guidelines to Interface Design. *Interacting with Computers, 12*(3), 225-243.

ISO/TC 159 Ergonomics. ISO TR 18529 - The Usability Maturity Model. .

Lewis, C., & Rieman, J. (1994). *Task-Centred User Interface Design*. Available: http://www.hcibib.org/tcuid/appx-m.html [2001]

Mayhew, D. J. (1999). *The Usability Engineering Lifecycle: A Practioner's Handbook for User Interface Design*: Morgan Kaufman Publishers.

Metzker, E., & Offergeld, M. (2001). An Interdisciplinary Approach for Successfully Integrating Human-Centered Design Methods Into Development Processes Practiced by Industrial Software Development Organizations. In R. Little & L. Nigay (Eds.), *Engineering for Human Computer Interaction: 8th IFIP International Conference, EHCI 2001 (EHCI'01)* (pp. 21-36). Toronto, Canada: Springer.

Norman, D. A. (1998). *The Invisible Computer*: MIT Press.

Seffah, A., & Rilling, J. (2001.). *Investigating the Relationship between Usability and Conceptual Gaps for Human-Centric CASE Tools*. Paper presented at the IEEE Symposium on Human-Centric Computing Languages and Environments, Stresa, Italy.

Welie, M. v. (1999). *Breaking Down Usability*. Paper presented at the IFIP TC.13 International Conference on Human-Computer Interaction (INTERACT), Endinburgh, UK.

# Delegation Systems:
## Staying in Charge of Highly Flexible Automation

*Christopher A. Miller,*
*Robert P. Goldman, Harry B. Funk*

Smart Information Flow Technologies
Minneapolis, MN. USA
{cmiller, hfunk, rgoldman}@siftech.com

*Raja Parasuraman*

Cognitive Science Laboratory
The Catholic University of America
Washington, DC USA
parasuraman@cua.edu

## Abstract

Delegation systems strive to provide the same level of flexibility in how and at what level automation is used as humans have in supervisory relationships with well-trained subordinates. This sort of interaction requires a vocabulary for talking about tasks and how to perform them, which can be provided by various task modeling formalisms for representing knowledge about the domain. It also requires an intelligent, automated planning capability able to understand that vocabulary. We present a general architecture for delegation systems and provide a detailed walkthrough of human interactions with one delegation system that uses the metaphor of a sports teams' Playbook.

## 1 Introduction

As systems become more complex, there is increasing temptation to control them via automation. While the study of human-automation interactions has traditionally not been synonymous with that of human-computer interaction (HCI), this is rapidly changing. Modern computer systems are increasingly, (one might even say 'ubiquitously') becoming automation systems in that they exert control over ever more important aspects of our daily lives—our bank accounts, automobile navigation, medication monitoring, etc. Hence, lessons learned from the design of human-automation systems take on increasing importance for how we should create HCI systems.

One such lesson is that automation can be a double-edged sword. Billings (1997) documents payoffs of increased automation in four key areas: safety, reliability, economy, and comfort. But automation has also been shown to pose novel problems for human operators in some circumstances: to increase workload and training requirements, to result in decreased situation awareness and, in specific instances, to cause accidents (e.g., Reason, 1997; Parasuraman & Riley, 1997).

An approach to human-computer relationships that retains the benefits of automation while minimizing its costs and hazards is needed. Such an approach requires that neither human nor automation be exclusively in charge of most tasks, but rather demands flexibility in the role and level of automation while placing control of that flexibility in the human operator's hands. This implies that, for most tasks, automation levels should be at neither end of a possible spectrum, but rather at some intermediate, and adjustable, point. Human operators need to be able to *delegate* tasks to automation, and receive feedback on their performance, in much the same way that delegation can be performed in successful human-human teams and organizations: at various levels of detail and granularity, and with various constraints, stipulations, contingencies, and alternatives.

## 2  Delegation and Supervisory Control

Sheridan (1987) is generally credited with defining "supervisory control"—a relationship where control automation allows the human to behave as if interacting with an intelligent, human subordinate. He described a spectrum of *levels of automation* whose endpoints are full control autonomy for the human (essentially no role for automation) and vice versa, with a progression of shared roles in between. While Sheridan's conception of supervisory control explicitly included the need for a supervisor to instruct, teach or program automation, this concept has rarely been implemented beyond initial programming or, perhaps, the human selection of automation 'modes.'

A critical requirement for human-automation delegation relationships is a vocabulary for communication about actions in the domain. This communication requires more than the relatively flat automation levels in Sheridan's spectrum. Sheridan's levels implicitly reference a task to be performed by some mix of human and automation. But tasks can be decomposed in various ways to represent various methods or steps in task performance. Various task decomposition methods exist including functional decompositions (e.g., Plan-Goal Graphs, Operator Functional Modelling, GOMS, etc.) and sequential process models (e.g., PERT charts, Petri nets, etc.). Each approach is inherently hierarchical and may proceed through any number of levels to some primitive, "stopping" level. Automation may be applied differently to each and every subtask that comprises the parent task. When identifying a level of automation for a system in Sheridan's sense, one identifies an average or modal level over the subtasks the system accomplishes. In practice, one could identify the *specific* subtasks to be performed and represent an automation level or role for each.

This representation of what is to be done, by whom and how, is precisely the vocabulary needed for delegating task performance. Delegation is a process of assigning specific roles and responsibilities for the subtasks of a parent task for which the supervisor retains authority (and responsibility). Furthermore, communication about intent to subordinates is frequently in terms of specific goals, methods and/or constraints on how, when and with what resources the subtasks should be accomplished (Klein 1998; Jones & Jasek 1997). Thus, it is essential that a model of tasks, roles and resources be at the core of a human-automation delegation system.

We are developing such delegation systems. Our goal is to more closely emulate delegation in human-human work relationships, allowing the operator to smoothly adjust the amount and level of automation used depending on such variables as time available, workload, decisions criticality, trust, etc. Such a system must avoid two problems: (1) requiring excessive workload for the task of delegating itself, and (2) permitting unsafe or ineffective overall behaviour. We have created a design metaphor and system architecture that addresses these concerns. Our particular approach to enabling, facilitating and ensuring correctness from a "delegation system", we call a *Playbook*—because it is based on the metaphor of a sports team's book of approved plays.

## 3  A Playbook Approach to Tasking Interfaces

Figure 1 presents our general architecture for delegation systems. The Playbook architecture has been described in prior publications (Miller, Pelican & Goldman 2000; Miller & Parasuraman submitted) and will be briefly summarized here.

The Playbook 'proper' consists of a User Interface (UI) and a constraint propagation planner known as the Mission Analysis Component (MAC) that communicate with each other and the operator via a Shared Task Model. The user gives instructions in the form of desired goals, tasks, partial plans or constraints, using the task structures of the shared task model. The MAC is an

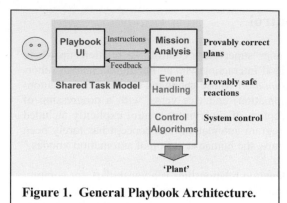

**Figure 1. General Playbook Architecture.**

automated planning system that understands these instructions and (a) evaluates them for feasibility and/or (b) expands them into fully executable plans. Outside of the delegation system, but essential to its use, are two additional components. An Event Handling component, itself a discrete control system capable of making momentary adjustments during execution, takes plans from the Playbook. These instructions are sent to control algorithms that effect the behaviours.

# 4   Usage Scenario

In work performed while the author were at Honeywell Laboratories, the Playbook concept described above was partially implemented for mission planning for Unmanned Combat Air Vehicles (UCAVs). To date, we have concentrated on a ground-based delegation to be used for a priori mission planning, but we intend this prototype as an illustration of the general concept and believe that, with appropriate modifications, this approach will be suited to in-flight delegation as well.

Figure 2 shows the five primary regions of the Playbook UI. The upper half of the screen is a Mission Composition Space that shows the plan composed thus far. The lower left corner is an Available Resource Space, currently presenting the aircraft available for use. The lower right contains an interactive Terrain Map, used to facilitate interactions with significant geographic content. The lower middle space is for Resources in Use—as resources (e.g., UCAVs, munitions, etc.) are selected for use, they are moved to this workspace, where they can be interacted with in more detail. Finally, the lower control buttons is always present. This includes options such as "Finish Plan" for handing the partial plan off to the MAC for completion and/or review and "Show Schedule" for obtaining a Gantt chart timeline of the activities planned for each actor, etc.

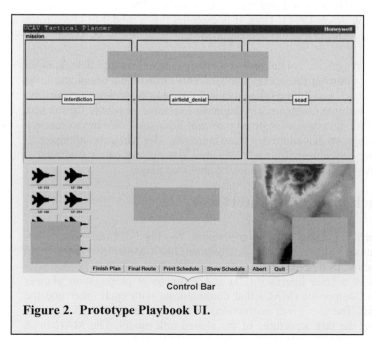

**Figure 2. Prototype Playbook UI.**

At startup, the Mission Composition Space presents the three top-level plays (or 'mission types') the system currently knows about: Interdiction, Airfield Denial, and Suppress Enemy Air Defences (SEAD). The mission leader would interact with the playbook to, first, declare that the overall mission "play" for the day was, say, "Airfield Denial." In principle, the user could define a new top-level play either by reference to existing play structures or completely from scratch, but this capability has not been implemented yet.

Clicking on "Airfield Denial" produces pop-up menu options to tell the MAC to "Plan this Task" (that is, develop a plan to accomplish it) or indicate that s/he will `Choose airfield denial' as a task that to flesh out further. The menu also contains a context-sensitive list of optional subtasks that the operator can choose to include under this task. This list is generated by the MAC with reference to the existing play structures in the play library, filtered for current feasibility.

At this point, having been told only that the task for the day is "Airfield Denial," a team of trained pilots would have a very good general picture of the mission they would fly. Similarly, the delegation interface (via the Shared Task Model) knows that a typical airfield denial plan consists of ingress, attack and egress phases and may also contain a suppress air defence task before or in parallel with attack. But just as a leader instructing a human flight team could not leave the delegation instructions at a simple 'Let's do an Airfield Denial mission today,' so the operator of the tasking interface is required to provide a bit more information. The supervisor must provide four additional items: a target, a homebase, a staging and a rendezvous point. Most of these activities are geographical in nature and users typically find it easier to perform them with reference to a terrain map. Hence, by selecting any of them from the pop up menu, the user enables direct interaction with the Terrain Map to designate an appropriate point. Since the Playbook knows what task and parameter the point is meant to indicate, appropriate semantics are preserved between user and system. As for all plans, the specific aircraft to be used may be selected by the user or left to the MAC. If the user wishes to make the selection, s/he views available aircraft in the Available Resource Space and chooses them by clicking and moving them to the Resources in Use Area.

A mission leader working with human pilots could, if time, mission complexity or degree of trust made it desirable, delegate the mission planning task at this point. The playbook operator can do this as well, enabling the MAC via the "Finish Plan" button. The leader might wish, however, to provide more detailed delegation instructions. S/he can do this by progressively interacting with the playbook UI to provide deeper layers of task selection, or to impose constraints or stipulations on the resources to be used, waypoints to be flown, etc.

To provide detailed instructions on Ingress, the user must select it, producing a "generic" Ingress template or "play". This is not a default method of doing Ingress but rather an uninstantiated template, corresponding to what a human expert knows about what constitutes an Ingress task and how it can or should be performed—i.e., that it can be done in formation or dispersed mode, that it must involve a "Take Off" subtask followed by "Fly to Location" subtasks, etc. Similarly, the operator can select from available options via context-sensitive, MAC-generated menus appropriate to each level of task decomposition.

The user can continue to specify and instantiate tasks down to the "primitive" level where the subtasks are behaviours the control algorithms (see Figure 1) can reliably execute. Alternatively, at any point after the initial selection of the top level mission task and its required parameters, the operator can hand the partly developed plan over to the MAC for completion and/or review. In extreme cases, a viable "Airfield Denial" plan could be created in our prototype with as few as five selections and more sophisticated planning capabilities could reduce this number further. If the MAC is incapable of developing a viable plan within the constraints imposed, (e.g., if the user has stipulated distant targets that exceed aircraft fuel supplies) it will inform the user of these.

# 5  Implications and Future Work

The Playbook described above enables a human supervisor to use varying levels of automation for the task of planning a UCAV mission similarly to delegating to a knowledgeable subordinate. To date, we have merely demonstrated the feasibility of this approach. Much work remains to be done. Specifically, playbook prototypes need to be evaluated to ascertain their usability and performance benefits. We believe significant payoffs will result especially because, by providing true supervisory control with all the flexibility of human supervisory roles, we should enable and encourage enhanced human engagement with automation in task performance. This in turn should provide (1) improved situation awareness relative interfaces with fixed levels of automation, (2) improved workload relative to interfaces providing less automation support, and (3) improved overall human + machine performance relative to either human alone or machine alone. Delegation is an extraordinarily powerful form of human interaction and, as such, has been adopted by military, government, and commercial organizations—not to mention sport teams—throughout history. But just as a poorly trained teammate can require more work than doing the job oneself, so a poorly designed delegation interface may make the task more difficult. We look forward to exploring, tuning and evaluating the delegation interaction format in future projects. Additional Playbook work in unmanned vehicle domains is currently ongoing with funding from the DARPA Information Exploitation Office's (IXO) Mixed Initiative Control of Automa-teams program and will be commencing soon with a DARPA-IXO Phase II Small Business Innovative Research Grant on control of Organic Air Vehicles.

## References

Billings, C. (1997). *Aviation Automation*. Mahwah, NJ: Lawrence Erlbaum.

Jones, P. & Jasek, C. (1997). Intelligent Support for Activity Management (ISAM). *IEEE Trans. on Sys. Man & Cyber.—Pt. A: Sys & Hum*, 27, 274-288.

Klein, G. (1998). *Sources of Power; How People Make Decisions.* Cambridge, MA; MIT Press.

Miller, C. & Parasuraman, P. (submitted). 'Tasking' Interfaces for Flexible Interaction with Automation: Playbooks for Supervisory Control. Submitted to *IEEE Transactions on Systems, Man and Cybernetics; Part A: Systems and Humans.*

Miller, C., Goldman, R. & Pelican, M., 'Tasking' Interfaces for Flexible Interaction with Automation: Keeping the Operator in Control. In *Proceedings of the Conference on Human Interaction with Complex Systems.* Urbana-Champaign, Ill. May, 2000.

Parasuraman, R. & Riley, V. (1997). Humans and automation: Use, misuse, disuse, abuse. *Human Factors, 39,* 230-253.

Reason, J. (1997). *Managing the risks of organizational accidents.* Aldershot, UK: Ashgate.

Sheridan, T. Supervisory Control. In G. Salvendy, (Ed.), *Handbook of Human Factors* (pp. 1244-1268). New York: John Wiley & Sons.

# Designing for Proficient Users:
## Drawing from the Realities of Practice

*Dimitris Nathanael*        *Nicolas Marmaras*        *Bill Papantoniou*

National Technical University of Athens, School of Mechanical Engineering
GR- 15780, Zografou, GREECE
dnathan@central.ntua.gr   marmaras@central.ntua.gr   billpapa2001@yahoo.com

## Abstract

The phenomenological perspective is discussed as a means to obtain a better understanding of how experienced computer users shape and are being shaped by the artefacts they use. The concept of embodiment is introduced to describe the combined mind-body skill of situational discrimination and seamless immediate action in the world. The importance of this concept is demonstrated through examples from real world practice. We stress that bringing forth embodiment is a necessary step in understanding proficient interaction, thus essential in design situations of new versions of professional software, new modules for existing applications or genuinely new applications that are developed for specific worlds of practice.

## 1   Introduction

New challenges for the field of Human-Computer Interaction arise as more and more contemporary applications come as replacements or enhancements to older ones. As a consequence, when designing new applications, designers need to respect existing practices as their future users have already embodied them as a kind of language or sign system that is applied across various applications. Moreover, when designing a computer application for a particular work setting, it is vital to consider how the applications previously used, have profoundly shaped the way the community of practice think - act upon the object of their work (Nardi, 1996). Also, the study of how these applications are actually being used (or shaped) as tools by their users should provide valuable insights on the real requirements for a new application. People get shaped by the technology they use as much as they shape this technology (Coyne, 1995, 1999).

Traditional HCI research and the derived design methods and guidelines have evolved under the reductionist paradigm, which is based on the dual split between body and mind and between an individual and the external environment. For example, human behaviour is analysed rationally, by identification of "the underlying cognitive schemes" it is based upon. This representationalist - symbolic view has given good service in providing structured and coherent models of human behaviour. However, most successful applications of this view, involve novel activities, rationally structured domains and/or "generic human subjects", where reflective reasoning dominates and meaning/ontology generation can be sufficiently restrained by the analyst. But although these models offer predictive power in controlled environments, their general and detached character, make them inappropriate for coping with the richness and immediacy of the real world. This is most evident if one considers the frequency of encounter of the term *"contextual factors"* to accommodate for all the shortcomings of applying such models in real world settings.

In the present paper, we discuss the relevance of the phenomenological perspective through the notion of embodiment as a means to obtain a better understanding of how experienced users interact with software artefacts. As people appropriate a software environment they tent to use it with remarkable creativity, some times in ways that were not anticipated by their designers (catachresis). We advocate that the acknowledgment of this appropriation process and the identification of the deviant creative uses are becoming essential for successful software design.

## 2    Body-mind-world as a whole

### 2.1    Perceiving –acting
The habituated scanning of my eyes and the movement of my hand when trying to accomplish an action in a specific computer application are important. Peripheral sight and hand trajectory over the mouse in order to print a file are important. It forms an integral part of *knowing* how to print a file. Hearing of the floppy disk noise is important because it always happens when saving a file. Response lags on a specific PC or while at a specific Internet site are important. Actually I usually neither measure them nor think about them, I just coordinate my hand and eyes movement with these lags to perform my actions with the least tension. All the above bodily patterns are an integral part of coping with the world that the body "learns".

### 2.2    Acting-thinking
Even when confronted with a new software tool, and under the condition of an existing motivation to do something more or less specific through it, I have expectations. These expectations are, to a large extent, the result of previous experience with all kinds of software (and other artefacts) I have used before. Expectations are manifested through the pre-reflective *"I know I should be able to do this"* which many times (if a breakdown occurs) may be reflectively transformed to *"I know it should be somewhere"* referring to some familiar type of software routine or command.

This pre-reflective knowledge, the *"I know I should be able to do this"* is rooted in everyday experience, it is primarily a feeling. When people are forced to explain why they think it can be done, i.e., to reflect upon their feeling, they may reason using some folk knowledge of software engineering or point to an analogous (for them) routine or behaviour of another software tool.
As Mingers (2001) suggests, much of what we "know", in the sense that we are able to undertake particular actions and activities, is essentially tacit, habitual, and only partially open to our consciousness. Such knowledge is always learnt through practice and habituation. Pre-reflective knowledge such as the example above may end up right or wrong in a particular situation, but the important thing is the claim that it is useful for the purposes of design, to consider it as such.

This stand, towards the analysis of human behaviour as a whole composed of the body, world and cognition, is gaining acceptance over the last decade. It is more or less evident in cognitive psychology work on Distributed Cognition (Hutchins, 1995), Situated Cognition (Suchman), Activity Theory (Nardi 1996) and on Artificial Intelligence through the work of Winograd & Flores (1987) and others.

### 2.3    Embodiment
The notion of embodiment can be defined as the mind-body skill of situational discrimination and seamless immediate action. It draws from the phenomenological tradition of European philosophy and particularly from the work of M. Heidegger with his contemplations on the nature of concern-full every-day activity, and in a more radical way from the work of Merleau-Ponty (Dreyfus 1991,

1996). According to Merleau-Ponty, in everyday, absorbed, skilful coping, acting is experienced as a steady flow of skilful activity in response to one's sense of the situation. Accordingly, human behaviour can neither be explained in a behaviourist way in terms of external causes, nor internally in terms of conscious intentionality. Rather, it has to be explained structurally in terms of the physical structures of the body and nervous system as they develop in a circular interplay *within the world*. The world does not determine our perception, nor does our perception constitute the world. As Merleau-Ponty (sited by Dreyfus, 1996) puts it *"The relations between the organism and his milieu are not relations of linear causality but of circular causality"*.

The above suggests that the representationalist consideration of *context* as a noise factor that disturbs some default rational ontologies of the human mind is misleading; context is instrumental in building interpretations of ones actions. The proficient user of a computer application is, to a large extent, immersed in the world of his skilful activity, and just sees what needs to be done based on mature and practiced situational discrimination. Also as expertise builds up the body "knows" how to achieve the goal (Dreyfus 1996). Embodiment develops through engaged action regardless of any explicit effort to support it.

## 2.4   Embodiment in HCI

In HCI the notion of embodiment is present in a number of current HCI theoretical and hands-on research efforts (Dourish, 2001). In most of them embodiment is often reduced to efforts to enhance the interaction through more "natural" manipulation. That is, an effort to try to mimic a target physical analogy. See for example the Bishop marble answering machine (Dourish 2001) or tilting and squeezing a palmtop (Fishkin et all 2000). Thus, the so entitled "embodied interfaces" are tangible interfaces, built with an explicit aim to support *readiness-to-hand*.

Tangible interfaces offer increased interaction means and materiality of IT technology. But although multimodality and materiality of the interface seem to support the development of transparent action, this lies more on the concrete history of skilful/concerned activity of the individual than on any surface feature. Take as an example the skill of writing text on the cell phone. Although the cell phone's interface was not designed for writing text and is generally regarded as bad design for this task, certain individuals communicate through SMS messages with an impressive speed and ease, and without even rupturing their involvement in another social activity. Embodiment develops primarily through engaged action, regardless of any explicit effort to support it.

The same can be observed with expert computer users working on command line interfaces. From a phenomenological perspective such individuals neither type commands nor read outputs from the computer screen. The object of their concern lays elsewhere; it is mirrored in spontaneous verbalisations such as *"I am trying to locate this file"* or *"there is some dll conflict"*. The speed and seamlessness of interaction with the operating system, in conjunction with simultaneous statements such as the above is an indication that the interface has been embodied to a large extent.

Embodiment as pre-reflective flawless action extends well beyond the physically observable. It tends to dissolve from reflective thinking even abstract concepts and metaphors. An expert AutoCAD draftsman, when designing, views the world in layers. While immersed in that world, he doesn't *reason in terms of*, but *acts upon* layers with his eyes and hands (Goel 1995).

## 2.5    Embodiment goes beyond familiarization

It is a trivial observation that many work domains that were early adopters of information technology (i.e. accounting, banking) still work on outdated often command based environments. The persistence of such software should not be solely understood in terms of an acquired familiarity with a particular interface. People nourished in such domains inevitably "see and act upon their work" through their embodied understanding of such artefacts undivided from the "semantics of their work". "Ctrl - F5" has particular meaning for a bank teller community, it is integral to their praxis.

Consider a particular case from banking. A small regional bank, in order to start a dedicated loans collection unit, employed a loans overdue collection expert that previously worked in a big multinational bank. Although the person had more than 10 years experience in loans overdue collection, she had tremendous difficulty in transferring this expertise to the new setting. While she was working with a management consultancy in order to set up standard procedures for the new unit, she kept giving what seemed "irrational explanations" for many parts of the process. It was latter realised that she was almost unable to differentiate between the "loans overdue collection task" *per se* and the use of the information system that she had been using as a medium for this work in her previous position in the multinational bank. Her meaning making had evolved inseparably from the particular information system. The above may be an extreme case, but this type of phenomena exists almost in every engaged activity.

The above examples point to the need to acknowledge that people progressively create meaning through engaged action in the world. Nowadays, more and more people have already established authentic ecologies containing particular software tools. They should not be viewed as mere humans but as experienced workers with an embodied understanding of their activities. In such cases, new designs should explicitly consider and support this embodiment. A good example illustrating acknowledgement of embodiment is MS Excel™ which, for many successive versions, provided an alternative Lotus 123™ interface (the first widely adopted electronic spreadsheet) or the preservation of the old shortcut keys in new graphical interfaces for bank tellers.

Through appropriation of particular software environments, people progressively pass to a stage where they perceive and exploit opportunities. They tend to use software features with remarkable creativity in ways for which they are not overtly designed. *Readiness-to-hand* opens a whole new *worldview*. In this sense, a search engine may become an ecological phrase checker, just by looking at the number of occurrences in the WWW, or a folder name on the desktop, may become a phone number memo, just by right click and type (Papantoniou, Nathanael & Marmaras, 2002). These deviant uses of software that are to a large extent are grounded to embodiment may be seen as *catachreses* (Béguin & Rabardel, 2000). Catachreses do not necessarily imply a misinterpretation by the designer of users needs. As we have pointed above, people are highly creative and the contexts they may find themselves in are infinite. Catachreses may well point to the inescapable co-evolution of people and artefacts. Identifying and explicitly considering them in may well provide fresh opportunities for improved design.

## 3    Epilogue

In this paper we advocate that successful design for proficient users should acknowledge embodiment and deviant creative uses of IT artefacts. The discussion and the examples provided suggest that universal-principle-based HCI design, detached from the historically evolved human practice, is insufficient for the support of every day skilful activity.

For the analyst, trying to bring forth embodied interaction requires involvement since it is experiential to a large extent. We are not yet in a position to communicate in a systematic way how one recognizes embodied features of specific worldviews or practices. Spontaneous verbalisations, catachreses, software logs seemingly irrational habits etc., provide valuable hinds. Further work needs to be done towards this direction.

In is important to not that, as the realities of software design demonstrate, embodiment is actually respected to a large extent. It is more so for widely used historically evolved applications and on the WWW as manifested in by the trend towards uniformity of Interaction platforms and the implicit or explicit "mimic" practice inside the design community.

## References

Béguin, P. & Rabardel, P. (2000). Designing for instrument mediated activity. *Scandinavian Journal of Information Systems*, vol.12.

Clark, A. (1997). Being there: Putting brain, body and world together again. Cambridge MA: MIT Press

Coyne, R. D. (1995): Designing Information Technology in the Postmodern Age., Cambridge MA: The MIT Press.

Coyne R. D. (1999). Technoromanticism: digital narrative, holism, and the romance of the real. Cambridge MA: The MIT Press.

Dreyfus, H. L. (1996). The current relevance of Merleau-Ponty's Phenomenology of Embodiment. *The Electronic Journal of Analytic Philosophy*, Issue 4.

Dreyfuss, H. L. (1991). Being in the World. Cambridge MA: The MIT Press.

Dourish, P. (2001). Where the Action Is: The Foundations of Embodied Interaction. Cambridge MA: MIT Press,

Fishkin, K., Gujar, A., Harrison, B., Moran, T. & Want, R. (2000). Embodied user interfaces for really direct manipulation. *Communications of the ACM*, 43 (9).

Goel, V. (1995). Sketches of thought. Cambridge MA: The MIT Press

Hutchins, E. L. (1995). Cognition in the Wild. Cambridge MA: The MIT Press.

Mingers, J. (2001). Embodying information systems: the contribution of phenomenology. *Information and Organization,* 11 (2), pp. 103-128.

Nardi, B. (1996). Context and Consciousness: Activity Theory and Human-Computer Interaction. Cambridge MA: The MIT Press.

Papantoniou, B., Nathanael, D., Marmaras, N. (2003). Moving Target: Designing for Evolving Practice. *HCI International 2003* (accepted for presentation/publication in proceedings).

Suchnan, L. (1987). Plans and situated actions: the problem of human/machine communication. Cambridge: Cambridge University Press.

Winograd, T. & Flores, F. (1987). Understanding Computers and Cognition. NY: Addison-Wesley.

# Mindtape – a Technique in Verbal Protocol Analysis

*Janni Nielsen, Nina Christiansen, Torkil Clemmensen, Carsten Yssing*

Copenhagen Business School, Department of Informatics
60 Howitzvej, DK - 2000 Frederiksberg, DENMARK
{janni.nielsen, nc.inf, tc.inf, cy.inf}@cbs.dk

## Abstract

This paper reports on the explorative use of video recordings in studies of distributed collaboration. The primary goal for the analyst is to acquire a better understanding of communication and interaction in CSCW. A technique called Mindtape is explored, the essence of which is to review a priori identified video sequences in a dialogue with the users. It is suggested that in the search for techniques, which may bring us closer to what goes on in people's mind, Mindtape brings us beyond the limitations of thinking aloud, and beyond the "experts as analysts approach". Mindtape allows research and design to cross over from objectification of the user to participation of the user, and brings us one step closer to knowing the workings of another mind.

## 1    Introduction

Video is used extensively in HCI work and has been applied in studies of control room work and as a tool for reflection (Suchmann & Trigg, 1991) to inform participatory design workshops and to serve as a basis for developing scenarios (Nielsen and Christiansen, 1996). The methods of quantitative and qualitative analysis of video recordings have been discussed (Kensing et al., 1997) and a framework for the analysis of complex work and learning situations has been developed (Jordan & Henderson, 1994). Different forms of notation in the analysis have been demonstrated, but it has been argued that there is a need for rethinking video analysis (Koschmann et al., 1998; Nielsen et al, 2001b). One investigated approach is to review videotapes together with the users. This allows research to cross over from objectification of the user to participation of the user, an approach to usability work that falls within the frames of Scandinavian Systems Development. In Denmark there is a tradition for dialogue with users in the design process (Nielsen et al., 2001a; Bagger & Buur, 1999) and video is used extensively (Dirckinck-Holmfeld, 1997). The Mindtape technique that we discuss in this paper falls within this tradition, but has its origin in psychology and hermeneutic phenomenology - not in computer science.

## 2    Mindtape

Reviewing videotapes together with users has been suggested as a way to acquire insight into the interaction (Koschmann et al., 1998; Jacobsen et al., 1998), to come closer to an understanding of what goes on in the users' minds and to test the validity of the researcher's interpretation. In our research, we have found that reviewing video together with users seems to serve as a mental trigger for their recall. When users hear and see the dialogue unfolding between themselves and their colleagues, their memory is triggered in a very special way. They seem to recall in extreme detail what they did, why, what they expected to happen, what they thought when a visual image appeared on the screen, why they juxtaposed another image, etc. They seem capable of making their internal thought processes explicit, almost as if a "Mindtape" of their tacit inferences is being replayed.

# 3    The Technique

The primary goal with the Mindtape technique was to acquire a better understanding of the communication and interaction between geographically distributed researchers collaborating through use of a CSCW system. Video was used to record the collaboration, capturing actions and verbal expressions at three different sites at the same time. The video camera was set to capture the user from the side, with part of the facial expression and upper torso, the screen, the space around the computer (sometimes capturing other researchers who were hanging around) and part of the office. It is the analysis of these videotapes that the Mindtape technique was developed for.

The Mindtape technique inscribes itself into the tradition of video analysis but moves beyond this by inviting the user to become participant in the analysis of the tapes. The steps in this group analysis are: First, view the whole tape with no interruption. Each analyst writes a small summary. In the second step, the tape is divided into ten-minute sequences or into events (Jordan & Henderson 1994). Again each analyst writes a small summary. Based on the group discussion, sequences from the tape are then identified for closer analysis. In praxis this identification means making verbatim transcripts of dialogue, descriptions of actions, etc., on basis of a schema developed for this purpose. We also note things that capture our mind's eye, but do not belong in any analytical categories identified a priori[1].

From our initial analysis, it was clear that the recordings of the actions and the voices speaking, though providing us with rich data, were not sufficient. Much more took place and our analysis often reached an impasse. E.g. somebody would enter the physical office at site X and interrupt the participant. On the VDUs at site Y, Z and W, the other participants would not know that something was taking place at X, though the event would be expressed through complete silence in dialogue and action from site X. Or participants would be engaged in a discussion and suddenly one participant at one site would move into action: He would get up from his chair, search for something in the book shelves, come back and sit down in front of his computer, start reading in the paper found, being very concentrated for quite a while, then start moving objects in a window, open a new window, integrate the figure from his private file into the shared work file - without saying a word. Only the shared work file on the screen was visible to the others – yet they asked no questions, but just accepted the change and continued their discussion. We needed to understand what was happening. What went on in the users' minds? Why did he do that, what was he thinking? Why did the others not comment when the shared document was changed. How did it influence the way, they proceeded? Did they perceive that though he was back in front of the screen and camera, he was not mentally present in virtual space? (Christiansen et al. 2003). Here the need for Mindtape sessions occurred. The actual participant is invited to watch the tape together with one of the analysts. A priori selected tape sequences are shown and a dialogue unfolds between the two; new information is brought into the light and interpretations may be confirmed or completely refuted.

---

[1] These are things which once seen, cannot be unseen. They are things which shoot out of the videotape and hit us – and stay with us, though we do not always understand why. "Punctum" is what Roland Barthes (1983) calls this dimension in his analysis of the "eigenart" of the photography. We register them because we know they influence our understanding and we use the registration to reflect critically on our analysis.

To get closer to what went on in the users´ minds in this kind of work, we obviously could not rely of the traditional HCI test: The Thinking Aloud technique, simply because the users cannot Think Aloud while they are engaged in a dialogue. However, the Thinking Aloud classification did seem to offer a foundation for our approach.

# 4    Thinking Aloud

Ericsson and Simon argue that to understand people's actions, even in very simple tasks, some kind of verbal reporting is necessary (1984, 1998). Their understanding is that a sentence is the verbal realisation of an idea, and verbs in a sentence can be used to identify different kinds of information and different cognitive processes. Their aim was to study task directed cognitive processes. They argued that if the participant performed a given task, only concurrent and certain kinds of retrospective verbalising would address the information employed. Their interest was to identify and analyse these verbalisations and they distinguished between three kinds:

- Level 1: Vocalisations of thoughts that are already encoded in the verbal form (*talk aloud*).
- Level 2: Verbalisation of a sequence of thought that are held in memory in some other form, e.g. visually (*think aloud*).
- Level 3: Other verbalisations (*retrospective reports on thoughts not held in memory*)

Thinking aloud occurs at level 1 and level 2: Vocalisations and verbalisations, where thinking can be verbalized without altering its sequence. However at Level 3 the subject does not Think Aloud, but describes and explains her/is thoughts, and hereby alters the sequence of thoughts (Ericsson & Simon 1998) [2]. To obtain TA protocols of the subjects' thoughts: 'The single most important precondition for successful direct expression of thinking is that the participants are allowed to maintain undisrupted focus on the completion of the presented tasks' (Ericsson & Simon 1998, p 181). But what is level 3 data? Ericsson and Simon suggest that participants, while describing and explaining their thinking, use and develop a framework of ideas and perceptually available objects to make their often complex thoughts understandable to others. The verbalization of level 3 data therefore includes (Ericsson 1998):

- Monitoring of own speech to ensure that it is understandable.
- Correction and explications of not understandable verbalizations.
- The use of more orderly strategies that are more easy to communicate

For example, during a CSCW session (video recorded) participant 'A' follows a presentation on the screen delivered by 'B', located in another country. The presentation obviously puts a heavy cognitive load on 'A'. He nods and says 'okay' repeatedly while he carefully lets his cursor follow 'B' s explanations. A third person 'C' is present in the room with 'A'. We can not know from the video observation and transcripts if A' is talking aloud, thinking aloud, or if 'A' is enlightening his less knowledgeable colleague 'C', by talking and by nodding and pointing to the graphs shown. The question is whether the "Okays" are to be categorised as concurrent verbal report of his understanding of what goes on in the presentation, or whether the verbal report and cursor actions

---

[2] With Mindtape we deliberately focus on retrospective reports – but at the same time, the sequence of thoughts are not organised by the user's memory, but structured and led by the sequence of events taking place on the tapes.

must be understood as "socially directed verbalizations", generated to communicate explanation for the benefit of 'C' (level 3 data). A Mindtape session with 'A' followed:

> Analyst: "Can you remember ..[]..there are some graphs on the screen...", (the analyst points towards the video where Participant A nods, says 'Okay' and lets his cursor follow the presenter's ('B') explanation.
> Participant A: "No, I cannot".
> The analyst plays the video sequence again.: "Is this ('A's explanations) for your own sake or for (participant C's) sake?"
> Participant A: (silent for a moment) "Oehh...( )...this is really funny...I didn't think about that".
> Analyst: "?" (asks again).
> Participant A: "This is just one of those things that you do...I don't believe that I thought about it" ('A' goes on and explains possible reasons to why he reacts in the way he does to 'B's presentation).

The retrospective report from the Mindtape session does not allow us to make inferences of the actual content in short term memory, at the time of the first session. But the replay does trigger a vivid memory recall in participant A, which enables participant and analyst to learn and understand. Hence Mindtape session provide opportunities for learning for all involved.

The study of level 3 verbalization in everyday work is therefore an important way to get insight into the critical thought processes that people use to learn individually and collectively (Ericsson 1998). With this distinction between thinking aloud and socially directed descriptions and explanations we can explain why the Mindtape technique may give us access to retrospective reports – without altering the sequence of thoughts.

# 5    Conclusion: Returning to Mindtape

Our initial experiences with the Mindtape technique seemed promising because the processes of insight that runs associatively while the user interacts with a computer system and other participants may become partly explicit during the reviewing session. The video images serve as triggers. When the events are shown on the tapes the user seem to recall in detail the thought processes that took place during the interaction, and the explication flows easy *with the actual sequence of events structuring the recalls*. It is not the user's memory that structures the recall hence the verbal reporting, but the actual sequence of events which took place.

Mindtape focuses on thoughts not currently held in memory – but thoughts which took place some time ago. Mindtape gives us access to level 3 data, the retrospective reports, understood as 'socially directed' verbalizations. However, Mindtape has procedures that we believe minimize the reactive effects of verbalisation of the sequence of thoughts:

- Video images serves as triggers for correction and explication of not understandable verbalizations
- Video clips preserve the sequence of thoughts otherwise altered by the subjects descriptions and explanation
- Instructions to subjects are always directed towards the video, e.g. "what you see on the video", "what did you do here". Hence the analyst stay in the background and the subject has a dialogue with him or her self about the performance showed on the video.

Furthermore, these procedures allow the Mindtape technique to produce Thinking Aloud – like data on types of performancec that can not be caught with ordinary TA procedures. As argued, in most CSCW and CMC situations the participants are certainly *not* allowed to maintain undisrupted focus on the completion of the presented tasks. Thus we cannot produce thinking aloud protocols with the procedures suggested in the classic approach, but must instead use Mindtapes. But also in individual testing, Mindtape may be an asset. The technique helps us get beyond some of the limitations of the Think Aloud method; e.g. when users complain that they think faster than they can speak, that their thought processes are much more complex than they can verbalise, and their Thinking Aloud interferes with their interaction with the interfaces and the task. Besides, Thinking Aloud does not come naturally to most people and other techniques should be explored within the HCI community to improve quality of testing.

## References

Bagger, K. & Buur, J., (1999). Replacing usability Testing with User Dialogue in *Communications of the ACM, May vol. 42, no. 5, p 63-67.*

Barthes, R. (1983). *Le Chambre Claire*. Politisk Revy.

Christiansen Nina & Kelly Maglaughlin (2003) Crossing from Physical Workplace to Virtual Workspace: be AWARE, accepted for *HCII2003, Crete Proceedings*

Dirckinck-Holmfeld, L. (1997) (ed). *Video observation*. Aalborg Universitets forlag.

Ericsson, K.A. & Simon, H.A. (1984) *Protocol Analysis, Verbal reports as data.* Cambridge, Massachusetts.

Ericsson, K. A. & Simon, H. A., (1998), How to study thinking in everyday life: Contrasting think-aloud protocols with descriptions and explanations of thinking. *Mind, Culture, & Activity, vol. 5, no. 3, pp. 178-186.*

Jacobsen, N. E., Hertzum M. & John, B.E., (1998). The Evaluator Effect in Usability Tests, *CHI 98, p. 255- 257.*

Jordan, B., Henderson A. (1994). *Interaction Analysis Foundation and Practice*, Xerox Palo Alto Research Center and Institute for Research on Learning, Palo Alto, California.

Kensing, F., J. Simonsen, K. Bødker, S. (1997). Designing for Cooperation at a Radio Station, In *Proceedings of ECSCW 1997, Lancaster, England, Kluwer Academic Publishers, 329-345.*

Koschmann T., Anderson A., Hall, R., Heath, C., LeBaron, C., Olson, J.& Suchmann, L. (1998). Six Reading of a Single Text: A Videoanalytic Session Panel, *Proceedings CSCW 98, Seattle.*

Nielsen Janni, Ørngreen R., Siggaard, S. J. & Christiansen, E. (2001a). Learning Happens - Rethinking video analysis, in Dirckinck-Holmfeld L. & B. Fibiger (eds) Learning in Virtual Environments, Samfundslitteraturen , Copenhagen, 310-341

Nielsen Janni, Dirckinck-Holmfeldt, L. & Danielsen, O. (2001b). Dialogue Design - with mutual learning as guiding principle. *Int. J. of Human Computer Interaction, 15, (1), 21-40.*

Nielsen, Janni, Clemmensen, T. & Yssing, C. (2002). Getting Access to what goes on in Peoples Head – Reflection on the think-aloud technique, *NORDICHI 2002 proceedings*, Aarhus.

Nielsen, Janni & Christiansen, N. (1996). Tacit knowledge in a distributed collaborative visualisation system. *Position paper for Workshop 8: Tacit knowledge, CSCW, Cambridge Mass. November 16-21.*

Suchmann, L. & Trigg, R. (1991). Understandings practice. Greenbaum, J. & Kyng, M. *Design at Work*. Lawrence Erlbaum.

# Development of GUI Design Consistency Auto-Scoring System

*Hidehiko Okada and Toshiyuki Asahi*

Internet Systems Research Laboratories, NEC Corporation
8916-47, Takayama-cho, Ikoma, Nara 630-0101, Japan
h-okada@cq.jp.nec.com, t-asahi@bx.jp.nec.com

## Abstract

This paper proposes a usability scoring method that auto-scores the design consistency of graphical user interfaces (GUIs). The essence of the proposed method is to use usability auto-checking results obtainable with a tool we previously developed. Because the checking results produced by the tool do not depend on the skills and experience of human evaluators, the proposed method can calculate more objective scores than other scoring methods. By using the tool, the number of consistent/inconsistent widgets in an application can be auto-counted. The method uses these widget counts. We describe our system employing the auto-scoring method and show example screenshots of score charts and tables produced by the system. The effectiveness of our system is confirmed in a case study in which the design consistency scores of five existing emailers were auto-calculated by the system.

## 1   Introduction

A method for quantitative usability scoring is very useful because it will enable quantitative usability comparisons among multiple systems, quantitative verification of UI redesigns/modifications in the iterative user interface (UI) design cycle, and quantitative usability goal setting in system development. Previous methods (Sugizaki, Kurosu & Matsuura, 1999; Wakizaka, Morimoto & Kurokawa, 2001; Yamada & Tsuchiya, 1996) calculate usability scores from usability evaluation results obtained by manual evaluation methods, such as the heuristic evaluation method (Nielsen, 1994). The evaluation results obtained by manual methods depend on the skills and experience of human evaluators. Thus, scores obtained by the previous methods will vary depending on the evaluators employed. We propose a scoring method that can calculate objective scores.

The essence of our method is to use auto-checking results obtainable by our tool "GUITESTER2" (Okada, Fukuzumi & Asahi, 1999; Okada & Asahi, 2002). The tool can auto-check graphical UI (GUI) design consistency within a single PC application. Because the results produced by GUITESTER2 do not depend on human evaluators, the method we propose in this paper can calculate scores more objectively than the previous methods.

## 2   GUI Design Consistency Auto-Checker "GUITESTER2"

Consistency is one of the usability heuristics - "users should not have to wonder whether different words, situations, or actions mean the same thing" (Nielsen, 1994). The auto-checking method of our tool is based on the following concept: Consistency means avoidance of unnecessary design variations. If most of the GUI widgets (e.g., buttons) in an application have the same design and the remaining small portion of widgets have different designs, those in the minority are possibly inconsistent with the majority. For example, suppose that most "OK" buttons in an application are located on the lower right area of windows and the remaining "OK" buttons are located in the upper right area. In this case, human evaluators will judge the majority (those located on the lower right area) as the basis for consistency and the minority (those located on the upper right area) as inconsistent with that basis. Thus, design consistency can be checked by 1) grouping widgets with the same design by design comparison, 2) counting the number of widgets for each widget group to find out which designs are major/minor in the application, and 3) judging which widgets are consistent/inconsistent based on the widget counts. Figure 1 illustrates this concept of our auto-checking method. Based on this concept, we developed consistency auto-checking functions in GUITESTER2. The tool can check consistency for ten criteria including terminology in widget labels, widget locations, widget sizes and key assignments for accessing functions/widgets.

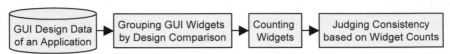

**Figure 1:** Concept of Consistency Auto-Checking Method by GUITESTER2

# 3   Design Consistency Auto-Scoring Method

We propose a method for auto-scoring GUI design consistency that uses the consistency auto-checking result produced by GUITESTER2. The method meets the following requirements because we consider such a score to be intuitive due to its analogy to a score in an academic test.

1.   The greater the number of inconsistent widgets is, the smaller the score is.
2.   The score ranges from 0 to 100.

GUITESTER2 judges the consistency of each widget as OK, NG or *unknown* - if some design variations exist and the number of widgets with each design is the same among the design variations, GUITESTER judges the widgets as unknown. As an example of such a case for key assignments, suppose 1) the same N widgets (e.g., "Help" buttons) are used in an application, 2) the assigned key for each of the N widgets is one of M different keys (e.g., H, L, P), and 3) the number of widgets for each of the M keys is the same (i.e., N/M). In this case, GUITESTER2 cannot decide which of the M keys is the basis for consistency because the tool judges major designs (i.e., designs whose numbers of widgets are relatively greater than those of others) as the bases for consistency. Thus, the consistency of key assignments for the N widgets is judged as unknown. In such a case, the consistency score should be smaller as more design variations exist (i.e., as M is greater). In addition, to modify the designs of those N widgets for consistency, UI designers will need to choose one of the M designs as the basis for consistency (e.g., the H key for the Help buttons) and modify the widgets that have other designs so that all N widgets have the same design. Therefore, N/M widgets out of the N widgets can be counted as OK and the other (N-N/M) widgets can be counted as NG.

Based on this consideration, the method proposed in this paper calculates consistency scores for the ten criteria. Let us denote the number of widgets judged as OK/NG for each criterion C[i] (i=1,2,...,10) as Num_OK[i] and Num_NG[i]. If widgets judged as unknown exist, those with one design are counted as OK and the other widgets are counted as NG. The proposed method calculates the consistency score S for the application by the following equation, where S[i] is the score for each criteria C[i], and w[i] is a weight corresponding to S[i]. In the simplest case where w[1]=w[2]=...=W[10]=0.1, S is equal to the average of S[1]-S[10].

S = w[1]*S[1] + w[2]*S[2] +...+ w[10]*S[10],
S[i] = 100*Num_OK[i]/(Num_OK[i] + Num_NG[i]), i=1,2,...,10,
w[1] + w[2] + ... + w[10] = 1.

# 4 Design Consistency Auto-Scoring System

This section describes our system for auto-scoring design consistency by the proposed method. Figure 2 shows the system configuration. The system calculates scores from the consistency auto-checking results produced by GUITESTER2. The calculated scores are saved in HTML files so that the scores can be easily browsed both locally and via the Web (usually the Intranet in an organization).

Figures 3-6 show example screenshots of consistency scores displayed in a Web browser. The chart in Figure 3 shows the total score S for each application. Such charts will help human evaluators compare the total scores among applications. "Application1"-"Application5" in Figure 3 denote application names (in actual use, application names are displayed instead of "Application1"-"Application5"). "Application1"-"Application5" in Figures 4-6 are the same. A human evaluator enters each application name when she/he makes GUITESTER2 auto-record the GUI design data of the application. The chart in Figure 4 shows consistency scores for widget sizes. Score charts for the other nine criteria are similar to that shown in Figure 4. Such charts will help human evaluators compare the scores of a specific criterion among applications. The chart in Figure 5 shows consistency scores for a single application (Application4 in this figure). Such charts will help human evaluators identify criteria for which scores are great/small in the

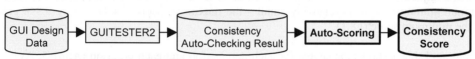

**Figure 2:** Configuration of Our Consistency Auto-Scoring System

**Consistency Total Score**

| Application1 | 74.4 |
| Application2 | 73.2 |
| Application3 | 71.9 |
| Application4 | 68.1 |
| Application5 | 71.7 |
| Average | 71.8 |

**Figure 3:** Example Screenshot of Total Score Chart for Each Application

**Consistency Score**

Criterion: Widget Sizes

| Application1 | 40.6 |
| Application2 | 62.7 |
| Application3 | 50.2 |
| Application4 | 57.2 |
| Application5 | 38.7 |
| Average | 49.5 |

**Figure 4:** Example Screenshot of Score Chart for the Same Criterion

**Consistency Score**

Application: Application4

| Combination of widgets on each window | 59.6 |
| Terminology in widget labels | 98.6 |
| Terminology in window titles and widget labels | 22.4 |
| Widget locations | 77.3 |
| Widget sizes | 57.2 |
| Window margins/margins between widgets | 39.6 |
| Key assignments for accessing functions/widgets | 97.4 |
| Usage of message windows | 40.6 |
| Methods for indicating selected options/currently-focused widgets | 98.7 |
| Rules for controlling user operations | 90.3 |
| Total | 68.1 |

**Figure 5:** Example Screenshot of Score Chart for a Single Application

**Consistency Score**

Application: Application4

| | OK | NG | Score |
|---|---|---|---|
| Combination of widgets on each window | 31 | 21 | 59.6 |
| Terminology in widget labels | 2021 | 28 | 98.6 |
| Terminology in window titles and widget labels | 22 | 76 | 22.4 |
| Widget locations | 215 | 63 | 77.3 |
| Widget sizes | 724 | 542 | 57.2 |
| Window margins/margins between widgets | 42 | 64 | 39.6 |
| Key assignments for accessing functions/widgets | 1336 | 36 | 97.4 |
| Usage of message windows | 41 | 60 | 40.6 |
| Methods for indicating selected options/currently-focused widgets | 152 | 2 | 98.7 |
| Rules for controlling user operations | 917 | 99 | 90.3 |
| Total | 5501 | 991 | 68.1 |

**Figure 6:** Example Screenshot of Score Table for a Single Application

application. The table in Figure 6 shows the scores for a single application (also Application4 in this figure) and the number of OK/NG widgets in the application for each criterion (i.e., Num_OK[i] and Num_NG[i]). Such tables will help human evaluators determine how many OK/NG widgets are included in each application. In actual web browser screens, human evaluators can jump from one of the charts/table to another by clicking on hyperlinks (the hyperlink area is not shown in Figures 3-6 to save space). For example, human evaluators can jump from the chart in Figure 3 to one of the charts/table in Figures 4-6 by clicking on hyperlinks located below the chart.

# 5 Case Study

As a case study, we selected five existing emailers as GUI applications to be scored. This is because 1) score comparison among applications in the same domain is the main intended purpose of this system, 2) emailers are widely used and many kinds of emailers are available, and 3) the five selected emailers are major ones from a user share viewpoint.

For each of the five emailers, GUITESTER2 auto-recorded the GUI design data and auto-checked the design consistency. Consistency scores were auto-calculated from the results obtained by GUITESTER2. Figures 3-6 are the charts/table obtained in this case study. Figure 3 establishes that the total score for each of the five emailers is around 70 and the difference between the maximum/minimum score is relatively small - the maximum score (74.4) is only 9.3% greater than the minimum score (68.1).

In addition, Table 1 shows the scores for each application and each criterion (App1-5, Ave., and SD denote Application1-5, average, and standard deviation, respectively). From this table, we found the following for the five emailers.

A) Although the total score range in the five emailers is small, the score ranges are relatively great for the following three criteria. (The values in ( ) are the standard deviations for each criterion.)

- Combination of widgets on each window (31.2)
- Widget sizes (20.7)
- Usage of message windows (36.1)

That is, the superiority of each emailer from the viewpoint of design consistency in the three criteria varies remarkably. For example, the score of Application5 for "Combination of widgets on

**Table 1:** Consistency Scores for Five Existing Emailers

| Criteria | App1 | App2 | App3 | App4 | App5 | Ave. | SD |
|---|---|---|---|---|---|---|---|
| Combination of widgets on each window | 68.3 | 55.6 | 39.6 | 59.6 | 81.2 | 63.0 | 31.2 |
| Terminology in widget labels | 98.9 | 99.4 | 98.7 | 98.6 | 98.1 | 98.7 | 0.95 |
| Terminology in window titles and widget labels | 32.6 | 32.7 | 30.2 | 22.4 | 31.4 | 29.5 | 8.62 |
| Widget locations | 87.3 | 76.8 | 82.7 | 77.3 | 76.1 | 80.7 | 9.77 |
| Widget sizes | 40.6 | 62.7 | 50.2 | 57.2 | 38.7 | 49.5 | 20.7 |
| Window margins/margins between widgets | 43.9 | 47.4 | 49.0 | 39.6 | 45.1 | 45.5 | 7.31 |
| Key assignments for accessing functions/widgets | 97.6 | 97.6 | 96.5 | 97.4 | 100.0 | 97.7 | 2.62 |
| Usage of message windows | 78.2 | 68.8 | 79.7 | 40.6 | 48.1 | 60.3 | 36.1 |
| Methods for indicating selected options/ currently-focused widgets | 99.0 | 96.0 | 96.2 | 98.7 | 98.4 | 97.8 | 2.90 |
| Rules for controlling user operations | 98.5 | 95.9 | 96.4 | 90.3 | 100.0 | 96.2 | 7.39 |
| Total (=Average) | 74.4 | 73.2 | 71.9 | 68.1 | 71.7 | 71.8 | 4.71 |

each window" is 205% greater than that of Application3 (81.2/39.6).

B) The average scores for the following three criteria are relatively small. (The values in ( ) are the average scores for each criterion.)

- Terminology in window titles and widget labels (29.5)
- Widget sizes (49.5)
- Window margins/margins between widgets (45.5)

We should further analyze to determine whether these findings are common, i.e., whether or not similar results are obtainable in the case of other applications.

## 6　Conclusion

In this paper, we proposed a design consistency auto-scoring method for GUI applications and reported on our system employing the method. The essence of the method is to calculate scores from the auto-checking results produced by our auto-checker. Because the consistency checking results produced by the tool do not depend on the skills and experience of human evaluators, the system can score consistency more objectively than previous methods. We further reported on a case study applying the system to five existing emailers.

This paper focuses on intra-application consistency (i.e., consistency within a single application). In addition to GUITESTER2, which auto-checks the intra-application consistency, we have developed an auto-checker for *inter*-application consistency, i.e., consistency of an application with other applications (Okada & Asahi, 2001). We plan to develop a method and system for auto-scoring inter-application consistency by using the auto-checker.

## References

Nielsen, J. (1994). Heuristic evaluation. In Nielsen, J. & Mack, R. L. (Ed.), *Usability inspection methods* (pp. 25-62). John Wiley & Sons.

Okada, H., Fukuzumi, S. & Asahi, T. (1999). GUITESTER2: an automatic consistency evaluation tool for graphical user interfaces. *Proc. INTERACT'99*, 519-526.

Okada, H. & Asahi, T. (2001). An automatic GUI design checking tool from the viewpoint of design standards. *Proc. INTERACT2001*, 504-511.

Okada, H. & Asahi, T. (2002). Effectiveness evaluation of GUI design consistency auto-checker GUITESTER2. *Journal of Human Interface Society*, 4(2), 35-42.

Sugizaki, M., Kurosu, M. & Matsuura, S. (1999). Attempt at absolute indexing of usability. *Proc. Human Interface Symposium'99*, 511-518 (in Japanese).

Wakizaka, Y., Morimoto, K. & Kurokawa, T. (2001). A proposal of absolute index of usability of information apparatus. *Proc. Human Interface Symposium 2001*, 425-428 (in Japanese).

Yamada, S. & Tsuchiya, K. (1996). A study of usability evaluation of PC - discussion of model on PCs. *Proc. 37th Conf. of Japan Ergonomics Society*, 350-351 (in Japanese).

# Automatic Generation of Interactive Systems: Why A Task Model is not Enough

*Philippe Palanque, Rémi Bastide, Marco Winckler*

LIIHS-IRIT Université Paul Sabatier
118, route de Narbonne, 31062 TOULOUSE cedex 4, France
[palanque, bastide, winckler]@irit.fr

## Abstract

In this paper we present a very simple example that shows that creativity must remains an explicit phase in the design process of interactive systems and the generation of interactive system from task models cannot lead to efficient applications. We try to show here that the generation of an interactive system from task models is not a valid model as either you need to add information in the task model in order to generate the interactive system from it and in that case this is not a task model anymore; or you can generate a basic interactive system that does not cover efficiently users activity. In order to exemplify that, we will present a study case for a Tic-Tac-Toe application that supports user's strategy in selecting the best number. We show and discuss that such strategic part of users' activity is not represented in the task model, as this kind of model is not meant to encompass this kind o information.

## 1 Introduction

When dealing with interactive systems it is widely agreed upon that user information has to be taken into account and that this must be done through task analysis and task modelling. During the design process of interactive systems, user goals have to be analysed as part of the specification phase, while task analysis is conducted during the design phase.

Classically formal notations are meant to guaranty certain quality of the models. For instance they are used to insure completeness or non-ambiguity of the descriptions. We try to demonstrate in the following that they can also be used to ensure consistency among the various models that are build in the various phases of the development process of an interactive system. As models built using formal notations are non-ambiguous they can be analysed automatically by inspection tools.

Even though formal notations are used for task modelling, it is not possible to generate, from the task models, the interactive application supporting those tasks. The justification of this claim comes from the following argument we develop hereafter: "it is impossible because there is not enough information in the task models to generate code". It is possible to extend the task models with additional information, but in that case we claim that the resulting model is not anymore a task model but a merging of both a task model and a model of the behaviour of the system. Another alternative is to use generic information about interactive application and to use them in order to generate the code of the final interactive system. In that case we claim that the application generated is very stereotyped and that this is only possible for a very small number of applications.

We take as an example the TRIDENT project (Bodart et al., 94) that was dedicated to the generation of form-based interactive applications from task models and data models. Even for this kind of "simple" interactive application a data model was mandatory and was actually the core of the generation process (the task model was only used for the dialogue part of the application and the structure of the various windows of the application). Instead of generating the code from a task model, we propose to use the task model as a mean for checking that the system model is compliant with it (Palanque & Bastide, 97).

## 2 Users' Tasks Descriptions

Tasks correspond to actions that are to be performed by a user in order to reach a goal. The relationships between tasks and goals are clear as described in Norman's theory of action (Norman, 86). Indeed, each time a task is performed the user has to compare the perceivable state of the system with respect to the goal.

Various notations are currently available for modelling tasks. They range from very abstract to concrete. The representation of an abstract task model requires a declarative abstract notation. The notation must be able to describe actions and qualitative temporal relationships among actions. A notation belonging to this category is Concur Task Tree (Lecerof & Paternò, 98). A concrete task model requires a procedural notation describing both quantitative and qualitative temporal relationships among actions but also to represent the data needed in order to perform the actions (see for instance (Palanque et al., 95)).

### 2.1 Example: The game of 15

We consider a simple (and somewhat silly) game that requires two players. The rule is as follows: The numbers from 1 to 9 are initially available to any of the two players. The players play in turn. At each turn, the player chooses one of the remaining numbers (thus making it unavailable). If the player possesses three numbers that add up to exactly 15, she wins the game. You might want to try this game with one of your colleagues, without the help of any external tool (paper, pencil or other). It turns out, rather unexpectedly, that this game is almost impossible to play this way, because the user task is extremely complicated, as we will show by constructing a task model for it. We will later show that, although you may not be aware of it, you have almost certainly played this game.

### 2.2 An Abstract Task Model

Both players share a common goal: "win the game", and sub-goal: "if I cannot win, don't let the other player win". The task of someone playing the game of 15 can be described as follows:
- I wait for my turn.
- When my turn comes, I choose a number among the remaining numbers (here, I must apply some form of strategy to decide which number to take). Within this activity the player might also check whether or not the other player is about to win.
- I then evaluate if I win the game. This operation is especially difficult, because I need to compute all combinations of 3 numbers among the numbers I have already chosen, and decide if this combination adds up to 15.
- If I do not win, I let the other player take her turn.
- To accomplish this task properly, I need to perform two "background activities":
- Remember the numbers I have already taken,
- Remember the remaining numbers (or alternatively remember the numbers taken by my opponent),
- Remember who is next.

These background activities are very demanding to the short-term memory of the player, and contribute to making it actually impossible to play without adequate support.

## 2.3 A Concrete Task Model

A concrete task model for playing this game is presented in Figure 1. Of course, several simpler models could be built for instance including random selection of numbers. We have already considered these aspects in a previous paper (Moher et al., 96) and we focus our discussion here on the more efficient task models.

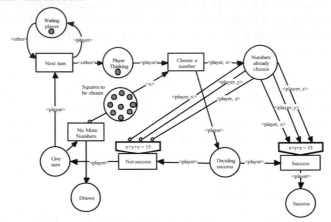

**Figure 1**: Petri net representation of the concrete task

In this Petri net model of tasks we can note that:
- Information is stored in places thus representing what the user has to remember,
- Actions are represented by transitions; as there is no support all the actions are to be performed directly by the user. Two transitions feature preconditions (namely transitions Not Success and Success), which means that the transition can occur only if the precondition is true. Transition Not Success is related to place Number Already Chosen by means of inhibitor arcs. This type of arcs have an effect on the precondition i.e. the transition Not Success can occur only if there is no combination in the place Number Already Chosen that matches the precondition.

## 3 A Classical System Model

Figure 2 presents an interactive system for playing the game. The behaviour of this system is modelled using the same formalism as for task model. However, in order to fully model an interactive system more information need to be represented such as activation (how the user can act on the interactors) and rendering (how the system renders information to the user).

**Figure 2**: Presentation part of a basic system.

Figure 3 only present here the behaviour part of the system. The proposed system fully supports all the actions that were left to the user in the unsupported game. For this reason, the behaviour of the system is very close to the concrete task model presented in Figure 1.

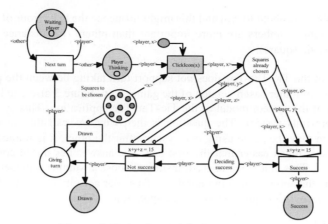

**Figure 3**: Behaviour of the basic system.

Some places are greyed-out, which means that there is some kind of rendering associated to them (not detailed here but for instance when a token enters place Success the player that won is displayed on the user interface). Transition ClickIcon(x) is greyed out, which means that this transition is related to the user interface and responsible for triggering transitions according to user's actions on the interactors.

## 4 The Improved System Model

Another system model for playing the game is presented in Figure 4-*right-hand side*. This system model is based on magic squares and tic-tac-toe as shown in the same figure 4 at the *middle* position. The left-handed part of Figure 4 presents a magic square. A magic square is made up of a set of cells each cell being filled in by a number. A 3x3 magic square is made up of 9 cells and the valid numbers are from 1 to 9. A specific feature of magic squares is that the sum of the numbers in each row equals the sum of the numbers in each column equals the sum of the numbers in each diagonal. In the case of a 3x3 magic square, the sum is always 15.

 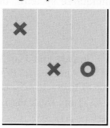 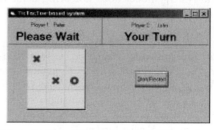

**Figure 4:** Example of the use of magic squares (*left-hand side*), TicTacToe (*middle*) and part of the system supporting for supporting user activity (*right-hand side*).

The user interface for playing the game is thus exactly the same as the one for playing the Tic-Tac-Toe (see Figure 4-*right-hand side*). This interface supports very well most actions in the task model:

- It shows what are the numbers still available (the empty cells)
- It shows the numbers already chosen for each player (for instance a cross for player 1 and a circle for player 2).
- It shows if a player has win (a row, a column or a diagonal is filled-up with the same symbol).
- From a strategy point of view, this user interface provides very relevant information:

201

- It shows if a player is about to win and this might influence the behaviour of the other player
- It shows that some numbers are more important than others (for instance number 5 in the centre of the magic square).

The paper version of the Tic-Tac-Toe does not support turn taking between the players but this is easily done using a computer-based version of the game. Now figure 3 also can be used to explain the behaviour part of the system model of the Tic-Tac-Toe application. This system model is the same as for the previous system. The question is then why we call this an improved system model? If you try to play the game using one system after the other, it is immediately noticeable that the second one is much easier that the first one. The answer lies in the concrete task model presented in Figure 1. This model features a transition Choose a Number representing the selection made a player from the set of remaining numbers. This user activity is highly demanding from a cognitive point of view, as it requires strategic computation.

# 5   Conclusion

In this position (and polemic) paper we have used a very simple example to show that creativity must remain an explicit phase in the design process of interactive systems and thus that generation of interactive systems from task models cannot lead to efficient interactive systems.

At the opposite of related work in the field such as (Paternò et al. 1999), we have tried to show here that the generation of an interactive system from task models is not a valid use of a task model as:

- Either you need to add information in the task model in order to be able to generate the interactive system from it and in that case this is not a task model anymore (this is done for instance in the work from (Bodart et al. 1995),
- Or you can generate a basic interactive system that does not cover efficiently users' activities. This has been shown on the example above where the Tic-Tac-Toe application supports users' strategy in selecting the best number. This strategic part of users' activity was not represented in the task model, as this kind of model is not meant to encompass this kind o information.

We also believe (but we have not shown it here for space reasons) that generating a task model from a system model does not bring a lot to interactive systems' design, as this is advocated in ((Lu et al. 1999).

# 6   References

Bodart, François, A-M. Hennebert, Leheureux J-M., I. Provot, and Jean Vanderdonckt. "A Model-Based Approach to Presentation: A Continuum from Task Analysis to Prototype." 1st. EUROGRAPHICS Workshop on Design Specification and Verification of Interactive System (DSV-IS'95), 1994.

Lecerof, A, and fabio Paternò. "Automatic Support for Usability Evaluation." IEEE Transactions on Software Engineering 24, no. 10 (1998): 863-88.

Lu, S., Paris, C., Vander Linden, K. Toward the automatic construction of task models from object-oriented diagrams. Engineering for Human-Computer Interaction. Kluwer Academic Publishers. 1999. 169-180.

Moher, Tom, Victor Dirda, Rémi Bastide, and Philippe Palanque. "Monolingual, Articulated Modeling of Users, Devices and Interfaces." Proceedings of the Eurographics Workshop on Design Specification and Verification of Interactive System, 312-29, Springer-Verlag, .Wien, 1996.

Norman, Donald. "Cognitive Engineering." User Centred System Design: New Perspectives on Human-Computer Interaction., 31-61.: Lawrence Erlbaum Associates, 1986.

Palanque, Philippe, and Rémi Bastide. "Synergistic Modelling of Tasks, Users and Systems Using Formal Specification Techniques." Interacting With Computers 9, no. 2, 129-53, 1997.

Paterno, F., Breedvelt-Schouten and N. de Koning. Deriving presentations from task models. In. Engineering for Human-Computer Interaction. Kluwer Academic Publishers. 1999, pp. 319-338.

# Loosing Reality in the Modeling Process

*Jenny Persson*

Department of Information Technology/Human Computer Interaction
Uppsala University, Box 337, SE-751 05 Uppsala, Sweden
jenny@hci.uu.se

## Abstract

Modeling is one crucial part in the software development process. I will in this paper discuss the concept of reality, its relation to metaphors as well as consequences of interpretations caused by the idea that reality can be understood as the sum of its fragments. By reflecting the contemporary dominating software development model, the Rational Unified Process, RUP, I will as well illustrate some of the problems that can appear in the modeling process with an empirical example chosen from a Swedish research project called the VERKA-project concerning two of the most important governmental authorities in Sweden. Both authorities use RUP and embody their own in-house systems development organizations. By modeling the state of work a picture of the work is created, a "model". This model can be more or less useful in the software development process, but it does not portray the entire or pure picture of reality. The understanding and awareness of the changes taking place in the interpretation of the work is crucial if the development of new computer systems is to be able to contribute to a healthy work environment that is as well characterized by efficiency, effectiveness and satisfaction.

## 1    Developing for Reality

Software developed for use in working life requires a genuine understanding of and respect for the work the software is going to be used in as well as for the users. In the VERKA-project[1] the software development projects mainly concern the work of case-handling in governmental authorities. One of the development projects studied is the development of a case-handling system built for the Swedish national registration. This computer system is primarily going to be used by approximately 700 people in their everyday work.

### 1.1    Understanding Reality

A custom-made system implies that the system is customized to fit a specific customer, acting in a specific (work) situation. The understanding of the work and its context in such a case is essential for the development process. The development of a computer system built for use in everyday work needs to match the work and the workers in an accurate way in order to promote positive effects in the work environment. This requires a profound understanding of the situation the

---

[1] The VERKA project started in the spring of year 2000 and ended in December 2002. VERKA is in Swedish an acronym for *Verksamhetsutveckling och Arbetsmiljö* (in English *Business Development and Work Environment*), *VERKA* means to *work; act.*

computer system is being built for. To understand the world, we use our preconceived notions to interpret what we perceive. Our interpretation and understanding of our environment is so to say affected through our use of language and through all experience embedded in ourselves. The French sociologist, Pierre Bourdieu's concept *habitus* describes the experience, knowledge and social possession embedded in a person (Bourdieu, P. 1979). According to Bourdieu we have the natural authority to fit into a social field, which by belonging to, we also contribute to, maintain and form. In the social field, our social skills and cultural capital is useful and understood in all its nuances, but if we want to change social field, it will become difficult, since our habitus consists of both the skills and the experience we can achieve, as well as of other factors, which we cannot control. All these attributes taken together give us authority to enter the social field.

In the kind of investigation, as the sociological investigation Bourdieu accomplishes, different social fields will appear. These do not only concern social classes alone, but will as well show differences between different work groups, workplaces, neighborhoods etcetera, in so far that they will have a place, a dominating social field on the sociologic map drawn. The differences in acceptance within the different fields cannot be explained with criteria or logic classes as education, skill, neighborhood, economy or other criteria possible to measure. This is why Bourdieu talks about habitus.

Habitus is the formative structure, which organizes the practices and the interpretation of the attributes, while acting as the structured structure as well, which in turn is the principle for the division into the different logical classifications that organize the way of assessing the social world, which itself is a product and a result of the way societies divide social classes, that will then exist in an embodied form. The social class is defined by the structure of all relevant characteristics. This is why we have to make clear that there is a web of secondary characteristics with which we more or less consciously present every time we deal with classes that are constructed on the basis of only one single criterion or a couple of criteria, as often seen in investigations or in modeling procedures.

A reduction of the number of criteria in an investigation of a phenomenon is common. We need to limit our investigations in order to be able to overview them. In this limiting process, our image of what the typical criteria are is normally picked to represent the phenomena as a whole. This affects the results, which in this way will become more uncertain. Reflecting Bourdieu's thoughts about habitus, this reduction leaves the formative structure outside the investigation in favor of the structured structure, namely the division into different logical classifications. Consequently this makes the investigation fractional and the results that follow are more or less biased towards the expected. A redundant approach does have a tendency to overestimate the interpretation of the results rather than to underestimate them.

## 1.2 An Empirical Example

The software project concerning the Swedish national registration can be used as an example of the relation between reality and the modeled reality from which the computer system is being built. Within this project, not only the development of the software, but as well the choice of structure within the development project organization is of interest. To value the construction of the project, the relations in the organization have to be understood. The Swedish National Tax Board, which is the authority governing the national registration, can be seen as consisting of three main sectors – the primary sector, which is set to discharge the duties of the Tax Authorities, the Tax Authorities themselves who are set to discharge the duties of the citizens and the companies that are active in Sweden, and finally there is the National Tax Board's own IT-department whose job is to develop and support the systems used in the authority. But at the same time there is a desire to work as if the relation between the discharging units and the IT-department is a normal client-supplier relationship.

The project manager for the development of the new computer system being built was from the start of the project an executive officer coming from the business sector of the National Tax Board. Next in the hierarchy, was the project manager of the software development, called IT project manager. The project group at that time encompassed mainly software developers, but there were as well a group of user representatives. This created a situation whereby the same party became both the customer and the supplier. The customer (the National Tax Board) who orders the system becomes by this construction responsible not only for the design of the work, but as well for the development of the system. This means that the National Tax Board in this way becomes responsible for the development of the system, for the requirements, for the ownership and for the support of the system. The system will as well be normative for the future work, which consequently means that the authority that is set to discharge the user organization also affect the design of the work through the design of the computer systems.

When the software development project first started in the spring of 2000, a lot of time was spent on learning Rational Unified Process, RUP, (Kruchten, 1998) which is a development model at that time, had never been used in the organization. Pressures of time forced the project to start with the modeling work before people really had gotten the hang of what the actual purpose with the use cases was. Every move was documented in the templates that followed with RUP. From an organizational point of view, the understanding of what business the system was being developed for, was the ultimate focus.

After spending almost a year modeling all kinds of situations, the project ended up in a jungle of use cases, which were to be transformed into system use cases. At this juncture, the question arose as to why the modeled situations were the ones chosen, and especially why the structure and content in the use cases were the way they were. What is their purpose and how are we going to use them? Everything had been documented, in accordance with RUP procedures, but at this point, the documentation had become almost as a project in itself.

## 1.3    Representations of Reality

The models drawn in the software project were mainly different work situations described as different flows. In the RUP model the processes of a business are defined as a number of different business use cases, each of which represents a specific workflow in the business. A business use case defines what should happen in the business when it is performed; it describes the performance of a sequence of actions that produces a valuable result to a particular business actor. A business process either generates value for the business or lessens costs to the business. (RUP, 2000)

A business use case shall, according to RUP describe "a sequence of actions performed in a business that produces a result of observable value to an individual actor of the business"(RUP, 2000). Hence, from an individual actor's perspective, a business use case defines the complete workflow that produces the desired results. This is similar to what is generally called a "business process", but "a business use case" has in RUP a more precise definition. The business use-case is a model of the business' intended functions. The business use-case model is used as an essential input to identify roles and deliverables in the organization, while the use-case model is defined as a model that describes a system's functional requirements in terms of use cases.

We often try to find a better way of understanding an organization by building descriptive models or metaphors. The metaphor is a magnificent descriptive tool, but the moment the metaphor becomes superior to reality, we will loose ourselves in sub-discussions. Or, to put it in another way, when the metaphor itself becomes the focus for decisions about strategies or changes, the risk of loosing control or diminishing the understanding of the actual organization increases dramatically. Awareness of what theoretical values or ideas that constitute the framework of the decisions we make, will increase the ability to find a beneficial way to reach our goals.

Metaphors are the pictures we choose to help us in our descriptions and understanding of certain situations, or more or less specific relations. In organization theory, as well as in managerial work situations, metaphors are often used. For instance we portray organizations in different ways depending on what we want to focus on.

There are a number of different metaphors with which we can analyze organizations described, for instance, by Morgan in his piece of work, *Images of Organization* (1986). These are metaphors such as machines, organisms, brains, cultures, political systems, psychic prisons, and flux and transformation. The metaphors are to be used to understand organizations in their complexity. Morgan emphasizes that organizational analysis must take into consideration the fact that organizations can be many things at the same time. The complexity, the ambiguous and the paradoxical basis, must be considered if one truly wants to understand an organization. Morgan also mentions the assumptions often made by organization theorists and managers. For instance, that they believe "that organizations are ultimately rational phenomena that must be understood with reference to their goals or objectives"(Morgan, 1986. p.322). He points out that this tendency to believe in a perspective that will override the complexity, will also get in the way of realistic analysis. In organizational analysis, Morgan's suggestion is a method in two steps. The first step is a diagnostic reading of the organization, where key aspects of the situation are highlighted with help from relevant metaphors. In the second step there is a critical evaluation of the different interpretations found in the first step. According to Morgan, these two steps will make it possible to explore the complexity of organizations in both a descriptive and a prescriptive manner.

The use of metaphors to picture different conditions in organizations has expanded. In trends today it is implied that metaphors that describe reality in terms of processes and flows. Everything is interpreted as a process or processes: the decision-making process, the work process, the life process, and the development process… As a tool for interpreting a phenomenon the process might be useful. But with this instrument as with other instruments the idea of a process implies something. Process thinking evolves from something that will color our understanding of the interpreted. The concept of process evolves in organizational analysis from the process industry, which implies that the business in the organization can be interpreted as a set of different flows, as in an industrial unit. In an organization other than a mechanized plant, the flow would consist of people's behavior in certain situations, which, if we use the process model, requires people to behave in the same way at a particular moment in a workflow. In software development, process thinking is predominant, but when interpreting businesses in its entirety, process thinking is too unsubtle a method of interpretation.

## 1.4 Loosing Reality

Understanding reality is as aforementioned one crucial matter in the development of custom-made software. But, what then is reality, and does reality actually get lost? The answer is not easy to give. Reality differs depending on the interpreter. In a software project, as in the example with the Swedish national registration, the reality in which the people performing the case-handling tasks was to be superior. The consequences from the modeling process though, showed that the modeling almost became an end in it self. The original focus was on the business, or to say the content or reality, but when the modeling process started, the focus shifted into the modeling itself, or to say towards its form. The *content* of the work became in this way transformed into the *form* of modeling. The targeted subject was in this way completely changed.

## 2    Conclusion

From the discussed matters in the previous paragraphs, some conditions that severely affect the understanding of reality can be concluded. Under pressure basic values often show clearer, however not necessarily clear to one self. Solutions for how to interpret reality, found in ready-made methods are grasped, and in the end, when time and money finally control the process, all the magnificent ideas of a system built for a better work environment have soon faded away. A lot of good work has gone to waste, as well the initial enthusiasm of many people. The national registration software project is now a mere shadow of its former ideal. With time and money as predominant factors, together with a shortsighted rational way of making decisions, many important values in the organization will risk getting lost. This loss reflects the basic values in the organization as well as it reflects the unawareness of the consequences of different strategic decisions. The basic values exposed in the preconceived notions show that the idea of the business understood as the sum of its workflows are dominating at the time. The workflow metaphor implies one kind of understanding of reality. This understanding is reduced to the conditions concerning processes or flows. A redundant approach will bias some conditions and leave other conditions of importance missing, in the interpretation of a specific situation. The redundant portray of "reality" may be usable for the understanding of the computer system, but will be far too shallow for the understanding of the business or the work and work environment. The modeling process will thereby itself increase the risk of loosing important aspects of reality. If we do not try to actually increase our awareness of our grounds for decisions, and in that sense not increase the awareness of our preconceived notions, we will ultimately end up loosing reality in the modeling process.

## References

Bourdieu, P. (1993). *Kultursociologiska texter.* Stockholm: Brutus Östling Bokförlag Symposion AB. (*Orig. La Distinction. Critique Sociale du Jugement* Paris: Éditions de Minuit 1979)

Kruchten, P.; (1998); *The Rational Unified Process – an Introduction*; Massachusetts, USA, Addison Wesley Longman, Inc, Reading.

Morgan, G. (1986). *Images of Organization.* London: Sage Publications Ltd.

Rational Unified Process, RUP, 2000

# Metaphors in Design – out of Date?

*Antti Pirhonen*

Department of Computer Science and Information Systems
P.O.Box 35
FIN-40014 University of Jyväskylä
Finland
pianta@cc.jyu.fi

## Abstract

The concept of metaphor is so widely used in the context of graphical user-interface that it is often seen as an outdated principle in design. However, we argue this is mainly due to the peculiar usage of the concept in design. We are defending the applicability of metaphors in design by reacting to central criticism against it, discussing its usage in design, and providing guidelines that help using metaphors in design in a way that they have been used in human language.

## 1    Introduction

Since the introduction of the Macintosh-style GUI (graphical user-interface), the concept of metaphor has been widely used as a guideline for design of human-computer interaction. Even if the empirical evidence about the superiority of the GUI compared to other alternatives is still lacking (Hazari & Reaves, 1994; Jones & Dumais, 1986; Petre & Green, 1990), the commercial success is clear, obviously indicating a certain kind of user preference.

It is quite easy to point out weaknesses in the GUI (Gentner & Nielson, 1996). Sometimes, the key problem with the GUI is argued to be the use of metaphors (Nardi & Zarmer, 1993), not the graphical information presentation itself. GUI is seen to be based on metaphors (e.g., Apple Computer, 1992), so it is no wonder that distrust against GUI is generating criticism against metaphors in human-computer interaction. However, the use of the term 'metaphor' seems to have changed in the context of user-interfaces from its origin in rhetoric.

In this paper we suggest that by creating metaphors in the way that is done in discourse, they still have power as a design principle. We first defend metaphors in design by reacting to two familiar arguments against them. In this way, we stress some essential properties of a metaphor as a means for supporting conceptualisation process. These properties have been neglected in the current usage of the term in the context of GUI design.

# 2 Usage of the Concept of Metaphor in Design

## 2.1 Criticism against Metaphors

There has been two central themes in the criticism against the use of metaphors in design:

1) A metaphor cannot cover the whole domain of its referent.

Using a metaphor is argued to limit functionality (Harrison, Fishkin, Gujar, Mochon, & Want, 1998; Johnson, 1987; Nardi & Zarmer, 1993). This argument immediately reveals the underlying conception of metaphor. In this conception, metaphor as a virtual entity parallels a real world entity as closely as possible. For example, if we create a virtual calculator, we try to make it resemble its real world counterpart as closely as possible with the technology available. We can easily find numerous examples of this kind in the short history of GUI. This, in most cases, forms a constraint concerning functionality. In the creation of a real calculator, there is its history, manufacturing processes and many other things that give rise to the form we are accustomed to. Even though we do not have the same physical constraints in the creation of a virtual calculator, we still adopt the same form. However, there are two serious problems in this conception of metaphor. First, there is a *motivational* problem. If we simply imitate a real world entity, its development process may suffer. As soon as our virtual calculator works exactly like its real-world counterpart, it often is deemed perfect. That is the absolute limit. In most cases, it is inferior to the original. For example, there have been attempts to create a virtual book by imitating a physical book (Harrison et al., 1998). However sophisticated technology we use, our UI cannot equal a real book. We call this a motivational problem, since trying to implement, in a virtual environment, something that has already been implemented in another kind of environment is mainly a technological challenge. It does not provide anything really new, only a virtual version of an existing thing – usually in worse quality – in a computerised environment.

The second problem in labelling imitation a metaphor is *theoretical*. Whether we refer to the classical, Aristotelian view of metaphor (Aristotle, 1984) or modern theory of metaphor (Lakoff, 1993; Lakoff & Johnson, 1980, 1999), the dissimilarity between a metaphor and its referent shouldn't be taken as a weakness of the metaphor but as its strength (Hamilton, 2000). We argue that mismatch between a metaphor and its referent is one of the core properties of a successful metaphor; it is the dissimilarities that make us aware of the similarity. This does not mean that we defend the dissimilarities in our previous examples (calculator and book). Rather, we argue that these examples were not examples of metaphors at all. If the aim is to imitate, the result is a simulation, not a metaphor. In simulation, all dissimilarities are weaknesses. In a metaphor, dissimilarities are intentional: a metaphor is a means of making the interpreter (user) focus attention to a *certain* property or properties of the referent item. The rationale of using simulation is quite different from the rationale of using metaphor.

2) The main rationale for the use of metaphor is in its power to enhance the learning of new concepts. So why use the metaphor after having learned to use the application (Gentner & Nielson, 1996; Johnson, 1987)?

Cooper (1993) analyses a concept of dead metaphor. By a dead metaphor he means something that once was a metaphor, evoked associations, and helped in the conceptualisation process of a new entity. Gradually, the connection to the origin of the metaphor weakens and finally vanishes. The metaphor turns into an idiom – it dies. Our point is that there is nothing wrong in this lifecycle of a metaphor: first it is born and finally it dies. Actually, a successful metaphor works for different users in a different way. For a novice, an expression may be metaphorical; whereas for an experienced user it may seem a mundane item, i.e., an idiom. As a user-interface, an everyday item such as a light switch, for example, has not much in common with its predecessors, in this case the lighting of oil or gas lamp. The idea of controlling the electric bulb with a tiny lever beside the doorway must have had an underlying metaphor. It is hard to say whether the primary metaphor concerned the hardware (the lever) or the controlling gesture. We do not need to think it anymore, because it is no more a metaphor. Rather, it is a part of our everyday life; it is an idiom. The peculiar thing in this example is that the underlying metaphor concerning the required gesture may have been different in Britain from that on the Continent, since the light switches work in opposite directions: on the Continent, gesture upwards means switching on, in the UK it is the other way round.

## 2.2  Non-verbal Metaphors

One difficulty in applying the concept of metaphor in the design of GUI is that GUI is largely non-verbal in nature whereas the traditional conception of metaphor is firmly related to language. The widely used definition of a metaphor as a figure of speech (Oxford English Dictionary, OED) is probably the clearest case. This reflects the tradition originating in the scripts of Aristotle (1984), who handled metaphor as a means of poetry and rhetoric. Metaphor theorists like Richards (1950) and Black (1962) have continued this tradition in the 20th century. Because the concept is directly related to a certain communication mode (language), it cannot be directly applied to another mode without re-defining it (see Wilks, 1995). A basis for understanding non-verbal metaphors could be modern metaphor theories, like Lakoff's (1993). In this theory, the essence of metaphor is seen in its power to enhance understanding of new concepts in terms of familiar ones. The other important point of Lakoff and Johnson is that metaphors are conceptual in nature and that metaphorical language is only one form of manifestation of those conceptual entities. From these ideas it is easy to derive an understanding of non-verbal metaphors, and we argue that this historical means of knowledge construction is still relevant in the age of multimedia.

## 3  Conclusions

The difference between the traditional metaphor and 'UI-metaphor' is so clear that it can be argued, like Gaver (1995), that these interface representations can be called metaphors only metaphorically. Of course, this is not necessarily the case; metaphors can be used in GUIs in their traditional sense as well, as 'real' metaphors. But the first step is to discard the idea that the GUI is based on metaphors. Only in this way are we able to reinstate metaphor in design, with its unique power.

The conception of metaphor proposed in this paper can be expressed more concretely in order to support design. So we summarise our findings in four points, explaining each one from the perspective of practical design.

A successful metaphor

- is an unexpected, unconventional mapping between two domains

    - In order to activate the user, we should not use parallels that are too obvious and conventional. With unconventional parallels we are able to affect even attitudes and emotions. So the combining property between a metaphor's source and its target doesn't need to be visual or anything else that is observable. Rather, an effective metaphor may rely on mental impressions. It may relate things that superficially don't seem to have anything to do with each other. The insight about the combining property makes that particular property extremely strong in the mind of the user.

- evokes insights into the core features of the target domain,

    - When creating a metaphor, we have to have a clear idea about the core features of the target of the metaphor. If we found a parallel that is fascinating and unexpected but directed the attention of the user to something unessential with it, the metaphor would hardly work. Therefore, we potentially have to compromise between the piquancy and the relevance of a metaphor.

- makes learning easier by utilising existing mental structures of the interpreter,

    - To support the construction of new concepts in terms of familiar ones is one of the main reasons for using metaphors. However, the designer has to keep in mind, that active knowledge construction process is a highly subjective process. So we can never be completely sure what kind of meanings our design evokes in the mind of the user. For the same reason, we are not able to push different meanings by the help of metaphors to the mind of the user and to construct knowledge on behalf of the user.
    - Since the aim is to support knowledge construction on the basis of existing mental structures, all information that might help to understand the user's perspective is valuable. A metaphor, which is found to work in one culture, for instance, does not necessarily work in another culture. Equally, metaphors designed for older adults might not come across for younger people.
    - Just as in all human communication, the creation of a working metaphor is highly dependent on the general communication skills of the designer of the metaphor. This includes the ability to see the world with the eyes of the user (i.e., understanding), and the ability to express meanings.

- may become an idiom in time.

    - Creating a user-interface element that works as a metaphor for a novice and as an idiom for an experienced user is the real challenge. The same expression (word, drawing, icon, sound etc.) should help the novice to understand the basic nature of something new, and at the same time, work as a simple symbol of something familiar for the experienced user. If the expression is too rich and illustrative (over-whelming use of presentation technology available), it might seem appropriate for the novice but irritate the experienced user who might be fed up with the use of all the effects. On the other hand, if the expression is minimal, it is probably adequate for the experienced user but does not say anything to the novice. The balance between the needs of the novice (using metaphor) and the experienced user (using idiom) is crucial.

# References

Apple Computer, I. (1992). Macintosh human interface guidelines. Reading (MA): Addison-Wesley.

Aristotle. (1984). The complete works of Aristotle: The revised Oxford translation / ed. by Jonathan Barnes (1 ed. Vol. 2). Princeton, N.J.: Princeton University Press.

Black, M. (1962). Models and metaphors. New York: Cornell University Press.

Cooper, P. A. (1993). Paradigm shifts in designed instruction: from behaviorism to cognitivism to constructivism. *Educational Technology*, 33(5), 12-19.

Gaver, W. W. (1995). Oh what a tangled web we weave: Metaphor and mapping in graphical interfaces. In *CHI'95, Conference companion on Human factors in computing systems* (pp. 270-271). ACM.

Gentner, D., & Nielson, J. (1996). The anti-mac interface. *Communications of the ACM*, 39(8), 70-82.

Hamilton, A. (2000). Metaphor in theory and practice: the influence of metaphors on expectations. *ACM Journal of Computer Documentation*, 24(4), 237-253.

Harrison, B. L., Fishkin, K. P., Gujar, A., Mochon, C., & Want, R. (1998). Squeeze me, hold me, tilt me! An exploration of manipulative user interfaces. In *CHI'98, Conference proceedings on Human factors in computing systems* (pp. 17-24). Los Angeles, CA: ACM Press.

Hazari, S. I., & Reaves, R. R. (1994). Student preferences toward microcomputer user interfaces. *Computers & Education*, 22(3), 225-229.

Johnson, J. (1987). How faithfully should the electronic office simulate the real one? *SIGCHI Bulletin*, 19(2), 21-25.

Jones, W. P., & Dumais, S. T. (1986). The spatial metaphor for user interfaces: experimental tests of reference by location versus name. *ACM Transactions on Information Systems*, 4(1), 42-63.

Lakoff, G. (1993). The contemporary theory of metaphor. In A. Ortony (Ed.), *Metaphor and thought* (2nd ed., pp. 202-251). Cambridge: Cambridge University Press.

Lakoff, G., & Johnson, M. (1980). Metaphors we live by. Chicago: The University of Chicago Press.

Lakoff, G., & Johnson, M. (1999). Philosophy in the flesh: the embodied mind and its challenge to western thought. New York: Basic Books.

Nardi, B. A., & Zarmer, C. L. (1993). Beyond models and metaphors: Visual formalisms in user interface design. *Journal of Visual Languages and Computing*, 4, 5-33.

Petre, M., & Green, T. R. G. (1990). Is graphical notation really superior to text, or just different? Some claims by logic designers about graphics in notation. In *Proceedings in the Fifth Conference on Cognitive Ergonomics*. Urbino, Italy, September 3 - 6, 1990.

Richards, I. A. (1950). The philosophy of rhetoric (2nd ed.). New York: Oxford University Press.

Wilks, Y. (1995). Language, vision and metaphor. *Artifial intelligence review*, 9(4-5), 273-289.

# A Theory of Information Scent

*Peter Pirolli*

PARC
3333 Coyote Hill Road, Palo Alto, CA 94304, U.S.A.
pirolli@parc.com

## Abstract

Information scent is a psychological theory of how people use perceptual cues, such as World Wide Web (WWW) links in order to make information-seeking decisions and to gain an overall sense of the contents of information collections. A spreading activation theory of information scent is presented, as well as a theory of learning categories (e.g., genres) of available information. The SNIF-ACT model incorporates information scent to model World Wide Web browsing. The InfoCLASS model simulates the learning of information categories from browsing.

## 1    Introduction

Information scent is a psychological theory of how people use perceptual cues, such as World Wide Web (WWW) links in order to make information-seeking decisions. Information scent cues play a role in guiding users to the information they seek, and they also play a role in providing users with an overall sense of the contents of information collections. This paper presents an overview of a theory and a set of computational models about the psychology of information scent.

## 2    The Role of Information Scent in Browsing

Figure 1 presents some examples of information scent, or mediating representations used to guide people to information. Figure 1a is a typical page generated by a WWW search engine in response to a user query. The page lists WWW search results summarizing documents that are predicted to be relevant to the query. Figure 1b illustrates, an alternative form of search result representation that is provided by *relevance enhanced thumbnails* (Woodruff, Rosenholtz, Morrison, Faulring, & Pirolli, 2002).

### 2.1    Effect of Information Scent on Navigation Costs

Figure 2 illustrates the, often dramatic, impact of the quality of information scent on the efficiency of user navigation through a WWW site, or any navigation environment with branching structure. Users often browse by a making a series of decisions involving the selection of a link from a presented set of links. These decisions are usually based on local information scent cues, such as the link representations in Figure 1. If the information scent cues are "perfect" then the user will make no navigation errors and will proceed directly to desired information. Completely uninformative information scent cues will lead to random choices at each decision point. Figure 2 illustrates navigation costs, in terms of number of WWW pages visited before finding a desired target page that is located at different depths from a starting page, for a hypothetical user searching through a WWW site that is 10 levels deep and that has an average of 10 alternative links per page. The curves were computed using a cost function equation described in Pirolli et al. (in press), and empirically determined *false alarm rates* (Woodruff et al., 2002) that characterize the likelihood of users choosing links they did not want to follow. The top curve characterizes the worst average false alarm rate for text link summaries (Figure 1a) and the bottom curve

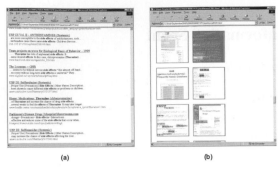

(a)                                    (b)

**Figure 1. Examples of information scent.**

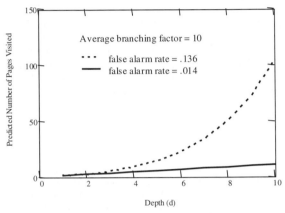

Depth (d)

**Figure 2. Cost due to information scent.**

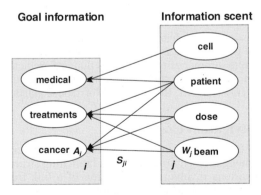

**Figure 3. Spreading activation network.**

characterizes the best average false alarm rate for relevance enhanced thumbnails (Figure 1b). Figure 2 illustrates that the costs associated with hierarchically organized navigation can be radically shifted by using improved information scent cues. Below I summarize a recent computational cognitive model, SNIF-ACT, that simulates the scent-based navigation behavior of users on the WWW.

## 2.2  Forming Concepts about the Information Environment

Over time, as users interact with information environments such as the WWW, they learn what types of information reside where, in what abundance, and having what utility. Studies (Pirolli, Schank, Hearst, & Diehl, 1996) of the Scatter/Gather browser (Cutting, Karger, & Pedersen, 1993) suggest that it is superior in helping users conceptualise the content of large document collections, in comparison to standard search engines: (1) Scatter/Gather users developed richer mental category structures to represent the contents of the document collection than users of a standard search engine (on the same repository and tasks), and (2) Scatter/Gather users showed greater convergence to one another in their category structure in comparison to the search engine users. Below, I summarize InfoCLASS, which is a computational cognitive model of users learning mental categories of information from their experience with the information scent cues presented by the Scatter/Gather browser.

## 3  A Theory of Information Scent

### 3.1  Navigation Choices Based a Spreading Activation

Our cognitive simulations use a *spreading activation* model of navigation based on information scent that extends the ACT-R theory (Anderson & Lebiere, 2000). Activation spreads from a set of cognitive structures that are the current focus of user attention through *associations* to other cognitive structures in memory. These cognitive structures are called *chunks* (Anderson & Lebiere, 2000). The basic idea is that a user's information goal activates a set of cognitive structures in a user's memory, and information scent cues in the world activates another set of cognitive structures. Activation spreads from these activated cognitive structures to associated structures in the spreading activation network. The amount of activation accumulating on the goal chunks provides an indicator of the likelihood of the goal based on the information scent cues (the expected utility of navigation choices).

Figure 3 assumes that a user has the goal of finding information about medical treatments for cancer and encounters a WWW link labelled with text that includes "cell", "patient", "dose", and "beam". The cognitive chunks representing information scent cues are presented on the right side of Figure 3 and the goal chunks are on the left. Figure 3 also shows that there are associations between the goal chunks and information scent chunks. These associations reflect past experiences in which chunks co-occurred. The *strength* of associations reflects the degree to which information scent cues predict the occurrence of goal chunks. For instance, the word "medical" and "patient" co-occur quite frequently and they would have a high strength of association. The stronger the associations, the greater the amount of activation flow. The activation of a chunk $i$ is

$$A_i = B_i + \sum_j W_j S_{ji} \tag{1}$$

where $B_i$ is the base-level activation of $i$, $S_{ji}$ is the association strength between an associated chunk $j$ and chunk $i$, and $W_j$ is reflects attention (*source activation*) on chunk $j$. One may interpret Equation 1 as reflection of a Bayesian prediction of the likelihood of one chunk in the context of other chunks. $A_i$ in Equation 1 is interpreted as reflecting the log posterior odds that $i$ is likely, $B_i$ reflects the log prior odds of $i$ being likely, and $S_{ji}$ reflects the log likelihood ratios that $i$ is likely given that it occurs in the context of chunk $j$. A particular action (e.g., the selection of a WWW link) may be evaluated based on those goal activation levels produced by the information scent of a particular set of cues (e.g., the words presented in the WWW link). It is possible to automatically construct large spreading activation networks from on-line text corpora. We employ a database that contains statistics relevant to setting the base-level activations of 200 million word tokens and the inter-word association strengths of 55 million word pairs, and we update this database (e.g., to track novel words entering the language) by using the WWW.

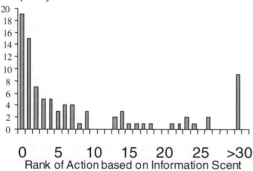

Frequency of Match to User Behavior

Rank of Action based on Information Scent

**Figure 4. Match of SNIF-ACT action evaluations to users.**

## 3.2 SNIF-ACT: A Cognitive Model of WWW Navigation

The spreading activation model of information scent has been incorporated in the ACT-IF cognitive model (Pirolli & Card, 1999) of user behavior with the Scatter/Gather browser, and a cognitive model of WWW browsing called SNIF-ACT (Pirolli & Fu, in press). User data collected in the laboratory using application logs have been fit to these models with a *user-tracing architecture* (Pirolli, Fu, Reeder, & Card, 2002). This architecture matches a cognitive simulation model to moment-by-moment user behavior (called a *user trace*). Figure 4 comes from the evaluation of the SNIF-ACT model by Pirolli and Fu (in press). Figure 4 plots data extracted from all the places where the SNIF-ACT simulation was compared against user data at the point when the user was just about to make a selection of a link on a WWW page (there are a total of 91 link selections in the data set, which comprise 48% of all the actions performed by four subjects on two tasks). Actions associated with following the links on a page were ranked in SNIF-ACT by their information scent utilities, as computed by spreading activation. The x-axis in Figure 4 plots SNIF-ACT's scent-based ranking of all the possible link-following actions available to the users at each decision point. The y-axis in Figure 4 plots the observed frequency with which the potential SNIF-ACT action matched a real user. Figure 4 shows that link choice is strongly related to the information scent values computed in SNIF-ACT. This result replicates a similar analysis made by Pirolli and Card (1999) concerning the ACT-IF model of the Scatter/Gather browser.

## 3.3    InfoCLASS: Learning Concepts about the Information Environment

InfoCLASS is a model of category formation that builds on Anderson's (1991) theory. The basic idea is that users elaborate their memory structure (e.g., Figure 3) with induced categories, such as Figure 5. In Figure 5, it is assumed, that the user has the goal of searching for new medical treatments for cancer and encounters a WWW link that activates the cognitive chunks representing the words "cell", "patient", "dose", and "beam". There are associations between the words and a category chunk that represents the "health" genre of documents. From that "health" chunk, there may be associations to the chunks that comprise the users goal. The category associations in Figure 5 co-exist with the memory-based associations in Figure 3. In Figure 5, activation spreads from the chunks associated with the information scent cues on the right, through the "health" chunk, to the goal chunks on the left. The strengths of the associations again reflect a Bayesian log odds computation: (1) the log odds that some array of information scent cues will lead to a particular category of information (reflected in the associations from the information scent cues to the category), and then (2) the log odds that that category of documents contains information that the user is seeking (reflected in the associations between the category and the goal information).

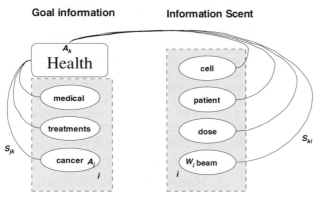

**Figure 5. A Category network.**

In InfoCLASS, the strengths in Figure 6 are learned by a Bayesian learning scheme proposed by Anderson (1991). This scheme specifies the learning of (a) the conditional probability that members of a category will have given features (represented as chunks) and (b) the conditional probability that an item with certain features is a member of a category. The InfoCLASS learning algorithm relies on a *coupling parameter* that determines the overall likelihood that any two items belong to the same category (which is used to determine if an item is novel enough to form a new category). In this algorithm, every set of information scent cues, that is perceived as a coherent object, such as a graphic or link text, is referred to generically as an *item*. As users encounter items, the items are categorized by the following process:

1.   If there are no existing categories (for instance this is the first time a user has visited a WWW site), then create a new category and assign the item as a member of the category, otherwise,
2.   Determine the probability that the item comes from a new category, and compare that to the maximum of the set of probabilities of the item belonging to an existing category,
     a.   Assign the item to a new category if that is more probable, otherwise,
     b.   Assign the item to the highest probability existing category.

An InfoCLASS category learning model was developed that was exposed to the exact same interface interactions as users ($N = 16$) in the Pirolli et al. (1996) study of Scatter/Gather. The goal of this category learning simulation was to predict a higher degree of conceptual coherence among the Scatter/Gather users as compared to users of a standard search engine called Similarity Search. For each user, the category learning simulation involved (a) parsing the user's interaction log (with either Scatter/Gather or Similarity Search) and extracting a sequence of items (link summary words) and (b) submitting the items, in the order they occurred in the user interaction log, to the category learning algorithm described above

An information-theoretic measure called the Average Divergence from the Average (ADA) was developed to measure the degree of conceptual coherence among Scatter/Gather vs Similarity Search simulations. ADA is an extension of a pairwise divergence metric called Total Divergence from the Average (Manning & Schuetze, 1999). The ADA is calculated by computing the average of all the categories formed by all user simulations in each group, and measuring the divergence of every category from the average (using an entropy measure). Low ADA values indicate greater category coherence. The Scatter/Gather user simulations showed greater coherence (ADA = .949 x $10^{-6}$ bits) than the Similarity Search simulations (ADA = 1.292 x $10^{-6}$ bits). The InfoCLASS simulations of Scatter/Gather users develop more coherent sets of categories than Similarity Search simulations.

# 4    General Discussion

Cognitive models such as SNIF-ACT and InfoCLASS can be used to develop cognitive engineering models to evaluate user interfaces (see also, Blackmon, Polson, Kitajima, & Lewis, 2002). The information scent theory has been used in applications that automatically evaluate WWW sites (Chi et al., 2003) and infer user goals (Chi, Rosien, & Heer, 2002). As the cognitive models evolve, we may expect to see them address richer interface content and off-the-desktop systems such as mobile devices.

## References

Anderson, J. R. (1991). The adaptive nature of human categorization. *Psychological Review, 98,* 409-429.

Anderson, J. R., & Lebiere, C. (2000). *The atomic components of thought.* Mahwah, NJ: Lawrence Erlbaum Associates.

Blackmon, M. H., Polson, P. G., Kitajima, M., & Lewis, C. (2002). *Cognitive Walkthrough for the Web.* Paper presented at the Human Factors in Computing Systems, CHI 2002, Minneapolis, MN.

Chi, E. H., Rosien, A., & Heer, J. (2002). *Lumberjack: Intelligent discovery and analysis of Web user traffic composition.* Paper presented at the ACM-SIGKIDD Workshop on Web mining for usage patterns and user profiles, WebKDD 2002, Edmonton, Canada.

Chi, E. H., Rosien, A., Suppattanasiri, G., Williams, A., Royer, C., Chow, C., Robles, E., Dalal, B., Chen, J., & Cousins, S. (2003). *The Bloodhound Project: Automating discovery of Web usability issues using the InfoScent simulator.* Paper presented at the Conference on Human Factors in Computing Systems, CHI 2003., Fort Lauderdale, FL.

Cutting, D. R., Karger, D. R., & Pedersen, J. O. (1993). *Constant interaction-time Scatter/Gather browsing of very large document collections.* Paper presented at the SIGIR '93, New York.

Manning, C. D., & Schuetze, H. (1999). *Foundations of statistical natural language processing.* Cambridge, MA: MIT Press.

Pirolli, P., & Card, S. K. (1999). Information foraging. *Psychological Review, 106,* 643-675.

Pirolli, P., Card, S. K., & Van Der Wege, M. M. (in press). The effects of information scent on visual search in the Hyperbolic Tree Browser.

Pirolli, P., & Fu, W. (in press). *SNIF-ACT: A model of information foraging on the World Wide Web.* Paper presented at the User Modeling, 2003, Johnstown, PA.

Pirolli, P., Fu, W., Reeder, R., & Card, S. K. (2002). *A user-tracing architecture for modeling interaction with the World Wide Web.* Paper presented at the Advanced Visual Interfaces, AVI 2002, Trento, Italy.

Pirolli, P., Schank, P., Hearst, M., & Diehl, C. (1996). *Scatter/Gather browsing communicates the topic structure of a very large text collection.* Paper presented at the Conference on Human Factors in Computing Systems, CHI '96, Vancouver, BC.

Woodruff, A., Rosenholtz, R., Morrison, J. B., Faulring, A., & Pirolli, P. (2002). A comparison of the use of text summaries, plain thumbnails, and enhanced thumbnails for Web search tasks. *Journal of the American Society for Information Science and Technology, 53,* 172-185.

# Social Network Analysis of a
# Participatory Designed Online Foreign Language Course

*Meenakshi Sundaram Rajasekaran*

*Panayiotis Zaphiris*

School of Informatics, City University,
London EC1V 0HB,
United Kingdom.
sundar_slm@rediffmail.com

The Centre for Human-Computer
Interaction Design,
School of Informatics, City University,
London EC1V 0HB,
United Kingdom.
zaphiri@soi.city.ac.uk

## Abstract

People intensive social systems like WBT (Web-Based Training) Systems are open natured and behave unpredictably with their environment. In other words, these systems are less responsive to well defined scientific methods. Hence a shift in focus from technical to non-technical approaches like Social network analysis (SNA) which is principally used in communication science, is necessitated. This study attempts to evaluate the applicability of Social Network Analysis (SNA) as an analytical and empirical method to analyse the interactions within the online community belonging to a Greek language for English speakers online course. The key objective of this study is to test a hypothesis that relates to the Participatory Design Methodology employed while developing this specific course.

## 1. Introduction

Some of the important concepts relating to this study are Web-based training systems (WBT), Online communities, Participatory Design (PD) methodology from the field of Human Computer Interaction (HCI), Social Network Analysis (SNA) and Netminer, the tool used for performing network analysis. A discussion of these topics is therefore important and the depth of discussion is curtailed pertaining to their association.

## 2. Web-Based Training System

Any application of web and communication technologies for the purpose of imparting knowledge or skill to someone is referred to as Web-based training. WBT enables one to obtain and deliver knowledge anytime, anywhere using a combination of the Internet and other communication technologies. The WBT system used in this study is 'Learn Greek Online' found at http://kypros.org/ Greek.

## 3. Online Communities

For the assessment of this WBT System, the online community belonging to the discussion forum at 'Learn Greek Online' system, is chosen as analysis. An online community can be defined as, a group of people with common interests who exchange ideas and words, or share emotions, predominantly over the Internet via electronic media like chat, email, online discussion boards etc.

# 4. Human – Computer Interaction (HCI)

The primary focus of HCI on human-computer based systems like WBT and discussion boards is to provide maximum usability to the ultimate users of a system. This, HCI aims to achieve by designing and developing a systems with a user-centered approach. HCI is a discipline concerned with the design, evaluation and implementation of interactive computing systems for human use and with the study of major phenomena surrounding them (Shneiderman, 1997).

## 4.1 Participatory Design (PD) in 'Learn Greek Online'

PD is a key technique from HCI that emphasizes active and seamless involvement of users at all stages of a system development. PD is the methodology underpinning design and development of 'Learn Greek Online' system at http://kypros.org/Greek. The rationale behind the usage of PD is paid to be motivation, commitment and satisfaction of users (Zaphiris & Zacharia, 2002b).

## 4.2 Hypothesis

> By involving users actively during the design and development phases of a system, a sense of ownership can be cultivated within the users, which consequently will influence them to play key roles in the system.

The extensive use of PD in design and development of 'Learn Greek Online' WBT System makes the hypothesis applicable in here (Zaphiris & Zacharia, 2002a). By testing the above hypothesis one can ensure that up fronting time and effort in Participatory Design (PD) can be significantly rewarding and consequently motivate system owners to implement PD, in the ongoing process of system improvement and maintenance.

# 5. Social Network Analysis (SNA)

Although Social network analysis (SNA) up to now has been principally used in communication science (Garton, 1999:76) for evaluating face to face communities, we think that its strengths can be of use in evaluating virtual communities too.

"Communities rely on relationships for their growth ..." (Preece, 2000). A community dwells on the notion of relationship and analysts have tried and established SNA as an empirical method to study social networks. In recent years, research has been carried out to test and demonstrate the applicability of SNA concepts in online communities (Preece, 2000:173).

Social network analysis (SNA) is the study of social relationships between a set of actors. SNA uses various concepts to evaluate different network properties like centrality, connectivity, cliques etc, each of which pertain to particular dimension of the network. This study has employed a few of these important concepts in testing the hypothesis and to reveal any useful implications over the system's performance.

### 5.1 Properties of social networks (Hanneman, 10-06-2002)

*5.1.1 Degree of an actor:* This is the number of connections an actor has with others in the network. This directly assists in establishing whether the actor forms a bottleneck in the flow of information within the system.

*5.1.2 Degree centrality:* This assesses the power of an actor based on number alters (other actors) that actor is directly connected to. If an actor receives many ties (aka relationships) or makes contact with many others then the actor is said to have high-prestige or high influence, respectively.

*5.1.3 Cliques:* A clique is a sub-set of actors in a network, who are more closely tied to each other than to other actors who are not part of the sub-set. Cliques focus on closely connected groups. They can aid in understanding how large social structures are built upon smaller and tighter components.

### 5.2 Netminer

Owing to voluminous calculations, this study has employed a GUI based computer tool called Netminer, to perform network analysis.

## 6. Methodology

In any online community there are two phases involved, namely *population phase* and *evolution phase*. The initial postings in the discussion board of Learn Greek online are associated with the initial population phase, which is rich in interactions and it is the set of postings that were subjected to analysis in this study.

The postings were carefully examined, categorized and tabulated with different details of posting made (e.g. originator and recipient of the post). From this table, information is transformed in the form of a data set, usable by Netminer for performing SNA. User's identity was masked with dummy identity to mitigate the concerns of privacy intrusion.

Examining these postings resulted in a list of 43 online community members who form the social network for this study of which 5 are from the Participatory Design Team (PDT). PDT refers to those involved in the design and development phases of 'Learn Greek Online' system.

## 7. Analyses and Results

A rigorous analysis has revealed many useful information from which constructive inferences and assertions can be made. A constraint matrix developed on this data reveals that 5 of the users impose most restriction on the other users, of which 3 belong to the participatory design team. Also a *'Reachability Matrix'* demonstrates another important property of the network, that every actor in the network is reachable by every other actor.

An in-degree visualization of the community at scale five, groups, 2 of the 5 PD team users at the centre (Figure 1, U17, U21), depicting that they have the highest in-degree amongst 2 others. Another PDT user (U19) is plotted at the next level showing high interaction level. As in the case of in-degree analysis, PDT-Users, U17, U19 and U21 are plotted as part of the most central actors.

Thus both in-degree and out-degree centrality analysis unmistakably denotes that PDT-Users have been playing key roles in the community.

## *Figure 1 : In-degree graph (Scale 5)*

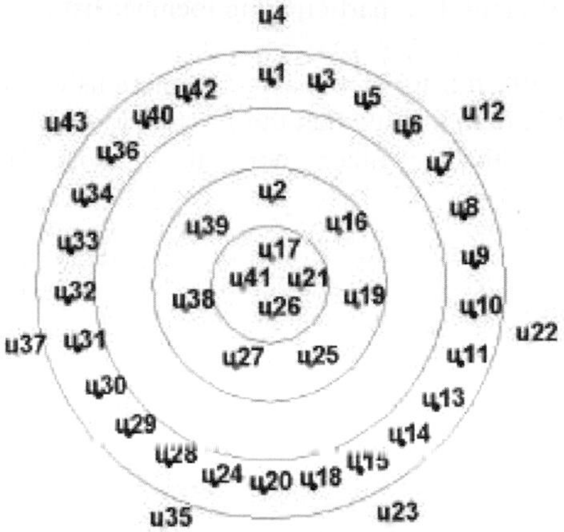

## *Table 1 :  Clique Analysis Report*

| Cliques | Members | Cliques | Members |
|---------|---------|---------|---------|
| K1  | (**u17, u19, u21,** u26, u27 , u12, u13) | | |
| K2  | (**u17, u19, u21,** u26, u27 , u12, u14) | K19 | (**u17, u19, u21,** u26, u27 , u25) |
| K3  | (**u17, u19, u21,** u26, u27 , u12, u15) | K20 | (**u17, u19, u21,** u26, u27 , u4) |
| K4  | (**u17, u19, u21,** u26, u27 , u7) | K21 | (**u17, u19, u21,** u26, u27 , u5) |
| K5  | (**u17, u19, u21,** u26, u27 , u8) | K22 | (**u17, u19, u21,** u26, u27 , u28) |
| K6  | (**u17, u19, u21,** u26, u27 , u9) | K23 | (**u17, u19, u21,** u26, u27 , u29) |
| K7  | (**u17, u19, u21,** u26, u27 , u10) | K24 | (**u17, u19, u21,** u26, u27 , u30, u43) |
| K8  | (**u17, u19, u21,** u26, u27 , u11) | K25 | (**u17, u19, u21,** u26, u27 , u31) |
| K9  | (**u17, u19, u21,** u26, u27 , u6) | K26 | (**u17, u19, u21,** u26, u27 , u33) |
| K10 | (**u17, u19, u21,** u26, u27 , u16) | K27 | (**u17, u19, u21,** u26, u27 , u34) |
| K11 | (**u17, u19, u21,** u26, u27 , u1) | K28 | (**u17, u19, u21,** u26, u27 , u35) |
| K12 | (**u17, u19, u21,** u26, u27 , u18) | K29 | (**u17, u19, u21,** u26, u27 , u36) |
| K13 | (**u17, u19, u21,** u26, u27 , u2) | K30 | (**u17, u19, u21,** u26, u27 , u37) |
| K14 | (**u17, u19, u21,** u26, u27 , u20) | K31 | (**u17, u19, u21,** u26, u27 , u38) |
| K15 | (**u17, u19, u21,** u26, u27 , u3) | K32 | (**u17, u19, u21,** u26, u27 , u39) |
| K16 | (**u17, u19, u21,** u26, u27 , u22) | K33 | (**u17, u19, u21,** u26, u27 , u40) |
| K17 | (**u17, u19, u21,** u26, u27 , u23, u32) | K34 | (**u17, u19, u21,** u26, u27 , u41) |
| K18 | (**u17, u19, u21,** u26, u27 , u24) | K35 | (**u17, u19, u21,** u26, u27 , u42) |

Sub-group analysis focuses on tightly knit components that satisfy certain criteria. Participation of network members in these closely-knit components in turn indicates a strong involvement of actors in the network. A subgroup analysis shows a group of 5, as those who are present in all of the 35 cliques formed in the network. This can be verified from the cliques table which lists all the 35 cliques (Table 1, K1 to K35) and their participating member list.

To strengthen the value of results, a filter of at least 3 interactions was imposed to the above clique analysis. And the results clearly demonstrate that the 5 actors (3 of whom are PD team members) have very strong relationships with every other actor in the network, forming crucial part of the network.

# 8. Conclusion

All of the above demonstrate that the PDT-Users who have been employed in the design and development phases of 'Learn Greek Online', have established long-term relationship, enduring interests, playing a vital role in the system. This signifies their sense of ownership over the system. This hence is in support of the initial hypothesis.

Hence by up-fronting effort and investment in the form of involving users (as in Participatory methodology) during the early phases of design and development of a system, can not only help in increasing the usability of a system, but also motivate them to have a long term relationship with the system.

# References

Garton L. et al (1999), Studying On-Line Social Networks in Jones S. (Eds) (1999), *Doing Internet Research,* Thousand Oaks : Sage Publications.

Hanneman R.A., http://faculty.ucr.edu/~hanneman/SOC157/TEXT/TextIndex.html, 10-06-2002, *Introduction to Social Network Methods.*

Preece J. (2000), *Online Communities – Designing Usability, Supporting Sociability* , Chichester : John Wiley & Sons.

Shneiderman, B. (1997), Designing the User Interface, Reading, MA: Addison-Wesley.

Zaphiris P.,Zacharia G. (2000), *Design Methodology of an Online Greek Language Course.* In Proceedings of CHI 2001 Conference on Human Factors in Computing Systems, March 31 – April 5. Seattle, WA.

Zaphiris P.,Zacharia G., *User-Centered Evaluation of an On-Line Modern Greek Language Course.* Proceedings of the WebNet 2000 Conference, October 23-27. Orlando, FL.

# A Comprehensive Process Model for Usable Information Architecture Systems: Integrating Top-down and Bottom-up Information Architecture

*Arno Reichenauer*

Siemens AG
Corporate Technology
Otto-Hahn-Ring 6
D-81730 Munich, Germany
Telephone: +49.89.636-40256
arno.reichenauer@web.de

*Tobias Komischke*

Siemens AG
Corporate Technology
Otto-Hahn-Ring 6
D-81730 Munich, Germany
Telephone: +49.89.636-41823
tobias.komischke@siemens.com

## Abstract

The term Information Architecture (IA) describes both a process and the resulting system. IA processes can be divided into top-down, bottom-up, and hybrid processes. Deficient IA systems are in many cases the cause for unusable web sites and intranets. A huge part of these system deficiencies can be traced back to deficiencies in the respective IA process. To remedy this root cause and to fully leverage the synergetic potential, we aim at a maximum integration of top-down and bottom-up activities in a comprehensive IA process model. The method for developing the process model and preliminary results are outlined.

## 1    Information Architecture: system and process view

The term Information Architecture (IA) can be understood in at least two ways: either describing specific elements of interactive systems (mainly web sites and intranets) related to their information organisation and navigation (= system view), or the underlying process yielding these elements (= process view). Still other denotations are in use (e.g. for the corresponding discipline / professional community), but are not relevant here.

### 1.1    System view

The main purpose of an IA system is to facilitate information finding and management. That way, it is at the same time both part of the end user's interface and part of the Content Manager's administration tools[1]. The building blocks of an IA system can be categorised in navigation, search, content, organisation and labelling components (Rhodes, 2000). Navigation and search components mainly support the end user in finding information. This can be a hierarchical menu to choose from or a search engine for keyword or full-text search. Components related to content and its organisation naturally address the Content Manager's needs. They include metadata

---

[1] A Content Manager can be defined as the person responsible for authoring, evaluating, organising, publishing, maintaining and storing content for a web site or intranet.

schemes, Controlled Vocabularies and taxonomies[2]. Labelling is the systematic application of terms used to represent larger chunks of information. Unlike the former components, this addresses the end user (e.g. as labels of navigation elements) just as much as the Content Manager (e.g. as a taxonomy's category labels). Of course, there is a vast number of intersections and interdependencies between these single building blocks, e.g. elements of a Controlled Vocabulary can be the basis for labels of a taxonomy which in turn can be used as a navigation system.

## 1.2   Process view

Obviously, any such categorisation will have limitations, and in fact this is only one of several discussed, each with different scope and connotations (see for example Rosenfeld & Morville, 2002). Similar to this heterogeneity in IA systems, IA processes also differ significantly, depending on the system view taken. A literature review suggests that they can be divided into three approaches:

(1) *Top-down IA processes* tend to be user centred, i.e., they focus mostly on user behaviour and needs to determine the web site's navigation and search. The resulting requirements define the navigation system, and the content is sorted likewise. Naturally, these processes resemble closely the processes of Usability Engineering or User-Centred Design, although User Interface Design in terms of developing a visual screen design is not always involved. However, most of the methods used here are known to Usability Engineers: interviews, workshops including card sorting exercises, persona and scenario development, paper prototyping, formative evaluations, etc.

(2) *Bottom-up IA processes* tend to be content centred, i.e. they focus mostly on how content can be described by characteristics inherent in each single information object in order to determine the organisation of the web site's content. This approach is rooted in disciplines such as Library Science and Content Management. The idea is to identify attributes which define or describe the information (i.e., metadata attributes; e.g. the name of the author). By assigning attribute values (e.g. "author name = Ms. Smith") to it, information is *classified*. Once information is classified, it also can be *categorised*, i.e. put in the appropriate position of a taxonomy. Unlike a top-down developed navigation system which is based on (maybe subjective but critical) user requirements, a taxonomy allows for an objective, unbiased organisation of information, based mostly on characteristics inherent in the information itself (i.e., "content requirements"). The methods include (manual) Content Inventories[3] and (automated) classification / categorisation tools[4], but for most of the work IA specialists have to rely on expert discussion and judgement.

(3) *Hybrid IA processes:* It can be assumed that in practice neither of both approaches presented above can be implemented alone. In fact, a minimum of both top-down and bottom-up approaches is indispensable (e.g., at least a high-level Content Inventory is needed to know what the content is, and to match it to the navigation system that was derived top-down). Hybrid IA processes formalise this notion, trying to explicitly balance their focus regarding bottom-up and top-down activities (see for example Veen & Fraser, 2001).

---

[2] Metadata is data about data and usually describes, summarises, or adds context to the data. A Controlled Vocabulary is a deliberately selected subset of natural language used to describe a specific domain; it is often used as the set of "valid" terms for a specific metadata element. Taxonomies are - broadly defined - (mostly hierarchical) structures of categories. Categories are labelled containers for content.
[3] A Content Inventory is a methodical review of a site's existing and future content.
[4] Automated classification and/or categorisation tools are mainly used in Content Management Systems.

## 2    The problem: unusable web sites and intranets

End users as well as Content Managers suffer from unusable web sites or intranets: end users cannot effectively find the content they look for, Content Managers cannot effectively manage the content (low effectiveness) or both of them only succeed with lots of effort (low efficiency). These problems escalate in collaborative portals, where virtually everybody is at the same time consumer (i.e., end user) and producer (i.e., Content Manager) of information. The consequences are manifold. To mention only some, business goals are not achieved; the costs for finding information (time, money & dissatisfaction) and *not* finding information (bad decisions, duplicated work, lost customers) increase; likewise, expenses for maintenance, support, training and hardware rise (Feldman & Sherman, 2001; Rosenfeld & Morville, 2002).

## 3    Cause for the problem: deficient IA systems

Our findings show that the problems end users and Content Managers experience day by day are in many cases due to deficient IA systems. If end users repeatedly have a hard time finding information on a specific web site which in fact is there, the IA system obviously fails to sufficiently take into account their needs and expectations regarding the web site's navigation and search systems, resulting in low effectiveness and/or low efficiency for the end user. But problems also arise if the information, once found, does not speak the users' language or cannot satisfy their informational needs. Even worse if essential information is not offered at all.

If Content Managers, on the other hand, cannot adequately describe information using the metadata values available, or if repeatedly there is no category that really fits, then the IA system obviously fails to support the adequate classification and categorisation of the web site's information, equalling low effectiveness and/or low efficiency for the Content Manager (both in turn weaken the system's usability for end users). But even before that, problems also arise if they have to compose content without understanding what content quality is desired. Even worse if they unknowingly have to manage information that the intended audience does not need at all.

## 4    Root cause: deficient IA processes

Our hypothesis is that a huge part of the problems an IA system holds can be traced back to deficiencies in the respective IA process used. We have identified three possible root causes:

(1) *Insufficient top-down IA:* unusable navigation and search systems are likely caused by an IA process that neglected the user's characteristics, needs and expectations.

(2) *Insufficient bottom-up IA:* if information cannot be properly classified and categorised, then the underlying IA process likely failed to fully leverage the potential of bottom-up IA.

(3) *Insufficient integration of top-down and bottom-up IA:* even if both a user-centred navigation/ search system and adequate classification / categorisation systems are accomplished, problems still can arise due to lack of sophisticated integration of the two single processes. If end user requirements and content requirements are not harmonized regarding content scope (redundant or missing content), terminology (end users and content do not speak the same language), or content quality (content is too superficial or too sophisticated), then the two processes failed to exchange information and base their work on the findings of each other.

# 5    Objective: an IA process model for usable IA systems

The (high-level) purpose of our efforts is to enable people to find and manage information effectively, efficiently and satisfactorily. To accomplish this for web sites and intranets with different business objectives, constraints, audiences and contents, our (immediate) objective is to develop a process model that incorporates our findings in a comprehensive process model. To remedy the root causes mentioned above and to fully leverage the synergetic potential, we aim at a maximum integration of top-down and bottom-up IA processes. Moreover, the characteristics of our typical IA projects demand that, to be efficient, the process model should be scalable (varying time and cost pressures), applicable in teamwork situations (distributed and multinational teams) and straightforward even for non-experts (team members are partly new to IA).

# 6    Method: result-driven development of an IA process model

The IA process model is developed and evaluated in close co-operation with project teams developing portals for a company ranked as one of Fortune's global top 30 (Fortune Global 500, 2002). The following sections present the steps that are taken:

(1) *Analytical investigation of Information Architecture concepts:* A literature review has been carried out to identify current IA systems with their building blocks and IA processes, process phases and single methods used. From these IA system and process instances we have induced general models of current IA systems and processes. An analysis of deficiencies of the reviewed IA systems is validated by usability testing with end users and interviewing Content Managers of the company's existing intranet for its more than 400.000 employees world-wide. The resulting model of deficient IA systems sets the stage for the to-be-developed process model twofold: First, from this deficient IA system model, a model of an optimised IA system can be derived. The resulting optimised IA system model shows which building blocks have to be addressed by the optimised IA process. Second, IA system deficiencies are traced back to their respective process deficiencies as outlined in Degen, Pedell & Schoen (2003), and these in turn have to be eliminated in the optimised process model.

(2) *Development of an evaluation instrument:* With the identified process and system deficiencies, we can define target criteria for the optimised process and devise an evaluation instrument regarding the quality of the resulting IA system. This will be a usability evaluation procedure for end users and Content Managers, combined with other methods like Card Sorting and its variants. The quality of the process itself is continuously evaluated with project team members using interviews and questionnaires. In that way, it is ensured that the process model not only yields effective and efficient IA systems (and thereby is an effective process), but that it is also an sufficiently efficient process itself according to the criteria defined in section 5.

(3) *Composing of the process model, execution and evaluation:* For each of the building blocks of the optimised IA system model, single process elements are introduced that output the building block. Next, input necessary for each process element can be defined. Based on the results of the analytical investigation in step (1) and on the ISO 13407 process model for Human-centred design processes (1999), we then can sketch the process flow as a flow of outputs and inputs. For each element of the process, an overview of applicable methods is given with the respective focus, costs and benefits. Depending on the characteristics of the actual projects, we will instantiate our process model with methods most suitable for the project at hand, and then execute, evaluate and iteratively improve our process model.

# 7 Preliminary results

The results of the analytical investigation have already been outlined in sections 1 to 3. For evaluating IA systems, we are currently using our established usability evaluation methodology. In future steps, we will incorporate other techniques as applicable. The first draft of the process model (see Figure 1) shows 5 distinct *phases* (column-wise, #1 to 5). In each phase, different process *elements* (#1.0 to 5.3) have to be performed regarding top-down, bottom-up and technical, implementation-related activities (row-wise). We call the sum of deliverables of the top-down process "Content Access System", while for the bottom-up process it is called "Content System". The first row basically shows a high-level view of an user-centred design process concentrating on navigation and search systems, while the second row focuses on content and its organisation systems. At the moment, we are detailing the single IA process elements and the interconnections between them. Further results will be published as soon as we have empirically validated the detailed process model.

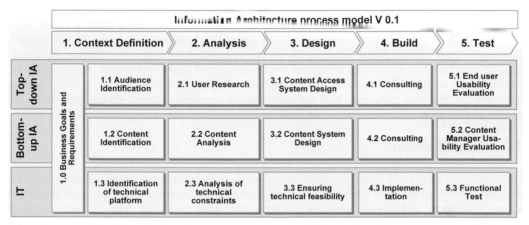

Figure 1: First draft of the Information Architecture process model

# References

Degen, H., Pedell, S. & Schoen, S. (2003). Configuring the Design Process – Applying the JIET Design Process Framework. *In this volume.*

Feldman, S. & Sherman, C. (2001). The High Cost of Not Finding Information. Retrieved July 31, 2002, from http://www.enfish.com/common/pdf/idc_white_paper.pdf

Fortune Global 500 (2002). Retrieved February 10, 2003, from http://www.fortune.com/fortune/global500/snapshot/0,15198,22,00.html

ISO 13407: Human-centered design processes for interactive systems. Berlin: Beuth

Rosenfeld, L. & Morville, P. (2002). Information Architecture for the World Wide Web (2nd ed.). Cambridge: O'Reilly & Associates.

Rhodes, J. S. (2000). The Intersection of Information Architecture and Usability. Retrieved August 20, 2002 from http://webword.com/interviews/head.html

Veen, J. & Fraser, J. (2001). Designing the Complete User Experience. Retrieved August 8, 2002, from http://www.adaptivepath.com/workshops/complete/

# A New Approach to Software Reuse Based on Interpretative Approach to Analogical Reasoning

*Nima Reyhani and Kambiz Badie*

Info-Society Dept., Iran Telecom Research Center
North Karegar Ave., Tehran, Iran
{nreyhani, k_badie@itrc.ac.ir}

## Abstract

With respect to the CBR approaches to software reuse, a number of difficulties may exists, out of which the necessity for high number of cases, the inability of a simple CBR in adapting the software solution to a new domain, as well as inability in transferring deep knowledge from a domain to another, are mentionable. To circumvent these problems, in this paper we propose a new approach to software reuse emphasizing on generating analysis and design diagrams for a new application domain based on an interpretative approach to analogical reasoning. It is seen that, in contrast with the classical approaches to analogical reasoning within which only the solutions belonging to the sources with predefined similarities are transferable, the proposed approach is capable of generating new solutions despite any lack of direct similarity between the source and the target.

## 1    Introduction

One important aspect of human computer interaction is to help the users create potential software solutions, which in ordinary cases call for an extensive effort from the user side. With respect to software analysis and design, reusable software has been proposed with a wide range of applications, mostly based on Object Oriented approaches, so as to utilize the essential requirements, diagrams, and codes of a successfully assessed software for a new problem. Analogical approaches (Gentner, 1983) in general, and Case Based Reasoning (Kolonder, 1993) in particular, have been successfully applied within this respect to take the responsibility of adapting the successfully assessed software solutions to new problem situations. With respect to the CBR approach to software reuse, a number of difficulties may exist, out of which the necessity for a high number of cases, the inability of a simple CBR in adapting the software solution to a new domain with a new characteristics, as well as the inability in transferring deep knowledge from a domain to another, are mentionable. These lead to the necessity of having an analogical approach, which can be creative in some manner, i.e., being able to produce competent solutions on the basis of only a few number of previously experienced cases. In this paper we propose a new approach to software reuse emphasizing on generating object oriented based design diagrams for a new application domain based on an interpretative approach to analogical reasoning (Badie, 2002).

## 2    An Overview on Some Existing Approaches to Software Reuse Through CBR and Analogical Reasoning

To help overcome conceptual complexity of software development processes, software designers should be equipped with intelligent tools to help them create new system using parts from previous ones (Gomes et al, 2002).

The main goal of software reuse is to shorten the development time of a software project improving its quality at the same time. Software development holds different levels of abstraction, each level dealing with several types of knowledge. To each of these levels corresponds a reuse level, varying from code reuse, to project requirements reuse, passing by software design reuse (Maiden & Sutcliffe, 1992). Most of the tools and systems developed for reuse purposes only provide help in retrieving software entities, like classes, functions or specifications, from repositories (Tautz & Althoff, 1998). However, reusing software involves also adaptation of the retrieved knowledge to the system being developed, which is actually left to the designer. Analogical reasoning (Gentner, 1983) appears as an efficient technique in this regard (Gomes et al, 2002). Within this respect, a computer aided software engineering tool called ReBuilder (Gomes et al, 2002) has been developed, which uses analogical reasoning to reuse software. Within this respect, an analogical engine helps the designer, provide mappings between the partial class diagram, described in terms of UML notation, of the system being developed, and class diagrams stored in a knowledge base of previous systems. It thus suggests new classes, attributes or methods to be added to the current class diagram. It seems that the above approach can suggest new entities to be added to the class diagram in the target, based on additional entities in the source class diagram, and thus can be regarded as a creative process for software design.

A CBR-oriented approach exists too, which works on the basis of the concept of experience factory; an organizational unit that supports reuse of experience and collective learning by developing, updating and providing, on request, clusters of competencies to be used by the project organization. Within this respect, definition of strategies or the way to join such an experience, should also be taken into account, where updating experience base is performed by employing a case-based reasoning strategy (Tautz & Althoff, 1998). The above approach makes use of semantic relationships among experience items to ensure the consistency of the experience base, to allow context- sensitive retrieval, and to facilitate the understanding of artefacts retrieved.

It should be noted that experience modification is a crucial stage with respect to the above approach, whose objective is to bridge the gap between the selected reuse candidates and given reuse requirements, where candidates can either be single, or multiple. It seems that any systems based on the above approach can help software designers be reminded of some methodology-oriented checkpoints approved by experience, which are essential to different stages of software development. This will turn into a sort of regularization or adjustment with respect to software development.

# 3    Creative Idea Generation

Classical approach to analogy is not sufficiently capable of suggesting a framework for creativity, mainly due to the following reasons:
- Creativity seems to be based on a sort of divergent thinking within which numerous alternatives are to be generated in the target space, whereas, in case of classical analogy, due to the monologue nature of structure mapping between the source and the target, such an objective seems to be hardly achievable.
- Creativity is a sort of clue-oriented process, within which new ideas are to be generated based on some clues from a previously structured relevant source(s), however classical analogy restricts itself to a predefined concept of similarity between the source and the target. In other words, in case of classical analogy, source has been assumed to be similar to the target in a predefined manner.

- In classical analogy, purpose of analogy is almost explicitly defined, whereas creative issues are instead concerned with a general intension or motivation of the creator.

To add the flavor of creativity to analogical reasoning, an intermediate space was proposed to be included between source and target, so that the inputs in the source can be interpreted in terms of some navigating concepts in the layer for further reinterpretation in the target space (Badie, 2002). This yields the possibility of producing numerous alternatives as the final ideas. Moreover, the co-working of the two interpretation and reinterpretation actions, one from the source into the intermediate space and the other from the intermediate space into the target space, eliminates the urgency for depending on inherent similarity between the source and the target. Within the above context, the interaction between the creator's intention and the intermediate space can yield generating particularly surprising ideas.

## 4    Our Proposed Approach

In our approach, we have made use of UML notation to present a suitable description of the processes as well as the results obtained in an OO process software analysis and design.

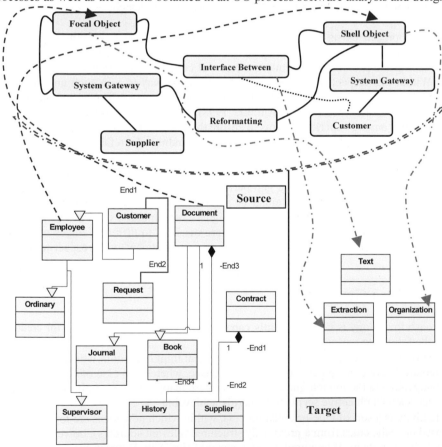

**Figure 1.** Illustrates an example within which the diagram developed for the purpose of the "Library Management" finally lead to generating a diagram for the purpose of "Text Mining"

Using the idea discussed in 3, our goal is to generate an appropriate design related diagram in the target, using a previously developed design diagram in the source, via interpreting the main objects of the source diagrams, in terms of some navigating concepts in the intermediate space, and reinterpreting these navigating concepts in terms of appropriate concepts in the target. In this respect, it is first necessary to define some appropriate concepts in the intermediate space, which can be regarded as interpretable forms of the "classes" (or "objects"), "inter-class relations", and "class attributes" as well as "class methods" of a object oriented diagram (Coad & Yordon, 1991). "Interclass Relation", is also in many cases interpretable in the same way defined in source diagram, and thus is to appear similarly in the target. Figure 1 illustrates a simple example within which the class diagram developed for the purpose of the "Library Management" will finally lead to generating a class diagram for the purpose of "Text Mining". As it seen from the figure, the class "entity", as a basic class for "Library Management", is first interpreted in terms of the navigating concept "focal object", and will then be reinterpreted in terms of the class "document" in the target. Interestingly, even if no alternative exists for interpreting some of the classes, as far as its relation with another class (or other classes) is (are) determined, the chance will exist to acquire its counterpart in the target. For instance, provided that there is no interpretable concept for the class "employee", as far as we know that it holds the relation "organizing" with the class "entity", which is interpreted in the same way in the intermediate space, the counterpart of the "employee" will finally be determined in the target through searching for a concept which holds the relation "organizing" towards the class "document". This, according to the knowledge in the target, can be decided to be "classification". Another interesting point is that such an approach to analogical reasoning provides the ability to generate extras ideas in the target through reinterpreting these concepts in the intermediate space, which do not correspond to specific concepts in the source, but are somehow connected to other concepts in the intermediate space. For instance, the interclass relation "interface-between" hasn't been originally cared in the source, but since it is indeed the relation between a "focal object" and the one "organizing" it, can be reinterpreted in terms of the concept "feature extractor" which has been defined to be the corresponding class-type relation between the classes "document" and "classification" in the target.

As mentioned earlier, creative issues are concerned with a general intension or motivation in the creator, which determines the way the essential structure consisting of the navigating concepts is to be shaped. Regarding this, for the final diagram in the target to be creative, the navigating concepts obtained through interpreting the source diagram should be subject to change based on the very intensions the creator is following within his/her design & analysis task. This means that the major items, which are significant to the creator, should somehow be reflected in the way the navigating structure is being shaped in the intermediate space, or reinterpreted in the target. A partial ontology of these items is illustrated in Figure 2.

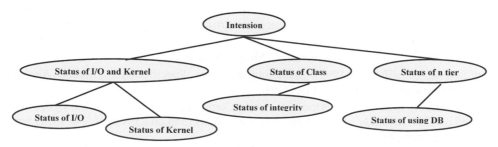

**Figure 2**. A partial ontology for the major items significant to the creator

It should be noticed that, based upon the node in the above ontology (as the intension of the creator), both the reinterpreted versions of the navigating concepts ("focal object", "interface-between", "reformatting", …) as well as their status of relations in the target will change. This provides a conducive medium for producing a well-tailored final analysis & design diagram in the target, based upon the analyzer/designer's intension.

## 5    Concluding Remarks

In this paper, we presented a framework for generating software solutions for a new problem based on an interpretative approach to analogical reasoning. A salient difference between this framework and the analogical framework proposed in (Gomes et al, 2002), is the ability of the proposed framework in handling those sources, which are not similar to the target in a particular way, and it is rather the co-working of the two processes of interpretation and reinterpretation which makes possible the transference of the crucial concepts from the source to the target. It is seen that any knowledge in the source diagram, which can be represented in terms of a semantic net and/or a frame, can be transferred to the target through this approach, if an appropriate intermediate space including ample of concepts can exist.

An important point in the proposed approach is the ability of using a non-software domain as sources for generating helpful patterns for designing software solutions. Example can be mentioned for designing an office automation system for a publishing company, based on the ideas related to the configuration of an industrial plant. Another important point in the proposed approach is the role of intension in guiding the creation process. In this view, one can think of a software creation process, within which the user's intention is being presented to the system in an interactive manner. The proposed approach can thus be helpful for complex design processes, which call for occasional interaction of human designer via presenting various components of their intention.

As a final conclusion, it is expected that the proposed framework can be used as a potential tool for intelligent computer aided design (ICAD) purposes.

## References

K. Badie (2002), "Creative Idea Generation via Interpretative Approach to Analogical Reasoning", *Kybernetes*, 31(9), Emerald Pub. Company, UK.

Coad, P. & Yordon, E.(1991), "Object Oriented Analysis", 2nd Edition, Yordon Press.

Gentner, D., "Structure Mapping: A Theoretical Framework for Analogy" (1983), *Cognitive Science*, 7(2), 155-170.

Gomes, P.,C. Pereira, F., Paiva, P., Seco,N., Carreiro, P., José L. Ferreira, & Bento, C. (2002). "Combining Case-Based Reasoning and Analogical Reasoning in Software Design". Proceedings of the 13th Irish Conference on Artificial Intelligence & Cognitive Science (AICS'02).

Kolonder, J. (1983), "Case-Based Reasoning", Morgan Kaufman.

Maiden, N. & Sutcliffe, A.(1992),"Exploiting Reusable Specifications Through Analogy", *Communications of the ACM*, 35(4), 55-64.

Tautz, C., Althoff, K.D. (1998), "Operationalizing Comprehensive Software Knowledge Reuse Based on CBR Methods", Proc. of the 6th German Workshop on CBR, Vol. 7 of IMIB Series, Berlin, Germany, 89-98.

# MenuSelector: Automated Generation of Dynamic Menus with Guidelines Support

*Jeremy Spoidenne, Jean Vanderdonckt*

Université catholique de Louvain (UCL)
School of Management (IAG) - Unit of Information Systems (ISYS)
Belgian Laboratory of Computer-Human Interaction (BCHI)
Placc des Doyens, 1 – B-1348 Louvain-la-Neuve (Belgium)
vanderdonckt@isys.ucl.ac.be

## Abstract

This paper presents a software tool that enables developers to quickly specify design options for a menu bar and related menus and sub-menus for any type of application. These design options are then stored in a universal user interface specification language (XIML) for later use. At any time, the current design of the menu can be checked against some basic usability guidelines, depending on currently being specified options. Once a menu is ready for full specification, it is then subject to automated generation of MacroMedia Flash code. Traditional menu abstractions are frequently based on the paradigm of Graphical User Interfaces. Here, we attempted to expand this paradigm to any type of menu interactor, with static or dynamic features.

## 1 Introduction

Incorporating guidelines in software tool to assist designers and developers has been for a long time a research and development issue that still remains not completely solved. Separating the user interface (UI) design phase from the evaluation phase has been criticized for a long time: not only both phases are decoupled, but the developer may become confused when the evaluation report, along with some feedback, is produced because she is no longer in a design context. This change of focus of attention may disrupt developers. To overcome this problem, we believe that producing evaluation feedback while designing a UI would be better appreciated by developers. While they are designing and developing a UI, they could at any time trigger an evaluation.

For this purpose, we developed a test-bed application, MenuSelector, that allows developers to design a menu bar with menus and sub-menus for any type of interactive application. Then, the developer can perform a guidelines checking by asking the software to directly assess the currently being designed menu. We choose the menu selection interaction style as many usability guidelines exist today to characterize whether a menu is usable or not, for instance (Norman, 1990). In particular, some guidelines (Kiger, 1994) (Norman & Chin, 1988) establish boundaries for keeping a usable trade-off between the breadth and depth. More specifically, even age differences of users can influence this trade-off (Zaphiris, 2001). To reach this goal, a new abstraction of the menu concept was required to specify not only traditional aspects (like menu items, breadth, font, size), but also non-traditional aspects (like presentation, animation effect). When designing a menu, the developer can specify appropriate values for any of these parameters. MenuSelector then stores this specification in XIML (exTensible Interface Mark-up Language) (Puerta & Eisenstein, 2002) for future use. For some combinations of these parameters, MenuSelector automatically generates MacroMedia Flash code in vector presentation style (SWG). Automated generation of menu from a conceptual schema has already been investigated and proved feasible (Vanderdonckt, 1998). However, incorporating support for guidelines evaluation has been underexplored so far (Scapin *et al.*, 2000).

## 2 The New Menu Abstraction

To support the specification of advanced menus, possibly with dynamic aspects, for traditional Graphical User Interfaces (GUIs) or for non-WIMP menus, possibly in three dimensions, a new abstraction of the menu artefact was needed. This abstraction consists in defining a series of parameters for each menu where appropriate. These parameters are summarized in Table 1.

| Menu | General organization | Full screen, localized |
|---|---|---|
| | Amount of spatial dimensions | 1D, 2D, 2D½, 3D (Vanderdonckt, 2003) |
| | Shape of main presentation axis | Linear, curve, polygonal |
| | Menu disposition | Horizontal, vertical, oblique, circular, polygonal, mixed |
| | Menu orientation | If vertical: left, right<br>If horizontal: top, bottom |
| | Modality types | Textual, iconic, bimodal |
| | Status | Compact, full |
| | Representation | Implicit, explicit |
| | Duration | Static, dynamic, persistent |
| | Apparition | Animated, not animated |
| | Navigation | Continuous, by selection |
| | Length | Fixed, variable |
| Sub-menu | Apparition | Expandable: drop-down, cascade, sub-menu, other<br>Not Expandable |
| | Reference to lower level | Upper level complete, by title, inexistent |
| Menu item | Type | Command item, dialog item, sub-menu item, toggle item, radio item |
| | Highlighting method | None, for "on-mouse over" state, for "on-mouse down" state, similar for both, different for both. |

**Table 1:** Parameters of the new Menu abstraction.

Figure 1 shows a menu with the following parameters: localized (the menu only consumes some part of the screen, not the whole screen), 3D (as the presentation of items are in three dimensions, linear (items are aligned on a horizontal axis), horizontal menu disposition, items are oriented from left to right. Each menu item is bimodal as icons are augmented with menu labels. The menu is full (no alternative presentation depending on the user for instance), the representation is explicit, their duration is static (items are presented only during the menu selection), they are animated with a 3D special effect, the navigation is continuous, and the length is fixed.

**Figure 1:** Example of non-traditional menu

# 3    Tool support

A prototypical tool has been implemented in Microsoft Visual Basic V6.0 that allows developers to specify any menu according to the parameters defined in the menu abstraction. The tool verifies that each menu specification is compliant with some basic guidelines. For instance, it is possible to check that the amount of items at the first level does not exceed the threshold prescribed by usability guidelines. Therefore, if the developer is attempting to introduce to many items at the first level with too few items on the subsequent levels, the tool may suggest re-ordering items.

Depending on previously entered specifications, the tool activates or deactivates further possible specifications (Fig. 2a). For instance, when choosing a design option is restricting the scope of other design decisions, only the possible choices are displayed and activated. For example, when the amount of spatial dimensions is specified as 2 (Fig. 2b), the developer can specify various shapes for the principal axis along which the menu items will be presented, here a curve. Due to this choice, the tool restricts the range of potential values for the other parameters.

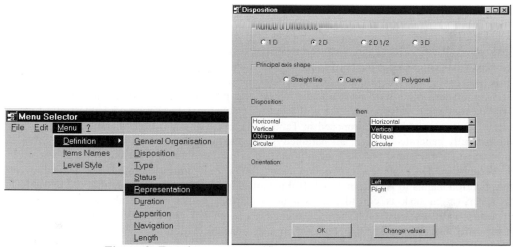

**Figure 2:** Entering menu specifications with MenuSelector.

It is important for developers not to feel forced to enter all parameters at the same time. Indeed, UI design process remains very flexible, largely incomplete or underspecified at the beginning. The early requirements may evolve with the rest of the interactive application during the traditional development life cycle. As this cycle is progressing, new requirements can be elicited that can be transformed into new or updated specifications for the menu. At any time, the developer can specify the hierarchy of menu, how many levels, what are the items (the text and the icon where appropriate), what kind of transitions should be supported in the menu selection. This can be updated at any time. Although the abstraction represented in Table 1 goes further than the usual menu definition, developers should be enabled to add any extra, possibly unsupported, definition of parameters.

Each menu specification is internally stored in XIML (exTensible Interface Markup Language) (Puerta & Eisenstein, 2002), which is a UI specification language supporting various models (i.e., task, domain, user, presentation, dialog, platform, and general purpose). Each model may contain attributes defined locally or globally. Between models or within models, relationships can be defined. For the basic parameters defined in table 1, a canonical representation has been defined in

XIML so as to express any combination of these parameters. For custom, possibly unsupported, parameters, the general purpose attribute definition is used. Therefore, each menu is modelled in a presentation model, which is recursively decomposed into presentation elements. At the first level (e.g., a menu bar), a first presentation element of the class "Menu Bar" is instantiated. To add pull-down menus, instances of the class "Pull-down menus" are created. Similarly, for any sub-menu or final menu item, each parameter can be specified where appropriate. This XIML specification can then be integrated with a larger UI specification, for instance the other models. Or the menu definition can itself become a particular presentation element of the global presentation elements hierarchy. At any time, the current specifications can be exported in a XIML-compliant file to be integrated in a larger UI design repository or for another tool to generate final code. A summary of the specifications can be displayed in XIML format on demand. As this XIML file is itself XML-compliant, it can be the source of any XSLT transformation so as to create other definitions, or XML compatible UI definitions.

MenuSelector is intended to automatically generate code corresponding to the menu in MacroMedia Flash format. This language allows much more flexibility in the presentation styles and dynamics than traditional mark-up languages (e.g., HTML) or programming languages (e.g., Visual Basic), unless special libraries are used (e.g. AfterEffect). Fig. 3 represents a 1D vertical menu where the first-level items can be deployed at any time and separately. Clicking on any coloured tab open or closes the corresponding sub-menu. Contrarily, fig. 4 presents the same menu items, but in a purely hierarchical menu, where only one level of menu can be open at a time. Clicking on any other tab automatically closes the previously opened menu. Although this variation may seem very limited, the tool allows the developer to change parameters at a higher level of abstraction (the one proposed in Table 1) than at the code level. MenuSelector does not automatically generate code for all parameters combinations that can be inferred from Table 1. Rather, some parameters, that have been identified as fundamental, are currently being supported. The exported XIML file can be edited manually or may initiate another UI generation or interpretation thanks to an external tool. If this tool supports the abstraction introduced in Table 1, the specification task is considered as a good starting point. Not all abstractions should be supported by all external tools. Some parameters can be left unsupported, such as the general purpose attributes that can be kept only for the sake of specifications.

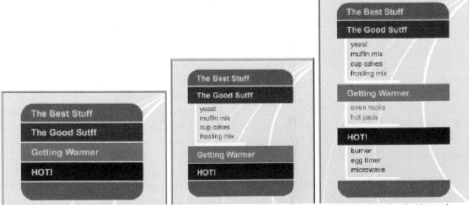

**Figure 3:** Example of a hierarchical menu with possible multiple levels deployed
(Source: http://www.freehandsource.com/_test/hierarchical_menus.html )

**Figure 4:** Example of a purely hierarchical menu

## Appendix

The web site of MenuSelector (Computer-Aided Design of Animated Web Menus in Flash) is accessible at http://www.isys.ucl.ac.be/bchi/research/selector.htm

## References

Jacko, J., & Salvendy, G. (1996). Hierarchical Menu Design: Breadth, Depth, and Task. *Complexity. Perceptual and Motor Skills*, 82, 1187-1201.

Kiger, J.I. (1984). The Depth/Breadth Tradeoff in the Design of Menu-Driven Interfaces. *International Journal of Man-Machine Studies*, 20, 201-213.

Norman, K.L., & Chin, J. (1988). The effect of tree structures on search in a hierarchical menu selection system. *Behaviour and Information Technology*, 7, 51-65.

Norman, K.L. (1990). The Psychology of Menu Selection: Designing Cognitive Control of the Human/Computer Interface. Norwood: Ablex Publishing Corporation.

Puerta, A., & Eisenstein, J. (2002). XIML: A Common Representation for Interaction Data. In Proceedings of 7th ACM International Conference on Intelligent User Interfaces IUI'2002 (San Francisco, January 13-16, 2002). New York: ACM Press. Accessible at http://www.iuiconf.org/02pdf/2002-002-0043.pdf

Scapin, D., Leulier, C., Vanderdonckt, J., Mariage, C., Bastien, Ch., Farenc, Ch., Palanque, Ph., & Bastide, R. (2000). A Framework for Organizing Web Usability Guidelines. In Kortum, Ph. & Kudzinger, E. (Eds.), Proceedings of 6th Conference on Human Factors and the Web HFWeb'2000 (Austin, 19 June 2000). Austin: University of Texas. Accessible at http://www.tri.sbc.com/hfweb/scapin/Scapin.html and http://www.isys.ucl.ac.be/bchi/publications/2000/Scapin-HFWeb2000.htm

Vanderdonckt, J. (1998). Computer-Aided Design of Menu Bar and Pull-Down Menus for Business Oriented Applications. In Martins, F.M. & Duke, D.J. (Eds.), *Proceedings of 6th International Eurographics Workshop on Design, Specification, Verification of Interactive Systems DSV-IS'99* (Braga, 2-4 June 1999). Vol. 2. Vienna: Springer-Verlag, pp. 73-88.

Vanderdonckt, J. (2003). Visual Design Methods in Interactive Applications. Chapter 7. In Albers, M. & Mazur, B. (Eds.), Content and Complexity: Information Design in Technical Communication. Mahwah: Lawrence Erlbaum Associates, pp. 187-203. Additional web page accessible at http://www.isys.ucl.ac.be/bchi/members/jva/pub/visual.htm

Zaphiris, P. (2001). Age Differences and the Depth-Breadth Tradeoff in Hierarchical Online Information Systems. In Stephanidis, C. (Ed.). Proceedings of 1st International Conference on Universal Access in HCI UAHCI'2001 (New Orleans, August 5-10, 2001). Mahwah: Lawrence Erlbaum Associates, pp. 540-544.

# From Scenarios to Interactive Workflow Specifications

*Chris Stary*

University of Linz, Department of Business Information Systems
Communications Engineering, Freistädterstraße 315, A-4040 Linz
Christian.Stary@jku.at

## Abstract

Scenario-based design has become a widely acknowledged approach in interactive-system as well as organizational development. However, from the methodological perspective when using work scenarios for workflow-driven design a bundle of issues is still open. One of the key questions still is: Which knowledge has to be represented and refined for workflow-based design? Tool support to that respect requires processing that knowledge in a workflow-oriented way. The paper deals with the ontology conform to work processes, and its implementation for application generation.

## 1    Introduction

In today's companies most of the workflows are usually supported by interactive information systems along business-process definitions. According to these specifications people work together to achieve some desired output to reach the business objectives of the company (Ould et al., 1995). Analyses of traditional workflow specifications have shown, e.g., Alonso et al. (1995), that the particular characteristics of work, such as context-sensitive collaboration, cannot be represented through traditional entities and relationships for modelling, and thus, cannot directly contribute to organizational or technical advancements. In addition, the interactive system community still (cf. Siewiorek, 2002, Stary, 2003) states a need for context-aware development methodologies to come up with work-conform solutions. Scenario-based design (Carroll, 1995, CACM, 1995) is a candidate to be part of such methodologies. However, most of the existing development procedures, although providing representations for tasks and/or workflows, e.g., Brusilovsky (2002), lack ontologies that actively support mapping work-task knowledge to interactive workflow specifications.

In this paper we first revisit the conceptual elements for representing workflows (section 2). We then propose a solution on how to represent, refine and process work items for workflow-driven application design (section 3). The ontology is being implemented in the portal developer KnowIt! (Stoiber et al., 2002) based on the results of the TADEUS project (Stary, 1996). The paper concludes with an appraisal of results and topics for further research (section 4).

## 2    Workflow Specifications based on Business Process Definitions

In general, a workflow is considered as a collection of activities for the accomplishment of tasks. Each sequence of activities has to be performed in accordance with the business processes of the organization at hand (IBM, 1994). The workflow is composed of work objects and related control information. Both are passed from one user to another for stepwise task accomplishment within the organization. In a workflow model the following minimal information should be associated with each activity (Stary, 1996): (R1) Pre- and post conditions (what conditions must be met before an activity starts, and when an activity is completed); (R2) Who has control over the activity; (R3) Which other activities are required to complete the activity; (R4) The input/output of the activity, i.e. the data and control information required for task accomplishment. To represent all the required information addressed above a traditional workflow model captures business intelligence through a variety of elements:

- *Organizational unit*: Describes departments, positions etc. within the organization structure of a company.
- *Role*: Describes employees. Staff is assigned to roles. A role can be captured from more than one employee and vice versa.
- *Activity*, also called *task* or *function*: Piece of work within a business process.
- *Data*: Contains all data relevant for the business process.
- *Event*: Describes an incident which might occur during the accomplishment of an activity.
- *Material*: Represents a device, a machine, a tool etc. which is required to accomplish a task.
- *Process*: Contains several activities in the order and the constraints under which activities have to be accomplished to attain a business goal of the company.

# 3　Mapping Work Scenarios to Interactive Workflow Specifications

**The Ontology.** According to the requirements for workflow specification, several conceptual entities have to be part of the scenarios used in the course of analysis, design and for the mapping to executable specifications: The *organization* and/or *organizational units* are required to provide a framework for the representation of application development. They might represent the set of departments the organization is composed of or hierarchical layers within organizations. *Positions* and *persons* capturing functional roles (meeting R2) are required to achieve a comprehensive representation of the organizational setting, and to model users. *Activities* are all actions that might be performed by persons or machines, as they occur in the course of task accomplishment (meeting R3). They are part of tasks or sub tasks. They might also depend on mutual temporal and causal constraints (meeting R1), thus being part of the workflow to be supported. *Material* can either be data or physical objects that have to be processed in order to accomplish a task. Usually, they are assigned to activities and manipulated in the course of task accomplishment, finally representing the results of work processes (meeting R4). *Events* are those points of reference that might have to be considered as being particular for the workflow to be supported. Events might lead to specific procedures, such as handling emergency cases. Events might also be described at several levels of detail, in order to understand the consequences when particular situations occur (meeting R1). Finally, *tools* and *IT-resources* are means for task accomplishment.

Several conceptual relationships link the conceptual entities listed above: *employs, has, is filled, is superior, is a, has part, handles, creates, informs, controls, requires, before, is based on*. They do not only support the structural perspective on work processes and the flow of work, but also the mapping of conceptual entities, such as work tasks to problem domain data through 'is based on'. The relationships are also used to integrate perspectives or views on the development knowledge, in the following termed models. Figure 1 shows an application of the 'handles' and 'creates' relationship denoted as links between entities (rectangles). It links a position to an activity and material being processed in the course of course-preparation by a professor. The ontology allows to map task specifications, such as 'Select Option' for a cinema visit in Figure 2, to problem domain data, such as 'Film' in the figure. The same way it allows the mapping to interaction styles, such as form-based interaction for acquiring options for the cinema visit, using 'is-presented' between task or data elements and interaction elements. Table 1 shows the major inter- and intra-relationships of models that are used beyond the traditional object-oriented relationships. For instance, 'before' is a typical relationship being used within a model (in this case, the Task Model), whereas 'handles' is a typical relationship to connect elements of different models (in this case the User and the Task model – a task is handled by a position or person). As shown above, the (declarative) relationships between ontology elements are used for (i) accurate domain specification, (ii) for tuning perspectives, and (iii) for mapping the Task Model to an executable Application Model. For each of the relationships the support environment has to provide algorithms that check their use.

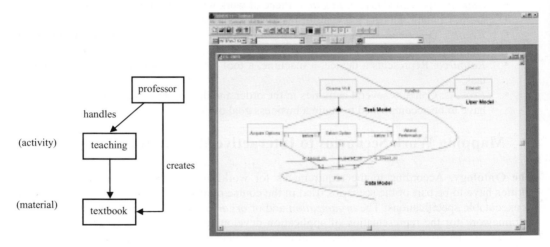

**Figure 1:** An Instance of the Ontology

**Figure 2:** Tool-based Model-Based Design

**Table 1:** Intra- and Inter-Relationships of Models to Tune Perspectives

|  | Task M. | User M. | Data M. | Inter. M. |
|---|---|---|---|---|
| **Task Model** | Before |  | Is-based-On | Is-presented |
| **User Model** | Handles Controls | Informs Controls | Creates |  |
| **Data Model** |  |  |  | Is-presented |

**Engineering an Application Prototype.** Using a diagrammatic notation for specification, such as UML, requires:

- Relationship Diagrams describing the structural relationships between ontology elements
- Behavior Diagrams describing the behavior (dynamics) of ontology elements
- Interaction Diagrams describing the interaction between life cycles of elements.

Through scenario definition those tasks are selected that are expected to be supported through interactive software. These tasks are refined and related to each other, according to their sequence of the business processes in a Relationship Diagram RD – the sequence is specified through the 'before' relationship. This relationship is mapped to BDs together with the involved tasks in the course of design. Initially, the tasks are specified in an object-oriented way in terms of abstract classes (containing only identifier and description), the structure and behavior relationships as links between the classes. For instance, the global task Cinema Visit might be decomposed into sub tasks like shown in Figure 2 (rectangles denote classes, links relationships, such as the black triangle 'has part'). This specification is the starting point for mapping to data and interaction model elements, leading to instances of application specifications. Each model consists of RDs and BDs that are linked to RDs and BDs of other models given the relationships in Table 1. Linking BDs constitute IDs. For each sub task at the end of an aggregation line of a global task, such as Select Option in Figure 2, the workflow to be followed has to be defined at the ID level. As a consequence, each RD of the task model is related to a set of BDs containing the workflow of the task.

The structure part of the User Model (i.e. again a set of RDs) comprises user (group) definitions. There are two ways to define users or user groups, namely the functional and the individual perspective. For instance, each department of the organization has a particular set of privileges, such as the right to manipulate film data in case of entertainment management (functional perspective). Each staff member also has a user profile based on individual skills and preferences, such as the use of button bars instead of menus. The integration of both perspectives has to be performed at the level of User-Model RDs. The integration has to propagate to the concerned problem domain data (in the Data-Model RDs) and the dialog elements (in the Interaction-Model RDs). Finally, the BDs of the involved models propagate the behavior to an executable ID.

Coupling the User Model with the Task Model requires the use of 'handles' relationships. The dynamic part of the User Model captures the work process from the perspective of a particular person. For instance, an employee might be involved in tasks that are processed by several departments, such as an accountant performing human resource management as well as sales calculations. Hence, from the task perspective, the dynamic part of the User Model is a synchronized combination of Task-Model BDs under the umbrella of a particular user role. The specification has to show the task-relevant, synchronized involvement of a particular user group in one or more subtasks. This step can be achieved through synchronizing BDs, i.e. specifying the passing of flow control between object life cycles.

Mapping tasks to problem domain data defines the resources required for task accomplishment. Identifiers, attributes, operations and relationships have to be provided. For instance, handling film requests requires to create an object of the class Film. Setting up a Data Model is also required, in order to provide information for the mapping of the data-related functionality to the interaction facilities later on (such as assigning input fields to data that are expected to be entered by the user). In order to ensure the integrity and completeness with respect to the tasks that are going to be supported, the elements of the structure part of the Task Model have to be mapped (i.e. directly linked) to the elements of the structure part of the Data Model at the level of RDs. This step is achieved through setting the 'is based on' relationship between (sub) tasks and data specifications. For instance, the task Select Option 'is based on' Film. Additionally, the declarative relationships to the User Model have to be mapped to the Data Model to checked whether the access permits given through the role specification in the User Model fit to the specified data-model elements, and vice versa, whether each of the data classes has been actually assigned to at least one functional role specified in the User Model. The behavior of the problem domain data has also to be specified through life cycle diagrams. For instance, the life cycle of Film has to be defined, according to the attributes and methods specified in the class Film. In case of multiple involvement of a data element in several tasks, such as Film for handling film requests (person: cinema visitor) and updating film offers (position: offer manager), the dynamic specification integrates different behavior specifications in a single representation. It captures all possible states of that data element. Finally, the life cycle has to be synchronized with one or more BDs of the dynamic part of the Task Model, since each of the transitions concerning data has to be performed in conformance to the tasks specified in the dynamic part of Task Model (BDs). The results of this synchronization process are again OIDs. Once the dynamic parts (BDs) of the Task, User, and Data Model have been synchronized with the Interaction Model, an interactive workflow specification (i.e. Application Model) has been completed, and can be executed by a platform, such as a GUI builder, or a workflow system.

# 4 Conclusion

Although scenario-based design has become a widely used acquisition and design approach in interactive system and organizational development, mapping task and user specifications to application specifications at the structural and behaviour level is still an open issue. We suggest an active ontology used throughout the different phases of development. We have developed a scheme for representing design knowledge that captures work scenarios and supports the refinement to interactive workflow specifications. The latter can be processed for application generation and user-interface prototyping.

# 5 References

Alonso G., Günthör R., Kamath M., Agrawal D., Abbadi A. El., Mohan C. (1995). Exotica/FMDC: Handling Disconnected Clients in a Workflow Management System, 3rd International Conference on *Cooperative Information Systems*.

Brusilovsky, P. (2002). Domain, Task, and User Models for an Adaptive Hypermedia Performance Support System, Proceedings *IUI'02*, 23-30.

CACM (1995). Requirements Gathering: The Human Factor. *Special Issue of the Communications of the ACM*, 38(5).

Carroll, J.M. (ed.) (1995). Scenario-based Design: Envisioning Work and Technology in System Design, New York: Wiley.

IBM (1994). Flowmark, Modeling Workflow, Release 1.1, SH19-8175-01.

Ould M.A. (1995). Business Processes - Modelling and Analysis for Re-engineering and Improvement, New York: Macmillan.

Siewiorek, D.P. (2002). New Frontiers of Application Design. *Communications of the ACM,* 45(12), 79-82.

Stary, Ch. (1996). Integrating Workflow Representations into User Interface Design Representations, *Software - Concepts and Tools*, 17, 173-187.

Stary, Ch. (2003). Shifting Knowledge from Task Analysis to Design – Requirements for Contextual User Interface Development, *Behavior and Information Technology*, 21(6), 425-440.

Stoiber, S. & Stary, Ch. (2002). Organizational Learning Online, in: Proceedings *HICSS-35,* Hawai'i International Conference on System Sciences, IEEE, on CD-ROM.

# Synthesising Creativity: Systems to support interactive human processes for aesthetic product design

*Modestos Stavrakis*
modestos@aegean.gr

*Thomas Spyrou*
tsp@aegean.gr

*John Darzentas*
idarz@aegean.gr

University of the Aegean
Department of Product and Systems Design Engineering
Ermoupolis, Syros,
GR 84110, Greece

## Abstract

Current technology systems, focusing in product design, do not provide a complete creative environment. This paper proposes the 'fusion' of artificial agents and human designers in a synthetic environment for aesthetic product design. Computational mo dels alone cannot fulfil the cycle of aesthetic product design because they are unable to evaluate variables related to aesthetic features of products/artefacts intended for humans. To overcome this limitation this paper proposes a system based on agents that act as a society and interact with designers in order to support their creative processes for aesthetic product design.

## 1 Introduction

Human creativity is both special and ordinary. Humans are beings known for their creativity their ability to make art and literature, and to produce their own artificial tools and environments (Smith, 1997). However, at present, tools do not support creativity in any significant way. "Creative design requires deeper rather than shallow analysis and the current CAD techniques are not necessarily promoting this" (Lawson, 1999). The aim of this research is to identify requirements of computer support for interactive creative work by exploring the potential of new technology for aesthetic product design. Fundamental to this work is an understanding of how creativity, in product designing, works. The objective is to enable designers to produce innovative artefacts and study the implications of future technologies in aesthetic design. In addition, a key area is to understand the requirements for support structures for creative practitioners.

Based on human centred designs that collaborate with new media, this research builds upon existing art forms and art practice to eventually lead to entirely new ways of working/designing, and very probably new genres of artefacts, that could be mostly digital.

There are great opportunities to expand the range of tools that support and amplify the creative processes. Current research looks at this by studying human creative actions/behaviour and identifies methods that can aid designers in producing novelty for product/artefact aesthetic design, by interacting with intelligent computational systems that act as the new creative medium.

The research described here is focusing on a society of artificial agents that observe human processes, explore the human-computer interactions that support the emergence of ideas and evaluate the capacity of novelty of the artefact; therefore, the evolution of the design.

This paper presents an approach to creating a support environment for creativity, which is based on a community of adaptive agents that act as a mechanism to motivate users and support them during the conceptual phase of the creative process of aesthetic product design, by observing their actions\behaviour and suggesting methods that extend or move the state space of potential designs. Furthermore the system's architecture will support the sharing of knowledge, among the society of heterogeneous designers, and assist designers' creativity to emerge, in a manner similar to that occurring in group design teams (Edmonds, 1994)(Candy, 2000-2002)(Mamykina, 2002).

## 2 Models of Creativity

Creativity is a term inadequately defined, although there are many definitions in psychological literature (Taylor, 1988). The fuzziness of the word is related to the variety of the domains to which the meaning of creativity refers.

There are many theoretical models proposed and many empirical studies introduced in order to provide detailed accounts of the processes involved in creativity (Saunders, 2002). Finally there is an attempt to merge all previous research by unified models that classify creativity in terms of theoretical frameworks. Boden outlines two different interpretations:

- P-creativity or personal/psychological creativity, where concepts or ideas are novel to the designer's mind from which the initial idea emerged
- H-creativity or historical creativity, where the idea is truly novel and have never been thought before; (Boden, 1990).

An important extension to Boden's model is the S-creativity model introduced by Gero:

- S-creativity or situated creativity occurs in designing when the design contains ideas which were not expected to be in the design when the design was commenced. Thus the design contains ideas that are not necessarily novel in any absolute sense or novel to the designer but that are novel in that particular design situation (Suwa, M., Gero, J. S. and Purcell, T., 2000).

In addition to these models (Shneiderman, 2000), identifies the human creative processes, when interacting with computer systems, and distinguishes four phases of action: Collect Relate, Create, and Donate. These are similar to the characterizations of (Couger, 1996): Preparation, Incubation, Illumination and Verification.

Based on these four processes Shneiderman introduces eight activities of human-computer interaction: Searching and browsing digital libraries, Consulting with peers and mentors, Visualizing data and processes, Thinking by free associations, Exploring solutions- What if tools, Composing artefacts and performances, Reviewing and replaying session histories and Disseminating results.

Finally, for creativity taking place in synthetic environments the research of Edmonds (Edmonds, 1999) considers that designer 'manipulates' his imaginative artefact in three different realities: Imagined Reality, Virtual Reality and Actual Reality.

Aesthetic design is probably the one of the most important tasks in creative design when introducing innovative products, and in many cases, novel. Existing design tools and systems tend to ignore this aspect of product design because of its complexity; it involves the aspect of creativity, mentioned earlier. In defining product design by taking into account situatedness, it can be treated in terms of variables conveniently categorised to define structure, behaviour and function (Gero, 1994).

## 3 Agents that Support Human Creativity

Design is a complex process closely related to creativity and creativity itself a process closely related to human processes still unknown. It is also believed that the interaction between designers and the environment, through the support of computational models, strongly supports their creative processes.

An important consideration when designing support systems is the introduction of adaptive (virtual) autonomous *agents*. Autonomous adaptive agents are dynamic entities and are designed to autonomously decide when to be re-designed to adapt to their environment. Their functionality can evolve during their lifetime (Brazier, 2002).

Fundamental arguments to creativity and agents are analogy and situatedness.

- *Analogy* is the novel combination of familiar ideas (Boden 1994). Analogy-based models, implemented in software, are usually seen as communities of agents.

- Additional to analogy is *situatedness* a notion recently introduced to the agent model by (Gero, 2002). Situatedness means that concepts related to designing change according to what designers see, which itself is a function of what they have done. This can be modelled as the interaction of three worlds: *external*, *interpreted* and *expected* world that exists inside interpreted world (see Figure 1).

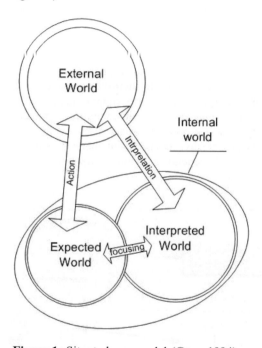

**Figure 1:** Situatedness model (Gero, 1994)

The introduction of agents that support human creativity have to ensure that the proposed computational models interact with designers and help them by suggesting, identifying and even evaluating differences between familiar ideas and novel ones (Boden, 1994) and in this case by considering the variables of the designed artefact. At a higher level this relates to the ability of these agents to decode and adapt the symbols that exist to their external world.

Finally, this community of agents must have the ability to observe human actions and behaviour, while the humans are interacting with the system, and performing operations to stimulate his designers' creativity. This is the user personalisation that adaptive agents often are used for and in extension the generation of a perspective view of the behaviour of a community of designers. Agents within a semantic network not only effect associations but also trace the associative pathways involved, which in itself might prompt the user to new insights (Boden, 1994).

## 4   Concept Architecture

In order to make more concrete this line of research, this paper introduces an architecture (see Figure 2) that unifies previous research and aims to support the human in aesthetic product design. The basic components are: the Human/Designer, the agents' society, the Virtual Design Tool and Digital Libraries.

Humans as creative individuals can be characterised by situatedness. They have internal and external world and in addition physical representation. Built in their inner world, in terms of sensory experiences, perceptions and concepts, is their interpreted world. With imagined actions, an expected world is produced which is identical to their imagined reality (focusing). Their external world is the world that is composed of representations outside them, everything they are interacting with.

To begin with, the humans construct an initial idea regarding their expectations in their imagined reality; this is the first step in their constructive memory. The next step is to interact with the system (combined VR/CAD) in order to produce an early representation of the artefact. This interactive process is observed by the agents' society. When agents, through interpretation of their external world (i.e. human actions in VR and actual behaviour, search in digital libraries), identify symbolic language known to them, (built into their own capacity), they plan their expected world and finally they perform appropriate actions to support human aesthetic decisions. Continuous cycle of this procedure causes situatedness for the evolution of the aesthetic design of the product (different values in every cycle for product structure, behaviour and function). Because humans and agents collaborate in this creative environment there is a possibility of novel conclusions, although the evaluation is mostly dependent on human judgments.

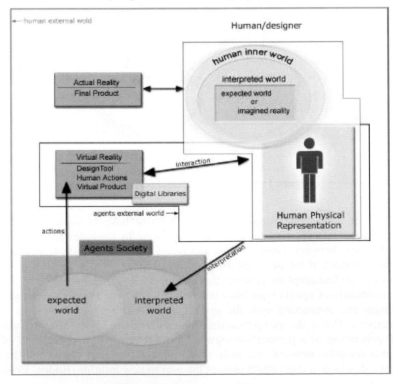

**Figure 2: Concept Architecture.**

# 5   Conclusion and future work

This paper identifies an approach for a system that analyses concepts that structure human creative perception which lead to the phenomenon of the emergence of novel ideas in a creative environment of multi-agent system that study human behaviour during progression towards novelty for aesthetic product design. Moreover, they construct a personal profile for each user/designer and support sharing of knowledge in the design domain. Future work includes a detailed description of the role of the agents that exist in the agent society and further evaluation of the endurance of the model in time.

# 6   References

Boden, M (1990). The creative mind, myths and mechanisms, London: Abacus.

Boden, M (1994). Agents and creativity, *Communications of the ACM*, 37(7), 117-121.

Brazier, F.M.T., Wijngaards, N.J.E., (2002). Designing Creativity. *Proceedings of the Learning and Creativity in Design workshop at AID'02*

Candy, L. (2000). Dimensions of Art-Technology Partnerships in Collaborative Creativity, *Collective Creativity Workshop*, Nara Institute of Science and Technology.

Candy, L., Edmonds, E.A. (2002). Modeling co-creativity in art and technology. *In Proceedings of the 4th International Conference on Creativity and Cognition, ACM*. 134- 141.

Mamykina, L. Candy, L. and Edmonds, E.A. (2002). Collaborative Creativity. *Communications of the ACM Special Section on Creativity and Interface*, 45(10), 96-99.

Edmonds, E. A., Candy, L., Jones, R., & Soufi, B. (1994). Support for collaborative design: Agents and emergence. *Communications of the ACM*, 37(7) 41-47.

Edmonds, E. A., Candy, L., (1999) Computation, Interaction and Imagination: Into Virtual Space and Back to Reality. *Computational Models of Creative Design IV*, 19-31.

Gero, J. S. (1994). Towards a model of exploration in computer-aided design, *in* J. S. Gero and E. Tyugu (eds) (1994). *Formal Design Methods for CAD*, 315-336

Gero, J. S. (1996). Creativity emergence and evolution in design: concepts and framework. *Knowledge-Based Systems, 9(7)*, 435-448.

Gero, J. S. (2000). Computational models of innovative and creative design processes. *Technological Forecasting and Social Change, 64(2-3)*, 183-196.

Gero, J S. (2002). Computational models of creative designing based on situated cognition. *Creativity and Cognition 4*, Loughborough University (to appear).

Lawson, B. (1999). 'Fake' and 'Real' Creativity using Computer Aided Design: Some Lessons from Herman Hertzberger. *In Proceedings Creativity & Cognition 9, ACM*, 174-180.

Shneiderman, B. (2000) Creating Creativity: User Interfaces for Supporting Innovation. *ACM Transactions on Computer-Human Interaction*, 7(1), 114-138.

Smith, S. M. (1997). The Machinery of Creative Thinking. Retrieved January, 2003, from http://www.winstonbrill.com/bril001/html/article_index/articles/251-300/article290_body.html

Suwa, M., Gero, J. S. and Purcell, T. (2000). Unexpected discoveries and s-inventions of design requirements: Important vehicles for a design process, *Design Studies* 21(6), 539-567.

Taylor, C. W. (1988). Various approaches to and definitions of creativity. In Sternberg, R. J. (Ed.). *The nature of creativity: Contemporary psychological perspectives* (pp. 99-121). Cambridge: Cambridge University Press.

# Task-Object Models for the Development of Interactive Web Sites

*Gerd Szwillus*                              *Birgit Bomsdorf*

Universität Paderborn                    Universität Hagen
D-33102 Paderborn, Fürstenallee 11    D-58084 Hagen, Universitätsstraße 1
szwillus@upb.de                    birgit.bomsdorf@fernuni-hagen.de

## Abstract

A systematic web site development process needs to start from a sound knowledge of the organization's activities and tasks to be supported. The necessary knowledge can be captured by means of expressive scenarios or refined use cases, covering the relevant aspects of these activities. In our approach we start web site development from such a collection to derive a formal task model focussing on the explicit specification of the task objects and their modifications as identified from the scenarios. An executable specification of this model, referred to as *Web Object Life Cycle Model (WOLM)*, serves as central content source for the resulting web site, thus becoming the core part of a task-based web engineering approach.

## 1   Scenarios in System Development

Scenarios are used within nearly each phase of the system development process with different goals, such as establishing requirements, describing design decisions, or supporting evaluation. The degree of formality and the way scenarios are used varies a lot within existing approaches (Caroll, 1995). In this paper we focus on scenarios as applied within the process of requirement collection, where scenarios are informal, descriptive narratives of how a person or a group performs activities in a given situation. They are used for two purposes: On one hand, they are used for analyzing the current work situation, referred to as problem scenarios in (Rosson & Caroll, 2002); on the other hand, they are used for designing how a user will interact in the future with the target system currently under development, referred to as design scenarios in (Rosson & Caroll, 2002). Scenarios can be created spontaneously and "easily", as no restrictive formalisms or techniques are being used. They can be understood by all members of the development team as well as by the future user of the system. Therefore, scenarios support problem detection and raise questions to find out about the objects and behavior of the system to be realized. With ongoing design they are combined with other models.

In the field of software engineering, scenarios are used in conjunction with use cases (Eriksson & Penker, 1998), which enumerate the system functions provided to the user. Their notation is somewhat more formal and their focus is on specifying system functionality as needed by different roles users may adopt while performing their tasks. If scenarios are written down for the purpose of requirement gathering, use cases are derived. On the other hand, scenarios are also used as refinements of use cases. Such a design scenario describes an instance of a use case representing an actual usage of the system. Details are added by means of additional formal models, such as activity diagrams, sequence diagrams, and state charts (Eriksson & Penker, 1998). Still, use cases are often claimed to be insufficient for adequately covering the user- and task-centered aspects of interactive systems. Therefore, they are increasingly combined with models developed in the field of HCI. In such modeling approaches (Vanderdonckt & Puerta, 1999), task and dialogue models are of particular importance to address user interface design problems. The task model represents the user's point of view, depicting the user's goals, activities, and the objects manipulated during task execution. This description is refined by the dialogue model, which specifies the behavior of the user interface in terms of user interactions and system feedback. Similar as in the field of software engineering, problem scenarios as well as design scenarios are in use, i.e. task models are derived from scenarios described during requirement analysis, e.g. (Paterno 2001), while design scenarios refine the task model thus supporting the specification of the dialogue model, e.g. (Bruins 98). In web site modeling approaches such as WebML (Ceri, Fraternali & Bongio, 2000), OOHDM (Vilain & Schwabe, 2002), and WSDM (De Troyer, 2001), scenarios and/or use cases are only utilized for requirement analysis in the beginning of the process. Based on these descriptions, the models specifying the relevant objects, the structure of the navigation space, and the abstract presentation are derived. The focus in web modeling traditionally lies on data-intensive web sites, the purpose of which is to provide information through which the user can navigate. Hence, the development process as well as the models used are strongly related to system-oriented aspects, such as the structure of web sites in terms of objects, navigation issues in terms of access rules and filtering mechanisms, as well as data retrieval and dynamic generation of web pages. Overall, similar as in software engineering, the focus is shifted early in the process from the task- and user-centered view to technical aspects. In some approaches, however, this deficit was recognized and more attention is paid to user tasks. In OOHDM and WSDM, for example, in addition to the model of the underlying information base, a user-related object model is introduced. This model describes those objects, properties, and semantic links to be shown to the user while performing web reading tasks. Furthermore, in OOHDM scenarios and use cases are refined by diagrams defining the data to be exchanged between the user and the web application, thus dealing with aspects of the dialogue between the user and the system. Still the approaches have to be considered as data-driven, as they do not care enough about the user and task-oriented view. This holds true especially, when the behavior of an interactive web application has to be modeled, such that the modifications of objects result from the processes of the business the web site represents.

In the following we propose a web site management approach which is based upon a task object model, derived from scenarios representing the user's view of the system. The core part of our approach is a formal task object model complemented with a declarative web site structure model.

## 2   Task-Object Based Web Management

As a first step towards the development of a web site, the future web site owner is interviewed by the analyst. The goal is to find out about typical processes in the business, which are described by means of scenarios. While collecting these stories, the analyst can identify the relevant tasks and involved objects of the business – what parts do they contain, what are their interrelations, and what are the changes they undergo in the context of the different tasks. The dynamic aspect covers creation, modifications, and destruction of the objects. The reason for these object transformations are the tasks performed in the business. Hence, we find out about the task objects of the underlying real happenings from the view of the people performing the tasks. Assume, for instance, a library loaning system which has to be "put on the web". While talking to the librarian, she might tell the story of the *customer Tom, who enters the library, identifies himself with a identity card, and then is told by the librarian which books he has currently loaned and which of these he can prolong.*

Process scenarios like this are collected, which capture information about the given ("old") way of performing the business' tasks. This scenario identifies important objects of the application, such as *customer, identification, book*, and *book loan*; in addition, object relations become visible *("Tom has lent a book")*, and states of objects show up *("a book can be prolonged or not")*. State transitions in this terminology correspond to task performance in the business. Likewise, the librarian together with a designer can specify design scenarios describing the future system which is to be created. Corresponding to the process scenario given above, the designer might describe the *internet user Tom who browses to the library login page, where he inputs account and password information, and pushes the login button. If the login information was valid Tom is presented a page with a list of all books he has currently lend and their current return dates.*

Old and new objects, object relations, and object states can be detected in this scenario, such as the *login information (account and password)*, the state of a *user* to be *logged in* or *not*, or the existence of some controlling instance doing the login check. As mentioned above, both types of scenarios, describing the old situation and the new envisioned situation, are useful to find out about objects, their relations, and dynamics. In addition, the scenarios capture information about which roles of users exist in the business, and what their respective duties and responsibilities are. Within our approach, all this information is used to create a formal system of the task objects.

We refer to this model as *Web Object Life Cycle Model (WOLM)*, as it describes the task performance of the business in terms of the modifications the relevant task objects undergo during their lifetime. Hence, WOLM is an object model, specifying the object structures, object relations, states, and state transitions with respect to task execution, rather than building upon a task hierarchy and temporal relations, as a classical task model would do. As the task objects originate from the scenarios mentioned above they have a clear semantic meaning in terms of the business. *In the library loaning example, the WOLM would contain objects of type user, book, loan, with certain substructures such as user.account and user.password, corresponding to the information to be stored within the objects.* By defining states and state transitions a state model is added to each object, with task execution resulting in state transitions, *such as the login operation transforming the user object from the state "loggedOut" to "loggedIn".* The model needs not specify all objects of the business, nor needs it to define the complete "business logic". Only the subpart relevant for the web site representation has to be specified within the model.

## 2.1 The Web Object Life Cycle Model

WOLM objects are described by means of a class specification, which defines the properties of a type of objects. The class specification contains attribute declarations, initialization information, and methods to be performed "on" the objects of the specified type. Attribute types are scalar types, such as `integer`, `boolean`, or `string`, references to other WOLM objects, and lists of values of these types. *In the library loaning example we have a class* `user` *with an* `integer` *attribute* `id` *and* `string` *attributes* `account, name,` *and* `password`. *As methods we might have, among others, a* `login()` *method and a* `logout()` *method.*

An essential concept of the model is the explicit modeling of web object states. The different states represent "situations" or "configurations" a web object can be in. Transitions between them describe the changes a web object undergoes during its life time through execution of tasks. To define states and state transitions, a WOLM class specification contains a definition of the state set of the objects, and rules governing the transition between these states. *In the case of the class* user *we need to distinguish between the state of a user to be logged in or not, hence we specify the state set as* {loggedIn, loggedOut}, *and define the behavior by the two rules:* loggedOut login → loggedIn, loggedIn logout → loggedOut

Without going into detail about these definitions, let us mention that the state set and rule definition follows the same concepts as defined for DSN (Curry & Monk, 1995) and DSN/2 (Szwillus & Kespohl, 1996). For a given state, the rules specify which state transitions are possible and what the resulting state is. A state transition triggered by an input event causes a method invocation defining the effects of the transition on the object's attribute values. *If a user is not logged in, then the event "*login*" is possible in the corresponding* user *object. If this actually happens, then the method "*login()*" is executed, which modifies the* user *object's attributes.*

The execution of methods, i.e. attribute modifications, can necessitate input parameters from the user of the WOLM, which are requested at run time. Such an input request is an abstract specification of the user system dialog, which is mapped onto concrete web site interaction elements later, when designing the actual web pages. *In our example we introduce a* `controller` *object of the library loaning system, which is the manager of the login-logout-process. The* `controller` *object owns, the method* `tryLogin(String account, String password)` *and the rule* `loginPossible tryLogin` → `loginSucceeded`. *As long as the object is in the state* `loginPossible`, *it is waiting for some user to log on. If a user tries to do so he has to provide account and password strings to the controller, as defined by the method* `tryLogin`.

The web object model specifies the modifications in the business from a strictly user centered point of view. When specifying the state transition possibilities and their effects on objects, it is not defined where triggers for state transitions originate. The model is explicitly concentrating on their effects and interrelations between the objects. *Dependent on the result of the check captured in the method* `tryLogin(...)` *the state of the* `controller` *object is changed to either* `loginSucceeded` *or* `loginFailed`. *In the first case, the* `user` *object performs its* `login()` *method.* A given state transition is only possible if the object is in an appropriate state, thus the model contains implicit checks of the semantics. Furthermore, as the model is independent from any web representation specification, the two design spaces "task object model" and "web representation" are clearly separated in this approach.

## 2.2 Abstract Web Presentation based on WOLM

The web representation is described by an additional model, which is derived from the WOLM by the definition of an abstract web site structure (AWS). In this creative design step decisions are made about the distribution of object information onto different parts of the web site, and how navigation is structured. The resulting AWS specifies the structure of the object information in terms of a hierarchy of web pages and page parts, down to single text or graphic elements, and links. This model is called abstract, as it does not include any page optics decisions, such as the use of graphics, geometric grouping, colours, or fonts. By means of a declarative specification we assign the task objects as defined in WOLM to the AWS. Hence, WOLM objects are related to web pages or page parts, and these in turn are defined dependent on the WOLM objects' attribute values and states. As a result, the AWS part corresponding to an object changes as soon as the object changes itself.

*The single line representation of a book on the user's homepage shows information about the book, and may change dependent on whether the loan period of the book can be prolonged or not. Hence, the AWS contains a part describing the representation of a* `Book` *object* `b` *as*

```
Field LoanedBook(Book b)
    Text RegisterNumber [b.regnumber]
    Text Title        [b.title]
    if (b inState reserved) Text "prolong impossible"
    else                    Text "prolong possible"
```

Interactivity of web sites is covered in our approach through interaction elements contained in the AWS. They are connected to WOLM object state transitions, thus enabling web site visitors to perform well-defined operations. We will not go into details about these concepts here, as they are treated elsewhere (Bomsdorf & Szwillus, 2003).

## 2.3 Managing the Living Web Site

The WOLM itself allows execution and validation of the happenings in the model against what happens in reality. The model, however, can not only be simulated, it can also serve as the driving instance of the living web site. The principle idea is that things occurring in the business are reflected in the web site, thus creating a mirror of reality. For this purpose, we generate the final web pages via the AWS, merged with concrete representation information (using standard techniques such as XSLT) from the current WOLM state. Updating the web site is then reduced to

251

updating that state (Szwillus & Bomsdorf, 2002). Furthermore, we have implemented a simple web interface, which enables the web site owner to inspect the current state of the WOLM objects, and to trigger the necessary state transitions. If additional information is needed to perform the state transition this is requested by the transition,as specified in the WOLM model. Once the modifications are done, the next web site visitor will see the new pages, as generated from the updated state of the WOLM. This update concept was enhanced further by adding timing information to the WOLM object state transitions, which can trigger automatic events at given times, or results in recommendations about the next login date. In the case of direct web site interaction with the visitor, the triggered state transitions result in immediate feedback of modified web pages or page parts, thus allowing the definition of sophisticated web interaction.

# 3 Conclusion

Our approach covers the full range of web site dynamics from purely static to highly interactive, dynamically generated web sites. By explicitly using the scenario-based task object model as the core model, the web site is derived systematically from the behavior of the business. All levels of the development, including the actual running and updating of the web site, directly reflect and utilize this strong connection between web representation and reality. Other than in web modeling approaches known from literature, behaviour is specified prior to web site structure, which is added as AWS complement to the kernel WOLM model. In addition, the behaviour specification originates from the tasks of the business, thus directly reflecting task execution and the objects' interrelations with business processes. Up to now, we have verified the general feasibility of the approach, by trying the ideas on small example web sites. We are currently working on a tool collection supporting the development approach, including a WOLM model editor and simulator, an AWS editor, and mechanisms for linking concrete web designs to the AWS.

# 4 References

Bomsdorf, B., Szwillus, G. (2003). User-Centered Modeling of Interactive Web Sites, Poster, WWW2003.

Bruins, A. (1998). The Value of Task Analysis in Interaction Design, http://www.upb.de/cs/ag-szwillus/chi98ws.

Caroll, J.M. (1995), Scenario-Based Design: Envisioning Work and Technology in System Development, John Wiley and Sons, Inc., New York.

Ceri, S., Fraternali, P., Bongio, A. (2000). Web Modeling Language (WebML): a modeling language for designing Web sites, in Proceedings of the WWW9 Conference.

Curry, M., Monk, A. (1995) Dialogue modelling of graphical user interfaces with a production system, Behaviour & Information Technology, Vol. 14, No. 1, 41-55.

De Troyer, O. (2001). Audience-driven Web Design, in: Information modelling in the new millennium, eds. Matt Rossi and Keng Siau, IDEA Group Publishing.

Eriksson, H.-E., Penker, M. (1998). UML Toolkit, John Wiley and Sons, Inc., New York.

Paterno, F., (2001). Towards a UML for Interactive Systems, Proceedings of EHCI 2001,175-185.

Rosson, M.B., Caroll, J.M. (2002), Usability Engineering: Scenario-Based Development of Human Computer Interaction, Morgan Kaufmann, New York.

Szwillus, G., Kespohl, K., (1996). Prototyping Technical Device Interfaces with DSN/2, CHI'96, Conference Companion, Vancouver, Canada.

Szwillus, G., Bomsdorf, B (2002). Models for Task-Object-Based Web Site Management, in Proceedings of DSV-IS 2002 (Rostock, June 2002), 267-281.

Vanderdonckt, J., Puerta A. (1999). Computer-Aided Design of User Interfaces II, Proceedings CADUI'99, Kluwer Academic Publishers.

Vilain, P., & Schwabe, D. (2002). Improving the Web Application Design Process with UIDs, 2nd International Workshop on Web-Oriented Software Technology

# Coordination Patterns –Towards Declarative Modeling of Coordination Requirements within Cooperative Work Arrangements

*Peter Thies*

Fraunhofer Institute for Industrial Engineering (IAO)
Nobelstr. 12, D-70569 Stuttgart, Germany.
Peter.Thies@iao.fhg.de

## Abstract

This article introduces Coordination Patterns[1] as an approach to coordinating cooperative work arrangements. As they frequently contain creative and problem-solving tasks, cooperative work arrangements are characterized by emergent processes. Hence they often feature indeterminism emergent processes cannot be procedurally modeled in advance. Thus, supporting them by means of workflow management systems is difficult. A solution is presented, enabling coordination of activities even in case of an a priori absent control flow. This is accomplished by using constraint-enabled ontologies for specifying consistency requirements of work arrangements. By means of dedicated tools, Coordination Patterns can be embedded into groupware environments, supporting e.g. the coordination of compiling complex content.

## 1    Introduction

Product development or customer interaction is often characterized by creative and problem-solving processes. As they often feature indeterminism, it is difficult to standardize these processes. As a result of its stability, the domain of materials management or accounting and finance can be widely standardized and supported by traditional workflow management systems (WFMS, Jablonski et al., 1997). In order to work properly, WFMS force the user to thoroughly specify process schemata comprising activities and control flow. In contrast to this, creative and problem-solving activities are frequently affected by varying conditions. This leads to volatile processes. Hence, they cannot be modeled in advance to a satisfying extent, one could call them emergent processes. They develop – at least partially – during runtime, thus requiring flexible and adaptive tools (Carlsen & Jørgensen, 1998; Klein et al. 2000). Here, conventional approaches of business processes modeling are not applicable.

Nonetheless, processes within product development normally comprise a priori definable as well as emergent sections. These processes can at least partially be modeled in advance by means of e.g. process chains featuring gaps or indeterminism in various places. It is possible to distinguish between horizontal, vertical and causal indeterminism. Horizontal indeterminism arises from missing links within process chains. Vertical indeterminism occurs in case of insufficiently elaborated actions. Causal indeterminism appears as incomplete control flow between actions. Indeterminism indicates vagueness (Herrmann & Loser, 1999), thus making coordination (Crowston 1994; Malone & Crowston 1994) by means of a priori defined process chains difficult.

---

[1] This work is supported by the German Ministry of Research and Education under Grant 01HG9985.

Additionally, focusing on procedural aspects of coordination such as control flow, conventional WFMS offer a strongly restricted view of structural and semantic features of information objects. These restrictions lead to control flow based on evaluating name-value pairs as they are known from forms within WFMS. In conclusion, applying WFMS requires knowledge of the overall process and data structures in advance. Therefore, those systems are not applicable to team-oriented work.

## 2  Conception of Coordination Patterns

In contrast to procedural mechanisms of WFMS, Coordination Patterns support emergent processes by providing a declarative approach to coordination. Here, cooperative work arrangements (Schmidt & Bannon, 1992) are modeled by using constraint-enhanced ontologies (Fridman & Hafner, 1997). This is accomplished by using ontologies for specifying arbitrary aspects of cooperative work arrangements. This approach goes far beyond conventional workflow since it provides much richer resources than simple action flows. An explicit control flow specification isn't necessary in advance even though it is not forbidden. In opposition to the workflow management approach, Coordination Patterns can be elaborated during runtime. Adapting the scope of a model according to the use-case thus becomes possible.

According to this approach, constraints are restricting possible actions a user might choose, resulting in implicit process control. It offers basic support every groupware could benefit from. Model elements are specified using a homogeneous modeling and query language. A major contribution of the method is the opportunity of supporting repeating similar tasks by reusable patterns (Alexander et al. 1977; Gamma et al. 1995).

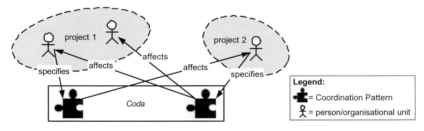

**Figure 1:** Exemplary use-case of Coordination Patterns

Coordination Patterns can be applied to various use-cases such as process planning, realizing awareness tools (Dourish & Belotti 1992) and managing large-scaled document structures. Figure 1 shows an exemplary use-case where several persons are working on two projects. They are supported by a pattern-based tool named Coda (ref. section 3) which is coordinating their activities. According to the system, users can play different roles. Some persons could be engaged in specifying Coordinations Patterns while others would only be affected by these patterns by using Coda-enhanced groupware.

**Definition (Coordination Pattern).** A Coordination Patterns $Cp=(O,K,M)$ is comprising an ontology $O$, a set of Coordinators $K$ and a set of subpatterns $M$. $O$ consists of concepts pertaining to some work arrangement and their interrelationships as well as a set of axioms $A$ for deriving implicit[2] knowledge contained within $O$. The set of Coordinators $K$ is consisting of pairs $k \in C{\times}V$, where $C{\subset}O$ is a set of constraints and $V$ is a set of rules for applying the elements of $C$.

---

[2] Not to be confused with tacit knowledge as known from Knowledge Management!

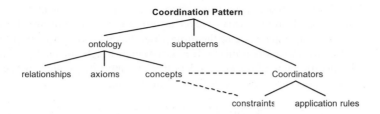

**Figure 2:** Structure of Coordination Patterns

Although some of an ontology's basic structures are presumed to be able to handle cooperative work arrangements, ontologies used within Coordination Patterns can be structured arbitrarily. Among others, supplementary basic elements of an ontology such as information objects, actions, organizational entities and technical systems are predefined. Possible states and state transitions of concepts can be arbitrarily defined. The states can be associated with intervals which themselves can be ordered by using interval relationships (Allen, 1983). Therefore it is possible to overcome the limitations of conventional WFMS which allow this in a very restricted manner (possible states are often stipulated) for activities only. Besides a pattern's ontology, Coordinators play a pertinent role within patterns. Concepts may be linked to Coordinators comprising a constraint and a dedicated application rule. Constraints specify boundary conditions within an ontology and, therefore, within the associated work arrangement. A Coordinator's application rule defines conditions under which the dedicated constraint should hold.

Coordination Patterns are represented by means of Frame Logic (Kifer et al. 1995). Frame Logic offers formally defined mechanisms[3] such as a homogeneous specification and query language and extensive deductive features. Deduction is of interest especially during runtime to analyze broken constraints.

**Example (Coordinator).** Assuming that there exists an ontology for modeling publicly funded research proposals, an exemplary constraint could state that the number of some letters of intent (LOI) to be gathered for proposal submission should be greater than 2:

`count{LOI|Instance[lettersOfIntent->>{LOI}]} > 2.`

To state that the constraint should be respected after December 12th, 2003, we could specify the following predicate as application rule:

`after(today,date[day->12, month->12, year->2003]).`

The application rule and the constraint together are forming a Coordinator, specifying that there should be more than two letters of intent after December 12th, 2003.

# 3    The Coda Software

Coda is a prototypical system for administering and verifying Coordination Patterns. The software is comprising a graphical user interface for maintaining patterns. Hence Coda is in prototypical status, it offers restricted functionality. However, the software could be enhanced for operational use. The presented version of Coda is not intended as a groupware client. As depicted in Figure 3a), upcoming versions will be widely integrated into existing groupware environments offering user-friendly interfaces for using and maintaining patterns.

Coda has been implemented using Java[4]. Although Coordination Patterns are handled as logical expressions, Coda is hiding the Frame Logic -based representation from the user as far as possible.

---

[3] These features would be unavailable in case of using the Unified Modeling Language (UML) in conjunction with the Object Constraint Language (OCL) (Warmer & Kleppe 1999; OMG 2000).
[4] http://java.sun.com

Figure 3b) displays Coda's graphical user interface. The software is organizing Coordination Patterns through hierarchical projects (tree widget in left frame of Figure 3b). Projects contain arbitrary numbers of patterns. Each node of the project tree has its own type-reflecting icon. Users can freely edit this tree for modeling projects. If a user selects a node of the tree, the right frame will display a node-specific view of the node's details. For supporting pattern usage each project can contain logical programs (using Prolog or Frame Logic syntax).

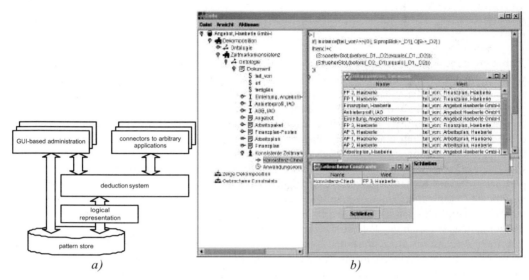

a)  b)

**Figure 3:** Coda architecture and user interface

Coda has been integrated with the deductive database system XSB[5], a Prolog-based system with a Frame Logic implementation called Flora -2. XSB is capable to interpret Coordination Patterns as well as answering arbitrarily complex queries according to the respective project. By means of logical programs, complex and repeating tasks and queries (e.g. analysing broken constraints) can be realized.

As already mentioned, future versions of Coda will be widely embeddable into arbitrary groupware. As a result of this, Coda could coordinate actions of those systems. The GUI presented here will then be restricted to administering Coordination Patterns. Awareness of patterns would be realized through elements of frontend software Coda will be integrated with. Actually pattern awareness is accomplished by dialogs based on logical programs being cyclically displayed or reloaded.

# 4  Conclusions

This contribution has introduced the concept of Coordination Patterns. These patterns are offering coordination support for cooperative work arrangements even in case of an a priori absent control flow. Coordination Patterns are based on constraint-enhanced ontologies offering support for specifying structure and consistency-specific conditions of work arrangements. The approach is generally applicable to every domain where WFMS requiring complete process specifications in advance fail. This is especially the case where emergent processes occur. Here, Coordination

---

[5] http://xsb.sourceforge.net oder http://www.cs.sunysb.edu

Patterns allow an incremental procedure of modeling and using a cooperative work arrangement's structure and its boundary conditions.

For a more extensive description of the method and the Coda prototype, the reader is referred to forthcoming publications, where among others several exemplary patterns such as interval patterns, temporal coupling, mutual exclusion and produced-consumer-patterns, will be presented.

# 5 References

Alexander C.; Ishikawa S.; Silverstein M.; Jacobson M.; Fiksdahl-King I.; Angel S. (1977): *A Pattern Language* .New York : Oxford University Press.

Allen, J.F. (1983): Maintaining knowledge about temporal intervals. In: *Communications of the ACM 26*, November, Nr.11, S.832–843.

Carlsen, S.; Jørgensen, H.D. (1998): *Emergent workflow: The AIS workware demonstrator.* CSCW Towards Adaptive Workflow Systems Workshop, Seattle, WA, November 1998. http://ccs.mit.edu/klein/cscw-ws html, accessed: 12/2000.

Crowston, K.(1994): *A Taxonomy of Organizational Dependencies and Coordination Mechanisms* MIT-Center for Coordination Science. http://ccs.mit.edu/CCSWP174.html, accessed: 08-05-1998, (174).–Tech. Rep. Cambridge, MA.

Dourish, P.; Belotti, V. (1992): Awareness and Coordination in Shared Workspaces. In: *Proc. ACM Conf. on Computer Supported Cooperative Work CSCW '92* (Toronto, Canada, Nov.1992). New York, ACM Press, S.107–114.

Dourish, P.; Holmes, J.; MacLean, A.; Marqvardsen, P.; Zbyslaw, A. (1996): Free?ow: Mediating between representation and action in work?ow systems. In: *Proc. ACM Conf. on Computer Supported Cooperative Work CSCW '96* (Boston, Nov.16-20. 1996). New York, ACM Press, S.190–198.

Fridman, N.; Hafner, C.D. (1997): The State of the Art in Ontology Design. In: *AI Magazine 18*, Nr.3, S.53–74.

Gamma, E.; Helm, R.; Johnson, R.; Vlissides, J. (1995): *Design Patterns: Elements of Reusable Object-Oriented Software*. Reading, MA: Addison-Wesley.

Herrmann, T.; Loser, K.-U. (1999): Vagueness in models of social-technical systems. In: *Behaviour & Information Technology*. Vol. 18. No. 5. S. 313-323. Taylor & Francis.

Jablonski, S.; Böhm, M.; Schulze, W. (1997): *Work?ow Management: Entwicklung von Anwendungen und Systemen. Facetten einer neuen Technologie*. Heidelberg: dpunkt-Verlag für digitale Technologie.

Kifer, M.; Lausen, G..; Wu, J. (1995): *Logical foundations of object-oriented and frame-based languages*. Journal of the ACM, 42:741-843, July 1995.

Klein, M.; Dellarocas, C.; Bernstein, A. (2000): *Adaptive Work?ow Systems*. Dordrecht, NL: Kluwer.

Malone, T.W.; Crowston, K. (1994): *Toward an interdisciplinary theory of coordination*. In: ACM Computing Surveys 26 (1994), Nr.1, S.87–119.

Object Management Group (2000): *Uni?ed Modeling Language Speci?cation Version 1.3*. http://www.omg.org, accessed: March 2000.

Schmidt, K.; Bannon, L. (1992): Taking CSCW Seriously. In: *Int. Journal on Computer Supported Cooperative Work 1* (1992), Nr.1-2, S.7 –39.

Warmer, J.; Kleppe, A. (1999): *The Object Constraint Language: Precise Modeling with UML*. Reading, MA: Addison-Wesley Longman, (Object Technology Series).

# System Development Influenced by Rituals and Taboos

*Karin Tweddell Levinsen*

Department of Informatics
Copenhagen Business School, Howitzvej 60, DK-2000 Frederiksberg
kle.inf@cbs.dk

## Abstract

This paper discusses the design process of an educational CD-Rom dealing with safety issues at the workplace - an ambiguous topic – within a professional production context. The learning objective was to change attitudes rather than training of skills. As the aim of the application was to touch upon people's actual attitudes and behavior, we wanted to involve the users actively in the design process. The ideal was to apply user centered and participatory design methods as far as time and money would allow. However, the introductory research revealed situations where the involvement of users and the access to sources of information had to be reconsidered for ethical reasons. The paper describes how we used the introductory research to uncover these special conditions and how they were dealt with, by combining and adjusting methods from Communication Studies such as target group analysis, modified future workshop, and the café-model.

## 1    Introduction

As part of a broad campaign focusing on safety issues, the task was to produce an educational CD-Rom aimed at new employees. On company -, social -, and individual levels, accidents had severe consequences involving heavy costs. Therefore companies, unions and other organizations were focusing on training employees to avoid accidents.

From the start of the project, we had the following information: 1) The learning objective was to introduce knowledge and awareness of safety on the job. 2) The CD-Rom should be audio-visual as many workers were reported to be poor readers owing to short or unsuccessful schooling or because Danish was not their first language. 3) The material should not be presented in a school-like manner as most workers carried within them negative learning experiences from school. 4) Genuine ignorance was not considered to be the cause of accidents as everybody entering a site was instructed in the rules and were obliged to wear safety gear. 5) As the aim of the application was to affect the workers' actual attitudes and behavior, the users should be actively involved in the design process. 7) We should pay special attention to a behavior called the "Tarzan Syndrome": employees who ignored safety and behaved dangerously in order to look cool.

## 2    Identifying Target Groups

At first, we wanted to understand all the complexities involved in this particular context. Therefore, we collected data from various sources: We talked with safety representatives who were also stakeholders in the project, read safety regulations and reports on accidents and safety problems, visited construction sites, engaged in informal conversations and made observations and interviews with workers. The investigation showed that the attitude to safety in workers who neglected to follow the rules could be described as either *rational* or *alienated*.

**Rational behavior**. Example: A worker took off his safety goggles. He explained that he could not see properly as his glasses got steamy. In such situations, workers were willing to discuss their choice. They were perfectly aware that they made a risk calculation, but refused to discuss the possibility or consequences of having an accident. We could approach this group through observation and interviews. But any talk about accidents and consequences was taboo. Also, they could only describe the "Tarzan Syndrome" in a very superficial language – in clichés.

**Alienated behavior**. In this group, we found two kinds of arguments involving either excuses or bragging. Examples: *1) Excuses*: A worker was standing on top of a high ladder without anybody to support the ladder. He explained that it was almost closing time and everybody was busy finishing his own tasks. Therefore, he did not like to bother them by asking for help. *2) Bragging*: A worker was using a grinding machine without wearing safety goggles. He claimed that it was no big deal having splinters removed from his eyes. It happened two or three times a year and gave him some days off work. He thought that was cool. In these cases, the workers were not communicative and would signal that this was the end of the conversation – period! They also refused to discuss any possibility or consequences of having an accident.

With the projects overall focus on safety, the alienated group became most important, but also very difficult to reach. They would not discuss their behavior, accidents or consequences. Some of them were not even aware that their behavior was dangerous. Such issues were taboo in these groups. The alienated behavior matched descriptions of the "Tarzan Syndrome" in reports and papers in organizational literature. Tarzan manifests himself in two ways: The Soft Tarzan corresponding to our excusing type and the He-man corresponding to our bragging type. *The Soft Tarzan* tries to meet expectations (imagined or real). He puts moral pressure on his colleagues. The *He-man* tries to match a mythological ideal of being omnipotent. The He-man leads his work-gang as a hierarchy with a strong pecking order. Both kinds of Tarzans ignore the symptoms of strain on the body and the signs of danger on the site, and they ignore safety rules, though risk behavior may affect both themselves and others.

Based on discussions with the safety representatives involved, it was decided to let these interpretations be the basis for outlining requirements for the design phase. Both groups shared the taboo against talking about accidents and consequences. As a result, workers *would not* talk about these things. Both groups could only talk about Tarzan as a cliché. Therefore, workers *did not possess* a language for discussing the "Tarzan Syndrome". Individuals and groups involved in the "Tarzan Syndrome" shared a social behavioral pattern, which we choose to characterize as a ritual[1]. Instead of a task-orientated and directed agency, the workers were alienated into behavioral patterns that might cause accidents. As a consequence, workers *could not* talk about these things.

In terms of the overall design and the HCI aspects of the application as such, our findings had two important implications: *A direct approach to these sensitive matters during the design phase should be avoided*. Otherwise, the users might reject the CD-Rom project. It also meant that the

---

[1] Rituals are interpreted as keepers of order and norms in a culture or sub-culture. A ritual is a sequence of acts that follow a certain pattern. They have an effect on people, because they involve strong emotions such as the feeling of belonging. A ritual is performed in silence. The point is that nothing must be said. The rules are inherent in the actors as a natural behaviour - like tacit knowledge (Sjørslev, 1995).

CD-Rom should *support the development and acquisition of a shared language and encourage* the workers to engage in discussions[2].

# 3 Ethical Questions

By choosing to understand the behavior as ritual and taboos, we became aware of a strong emotional perspective, which raised ethical questions to the use of participatory design, direct observation and interview as suitable approaches. The background material told us that the "Tarzan syndrome" is about lack of self-confidence. *"Many men having a Tarzan syndrome are broken when they can no longer work. Even if their wife supports them and adapt, they do not want to accept their new role"* (HK Industri, 2001; authors translation). Often these men lose their whole social network. From reports on accidents, we knew that many workers had encountered accidents and serious consequences. They had suppressed both the experience and the loss of a college or friend who did not die, but vanished into a social deroute.

Thus, both these aspects touched upon deep emotions in the involved individuals, and moving in too close, would be serious business. We were not professionals in emotional matters; therefore we had to consider how far we dared push a direct approach. We could choose hidden agenda, which is normally used when studying extreme or criminal organizations. This was not the case on a construction site. Thus, the hidden-agenda-approach was considered unethical. The answer to this ethical discussion was that we would have to gain information from other sources.

# 4 Future Workshop

To provide the process with a time schedule and structure, we decided to use a modified two-day *Future Workshop* (Jungk & Müllert, 1987). A future workshop consists of three phases: Critique, Fantasy and Implementation. We invited relevant participants: The *board members* who had an overall knowledge of the area, and the *safety representatives* who had the day-to-day experience. Besides, they were going to be the future instructors. As they belonged to the workers group, they shared characteristics with our target-group. Thus, we could indirectly collect information about the target-group by observing the safety representatives during the workshop. In addition, the *design team* participated. The first workshop covered the Critique and Fantasy phase. The second workshop covered the Implementation phase.

In the **Critique phase**, the participants lined up the negative responses to their initiatives. Then, they had a brainstorm on how to change the approach to safety behavior. The ideas might be wild and unrealistic, but they could not be questioned and did not have to be defended. All the ideas were presented in big letters on posters and taped to the wall.

In the **Fantasy phase**, the results of the Critique phase were turned into positive and more concrete ideas. The task was to imagine how it should feel to use the CD-Rom, and which response it should produce in the users. The participants worked in small interdisciplinary groups. We used a set-up called the *café-model* (Hansen, 2000), which helps to clarify the purpose and perspective shared in larger groups. The model implies pre-prepared tasks and a strict procedure controlled by a facilitator, and it is a useful tool for exposing the distinction between discussion

---

[2] For further reading on dealing with these aspects, see (Levinsen, 2002, pp. 159 - 165)

and dialog. Furthermore the model supports creativity as iterative rotation among the groups develops synergy.

First, each group defined the five most important aspects of the given task. Now, one member in each group remained as a host while the other members moved to other groups as guests – one to each group. In the new groups, discussions were not allowed. The host told the guests about the group's work, while the guests in turn asked questions to the host. The members returned to their home group and shared their experiences: Did we learn something from the other groups? Did somebody ask unexpected questions about our own ideas? The rotation was repeated while the focus of the task was narrowed down. Scenarios and lists of aspects and ideas were formulated in a representational form and drawn on posters. Again, there was no limit in terms of realism or resources.

In the second workshop, the **Implementation phase**, all participants took part in sorting the ideas into categories and listing priorities (importance, level of difficulty, etc.) on the sublevels. The aspects were written on large Post IT labels and put on the wall where they were moved around while discussing the movements. After the workshop, the design team processed the material from the two workshops into an illustrated pedagogical specification report providing a framework for the succeeding design process.

## 5    Designing by Exploring Constraints

During the workshops, we had learned more about the reaction to the taboos and the lack of language surrounding Tarzan by observing the safety representatives. We now knew that we would have to provoke a response and test the emotional limits indirectly. By closely studying the reports on accidents that had occurred within the preceding year, all the accidents were sorted into categories such as: accidents that may occur anywhere or happen to anyone, accidents that relate to specific places or job types, who initiated the accident, type of injury and seriousness, etc. These categories were turned into illustrated exemplary stories.

We deliberately created superficial stories hoping that the safety representatives would be dissatisfied and try to improve our formulations. And they *did* improve the stories making them more realistic. The stories were sent to the safety representatives for verification and comments – a request that was met on all levels from narratives, lines of speech, to specific details and usability issues. The comments were given over the phone as the safety representatives found it difficult to write. The entire process of forming the content went through this procedure. When necessary, it was repeated a couple of times. The approach turned out to be very engaging and creative for the safety representatives – and successful for the design team. In this way, we could test the constraints of our pedagogical proposals and thus avoid rejection. Gradually we could develop solutions that supported an explorative approach towards the sensitive matters among the end-users.

Verification was ensured based on the criteria for communicative validity recommended in qualitative interviews (Kvale, 1996). The safety representatives tested internal validity while members of the board and external experts tested external validity.

## 5.1   Simple Prototype Testing

The inscription and translation[3] process was difficult. How could we be sure that the target group would understand the overall purpose? For an untrained person it might be difficult to imagine the realization of a manuscript as an interactive application. Therefore, we decided to perform tests by using simple prototypes. In this way, we could also test the traditional usability issues. Again, the safety representatives were the test users - the second best solution - as the real end-users were out of the question.

In the production industry there is often a resistance against prototypes or models that are too rudimentary. It is argued that users cannot relate to rudimentary test objects. This is not our experience. However, the point is to make the model clearly rudimentary. This was done by making the esthetics of the model recognizable, but absolutely ugly while presenting or simulating full functionality. This made the test users accept the model for what it really was – *a model that needed help*. To help develop the model into something real and useful the test users came with expert comments. When necessary, the process repeated itself a couple of times.

## 6   Conclusion

Our research on attitudes towards safety showed that we were dealing with a wide range of factors of which taboos, rituals, alienation and strong, suppressed emotions played important roles. To ignore this might provoke unnecessary emotional reactions, complicate cooperation with users during design, and provoke rejection of the final CD-rom by the end-users. Our method suggests a way to approach the users indirectly during the design process. The users were still active as they contributed with their knowledge and experiences. The sensitive aspects were interpreted *indirectly* by observing the security representatives, who shared traits with the target group. All design processes are unique. Thus, our method is a *guideline* for approach - not a recipe - in cases, where introductory research uncovers conditions that exclude end users from the design process.

## References

Hansen, F.T. (2000). *Den sokratiske dialoggruppe - et værktøj til værdiafklaring*. (The Socratic dialog-group – a tool for value clarification). Copenhagen: Gyldendal.

*HK Industri, 12*, december 2001. Ægteskabet testes (Testing the marriage),

Jungk, R. & Müllert, N. (1987). *Future workshops: How to create desirable future*. London.

Kvale, S. (1996). *InterViews: An Introduction to Qualitative Research Interviewing*. Sage.

Levinsen, K. (2002). When narrative becomes an obstacle, in Danielsen, Nielsen & Holm Sørensen (eds.), *Learning and Narrativity in Digital Media* (pp. 141-165). Copenhagen: Samfundslitteratur.

Sjørslev, I. (1995). *Gudernes rum. En beretning om ritualer og tro i Brasilien*. (The Gods space. A tale of rituals and belief in Brazil). Copenhagen: Gyldendal.

---

3 Inscription and translation refer to the communication studies' view that any application is a semiotic system implying that all participants in a communicative interaction process construct their own meaning. This applies to both the one who express something (designer, sender) and the one who perceive (user, receiver).

# TOMBOLA: Simulation and User-Specific Presentation of Executable Task Models

*Holger Uhr*

University of Paderborn

Warburger Str. 100, D-33095 Paderborn

huhr@uni-paderborn.de

## Abstract

Task models are an important contribution to the design of user-oriented applications, because they provide a method to communicate user interface requirements from interface experts to designers. While several formal presentations for task models exist, none of them is suitable to be understood by users. This paper presents the TOMBOLA task simulator which allows to present task model execution in a user specific way, thereby helping the user to understand and verify the task model.

## 1    The Role of Task Models in the Design Process

The analysis of a user's task can provide valuable information for designing an application. Tasks are of special interest to developers because "Computer systems are designed to help people carry out tasks more efficiently and to carry out tasks that were previously not possible." (Johnson, 1992) In general, task analysis provides a common ground between designers and users. "It allows designers to understand users' work and users to understand designs." (Johnson, Wilson, & Johnson, 1997). It is also said "that understanding the parts and their structural relationship can be instrumental in designing performable and learnable tasks, artifacts to faciliate these tasks, prerequisite instruction etc." (Anderson, Carroll, Grudin, McGrew, & Scapin, 1990)

## 2    Presenting Task Models to the User

There exist many methods to formulate task models, such as GOMS (Card, Moran, & Newell, 1983), Diane+ (Tarby & Barthet, 1996), or ConcurTaskTrees (Paternò, Mancini, & Meniconi, 1997). It is a common problem when building a task model from user observations (after gathering data in a task analysis process) that the user is unable to verify the correctness of the task model because she doesn't understand the formal model description language. In addition, most methods offer only a static display of the task model without the possibiliy to simulate the tasks and explore the dynamic aspects of the model.

TOMBOLA is a task simulator that enables the developer to build an executable combined task and data model and to formulate display rules for the presentation of the current state of the execution. A specifically customized view of the model is presented to the user, where she finds elements of her familiar task world. Thus she can check the correctness of the model and point out errors in the model introduced by the developer.

# 3 Building Executable Task Models in TOMBOLA

The developer enters a textual description of the combined task and data model into the TOMBOLA simulator. She also defines a set of display rules that map specific task states, data objects and their values to features such as images or colors.

## 3.1 TOMBOLA's Task Model

TOMBOLA provides a hierarchic task structure where, starting from the top task, each non-atomic task can have an arbitrary number of subtasks. Each task controls the temporal relation of its subtasks. The temporal relations provided are

*Sequential:* The subtasks must execute in a fixed sequence.
*Serial:* The subtasks must execute in an arbitrary sequence.
*Parallel:* The subtasks can start and end at random.
*Simultaneous:* All subtasks have to start in an arbitrary sequence before any task can end. Therefore at least one moment exists where all subtasks are running simultaneously.
*Alternative:* Exactly one randomly selected subtask can execute.
*Optional:* One or no subtask at all can execute.
*Loop:* A single task is repeatedly executed while a given condition is true.

By combining several temporal relations in super/subtasks, a rich set of control flows can be generated to model the user's task flow.

TOMBOLA provides basically two different kinds of task:

*External tasks* are used to describe bigger task units that can be executed "on their own" without being directly dependent on their supertask. They are somehow comparable to functions in a language like "C" in that they can be called from several points and establish a new variable scope of their own. A task that has an external task as subtask, can communicate with this task via parameter values.

***Internal tasks*** build small substructures inside external tasks, like blocks inside functions in "C". They depend directly on their supertask and therefore cannot be used outside their context. Internal tasks inherit the variable scope of their supertask.

A task may carry the attributes "autostart" or "autostop", in which case it can start or stop on its own without any user interaction. This helps modelling system tasks that are executed by the application alone.

## 3.2 TOMBOLA's data model

To build the data model, the developer defines data variables to be used in the scope of a given task. These variables (possible types include integer, boolean, string, class types and arrays), can be used as task attributes. They are initialized when a task starts and can be changed in the postcondition of a task and queried in the precondition of a task.

Developers can declare their own class types and use inheritance to build subclasses. The membership of an object to a class can be tested by an "instanceof" operator (similar to "Java") in logical expressions.

TOMBOLA offers the declaration of object arrays and the instantiation of subtasks by iterating over the array. It is thus possible to spawn a subtask for every object in the array. (The number of objects in the array is defined dynamically during the simulation.)

Each task can have a precondition, an expression involving object variables, to be fulfilled before the task can start, and a postcondition, a list of assignments of expressions to variables that are executed when the task has been finished. In particular, TOMBOLA provides interactive input expressions that can read in integer, string or boolean values entered by the user during simulation. The simulation can therefore dynamically react to user decisions. In conjunction with iteration over arrays as mentioned above, the user can thus interactively control the number of subtasks for a given task by chosing the number of elements in an array.

## 3.3 Visualization and execution of tasks

When the task model is simulated by TOMBOLA, the current state of the execution is internally expressed as an XML model after each step. This includes all existing tasks, their attributes and states and all data objects. The XML model is then transformed by an XSL stylesheet into a set of HTML pages, which can be inspected by the user through a standard web browser. By defining special transformation rules in the stylesheet, the developer can build a user specific presentation of the state of the model execution, to help him understand the model. For example, specific presentations can be used for

- the various attributes of a task (temporal relation, autostart and autostop flags, pre- and postconditions),
- the current state of a task (not ready, ready, executing, etc.),

- the relations between tasks (subtask, supertask, sibling), and
- the data objects with their attributes. (Attribute types include integer, boolean, string, class types and arrays.)

The developer can use all the elements available in an HTML page, such as icons (showing for example real-world objects that are used by the user), colors, font attributes etc. She can set up rules that place the presentation of certain tasks on seperate (linked) pages or that make them invisible according to their state.

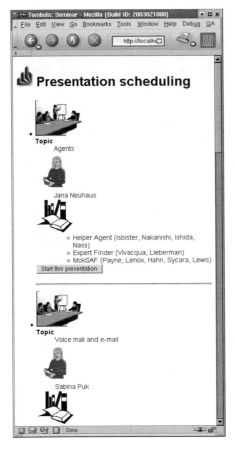

Figure 1: Screenshot of the seminar visualization example

When a task can start or stop in a given situation, the developer can provide the visualization with an according button. Once the user presses the button, the simulator is notified, starts or stops the specific task and updates the HTML pages to present the new state of the task model. Similarly, tasks which are waiting for interactive input are displayed with a HTML form where the user can input the according value which is then flowing into the task model.

By using TOMBOLA, the developer can build an executable version of the task model with a presentation that is specifically adapted to the user, using the flexibility of HTML design elements. The user can understand and inspect the task model and point out possible errors in the model.

## 4    Current Usage

A default stylesheet exists that displays all the information of the current state of execution of the task model in a textual format. This stylesheet serves as a starting point for developers of user specific visualization and helps to overcome the difficulties emerging when developers start to program in XSL.

Currently a specific visualization is developed to model web site content management tasks. A task model for organizing seminars at a university exists, for which a visualization is currently developed (see Figure 1). Here icons are used to symbolyze the different tasks and their attributes.

# 5    Future plans

To help the developer formulating her task model in TOMBOLA, a graphical editor will be developed that can be used as an alternative to the textual input language. The task model developer will be able to use visualizations of tasks and variables while specifying the task model.

By using a web browser as a "display engine" for the specific visualization of task models, TOMBOLA gets some useful features:

- A rich set of highly configurable display elements like text, pictures, animations etc.,
- an easy way to navigate different views by hyperlinks,
- interactive elements to enter data values and send signals to the simulator by using forms, and
- a textual language describing the different views and the elements they contain that is sufficient to create the views for the user.

Currently we explore the possibility of coupling TOMBOLA with different display engines that provide similar features for displaying views and interacting with the simulator. The Mozilla platform (Oeschger, Murphy, King, Collins, & Boswell, 2002) e. g., which can be programmed with the XML-based XUL user interface language, seems to be a suitable candidate.

# References

Anderson, R. I., Carroll, J. M., Grudin, J., McGrew, J. F., & Scapin, D. L. (1990). Task analysis: The oft missing step in the development of computer-human interfaces; its desirable nature, value, and role. In D. Diaper, G. Cockton, D. Gilmore, & B. Shackel (Eds.), *Human computer interaction – Interact '90* (pp. 1051–1054). Cambridge, UK: North Holland, Amsterdam. (Panel Session)

Card, S. P., Moran, T. P., & Newell, A. (1983). *The psychology of human-computer interaction.* Hillsdale, New Jersey: Lawrence Erlbaum Associates.

Johnson, P. (1992). *Human computer interaction: Psychology, task analysis, and software engineering.* Maidenhead, England: McGraw-Hill Book Company Europe.

Johnson, P., Wilson, S., & Johnson, H. (1997). Designing user interfaces from analyses of users' tasks. In S. Pemberton (Ed.), *CHI 97 electronic publications.* Atlanta, Georgia: acm PRESS.

Oeschger, I., Murphy, E., King, B., Collins, P., & Boswell, D. (2002). *Creating applications with Mozilla.* Sebastopol, CA: O'Reilly & Associates.

Paternò, F., Mancini, C., & Meniconi, S. (1997). ConcurTaskTrees: A diagrammatic notation for specifying task models. In S. Howard, J. Hammond, & G. Lindgaard (Eds.), *Human-computer interaction interact '97* (pp. 362–369). Sidney, Australia: Chapman & Hall, London.

Tarby, J.-C., & Barthet, M.-F. (1996). The Diane+ method. In J. Vanderdonckt (Ed.), *Proceedings of the 2nd international workshop on computer-aided design of user interfaces (CADUI'96)* (Vol. 15, pp. 95–120). Namur: Presses Universitaires de Namur, Namur.

# Development of a Crew Station Design Tool

*Brett Walters, Susan Archer and Shannon Pray*

Micro Analysis and Design
Boulder, CO 80301 U.S.A.
bwalters@maad.com

## Abstract

Designing the layout of any type of crew station is a complex and difficult process. Engineers have to consider the size of the crew station, the physical attributes of the crew station operators, the tasks the operators have to perform, and the method (modality) in which the operators perform the tasks. In addition, a number of different controls and displays are available that can be used to perform the same task. Although standards exist for the type and location of controls and displays for some types of crew stations (e.g., the cockpit of fixed winged aircraft), the layout of most crew stations is often left up to engineers. The recent increase in multi-modal input and output technologies only adds to the complexity of the engineer's job. In order to simplify and structure the process of designing a crew station, a software tool is being developed that will automatically determine the optimum arrangement of controls and displays based upon sound human engineering and ergonomics principles. The Crew Station Design Tool (CSDT) utilizes task network modeling to help identify conflicts among operator, task, and control groupings and a 3-dimensional environment that allows users to visualize prototype layouts of their crew stations. This work is being funded by the U.S. Army Research Laboratory, Human Research and Engineering Directorate.

## 1    Introduction

The increasing complexity of weapons systems, coupled with budget pressure and manpower limitations, has fostered a deep and sincere appreciation for ensuring that soldiers will be able to effectively and efficiently operate and maintain new systems when they are fielded. However, designing the layout of a crew station is a complex and difficult process. Engineers have to consider several factors, such as the size of the crew station, the physical attributes of the soldiers, the tasks the soldiers have to perform, and the method (modality) in which the soldiers perform the tasks. In addition, a number of different controls and displays are available that can be used to perform the same task. Although standards exist for the type and location of controls and displays for some types of crew stations, the layout of most crew stations is often left up to engineers. The recent increase in multi-modal input and output technologies only adds to the complexity of the engineer's job. This has fueled the hunger for design visualization techniques and tools.

The primary objective of this work was to define a practical solution that will enable crew station designers to visualize and optimize their choices of controls and displays, and the position of those elements in the workstation. Figure 1 shows the process flow we have developed for the Crew Station Design Tool (CSDT). Each element of the flow is described in a section of this report. The black triangles in the figure indicate three different "points of entry" into the CSDT.

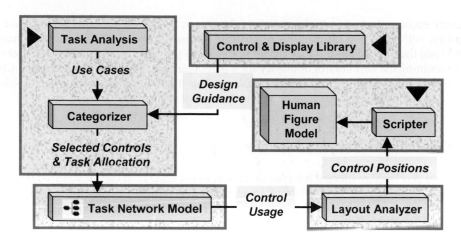

**Figure 1:** Overview of the CSDT process

## 2   Task Analysis

The CSDT starts by guiding the user through a requirement-driven task analysis for the jobs each operator will perform in the crew station.  Users begin at a high level by first identifying their system requirements.  The user must then break each requirement down into functions.  To support the user through this process, we have populated the CSDT with a database containing default requirements and functions for the following crew stations: armoured vehicles, fixed and rotary wing aircraft, ground control stations, motorcycles, passenger vehicles, and trucks.

The final step in the task analysis is to decompose the previously defined functions into use cases. Use cases can be thought of as tightly defined tasks composed of actions and objects (e.g., Increase altitude").  Actions are selected from a list that is stored in the CSDT database.  This list provides later links to human resources, or modalities, and also provides the basis for helping the user select the correct control or display for the crew station.  Objects are information elements. Objects can also be selected from a list stored in the CSDT database, or the user can simply enter a new object. The outputs of the task analysis are the use cases and resulting actions.

## 3   Categorizer

The Categorizer allows users to select a control or display for each use case (i.e., task) the crewmembers will perform.  The inputs to this part of the tool are the use cases from the task analysis.  It is through the Categorizer that a user will have access to the information stored in the Control and Display Library (described in Section 4) and the query system stored in the library.

In the Categorizer, actions (from the use case statements) are assigned to controls and displays in a two-step process.  First, they are automatically categorized into a primary modality.  Some actions, particularly those assigned to motor modalities, may require a secondary modality.  For example, a soldier may have to look at something they are attempting to reach, or they are listening for feedback that a motor control has been properly activated.  Based on the primary and secondary modalities, a list of possible controls and displays will be presented to the user to

269

effectively perform the task in the crew station. Users can change the modality for each use case. This action will result in a different list of possible controls to be selected. Users also have the option to add their own control if they cannot find it in the possible list of controls. If a user needs help determining the best control or display for a particular task, he or she can access the Control and Display Library. When the user has finished using the Categorizer, each action (excluding cognitive and automated actions) will be assigned a control or display.

# 4   Control and Display Library

The Control and Display Library is a critical element of the CSDT. It can be accessed through the Categorizer, as users need help selecting an appropriate control or display for a particular task. The library is also a stand-alone module, in that it can be accessed without performing the function analysis and use case language.

Within the Control and Display Library, a comprehensive list of controls and displays is separated into categories that are aligned to one or more sensory modalities (e.g., visual, auditory, motor), and detailed parameters (e.g., size and visual requirements). These parameters were carefully selected based on their correspondence to existing design guidelines and standards. This process resulted in a decision tree structure, in which user selections for attributes of their use cases will drive the identification of appropriate controls and displays. The process uses a series of tabs and checkboxes to help the user decide from all the possible choices of controls and displays for that action and modality. As users select different parameters, the list of possible controls is adjusted accordingly. Finally, the Control and Display Library stores and indexes relevant design guideline information for each control and display in a relational database. It is open-ended so that new control and display technology characteristics can be added as they develop.

# 5   Task Network Model

This element of the tool determines whether the initial assignment of tasks to controls and displays is likely to cause any performance-based problems or resource channel-based workload conflicts. The first step in this process is to import the use case information into the task network modeling tool (e.g., Micro Saint), sequence the use cases, make an initial assignment of use cases to crewmembers, and generate a human performance model that will provide a variety of outputs.

In order for the model to calculate a moment-to-moment overall mental workload value for each of the crewmembers defined in the analysis, each task must be assigned to resource/interface channels. These are pairs of the mental resources (e.g., visual, auditory, motor, speech, cognitive) that the crewmembers will be using to operate the selected controls and displays in order to perform each task. The output from the Categorizer will provide the information needed for this step. For each resource/interface channel assigned to a task, the user will select a workload value from a table of benchmarked values representing a numeric demand for performing the task.

Many outputs are provided to the user from the task network-modeling element in order to support a comprehensive understanding of the performance implications of the current assignment of tasks to crewmembers and to controls and displays. These outputs consist of:

- Frequency of use – to assist in the consolidation of tasks, controls and displays

- Concurrencies of use – to ensure that conflicts between resources (i.e., modalities) across tasks are resolved
- Crewmember utilization – to ensure that the tasks are allocated in a balanced manner
- Workload by human resource (or channel) – to assist in task allocation and to ensure that the tasks are appropriately supported by the choice of control and display

The task network model is used in an iterative manner with the preceding elements of the analysis. The output of the task network model is a workable assignment of controls and displays to use cases from a task performance and workload perspective.

**Figure 2:** Sample crew station designed in Open Inventor

# 6    Layout Analyzer

Once the controls and displays have been chosen to support each task, they must be positioned in the crew station. The Layout Analyzer will be used to make the first pass at identifying a workable physical location for each control and display. Users must first enter the size of the crew station. They are then asked to position any unusable space into the crew station (e.g., air vents, doorways, and windows). Once the user has answered these questions, he or she will have the opportunity to select which conventions (if any) they would like to be used to arrange their controls and displays. Design conventions (also known as stereotypes) are properties or characteristics that are *typically* associated with an object (e.g., how it is used and its location). Over 250 conventions were obtained from Military Standards and Federal Motor Vehicle Safety Standards and stored in the CSDT's database.

After the user has selected their conventions, a three-dimensional workspace will be presented with the dimensions entered by the user. Next, the Layout Analyzer will position the controls and displays in this scene in the optimal spot for each operator. This will be done using a combination of anthropometric and ergonomic data as well as a series of priorities (e.g., conventions, accessibility, frequency of use, and sequence of use). Users will have the option to select which priorities they want the CSDT to use to place their controls and displays. Figure 2 shows an example of a passenger vehicle crew station. We have used the commercial software Open Inventor (developed by TGS, Inc.) to show the output of the Layout Analyzer in a three-dimensional environment. The output of the Layout Analyzer will be a determination of the position of the controls and displays in three-dimensional space, and the reporting of that position.

# 7 Human Figure Model

The final step in the CSDT is the human figure-modeling element of the tool. We chose the software JACK® as the human figure model. [JACK® is currently owned and maintained by Electronic Data Systems (EDS)]. The input to this element, provided by the Open Environment project, will consist of a computer-aided design (CAD) image of the crew station shell, an initial position of each control and display to be placed in that crew station, and a script that will animate the human figure to use the controls and displays to perform the use cases.

Human figure models simulate the physical behavior of humans interacting in a three-dimensional environment. They allow designers to create virtual environments by importing CAD data, populate the environment with biomechanically accurate human figures, assign tasks to these virtual humans, and obtain valuable information about their behavior. This type of software also allows designers to visualize the feasibility of certain tasks (i.e., can a human see and actuate a control within the specified environment), obtain information on strength capability, and generate "possible" and "comfortable" reach envelopes. When the user has finished viewing the JACK® model, they can return to any part of the CSDT to make adjustments in their design.

# 8 Discussion

Because of the CSDT's modular approach, not all of the elements are required for the tool to function. For example, the Control and Display Library is a stand-alone component of the CSDT. Our vision is that this application could reside on a designer's desktop as a tool that will provide human factors guidance and data, without performing the function analysis and use case language. The human figure modeling is also an optional element of the tool: the Layout Analyzer and Open Inventor portions of the CSDT can provide a detailed first-glance of the crew station.

The approach used within the CSDT is innovative in several ways. First, the task network model determines areas of conflicting resources for tasks performed simultaneously and the overuse of controls and displays. Second, the Control and Display Library helps the user select the appropriate control and display for each task through the use of a query system and a relational database linking controls and displays to a wide variety of characteristics. Finally, the three-dimensional environments allow users to visualize how operators will perform tasks within their design. The CSDT will be completed in January 2004.

# Acknowledgements

This work was conducted under contract DAAD17-02-C-0019 for the U.S. Army Research Laboratory. The opinions and assertions contained herein are those of the authors and are not to be construed as official or reflecting the views of the U.S. Army Research Laboratory.

# A Methodology for the Component-Based Development of Web Applications

*Michael Wissen, Juergen Ziegler*

Fraunhofer IAO
Nobelstrasse 12
Stuttgart, Germany
{Michael.Wissen, Juergen.Ziegler}@iao.fhg.de

## Abstract

This paper presents a universal and integrated methodology for the component-based design of web applications. Modelling methods for design areas relevant in the web context are explicated which are enhancing the concepts of existing design methods by aspects of metadata modelling as well as of view and navigation modelling. Core of the approach is the application of metamodels and modelling techniques as well as tools for prototypical implementation of developed models in a run-capable web application.

## 1 Introduction

The development of web-based information systems has been technologically supported for some years by a multiplicity of different systems, such as web editors or content management systems. The focus of these technologies lies on the design and flexible development of graphic representations of contents, placing and administering of contents within an editing process as well as the connection to data bases and existing applications. Contrary to conventional applications, web-based applications are often subject to continuous revision regarding their structures as well as their contents which entails an extensive new design of the application when using current procedures and systems. Typical characteristics of current web applications are a mixture of information objects of varying structuring degrees, high modification dynamics and inclusion of external information components and services. Design techniques and development processes have to answer the questions connected with this changed situation and, at the same time, meet the requirements of the widely expanded spectra of diverse designer and developer roles.

## 2 State of the Art

The use of systematic design methods is yet uncommon in the development of web applications. Studies like Barry & Lang (Barry & Lang, 2001) show that specific methods are hardly used, particularly in the media field, and if so, they have a low standardization degree. Even standard methods like UML are applied as yet only rarely in the web area. Several approaches of methods were proposed for specific aspects of the web development. Some of these methods refer to the modelling of the hypermedial structure of web sites, such as HDM (Garzotti et al., 1993), RMM (Isakowitz & Stohr, 1995), OOHDM (Schwabe & Rossi, 1994) as well as the UML-extension for the modelling of hypermedial systems (Baumeister et al., 1999). However, these methods usually are inadequately integrable with other development perspectives and do not possess sufficient tool support. Moreover, these methods lack, on one hand, the possibility for timely evaluation of efficiency, navigationability and representation of a web application with a generated prototype and, on the other hand, the conceptual support of dynamically changing scope of concepts and contents.

# 3 Ontology-based Approach for Modelling Web Applications

This paper presents methods and metamodels for the systemization of the construction of web applications and support of the entire development process up to service and maintenance of web applications. Basis of the methodology is a method compound containing concepts and technologies for significant modelling aspects. The concept of the method compound enables software-based tools to support the generation process semiautomatically so that web models generated by the user can be represented prototypically. This permits initial application of mechanisms for efficiency evaluation of structure, navigation, and representation. Anchor point is an ontological description (Lenat & Guha, 1990) of the information base which serves as basis for the systematic design of a navigation and view structure. Based on this, so-called navigational classes enable the automated classification of information objects in individual subjects of the navigation structure with consideration to the different views. Moreover, distributed services available, such as web services, can be integrated in the form of functional components.Construction of a web application site and navigation structure is accomplished based on various components. The web structure is generated with the aid of containers and partitions which have the capacity to be embedded into one another for representation purposes so that also nested web pages can be represented. The method compound is expanded by procedures which provide the developer of a web application with possibilities for modelling and integration of context-dependent information offers. So-called view classes are made available which enable a role-dependent, respectively location-dependent.
The design process of a web application is composed of the following steps: design of metadata model (topic structure), generation of a navigation model, formal analysis of user requirements, derivation of view structure, configuration of external services and definition of graphic representation and interaction.

## 2.1 Conceptual Model

The resource model serves as basis for the construction of the topic structure. It contains concepts as well as individual information objects in the form of documents, tasks, processes as instances as well as their associations to other instances. This, however, is not sufficient for a defined navigation in the subsequent web application, since it is not possible at this point to access information components directly. But with the aid of an ontology, explorative accesses within the web application can be realized and, moreover, navigation possibilities can be expanded to dynamic content.

## 2.2 Component Design and Structuring

In this step, essential elements of a web application are described in components, which are composed into the form of component models. Structure and construction can be modeled using containers and partitions which can be realized, e.g., by frames in the subsequent web application. *Atomic components* are concrete objects which cannot be divided further. This includes, e.g., texts, images, multimedia objects, external references, etc. Atomic components are considered instances of their respective atomic class within the conceptual model and, primarily, serve the purpose of unitary viewing of different objects of the navigation model.

274

The elements describing individual web sites are kept in so-called *containers* which can be embedded into one another so that nested web sites can also be represented. The navigation model, which is the basis for the subsequent representation of the web application, is based on a HiGraph-like approach (Harel et al., 1987). Several subcontainers within a container, which are on the same level, are represented in accordance with an XOR link, i.e., in this case, always exactly one subcontainer is visible. For a parallel representation of different containers, containers are divided into individual *partitions* (cf. Fig. 1).

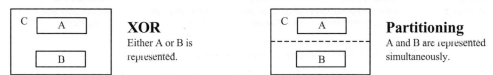

Fig. 1: Containers and Partitions

Atomic classes are considered independent containers which, however, cannot be divided further, If there are several subcontainers on the same level of a container, it needs to be determined which container is visible at first. To do so, a initial state is defined, pictured by an arrow starting from a dot. Containers as well as partitions define the destination area where the contents are to be represented later. With the aid of place holders (dashed component), partial areas of the web application can be represented schematically for better clarity. Detailed modelling of the area coarsely represented by the placeholder, can be done elsewhere.

An alteration of the contents represented on a web page is induced by an *event* which is typically initiated by interaction by a user or caused automatically (e.g., by lapse of time or previous events). Processing the event is performed by a *link* which will prompt an altered representation of a web application. The destination area of a link defines the container, respectively the partition, where changes are to be made. The current context is carried along in a link in form of the source instance. The source instance is the instance of a class represented by the link. Transfer of the context is necessary in order to serve dependencies of the information to be built with the current context. Since the destination area cannot be identified by an atomic component, in addition, a destination class must be known to the link. Based on this and the source instance, it is possible to detect the instance, respectively the set of instances. If the destination class is an atomic class, the source instance is the component in which the link was defined. This component can be either a container or a partition. It is, therefore, possible to refer to arbitrary objects within a web application which are not contained in the ontology. In the graphic representation of the navigation model, links define the transition from one component to another.

In a destination area, either an individual instance of a class can be represented or, if several instances are linked to one class, an overview on the instances available can be displayed with the possibility to view the instances individually at a defined location. For this purpose, *navigational classes* are defined which enable navigation of a set of instances with the aid of an overview. The manner in which this overview is represented, can be determined either by the navigational class itself or has already been determined in the underlying navigation model. In the latter case, a representation type is selected from a set of predefined navigation models. This set includes, according to Fig. 2, types such as tree, index, sequence, etc. (cf. Ziegler, 1997). If the representation type of the navigational class is detected dynamically, the cardinality of the instance set to be represented, is applied. The set of the instances to be represented can be described with the aid of the instance of the original class and the original relation. The selection of attributes represented is defined in the attribute set of the destination instance and arranged according to a sorting function.

The graphic representation of navigational classes consists of two components which are linked by a navigation relation. The navigation relation is a parameterizable link which is initialized by the individual instances in the navigational class when called. It is marked by an open arrowhead.

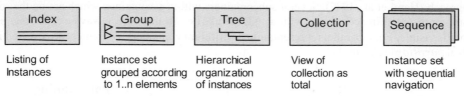

**Fig. 2:** Selection of different types of navigational classes

The *composition model* defines the compilation of arbitrary components into individual web sites. It specifies, with the aid of containers and partitions, the structure of a web site and the site-inherent navigation. The *navigation model* is composed of the composition models of the individual web sites and an additional higher-ranking navigation structure. This defines the navigation between the individual web sites. Fig. 3 is an example of the navigation model of a portal page. In the entire web application, a header line is to be found in the upper area, directly beneath it is a menu bar with a search field attached to the right in form of an application. Beneath these two components, there is a further partition. It contains in the start condition at first a login field on the left. Upon successful login, it is replaced by a vertical menu with an overlying image. On the right, nothing is represented at first, only upon finalization of the login, the ticker is activated. The lower area represents the main area of the application. It is, at the same time, the destination area for both menu bars and the search field. The different components, which are represented in this area, can now be specified more precisely at another location.

## 2.3 View Model

Different views on existing information are created with the aid of view classes, with the objective of generating an information supply customized to user requirements. Their mission is representing defined characteristics of an instance of a class under consideration of constraints. A resource class can encompass several view classes with which the different views on the underlying contents can be realized. Similar to OOHDM, a SQL-like query language would be the obvious choice for the definition of the view classes.

## 2.4 Representation Model

The representation model serves the transformation of the conceptual schema on a lower-level representation and basically contains the information necessary for graphic representation. This includes exact information on position and graphic design of the contents and interaction interfaces to be represented. These information items are assigned to the individual view classes using style sheets and retrieved at runtime.

## 3 Implementation and Future Development

With the aid of parameterizable components, which generate, based on defined metadata and navigational models on runtime, dynamic and context-sensitive navigation structures, models generated by the user can be realized in a runtime environment for test and verification possibilities. For this, processes for automatic generation of navigation structures based on the developed models are designed and implemented. Their high reusability rate enables the implementation of a modeled web application in a run-capable system.

This paper describes the procedure for systematic and universal development of web applications. Special attention has been paid to a most easy perceivability of the models and especially to the support possibilities by software-based tools which are to relieve the user in generation and verification of the models to the greatest possible extent. In a further step, the component-oriented approach could be extended to model components available in the web which are characterized by an integrative description. It would offer the developer of a web application the advantage of being able to access already existing self-generated or external solutions for certain components of his application.

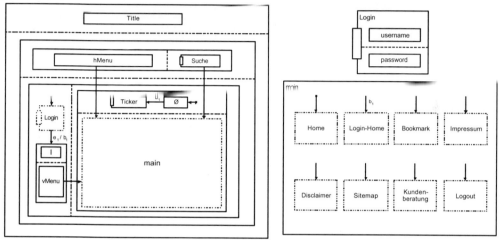

**Fig. 3:** Navigation model (Excerpt from a portal)

# References

Barry, C.; Lang, M. (2001): A Survey of Multimedia and Web Development Techniques and Methodology Usage. IEEE Multimedia. 8(3), 52-60

Baumeister, H.; Koch, N.; Mandel, L. (1999): Towards a UML extension for hypermedia design. In UML'99, Fort Collins, USA

Garzotti, F.; Paolini, P.; Schwabe, D. (1993): HDM – A model-based approach to hypermedia application design. ACM-Transactions on Information Systems, 11/1

Harel, D.; Pnueli, A.; Schmidt, J. P.; Sherman, R. (1987): On the formal semantics of statecharts. Proc. 2nd. IEEE Symposium on Logic in Computer Science, Ithaca, N.Y.

Isakowitz, T.; Stohr, E. A.; Balasubramanian, P. (1995): RMM: A Methodology for Structured Hypermedia Design. Communications of the ACM, 38/8:34-44

Lenat, D. B.; Guha, R. V. (1990): Building large knowledge-based systems. Addison-Wesley

Schwabe, D.; Rossi, G. (1994): From Domain Models to Hypermedia Applications: An Object-Oriented Approach. International Workshop on Methodologies for Designing and Developing Hypermedia Applications, Edinburgh

Ziegler, J. E. (1997): ViewNet - Conceptual design and modelling of navigation. In S. Howard, J. Hammon & G. Lingaard (Eds.), Human-Computer Interaction: Interact'97, London: Chapman & Hall

# Section 2

# HCI Methodologies and Perspectives

# Eighteen Classes of Functionality:
# The D.EU.PS. Model of Information Systems Use

*Pär J. Ågerfalk*

Dept. of Informatics (ESI)
Örebro University
SE-701 82 Örebro, Sweden
pak@esi.oru.se

*Emma Eliason*

Dept. of Informatics (ESI)
Örebro University
SE-701 82 Örebro, Sweden
eaen@esi.oru.se

## Abstract

Five high-level categories of information systems functionality (desired, existing, utilized, perceived, and satisfactory), which combine to 18 classes, are presented. The resulting model is referred to as the D.EU.PS. model of information systems use (pronounced 'dupes'). Along with the presentation of the classes their validity is argued by use of examples from three case studies. In doing so, the usefulness of the model as a tool for data collection and analysis within different use-contexts is exemplified.

## 1 Introduction

Usability, according to ISO 9241-11 (1998), is concerned with the effectiveness, efficiency and satisfaction with which users can achieve specified goals in specified contexts of use. Preece, Rogers & Sharp (2002, p. 14) explain the term effectiveness as 'how good a system is at doing what it is supposed to do'. That is, 'effectiveness' suggests that specified goals are to be achieved with accuracy and completeness (ISO 9241-11, 1998). Effectiveness is thus related to a system's de-sired functionality, what users are supposed to use the system for. The term efficiency suggests that specified goals are to be achieved with as little expenditure of resources as possible (ISO 9241-11, 1998). Bevan & Macleod (1994), for example, suggest that (temporal) efficiency equals 'effectiveness / task time'. In this view, efficiency is related to how instrumentally efficient the desired functionality can be used by specified users. The third criterion, satisfaction, suggests that users should feel comfortable with and have positive attitudes towards the use of the system (ISO 9241-11, 1998). Satisfaction is thus related to how subjectively pleasing desired functionality is in the eyes of the users.

If applying these categories unreflectively, an implicit assumption is made that information systems consist of desired functionality that to varying extent is efficient and satisfactory. That is, there are two primary overlapping classes of functionality: the desired and the satisfactory, which may or may not be efficient. In this paper we argue that to fully understand the use of an information system (IS) and to assess its usability there is a need to introduce further categories of IS functionality. In the remainder of the paper we present five categories, which combine to 18 classes of IS functionality. Along with the classes we present examples from an ongoing project involving evaluation of an Internet banking system (the Internet Bank), an IS used to book rooms, teachers, students and extra equipment at a university (the Booking System), and a corporate intranet (the Intranet). In these cases, the resulting model has proved useful in directing evaluators' attention during data collection and in functioning as a tool for analysis of user-system interaction within different business contexts (see Ågerfalk, Sjöström, Eliason, Cronholm & Goldkuhl, 2002).

## 2    The D.EU.PS Model of Information Systems Use

The reason for designing an IS in the first place is probably a belief that there are some tasks that the system may support. Hence, there is a desired functionality that the IS is supposed to implement and deliver to its users. We may distinguish between this desired functionality from the functionality that does eventually exist in the system. These two categories may not overlap completely, which is why there may be a need for further development and implementation of new non-existing but desired functionality. It may also be the case that there is undesired existing functionality, which may as well be removed. Furthermore, we may distinguish existing functionality from functionality users perceive exist. According to Norman (1999), the art of the designer is to ensure that desired, relevant actions are readily perceivable. Therefore it is equally important to recognize not only what functions an IS implements but also actions that users perceive as possible to perform. Furthermore, it is a well-known fact that many systems or parts thereof are not used to the extent anticipated during design (e.g., Davis, 1989). Therefore we may additionally distinguish between what exists, perceived or not, and what is actually utilized. Finally, as suggested by the ISO 9241 usability definition, functionality perceived by users may be perceived as satisfactory or unsatisfactory; that is without discomfort and with positive attitudes towards the use of the product. Davis (1989) points out that in addition to desired functionality (he refer to this as usefulness), utilization is influenced by perceived ease of use (as in satisfaction), and the utilization of existing functionality depends on what functionality a person believes enhances his or her job performance. It is important not to confuse desired functionality with satisfactory. Davis (1989) showed that even if users believe an application to be useful (as in desired), they might still consider it too hard to use (as in unsatisfactory). On the other hand, people may regard an undesired function as satisfactory: 'sure, it works fine, but I don't need it'. Satisfaction is thus not restricted to ease of use alone. Satisfaction relates to the contentment of the actual implementation of a function while desired functionality more holistically describes the desirability of a function.

Altogether this gives us five high-level categories of desired, existing, utilized, perceived, and satisfactory functionality. Desired functionality is functionality that is believed to enable a user to accomplish intended business effects. Existing functionality is functionality that is implemented and accessible in the system. Utilized functionality is existing functionality which is actually being used. Perceived functionality is functionality that users perceive to exist. Finally, satis-factory functionality is perceived functionality which can be used with satisfaction. These five categories combine to 18 classes of IS functionality. We refer to the resulting model as the D.EU.PS. model of IS use (pronounced 'dupes'), first introduced by Ågerfalk et al. (2002). The model is depicted in Figure 1 (next page). Note that the respective sizes of the classes in Figure 1 do not correspond to any quantitative empirical measure. They have been chosen simply to facilitate presentation.

Note that in the model, utilized is regarded a subset of existing, and satisfactory a subclass of perceived (something that is utilized always exist, and something that is satisfactory is always perceived). This is the reason for calling the model D.EU.PS. This is also the reason why not all five letters are used in all class names below; they can be uniquely identified anyway.

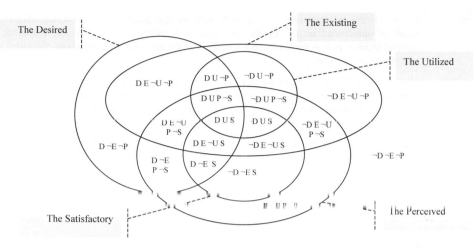

**Figure 1**: The D.EU.PS. model of information systems use.

In Table 1 we present these combinations and give examples of each as derived from the three performed information systems evaluations in which the model has been used as an analytic tool. A note should be made regarding functionality that is undesired (¬D), non-existing (¬E), and unperceived (¬P). This would, for example, be the case if a new feature were proposed in response to identified problems during evaluation and we were to decide whether to implement it or not. That is, to decide, for example, if it is desired or not. In set-theoretical terms, we can think of such a class as the complement of the union of desired, existing and perceived functionality.

**Table 1**: Classes of the D.EU.PS. model.

| Class | Description and Example |
|---|---|
| ¬D ¬E ¬P | An undesired function that does not exist in the system, everyone agrees to this, and there is no reason for implementing it either. |
| ¬D E ¬U ¬P | This function is neither desired nor used, and not even recognized. It can probably be removed. Example: An Internet Bank user did not know that it was possible to categorize pre-registered receivers, and he did not want that function anyway. |
| ¬D U ¬P | This function is used even though users are not aware of its existence, and they do not want it. This may be a result of unattended action (e.g., trial and error) and should, in that case, be avoided. It may also be the case that the system does something automatically without the user's knowledge and approval. Example: The Internet Bank showed a list of 'new offers' whenever someone logged on to the system, which may be more annoying than informative. The same goes for logging of user actions by use of cookies that the user may not be aware of. |
| D ¬E ¬P | This should really be there. A missing user requirement. Example: An Internet Bank user wanted a function showing how much money that would be left on his account should he fulfil the current payment. |
| D E ¬U ¬P | Users do not perceive this function so they do not use it, but they would have used it if they understood it. Example: An Internet Bank user did not know how to change a partially completed payment order. The function was 'hidden' in the design since it had the same visual appearance as help texts and therefore was perceived as such. |

| | |
|---|---|
| D U ¬P | Users perform this function even though they are not aware of its existence. Probably a result of unattended action. Relates probably to learnability or education. Example: An Internet Bank user was not aware that when a date for payment was not given, the today date was automatically used. |
| ¬D ¬E P ¬S | Users believe that this undesired function exists but that it could be better, but it does not exist at all, and we better leave it that way. This is probably related to learnability or misleading instructions/education. Example: The Intranet did not log user activities to the extent feared by some users. |
| ¬D ¬E S | Users believe this undesired function exists and that it works fine, but it does not, and we better leave it that way. This is probably related to learnability or misleading instructions/education. Example: Some Intranet users did not care that the system logged their activities (which it did not). |
| ¬D E ¬U S | Sure it works fine but users do not need the functionality. Example: In the Booking System it was possible to create diagrams showing bookings made by a teacher. A function that users did not want and did not use. |
| ¬D E ¬U P ¬S | An undesired function that the users do not need and think is unsatisfactory. Example: The function in the previous example (¬D E ¬U S) was not only undesired but was also considered hard to use by some users. |
| ¬D U S | It works well even though users do not really want it. Example: Some Internet Bank users thought that the system required too many safety codes that did not promote trust, yet they accepted it and did not think it was a big deal. |
| ¬D U P ¬S | Not only forced to use it, it is unpleasant as well. Example: Some Internet Bank users regarded the repeated entering of security codes as a waste of time. |
| D ¬E S | Users want this function and believe that it exists and looks good, but obviously they do something else than what they think they do. Example: A Booking System user was very happy that he had booked a classroom, but when the class started another teacher claimed the room. The user was not aware that the system used the wrong database for storing the bookings of that particular room. |
| D ¬E P ¬S | Users want this function and believe that it exists even though it could be better, but obviously they do something else than what they think they do. Example: When discovering that the Booking System did not book the room as first believed, the user was no longer that happy with the function. |
| D E ¬U P ¬S | Its so bad users cannot really use it even though they would like to. Example: A user thought that the Internet Bank's online help did not help at all. As a consequence, he did not use it even though in need for help. |
| D E ¬U S | Sure, but another way works even better. Example: In the Internet Bank there were two different ways to input the receiver of a payment; the other way was preferred. |
| D U P ¬S | This is good but could be better. Example: An Internet Bank user would rather have registered new receivers in a separate window. |
| D U S | This is good! This is they way all systems should be like! Example: An Internet Bank user was happy that she could check when the money had been withdrawn from her account. |

The D.EU.PS. model captures important aspects to consider when studying user interaction with information systems. The model makes it possible to discuss the functionality of a system in terms of what is desired and what is not, what exists and what is missing, what is actually utilized and what is needless, what is believed to exist, and what can be used with satisfaction? This in turn makes it possible to identify and discuss different users' perceptions of and attitudes towards a system, and how these perceptions and attitudes change over time, as evident from the examples in Table 1. It would, for example, not be very effective to re-design a part of a system if that part is unlikely to be used anyway – often the user interface is not the problem at all (Mathieson & Keil, 1998).

The D.EU.PS. model represents an operationalization of usability that provides an analytic flexibility, encompassing not only more or less efficient and satisfactory desired (effective) functionality, by putting the usability categories in a broader situated and evolving context. Taking these aspects into account is important in order to pinpoint misunderstandings and focus on the real usability problems.

# 3 Acknowledgements

The D.EU.PS. model has been developed within a research project funded by The Swedish Agency for Innovation Systems (VINNOVA). We would like to thank the other project members who have all contributed to the development of the D.EU.PS. model: Stefan Cronholm, Göran Goldkuhl and Jonas Sjöström (a special thanks to Jonas for performing the evaluation of the Booking System and providing us with material).

# 4 References

Ågerfalk, P. J., Sjöström, J., Eliason, E., Cronholm, S., & Goldkuhl, G. (2002). Setting the Scene for Actability Evaluation: Understanding Information Systems in Context, *Proceedings of ECITE 2002* (pp. 1–9), 15–16 July 2002, Paris, France.

Bevan, N., & Macleod, M. (1994). Usability Measurement in Context, *Behaviour & Information Technology*, 13(1/2), 132–145.

Davis, F. D. (1989). Perceived Usefulness, Perceived Ease of Use, and User Acceptance of Information Technology, *MIS Quarterly*, 13(3), 319–340.

ISO 9241-11 (1998). Ergonomic Requirements for Office Work with Visual Display Terminals (VDTs), Part 11: Guidance on Usability, 1st ed., 1998-03-15, Geneva: International Organization for Standardization.

Mathieson, K., & Keil, M. (1998). Beyond the Interface: Ease of Use and Task/Technology Fit, *Information & Management*, 34(4), 221–230.

Norman, D. A. (1999). Affordance, Conventions, and Design, *Interactions*, 6(3), 38–43.

Preece, J., Rogers, Y., & Sharp, H. (2002). Interaction Design: Beyond Human-Computer Interaction, New York, NY: John Wiley & Sons, Inc.

# Use Cases and User Interface Artefacts

*Tricia Balfe*

Motorola
Mahon Industrial Estate, Cork, Ireland
tricia.balfe@motorola.com

*Frank O'Connor*

Motorola
Mahon Industrial Estate, Cork, Ireland
foconnor@motorola.com

## Abstract

Use cases are *prima facie* user-centric. Some common assumptions that people make as a result of this are that a use case model provides a good user-centred system description and can provide a good basis for maintaining a user-centred focus in a project, and that the use cases provide a good structure which should be reflected in the structure of the artefacts that define the user interface.

We argue that use cases become too complex to serve as an adequate system description from the user's perspective, and that the use case model is not holistic enough to be the centrepiece around which the user interface definition should be structured.

Adopting use cases does not guarantee a user-centred approach. We believe it is essential to explore and communicate the design of the system from an end-user's perspective in a concrete way through the production of prototypes and precise definitions of the user interface. If the user interface artefacts are embedded in the use case model, rather than maintained as an integral unit outside the model, then you lose a valuable, intuitive, and user-centred view of the product. These artefacts should be maintained separately and in parallel to the model. Moreover, care should be taken to ensure that these user interface artefacts are not sidelined by the use case model.

## 1 Introduction

We discuss some of the shortcomings of a use case based approach for user-centred design. We argue that, rather than extending the use case model to improve its support for user interface design, the emphasis should be placed on producing separate deliverables whose main content is pictures of screens, hand drawn or otherwise, with supplementary notes.

The authors of this paper work in a small usability group. Our comments are based on our experiences of working on the usability of products within a development organisation that has employed a use case based requirements methodology for the past number of years.

## 2 The Complexity of Use Cases

In order to maintain a user focus in a project, it is important to create artefacts that provide a clear description of the system behaviour from an end-user's perspective. This system description should be understandable to end-users, and should help the project team to keep a focus on end-users right through the project. Use cases are too complex to provide such a system description.

In our organisation, the initial work on use cases for the system was very positive from the perspective of user-centred design. When the use cases were being identified and named and the overall structure of the use case model was being drawn up, we found that there were plenty of discussions about user goals and user tasks. The first drafts of the use cases were quite simple, user-centric and easy to follow. The introduction of a use case based approach encouraged a user-centric process at the early stage in the project.

However, as the analysis and requirements work proceeded, the use cases became increasingly complex. The use cases that were eventually drawn up documented far more than a simple, slightly abstracted description of how users interact with the system. They described the interactions with other actors outside the system boundary. They defined all but the non-functional requirements in the system. They referenced the architecturally significant interfaces within the system, about which the end-user need know nothing. They were used as test cases by the test organisation, and so the use cases documented aspects of the system that were deemed to be testable but that did not relate to user interaction.

All of this placed a huge burden on the fuzzy, English text use case, and resulted in a system description that no longer provided an intuitive picture of how the system works from the end-user's perspective.

The complexity of the use cases also made it more difficult for the project team to maintain a clear user focus. This loss of user focus was understandable, since the use cases were designed to facilitate the needs of other stakeholders. As the use cases were filled out, the emphasis shifted to providing a system specification suitable for developers and testers, and to reducing the maintenance overhead associated with keeping the system specification up-to-date.

One could argue that, in our organisation, we misused use cases. The authors of the Rational Unified Process recommend that use cases should be simple and clear enough to be understandable by a wide range of stakeholders (Kruchten, 2000). However, there is no absolute definition of what a use case should be, or how to write it. As Constantine and Lockwood write, use cases suffer from "a certain imprecision in definition" and as a result "almost anything may be called a use case despite the enormous variability in detail, focus, format, structure, style and content" (Constantine & Lockwood, 2001, p.245). Our experience has been that use cases turned out to be unintuitive and difficult. This experience of increasing complexity in the use cases is backed up in the literature. For example, Gulliksen, Göransson and Lif write that use cases very quickly become too complex to give a clear picture of what users can expect from a system (Gulliksen, Göransson and Lif, 2001).

We also note that a number of improvements and extensions to the use case model are suggested in the literature that might mitigate the problem of the complexity that we found. For example, Constantine and Lockwood argue that, to properly support user-centred design, you should produce a subset of use cases relating to the user tasks. They call these 'structured, essential use cases': 'structured' because the user's activities are clearly separated, using a table format, from the system's activities; 'essential' because they relate to end-user tasks, and are written in "abstract, technology-free, implementation-independent terms using the language of the application domain and of external users in role" (Constantine & Lockwood, 2001, p.249). Along a similar vein, it could be argued that the early sketchy versions of the use cases for a project should be saved and used as the description of the system from the end-user's perspective.

However, there is an overhead associated with writing a new set of essential use cases, and there is an overhead associated with reaching agreement within the project on the initial set of high level use cases to be saved as the system description. In addition, these sets of use cases need to be maintained as the more detailed downstream model changes. Given the iterative nature of the modelling process, we expect the downstream model to change frequently. In our experience, only artefacts on which people really and truly rely are maintained. Invariably, these are the artefacts that contain plenty of detailed information.

Quite apart from the difficulty of creating and maintaining additional sets of use cases, it is not entirely clear that even these sets of use cases provide a good description of the system from the end-user's perspective. We question the value of including yet more textual-based descriptions in the model, and would prefer to focus on what is generally agreed to be a key component of user-centred design, prototyping. Effort is best spent on producing pictorial representations of the system.

Moreover, it is unlikely that a very similar, more abstract model would ever be used by the project team as a means of maintaining focus on system behaviour for the end-user. The creation of a clear and simple document, that describes pictorially, and at a high level, what is proposed for the user interface, is a far more useful description for this purpose.

## 3    A Holistic or Fragmented Product View

There is a tendency to believe, since the use cases are understood to be user-centric, that the system should be understood primarily in terms of use case descriptions. Understanding can be 'filled out', or concretised, by viewing the user interfaces that realise a particular use case. It is easy to believe that the design and documentation of the user interface should flow from the use case model.

For example, it has been recommended to us, at a UML course certified by Rational, that prototype screens should be embedded into use cases. The authors of the Rational Unified Process recommend a slightly different approach. They suggest that use case storyboards be created for each use case that relates to a user interface. These storyboards detail how a user will achieve the goals outlined in the use case via the user interface (Kruchten, 2000). The thread by which you understand the system is from use case, to use case storyboard, to user interface specification.

Within our own organisation, we have been asked to associate an individual document containing pictorial representations of windows with each use case that relates to end-users. Or even more extreme, we have been asked to embed pictorial representations into individual steps in a use case flow to indicate how the step is 'concretely realised'.

There are some practical problems with 'filling out' the use cases with descriptions of the user interface. Quite clearly the notion of embedding the actual user interface into use cases is problematic. Susan Lilly warns against it in her paper "How to Avoid Use-Case Pitfalls". She claims that it will result in the use cases never being completed, since user interfaces change too frequently (Lilly, 2000).

In addition, the use case model is not always well structured. For example, use case authors often overuse the 'extends' and 'uses' modelling conventions in order to try and avoid duplication of information, and ease the burden of maintenance. This results in an overly complex structure, and in abstruse use case descriptions. There is also a tendency for software engineers to prefer small use cases (Gulliksen et al, 2001), again resulting in an awkward model structure. But perhaps it could be argued that, with experience, we could build a well-structured model.

There is another problem. The use case model is made up of use cases, one per major user task. This is a fragmented view. Following the use case model too closely can lead to the design of a similarly fragmented user interface. For example, Gulliksen, Göransson and Lif claim that they have encountered situations where a use case based approach has lead to the design of a fragmented interface where one use case relates to one window (Gulliksen et al, 2001). This design strategy is ill conceived. For example, where a number of tasks relate to the same information, it may be useful to design a single window that supports multiple tasks. Or where a number of use cases share a common set of sub-steps, it may be useful to design the system so that the user navigates to another window for these sub-steps. As William Hudson states "one of the greatest failings of object-oriented design methods is that although they take a holistic approach to the architecture of components...they generally encourage piecemeal development of the UI." (Hudson, 2002, p.96). But one could argue that it is the job of the user interface designers to be aware of and avoid these pitfalls.

There is, however, a more fundamental problem. A view of the user interface is essential for communicating the idea of the product among the project team. The use case model does not provide such an overall view. If too much emphasis is placed on the use case model as the centrepiece of the system description, and the user interface is understood in terms of this central model, then your understanding of the system's behaviour from a user's perspective will be fragmented. The more holistic and very valuable view of a product that you get from a high-level, pictorial representation of the entire product's user interface is sidelined, under-used and possibly lost.

# 4    Our Approach

Our preferred approach is to produce a fairly loose 'storyboard-like' document that includes a simple task list. This is a separate and self-contained artefact that sits alongside the use-case model. At an early stage in the project, we produce a version of the document that includes rough sketches and suggestions for the user interface. At a later stage, we produce a version that contains precise definitions of the windows that can be used to complete these tasks, and a list of all the menus that can be used to navigate between the windows. It is quite easy to trace through this type of system description in an informal way and understand the user interactions with the system. The system description can also be used to validate that the use case flows relating to user interactions are well supported. This artefact provides an excellent view of the system from an end-user's perspective, and is a very useful means of communicating an intuitive and common vision of the product to all the members of the project.

Apart from the emphasis that we place on the production of this artefact, our work to maintain the focus on user-centred design closely follows the recommendations made by the authors of the Rational Unified Process. We work in parallel with the use case author to design the system's user interface. We prototype. We produce concrete specifications of the user interface. We hold reviews of these prototypes and concrete specifications of the user interface with the same people who write and review the use case model (Kruchten, 2000). Having explored the user tasks through prototyping, we review the use cases to ensure that key steps in the flows are included. We write terse, hierarchical task lists that are similar to the table of contents for the use case model, but that include additional tasks not drawn out in the use case model such as printing, saving, sorting, filtering and finding data.

## 5 Conclusion

We have argued that the adoption of use cases does not guarantee a user-centred approach. The early work on identifying and creating outline use cases can certainly increase focus on end-users. Furthermore, the emphasis on the definition and maintenance of a system model gives the usability team early visibility of the system specification. This allows us to review and critique the system, and to ensure that important user requirements are included. However, there are also some problems. Use cases become complex, and as a result, do not provide a good user-centred view of the system. In addition, the use case model is necessarily a fragmented model of the system. We argue that it is important to create a separate artefact that documents, at a high level, the proposed user interface of the product. This serves the purpose of providing an intuitive and holistic view of the system that is very valuable in increasing the awareness of end-user requirements among the whole project team.

## 6 References

Constantine, L. L., & Lockwood, L. A. D., (2001). Structure and Style in Use Cases for User Interface Design. In Van Harleman, M. (Ed.), *Object Modelling and User Interface Design: Designing Interactive Systems* (pp. 245-280). New York: Addison Wesley.

Gulliksen J., Göransson B., & Lif, M. (2001). A User-Centred Approach to Object-Oriented User Interface Design. In Van Harleman, M. (Ed.), *Object Modelling and User Interface Design: Designing Interactive Systems* (pp. 283-312). New York: Addison Wesley.

Hudson, W. (2002). Industry Briefs: Syntagm. *Interactions*, 9(2), 95-98.

Kruchten, P. (2000). The Rational Unified Process: An Introduction (2nd Ed). New York: Addison Wesley.

Kruchten, P., Ahlquist, S., & Bylund, S. (2001). User Interface Design in the Rational Unified Process. In Van Harleman, M. (Ed.), *Object Modelling and User Interface Design: Designing Interactive Systems* (pp. 159 - 196). New York: Addison Wesley.

Lilly, S. (2000). How to Avoid Use-Case Pitfalls. *Software Development*, January. Retrieved February 11th, 2003, from http://www.sdmagazine.com.

# Δ: Modelling Cognitive Performance

*Laurent Bayssié and Laurent Chaudron*
Onera-Cert
2, avenue E. Belin
31055 Toulouse Cedex 4
{bayssie,chaudron@cert.fr}

### Abstract

The study of the relations between the cognitive performance of a human operator and his/her task is a fundamental issue as far as the improvement of the performances of socio-technical systems is concerned. In this paper, we propose a closed loop methodology, called Δ, for the evaluation of evolving interaction systems based on the study of the causal relations between the variations of the cognitive performance of the operator $P$ and the variations of the external conditions of his/her task $T$, *i.e.* $dP/dT$. The methodology relies on symbolic data and qualitative analyses. Our contribution consists in the description of the Δ method: - a conceptual model, - a pragmatic methodology, - a formal framework and, - an aeronautical application.

## 1   Δ Method

### 1.1   Approach

Controlling complex systems (*e.g.* aircraft piloting..) generates specific problems in which the quality of the interactions between the operator and his/her system is critical. Thus, the conjoint improvement of man-machine interactions systems performances leads to search for some validated modifications of these interaction systems (*i.e.* ergonomy, procedures, training). The crucial problem is the evaluation of the effective result of an improvement of the man-machine interactions on the actual performance of the operator. In this context, our approach aims at giving a framework and a methodology for the objective assessment of the effects of evolving interaction systems on the performances reached by the operator. Thus, the man-machine interaction evaluation is studied through the causal relations between the variations of the parameters of the task $T$ and the effective variations of the cognitive performance $P$ of the operator, measured through his/her activity. Finally, the Δ method consists in the analysis of the differential functional relation: $f(dP/dT)$.

### 1.2   Cognitive performance variability

Whatever the domain (*e.g.* aeronautics..), the conception of cooperative systems faces with the high variability of the operator's activity. This variability is a natural element of human activity which participates directly to the regulation of the activity. Thus, an approach aiming at improving the performances of such systems by the design of new interfaces (*e.g.* instructions, ergonomics..) should take into account this variability. The variability of the performance comprises two components: -

the inter-individual differences between operators, and - the intra-individual differences in actions performed during a task. In this context, the lack of operative operator's model leads us to modelize his/her cognitive performance. The classical tools analyse deviations from an external reference norm (*e.g.* a prescribed task..) so that they only focus on negative events and prevent taking into account the various possible ways of achieving the goal's task. Hence, the adressed problem is the capture of the deviations and the positive adaptions of the operator faced with his/her various task and so, the definition of a reference norm. In this context, the $\Delta$ approach consists in the qualitative analysis of the variability of the operator's activity near a so-called reference activity[1] [1].

## 1.3   Knowledge representation

The operator's activity is classically analysed by two axis: - 1) the estimation of the conditions of the task's realization which contributes to the comfort and easyness of the work, - 2) the performances of the operator which are actually measurable. Frequently, the only parameters of type 1) that are considered for the evaluation of the conditions of the task's realization by the operator are subjective ones (*e.g.* inner feeling of workload such as Cooper-Harper, Nasa TLX.., operator evaluation of new interaction systems..). Unfortunately, different types of bias appear in the exclusive use of subjective parameters: - the invasive part of the on line protocols ( e.g. increase of the degree's feeling..) - the self knowledge bias ( e.g. rationalisation *a posteriori*..) [2]. As the essential point for the aeronautical conception is to guarantee the results' fiability as well as in terms of performances effectively measured as their acceptation and their conscious level by the operators, the performance of the operator is defined both in terms of objective and subjective data. This leads to establish three classes of parameters in the $\Delta$ methodology:

1) the subject's **actual-performance** $A$,
2) the subject's **feeling** relative to the task $F$,
3) the **self-evaluation** of the effective task $S$.

Thus, in the $\Delta$ method, the global performance is defined as a triple $P = (A, F, S)$ which is analysed combined with the task $T$ (see Table 1).

| | Objective | Subjective | Examples |
|---|:---:|:---:|:---:|
| $T$ Task | × | | mission, reasoning test ... |
| $A$ Actual-performance | × | | flight parameters, answer vector ... |
| $F$ Feeling | | × | TLX scale, free text ... |
| $S$ Self-evaluation | | × | answer sheet, ... free text ... |

Table 1. The quadruple $(T, A, F, S)$.

## 1.4   Pragmatic methodology

The main topic of the $\Delta$ method is to analyse the relations between the task (and its variation) and the global performance (and its variations), considering a corpus of experiments. This implies to take into account a set of operators or (ergodic hypothesis) and a set of instances of operators' performances. The methodology is summarized as follows:

---

[1]The reference plays the role of a gauge value and is not considered as a nominal one.

(a) **Representation step**: definition of the quadruple $(T, A, F, S)$,

(b) **Preparation step**: determination of the set of operators: $\{O_i\}_{i \in [1,n]}$ and the controlled variations of the task $dT$,

(c) **Experimental step**: for all $O_i$, effective experiments of the tasks $T$ and $T + dT$, recording of $(dT, A_i, F_i, S_i)$ in a database,

(d) **Qualitative analysis step**: the database is formally analysed so as to produce: **clusters** and **rules**.

The qualitative analysis tool is the Generalized Formal Analysis technique described in [3]. The mathematical groundings are described in an algebraic frame in which a lattice structure, the Cubes model, is defined for cognitive performance modelling. Rules induction techniques are also used so as to produce causal relations.

# 2 Aeronautical application

Involved in the study of the noise annoyance [4], an experiment - conducted during Paris Air Show 2001 - consisted in assessing the cognitive performance in a set of aeronautical relevant cognitive tasks for which we included a comfort parameter in the environment (aircraft's noise vs calm).

## 2.1 Experiments

Given the crucial role of the operator's reasoning in man-machine interactions and the supposed variability of the cognitive performance associated with noisy conditions, the chosen task $T$ was deductive reasoning in flight procedures evaluation[2]. Hence, the problem is to decide wether a particular mission's situation (conclusion) necessary follows from a general flight's procedure (premises) as follows: given the procedure: "When a fighter pilot patrols in an enemy territory, he has to flight at a low altitude", the observation is: "During a mission, if a fighter pilot patrols in an enemy territory and if the climatic conditions are unfavourable then he has to flight at a low altitude".
The measures of the actual performances $A$ are: - the logic performance, *i.e.* the discrepancies between the subjects' answers and the reasoning logic model based upon the theory of formal rules of inferences (see Table 2.) [5] and, - the answer's time in seconds. A questionnaire is distributed to each person in order to express his/her feeling $F$ of the conditions of the task (aircraft's noise vs calm) and also his/her self-evaluation $S$ of the performance.

| Rules | Form | | Truth Value |
|---|---|---|---|
| | Premise | Conclusion | |
| 1 | $a \rightarrow b$ | $a \wedge c \rightarrow b$ | True |
| 2 | $a \rightarrow b \wedge c$ | $a \rightarrow b$ | True |
| 3 | $a \rightarrow b$ | $a \rightarrow b \wedge c$ | False |
| 4 | $a \wedge c \rightarrow b$ | $a \rightarrow b$ | False |
| 5 | $a \rightarrow b$ | $\neg a \rightarrow b$ | False |
| 6 | $a \rightarrow b$ | $b \rightarrow a$ | False |
| 7 | $a \rightarrow b$ | $\neg a \rightarrow \neg b$ | False |
| 8 | $a \rightarrow b$ | $\neg b \rightarrow \neg a$ | True |

Table 2. Reasoning task.

54 voluntary persons - visitors or professionals of the Paris Air Show - participated to the experiment. The performances are assessed both in noisy and in calm conditions. Simplifying the data

---

[2]This situation is representative of the operator's activity when most of the time he/she must check the adequacy between the known procedures and new circumstances. Despite that the problems are relative to the aeronautical domain, the understanding of the arguments do not require an expert knowledge.

features as: - "no-noise/noise" or $dT(+,-)$, - quality/time and positive/equal/negative actual performances $dA(qual,+/=/-)$ and $dA(time,+/=/-)$, - quality/time over/equal/under estimation of performances $dS(qual,+/=/-)$ and $dS(time,+/=/-)$, - "annoyance/no-annoyance" $dF(+/-)$, the performance vector can be summarized as:

| $O_i$ | $dA$ | | | | | | $dS$ | | | | | | $dF$ | |
|---|---|---|---|---|---|---|---|---|---|---|---|---|---|---|
| | $dA(qual)$ | | | $dA(time)$ | | | $dS(qual)$ | | | $dS(time)$ | | | $dF(-)$ | $dF(+)$ |
| n | $-$ | $=$ | $+$ | $-$ | $=$ | $+$ | $-$ | $=$ | $+$ | $-$ | $=$ | $+$ | | |

For all subject $s$, the vector $(T, A, F, S)$ is recorded in a logic database as:

```
subject(name,<task,actual-performance,feeling,self-evaluation>),
```

which is represented in a global lattice struture (the cube model).

## 2.2  Formal Analysis

The results are analysed through a qualitative analysis model - called "Generalized Formal Analysis" [6], which allows qualitative clustering and rules inductions capabilites. The global database of the 54 subjects with their characteristics can be analysed and the clusters: (groups of subjects, common $(T, A, F, R)$ features) can be computed so as to form a global lattice structure (see Fig. 2.):

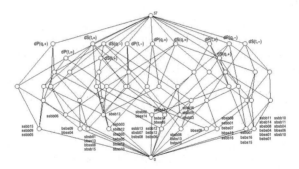

Fig. 2. The qualitative clusters lattice.

The "Generalized Formal Analysis" [3] allows the induction of symbolic rules which can be gradated according to their degree of plausibility, *i.e.* the support which is the ratio of interest of the rule among the data-base and confidence, *i.e.* the ratio of validity [7].

The main characteristics of the results obtained thanks to the $\Delta$ method can be summarized as follows:

- for rule(ident,<ant>,cons,conf>0.5,0.5<supp>1)
- $dF(+) \rightarrow dT(+)$,conf=1, supp=0.98
- $dS(time,-) \rightarrow dT(+)$,conf=1, supp=0.54
- $dA(time,+) \rightarrow dT(+)$,conf=1, supp=0.56
- $dS(time,-) \rightarrow dF(+)$,conf=1, supp=0.54
- $dS(time,+) \rightarrow dF(+)$,conf=1, supp=0.56
- $dS(time,+) \rightarrow dA(+)$,conf=1, supp=0.52

The last result is interesting as far as for a significant (52%) proportion of the subjects, in a noisy environment, they did improve their performances while they thought they degraded them.

The formal analysis of the data revealed significant differences between the following dimensions: - the Task $T$, - the actual Activity of the operator $A$, - his/her Feelings $F$ of the conditions of the task, - his/her Self evaluation of the performance $S$. From a formal point of view, the functional relation between the partially ordered task space T and the objective/subjective performance P is *not* monotonic.

The study of the properties of the morphisms between those spaces is the current purpose of our researches.

# 3   Conclusions

The contribution of this article is an evaluation methodology of interaction systems. Preferentially devoted to the applications in which the operator's variability is high, the method $\Delta$ allows a formal analysis of the vectors $(T, A, F, S)$. Thus, it is possible to assess in a qualitative and comparative manner the subjective and objective effects on the cognitive performance of the operator.

After a description of the adopted method $\Delta$, a first set of general experiences and results formally analysed is presented through a protocol of reasoning tasks with an external perturbation.

The formal analysis of the data underlined significant differences between the dimensions of the performance previously defined, *i.e.* $P = A, F, S$). The current study is devoted to the *Eucalepic effect*, *i.e.* the conjecture in which a "greater" cognitive task can lead to an "easier" performance (*e.g.* instructions, sub-tasks..). The "Continuous Inference" model is used so as to track cognitive performance and also offers learning capabilities for a human cognitive agent [3].

# References

[1] L. Bayssié, L. Chaudron, P. Leblaye, N. Maille, and S. Sadok. Pilot activity modelling for flight analysis. In *Digital Human Modeling for Design and Engineering*, Montreal, Canada, June, 16-19 2003. DHM.

[2] J. St. B. T. Evans. *Biais in human reasoning: causes and consequences*. Lawrence Erlbaum Associates, Hove and London (UK), Hillsdale (USA), 1990.

[3] L. Chaudron, N. Maille, and M. Boyer. The CUBE lattice model and its applications. *Applied Artificial Intelligence*, 2003.

[4] M. Boyer. *Induction de régularités dans une base de connaissances, application au phénomène bruit/gêne et ses extensions*. PhD thesis, Supaèro, décembre 2001.

[5] L.J. Rips and G. Conrad. Individual differences in deduction. *Cognition and brain theory*, 1983. Lawrence Erlbaum Associates.

[6] L. Chaudron, N. Maille, and M. Boyer. Rule Induction based on Cubical-FCA. In *ECAI'2002 15th European Conference on Artificial Intelligence*, Lyon, 21 - 26 July 2002. Workshop ECAI'2002 "Advances in Formal Concept Analysis for Knowledge Discovery in Databases.

[7] Christian Borgelt and Rudolf Kruse. Induction of association rules: Apriori implementation. In *Proc. 15th Conf. on Computational Statistics (Compstat 2002, Berlin, Germany)*, Heidelberg, Germany, 2002. Physika Verlag.

# Expanding HCI Methodologies To Incorporate Motivational Evaluation

Winslow Burleson
MIT Media Lab
20 Ames St., Cambridge, MA, 02139
win@media.mit.edu

## Abstract

Traditional Human Computer Interaction (HCI) methodologies have ignored motivational elements. A schema for expanding the scope and application of existing HCI disciplines is developed and the benefit of incorporating analysis of motivation into interaction design and evaluation is presented. A hybrid research strategy comprised of rich and diverse HCI methodological approaches (Field, Experimental, Respondent, and Theoretical) is proposed.

## 1 Motivation and Traditional HCI

"Usability as a concept does not seem to include [positive] feeling such as pride, excitement or surprise"; embracing positive emotions is "not reflected in traditional human factor practices" (Jordan, 1998; Monk, 2002). Put another way, when "fun and enjoyment are an important 'candidate software requirement'…it has become clear that our current understanding of user concerns, derived from the world of work, is simply not adequate to this new design challenge (Monk, 2002)." There is some emerging interest in the HCI community to address this deficit: Thomas Malone's early work on heuristics (usability guidelines) for designing enjoyable user interfaces; Andrew Monk's *Computer's and Fun*, York, December, 2000 (Monk, 2000); the *International Conference for Affective Human Factors Design* in Singapore, in June, 2001; and a few research efforts by Clifford Nass and Byron Reeves of Stanford University's project on Social Responses to Communication Technologies, Rosalind Picard's Affective Computing at the MIT Media Lab, BJ Fogg's Captology at Stanford University, Sweden's PLAY Interactive Institute, Carnegie Mellon's Entertainment Technology Program, and Bill Gaver's Ludic design at the Royal College of Art.

Motivation is the psychological feature arousing an organism to action. Fun, reward, enjoyment, beliefs, needs, opinions, challenge, pleasure, creativity, play, and Flow are *motivational elements* (Ford, 1992; Malone, 1981). Traditional HCI practices have relied on usability heuristics highlighting: "visibility", "consistency", "error prevention", "recognition", "minimalist design", "diagnose and recover", and "help and document", as their main parameters (Tognazzini,

2001; Molich, Nielsen, 1990; Nielsen, 1993, 1994a, 1994b). Not only do these existing rubrics lack the incorporation of motivational elements they endanger motivational elements which are not heuristically protected. For example, "interface minimalism" may call for the removal of a motivational element such as sound, animation, or an ethereal glow in the name of productivity. Both productivity and motivation should be considered. HCI often "[undervalues] what people experience and report", (Nielsen, 1993; Monk 2002). These subjective elements are most accessible through "Field" (observation under natural conditions) and "Respondent" (responses of participants) research strategies. These strategies and their data are often seen as too "fuzzy, maybe too 'contradictive', 'irrational' and 'non-predictive', to seriously ground design decisions",(Monk, 2002). There exists the danger that significant elements of the HCI, particularly those that manifest themselves primarily in this subjective realm, are not being attended to "productively". Without correction it may be difficult to rely on the existing HCI methodologies to provide greater advances. Still, there can be some comfort taken in the realization that there are numerous well-founded methods drawn from the diverse disciplines of anthropology, sociology, psychology, cognitive science, philosophy, design engineering, and computer science that comprise the HCI Community (Rosson, Carroll, 20002). Joseph McGrath sensibly argues support for research that draws upon the combined strengths of "Field", "Experimental", "Respondent", and "Theoretical" approaches (McGrath, 1995). Embracing these methods as a hybrid approach to the design, development and evaluation process should lead to greater acceptance of new and relevant research directions; embracing motivational elements and their benefits will provide substantial contributions to HCI.

## 2 New Opportunities for HCI

Interdisciplinary research combining the strengths of methods ranging from experimental design to ethnographic study is increasing our understanding of HCI. Germane examples from the fields of Psychology and Creativity Research include the hybrid research strategies of Mihaly Csikszentmihalyi and Teresa Amabile. Each supports the development of holistic approaches combining field and case studies, surveys, and interviews, "Field" and "Respondent" approaches in concert with "Experimental" strategies. Csikszentmihalyi has applied these toward a refinement of research on Flow. Each has applied these strategies to develop "Theoretical" constructs as well, Csikszentmihalyi's systems model of creativity and Amabile's componential model of creativity. Both rely heavily upon motivational elements (Csikszentmihalyi, 1990, 1996; Amabile, Mueller, 2002). Psychologist Martin Ford makes a substantive case for techniques that encourage implementation of multiple motivations (Ford, 1992). Supporting fun, reward, enjoyment, etc… should enhance user's subjective experience and creativity (Burleson, 2003); and particularly in under-motivated circumstances, should benefit traditional HCI productivity goals as well. Ben Shneiderman calls for tools that facilitate the creativity of acquiring new knowledge, relating it to other knowledge, and donating the product to others (Relate-Create-Donate). He presents a philosophy that is a motivational cycle of social, collaborative, and creative production (Shneiderman, 1997; Csikszentmihalyi 1996, Amabile, Mueller, 2002).

## 3    Embracing Diverse HCI Methodologies to Study Motivation

HCI "Field" strategies including: ethnographic observation, case studies, and participatory design; elucidate the spectrum and mechanisms by which motivation operates. "Experimental" strategies including: Wizard of OZ techniques, user-testing, and controlled experiments; quantify and evaluate motivational elements in specific conditions. "Respondent" strategies including: case studies, participatory design, Walk Through evaluations, Active Talk (Amabile, Mueller, 2002), surveys, evaluation forms, and focus groups; gather qualitative and quantitative responses from users. Respondent strategies are applied in ecological and laboratory settings, often in conjunction with "Experimental" strategies, to elicit and record subjective experience, confirm "Field" findings, and assess "Theoretical" models. "Theoretical" strategies include: expert and heuristic evaluations, models of motivation, cognitive models, and user interaction and perception models. "Theoretical" constructs are formulated through each of the other research strategies as a culmination and aggregation of research into unified higher level constructs. Theory serves as a philosophical resource providing guidance for hypothesis which will validate, expand, or discredit the "Theoretical" constructs.

## 4    Research Examples

In Teresa Amabile's "Field" research, extensive ethnographic work did not reveal the physical work environment as a salient contributor to the creative process (subsequently two doctoral thesis also yielded poor results on this topic) (Amabile, 2001). This is the kind of powerful statement that strong ethnographic research can provide: a scope for initial and ongoing salient explorations. It is important to look at both the highly positive as well as the depressingly negative events to obtain holistic findings. A broad scope strategy will be needed as the interaction of technology and motivation is defined, refined, and studied. In related research, Csikszentmihalyi applied the Experience Sampling Method and the subsequent pager based implementation to measure Flow States during activity engagement. A brief interruption of normal activity at a random interval was followed by a survey instrument used to measure elements of Flow: optimal experience, challenge, skill, anxiety, apathy, boredom, etc… (Csikszentmihalyi, 1990). This "Respondent" strategy proved quite successful, supporting the systems model of creativity. Separately, Chet Hedden's study of interactive adventures, notes the difficulty of doing in-sitiu evaluation of computer users and its tendency to disrupt Flow (Hedden, 1993). In developing heuristics, Malone circumvented this "in-sitiu difficulty" in his empirical studies by administering pre and post user response survey instruments (Malone, 1981). In developing in-sitiu "Respondent" techniques perhaps the most promising is the Active Talk method utilized by Amabile (Amabile, Mueller, 2002). She has found that Active Talk methods, in which subjects vocalize their thoughts as they engage in tasks, have not significantly impacted either task performance or the creative process. Since these creative processes are highly correlated with motivation it is likely that the Active Talk methods may prove to be a cornerstone of the solution to these methodological issues within HCI evaluation. It is important to note the rigorous coding protocols implemented in Amabile's analysis of the vocal data. These protocols are critical in transforming otherwise subjective experiences into quantitative experimental data. A similar, less formal, technique is utilized in Walk Through assessment (Shneiderman, 1997a). Picard's Affective Computing research has developed rigorous coding methods but has the distinction, that rather than requiring subject vocalizations she uses the interaction technology as a sensing system to gather information which is then coded into representations of affective, context, and physiological states. This information obtained and utilized during natural user interactions, can then be correlated to other more traditional research measurement instruments to insure validity (Picard, 1997; Burleson, 2003). Some of HCI's most notable practitioners are calling for greater attention to motivation elements; Don Norman in *Learning From the Success of Computer Games* and Jakob Nielsen in his study of website usability for children call for animation, sound effects, and exploratory interactions (Norman, 2002, 2002a; Nielsen, 2002). It is important to unite the HCI community drawing on the merits of each of the sub-disciplines and their cross fertilization.

## 7 Conclusions

As the HCI field emerges the salient focus must migrate. What was once "most important" is now well addressed leaving exposed new and more relevant "most important" issues (Thomas, 2002). Attention to subjective user experience is particularly lacking. Emerging research efforts are beginning to combine "Field", "Experimental", "Respondent", and "Theoretical" techniques across HCI disciplines, presenting findings that are increasingly supportive of both the incorporation of motivational elements and the hybrid research strategy itself. These benefits will not only be realized for subjective experiences of users but will greatly enhance the traditional HCI goals of productivity and usability.

# 8 References

Amabile, T.M.; Presentation MIT Media Lab, June 2001.

Amabile, T.M.; Mueller, J.S. (2002) Assessing Creativity and Its Antecedents: An Exploration of the Componential Theory of Creativity; Handbook of Organizational Creativity, edited by Cameron Ford. Mahwah, NJ: Lawrence Erlbaum Associates.

Burleson, W.; Developing a Framework for HCI Influences on Creativity; HCI International 2003, the 10th International Conference on Human-Computer Interaction. Crete, Greece, June, 2003.

Csikszentmihalyi, M. (1990). *Flow: the psychology of optimal experience*. Harper Perennial, New York, NY.

Csikszentmihalyi, M. (1996) Creativity: Flow and the Psychology of Discovery and Invention; Harper Collins Publishers, New York, NY.

Ford, M.E. (1992) Motivating humans: Goals, emotions, and personal agency beliefs. Sage Publications, Newbury Park, CA.

Hedden, C. (1993) Challenge and Motivation in Interactive Adventures, College of Education Research/Inquiry Presentation Autumn 1993, University of Washington. Jordan, P. (1998) Human Factors for pleasure in product use. Applied Ergonomics, 29[1], p. 25-33.

Jordan,P.(1998). Human factors for pleasure in product use. *Applied Ergonomics*, 29[1],p.25-33.

Malone, T. (1981) Heuristics for Designing Enjoyable User Interfaces: Lessons from Computer Games, Association for Computing Machinery.

McGrath, J. (1995) "Methodology Matters: doing research in the behavioural and social sciences" Readings in human-computer interaction: Towards the Year 2000. Eds. Ronald Baecker, Jonathan Grudin, William Buxton, Saul Greenberg. Morgan Kaufmann.

Molich, R., and Nielsen, J. (1990). Improving a human-computer dialogue, *Communications of the ACM* 33, 3 (March), 338-348.

Monk, A. (2000) Computer's and Fun, York, December 2000.

Monk, A. (2002) Funology: designing enjoyment, CHI 2002 Workshop Abstract.

Nielsen, J. (1993) Usability Engineering, Academic Press, Inc. New York, NY.

Nielsen, J. (1994a). Enhancing the explanatory power of usability heuristics. *Proc. ACM CHI'94 Conf.* (Boston, MA, April 24-28), 152-158.

Nielsen, J. (1994b). Heuristic evaluation. In Nielsen, J., and Mack, R.L. (Eds.), *Usability Inspection Methods*, John Wiley & Sons, New York, NY.

Nielsen, J. (2002) Kids' Corner:Website Usability for Children, http://www.useit.com/alertbox/20020414.html.

Norman, D. (2002) www.jnd.org/dn.mss/ComputerGames.html

Norman, D. (2002a) www.jnd/dn.mss/Northwestern Commencement.html

Picard, R. (1997) Affective Computing, MIT Press, Cambridge, MA.

Rosson, M.; Carroll, J. (2002) Usability Engineering: Scenario-based development of human-computer interaction. Morgan Kaufmann Publishers, Tokyo, Japan.

Shneiderman, B. (1997) Relate-Create-Donate: An educational philosophy for the cyber-generation Computers & Education 31, 1 (1998), 25-39.

Shneiderman, B. (1997a) Designing the user interface: strategies for effective human-computer interaction, Addison Wesley.

Thomas, J.C.; Personal Communication, July 2002.

Tognazzini, B. (2001)www.asktog.com/basics/firstPrinciples.html.

# Event Cycle and Knowledge Development in NASA Mission Control Center

*Barrett Caldwell*

Purdue University
West Lafayette, Indiana
bcaldwel@ecn.purdue.edu

*Enlie Wang*

Purdue University
West Lafayette, Indiana
ewang1@purdue.edu

## Abstract

The objective of this paper is to study the relationship between event cycles and operational knowledge development in the NASA Mission Control Center (MCC) environment. This work is based on preliminary analysis of communication data from mission controllers and crewmembers during MCC simulation exercises. MCC is a prototypical operation-critical and time-constrained collaboration environment because any mistake or unnecessary delay during the operation can lead to severe outcomes. In such collaboration environments, reliability and accuracy must be maintained to assure the success of mission. To achieve high reliability and accuracy, improving the hardware and technology is not enough; we also need to improve human performance and task coordination as well. In this case, it is very important to how and within what time scales the knowledge is created, shared and utilized by the mission controllers and crewmembers.

## 1    Introduction

Knowledge is usually classified in the literature by the criterion of verbalization, such as tacit vs. explicit knowledge (Nonaka & Takeuchi, 1995) and procedural vs. declarative knowledge (Proctor & VanZandt, 1994). However, our research highlights the development and interaction of two major types of knowledge in the MCC environment: operational and reference knowledge (Garrett & Caldwell, 2002). Operational knowledge includes the physical actions as well as the understanding of what actions to perform to achieve a desired result. Reference knowledge refers to the documented results of operational knowledge experience, event referencing, and context description archived for future use. In this paper, we will focus on operational knowledge event cycles.

## 2    Knowledge Development Cycles

Each knowledge development cycle depends on related events and their timescales. The basic unit of operational knowledge development in MCC environment is the operational event. There are active and latent phase aspects of each event cycle, representing externally recognized activity and interim processes or pauses between identifiable activities.

There are seven types of events in the Mission Control environment (Garrett & Caldwell, 2001):
- *Spacecraft* - what is happening on the vehicle (system state change)
- *Software* - two main types

- sensory data capture, storage and transmission

- data processing and model usage

- *Network* - data and telemetry exchange between MCC and vehicle

- *Command* - an algorithm execution between machines (i.e. software on a computer causing a switch to close)

- *Cognition* - sensemaking issues, process of comprehending information and making it usable

- *Control* - humans directly interact with machines

- *Collaboration* - humans interact with each other

These seven events can be categorized in three levels with different timescales:

> Level 1, Machine-Machine Interaction (MMI): spacecraft, software, network and command events ranging from 10 ms (or less) to 10s;
> Level 2, Human-Machine Interaction (HMI): cognition and control events between 1 and 30s;
> Level 3, Human-Human Interaction (HHI): collaboration event in the order of seconds to minutes.

According to Newell's theory, human's action can be categorized into four bands. Each band has its own time scale. The actions in biological band such as transmission between neurons usually take about $10^{-4}$ to $10^{-2}$ seconds, while unit tasks, operations and deliberate acts in the cognitive band require from $10^{-1}$ to $10^{1}$ seconds. Above the cognitive band lies the rational bands, which is of the order of minutes to hours ($10^{2}$ to $10^{4}$ seconds); and the highest band is social band with the time scale of $10^{5}$ to $10^{7}$ seconds (days to months). Although this classification is based on single human action, it also can be mapped to the timescales of HHI interaction events (see table 1).

**Table1:** Timescale mapping between single human action and interaction level

| Time scale | Single human action | Interaction level |
|---|---|---|
| $10^{-4}$ to $10^{-2}$ seconds | Biological band | MMI |
| $10^{-1}$ to $10^{1}$ seconds | Cognitive band | Computer mediated HMI (e.g. press the right key) |
| $10^{2}$ to $10^{4}$ seconds | Rational band | Computer mediated HMI and computer mediated HHI (with reasonable interaction delay) |
| $10^{5}$ to $10^{7}$ seconds | Social band | Computer mediated HHI (with excessive interaction delays like mission to mars) |

In addition, different groups of MCC controllers have responsibilities for managing and responding to different types of operational events at different time scales. This requirement for task coordination and information exchange requires controllers to be aware of, and utilize information technologies that support, events at multiple time scales. In addition, information technologies must effectively support transitions from operations to reference knowledge with additional requirements for knowledge synchronization among controllers and documentation sources. The MCC simulation exercises studied in this work are the primary sources for the initiation of many operations – reference knowledge development cycles.

# 3 Event Cycles and Delay Tolerance

Our previous studies show that event cycle directly affects user's delay tolerance. Caldwell and Paradkar (1995) reported that the voicemail server-local phone delays could significantly influence user's tolerance of system performance in their study of Voice Mail message transmission. The voicemail server traffic load reflects possible sources of transmission delays during machine-machine interaction. In some environments like mission control – ground communication for distant exploration missions, machine-machine interaction could take as longer as hours or days depending on distance and transmission frequency. For example, notifying changes of procedure to the next shift of controllers may take up to 8 hours because the receiver is not available in current shift. In the MCC environment, multiple events might, however, be processed simultaneously if each event requires different resources (visual or audio modalities) or their active phases do not directly overlap in time or resource requirements.

In a series of studies examining users' performance in human-computer interaction with delayed information flow (Wang, 2002, Wang, Caldwell & Zhang, 2002), we also studied event cycle effects on user's delay tolerance in level 3 (HHI) interactions. In these experiments, we manipulated event cycles by changing the task type and difficulty level. Users were able to actively control the delay they are willing to accept while maintaining normal performance. The results show that task type and difficulty level can significantly affect the delay tolerance during interaction. Users expected more time delays working on calculation tasks than working on search tasks; they also tolerated more time delays when the difficulty level increases (see table 2).

**Table 2:** Acceptable Time Delay in Different Treatments*

| Treatments | Male | Female | Total |
|---|---|---|---|
| Search(easy) | 4.2 (1.8) | 4.6 (1.3) | 4.4 (1.5) |
| Search(hard) | 9.3 (4.5) | 10.5 (6.2) | 9.9 (5.3) |
| Calculation(easy) | 15.7 (6.9) | 14.8 (6.4) | 15.3 (6.5) |
| Calculation(hard) | 24.6 (11.3) | 29.2 (11.2) | 27.0 (11.3) |

* ATD is measured in seconds. Corresponding SD is in the bracket.

The results of the first experiment also help us determine the delay levels, which were used as independent variables in two later experiments.

Five factors (gender, Type-A personality, task type, difficulty level and time delay) were investigated in both experiments. Users experienced passive delays as an experimental control during interaction with their "partner". Perceived delay and tolerance rating were collected during collaboration tasks using pop-up window. The only difference between the experiments was the partner type. In experiment 2, users were convinced that their partner during collaboration tasks was a real human; in experiment 3, they were convinced that their partner was a software agent. In fact, the partner was a software agent in both experiments. In both experiments, only delays of 2-6 seconds were estimated accurately by users, in support of Vierordt's Law (Woodrow, 1951).

Actual delay and perceived showed a significantly positive correlation. One interesting finding was that users tended to report there was delay even if there was not. This finding suggests that humans are always expecting some amount of delays during collaboration tasks. Comparisons of delay tolerance in experiment 2 and experiment 3, yielded an unexpected finding (see table 3 and figure 1). We found that the perceived partner type (Human or software agent) can significantly affect user's delay tolerance in calculation tasks. They will expect fewer delays when working with software agent than working with humans in calculation tasks. In search tasks, users will tolerate some amount of delays regardless of their partner type. Although the software agent is able to complete all the tasks within one second, users were still using human scale event cycles to predict the performance and expected delay of software agents.

**Table 3:** Comparison of the expected delay *

|  | Human Partner | | Software Agent | |
|---|---|---|---|---|
|  | Physical delay | SD | Physical delay | SD |
| Easy search | 7.11 | 6.10 | 7.05 | 4.41 |
| Hard search | 12.63 | 8.05 | 13.60 | 6.21 |
| Easy calculation | 11.37 | 7.85 | 15.45 | 7.55 |
| Hard calculation | 12.91 | 9.04 | 24.70 | 10.47 |

*The expected delay is the physical delay, which is rated as point 5 in the 9-point scale.

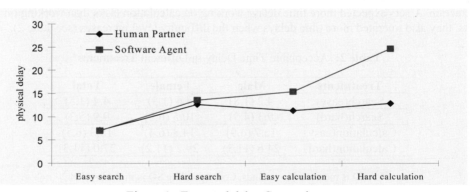

Figure 1. Expected delay Comparison

We also found that the partner type (Human or software agent) can significantly affect user's delay tolerance in calculation tasks. They will expect fewer delays when working with software agent than working with humans. In searching tasks, users will tolerate some amount of delays regardless of their partner type. Although the software agent is able to complete all the tasks within one second, users are still treating the software agent as human and use human's performance to predict agent's performance. This suggested that humans would expect a partner to keep a working speed in synchrony with their preferred event cycles. Previous studies (Billard & Pasquale, 1993) indicate that excessive delay of information sharing will deteriorate the quality of distributed decision making and cause emotional arousal (e.g. annoyance and anger); while working without any delay will lead to stress and increased mental workload to users. Humans' capacity for real-time information processing is limited (Newell 1990); task cycles must be attuned to the interaction speed at the level that the user can perform the task with minimal desynchronization.

**Postscript:**

As of 14 Feb 2003, investigation of the events leading to the loss of the Space Shuttle Columbia (STS-107) has been underway in the US. Ironically and sadly, the analysis of those events presents a clear example of the challenges and requirements of group-level operational knowledge development, and sensitivity to the range of event cycles, operating in the MCC environment. The investigation to date has highlighted several distinct aspects of the process of distributed supervisory coordination and MMI, HMI and HHI information flows. Rather than suggest fault or place blame, the authors insist that the issues highlighted in this paper underscore the level of professional competence, dedication, and skill exhibited by flight controllers in an extremely complex technological environment. Although improvements in information and communication technology systems may provide enhanced event referencing and archiving capabilities, the effectiveness of the MCC model still relies on the coordination skill, knowledge base, and shared awareness that is the emphasis of the shared teamwork, taskwork, and pathwork expertise among members of the MCC flight control team.

# 4   References

Billard, E.A. and Pasquale, J.C.(1993), Effects of Periodic Communication on Distributed Decision-Making, *IEEE International Conference on Systems, Man, and Cybernetics*, New York.

Caldwell, B.S. and Paradkar P. (1995), Factors Affecting User Tolerance for Voice Mail Message Transmission Delays, *International Journal of Human-Computer Interaction*, Vol 7, 3, 235-248.

Garrett, S. K., Caldwell, B. S. (2001) Information Sharing and Knowledge Management In MCC System Evolution. *NASA / ASEE Summer Faculty Fellowship Program Reports*. Houston, TX: NASA Johnson Space Center.

Garrett, S. K., and Caldwell, B. S. (2002). Describing functional requirements for knowledge sharing communities. *Behaviour and Information Technology*, vol **21** (5), 359-364.

Newell, A. (1990), *Unified Theories of Cognition*, Harvard University Press, Cambridge, Mass.

Wang, E. (2002), User's Delay Perception and Tolerance in Human-Computer Interaction. Proceedings of the Human Factors and Ergonomics Society 46th Annual meeting, 651-655.

Wang, E., & Caldwell, B.S. & Zhang K. (2002). Time delay tolerance in Computer Supported Cooperative Work. *Proceeding of 6th International Scientific Conference on Work With Display Unit*, 199-201.

Woodrow, H. (1951). Time Perception. In Stevens S. S. (Ed.), *Handbook of experimental psychology*. New York: Wiley.

# Need for Action Oriented Design and Evaluation of Information Systems

*Stefan Cronholm*

Dept. of Computer and Information Science
Linköping University, SE-581 83 Linköping, Sweden,
E-mail: stecr@ida.liu.se

**Abstract**
The problem we are approaching is that the actions offered by computerized information systems (IS) often seem to disharmonize with the actions performed in work practice. In this paper, we have analysed a representative usability-oriented checklist supporting design and evaluation of computer-based systems. The analysis has been made from an action perspective. One result is that the checklist is more oriented towards cognitive aspects. Action oriented improvements are suggested.

## 1 Introduction

The problem we are approaching is that the actions offered by computerized information systems (IS) often seem to disharmonise with the actions performed in work practice. Several researchers report faults in IS use. For example, Henderson & Kyng (1994) claim that there is a discrepancy between the design of IS and the work situation. Bannon (1994) claims that there is need for a better understanding among researchers and system designers about users and their work settings. We need to understand people as actors with a set of skills and shared practices based on work experience (ibid.)

There are many different philosophies, methodologies or checklists (such as object-oriented, traditional, participatory design, prototyping) aiming at supporting IS development or evaluation. In the Human Computer Interaction area the concept of usability and IS are focused. In this area we can find checklists such as Nielsen's (1994) ten usability heuristics, design principles such as Shneiderman's (1998) eight golden rules and usability models presented by Shackel (1984). What they all seem to miss or at least not have in focus is the action character of the IS.

In this paper, we adopt an action perspective on information systems (see section 2). The action perspective emphasises what users do while communicating through an IS. In Cronholm & Goldkuhl (2002) several actable quality ideals have been suggested. These quality ideals are derived from action theory (Mead, 1938; Goldkuhl, 2001) and empirical findings. The aim of the quality ideals is to support design and evaluation of IS. The aim of this paper is to show that the action character is not focused on in a representative checklist developed from a usability perspective. We have chosen Nielsen's (1994) ten usability heuristics since they are well known and frequently used. Finally, we will suggest some complementary quality ideals.

## 2 An Action Perspective on Information Systems

We adopt an action perspective on information systems. Actions are humans´ intentional way of changing the world. Humans intervene in the external world. These intervening actions are overt actions, which can be communicative or material. Actions can also be covert. In such a situation a human tries to make sense of something external. He performs an interpretative action. He is not changing something externally as in intervening actions. He is instead trying to change his inner world, his knowledge of the external world. Besides interpretative actions, there are other covert actions. When a human is intentionally trying to solve a problem mentally through reflection this can be seen as a covert action. This action view is inspired by American pragmatism (e.g. Mead, 1938), social phenomenology (e.g Schutz, 1962) and language action theories (e.g. Austin, 1962). See also Goldkuhl (2001) and Goldkuhl & Ågerfalk (2000).

From an action-oriented perspective, IS are viewed as communication systems, as distinct from strict representational views of information. A representational view of information means that designers try to create an 'image' of the reality in order to have the analysed piece of reality properly represented in the systems database. This strict representational view can be challenged, which an action perspective certainly does (e.g. Goldkuhl & Lyytinen, 1982). In the action-oriented perspective, IS are not considered as "containers of facts" or "instruments for information transmission" (Goldkuhl & Ågerfalk, 2000). The action-oriented perspective emphasises what users do while communicating through an IS (ibid.). IS are systems for action in work practices, and such actions are the means by which work practice relations are created.

IS have action ability. We call this IS actability. We define actability as an IS ability to perform actions and to permit, promote and facilitate users to perform their actions both through the system and based on messages from the system, in some work practice context (ibid.). Within the actability perspective the notion of IS can be defined in the following way: An IS consists of
- action potential (a predefined and regulated repertoire of actions)
- actions performed interactively by the user and the system and/or automatically by the system.
- action memory (a memory of earlier actions and including other prerequisites for action)
- documents (as action conditions, action media, action results)
- a contained structured work practice language (giving frames for actions, action memory and documents)

Designing an IS means suggesting and establishing an action potential. An action potential both enables and delimits actions. It entails a repertoire of actions and a related vocabulary. The vocabulary consists of concepts related to the work practice language. An IS must also offer a record of actions performed. Information about these performed actions can normally be found in the IS database. We call it an action memory, which is part of an organizational memory.

# 3 Analysis Model and Categorization

In order to analyse the action character of the usability heuristics we have used the elementary interaction loop model (Cronholm & Goldkuhl, 2002). The model is inspired by Mead's (1938) stages of an act. The reason for using this model is that it in a generic sense describes the interaction between a user and a computerised IS. The interaction is divided into four phases within the loop: informing, execution, IS action and interpretation (see figure 1).

In the informing phase the user has to be informed from the screen document about what can be done. He/she must have knowledge about which possible actions can be carried out. After being informed, the user executes an action (for example by clicking on a button on the screen document). The IS reacts by performing its corresponding IS action. When the IS action is performed the user interprets what the IS has done.

The screen document plays important roles in the interaction. One can say that the screen document is multifunctional. It contains *information about the action possibilities*. In this sense it is used in the informing phase when the user is reading the screen figuring out what to do.

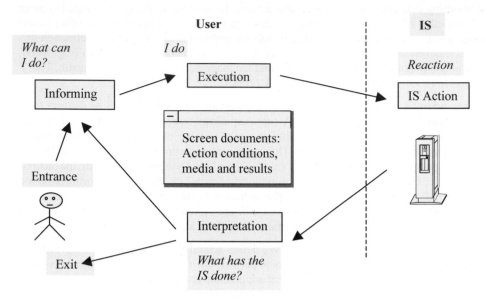

**Figure 1.** The elementary interaction loop (EIAL), (Cronholm, Goldkuhl, 2002)

The screen document also functions as an *action media* in the execution phase when the user performs an action (for example clicking on a button). The user can also (in the execution phase) type some information in a field and the screen document consists in this sense of *action results* of the user execution action. The IS action can result in changes of the screen document (as a feedback to the user). This means that it contains *results of the IS action*.

In order to analyse the action character of the Nielsen´s ten heuristics we have categorized the heuristics according to the model phases informing, execution, IS action and interpretation (see table 1).

Table 1. Categorization of the usability heuristics

| Usability Heuristics (Nielsen, 1994) | Categori- zation |
|---|---|
| *Visibility of system status*. The system should always keep users informed about what is going on, through appropriate feedback within a reasonable time. | Informing IS action Interpretation |
| *Match between system and the real world*. The system should speak the users' language, with words, phrases and concepts familiar to the user, rather than system-oriented terms. Follow real-world conventions, making information appear in a natural and logical order. | Informing IS action Interpretation |
| *User control and freedom*. Users often choose system functions by mistake and will need a clearly marked "emergency exit" to leave the unwanted state without having to go through an extended dialogue. Support, undo and redo. | Informing Execution |
| *Consistency and standards*. Users should not have to wonder whether different words, situations, or actions mean the same thing. Follow platform conventions. | Informing |

| Usability Heuristics (Nielsen, 1994) | Categori-zation |
|---|---|
| *Error prevention.* Even better than good error messages is a careful design which prevents a problem from occurring in the first place. | Informing Execution |
| *Recognition rather than recall.* Make objects, actions, and options visible. The user should not have to remember information from one part of the dialogue to another. Instructions for use of the system should be visible or easily retrievable whenever appropriate. | Informing |
| *Flexibility and efficiency of use.* Accelerators -- unseen by the novice user -- may often speed up the interaction for the expert user such that the system can cater to both inexperienced and experienced users. Allow users to tailor frequent actions. | Execution |
| *Aesthetic and minimalist design.* Dialogues should not contain information which is irrelevant or rarely needed. Every extra unit of information in a dialogue competes with the relevant units of information and diminishes their relative visibility. | Informing |
| *Help users recognize, diagnose, and recover from errors.* Error messages should be expressed in plain language (no codes), precisely indicate the problem, and constructively suggest a solution. | IS action Interpretation |
| *Help and documentation.* Even though it is better if the system can be used without documentation, it may be necessary to provide help and documentation. Any such information should be easy to search, focused on the user's task, list concrete steps to be carried out. | Informing |

# 4 Conclusion

It is clear that some of Nielsen's heuristics have an action character such as the informing-related heuristics "recognition rather than recall" and "visibility of the systems status". However, most of the heuristics are more oriented towards cognitive aspects. When comparing the analysis results with Shackel's (1984) classical model consisting of the related components user, tool, task and environment it is clear that the heuristics are more user- and tool-oriented. From an action perspective, IS are viewed as communication systems. The action perspective emphasises the actors (communicators, receivers) and the actions (what the actors do while communicating through an IS). We claim that the heuristics should also include action oriented quality ideals such as "understanding the screen document" and "action memory – easy accessible and personalized". (For a exhaustive description of actable qualitative ideals, see Cronholm & Goldkuhl, 2002.)

"Understanding the screen document" means that the contents of the screen document should offer good conditions for performing actions both within and outside the IS. Information presented must be easily understandable. Relations between IS-supported actions must be visualised in a way that the users easily understand if there is a specific order among the offered actions. "Action memory – easy accessible and personalised means that information stored earlier about previous actions should be easy to access. The action memory can consist of both historical information (actions that have been performed) and expected actions (actions that should be performed). It should also be clear who is responsible for the content of information communicated. Information about "who has said what" should be stored in the IS as part of the action memory. This quality ideal can be seen as an exhortation to avoid anonymity in information systems.

Further, we think that Nielsen's heuristics can to high degree be seen as a checklist for user-interface design. One fault is that the heuristics are presented in a sequential list without any explicit order and that they are not explicitly or theoretically grounded. In order to analyse the action character, we have categorized the heuristics according to a theoretical model that in a generic sense describes the interaction between a user and a computerised IS.

We claim that if one includes action oriented quality ideals in a design or evaluation situation, one will have a high degree of probability to reach an actable information system. Of course there are other ways to arrive at actable systems. Following other approaches does not exclude the possibility to create an actable system. For example following object-oriented approaches (e.g Kruchten, 1999) or Nielsen´s usability heuristics might well lead to actable systems although those approaches do not contain explicit criteria for actability design. In such cases actable systems are created *by chance*. In the case of using the actability quality ideals, actable information systems are created by *intentional and conscious design*. The main contribution of this paper is to present the elementary interaction loop and argue for that an action perspective can complement established criteria/heuristics developed from a usability perspective.

# References

Austin J L (1962). *How to do things with words*. Oxford University press

Bannon L (1994). From Human Factors to Human Actors, in *Design at Work* (Greenbaum J & Kyng M eds), Lawrence Erlbaum Associates. Hillsdale, New Jersey.

Cronholm S & Goldkuhl G (2002). Actable Information Systems - Quality Ideals Put Into Practice. *Accepted to the Eleventh Conference On Information Systems (ISD 2002)*. Riga, Latvia.

Goldkuhl G (2001). Communicative vs material actions: Instrumentality, sociality and comprehensibility, in Schoop M, Taylor J (Eds, 2001), *Proocedings of the 6th Int Workshop on the Language Action Perspective (LAP 2001)*. RWTH, Aachen.

Goldkuhl G & Lyytinen K (1982). A language action view of information systems, *in proceedings of third Intl. Conference on Information systems*, Ginzberg & Ross (eds). Ann Arbor

Goldkuhl G, Ågerfalk PJ (2000) Actability: A way to understand information systems pragmatics, accepted to the 3rd International Workshop on organisational semiotics, Staffordshire University.

Henderson A & Kyng M (1994). There's No Place Like Home, in *Design at Work* (Greenbaum J & Kyng M eds), Lawrence Erlbaum Associates. Hillsdale, New Jersey.

Kruchten P (1999). *The Rational Unified Process: an Introduction*. Addison Wesley Inc.

Mead G H (1938). *Philosophy of the act*. The University of Chicago Press

Nielsen J (1994). *How to Conduct a Heuristic Evaluation*. http://useit.com/papers/heuristic/heuristic_evaluation.html, retrieved 2003-02-10

Schutz A (1962) *Collected papers I*, Martinus Nijhoff, Haag

Shackel B (1984). The Concept of Usability, in *Visual Display Terminals: Usabiblity Issues and Health Concerns* (Bennet, Case, Sandelin and Smith eds), Prentice Hall, Englewood Cliffs, NJ

Shneiderman B (1998). *Deigning the User Interface: Strategies for Effective Human-Computer Interaction*. Addison Wesley.

# Interaction Literacy:
# Form, Function and Fitness at the Interface

*Elisabeth Davenport*

Napier University School of Computing
10 Colinton Road Edinburgh EH10 5DT UK
e.davenport@napier.ac.uk

## Abstract

The presentation develops the concept of 'interaction literacy' as a way of extending current notions of usability and design in HCI. In many installed systems, designers and usability engineers have simplified the user interface in an attempt to minimize user effort to complete tasks. Literacy in such circumstances is construed primarily as training for navigation and this can influence ways in which designers support interaction. Such an approach obscures the rationale, legitimacy and consequences of interaction both inside and outside a given system. Though a 'black box' may be appropriate in some cases, in others it is not. An alternative approach to interface design is proposed that provides insight into the locations, trajectories and destinations of manipulated data.

## 1    Introduction

Dourish (2000) has proposed an approach to HCI design that takes account of embodiment and embeddedness. His rationale is that those who interact with computational devices do so from a position of physical presence in a situation or context. Dourish discusses two main research agendas ("tangible" and "social" computing) and makes the case for "reflective" interaction design, or interfaces that "account" for the systems that support them. Most interface design, he claims, aims to hide such accounts to the detriment of an interactor's comfort and security. Where the workings of a device are transparent, expectations of what may be achieved in an interaction are likely to be more realistic. Interfaces and systems that support online interaction thus need to be "accountable" or interactors will have inadequate understanding of the consequences of their moves, and may, as physical actors, have insufficient confidence in the context that envelops them. Such confidence has been labelled "systems trust" by Baier (1986). Where it is lacking, users' inclination to fully engage with a system may be inhibited.

To show how  problems  may arise where interaction is interpreted in a parsimonious way, a case study is presented of a small group of computer-supported adult learners in the tourism sector. A recent European Community project, Net Quality, explored the design of a learning platform that can assist micro-organizations to work together in the interests of local or regional development. The learning platform was intended as a probe that would allow a research team to evaluate the role of communities in organizational learning, and the role of computational media in that process. Interaction literacy was explored from two perspectives. Firstly, a social perspective, as the project required independent proprietors of small enterprises to interact as a collective "virtual enterprise." Secondly, a technological perspective, as novices had to attain a requisite level of comfort with online interaction.

## 2 The classroom case study: networked adult learners

A small target cohort of local entrepreneurs worked with the developers over a six-month period on a pilot application to engineer a community that would foster both online and "domain" (or "destination management") competence. A learning shell based on five phased modules was constructed by a team that included participants with experience of instructional design. The design of this shell was influenced by prior work in management education on structured interaction and situated learning (Gherardi et al., 1998). The team also drew heavily on theories of communities of practice (CoPs) and their role in organizational learning, specifically on a "learning framework" developed at Napier. (Davenport and Hall, 2001). The framework assumes that CoPs, supported by an appropriate "social infrastructure" (SI), are sites for three types of learning: situated learning (SL) (Lave and Wenger, 1991), situated action (SA) (Suchman, 1987) and distributed cognition (DC) (Hutchins, 1991). Interaction exercises were designed for each module to support these. Details of the modules can be found in Davenport and Taylor (2000).

The CoP template was mapped on to the learning shell in the following way. Phase One of each unit established an SI environment, by capturing the expectations and motivations of participants, and introducing them to the "ways" of the unit: the tools, processes and the etiquette that shape the online learning space. Participants were required to use narratives of their own experience (SA) as a starting point in addressing a number of questions relevant to their workplace. The role of the "broker" or tutor was carefully constructed. Tutors are responsible for managing DC by synthesizing, shaping, and archiving the insights that emerge as learners work first in pairs, then in larger groups that provide suggestions and insights on the problem areas using appropriate artefacts (such as discussion lists and news groups). At each stage, the output from groups was to be archived for 'social learning' (SI) purposes.

### 2.1 The interaction design approach

Designers of online learning platforms must accommodate a Doppelganger effect: bodies are embedded in two settings, that of the machine and that of the domain that is the focus of learning. Though aware of potential problems arising from the dual nature of design for online instruction, the Net Quality team focused on contextual factors such as social learning for business processes (the domain) as the primary challenges for the SMEs involved in the project. As a result, the learning activities in the modules featured problem solving by practitioner communities supported by appropriate media/artefacts and the solving of work-based problems. The problems ranged from simple puzzle solving, to designing a common information architecture for a regional tourism destination that learners might wish to manage together. These problem solving activities, undertaken and captured online, were intended to function as aids to reflective practice, by capturing material to prompt self-evaluation, and be the basis of "virtual apprenticeship" (Davenport and Cronin, 1991) or knowledge of how to behave in a specific online community. We also believed that we would help participating entrepreneurs understand, by anatomizing the process, how organizational learning might happen in the "virtual enterprises" promoted by the project. Three modules of the learning platform were tested in May 2001 with a small cohort of ten distance learners, either involved in micro-enterprise or intending to start up in business in the near future.

# 3     Findings: form and function at the interface

Given the design team's inattention to the dual nature of design for online learning, and their over-emphasis on the domain setting at the expense of the technology context, it is not surprising that a number of technical problems were raised by users. This was largely due to their perplexity over the consequences of basic physical actions such as unexpected screen loss in response to a keystroke, or fear of submitting a photograph for scanning as they did not know "where it might end up being used." Our learners were relatively inexperienced (many of them were comparative novices to internet-based work, and one was a total computing naïf). They were in no position to exploit the affordance of the off-the-peg systems that had been combined to form the platform (such as Microsoft Office and Yahoo! Groups).

Problems arising from user naiveté were compounded by local system incompatibilities, and much of the early part of the project involved "repair" (social and technical) work by designers and tutors. After the pilot, intensive one-to-one evaluation-and-help sessions were arranged to explain what happens behind the screen when different moves are made, and to help users to form a mental model of where information is held and processed across an assembly of local and remote servers. But some learners failed to become competent in any but the most basic functions: though they could work at a distance in pairs, as the elements of an e-mail interface are not difficult to grasp, working with a broader range of interactive genres (discussion list, e-mail, archive, dialogue box) raised a number of problems such as perplexity over private and public space, and how to exploit backchannels, a phenomenon that has been observed in learning communities in other contexts (Haythornthwaite, 1999).

With hindsight, it is clear that technological trust (trust in the machines that support a task), a key component in interaction literacy, should have been an issue in design. The team, as noted above, initially focused on interpersonal trust (Lewicki and Bunker, 1996) as a critical factor in online education, as there can be little social learning where sources (human and textual) are not recognized as credible or legitimate.

# 4     Lessons learned

The design team has learned a number of lessons from the Net Quality project that have implications for interaction design and interaction literacy. The first finding is that there is a need for hybrid tutors, who can solve problems that require both technical and domain expertise. Though the project started with a strong focus on the "learner," the consortium quickly recognized that the tutor role needed equal attention, as the profile that is required is very different from the IT "trainers" that have characterized investment in interactive training to date. The second finding relates to the emphasis of domain at the expense of interaction. In attempting to design a learning platform that was innovative in its emphasis on social learning, and that would compensate for the narrow focus of much existing "ICT" training, the team over-emphasized the more abstract context of the learning domain ("embedding"), and neglected the ("embodied") context to hand – the computer mediated workspace. A third finding concerned the nature of the interaction problems experienced by learners. These were not merely ergonomic. When learners express concern that they do not understand understanding the consequences of a "physical" move (such as keystroking or scanning an image), they are addressing a set of non-trivial issues about transparency, accountability, privacy and control.

Comparable issues have been discussed by a number of analysts in recent years (Johnson and Nissenbaum, 1995; Lessig, 1999; Castells, 2001) who have expressed concern that users, or consumers or citizens, have little or no access to the pathways and "internal" interfaces that shape the transformations and transitions of personal data in human computer interactions. The problem is both structural and political. Transparency is difficult to achieve in this complex situation (what Nissenbaum calls the "many hands" problem), but it may also not be in the interests of producers to reveal the input and output paths of personal data. Users may then contest the status of this valuable proprietary resource, and challenge the appropriation of property rights. For many producers opacity (it may even be described as the "tacit knowledge" of the machine) is a desideratum in networked interaction.

## 5    What do the "lessons learned" offer HCI education?

The findings from the adult learner case study are comparable to those of other studies by the Napier group, notably in the area of the design of online consultation platforms to support e-democracy (Whyte and Macintosh, 2001). These suggest that a more flexible approach to HCI education may be called for, that liberates the domain from the curricular silos that characaterize instruction in many institutions. An extended curriculum may address areas such as HCI for collective, as well as stand-alone, HCI that embraces, rather than hides, complexity and HCI that uses black boxes with due attention to fitness.

The last point is important. By attempting to "black box," or hide the deep details of technology, the team committed an error that Dourish suggests is committed by many professional suppliers of infrastructure who ignore the need for sense-making and provide "one step solutions" that confuse local practitioners.  As is noted above, Dourish makes the case for "reflective" computing, or interfaces that "account" for the systems that support them.   This complements "reflective practice," and fully-fledged members of such communities may be seen "double agents," at ease in both online and domain worlds and making no distinction between the two. To attain this level of competence, interactive users will require both educational and infrastructural support. The latter involves designers, who need to be trained to see their clients in a different way. Rather than patronizing users by "hiding" detail (the "Don't make me think" syndrome (Krug & Black, 2000)) that might act as a scaffold, they can work with them on extensible infrastructure that provides users with whatever scaffolds they require.

Such an approach would entail a different pedagogy, premised on complexity rather than simplicity. A different type of black box may be pertinent here; the teaching technique described by Haberman and Kolikaut (n.d.), that builds on direct observation to establish what models novices exploit when faced with  new tasks. Such an approach can assist designers to establish what kinds of scaffold are required by which users in what situations. The paradoxical outcome may be interfaces for novices that not only make them think, but help them pursue concerns and clarify uncertainties that inhibit interaction.

## Acknowledgments

Net Quality was funded by DGXXIII of the EC (DGXXIII 98006361/IT-9). This paper presents the views of the author, not the overall consortium.

# References

Baier, A. (1986). Trust and Antitrust. *Ethics* 96, 231-260.

Castells, M. (2001) The Internet Galaxy.: reflections on the Internet, business and society. Oxford: Oxford University Press.

Davenport, E. (2000). Localisation, globalisation and SMEs in European tourism: the "virtual enterprise" model of intervention. In N D. Kraft (Ed.) *ASIS 2000: Proceedings of the 63rd ASIS Annual Meeting Chicago, IL, November 12 – 16, 2000*. Vol. 37. Medford NJ: Information Today Inc, 309 – 319.

Davenport, E. and Hall, H. (2001). New knowledge and micro-level online organization: "communities of practice" as a development framework. In *Proceedings of HICSS-34*. Los Alamitos: IEEE (CD ROM)

Dourish, P. (2001). *Where the action is: the foundations of embodied interaction.* Cambridge, MA: MIT Press.

Dourish, P. (2001). Seeking a foundation for context-aware computing. *Human-computer Interaction*, 16, 229-241.

Gherardi, S. Nicolini, D. and Odella, F. (1998). Toward a social understanding of how people learn in organizations: the notion of situated curriculum. *Management Learning*, 29(3), 273 – 298.

Haberman, B. and Kolikaut, Y. Ben-David. Acitvating "black boxes" instead of opening "zippers" – a method of teaching novices basic concepts in computation. Accessed 23 October 2002 http://stwww.weizmann.ac.il/g-cs/yifat/papers/zipper/zzzipper.html.

Hutchins, E. (1991). The social organization of distributed cognition. In L.R. Resnick et al., eds. *Perspectives on socially shared cognition*. Washington DC: American Psychological Association, 284 - 307.

Johnson, D.G. and Nissenbaum, H. (1995). *Computers, ethics and human values*. Upper Saddle River: Prentice-Hall.

Krug, S. and Black, R. (2000). *Don't make me think: a common sense approach to web usability.* Que.

Lave, J. and Wenger, E. (1991). *Situated learning: legitimate peripheral participation.* Cambridge: Cambridge University Press.

Lessig, L. (1999). *Code and Other Laws of Cyberspace.* New York: Basic Books.

Lewicki, R.J. and Bunker, B.B. (1996). Developing and maintaining trust in working relationships. In Kramer, R.K. and Tyler, T.R., eds. *Trust in organizations*. London: Sage, 114 - 139.

Johnston, D.G. and Nissenbaum, H. (1995). *Computing, ethics and human values*. Prentice Hall.

Suchman, L. (1987). *Plans and situated actions: the problem of human-machine communication.* Cambridge: Cambridge University Press.

Wenger, E. (1998). *Communities of practice: learning, meaning, and identity*. New York: Cambridge University Press.

Whyte, A. and Macintosh, A. (2001). Transparency and teledemocracy: issues from an e-consultation. *Journal of Information Science*, 27(4), 187-198

# Web Design, Interface Design, Usability, Software Ergonomics: Mixing All Those Approaches

*Anamaria de Moraes,*
*Robson Santos,*

LEUI – Laboratory of Ergonomics and Usability of Human-Technology Systems
Interfaces
PUC-Rio  Pontifical Catholic University of Rio de Janeiro
Rua Voluntários da Pátria, 98, apto 601    22270-010   Rio de Janeiro, RJ – Brazil
ergonana@terra.com.br; robson.s@terra.com.br

## Abstract

This paper presents the objectives, the content, the methodology, the results of the first course taught about interface usability from the point of view of the software ergonomics. The focus was the usability evaluation considering the user satisfaction. Some problems had to be faced because the different backgrounds and expectances of the students. So, three versions of the course were experienced to reach a final one that was considered good after the results obtained in the proposed practical exercises.

## 1    The Problem

It is an important experience that has been carried on by LEUI: Laboratory of Ergonomics and Usability of Interfaces in Human-Technology Systems. Firstly we began with a one year course with the title of "Design of Interfaces". It was difficult to integrate the different backgrounds and objectives of the students.

One group (web designers) wanted to learn the latest versions of the multimedia authoring tools and the web design softwares and use it in the "design" of an interface. Others (people from informatics) expected to learn how to "design" a beautiful interface, considering typography, colors, icons. Few of them understood that we intended to teach something new: how to evaluate the usability of the interfaces of softwares applications and web sites, considering principally the user and its satisfaction, easy of learning and easy of use.

What they should learn were the fundamental concepts about human computer-interaction, usability evaluation, guidelines, golden rules, heuristics, hierarchical task analysis, cooperative evaluation, prototyping and tests of new proposals.

In the first version of the course those divergences caused a great number of abandon. In the second version the coordinator emphasized the objectives and took off the syllabus any discipline that taught softwares. In the third version all the students knew since the beginning that our aim was interface design and usability evaluation. Even so we had problems – some students complained about the theoretical approach of the disciplines. To avoid loosing students we put forward a practical exercise that we divided in two phases.

## 2    The Objectives

We had as target public web designers, informational designers, ergonomists, architects, and people from informatics. To all of them we defined and emphasized, since the beginning, the objectives of the course:

- Provide to users more efficient sites developed considering the user-centered design.

- Understand the difference between designer mental model and user mental model.

- Teach some techniques of knowledge elicitation.

- Learn and apply techniques of usability engineering as heuristic evaluation, cooperative evaluation, focus group, and task analysis.

- Develop and test a low technique prototype of proposals to solve some usability problems detected in the analyzed interfaces.

## 3    The Curriculum

1. SOME BASIC CONCEPTS: Human-computer interaction; Usability; Interface design; Screen design

2. USERS PROFILE AND VARIABILITY: Cognitive and mental characteristics of the users; Mental processing of information – perception, cognition, signals, signs and symbols; skills, rules and knowledge; mental models.

3. SPECIFIC APPLICATIONS: shopping, learning, information, identity and information web sites, e-government and intranets

4. INFORMATION ARCHITECTURE: information organization; navigation; information hierarchy; navigation aids, card sorting

5. SOME RECOMMENDATIONS FOR INTERFACE DESIGN: guidelines (Nielsen, Bastien & Scapin) and golden rules (Shneiderman); user centered design; screen design

6. TASK ANALYSIS: hierarchic task analysis; cognitive task analysis; scenarios, TAKD; GOMS, TAG

7. USABILITY EVALUATION: Cognitive walkthrough; Questionnaire of user satisfaction (Schneiderman); heuristic evaluation (Nielsen); cooperative evaluation (Monk); field studies

8. PROTOTYPING: levels of prototyping; low-tech prototyping (Sade); some dirty techniques; quick prototyping

9. WEB CONTENT MANAGEMENT: problem definition; CMS; concepts and principles; workflow; content deployment; designing and implementation

# 4      Methodology

- To present the topics of the syllabus discuss the concepts and techniques introduced considering the technical literature from 1990.

- Exercises to evaluate usability of existing specific interfaces chosen by the students (shopping, learning, information, identity and information web sites, e-government and intranets)

- Exercise to develop conceptual models and prototypes of proposals to correct some of the detected usability problems

Some of the author considered in the development of course content were: Andrew Monk, Aaron Marcus, Mark Lansdale, Brenda Laurel, Deborah Mayhew, Ben Shneiderman, Jakob Nielsen, Jeffrey Rubin, Joseph Dumas, Janice Redish, Jeniffer Fleming, Louis Rosenfeld and Peter Morville, Jenny Preece and others.

# 5      Practical Exercise

The practical exercise was divided in two phases. The first one included: select two web sites from an specific application; describe the interface and its features; apply the heuristic evaluation and the cooperative evaluation to discover some usability problems. The appraisal of the interface was finished with the presentation of problems discovered and proposals of improvements.

The second phase included: the definition of the requirements to solve the problems; the definition of elements for a new interface and features; prototyping the new proposal; conduct usability testing to evaluate the prototype with the solutions proposed.

# 6      Examples Of Works Presented By The Students

## 6.1    Evaluation and proposals for Zona Sul supermarket e-commerce web site

Some problems discovered: there is no "back" button in the all pages; the main information is presented in a second window; bad indexing for the search system; it's not possible to see the information about the products included in the shopping cart;

Some proposals: to present the main content in the first browser window; modify the navigational system; to improve the search by using a more dynamic and intelligent dictionary; to show thumbnails of the products in the shopping cart.

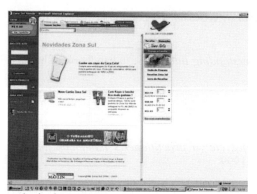

web site home page            The proposed home page interfce

## 6.2     Evaluation and proposals for Companhia Siderurgica Nacional corporate web site

<u>Some problems discovered:</u> lack of consistency between links and heading of the pages; no error prevention; difficulty to interpret the existent icons; inconsistency in the presentation of the links; there is no site map or other aid for helping the user.

<u>Some proposals:</u> to redesign the site icons, with the inclusion of labels; inclusion of a visible search field in all pages; to redesign the entire navigation system.

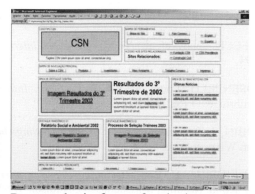

Homepage evaluated            Proposed wireframe for the home page.

## 7     Conclusions

The results presented by the students in each phase of the exercise show that they understood the objectives of the course, the concept of usability and applied correctly the methods introduced.

# References

DUMAS, Joseph S.; REDISH, Janice C. *A practical guide to usability testing*. rev ed, New Jersey, Ablex Publishing Corporation, 1999.

HELANDER, M. *Handbook of human-computer interaction*. New York, North Holland, 1988.

KIRWAN, Barry. *A guide to practical human reliability assessment*. London, Taylor & Francis, 1994.

LANSDALE, Mark W., ORMEROD, Thomas. *Understanding interfaces*: a handbook of human-computer dialog. London: Academic Press, 1994.

LAUREL, Brenda (Ed.). *The art of human-computer interface design*. Reading, Massachusetts, Addison-Wesley, 1990.

MAYHEW, Deborah. *The usability engineering lifecycle: a practitioner's handbook for user interface design*. San Francisco: Morgan Kaufmann, 1999

MAYHEW, Deborah. *Principles and guidelines in software user interface design*. New Jersey: Prentice-Hall, 1992.

MARCUS, Aaron. *Graphic design for electronic document and user interface*. New York: ACM, 1992.

MONK, Andrew; WRIGHT, Peter; HABER, Jeanne; DAVENPORT, Lora. *Improving your human-computer interface:* a practical technique. New York, Prentice Hall, 1993.

NIELSEN, Jakob, THAIR, Marie. *Home page usability*: 50 web sites deconstructed. Indianapolis, Indiana: New Riders, 2001.

NIELSEN, Jakob. *Designing web usability*. Indianapolis, Indiana: New Riders, 2001.

NIELSEN, Jakob; MACK, Robert L. *Usability inspection methods*. John Wiley & Sons, 1994.

PREECE, Jenny et al. *Human-computer interaction*. Wokingham, England, Addison-Wesley, 1994.

ROSENFELD, Louis, MORVILLE, Peter. *Information architecture for the world wide web*. Cambridge: O'Reilly, 2002. 2nd ed.

RUBIN, Jeffery. *Handbook of usability testing: how to plan, design and conduct effective tests*. John Wiley & Sons, 1994.

SHNEIDERMAN, Ben. *Designing the user interface*; strategies for effective human-computer interaction. Reading, Massachusetts, Addison-Wesley, 1992. 2nd ed.

# Integrating Data Analysis, Navigation and Knowledge Transfer by Visualizing Conceptual Models

*Maximilian Eibl*

GESIS
Schiffbauerdamm 19
10117 Berlin, Germany
eibl@berlin.iz-soz.de

## Abstract

Information visualization is usually used in three areas of application: data analysis, navigation, and knowledge transfer. Here, totally different kinds of visualizations are developed. This article describes an example that integrates the three areas into one single graphical representation. The application context is a sociological model of electoral behaviour. The here described visualization is part of the Internet journal sowinet, which provides social sciences information and is maintained by GESIS (http://www.gesis.org).

## 1    Introduction

The usual context of information visualization is data analysis. Here even simple graphical presentations like bar or line charts allow to see trends, patterns or peculiarities (Tufte 1983). More elaborated applications allow dynamic interaction with these diagrams (Wolff 1996). In information retrieval a lot of research has been done in visualizing the several aspects of the retrieval process (Eibl2000). Another context of information visualization is navigation. Here, tree-based visualizations are used in order to facilitate navigation in hierarchical data (Eibl 2000). Web sites frequently use those visualizations for site maps. Another navigational visualization is the topographic map which enables the user to browse a data space visually. Here, no hierarchical system is required. Last but not least, visualization is frequently used as a method of knowledge transfer. Learning by visual elements is much easier than learning by plain text. The here described visualization integrates all three aspects of visualization.

The conceptual base of the visualization is the sociological model of electoral behaviour. This model describes the main factors and effects responsible for individual voting decisions. The main problem of working with this model is its non-deterministic nature. Easy to handle diagrams like rule based flow-charts are not applicable. The rather fuzzy conception needs a less mechanic and more vague visualization.

A more thorough discussion of the model is provided by Eibl & Quandt 2002 and Bürklin & Klein 1998. Here, only the main points shell be listed:

- *Internal factors:* Changeable psychological factors like: preferences of parties and candidates, influences of current debates. Next to those short-term factors also long-term psychological factors are to be displayed like: political world view or individual values.

- *External factors:* Factors which have an outside effect on the individual decision like economic situation, electoral system, Media, etc.
- *Mixed factors:* Those are somewhat in between internal and external factors. Mixed factors are typically socio-demographic characteristics like religion, sex, occupation, etc.
- *Effects:* All these factors can immediately affect the vote or can affect other factors.

The visualization has to display these elements as well as their relationships. Additionally it has to emphasize their vagueness.

## 2    The Model

A first realization of the model contains two encapsulated clouds – the inner cloud representing the internal factors and the outer cloud representing the external factors (see Figure 1). The inner cloud also contains a thin line separating long-term factors from short-term factors. The mixed factors are visualized by a thick white line separating the two clouds. Left of the actual model a text box provides additional information to the single factors. The content of this box is changed by clicking on the factors of the model.

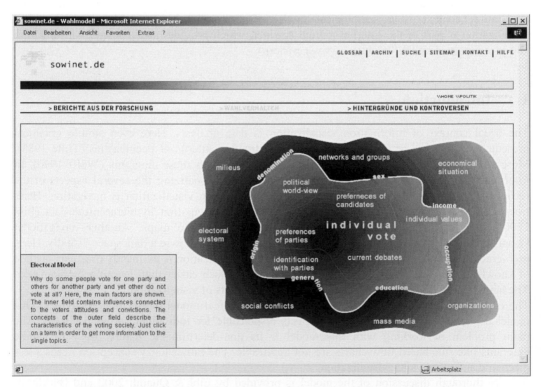

**Figure 1:** Visualization of the factors of the model of electoral behaviour

By mouse-interaction additional information to the single factors can be viewed as well as the effects, represented by appearing/disappearing arrows. Here again additional information to the effects can be displayed. Figure 2 shows the visualization after clicking on the factor *origin*. The text box on the left now contains information about the factor *origin* and three additional arrows appear to show the effects of this factor to three other factors *political world-view*, *individual values* and *identification with parties*. The mouse pointer is positioned above the arrow from *origin* to *individual values* causing an additional text box containing information to this effect to appear.

At this stage the visualization is suitable for gaining a quick overview and understanding the relatively complex electoral model. Additional information displayed depending on user interaction provides a first entry to a deeper understanding.

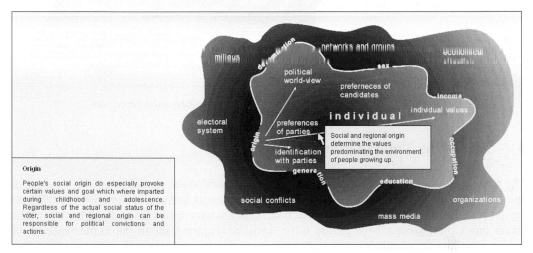

**Figure 2:** Visualizing the effects of the model of electoral behaviour

# 3    Navigating by the Model

In a next step the model was enriched by links to further information sources. Originally this possibility was introduced in order to set up the model as the central navigational aid for the several texts of sowinet. The aim was to provide a visual concept of the different articles as well as a starting point for navigating these articles. Next to this intra-sowinet navigation the visualization is also connected to content of the GESIS site www.gesis.org and other Internet sources. The grouping of the available links is twofold. First, the links are grouped into links to sowinet, links to GESIS, and links to the Internet. Second, the links shown always concern the factor actually activated. Thus, the visualization can be used as a navigational tool as well as a starting point to gain further information about scientific background of elections. Figure 3 shows links to sowinet displayed in an additional box. Further links to the GESIS site and the Internet can be selected.

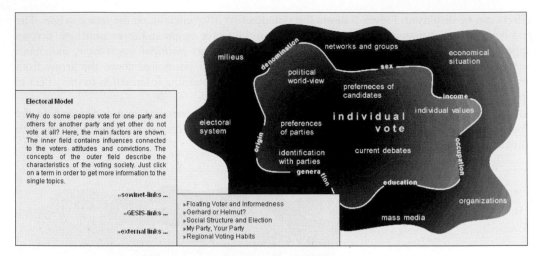

**Figure 3:** Linking the visualization to further information resources

# 4    Analyzing Voters Groups

The next step demonstrates the integration of empirical data to the visualization. On the base of opinion polls we conducted a CHAID (chi squared automated interaction detection) analysis on the factors of the last German national election. This analysis results in a categorisation of the voters into typical groups. Nine groups were identified. Figure 4 shows the implementation of these results into the visualization. An additional selection box is placed left to the visualization. In Figure 4 the group *CDU-voters amongst farmers and self-employed* is selected. A bar chart shows how many people of this group voted for a specific party. Here 54-70% voted for the German Christian Democrats (CDU). The little arrow below the chart signifies the percentage the CDU got all in all. The characteristics of this group are integrated into the visualization.

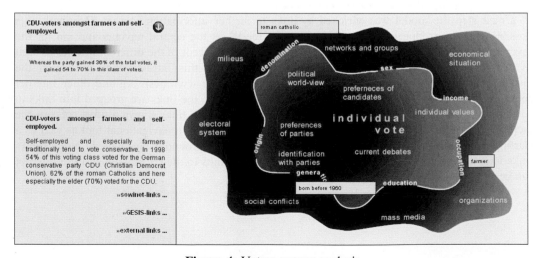

**Figure 4:** Voters groups analysis

# 5    Final Remark

The visualization is part of the Internet journal sowinet.de (http://www.sowinet.de), a project sponsored by the German Ministry for Education and Research until April 2002. Some parts of the journal are still available online as is the here described visualization: http://www.sowinet.de/wahlmodell/modell.html. The results of the project are described in Eibl 2001 and Eibl & Quandt 2002.

# 6    References

Bürklin Wilhelm· Klein, Markus (1998). Wahlen und Wahlverhalten. Eine Einführung. Opladen.

Eibl, Maximilian (2000). Visualisierung im Document Retrieval – Theoretische und praktische Zusammenführung von Softwareergonomie und Graphik Design. IZ-Forschungsberichte Bd. 3, Bonn.

Eibl, Maximilian (2001). Interaction through Multimedia. In: Abridged Proceedings of the 9th HCI International Poster Sessions, 130-132.

Eibl, Maximilian; Quandt, Markus (2002). Modellbildung und Visualisierung: Das Sowinet.de-Wahlmodell. In: Proceedings of the 8th International Symposium on Information Science. Pp 131-146.

Mandl, Thomas; Eibl, Maximilian (2001). Evaluating Visualizations: A Method for Comparing 2D Maps. In: Proceedings of the 9th HCI International, New Orleans, August 5-10, 2001, 1, 1145-1149.

Tufte, Edward (1983). The Visual Display of Quantitative Information. Cheshire Connecticut.

Wolff, Christian (1996). Graphisches Faktenretrieval mit Liniendiagrammen: Gestaltung und Evaluierung eines experimentellen Rechercheverfahrens auf Grundlage kognitiver Theorien der Graphenwahrnehmung. Schriften zur Informationswissenschaft Bd.24, Konstanz.

# The Copernican Shift:
# HCI Education & the Design Enterprise

*Anthony Faiola*

Indiana University – Purdue University
School of Informatics, Indianapolis, IN USA
afaiola@iupui.edu

## Abstract

The field of human-computer interaction (HCI) has brought us to a new plateau of reflection about its positive development for the future (Preece, Rogers, & Sharp, 2002; Carroll, 2001). The scope of HCI continues to migrate into new areas of study as technology becomes even more pervasive in its effect on the users' personal and community interactive experience. Similarly design is taking an even greater role and responsibility for shaping product look, feel, and strategy (Norman 1998; Norman, 2002). Designers are being consulted and depended upon to provide an insight rarely observed in the technology domain (Olsen, 2002; Faiola, 1998). Peer acceptance is gaining momentum in a discipline once dominated by computing professionals. This change is due to an increasing realization that technologists and social scientists provide a necessary, yet marginal component to transforming information and research findings into tangible design strategies. Beginning with a historical overview of HCI, this paper addresses the gradual shift to the central role that design is playing in product modeling and the integration of this view into HCI education, with a proposal outlining a new HCI curriculum framework at Indiana University's School of Informatics that embraces HCI as a "design enterprise."

## 1    Disciplinary Integration  - Good-bye to the Ptolemaic View

For over 20 years computer science (CS) programs have steadily evolved (Myers, 1998) in developing internal HCI course content and, to eventually become one of its nine "core" areas (Denning, et al, 1988). At its inception, HCI applied the earliest understanding of user engineering with breakthroughs in the engineering model of human performance greatly contributing to the "model human processor" (Myers, Hollan, & Cruz, 1996). As CS and cognitive psychology formed a closer bond, HCI moved toward the next stage of curriculum development, yet was still overshadowed by a CS agenda that was slow to recognize the importance of user interface design and HCI issues in software development (Douglas, et al, 2002). Since the 1990s, an increasing number of HCI, information technology (IT), and CS programs have initiated collaborative relationships with those departments that teach applied arts and design, e.g. graphic/design communication, industrial design, and interactive media, etc. By pursuing such partnerships, they provided students with multiple course options, with an opportunity to interact with faculty and students from other disciplines, and a dynamic class atmosphere where multidisciplinary team projects became more closely attuned to a real-world scenario that could generate a new synergy in problem-solving and innovation. Nevertheless, as team members juxtapose their orientation of product development in a collaborative environment, the potential for problem-solving around the informality and unpredictability of human behaviour has been difficult, largely due to technology's historical resistance to a user-centered paradigm in product design (Arias, et al., 1999; Snow, 1993). Shneiderman (2002, p. 71) suggests that this issue is related to "technology-centered researchers who value mathematical formalism more than psychological experimentation," and their preference to "work on problems in isolation" makes dealing with users quite troublesome. A greater concern is "their perception is that user experience issues are the paint you put on when the building is done," rather than the user experience being "the steel structure that frames the building (Shneiderman, 2002, p. 72)." Moreover, Shneiderman (2002, pp. 12-13) states that we are now in the second transformation of computing, where "the shift from machine centered automation to user-centered services and tools," are enabling users to be "more creative." This evolutionary transformation, which he refers to as the "Copernican shift," is an emerging focus "not on what machines can do, but on what users can do," being supported, enhanced, empowered, and enabled to extend their existing creative and cognitive abilities to fulfill their personal or community related desires.

## 2   User-Centricity – Hello to the "Copernican Shift"

Copernicus (Figure 1.) calculated that the sun did not revolve around the earth as purported by ancient astronomers using the Ptolemaic View, but rather that any point has the potential to be designated the center of the universe. Hence, the Copernican Shift, as the base of reference came with a radical change in the interpretation of observed phenomena (Crowe, 1990). By 1920, Harlow Shapley was able to calculate the distance of our galaxy and determine that our sun is far from the galaxy's center. By so doing, he proved that the earth and our solar system were not at the center of the universe.

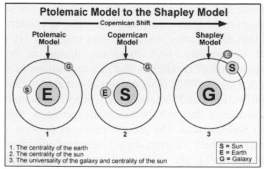 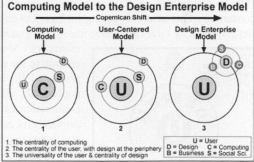

**Figure 1:** A shift from Ptolemaic to Shapley.     **Figure 2:** A shift from computing to design.

Technologists, confronted with the "new reality" of the Copernican shift, are slowly adapting to the gravitational pull of user-centricity over computing as the center of product modeling (Figure 2). Research and professional best practices are increasingly establishing a more viable model for product design. By the latter part of the 1990s there was a gradual acceptance of the user-centered model of teaching HCI within most programs in the U.S. and abroad. This model, however, only provided an incremental shift away from computing, as well as a marginal theoretical infrastructure on which to build a formal pedagogical framework or cohesive methodology, both of which are critical to dealing with the "problem space" of emerging and complex systems. Supportive of this view, Rogers (2001) outlines the dynamic shift away from the pure science of cognitive psychology's theorizing and experimentation with product concept modeling and toward the "boundless domain" (Barnard, et al, 2000, p. 221). Barnard, et al (2000) states that "Everything is in a state of flux: the theory driving the research is changing, many new concepts are emerging, the domains and type of users being studied are diversifying, many of the ways of doing design are new and much of what is being designed is significantly different."

Barnard, et al (2000) also asserts that what was originally a confined problem space with a clear focus and small set of usability methods ("that of designing computer systems to make them more easy and efficient to use by a single user"), is much broader in scope, i.e., it is a "vast problem space, with a much less clear focus and a bricolage of methods and theories being thrown at it." Building on Barnard's view, the author proposes that Rogers' thesis of "knowledge transfer" be expanded to include the "enterprise of design," not as a peripheral activity (as in the past), but an integral and universal mechanism for product design that is far more inclusive than most design traditionalists might affirm. As Rogers states, "Innovative design methods unheard of in the 80s, have been imported and adapted from far a field to study and investigate what people do in diverse settings. Ethnography, information design, cultural probes and scenario-based design are examples of these. ...Alongside these methodological and conceptual developments, has been a major rethink of whether, how and what kinds of theory can be of value in contributing to the design of new technologies." No doubt, as Arias, et al (2000, pp. 85-87) outline, the history of HCI has brought about new challenges that center on the long-term theories of design, system, technology, and media. Artefact creation, however, must be supported with knowledge of design processes and the need to manage large amounts of information relevant to the design task in the context of different (disciplinary) perspectives of the problem.

Despite a wealth of course content on computing, cognitive theory, and interface design, HCI students still lack a comprehensive understanding of information design and human-centered design as an enterprise of knowledge building and management, i.e., shaping a product's conceptual model to include a well-defined business strategy that not only leverages information from a social and cultural context of target users, but also contextualizes existing business conditions. This means focusing on business decision-making as giving tangible value to product development, a vital dimension rarely integrated into the iterative design process taught by HCI educators. More importantly, HCI must use design as an enterprise for exploration of conditions and presuppositions of a problem (Blevis & Siegel, 2002). This process is initiated with a contextual identification and analysis of people-to-people communication mediated through human-made artefacts that provide processes and services. The process proceeds through a structured sequence in which information is researched and ideas and systems are explored and evaluated until an optimum solution is devised that provides a new frame of reference in the design of information products – a discipline that the author refers to as HCI Design (HCID) (see Figure 3).

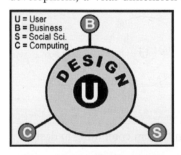

**Figure 3:** Design Enterprise

Given the perpetual changes of HCI theory, are HCI students being adequately prepared beyond a traditional framework of education? Are emerging online, wireless, and community-based technologies conceiving a new kind of designer of interactive products? Will HCI professionals take on a greater leadership role in product strategic planning and design, leaving product implementation to computing specialists, as building contractors might execute the design of the architect (Kapor, 1996)? Given Sutcliffe's (2000) statement that a fundamental mission of HCI is to bring the social sciences to bear upon design, the author proposes a concise philosophical shift of what might constitute a further metamorphosis in domain knowledge for HCID students (Faiola, 2002).

# 3 HCID & the Design Enterprise at Indiana University

Working under the banner mission of Indiana University's School of Informatics, the newly formed HCID Graduate Program addresses a long-term curriculum goal to provide students with an understanding of product development as a design-driven process comprised of multiple integrated models. From this perspective, students are equipped in the context of a field that is increasingly inclusive of many disciplines. The M.S. degree focuses students on a plan of study in design, research, development, and design evaluation of emerging technologies relative to an evolving world of interactive systems (see Table 1 as an example).

**Table 1:** One typical scenario and plan of study with core, flexible core, and electives.

| COURSE | SCHOOL | COURSE NAME | COURSE SUMMARY |
|---|---|---|---|
| I501 [1] I502 [1] | Informatics (2 Courses) (6 Credits) | 1) **Introduction to Informatics** 2) **Information Management** | 1) Economics of information sharing, i.e., representation, processing, searching and organization, as well as evaluation and analysis. 2) Information ... management i.e. database structures/models, access strategies, and indexing, analysis and representation of structured data sets. |
| I596 [1] I496 [1] | Informatics (2 Courses) (6 Credits) | 1) **HCID I** 2) **HCID II** | Both courses provide a broad range of theory and best practices that encompass the "design enterprise": 1. **THEORY**: From the human perspective, emphasis is placed on user-behaviour, information processing, and cognitive modeling; from the technology perspective, emphasis is placed on the design and modeling of interfaces, system design, and information architecture; and from a business perspective, emphasis is place on competitive markets, product management, and product analysis. 2. **PRACTICE**: Product management, planning, design, system creation, ideation, prototyping/sketching, qualitative/quantitative product testing techniques, document/report design. |
| I694 [1] | Informatics (6 Credits) | 1) Thesis: Report & Proposal 2) Project: Artefact Development | Final capstone project, including product proposal / theory-based thesis and product design and implementation and (market/user) testing. |
| N450 [2] | Informatics | Usability Principles for New Media | Principles of HCI and user experience modeling; study of theory and application design, usability, and usability testing in the context of new media product development. |
| L503 [2] | Lib. Info. Sc. | User Needs & Behaviour | Analysis and assessment of user information, from a human (socio/psycho) perspective, focus on theoretical models and practical techniques that underpin the field. |
| L579 [2] | Lib. Info. Sc. | Information Visualization | Data analysis and processes to extract relationships and interaction techniques to produce the perceptual basis of effective visualizations. |
| CS538 [2] | Computer Sc. | Design of Interactive Systems | Concepts / tools for designing interfaces for human-machine interaction, including: design problem/concepts, cognitive and predictive models, & experimental design and analysis |
| J522 [3] | Business | Strategic Management of Tech. Innovation | Examines interrelationship of technology, innovation, and strategy concepts and explores how managers employ the development of technology to improve bus. strategic position. |
| L575 [3] | Lib. Info. Sc. | Interface Design for Collaborative Info. Spaces | 2-dimensional and 3-dimensional interface design, including task and user analysis, interface goals and design methods, and empirical evaluation. |
| [1] = Core / [2] = Flexible Core / [3] = Elective | | HCID = Human computer interaction design | |

# 4 Conclusion

By providing theory, tools and the vision of a new HCI paradigm, students expand their learning potential by building upon their undergraduate degree with HCID fundamentals. Course content is applied to team projects, where students: (1) obtain skills in tangible ways that demonstrate the value of design innovation fused with human-centered product strategies and (2) enhance their leadership skills with the power and influence to ensure that the "design enterprise" should both drive and lead product development in a balanced, collaborative atmosphere. Under this plan, students learn how to configure a product's cognitive model through contextual discovery and analysis of information, problem space identification, design/concept building, prototyping, and a product implementation that insures the integrity of the original business model.

# References

Arias, E., Eden, H., Fischer, G., Gorman, A., and Scharff, E., (1999). Transcending the individual human mind creating shared understanding through collaborative design. In *Proc. ACM Transactions on Computer-Human interaction, Vol. 7, No.1, March 2000, 84-113.*

Blevis, E. & Siegel, M. (2003). (in manuscript) *Constructing Design Arguments*. Indiana University.

Barnard, P. (1991). Bridging between basic theories and the artifacts of human-computer interaction. In *Designing Interaction: Psychology at the Human-Computer Interface.* Carroll, J. (Ed.), New York: Cambridge University Press, 103-127.

Carroll, J. M. (Ed.) (2001). *Human-Computer Interaction in the New Millennium*. Addison-Wesley, Boston.

Crowe, M.J. (1990). *Theories of the World from Antiquity to the Copernican Revolution*, Dover: New York.

Denning, P. J., D. E. Comer, D. Gries, M. C. Mulder, A. Tucker, A. J. Turner, and P. R. Young. (1988). *Computing as a Discipline: Report of the ACM Task Force on the Core of Computer Science*. New York: ACM Press.

Douglas, S., Tremaine, M., Leventhal, L., Wills, C.E., Manaris, B. (2002). Incorporating human-computer interaction into the undergraduate computer science curriculum. In *Proceedings of the SIGCSE 2002 Conference (pp. 211-212), Covington, Kentucky: ACM Press.*

Faiola, A. (2002). New media usability: HCI curriculum focus in the School of Informatics, IUPUI. *ACM Interactions – New Visions of Human-Computer, ACM Interaction.* 9(2), 25-27.

Faiola, A. (1999). Re-Designing graphic arts education: A closer look at strategies for a New Millennium of Digital Communication and Globalization. *The Journal of Technology Studies. 25(2), 47-50.*

Kaplor, M. (1996). A software design manifesto. In *Bringing Design to Software*, Winograd, T., Bennett, J., DeYoung, L. & Hartfield B. (eds.), New York: Addison Wesley, 1-16.

Myers, B. A., Hollan, J., Cruz, I., et. al. (1996). Strategic directions in human computer interaction. *ACM Computing Surveys 28*, 4.

Myers, B. A. (1998). *A brief history of the human computer interaction technology*. ACM interactions. 5(2) 44-54.

Norman, D. A. (2002). Emotion and design: Attractive things work better. *Interactions Magazine*, ix (4), 36-42.

Norman, D. A. (1998). *The Invisible Computer: Why good products can fail, the personal computer is so complex, and information appliances are the solution*. Cambridge, MA: MIT Press.

Olsen, G. (2002) The emperor has no lab coat. *ACM Interactions*, 9(4), pp. 13--17.

Preece, J., Rogers, Y., & Sharp, H. (2002). *Interaction design: Human-computer interaction*. New York: John Wiley & Sons, Inc.

Rogers, Y. (2002). Knowledge transfer in a rapidly changing field: What can new theoretical approaches offer HCI? Retrieved September 8, 2002, from http://www.cogs.susx.ac.uk/interact/papers/

Shneiderman, B. 2002. *Leonardo's Laptop: Human Needs and the New Computing Technologies*. Cambridge, MA: MIT Press.

Snow, C.P. (1993). *The Two Cultures*. New York: Cambridge University Press.

Sutcliffe, A. (2000). On the effective use and re-use of HCI knowledge. *ACM Transactions on Computer-Human Interaction*, 7(2), 197-221.

# Conceptualising an Experience Framework for HCI

*Salvatore G. Fiore*

Northumbria University,
School of Informatics, Ellison Place, Newcastle upon Tyne, NE1 8ST, U.K
salvatore.fiore@unn.ac.uk

## Abstract

In this paper, some emerging frames of reference for HCI and recent advances in the study of experience are discussed. Key works contributing to a conceptualisation of human experience, drawing from aesthetic disciplines outside the traditional HCI domain are then summarised. Based on this, an initial framework for designing and evaluating artefacts as the focus of electronically mediated user experience, emphasizing emotional, sensory and other dimensions is presented.

## 1    Introduction

The study of experience has so far received little consideration in HCI. This is perhaps due to the modernist origins of HCI and a resulting tendency to privilege values of information and efficiency over community, creativity or pleasure (Muller, 1995). Prevailing approaches in HCI have largely failed to address user emotion or account for the social, cultural and historical context in which people live. Thus, existing frameworks, many based on the representation of tasks (Wright et al, 2000) are limited in their applicability to design when the goal is to support user experience. People are now using computers for activities for which deconstruction by task is difficult (e.g. gaming) and this raises a dichotomy between designing for *usability* versus *experience*. Preece, Rogers and Sharp, introduce user experience goals as "what the system *feels* like to the users" and note that they differ from usability goals "…in that they are concerned with how users experience an interactive product from their perspective" (Preece et al, 2002, p19). Designing a system to account for individual psychological phenomena and meet experience goals (e.g. satisfying, fun, etc.) may mean that the system is purposefully 'non-easy' to use. What these experience goals do not account for, however, are computer supported activities that are, for example, bewildering, frightening, frustrating or upsetting, that are nonetheless meaningful experiences. Understanding the nature of experience and how it evolves within a specific context, is a necessary basis for supporting experiences through design. In this direction, HCI researchers have recently turned to emerging frames of reference for HCI in an uncoordinated attempt to re-frame the discipline and account for extra-usability issues in design.

## 2    New foundations: vocalising context and collaboration in HCI

Before talking about how people perceive and live though activities and thus experience them, it is important to understand user activity in context. Activity Theory (AT) (Bødker, 1991) has been adapted as an emerging frame of reference for HCI, functioning as a philosophical framework for analysing activity. AT enables the cultural, social and historical context of an activity to be considered integrally when designing a system to support it. It enables designers to interpret the use of physical and psychological tools used by people in working towards a goal and although it cannot model experience, it can help establish how experience evolves through activity. Adopting an AT approach to a design challenge involves accessing the actual context of use and the participation of the users of the system in a dialogue between designer and users, to jointly decide the future of the artefact. Extending this dialogue to actively involve all stakeholders in the design, is a key concern of ethnocriticism. The approach, from literary criticism, is concerned with diversity and collaboration across differences in culture and power. That ethnocriticism is being embraced within HCI at the time when emerging technologies such as the Internet, which cross socio-cultural borders, are becoming increasingly pervasive, is interesting. Muller's Ethnocritical Heuristics (Muller, 1995), based on the position that meaning is a product of interaction, interrelationship and argumentation, are a case in point. The key to the heuristics, is the 'polyvocal polity' that characterises them: the involvement and collective decision making by all who will be affected by the new system. In this way, design becomes an occasion to enrich the experiences of all involved, in positioning the collective team of stakeholders, designers and users at the frontier between the work practices of each. Also relating to the notion of a collaboratively and contextually organised world of artefacts and actions, Distributed Cognition views thinking as occurring not only as an activity within a person's head, but in external relationships with objects and people. Central is the *functional system*: the actor(s) and the system/technology/tools as they are related to one another in the environment of their situation. Analysing how these components are coordinated and organised externally and cognitively, and why such coordination may fail, is the concern of Distributed Cognition. (Wright, Fields & Harrison, 2000). Here, in particular, actions both depend upon events and are affected by context. Nonetheless, despite drawing attention to the human perspective in HCI, each of these approaches is limited, in failing to allow for the complex implications of human *experience* as a focus for design.

## 3 Focussing towards designing for experience

Recent developments are fostering an area of HCI targeted specifically towards designing for user experience. Such approaches, drawing attention to the complexities of experience, have focussed on thinking beyond usability to user experience needs and positioning user experience as the focus of design ideation, reflecting changes in users themselves, their relationships and experiences and establishing empathy with them through collaboration (Sanders, 2000). Additionally, the importance of involving and building relationships with users throughout design has come to the fore, acknowledging human experience itself as a source of new products (Margolin, 1997). However, work to inform design for experiencing has tended to aim explicitly for rich or immersive experiences, based on the notion that, when all channels and dimensions for experiencing are taken into account, there is potential for such highly involving experiences (Forlizzi, 2002). Researchers have also begun to look outside of the traditional HCI matrix for inspiration on how to design for experience. Laurel, for example, sees the issue through the lens of drama theory (Laurel, 1993). Wright and McCarthy instead, appeal to literary studies and philosophy in developing a framework to analyse experience with emerging technologies (Wright & McCarthy, 2003).

## 4 Informing the construction of an experience framework

In constructing a holistic experience framework for HCI design work, some key contributions aid an understanding of what experience is and how it manifests in human activity. Csikszentmihalyi provides a cornerstone to an understanding of peak, emotionally rewarding experiences, which he calls *flow* (Csikszentmihalyi, 1988). He identifies a precondition for flow: "a balance between the challenges perceived in a given situation and the skills a person brings to it…any possibility for action to which a skill corresponds can produce an autotelic experience" (ibid, p.30). In addition, certain conditions characterise flow, namely clear goals, rapid and unambiguous feedback, focussed concentration on the activity, perceived control over activity outcomes, a distorted sense of time and loss of self-consciousness. A more comprehensive and holistic conceptualisation of human experience to emphasise the meaningful and aesthetic in experience comes from Dewey (Dewey, 1958). For Dewey, experience is *continuous*, undergoing change, with events being temporally related and connected, interacting and transforming one another. It is also *eventful* with a the unity between experiencing subject and experienced object changeable, and *historical:* there are no ends for it, as individual experiences are connected to one another through a continuous flow each carrying the memory of past experience. Experience is also *qualitative*, so that each experience is pervaded by a quality that gives it it's individuality (e.g. attending *that* meeting) and unity. This experiential quality belongs equally to the experiencing subject and the objects experienced. Dewey also maintains that, as experience is controlled through attention to it and it's relation to other things, the quality of an experience may suggest the presence of other conditions which may then be altered to rid of the quality if desired. In this sense, knowing is *instrumental* and experience is *experimental*. Experience is also *meaningful* and pervaded by communication: the mind creates meaning, observes and comes to believe through social processes so that experience is fundamentally *social*. Fundamentally, Dewey distinguishes between *experiencing* and having *an experience*, emphasising that no experience lives and dies of itself. Each individually consummated and whole experience exists within an *experiential continuum* and occurs only when the material experienced reaches fulfilment. *An experience* has it's own inception, rhythmic movement and closure, pervaded by a unique quality, while an aesthetic experience is additionally appreciative, perceiving and usually involves some element of undergoing, which may or may not be painful depending on conditions. The emotional phase of experience binds it as a whole, the intellectual phase labels it with meaning, while the practical phase represents an organism interacting with surrounding events and objects. For an experience to be aesthetic, these phases must be linked and move together toward an inclusive and fulfilling close. So with this description of experiencing as a base, Bakhtin provides a final link to explain how people interpret meaning from their experiences. For Bakhtin, there is an intersubjectivity of meaning: that it "…is always found in the space between expression and understanding, and that this space – the 'inter' separating subjects – is not a limitation but the very condition of meaningful utterance" (Hirschkop, 1999, p5). So, an experience is made sense of by the individual as they relate to others and to history, and not in isolation. Communication of meaning then occurs as dialogue, the consummating relationship leading to which, Bakhtin calls Dialogism: that which is experienced meaningfully is experienced dialogically (Hirschkop, 1999).

## 5    The Framework

The framework summarised next, is founded on knowledge about experience obtained from Csikszentmihalyi, Dewey and Bakhtin and is intended to guide designers in supporting user aesthetic experience. It is composed of a main conceptual tool, to be used for analysing an experience according to dimensions common to all aesthetic experiences. These characteristics may be loosely applied as criteria in assessing the experience or evaluating experiential outcomes of a design process. Additional optional dimensions are later suggested to help the designer achieve a more detailed description of the experience of an individual user. Firstly, experience is *educative and memorable*. As the quality of every experience is affected by previous ones, so it affects the next, altering the experiencing subject and the objective possibilities open to future experiences. An artefact must be able to keep up with and support the evolution of experience. User experience can also be described as *whole, unique and non-reduplicable*. This means that during interaction, there are possibilities (e.g. interface affordances) and limitations available to be explored, acted on and understood. Users have expectations and will generate responses, reflecting, relating, recounting and making meaning of the experience. The *historical* dimension of user experiences derives from the shaping of activities by cultural-historical developments, social factors and past experiences. Additionally, the user may have previous experience of the artefact or similar ones for the same activity. Existing knowledge thus enables the user to join action and expected outcome in their perception, increasing the depth of the experience. Thus, the *meaningful* dimension in aesthetic experiences comes from an active interplay between action and perceiving and undergoing. Users participate actively with objects to build their emotional responses in a situation, whereby understanding and action may be motivated by emotion or by intellectual reasoning of possible actions and their consequences. Therefore, the range of possibilities for involving the *qualitative* dimension of experience are vast. Knowledge held by a user is instrumental to altering this quality, however, the designer and user each have roles to fulfill and bring meaning to the interaction. Particularly, the designer can be conceptualized in two roles: educator and artist. They seek to engage the user in learning the potential for current and future interactions. They must arrange for experiences that engage the activities of the user, judging which are detrimental or conducive to growth, while understanding users as individuals. Additionally, an aesthetic experience with a system is connected with the experience of making; the designer constructs the artefact while at the same time experiencing it and empathizing with the user, much like an artist. The user then perceives the 'artwork', creating their experience of it, inseparable from that of the designer. These collaborative roles mean that an experience is necessarily *instrumental*, *constructed* and *intersubjective*. It is also *contextual*. Specifically, an experience cannot be detached from the socio-cultural context in which it is situated where objects and artifacts within the experience are conceived, used and related to. Indeed, experience always involves a *physical or sensory* dimension where the user interacts with other people and objects around them, relying on physical instruments (hand, eye, ears, etc.) throughout. Similarly, this sensory dimension may be *embodied* as knowledge where, for example, a particular colour, smell or sound has immediate meaning (e.g. darkness may be felt as vulnerability). Finally, an experience has a *spatio-temporally situated* dimension, whereby events occurring in an experience are temporally related to one another. Time may appear to pass quickly or slowly and a space (physical or virtual) may hold relevance, be comfortable, private or public for the user. Metaphors may also represent space that is not real (e.g. 'shelves' in an on-line book store). Perception of time, on the other hand, might help distinguish between events or direct action. Some additional considerations for designers may be used to achieve a more detailed description of user experience and needs. These extra dimensions relate predominantly to the activity of the experience, including whether it is driven by any *goal* or is *autotelic*, if it is related to *work* or *leisure*, if the activity is *novel or familiar* for the user and if it has a *real-world equivalent* (e.g. on-line shopping). These bear important implications in themselves and particularly the goal and work/play distinction; Dewey suggests that work activity is set apart from play in the sense that it is enriched by the perception that it leads to something. However, both work and play may be autotelic, moving away from the traditional focus in HCI on goal driven work activities.

# 6    Conclusions

This paper has attempted to draw attention to frameworks and design tools which help understand the process of design for user experiencing. An initial framework has been proposed as a tool for understanding and evaluating user experience in computer mediated activity aiming for an holistic approach that draws from aesthetic disciplines to re-examine the roles and relationships of designers and users in constructing experience. It is intended to help designers account for the lack of sensory dimension within electronically mediated experiences and the evolution of experience, due to user learning. The framework may offer benefits to designers to prioritise user experience, and interpret its dimensions, including emotional responses, historical, social and cultural context and perceptions of the activity and artefact. It is intended to be flexible and general enough to allow for specification to a design case. Clearly, practical application of the framework is needed to validate it and determine best methods for it's application through methods where user-designer dialogue is central. The various theoretical dimensions of the proposed framework must be validated appropriately through evaluation of actual user experiences. Perhaps the most important contribution and innovation of this work however, in drawing on philosophical, psychological and literary approaches outside the traditional HCI matrix, is the support it lends to the potential for the enrichment of HCI in the direction of aesthetic and humanist concerns.

# Acknowledgement

Thanks to Dr Paul Vickers @ Northumbria University for his helpful comments and suggestions.

# References

Bødker, S. (1991) *Through the Interface*. USA: Erlbaum.

Csikszentmihalyi, M. (1988) The Flow experience and its significance for human psychology. In Csikszentmihalyi, M. & Csikszentmihalyi, I. S. (Eds.). *Optimal experience: psychological studies of flow in consciousness*. USA: Cambridge University press. pp. 15-35.

Dewey, J. (1938). *Experience and Education*. USA: Macmillan.

Dewey, J. (1958). *Art as Experience*. USA: Capricorn Books.

Forlizzi, J. (2002). Towards a Framework of Interaction and Experience As It Relates to Product Design, Retrieved from http://goodgestreet.com/experience/home.html

Forlizzi, J. & Ford, S. (2000). The Building Blocks of Experience: An Early Framework for Interaction Designers. *Proceedings of the DIS00 Conference*.

Hirschkop, K. (1999). *Mikhail Bakhtin: an aesthetic for democracy*. UK: Oxford University Press.

Laurel, B. (1993). *Computers as theatre*. USA: Addison-Wesley.

Margolin, V. (1997). Getting to know the user. *Design Studies*. 18(3): pp.227-236

Muller, M. J. (1995). Ethnocritical Heuristics for HCI Work with Users and Other Stakeholders. Retrieved from http://iris.informatik.gu.se/conference/iris18/iris1844.htm#E21E44

Preece, J., Rogers, Y. & Sharp, H. (2002). *Interaction design: beyond human-computer interaction*. USA: John Wiley & Sons.

Sanders, E. B-N. (2000). From User-Centered to Participatory Design Approaches, In Frascara, J. (Ed.), *Design and the Social Sciences: Making Connections* (pp. 1-8). UK: Taylor & Francis.

Wright, P., Fields, R., Harrison. M. (2000). Analysing Human-Computer Interaction As Distributed Cognition: The Resources Model. *Human Computer Interaction*. 51(1): pp.1-41.

Wright, P.C. & McCarthy, J. .M. (2003) A framework for analysing user experience. In Blythe, M., Monk, A., Wright, P.C. and Overbeeke, C. (Eds.) *Funology: From Usability to user enjoyment*. UK: Kluwer.

# Improving Knowledge Transfer Through Ubiquitous Multimedia Applications

*Stephen Giff*

Microsoft Corporation
One Microsoft Way
Redmond, WA, 98052-6399, USA
sgiff@microsoft.com

## Abstract

The objective of this paper is to demonstrate that effective HCI intervention can lead to better learning in a multimedia training environment. Innovative HCI methodology, leveraged from cognitive psychology (Paas & Van Merriënboer, 1994) was employed to prove that explicit navigation and context are key factors in successful multimedia training and that a poorly designed interface can lead to an elevation of Extraneous Cognitive Load (ECL). A new form of menu design that addresses the problem of elevated ECL, usability, and screen real estate is also proposed.

## 1    The Problem of Instructional Design in Multimedia

The problem of Instructional Design is to ensure an effective transfer of knowledge to a user which enables that knowledge to be applied to real world scenarios. Cognitive Load Theory (CLT) (Sweller, 1994; Kirschner, 2002) posits that human working memory capacity is limited and in order for a positive effect of learning to be achieved a combination of appropriate Intrinsic Cognitive Load, low Extraneous Cognitive Load, and high Germane Cognitive Load are required. Multimedia (Mayer, 2001) offers a partial solution to these fundamental problems. However, an ineffective or poorly designed multimedia interface may in fact lead to an elevation of ECL and consequently a reduction in cognitive processing ability (Feinberg & Murphy, 2000). This decrease in available cognitive processes is caused by such factors as the user needing to continually learn and relearn the interface, poor navigation, and also inadequate context.

## 2    The Problem of HCI in Multimedia

The fundamental problem of HCI is to ensure that humans and computers effectively interact to perform work at a desired level of performance (e.g. Dowell & Long, 1989). Work in this case refers to the effective transfer of knowledge. The goal of HCI therefore should be to ensure that multimedia applications are highly usable and easily learned. These applications should be 'invisible' (ubiquitous) and should not unnecessarily elevate ECL. The applications should also present an explicit and persistent navigation scheme that enables the user to rapidly develop a clear mental representation of that scheme. In addition, contextual cues should be clear and readily available so that users will not waste valuable cognitive processes. Similarly, the Coherence Principle posited by Mayer and his colleagues (Harp & Mayer, 1997; Moreno & Mayer, 2000;

Mayer, 2001) has demonstrated that superfluous content can affect learning due to the competition for cognitive resources caused by the extraneous material. It is likely then that unrelated interface elements (e.g. a persistent Table of Contents (TOC)) will be distracting to users and will therefore also be detrimental to learning. This will be discussed in the following section.

## 3 Table of Contents: Context, Navigation, and ECL

A series of usability studies have taken place on the Microsoft Redmond campus to investigate the importance of context and navigation in multimedia training and learning effects associated with a new style of Table of Contents (TOC). The studies centred on Microsoft® Enterprise Learning Library (MELL), which is a Web-based training suite that offers interactive training for Microsoft software. The training features a series of step by step simulations and demonstrations.

The Table of Contents is an essential element of any online training, ensuring context, major navigation, and structure. An early version of MELL with a persistent Table of Contents is displayed in Figure 1. It can be seen that in this version, users were forced to continually scroll horizontally or resize the TOC to view the lessons and topics. These actions will, by their very nature, lead to an elevation of cognitive processes and physical activity.

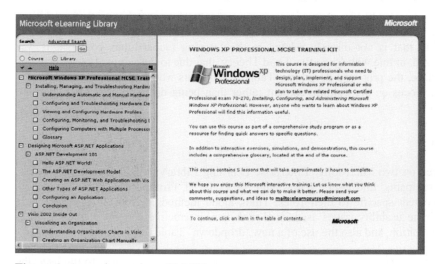

**Figure 1.** An early version of MELL

In order to completely solve this problem, the TOC would either need to take up most of the content screen to avoid horizontal scrolling, or the text would need to be wrapped to such a degree that the lesson/topic titles would appear over several lines - causing the TOC to be extremely long and therefore requiring excessive vertical scrolling. This of course would not solve the potential problem of persistency. Figure 2 shows the proposed solution that addresses all of these problems. Notice from the screenshot that the Table of Contents (TOC) acts in the same way as a dropdown file menu, 'masking' the content in the window behind. In more recent iterations, we have begun implementing a 'transparent' or see through menu, further reducing distraction from the content window.

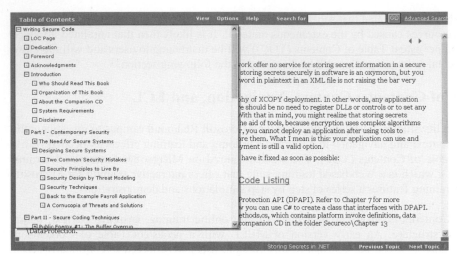

**Figure 2.** A later version of MELL with the proposed dropdown Table of Contents

Would this type of menu lead to a reduction in cognitive load? The extensive work from Mayer and his colleagues would suggest that a persistent Table of Contents would increase Cognitive load as it is content that is not directly related to the learning (superfluous content). It was therefore hypothesized, in line with Cognitive Load Theory, that due to the removal of superfluous stimuli in the interface, the proposed dropdown Table of Contents would indeed reduce ECL by allowing the user to focus on the training content rather than being distracted by this important, but unrelated stimuli.

## 4 Methods

Studies 1&2 focused on two Microsoft® Enterprise Learning Library (MELL) training viewers. Four groups of participants, End Users, Knowledge Workers, IT Professionals, and Developers were tested with content specific to their roles. Think-aloud methods (Protocol Analysis) were employed to determine usability issues associated with new interactive content elements, intra-topic and global navigation, and also the use of a new 'dropdown' Table of Contents.

Study 3 examined the necessity of displaying topic level context ($4^{th}$ level topics) to the user, the positioning of these topics, and the importance of explicit navigational elements. Two viewers were compared to determine whether differences in the interface – navigation, context, and Table of Contents - affect transfer of knowledge. Innovative methodology, leveraged from cognitive psychology, was tested whereby participants (IT Professionals and Developers) were presented with content in one of two viewers. After each lesson, multiple choice questions tested the recall of the content just learned. At the end of each session subjective questions relating to the interface and perceived cognitive load (mental effort) were asked. To test transfer of knowledge, questions were asked that required the participant to apply the 'type' of knowledge they had just learned to a novel situation.

Study 4 again utilized the new methodology mentioned above to investigate the effectiveness of the dropdown TOC on the transfer of knowledge by directly comparing this new design with a persistent TOC. In addition, effects associated with content were also tested (not discussed here).

# 5 Results

The results indicate that the interface elements can indeed affect ECL. A poorly designed interface can lead to a fall in performance and elevate perceived cognitive load in a training environment.

## 5.1 The Importance of Context and Navigation

The importance of context was investigated by presenting content to participants in one of two viewers. In the first viewer (Viewer 1), level 4 topics were presented in a Table of Contents. In the second viewer (Viewer 2), level 4 topics were displayed in a dropdown topic menu within the content window (see Figure 3). Participants did not clearly understand the purpose of the additional dropdown topic menu which consequently led to the menu not being utilized to navigate between topics. The context provided by this type of navigation was ineffective as it added an element of complexity to the overall navigation scheme.

Similarly, the importance of major navigational elements was tested by comparing the navigation between topics, utilizing either a single major navigational element (Viewer 1 with 'previous' and 'next' buttons only) or multiple navigational schemes (Viewer 2 with conflicting topic/lesson 'previous' and 'next' buttons). It was found that the use of multiple navigation schemes caused several usability problems and introduced ambiguity into the interface.

Participants performed better and reported a lower rate of cognitive load for Viewer 1, which included level 4 topics in the Table of Contents and did not utilize conflicting navigation.

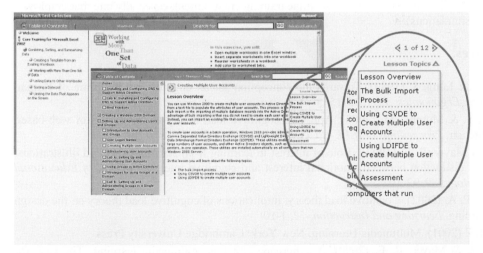

**Figure 3.** Viewers 1 and 2; in Viewer 2 topics are displayed in a dropdown menu

## 5.2 Table of Contents

The proposed dropdown Table of Contents (TOC) was effectively utilized to complete tasks. In addition, behavioural and verbatim evidence indicate that the proposed design was favoured. There were no significant differences between the groups in the number of transfer and recall questions answered. Therefore, no learning effects were found in the comparison of the persistent, and the new dropdown TOC. It is important to note however that in MELL, most of the actual training (the simulations) occurs within a separate pop up browser. It is expected that in future versions, simulations will be integrated into the viewer itself; a method employed by many online training organizations. It is predicted that when this occurs, the proposed TOC will address the problem of elevated Extraneous Cognitive Load, leading to better learning of the content. Further studies are planned.

# 6    Conclusion

These findings show that effective HCI intervention can lead to better learning in a Multimedia Training environment. This is achieved by ensuring a transparent or ubiquitous interface that offers an appropriate level of context, clear navigation, and a reduction in the superfluous content and interface elements that may be detrimental to learning. New methodology has been proposed, aimed at testing the learning effects associated with these problems.

It is concluded that:
- Navigational elements should be persistent, explicit, and clearly associated with the Table of Contents
- Topic level items should be presented as a parent/child relationship in the TOC
- Conflicting local (intra topic) navigation causes confusion and should be avoided – global navigation should take the user to topic level
- The proposed dropdown TOC can effectively reduce problems associated with elevated ECL. In addition, the proposed design addresses the issue of limited screen real estate (a valuable commodity in an online training environment, especially one that employs simulations).

# References

Dowell, J. & Long, J. (1989). Towards the conception for an engineering discipline of Human Factors. *Ergonomics, 32,* 1513-1535

Feinberg, S. & Murphy, M. (2000). Applying cognitive load theory to the design of web-based instruction. *Proceedings of IEEE-IPCC*. 353-360

Harp, S. & Mayer, R. E. (1997). Role of interest in learning from scientific text and illustrations: On the distinction between emotional interest and cognitive interest. *Journal of Educational Psychology, 89*; 92-102

Kirschner, P. A. (2002). Cognitive load theory: implications of cognitive load theory on the design of learning. *Learning and Instruction, 12,* 1-10

Mayer, R. E (2001). Multimedia Learning. New York: Cambridge University Press

Moreno, R. & Mayer, R. E. (2000). A coherence effect in multimedia learning: The case for minimizing irrelevant sounds in the design of multimedia instructional messages. *Journal of Educational Psychology, 92,* 117-125

Paas, G. W. C. & Van Merriënboer, J. G. (1994). Variability of worked examples and transfer of geometrical problem-solving skills: A cognitive-load approach. *Journal of Educational Psychology, 86,* 122-133

Sweller, J. (1994). Cognitive load theory, learning difficulty and instructional design. *Learning and Instruction, 4, 295*-312

# The New Demographic: Transforming the HCI Curriculum

*William M. Gribbons*

Bentley College
Waltham, MA 02452
wgribbons@bentley.edu

## Abstract

This paper examines the ever-changing demographic profile found in the technology marketplace and the implications of these changes for the education of today's HCI professional. The paper concludes that the vast majority of accommodations made for one profile of users will benefit the entire user community.

## 1.    Introduction

As any human-computer interaction (HCI) professional can tell you, the qualities that characterize a usable product are an ever-shifting target. While one group of users might find a product highly usable and valuable, a different group of users with different needs or abilities might very well reject it. Gone are the days when we designed systems exclusively for users that are technologically capable and highly motivated. Most significant for the topic of this paper is the fact that the field of HCI emerged during a period when this demographic profile dominated the marketplace. Quite appropriately, the "model" adopted in HCI education during this same time mirrored the needs and requirements of this user profile and the complex systems they used.

Times have changed and so too must our concept of HCI. The field is no longer focused exclusively on complex systems. Increasingly, the products include consumer electronics, web-enabled services, e-learning products, public kiosks, and factory automation controls. The users of this ever-expanding array of technology products are no longer a homogeneous collection of tolerant, adaptive users. In its place we find a rapidly changing demographic characterized by a general lack of tolerance, wide-ranging technical and literacy skills, and an unwillingness or inability to adapt to incompatible systems. Who are these new users? One subset are the users of the past who now find themselves overwhelmed by a never-ending stream of new technologies. The "new user" is functionally illiterate, a class of users traditionally marginalized by our industry as too expensive to accommodate and support. The "new user" is the international user who now makes up 65% of the revenues for the largest software producers in this country. The "new user" is the rapidly growing aged and aging population. And finally, the "new user" are children, a group that will dominate the marketplace for the next four decades. In the final analysis, it is the marketplace that defines product acceptance, not the HCI professional with some vague notion of what is right or the engineer driven by what is easiest or possible.

## 2.    The Existing User

All technology carries a learning and use burden, a burden that must be allocated to either the system or the user. Traditionally, the user has assumed the largest portion of the burden related to using technology and they did so with little or no complaining. Some of this tolerance was the product of our human ability to adapt as long as resources were available. Some of this tolerance was a token of our gratitude for the value offered by the new technology. It seems users were willing to endure the burden inherent in the new technology if it improved performance elsewhere in their lives.

Unfortunately, the model that drove the initial development of our industry will not sustain growth in the years ahead (Rockwell, 1999; Boehm and Sullivan, 2000, Zajicek, 2001). The ability of users to continually adapt assumes an adequate level of human resource. Consider some data the author collected recently for a client. In 1989, the average user of this client's product used 3-4 software packages to meet the requirements of their job. A decade later, this same user, on average, is now using 11-12 software packages. Consider also that today's software is updated far more frequently and offers considerably greater functionality. Consider all of these changes and it is easy to imagine that we are rapidly pushing users to the limits of their ability and willingness to learn new technology.

Today's HCI professional must target and mine the untapped usability potential in existing products. By making existing products more usable we free up human resources to enable existing users to purchase new technologies. Given these conditions, the HCI curriculum must redouble its efforts to examine means of reducing the workload imposed by technology and aligning the technology more closely with highly valued user goals.

# 3.     Lower Literacy Groups

Traditionally, lower literacy groups were viewed as a market with little or no value for technology products. This position was not grounded in market size but in the resources required to achieve the ease-of-use required by this group. A 1992 study of literacy among adults in the United States reported that 43% of the labor force scored at the two lowest levels of literacy proficiency (as measured by prose, document, and quantitative proficiency), while only one out of four labor force participants scored at the highest levels of proficiency (Sum, 1999, XIV). At the same time, literacy requirements in the workplace are constantly increasing due to the rapid infusion of computer technologies in all aspects of work. In 1997 half of the employed adults in the United States used a computer in their work (US Census Bureau, 1999). As we seek new markets for technology, we are now viewing this once ignored profile as an opportunity for market expansion.

The functionally illiterate profile presents many challenges. Most significantly among these challenges is the illiterate's inability to assume the burden for learning the system. The functionally illiterate often lack the cognitive flexibility to adapt to a system that does not align with their needs and abilities. The challenges posed by this profile are well documented:
- A comprehension deficit while processing written prose
- Limited number of text processing strategies
- An inability to distinguish levels of importance in text
- A limited short-term-memory
- A slower verbal coding speed (Smiley et al., 1977; Jonassen, 1985; Potter 1991; Perfetti and Lesgold, 1977)

While there is little research in this area in the HCI literature, guidance can be found in the instructional design literature suggesting how to support these shortcomings. Given the size and value of the illiterate profile, academic programs must expand their curricula to include instructional design theory aligned with the needs of less literate populations.

# 4.    International Users

A few years ago an important milestone was reached in the United States where the one hundred leading software producers derived 65% of their revenues from international sales. Unfortunately, our early efforts to accommodate cultural differences in system design have been less-than-adequate. While intercultural research notes that as much as 90% of the interaction with another culture is non-verbal, most development organizations limit their localization efforts to translation. Fortunately, the emerging field of cultural psychology provides valuable insight into the deeper psychological issues that affect interaction design for international users. (Matsumoto, 2000).

In our early localization efforts we can again witness the remarkable resilience of humans. Faced with little or no choice in the global marketplace, the international user has assumed the burden for learning and using a product that mirrored U.S. domestic behaviors and requirements. The author would caution our industry of the dangers of complacency and misinterpreting this early experience as a global willingness to accept a U.S. model. Instead, this early acceptance was shaped by the lack of choice, a factor that is likely to diminish as localization savvy competitors tailor their products to local needs. The field of intercultural communication identifies many areas where we must accommodate the international user. Most significant are the following:

- Reading patterns
- Iconic symbolism
- Design tradition
- Concept of time
- Concept of information category width
- Communication patterns  (del Galdo and Nielsen, 1996 and  Matsumoto, 2000).

Academic programs in HCI must move beyond a foundation grounded in a western perspective on cognitive psychology and embrace the emerging field of cultural psychology. This realignment is critical if we are to remain competitive in the rapidly expanding global marketplace.

# 5.    The Aging Workforce

The dramatic growth of the aging population suggests that older people will be one of the fastest growing user profiles in the decades ahead (Brouwer-Janse et al., 1997). Czaja and Lee (2002) report that in the year 2000 "persons 65 years or older represent 13% of the population, and it is estimated that people in this age group will represent 20% of the population by 2030" (414). These statistics would not be so troubling if it were not for the fact that young developers functioning at the peak of their cognitive and perceptual abilities typically design most computer technologies. If we are to capitalize on the pool of aged users, we must align our design practices with the requirements of the aging population.

Fortunately there is ample research to guide our efforts to accommodate the aging workforce. Cognitively, the older worker experiences an ever decreasing working memory resource; a deficit that affects the ability to formulate conceptual models of a system and the ability to navigate virtual environments. Systems that place a heavy load on this declining resource will be subject to usability problems. Product designs must support and extend this dwindling resource through mnemonics, cues, performance support, and visualization strategies (Brouwer-Janse et al., 1997; Czaja and Lee, 2002; Zajicek, 2001). Physiologically, the older user experiences problems associated with a loss of visual acuity, a decreased ability to detect color variations, a loss of contrast sensitivity, and a heightened sensitivity to surface glare. The aging user also experiences a decline of fine motor controls, which will cause grave problems for input devices such as the mouse or interacting with small targets (Czaja and Lee 2002; Zajicek, 2001). The decline in these abilities is not fixed and varies significantly across aging individuals. This "dynamic diversity" poses additional design challenges as we attempt to accommodate the varied and constantly changing needs of this population.

Failure to accommodate these differences will result in insurmountable usability problems for the older user. Clearly, academic programs in HCI must now infuse the study of gerontology and the emerging field of gerontechnology in the design of their curricula.

# 6.    Children

For anyone who has ever watched a seven year old navigate the web, play a hand-held video game, interact with computer-assisted learning aids or download software off the internet, you realize the users we will face in the years ahead bring radically different abilities and expectations to the table. They are generally unwilling to modify their behaviors, they work in social groups, and they live in a world of choice where they move quickly to a new product that conforms more closely to their needs and abilities.

A growing body of research is examining young users and their interaction with technology (Brouwer-Janse et al., 1997). That combined with research from the field of developmental psychology offers valuable guidance for those who design technology for children. This research suggests that children "learn" a system through their interaction with the system. They expect a multi-sensory interaction that moves beyond our current reliance on verbal and graphical controls. Finally, children are far less likely to rely on external support systems such as documentation and will learn most effectively through interaction and imbedded support.

Children also force us to make accommodations in our requirements gathering and testing methodologies. Due to their varied abilities or willingness to articulate their needs, HCI professionals must adapt traditional methods and adopt a more assertive role in their interactions with children while still relying heavily on keen observation of demonstrated behaviors. The social and contextual needs of children require HCI professionals to move requirements gathering and testing outside of the traditional lab and into the classroom, living room or other appropriate social settings (Brouwer-Janse et al., 1997; Hanna, Risden, and Alexander, 1997; and Druin, 1999). Once again, we see cause for academic programs expanding their curricula beyond the traditional focus on adult learning psychology and considering the developmental needs of children.

# 7.    Conclusion

Gone are the days of homogeneous populations of capable, tolerant, adaptive users. In their place we find a changing demographic composed of existing users with strained resources, functionally illiterate users, international users, older users, and children. A dramatic shift in the demographic requires an equally dramatic change in the way we educate HCI professionals.

The current debate in HCI education centers on whether these profiles should be addressed through stand alone modules focused on the interaction needs of the group or whether the theories and practices that address these specialized needs are woven seamlessly into the larger discussion of human-computer interaction. I would argue for the latter if we are to avoid the trap of pitting one group against the other and accept the tenet that all users ultimately benefit from the majority of the accommodations we make for individual profiles. Regardless of the approach we adopt, one thing is clear and that is the contemporary HCI curriculum must now expand to include the instructional design and support needs of lower literacy groups, embrace the emerging field of cultural psychology, include the theory of child growth and development, and consider the decline of abilities associated with aging. These accommodations are critical if we are to fully realize the social and economic benefits of newly emerging technologies.

## References

Boehm, B. and Sullivan, K. (2000). Software Economics: A Road Map. *Proceedings of the Conference on the Future of Software Engineering*. (pp. 319-343). Limerick, Ireland.

Brouwer-Janse, M., Suri, J., Yawitz, M., Vries, G., Fozart, J., and Coleman, R. (1997). User Interfaces for Young and Old. *Interactions*. 1 (2), 34 16.

Czaja, S. and Lee, C. (2003). Designing Computer Systems for Older Adults. In J. Jacko and A. Sears (Eds.) *The Human-Computer Interaction Handbook: Fundamentals, Evolving Technologies, and Emerging Applications*. (pp. 413-427). Hillsdale, NJ:Lawrence Erlbaum Associates.

del Galdo. E. and Nielsen, J. (1996) *International User Interfaces*. New York: John Wiley & Sons.

Druin, A. (1999). Cooperative Inquiry: Developing New Technologies for Children with Children. In *Proceedings of the CHI 99 Conference on Human Factors in Computing Systems: The CHI is The Limit*. (pp. 592 – 599). New York: ACM Press.

Hanna, L. Risden, K. and Alexander, K. (1997) Guidelines for Usability Testing With Children. *Interactions*. 4 (5), 9-14.

Jonassen, D. (1985). *The Technology of Text*. Englewood Cliffs, NJ: Educational Technology Publications.

Matsumoto, D. (2000). *Culture and Psychology*. Stanford, CT: Wadsworth.

Perfetti, C. and Lesgold, A. (1977). Discourse Comprehension of Individual Differences. In A. Just and P. Carpenter (Eds.), *Cognitive Processes in Comprehension*. (pp. 141-183) Hillsdale, NJ: Lawrence Erbaum Associates.

Potter, F. (1991). The Use Made of the Succeeding Linguistic Context by Good and Poor Readers. In J. Chapman (Ed.) The Reader and Text (pp. 106-114). London: Hunemann Educational Books Ltd.

Rockwell. C. (1999). Customer Connection Creates a Winning Product: Building Success with Contextual Techniques. *Interactions*. 6 (1) 50-57.

Smiley, S., Oakley, D., Worthen, D., Campione, J., and Brown, A. (1977). Recall of Thematically Relevant Material by Adolescent Good and Poor Readers as a Function of Written and Oral Presentation. *Technical Report #23*. Center for the Study of Reading: University of Illinois.

Sum, A (1999). *Literacy in the Workforce: Results from the Adult Literacy Survey*. Washington, DC: U.S. Department of Labor.

US Census Bureau. (1999). *Computer Use in the United States*. Washington DC.

Zajicek, M. (2001). Interface Design for Older Adults. In *Proceedings of the 2001 EC/NSF Workshop on Universal Accessibility of Ubiquitous Computing: Providing for the Ederly*. (pp. 60-65). Alcácer do Sal, Portugal

# A Tentative Model for Procuring Usable Systems

*Stefan Holmlid and Henrik Artman*

Interaction and Presentation Lab, Royal Institute of Technology
100 44 STOCKHOLM, Sweden
{holmlid, artman}@nada.kth.se

## Abstract

In this paper a proposal for a system development model for procurers is presented. The model is aimed at increasing the usability of supplied systems. It is meant to be a supportive tool for any procurer, and is developed from experiences and documented problems in system development. The model is based on three field studies of procurement organizations and their role in the system development process. The model comprises planning, communication, monitoring and evaluation as parallel activities by procurers within any system development project.

## 1 Introduction

After a case description in a book about system development skill, the conclusion comes down to one problem:

> Through the ambition that everything should be specified exactly, one accomplished an extensive, and probably correct, foundation. But, by *only* doing an exact reproduction, both on high and low levels of abstraction, this thorough work created no understanding by the customer. [Hoberg, 1998, p. 34, our translation]

It is indeed a problem that suppliers and procurers do not share a common and shared view of the development process. Moreover, the quote represent an asymmetry often assumed in system development, that the procurer should understand, or more specifically, adapt to the supplier. We think that this assumed asymmetry is the problem, rather than the symptom. Therefore, we believe it is essential to understand the *procurer's* role in system development, in particular in relation to usability issues.

## 2 Theoretical background

Except for in a few discussions (Grudin, 1995; Winkler & Buie, 1995), the issue of procurement has attracted little attention of the Human Computer Interaction (HCI) community. There are virtually no empirical studies of how procurers reason about usability and user involvement as requirements for a system. Some aspects can be found in (Keil & Carmel, 1995; Ives & Olson, 1984; Forsgren, 1996), who emphasize the importance of user involvement and usability. Reviewing the general literature of HCI, one discovers that focus is on the production of the user interface in product and in house development (Helander, Landauer, & Prabhu, 1997; Preece, Rogers, Sharp & Benyon, 1996; Shneiderman, 1998; Dix, Finlay, Abowd & Beale, 1998; Cooper, 1999). System development methods and models for user-centred system development have also mainly focused on the developmental side of the contract (see Newman & Lamming, 1995;

Greenbaum & Kyng, 1991; Constantine & Lockwood, 1999). The implicit reasoning seems to be that it is the contractor's responsibility to ask the right questions and produce the right design (see Pohl, 1993; Dorfman & Thayer, 1990; Macaulay, 1996). This means that procurers have only a passive voice in the development of usable systems.

The procurer, to us, is the person or the group of persons who have a sketchy idea of the system, who are responsible for its fulfilment, who have the power to distribute resources during the completion of the project, and who have the authority to sign off payments and contracts. In our view, high procurement competence must include an awareness of usability issues as well as organisational goals. It must also include the ability to plan, communicate, monitor, and evaluate the process of reaching these goals.

# 3    Case review

In an open exploratory manner we have been working with cases in order to understand procurement as it occurs in real-life procurement situations. The three cases span a range of procurement situations; one small company procuring a web-based system from a small company, a large governmental body procuring a administrative system from the internal development department, and a large bank making a proposal for tender for a content management system.

## 3.1    Contracting negotiations and emergent requirements

In one of our case studies in contract development of an Internet application for working environment evaluation (Artman, 2002) we elucidated several conflicts between the procurer and the contractor. First we found that the procurer project leader and the contractor project leader had different views on usability and user involvement. The procurer project leader was more process-oriented, while the contractor viewed usability solely as product properties (see also Holmlid, 2002). They also had different views on how detailed the project was on the outset. The procurer found the project to be general and open-ended regarding its implementation, while the contractor found it to be very closed and detailed. During the design phase of the project several controversies became apparent. Iterations of the design sketches were not as rapid as had been anticipated, and did not take into account the emergent requirements from the formative evaluations. The procurer project leader did, after some iterations, prototype detailing user requirements and a suggestion for interaction design. Much of the controversies can be ascribed to different ideas of the motive for iterations. The procurer project leader expected rough sketches as suggestions and evaluative material, while the contractor designers saw their sketches as more firm suggestions. The contractor project leader, working as a mediator, did not want the designer to come in direct contact with users, which of course made the communication even more problematic (see Keil & Carmel, 1995). The project was not completed within time and budget. Interestingly, the contractor project leader today works for the procurer and is very focused on being platform and supplier independent. The case points out the importance of agreed upon definitions, process-oriented requirements, synchronised motives of process and product as well as an active dialogue bridging asymmetries.

## 3.2    Power and management

Our case study with the Swedish Tax Government (RSV) was a retrospective interview study of a project that was meant to be a best-practice project when it comes to user centred design and usability. The study elucidates several interesting problems. At the start of the project strong emphasis was put on usability and user-centred design, but gradually, the results of those efforts were not recognized as much as expected. Our analysis suggest that usability requirements was not recognized as much because of two major factors; the steer group did not have the competence and interest to evaluate the results, and RSV issued new directives for platform use. Further analyses suggest that the steer group mainly reacted to obvious project problems rather than taking a strategic perspective and steer the project in the directions of the initial directives. The new directives forced the developers to investigate new technologies and made some usability requirements (e.g. short commands) impossible. This in turn made power conflicts between the usability professionals and the programmers within the project apparent. As a consequence, one of the usability professionals skipped the project, as he could not see that it would fulfil the usability goals. Instead of being a best-practice project of usability, where user requirements and needs directed technology, the project took a traditional approach where technology directed user adaptations. Later investigations revealed that most of the requirements within the initial directive could have been taken care of, although not within time and budget. The case interviews suggest that the procurer steer group could have acted more proactively and strongly with continuously analysing requirements and the technology supporting these goals. RSV is now revising its procurement organisation and will take a more strategic account for usability.

## 3.3 Usability activities and conceptual models of work

Together with a large Swedish bank we followed the procurement of a Content Management system (see Markensten, this volume). The department handling the procurement had no formal education in usability or user-centred design, working more from out of common sense. Our presence, with a focus on usability activities, increased discussion on such issues. Markensten's analysis suggests that the procurers were influenced more by rational and technology centred models of the work process than the actual usage presented by the users. As an effect of that, they had problems listening to the descriptions of work procedures articulated by the users. As Markensten shows the interview became an arena for persuading the user how rational and consequent a certain work procedure would become. This in turn made the articulation of adequate user requirements problematic. The case points out how important it is for those articulating procurement and user requirements to have a conscious idea of usability and user-centred activities when collecting information to articulate user requirements.

## 4 A tentative procurement model

The framework model consists of four general activities throughout a system development process. Perspectives to be taken into account while dealing with these activities are roles, process issues, power-relationship and goals (see table 1).

**Table 1.** The structure of the proposed model

| Activity / Perspective | Planning/ Organization | Communication | Monitoring | Evaluation/ Assessment |
|---|---|---|---|---|
| Goals | Fit between business goals and usability goals | Requirements formulation | Secure the projects development to planned goals | Fulfilment of technology, usability and business goals |
| Process | Stakeholder involvement Assess motives | Prototyping Conceptual modelling | Assess progression Usability tests | Evaluation |
| Role | Analyst Steer-group | Designer | Procurer project leader Steer-group | Evaluator |
| Power-relationship | Conflicting goals | Asymmetrical motives and knowledge | Usability activities vs. technology implementation | Resistance to change Conflicts of responsibility |

During *planning and organisation* procurers must become aware of the fact that users are the ones who will contribute the most to meet the business goals, as well as acknowledge how this should be done. The problem a procurer must take care of during planning is that most system development processes is supplier focused, where phases of the projects are defined more on the terms of the production, than the needs and phases of the procurer organisation. They must also become aware of the asymmetry of knowledge of and differing motives for the procurer and contractor to participate. This is essential to be able to monitor and handle upcoming misunderstandings. Costs are one of the main aspects that procurers acknowledge, but as we know from other studies few projects finish on time and budget. Procurers should be more focused on quality aspects, as these later will appear in evaluations. A common problem is that procurers order specific technology often based on shallow knowledge and analyses. Within the model there is also room for perspectives and definitions on technology development, as means for organisational change, competence development or as plain information processes.

The *communication process* can be divided into several parts, the communication between the procurer project leader and the users, the communication between procurer and contractor, etc. In the first relation we have seen how project leaders try to impose certain information structures and requirements on the users. In the second, we have found that the procurer should not assimilate and adopt the jargon of the developers / contractors, as labels might hide the mismatch for each actual understanding. The goals of the procurer, users and the developer might be only loosely fit. The problem facing the procurers deals here with that of false belief of shared motives and references to concepts. The procurer must take charge of articulation of motives, requirements and definitions of concepts. That is, the procurer organisation must take a proactive role in respect to the supplier in order to balance the asymmetries of jargon and systematic knowledge, as well as being one step ahead of the project.

*Monitoring* is what we would like to think of as the actual change project management. In our studies we have seen that the procurers is marginalized as soon as the development project starts. The project becomes a black box for the procurer, and only appears full-fledged when there are problems. This marginalizes the procurer to reactive thinking and action rather than being proactive and the one who leads and decides. First of all, goals, evaluation measures and criteria must be defined early. Secondly, the procurer must be secure in that it is her appropriate knowledge of the work that is in the forefront of the criteria. Emergent requirements and partly dependent projects must be monitored and evaluated. Project processes of usability must be monitored, for example by probing project member of the users needs and work processes.

*Evaluation and assessment* should be evident, but is not always. We see the problems and solutions of evaluation in relation to the above. Especially, the focus on the relation between usability goals and organisational goals seem to be important. Still many organisational goals may not be evident at the completion of the project. Usability is sometimes seen as self-contained rather than connected to organisational goals. Another issue concerning evaluations is when and by whom they should be done. An independent evaluator can take the criteria and evaluate without any pre-understanding of project progression. On the other hand, independent evaluators may not have domain knowledge. Evaluations and criteria for acceptance must be set early in the project, and these must be monitored and secured during planning and contracting.

To summarize, we believe that the procurer must formulate the "what and how" of users work situation before actual procurement. What activities and tasks should the user perform and in what way are these activities supporting the organisation's need for change? Secondly, how should the user perform these activities? These two questions can then direct the contractor – what must be done and how it should be done in order to conform to the goals of the procurer organisation.

# 5 Future work

We will continue to develop the details of the model, especially with emphasis on how procurers can streamline organisational and usability goals as well as how procurers can monitor actual development projects. We will do this through case studies of actual procurement and through more active participation in development projects together with different stakeholders. We also organise workshops with usability professionals and researchers. Another line of future work is to organise reflection seminars together with procurement practitioners were we present and discuss research and actual development projects. We also expect that education and focus will change accordingly in order to meet the demands of future usability procurement.

# 6 Acknowledgements

We are grateful for the energy and enthusiasm Emma Borgström, Erik Markensten and Jon Svensson has put into the work with the cases. We wish to extend our gratitude to the participating companies and organizations. This research project is supported by a research grant from Vinnova, the Swedish Agency for Innovation Systems.

# References

Artman, H. (2002). Procurer Usability Requirements: Negotiations in contract development. Proceedings of NORDICHI 02. pp 61-70.

Constantine, L.L., Lockwood, L.A.D. (1999) *Software for Use*. New York: Addison-Wesley.

Cooper, A. (1999). *The Inmates are Running the Asylum*. New York: SAMS.

Dix, A., Finlay, J., Abowd, G., Beale, R. (1997). *Human Computer Interaction*. Harlow: Prentice Hall.

Dorfman, M., Thayer, R., H. (1993). *Standards, Guidelines and Examples of System and Software Requirements Engineering*. International Thompson Publishing.

Forsgren, P. (1996). *Management of Industrial IT Procurement*, Doctoral Thesis. Industrial Control Systems, Royal Institute of Technology, Stockholm, Sweden.

Greenbaum, J. & Kyng M. (eds.), (1991), *Design at Work: Cooperative Design of Computer Systems*. Hillsdale, NJ: Lawrence Erlbaum Associates.

Grudin, J., 1991. *The Development of Interactive Systems: Bridging the Gaps Between Developers and Users. IEEE Computer*, 24, 4, 59 –69.

Helander, M., Landauer, T., Prabhu, P. (1997) *Handbook of Human Computer Interaction*. Amsterdam: Elsevier Science.

Hoberg, C. (1998). Precision och improvisation [Precision and improvisation]. Stockholm: Combitech Software and Dialoger. [in Swedish].

Holmlid, S. (2002). *Adapting users: Towards a theory of use quality*. Linköping Studies in Science and Technology, Diss. No. 765. Linköpings universitet, Sweden.

Keil, M. and Carmel, E. (1995) Customer Developer Links in Software Development. *Communications of the ACM*, Vol. 38, Num. 5, May, pp.33 –44.

Macaulay, L., A. (1996). *Requirements Engineering*. London: Springer.

Newman, W., Lamming, M (1995). *Interactive Systems Design*. Harlow:Addison-Wesley.

Pohl, K. (1993). *The Three Dimensions of Requirements Engineering*. In Rolland, C., Bodart, F. And Cauvet, C (eds.) Proc. CAISE'93. Paris: Springer, 175 –292.

Preece, J., Rogers, Y., Sharp, H., Benyon, D., (1995). *Human-Computer Interaction*. Addison-Wesley.

Shneiderman, B. (1998). *Designing the User Interface* (3ed). Addison-Wesley.

Winkler, I., Buie, E. (1995) HCI Challenges in Government Contracting. *ACM SIGCHI Bulletin, v.27 n.4 p.35-37*.

# Conceptual Modeling for Interaction Design

*Qingyi Hua, Hui Wang, and Matthias Hemmje*

Fraunhofer – IPSI
Dolivostr. 15, D-64293 Darmstadt, Germany
{hua, hwang, hemmje}@ipsi.fhg.de

## Abstract

A usable interactive system provides its users with presentation and manipulation of useful concepts for solving their problems at hand without becoming bogged down in accidental features of user interface. It implicates that interaction design is directed by the acquisition and representation of knowledge about the context of use in a way that can be traced back to the users' problem-solving activity. In this paper we focus on conceptualization of that activity. The conceptualization is characterized by a set of ontological terms that capture a continuum between user tasks and problem domain in a declarative way, and by a framework of conceptual models that describe a system on a very abstract level without being limited to particular design models.

## 1    Introduction

In HCI a task is defined as an activity performed to achieve a goal (Welie, Veer & Elens, 1998). However, there is at least one other more or less parallel activity during the performance, that is, the mental representation and processing of the problem to be solved at hand. Based on the beliefs on the problem domain, the problem-solving activity generates the goal and intentions performing the task in terms of Norman's action theory (Norman, 1986). Users could distract their mental effort from their working situations to the manipulation of the user interface if a system fails to provide the relevant concepts to the problem-solving context. As a result, usability relies on that the system provides appropriate content at appropriate time during the process of problem solving. The problem-solving activity relies on the knowledge structures supporting understanding (schemas) and the mechanisms used to organize that knowledge (plans) (see (Robilland, 1999) for a cognitive discuss). In AI terms, a problem solver requires two types of knowledge (Chandrasekaran, Josephson, & Benjamins, 1998): domain factual knowledge (objects, events, relations, processes, etc) and problem-solving knowledge about how to achieve various goals, such as problem-solving methods. However, mental processing and representation differs from a system description in many ways. For example, human can have semantically much richer, and more complex schemas than the system, whereas the amount of knowledge that can be handled by a human mind at any given time is limited to 5±2 chunks according to psychologists. As a result, it is fundamental to develop a shared understanding between the users and designers according to the context of use, such as the intended users, their tasks and environments.

Contextual development has been recognized to be crucial for meeting the demands of user-centred systems design (Stary, 2000). Two categories of approaches can be found in HCI: task-based and context-oriented. Task-based approaches, e.g. (Welie, et al., 1998), capture only the hierarchical properties of tasks without memory reminding why the tasks are executed or decomposed, whereas context-oriented ones, e.g. (Maguire, 2001), describe working situations without mental abstraction. However, users guide their tasks to be done by making plans. The properties of plans are anticipation and simplification (Robilland, 1999): a heuristic nature without a detail analysis of the situation, optimal use of memory by keeping only critical properties of objects and events, and higher control level without details of the activity being processed.

In this paper, we present our initial results on conceptualization of knowledge about the context of use. The purpose of the conceptualization is to develop a shared understanding and representation of that knowledge between stakeholders. Section 2 introduces a set of ontological terms that capture the conceptualization and that act as meta-knowledge for the knowledge acquisition from a user's perspective. Section 3 describes the conceptual models and demonstrates a case study. The description of the system is done with the conceptual models on a very abstract level. On this level, we talk of mental states instead of system states, of intentions instead of task decompositions. Consequently, this level allows us to represent knowledge about the context of use in a way that can be traced back to the users' problem-solving ability. On the other hand, it makes possible provide opportunities and constraints for interaction and system design without being limited to a special set of design models. Finally the conclusions are presented in section 4.

## 2 Ontological assumptions of problem-solving activity

The notion of ontology represents knowledge shared and reused in some domain of interest. The essence of the notion is in (Gruber, 1993): an ontology is an explicit specification of a conceptualization. A conceptualization makes some assumptions about concepts and their relationships in the domain to be modelled. In AI and software engineering, for instance, objects, events, processes, and relations are general terms for modelling domain knowledge. Recently ontological assumptions have been also proposed for modelling problem-solving methods in AI. For instance, (Chandrasekaran, et al., 1998) defines an ontology of problem solver and an ontology of methods for establishing assumptions of the knowledge structures used by a problem solver and of the control mechanisms to use those structures.

However, the difference between the human mind and the system lies in the capability of knowledge processing and representation. To make the system usable, the phenomenon of knowledge chunking has to be taken into account. Chunks are general and do not refer to the information content of knowledge. This implicates that they are a measure of the unrelated knowledge that can be processed naturally (Robilland, 1999). The phenomenon illustrates that the problem-solving knowledge used by the human mind is organized on a more abstract level than the level of domain knowledge. For the mediation of the two different levels, the system has to organize contextual information about task-performing states, or domain states to meet the users' needs. In the remainder of this section, we define a set of primitive terms in order to match the elicitation and representation of the needs.

*Term 2.1. Domain object/action/state.* We apply the traditional OOA definition for the three terms in general. In particular, we define that a (problem) domain state is a set of values of state variables representing objects in the problem domain.

*Term 2.2. Task object.* A task object is supposed to match a mental representation describing a problem domain state that a user believes or pursues. Task objects can have the same abstract mechanism and form as domain objects. For example, a task object can be specialized as an entity, relationship, or event depending on whether it is autonomous, subordinate, or instantaneous, respectively. Task objects are essential for the problem-solving activity because they are a way of representing the user's knowledge about the problem domain from a view of task. In other words, they represent the content of tasks to be undertaken. In the domain of a hotel, for example, domain objects can be room, guest list and so on, whereas a guest may have particular concepts about the domain states, such as availability, preference and so on, when she wants to make a reservation. It

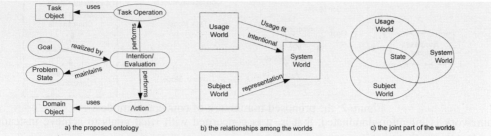

| a) the proposed ontology | b) the relationships among the worlds | c) the joint part of the worlds |

Figure 1: the proposed ontology and its relationship with the three worlds

is also desirable to anticipate possibilities, to determine the current situation, and to remember working history by use of task objects. As a result, task objects provide information about how to represent knowledge about problem domain on the mental level.

**Term 2.3. Task operation.** A task operation is supposed to match a mental operation representing an input-output relation over task objects. Task operation applications determine state transitions in the domain of task objects. Like a domain action, task operations can be characterized by pre/post conditions, and trigger conditions that capture the elementary state transitions. A task operation can be specialized as *required* or *requested* dependent on if it changes domain states of task objects and generates an event or only queries the task domain, respectively.

**Term 2.4. Problem state.** The process of problem solving creates, uses and changes a number of task objects referring the states of the affairs. A problem state is a set of values of state variables (e.g. knowledge chunks) representing these task objects. Problem states include information about current goals. Problem states also include all information generated during the process of problem solving, such as beliefs, desires and so on.

**Term2.5. Goal.** A goal is supposed to match a mental representation that the user has an attitude describing an expected problem state (e.g. 'Make reservation' is the goal in the above example). Goals are realized by intentions and evaluations. The important point is that a goal is some desired end state to be wished by the user, rather than a state to be reached after successful execution of a task in traditional task analysis. As a result, traditional analysis cannot answer if a goal can be achieved, and how failures can be recovered.

**Term 2.6. Intention/Evaluation.** An intention (or evaluation) is supposed to match a mental thread of problem solving and of maintaining problem states. A mental thread performs a set of required or requested task operations dependent on whether it is an intention realizing a current goal or an evaluation establishing a belief of current domain state, respectively. In general, mental threads change problem states, which in turn, invoke a set of domain actions to complete intentions or evaluations. Intentions (or evaluations) are characterized by trigger and stop conditions. As shown in Fig. 1a, intention/evaluation is significant in that it represents a basic unit of problem-solving knowledge, and establishes a continuum between the different views of task and domain.

(Jarke, Pohl & Rolland, 1994) identifies three worlds and the possible relationships that need to be understood and modelled (Fig. 1b). For the sake of usability, we argue that the joint part of the three worlds has to be taken into account, that is, the states of affairs among the worlds (Fig. 1c). We have identified the joint part between the usage world and the subject world by the definition of the term *problem state*.

## 3  Conceptual models on the knowledge level

A conceptual model is a set of concepts and their relationships, which embodies the view captured by a set of ontological terms with respect to some domain of interest. The view of the proposed

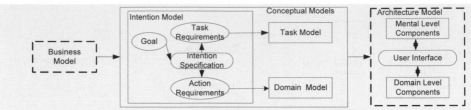
Figure 2: the proposed framework of conceptual models

framework is intention-dominated, that is, it is concerned with what goals to achieve instead of how to achieve these goals.

As shown in Fig. 2, the business and architectural models (the dashed-line boxes) do not belong to the framework. The business process model provides contextual information for the conceptualization, whereas the architecture model is used for the realization of the conceptual modelling. Because of the limited space, in the remainder we will not mention them. The models in the framework are described as follows.

***Model 3.1. Intention model.*** An intentional model consist of a goal the model intends to achieve and three interrelated parts:

- Intention specification. Intention specification is a cluster of intentions/evaluations without sequencing, in which intentions and evaluations realize the goal, and validate whether the goal is achieved, respectively. For interaction design, it provides information how to organize dialogue and presentation of user interface towards the current goal it realized.
- Task requirements. Task requirements are a collection of task operations invoked by the intention specification without sequencing. Task sequencing is determined by intention specification. For interaction design, task requirements determine the actions that a user may initiate.
- Domain requirements. Domain requirements are a collection of actions invoked by the intention specification without sequencing. Domain requirements in general are independent of task requirements, although they are indispensable for system design.

***Model 3.2. Task model.*** A task model is a collection of task objects that represent the content of tasks, and that represent all information that needs to be presented. Computationally, the model represents a working memory that could be seen as the extension of the users' working memory.

***Model 3.3. Domain mode.*** A domain model is a collection of domain objects that represent domain factual knowledge, and it is therefore independent of the task model. In our framework, the model represents only the underlying information that a system should maintain for the purpose of functionality, and it is therefore, usually unperceivable by the users.

In remainder of this section we demonstrate a case study of hotel reservation. For simplicity, we assume that the users' needs are to make a room reservation. The modelling process starts from the artefacts of business process modelling. In general, the artefacts include a domain process model and a domain model. For the sake of the limited space, only the domain model in our example is show in Fig. 3a.

The modelling process includes two complementary steps of intention/evaluation analysis and task elicitation by analysis of the artefacts of business modelling and by asking the users questions repeatedly: *What* do you *intend* to *do* when doing this thing? *What* do you *expect* to *get* from doing this thing? The *what* in general means the content of task (i.e. task objects) and the *do* (and the *get*) means task operations. The users can usually answer such questions because these questions are on the same level as their plans of guiding their tasks to be done. For example, a user may answer that 'I want to know if a room is available in this period as I am making a call to this

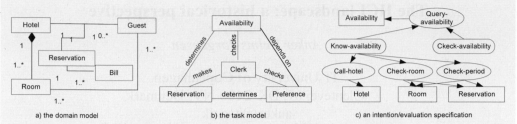

Figure 3: a part of conceptual models in the case study

hotel'. From this answer we can find a task object 'availability' with properties 'room', 'hotel' and 'period', and an operation 'query-availability'. The properties represent a set of state variables relevant to the domain knowledge in Fig. 3a. It is notable that we aim at eliciting which intentions should be realized, rather than exploring how these intentions are realized by control knowledge as it does in traditional task analysis.

Fig. 3b depicts all the task objects and their relationships by repeating the steps of the process. It is not too difficult to establish the relationships between the identified intentions/evaluations (and the operations), and domain actions because each task object, more precisely, its properties make explicit references to domain objects, or their properties. For example, fig. 3c shows a specification for the intention of knowing availability and the evaluation of checking availability.

## 4    Conclusions

We are not developing a brand-new method, but reusing the traditional OOA technology on the knowledge level. The ontological terms we defined in this paper emphasize acquisition of knowledge about the content of task domain, rather than about control mechanisms in that domain. The continuum defined by these terms represents a declarative specification to mediate activities of problem solving and task performing. On the knowledge level, the proposed conceptual models can represent the requirements for a system in the same state space as the task knowledge, which makes possible build the system to delegate user tasks on the same level.

## References

Chandrasekaran, B. Josephson, J. & Benjamins, R. (1998). Ontology of Tasks and Methods. In the Proceedings of KAW'98, Voyager Inn, Banff, Alberta, Canada

Gruber, T. R., (1993). A translation approach to portable ontology specification. Knowledge Acquisition, 6(2), 199-221

Jarke, M. Pohl, K. and Rolland, C. (1993). Establishing visions in context: towards a model of requirements processes, Proc. 12th Intl. Conf. Information Systems, Orlando, Fl.

Maguire, M. (2001). Methods to support human-centred design. Int. J. Human-Computer Studies, 55, 587- 634

Norman, D. (1986). Cognitive engineering. In: Norman, D. and Draper, S (Ed.): User centered system design (pp. 31-61), Lawrence Erlbaum Associates, Hillsdale, NJ.

Robilland, N. (1999). The role of knowledge in software development. Communication of the ACM, 42 (1), 87-92

Stary, C. (2000). Contextual prototyping of user interfaces. In the Proceedings of ACM DIS'00, 388-395

Welie, M., Veer, G., & Elens, A. (1998). An ontology for task world models. In the proceedings of DSV-IS'98, Springer Verlag, 57-70

# The HCI landscape: a historical perspective

*Anker Helms Jørgensen*

IT University of Copenhagen
Glentevej 67, DK-2400 NV, Denmark
anker@it-c.dk

## Abstract

The computer is now a universal and pervasive device that must be viewed as a cultural and social phenomenon in order to be understood and exploited fully. Initially the computer belonged to the technical realm, used by individual technical specialists, but along with the development a number of novel academic disciplines emerged, e.g., Human-Computer Interaction (HCI) and Participatory Design (PD), drawing upon an even greater number of supporting academic disciplines (e.g., Sociology and Art History). This paper presents a historical overview of the developments in supporting academic disciplines as seen from a HCI perspective. The overview takes shape as a table with seven entries: applications of computers; technologies; paradigms; interaction forms; forms of HCI knowledge; novel emerging disciplines; and supporting disciplines.

## 1    Introduction

Some years ago I had a rewarding experience: I came across Steven Johnson's book *Interface Culture* (1997). Alas, the topic I had been working with for two decades had finally become a cultural phenomenon! The book gives a cultural perspective on recent developments of the computer, centered around the user interface. The publication of this book witnesses the incredible changes that have taken place in the last 50 years with the computer: from a calculator as used by engineers and astronomers to a social agent being used by millions of ordinary people in everyday activities. This paper continues this line of thinking by presenting a historical overview of the developments in accompanying academic disciplines as seen from a HCI perspective. The paper has three aims. Firstly, to support newcomers in the field to get a grasp the origins and history of HCI. Secondly, to provide grounds for an understanding of HCI as a field - interdisciplinary or a field in it's own right. Thirdly, to provide grounds for a discussion of the field of usability that is being challenged by the social and cultural penetration (Jørgensen, 2002). This exposition is by no means an attempt to tell the complete history – this has been attempted by others, e.g., by Wurster (2002) addressing the history of computers, by Myers (1998) addressing the history of human-computer interaction technology, and by Goldberg (1988) addressing the history of personal workstations. In addition, an elaborated analogy is found in the field of Human Factors and Ergonomics described by Meister (1999). As to method, the paper is a personal account, not based on any systematic analysis of the literature.

## 2    The Landscape

The landscape takes shape as a table with seven entries (see Table 1): applications of computers (e.g., administration); technologies (e.g., networks); paradigms (e.g., technical); interaction forms (e.g., speech); forms of HCI evidence (e.g., theories); novel emerging academic disciplines (e.g., Human-Compuer Interaction); supporting disciplines (e.g., ethnography). The table and the presentation do not include timelines as fairly precise timing is hard and in many cases not meaningful as phases often overlap.

## 2.1 Applications of computers and technologies

The first *application areas* where computers were applied included largely military, technical, and research applications, such as calculations of planet orbits and missile trajectories. Indeed, the name of the first major programming language FORTRAN reflected this type of applications: FORmula TRANslator. Soon the potential of the computer for handling and storing large amounts of data was realized, leading to applications in business and administration. The next major programming language reflects this development: COBOL is short for COmmon Business Oriented Language. Gradually novel application areas were developed, e.g., *communication* and *education*. With the miniaturisation, computers became increasingly embedded in mechanical and electronic devices, firstly in machinery and later in everyday devices such as convers and cameras. The most recent areas of application is *art, design,* and *entertainment*. An important area here is computer games, where technology, culture, and economy merge on a huge scene, economically comparable to the Hollywood film industry.

The first computers were developed in the labs of early *pioneers*, for example the Z1 by Konrad Zuse in Berlin in 1938, and the ENIAC by Eckert and Mauchly in 1946 in the US. These computers were extremy small in terms of capacity and extremely large in terms of sheer physical appearance. Computer manufacturing companies emerged and existing enterprises like IBM took up computer development and production. Gradually the capacity of the computers increased while the size of the processors and storage units decrased; mind you, a hard disc drive was the size of a washing machine in the early 1970's. This miniturization development included some leaps: from the radio tube over the transistor with hundreds of units on one board to the integrated circuit with ten of thousands of components on small chips.

Two trends appeared: towards larger and smaller computers. The large computers grew steadily, known as *mainframes*, typically used by large corporations like banks and insurance companies for administrative purposes. Among the best known mainframes is the IBM series 360 and 370 that were sold in thousands in the 1970. This trend has continued into today's supercomputers with huge processing capacity, primarily being used in research. The trend towards smaller computers firstly manifested in so-called *mini-computers* as local needs for computing power emerged, such as in departments of large companies. Among the leaders here were Digital Equipment Corporation's PDP 8 and PDP 11 that were also sold in huge numbers. This development has continued ever since, leading to the *personal computer* on the desktop and to extremely small integrated circuits that are built into all sorts of devices – now known as *ubicomp* (ubiquitous computing) and pervasive computing. Other important technological developments is the *network* with the military ARPA/DARPA networks and Xerox's Ethernet as "founding fathers", leading to the Internet with World Wide Web, email, chat rooms, etc. On the fringe, we have also seen developments in *Virtual Reality* and *robot* technologies, employing computer vision based on miniature video cameras.

## 2.2 Paradigms

There have been a number of shifts in the way we see the role of computers and computing, accompanying the technological developments. Not surprisingly, early on computers were seen as associated with technology and research. As they were moved out of the labs and used for ordinary tasks by human beings who were not engineers, it became clear that the interaction between the user and the computer had to be attended to, leading to design of user interfaces and also selection and training of computer personnel – a focus on the *individual* – typically the white collar worker in the offices. As computers were increasingly connected in and between organisations, they became a vehicle for collaboration and communication between individuals, thereby changing the focus from the individual perspective to the *organisational* context. The next step was the move out of the offices into the *social context*, i.e., into people's homes and everyday lives, enabling changes in the ways citizens communicate, consume and produce information, and perceive themselves in the modern society: the information society and the network society (Castells, 1996). This has huge *social*, *cultural*, *democratic*, *economical* implications – indicating another shift in paradigm towards globalization.

## 2.3    Interaction forms and forms of HCI evidence

There is an odd dichotomy in the ways humans interact with the computers. On the one hand, there has been almost no change in the way human users provide input to the computer in the last 50 years: the largely unchanged QWERTY *keyboard* is still the primary input medium – however now being supplemented by the mouse for selection. On the other hand, there has been a tremendous development on the output side, in particular the graphical user interface and multimedia capabilities. The predominant early forms – firstly *knobs and dials*, later *command languages* in the 1960's - were gradually complemented by visual/graphical forms, firstly by the traditional *menu-driven* systems of the 1980's (supporting recognition rather than recall) and later by the so-called *graphical user interfaces* (a.k.a. GUIs) and *multimodal* interfaces (allowing direct manipulation of interface representations of data objects). Interaction territories largely so far unchartered are *gestures* and *speech* input/output, although being potentially very useful. As an aside, with 20/20 hindsight it is striking to see how impoverished the great achievement of the 1980's - the "graphical" GUI interfaces with their rectangles and boxes - appear compared to the visual explosion in Web-interfaces - for the good and bad!

Along with the emergence of technology and interaction forms, it is interesting to observe the forms of evidence that have gelled in the field of Human-Computer Interaction. The first manifest principles were based on experience obtained by individuals who had developed dialogue systems (as interactive systems were called in the 1970's), expressed as *guidelines* such as "*Provide specific error messages*". Guidelines by the dozen emerged from all quarters, calling for systematizing and compilation. Smith and Mosier poineered this approach in the 1980's by compiling a suite of three reports, the last containing no less than 944 guidelines covering all aspects of the user interface (Smith and Mosier, 1986). It soon became clear that this approach was a dead end. Instead of focussing on features of the user interface, it was necccessary to address the development process. Here Gould and Lewis (1985) coined three development principles in the 1980's for developing usable systems: *early and contiuous focus on users*; *iterative development*; and *continuous user testing*. These principles still stand out as landmarks. But gradually models, concepts, and theories emerged in the 1980's, such as *affordances* and *gaps of execution and evaluation* (Norman, 1988) that provide a deeper understanding of aspects of the interaction between the human and the computer, rooted in cognitive psychology. The 1990's saw a lot of focus on *evaluation methods*, such as user testing, thinking-out loud, cognitive walkthrough, and heuristic evaluation, see Nielsen (1993), including comparisons of the strengths and weaknesses of the methods. The final focal point in forms of HCI evidence is *standards* that evolved around the GUI systems, ensuring consistency across systems. Apple was championing this approach in 1984 and followed later by other developing organisations. As a sad fact of life, the widespread standardization that has been achieved in GUI platforms has been completely undermined by the Web due to its anarchistic nature. It seems that the only item being fully standardized on the Web is how to click on clickable interface objects.

## 2.4    Novel and supporting academic disciplines

Undoubtedly, the most prominent academic discpline associated with the computer is *computer science*, starting more than 50 years ago with mathematics as theoretical basis and electronic enginccring as basis. An early emerging field was AI: *Artificial Intelligenc*e, growing out of *Cybernetics* and supported by *Philosophy*. AI had great promise for the future – a future that has only to a limited extent been realized. Other emerging disciplines were *Human-Computer Interaction*, taking shape in the 1970's with research lab developments and practical development experience as starting points. The supporting disciplines were primarily *cognitive psychology, linguistics,* and *ergonomics*. As the computer grew graphical and displayed symbols in addition to characters, the field of *semiotics* became relevant as did. With the invasion of the computer in organisations and the subsequent ever increasing close linking between organisation and technology, *organisational theory* became a cornerstone of HCI. As communication and collaboration became more prevalent, the disciplines *CMC: Computer Mediated Communication* and *CSCW: Computer Supported Collaborative Work* emerged with *communication studies* and *sociology/ethnography* as supporting displines, respectively. With the digital penetration of the social and cultural spheres, *Digital Aesthetics* may be the next academic discipline to take shape, supported by *Arts History, Cultural Studies, Media Studies,* and *Literary Theory*.

## 3    References

Castells, M. (1996): The information age: Economy, society and culture, vol. 1. The Rise of the Network Society. Blackwell.

Goldberg, A. (Ed) (1988): A History of Personal Workstations. New York: Addison-Wesley.

Gould and Lewis (1985): Designing for usability: Key principles and what designers think. *Comm. ACM*, 28 (3), 300-311.

Johnson, S. (1997): Interface Culture. Basic Books.

Jørgensen, A.H. (2002): The Concept of Usability: Challenges from Mobility and the Internet Proc. Work With Display Units, Berchtesgarden, Germany, May 22-25th, 2002, pp. 131-132.

Meister, D. (1999): History of Human Factors and Ergonomics. Lawrence Erlbaum.

Myers, B.A. (1998): A brief history of human-computer interaction technology. *ACM Interactions*, 5(2), 44-54.

Nielsen, J. (1993). Usability Engineering. Academic Press

Norman, D. A. (1988): The Psychology of Everyday Things. Basic Books.

Smith, S.L. and Mosier, J.N. (1986). Guidelines for designing user interface software. Mitre Corporation Report MTR-9420, Mitre Corporation.

Wurster, C. (2002): Computers: An Illustrated History. Benedikt Taschen Verlag.

**Table 1:** Overview of the HCI landscape

| Applications | Technologies | Paradigms | Interaction forms | Forms of HCI Evidence | Novel Disciplines | Supporting Disciplines |
|---|---|---|---|---|---|---|
| Calculations | Pioneers' works | Technical | Knobs & dials | Experience | Computer Science | Electronics |
| Administration | Mainframes | The Individual | Commands | Guidelines | Cognitive Ergonomics | Ergonomics |
| Everyday things | Minicomputers | | Menus & forms | Cognitive Experiments | | Cognitive Psychology |
| Communication | Personal computers | Organisational context | Graphical User Interfaces | Concepts | Human-Computer Interaction | Linguistics |
| Education | Networks | Social Context | Speech I/O | Theories | Computer-Supported Cooperative Work | Semiotics |
| Entertainment | Virtual Reality | Identity | Multimedia | Development Methods | Participatory Design | Philosophy |
| Design | Robots | Democracy | Gestures | Standards | Computer-Supported Cooperative Learning | Pedagogics |
| Art | Ubicomp | Economy | Physiology | Evaluation Methods | Computer-Mediated Communication | Communication |
| Culture | | | Touch | | | Media Studies |
| | | | | | | Organisational Theory |
| | | | | | | Design |
| | | | | | | Sociology |
| | | | | | | Etnography |
| | | | | | | Cultural Studies |
| | | | | | | Aesthetics |
| | | | | | | Litterary Theory |
| | | | | | | Art History |
| | | | | | | Drama |
| | | | | | | Occupational Health |

# Interaction and Distance Education

*Stefano Levialdi*

University of Rome (La Sapienza)
Department of Informatics, Rome,
Italy
levialdi@dsi.uniroma1.it

*Maria De Marsico*

University of Rome (La Sapienza)
Department of Informatics, Rome,
Italy
demarsico@dsi. uniroma1.it

## Abstract

There are many approaches to distance education that either try to simulate classroom activities or follow the augmented reality strategy. We have chosen this second one believing in off-line teaching (the student may always choose the time and place for his study session) in cooperation with face-to-face learning. Students may download the lecture before or even after the normal class so allowing them to prepare for the specific concepts to be taught or to revise them after the class. A number of extra services are provided in our Java-based platform called MultiCom II (possibility to link parts of the course with other parts or even other courses, communicate with the university community). In the near future we will add tools to annotate text and to manage multiple-choice tests.

## 1 Introduction

Research in the field of teaching/learning theories and pedagogical principles has received new impetus from emerging technologies. The potential of computer-supported learning (CSL) is coupled with increased use of the Internet and the World Wide Web and with remarkable technological innovations (Kearsley, 2000). This scenario provides a powerful framework to optimize usage of educational resources. Basically, any kind of technological approach must rely on a tested educational methodology; we may consider a first course distinction between objectivistic vs. constructivistic learning.

The objectivistic philosophy is based on the assumption that one learns by being told, that knowledge is objective and identical for all. In this approach both teaching and learning are explicit and declarative (Guttormsen and Krueger, 2000). On the contrary, the constructivistic philosophy relies on the learn-by-doing principle, i.e. on constructing knowledge by reorganizing cognitive experiences within a theoretical framework (Kintsch, Franzke, Kintsch, 1996). In this way, learning is mainly an intuitive process, while teaching should provide tools for the reorganization process.

In the first case, the corresponding system will be a tutoring system, i.e. a system in which the learning material is organized in a strict sequence by the educator: the student has to follow the predefined cognitive path in learning. Conversely, in the constructivistic case, the system will provide learn-by-doing and free navigation facilities, eventually mediated by the educator. As a consequence, the hierarchical model implicit in the objectivistic approach is substituted by a cooperative model, where educators and students collaborate to reach the cognitive goal. This implies that students move from a passive to an active role. Collaboration with classmates and educators provides a way to externalize individual acquired knowledge, in order to validate it and to support mutual understanding. Social interactions initiate and reinforce mental activity thus fixing knowledge. Different learning styles and/or different knowledge and skill about a subject at hand can be accommodated for by a flexible system allowing to choose a learning mode at a certain moment and for a given cognitive task (Carswell and Venkatesh, 2002).

A further choice is either to substitute or integrate a classroom mode. In the first case, the whole teaching activity is performed through a CSL system, eventually at a distance (video conference). The main drawback of this approach is the lack of personal interaction between students and educators. Even if the quality of personal interaction can be affected by individual teaching and learning styles, nevertheless it represents, in many circumstances, an important factor in the education process. In the second approach, the distance learning system integrates the classroom, providing further material, giving the opportunity to preview a lesson while also extending teaching material requested for full comprehension. Moreover, this approach gives the opportunity to reconsider a lesson to study and further explore its contents, eventually connecting it with external learning material. The classroom is substituted by a virtual global education environment so becoming a substantial extension instead of simply mimicking a face-to-face lesson. The main possible approaches we discussed are summarized in Table 1.

**Table 1:**

Approaches to teaching-learning environments

❑ **Objectivistic vs. Constructivistic learning**
  ○ Tutoring systems vs. free navigation
    ▪ Hierarchical model vs. cooperative model
    ▪ Passive vs. active students
❑ **Substitute classroom mode vs. support classroom mode**
  ○ Reproduction vs. extension
  ○ Classroom metaphor vs. global education environment

The spatial separation between students and an educator may or may not be accompanied by temporal separation (asynchronous vs. synchronous), as shown in Figure 1.

| | | |
|---|---|---|
| **Different Time** | **Learning point** | Asynchronous Distance Education |
| **Same Time** | **Classroom** | Synchronous Distance Education |
| | Same Place | Different Place |

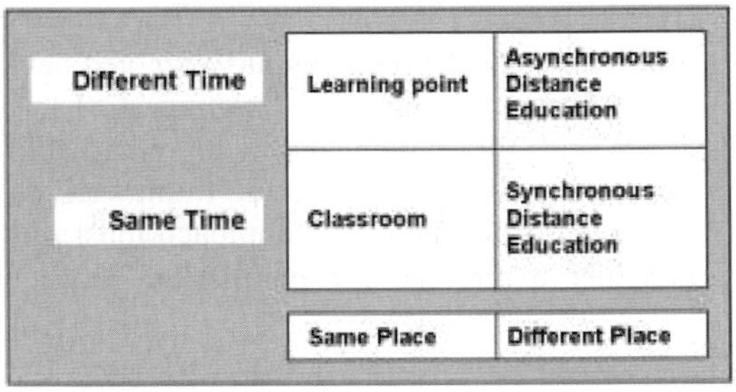

**Figure 1:** Space/time arrangements

In distance education, the interaction between student, instructor and content is essentially mediated by technology, i.e. facilitated by a medium, or better by a number of media (multimedia) (Guttormsen and Krueger, 2000). Within this mediation there must be a mutual adaptation between the learning material and the communication medium. All contents must be adapted to the new teaching paradigm in order to be effective, and adequate media must be selected so that information is presented according to the learning material and to the students' needs (Jonassen, Peck and Wilson, 1999).

## 2  MultiCom II

MultiCom II is a system for distance-supported education developed by our research group, based on a client-server architecture. We next summarize our educational assumptions, and then show the derived system architecture and functionalities.

### 2.1  Educational Approach

We chose to follow a constructivist approach, because we believe that learning is an active process, performed by the learner according to his/her personal profile, aiming to reorganize new concepts through previous knowledge. Accepting the idea of active learning shifts the main focus of educational activities from teaching to learning, where the student plays the demiurgic role in the cognitive process. The educator becomes the mediator of this process, providing guidance and assistance in constructing meaning from the learning material, allowing and suggesting personal cognitive paths. These paths are not limited to a single course. They can eventually span a complex network of linked chunks from different courses, when these are believed useful to clarify the lesson. Typically, during natural studying, a student often uses external material to improve his/her comprehension or to integrate the lesson.

Our aim is to facilitate this natural way of learning. For this reason, differently from existing e-learning platforms that tend to isolate courses among them, we provide a unified view for the learning material of all courses. Moreover, we allow to establish personal and permanent links among documents of the same course as well as of different courses. The educator can also provide this kind of (shared) link in his/her material, suggesting a diversion from the lesson to enhance student comprehension. Based on his/her knowledge and present cognitive goal, the student can choose either to follow these links or not. These chunks of information can be considered as the "learning objects" (Wiley, 2000) that the student builds up to construct his/her cognitive path. The learning material can also be integrated by the student with personal material, annotations and notes that can eventually be published under educator mediation on the central repository.

The main drawback in using hypertext and links is the possibility of becoming lost or confused in an information jungle. Furthermore, it is a fact that beginner students of a certain subject prefer to have a strict learning sequence to follow. For this reason we designed an intuitive interface for educators to organize their course material in a hierarchical structure, using the folder-subfolder model, so that the learning material has a given sequential pathway. Notice that uploading of documents is fast and intuitive and neither requires intermediate intervention by a publisher nor the need to remember the specific place in the repository reserved for the course. The student will use the same interface to browse the learning material. This interface allows nonlinear learner-controlled sequences for more experienced students. It can be considered a kind of augmented reality environment, where annotations of various kinds are superimposed to the lesson.

Collaborative learning implies interaction with classmates and educators and cooperation in solving cognitive tasks as a key issue for effective learning. The system includes exchange of messages (asynchronous communication), as well as chat management for each course (synchronous communication). Moreover, course notice boards allow fast communication of important and urgent news. Forums about general course content or about specific topics complete the communication scenario. Last but not least, specific information can be obtained about courses and educators. Administrative personnel also communicate with educators and students in order to answer questions on exams, etc. The integration of the three communities (educators, students and administrators) is aided by the fact that the system does not rely on any external mail, chat or web services (even if web pages can be directly linked to the learning material), so that it is self-consistent. This feature also strengthens the psychological feeling of belonging to a group.

## 2.2   System Architecture and Functionalities

MultiCom II is a Java client-server system, so that both client and server sides are portable on any operating system supported by JDK1.4.1. Only a Java Virtual Machine is needed for the application to run. In other present platforms, the server side is operating system dependent. Moreover, the modularity of the system allows easy integration of more functionalities (see Figure

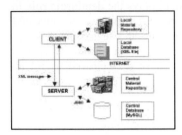

2). The server side of the application manages all information and procedures related to users, classified in the three communities, and to courses and connected services, i.e. learning material, chats, notice boards and forums. The information is gathered in a MySQL database, managed and queried by the server using the JDBC Java package. Learning material of all courses is collected on the server, in a way completely transparent to users, to whom it is shown as a single tree. The server also allows mail-like messages among users. Services are requested by the client side of the application using XML messages, sent over TCP/IP sockets. Incoming messages are parsed by the server according to a DTD file. The

**Figure 2:** MultiCom II system

same happens for incoming messages on the client side.

**Figure 3:** MultiCom II toolbar

The client maintains a local copy of the documents downloaded by the user that can also be read offline. Moreover, an XML file maintains information and messages related to the user, plus a number of personal settings, such as personal links among documents, with the same fields defined for the central database. The client interface is composed of a series of stanzas (see Figure 3). The toolbar is integrated by a couple of status/action buttons that switch between online and offline status.

The first stanza that the user enters is the learning material stanza: the user can visualize the tree of documents (Figure 4), visualize selected documents, and create his/her personal links, or update the material online and download new documents. In the same stanza, a number of buttons, not shown to students, allow the lecturer to add or update sections and documents to his/her course. The lecturer can also include suggested links among documents, even of different courses. At present, learning documents directly visible by the client interface are text, rtf and standard html documents. Furthermore extra files of any format (e.g. pdf) can be provided, requiring external viewers or applications. This additional material is listed in the first node of the tree that corresponds to the course. The next stanza is for newsgroups that share the tree structure of learning material. A stanza is allocated to notice boards. A general notice board is associated with each course upon its creation. More notice boards can be added to a course for specific topics.

The messages stanza provides the list of received messages and a contact list, where the user can include frequently messaged MultiCom users. A pop-up window allows full list management. As for notice boards, a general chat in the chat stanza is created for each new course. More chats can be created for specific discussion groups. The next two stanzas provide public information about lecturers, their courses, and information about single courses, including their lecturers. Additional buttons, shown only to administrative staff, allow creation of new courses, insertion of new lecturers, and update of information. The last two stanzas, present in the administrative staff interface only, allow the management of students registration and capabilities for a number of operations, that can be granted to certain students under lecturer authorization.

## 3 Future Work

The next steps are to implement this approach in a new media curriculum and conduct evaluation research to assess the learning outcomes because research "has been mixed about the effectiveness of PBL, although most of this work has focused on its use in medical education" (Rosenbaum, 2000). Vernon and Blake (1993) find that it is a markedly superior approach to traditional learning while Fenwick and Parsons (1998) are critical of the approach. Having the opportunity to conduct comparative research comparing a class using the approach described above with a class using traditional approach would be useful in attempting to settle this debate. We are also working on a version of this approach that will be appropriate for use in doctoral curricula and hope to report on our work in the near future.

# References

Abdullah, M.H. (1998) Problem-Based Learning in Language Instruction: A Constructivist Model. *Eric Digest*. ERIC Clearinghouse, Bloomington, IN.

Barrows, H.S. (1998). The Essentials of Problem-Based Learning. *Journal of Dental Education*. 62(9). 630-633.

Bentley, J., Sandy, G., and Lowry, G. (2002). Problem-based learning in information systems analysis and design. In Cohen, E. (ed). *Challenges of Information Technology Education in the 21st Century*. Hershey, PA: Idea Group Publishing. p. 100-124.

Edens, K.M. (2000). Preparing Problem Solvers for the 21st Century through Problem-Based Learning. *College Teaching*, Spring, 48(2), pp.55-61

Faiola, A. & Rosenbaum, H. (2002). Challenges in teaching usability theory and testing for new media curricula. ACM SIGCHI. Annual conference. April 23. Minneapolis. Minn.

Fenwick, T. & Parsons, J. (1998). Boldly solving the world: A critical analysis of problem-based learning as a method of Professional Education. *Studies in the Education of Adults*, 3(1): 53-65.

Hewett, Baecker, Card, Carey, Gasen, Mantei, Perlman, Strong & Verplank. (1996). ACM SIGCHI Curricula for Human-Computer Interaction. ACM Special Interest Group on Computer-Human Interaction Curriculum Development Group http://sigchi.org/cdg/

Meyers, B., Holland, J., & Cruz, I. (1996). Strategic Directions in Human Computer Interaction. *ACM Computing Surveys* 28(4). http://www-2.cs.cmu.edu/~bam/nsfworkshop/hcireport.html

Moiao, Y. (2000). Supporting Self-directed Learning Processes in a Virtual Collaborative Problem Based Learning Environment. In Chung, M.H. (Ed.). *Proceedings of the 6th America's Conference on Information Systems. Association for Information Systems*. 1784-1790

Rosenbaum, H. (2000). Teaching Electronic Commerce: Problem-based Learning in a Virtual Economy. *Journal of Informatics Education and Research*. 2(2). 45-58.

Rossen, M.B. & Carroll, J.M. (2002). *Usability Engineering: Scenario-Based Development of Human-Computer Interaction*. Morgan Kaufmann.

Savery, J.R. & Duffy, T.M. (1995). Problem based learning: An instructional model and its constructivist framework. *Educational Technology*. 35(5) 35-38.

Shanley, D.B. & Kelly, M. (1994). Why Problem-Based Learning?

http://www.odont.lu.se/projects/ADEE/shanley.html

Vat, K.H. (2001). Teaching HCI with scenario-based design: the constructivist's synthesis. Proceedings of the 6th annual conference on Innovation and Technology in Computer Science Education. Canterbury, UK.

# The Etiquette Perspective on Human-Computer Interaction

*Christopher A. Miller*

Smart Information Flow Technologies
Minneapolis, MN U.S.A.
cmiller@siftech.com

## Abstract

As computer systems become more autonomous, intelligent and ubiquitous, it becomes more important to define roles for both the human and the computer that both can live with. There is evidence (extensively from Reeves and Nass, 1996) that we, by default, expect intelligent automated agents to adhere to rules of good conduct that human agents in our experience exhibit. The rules that govern such relationships are "etiquette" rules. In this paper, we define two senses in which the term is appropriately applied to the design of computer systems and review work from both perspectives presented at the AAAI Fall Symposium on Etiquette for Human-Computer Work.

## 1   Introduction

As computers become more ubiquitous in our daily lives, as they become smarter and more capable, and as we allow them to take on autonomous or semi-autonomous control of more critical functions, it becomes more important to define roles for both the human and the computer that each side can live with. The rules that govern such relationships are, we claim, etiquette rules. By 'etiquette', we mean the defined roles and acceptable behaviors and interaction moves of each participant in a common 'social' setting—that is, one that involves more than one intelligent agent. Etiquette rules create an informal contract between participants in a social interaction, allowing expectations to be formed and used about the behavior of other parties, and defining what counts as good behavior.

Each of us encounters and uses a host of different etiquettes in our daily lives. Etiquette in human-human relations is not just about which fork to use, but also about who speaks first when ordering a meal, how long a pause in speech or action must be before an onlooker concludes the actor has missed a cue, whether or not one can swear in a poker game vs. a church, who's in charge of deploying the landing gear in flight, etc. We establish etiquettes to smooth interactions and make them more efficient and comfortable, yet etiquette may be may be more or less effective—it frequently facilitates communication, but can sometimes obstruct it. Etiquette expectations may be violated—sometimes to potent and useful effect, but more frequently resulting in confusion and/or frustration.

Concerns about human-computer etiquette are not exactly synonymous with User Interface generation—since they say little about the 'look' part of look and feel. Etiquette can and does exist in a wide range of interface implementations (R2D2 had a more effective work etiquette than HAL, though HAL won out on the natural language front, for example). Etiquette concerns are not synonymous with human-centered design approaches, which are a much broader category of which etiquette would be a subset. Etiquette should be an aspect of the design of sophisticated, adaptive automation systems and should concern itself with how human and automation need to, and want to, interact for effective and acceptable relationships.

## 2 Why Etiquette?

Why use the term "etiquette" as opposed to other, related terms that denote other, related fields of study (e.g., social computing (Preece, 2002), embodied or personified agents (Cassell, Sullivan, Prevost & Churchill, 2000), human-centred design (Billings, 1997), etc.? How does a focus on etiquette differ from these other fields? "Etiquette" certainly works as an extended metaphor implying that sophisticated computer systems ought to behave in a more polite and sophisticated manner— graceful and suave. It, of course, also has the

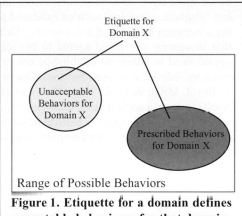

Figure 1. Etiquette for a domain defines acceptable behaviours for that domain.

benefit of being somewhat provocative—especially among those who are used to thinking of computers as either unintelligent tools that must be forced to perform desired tasks, or as demons actively trying to thwart successful human endeavours.

Reeves and Nass (1996) have shown, however, that media will tend to be regarded with the same set of expectations and assumptions that we use to interact with human actors. This is, perhaps, more true to the degree that the media exhibit dynamic, adaptive, autonomous action. Thus, it is perhaps not surprising that human-human "etiquette" serves as more than a metaphor for human-computer relationships and, instead, provides a reasonable first guide to shaping the desirable behaviors of sophisticated automation systems.

The American Heritage Dictionary defines "etiquette" as follows: "(1) the body of prescribed social usages. (2) Any special code of behaviour or courtesy: "In the code of military etiquette, silence and fixity are forms of deference" (Ambrose Bierce). ... Synonyms: etiquette, propriety, decorum, protocol. These nouns refer to codes governing correct behaviour." (Morris, 1979). Of the two definitions offered, the second is closer to the definition offered in the second paragraph above and, perhaps, of more utility for HCI systems designed to engage in working relationships with human users. Etiquette, in this sense, need have little to do with politeness or the "social niceties," and specific environments, specific work "cultures," even individual teams of humans will all have their unique sets of expected behavioural norms or etiquettes. Essentially, an etiquette for a specific domain is formed by prescribing some subset of the range of possible human behaviours as appropriate or inappropriate, expected or unexpected for those who participate in that domain (see Figure 1). These behaviours may pertain to speech, dress, movement, etc., or to more specific protocol behaviours (e.g., 'in this plant, we always empty a vessel when it's not in use'). As such, these are the behaviours that any human or automation agent should strive to adhere to if it wants to be accepted into that milieu.

At the recent AAAI Fall Symposium on Human Computer Etiquette (cf. Miller, 2002 for papers presented by the authors listed below), several authors reported work that did or could take this perspective. Susan Hahn reported work indicating that when automation took upon itself a more comprehensive role vis a vis the human operator (e.g., that of a decision aiding 'oracle' as opposed to an information providing table or list) operators chose to use it more frequently even if it produced more errors and made those errors harder to uncover. John Lee reviewed variables, including "etiquette variables" such as exhibited personality trait similarities, which affect trust calibrations between human and automation. Debra Schreckenghost reported work at NASA showing that designing automated agents to participate in distributed control of space operations, such agents need to adhere to established norms of alerting and notification—not so much for acceptance by individual humans but for efficient group coordination and functioning. Finally, my own work (cf. Miller & Funk, 2001) shows greater user acceptance and significant workload reduction resulting from a sophisticated decision aiding and information management in a military rotorcraft when that aid is designed to participate in the "etiquette" of cockpit crew coordination—even if its capabilities for doing so are comparatively primitive.

Even so, the first definition provided above applies equally well. Indeed, several papers presented at the symposium took the perspective of etiquette as good manners, though they investigated different aspects of what that might mean. This is hardly problematic. Just as we encounter multiple etiquettes in our daily lives (what counts as appropriate, expected behaviour is *different* at work, in a bar, in church, etc.), so computers designed with etiquette in mind should be expected to exhibit different kinds of etiquette for the different kinds of contexts they are used in.

For example, Lewis Johnson of the University of Southern California reported work illustrating that making pedagogical 'guidebots' (personified agents with a teaching role) more socially intelligent—making them respond to users in more socially appropriate ways (both positive and negative)—promotes user learning. Max Louwerse goes further and points to very specific mixed initiative dialogue and turn taking strategies in tutorial interactions and shows that their violation can have adverse effects on learning. Finally, Barry Kort and John Davis both presented separate work on the roles of technology in fostering civility in on-line interactions—Davis stressing the role of increased "individuation" through reputation tracking, personal profiles and voice; while Kort stressed the importance of a mutually agreed-upon and tracked "social contract".

The relationship between the "good manners" and the "established codes of behaviour" meanings of the term "etiquette" seems to be a simple one of general vs. specialized application. The etiquette rules of good manners are general, at least in the sense that they are intended to apply in situations where no more specific set of rules is known to apply. For example, the use of "please" and "thank you" and the polite forms of address can frequently be dispensed with, but they are a good way to start an interaction with someone who is unknown to us. In fact, Brown& Levinson (1987) claims that most of the polite forms in conversation, regardless of their specific linguistic or cultural manifestation, are intended precisely to ensure that the least offensive interpretation possible is applied to our utterances and actions.

By contrast, the etiquette rules of specialized domains may be less formal and may involve substantial deviation from the "polite" forms. Work domains are a common sphere in which specialized etiquettes prevail—etiquettes that differ from formal and polite norms. This is true for forms of address and the use of "please" and "thank you," but it is true in more interesting ways as well. For example, (as Reeves and Nass, 1996, demonstrate) we take it as inappropriate for a conversational partner in social interaction not to maintain a certain amount of small movements— indicative of a lack of intelligence or attention, or of artificiality. Nevertheless, an assistant in a work setting who is always moving about the office may be seen as intrusive or at least fidgety. It is also true that many work domains have evolved highly specialized etiquettes designed to smooth operations and/or ensure safety. The conversational conventions used between pilots and Air Traffic Controllers is one particularly formal example. When such conventions exist, whether formally documented and trained (as in the air traffic control case) or not, it will be important for computers which wish to perform actions similar to humans in the domain to perform in a fashion that preserves the expected etiquette which those humans would adhere to.

In fact, work at the intersection of the two definitions may be the most interesting. Raja Parasuraman reported research demonstrating that polite etiquette from machine automation can affect human + machine performance in both good and bad ways—even in a highly structured work domain in which 'politeness' is not normally regarded as important. For example, in a simulated aircraft piloting task, Parasurman contrasted a "rude" decision aid (one which made intrusive recommendations and impatient queries) vs. a "polite" in the domain of detecting, diagnosing and repairing engine faults. This politeness dimension was crossed with the accuracy with which the aid provided its recommendations: 60% vs. 80%. Parasuraman's results showed that the difference between a polite and an impolite delivery style is as large as the 20% difference in accuracy rates in terms of users' rated trust and observed acceptance of those recommendations. That is, a polite but inaccurate aid was used and trusted about as a much as a rude but accurate one. The implication may be that a 'sweet-tongued' decision aid may gain more user acceptance than it deserves, while a brusque one may turn users off even when they should attend to its advice.

So, in short, both senses of the term "etiquette" seem relevant for human-computer interaction—though perhaps in different settings. Politeness and adherence to social codes will likely be useful in exactly those situations where it is important in human-human interactions: in general situations, between strangers, when the specific context or relationship with the conversant is unknown, when it is important not to give offence, etc. On the other hand, especially given that computers are frequently (though certainly not exclusively) used as tools in the pursuit of work goals in specific work domains, the specialized codes of etiquette which are familiar and expected in those domains will be important for the computer to use. As a simple example: workers in an oil refinery are generally not accustomed to addressing each other with honorifics; therefore it is likely that a computer system that does so will be seen as out of place.

# 3   Is Etiquette 'just' good interface design?

The notion of etiquette does, I believe, force us to consider several aspects of human-computer relationships that traditional design concerns do not. By placing the system to be designed in the role of a well-behaved human collaborator we gain insights into how users might like or expect a system to act and how its actions might be interpreted *as a social being* that rarely come from any other source of design (with the possible exception of usability reviews for an already-designed system). I find it instructive to ask, for example, "if this system were replaced by an ideal human assistant, how would that assistant behave?" and, alternatively, "if a human assistant were to provide this information/decision /recommendation in this way, how would s/he be perceived by colleagues?" To pick, perhaps unfairly, on a well-known example: how would I regard a human office assistant who, several times a day, interrupted my work to offer to help me write a letter?

Would traditional approaches to human-computer interface design provide similar insights? Perhaps, but a growing set of criticisms claim otherwise. Traditional HCI design approaches have been criticized for being too focused on a single individual and his or her computer, rather than the role that that individual (and technology) play in a greater social/organizational setting (McNeese, Salas, & Endsley, 2001). Similarly, one of the goals of the "social computing" movement is to design HCIs and computer-mediated human-human interactions that foster richer communication and the building of more "social capital" than have traditionally been the case (Preece, 2001). Social capital comes, in part, from interactions that preserve the richer roles and cues of human-

human interaction. Finally, Don Norman (2002) has augmented his prior calls to usability by claiming that things we regard as beautiful will be used better—not just because we like them (though that effect cannot be overestimated), but also because our liking of them produces cognitive effects including greater creativity and an ability to be less distracted by inefficiencies in either the design or the problem context itself. The relationship between etiquette and aesthetics is far from clear, but insofar as adherence to proper etiquette produces more "pleasure" than violations of etiquette, Norman's claims should hold for good HCI etiquette as well.

In the end, an etiquette perspective on human-computer relationships may not so much represent a qualitative difference over traditional approaches as rather another perspective that can afford new insights. After all, designing a system for "usability" *should* include designing for user affect, perceived competence and trust, etc. The calls cited above for new approaches are *de facto* evidence that traditional approaches have not been completely successful in providing those foci. As our automation becomes more ubiquitous and autonomous, we increasingly need to attend to its manners if we are live comfortably and successfully with it. Taking an etiquette perspective in design is one way to accomplish that goal.

# 4 References

Billings, C. (1997). *Aviation Automation: The search for a human-centered approach*. Erlbaum. Mahwah, NJ.

Brown, P. & Levinson, S. (1987). *Politeness: Some universals in language usage*. Cambridge University Press; Cambridge, UK.

Cassell, J., Sullivan, J., Prevost, S. and Churchill, E. (Eds.) (2000). *Embodied Conversational Agents*. MIT Press, Cambridge, MA.

McNeese, M., Salas, E. & Endsley, M. (2001). *New trends in cooperative activities: understanding system dynamics in complex environments*. HFES. Santa Monica, CA.

Miller, C. A. (Ed.) (2002). *Working Notes of the AAAI Fall Symposium on Etiquette for Human-Computer Work*. November 15-17; N. Falmouth, MA.

Miller, C. & Funk, H. (2001). "Associates with Etiquette: Meta-Communication to Make Human-Automation Interaction more Natural, Productive and Polite". In *Proceedings of the 8th European Conference on Cognitive Science Approaches to Process Control*. September 24-26, 2001; Munich. 329-338.

Morris, W. (Ed.) (1978). *The American Heritage Dictionary of the English Language*. Houghton Mifflin, Boston. 451.

Norman, D. (2002). Emotion and Design: Attractive things work better. *Interactions*. July/August. 36-4

Preece, J. (Ed.) (2002). Supporting Community and Building Social Capital. In *Communications of the ACM*, 4(4), 36-73.

Reeves, B. & Nass, C. (1996). The Media Equation. Cambridge University Press, Cambridge, UK.

# The Imaginative Powers of the User´s Mind
## - a prerequisite in Human-Computer Interaction

*Janni Nielsen*

Copenhagen Business School, Department of Informatics
Howitzvej 60, DK 2000 Frederiksberg, Denmark
janni.nielsen@cbs.dk

## Abstract

The interface is visual and designed to communicate, however, it is interaction that distinguishes digital interfaces from the more traditional media. With the aim of pursuing a multi disciplinary theoretical framework for design of interactive interfaces different perspectives on visual communication: communication studies, film and media and architecture are discussed. It is argued that without the imaginative power of the user's mind, visual communication will not open for interaction. Hence, it is necessary to enhance the theoretical framework and the theory of tacit inferences, the imaginative powers of the user's mind is discussed arguing that the basis for making sense out of fragmented visual information is mental immersion.

## 1    Steps To A Theoretical Frame For Interface Design

Creative disciplines such as graphic design, media and film making focus on visual communication. The development in the IT-industry indicates that design of pleasurable and aesthetic interfaces are a competitive advantage[1], and different research traditions bring different conceptualisations. However, the field has yet to establish a multi disciplinary foundation.

Only few theoretical models have been developed though visual communication is also fundamental for design of interactive interfaces. However, Thorlacius´ theoretical work (2001, 2002) offers an approach to analysis and design of visual communication. The author introduces the concept; the emotive function which is ascribed to both the addresser and addressee: the addresser may possess/may not possess expressions of emotions that s/he evokes in the user, and the addresser may unintentionally evoke expressions of emotions in the user. The emotive is closely related to the aesthetic aspect, which is differentiated into the expressible – and the inexpressible aesthetic aspects, the origin of which is ascribed to senses and feelings. The inexpressible is the ability of visual language to communicate that which cannot be classified, and Thorlacius´ understanding of the inexpressible aspects extends to a user. It is a mutual experience (the designer´s as well as the user´s feelings of inexpressibility), of being unable to express what s/he perceives.

By focusing on the emotive and the aesthetics, Thorlacius brings the user's experience (of pleasure, beauty, ugliness) into design and research and points to the psychological powers at work. However, the model does not leave room for the subjective understanding that the individual brings to the interaction, nor does the design approach deal with the interactivity of the

---

[1] Navision – personal communication

interface. Objects on the screen move, objects emerge, windows appear, click on objects lead to new digital pages etc. Hence interfaces are dynamic not static.

Filmmaking is the creation of dynamic pictures, and it is not showing, but constructing an aspect of vision that the film-viewer needs to make sense of the film (Davis 2002). The filmmaker does so by offering, "fragmentary evidence, organised with a view to affording certain assumptions and interpretations,..", and Davis continue: ". and the film-viewer (partly on the basis of shared conventions) duly makes those interpretations". He suggests that ideally the design should make users "believe that they are interactive observers of a world" and this requires psychological immersion in the spatial design Davis argues that interactive interfaces are constraint in their spatiality and suggests exploring the issue by focusing on the "make believe", found in the "spatial maturity" in filmmaking. Davis´ concept of the participant observer opens for further analysis of the psychological – or mental – processes of the user. However, in the world interactive interfaces, it is not sufficient to observe, to participate in sense making and to feel immersed. The user is required to act, to engage in mental visual interactions, to become an active user and contributor to the interaction on the basis of shared conventions, which cannot be separated from the subjective powers of human imagination.

In architecture visual communication is understood as embodied spaces that invites interaction, and this opens for the actor's perspective. The essence of immersion in architectural prototypes is physical - body moving into, and Luescher (2002) assumes that it is the same physical immersion which takes place in relation to panoramic paintings: bodies and heads moving along, and eyes roam. He uses this understanding to question the claimed revolutionary potential of the computer in terms of imaging and the promise of immersion. He argues that this exactly the limitation of the computer because computer representations remain screen-bound, small-scale and impenetrable. Hence, engagement in a computer based design will never result in "..bodily, sensuous, emotional and intellectual understanding of space". But how can it be separated from the psychological immersion in the image? Physical interaction – as the body moves alongside a painting or into a man size architectural construction - is also a mental (psychological) interaction. The construction is a simulation, which requires the creative powers of the human mind; to imagine the experience of moving around in an imagined space while moving in its simulation. This is one of the incredible forces of human life. We are constantly engaged in making sense whether we interact with people, watch a movie, surf on the net or use an administrative program. It is the imaginative powers of the mind that are at play here. In the following, a model of human perception and sense giving is introduced in order to contribute to the multi disciplinary theoretical framework.

## 2    The Imaginative Powers of the Mind

Michael Polanyi (1968) argues that perception fundamentally rests upon indeterminacy[2] suggesting that "explanations must be understood as a particular form of insight". This implies that there are other kinds of insights, and is these other (psychological) forms, which has his interest. Their essence is tacit inferences and tacit knowledge based on "indefinable powers of thought". As such perception is beyond language, a psychological process that established an observation of external facts "without formal argument and even without explicitly stating the results". The foundation for this process is focal awareness and subsidiary awareness. One can be consciously aware and focus on something specific e.g. the meaning of the heading above; this is focal awareness. At the same time, other factors in the borderline of one's conscious awareness are

---

[2] in content, in the connections we see and in the data upon which we base our results. He

interacting and supporting the focal awareness, e.g. each letter, the spaces etc., this is the subsidiary awareness. Meaning, comes into being through an act of sense giving: we come to know, a tacit inference, a tacit act which is irreversible, done through a process of integration, and we do it by projecting our senses out into the world.

The process of projecting ourselves out into the world is also described as "pouring one's body into" or interiorising it. To make "something function as subsidiary is to interiorise it, or else to pour one's body into it". Polanyi calls the ability to inhabit the artefacts around us for "indwelling". He considers this cognitive and emotive qualification essential for the construction of knowledge and meaning. Our senses point into external space just as our actions are projected outward and the objects of our conscious attention lie predominantly away from us. Perception is a very complicated cognitive process through which we, in the act of giving sense come to understand our world. Hence, knowing is not the same as verbalisation. On the contrary, it is much more than what we can verbalise and it involves the imaginative powers of the mind

# 3 Human Computer Interaction

In a design process, one moves through many drafts and prototypes of the interface design. During this process, design elements come together in incidental and unplanned ways. Some of these "coming together" incidents may speak so clearly for themselves that the designer includes them in the design – without realizing that they have "constituted themselves". Hence, the design elements themselves may contribute to the unintentional, but expressible aspects that cannot be classified. However, my interest is a psychological frame that may enhance the understanding of the user. In this connection, the aesthetic function – the inexpressible aspect that the user perceives although it cannot be explicitly verbalised - becomes interesting. This step in the analysis of the visual communication takes us close to the psychological forces working in the subject. But it stops short of interaction – and it is not able to explain the psychological subject or what the user brings to the understanding of visual communication. However, it is a prerequisite that somebody interacts psychologically with the site, reads the screen, makes sense of the information, navigates, etc. Interaction with an interface is more mental than physical. It is not sufficient for a hand to click on the mouse to make communication work. The computer is a symbolic processing machine, presenting graphics on the screen that meet the eye and interact directly with our mind, and visual communication cannot be separated from mental interaction.

## 3.1 Giving sense to fragmentary life
Fragmentary evidence is organised to enhance certain assumptions and interpretations, and on this basis, we may look at the viewer as participating in creating understandings. Here, Davis (2002) indicates the psychological activity required, and as such he makes room for mental processes. But in traditional films, the viewer is an observer, because s/he has no influence on the story unfolding – the filmmaker has designed its course. However, it is the assumptions about the human mind embedded in Davis´ theory that I find interesting. Davis believes that a human being is able to make sense out of fragmented evidence. It is a general human competence of meaning construction (Bruner, Jerome 1990). Human beings live a life that is fragmented – seen from the outside. Fragmented information, fragmented interaction and fragmented communication fragment our actions, etc. However, seen from the inside of an individual human being, his/her life is coherent because of his/her ability to make sense – it is a continuous (life)story. We are engaged in a constant process of sense making, seeing coherence where there is no coherence and constructing life as meaningful. Life becomes meaningful because we are not observers; we live

life and understand it from an embodied point of view. We are immersed in life, whatever actions we engage in – we are not outside it. (Winograd Terry and Fernando Flores, 1984).

Fragmentary evidence only appears as fragments on the background of nothingness and the black holes between the fragments. Making sense of fragmentary evidence and the background is an active process, in which the elements must be constructed into coherent stories. Such a process requires mental immersion and not just observations. It is necessary to project our senses out into nothingness, fragments and incoherences. Only by dwelling in may coherence come into being, and here the fragmentary evidence and the black holes function in the subsidiary awareness of our focus, bearing down on our focal awareness. In a tacit process of from-to-knowing, the two elements are connected to a whole, coherence is seen and sense is given. This is a psychological process. Just think of the process of surfing on the web, trying to find information that is not specified clearly – or think of reading and writing a paper with fragmented sentences, fragmented understandings and fragmented knowing, which the surfer or writer brings into mental interaction.

## 3.2   Psychological immersion in virtual spaces

The user's experience of psychological immersion depends upon the representation of spatiality, which will induce users to believe that they are participant observers of the world. As such, the make-believe of filmmaking offers a possibility for psychological immersion. Ideally, I (the film viewer) must feel that I am up on the movie screen living among the people, I must mentally project my senses, indeed myself on to the screen and move around in the filmic space. I must forget my physical body and project my self onto the screen – I must feel psychologically immersed and believe that I am a participant observer. In this way, we may understand how the participant observer engages in a mental process where s/he mentally connects the fragments into coherence.

But in interactive media, visual communication must also be understood in relation to interaction, the user's visual interaction with the media. Interaction is not just based on "make believe". To some extent, we might say that there is no need for make believe in interactive media. The media requires that the user acts and interacts with the visual elements, changes, reconstructs or deconstructs, navigates into, moves on etc.[3] Here, immersion may be understood based on our ability to be in the world and to project our senses out into the world. We may speak of our ability to pour our bodies into the visual, dynamic graphics unfolding before our very eyes. We are only able to make sense of images, interact with them and work with them through the extension of our senses into the visual representation. The process kind of involves moving from inside-out and from outside–in, through which an internalisation takes place; we become one with the sense-giving and meaning construction.

This mental process does not differ from immersion in spatial images represented in solid material. Moving bodies along and into – is also an embodied perceptual experience. It is a physical immersion that can only be experienced by projecting our bodies into something, and what we feel is not the man size world, but the place where our senses encounter the world. In a phenomenal transformation, we give sense to our perceptions. Luescher  (2002) questions immersion in two-dimensional screen bound interfaces as opposed to immersion in panoramic paintings where our eyes must move around in and our body move alongside the painting.

---

[3] Obviously, the user's control over the information, interface and interaction depends upon the specific system. For the purpose of a theoretical discussion on visual interaction, I work from an ideal model of users' ownership of information and control of the interface.

However, immersion is not tied to and restricted by movement of the body. Besides, physical immersion in panoramic paintings cannot be separated from psychological immersion. As we move along the painting, our senses are assaulted by the traces left by the brush on the canvas, the colours in fat oily strokes, thousands of points, shapes, images, smell, etc.

Design is not just a question of offering fragmentary evidence that affords interpretations by the user. Interactive interfaces offer more than films and paintings. The space is visual as in paintings or films. But it offers fragmentary evidence that entices the user to act – to interact with the interface. It offers interaction that may open for exploration of representations, psychological involvement and result in bodily, sensual, emotional and intellectual understandings, even the experience of bodily and psychological immersion. And the activity becomes meaningful because the users are engaged in interaction and understand it from an embodied point of view. Users are immersed, also when they engage in actions such as interacting with an interface.

Computers offer representations, dynamic representations that embed possibilities of sensual, emotional and intellectual engagement; even possibilities of virtual exploration of materials and the electronic images and animations of designers' imagined vision. These representations require the user's participation, and the experience may enhance intellectual, sensual, emotional and even bodily understandings.

## References

Bruner, Jerome (1990. *Acts of Meaning*, Harvard University Press.
Davis, Steven Boyd (2002). Interacting with Pictures: film, narrative and interaction, *Digital Creativity*, vol. 13 no. 2, 71-85
Luescher, Andreas (2002) The Physical Trace, *Digital Creativity*, vol.13, no.2, 99-108
Nielsen J. Clemmensen T. and C. Yssing(2002) People's Heads, People's Minds? Theoretical reflections on Thinking aloud, in Dai Guozhong (ed) Proceedings of the 5th Asian Pacific Conference on computer Human Interaction:, vol. 2, 897-910
Nielsen, Janni(2002)Visual Cognition and Multimedia Artefacts in Danielsen O., Nielsen J. and B.H. Sørensen (eds) *Learning and Narrativity in Digital Media*, Samfundslitteraturen, 113-123
Nielsen, Janni: The Role of Text and Dynamic Graphics in Knowledge Acquisition, in McDouglas A. and C. Dowling (eds.) *Computers in Education*, Elsevier Science Publ. IFIP, 1990, pp. 107-115
Polanyi, Michael (1968) Logic and Psychology, *American Psychologist*, 23, 27-43
Thorlacius, Lisbeth (2001). *Model til analyse af lexi-visuel, æstetisk kommunikation – med et særligt henblik på websites*, Roskilde University, Denmark
Thorlacius ,Lisbeth (2002). A model of visual, aesthetic communication – focusing on web, *Digital Creativity*, vol. 13, no. 2, 85-98
Winograd Terry and Fernando Flores: *Understanding Computer and Cognition*. Addison Wesley, 1987.

# CII: A Taxonomic Model of Innovations in Human-Computer Interaction

*Timo Partala*

Tampere Unit for Human-Computer Interaction,
Department of Computer and Information Sciences,
FIN-33014 University of Tampere, Finland
tpa@cs.uta.fi

## Abstract

In the current paper, a new taxonomic model of innovations in human-computer interaction (HCI) is proposed. The model consists of three basic components of innovations: *change, improvement and indicator* (CII). The change component of the proposed model is divided into five categories: input-physical, input-interpretation, information processing, output-information presentation, and output-physical. The improvement component is divided into seven different subcategories: spatial, temporal, associative, coordinational, social, affective, and motivational. The indicator component is divided into five categories: performance measures, subjective ratings, and different cognitive, behavioral, and physiological measures. It is suggested that the CII model could be useful in both designing new HCI innovations and categorizing existing innovations.

## 1    Introduction

During the past few decades, a great number of innovations have been presented in the field of HCI. These have included a range of innovations from small improvements within software products to heavy-scale physical input device designs. Despite the large number of innovations, there have not been very many attempts at systematically structuring the design space of innovations related to HCI. Thus, it seems that there is room for new models related to HCI innovations.

While a great number of different models have been suggested in the field of human-computer interaction, many models have only focused on one aspect of human-computer interaction. For example, common model types include models of the user, the task, the computer system, or the surrounding world, in which the user performs the task (Nielsen, 1990). Consequently, it seems that most of the models in the field of HCI are quite narrow in scope. On the other hand, there are also some very complex models, consisting of hundreds of elements, for example, the model by Gavron, Drury, Czaja & Wilkins (1989), which is a taxonomy of variables affecting the user's performance in HCI.

Some of the classic successful models in human-computer interaction include the human processor model by Card, Moran & Newell (1983), the GOMS (goals, operations, methods, and selection rules) model and the family of related user interface analysis techniques (see e.g. John & Kieras, 1996), and the model of the design space of input devices by Card, MacKinlay & Robertson (1990). Later, the focus of suggested models shifted to, for example, models attempting to structure aspects of multi-modal HCI (e.g. Bernsen, 1994; Nigay & Coutaz, 1993).

The empirical evaluation of innovations is an integral part of research in HCI. That is why the models that have concentrated on the evaluation of human-computer interaction (e.g. Nielsen, 1993; Shneiderman, 1998; Sweeney, Maguire & Shackel, 1993) are also potential contributors to models of HCI innovations. The model presented in the current paper is partially based on the model by Sweeney et al. (1993). They distinguished between theory-based, expert-based, and user-based usability evaluation. Importantly, they defined six classes of usability indicators in user-based evaluation: performance (e.g. task times), non-verbal behavior (e.g. documentation access), attitude (e.g. questionnaire responses), cognition (e.g. verbal protocols), stress (e.g. galvanic skin response), and motivation (e.g. enthusiasm).

In the current paper, a new model of HCI innovations called CII is presented. The model is based on an analysis of recent HCI innovations as well as on the model by Sweeney et al. (1993). As opposed to many existing models, the present aim was to create a model that is broad in scope, but simple enough to be described briefly.

## 2 The CII Model

### 2.1 The Model Development Process

The CII model was created in the following way. First, a set of about 50 recent HCI innovations was selected from the proceedings of the ACM CHI conference series (years 1998-2002). All selected innovations had a clear HCI-related improvement, which was also validated in user evaluation. Second, the key components (i.e. changes, improvements, and indicators) related to each innovation were identified. Third, all the different changes, improvements, and indicators were organized into categories, and all the perceived overlaps were removed. Fourth, the categories were compared to those presented in the previous research (Sweeney et al., 1993) and the model was slightly revised. Finally, it was checked that the originally selected innovations still fell nicely into the categories of the updated model.

### 2.2 Description of the Model

The proposed model consists of three basic components of HCI innovations: *change, improvement and indicator* (see table 1). In the model, change refers to the technical novelty, which constitutes the practical implementation of the innovation in computer software and/or hardware. Improvement refers to the nature of the intended improvement on the user side, while indicator refers to the measurement indicators, which can be used to validate the improvement.

**Table 1:** Illustration of the CII Model

| CHANGE | Input-physical | Input-interpretation | | Information processing | Output– information presentation | | Output – physical |
|---|---|---|---|---|---|---|---|
| IMPROVEMENT | Spatial | Temporal | Associative | Coordinational | Social | Affective | Motivational |
| INDICATOR | Performance | Subjective | | Cognitive | Behavioral | | Physiological |

In the CII model, the change component is divided into five categories: input-physical, input-interpretation, information processing, output-information presentation, and output-physical. The innovations in the input-physical and output-physical categories are related to the design of physically novel input or output technologies, for example, a novel 3D input device or a new kind of input device with tactile feedback. The innovations in the input-interpretation and output-information presentation categories are related to the software processing of user input (e.g. mouse acceleration) and information presentation in output (e.g. a new information presentation method for audio interfaces). The information processing category refers to changes in software that are not directly related to input or output. For example, a more sophisticated artificial intelligence model could improve both input and output handling in conversational interfaces.

The improvement component of the proposed model is divided into seven different categories: spatial, temporal, associative, coordinational, social, affective, and motivational. The improvements in the spatial category could include, for example, more efficient use of screen space or physical space. An example of a temporal improvement could be caused by any change in the user interface, which enables the user to carry out a given task in a shorter time. The associative improvements could include, for example, better metaphors in the user interface or tools for supporting the user's creativity. An example of a coordinational improvement could be a tool that enables improved hand-eye coordination in a given task. Social improvements might include improved communications with other computer users, for example, while collaborating in a groupware application. Possible affective and motivational improvements include, for example, a system, which provides affective support to the user, and a learning environment, which aims at enhancing the user's learning motivation.

Finally, the indicator component can be divided into five categories: performance measures (e.g. task times and error rates), subjective ratings (e.g. subjective satisfaction, attitude towards the system, and ratings of affective state), as well as different cognitive, behavioral, and physiological measures (e.g. verbal reports, eye movements, and heart rate, respectively).

Obviously, it is possible that innovations may fall into more than one category inside each component of the CII model. For example, a new 3D input device could differ from existing devices in physical appearance and it could also use sophisticated mathematical models in the input interpretation. The improvement of such device could be both spatial and temporal, possibly also coordinational, for example, if two hands could be used effectively in co-operation to control the device. The possible indicators of the improvements could be performance (decreased task times and error rates), and possibly more positive ratings of subjective satisfaction, too.

When using the CII model, one has to realize that the improvement component of the model is in most cases relative to the product, to which the new innovation is compared in empirical evaluations. Though the researcher may choose, to which product the new innovation is compared, it is common to use the best prevailing empirically validated innovation as a point of comparison. However, many new products are quite large and may contain many improvements. For this reason, it can be presumed that in some cases the user of the CII model is only interested in the primary improvement and the primary indicator of a certain change.

# 3 Discussion

In the current paper, a new taxomonic model of innovations in human-computer interaction is suggested. The model is based on an analysis of a set of recent innovations and their empirical evaluations, as well as on an existing model (Sweeney et al., 1993). However, it is suggested that the model reorganizes the design space of HCI innovations in a new way by bringing the technical change, the nature of the improvement, and the indicators into a single model.

In the current model, the categories related to the change and improvement components were discovered based on the analysis of existing innovations. In contrast, the indicator category was based on the work of Sweeney et al. (1993) and it was only slightly modified. It is worth noticing that in the current model, motivation is considered as a category of improvement, while Sweeney et al. (1993) considered it as a usability indicator. In the CII model, the indicators of motivational improvements (e.g. subjective ratings of willingness or enthusiasm) would supposedly fall into the subjective indicator category. Another difference between the indicator component of the current model and the model by Sweeney et al. (1993) is the label for the physiological indicator category, which was updated from 'stress' to 'physiological improvements'. This broadens the scope of the category so that other than stress-related physiological measurements, for example, affective physiological measurements can be included, too. Indeed, in the field of HCI, there are recent efforts towards technology that supports the user on an affective level (Partala & Surakka, accepted). This aspect, missing from most HCI-related models, has only got major research interest in the field of HCI during the past few years.

In the development of the CII model, the emphasis has been in categorizing innovations as final products rather than their development process. For example, a new tool for user interface development could be regarded as a HCI innovation. However, the current model has been designed to categorize ready-to-use HCI innovations rather than their design tools. On the other hand, the model could be used in categorizing the innovation in the user interface development tool itself. For example, an improved information presentation in the user interface development tool could lead to temporally improved usage indicated by, for example, faster performance in development and enhanced subjective attitudes towards the tool. Similarly, the CII model is not especially suitable for analyzing evaluation methods, even though most researchers would probably agree that, for example, a new usability evaluation method could be characterized as a HCI innovation.

The CII model has been primarily designed so that the potential users of the model would be HCI researchers, the developers of new HCI innovations, and HCI students. The model is potentially useful in both designing new innovations and categorizing existing innovations. In innovation design, the model can be used as a checklist that displays possible types of changes, improvements, and indicators. The CII model can also be useful in categorizing existing innovations and thus bringing some order to the ever-growing mass of all HCI innovations. Because the goal was to keep the model simple enough to be easily understood and remembered, the model might also be useful in HCI education.

In order to fully understand the whole design space of innovations in human-computer interaction, and all the possible research aspects, further research is needed. The author suggests that the CII model presented in this paper can serve as one possible starting point in the further modeling of the human-computer interaction research field.

# 4 Acknowledgements

The author would like to thank University of Tampere and Tampere Graduate School in Information Science and Engineering for support.

# 5 References

Bernsen, N. O. (1994). Foundations of multimodal representations. A taxonomy of representational modalities. *Interacting with Computers* 6 (4), 347-71.

Card, S. K., Moran T. P. & Newell, A. (1983). *The Psychology of Human-Computer Interaction.*

Card, S. K., MacKinlay, J. D. & Robertson, G. G. (1990). The Design Space of Input Devices. In J. C. Chew & J. Whiteside (Eds.), *Proceedings of CHI 1990* (pp. 117-124). New York: ACM Press.

Gawron, V. J., Drury, C. G., Czaja, S. J. & Wilkins, D. M. (1989). A Taxonomy of Independent Variables Affecting Human Performance. *International Journal of Man-Machine Studies* 31 (6), 643-671.

John, B. E. & Kieras, D. E. (1996). The GOMS family of user interface analysis techniques: comparison and contrasts. *ACM Transactions on Computer-Human Interaction* 3 (4), 320-351.

Nielsen, J. (1990). A Meta-Model for Interacting with Computers. *Interacting with Computers,* 2 (2), 147-160.

Nielsen, J. (1993). Usability engineering. London: Academic Press.

Nigay, L. & Coutaz, J. (1993). A design space for multimodal systems: concurrent processing and data fusion. In *Proceedings of INTERCHI'93* (pp. 172-178). New York: ACM Press.

Partala, T. & Surakka, V. (accepted). Pupil size as an indication of affective processing. Accepted for publication in *International Journal of Human-Computer Studies.*

Shneiderman, B. (1998). Designing the user interface. Boston: Addison-Wesley.

Sweeney, M., Maguire, M. & Shackel, B. (1993). Evaluating user-computer interaction: a framework. *International Journal of Man-Machine Studies*, 38, 689-711.

# End-User Requirements for Seamless and Transparent Middleware

*Päivi Pöyry & Lauri Repokari*

Helsinki University of Technology
Software Business and Engineering Institute
P.O.Box 9600, FIN-02015 HUT, Finland
Paivi.T.Poyry@hut.fi, Lauri.Repokari@hut.fi

## Abstract

Middleware presents a challenge to those defining user requirements, because there may be many different kinds of users of the technology, and the technology itself may be invisible for some users while being visible to others. The purpose of this paper is to introduce a methodology of data collection and analysis to meet the requirements of end-users who are developing middleware technology for seamless and transparent services. The context of the study consists of combining two major research methodologies in an iterative manner. As a result of the study, conducted as a part of the NOMAD project, we will introduce the use trait model for seamless and transparent mobile services. The main findings of the study also include the possibility of categorizing reliably user needs based on statistical and qualitative research methods. However, the factors affecting the reliability and validity of the research must be considered carefully.

## 1 Introduction

This study was conducted as a part of the NOMAD (http://www.temagon.gr/nomad) project that is funded by the European Commission's IST programme in the 5th Framework Programme. The aim of the NOMAD project is to develop and demonstrate middleware capable of seamlessly integrating available and future technologies, i.e. UMTS and WLAN, as well as IP compatible multi-hop ad-hoc networks, into a single integrated NOMAD platform, employing new algorithms for the parallel use of multiple access interfaces. The product that will be developed in this project is seamless and transparent middleware that gives network and value added service providers, for example, the opportunity to carry out different business transactions.

Middleware presents a challenge to those defining user requirements because there may be many different kinds of users of the technology. For example, NOMAD will have end-users such as consumers of the mobile services enabled by NOMAD technology, but the service providers are NOMAD users as well. The NOMAD technology itself may be rather invisible for the 'consumer end-user', whilst the 'service provider-user' may be very well acquainted with the technology. In this study, the focus is on the consumer end-users and their needs and requirements for the middleware and services enabled by the platform.

In the NOMAD project, a *methodology* for defining end-user requirements especially for middleware was developed and tried out. In this study, this methodology was found suitable for categorizing the *use traits* that will be used in the middleware product development process. Both qualitative and quantitative research methods were used in the user requirements definition study. The potential end-users of the services enabled by the NOMAD technology were observed and interviewed and their opinions and attitudes towards technology were investigated with a questionnaire developed especially for this purpose. To sum up, the context of this study consists of combining two major research methodologies in an iterative manner. In this process, three major methods were used: observation, interviewing and survey-questionnaire.

The purpose of this paper is to introduce a methodology for data collection and analysis of end-user requirements when developing middleware technology for seamless and transparent services. As a result of the study, the use trait model developed with the methodology will be introduced. In a central role in this methodology is the data gathering and analysis process according to which the research progresses. Using this methodology it is possible to define factors that represent the user needs and use traits of the product to be developed. From these factors, or use traits, it is possible to derive the user requirements to be utilized in the product development process. However, the subjects of the study must be chosen carefully so that they are representative of the whole end-user group.

## 2 User Requirements in Product Development

A *requirement* can be defined as a function, constraint or other property that the system must provide in order to meet the needs of the users (Faulk 1997). A *user need* is defined as a task or a goal the user wants to perform or accomplish with the help of a product (Kotonya et al. 1998). The requirements of a product to be developed usually include technical requirements related to the system and the user requirements concerned with the users' tasks, needs and expectations (Kauppinen 2002). The user-centred analysis used in this study aims to capture the requirements from the users' point of view (Rumbaugh 1994). It is recommended that information be gathered on the current practices and behaviour of the end-users in order to gain insight into future needs, because the problems and needs of today are likely to continue into the future too (Patnaik et al. 1999). Requirements engineering is a systematic way of defining, managing and testing the requirements of a system (Kauppinen et al. 2002). User requirements are used to create the system concept and specification; they also are used as evaluation criteria in the later phases of the product development process (Mäkelä et al. 1999).

## 3 The Methodology for Gathering User Requirements

The methodology used in the NOMAD project for gathering the end-users' requirements follows the main principles of requirements engineering. In this process, both qualitative and quantitative research methods were used successfully. Data gathering by combining multiple methods aims to capture and define reliably the user needs and user requirements of the end-users of the NOMAD technology. Respectively, both the qualitative and the quantitative methods of analysis were used to extract the use traits from the data and to balance the weaknesses of one methodology with the strengths of the other. (Sudweeks et al. 1999) Qualitative content analysis was used for the interviews and observations. Statistical methods, such as factor analysis and analysis of variance, were used for the questionnaire data analysis. As a result of the analyses, a model of the use traits of mobile ICT devices was defined; on the basis of this, user requirements could be inferred.

In the initial phase of the study, potential users were observed and the information gathered used as a basis for the end-user interviews. Both the observations and the interviews were performed in the setting of an international airport, where people, most of whom are frequent travellers, use their mobile information and communication devices. The airport was selected because the end-users should be interviewed in their own environment, i.e. in the real context of use, in order to capture richer information about the user (Millen 2000). It was on this basis that the initial interview framework was formulated.

On the basis of the data gathered in the first semi-structured interviews, a new interview framework was developed in order to gather information from the users on their attitudes towards the use of new technology. This interview framework was again semi-structured, but, instead of questions, it consisted of statements that served as a basis for the conversation between the interviewer and the interviewees. The statements were used for eliciting the users' opinions and attitudes towards the new technology, as well as the user needs and requirements. It is known that users find it very difficult to define their needs and requirements if asked with direct questions. Instead, questions concerning the user needs and requirements should be embedded in realistic and concrete use-contexts if they are to elicit users' answers successfully. (Beyer et al. 1998.)

The survey questionnaire for end-users was formulated on the basis of the interviews and user scenarios of the NOMAD end-users. The questionnaire consisted of multiple-choice questions. Firstly, there were questions concerning the background information of the users, such as their age group, gender, as well as the use of the cell phone of the respondent and his or her geographical region. The second part of the questionnaire consisted of questions, or statements, that the respondent evaluated on a scale of 1 to 3, with 1 standing for "Agree", 2 for "Partly agree", and 3 for "Disagree". Our data collection effort focused on end-users and business users in the subject area of mobile connectivity at the CeBIT held in Hannover, 2002. We were able to collect 58 responses to the end-user questionnaire.

# 4    The Analysis and Results of the Study

The end-user questionnaires were analysed by using statistical methods with the help of SPSS-program. Frequencies and other descriptive statistics were calculated in order to get an overall picture of the data. Explorative factor analysis was used for extracting the user needs and traits. Analysis of variance (ANOVA) was used for comparing means between different end-user groups. Finally, correlations for finding connections between the factors were calculated.

Altogether 58 persons filled the end-user questionnaire at CeBIT. Of the respondents, 34,5% was female and 65,5 % male. The majority of the end-users were between 31-40 years of age. While most came from Europe, some came from Asia and North America as well. The end-users used the cell phone frequently; more than 70% used the cell phone daily, while less than 20% used it less frequently than once a week. The most popular mobile device (other than the cell phone) was the laptop; more than 50% had one. Approximately 30% had a Palm. Other mobile devices were quite rare. It was noticeable that almost 30% had no mobile device at all.

An explorative factor analysis was performed, and the Principal Axis Factoring method used to extract the factors. The Varimax rotation method was used for rotating the initial solution. Six factors with Eigenvalues >2 were extracted from the data. (Munro 2001; Coolican 1994.) Reliability analysis was performed, i.e. the Cronbach's alpha was calculated to the sum variables, which were formed on the basis of the factors' content variables. The sum variables were calculated for enabling further analysis on the data (e.g. means comparison, correlations between the factors) (Munro 2001.) As a result of the factor analysis, a model of user needs as well as traits of using the mobile IT devices was constructed.

One-way analysis of variance (ANOVA) was used for comparing the means of the sum variables formed on the basis of the factors. In this analysis, the means of the factors were compared in order to find out whether there were significant differences between the groups. The groups used in the comparison were formed according to the following criteria: age group, gender, use of cell phone and geographic region. The significance level of 0.05 (at least) was used in this analysis (Munro 2001.) The result was that no significant differences were found in the means between any of the groups. This indicates that the end-users form a rather uniform group with regard to user needs and use traits.

According to the analysis of the questionnaire data, there are six different factors or use traits, i.e. ways to work with Mobile IT devices. These are the basis for the user requirements of NOMAD technology. The user needs and users' requirements are, according to our analysis, mainly related to security and privacy, new technology and lifestyle, lead users, daily IT use, the (in)convenience of new technology and shopping on the Web. The factors and their reliability coefficients (Cronbach's Alpha) are presented in Table 1. The Alphas' values exceeding .60 indicate that the results of the study are reliable, i.e. internally consistent and stable (Munro 2001; Coolican 1994).

**Table 1.** Description and reliability of use traits. Bullets: first interpretations of user requirements

| |
|---|
| **Use Trait & Requirements related to security: [Alpha 0.73]** Security is experienced as a problem, which means that the user must have a feeling of security when using the mobile devices. Moreover, to be able to "work" with devices the users need the possibility of privacy. <br> ➢ User must be able to easily verify the security of the service he/she is using. <br> ➢ System must give an easy way of protecting against intruders and possible viruses. <br> ➢ Security should not be too complicated, but auto-configurative rather than configurable. |
| **Use Trait & Requirements related to technology and lifestyle: [Alpha 0.75]** Technology and mobile IT devices are seen as an indicator of a modern lifestyle – so technology must be "cool". The user must also be able to get information about the technology (e.g. how fast it is, how efficient it is, how powerful it is). <br> ➢ Devices must be designed to indicate the lifestyle of the user – (mass) design products. <br> ➢ Users (and others) must be able to recognize the efficiency of the product easily (by design or by other factors). |
| **Use Trait & Requirements related to lead users: [Alpha 0.72]** It is important for the more advanced users to have a feeling that "I am using the latest technology". <br> ➢ Users must have an easy way to get information about the technology they are using and about upcoming technology. |
| **Use Trait & Requirements related to Daily IT-use: [Alpha 0.72]** Ease of use is an important issue. <br> ➢ Middleware solutions must be able to be used with PDA. <br> ➢ E-calendar is in wide use, which means that the NOMAD technology should support it. <br> ➢ Solutions supporting e-calendar or timed services are possible for the daily users of IT devices. The only thing is that they must be easy to use. |
| **Use Trait & Requirements related to "the Inconveniency" of new technology: [Alpha 0.64]** Technology can be very inconvenient to the users. <br> ➢ NOMAD technology must have an easy way to switch it off or an easy way to customize the user profile so that the user can have privacy when needed (cooperation with E-calendar). |
| **Use Trait & Requirements related to shopping on the Web: [Alpha 0.66]** subjects were willing to do shopping with mobile devices. <br> ➢ NOMAD technology must support shopping on the Web. <br> ➢ If NOMAD technology supports advertisements, there must be an easy way to access and deny advertisements. <br> ➢ When shopping, there must be a secure way to make money transactions. <br> ➢ Nomad technology must give an instant way to accept a deal, e.g. if the customer gets (location based) advertisements "buy now and you will get a reduction". |

# 5    Conclusions

The results collected in this project are comparable to results in the other projects in the area. This fact can be interpreted as a validation of each other's results, even though the focus of the studies is not exactly the same. The main findings show the possibility of categorizing potential users' needs based on a combination of statistical and qualitative research methods. In addition to data coinciding with results from previous studies, we got valid information about end-user requirements for seamless and transparent services. As the type of services made possible by NOMAD is not currently available for common use, the results of our user studies are important and will be the basis for NOMAD technology functional specification in NOMAD product development.

In this study, the statistical methods were found to be suitable for the definition of user requirements. However, these methods are applicable only if the presuppositions and assumptions are correctly set so that the users participating in the study really represent the target group. The weakness of using this statistical method is the validity: the results are only statistical "truths" that must be validated as a part of the product development process. Future research is planned for testing the use traits and user requirements. The testing and validation of user requirements will be integrated into the field trials of the NOMAD platform during the fall of 2003.

# References

Beyer, H. & Holtzblatt, K. 1998. *Contextual Design – Defining Customer-Centered Systems.* Morgan Kaufmann Publishers, San Francisco.

Coolican, H. 1994. *Research Methods and Statistics in Psychology.* 2nd edition. Hodder & Stoughton.

Faulk, S. 1997. *Software Requirements: A Tutorial.* Software Requirements Engineering, 2nd edition. Edited by H. Thayer et al. IEEE Computer Society Press.

Kauppinen, Marjo. 2002. *Requirements Engineering.* Qure project. SoberIT, HUT.

Kauppinen, M., Kujala, S., Aaltio, T. & Lehtola, L. 2002. *Introducing Requirements Engineering: How to Make a Cultural Change Happen in Practice.* IEEE Joint International Conference on Requirements Engineering, Essen, Germany, September 2002.

Kotonya, G. & Sommerville, I. 1998. *Requirements Engineering – Process and Techniques.* John Wiley & Sons Ltd. England.

Millen, D.R. 2000. *Rapid Ethnography: Time Deepening Strategies for HCI Field Research.* DIS '00, New York.

Munro, B.H. 2001. *Statistical Methods for Health Care Research.* 4th edition. Lippincott.

Mäkelä, A. & Battarbee, K. 1999. *Applying Usability Methods to Concept Development of a Future Wireless Communication Device – Case in Maypole.* Proceedings of 17th International Symposium on Human Factors in Telecommunications. Copenhagen, Denmark, May 4-7.1999. Pp. 291-298.

Patnaik, D. & Beeker, R. 1999. *Needfinding: The Why and How of Uncovering the People's Needs.* Design Management Journal 1999.

Rumbaugh, J. 1994. *Getting started – Using use cases to capture requirements.* Journal of Object Oriented Programming, Vol. 11, No.1 September 1994.

Sudweeks, F. & Simoff, S.J, 1999. *Complementary Explorative Data Analysis, The Reconciliation of Quantitative and Qualitative Principles.* Jones, S. (ed) 1999. Doing Internet Research. Critical Issues and Methods for Examining the Net.

# The Role of Voluntary Attention in HCI

*Gabriella Pravettoni*

University of the Studies of Milan
Faculty of Political Sciences
gabriella.pravettoni@unimi.it

*Sebastiano Bagnara*

Politecnico of Milan

bagnara@unisi.it

## Abstract

It was generally accepted that attention is automatically attracted by particular characteristics of the external environment. Many researchers have provided empirical evidence about which critical stimulus properties automatically draw attention. A large number of studies have shown that changes in stimuli over time are likely to capture visual attention. The majority of studies emphasize the properties of stimuli that attract attention, and do not make much reference to an individual's cognitive goals. Clearly, individuals' goals were assumed to play a role in the control of attention, but it was generally accepted that voluntary attention is disrupted by the appearance of stimuli with properties to attract. Recent studies have shown that stimulus attributes which are generally thought to have a wired-in ability to capture attention automatically, have been shown to capture attention only as a consequence of voluntarily adopted mental sets. (Pashler, 2001). This view openly challenges the implicit theoretical assumptions underlying the actual practice adopted in designing HCI contexts. Given the economic importance of use of attention (Davenport, Beck, 2001), it may be of some interest to carry out research into the role of involuntary attention, that is of stimuli which automatically receive attention. The present exploratory study has been designed to investigate this. It aims to investigate to what extent attention may be disrupted when people are engaged in a challenging task, such as learning. 48 students took part in the experiment. They were presented with an online lesson with distractors. The results demonstrated that the distractors were well-recalled, while the number was not. There was no significant interaction. The distractors had a significant effect on learning performance. The preliminary results of this exploratory study indicate that, unlike when they are presented in isolation, these types of distractors did not influence recall. Stimuli attract attention when people adopt a purely exploratory mindset. When people are engaged in a challenging and motivating activity, such stimuli do not play any differential role in attracting attention. In an HCI context, attention should be considered to a much greater extent in terms of internal guidance factors, rather than of environmental attractiveness. Involuntary attention plays a role only when people adopt a purely exploratory mindset.

## 1 Introduction

It is generally accepted that attention plays a fundamental role in most cognitive activity, irrespective of the domain of application. Most HCI environments are designed on the basis of the principle that individuals enter and stay in a HCI context because their attention is attracted and oriented by the characteristics of the material presented. Much less consideration is given to individual motivation and cognitive settings. In terms of theories of attention, this means that HCI contexts are designed with reference to characteristics of involuntary, stimulus driven, rather than of voluntary, intention-guided attention.

Over the last decade, understanding of the notion of attention has undergone a transformation. Previously, it was generally accepted that attention, notwithstanding personal mindset, is automatically attracted by particular characteristics of the external environment. Both Titchener (1908) and James (1890) found that any sudden change or movement, including change in pitch, appear to distract subjects trying to concentrate on something else.

Since then, many researchers have provided empirical evidence about which critical stimulus properties automatically draw attention. A large number of studies have shown that changes in stimuli over time (transients) are likely to capture visual attention, especially if they occur rapidly. Yantis and Jonides (1984, 1990) maintained that stimuli with abrupt onsets have a special propensity to attract attention. Other studies have found that attention is powerfully attracted to "singletons", stimuli whose properties differ strongly from those of their immediate neighbours (Wolfe, 1992; Wolfe et al., 1989; Theeuwes, 1991). Other researchers have hypothesized that all stimuli have at least some tendency to attract attention. The strength of this tendency varies; stimuli with the highest relative salience are most likely to attract attention (Lee et al., 1999; Braun, 1999).

Overall, the majority of studies emphasize the properties of stimuli that attract attention, and do not make much reference to an individual's cognitive goals. Clearly, individuals' goals were assumed to play a role in the control of attention, but it was generally accepted that voluntary attention is disrupted by the appearance of stimuli with properties to attract. This view is implicitly shared by most HCI context designers.

However, recent studies have shown that stimulus attributes (e.g. onsets, transients), which are generally thought to have a wired-in ability to capture attention automatically, have been shown to capture attention only as a consequence of voluntarily adopted mental sets (Pashler et al, 2001; Pasher, 2001; Pashler & Harris, 2001). This means that attention is generally guided by internal, intentional, motivational factors, rather by external, stimulus related properties (Pashler, 2001). Moreover, it is maintained that stimulus related properties play a role in attracting attention only if individuals are mentally set to receive them. Viceversa, if the individual is mentally set to carry out an internally driven task, possible distractors would not play any role.

This view openly challenges the implicit theoretical assumptions underlying the actual practice adopted in designing HCI contexts. Given the economic importance of use of attention (Davenport & Beck, 2001), it may be of some interest to carry out research into the role of involuntary attention, that is of automatically receiving attention stimuli.

The present exploratory study has been designed to investigate this. It aims to investigate to what extent attention may be disrupted when people are engaged in a challenging task, such as learning.

## 2 Experiment

More than two decades ago, Schanke, Hirst, Neisser (1976) ran an astonishing experiment that revealed that people intent on watching a basketball game did not report having seen an oddly-dressed person walking around the field. The experiment to be described is based on the same idea: having people involved in a learning performance task exposed randomly to stimuli whose ability to differentially attract attention has been already tested. After completion of the task their performance was tested. They were then asked to return to the experimental setting after 48 hours to perform a similar task; instead they were requested to remember the type and the number of automatically attention-attracting stimuli they had been exposed to. The hypothesis to be explored was that their performance should not be differentially affected by the type of stimuli and that, given that the stimuli should not have attracted attention differentially, no difference should be found in recall performance.

## 2.1 Subjects

48 students, aged 19-23, took part in the experiment. Half were male and half female. Half of them were highly competent in interacting with computers, and half had low competence. The level of competence was established on the basis of their score on the test for the European Computer Driving Licence. The test consists of four topic areas, each with ten questions: hardware and operating systems, word processing, internet communication, spreadsheets and databases.

Task. The subjects were informed that they would be asked to carry out a learning task on computer use, that would last 30 minutes. They were also informed that later they would be tested on their learning.

They were then presented with an online lesson, consisting of 53 web pages. Only four high-competence subjects completed the task before the time limit.

When the experimental session finished, the subjects were asked to return to the experimental setting to be tested on their learning of the course content. The test consisted of a questionnaire of ten items, each having four possible answers. When they had completed the learning test, they were asked to try to remember the number and type of distractors that had appeared during the lesson.

## 2.2 Material

The web pages had border and background in blue, and texts in white (Figure 1).

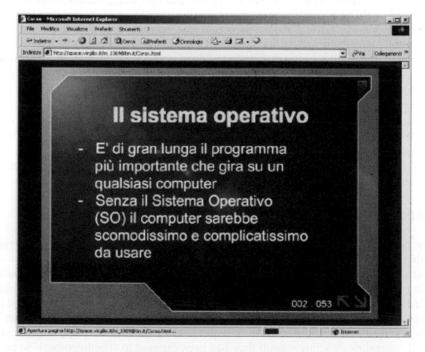

**Figure 1:** a lesson screen capture

Three series of distractors appeared randomly on the web pages, moving slowly across the screen: letters of the alphabet, horizontal rectangles and vertical rectangles. They were transparent and varied in color (black, light blue, grey, orange, red and yellow) (Figure 2).

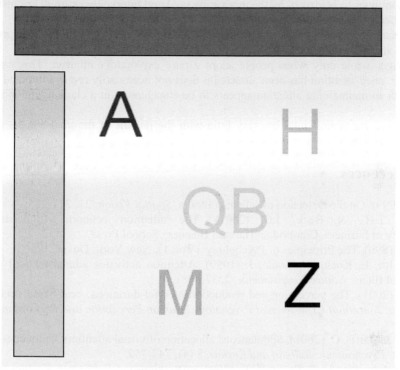

**Figure 2:** the distractors

## 2.3    Results

Two types of data were analysed: the outcome in the learning test (the possible scores ranged from 0 to 100, since a value of 10 was given for each correct answer), and the recall of type and number of distractors.

The types of distractors were well-recalled, while the number was not. There was no significant interaction. As expected, the subjects with high competence did significantly (.001) better than those with low competence. The distractors had significant effect of on learning performance.

## 3    Discussion and conclusion

The preliminary results of this exploratory study indicate that, unlike when they are presented in isolation, these type of distractors did not influence recall. Moreover, distractors do not seem to differentially affect the performance of the subjects when they are engaged in a meaningful activity.

It can be reasoned that differences in the capacity to attract attention of these stimuli can be found only when people adopt a specific mind set: to search for information in the external environment. Such a search should not be aimed at finding information in order to pursue an internal goal, but to discover novel features in the environment. In other words, stimuli attract attention when people adopt a purely exploratory mindset. When people are engaged in a challenging and motivating activity, such stimuli do not play any differential role in attracting attention: they are simply noted. Attention continues to be driven by the motivating goal and intervening stimuli are just monitored and do no succeed in attracting attention.

It can be concluded that in a HCI context, attention should be considered to a much greater extent in terms of internal guidance factors, rather than of environmental attractiveness. Involuntary attention plays a role only when people adopt a pure exploratory mindset. This rarely occurs. Furthermore, once attention has been attracted it does not necessarily remain attracted. Again, the crucial factor in maintaining attention appears to be engagement in a challenging and motivating goal.

Attention-attracting stimuli interfere very little with the pursuit of the goal and its guidance of attentional mechanisms.

# 4    References

Braun, J. (1999). On the Detection of salient contours. *Spatial Vision*, 21, 211-225.

Davenport, T.H., & Beck, J.C. (2001). The attention economy: understanding the new currency of business. Cambridge: Harvard Business School Press.

James, W. (1890). The Principles of Psychology (Vol. 1). New York: Dover, 1950.

Lee, D.K., Itti, L, Koch, C., Braun, J. (1999). Attention activates winner-take-all competition among visual filters. *Nature Neuroscience*, 2, 375-381.

Pashler, H. (2001). The perception and production of brief durations: beat-based versus interval-based timing. *Journal of Experimental Psychology: Human Perception and Performance*, 27, 485-493.

Pashler, H., & Harris, C. (2001). Spontaneous allocation of visual attention: Uniqueness dominates abrupt onset. *Psychonomic Bulletin and Review*, 8 (4), 747-752.

Pashler, H., Johnston, J., & Ruthruff, E. (2001). Attention and Performance. *Annual Review of Psychology*, 52, 629-651. Academic Press.

Pashler H. (2001). Involuntary orienting to flashing distractors in delayed search. In C.L. Folk and Bradley Gibson, eds., Attraction, Distraction, and Action: multiple perspectives on attentional capture. Advances in Psychology. Elsevier Science.

Schanke, E.S., Hirst, W., Neisser, U. (1976). Skills of divided attention. *Cognition*, 4, 215-230.

Theeuwes, J. (1991). Exogenous and endogenous control of attention: the effect of visual onsets and offsets. *Perception & Psychophysics*, 49, 83-90.

Titchener, E.B. (1908). Lectures on the elementary psychology of feeling and attention. New York: Macmillan.

Wolfe, J.M. (1992). The parallel guidance of visual attention. *Current Directions in Psychological Science*, 1, 124-128.

Wolfe, J.M., Cave, K.R., Franzel, S.L. (1989). Guided search: an alternative to the feature integration model of visual search. *Journal of Experimental Psychology: Human Perception & Performance*, 15, 419-433.

Yantis, S., & Jonides, J. (1984). Abrupt visual onsets and selective attention: evidence from visual search. *Journal of Experimental Psychology: Human Perception and Performance*, 10, 601-621.

Yantis, S., & Jonides, J. (1990). Abrupt visual onsets and selective attention: voluntary versus automatic allocation. *Journal of Experimental Psychology: Human Perception and Performance*, 16, 121-134.

# The Challenges of Teaching HCI Online: It's Mostly About Creating Community

*Jenny Preece*

Information Systems Department
University of Maryland, BC
Baltimore, MD 21250, USA
preece@umbc.edu

*Chadia Abras*

Information Systems Department
University of Maryland, BC
Baltimore, MD 21250, USA
abras@umbc.edu

## Abstract

Several studies have attempted to compare students' performance in online classes with face-to-face classes. Less attention has focused on professors' experiences of and attitudes towards teaching online. While many professors, including human-computer interaction (HCI) educators typically supplement their face-to-face classes with an online component, few HCI classes are taught totally via the Internet. In this paper we identify challenges for teaching HCI online; these include: (i) developing relationships; (ii) showing enthusiasm; (iii) balancing time versus activities; (iv) and creating and managing meaningful design projects. We suggest some possible solutions based on our experience of online teaching and our research on online communities.

## 1    Introduction

Interest in human-computer interaction (HCI) has grown dramatically in recent years. During the dot.com era of the mid-1990s 'ease of use', 'user friendly' and 'usability' became common terms used by developers. Informed web developers understood that they were designing for people whose experiences with computers differed widely. They also knew that users differed in age, educational attainments, and culture. In order to produce successful designs for this broad audience, developers had to pay attention to usability. The need to design for usability in turn created a demand for HCI courses at all levels.

At the same time the promise of reaching new student markets via the Internet created a strong interest in online education. Educators and educational administrators sought new student markets by creating online distance education courses in many disciplines. For students who lived far away from a university campus, or who worked full time and wanted to learn new skills at their own convenience, these courses were attractive. While many courses failed, including some proposed by top brand-name universities in the USA, the small fraction that met a real need succeeded after the initial hype ended. One of these programs is the Information Systems Online Master's program run by the University of Maryland Baltimore County (UMBC).

# 2    The UMBC online HCI class

This UMBC program requires students to study for 120 credits at the Master's level. HCI is a third level elective class. Students can opt to take this class after they have already completed classes in: information and communications technology, management information systems, networks, database management systems, and structured systems analysis and design. The class focuses on interaction design. It takes a process-oriented approach to design and goes beyond traditional HCI in that it introduces a range of design problems involving computers. Many of these applications involve mobile devices or computers that are embedded in other things. For example, computerized toys rely on computers for creating movement and speech but the casing of the toy, its size, shape and colour are also central design features essential for the success of the toy. The course is built around the text: *Interaction Design: Beyond Human-Computer Interaction* (Preece, Rogers & Sharp, 2002). Students read the fifteen chapters of the text and complete three assignments. The first five chapters provide an introduction to interaction design, users, collaboration and communication. The next four chapters discuss techniques for interaction design and the last six chapters discuss evaluation. A website (www.id-book.com) provides additional material, activities, examples and other material to support the book. Having acquired the concepts and practiced the skills taught in the book on small problems, the students then have the opportunity to bring their knowledge and skills together in two projects: an individual project and a collaborative team project.

The teaching challenges discussed below arise from the experience of teaching this course for the first time totally online.

# 3    Teaching challenges

There are many issues to take into account when teaching a totally online HCI class such as the one described in section 2; these include: (i) developing relationships; (ii) showing enthusiasm; (iii) balancing time versus activities; and (iv) creating and managing meaningful design projects.

## 3.1    Developing relationships

Online teachers are concerned about students feeling isolated. There is also a problem for professors, who want to know: Who is in this class? What are the students like as individuals and as groups – are there some extroverts who may help to generate conversation or even dominate it, do the students know each other, what are their skills? How are the students feeling about this or that? What do they expect from the course, from me, and from each other?

When the students have not and will not be able to meet each other, there is inevitably a short period of time needed to get to know the students. During this time the professor may feel isolated as if one is talking to a void. You have no idea what the answers to these questions are.

*Solution*: Develop 'getting acquainted exercises'. For example, invite the students to present themselves on the discussion board and point to a short web page about themselves. As well as giving obvious details such as name, occupation, reason for taking the class, what they hope to get from it, throw in some fun items too. Ask them to talk about personal interests such as a hobby, their favourite foods, their family, and the person in the world whom they most admire and say why.

## 3.2 Showing enthusiasm

Professors who are expressive in face-to-face classes may feel that it is hard to convey this enthusiasm meaningfully online. The body language, voice intonation, spontaneity and humour that helped them achieve years of successful teaching in the classroom, can no longer be relied upon.

*Solution:* HCI is an inherently interesting topic that changes fast and students respond well to pointers to newspaper articles, controversial comments on the web, and interesting designs that are pointed out. It also helps to broaden the students' experience of design possibilities and they respond by posting things that they find.

## 3.3 Balancing time and demands

One of the biggest challenges is time management both for professors and students. Professional students in Masters Programs such as the one at UMBC are often good time managers. They work out what they need to do to successfully complete the class and they cut out everything else. This means that the instructor needs to make additional activities accessible. So, forget the idea of holding many synchronous discussions because some students will not be able to attend and will become stressed. It is better to hold just one or two well-publicised sessions. Instructors also need to balance the trade-off of giving students what they want versus doing their own research.

*Solution:* Avoid the temptation of being too responsive on email. Train students not to expect immediate responses. Also work on creating a collaborative environment in which students will help each other rather than relying on the instructor all the time. To do this, the instructor needs to provide incentives and privacy. Many professors allocate some portion of the grade for collaboration but intrinsically motivating activities are preferable. For example, set challenging problems that require students to collaborate, review and comment on each other's work. Encouraging open and free discussion in which everyone feels valued is important. Also make it easy for students to talk in groups in private. Environments like Blackboard, for example, offer chat and discussion board facilities that can be made private.

## 3.4 Creating and managing meaningful design projects

A meaningful project is one that is needed in the real world and will live on after the end of the semester. These projects enable students to demonstrate the skills taught in the class and to work together. These can be daunting requirements. Professors, whether they teach online or in the classroom are often anxious about managing group projects. They believe that students should work individually and collaborating is tantamount to cheating. Even those who believe in group projects may be daunted by the prospect of managing and grading group projects in online classes.

In the HCI class discussed above the students completed an individual project and a group project. The individual project did not meet all the criteria just mentioned but the group project did. Even though the professor was experienced in managing group projects, she was apprehensive about how group projects would work in a totally online class but she need not have worried.

*Solution:* There seem to be three reasons for this success. *Reason 1:* many of the students knew each other and had worked together in previous classes. Collaboration is more common in information systems than in many disciplines. Furthermore, these students were all working professionals and many were trained in time management. *Reason 2:* a strong ethos of collaboration was developed that started with the first class and was clearly articulated as a goal for the class. The project teams each had their own private discussion areas. *Reason 3:* every student had a specific role that contributed to the project and there was a student team manager.

## 4    Conclusions: Community is the key

Many of the solutions that we have suggested point in the same direction – the need for creating a community of learners, which in this case means an online community. 'Community' is elusive to define and in this paper we use the term 'online community' broadly to refer to all communities that exist predominantly online. Online communities vary depending on: (i) whether they have physical as well as virtual (i.e., networked, physi-virtual) presence (Lazar, Tsao & Preece, 1999); (ii) purpose (e.g., health support, education, business, neighbourhood activities); (iii) the software environment supporting them (e.g., listserver, bulletin board, chat, instant messaging, or more often these days – some combination); (iv) size (small communities of fifty people are different from those of 5000 or 50,000); (v) duration of their existence; (vi) stage in their life-cycle; (vii) culture of their members (e.g., international, national, local and influences that may be related to politics, religion, gender, professional norms, etc.); (viii) governance structure (e.g., the kind of governance structure that develops and the types of norms and rules associated with it) (Preece, 2000.)

A learning online community usually has a short life of up to 16 weeks depending on the length of the class, which presents a challenge to the instructor on how to create a harmonious environment with some form of ties no matter how weak.  The short-lived nature of such communities can also make them more intense and harder to manage, but on the positive side the educator has a captive audience; the students have to participate in order to succeed, unlike other online communities where participation is voluntary.  However, the teacher still has to contend with students' motivation and satisfaction in order to keep them coming to the community.

In contrast to the case of the online HCI class presented above, the other researcher taught an online class for undergraduates, where motivation was not as high as with the graduate professional group.  In this case feedback was of the utmost importance. The results of several questionnaires given over a period of four semesters showed that the majority of undergraduates preferred face-to-face classes, unlike the graduate student professionals, who for the most part preferred the freedom of the asynchronous online class.

In another example both researchers created an online community to support graduate students in a humanities program. This community involved voluntary participation. 25.6% of the members participated by posting; however, we had a high number of readers who said they felt part of the community even though they do not participate (Abras, Maloney-Krichmar & Preece, 2003.)
Learning communities can be difficult to create because the learning climate encourages competition and a thriving community needs reciprocity and collaboration. But, as we have suggested, it is possible to promote community through carefully planned activities, and group work that motivates students to collaborate. Weak ties, and some strong ties, do form in such environments that last the duration of the course, and students report a sense of belonging. However, online communities may be more appropriate for professional graduate students than for undergraduates.

# 5    References

Abras, C., Maloney-Krichmar, D., Preece, J. (2003). *Emergence of a weak-tie online academic Community: Indicators of success.* (available from the authors)

Lazar, J. R., Tsao, R. and Preece, J. (1999). One foot in cyberspace and the other on the ground: A case study of analysis and design issues in a hybrid virtual and physical community. *WebNet Journal: Internet Technologies, Applications and Issues 1*(3), 49-57.

Preece, J. (2000) *Online communities: Designing usability, supporting sociability.* NY: John Wiley & Sons.

Preece, J., Rogers, Y. & Sharp, H. (2002). *Interaction design: Beyond human-computer interaction.* NY: John Wiley & Sons.

# "A Role With No Edges": The Work Practices of Information Architects

*Toni Robertson, Cindy Hewlett, Sam Harvey and Jenny Edwards*

Faculty of Information Technology, University of Technology, Sydney
PO Box 123 Broadway NSW AUSTRALIA 2007
<toni><chewlett>><sam><jenny>@it.uts.edu.au

## Abstract

This paper reports the analysis of a range of workplace interviews with people who define themselves as *information architects*. This is a recent job title that is commonly used to describe people who are involved in the design and development of web-based applications, including those designed to be used on mobile devices. Our research aimed to identify and understand the work practices that define the position of information architects. Our major finding was that the work that information architects actually do in practice varies considerably and demands expertise in a range of diverse fields. Moreover, this work shifts and expands to fill the gaps in expertise and user involvement within the specific project that are working on at a particular time. The results of our study open a space to reflect on the evolution of the processes, training, tools and techniques that are defining of Human-Computer Interaction as a field of research and practice.

## 1    Introduction

*Information Architect: 1) the individual who organises the patterns inherent in data, making the complex clear; 2) a person who creates the structure or map of information which allows others to find their personal paths to knowledge; 3) the emerging 21st century professional occupation addressing the needs of the age focused upon clarity, human understanding and the science of the organisation of information.*
Richard Wurman (1996) *Information Architects*, Graphic Press Corp, Switzerland, p. 9.

*Information Architecture: structural design of the information space to facilitate intuitive access to content.*
Jesse James Garret (2000) *The Elements of User Experience* http://www.jjg.net/ia/

The recent growth in the use of *information architect* as a job description implies an established and shared understanding of the meaning of the term; yet this meaning has been remarkably contested. As one of the information architects interviewed for this research remarked: 'It's just a source of endless idle argument on the newsgroups when the profession could be getting on with better things'. So our concern in this paper is to consider the skills and backgrounds of people who are called information architects and to understand the various processes and practices involved in their work. Of particular interest is how design processes defined as user-centred are being used by information architects in their work.

The research reported here is based on twenty-six intensive, loosely-structured, workplace interviews with information architects. Most of these are currently working in Sydney and Melbourne, in Australia, but five are based in Europe one in Hong Kong and one from the US. This research is the initial, scoping stage of a larger project that includes long term field studies of a range of work practices surrounding the design and use of web-based applications. In the interviews we sought to identify the common issues affecting the work of information architects as well as those issues that were highly situated and domain specific in their effect on different individuals. We wanted to ensure that the issues we pursued in our ongoing research were grounded in actual technology design practice.

We contacted eight of the interview participants directly and the others replied to a single request for participation posted on the listserv hosted by the Computer-Human Interaction Special Interest Group (CHISIG) of the Ergonomic Society of Australia. This is the Australian professional organisation for academics, researchers and practitioners in Human-Computer Interaction (HCI) and related fields. We were particularly interested in interviewing information architects who had a familiarity with, and commitment to, user-centred design methods.

We asked each person interviewed to situate their answers in their most recently completed project. This provided us with twenty-six examples of genuine work practice to represent what information architects have been doing. The interviews were transcribed and then independently analysed by the authors to identify the issues and categorisations relevant to the aims of the research. In turn these issues and categorisations were compared, sorted, iteratively re-evaluated and then further defined against the transcripts, interview notes and other field data before being used to structure our analysis.

Because our space here is limited, we cannot give a full account of our findings. Instead, in the following section we use a broad brush to represent the skills and backgrounds information architects bring to their work. From there we summarise the range of projects information architects contribute to and discuss the different kinds of work that the information architects themselves defined as part of their practices. Finally we reflect, briefly, on the challenges and opportunities posed by our study to the evolution of HCI.

## 2 The Actors

The ages of the participants ranged between approximately twenty-five to forty-five. Seven had worked as information architects or in closely related fields for more than five years, eight for three to five years and the remaining eleven for less than three years. Only one participant had no university qualification. Three had no undergraduate degree but had gained post-graduate qualifications. Of those with undergraduate degrees, eight had first degrees in the humanities, eight in architecture/design (including one each from industrial design, graphic design and ergonomics); two had first degrees in mathematics, two in new media (digital design etc) and one in economics. None had undergraduate degrees in information technology or information science. Fifteen had, or were completing, post graduate qualifications; five in information technology, three in new media and three had a specialist qualification in HCI. Two had postgraduate qualifications in education and two in the humanities. Interestingly, given its traditional contribution to information design as a discipline, none of our participants had studied information science. Their professional backgrounds, prior to working as information architects, were even more diverse and included: academia and research (five), web development (three), two each from

HCI, instructional design, visual design, theatre, publishing and marketing; and one each from school teaching, fine arts, communication, industrial design, consulting and public relations.

While it is common for those who work in IT design environments to come from other areas, we were both surprised and impressed by the diversity of qualifications and professional backgrounds among our participants. Some kind of design and/or research training and/or professional background was in fact common to most; it was just not the same kind of design training or background. The important point here is that the field is too recent for there to be a basic, shared body of professional knowledge and skills that can be assumed for people working as information architects.

In practice most used mailing lists and specialist websites to sustain their professional development in information architecture and related areas. Round half attended industry conferences or short courses, a third regularly bought popular industry publications through Amazon.com and only five read academic research papers. It was interesting to note that while many of our participants cited Garrett's (2000, 2002) definition of information architecture (quoted at the beginning of this paper) none of their own work practices were this specialised.

## 3    Their Roles

We cannot do justice here to the very rich and complex descriptions in our data of the various kinds of work each of our participants did within the design and development processes in their different organisations. But we found that that all the kinds of work mentioned in the interviews fitted within the broad categories of Research, Focused Designing, Evaluation, Coordinating Internal and External Stakeholders and Management. We emphasise that these categories were not imposed on the data but emerged from the iterative analysis of interview transcriptions and then validated against additional transcriptions. To give our readers some understanding of the extent of reduction in this categorisation, the work of Focused Designing includes practices as varied as requirements development, preparing customer 'pitches', defining scenarios and personas, developing sitemaps and navigational models, producing wireframes, designing the interface and the interaction, and specifying content. We labelled this category Focused Designing, rather than just Designing because we believe that actual technology design practice includes the full range of work summarised in the table (Robertson, 1996).

The other categories of work, included in the table, are at a similar level of abstraction from the very varied kinds of actual processes represented in our data. Research includes user, process and domain research, research of the existing application and its usage statistics. Evaluation included heuristic evaluation, card sorting, prototyping, user testing of various kinds, implementation reviews and project process reviews. Coordinating external and external stakeholders included working as the communication facilitator in a project, liaising with content providers, liaising with external providers, preparing presentations and the range of activities usually covered by 'attending meetings'. Finally, Management work included defining the overall process, managing that process, managing the project itself and/or the people working on it. A detailed discussion of the relations between all these activities and the project itself is beyond the scope of this paper.

The top section of the table below maps the work practices of each participant (columns labelled **a** to **z**) to these categories of work (rows). The bottom row of the table records whether the individual information architect indicated a familiarity with user-centred design methods.

398

**Table 1:** Summary of the Work Done By Information Architects[1]

| | a | b | c | d | e | f | g | h | i | j | k | l | m | n | o | p | q | r | s | t | u | v | w | x | y | z |
|---|---|---|---|---|---|---|---|---|---|---|---|---|---|---|---|---|---|---|---|---|---|---|---|---|---|---|
| **Research** | X | U | U | X | X | X | X | X | U | X | U | U | | * | U | X | X | U | U | X | * | U | U | * | U | U |
| **Focused Designing** | X | X | X | X | X | X | * | * | X | X | * | X | | * | X | X | X | X | X | X | X | X | X | X | X | * |
| **Evaluation** | | X | | *U | U | *U | | *U | | | * | X | | U | *X | | | X | * | X | | | | *U | *U | U |
| **Coordinating stakeholders** | X | X | X | X | X | X | X | X | X | X | X | X | X | X | X | * | X | X | X | X | * | X | X | X | X | X |
| **Management** | * | X | * | * | * | X | * | X | * | * | * | X | X | X | X | * | * | * | * | * | * | X | * | X | * | X |
| **User-Centred Design?** | X | X | | * | X | X | - | X | X | X | X | X | - | X | X | - | - | X | X | - | - | X | X | - | X | X |

The information architect represented in column **u** was a junior member of a task-based organisation and is the only person whose work was restricted to processes we have categorised as Focused Designing. Information architects were as involved in various kinds of research about the developing product as they were in specific design processes. Our data support the importance of research-based design to the design of any software product including those built for the web.

Our participants were less likely to be involved in the evaluation of the product (only seven of the twenty-six). Our data provide two explanations. The first is that some organisations either had people dedicated to usability evaluation in the design team or routinely outsourced those parts of the process to a specialist provider. The second is that in eleven of the twenty six projects (forty-two percent) no evaluation was done at all. Just over half the products that were evaluated involved users in that process (round thirty percent of the total).

Six information architects were not directly involved in the specific design process at all. Four of these were responsible for the management aspects of their projects. Of these one (column **m**) had chosen the title of information architect for strategic reasons and was rarely involved in the actual design of specific products. The positions of the information architects represented in columns **g** and **k** were defined within their organisations to facilitate their liaison with information architects in the specialist providers where development of products had been outsourced.

Of the two information architects who were not involved in stakeholder coordination, one is the junior staff member referred to above and the other (column **p**) was building an already specified application to work on an existing web-site. Our findings demonstrate the centrality of the coordination of stakeholders, to the work practices of information architects and confirm Robertson's (1996) and others earlier findings that the technology design process relies on the communication work done by designers to coordinate their work with the work of others. This coordination work is also where the politics of the design process is played out, and where usability gains are won and lost.

---

[1] A * indicates that the work for that row was done on the project but not by the information architect.
**X** indicates the individual architect was involved in these work practices in their last project. **U** indicates users were also involved. **\*U** means someone else did this work but users were involved.
A space in the first four rows means that the work was not done on the project at all.
In the bottom rows a - signifies a negative answer.

In the final row of the table, two thirds of the information architects noted an understanding of, and experience in, developing design concepts using direct user involvement. Yet only twelve of the twenty-six projects included user research of some kind and only eight involved users in the product evaluations. Some of the information architects explicitly noted that user involvement was the first item to be removed from the budget when things got tight. The resulting product is a design based on assumptions about the user, non on genuine user participation.

# 4    Some Reflections

In their findings from an extensive survey of user-centred design practices, Vredenburg, Mao, Smith and Carey (2002) wrote: 'Some common characteristics of an ideal user-centred design process were not found to be used in practice, namely focusing on the total user experience, end-to-end user involvement in the development process, and tracking customer satisfaction' (p. 478). We found similar issues in our study of the work of information architects. In fact one third of our participants either did not appear to be familiar with user-centred design processes at all, or while they may, in fact, have involved their users, they were unable to situate this involvement within the wider, established design methods for user-involvement. An interesting consequence for these information architects was a relative lack of control over the design priorities in the products they were building.

A clear issue in all our interviews was the constant efforts of almost all of our participants to develop the most usable product that they could. However, only those involved in managing the design process, and those protected by a manager, who defended the importance of usability and the involvement of users in that process, were able to routinely exploit user-centred design methods. Even so, those who were able to control their design processes rarely had control over the allocation of project budgets. This meant an increasing reliance on discount usability methods and the use of information architects as user-representatives in the design process.

The work information architects do shifts and expands to fill the gaps in expertise and user involvement within the specific project that are working on at a particular time. For those of us working within HCI the implications are sobering. Industry practitioners of our discipline are increasingly required to work as generalists in the technology design process where they must create and drive, rather than assume, opportunities for incorporating user-centred design methods in this process. Researchers in our discipline need to contribute a literature that can ground the practice of information architects in user-centred design methods that are sensitive to the constraints of their practice.

# References
Garrett, J. (2000). *The Elements of User Experience*. Diagram retrieved April 20, 2001, from http://www.jjg.net/ia/
Garrett, J. (2002). *The Elements of User Experience,* USA: New riders Publishing.
Krug, S. (2000). *Don't Make Me Think! A Common Sense Approach to Web Usability*, USA: Macmillan.
Robertson, T. (1996) Embodied Actions in Time and Place: The Cooperative Design of a Multimedia, Educational Computer Game. *Computer Supported Cooperative Work: Journal of Collaborative Computing*, 5, (4) 341-367.
Vredenburg, K., Mao, J., Smith, P. and Carey, T. (2002). A Survey of User-Centred Design Practice. In *CHI Letters*, vol. 4, No. 1, 471-478

# Problem-Based Learning in New Media Education: The Case for Human-Computer Interaction

*Howard Rosenbaum*

Indiana University
School of Library & Information
Science
Bloomington, IN USA
hrosenba@indiana.edu

*Margaret B. Swan*

Indiana University
School of Library & Information
Science
Bloomington, IN USA
mbswan@indiana.edu

## Abstract

In this paper we pose and answer two questions that we believe are central to the development of curriculum in new media programs. First, why should new media students learn about human-computer interaction (HCI)? We begin by arguing for the importance HCI for students learning to design a variety of digital media. Second, what is the best way to teach students HCI? To answer this question we present problem based learning (PBL) a pedagogical approach that is well suited to developing HCI curricula. The utility of PBL is illustrated with a problem-based scenario that could be used in a new media class to help students learn about HCI.

## 1 Introduction

The working definition of human-computer interaction (HCI) used here is that proposed by the Association of Computing Machinery's (ACM) Special Interest Group on Computer-Human Interaction Curriculum Development Group (Hewett et al., 1996); HCI "is a discipline concerned with the design, evaluation and implementation of interactive computing systems for human use and with the study of major phenomena surrounding them."

There is growing interest in HCI among new media faculty as was evident by the attendance at a session on the topic at SIGCHI'02 (Faiola and Rosenbaum, 2002). In this paper, we do not engage the debate about the proper location for HCI studies and assume that there is a need to incorporate HCI into new media curricula. We pose two questions that we believe are central to teaching HCI in these programs. First, why should new media students learn about HCI? To answer this question, we argue for the importance of this domain for students who are learning to design a variety of digital media. Second, what is an effective way to teach students HCI? Here we present a pedagogical approach, problem based learning (PBL) that is particularly apposite when developing HCI curricula. The utility of PBL is illustrated with a sample problem-based scenario that can be used in a new media class to help students learn about HCI.

## 2 Why New Media Students Should Learn About HCI

New media programs prepare students to work in digital environments. They learn how to design a wide variety of digital products, including web and database interfaces and multimedia. It is equally important for them to understand the social and organizational context of design. More

401

specifically, they should know how to evaluate the digital products they design in terms of efficiency, effectiveness, or other relevant criteria. They should have an appreciation for the role of the user in the design and use of these products and understand the value of user-centered and participatory design. We argue that the best way for new media faculty to help students learn about the context of design is by taking advantage of HCI research, by incorporating HCI modules into existing courses or by developing stand-alone HCI courses.

HCI research can aid in the design, development, and implementation of information and communication technologies (ICTs) and can improve students' abilities to design interfaces to these technologies (Hewett et al., 1996). This work can have serious consequences because "well-known catastrophes … have resulted from not paying enough attention to the human-computer interface" (Meyers, Holland, and Cruz, 1996). We are not arguing that the design work that lies ahead of new media students will typically have such serious implications, but we do believe that their design work can be improved by exposing them to the techniques and findings of HCI.

Four main themes should be a part of HCI classes in new media programs. Students should:
- Understand that HCI is a rigorous, methodologically sound, and theoretically grounded procedure for evaluating and improving the design, and use of ICTs.
- Understand that HCI research has practical applications in the systems development lifecycle that have significant impacts on the organization's bottom line.
- Be familiar with the best practices as revealed by the current research findings.
- Be able to apply their knowledge of HCI to a wide variety of situations in their work.

What then is an effective way to teach HCI to new media students?

# 3   Using Problem-based Learning to Teach HCI

In this section we describe problem-based learning (PBL), a user-centered approach that places students at the center of the process of inquiry. When coupled with scenario-based instruction, PBL provides an effective pedagogical strategy for learning about HCI.

## 3.1 Problem-based Learning and Scenario-based Instruction

Problem-based learning is a pedagogical strategy that assumes that learning occurs as students work through complex problems drawn from and approximating situations typical of professional practice (Abdullah, 1998; Savery and Duffy; 1995). In general, PBL is a good fit for new media programs since many of the career paths that students will follow are problem-centered. The focus on realistic problems is an important feature of PBL because it helps "students to take ownership of the learning experience" (Bentley, Sandy, and Lowry 2002: 107). PBL's origins in medical education (Barrows, 1998) and its presence in other disciplines are beyond the scope of this paper and are discussed elsewhere (Rosenbaum, 2000).

In a typical implementation, students are presented with a complex and ill-structured problem. Working in small groups, they decompose the problem into its constituent parts and begin an investigation "from which subject matter and instruction emerge" (Eden, 2000; 56). They collaboratively suggest informal hypotheses or research questions, propose and test solutions, and evaluate their results (Moiao, 2000). As they resolve each component part, they incorporate what they have learned from their reading and experimentation, developing pragmatic, group-based problem solving skills. Throughout the semester, the faculty member meets regularly with the teams to review their progress and assist them with problems that arise in their work.

PBL changes the role of faculty in the learning process. Rather than being at the center of the learning experience, the faculty member takes on the role of the guide and facilitator with the goal of helping students work towards the resolution of the problem. Teaching becomes a process of providing suggestions as students discuss their progress, directing them towards fruitful lines of inquiry and not a matter of dominating interactions with students (Shanley and Kelly, 1994).

To formulate the problem that serves as a focus of PBL, we suggest using scenario-based instruction (Rossen and Carroll, 2002). This pedagogical strategy is useful because it can "deepen the idea that HCI is concerned with understanding, designing, evaluating and implementing interactive computer systems to match the needs of people" (Vat, 2001; 9). Similar to a case analysis, a scenario is a narrative of a problematic situation adapted from professional practice (Rossen and Carroll, 2002). It can focus on three different components:

- The problem: students define the problem, determine its boundaries, and identify and consider the interests of the stakeholders involved in the design and development of a digital product.
- The interactions: they develop profiles of the people who will use the digital product.
- The activities: they decide how to resolve the problem, determine the appropriate technologies, divide the labor, and begin working on the design solution.

How then can PBL and scenario-based instruction be used as the basis upon which new media students can learn about HCI?

## 3.2 Using PBL to Teach HCI

At the beginning of the semester students are presented with a scenario. Their initial task is to analyze the scenario, an activity that can be carried out with the entire class or in small groups of four or five. The faculty role in this stage is to assist in the definition of the problem embedded in the scenario and the decomposition of the problem into smaller and manageable components. As students work through this stage, the instructor helps them evaluate the potential resolutions they suggest based on what they already know. With the faculty member's assistance, students then describe what they need to know, clarify key learning issues, and develop an action plan.

During the information gathering stage, which may take several weeks, students explore relevant literatures to determine what they need to do to resolve the various components of the problem. This involves focusing on the problem, activities, and interactions, as described above. Depending on the nature of the scenario, students may complete their information gathering in one period of time. However, each component of the problem may require its own round of research.

During this stage, students assume greater control of their educational experience and learning activities, although the teams continue to meet regularly with the instructor. At these meetings, teams report their progress, discuss problems they are facing and what they will do to resolve them, and describe what they will do next. The instructor listens and comments on the quality of their proposed resolutions and intervenes when necessary, preventing teams from exploring blind alleys. Regular meetings also allow the instructor to pay attention to team dynamics.

During the next stage, students engage in a process of synthesis. This involves a form of critical thinking where students abstract what they have learned from their reading and apply the relevant insights and findings to the problem and its components. The end product of this stage is a design strategy. The instructor works closely with the teams to ensure that their strategies are reasonable and can be completed with the time remaining in the term

Typically, teams present their work in class at the end of the semester. In addition to the instructor's evaluation of their work, students review their experiences by evaluating their performances and the work done by their team members. This allows them to explain what they have learned during the semester and gives the instructor an indication of the extent to which individual students have made progress and of each member's contribution to the team.

### 3.2.1  A PBL-based Scenario: Educational Product Design

A well-established company that designs digital, interactive products such as computer games and virtual comic books has hired you as a media designer. The company has recently decided to expand their product line to include educational tools. Because of your new media degree, your boss has made you the lead designer for the first new educational product, as yet unnamed. The product is geared toward 7-10 yr. old girls and is planned to be an interactive tool that teaches decision-making skills through storytelling.

You meet with your boss to decide how to proceed. The head graphic designer and marketing manager are also in attendance. Numerous ideas are tossed around concerning the end product content, "look and feel," and design medium. However, you feel that something is missing. Because of your HCI experience, you know that designing with the user in mind should also be a top priority. You mention this and the group half-heartedly agrees. Your boss is willing to give this a try, but other than focus groups, she is unsure of the process to take to ensure user-centered design (UCD) and wary of the costs associated with extra steps to product deployment.

After the meeting, you begin to think about how to explain the importance of UCD to your project colleagues. You must come up with a framework for the project that incorporates UCD and present it to your boss, colleagues, the marketing head and lead graphic designer. To make the plan airtight, you decide to also mock up a prototype of an end product and include it in your presentation. There is a time constraint; your boss expects a presentation in three weeks.

Your assignment is to assemble and present a solid design argument for a digital, interactive educational product that will be created with UCD. The design argument should include background information and rationale, user observations, content samples, use scenarios, design or system concepts, and interface prototypes. Be prepared to demonstrate the interface that you have developed for your working prototype.

## 4  Conclusions

We have argued in this brief paper that it is important for new media students to have an understanding of the research findings and techniques of HCI because this knowledge will improve their design and building skills. We presented a combination of PBL and scenario-based instruction as an effective pedagogical strategy for teaching new media students about HCI. This was followed by a brief scenario and suggestions for how it could be used in a problem-based design class as the basis for a semester long team project.

The next steps are to implement this approach in a new media curriculum and conduct evaluation research to assess the learning outcomes because research "has been mixed about the effectiveness of PBL, although most of this work has focused on its use in medical education" (Rosenbaum, 2000). Vernon and Blake (1993) find that it is a markedly superior approach to traditional learning while Fenwick and Parsons (1998) are critical of the approach. Having the opportunity to conduct comparative research comparing a class using the approach described above with a class using traditional approach would be useful in attempting to settle this debate. We are also working on a version of this approach that will be appropriate for use in doctoral curricula and hope to report on our work in the near future.

## References

Abdullah, M.H. (1998). Problem-Based Learning in Language Instruction: A Constructivist Model. *Eric Digest*. ERIC Clearinghouse, Bloomington, IN.

Barrows, H.S. (1998). The Essentials of Problem-Based Learning. *Journal of Dental Education*. 62(9). 630-633.

Bentley, J., Sandy, G., and Lowry, G. (2002). Problem-based learning in information systems analysis and design. In Cohen, E. (ed). *Challenges of Information Technology Education in the 21$^{st}$ Century*. Hershey, PA: Idea Group Publishing. p. 100-124.

Edens, K.M. (2000). Preparing Problem Solvers for the 21st Century through Problem-Based Learning. *College Teaching*, Spring, 48(2), pp.55-61

Faiola, A. & Rosenbaum, H. (2002). Challenges in teaching usability theory and testing for new media curricula. ACM SIGCHI. Annual conference. April 23. Minneapolis. Minn.

Fenwick, T. & Parsons, J. (1998). Boldly solving the world: A critical analysis of problem-based learning as a method of Professional Education. *Studies in the Education of Adults*, 3(1): 53-65.

Hewett, Baecker, Card, Carey, Gasen, Mantei, Perlman, Strong & Verplank. (1996). ACM SIGCHI Curricula for Human-Computer Interaction. ACM Special Interest Group on Computer-Human Interaction Curriculum Development Group http://sigchi.org/cdg/

Meyers, B., Holland, J., & Cruz, I. (1996). Strategic Directions in Human Computer Interaction. *ACM Computing Surveys* 28(4). http://www-2.cs.cmu.edu/~bam/nsfworkshop/hcireport.html

Moiao, Y. (2000). Supporting Self-directed Learning Processes in a Virtual Collaborative Problem Based Learning Environment. In Chung, M.H. (Ed.). *Proceedings of the 6th America's Conference on Information Systems. Association for Information Systems*. 1784-1790

Rosenbaum, H. (2000). Teaching Electronic Commerce: Problem-based Learning in a Virtual Economy. *Journal of Informatics Education and Research*. 2(2). 45-58.

Rossen, M.B. & Carroll, J.M. (2002). *Usability Engineering: Scenario-Based Development of Human-Computer Interaction*. Morgan Kaufmann.

Savery, J.R. & Duffy, T.M. (1995). Problem based learning: An instructional model and its constructivist framework. *Educational Technology*. 35(5) 35-38.

Shanley, D.B. & Kelly, M. (1994). Why Problem-Based Learning?

http://www.odont.lu.se/projects/ADEE/shanley.html

Vat, K.H. (2001). Teaching HCI with scenario-based design: the constructivist's synthesis. Proceedings of the 6th annual conference on Innovation and Technology in Computer Science Education. Canterbury, UK.

Vernon, D.T.A. & Blake, R.L. (1993). Does Problem-Based Learning Work - A Meta-Analysis of Evaluative Research. *Academic Medicine*, 68(7): 550-563.

# Human-system Interaction Container Paradigm

*Célestin Sedogbo, Pascal Bisson, Olivier Grisvard & Thierry Poibeau*

THALES Research & Technology France
Domaine de Corbeville – 91404 Orsay Cedex – France
[celestin.sedogbo,pascal.bisson,olivier.grisvard,thierry.poibeau]@thalesgroup.com

## Abstract

In this paper we present the concept of Human Interaction Container (HIC) which introduces an important shift in the field of Human-Computer Interaction, moving from an application-centric to user-centric perspective, through the adoption of a service-oriented view of application and user interface capabilities. The HIC has been designed as to propose an interaction infrastructure offering the necessary decoupling between application, interaction and presentation logics in order to enable intelligent adaptive interaction and easy integration of new interaction modalities and appliances. The HIC approach is in the process of implementation and validation on several THALES business cases such as Collaborative Decision Making for Air Traffic Management.

## 1    Introduction

Due to the increasing diversity and complexity of software systems and means, the scope of Human-Computer Interaction (HCI) extends far beyond the simple issue of providing human beings with means to use a system. First, HCI must also enable the users to access the system anywhere, anytime and anyhow, that is, more generally speaking, in any context of use. Second, it must provide the users with support and assistance in order for them to perform their task or carry out their mission.

Nevertheless, most existing systems are not directed towards the satisfaction of the users' needs, but rather merely designed as to propose an interface to a set of application functions. Moreover, these functions are often used through the prismatic view of the Graphical User Interface (GUI) and most GUIs do not take into account some key contextual information such as the users' task and behavior and do not exploit all the possibilities offered by the appliances at hands. Therefore, the usability and utility of the User Interface (UI), although being a key acceptance factor for nowadays systems, is often neglected to the detriment of the usability of the whole system.

One of the main stumbling blocks to the design of user-centered systems is the difficulty to clearly identify the interaction logic, that heavily depends on contextual information at several levels (application domain, user tasks, user profiles and preferences, hardware environment and interaction history), as opposed to the business application logic, which should be independent of any context of use, and the presentation logic, which is necessarily specific to a given UI on a given terminal. A consequence of this lack of proper separation between application, interaction and presentation services is that system designers and developers have to face important costs in terms of system deployment, upgrade and maintenance. Indeed, any evolution of the users' interaction needs, for instance when the physical or logical context of use changes, impacts both the applications and their interfaces. We argue that changing this situation in order to open the way to the design of highly interactive systems requires a move from an application-centric perspective, which is the characteristic of most existing systems, to a user-centric perspective.

Moreover, as today interaction means become more and more various and sophisticated, interaction demands the integration of heterogeneous modalities, such as voice, gesture, graphics and animation, as well as appliances, such as classical laptop and desktop workstations, mobile phones, Personal Digital Assistants (PDA), PC tablets, etc. Therefore, HCI systems must undergo an important mutation, moving from a one-to-one to a many-to-many scheme in terms of application to UI pairing, and become able to dynamically adapt or even create the interface according to the users' role and environment, without changing the application.

## 2    The Human Interaction Container paradigm

A solution for providing users with wider access to existing systems and for enhancing user friendliness of existing interaction means is to design intelligent interaction systems that dynamically adapt to the interaction environment and react appropriately in various contexts of use, without implying any modification to the core application. In order to achieve this, we propose a new HCI architecture, offering the necessary decoupling between application, interaction and presentation logics in order to implement adaptive interaction. This architecture is based on the key concept of Human Interaction Container (HIC).

The HIC aims at encapsulating all software components dedicated to user-system interaction management into a context-aware and context-sensitive container enacting as a mediator between the application services and the presentation services. As such, this container is designed so to ease the logic separation, between application, interaction and presentation, and handle all the interaction processes enabling an application and its various user interfaces to communicate with each other. It offers application-independent and interface-independent interaction services which support intelligent adaptive interaction. These generic interaction services include dialogue processing, task and activity planning, user adaptation, multi-modality input and output management, multimedia presentation generation and terminal adaptation.

The HIC also aims at providing application and UI designers with an open framework, independent from platforms, networks and appliances, and enabling them to design at compile-time interaction facilities that meet the operator and task requirements and use them at runtime without disrupting the realism of interaction and the user performance. As such, the interaction container can be seen as an attempt to exploit middleware facilities in order to satisfy the interaction demand at the application level. Therefore, from a middleware perspective, the HIC constitutes a set of additional layers, the *interaction layers*, on top of an existing system middleware, this assembly being called an *interaction middleware*. For multi-tier architectures, the introduction of the HIC implies a migration from three tier architectures to four tier architectures, with the HIC as a new tier, inserted between the application server tier and the client tier.

## 3    Architectures for human-computer interaction

Agent-based frameworks such as the Open Agent Architecture (Cheyer & Martin, 2001) or Ivy (Chatty, 2002) may constitute a first step towards the implementation of the interaction infrastructure we advocate for, as they can be used in order to encapsulate interaction components as agents into an interaction platform independent of either application and interface. Some interaction platforms like the iROS software/middleware system (Ponnekanti, Johanson, Kiciman

& Fox 2003) prefigure the shift towards the new kind of interaction architecture we propose. iROS uses an *event heap* (Johanson & Fox, 2002) as a coordination infrastructure for interactive workspaces. Based on a *tuplespace* model, it offers a satisfactory infrastructure for coordinating and assembling distributed components, even if it lacks several functionalities in terms of event management, especially regarding event life cycle and timestamp management. Moreover, the ICrafter system, (Ponnekanti, Lee, Fox, Hanrahan & Winograd, 2001) designed as a service framework for ubiquitous computing environments, is even closer to our proposal as it is designed as to let users interact with workspace services using a variety of modalities and appliances. The ICrafter architecture is built upon the event-based communication system of iROS and a context memory component used for storing workspace context information.

## 4    The Human Interaction Container architecture

The architecture we have designed for the HIC results from our background acquired in the field of HCI. It can be seen as a generalization of the approach we pursued in the past when addressing the problem of (semi-)automatic production of speech-based interfaces. This approach consisted in setting-up an agent-based HCI infrastructure where specialized agents devoted to interaction management were communicating through a supervisor. The implementation was done using OAA (cf. section 3). As can be seen on Figure 1, the HIC internal flow processing architecture is also based on a community of agents dedicated to the processing of interaction and implementing the various interaction services, as well as on a resource management facility which ensures real-time access to interaction resources such as activity state, dialogue history or UI contexts. This architecture is supported by the several other layers of the whole interaction middleware which offer the various services required in order to implement high-level interaction processes.

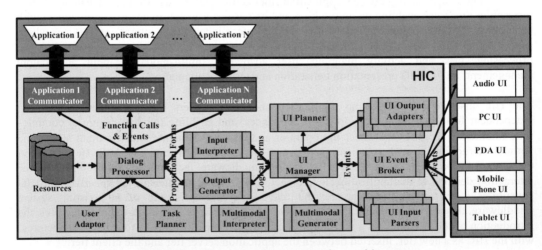

**Figure 1:** HIC internal flow processing architecture

## 5    The concept of Interaction Middleware

As stated above (cf. section 2), the HIC can be seen as an interaction middleware whose purpose is to ease the development of HCI systems where people have to interact and possibly collaborate through a heterogeneous set of devices ranging from mobile phones and PDAs to laptop and desktop PCs. The interaction middleware is organized as a set of additional layers on top of a classical system middleware. Such a middleware provides some core services such as object life cycle management, time management and persistence[1]. In our approach, these cores services are completed by two additional layers, one devoted to *technical services* and the other to *interaction services*. As shown on Figure 2, technical services support both the application container and the interaction container while interaction services provide the basis for the implementation of the HIC.

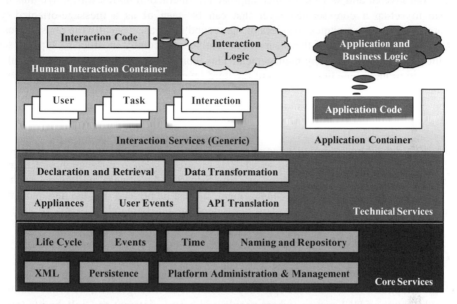

**Figure 2:** Architecture of an interaction middleware

Technical services are not directly implementing interaction management. They are primary specified and developed in order to be called and used by the upper layer as tools for interaction service implementation. The *declaration and retrieval service* is a facility for the management of the other services (addition, retrieval, removal, etc.). The *API translation service* translates an application API into a set of application services accessible through the interaction middleware in order to ensure compatibility with existing applications. The *data transformation service* offers tools for converting some data from a given format into another. The *user events service* manages user events at different levels of granularity, ranging from elementary events such as mouse moves or key strokes to complex ones reflecting the semantics of user actions such as selection in a list or connection to a device. Finally, the *appliances service* stores the characteristics of the various appliances which are used by some services of the upper layer in order for them to account for appliance variation and react accordingly.

Interaction services are the basis for the implementation of the intelligent interaction capabilities shown on Figure 1 above. These services are concerned with the management of the user task descriptions, the acquisition and update of the user activity state descriptions at runtime, the management of user class and user specific profiles and preferences as well as the management of the interaction (or *dialogue*) state and history.

---

[1] As core services are common to most existing system middlewares, we will not describe them in details.

# 6 Implementation and applications

The HIC concept is under a process of implementation by way of an integrated approach that mixes technological development together with operational application. Our first implementation of the HIC technology relies on the iROS middleware (cf. section 3) as a good candidate to experiment our layered and service oriented approach to interaction middlewares. We thus use the event heap to set-up a core service layer that can be seen of as a message-oriented system middleware. A few of the technical services of the second layer are basically the ones proposed by ICrafter but we develop most of these services ourselves, especially in order to support service classification and service hierarchy management. This work is now under completion. Finally, the implementation of the interaction services of the third level as well as the interaction processes of the HIC core will be done on the basis of a model-based approach which enables derivation of interaction model from application model and user task model at-compile-time and management of business interaction patterns and rules at runtime.

While the HIC may address a broader range of applications, its implementation is mainly driven by a set of THALES business cases in which it is meant to support tasks such as Control and Command, Collaborative Decision-Making or Team Situation Awareness, in domains such as Air Traffic Management or Naval Combat Management. It must be noticed that the deployment of the resulting technology, while being strongly dependent on the demand in terms of interaction services, also leads us to design specific architectural patterns per domain and business.

# 7 Conclusion

Through the adoption of a user-centric approach to interaction management, we argue that the HIC prefigures the future of HCI systems where people will have to interact and collaborate through a heterogeneous set of modalities and appliances. In terms of software development and management, the HIC as an interaction middleware offers the HCI designers an open framework enabling them to design an interaction system that meets business requirements such as enhanced user support and easy integration of new devices, without implying important changes to the application, which is a major requirement of THALES system designers and developers.

## References

Chatty, S. (2002). The Ivy software bus – a white paper. CENA Technical Note NT02-816, from http://www.tls.cena.fr/products/ivy/

Cheyer, A., & Martin, D. (2001). The Open Agent Architecture. *Journal of Autonomous Agents and Multi-Agent Systems*, 4 (1), 143-148.

Johanson, B., & Fox, A. (2002). The Event Heap: A Coordination Infrastructure for Interactive Workspaces. *Proceedings of the 4th IEEE Workshop on Mobile Computing Systems and Applications (WMCSA 2002)*.

Ponnekanti, S. R., Lee, B., Fox, A., Hanrahan, P., & Winograd, T. (2001). ICrafter: A Service Framework for Ubiquitous Computing Environments, *Proceedings of the Ubiquitous Computing Conference (UBICOMP 2001)*.

Ponnekanti, S. R., Johanson, B., Kiciman, E., & Fox A. (2003). Portability, Extensibility and Robustness in iROS. *Proceedings of the IEEE International Conference on Pervasive Computing and Communications (Percom 2003)*.

# Measuring the Immeasurable:
# System Usability, User Satisfaction and Quality Management

*Marcin Sikorski*

Gdansk University of Technology
Faculty of Management and Economics, Ergonomics Dept.
ul. Narutowicza 11/12, 80-950 Gdansk, Poland
Marcin.Sikorski@zie.pg.gda.pl

Abstract

This paper discusses opportunities for development of a joint approach, which combines usability studies, consumer research and quality management techniques. Consumer research perspective is presented as especially promising in the context of today's blooming market of interactive products and in the context of future developments for human-computer interaction (HCI) as a scientific discipline.

## 1    Introduction

HCI research has not yet produced a generic model of interaction between the user and the system. User-based testing and evaluation in recent years tends to prevail in most of usability studies, making them a sort of qualitative research, related to product quality improvement cycle. Newly latest interactive products are often aimed for mobile use and are subject to free consumer decisions, what seems to force broadening the perspective of current HCI and usability research. Including customer research and quality management tools into development process for contemporary interactive products seems to be a close perspective and a market demand for HCI.

## 2    Research perspectives for interactive products

For interactive products following main research perspectives are observed:
- **User-system interaction design perspective**
  - User-system interaction up to recently seems to be the dominant approach applied in HCI and in usability research; it relies on investigating the interaction between the user and the system in the specific context of use. Main variables, affecting the quality of user-system interaction can be identified as user interface characteristics, or as environmental factors, known as the context of use, which includes also characteristics of target users.
  - In user-system interaction design and research primarily experimental techniques are applied, aimed to identify factors affecting goal accomplishment, mental workload and satisfaction of the user. Extensive use of user interviews, questionnaires or web surveys suggests that interaction-oriented research today is dominantly user-based, not model-based – because predictive power of available user-system interaction models is still too weak (Rauterberg, 2000).

411

- Many available HCI techniques have been formerly developed for optimising usability of desktop software applications, and they are today sometimes hardly adaptable to investigation of new products like digital media or handheld mobile appliances. Therefore, a classical interaction design perspective to system usability is in many up-to-date cases too limited, also because more and more interactive products can be used also for fun and entertainment.

- **Quality management perspective**
  - A quality management perspective in HCI and usability studies focuses on providing user/customer with lasting quality of an interactive product across its lifecycle. Broad availability of consumer electronics products, which are shaped by subjective user preferences, not only by documented requirements, resulted recently in higher interest towards customer needs and attempts to include them in the user-centred software development process.
  - In contradiction to traditional manufacturing, contemporary quality management for interactive systems did not succeed to apply statistical control approach: extensive inspections and participatory techniques, like user testing are still dominant methods. Also Total Quality Management (TQM) approach gains a growing popularity in information technology industry (Woodal et al., 1997). In the quality management approach product usability is measured partly by a set of technical parameters, partly by users' ratings and partly also using benchmarking references to competitive products.
  - Primary focus in the quality management perspective is set on improving usability and attractiveness of the whole product by providing appropriate system functionality and lasting satisfaction for the users. Continuous improvement in the development process, aimed on transforming recognized user requirements into required product features, is perceived as the key success factor for satisfying target users of interactive products.

- **Contextual design perspective**
  - In contextual design approach, introduced by Beyer and Holtzblatt (1998) primary focus is set on observing behaviour of users when they are using the product, and on identifying user preferences, crucial for product acceptance in a real task environment. In this approach identifying social, organisational and cultural factors, which act as productivity stimulators (or as design limitations), is especially important for developing a interactive system, which would be highly usable and widely accepted by users.
  - Observatory and participatory techniques are central in this approach, together with cooperative task execution and testing prototypes to gather users' opinions that can be used for improving the system towards better fit to actual task environment.
  - Gathering quantitative data in not the primary goal in this perspective: most of valuable information remains in descriptive form, but the insight into actual tasks environment allows identifying factors critical for improving users' performance in computer-supported tasks.

- **Consumer research perspective**
  - In consumer research perspective primary focus is set on analyzing reactions of customers when they are confronted with the interactive product, and on understanding complex relationships user-product also beyond the user interface – touching also emotional spheres, lifestyles and mechanisms of buying decisions. Consumer behaviour is observed in a dynamic environment in variety of contexts, when user has access and to unlimited various sources of marketing information.
  - Methods and techniques used in consumer research are mostly based on surveys (questionnaires, interviews and focus groups) and on observation of users/customers during product tests. Statistical techniques (e.g. conjoint analysis, broadly used in

marketing research) can be also helpful in analysing customer behaviour and decision making in common interactive activities, like internet shopping.

- o  Moreover, the end user of an interactive product is treated as a typical consumer, who takes decisions in order to optimise his/her span between two goals: achieving required performance in various application scenarios and respecting specific budget limitations.

Customer research perspective adds to HCI and usability studies important new elements, which nowadays strongly shape actual customer decisions:

- total costs of ownership,
- integration with task environment,
- adaptability to different contexts of use,
- wide accessibility: anywhere, anytime.

Akoumianakis and Stephanidis (2001) have discussed recently observed shift in usability design goals from pure desktop applications usability, towards providing full accessibility for all users of interactive systems Moreover, consumer research perspective treats a *service*, offered through an interactive system, as a desired object, which can be *independent from the device*. Therefore it is claimed that the perceived quality of software-based, *interactive service* is more important for the business success and for the customer loyalty than the quality of the relevant interactive device.

# 3    User-centred design and the "new usability"

Nowadays availability of mobile devices with different add-ons, which can change the device functionality and download different digital contents, caused *interactive services* to be the central point of contemporary studies relevant to human life and work in the networked environment. As a result, an interactive product with its user interface is often treated as a device-based *enabler* for a particular service. However, not the devices itself, but the characteristics of the services and subjective reaction of users seem to be the key factor for the market success. Moreover, the terminal device (like a cell phone, a PC or a TV set) is usually owned by the end user, but the service content is usually delivered on the subscription basis.

New forms of interactions like ubiquitous computing, mobile devices, e-business, m-services, led to changes in lifestyles: users are almost always on-line, living in the information space and expecting accessibility to information services anytime, anywhere. Technological developments and transitions in users' requirements and lifestyles led to important changes in general approach how the interactive systems are currently developed:

- general shift of interest from interactive products to interactive services, which are operated by end user/customer on self-service basis,
- integration of services in subsequent steps: from dedicated software applications, web sites to web services, and ultimately, to ubiquitous net services,
- development of multi-purpose devices, which can deliver contents, perform operations, support decisions, solve specific problems or facilitate communication,
- treating usability as a competitive edge:
  - o  user is a consumer and decision-maker when choosing the device, service and its provider, while relationships with subscription-based service provider are often more strong and more lasting than with a device vendor,
  - o  consumer's choice can be guided by addressing users self-esteem, provider's empathy and other promotional techniques, addressing user's cultural values,
  - o  perceived trade-off between perceived product usability and actual costs of ownership has become  important factor in consumer decisions.

Commonly observed trend towards designing a positive "user experience" and emotional involvement (very important especially in web sites, web services and edutainment) seems to have its roots in the model of experiential learning (Kolb, 1984). By sequentially addressing users' emotional, analytic, reflective and activity spheres, user experience in modern interactive services addresses also customer values in the emotional (experience-related) and functional (usability-related) dimensions.

The trends discussed above seem to lead to new stream in HCI research and development, possibly reorienting paradigm shift in user-centered design from interaction design and optimizing the user interface to **optimizing user relationship with interactive e-service** and its business provider. The "new usability" approach attempts to address these issues, and to adopt some research techniques from consumer research to developing interactive services – in addition to optimizing user interfaces of interactive devices.

# 4    Quality of interactive services

In the light of presented trends, broadening of the HCI and usability research towards customer-oriented perspective seems to be a natural step suitable for building closer links between research and application practice. Expected shift in HCI and usability research will probably touch the basic focus of investigation from user's interaction with tangible product to user's relationship with intangible service and its provider. This shift is likely not only to extend the scope of the HCI and usability research towards the whole product lifecycle, but consequently, it will also include new factors like lifestyles, fashion, cultural issues in addition to task-oriented, business-related perspective of current HCI studies.

Despite of evident differences among the ways in which people use desktop products, mobile products, websites and net services, identifying key components of user satisfaction will probably remain as the common issue in the future HCI research. Moreover, the primary object of evaluation will be no longer the interactive product and its user interface, but the service delivered with the help of the specific product (device) - primarily the relationships between the customer and service provider. The issues like gaining consumer loyalty, building solid relationship with a customer base and building value chains for customers (Triekenens and Venendaal, 1997) with interactive services will be probably the novel issues reported not only at future HCI conferences.

Expected shift in HCI research focus - from software process improvement to building user-service relationship - will surely result in **a new model for development of interactive products and services**. Possibly, it will also produce new methods for evaluating and measuring interactive product quality, including currently marginal factors like customer satisfaction trends, loyalty assessment and provider-customer relationship studies. These factors have been so far often avoided in classical HCI research, and were considered as almost immeasurable. Some early notice of these trends can be observed in the work of Kettinger and Lee (1999), who applied a new SERVQAL as a measure of service quality evaluated by users and providers. Also recent adoptions of Quality Function Deployment (QFD) technique for intangible products like software (Fehin, 1997) or web services (Sikorski, 2002) attempt to capture "customer voice" for the user-centred quality assurance in the development cycle for an interactive product.

# 5    Conclusions and summary

Prospective enhancements the HCI scope and methodology, may yet seem to be quite hypothetical, despite some early notice of changes discussed here can be observed in research reports and in industrial practice. Expanding research methodology not always needs to strengthen maturity of HCI as a scientific discipline; HCI methodology should be original, not adopted in a big part from other disciplines. Therefore adopting tools and methods from perspectives discussed here should be careful and purposeful.

As a final remark, it can be noted here, that some dozens years ago similar search for status and original identity was performed among early researchers in marketing sciences or management. Today the status of marketing and management sciences as research disciplines is evidently more clear that this of HCI, since proving practical validity of marketing or management studies has solved most of the doubts. Because similar problems stand today before HCI community, revealing relationships between usability studies, consumer research and quality management may bring a stimulating influence also for future HCI research and its status among other disciplines.

# References

Akoumianakis, D. and Stephanidis, C. (2001). Re-thinking HCI in terms of universal design. In: Stephanidis C.: Universal Access in HCI: Towards and Information Society for All. Proceedings of HCI International 2001. Vol. 3. Lawrence Erlbaum, London.8-12

Beyer, H. and Holtzblatt, K. Contextual Design: Defining Customer-Centered Systems. Morgan Kaufmann: San Francisco, CA, 1998.

Blackwell, R.D., Miniard, P.W. and Engel, J.F. (2001). Consumer behavior. Hartcourt, London.

Fehin, P. (1997). Using QFD as a framework fro a user-driven design process. In: van Veenedaal, and McMullan, J. (eds): Achieving Software Quality. UTN Publishers, Den Bosch, The Netherlands, 69-82..

Kettinger, W.J., and Lee, C.C. (1999). Replication of measures in information systems research: The case of IS SERVQUAL. *Decision Science*, 30,33, 893-899.

Kolb, J. (1984). Experiential learning. Prentice Hall, New York.

Rauterberg, M. (2000). How to characterize a research line for user-system interaction. IPO Annual Report '2000, TU Eindhoven. 66-86.

Sikorski, M. (2002). Zastosowanie metody QFD do doskonalenia jakosci uzytkowej serwisow WWW. (Application of QFD Method for Improving Websites Usability, in Polish) In: Research Papers (Zeszyty Naukowe) of Poznan University of Technology. Series: Organizacja i Zarzadzanie, 35, 13-24.

Trienekens, J. and van Venendaal, E. (1997). Software quality from a business perspective. Kluwer, Deventer, The Netherlands.

Woodal, J., Rebuck, D.K., and Voehl, F. (1997). Total quality in information systems and technology. St. Lucie Press, Delray Beach, Florida.

# Section 3

## Usability

# Coping with Increasing SW Complexity - Stepwise Feature Introduction and User-Centred Design

*Heikki Anttila\*, Ralph-Johan Back\*\*, Pekka Ketola\*, Katja Konkka\*, Jyrki Leskelä\* and Erkki Rysä\**

\* Nokia Mobile Phones \*\* Abo Akademi University and TUCS

pekka.ketola@nokia.com

## Abstract:

We present the ladder development process, which combines the stepwise SW feature introduction method with the user-centred design method for system specification. The ladder process is tried out and refined through a case study, where we construct an application for a mobile terminal, the Teenage Girl Diary. We describe our experiences of using the ladder process, and also propose some enhancements, in particular how to incorporate Extreme Programming in the ladder process.

## 1. Introduction

Today's software construction takes place in a turbulent environment, where both the underlying technology and user needs are evolving. We need methods for building SW that rapidly responds to new user needs and market opportunities. Rather than looking at system as a monolithic entity, which is completely specified at the beginning of the project, it seems realistic to look at SW as an artefact that continuously adapts to changing user needs. This requires that we have a SW architecture that allows the evolution to take place, that we have a user driven planning process that shows in which way the SW should evolve, and we have a SW process supporting the evolution. We will propose a way to achieve these goals [1], based on combining two existing but recent techniques: *stepwise feature introduction* [2] and *user-centred design* (UCD) [6]. We refer to the combination of these as the *ladder process* for SW development.

Stepwise feature introduction provides architecture for constructing layered SW. A layer introduces a new feature into the system. Each layer (with the underlying layers) forms a release, which functions as a complete application. UCD is an iterative approach for designing a system. The design is produced through a cycle where one first studies the user's requirements, environment and tasks. One then defines the success criteria for the product, in terms of user tasks. The design is evaluated against user tasks and their success criteria. If the evaluation shows a need for improvement, the process is repeated, until a satisfying design is produced.

Our aim is to define a development process that allows the system to be built incrementally based on the continuous user feedback and new feature needs. Key features of this process would be: flexible product architecture and easily maintainable SW, increased reliability through incremental testing, easier product variation, fast and flexible response to user-needs, and better co-operation between people. We evaluate and refine the ladder process through a case study. We are interested in application SW for mobile terminals. The application that we have been looking at is a teenage girl diary on a communicator platform. A team of six persons built the application. Two of the team members were specialized in UI design, and four were specialized in SW construction. We will first give overviews of the two techniques (stepwise feature introduction and UCD), and how these are combined in the ladder process. Then we will study how the ladder process works in practice. We will evaluate the ladder process, based on our experience. We compare our approach with some other SW construction processes, in particular with Extreme Programming (XP).

## 2. Software design and implementation processes

An object-oriented software system is built out of a collection of components. A new system *feature* is realized as extensions of components. Stepwise feature introduction is based on incrementally extending the system with new *features*. A collection of features makes up a *layer*. A layer together with the layers below it constitutes a *release*. Because the system is being built bottom-up, there is a danger that we get into a blind alley: a new feature cannot be introduced in a proper way because of the way earlier features have been built. Therefore, we need to be able to change the system structure when needed, i.e. *refactor* the SW (Fowler [5]).

ISO 13407 [6] represents the prevailing understanding for creating usable systems. UCD works best in the early development, but there are challenges in applying UCD in traditional sequenced development due to timetable pressures (design iteration is not always possible or limited), confidentiality limitations (it is not always possible to involve real users), increasing system complexity (many functions are designed and implemented in one shot) and increasingly complex and evolving user requirements. Stepwise feature introduction is an opportunity for UCD because: design iteration is easier because the set of features to be iterated is small, iteration (in the form of user testing) is a natural activity *between* releases, system complexity can be managed because the set of features to be introduced is small, and user requirements can be checked between releases.

### 2.3 The Ladder Process and two teams

We combine UCD with stepwise feature introduction in a process where the system is specified and implemented in layers. The system is developed by two teams. One concentrates on analysing the requirements and creating the system specifications. The other concentrates on implementing the system. The system specifications are created as a sequence of releases. Each release defines the new features to be added, and how they interact with the existing features. The *release specification* is handed over to the implementation team, which introduces the required features by adding a layer on top of the existing layer. This new layer, together with the underlying layers, is a *release implementation*. The implementation is handed back to the UI team, which evaluates the implementation for conformance with the requirements. Based on the evaluation, a new updated specification may be given to the implementation team (together with a list of detected errors), which is then used to create an updated implementation. This exchange is repeated until both teams are satisfied with the release. The process is repeated, a step higher up in the ladder (Fig. 1).

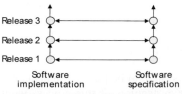

Figure 1: The ladder process

The advantage of this process is that the design of the whole system is broken up into smaller, manageable design problems, where only few features are introduced each time. The team will get early feedback on how well the feature works in practice and how it interacts with the previous features.

### 2.4 The Release Specification Process

We have identified the *core activities* necessary for UCD in stepwise feature introduction:
- Understanding the user needs regarding the release to be built
- Composing stories, use scenarios, describing the use of the release
- Analysing the steps that are required in performing a certain activity
- Documenting the activity as a structured use-case description
- Composing and analysing requirements based on the use case description
- Turning the requirements to concrete user interface specifications.

420

UCD process requires co-ordination: *Long-term planning* defines the overall development goals of each release, and *short term planning* consists of the core activities. A core activity produces a *release specification* with requirements to guide the development of the *next* release. User testing after release implementation functions as follow-up activity. It validates the completed functionality and produces improvement requirements.

## 2.5 The Release Specification and Implementation Process

UCD is combined with stepwise feature introduction in the *release specification process* (right side, Fig. 2. The development starts by studying user needs. Needs are translated to specification through core activities. The specification is implemented and tested (technical tests, specification compliance check and usability test). The tests produce change requests and feature proposals. The specification release is open to changes until it is fully implemented and tested. The implementation process (left side, Fig. 2) implements the specification.

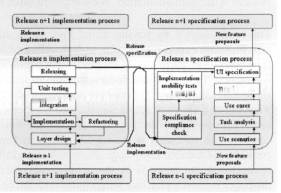

Figure 2. Release process, one ladder step

## 3. Case: Teenage Girl Diary

*Teenage girl diary* is a time-management application for teenage girls. The users, school girls, store term-specific school timetables in addition to normal calendar events. A school term is the major factor in organizing appointments with friends and hobbies – not the annual calendar. According to user studies attaching pictures and notes in the paper-calendars and diary pages is a daily activity. Visuality and personality are important issues. Our vision of the application was to support this kind of basic use, and to add functionality as the requirements are known better and new ideas are generated. An example screen of the diary is presented in Figure. 3. The purpose of the case study was to check out how the ladder process works in practice, and to refine the process. The description of the ladder process is thus a result of the case study described here, so the case study cannot be considered a test of the ladder process, but is an illustration of the process. Further studies are needed to test the ladder process in practical development.

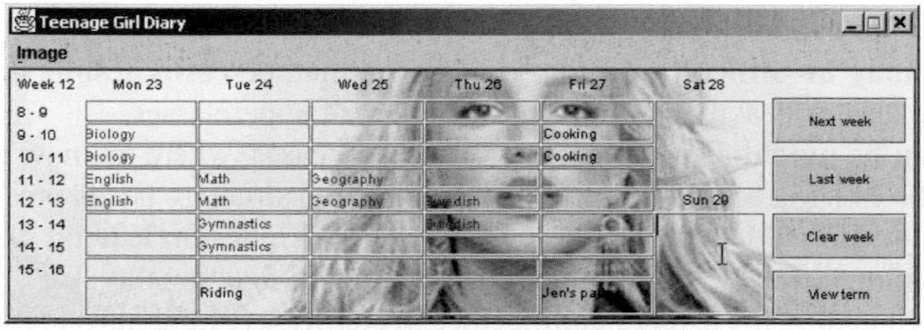

Figure 3. Screen shot of Teenage Girl Diary

We had a vision of the final system to be developed and a clear idea of the user needs. We created a long-term UI-plan for introducing the different SW functions in different *releases*. Each release was planned to be a stand-alone product. For really trying out how emerging user needs and features that are not known beforehand can be implemented in the *next* release, this long-term plan was only visible to specification team and not to implementation team. We tried to maximize opportunities for re-factoring, feature introduction and releases. Between releases we did short-term planning, which consisted of studying the current state of the implementation, and creating the next iteration of the specification through core activities. The output of the short-term planning was a document describing use cases, requirements and specification. It described the functions of the release and a list of corrections to the previous iteration of the release. Based on this document the implementation for the basic school timetable started. In the next phase the specification team evaluated the implementation. We completed three releases.

The work started by having initial knowledge of *user needs*. While going through the core activities and expert reviews the understanding of user needs increased and was used for creating *scenarios*. In order to clarify the motivation for new functionality we created *stories* describing the use of new functions. We documented the *exact steps* for performing the activity in our system. By identifying the needed steps we understood the required user and system interaction. Use cases were defined for each set of functions to be specified. Based on use case description we defined *user requirements*. The use cases alone were not always sufficient information, but we had to consider the scenarios also with SW platform functionality and consistency. The requirements could be turned into *UI specifications* in a straightforward way. The design process is simple and clear enough to be used in the iterative development. It is efficient, because the data that it produces is mostly platform independent and can be reused in further design releases.

## 4. Evaluating the Ladder Process – Pros and cons

We started out on the case study with a rough idea of the ladder process, essentially only knowing that we wanted to combine stepwise feature introduction with UCD. Our experiences after the case study are positive. The two techniques turned out to be easy to combine, and the division into specification and implementation teams felt natural, and did not cause overlapping work or redundant communication. The method worked in the case study, and the results encourage further experimentation. Below are some highlights on the experiences collected during the study:

+ **Parallel user testing:** Stepwise feature introduction with user feedback and parallel testing is supported by the layered approach.

+ **Reduced risk:** Having the previous application layers runnable at any time reduces risks. It is possible to revert back to using the previously added layer if needed.

+ **Simple to follow:** The idea of introducing new features as layers of objects built on top of existing ones was considered to be easy to learn and understand.

+ **No need for system simulation**: The system is functional all the time, so there is no need for to prepare additional simulations for user testing.

- **Resulting code may be difficult to understand:** The method easily results in deep inheritance hierarchies, where classes heavily redefine methods in their superclasses.

- **Refactorings can be potentially quite large:** There are cases, where the new functionality is so conflicting with the existing SW and its architecture, that the structure needs profound changes.

- **Not optimised for embedded SW:** The approach heavily relies on leaving existing code intact and mostly just adding new code that often overrides existing functionality.

We concentrated on getting the layered architecture in stepwise feature introduction correct, as this was a new thing of which the implementation team had no prior experience. In a future project a combination of stepwise feature introduction and XP would be most fruitful. The stepwise feature introduction method does not fit well with the traditional SW process, with its strong separation of requirements analysis, design, coding, testing and maintenance. However, it does fit well with the XP approach (Beck [4], Back and others [3]). Among the practices advocated are short iteration cycles, concentration on producing code rather than documentation, striving for simplicity, and frequent testing and integration during development. It also encourages practices like pair-programming and on-site customers, in order to create the right team dynamics.

XP emphasizes incremental SW construction. It lets the customer prioritise the features, while the programming team estimates the resources needed for implementation. The next iteration cycle is carried out in a fixed time with those features the customer has selected. The ladder process complements XP by giving explicit guidelines in places that XP leaves open. Stepwise feature introduction describes precisely how to structure the SW and build the components that are developed. UCD shows in more detail how the customer works in order to create the product specifications and how the customer should evaluate the releases that are produced by the XP team. XP defines "customer", which refers to an internal team member that represents users. To our view, for mobile phone development this approach is insufficient in cases where the SW is being developed for customers in several market areas. There is a need and possibility to improve XP by adding UCD activities in the process. XP has been seen as a new coming of hackerism, as it de-emphasizes documentation and careful design of the SW, and emphasizes the SW coding. The ladder process could, in combination with XP, provide the SW architecture, the guidelines and the long term planning necessary for keeping SW development on track.

# 5. Conclusions

We studied how to combine stepwise feature introduction with UCD, with the purpose of finding a process that supports SW evolution in response to changing user needs. We defined a development process, the ladder process, which combines these methods, and studied how this process works in practice through a case study. The case study shows that the method works, and provides clean SW architecture, and simplifies the UI specification. It seems that the ladder process could profitably be combined with XP. We were able to apply the whole UCD process during this experiment. This information is valuable because there is little real-life evidence about applying UCD in the development of mobile terminal SW. In our view, stepwise feature introduction and XP provide useful elements to implement UCD. Especially the possibility to iterate the design between releases is an asset. This could lead to cleaner SW and functionality in the long run. A more thorough investigation into using the ladder process for practical SW development is needed in order to understand the benefits and disadvantages of this approach. Our initial experiences have, in spite of limited time and resources, been encouraging.

# References

[1] Anttila, H., Back, R.-J., Ketola, P., Konkka, K., Leskelä, J. and Rysä, E.: Combining Stepwise Feature Introduction with User-Centric Design. TUCS Tech. Report 495, 2002, /www.tucs.fi/Research/Series/techreports.

[2] Back, R.J.R.: Software Construction by Stepwise Feature Introduction. In *ZB2002: Formal Specification and Development.* In Z and B (eds. D. Bert, J. Bowen, M. Henson and K. Robinsons), pp. 162-183. Springer Lecture Notes in Computer Science 2272, January 2002.

[3] Back, R.-J. and Milovanov, L. and Porres, I. and Preoteasa, V. An Experiment on Extreme Programming and Stepwise Feature Introduction. TUCS Tech. Report **495**, 2002, /www.tucs.fi/Research/Series/techreports.

[4] Beck, K.: *Extreme Programming explained.* Addison Wesley 1999.

[5] Fowler, M.: *Refactoring: Improving the Design of Existing Code.* Addison Wesley, 1999.

[6] ISO13407: *Human-centred Design Processes for Interactive Systems.* International Organisation for Standardization, Geneva, Switzerland. 1999.

# User-Centered Software Design and Development: Ensuring Customer Satisfaction

*Nuray Aykin*

Department Head
User Interface Design Center
Siemens Corporate Research
755 College Road East
Princeton NJ 08540
nuray.aykin@scr.siemens.com

## Abstract

In highly competitive global markets, meeting customer expectations becomes a key to success. In order to anticipate and respond to the challenges of today's markets and stiff competition we need to think about "more" not in terms of additional features, but in terms of meeting customer expectations. There is a strong need to connect and integrate information, people, processes and systems to increase customer satisfaction. Companies today are becoming more and more "customer-centric". Competitive software solutions must be designed for reusability, portability and maintainability to keep pace with customers' changing needs and to reduce product life-cycle costs. The development process should integrate user interface design, architecture design and quality management with the entire product development cycle. In response to these challenges, we must provide an integrated user-centered software design and development process, a main factor for the successful implementation of any product development project. The development of solutions built to meet or even exceed customer expectations is the mission of a user-centered software development process.

In this paper, we describe how two separate processes can be combined, and how we could convince product managers that user-centered design is not a luxury, but a necessity. This combined process brings the UI design and software development steps into a single structure. The key is in viewing each process step as a pair of interlocking puzzle pieces. One piece represents user-centered design activities, while the other represents software engineering activities. When snapped together, these pieces form a remarkably effective, collaborative process.

## 1    User-Centered Design Process

Any software development project involves a process. And, any user interface design also involves a process. The goal of both these processes is to create a successful product. A successful product is one that provides user satisfaction *and* has been developed within its planned budget and schedule. In the last ten or fifteen years, there have been many books written on software processes and user-centered design, but many of them have described their processes in their own terms, and created duplicate terminology for the same thing. Different people handled these two processes as separate phases in the product development lifecycle, sometimes causing unnecessary overlaps or different milestones. Many times what we see is software engineers working on requirements analysis, while user interface designers are collecting data from users in order to identify the similar requirements from a different perspective. Bringing these two forces

424

together is definitely the right methodology, but somehow not very much has been done about it until recently. Finally, in the past few years, we see a trend towards bringing these two processes and proven methodologies together with a goal of creating a "successful product" (Constantine & Lockwood, 1999; Mayhew, 1999).

However, we still see the readers of the similar books and articles on user-centered design and the software development lifecycle staying within their own territory. And, they learn the terminology of their own profession. For example, in the user-centered design process literature, we never hear the terms "actor" or "stage" as in Object-Oriented Software Engineering (OOSE) (Jacobson et al., 1992), and they never hear the terms "contextual inquiry" or "storyboarding". The danger has been that the two processes would end up following their own paths, without thinking about crossing, learning from each other, and eliminating duplication of efforts. Mayhew (1999) gives a great insight on how the user-centered design lifecycle could relate to the software development lifecycle. In her book, she follows very detailed user interface design lifecycle steps, and relates each step to an OOSE (Object Oriented Software Engineering) lifecycle as defined in Jacobson et al. (1992).

This paper is not intended to provide all the details about bringing the two processes together - although that is the ultimate goal - but it is about presenting product development as one single process at a very high level of integration. It is making the line between software engineers and user interface designers blurrier, while making their ties stronger. We need to show how we describe these processes as parts of a puzzle, step by step, with puzzle pieces representing the matching steps in software development and user-centered design processes. This paper describes our initial attempts at Siemens Corporate Research (SCR) to combine these two processes. We began understanding each other's methodology and terminologies, so that we can integrate these two processes in a seamless fashion. So far, our main success regarding bringing these two processes together was reflected in our joint internal project acquisitions. Specifically, the User Interface Design Center (UIDC) and the Software Engineering (SE) Departments have been promoting this single process to our internal customers together, rather than just promoting our own areas of work. Our internal customers start to see us as a "one-stop-shop" to help them produce high quality products.

## 2    Product Development Process – Pieces of a Puzzle

The main goal of creating the following puzzle is to merge terminologies used in the two separate processes and to create a common terminology set. We realize that although there are steps within each process that are highly common, using different terminologies is creating an artificial gap between these two processes. And, as you can imagine, the experts in each field (software engineers and user interface designers) are trying hard to convince product development that their process would solve the same problem of creating a "successful product", making customers happy, and saving on development costs. Bringing these processes together strengthens the case of making products successful.

At SCR's User Interface Design Center, we decided to tie our processes to the software development processes promoted by the SE Department. We believe this would impact our customers' understanding of our value as a whole, rather than two separate pieces. As a first step, we looked at the various terminologies being used in defining these processes. From what we found, we realized that we could come up with a terminology, that is the "Least Common Denominator (LCD)" which could describe these processes as a single, integrated process. The task is not that easy, if you consider all the different terminologies being used in academics and industry. A small sample of these terms include: requirements analysis, market analysis, use cases, usage scenarios, task flows, information architecture, software architecture, quality assurance, prototyping, iterative usability evaluation, etc.

Before we really could define a set of terminology that we both could understand, we decided that we would look into the phases of these two processes and see if there are phases that can be performed in parallel.

We came up with the following puzzle-process (shown in Figure 1) that describes how these two processes had their puzzle pieces (phases) together – as one streamlined process. In order to create an equivalent user-centered design process for every step in the software development process, we came up with common general steps that exist in both processes. We tried to make these steps complimentary, not overlapping, in nature. Note that the process is not yet validated, and one can argue that it is an oversimplification. But the diagram shown in Figure 1 was successful in one important way: it helped our internal Siemens customers to see how we fit together. It became a great marketing tool we both started to use in our meetings with the clients.

As we all know, user interface design specialists always complain about being brought in late in the product development cycle. One of the main reasons for this is that it is hard to relate one process as an outsider (i.e. user-centered design process) to the one accepted more widely (i.e. software development process). With this puzzle-process we are making a statement that for every step in a software development process, there is a complementary user-centered design process, and vice versa. For example, what user interface designers collect at the Requirements Analysis phase is not the same as the requirements collected by the software developers/architects. You need both in order to perform a comprehensive requirements analysis. With this approach, you are covering all the bases to understand what the requirements are. The following section summarizes the methodology and the deliverables for each step in the process. This is not intended to be inclusive of the entire methodology in either process, but rather a start to bring some terminologies and complementing steps and deliverables together.

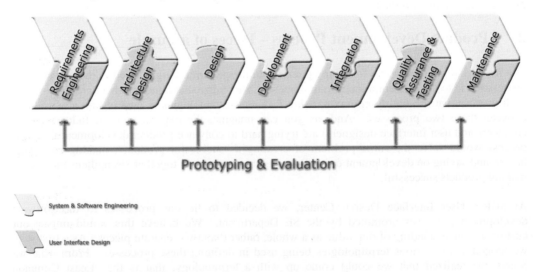

**Figure 1.** User Centered Software Development: An integrated process

426

# 3 Process Steps

In each step described below, we still list the software design and user interface design processes separately. Currently we are working with the SE Department to understand how the contents of each phase could be brought together to complete our puzzle. The work includes finding common terminologies, identifying overlapping and complementary activities so that we can work together in an integrated fashion.

- **Requirements Analysis**

  Eliciting marketing, user and software requirements are a critical foundation for creating the appropriate software product to meet a market need
  - Software design process: Competitive analysis, Market requirements, business modeling using UML and design tools
  - User interface design process: Gather user requirements via interviews focus groups, customer workshops, surveys, feedback, brainstorming on task scenarios, affinity diagrams, work models.

- **Architecture Definition**

  Software architecture and user interface design ensures flexibility and integrity throughout a product's life cycle and provides a competitive advantage.
  - Software design process: Architecture-Centric Software Project Planning (ACSPP), Product Line Architecture (PLA) Planning (Paulish, 2001)
  - User interface design process: UI prototyping, conceptual modelling, high-level design, storyboarding, and style guide

- **Design, Development and Integration**

  Software design, development and integration is where everything merges into a product. The product will reflect the quality of the design and the time-to-market and the cost of the product will reflect how well the processes are implemented.
  - Software design process: Detailed software design, global development
  - User Interface design process: Detailed UI design, iterative UI design evaluation

- **Iterative Prototyping and Evaluation**

  Prototyping the user interface at an early stage of development is a cost-effective and efficient method to collect customer input during the design phase. Prototyping critical components to validate a design can also reduce the risk to schedules and budgets. This step is repeated throughout the design and development process.
  - Software design process: Architecture Tradeoff Analysis Method (ATAM), Technical Architecture Prototypes
  - User Interface design process: UI prototypes, results of usability evaluation

- **Quality Assurance and Testing**

  Evaluation of the software and the usability of products at all stages of development, ensuring that the solutions we offer are easy to use
  - Software design process: System architecture testing, performance tests
  - User Interface design process: UI style guide compliance testing, usability testing

- **Maintenance**

Evaluation of the software and the usability of products after the development ensuring that the product meets the business and market needs, and it is usable.
- Software design process: Bug fixes, new feature identification for next releases
- User Interface design process: user feedback on product via usability evaluation, product feedback, customer care input

# 4    Summary

This paper describes how we at the SCR User Interface Design Center integrated our processes with those of a companion SE Department. This paper describes at a very high level how the terminology and process integration works. Although our integrated process is not detailed out to explain all phases and joint deliverables, the greatest benefit we have gained so far is that our internal customers see us as an integrated approach for product development, and we end up working on projects together with SE Department much earlier in the development cycle. We still have a lot to accomplish to make this integrated approach successful. Our next goal is to work together with the SE Department on identifying where we complement each other and where we can have joint work items and deliverables. This will help us to move forward towards a fully-integrated process where we will work together more efficiently.

## References

Constantine, L.L., Lockwood L.A.D. *Software For Use: A Practical Guide to the Models and Methods of Usage-Centered Design.* 1999. Reading, Mass.: ACM Press/Addison-Wesley.

Jacobson, I., M. Christerson, P. Johnson, and G. Overgaard. 1992. *Object-Oriented Software Engineering: A Use Case Driven Approach.* Reading, Mass.: ACM Press/Addison-Wesley.

Mayhew, D.J. *The Usability Engineering Lifecycle: A Practitioner's Handbook for User Interface Design.* 1992. San Francisco, Ca: Morgan Kaufmann Publishers, Inc.

Paulish, D. *Architecture-Centric Software Project Management: A Practical Guide.* 2001. Boston, Mass: Addison Wesley.

# Accreditation of Usability Professionals

*Nigel Bevan*

Serco Usability Services
22 Hand Court, London WC1V 6JF, UK
nbevan@usability.serco.com

## Abstract

There is increasing demand for professional recognition of usability expertise, to give status to professionals in the area, and to enable potential employers and clients to identify people who are competent in the field. Proposals for usability accreditation were first developed by the EU INUSE project in 1997, and these formed the basis of subsequent UK-based initiatives to establish usability accreditation. The EU UsabilityNet project worked with an international group to develop the proposals into an international scheme that was announced in April 2002. Feedback was collected through a web questionnaire and sessions at several international conferences. The web survey showed that while 77% of people new to the field would seek certification, only 39% of the most experienced would do so. At this point, the major US sponsor withdrew for financial reasons, and the Usability Professionals Association that was supporting the project concluded that it did not have the resources to go it alone with a controversial accreditation programme. Support for usability accreditation remains strong in Europe, and the preferred European approach would be to develop usability accreditation as part of a professional development scheme.

## 1    Why usability accreditation?

The possibility of usability accreditation has been discussed for several years. The potential advantages and disadvantages for potential stakeholders were identified at an international workshop in 2001 (Day & Bevan 2002).

### 1.1    Purchasers of usability services

Advantages: Provides criteria to choose a service provider, requires less expertise on part of purchaser to make decision, helps justify decisions to management, likelier to get better services resulting in a higher quality product.
Disadvantages: Needs to be properly advertised and managed, a certified person will cost more, could stifle innovation, guarantees mediocrity.

### 1.2    Usability professionals

Advantages: Provides status and a level of authority, promotes accepted values within the profession, makes moving job easier, will command a higher salary, helps define a career path
Disadvantages: Does not distinguish between the newly qualified and the very experienced, will need time, effort and money, will not cater for the specialist, could homogenise the profession.

### 1.3    Usability aware employer

Advantages: Encourages development of less experienced employees, provides criteria to select a new employee, helps justify decisions about hires, lowers the risk of selecting an inappropriate employee, a certified employee will need less local training, easier to identify training requirements, can be part of a reward/compensation package, provides a basis for differentiating employee expertise, provides guidance for professional development.

Disadvantages: certified employees will expect a higher salary, the employer may be expected to pay for certification and training, leaves less latitude for personal judgment, may focus only on certification and not on other relevant skills.

### 1.4 Entry level practitioners

Advantages: Provides something to aspire to and guidance on skills required, provides self-evaluation standard, gives status, makes the subject more meaningful, helps decide "should I go into this field?"

Disadvantages: Costly in money and time, the employer not as likely to pay, provides a barrier to entry to field

### 1.5 Usability consultancies

Advantages: easier to gain credibility, provides differentiation to competitors

Disadvantages: cost to get employees certified, may result in higher staff turnover, more difficult to retain staff, consultancy fees will need to go up.

### 1.6 Training organizations

Advantages: easier to get students (marketing), may increase business of certified courses

Disadvantages: must meet standards (could be costly), cost of certification

Pos/Neg: defines a program for training

### 1.7 Lower priority stakeholders

Other potential stakeholders include end users, project managers, executives sponsors of usability professionals, academic teachers, professional organizations, industry usability trainers, students in academic programs, project team, and legal professionals (liability).

## 2 European and UK initiatives

### 2.1 INUSE scheme

One of the outputs of the EC INUSE project was a proposed scheme for accreditation of usability support providers (Bevan et al 1998). It included assessment criteria derived from the usability maturity model proposed by Earthy (1999) (this later formed the basis for ISO TR 18529). The scheme was validated by assessing the capability of the INUSE project partners, but no further use was made of the scheme at that time.

### 2.2 British HCI Group scheme

Jonathan Earthy then worked with the British HCI Group to develop a proposed lightweight scheme of self-certification (BHCIG 2001) that received significant support from British HCI Group members. Companies or individuals who feel they meet the criteria would attest competence in one or more of seven areas: usability consultancy, planning user centred design, evaluation and testing; and the optional specialist competencies: requirements engineering, product design support, training courses and technology transfer. When accredited the company or individual could claim these competencies in communications with customers.

The British HCI Group is currently investigating whether HCI competence can be incorporated into the BCS curriculum and professional development and professional membership schemes.

## 2.3 UK UPA scheme

Another proposed scheme was developed, based on the same INUSE competencies, but with procedures modelled on those used by the British Computer Society to assess candidates for professional membership. The UK UPA also surveyed its members, and found that 77% of those replying were in favour of an accreditation scheme, and the majority were in favour of a formal international scheme rather than the self-certification proposed by the British HCI Group.

# 3 International workshop

In the USA, Surgeworks, which has participated in a DSDM certification scheme, hosted an international workshop jointly sponsored with the UPA to discuss setting up a usability certification scheme. There were 13 participants at the workshop, including Nigel Bevan and Jonathan Earthy from the UK, Masaaki Kurosu from Japan and 10 US representatives from industry, government and academia. The UK submitted an agreed position on accreditation based on the outcome of a meeting between representatives of the British HCI Group, UK UPA, Ergonomics Society, IEE and with European input from the EU UsabilityNet project (www.usabilitynet.org).

## 3.1 Workshop conclusions

The workshop accepted all the European inputs, and reached the following conclusions.

### 3.1.1 Criteria for certification

The preferred approach is to define core competency complemented by elective specialities. The scope of assessment will be user centred design, and the assessment criteria will be derived from the "Technical competencies for User Centred Design professionals" document (Earthy et al 2002) submitted by the UK. The core competencies are expected to be:
- Plan and manage the human-centred design process
- Understand and specify user and organisational requirements and context of use
- Produce design solutions
- Evaluate designs against requirements

Jonathan Earthy and Nigel Bevan (with support from UsabilityNet) will be responsible for developing a strawman model including pass and fail criteria for each competency item for typical personas seeking certification.

### 3.1.2 Certification process

The suggested elements are:
- a points system to assess eligibility based on education and experience (there was little enthusiasm to accredit training courses in the US in the first phase)
- submission of material describing use of UCD on a project. The DSDM model of a 2000 word explanation of how and why UCD principles has (or has not) been applied was favoured.
- submission of structured peer references
- a possible written exam composed of problem-solving questions
- a structured interview

Assessment should be operated by a not-for-profit consortium including representatives of professional bodies and major companies, and the scheme should be operated internationally.

# 4    Consultation

UsabilityNet then organised several workshops in London to further refine and develop the definition of competencies, in conjunction with the US group. The proposed scheme was announced in April 2002, and feedback was collected through a web questionnaire and sessions at several international conferences.

The web survey (UPA 2002) showed that while 77% of people new to the field would seek certification, only 39% of the most experienced would do so. Strong opinions, both in favour and against certification were expressed on a number of themes (Table 1).

| | |
|---|---|
| Theme: Value to Customers<br>• Low value to managers<br>• Value to consultants<br>• Acceptance affects value | Theme: Value in Hiring and Self-Promotion<br>• Extra credentials<br>• Comparing credentials<br>• Low value for academics |
| Theme: Value of Certification to the Profession<br>• Builds legitimacy and credibility<br>• Raise standards in profession<br>• No guarantee of quality<br>• No value or benefit<br>• Potentially harmful or divisive | Theme: The Certification Process<br>• Costs<br>• International Recognition<br>• Code of Conduct |
| Theme: Value for Those Entering the Field<br>• Barrier to entry<br>• Alternate route to establish credentials<br>• Value is in self-assessment<br>• Must not be too easy | Theme: Education vs Experience<br>• Need links between certification and degree programs<br>• Degrees more important than certification or experience<br>• Experience more important than degrees |
| Theme: Defining the Field<br>• Need to define core skills<br>• Field too undefined at this point<br>• Could restrict creativity<br>• Field too broad | Theme: Project and Process<br>• Don't compete with existing schemes<br>• Don't emulate existing certifications<br>• Governance and control<br>• UPA's involvement in this project<br>• Comments on the Survey |

**Table 1:** Survey comments

Common fears expressed were that: accreditation would be a barrier to entry to the profession, would exclude people who had been in the field for years, would cause a rift between the haves and the have nots, would be a barrier to innovation and would stipulate the usability methods to be used. Experienced professionals were the most voluble in their criticism of certification in responses to the survey, and in email discussion groups. They were particularly concerned that it might challenge their established expertise.

Many of the concerns were a misunderstanding of the proposed scheme, but at this point, the major US sponsor, SurgeWorks, withdrew for financial reasons, and the Usability Professionals Association concluded that it had neither the financial nor people resources to go it alone with a controversial certification programme.

# 5  The future: a professional development scheme

Usability accreditation is on the agenda of the European Usability Forum (Tscheligi et al 2003) set up by UsabilityNet. In the mean time other organisations are taking the initiative with their own narrower schemes, e.g. usability certification awarded by HFI (www.humanfactors.com) in the USA, and the SFIA (www.e-skills.com) definition of system ergonomics in the UK.

Many people regard professional usability accreditation as inevitable in the long run, but it is likely to be more acceptable in Europe if it forms part of a professional development scheme. As an individual progresses in their career they need to acquire two complementary types of skills:

1. technical capability, for example to apply user centred design methods
2. general professional skills that are acquired by practitioners with increasing responsibility

The technical skills for a usability professional are generic, and are defined by a set of levels, for example;

1. Assistant practitioner undergoing training and becoming familiar with the scope of their tasks
2. Trained practitioner familiar with the scope of their tasks, working under supervision with specific instructions
3. Professional. Selects appropriately from applicable standards, methods and tools. Supervise others. Decisions influence success of projects and team objectives.
4. Senior practitioner. Full accountability for their own technical work or project/supervisory responsibilities. Receives assignments in the form of objectives. Establishes own milestones, team objectives and delegates assignments.

Professional development can then be defined as a combination of:

1. Acquiring the necessary technical skills (through education, training and experience)
2. Gaining the professional skills (through training and experience) to apply the technical skills at increasing degrees of responsibility.

A professional development scheme for usability professionals would enable an individual and employer to track the individual's progress in gaining both technical and professional skills, with the objective of becoming an accredited professional.

# References

Bevan, N., Claridge, N., Earthy, J., Kirakowski, J. (1998) INUSE Deliverable D5.2.3: Proposed usability engineering assurance scheme, v2. www.lboro.ac.uk/eusc/index_r_assurance.html

BHCIG (2001) Draft HCI accreditation. http ://www.bcs-hci.org.uk/accreditation.html

Day, D., Bevan, N. (2002) Certifying usability (professionals): a scheme to qualify practitioners. Interactions, January-February 2002, page 7-9.

Earthy, J. (1999) INUSE Deliverable D5.1.4(p) Usability Maturity Model: Processes, v2.2 www.lboro.ac.uk/eusc/index_r_assurance.html

Earthy, J., Bevan, N., Philip, R. (2002). Technical competencies for Human Centred Design professionals, v0.7. www.upcertification.org/report_drafts.html

ISO 13407 (1998) User centred design process for interactive systems.

ISO TR 18529 (2000) Human-centred lifecycle process descriptions

Tscheligi, M., Giller, V., Frohlich, P. (2003) European usability forum. Proc. HCI International 2003

UPA (2002) Certification of Usability/UCD Professionals. www.upcertification.org

# UsabilityNet Methods for User Centred Design

*Nigel Bevan*

Serco Usability Services
22 Hand Court, London WC1V 6JF, UK
nbevan@usability.serco.com

## Abstract

There is a wide variety of conflicting guidance available on appropriate methods to use to support user centred design. The ISO 13407 standard provides a framework for applying user centred design, without stipulating which methods should be used. Based on wide experience of EC and commercial projects, the TRUMP and UsabilityNet projects selected sets of methods to support ISO 13407 that have been found to be cost-effective in commercial application. The paper compares these methods with those found in textbooks, and discusses the most effective way to present them through a web site.

## 1    User centred design process: ISO 13407

ISO 13407 provides guidance on achieving usability by incorporating user centred design activities throughout the life cycle of interactive computer-based systems.  It describes user centred design as a multi-disciplinary activity.

The standard describes four user centred design activities that need to start at the earliest stages of a project.  These are to:

- understand and specify the context of use

- specify the user and organisational requirements

- produce design solutions

- evaluate designs against requirements.

The iterative nature of these activities is illustrated in Figure 1.

The process involves iterating until the objectives are satisfied. ISO 13407 describes the basic principles, but does not stipulate specific methods. The sequence in which the activities are performed and the level of effort and detail that is appropriate varies depending on the design environment and the stage of the design process.

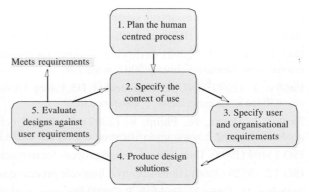

**Figure 1:** ISO 13407 activities

## 2 TRUMP

The EC-funded TRUMP project (Bevan et al, 2000) trialled use of user centred design methods based on ISO 13407 in two contrasting application areas: office applications in the Inland Revenue/EDS, and avionics systems in Israel Aircraft Industries. The methods used were selected to be simple to plan and apply, and easy to learn by development teams. From the common experience of these trials, 10 methods were selected as generally applicable across a wide range of development environments. Figure 2 shows how each of the recommended methods relates to the lifecycle stages and the processes described in ISO 13407.

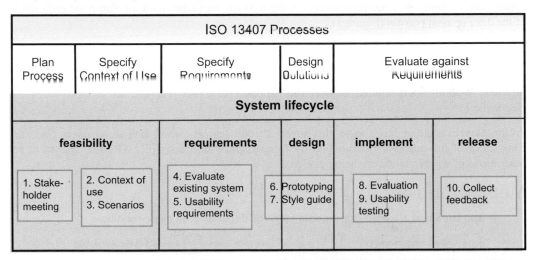

**Figure 2:** TRUMP methods

*1. Stakeholder meeting* A half-day meeting to identify and agree on the role of usability, broadly identifying the intended context of use and usability goals, and how these relate to the business objectives and success criteria for the system.

*2. Context of use* A half-day workshop to collect and agree detailed information about the intended users, their tasks, and the technical and environmental constraints.

*3. Scenarios of use* A half day workshop to document examples of how users are expected carry out key tasks in a specified contexts, to provide an input to design and a basis for usability testing.

*4. Evaluate an existing system* Evaluate an earlier version or competitor system to identify usability problems and obtain measures of usability as an input to usability requirements.

*5. Usability requirements* A half-day workshop to establish usability requirements for the user groups and tasks identified in the context of use analysis and in the scenarios.

*6. Paper prototyping* Evaluation by users of quick low fidelity prototypes (using paper or other materials) to clarify requirements and enable draft designs to be rapidly simulated and tested.

*7. Style guide* Identify, document and adhere to industry, corporate or project conventions for screen and page design.

*8. Evaluation of machine prototypes* Informal usability testing with 3-5 representative users carrying out key tasks to provide rapid feedback on the usability of prototypes.

*9. Usability testing* Formal usability testing with at least 8 users carrying out key tasks to identify any remaining usability problems and evaluate whether usability objectives have been achieved.

*10. Collect feedback from users* Collect information from sources such as usability surveys, help lines and support services to identify any problems that should be fixed in future versions.

# 3    UsabilityNet

One of the objectives of the EC UsabilityNet project (Bevan et al 2002) has been to provide usability professionals with an authoritative website of resources, including recommended methods for user centred design. UsabilityNet partners reviewed a wide range of methods, and based on the partners' experience in EC and commercial projects, 35 methods was selected that had a track record of cost-effective application in a commercial environment. These were categorised into the same stages of the development process as in TRUMP, except that testing and measuring was identified as a separate activity at the end of implementation. To help users select appropriate methods, they are represented on the web site in a table with a column for each stage of the development process (Figure 3).

A description of the method can be obtained by clicking the appropriate cell. The methods can also be filtered based on three criteria: limited time or resources, no direct access to users or limited skills or expertise. Inappropriate methods are greyed out depending on the criteria selected. With all filters applied, eight remaining basic early lifecycle methods are recommended (see Table 1).

Preliminary evaluation has shown this to be a good learning tool, but some users find the complexity of the interface intimidating by comparison with the simpler TRUMP approach, so other forms of representation are being explored.

## Methods table

you can select the most appropriate methods depending on three conditions

☐ limited time/resources    ☐ No direct access to users    ☐ Limited skills/expertise

| Planning & Feasibility | Requirements | Design | Implementation | Test & Measure | Post Release |
|---|---|---|---|---|---|
| Getting started | User Surveys | Design guidelines | Style guides | Diagnostic evaluation | Post release testing |
| Stakeholder meeting | Interviews | Paper prototyping | Rapid prototyping | Performance testing | Subjective assessment |
| Analyse content | Contextual inquiry | Heuristic evaluation | | Subjective evaluation | User surveys |
| ISO 13407 | User Observation | Parallel design | | Heuristic evaluation | Remote evaluation |
| Planning | Context | Storyboarding | | Critical Incidence Technique | |
| Competitor Analysis | Focus Groups | Evaluate prototype | | Pleasure | |
| | Brainstorming | Wizard of Oz | | | |
| | Evaluting existing systems | Interface design patterns | | | |
| | Card Sorting | | | | |
| | Affinity diagramming | | | | |
| | Scenarios of use | | | | |
| | Task Anaysis | | | | |
| | Requirements meeting | | | | |

**Figure 3:** UsabilityNet methods

| Method | UN | Nielsen | Mayhew | Vred. |
|---|---|---|---|---|
| ***Planning & feasibility*** | | | | |
| Stakeholder meeting | √ | | | |
| Planning | √ | √ | √ | √ |
| Cost benefit analysis | √ | √ | √ | √ |
| Competitor analysis | √ | √ | | |
| ***Requirements*** | | | | |
| User survey questionnaire | √ | | √ | |
| Interviews | √ | √ | √ | |
| Contextual inquiry/interview | √ | | √ | √ |
| User observation/field study | √ | √ | √ | |
| Analyse context of use | √ | √ | √ | √ |
| Focus group (requirements) | √ | √ | | |
| Brainstorming | √ | | | |
| Evaluate existing system | √ | | | √ |
| Card sorting | √ | √ | | √ |
| Affinity diagramming | √ | | √ | √ |
| Scenarios of use/use cases | √ | √ | √ | √ |
| Task analysis (analytical) | √ | √ | | √ |
| Set usability goals | √ | √ | √ | √ |
| Design patterns | √ | | | |
| ***Design*** | | | | |
| Design guidelines | √ | √ | √ | |
| Paper prototyping | √ | √ | √ | √ |
| Heuristic/expert evaluation | √ | √ | √ | √ |
| Parallel design | √ | √ | | |
| Storyboarding | √ | | | |
| Evaluate prototype | √ | √ | √ | √ |
| Wizard of Oz | √ | | | |
| Conceptual models | | | √ | |
| Participatory design | | √ | √ | |
| Design walkthrough | | | | √ |
| ***Implementation*** | | | | |
| Style guides | √ | √ | √ | |
| ***Test & Measure*** | | | | |
| Diagnostic evaluation | √ | √ | √ | √ |
| Performance testing | √ | √ | √ | √ |
| Subjective evaluation | √ | √ | | √ |
| Critical Incident Technique | √ | √ | | |
| Pleasure | √ | | | |
| ***Post-release*** | | | | |
| Testing and measurement | √ | √ | | √ |
| Subjective assessment | √ | √ | √ | |
| User survey questionnaire | √ | √ | √ | √ |
| Remote evaluation | √ | | | |
| Logging | | √ | | |
| Field study | | √ | √ | |

**Table 1:** Recommended methods

**Key**
UN: UsabilityNet
Nielsen: Nielsen (1993)
Mayhew: Mayhew (1999)
Vred.: Vredenburg et al (2002)

√: described
√: mentioned

⬚ √ : basic or recommended

**Note**
The methods listed in the table exclude general advice and duplications

# 4    Comparison

The methods recommended by UsabilityNet have been compared with those recommended in three textbooks (Table 1).  UsabilityNet (2002) recommends 35 methods, of which seven are core methods.  Nielsen (1993) describes 24 methods (and mentions four more), of which five are rated as having the biggest impact on usability.  Mayhew (1999) describes 18 of the methods and mentions four more, Vredenburg et al (2002) describe 18 and mention 2 more.

The difference in methods recommended by different sources can partly be explained by implicit assumptions about the development environment in which user centred design is expected to be applied.  More guidance is needed on the appropriateness of the methods in different contexts of use.  For example:

- Consultancy: The *stakeholder meeting* is an essential activity for consultancies (and in-house usability groups that act in consultancy mode) to establish which usability methods will support the particular business and marketing priorities.
- In-house development: *Participatory design* is much easier to achieve when a system is developed for in-house users.
- Web development: Some techniques such as *card sorting* are particularly appropriate when developing web sites.

Some other methods are only required in specialised circumstances, for example *focus groups*, *brainstorming*, *parallel design*, *storyboarding*, *wizard of oz*, *remote evaluation* and *logging*. Other methods are relatively new, and not yet widely adopted, such as: *design patterns*, *affinity diagramming*, *critical incident technique* and *pleasure*.

Most surprising is the low profile of *questionnaires* and *subjective assessment*.  This seems to reflect the opinion prevalent in some American organisations that what matters is whether a user can achieve a task, rather than the user's attitude to the product.  However the priorities in industry are changing with a greater appreciation of the importance of user satisfaction for web sites and consumer technology, and growing emphasis not only on preference but also on engagement and *pleasure* (Green and Jordan 2002).

Another noticeable difference between UsabilityNet and the other sources is the terminology used by UsabilityNet, derived from ISO 9241-11 and ISO 13407, where usability is defined as: the extent to which a product can be used by specified users to achieve specified goals with effectiveness, efficiency and satisfaction in a specified context of use.  This highlights the importance of satisfaction, and uses the term *context of use* to refer to the users, tasks and environments of use.  The term context of use is gaining acceptance in Europe, but in the USA the terms user analysis and task analysis are more commonly used.

# References

Bevan, N, Bogomolni, I, Ryan, N (2000) TRUMP. www.usability.serco.com/trump

Bevan, N., Claridge, N., Fröhlich, P., Granlund, A., Kirakowski, J., Tscheligi, M. (2002). UsabilityNet: Usability support network. www.usabilitynet.org

Green, W.S., Jordan, P.W. (2002) Pleasure with products: Beyond usability. Taylor and Francis.

Mayhew, D.J. (1999) The usability engineering lifecycle. Morgan Kaufmann.

Nielsen, J. (1993). Usability engineering. Academic Press.

Vredenburg, K., Isensee, S., Righi, C. (2002) User-centered design.  Prentice Hall.

# Usability Design for the Home Media Station

*Konstantinos Chorianopoulos*                    *Diomidis Spinellis*


ELTRUN, Athens University of Economics & Business
http://itv.eltrun.aueb.gr
chk@aueb.gr, dds@aueb.gr

## Abstract

A different usability design approach is needed for the emerging class of home infotainment appliances, collectively referred to as the home media station (HMS). Mass-media theory, consumer electronics engineering, content creation and content distribution play a major role towards the human-centered design of home media appliances. Different audience behavior factors, such as the attention span, and group watching, affect the design of the HMS. We have employed our approach in the design of a system offering dynamic synthesis of the advertising-break at each television set-top box.

## 1    Introduction

There is growing evidence from the marketplace that information technology is migrating from the PC in the office to the set-top box in the living room. The HMS category encompasses stand-alone or networked devices that range from video game consoles (Sony PS2, Microsoft XBox), MP3 juke boxes (HP dec100), digital tuners (Nokia Mediamaster), digital video recorders (TiVo, ReplayTV), as far as combinations of the above (Digeo Moxi Media Center). User access to the HMS is currently done according to the manufacturer's idiosyncrasy or by applying rules from the desktop and Internet experience (Carey 1997; Wallich 2002).

Most notable among the recent findings regarding the HMS class of devices is the realization that users' subjective satisfaction is at odds with performance metrics. In a usability test of three video skipping interfaces, users preferred the interface that required more time, more clicks and had the highest error rate (Drucker, Glatzer, Mar, and Wong 2002). Users reasoned their choice on the basis of how fun and relaxing an interface was.

It has also been widely evident, in our literature review, that the approach followed by the majority of scientific publications, regarding entertainment computing in the home, is mainly PC-centric. Since traditional HCI principles assume a task-oriented approach, where the human interacts with an application to accomplish a particular goal, computer-mediated leisure applications require a fresh view of the current paradigms.

## 2    Dimensions of Usability Design for Home Infotainment Appliances

The field of HCI has been benefited by a multidisciplinary approach to design problems (Marcus 2002). Successful user interfaces, apart from proven methodologies and multiple design iterations, draw from a diverse array of design disciplines. For the case of the HMS we have identified: 1) Mass-media theory, 2) content creation and content distribution, and 3) broadcasting and consumer electronics engineering. The role of each discipline to the design of HMS applications is summarized in a table and each one of the factors is described with a reference to original source.

## 2.1 Mass-Media Theory

Useful insights can be gained from studying previous research, regarding the use of traditional television, published in advertising, psychology and sociology journals. Lee and Lee (1995) have identified a variety of fundamental uses and gratifications that people seek from television watching. For example, there are four levels of attention to the television set —from background noise to full concentration— which contrasts 'to the image of the highly interactive viewer intently engaged with the television set that is often summoned up in talking about new possibilities.'

Studies of media use in the home indicate that there is an important technology-driven shift in the household's media consumption patterns every decade or so. In the 80's there was the PC (Vitalari, Venkatesh and Gronhaug 1985), in the 90's there was the Internet (Kraut, Mukhopadhyay, Szczypula, Kiesler and Scherlis 1998). The current trend towards digital television transmission, local storage and manipulation of media content through home media networks (Bell and Gemmell 2002) is already apparent and has been studied by ethnographers for the case of the digital set-top box (O'Brien, Rodden, Rouncefield and J. Hughes 1999). Table 1 summarizes the most important and relevant to the case of the HMS design factors that regard the user from a media consumption point of view.

**Table 1** Audience behavior factors that affect the design of the HMS

| Design Factor | Description |
| --- | --- |
| Attention to the TV set | There are multiple levels of attention to television. One can watch television or leave it open as an electric light (Lee and Lee 1995) |
| Group Watching | There is either group or solitary use of television centric appliances (O'Brien et al. 1999) |
| Impulse Program Selection | Viewers are loyal to a small number of programs, but now they are faced with an increased number of channels and ways of viewing their favorite programs (Lee and Lee 1995) |
| Distribution of Functionality | Concentration of functionality on a single device (server) is incompatible with the distributed nature of home life (O'Brien et al. 1999) |
| Automation of Tasks | Automation is usually synonymous with relaxed use, but sometimes users prefer the hassle of control, in cases such as their privacy (O'Brien et al. 1999) |

## 2.2 Media Content Creation and Content Distribution

The introduction and wide adoption of the Web has been promoted and attributed to the interactive content of the new medium. It often goes without much thought, that if something is interactive then it is also better and it will be preferable. However, the passive uses of the broadcasted media are either desirable, or have an implicit interactive dimension that takes place outside the medium itself (Lee and Lee 1995). An example of the latter is the social interaction that takes place in groups of TV viewers, or the virtual-competition with the televised players of quiz programs.

Computer-like menus, pages and navigation look irrelevant on a TV screen, even when used by experienced computer users (Lekakos, Chorianopoulos and Spinellis 2001). Therefore, there is a need for television-values information design and story driven content (Jaaskelainen 2001). Furthermore, interactivity and informational elements should be used to augment entertainment content (Livaditi, Vassilopoulou, Lougos and Chorianopoulos 2003). Table 2 summarizes the factors that affect the design of the IIMS from the perspective of media content creation.

**Table 2** Content production and distribution factors that affect the design of the HMS

| Design Factor | Description |
|---|---|
| Infotainment | Both information and entertainment should be offered in a relaxed way (Livaditi et al. 2003) |
| Interactivity | Current television patterns of use are passive, but interactivity can improve certain television experiences (Lee and Lee 1995) |
| Content Patterns | Television viewers are accustomed to stories and characters in contrast to computer users who prefer objects and actions (Lekakos et al. 2001) |
| Personalization | Television provides shared experiences that people can talk about, which contrasts with the effort for personalization (Lee and Lee 1995) |
| User control | Television experience has been linear and story driven so far, but the video-games industry has invented ways of adding user control within a story (Jaaskelainen 2001) |

## 2.3   Broadcasting and Consumer Electronics Engineering

The broadcasting model of computing encompasses a radical shift in the mentality of application development process and tools. Milenkovic (1998) highlights the differences with the client-server mentality, describes the concept of the carousel and explains why the characteristics of the networking infrastructure are an important factor in the type of feasibly deployed applications. Digital local storage technology (Whittingham 2001) takes viewer control from simple channel selection with the remote to non-linear local programming (Chorianopoulos and Spinellis 2002). However, storing copyrighted content locally is against the interests of media owners (Bell and Gemmell 2002). Apart from the rise of digital rights management issues networking and transferring content between devices in the home poses a number of significant maintainability issues (Spinellis 2002). Table 3 summarizes the design factors that affect HMS use from the perspective of technology.

**Table 3** Technological factors that affect the design of the HMS

| Design Factor | Description |
|---|---|
| TiVo content | Both stored and broadcasted programming should be available and complementary to each other (Whittingham 2001; Milenkovic 1998) |
| Networked Storage | Local storage and networking at the home is a favorite, but copyright holders are concerned (Bell and Gemmell 2002) |

| | |
|---|---|
| Maintainability | There are numerous opportunities for networking between diverse home appliances, which come at the cost of maintainability (Spinellis 2002) |

## 3 Usability Design for Personalized Television Advertising

The dynamic insertion of advertising, during the play-out of copyrighted media content in the home, is a form of substitution for the royalty rights that have to been paid to media owners (Bell and Gemmel 2002). Moreover, the television advertising-break has a fixed duration, small hard-disk storage requirements and is relatively simple to integrate with real-time broadcasts. Next, we describe the iMEDIA system that offers dynamic construction of the advertising-break at the television set-top box for each home. The iMEDIA system has been designed and tested with the purpose of replacing the broadcast television's advertising-break, but can be extended to handle similar cases of dynamic advertising insertion, such as television programming stored on a hard-disk or MP3 music. The design of the iMEDIA system is described in Table 4 on the basis of the factors that affect user interface design for home media applications and the respective resolution strategy that was followed.

**Table 4** The resolution strategy for each design factor for the case of the iMEDIA system

| Design Factor | Resolution Strategy |
|---|---|
| TiVo content | The advertising-break is dynamically created for each set-top box. The overall experience is seamless for the viewer |
| Group Watching | The break may be personalized on the group or the individual level |
| Interactivity | Some advertisement spots may have additional interactive content available for later browsing |
| Infotainment | Interactive content of a commercial offers opportunities for informative material in the form of a micro-web-site |
| User Control | The users may choose to opt-out from receiving personalized advertising-breaks |
| Personalization | The advertising-break is targeted to household viewers, while some of the spots within it may be the same across different households |

## 4 Further Research

Additional case studies of important television programming include news and music video-clips. Music video-clip programming from channels such as MTV may offer a personalized 'top 20' for each week. Moreover, receiving trivial and lyrics —through Internet resources— offers personalization of the overlaid information. While there have been various treatments to adapt multimedia content to the user (e.g. content and collaborative filtering), the results of further research should address prominent HCI questions such as the feasibility of an agent-mediated (using MS Agent) dialog interface for leisure applications and how to evaluate a user interface when the goal is to be relaxing and fun to use.

# 5    References

G. Bell and J. Gemmell. A call for the home media network. *Communications of the ACM*, 45(7):71–75, 2002.

J. Carey. Interactive television trials and marketplace experiences. *Multimedia Tools and Applications*, 5(2):207–216, 1997.

J. Carey. Content and services for the new digital tv environment. In D. Gerbarg, editor, *The Economics, Technology and Content of Digital TV*, pages 88–102. Kluwer Academic Publishers, 1999.

K. Chorianopoulos and D. Spinellis. A metaphor for personalized television programming. In N. Carbonelle and C. Staphanidis, editors, *Proceedings of the 7th ERCIM Workshop User Interfaces for All*, pages 138–144. Springer-Verlag, 2002. Lecture Notes in Computer Science: State-of-the-Art Surveys.

S. M. Drucker, A. Glatzer, S. D. Mai, and C. Wong. Smartskip: consumer level browsing and skipping of digital video content. In *Proceedings of the SIGCHI conference on Human factors in computing systems*, pages 219–226. ACM Press, 2002.

K. Jaaskelainen. *Strategic Questions in the Development of Interactive Television Programs*. PhD thesis, University of Art and Design Helsinki, 2001.

R. Kraut, T. Mukhopadhyay, J. Szczypula, S. Kiesler, and W. Scherlis. Communication and information alternative uses of the Internet in households. In *Proceedings of the Conference on Human Factors in Computing Systems (CHI-98) : Making the Impossible Possible*, pages 368–375, New York, Apr.~18--23 1998. ACM Press.

B. Lee and R. S. Lee. How and why people watch tv: Implications for the future of interactive television. *Journal of Advertising Research*, 35(6), 1995.

G. Lekakos, K. Chorianopoulos, and D. Spinellis. Information systems in the living room: A case study of personalized interactive TV design. In *Proceedings of the 9th European Conference on Information Systems*, Bled, Slovenia, June 2001.

J. Livaditi, K. Vassilopoulou, C. Lougos, and K. Chorianopoulos. Needs and gratifications for interactive tv applications: Implications for designers. In *Proceeding of Hawaii International Conference on System Sciences 2003*. IEEE, 2003.

A. Marcus. Chi as a cross-tribal community. *interactions*, 9(4):31–35, 2002.

M. Milenkovic. Delivering interactive services via a digital TV infrastructure. *IEEE MultiMedia*, 5(4):34–43, Oct.--Dec. 1998.

J. O'Brien, T. Rodden, M. Rouncefield, and J. Hughes. At home with the technology: an ethnographic study of a set-top-box trial. *ACM Transactions on Computer-Human Interaction (TOCHI)*, 6(3):282– 308, 1999.

D. Spinellis. The information furnace: User-friendly home control. In *Proceedings of the 3rd International System Administration and Networking Conference SANE 2002*, pages 145–174, Maastricht, The Netherlands, May 2002.

N. P. Vitalari, A. Venkatesh, and K. Gronhaug. Computing in the home: shifts in the time allocation patterns of households. *Communications of the ACM*, 28(5):512–522, May 1985.

P. Wallich. Digital hubbub. *IEEE Spectrum*, pages 26–31, july 2002.

Whittingham. Digital local storage. Technical report, Durlacher, May 2001.

# Usability Support for Managers

*Nigel Claridge*

Scandinavian Usability Associates
Angantyrvägen 11, 182 54 Djursholm, Sweden
nigel.claridge@wammi.com

## Abstract

Much current usability and user-centred design material is not presented in a form suitable to managers. Managers respond more positively to storytelling where the message is increase profitability based on success stories. Based on a survey of managers' requirements, UsabilityNet created support material with a structure, form and content that matched the needs of different groups of managers. Much of the content focuses on the business cases for usability and is supported by up-to-date examples and case studies. According to initial web site traffic analysis and an evaluation study by thirty managers, other successful content includes: a list of individuals and organisations providing usability consultancy, a section providing a high level overview of usability and user centred design, and an introduction to topics such as relevant international standards, assessing usability capability and usability in procurement.

## 1 Managers hold the key

The successful uptake of usability and user centred design is dependent on the degree of awareness and understanding possessed by managers and procurers. For example, a project manager who understands the benefits of good usability will plan and budget for appropriate user-centred resources and activities early in the development process

Managers who are unable to perceive the benefits of usability and user centred design, in particular in terms of profitability, will not make decisions that will influence the efficient uptake of usability and user centred design in development and procurement. Despite increasing attention given to usability, there still seems to be an alarming lack of understanding of usability within many organizations. Usability practitioners are under constant pressure to deliver results in a form that are comprehensible not only to them and their teams, but also to their managers (Trenner 1998). This affects the quality of the final product and may result in an inefficient use of invested capital.

Much current usability and user-centred design material is not presented in a form that suits the needs of managers. It is often too academic, too detailed or too technical and seldom addresses fundamental management issues such as how to help managers do their job better, and increase profitability and Return on Investment.

Managers tend to be more receptive to examples rather than theory and principle, in particular if the examples are relevant to the organisation in which they are being told and provide a compelling example of success. (Tyler et al 2002). This supports the need for case studies and examples presenting the business case for usability, especially where the balance sheet is affected in a positive way. UsabilityNet identified this after very early requirements gathering activities.

# 2 Objectives

One of the main tasks in the second stage of UsabilityNet has been to address the needs of managers. The aim was to specify the structure, form, and content of support material and information in order to increase their awareness and understanding of usability and user interface design issues of middle management.

In this context, middle management is defined as project managers, line managers, product managers, IT managers and procurers. Top management were considered but no activity was carried out with this group due to limitations in resources. Usability managers were not specifically included as they were regarded as having a dual interest as managers and usability professionals who had already been catered for.

# 3 Requirements activities

In order to obtain a clear understanding of what managers wanted, a requirements survey was carried out. Twenty managers (five line managers, four project managers, two purchase managers, three product managers and six other kinds of manager) participated in the survey from Sweden, UK, South Africa, China and Spain.

The following activities were carried out:

1. Telephone interviews with selected managers in Sweden and the UK. Interviews focused on the type of information/support required for managers to do their job better, and circumstances under which information was required.

2. Electronic questionnaires sent out to managers in Spain, China and South Africa via the project associate partners. The focus was on current needs given current work/position.

# 4 Activities and Results in summary

Over 80% of the managers who participated in this study stated that usability and user centred design was either important or very important. 65% stated that their working knowledge of the subject area was good or very good.

The results indicated that managers obtained information from a wide range of sources, including usability experts (often in-house specialists), an assortment of books, guides and handbooks, and a number of web sites. Some had attended seminars and conferences, which covered the area of user interface design and usability. However, over 70% of managers felt that these sources were very spread out and would prefer to have a more centralised single location, which would provide them with a broad amount of fundamental information, and then access or links to further sources.

A large proportion of managers (75%) wanted information that was updated regularly. "Fresh, up-to-date information" combined with a preference for information/material or guides not in book form. This supported the case for information to be available in electronic form. (65% of managers preferred not to have a book or guide in printed form). In order to be informed about new updates, some managers (20%) proposed a mailing list be set up to act as a reminder.

Managers (85%) stated that content should focus on the business cases for usability, supported by up-to-date examples and case studies. It was important for managers to be able to relate the information to their own projects. Cost justification (65%) was required, not so much for estimates within their own projects, but for promoting the advantages of usability and user centred design within their organisation. A large number of managers (75%) stated a general lack of understanding of usability within their organisation. It was considered as an add-on, and something that could be cut during hard times. Cost justification examples should provide sound arguments to prevent this from happening.

35% of managers and procurers expressed an interest in a list of national/international usability expert resources, preferably with some form of classification or statement of experience. This list should be combined with details about competences required during different stages of development and which experts had the correct competence for any given stage.

Those managers working closer to usability were less in need of an overview of the process. They were more interested in details of different tools and methods: when and how to use them, the pitfalls, and examples/cases where they have been used. A feature requested by a number of managers (15%) was a forum where managers could share experiences and learn from each other.

## 5 Structure and content of managers support material

As a result of the requirement activities, content with the structure shown below was created and located at www.usabilitynet.org/management.htm.

Commercial advantages
- Quotes
- Business case for usability
- Business value of enhancing your user's experience
- Cost justifying usability
- Calculate the benefits of usability - for your business

Usability resources
- Consultancy list: individuals and organisations providing usability consultancy
- Job descriptions of usability competencies required during development
- Other resources can be found in other parts of this site

Manager forum
- FAQs
- Ask the experts
- Discussion list for managers

Basics of usability
- Overview of usability and the user centred design process
- How to promote usability in an organisation
- Design principles to support user centred design
- Top ten mistakes in web design
- What is usability: how is usability defined?
- International standards for usability
- Planning user-centred design: enforcing UCD during development
- Usability in procurement

*Commercial advantages* – this section provides quotes, statistics and information about the business case for usability and user centred design, with statistics, examples and case studies. The success of this section is dependent on receiving case studies from colleagues around the world and across sectors. It is important not to focus only on software and internet solutions, but consider also other areas such as consumer products, Interactive TV and mobile applications. The section provides details about cost justification of usability and guides managers through the jungle of calculating the benefits.

*Statistics example*: The business case for usability can be summarised by saying that good usability leads to satisfied, purchasing, and returning customers. The Gartner group found in 1992 that a user-centred approach raised customer satisfaction by 40%. Bad usability on the other hand, leads to angry customers and loss in sales. AT Kearney in 2001 reported that 82% of users they surveyed attempted to purchase something but gave up because of poor design and usability. A useful metric of your web site's success is not traffic but conversion rate: the number of visitors who are 'converted' into buyers.

E-tailers could miss out on as much as $10 billion this holiday season because of Web site usability defects that prevent shoppers from buying online. Improving usability can raise sales and keep customers coming back. (Gartner group, 2001).

*Quotes example:* "The benefits of usable technology include reduced training costs, limited user risk and enhanced performance ... American industry and government will become even more productive if they take advantage of usability engineering techniques". Vice president Al Gore, 1998

*Business case for usability example:* The benefits of usability engineering extend beyond improving the user interface and end user productivity: its beneficiaries include not only end users but also developers and their companies. User centred design can reduce software and e-commerce costs (including development, support, training, documentation and maintenance costs), shorten development time and improve marketability.

*Business value of enhancing your user's experience example:* If your user can't use it, it doesn't work! Nowadays, people expect things to simply work - no prior reading and certainly no training. Either they can gain immediate value, or they will move on. So it's win or lose, based on the initial user experience. For many organizations, this user experience is directly mapped to business success.

*Cost justifying usability.* Provides a number of case study summaries from multiple authoritative sources based on numerous research projects.

**Usability resources** – this section presents a list of independent external consultants and organisations providing usability consultancy by country. It also presents a list of job descriptions of usability competencies required during development.

*Consultancy list: individuals and organisations providing usability consultancy.* The list is provided to assist managers who would like to know what usability expertise is available in their own country, or who require access to usability services in other countries. Any organisation advertising usability services on a web site can be included on the list, although UsabilityNet reserves the right to decline or remove listings, and may request references.

*Job descriptions of usability competencies required during development.* Managers often wish to know what competencies are needed at different stages of the development process. Usability competencies are grouped into five main areas of activities.

*Other resources can be found in other parts of this site.* These consists of links to topics on other parts of the site such as usability discussion lists, annual conferences, usability booklist links, relevant journals and magazines, university and commercial courses, tools and methods and more detailed information about international standards for HCI and usability.

***Manager's forum*** – the manager's forum enables managers to share experiences and learn from each other. It also allows managers to "quiz" a usability panel of experts. This panel changes each month. The forum also offers a number of frequently and seldom asked questions, including the questions posed to the panel of experts.

***Basics of usability*** – this section provides a high level overview of usability and user centred design and an introduction to basic topics of interest to managers.

# 6    Evaluation of structure and content

The Managers Guide to Usability was published on the UsabilityNet web site in late September 2002. An evaluation activity was initiated during December 2002, where thirty managers from different European countries, South Africa and the Far East were asked to assess the content based on their current needs. Preliminary results were positive (30% response rate at time of writing). The 'managers' section of the web site has been one of the most visited according to web traffic analysis statistics.

All respondents stated that the guide material supported them in their work and was easy to use.  In particular, they found the basics of usability, the case studies, the cost benefit methods and the expert resources most useful.

In order to improve the content, they stated that they wanted more case studies and cost benefit application examples, in particular showing how profitability can be improved. The scope of the content could be expanded to include areas other than software and the Internet.

Although initial results indicate a positive response from managers (all respondents would recommend the site to colleagues), there are very clearly some challenges ahead. Our aim is for managers to use the site regularly, in particular the managers' forum. This is more likely if we can maintain a high degree of "freshness" by keeping the site up-to-date with new information and case studies.

# References

A T Kearney (2001) "Satisfying the experienced online shopper" Global e-shopping survey. www.atkearney.de/veroeffentlichungen/doc/E-shopping_survey.pdf

Bevan, N., Claridge, N., Frohlich, P., Granlund, A., Kirakowski, J., Tscheligi, M. (2002). UsabilityNet: Usability for managers.  www.usabilitynet.org/management.htm

Gartner Group (2001) Better Web Usability Can Boost Holiday Sales by $10 Billion www.gartner.com/DisplayDocument?doc_cd=102554

Tyler,S.,Followell, D,. Geelhoed,E,. (2002) Usability managers' forum: A commercial perspective,

Trenner, L. and Bawa, J. (eds) The Politics of Usability, page xiv.  Springer-Verlag (1998).

# Usage-Centered Design:
# Scalability and Integration with Software Engineering

*Larry Constantine*

University of Technology, Sydney
Constantine & Lockwood, Ltd.
58 Kathleen Circle, Rowley, MA 01969
lconstantine@foruse.com

*Helmut Windl*

Siemens AG
A&D AS RD 221, Gleiwitzer Str. 555
90475 Nuremberg, Germany
helmut.windl@siemens.com

## Abstract

Usage-centered design is described and compared with traditional user-centered approaches in relation to software engineering and development. Experiences in integrating usage-centered design with software engineering across a spectrum of process and project models are reviewed. Successful techniques for mitigating the effects on time and resource are reviewed, including overlapping early-cycle activities, splitting support of User Actors from System Actors, recycling of usage-centered models, and newly developed agile modeling and design techniques.

## 1   Introduction

Usage-centered design (Constantine and Lockwood, 1999; 2002) is a systematic, model-driven approach to visual and interaction design for user interfaces in software and Web-based applications. On projects of widely varying scope and scale in a variety of application areas (Anderson et al., 2001; Constantine and Lockwood, 2002; Patton, 2002; Strope, 2002; Windl, 2002b), usage-centered design has proved capable of expeditious delivery of superior designs (Windl and Constantine, 2001). As the name suggests, usage-centered design differs from conventional user-centered approaches primarily in a shift of focus from users *per se* to usage, that is, to the tasks to be accomplished by users. This difference in emphasis is reflected in differing practices with a significant impact on the development life cycle and integration with software engineering and development. Most importantly, usage-centered design eschews the repetitive cycles of trial design and user testing common to traditional user-centered approaches in favor of a design process in which final solutions may be derived more or less directly from robust and highly refined models reflecting genuine user needs. The goal is an initial design requiring only limited usability testing and minimal refinement. Table 1 summarizes salient differences.

Whereas user-centered design emerged from a separate profession and evolved largely independent of software engineering, usage-centered design is grounded in a strong engineering orientation reflecting the background of its co-developers, including one of the early pioneer software methodologists. Usage-centered design integrates readily with software engineering precisely because it was developed from the outset to be compatible with object-oriented software engineering, using extensions and refinements to well established models and techniques, such as actors and use cases (Constantine and Lockwood, 2001).

**Table 1:** Salient Differences between Usage-Centered and User-Centered Design

| User-Centered Design | Usage-Centered Design |
|---|---|
| Focus on users:<br>  user experience, user satisfaction | Focus on usage:<br>  improved tools supporting task accomplishment |
| Driven by user input | Driven by models |
| Substantial user involvement:<br>  user studies, participatory design, user feedback, user testing | Selective user involvement<br>  exploratory modeling, model validation, structured usability inspections |
| Descriptions of users, user characteristics | Models of user relationships with system |
| Design by iterative prototyping | Design by modeling |
| Varied, often informal or unspecified processes | Systematic, fully specified process |
| Evolution through trial-and-error | Derivation through engineering |

## 2    Essential Models and Model-Driven Design

Usage-centered design employs three closely related abstract models: a role model capturing salient characteristics of relationships between users and a system, a task model representing the fine structure of work users need to accomplish with a system, and an interface model representing the contents and organization of the user interface needed to support the identified tasks. Although these correspond to and may superficially resemble more traditional user and task models and paper prototypes, they differ in important ways. For example, rather than modelling users and their characteristics, user roles model relationships between users and systems, which have a more direct and immediate impact on user interface design.

Work is modelled as task cases or "essential use cases" (Constantine, 1995; Constantine & Lockwood, 1999), a specialized and highly abstract form of use cases (Jacobson et al., 1992). Use cases as employed in software engineering model functional aspects of systems: "sequences of actions, including variant sequences and error sequences, that a system, subsystem, or class can perform by interacting with outside actors" (Rumbaugh et al, 1999: 488). Task cases, rather than describing interactions with systems, model the discrete intentions of users playing roles in relation to a system, taking the form of an interrelated collection of highly simplified narratives that are abstract, implementation independent, and devoid of technological assumptions.

A usage-centered task model differs from both use cases and ubiquitous scenarios in providing a much simpler and finer-grained representation of the elemental structure of work being accomplished by users. This simplicity and granularity enables more comprehensive modelling, promotes reuse of design elements and components, and facilitates validation of models.

It is only possible to offer a bare outline of the complete process here; details are available elsewhere (Constantine and Lockwood, 1999; 2002a). The entire process is model-driven and follows a straightforward logical derivation. Even initial inquiry and field investigation is guided by exploratory models of user roles and tasks (Windl, 2002a). As in conventional object-oriented software engineering (Jacobson et al., 1992), actors are identified, but user actors are distinguished from system actors (non-humans). A role model is constructed consisting of focused descriptions of the roles played by direct users and a map of the interrelationships among roles. A task model represents all the discrete task cases needed to support the user roles along with a map of the relationships among task cases. An interface model represents the abstract contents of the various parts of the user interface along with a navigation map of the interconnections among these parts. The final visual and interaction design—or implementation model—derives more or less directly from the interface model, particularly when the latter is expressed in canonical form (Constantine, Windl, Noble, and Lockwood, 2000). Other models are developed in parallel with the three core models, most importantly a domain model in the form of a glossary, data dictionary, or domain object model. Models from the usage-centered design process also feed the software engineering process, either directly or in altered form.

# 3   Impact on Software Engineering and Development

The tensions between traditional user-centered methods and software engineering and development are widely acknowledged (McCoy, 2002). An emphasis on thorough investigation and extensive field observation, particularly in ethnographic approaches, combined with repeated cycles of prototyping and user feedback, leads to a substantial, often unpredictable, analysis and design commitment that typically must precede software design and development. A cornerstone of traditional methods is usability testing with working prototypes or systems, which necessarily comes late in the development cycle, typically after an alpha (internal) or beta release. Results of usability testing, because they arrive so late in the process, are all too often ignored or deferred for later releases. When testing uncovers architectural defects or problems in the basic organization of the user interface, which is not uncommon, little if anything can be done.

Usage-centered design also imposes a substantial responsibility for up-front design, but the models themselves by nature are readily partitioned and lend themselves to iterative refinement as well as to simplification and shortcuts (Constantine, 2000). Moreover, because the entire process is inherently streamlined and simplified, the actual commitment for even the most thorough and in-depth design can itself be quite modest. For example, on the STEP 7 Lite project, a 50-person-year effort of more than two years duration, the complete usage-centered design of a user interface supporting over 300 task cases took only 14 weeks (Windl, 2002b).

# 4   Scalability

Usage-centered design was devised from the start to be a scalable process that could yield first-quality designs on projects ranging from a few weeks to multi-year efforts. It scales well for several reasons. Both highly simplified and fully elaborated forms of the models have been fully worked out and refined through practical application. At the one end of the spectrum are structured role models and structured task cases (Constantine and Lockwood, 2001) based on multi-faceted templates. Such structured models lend themselves to large, complex, and mission-critical applications, such as the STEP 7 Lite project cited above, and to highly organized methods, such as the unified process (Jacobson et al., 1999). At the other extreme are simple inventories that merely identify all roles or tasks. These have proved sufficient for driving effective UI design on accelerated "crunch mode" projects (Constantine, 2000; Constantine and Lockwood, 2002) and for projects using the emerging "agile" methods (Cockburn, 2002). Simple role and task inventories can be elaborated into more detailed models at any time as needed, including piecemeal or iteratively in successive short release cycles.

# 5   Concurrency and Integration

A range of actual projects exemplify current approaches to integration of usage-centered design with software engineering. At one extreme are complex, large scale systems developed through formal or highly structured engineering processes. Such a project was STEP 7 Lite, a complete integrated development environment (IDE) for PLC programming of automation applications

(Windl, 2002b). At the other end are small to intermediate applications developed on accelerated time frames, such as a classroom information management system for K-12 teachers (Constantine and Lockwood, 2002). In the first instance, task cases in the form of essential narratives were translated into conventional "concrete" use cases for the handoff to software engineering. Thus, although expressed in different narrative idioms, use cases became a common thread connecting the entire development process, including documentation and help system design. In addition to refinements to the object model, annotated detailed drawings of 34 user interface contexts supplemented by some 200 pages of added design documentation were delivered to the software engineers. For the educational application, developed in an accelerated XP-style iterative process, task cases were used only by the UI designers and the software engineering was guided by annotated visual designs supplemented by engineering notes and in-person communication.

Creative hybrids have also been employed with success, including in a medical informatics project in which a thorough and somewhat extended usage-centered design was followed by an agile implementation process based on XP. In this project, UI designs and task cases became the basis for so-called user stories that drove successive implementation cycles.

The STEP 7 Lite project also illustrates how concurrent engineering can reduce early-stage delays or bottlenecks. Selected internal subsystems with little or no dependence on the UI design were identified for development in parallel with the UI design. Software engineering can proceed in parallel with usage-centered design. Isolation of system actors from user actors permits independent design of system interfaces and supporting internals for all system actors. A robust architecture that insulates views and presentation layers from data structures and algorithms permits additional concurrent design and development.

Extreme programming (Beck, 2000) and other agile methods represent a new direction in highly disciplined programming practices, such as pair programming, are used within very short release cycles to deliver successively refined versions of reliable software. Card-based modelling employing ordinary index cards is a common technique used in such methods to yield simple models very quickly (Jeffries, 2001). Usage-centered models lend themselves to card-based modelling, and usable role and task models can be constructed and prioritized very rapidly (Constantine, 2002; Constantine and Lockwood, 2002).

Yet another approach, pioneered by Biddle, Noble, and Tempero (2002), is for task cases to serve without modification or translation as a direct input to the object design process. In this technique, a variant of responsibility-driven design (Wirsf-Brock and McKean, 2002) identifies objects and allocates responsibilities to objects based on the simplified abstract narratives that define each task case. It has been used successfully on a number of projects and in teaching.

## References

Anderson, J., Fleek, F., Garrity, K., and Drake, F. (2001) Integrating usability techniques into software development. *IEEE Software,* 18 (1), January/February.

Beck, K. (2000). Ext reme programming explained. Reading, MA: Addison-Wesley.

Biddle, R., Noble, J., and Tempero, E. From essential use cases to objects. In L. Constantine (Ed.), *forUSE 2002: Proceedings of the first international conference on usage-centered, task-centered, and performance-centered design.* Rowley, MA: Ampersand Press.

Cockburn, A. (2002). Agile software development. Boston: Addison-Wesley.

Constantine, L. L. (1995) "Essential Modeling: Use Cases for User Interfaces," *interactions 2* (2).

Constantine, L. L. (2000). Cutting corners: Shortcuts in model-driven web design. *Software Development,* 8 (2). Reprinted in L. Constantine (Ed.), *Beyond chaos: The expert edge in managing software development.* Boston: Addison-Wesley, 2001.

Constantine, L. L. (2002). Process agility and software usability: Toward lightweight usage-centered design. *Information Age,* 8 (2). Reprinted in L. Constantine (Ed.), *Beyond chaos: The expert edge in managing software development.* Boston: Addison-Wesley, 2001.

Constantine, L. L., and Lockwood, L. A. D. (1999) Software for Use: A Practical Guide to the Models and Methods of Usage-Centered Design. Reading, MA: Addison-Wesley

Constantine, L. L., and Lockwood, L. A. D. (2001) "Structure and Style in Use Cases for User Interface Design." In M. van Harmelan (Ed.), *Object Modeling an User Interface Design.* Boston: Addison-Wesley.

Constantine, L. L., and Lockwood, L. A. D. (2002) Usage-centered engineering for Web applications. *IEEE Software,* 19 (2), March/April.

Constantine, L. L., Windl, H., Noble, J., and Lockwood, L. A. D. (2000) From abstraction to realization in user interface design: Abstract prototypes based on canonical abstract components." http://www.foruse.com/articles/canonical.pdf

Jacobson, I., Christerson, M., Jonsson, P., and Övergaard, G (1992) Object-Oriented Software Engineering: A Use Case Driven Approach. Reading, MA: Addison-Wesley, 1992.

Jacobson, I., Booch, E. G., and Rumbaugh, J. (1999) The unified software development process. Reading, MA: Addison-Wesley.

Jeffries, R. (2001). Card magic for managers. *Software Development,* 8 (12), December. Reprinted in L. Constantine (Ed.), *Beyond chaos: The expert edge in managing software development.* Boston: Addison-Wesley, 2001.

Jeffries, R., Anderson, A. Hendrickson, C. (2001) Extreme Programming Installed. Boston: Addison-Wesley.

McCoy, T. (2002) Letter from the dark side: confessions of an applications developer. *interactions* 9 (6): 11 - 15.

Patton, J. Extreme design: Usage-centered design in XP and agile development. In L. Constantine (Ed.), *forUSE 2002: Proceedings of the first international conference on usage-centered, task-centered, and performance-centered design.* Rowley, MA: Ampersand Press.

Rumbaugh, J., Jacobson, I., and Booch, E. G. (1999) The Unified Modeling Language Reference Manual. Reading, MA: Addison-Wesley.

Strope, J. (2002) Putting usage-centered design to work: Clinical applications. In L. Constantine (Ed.), *forUSE 2002: Proceedings of the first international conference on usage-centered, task-centered, and performance-centered design.* Rowley, MA: Ampersand Press.

Windl, H., and Constantine, L. (2001) "Performance-Centered Design: STEP 7 Lite." Winning submission, Performance-Centered Design 2001, http://foruse.com/pcd/

Windl, H. (2002a) Usage-Centered Exploration: Speeding the Initial Design Process. In L. Constantine (Ed.), *forUSE 2002: Proceedings of the first international conference on usage-centered, task-centered, and performance-centered design.* Rowley, MA: Ampersand Press.

Windl, H. (2002b) Designing a Winner: Creating STEP 7 Lite with Usage-Centered Design. In L. Constantine (Ed.), *forUSE 2002: Proceedings of the first international conference on usage-centered, task-centered, and performance-centered design.* Rowley, MA: Ampersand Press.

Wirfs-Brock, R. J., and McKean, A. (2002). Object design: Roles, responsibilities, and collaborations. Boston: Addison-Wesley.

# User Research at Adobe: Establishing a User-Centered Culture

*Sheryl Ehrlich*

Adobe Systems
321 Park Avenue, San Jose, CA 95110
sehrlich@adobe.com

## Abstract

Adobe Systems has been creating software for over 20 years. However, it was not until mid-2000 that the company began a systematic approach to user research. Since that time, Adobe's User Research Team has been pushing to create a user-centered culture at the company. This paper describes the establishment of the User Research Team. It discusses how the group was founded and developed, issues that were encountered and the approach that was taken to address them, where the team is today, and the future of the team and user-centered culture at Adobe.

## 1  Laying the Foundation

Adobe's User Research Team was founded in the summer of 2000. Although some usability studies had been outsourced sporadically before that point, there was no systematic focus on user research. It was through the desire and lobbying of two individuals—one in the Product Design Group and one in the Advanced Technology Group—that the initial research position was established and I was brought in to begin developing a team.[1] From the start, we sought to instill a user-centered culture in the company.

When the program first started, the most in-demand type of user research was usability studies—the one type of research with which most people were familiar. Conducting these high visibility, in-demand studies was the most tangible way to expose a diverse set of product teams to the value of research first-hand, so I proceeded with conducting one-off studies across numerous products.

All of this was going on as the economy boomed in Silicon Valley. Within Adobe, we created a small "Design Seed" team, led by the head of the former Product Design Group. Design Seed sought to initiate strategic long-term design programs and ideas within the company; our first project was to develop the young User Research Team. This involved several key areas: providing education, advocacy, and financial justification; setting up infrastructure; and establishing a research program and relationships with other groups within the company. Having Design Seed focus on creating the team gave us the critical mass and bandwidth to put these pieces in place while still conducting research so that we were not in a position of preaching without proof.

### 1.1  Education

As we started to build the team and infrastructure, there was a need for education throughout the

---

[1] It is interesting to note that Adobe has traditionally been a very engineering-driven company. This comes as a surprise to many people given that one of the core business areas is in software for creative professionals (e.g., graphic designers). However, it was not until about 5 years ago that the first interaction designers arrived at Adobe and began to establish what is now a pervasive, systematic approach to user interface design across the company.

company across all disciplines and levels. Therefore, individually-tailored presentations were crafted to ensure understanding all the way from the VPs to individual engineers. The main goals of the educational campaign were to provide a financial argument explaining the return-on-investment of such a program and to educate product teams on the value of integrating research.

The executives and financial controllers needed to understand the short- and long-term financial value that user research could provide to Adobe. Presentations covered topics such as the return-on-investment of a user research program, why it was necessary at this point in time, the value in different types of research throughout the product development process, resource needs in building a team, and how our resources compared to those of other software companies.

In addition to the executives, the product teams needed to understand the impact that different types of user research could have on their products. Adobe had prospered as a company for almost 20 years without user research, so why was it so important to pursue at this point in time? The arguments for initiating a systematic research program included the following: although we had enjoyed virtually inclusive market share in some areas (e.g., the creative professional market with Photoshop), there was increasing competition in terms of products and alternatives for our users; having usable products would provide a competitive advantage; web products—in which usability is often the leading reason for success or failure—were becoming widespread; we were seeking to expand our user base to new, non-professional user groups including enterprise and consumers (where usability is a key factor in product reviews and purchases).

Although usability studies were in demand, other types of research (e.g., site visits, heuristic evaluations, workflow analyses, prototype studies) were not as well understood. In addition to explanatory presentations, the researchers provided much of this product-level education both by example and through the strong relationships formed within their product teams (with design, marketing, engineering, and quality engineering). We also created an online research document database so the entire company could easily access results from the various types of research studies. To further help with this aspect of the education process, we made a conscious decision to call the group the "User Research Team" instead of using the more well-recognized term "Usability." This was intended to help convey the fact that we do much more than usability studies, incorporating a wide variety of methods throughout all stages of product development.

## 1.2    Infrastructure

Rather than insisting that the company allocate space in an overflowing building to create a fully-outfitted lab large enough for the future team's size, we "collected" space piece-by-piece. Starting with two offices set up as the informal lab and observation room, we went through several transformations of the space. We also acquired equipment in a similar manner, always capitalizing on surplus items within the company and purchasing a small amount of new equipment on an on-going basis each quarter. Although it was time-consuming to continually change the set-up, it was best way to keep the lab growing in-step with the team.

Having research team members willing to pitch in on all fronts was critical. We relied heavily on each individual's unique skill set, whether it was in setting up a server or database or wiring lab rooms and installing audio-visual equipment during seemingly continuous moves. Through a true team effort we were able to put in place key pieces of infrastructure, including labs and observation rooms with digital video capabilities, a team web site, and participant database.

One issue that all companies need to address as they establish research programs is finding users to participate in research studies. We began with the "friends and family" program in which we would enlist the help of our friends and family, their friends, and friends of people around the company. However, our needs quickly outgrew this population. We did not outsource recruiting to external agencies primarily due to cost, but also because a lot of our products have very specific target users (e.g., video professionals who use Premiere) to whom such firms did not have access.

Our products address widely varying user groups—creative professionals, consumers, enterprise— so we needed mechanisms to attract all of these different demographics. We began to recruit users in diverse ways—inviting registered product users, posting on a popular local website, hanging flyers at local schools and stores, etc. As the number of users grew, we quickly realized that we needed a centralized database to maintain this information. Working with the Adobe.com team, we created a usability registration page that is now linked from multiple places on our external website. This information is then fed into a database which each researcher can access to search by user type, applications used, location, etc. Around the same time, a more formal American Express Gift Cheque compensation system was put in place, and we worked with our legal department to craft Orientation, Informed Consent, and NDA forms.

## 1.3   Research and Relationships

One of the most exciting parts of developing the User Research program at Adobe has been creating the research process. The team's strong and diverse research backgrounds (e.g., human-computer interaction, cognitive and social psychology, sociology) have been key in allowing us to create a wide variety of methods that we can draw upon to best meet the product team's needs. We are continuing to develop and introduce new methods (both standard and non-standard in the field) into our "toolbox" and leverage the strengths of the team in creating optimal methods. Although many other groups have set processes and deliverables at each development milestone, a critical part of the success of the research program has been each researcher's flexibility in determining the most appropriate methods at any given point in the product cycle.

We have focused on creating key relationships both within the product teams as well as across the company. A large part of our success has been our ability to work closely not only with design, but also with marketing, engineering, and quality engineering. Activities include brainstorming with designers and providing continual feedback, soliciting input from marketing on target users and research priorities, getting engineering's feedback on studies, and debriefing with the entire product team. Although these relationships take a lot of time to form and can be quite complex, they are critical. We have been careful to keep the product team members informed of our research and to be as inclusive as we can in creating our research plans to ensure that we are in synch and receive buy-in. We also encourage the entire team to observe studies and participate in site visits whenever possible. To further instill a user-centered mindset on the product teams, as the research team grew we made a conscious decision to abandon our original shotgun approach of one-off usability studies across many different products in favor of a model where each researcher addressed no more than two products. Rather than jumping in at the end to test what had been developed, the team's focus on user research could start before development and continue throughout the process.

Although the company has been extremely receptive to research and its value, the recent downturn in the economy has led to a constant struggle for resources to build the team to cover even the top priority projects. Thus, creative ways of securing resources were implemented, including continual pitching across different product teams explaining how our work could address their needs. In addition to building up the team via the product teams' support, we have and continue to also seek resources from within our larger department, the cross-company engineering division.

## 2 Where We Are Today

Today Adobe's User Research Team has eight members with backgrounds in various disciplines such as human-computer-interaction, cognitive and social psychology, and sociology, in addition to a wealth of industry experience. The research team addresses all three of Adobe's customer segments—creative professionals, enterprise, and consumers—focusing on the top priority products within each area. We occasionally shift coverage to ensure that we maintain support on the top corporate priorities, also considering other variables such as type of user and scope of changes. We are continuing to expand the breadth and depth of research conducted on the subset of products we address, decreasing the number of products each researcher covers.

We are now part of the User Interface Team, comprised of Interaction Designers, Visual Designers, Technical Writers, and User Researcher Specialists. We work extremely closely with the designers in the creation and iteration of their designs. We also establish close product-team relationships outside of the User Interface Team, with marketing to input into the earliest stages of the product concept, and with engineering and quality engineering to understand the technical limitations and drive change. We seek to help the product teams develop an understanding of users, typical patterns of interaction with products, usability issues and possible solutions, and pain points, workflows, goals, and needs. We do this by integrating research throughout the entire product development cycle, emphasizing early research using a wide variety of methods.

Adobe's product development process lends itself to three phases of research. In the first phase, we conduct formative "background research." During this phase, we work closely with marketing using methods such as site visits, user profiling, workflow analyses, and surveys. Data from these studies feeds directly into the concept proposal for the product. Once the concept has been accepted, the designers begin creating mock-ups of potential features and interfaces. During this "idea evaluation" phase, research methods include iterative mock-up and prototype studies, participatory design sessions (with both low and high-tech methods), and competitive usability analyses. Finally, as we begin to have stable builds of the product incorporating new features at alpha and beta, we progress to the "product evaluation" stage, conducting lab studies with functioning software, informal heuristic evaluations, and structured activities with the pre-release teams. Once the product has shipped, we conduct a final lab study on the application to assess the product as we begin to think about the next version of the product.

Although we have a systematic product development process across all of our products, each product still has its own personality and research needs. Therefore, we have been careful to maintain flexibility and empower each researcher to determine the needs and best approach for each product on an individual basis, rather than prescriptive, required cross-product milestones and deliverables. There are also different approaches to incorporating user feedback, from tight feedback loops with designers to more formal "bug" tracking systems with quality engineering. We continue to introduce various types of research into our team's toolbox, as well as issue tracking mechanisms, learning from the experiences of each member of the team. In addition to our product work, the User Research Team also leads large cross-product research initiatives exploring areas of interest to the company and serves as "research consultants" on a wide variety of smaller internal and external projects.

As for infrastructure, we now have four labs with live web-casting. Our real-time web-casting system has been a great way to ensure that the remote product teams (located throughout the United States and around the world) are part of the research process. We also rely heavily on videoconferences to discuss findings, as well as testing at remote offices to promote engagement of the entire product team. In our participant database, we have over 3,000 local users and nearly 10,000 worldwide. We continually drive the user recruiting campaign–from distributing sticky pads and pens displaying the registration URL to attending user meetings and tradeshows.

# 3    Future Directions

Up to this point, we have continually expanded our labs and resources as the team grew rather than all at once. This has allowed for small, but continual allocations in quarterly budgets. However, looking toward future, we now have a unique opportunity to create a large lab, as the company is building a new tower that relieves all space pressures. We are working to ensure that we can establish a permanent lab that will be large enough for the team as it continues to grow.

Realizing that creating products for a global audience rather than just focusing on users in the United States is critical, we have begun to lay the foundation for international research. We are starting to take advantage of the infrastructure we have put in place, using our worldwide database to find users for international usability studies, site visits, diary studies, and surveys. We have also recently begun to conduct remote usability sessions via the internet. Early in the team's development, the User Interface Team's International Solutions Manager and I went to Japan to learn more about the state of user research within ten Japanese companies and to conduct Adobe's first international usability study. This trip was invaluable in that it allowed us to expose Adobe's Tokyo office to user research first-hand, while providing significant information to bring home to help educate headquarters in the United States. Our efforts to build up an international research program are ongoing, and we hope to establish a team in Tokyo in the near future.

We have also begun to expand our interactions with other teams across the company. We have established regular meetings with Market Research to help drive the global research picture across the teams. With Instructional Communications, we have not only begun to incorporate testing of their user documentation and help systems in our product testing, but we are also involved in more exploratory research to inform their future directions. In recent conversations with Technical Support, we have begun to explore ways to leverage the information they collect and store in their customer support databases. Such collaborations will help ensure that the user-centered approach spans across the entire company and impacts the end-to-end user experience as we move forward.

We are also becoming more engaged in future-looking research endeavours, working with the company's "seed" projects to explore potential new business opportunities. Rather than having new products be driven primarily by technology creation, we feel this will help focus Adobe on new products that are driven by customer needs and desires. This will further help us establish a user-centered culture not only for the development of today's products, but also tomorrow's.

## Acknowledgements

Thanks to the User Research Team, Design Seed, Philippe Cailloux, and Katja Rimmi whose hard work and support have made the move toward a user-centered company a reality.

# Usability Evaluation as a Component of the OPEN Development Framework

*John Eklund*

Access Testing Centre and The
University of Sydney
112 Alexander Street, Crows Nest
NSW 2065 Australia
johne@testingcentre.com

*Matthew Baker, David Lowe*

University of Technology, Sydney
P.O. Box 123 Broadway
NSW 2007 Australia
mattyb@bigpond.net.au
david.lowe@uts.edu.au

## Abstract

Considerable work has been done in the area of formative usability evaluation of websites, and usability is now an accepted part of the development process. However, a significant portion of the literature on usability does not examine the broader aspects of the product as it evolves and focuses instead on user interactions. In practice, usability often overlooks the underlying architecture and the business model in developing the design of a website and has a clear focus on interface design. In this paper we report on a collaborative research project that examines the role of usability evaluation within a broader development framework – a process model called the Object Oriented Process Environment and Notation (OPEN), and in particular its application in Web OPEN. The paper considers practical methods for the integration of usability evaluation into the framework through a case study of a practicing usability group at Access Testing Centre in Sydney, Australia. This research showed that Web OPEN, with some extensions, is an appropriate framework to describe and support usability activities within current Web development practices. This work serves three purposes: It validates the Web OPEN process model against current industry practice; it identifies areas of the process model that require refinement to describe usability activities in web development; and it places our activities as Human Computer Interaction (HCI) practitioners in the broader context of a Web development framework.

## 1 Introduction

Website development is in a state of rapid change and rapid prototyping tools and shortened timeframes have contributed its relative immaturity compared to traditional software development. In response, much work has been undertaken to introduce more formal methods for developing websites. One such framework is that of Web OPEN. Web OPEN is an extension of the Object-oriented Process Environment and Notation designed to provide guidance for website development projects. As the Web OPEN process framework is only relatively new, the goal of this project was to further develop and detail the framework in the area of website testing and evaluation, with an emphasis on the role that usability testing and HCI plays in the development process. A part of developing and detailing this framework has been to undertake a case study with Access Testing Centre (ATC), a leading Australian website testing company that conducts usability evaluation and consultancy. This paper provides an overview of our research and details initial findings.

459

# 2    Overview of the OPEN framework

OPEN is a third generation Object Oriented development framework, which in recent years has been enhanced to account for the unique nature of Web development.   As OPEN is in the form of a process framework rather than a fixed development methodology, it readily allows for feedback and changes to the development process as tools and techniques are further refined, or as components of the development processes are improved and approaches mature.

In the OPEN Process specification (Graham, Henderson-Sellers & Younessi 1997), the OPEN process model, and consequently Web OPEN from which it is derived, have the following essential elements:

- *Producers* – people in the process model who perform work units, which leads the production of work products.
- *Work Units* – the elements of work that are used to produce a work product.
- *Activities* – major work units made up of a series of related work products. Activities are in turn accomplished using a combination of various tasks (individual elements to be completed in accomplishing an activity) and techniques (the methods used to complete both tasks and activities).
- *Work Products* – the tangible outputs of the work units.

Additional elements of the framework come in the form of:

- *Stages* – used to provide higher-level organisation of the various work units and work products by grouping them together.
- *Languages* – used to document work products.   They can take many forms such as modelling languages, implementation (programming) languages or natural languages, such as English.

Web OPEN is a theoretical framework that does not explicitly account for HCI activities. Our research was based on a desire to benchmark the framework with industry practice, and identify areas that require enhancement to account for the application of HCI services.

# 3    HCI practice and business needs

In completing this research project, many observations were made about the current status of both website development as well as website evaluation. One  particular area of note in the testing process is that of acceptance criteria and the requirements from which they are derived.   In developing requirements which detail the client's needs, as well as the purpose and scope of the website, the development team is then able to work with a customer in selecting metrics and target values in order to enable the client and the development team to negotiate acceptance levels for the website. While requirements are an important part of evaluation they also help to  define the direction of the project at the initial stages, as well as providing a measure of project success.

In defining these requirements, some important questions need to be answered, such as who the users are and what are their tasks, how will they use the system and how do these requirements fit in with the needs and objectives of the business that wishes to have the site developed.   Problems that can result from not having properly defined these requirements include websites that are unsuitable for their intended function, un-maintainable, and cost more due to the need for multiple fixes if the scope and the requirements begin to creep.

Another aspect of web evaluation that is not as frequently considered in HCI work is the business case or business needs of a website. In other words, ensuring the website meets business goals and user acceptance levels. Ensuring users see the value of a website is more important than its usability. This has important implications for HCI practice, where market acceptance of a design precedes usability. E-commerce is relatively new and there is also a limited amount of experience to guide people in building websites that work to fit a set of business needs. While some ideas for websites may provide value only to the business, others may provide value only to the customer. It is this area where these two sets of goals interact that provides the true benefits of having a website.

In addition to providing value to both the customer and the business simultaneously, there may be other intangible aspects or values to a particular business which might drive them towards developing a website, such as status, or a desire to stay ahead of competitors. The goal of many websites may be simply usage goals, expressed in user sessions. Real meaning can only come from these goals when they are targeted and specified appropriately, which can be difficult to achieve as clients frequently do not understand their needs, or the way that technology can assist them. To make matters more complex, it is often difficult to derive requirements arising from different professions and practices within a business, as each will often have their own different priorities and language used to express those needs and priorities.

## 4    Commercial practice

Based on the results of the case study at Access Testing Centre, it was clear that many organisations frequently failed to understand their needs from a business perspective, and thus how technology could help them meet their business goals. This meant that the end of a project, user acceptance levels were low; hence the usability of the site was of less consequence. This is echoed by the experiences of some of ATC's staff members during the "Dot Com" boom. During this period, it was found that many companies had very unusual ideas for websites or web applications which were fundamentally unsound, as studies revealed that virtually no-one had a need for the product that a company had devised, and that even if there was a demand for the particular product, the site that was developed to meet this perceived need was very difficult to use. As a result, users would often stay away, and the results would be disastrous for the organisation backing the website.

More recently there has been a shift in that customers now may have a goals defined, but at the same time there is still a lack of understanding on how to best design the website to achieve these goals. This means that the initial definition of requirements can still be quite difficult. These problems vary greatly as some organisations are able to articulate business goals very clearly, and understand how the web can help them achieve their goals.

Another observation made was the nature of usability testing as a part of the development process. Presently, many clients appear to come to ATC in a state of crisis, thus resulting in ATC's services being used to "fight fires" in a reactive way, dealing with problems on a last minute basis just prior to launch. Customers in these circumstances may frequently approach the organisation with a view to fixing problems that they believe could exist with their website or that they know exist but don't know how to fix. However, they most likely will not have conducted a large amount of testing during the development process, instead preferring to leave the majority of testing until the last stage of development.

Looking forward, ATC is working with its clients to change this attitude, to a model where testing is done at different points during development, at different project milestones during the course of the project. This change is being facilitated by educating clients about the benefits of iterative testing, through activities such as quality advocacy. The benefit of this new approach to testing is that in can provide a much more value added and effective way to conduct testing for ATC's clients, as performing activities such as last minute system or usability testing makes it difficult to add a large amount of value to the client's business. Having said this, some organisations already adopt this more proactive approach, as a means of good software development practice.

Another means of facilitating the changes mentioned above is by ATC moving away from working project by project with developers, and moving more towards a model of partnering with the business. In this case ATC is their "Quality partner". This means that ATC can get involved in the project earlier with a view to defining acceptance criteria, and helping to ensure that these criteria are met through iterative evaluation. An added benefit of this approach is that a higher level of impartiality can be achieved between ATC and its clients.

# 5   Relationship to OPEN

In undertaking this work on the Web OPEN framework, several interesting observations were made in relation to the process framework, as well as website testing and evaluation. With respect to Web OPEN in the broader sense, the main observation made was that it appears that the process framework is still largely relevant and reflective of current industry practice in general terms. Although the Web OPEN framework was not specifically evaluated in the broader development context, evidence obtained during the case study suggests that the framework is still very useful in for website development.

Website development and specifically website testing and evaluation, are still rapidly developing areas technologically. At the same time however the level of maturity with respect to testing and evaluation can vary widely. This was further reinforced by the case study, which found a large number of Access Testing Centre's clients still undertaking a large single stage of testing just before release in order to identify and correct any problems. Fortunately there appears to be a trend away from this model, to one that encompasses a more continuous stream of testing throughout the development process, albeit on a smaller scale at any one given time. This should help to ensure that testing and evaluation efforts continue to become more effective in terms of the feedback into the development process and the mitigation of potential problems within the finished product.

In terms of the manner in which testing is conducted, it is also interesting to note how new technologies are being employed to further enhance and refine existing testing and evaluation techniques. Further to this, the reuse of some existing software engineering techniques with regards to website testing and evaluation were also observed – hence the inclusion of some additional OPEN tasks and techniques into Web OPEN to help reflect this trend. Thus while website development is still a relatively new field within the context of software development, overall there appears to be a move towards the use of more formal methods and a more mature attitude towards website testing and evaluation.

# 6 Conclusions and further research

While much work has been done on website development in previous years, much still remains to be done to ensure that website development reaches a level similar of process maturity to that of traditional software development. That the available tools and methods allow websites to be rapidly developed with throwaway prototypes does not mean that a process framework for web development is not required. Our research has shown that Web OPEN is an appropriate model that conforms to current practice in web development and iterative evaluation. It has however highlighted the need for Web OPEN to be extended to further consider business needs. As OPEN includes support for business re-engineering and modelling, this is one area which could be investigated and developed further within the area of Web OPEN, with a view to clarifying many of the questions that appear to be unanswered with respect to business needs and HCI practice.

Looking to the future, it appears that usability testing is beginning to become more commoditised, due to the increasing prevalence of literature on the topic and a better understanding of its value to economic decision makers. As a result, there appears to be a shift towards ensuring quality and maturity of process websites during development rather than just the quality of the end product. In other words, organisations can often gain large benefits out of improving their processes, which in turn will help to ensure product improvement including the usability of their website.

# 7 References

Barnum, C (2002) The "Magic Number 5": Is it Enough for Testing? In M Maguire & K Adeboye (Eds.) European Usability Professionals Association Conference, Proceedings Volume 3, p.45-69. British Computer Society.

Firesmith, D., and Henderson-Sellers, B., (2002), *The OPEN Process Framework*, Addison-Wesley.

Graham, I., Henderson-Sellers, B., and Younessi, H., (1997), *The OPEN Process Specification*, Addison-Wesley.

Haire, B., (2000), "Web OPEN: An Extension to the OPEN Framework", Unpublished

Henderson-Sellers, B., Lowe, D., & Haire, B. (2002). OPEN Process Support for Web Development. *Annals of Software Engineering*, 13, 163-202.

Lowe D., (2000b), "A Framework for Defining Acceptance Criteria for Web Development Projects", *Second ICSE Workshop on Web Engineering*, June 2000; Limerick, Ireland.

Lowe, D., & Eklund, J. (2002, 7-11 May 2002). Client Needs and the Design Process in Web Projects. Paper presented at the *WWW'2002: The Eleventh International World Wide Web Conference*, Hawaii, USA.

Lowe D and Henderson-Sellers B., (2001), "Web Development: Addressing Process Differences", *Cutter IT Journal*, Volume 14, July.

Nielsen, J., (2000), *Designing Web Usability: The Practice of Simplicity*, New Riders.

# How to Integrate Usability and Functional Requirements: A Usability Requirements Model

*Johan Fransson, Emma Bosson, Erika Svensson*

Swedish Defence Research Agency
PO Box 1165, 581 11 Linköping
johan.fransson@foi.se; emma.bosson@foi.se; erika.svensson@foi.se

## Abstract

Proper management of user requirements is vital to develop usable information systems. Methods exist which handle functional requirements, but those methods often lack explicit consideration of usability requirements. Similarly, methods focusing on usability often lack support to handle utility. Integrating usability requirements and functional requirements is of major concern for enhancing system development processes. A hierarchical model where usability requirements could be described at several different levels, integrated with functional requirements would elicit user context knowledge to system developers. The usability requirements model consists of five levels: analysis of context, system goals, system requirements, interface requirements and technical requirements. At the different levels, usability requirements could be described from abstract requirements to detailed requirements, and from guideline requirements to specific technical requirements.

## 1    Introduction

One of the most common and severe problem associated with software development is related to requirements, i.e. to be able to determine a correct and complete set of requirements (Leffingwell & Widrig, 2000). Although requirements management is a problematic area, it is reconsidered a major component for successful development of software systems (Hofmann & Lehner, 2001).

The field of human–computer interaction strives to achieve methods, techniques, and guidelines for use in development of usable system where overall business effectiveness is enhanced. In the domain of HCI many methods and techniques have been presented where usability aspects are considered. Although methods exist they have not been in extensive use by manufacturing companies, which can be traced to that the methods and techniques have been developed as stand-alone products as they leave out other necessary aspects of system development. To be able to use usability requirements models, those models must take functional requirements into consideration, as those also are an important when developing systems. In this paper, problems of usability will be addressed considering the early phases in software development processes, where user requirements are stated. The objective is to explore a model for managing usability requirements, which focuses on integration of usability aspects and usability requirements with traditional requirements management to ensure that usability aspects are treated as importantly as other requirements.

# 2 Background

The development of a usability requirements model requires analysis of usability and factors which make systems usable as well as understanding what requirements management is. The following questions give a structure from which usability aspects and attributes can be further elaborated. How is usability defined? How can usability be measured? How are usability requirements used? How do usability requirements relate to other requirements? Knowledge regarding usability requirements is accessed from the HCI domain and knowledge regarding functional requirement is accessed from the requirements engineering field.

## 2.1 Requirements engineering

In requirements engineering research has been done to enhance activities concerning better and more effective management of requirements. Requirements engineering can be stated as containing the "activities involved in discovering, documenting and maintaining a set of requirements for a computer-based system" (Sommerville & Sawyer, 1997, p. 5). Requirements engineering is developed with an internal motivation to make software development projects more effective, hence it does not extensively treat usability aspects, because usability concerns external relations, i.e. developer vs. costumer. Focus in the area has been devoted to management of functional requirements. Functional requirements state what a system would be able to do (Sommerville, 1998; Kruchten, 2000) and all other requirements are categorized as non-functional requirements. Those non-functional requirements are treated as limitations in design space for the design of functional requirements. It is also explicitly stated that design requirements should not be integrated in requirements specifications. In software engineering, design issues are addressed during later phases of the development process, i.e. after the requirements specification phase. In practical development work, requirement assessment and design are intertwined (Leffingwell & Widrig, 2000), which points to that it is necessary to consider design aspects while assessing functional requirements.

## 2.2 Usability requirements

User requirements can be differentiated on the basis of whether they concern usefulness or usability (Nielsen, 1993; Nickerson, & Landauer, 1996). A system can be useful if the set of requirements includes as many necessary features as possible that help users to complete their tasks, but the system is usable only if those requirements are designed and implemented into tools that support efficient use. Usability requirements exist at different levels, from abstract requirements to detailed requirements and from guidelines to technical requirements. One hierarchical classification of HCI standards is (Cakir & Dzida, 1997):

1: Basic principles of interactions
2: Principles of HCI
3: Standard requirements of HCI
4: Standard technical specifications of HCI

Level four regards standards where design solutions are stated. The higher levels of standards regard principles of interaction, principles of HCI and ergonomics requirements. All those levels must be possible to define in a usability requirements model. It is commonly known in the area of HCI that usability is highly context dependent, i.e. usability depends on users, tasks, equipment and domain (ISO 9241-11). Only addressing what the system ought to perform and neglect how functions are designed is to say that the context, with its users, does not matter.

# 3 The usability requirements model

The model is described as a static structure (Bosson & Svensson, 2001) with relations between different requirement components. The model takes a top–down development perspective where high level goals are stated first and then further decomposed into different types of requirements. The model is divided into five levels: (1) Analysis of context, (2) System goals, (3) System requirements, (4) Interface requirements and (5) Technical requirements, see Figure 1. Each level describes requirements except the first level, Analysis of context, where context of use is described and no requirements are specified. The second level describes general system goals. The third level consists of functional requirements and interaction requirements, which are requirements describing both what the system should be able to do and how it will be done. The fourth level states requirements of how information should be represented, i.e. representational requirements. The last level consists of requirements which are detailed enough to be directly implemented, i.e. implementation requirements.

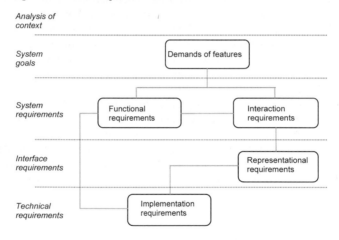

**Figure 1:** The usability requirements model

The model does not explicitly contain process descriptions for requirements elicitation. However, the model gives two alternative ways to achieve implementation requirements from functional requirements depending on whether or not user interaction is involved. Requirements that do not involve user interaction can be analysed and describe as implementation requirements directly, while those where users are involved must be further analysed.

## 3.1 Analysis of context

All development starts with needs to solve problems for some work situation. The work domain is analysed and documented in relation to four areas which are: tasks, users, environment and organizational goals. Of special interest are those factors in the user domain that may influence the system that will be developed. The documentation of the context will serve as a framework which all future analysis, development and design must consider.

## 3.2 System goals

At the system goals level the system features are defined. The demands of features state what the system should accomplish, what problems it should solve and what tasks it should support. Features are stated in user domain language because the development of the system starts at this

466

level and goals must be oriented towards the users. Usability requirements consider high level usability aspects which in the model include: interaction style requirements, training requirements, and navigation requirements.

Table 1: Example of usability requirements at the system goal level

| Interaction style | Direct manipulation should be used |
|---|---|
| Training | Training time for the system should not exceed 8 hours |
| Navigation | All functions should be accessible from menus |

## 3.3 System requirements

At the level of system requirements, functional and interaction requirements are specified. The functional requirements are to attain utility and the interaction requirements are to attain usability. Functional requirements can be specified as software requirements, which are the programming requirements to provide features, and hardware requirements to execute the functionality. Interaction requirements concern communication between the user and the system, which can be divided in operations and information. Operation regards actions which users or the system perform, while information states which kind of information is in use during the operation. Both operations and information can be divided into output and input. Operation output requirements state how systems respond and present information and operation input how users will act when using functions. Software requirements are stated in software domain language as it is more oriented towards the system development domain, than at system goals level.

Table 2: Example of usability requirements at the system requirements level

| Operation:Input | 1. Choose function<br>2. Click in map or use software buttons |
|---|---|
| Operation:Output | 1. Show active function<br>2. Move map |
| Information:Input | New centre point for map. Present scale will be used |
| Information:Output | Geographical information; e.g. roads, rivers or towns. |

Every interaction requirement also consists of measurements as defined in ISO, where usability is measured in effectiveness, efficiency and satisfaction (ISO 9241-11, 1998). Well proven usability principles of interaction are mapped into interaction requirements so that those principles can guide specification of every interaction requirement.

## 3.4 Interface requirements

Interface requirements define the representational design of the graphical user interface. Both visualization of information in the system and visualization of the graphical user interface are specified and they can be divided, as with interaction requirements, into output and input. Requirements regarding operation are definitions of graphical user interface components.

Table 3: Example of usability requirements at the system goal level

| Operation:Input | The button for activation of moving map function should use a sign of a magnifying glass with a plus sign inside. |
|---|---|
| Operation:Output | The cursor should use a plus sign when active in the moving map function. |
| Information:Input | The point which the user has chosen in the map should be represented with a black circle. |
| Information:Output | The digital map should be consistent with traditional paper map representations, e.g. roads in black and rivers in blue. |

Usability principles of representation exist at the interface requirements level, in the same way as general interaction principles in the level above.

## 3.5 Technical requirements

At the last level all requirements terminate as implementation requirements. Implementation requirements are requirements that are directly possible to implement. By gathering all requirements at the end as implementation requirements, all requirements can be seen as functional requirements, but the usability aspect has influenced and affected the requirements.

## 3.6 Usability requirements

Usability can be found on all levels in the model. Already at the System goals level usability can be found in requirements on interaction style, training requirements and navigation. The decisions made at that level shape the rest of the requirements specification and serve as guidelines. At system requirements level, usability is found in requirements on operation, information, and general principles for interaction. Moreover, interface requirements take usability into consideration in operation, information and general principles for design.

## 4 Conclusion

The model gives the possibility to elicit requirements from different perspectives, such as from purpose of the system to functionality and graphical user interfaces. Users, which focus on external properties of the system, can define interaction requirements and representation requirements while system developers can state functional requirements, as they focus on internal properties.

## References

Bosson, Emma & Svensson, Erika (2001). Usability Requirements – How to relate usability to functionality in requirements specification (in Swedish), FOI Base data report, FOI-R--0252--SE, Linköping

Hofmann, Hubert F. & Lehner, Franz (2001). Requirements Engineering as a Success Factor in Software Projects, IEEE Software, July/August.

ISO (1998). ISO 9241-11 Ergonomic Requirements For Office Work With Visual Display Terminals (VDTs) - Part 11: Guidance on Usability. Geneva: International Organization for Standardization.

Kruchten, Philippe (2000). The Rational Unified Process An Introduction (2nd ed.). Boston: Addison-Wesley.

Leffingwell, Dean &b Widrig, Don (2000). Managing software requirements A unified approach. New York: Addison Wesley.

Mayhew, Deborah. J. (1999). The Usability Engineering Lifecycle - A Practitioner's Handbook For User Interface Design. San Francisco: Morgan Kaufmann.

Nickerson, Raymond S. & Landauer, Thomas K. (1997) Human-Computer Interaction: Background and Issues in Human Computer Interaction in Helander et al. Handbook of Handbook of Human Computer Interaction, second edition. New York: Elsevier Science B.V.

Nielsen, J. (1993). Usability Engineering. San Diego: Academic Press.

Sommerville, Ian & Sawyer, Pete (1997). Requirements engineering – A practice guide. New York: Wiley.

Sommerville, I. (1998). Software Engineering (5th ed.). Harlow: Addison-Wesley.

# ObSys – a Tool for Visualizing Usability Evaluation Patterns with Mousemaps

*Michael Gellner*

University of Rostock
Department of Computer Science
Software Engineering Group
Albert-Einstein-Str. 21
18051 Rostock, Germany

*Peter Forbrig*

University of Rostock
Department of Computer Science
Software Engineering Group
Albert-Einstein-Str. 21
18051 Rostock, Germany

## Abstract

This paper presents *mousemaps,* multidimensional visualizations of user interactions with software systems. The source for that visualizations are data that are recorded with the event recorder *ObSys*. On the one hand mousemaps are proper for enhancing the effectiveness of video analyses, on the other hand it is possible to empower the efficiency of video analyses by focusing automatically critical situations. Furthermore mousemaps can substitute video analyses because the visualization offers most of the interaction information on a two-dimensional view without the need to watch video material. The low level data recorded by *ObSys* allow the recognition of »mouse gestures« that are characteristically for certain usability problems. Automating the analyses of usability testing sessions by applying usability evaluation patterns without the need of modeling information as required for other automating tools becomes possible.

## 1    Introduction – our work and what was there before

Usability testing with testing persons is of prime importance to get information about how users really can work with a system. Since it is impossible to analyse sessions on the fly or in between testing sessions are recorded to be analysed afterwards. For recording such a session there are different technologies, which offer features but disadvantages otherwise.

Often applied is video recording, since it allows to save the actions of the user, the behaviour of the system and especially the mimic and gesture information. But video recording is not efficient: For every hour of material one has to calculate around six to eight hours watching for recognizing problems. Recording from different perspectives multiplies the time of the sessions with the time of the analysing. As a consequence often taped material stays unevaluated (Mayhew, 1999) (Preece, 1994). There are further factors that make video recording inefficient like high hardware requirements, the necessity to hide the intimidating technology and personal needs. Although video recordings contain effective information for an human evaluator it is nearly impossible to analyse such data by software automatically. A similar situation arises if the screen capturing method is used: one does not need to fulfil high requirements in hardware and human resources but the results after the sessions are videos that have to be analysed by humans again.

Mousemaps are based on event recording. Every user input with mice, keyboards, joysticks or other devices send signals to the operating systems, so called *messages*. The event recorder *ObSys* captures every message from input devices that MS Windows operating systems put in the message queues. So far this method seems to be similar to those from tools like Kaldi (Al-Qaimari & McRosstie, 1999), GUITESTER (Okada & Asahi 1999) and others. In contradiction to those systems *ObSys* offers synchronisation to video files from cameras, screen capturing tools and event logging (input of manually typed comments by predefined short cuts with timestamps). In difference to the other published tools *ObSys* can play back the captured events. With that function *ObSys* can substitute the complete screen capturing technology with the advantage of getting 150 times smaller files[1].

The availability of this variety of methods enables studies which compare the abilities of the different methods. Although there are some papers that offer method comparisons, they are predominantly literature reviews; mostly results are compared based on estimations and derivations from different studies with different subjects, conducted by different evaluators with different testing persons, see for example (Gray & Salzman, 1998), (Hilbert & Redmiles, 1999) or (Ivory & Hearst, 2001).

## 2    Mousemaps – what they are, how do they look

Although mousemaps are projected on two dimensional windows they give multidimensional views. Beneath lines for visualising the mouse movements colours, thickness of lines and dots are used for further visual coding: The direction of each mouse move is indicated, the mouse clicks are given and the speed of the mouse moving is shown. For demonstrating this, we take the following minimal scenario: A small window is created with a button which has to be pressed (for isolating this visually another program is running that creates only a white background). Figure 1 shows the demonstration window in two testing series. On the left hand side the mouse is moved extremely slow (constantly), whereas the right screenshot shows slower and faster movements alternating.

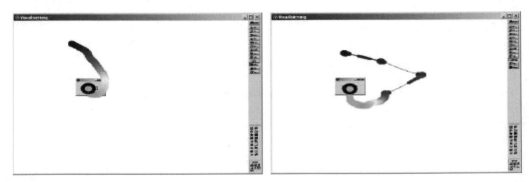

**Figure 1: Visualisation of the speed of mouse moves (left slow, right mixed)**

Figure 1 shows further that clicks are visualized as dots. As can be seen it can happen that after a certain number of clicks there is nothing more to recognize. For that reason *ObSys* allows to configure a range for the diameter of the clicks and for the width of the lines. The scenario in Figure 2 consists again of a minimal application: there are three buttons that have to be clicked with the left mouse button. Whereas on the left hand side there is hardly to see what lays under the click symbols the mousemap on the right gives a clear view what happened and on which positions that was (in case there are problems in finding details please consider the downscaling from originally 1280 × 1024 pixel).

---

[1] With one precondition that has to be considered by developing scenarios: All operations on windows (like sizing, moving, reconfiguring sub windows) and GUI elements must be reversible.

**Figure 2: Visualising clicks – keeping track of the things that happened by customizing views**

Another example demonstrates the visualisation of the direction to that the mouse was moved. In Figure 3 a paragraph was marked for applying any operation to that part. The first time the marking started from the left hand side and was moved to lower right, the second time the marking begun right and ended left above.

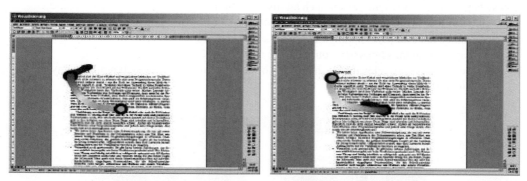

**Figure 3: Marking text beginning from the left hand side (left screenshot) and beginning from the right hand side (right screenshot)**

## 3    What do mousemaps tell us?

One of the most interesting aspects is that mousemaps give the tester a small insight into unconscious aspects of users behaviour. Quite a lot of actions seem to be accompanied by a mouse based pendant of body language. As can be seen in the figures above in general users do not take direct ways for their actions. Instead along linear paths the mouse is moved in curves similar to quarter segments of circles. Figure 4 shows the activities of an user by looking for information about publishing companies in Google (left). As the dots show only few manipulations took place.[2] Most of the movements supported the reading. *ObSys* allows to generate a view that considers only the activating actions (this excludes moves without activating further events). In this scenario moving causes no further actions like e.g. dragging. The figure on the right hand side shows the few lines that were necessary to activate the windows the user wanted. Errors of usage that would make those further acting necessary did not happen.

---

[2] some links were followed and some were opened in a separate window

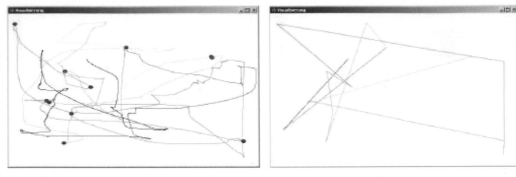

**Figure 4: Mousemap with a view on all activities (left) and a click to click reduction**

Contradictory explanations are possible: (I) The curvy behaviour is shown for physiological reasons. (II) As mentioned above: Similar to body language, mimic and gestures a mirror of our unconscious minds is given by using the mouse. To get more in detail at this point something that forces more precise placements of the mouse clicks has to be applied. The Windows game Minesweeper offers some features for such an investigation:

- it forces precise placements of mouse clicks
- since the user plays again time a little stress appears
- apart from the clicks the mouse can be moved free
- only clicks are relevant for that task, so the click to click view will not hide any important information

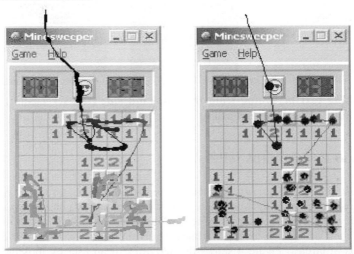

**Figure 5: Mousemap for playing Minesweeper (left: all actions, right: click to click**

Mousemaps for playing Minesweeper are shown in Figure 5. If the swinging movements would be necessary for physiological reasons to place the mouse on a matching position the left mousemap had to show a lot of curves. Instead of this there are nearly linear movements. The right mousemap shows the same testing series but reduced to the relevant click to click lines. In difference to Figure 4 the click to click shows mostly similarities to the normal view on the left hand side for that scenario. Based on that few data the physiological aspects seem to be not dominantly.

# 4    Conclusions and Perspectives

As (Smith et al., 2000) published based on eye tracking studies patterns of interaction between eye and mouse can be found. As outlined it is possible that acting with the mouse is one more way for communicating unconsciously. At the moment we tend to interpret the curvy mouse moving as a »normal« behaviour. *Normal* in that context means: without stress or forces from outside (like matching lines for assigning colours to them). Automating usability evaluations based on mousemaps were possible if one finds further indicators for the inner state of the user. This is our approach for further work.

This paper expresses our first experiences with our tool *ObSys* and the implemented multidimensional visualisation technology. On the one hand there are many aspects that we want to have implemented for further evaluations: The mouse event to key event ratio, the workflow jitter by using mouse and keyboard, an indicator for the time an input device is used and lots more. On the other hand a high number of evaluations with many scenarios and types of users are necessary to contain more facts with a high degree of clarity.

# References

Al-Qaimari, G. & McRosstie (1999). *KALDI: A Computer-Aided Usability Engineering Tool for Supporting Testing and Analysis of Human-Computer Interaction*. In: Vanderdonckt, J.; Puerta A. (Ed.), Computer Aided Desgin of User Interfaces II (Proceedings of the Third International Conference on Computer-Aided Design of User Interfaces), Kluwer Academic Publishers, Lovain-la-Neuve, pp. 337 - 355.

Gray, W. D. & Salzman, M. C. (1998). *Damaged Merchandise? A Review of Experiments That Compare Usability Evaluation Methods.* In: Moran, T.D. (Ed.): *Human-Computer Interaction, A Journal of Theoretical, Empirical and Methodological Issues of User Science and of System Design*, Vol. 13, (3), pp. 203 - 261.

Hilbert, D.M. & Redmiles D.F. (1999). *Extracting Usability information from user interface events*. Technical Report UCI-ICS-99-40, Irvine: Department of Information and Computer Science, University of California.

Ivory, M. Y. & Hearst, M. A. (2001). *The State of the Art in Automating Usability Evaluation of User Interfaces*. In: ACM Computer Surveys, Vol. 33, No. 4, pp. 470 - 516.

Mayhew, D. J. (1999). *The Usability Engineering Lifecycle*. San Francisco: Morgan Kaufmann Publishers, Inc.

Preece, J. (1994). *Human-Computer-Interaction*. Harlow: Addison-Wesley.

Okada, H. & Asahi, T. (1999). *GUITESTER: A log-based usability testing tool for Graphical User Interfaces*. In: IEICE Transactions on information and systems. Volume E82, pp. 1030 - 1041.

Smith, B. A., Ho, J., Ark, W. & Zhai, S. (2000). *Hand eye coordination patterns in target selection*. In: Proceedings of the symposium on Eye tracking research & applications 2000, ACM Press, pp. 117 - 122.

# Lightweight Usability Engineering
# Scaling Usability-Evaluation to a Minimum?

*Ronald Hartwig, Cristina Darolti, Michael Herczeg*
Institute for Multimedia und Interactive Systems - University of Luebeck
Willy Brandt-Allee 31a; D-23554 Luebeck, Germany

Hartwig|Darolti|Herczeg@imis.uni-luebeck.de

## Abstract

Today there are several approaches available to develop usable software, but many products still lack a minimum of usability. One reason may be the lack of usability in some "heavyweight" usability engineering and evaluation methods themselves. It seems that many development teams think such elaborate process models are too complex to be applied, so they just don't do usability engineering. This paper offers a scalable "lightweight" approach to usability engineering. It starts from the idea that an easy and imperfect but used method is better than a complex unused method. The paper gives a short overview on the intentions, the basic ideas, the implementation, as well as experiences in using the method and discusses their possible application in product certification.

## 1    Introduction

*"Nothing will ever be attempted if all possible objections must first be overcome."* Samuel *Johnson (Gediga & Furness 1993)*

The European directive for "Work with Video Display Units" (referenced as VDU-directive, EU/90/270) forces the employers to apply usability principles to software based working systems. Because of this, software companies are willing or impelled to ensure a certain grade of usability for their products. The following approach is a rethinking of some well known and often costly and difficult-to-apply usability engineering methods. (Dicks 2002) states, "most product managers are just not willing to pay the price in time and money" for supporting traditional usability engineering in their company, especially since the results are not necessarily immediate or direct. They are just unsure about the benefit of following in detail each process step as many elaborate process models (like the ISO 9000-x) propose; the standards are perceived as too general or abstract. Similar to their "quick and easy" programming, developers, who still hold the power, want a simple way for usability. The recent summary of practitioners' feedback (Rosenbaum et al. 2000) supports this affirmation. It can be said that there is a need for usability of usability methods, or to cite (Nielsen 1995): "User interface professionals ought to take their own medicine some more."

Before discussing the usability of small scale usability engineering methods it is necessary to define a common ground for the scope of usability itself. The international standard ISO 9241 (ISO 1996) and its software related parts 10 to 17 seem currently to be the commonly accepted base, but still there are known critics and disadvantages: Sometimes this standard is very detailed;

sometimes only principles and examples are given. Therefore we focused on the definitions from ISO 9241 part 11 (defining usability as the product of effectiveness, efficiency and satisfaction) and the principles of part 10 (dialogue principles) and part 12 (presentation of information).

## 2 The Method

The idea of scaling and combining different approaches was the conclusion of practical experiences with more elaborate methods. The first indicator was found in a master thesis (Hartwig 1997). The study showed that the use of static usability criteria lists did not justify the investment. In the study a software system was analyzed using heuristics, user tests and a criteria list derived from ISO 9241 called EVADIS (Oppermann et al. 1992). In this experiment the combination of heuristics and usability testing (4 tests) achieved an overall performance of detecting 90% of 48 problems (100% of all fatal problems, 89% severe, 71% medium, 100% simple) while the item list only helped finding 32% of all problems (20% fatal, 67% severe, 41% medium, 0% simple) Questionnaires were not used because the user sample was too small and assuming a return rate of 10 to 30% this would not have been reliable enough in terms of the statistical requirements of testing theory. The initial finding was that in spite of the low performance of the criteria list, the effort of using it was immense.

During the usability support of the multimedia production project "VFH" it got clear that the formative use of questionnaires did not support the constructive consulting work very much but again the effort to do it was high (Hartwig et al. 2002). No fatal or severe problem was detected using questionnaires, but the effort to organize and accomplish the inquiry took several days and involved many organizational resources. A third cause to doubt the usability of (too) heavyweight methods were experiences made in the development of a general standardized usability evaluation standard called "ErgoNorm" for the german accreditation office "DATech" (see (Dzida et al. 2001)) which tries to constitute a common ground for usability (product or process) certification. The authors think that the very elaborate way to conduct the engineering process as proposed by ISO 9000 derivates seems to deter many practitioners. But resources and the question at hand should match (Wixon et al. 2002). This impression is supported by the feedback from several ACM workshops described in (Rosenbaum et al. 2000).

The drawbacks described above determined the basic, maybe self evident, idea of this method: to allow downscaling to available resources, project size and required quality. Compared to the heavyweight approaches, where completeness is mandatory, we allow for scaling down the investment in usability engineering to whatever the project offers. The basic methods used for defining users, goals and usability requirements, for designing, testing and iterating are the classical, well known ones, (as described in approaches like "scenario based design" (Rosson, Carol 2002) or "contextual design" (Holtzblatt, Beyer, 1996)) but they are applied at a different level of detail and accuracy. The important difference to naively cutting down all efforts is keeping the central idea of quality management in mind: All findings must be comprehendible. If during the usability testing a potential problem appears it is checked against the context of use (Herczeg 1994) first. Only if the context of use and task and user attributes allow it, we call it a deficiency. Issues are justified using heuristic methods and expert usability knowledge. Only those issues that really affect the usability criteria and are relevant for the analysed task are rated as a problem. This ensures a minimum level of validity. Using the ideas of scenario based design (Rosson&Carroll 2002) ensures that at least the main tasks of the user are actually tested and reviewed. But in contrast to heavyweight approaches it is seen as sufficient to have one or two rough and general scenarios and to derive a small set of requirements instead of doing a full featured version. The next subchapter gives an idea on how and where heavyweight and lightweight process versions differ.

Decisions about the level of accuracy i.e. the testing quality (lightweight vs. heavyweight) have to be a part of the general process and product quality goal definition process. Analogue to the development process this should take place as a "claims analysis" as described by (Rosson&Carroll, 2002). This allows for a legitimate declaration of quality and, thereby, for possible certification activities.

**Table 1**: Lightweight usability engineering in practice of the VFH-project ("lightweight") in comparison to the more elaborate DATech-Process ("Heavyweight")

| Process activity | Heavyweight | Lightweight | Reasoning |
|---|---|---|---|
| Context of use documentation | Several use scenarios observed in real life context with iterative re-validation | Generic scenario derived from observations during first user tests | Being on site at the users workplaces for usability testing was a good time to document usage scenarios as well. For the first tests a hypothetical scenario was assumed and then adopted after testing. |
| Requirements engineering | The scenarios are analyzed and all requirements that evolve are documented. Test criteria are derived from these requirements | Critical incidents from user testing and reviews as well as questions from developers are used as indicators for clarification needs. Mainly these "open questions" are taken as starting points for scenario based requirements engineering. | Many requirements are obvious enough or fulfilled anyway so they don't have to be documented at all. Instead of a complete top-down approach, potential problems are used as actuators if they were confirmed to be relevant. Only these are target to requirements engineering. This allows shortening the documentation, communication and evaluation of requirements to an absolute minimum. |
| Evaluation | Expert reviews, usability tests and questionnaires are conducted based on the criteria derived from the scenarios. | Expert reviews are done based on the smaller set of criteria from above. Usability testing is combined with the context of use analysis. Questionnaires are only used on a more general level and mainly for summative purposes. | The main goal of the evaluation is formative, so it mainly relies on expertise and user testing. Both are loosely connected to the criteria developed so far. If critical incidents appear, the context of use documentation and the requirements engineering are refined for this part. The more summative questionnaires are used to back up the assumptions of the overall quality |

Because of the limited space the implementation of the process is only denoted here (see Table 1) to give an idea on how this lightweight process looked like in practice of the VFH-project ("lightweight") in comparison to the more elaborate DATech-Process ("Heavyweight"). A more detailed description can be found in (Hartwig et al. 2002).

# 3    Results

It is difficult to quantify the success of the described lightweight method. In practice it was impossible to do the same project twice in order to compare the usability of this minimal approach to the complete heavyweight approach. But there are hints from practical use which support this more flexible way of applying usability engineering methods. The described method was used to assure the usability of 35 multimedia learning modules in a 5 year production process. About 120 persons were involved in the production process; these were spread over 5 major development laboratories all over northern Germany. About 7 man-years of usability engineering were available from two persons. 91 users were polled using standardized questionnaires after about 15 of the modules had been released and were used in daily business. Users were asked to answer how important a usability item is for their personal use and how satisfied they are with respect to this item. The answers were combined to a stress/relief-rating. Items which were rated as important influenced the rating more than those which users thought to be "nice to have" (see (Hartwig et al. 2002) for details). The analysis showed that those modules that were part of this lightweight usability engineering process were rated significantly better than those which where not affected. Also the overall average user satisfaction indicated a preference for those modules which had been produced using the described process model. Effectiveness and efficiency were rated based on the user tests where obstructions and impediments got obvious. Again, those modules which were part of the process had less fatal or severe problems. Compared, at least, to "no usability engineering at all" this method seems to support a better overall usability. The comparison to more elaborate methods seems self evident: It was simply impossible to allocate more resources to the usability engineering task so insisting on a possibly more exact and in terms of testing theory "better" engineering and evaluation approach would have stopped the project or at least all usability engineering activities.

# 4    Discussion

First of all the described results are specific to the project VFH and may not directly be taken as evidence for other contexts. But the indicators show that our minimal approach is appropriate for many other development projects as well, although a number of problems arise: Scalability, which gives way for customization, also gives way for possible poor application of the method; so the success of the method depends on the way it is applied to the context. The authors think that in practice an easy method which introduces errors is better than no usability engineering at all. Problems with small test samples may by overcome using qualitative usability methods which are independent of the number of test-users. Concentrating on the main goal to make tools helpful for accomplishing tasks is the key to sensibly allocate scarce resources: "A lot can be accomplished quickly, easily, with little time, effort, or expense. The main secret is to observe real people doing real tasks." (Norman 2000). Relying on user testing has shortcomings because (especially laboratory) testing is always an artificial situation, participants are rarely representative of the whole population and the results, even if positive, cannot ensure that the product works. Testing focuses on tasks and therefore may not uncover larger problems like problems in overall system conceptual model (Dicks 2002). But both, "heavyweight" and "lightweight" methods suffer of this. In the end we mention the idea of "lightweight" quality certification by this method. It may be doubtful if it is fair to give away certificates based on such a method. Since quality may also be seen as scalable and controllable, the customers must be told about the limitations of the certificates. Any certificate must not be used as a 100% usability guarantee. Usability certificate should be seen as one minimal criterion among others (functionality, price, strategic decisions) when buying or planning software and its usage. It shows that at least the main tasks can be done with this software without too much harm.

# References

Dicks, R. S. (2002). Mis -Usability: On the Uses and Misuses of Usability Testing. Proceedings of the 20th annual international conference on Computer documentation, ACM Press.

Dzida, W., & Freitag, R. (2001). Usability Testing - The DATech Standard. In Wieczorek, Meyerhoff (Editor): Software Quality - State of the Art in Management, Testing And Tools. Springer, pp. 160-177.

Gediga, & Furness, T. (1993). Virtual Worlds – Why and When? User Interface Strategies a live satellite TV Broadcast, University of Maryland, College Park

Hartwig, R. (1997). Gestaltung und Bewertung der Gebrauchstauglichkeit von Software - Überprüfung der Machbarkeit am Beispiel der Benutzungsoberfläche eines Moduls eines Schichtplangestaltungssystems (Thesis); Universität Oldenburg FB Informatik, Oldenburg

Hartwig, R.; Triebe, J.K.; Herczeg, M. (2002): Usability Engineering as an Important Part of Quality Management for a Virtual University. In: Proceedings of Networked Learning 2002 - Challenges and Solutions for Virtual Education. Technical University of Berlin, Germany, ICSC-NAISO Academic Press, Canada.

Herczeg, M. (1994). Software-Ergonomie – Grundlagen der Mensch-Computer-Kommunikation, Addison-Wesley, Bonn; Paris ; Reading (Mass.)

Holtzblatt, K., & Beyer, H. (1996). Contextual Design: Principles and Practice. Field Methods for Software and Systems Design. D. Wixon and J. Ramey (Eds.), New York: John Wiley & Sons, Inc..

ISO (1996): ISO 9241 – Ergonomic requirements for office work with visual display terminals (VDTS) – International Standard

Nielsen, J. (1995): Getting usability used, in Human-Computer Interaction – Interact 95, Chapman&Hall, London

Norman, D. (2000). Usability Is Not A Luxury. Retrieved November, 2002, from http://www.jnd.org/dn.mss/UsabilityIsNotLuxury.html

Oppermann, R., Murchner, B., Reiterer, H., Koch, M. (1992): Softwareergonomische Evaluation - Der Leifaden EVADIS II (2. Auflage) --Berlin; New York: Walter de Gruyter

Rosenbaum, S., Rohn, J.A., & Humburg, J. (2000). A toolkit for strategic usability: results from workshops, panels, and surveys. Proceedings of the CHI 2000, ACM Press, New York.

Rosson, M. B., & Carroll, J.M. (2002) Usability Engineering – Scenario based development of human-computer interaction, Morgan Kaufmann Pub.; San Francisco.

Wixon, D. R., Ramey, J., Holtzblatt, K., Beyer, H., Hackos, J., Rosenbaum, S., Page, C., Laakso, S., & Laakso, K. (2002). Usability in practice: field methods evolution and revolution , in Proceedings of the CHI 2002, ACM Press, New York, 2002

# State of the Art: Approaches to Behaviour Coding in Usability Laboratories in German-Speaking Countries

*Britta Hofmann, Marc Hümmer & Peter Blachani*

Fraunhofer Institute for Applied Information Technology (FhG-FIT)
Schloss Birlinghoven, 53754 Sankt Augustin, Germany
Britta.Hofmann@fit.fhg.de, Marc.Huemmer@fit.fhg.de, Peter.Blachani@fit.fhg.de

## Abstract

Behaviour coding is a systematic methodological approach to use observational data gained from laboratories regarding usability issues. Several theoretical approaches are reviewed, including User Concerns (Robson & Carrol, 1995), Task Intention Coding (Hofmann et al. 2000), Task Action Grammar (Payne & Green, 1986) and GOMS (Card et al., 1983). A survey among sixteen usability laboratories in German-speaking countries was conducted to obtain an overview of the working practice in behaviour coding. Ten laboratories reported to use coding schemes. Only three laboratories use generic and domain-independent coding schemes. The general trend is towards a minimalist approach with a generic framework that is individually adjusted to each specific application domain. Generally, the feasibility of generic and domain-independent coding schemes was doubted. It is concluded that there is a need for a solid theoretical foundation for coding schemes that can be developed for domain-specific applications (e.g.: e-shops).

## 1 Introduction

The use of laboratories can support the general usability evaluation process by providing a wealth of observational data. These can be obtained by video, audio and screen recordings of the user's interaction with the system. Subsequently, the data can be utilized for three main purposes: documentation, categorization and analysis. While documentation and categorisation are purely descriptive, an analysis requires assumptions about the user's inner state and is hence interpretive. However, there is a need for sound methodological approaches to evaluate the observational data (Freitag & Dzida, 2002). Here, coding schemes for user behaviour are of immense value.

Human computer interaction can be regarded as a behaviour sequence that can be described in terms of user actions and systems reactions (Hofmann, Freitag & Dzida, 2000). Thus, it should be possible to derive coding schemes to capture user behaviour in a usability context. However, behaviour coding should not be regarded as a stand-alone method for usability testing. Nonetheless, it is considered to be a valuable and beneficial supplement to other usability evaluation techniques.

Our motivation for this paper was to obtain an overview of the various approaches towards behaviour coding. Therefore, we have considered some theoretical work on behaviour and interaction coding and conducted a survey about the working practice of usability laboratories.

## 2 Different theoretical approaches to behaviour coding and categorization

## 2.1   User concerns

Rosson and Carrol (1995) introduced a typology of generic usage situations that can be used as a coding scheme. The main concepts in their approach are the "user concerns". User concerns occur whenever users face difficulties of some kind in their goal pursuit. Rosson and Carrol (1995) distinguish between six concerns: orienting to goals, opportunistic interaction, searching for information, carrying out procedures, making sense and reflecting on skills. Each of these concerns deals with a specific usage situation. Thus, the interaction process can be coded along these concerns.

## 2.2   Task Intention Coding (TIC)

This approach was developed by Hofmann at al. (2000) and has been further refined by Freitag and Dzida (2002). Central to the TIC approach is the concept of a "mismatch". A mismatch occurs whenever users get stuck in their workflow as a consequence of differences between the intentions of the user and the system's designer. These situations are also referred to as "critical incidents". The intentions of users can be inferred from their behaviour and can subsequently be contrasted to the reactions of the system, which reflect the intentions of the designer.

The TIC is intended to be a generic and domain independent coding scheme that requires only minor adjustments for each particular application. The foundation of the TIC is a generic task structure. The behaviour of the user is evaluated in a combination of a sequential and a structural task model.

The sequential model distinguishes between three phases: task preparation, command execution and result evaluation. In the structural model, the existence of three components is supposed: a tool, an object and a parameter. The user performs an action with these components at different stages during the sequential model and receives as an outcome a result with certain attributes.

Within this workflow, mismatches can occur. Freitag and Dzida (2002) assume three mismatch management strategies: opportunistic behaviour, exploration and error management. Opportunistic behaviour characterises users who use a trial and error method to resolve their problem. Exploring occurs when users are trying to discover aspects of the system that they have not previously taken into account. Finally, error management is concerned with the steps that users undertake in order to reverse errors.

Based on the underlying task model and mismatch management strategies, Freitag and Dzida (2002) have developed a generic coding scheme that can be adjusted to the desired level of task analysis. In order to minimize resources, it is suggested that only critical incidents should be coded in detail. As a last step, the user and the system designer are asked to validate the coding protocol. Thus a clear understanding of the intentions of both sides can be obtained. Unfortunately, the TIC has not been extensively tested in practice so far, so no information about its feasibility can be given.

## 2.3   Task-Action Grammar (TAG)

Payne and Green (1986) established a rather different approach. They were inspired by linguistic theory and developed on these grounds the TAG. The TAG uses the principles of generative grammars to rewrite basic tasks into action specifications. In order to build a TAG of a specific task, specifications of the basic tasks and the transformation rules have to be provided. By considering the number of rules needed to build a TAG, a measurement about the relative learnability of the task can be obtained. Furthermore, resemblance between certain task-action mappings can be used to optimise the task language. The TAG is a very powerful and appealing theoretical device. However, because of its complexity and the resulting overhead, it is an impractical tool to use in practice.

## 2.4    Goals, Operators, Methods and Selection Rules (GOMS)

Probably the most widely known theoretical construct for interaction coding is the GOMS model. It was introduced by Card, Moran and Newell (1983) as a cognitive model of users' procedural knowledge. The acronym GOMS stands for goals, operators, methods and selection rules. GOMS models can be used to make predictions about skilled-performance time, method-learning time and likelihood of memory errors (John, 1995). There are several variations of the original GOMS model, such as the Keystroke Level Model (KLM) (Card et al. 1983) which can be regarded as a generic and domain independent coding scheme.

Although GOMS models have been successfully applied to a variety of real world application in published papers, including a KLM analysis by Nielsen (1995) for websites, they have never reached the breakthrough into industrial practice. This can be attributed to the implicit restriction: GOMS models are only valid for skilled users. Therefore, GOMS is not appropriate for many usage situations, e.g. prototyping or novel users. Moreover, the development of GOMS models is time and resource consuming and is therefore in many situations uneconomic.

## 3    Practice in Usability Laboratories

As part of this paper, we have conducted a survey about the working practice in usability laboratories in German-speaking countries. Our focus was on the approaches towards behaviour coding. Thus, we developed a question catalogue based on six questions, which will be treated in detail later on. Overall, we have received feedback from 16 laboratories. Here, we would like to thank all participating laboratories for their help.

*Question one: "Do you use video and screen recordings of the user?"*

All but one laboratories reported to use video recordings of the user. The majority of laboratories use two or more cameras to record different aspects. All laboratories use screen recordings of the testing sessions.

There was a notable difference in the utilization of the recorded material. One group of laboratories maintained to give prime importance to users' bodily behaviour, such as facial appearance, gestures and body posture, whereas another group regarded recordings of screen events as a richer and more reliable source. However, it was argued that the specific evaluation criteria are usually set by the domain and customer demands.

In addition, nearly all laboratories mentioned audio recording as further tool, especially in connection with thinking-aloud protocols. Seven laboratories also use eye-tracking systems in their usability evaluation.

*Question two: "Do you use specialised software to assist your evaluation?"*

Although the majority of nine laboratories employed specialised software, three of them use the software exclusively to record and visualize the data from the eye-tracker. Furthermore, the general trend is that the software is primarily used for automatic processing of the data. Four laboratories use their own software for the analysis, whereas the remaining five use third-party software. Three of them use the Noldus Observer Pro.

Most laboratories that do not use specialised software utilize common office applications, such as MS Excel or MS Access to record their observations. The left over laboratories use the traditional pen-and-paper method.

*Question three: "Do you use a coding scheme for the user's behaviour?"*

Ten laboratories reported to use some kind of coding scheme. However, there is an immense disparity between the various approaches. As a trend, there seems to be two general approaches: descriptive and interpretive observations. The descriptive approach notes solely the objectively observable behaviour, such as error occurrences or task completion time, whereas the interpretive approach makes assumption about the user's inner mental state.

The remaining six laboratories do not use coding schemes. Nevertheless, some explained to note users' behaviour in an informal way by logging general impressions, such as surprise or anger. Two laboratories stated no need for a coding scheme that they argued to be obsolete for qualitative analysis.

*Question four: "Do you use general or project-specific coding schemes?"*

This question is only applicable to those ten laboratories that use coding schemes. Only three laboratories reported to use a general coding scheme. The residual seven laboratories stated to employ project-specific coding schemes that are developed for each application. This was explained as being necessary because of the differences in the application domains. However, there is a trend towards a minimalist approach. Most of the seven laboratories use some kind of generic framework that is individually adjusted and modified according to the requirements of the specific application domain. This approach was argued to have the advantage of being flexible and economical in its resource needs. Only one laboratory stated to develop a completely new coding scheme for each project.

One laboratory asserted to have experimented with different general coding schemes but finally refrained from the general to the minimalist approach. This was attributed to problems with the granularity level of the schemes. Whereas a fine level of granularity was desired for some applications, it proved to be awkward and unfeasible for other applications. A gross level of granularity that could equally be applied to all applications was criticised for not containing sufficient valuable information that would justify the effort. In general, most laboratories doubted the feasibility of a generic coding scheme because of the differences in the application domains.

*Question five: "Do you use a theoretical background for your coding scheme?"*

Here, it was difficult to obtain a yes-or-no answer. All laboratories did not specify a single formal theory as the origin of their coding schemes. All laboratories said to give prime importance to practical experiences and operative knowledge, although none denied the incorporation of theoretical principles from various source (e.g.: perceptual theory). Nevertheless, nearly all laboratories reported to rely on the ISO standards, especially on ISO 9241-10 and 9241-11, as the main influence and guidance in their usability evaluation.

*Question seven: "Do you have practical experience in using GOMS, TAG or TIC?"*

All questioned staff reported to be familiar with GOMS through their academic education. However, only a few were familiar with TAG and none with TIC. Despite the relative awareness of GOMS, all staff reported to have never used a GOMS or a TAG model in industrial practice. Most staff argued that GOMS models are unfeasible, especially concerning their cost-benefit relation. This was attributed to the complexity and resource-consuming development of GOMS models. Another factor that was stated against GOMS models was their limited usability measurements, since they exclusively produces quantitative measurements. Overall, the general trend is that GOMS is considered to be an interesting tool in theory but unfeasible and inadequate in practice.

# 4    Conclusion

There are surprisingly few theoretical constructs and approaches for behaviour coding in usability context. Thus, there is an apparent need for more elaborated and feasible approaches that have been extensively tested and adjusted in working practice. Here, it was indicated by the survey that the ISO standards 9241-10 and 9241-11 could serve as a constructive foundation.

However, the development of a generic domain-independent coding faces severe difficulties as revealed by the survey. The differences in the application domains pose the main problem. Therefore, it can be doubted whether generic domain-independent coding schemes can capture important and essential aspects of an application. Furthermore, the specification of the target group is also essential for the usability testing process and thus restricts the use of a generic coding scheme. Overall, it can be said that the usage contexts for different application are subject to vast variations and hence do not permit the use of a generic domain-independent coding scheme.

Nevertheless, it could be workable to develop generic coding schemes for various applications within the same domain. E-shops could be an example for such a domain. Since most e-shops have a similar structure (e.g.: shopping-cart) and since the tasks (e.g.: search & select product) and goals (e.g.: buy an item) of the user can be regarded as mutual, a sophisticated coding scheme could have validity for a wide range of different e-shop applications. Such a coding scheme would not only allow for improved and standardised comparability between different e-shop solutions but would additionally permit cross-validation of usability evaluations through different laboratories.

## References

Card, S.K.; Moran, T.P. & Newell, A. (1983). The Psychology of Human-Computer Interaction. Hillsdale: Lawrence Erlbaum Associates.

Freitag, R. & Dzida, W. (2002). The Noldus Observer: An action guiding user manual for assisting the usage of NOLDIS Observer in Usability-Laboratories. (in press), pp, 61 – 63.

Hofmann, B.; Freitag, R. & Dzida, W. (2000). Coding Behaviour of Human and Computer in Terms of Intentions. *Measuring Behaviour 2000. Proceedings to the 3$^{rd}$ International Conference on Methods and Techniques in Behavioural Research*, pp. 142 – 143.

ISO 9241 (1998): Ergonomic requirements for office work with visual display terminals (VDTs).

John, B.E. (1995). Why GOMS?. *Interactions*, 2 (4), pp. 80 – 89.

Nielsen, J. (1995). Using a Version of the Keystroke-Level Model. *Interactions*, 2 (4), p. 86.

Payne, S.J. & Green, T.R.G. (1986). Task Action Grammars: A Model of the Mental Representation of Task Languages. *Human-Computer Interaction*, 2, pp. 99 – 135.

Rosson, M.B. & Carrol, J.M. (1995). Narrowing the Specification-Implementation Gap on Scenario-Based Design. In: Carroll, J.M. (ed.). Scenario-Based Design. New York: John Wiley & Sons, pp 247 – 278.

# Communication across the HCI/SE divide: ISO 13407 and the Rational Unified Process®[1]

Bonnie E. John

HCI Institute
Carnegie Mellon
Pittsburgh, PA 15213 USA
bej@cs.cmu.edu

Len Bass

Software Engineering Institute
Carnegie Mellon
Pittsburgh, PA 15213 USA
ljb@sei.cmu.edu

Rob J. Adams

HCI Institute
Carnegie Mellon
Pittsburgh, PA 15213 US
rjadams@cs.cmu.edu

## Abstract

Human-computer interaction practitioners and software engineers must work together to develop useful and usable systems. Although both groups learn techniques for product development activities like requirements generation, high-level and detailed design, and testing, the traditional outputs of these techniques may not make effective contact with the results of the other group or the needs of the project as a whole. We examine the products of different activities in software development to explore whether information flow from one group to the other has natural points of contact or if gaps exist. We use ISO 13407 as an example of a human-centred development process and the Rational Unified Process® as an example of a software engineering process.

## 1    Introduction

Software development processes differ in their formality, rigidity, scale and scope, but each share the goal of meeting the customers' needs at the end of the process. Some processes arise from a software engineering (SE) tradition, like the Rational Unified Process® (RUP®, 2001) or Extreme Programming (Beck, 1999). Others arise from human-centred design (HCD) or human-computer interaction (HCI), like Participatory Design (Muller, et. al. 1997), or the ISO standard 13407. Some processes have clear points of formal communication between the participating SE and HCI practitioners and/or users, and others rely on naturally occurring communication through team membership and collocation.

It is our hypothesis that identified points of communication, and the format of that communication, are critical to the success of multi-disciplinary teams, and that currently this communication is successful mainly through particular personalities rather than codified into the processes. To begin to test that hypothesis, we examine the ISO 13407 and the RUP®, looking for the points of communication mentioned in published summaries of those processes. We are looking for points of communication explicitly mentioned since those not mentioned occur as a result of personal discretion rather than process guided. We also looked for asymmetries in these exemplars of SE and HCI processes, where one side considers an activity and the communication of its results important and the other side doesn't mention it.

## 2    The basis of our comparison of ISO 13407 and RUP®

We chose two documents to compare. The *ISO 13407 Human-centred design processes for interactive systems* (1999) is a 26-page document describing the HCD process. The *Rational*

---

[1] This work supported by the U.S. Department of Defense and NASA

*Unified Process: Best practices for software development teams* (2001) is a 21 page white paper from the Rational Software Corporation. Both documents summarize their processes rather than fully describing them, and both documents make reference to other publications for more detail. Thus, they are similar in length and detail, making it reasonable to use them to look for points of communication, commonalities, and asymmetries.

ISO 13407 identifies four human-centred design activities that should take place during system development: (1) understand and specify the context of use, (2) specify the user and organizational requirements, (3) produce design solutions, and (4) evaluate designs against requirements. Each activity is expanded to include subactivities, outcomes, and some discussion of applicable HCD methods. The standard also includes a sample checklist for assessing whether an organization's development process conforms to this standard. We used this checklist to compare to the RUP® document, augmented by a few items that appear in the main prose but not in the sample checklist.

The RUP® white paper identifies four phases of system development: Inception, Elaboration, Construction and Transition. In addition, it identifies six core engineering workflows (business modeling, requirements, analysis & design, implementation, test, and deployment) and three supporting workflows (project management, configuration and change management, and environment of the development team) that occur during the phases. Each phase is presented with its outcomes and criteria for success. Each workflow gives slightly more detail about the outcomes to which it contributes. The white paper also includes a description of the RUP® Product - a suite of tools that support the workflows. The list of outcomes from each phase, augmented with other artifacts mentioned in the workflows, were used to compare to ISO 13407.

We look first at similarities between the two processes and then at differences where one process mentions activities or artifacts and the other does not.

# 3    Ostensible similarities in ISO 13407 and RUP®

We find several similarities in what the authors of these summaries found important to include: project planning, requirements generation and specification, production of design solutions and iteration, and testing against requirements. However, within some of these similarities lurk subtle differences that may hinder communication between HCI and SE team members.

## 3.1    Project planning

Both ISO 13407 and RUP® talk about project planning. In ISO 13407, the existence of a project plan is the first criteria for conformance to the standard. There must be a list of the HCD activities, assignment of responsibility, suitable timing of the activities, and mechanisms for ensuring that information from HCD activities feed into the system design. In RUP®, a project plan, showing phases and iterations, is an outcome of the inception phase. A more detailed development plan, showing iterations and the criteria for each iteration, and a development case, specifying processes to be used, are outcomes of the elaboration phase. In addition, project management is one of the supporting workflows of RUP® and it is expected to occur about equally in all phases of the development. Thus, both ISO 13407 and RUP® see project planning as an important step. ISO 13407 explicitly recognizes that HCD activities must be coordinated with other development activities, whereas RUP® treats the project as a whole, not mentioning HCD or other activities.

## 3.2    User requirements generation and specification

User requirements generation and specification are important to both ISO 13407 and RUP®. ISO 13407 requires a specification of the range of intended users (e.g., end users, maintainers, installers, etc.), the tasks to be supported, the environment in which the system will be used, and the organizational requirements. RUP® includes several documents that encode such requirements. In the inception phase, four preliminary requirements documents are produced. A vision document includes the core requirements, key features and main constraints. A business case and business model document organizational requirements (e.g., revenue projection, market recognition, etc.). An initial use-case model should be 10-20% complete at the end of the inception phase, and at least 80% complete at the end of the elaboration phase. The elaboration phase also produces a supplementary requirements document that captures any non-functional requirements.

User requirements are documented in use cases in RUP®. They record all the expected uses of the system. Although use cases usually involve exercising core functionality of the system, they could be expanded to record the types of specifications recommended or implied by ISO 13407. Use cases could include all users of the system including installers, maintainers, and help-desk personnel. They could record situations where users make errors, or where the product supports an organizational goal. They could include environmental constraints, like having to recover from interruptions, or survive hazardous physical environments. This may be placing a heavier burden on use cases than RUP® intended, but, in principle, use cases are the most likely vehicle for injecting HCD information into the system design process and maintaining its influence. The supplementary requirements document can be the repository of those usability requirements not easily captured as use cases.

## 3.3 Production of design solutions and iteration

The production of design solutions is, of course, the core of system development and appears in both processes. ISO 13407 talks in terms of prototypes, simulations, models and mockups that can be assessed by users at different times in the design. It emphasizes the sources of good design, including basic literature in psychology, product design, and related fields, standards and style guides, marketing information, and user testing. RUP® lists one or several prototypes as an outcome of the inception phase, an executable architectural prototype as an outcome of the elaboration phase, a software product fully integrated on appropriate platforms as an outcome of the construction phase, and deployable software as an outcome of the construction phase. RUP® does not elaborate on sources of design solutions. Also, RUP® seems to jump from architectural prototype to fully-functional software without discussing detailed design of the interface.

Both processes emphasizes the importance of iteration on design solutions throughout the development process. ISO 13407 speaks exclusively about user testing and iteration of the design based on that testing. Thus, ISO 13407's notion of iteration is primarily changing the interface to the product. RUP® discusses iteration within every phase of the development, and on every artifact. However, RUP®'s notion of iteration is primarily one of successive deepening or coverage; the system "grows incrementally from iteration to iteration". Thus, *iteration* means different things in these processes, leaving the door open for miscommunication between groups.

## 3.4 Testing against requirements

Both ISO 13407 and RUP® test the system against requirements. ISO 13407 differentiates between testing against design standards and user testing to ensure that the system meets the stated usefulness and usability requirements. RUP® tests against all requirements that are recorded in the use cases and supplementary specifications. Assuming all usability requirements have been encoded in those RUP® artifacts, then RUP® encompasses ISO 13407's notion of testing.

Both processes talk about different types of testing. ISO 13407 focuses on user testing in what RUP® would call the elaboration phase, when prototypes are used to establish the detailed design. RUP® focuses on functional and interoperability testing through the elaboration to transition phases, advocating automatic testing where possible. Both processes include field testing - called beta-testing in RUP® - toward the end of the development. ISO 13407 also advocates long-term monitoring of the system's performance and its effects on the workplace.

# 4  Asymmetries between ISO 13407 and RUP®

We also find several differences in what the authors of our two source documents found important enough to include in the summaries. ISO 13407 values design rationale and assessing the competence of staffing and methods used. RUP® values a business case, risk analysis, system architecture, and producing a deployable system. This is not to say that more lengthy descriptions of each process does not include a discussion of what the other process values, or that practitioners do not include these topics in practice; we cannot know from these documents alone. We do believe, however, that these differences are a preliminary indication of a fundamental difference in values between the HCI and the SE communities.

## 4.1  Recording design rationale

ISO 13407 talks about documenting design rationale at many stages. It calls for recording the source of information about context of use, legislative requirements, design standards used and ignored (and why), sources of evaluation feedback, etc. RUP® does not mention design rationale at all.

## 4.2  Assuring competency of staff and proper methods

ISO 13407 calls for assurance that competent staff are employed to carry out the required activities, and that proper methods are used. Although this summary does not explicitly dictate the staff's background and training, nor the specific methods to use, it does insist that an organization record the staffing and the methods. It also specifically requires documentation that user requirements have been confirmed by users. Its insistence on such documentation hints that unqualified staff or improperly chosen or conducted methods have been a problem with implementing HCD in the past. RUP® makes no such checks on staff or methods, assuming, perhaps, that all development teams are qualified to perform all necessary activities. Alternatively, the assessment of whether code fulfils a functional or performance requirement can be more objective than the assessment of whether usability input is valid. Tests of software are constructed to be pass/fail unambiguously. Invalid user data or analyses are not obvious on inspection and can lead the project in a wrong direction, thereby motivating the competency check.

## 4.3  Business case & risk analysis

RUP® includes an initial business case and risk analysis as outcomes of the inception phase and a revision of both as outcomes of the elaboration phase. ISO 13407 does not mention such documents. ISO 13407 seems to be smaller in scope than RUP®, not extending to business concerns. Risk analysis associated with user requirements is an important omission.

## 4.4 System architecture

RUP® lists a software architecture description as an outcome of the elaboration phase, but ISO 13407 does not mention software architecture. We have discussed elsewhere (Bass & John, 2001) that usability requirements have an impact on architecture design and that architectural decisions can preclude delivery of a usable system. The disregard for architecture in ISO 13407 exacerbates the problem by ignoring the interdependencies between HCD requirements and software architecture. We believe the inclusion of architecturally sensitive usability scenarios as use cases could integrate these concerns into RUP®.

## 4.5 Deliverable system

Surprisingly, only RUP® seems to acknowledge the need to deliver a fully running system. ISO 13407 discusses constructing prototypes and mock-ups, but only implies the construction of the deployable system by reference to field tests and long-term monitoring. There is considerable development effort in going from prototypes to deployment, and failure to discuss that effort and how it interacts with iterative design, is a shortcoming of the ISO process.

## 5 Summary and preliminary conclusions

The result of our preliminary analysis of an HCD and SE development process has revealed that there are both considerable commonalities and major asymmetries. Probing the similarities slightly, we find that the two can fit together, but only if assumptions are made about the content of "requirements", "design", and "testing". We also find differences in terminology that may contribute to difficulties in communication. The asymmetries highlight differences in assumptions about staff and methods, and gaps in the scope of HCD. This preliminary analysis gives us more focused hypotheses, which need to be checked against more detailed descriptions of these processes, against other documented processes, and against the processes as conducted in practice.

## 6 References

Bass, L. J. & John, B. E. (2001) Supporting usability through software architecture. *IEEE Computer*, 113-115.

Beck, K. (1999) *Extreme Programming Explained: Embrace Change.* Boston: Addison Wesley.

ISO 13407:1999, *Human-centred design processes for interactive system teams.*

Muller, M. J., Haslwanter, J.H., Dayton, T. (1997). Participatory practices in the software lifecycle. In M. Helander, T. K. Landauer, & P. V. Prabhu (Eds.), *Handbook of Human-Computer Interaction* (pp. 255-313). Amsterdam: Elsevier Science.

*Rational Unified Process: Best practices for software development teams.* (2001, Nov.) Retrieved 20 January 2003, from http://www.rational.com/media/whitepapers/rup_bestpractices.pdf

# Systematic Determination of Quantitative Usability Requirements

*Timo Jokela*

University of Oulu
P.O.Box 3000, FIN-90014 Oulu,
Finland
timo.jokela@oulu.fi

*Netta Iivari*

University of Oulu
P.O.Box 3000, FIN-90014 Oulu,
Finland
netta.iivari@oulu.fi

## Abstract

Quantitative usability requirements are often perceived difficult to determine. We propose a step-by-step process for determining quantitative usability requirements, developed in a series of field studies. The process was perceived useful and it helped the project teams get committed to usability activities. On the other hand, determining usability requirements was found such a complex task that further work is required for gaining concrete guidance to all steps of the process.

## 1  Introduction

A typical project management practice is to use quantitative requirements for product quality in project development projects. Quantitative, measurable quality requirements provide a clear direction of work and acceptance criteria for a development project. Usability requirements, however, are quite seldom among those quantitative requirements. One of the consequences of not defining usability requirements is that other objectives dominate and usability is considered only as a secondary objective of a project. The obvious consequence is a product with usability problems.

Many well-known usability books, such as (Nielsen 1993), (Hackos & Redish 1998)and (Rosson & Carroll 2002) address the issue of determining quantitative usability requirements. The topic is discussed by an article (Wixon & Wilson 1997) and case studies presented in a research report of the European PRUE project (Bevan et al. 2002). Related standards are ISO 9241, (ISO/IEC 1998) ISO 13407 (ISO/IEC 1999) and ISO 9126 (ISO/IEC 2000).

Literature, however, mainly focuses on presenting the characteristics of usability requirements. There do not exist many practical guidelines on how to derive the 'right set' of usability requirements; i.e. a set of requirements that really depict a usable product. Wixon & Wilson define a six-step process. The standard ISO 13407 identifies a process 'Specify the user and organizational requirements'. These guidelines, however, are at rather high level of abstraction. Context of use analysis (Thomas & Bevan 1996) is a stepwise method. It, however, is perhaps even too detailed to be practical. A practitioner needs concrete but practical guidelines.

The objective of this study was to learn how usability requirements could be practically determined with a systematic, stepwise process. We carried out a number of field studies in development projects. After each study, we gathered feedback from the project team. In the subsequent studies, we tried to do the things better, based on the lessons learnt.

We chose the definition of usability of ISO 9241-11(ISO/IEC 1998) as a basis for our research. According to ISO 9241-11, usability is "the extent to which a product can be used by specified users to achieve specified goals with effectiveness, efficiency and satisfaction in a specified context of use". Our strategy was to carry systematically out context of use analysis at a practical level of detail and specify quantitative usability requirements related to specific users and their specified goals.

In the next section, we describe the process that is the result of the field studies. In the last section, we draw conclusions and discuss the results.

## 2    The Steps of the Process

We determine quantitative usability requirements in workshops that last from one to two days. We form a set of working teams – each 3 to 4 persons – from the project staff, and use various teamwork techniques, such as post-it notes, board walking, and voting.

### 2.1    Objectives of the workshop

We brief the participants about the objectives and contents of the workshop. We want to make everyone understand what product or system is the object of the workshop. Unlike some other approaches, e.g. the stakeholder meeting (UsabilityNet 2003), we do not try to determine the business success factors of the project at this stage. Vice versa, we find that usability related business success factors can be determined based on the results of the workshop.

### 2.2    Identification of the user groups

We identify user groups by the *job role* of the users, using brainstorming. We do not try to achieve the final truth at this stage – our experience is that the following steps of the process clarify what is the appropriate set of users.

To keep to process manageable, we do not identify the user groups based on the experience of the users (novices, intermediate, experts etc.) nor on the cultural aspects (international users). These issues are relevant, but they can be taken into account later when usability evaluations are planned.

### 2.3    Prioritising of the user groups

We prioritize the user groups in order to determine the working order. If there are very many user groups, another workshop is possibly required to analyze the lower priority user groups.

### 2.4    Identification of the user accomplishments

We start with the most important user groups, and assign one user group for each working team. Then we ask the teams to identify the different accomplishments (goals) that users would want to gain with the product under development.

We have found this to be one of the most challenging parts of the process to the designers. It seems to be so much easier to think of tasks than the accomplishments. In the end, however, the participants usually find this kind of thinking most useful. - The facilitator's role at this stage is critical. The facilitator should challenge the teams and constructively 'not accept' too simple answers, such as "There is an only one user accomplishment: testing".

We have often found necessary to go back and check whether refinements in the set of user groups are required. When determining accomplishments, one often realizes that the original set of user groups was not exactly the right one.

## 2.5    Prioritising accomplishments

We ask the teams to prioritise the accomplishments based on a list of task attributes (frequency, time criticality, error criticality).

This stage, again, may lead to iteration. One is often able to refine the set of user accomplishments when the task attributes are brainstormed. For example, one may realize at there are different accomplishments related to 'testing' if one kind of testing is done frequently while another kind of testing is carried out only quite seldom.

## 2.6    Identifying critical accomplishments

A user-task matrix - such as proposed e.g. in (Hackos and Redish 1998) - is created. It shows whether many user groups, or one user group only perform a task. All the accomplishments of all the user groups are consolidated into a single table.

We ask the teams to check the priorities of the accomplishments of all the user groups. This outcome, the priorities of all accomplishments of all user groups, is a very central result of the workshop. We now know which accomplishments by which user groups form the basis of the criteria for usability of the product.

## 2.7    Consolidating requirements

At this phase, we may have quite a large number of accomplishments that may be  prioritised critical. While too many usability goals are impractical, one should plan how to consolidate those accomplishments into a reasonable number of usability requirements. Different approaches may be used. For example, an approach where the criteria are determined with the average performance of tasks is proposed in (Jokela and Pirkola 1999).

## 2.8    Qualitative descriptions of requirements

We distribute the most important accomplishments to the working teams, and ask them to define descriptive statements about how the requirements are successfully achieved. The statements should reflect the critical attributes of the accomplishments. For example, a task that is performed frequently, would probably lead to a statement that reads 'quickly'. Another example: "This task is a 'one-shot' tasks and it is utmost important that the outcome of the task is correct every time. On the other hand, the task is not very time-critical".

## 2.9    Quantitative measures for the requirements

The qualitative goals are transformed into quantitative ones. This is another step that is typically perceived challenging. Generally, we recommend setting goals in relation to the old version of the product (or a competitive product). This is where we have often found lack of information: the members of the project team do not know about the performance of the existing product.

## 2.10 Final reality check

When the final set of goals exists, we take a step backwards and make a final reality check. We especially remind that much of the results of this kind of work are based on the knowledge of the participants. In this case one should consider to which extent do we really know the world of user – how valid is the data that we derived?

## 2.11 Summary

The steps, with outcomes and comments, are summarised in Table 1.

**Table 1:** Summary of the Steps

| No | Step | Outcome | Comment |
|----|------|---------|---------|
| 1 | Workshop objectives | Definition of the scope of the workshop (what is the product) | The project manager has a critical role |
| 2 | Identification of user groups | A set of user groups: names, brief descriptions | Identified through job role |
| 3 | Prioritizing user groups | An ordered list of user groups | Based on the size or criticality of the user group |
| 4 | Identification of user accomplishments | A set of user accomplishments per user group  Refinements in the set of user groups (typically more that originally) | Describe the accomplishment (not the performance). Perceived challenging but useful. The role of the facilitator critical. |
| 5 | Prioritizing accomplishments | An ordered list of accomplishments  Refinements in the set of accomplishments (typically more than identified originally) | Frequency, time criticality, error criticality used as guiding factors. |
| 6 | Identifying critical accomplishments | A list of critical accomplishments of all user groups | A core intermediate result |
| 7 | Consolidating requirements | Baseline for goal setting | Challenge to cope if a large number of critical accomplishments |
| 8 | Qualitative descriptions | Qualitative descriptions of achieving goals successfully | Describe successful performance, not the accomplishment |
| 9 | Quantitative measures | Transform qualitative goals into quantitative ones. | Challenging to determine the measures. Even more challenging to determine the 'right' requirement values. |
| 10 | Final reality check | Potentially refinements in the quantitative goals. | Taking a step backwards, taking an overview of the results. |

## 3    Conclusion and Discussion

Quantitative usability requirements are often perceived difficult to determine. We propose a step-by-step process for determining quantitative usability requirements, developed in a series of field studies. The process was perceived useful and helped the project team get committed to usability activities.

On the other hand, we found that the process cannot be applied mechanically. Especially with systems with many different user groups and a large number of tasks, one has to tackle with 'space explosion' all the time (the number of items, and their combinations becomes so large). Prioritizing of different issues – e.g. the different user groups - is therefore essential throughout the process.

In most of the cases, we were able to carry the process through (i.e. we were able to determine a set of quantitative usability requirements). In some cases we were able to cover the requirements only partially: we did not have time in the given time frame to process those issues that we prioritized less important.

We found iteration - one principle of user-centred design – be a natural part of the process. For example, when working with user goals, we often found it sensible to take a step back and redefine the set of user groups. Another finding was that the role of the facilitator was a critical one; she/he needed to guide the participants towards the 'right track'.

We conclude that determining quantitative usability requirements based on a systematic context of use analysis is a useful but not a simple activity. Many early phases of usability engineering can be efficiently processed in the workshops. The process is effective training of usability issues to the designers, and it helps the development team to focus on the relevant usability issues. The challenge is in the management of complexity where further research is needed.

# 4   References

Bevan, N., N. Claridge, et al. (2002). Guide to specifying and evaluating usability as part of a contract, version1.0. PRUE project. London, Serco Usability Services: 47.

Hackos, J. T. and J. C. Redish (1998). User and Task Analysis for Interface Design, Wiley Computer Publishing.

ISO/IEC (1998). 9241-11 Ergonomic requirements for office work with visual display terminals (VDT)s - Part 11 Guidance on usability. ISO/IEC 9241-11: 1998 (E).

ISO/IEC (1999). 13407 Human-Centred Design Processes for Interactive Systems. ISO/IEC 13407: 1999 (E).

ISO/IEC (2000). 9126 Software Product Quality - Quality Model. ISO/IEC 9126: 2000 (E).

Jokela, T. and J. Pirkola (1999). Using Quantitative Usability Goals in the Design of a User Interface for Cellular Phones. INTERACT '99 (Volume II), Edinborough, UK, British Computer Society, Wiltshire, UK.

Nielsen, J. (1993). Usability Engineering. San Diego, Academic Press, Inc.

Rosson, M. B. and J. M. Carroll (2002). Usability Engineering. Scenario-Based Development of Human-Computer Interaction, Morgan Kaufmann Publishers.

Thomas, C. and N. Bevan (1996). Usability Context Analysis: A Practical Guide. Version 4.04. Teddington, National Physical Laboratory.

UsabilityNet (2003). Stakeholder meeting, Retrieved February 10, 2003, from www.usabilitynet.org

Wixon, D. and C. Wilson (1997). The Usability Engineering Framework for Product Design and Evaluation. Handbook of Human-Computer Interaction. M. Helander, T. Landauer and P. Prabhu. Amsterdam, Elsevier Science B.V: 653-688.

# Diary as a Usability Testing Method

*Kangas Eeva*

Digia
Elektroniikkatie 2,
90570 Oulu, Finland
eeva.kangas@digia.com

*Sinisammal Janne*

University of Oulu,
P.O. Box 4300, 90014
University of Oulu,
Finland
janne.Sinisammal@oulu.fi

*Paihonen Sami*

Nokia
Elektroniikkatie 13,
90571 Oulu, Finland
sami.paihonen@nokia.com

## Abstract

A diary is useful as a usability method when "traditional" usability tests conducted in a laboratory are not sufficient. The diary method is especially suitable for studying mobile devices, as it makes it possible to test usability in a natural context rather than in a laboratory environment. Different variations of diary studies can give much important information for developing a product. The importance of motivation should not be underestimated at any stage of the study. The form of the diary also requires careful planning. It is necessary to visit users to help maintain and bolster their efforts in completing the diary. When the test period is over, interviews and questionnaires give more information on the study.

## 1   Introduction

It is not sufficient to deliver products, which only have technical excellence – products should be made for the users. There are four user-centred activities that need to be taken into account in all stages of product development. These are: understanding and specifying the context of use, specifying the user and organizational requirements, producing designs solutions and evaluating the design against requirements (Bevan, 1999). The characteristics of the context (users, tasks and environment) may be as important in determining usability as the characteristics of the product itself. Changing any relevant aspect of the context of use may change the usability of the product. When usability is measured, it is important that the conditions for a test represent important aspects of the overall context of use (Bevan, 1994).

Several studies indicate the effectiveness of user testing in developing systems. This does not, however, mean that every kind of user testing fits every situation equally well. Hertzum (1999) found out that 76% of laboratory test results concerned the usability of the system, but field tests uncovered a broader mix of problems. Laboratory tests were biased toward how tasks were performed with the system, at the expense of what tasks could be performed. That is why laboratory tests are less suitable early in the development process, where the applicability of the system, i.e., what the system can do, was the major design concern.

With "traditional" usability testing it is almost impossible to conduct a test that reveals such issues as efficiency, context of real use and pleasantness of use over the long term.  Efficiency in particular is often a relatively hard issue to tackle.

## 2 The diary method

Rieman (1993) presents the diary study method as a middle-ground solution to the opposing limitations of laboratory studies and field studies. Diary studies can impose useful experimental constraints while maintaining ecological validity because they are conducted in natural settings but retain some level of researcher control. A diary is useful when studying user behaviour over a period of time. Diaries may vary from open-ended to highly structured tick-box forms. In open-ended diaries users comment with their own words and in tick-box diaries users answer simple multiple choice or yes/no questions. The results of diary studies depend on the context of the diary, varying from qualitative data on videotape to quantitative data on a questionnaire. Because the content of a diary can vary considerably, the method can be used in different phases of product development.

As the diary is always with the user, he can write ad-hoc comments about the product. If the users don't keep a diary after using the product, many important things are forgotten. The diary can contain questions like "Was there something you didn't like?" "Was it easy to use this feature?" "Did you have any problems today?" These kinds of questions make it easier to remember what happened when using the product. If they are not asked, users will not write them down and later they will not remember the situation correctly.

It is important that the users also have an opportunity to contact the company during the test period. Personal interactions are a key part of a successful study. Participants must be convinced to make a considerable effort to record their activities over the time-span of the diary (Rieman, 1993). Co-operation is easier if the contact person knows all the test users.

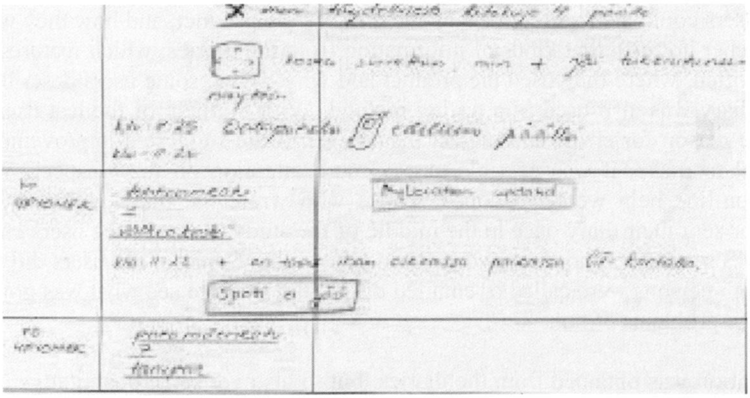

**Figure 1: It's easy to draw and use different colours when writing a diary by hand.**

With paper diaries (see Figure 1), users must stop their activity and manually record it; this is not as troublesome when users work at a desk and the activities of interest occur there, but it becomes problematic when the participants are mobile. Although using a mobile phone to call into a voice-mail line also means the subjects must stop their activity, they are often able to suspend and resume activities more quickly when using a device they would likely carry anyway, and often provide a richer description than if they were to make notes on paper (Palen et al, 2002).

# 3    Using the diary method: Genimap®Navigator study

Genimap®Navigator is an application developed for Nokia 9200 Communicator Series users. With Genimap®Navigator users can load maps into their Communicator and find addresses and services there. They can also use GPS when navigating with the product. With this kind of application, using only laboratory tests is not enough.

When developing Genimap®Navigator, a diary study was conducted after paper prototype tests to get more information from the users. Twenty users were asked to use Genimap®Navigator and keep a diary of their use for three weeks. Fourteen of the test users lived in Helsinki and six in Oulu. Eighteen of the test users were men (90%) and two were women. Their average age was 34 years. The oldest was 52 years old and youngest was 27. The user sample was gathered so that it included both novice and expert users related to map usage, GPS know-how and Communicator usage.

Before the study, everyone was asked to come to a meeting, where the users were told what they should write in their diaries and where they could also start using the Genimap®Navigator. They also met the contact person, whom they could contact whenever they needed during the study. Thirteen of the users had their own Communicator, and seven were given a Communicator for the study. Twelve of the users also received a GPS device.

The users had a choice of writing their diaries by pen, with a PC or with the Communicator. Only one of the users wrote his diary with the Communicator, one didn't write the diary at all, nine wrote it with a PC and nine used a pen and paper. In the diary they wrote the time when they used the application, what information they tried to get from it and what kinds of problems they had with it. The users could also write how they liked to use the product, and how they would improve it. The researcher got different kinds of information from the diaries: which features the users did use and how often, where they used the product and why. When some users described errors step by step, the diary was also used as a testing method. With the help of the test users, developers could find the reason for errors and correct them. Even if the suggested improvements proved to be impractical to make, it was possible to pay more attention on these aspects when the user manual and on-line help were developed. Those who wrote the diary with a PC or with the Communicator sent their diary once in the middle of the study. Some of the users emailed, called, or sent an SMS message almost every day during the study. Some of the users did not contact at all. The "silent" persons were called or emailed during the study to see what was going on and did they have some problems or not.

Much information was obtained from the diaries, but to also get some quantitative data, the users were also interviewed after the study, and 13 users filled in the system usability scale questionnaire (SUS). The users returned the Communicators, GPS devices and diaries at the same final meeting.

There was considerable variation in how much the users wrote in their diary. If a user did not write in the diary frequently, he forgot which kinds of problems he had with the product. He was able to remember the problem, but did not remember what he did before the problem occurred. Although interview data alone is relatively soft and unreliable, the participant's statements are "hardened" by relating them to the diary records (Rieman, 1993). For example, if an interviewee says, "I used the possibility to navigate with the GPS a lot" and a three-week diary confirms this, the evidence is much stronger than an interview statement without further support.

# 4    What to remember when conducting a diary study?

Certain issues have been identified when conducting a diary study as a usability testing method. First, the diary requires a strong commitment from the test users. It is very easy to neglect filling out the diary Second, when designing a diary-based test case, the researcher should remember that completing a diary always takes time, even though the event to be reported might not be of great importance.

Before doing a diary study, there are some things to remember. First, decide on what sort of diary should be used. Free-form diaries allow persons to express themselves freely (See Figure 2). Structured diaries with fixed response formats are easier to analyse. If a structured method is used, a careful selection of question and response categories must be produced, and it should be decided how often the respondents should make entries in the diary. This will be determined by the nature of the data that needs to be captured and the tasks being carried out.

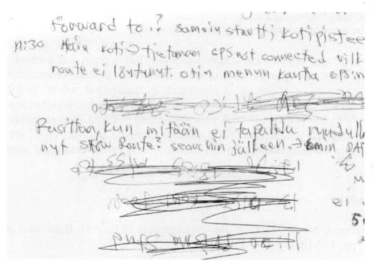

**Figure 2: Hand-written diaries are sometimes hard to read.**

Semi- or highly structured diaries are useful for a larger sample population. Instructions on keeping the diary have to be clear enough so the test user does not need to think about the test itself, but can focus on the real issue, filling out the diary.

Before the study it is good to provide a means (e.g., a telephone number) whereby the respondent can ask for help and the researcher can check on the diary keeping procedure. Co-operation between the test users and the researcher is very important. The test users should know what they should do, when and why. If the testing period is long, the users should be contacted regularly by phone or by email to help them overcome potential problems encountered in completing the diaries or using the product. Contacting the users also provides a valuable opportunity to reflect and empower them to carry on with the task

Users may forget to fill out their diary or if insufficient instructions are given, may fail to fill it out properly. Different users can produce totally different information due to the open nature of the data collection. In addition, people are quite poor at tracings their own actions; it's just not their normal, daily behaviour. Users are also quite quick in adapting to coming situations, this adaptation sometimes happens in just a fraction of a second, so it's quite hard to identify and even harder to remember this kind of "error cases".

When the test period is over, interviewing the users after the test is a necessary task. This can be called a walk through the diary, where the test users are asked to go through the events described in the diary and explains them in their own words. These sessions can be video-recorded or tape-recorded, so that researchers can analyse the session afterwards. In this interview session it is always important to let the test user proceed at the speed he wants when going through the data in the diary. The interviewer needs to ensure that the interviewee keeps focused on the issue. Sometimes, a group discussion is held after the test period so the test users can share their experiences of keeping the diary.

The importance of motivation cannot be underestimated at any stage of the test; even after the test. When setting up the test, it should be always emphasized how important the test is and that the test users are helping the company to create a more user-friendly product. Normally there is some kind of reward for participating in a test.

Finally, diary keeping is still more a research method than a plain test method. It can provide useful data about the way of using certain devices without the researcher interrupting the user in the middle of an event or action. It is important to understand where in the research and development (R&D) cycle this test is conducted, to which problems or ideas it is searching for answers. If diary keeping is used to check how something is being used (utilization rate, way of using), these test findings can provide useful data for immediate correction. If the R&D is in the visioning stage, then these findings are used for validating or correcting the strategy towards a certain vision.

# 5 References

Bevan, N. & Macleod, M. (1994), Usability measurement in context, Behaviour and Information Technology, 13, pp. 132-145.

Bevan, N. (1999), Quality in use: meeting user needs for quality, Journal of systems and software.

Hertzum, M. (1999), User testing in industry: a case study of laboratory, workshop, and field tests.

Palen, L. & Salzman, M. (2002), Voice-mail diaries for naturalistic data capture under Mobile conditions, Proceedings of the ACM conference on computer supported cooperative work, CSCW '02, New Orleans, LA.

Rieman, J. (1993), The Diary Study: A Workplace-Oriented Research Tool to Guide Laboratory Efforts, in S. Ashlund, K. Mullet, A. Henderson, E. Hollnagel, & T. White (eds.), Proceedings of INTERCHI'93 Conference on Human Factors in Computing Systems, ACM Press, New York, pp. 321-326.

# Usability of Ergonomics Softwares in the Design Process

*Sultan Kaygin*

Middle East Tech. Univ.
Ankara, Turkey
kaygin@metu.edu.tr

*Cigdem Erbug*

Middle East Tech. Univ.
Ankara, Turkey
erbug@metu.edu.tr

*Murat Alibaba*

Middle East Tech. Univ.
Ankara, Turkey
alibaba.m@lycos.com

## Abstract

This paper will discuss the integration of ergonomics knowledge into the design education by using ergonomics softwares. Research is mainly focused on designers' information requirements, ergonomics education and usability of ergonomics softwares.

A test on two ergonomics softwares, namely, MannequinPro and ErgoEaser will be presented, to make comparisons that highlight the positive and negative aspects of each program and to state their benefits and deficiencies in design process.

## 1  Introduction

Design students might be successful in theoretical courses but the integration of this knowledge into their product designs is missing. For example, students do not consider the usability or safety of the product as much as the aesthetical or functional aspects. In this context, there is a continuous debate between the educators to find solutions for the integration of the theoretical knowledge (ergonomics, structure, materials, etc.) to the students' design process. Among these courses the integration of ergonomics knowledge into the design process is the main issue of this paper.

Ergonomics information and education might be given by computers as well as by human teachers. It is evident that computers are not substitutes for ergonomics specialists but on the other hand, there is a need for using computer techniques to retrieve quick and easy information about the specific problems of design. To deny the human teacher, and to use computers for tutoring is not the main objective of computer usage in ergonomics. However, the computers are mostly suggested as aids to the students or to the ergonomics specialists. In this respect, computers are used as tools to apply simulation techniques, for data analysis, to have accesses to certain data banks, etc. (Erbug, 1999)

## 2    Design Process and Ergonomics

Nickerson (1999, p.609) pointed out that, "...designers are not full-fledged human factors practitioners. They are team leaders who must consider many additional factors during the design process". As it is also stated by different authors (Feyen, Liu, Chaffin, Jimmerson & Joseph, 1998, p.1), "Designers often have difficulty in incorporating ergonomics information about the human operator into their designs because of usability problems". And Pheasant (1986) has shown that, "if ergonomics information is not easily available, designers will base their design on themselves and their own experience". (as cited in Woodcock & Flyte, 1997, p.124)

Ergonomics should be made more concrete for designers and more applicable to the design process. Such an outlook can be achieved by educating designers on the usage of ergonomics data by explaining the role of ergonomics in design. The designer should be able to identify strategic stages where ergonomics expertise or ergonomics data is required without interfering with the cumulative strategy for design. The designer, being aware of the available ergonomics data and the appropriate stages to use these data, will be able to manipulate some of the constraints during the design process" (Erbug, 1999).

# 3 Ergonomics Software Tools

Ergonomic software tools are expected to be used as design aids. Deitz (1995) states, "When you simulate a precise human model moving within a solid model of the product, you see ways of solving the design problem that you just wouldn't think of if you looked at a table full of numbers". Research has also shown that "More and more people are looking for human-factors simulations, not just the human factors experts" (Robertson, B., 1997, p.34). The advantages of using computer techniques for the specific problems of design are:

1. Shorten the time required to retrieve specific information.
2. Shorten the time required for simulations.
3. Motivate designers to apply their ergonomics skills more broadly.
4. Save time during design.
5. Allow time for more sophisticated discussions.

The survey performed by the authors to discover the current ergonomics softwares available for the purposes of product design covers the programs such as: MannequinPRO, MannequinOnsite, ErgoEaser,ErgoMaster,ErgoIntelligence, Transom Jack, MDHMS, Safework, Combiman, Ramsis, People Size 2000 and WIN EDS. Unfortunately, the survey is limited to the world wide web pages of the programs as the original versions - except PeopleSize 2000 - were not available. These programs were limited full-versions for 30-day trial. Among these programs it is decided to exemplify one 2D and one 3D program. Comparison tables were prepared to analyze the contents of the programs and some of the programs were eliminated and one is left for each category: MannequinPro and ErgoEaser . MannequinPro represented 3D programs ErgoEaser 2D. The main differences of these two softwares are presented in Table 1.

**Table 1:** Main differences of ErgoEaser and MannequinPro

| ERGOEASER | MANNEQUINPRO |
|---|---|
| It is a PC based 2D program. | It is a PC based 3D program. |
| It provides training and analysis tools on potential ergonomic hazards. | It provides human body dynamics in the design of a wide range of products, equipment and facilities. |
| It does not have drawing or modelling capabilities. | It has 2D and 3D drawing and editing tools, 3D primitive solid creation extrusion and revolution of 2D objects into 3D abilities. |
| It seems to be useful mostly in the early stages of design by providing anthropometrics information. | It looks more useful in the further stages of the design process while testing the object. |

# 4 The Test

The aim of the test was to analyse the expectations of Industrial Design students from ergonomics softwares and to track the usability problems. 24 test participants were chosen among 3$^{rd}$ year, 4$^{th}$

year and graduate students from the department of Industrial Design, at METU. They have taken an ergonomics course in the second year of Industrial Design education in which they were introduced the basics of the discipline. They also have the necessary computer and modeling software knowledge. The test consisted of 4 main parts: pre-test interview, software demonstration, task observation, post-test interview.

### 4.1.1.  Pre-test Interview

The aim of this interview was to obtain information about the users' level of awareness in integrating ergonomics into design process and to understand the process they go through while designing. Before the sample user was introduced to the software, he/she was asked some questions mainly focusing on his/her expectations from an ergonomics program.

It is seen that none of the sample users had used an ergonomics software before. 54% of them never heard of such programs while 25% declared that they did not feel the need to use it. They are asked about the software they used at present to complete their projects. Nearly all of them (91%) use Rhinoceros, 3D Studio Max, Photoshop, and Poser during the design process. They expected an ergonomics software to provide anthropometrics data (databases) for different user groups. 21% of the users expected an ergonomics software to be 3D so that he/she could import his/her model into the program and "check whether it is ergonomic or not…".

### 4.1.2.  Software demonstration

Demonstration of sequential screenshots for each program is prepared by using Macromedia Flash. The aim of this was to give clues of the interface of each program and show very briefly how it worked. In this way students became familiar with the contents of the two programs.

### 4.1.3.  Task observation

Tasks were designed separately for each program since the programs operate different in nature. While the test participant was performing the task, the major problems observed and elapsed time for each task are recorded. A VDT Workstation (visual display terminal) analysis is carried out by the test participants by giving a starting file for each software.

### 4.1.4.  Post-test Interview

After the tasks were completed, a post-test interview is carried with the participants to investigate the first impressions of the software on them, difficulties faced in each program and the expectations raised.

## 5.  Observations & Recommendations

After using these softwares, most of the test participants stated that these softwares could be very useful for them if they were easier to use. The problems detected in analysis are presented in Table 2.

**Table 2:** Main problems detected in the analysis

**Table 2:** Main problems detected in the analysis

| | | ERGOEASER | MANNEQUINPRO |
|---|---|---|---|
| **ANALOGY** | Compatibility with other programs | It does not have import or export options. | It has a common file format. It allows exporting the model that is created to other softwares. |
| | Flexibility | It consists of three interactive components. It enables users to input key dimensions and variables then adjust the values and view results in a simulated display. However it does not have a flexibility of allowing user to make analysis different from the ones offered by the program. | It allows the user to create his/her own model in the program or import a model (human, object or environment) from any compatible modelling program. It gives flexibility to the user. |
| **FEEDBACK** | Feedback | Most of the users couldn't realize that the red crosses on the interactive schema gave feedback of the problem occurred and offered a solution for that problem when clicked on. Some of the users who read the offered solution still could not understand how to solve the problem. | The icons on the toolbar are not identifiable. When the mouse is kept on the icon, it is expected to give popup explanation but it does not. There is no feedback when command buttons are pressed and activated. |
| **VISUALISATION** | Graphic quality | It seems more usable than MannequinPro in terms of interface however it still has the potential to be improved. | The interface graphically needs to be improved. It has very different and unique icons and commands. |
| | | Only a few participants could realise (some of them unconsciously) that an automatic "make adjustments" button exists and fixes all the problems at once. Most of the participants couldn't realize the "hints" button. They are both weak to tell their function. | The command line at the bottom of the screen, which could be very helpful to the user, is not recognizable. |
| | Ease of use | ErgoEaser employs a graphical user interface that is easy to learn and use. | Its analytical tools are quite sophisticated for a new user. |
| | | Most people click on the 2D representational drawing to change the measurements of a specific part. | Finding the icons of certain commands (move, rotate, insert mannequin, reach) is a hard issue. Especially rotating precisely (90 degrees etc...) is difficult. |
| | | Nearly all of the users could not realize for a long time that other menu items (job, chair, inputs and display) exist. They got stuck on human. | The users got stuck on a command. They hit "ESC" to exit the command, but it did not work. "Undo" command does not work for more than a single step. Most of the users restarted their task. |

Participants commented that if ErgoEaser was more flexible in terms of setting up your own environment and importing your own product, it would be a very useful program for designers. MannequinPro, although providing this flexibility had some usability problems. Though it has similarities with other 3D softwares designers use for modelling and rendering purposes and shares mainly the same logic, it has very different and unique icons and commands. Its property of having the ability of importing your own drawing into MannequinPro was found to be the most positive aspect of the software.

ErgoEaser has the ability of providing interactive anthropometric data of various body parts and teaching the subject matter to the user by providing figures showing body part and its technical name in the literature (e.g. popliteal height). It also gives information about possible strains the human body will face under the circumstances given.
MannequinPro is more flexible and seems more useful in product development, whereas ErgoEaser seems to be useful mostly in the early stages of design by providing information on anthropometry. Therefore, it seems that a combination of these programs can be an effective tool to use in different stages of the design process.

## 5.1.Conclusion

2D and 3D computer models are being evaluated for some years but they still need further development. The observations and interviews revealed that designers need flexible and usable softwares which can be used in different stages of design. The reason is that ergonomics information required for design is dynamic, rather than static. It involves morphological study, criticism of existing products, analysis of standards and regulations, accident analysis, user profile, anthropometric study, biomechanical study, and cognitive study (Erbug, 1999). The softwares analysed in this study doesn't go directly from concept to final design in the same environment. Designers have to use separate tools for different analysis. In addition they don't use the same visual language for their icons, menu, etc.

As it is also stated by Robertson (1997), one of the major problems in ergonomics softwares is that you can carry any kind of analysis with these softwares in theory, but in real you can not, since they are typically designed with specific tasks in mind. They are not flexible enough to be used at different stages of design process and/or for specific requirements of the designer. In addition, they do not provide an efficient usage in conjunction with CAD systems. In this regard, they are neither very successful in motivating designers to apply their ergonomics skills more broadly, nor time saving.

## References

Deitz, D. (1995). Human-integrated design. *Mechanical Engineering,* 117 (8), 92-97
Erbug, C. (1999). Use of computers to teach ergonomics to designers (CD ROM). In Mondelo, M. Mattila & W. Karwowski (Eds.) *International Conference on Computer-aided Ergonomics & Safety*. Barcelona: Universitat Politecnica de Catalunya.
Feyen, R., Liu, Y., Chaffin, D., Jimmerson, G. & Joseph, B. (1999). Computer-aided ergonomics: a case study of incorporating analysis into workplace design. *Applied Ergonomics* 31, 291-300.
Nickerson, H. (1999). A framework for human factors education in the design disciplines. *Proceedings of The Second International Cyberspace Conference on Ergonomics.* (pp.609-620) Perth, Australia: Curtin University of Technology.
Pheasant, S. (1986). Bodyspace. Anthropometry, ergonomics and Design. London: Taylor and Francis.
Robertson, B. (July 1997). Virtual humans at work. *Computer Graphics World,* 20 (7), 33-39
Woodcock, A. & Galer Flyte, M., D. (1997). The development of computer based tools to support the use of ergonomics in automotive design. *Proceedings of 13th triennial congress of the international ergonomics association* (pp. 124-126)

# Why Can't Software Engineers and HCI Practitioners Work Together?

*Rick Kazman*

Software Engineering Institute,
Carnegie Mellon University and
ITM Department,
University of Hawaii
kazman@sei.cmu.edu

*Junius Gunaratne, Bill Jerome*

Human Computer Interaction Institute
Carnegie Mellon University
{jgunarat, wjj}@cs.cmu.edu

## Abstract

In this paper, we examine the state of Software Engineering and Human-Computer Interaction in practice by presenting the results of a survey that investigates how HCI practitioners and SEs interact. The main findings of the survey are disturbing: there is a substantial lack of mutual understanding among SEs and HCI specialists. Furthermore, there appear to be important differences in how SEs and HCI practitioners view their interaction in the software engineering life cycle. The final, and perhaps most serious, finding of this paper is that SEs and HCI practitioners tend to interact and communicate with each other late in the software life cycle; too late to fix the most fundamental usability problems.

## 1   Introduction

Almost half of the software in systems being developed today and thirty-seven to fifty percent of efforts throughout the software life cycle are related to the system's user interface (Myers & Rosson, 1992). For this reason the issues, methods and practices from the field of human-computer interaction (HCI) affect the overall process of software engineering (SE) tremendously. Given the enormous efforts that go in to implementing the user interface portion of a system, one would naturally assume that SEs (hereafter SEs)—typically in charge of the overall system development—would need to work closely with and interact early and often with HCI experts. One would be wrong in this assumption. Despite the need to practice and apply effective HCI methods and integrate these smoothly into the overall product development life cycle, there still exist major gaps of understanding between suggested practice, primarily coming from academic communities, and how software is developed in industry. And, not surprisingly, there are major gaps of communication between HCI and SE groups within software development organizations.

More specifically, the application of HCI methods continues to be an afterthought in the development of software despite many suggested practices of tightly intertwining HCI methods with software development processes (Artim, 1998; Hefley et al, 1994; Jambon et al, 2001; Metzker & Offergeld, 2001). There is an apparent lack of overlap between current SE processes and HCI methods, which cannot be directly mapped to corresponding processes and methods in HCI. SEs' and HCI practitioners' misconceptions about each others' fields further exacerbate the problems created by misalignments between SE processes and HCI methods.

This paper examines how SE and HCI professionals interact in practice. Data for our conclusions are drawn from the results of a survey that we administered to 96 SE and HCI professionals—63 HCI practitioners and 33 SEs.

## 2 The State of the Practice

We electronically distributed a survey to groups of software professionals from both disciplines. The survey was distributed in two forms to the two different audiences; one survey was aimed at SEs, and one was aimed at HCI practitioners. In one survey we asked SEs about their knowledge of HCI, and in the other we asked HCI practitioners about their knowledge of SE. 96 software professionals completed the survey in total (63 HCI practitioners and 33 SEs), which was delivered via a web site.[1] We solicited involvement in the survey via electronic mailing lists to the SE and HCI practitioner communities such as the British HCI Group, various SIGCHI chapters, the ISERN mailing list, CMU's HCI and SE alumni mailing lists, and ICMC mailing lists. Participation was voluntary and unpaid. Each survey consisted of twenty questions, eighteen of which contained a multiple choice component. Most questions were single-select options, although some allowed users to select multiple responses. Fifteen of the questions contained at least one text field for free responses to a question. Respondents were asked to identify their field of work and were asked about their knowledge of HCI or SE principles. SEs were asked primarily about HCI principles and vice-versa. Following these questions, respondents were asked how their organizations involved SEs and HCI specialists on projects (if at all) via a series of questions regarding product life cycles and the frequency of interactions with the other group.

What we found from the surveys was disturbing: the SE and HCI fields are still relative islands. The usability engineering life cycle and the software engineering life cycle are not aligned—sometimes they even use different names for the same activities. And the two groups of practitioners even have differing perceptions regarding *how often* they communicate.

### 2.1 Knowledge of SE and HCI

An overwhelming majority of these professionals learn about each other's field not through taking courses or through reading published material, but rather through personal contact with other professionals. 35 of 63 HCI practitioners indicated they learned about SE entirely through informal processes—interacting with their colleagues. More significantly, 12 of the 54 HCI practitioners answered they didn't keep up with SE at all! In fact, only 13 of the 63 practitioners claim to have a degree in HCI or a related field, and only 5 others have claimed to have taken classes in HCI methods. Thus, the vast majority of our HCI practitioners in this survey are self-taught. The HCI practitioners' knowledge of software engineering follows much the same pattern: 35 of 62 respondents claim that their knowledge stems from interactions with software professionals, 5 have a degree in software engineering, and 6 have taken classes in the field.

Similarly 18 of 26 SEs[2] who answered the question said they learned about HCI entirely through informal processes. A mere 6 had a degree in software engineering and a further 2 claimed to have taken classes in the field. Their knowledge of HCI was also relatively informally grounded: 10 of 31 claimed that they learned about HCI through interaction with HCI professionals (who, as we have seen from the previous discussion are themselves seldom formally trained in the field!), only 3 SEs, fewer than 10%, reported having taken a class in HCI, and the remainder learned from books and "other".

We also asked our two groups of survey respondents "What methods or channels of communication do you use to keep abreast of research developments in [the other field]?" The results were, from our perspective, not encouraging. 33% of SEs (8 of 24) relied on ad hoc means,

---

[1] The full text of the surveys can be found at: http://www.andrew.cmu.edu/~wjj/hci_practitioner_survey.html and http://www.andrew.cmu.edu/~wjj/software_engineer_survey.html.
[2] Note that not all questions were answered by all survey participants. This question, for example, was only answered by 26 of the 33 SE survey respondents.

to keep up with HCI: "Personal communication with field specialists". A full 38% (9 of 24) replied "none"; they did not keep up at all with HCI. The HCI practitioners answered similarly, saying that they kept abreast of developments in SE primarily through personal communications (28 of 54 or 52%) or "none" (12 of 54 or 22%). Only a small minority of practitioners in both groups read journals or mailing lists, or attended conferences.

## 2.2 Working Together in the Workplace

The majority of our survey respondents reported a distinct separation between the roles of SEs and HCI practitioners. Best practices in SE and HCI advise that professionals from both fields work closely together in the design and implementation of software. Yet, according to our survey results, most professionals *do not* closely collaborate with other professionals outside of their area. For example, we have evidence that key design decisions that affect the user interface are made by SEs without consulting HCI practitioners (13 of 19, or 68% indicated that this was the case). An even greater percentage, 91% (52 of 57), of HCI practitioners believed that SEs were making crucial design decisions without consulting the HCI practitioners. Why did they do such things? Several respondents claimed that time constraints prevent SEs from waiting for HCI data to be collected—and that HCI data could potentially affect the product's architecture. One respondent said that the reason some HCI recommendations were not implemented was that "… they were an issue of time to implement, not ability to implement." There are two implications that arise from the responses to this question: 1) there is, once again, a large difference in the perceptions of SEs and HCI practitioners regarding their shared development process; and 2) a substantial number of user interface design decisions are made primarily by SEs, and these decisions are made without the benefit of usability data.

There is also a large difference of opinion regarding how internal testing of the usability of software is conducted. 50% of SEs (14 of 28) answered that "Quality assurance handles usability". A further 10 of 28 answered that SEs conduct the usability testing, and 4 of 28 responded that "other departments suggest usability changes". This is in stark contract to the HCI practitioners who responded overwhelmingly that *they* conduct the usability testing (48 of 53, or 91%). The remaining 9% is split between "quality assurance" (2 or 53 respondents) and "outsourced quality assurance" (3 of 53 respondents). SEs did not rate even a single mention by the HCI practitioners as internal usability testers. There two views of the world are obviously irreconcilable.

## 2.3 Software Process Interactions

Another difference of opinion lies in the two group's beliefs about how often they interact. HCI practitioners have the perception that they have frequent contact and correspondence with SEs. SEs, on the other hand, are more likely to believe they have little or no contact with HCI practitioners. 40% of HCI practitioners (24 of 60) say that this contact happens "very frequently" and another 43% (26 of 60) of them said "occasionally" when asked how often they correspond with SEs. Thus the vast majority of HCI practitioners thought that they had at least occasional contact with their SE colleagues. When we turn to the SEs, we once again see a different view of the world. Their responses were split down the middle: 20% (4 of 20) felt that they "correspond with each other in the software development process" "very frequently", with 30% (6 of 20) saying that this happens "occasionally". On the other hand another 20% (4 of 20) said that this happens "rarely", and a full 30% (6 of 20) said that it "never" happens. One possible explanation is found in this HCI practitioner's comment: "I think there is more reluctance for SEs to adopt HCI processes than for HCI people to 'fit in' with SEs."

The received wisdom is that the earlier a problem can be eliminated in the software process, the less costly that problem will be. So we would expect that SEs and HCI practitioners correspond with each other often, particularly in the early phases of the software life cycle. Just the opposite is, unfortunately, the case. 29% of SEs (6 of 21) said they corresponded with HCI practitioners during the coding phase. 33% of SEs said they corresponded with HCI practitioners after software development had been completed (during the testing or release phases) and 24% (5 of 21) indicate that they have no correspondence at all with their HCI "colleagues". Only 1 SE claimed to correspond with HCI practitioners in the gathering and writing of software specifications. The HCI practitioners reported slightly different numbers, but these were similarly bleak in their implication. 78% (47 of 60) indicated that they corresponded with the SEs during the testing or release phases of the software—i.e. far too late to fix usability problems economically and with minimal user impact. A mere 3% (2 of 60) claimed that they worked with SEs during the specification phase of the project. This is consistent with the numbers reported by the SEs.

We also questioned both groups about when HCI methods are used in the software development process. The vast majority of HCI specialists (43 of 61 or 70%) indicated that these methods were used when the software was "already in production", and a mere 8% (5 of 61) said that these methods were used at the requirements stage. One practitioner commented: "In extreme cases, products sometimes need to be re-architected to improve consistency or usability." Another response includes the comment "we work closely with the product team to influence the design decisions. There are also cases we cannot change most of the UI because we get involved late in the product development cycle." The pattern of the SEs is similar, but not as pronounced. 30% of the responses (6 of 20) indicated that the SEs believe that HCI methods are used when the software was in development and 25% (5 of 20) claim that this occurs when the software is in production. In contrast, just 20% (4 of 20) believe that such methods were used at the requirements phase. The magnitudes of the discrepancies are different among the SEs and the HCI practitioners, but the results are still demoralizing: HCI methods are being used too late in the life cycle to be truly cost and time efficient. One software engineer noting this situation stated "HCI and design occur simultaneously to reduce development time. Sometimes this requires later re-design." Another pragmatic response talks about a "web interface in which the architecture of the design is often mixed with the User Interface. Not that it is right, but it happens."

## 2.4   Implications of the Process Issues

When you consider the last few results, it becomes clear that there is little collaboration happening between SEs and HCI practitioners and what little collaboration there is, is occurring too late in the life cycle to be effective. A lack of collaboration between SEs and HCI practitioners, coupled with different perceptions about how and when collaboration occurs, suggests that HCI practitioners are less involved in the design of software than they think. In fact, few practicing HCI professionals are aware when changes occur in the software processes that are used by their organization. The same is true with regards to SEs' knowledge of when new HCI methods are adopted at their organizations. Only about 40% of survey respondents could name an exact type of process or method, outside of their own field, that was being adopted at their organization. 52% of HCI respondents could not even give a time frame as to approximately when such processes or methods were adopted.

SEs and HCI practitioners are not keeping informed about changes being made in each other's development processes. This could explain why SEs are often forced to make design decisions that affect the user interface without consulting HCI practitioners. One practitioner, in a comment field, said "The constraints are not on the part of the HCI, but of the SE methods; in general SE methods do not take the user into account much, or if they do, when the testing is done mostly programmers and the designer team run the walkthroughs, which in some cases leaves out most of the target audience who have different expectations than programmers and designers." A software engineer signalled agreement by saying "Many times the current architecture or toolkit doesn't allow us to do what we intend, and we are blocked by architecture redesign. For instance, we would like to use a single-field control for entering dates with a spin button and masked controls, but our toolkit doesn't provide one, and we need to work on the toolkit so that it does." One HCI practitioner said "Designers have to work around the architectural decisions of the SW Engineers". Another respondent, commenting on compromises between SE and HCI professionals said "None. [Software engineering] always wins."

## 3    Conclusions/Future Work

There is a strong strain of research in the areas of HCI and SE that reflects a cross breeding of ideas between the two fields and a genuine concern to work on shared problems. For example, the inclusion of usability considerations into UML (Paterno, 2001) and the creation of maturity models for usability indicate a growing awareness of the need for the fields of software engineering and human-computer interaction to interact. Yet despite these growing trends in research that are steering organizations towards using an amalgamation of SE and HCI activities, industry professionals have yet to follow suggestions from the research. For the most part, SEs and HCI practitioners continue to work separately. While collaboration between the two groups does occur (as it *must*, since products do eventually get produced and these products frequently have a user interface), the collaboration does not happen frequently enough or early enough in the software life cycle. Infrequent contact leads to misperceptions about what is happening in the software development process. We have certainly seen such misperceptions and miscommunications as indicated by the responses to our pair of surveys.

HCI methods should be used to design usable software from the ground up. HCI methods should not be applied as patches to software after major development has already occurred; nevertheless that is exactly what is continuing to occur in industry today. To ensure successful system development it is necessary to increase emphasis on defining appropriate HCI processes and to integrate such processes with existing system and software development processes. We, as a community, need to be advocates for both education and industry, to ensure that this happens.

For the future, we would like to extend this surveying effort, looking more closely at the interaction among SEs and HCI practitioners. In particular, we would like to focus on surveying individual companies, to see how corporate culture affects their interaction.

## 4    REFERENCES

Artim, J., et. al., (1998). Incorporating Work, Process and Task Analysis Into Commercial and Industrial Object-Oriented System Development, *SIGCHI Bulletin*, 30(4).

Hefley, W., et al. (1994). Integrating Human Factors with Software Engineering Practices. *Human-Computer Interaction Institute Technical Report*. CMU-CHII-94-103.

Jambon, F., Girard, P., & Ait-ameur, Y. (2001). Interactive System Safety and Usability Enforced with the Development Process. *Engineering for Human-Computer Interaction*. Berlin: Springer Verlag, 39-52.

Myers, B. & Rosson, M. (1992). Survey on User Interface Programming. *Proceedings CHI'92*. New York: ACM, 195-202.

Metzker, E. & Offergeld, M. (2001). An Interdisciplinary Approach for Successfully Integrating Human-Centered Design Methods into Development Processes Practiced by Industrial Software Development Organizations. *Engineering for Human-Computer Interaction*. Berlin: Springer Verlag, 2001, 19-31.

Paterno, F. (2001) Towards a UML for Interactive Systems. *Engineering for Human-Computer Interaction*. Berlin: Springer Verlag, 7-17.

# Usability Support for EU Projects
## Experiences and Actions

*Jurek Kirakowski[1], Manfred Tscheligi[2], Verena Giller[2] and Peter Fröhlich[2]*

[1]Human Factors Research Group
University College Cork, Ireland
hfrg.ucc.ie

[2]Center for Usability Research and Engineering
Hauffgasse 3-5, 1110 Wien, Austria
www.cure.at

## Abstract

In order to meet the vision for IST programmes, a considerable investment in usability engineering technology is required. Lessons drawn from previous support notions suggest that information about strategically selected methods and concepts is critical. Service should be provided to projects based on the need of the individual project . UsabilityNet has undertaken a set of actions in the realm of disseminating information and advice about best practice.

# 1 The Vision: User-centred IST

The IST programme – in both EU framework programmes 5 and 6 - is strongly oriented to the concept of User Centred Design (UCD), as can be seen in the following quotation:

> *"Research will focus on the future generation of technologies in which computers and networks will be integrated into the everyday environment, rendering accessible a multitude of services and applications through **easy-to-use human interfaces**. This vision of "ambient intelligence" places **the user, the individual, at the centre of future developments** for an inclusive knowledge-based society for all.*
>
> *(Decision No 1513/2002/EC of the European Parliament and of the council of 27 June 2002)*

The three areas of "core technologies" that are to be developed within the IST (FP6) should be mobile communication infrastructures and computing technologies, intelligent user-friendly interfaces and the optimisation of micro-system components. All these technologies have to be focused on the users' needs and expectations in order to facilitate success.

# 2 The goal: user-centred IST projects

In order to realize this vision of user-centeredness, projects should take account of findings and results from Usability Engineering, Human Computer Interaction and Human Factors.

The major principles of User-Centred Design are as follows (and are, of course, clearly applicable to IST projects):

- Understand the context of use: Specify the characteristics of the users, their tasks, and their environment.
- Involve users in the design process throughout: Conduct user based requirements specification and evaluation (i.e. usability tests).
- Iterate design solutions: Refine designs and prototypes step by step, based on user feedback
- Integrate UCD into all development steps: Ensure that user centred design is accepted and integrated appropriately in development.

- Use existing knowledge: Apply existing guidelines and standards as appropriate.

What has been the actual experience of adoption of these principles?

# 3  Experience from project support

The Baseline project (IE 2013) was a support action which ran from the start of 1996 till the end of 1999, and was intended to support EC projects in the Telematics Application area. One of its important activities was a help desk from which a list of frequently asked questions was compiled. The list has been useful to other support action projects which followed Baseline and indeed applications projects in Framework V and VI.

This FAQ is a good indicator of the areas of concern of EC project with regard to usability, and the major questions raised in queries to the help desk will be summarised in this section. The answers, of course, are currently best given by looking at the UsabilityNet web site, especially the EU project support page (UsabilityNet, 2003).

## 3.1  Planning for user centred design

Although many projects are aware of the need for investing time and effort into the planning of the user centred activities, definitional problems were frequently raised: for instance, what is the difference between usability, ease of use, and user friendliness? Although many projects were eager to embrace the concept, quite what needed to be embraced was not always clear. An additional question strongly bearing on planning was, how do we involve end users in the activities? Although some projects had end user representatives in the project structure, it was not always clear how best to utilise their input, especially since the technical processes seemed to be so recondite that even an informed end user could not easily contribute to the discussion.

To this end, collections of usability methods began to be compiled and published which showed projects how end users could be meaningfully involved right from the very beginning: the importance of paper prototyping for instance began to be increasingly realised.

In general, approximately 10% to 15% of project effort should be spent directly on user centred activities: however, we found that the most successful projects were informed by a pervading sense of 'usability culture.' which goes beyond the adoption of methods for specific work tasks.

## 3.2  Who are the real users?

A baffling aspect of user centred design seemed to be the need to focus on specific users carrying out specific tasks (see the usability definition in ISO 9241 part 11 - ISO 1998). Many projects demurred, maintaining that the results of their project were generally applicable; or applicable to so wide a range of users over so wide a range of tasks that user-centred design was impossible *a priori*. Essentially, the Context of Use method (Kirakowski and Cierlik, 1999) takes a decompositional approach to the problem of characterising users and their associated tasks and makes what looks like a seemingly large and intractable problem into a series of much smaller, soluble ones. Unless real users can be identified, and their feedback incorporated into a project, user centred design cannot start.

## 3.3  The importance of users with special needs

Special needs users - such as the elderly, the very young, the handicapped - cannot be excluded from the potential benefits of investment in IT, and yet often projects were unable to get to grips with the issue of how to begin to consider the needs of such groups.

## 3.4 Context of Use and its consequences

A context of use analysis informs the entire project: firstly by suggesting what end users groups may be associated with the project, and what should be the main activities to focus on; and later, by indicating how the project may be assessed to show whether it actually has achieved its user centred objectives. Context of use is not a static issue: the initial analysis, no matter how thorough, will always need to be modified as the project progresses and greater understanding of the technology being developed is gained. A certain amount of fluidity in the project administration is needed to harness this concept in its most beneficial way, and this is not always possible.

## 3.5 Benchmarking

How does one know when a project has attained its objectives? The issue of setting real, objectively verifiable goals is a difficult one which projects continually face. The most common reaction to this question is to set performance goals. that is, to attempt to quantify how *effective* the project solution is. However, the *efficiency* with which the project solution works is often un-explored and rarely a feature of the assessment, still less the end user *satisfaction*. Metrics for all three aspects of quality of use are available, albeit less well developed in some areas than in others. User satisfaction, surprisingly, is a well understood area with large available databases which can be used for the benchmarking process. Efficiency is also moderately well represented, but effectiveness, despite its importance, remains a quality for which reliable metrics are scarce.

## 3.6 Usability Methods

A measure of the maturity of a discipline is the extent to which it employs standardised methods and procedures. The Baseline project over the years of activity monitored the kinds of tools and methods employed by the projects it was supporting. These data are summarised in the final report (Kirakowski, 1998). Although there were signs of improvement over the years, the concluding comment is:

> *It will be seen ... that there is still a heavy reliance on 'soft' methods such as in-house questionnaires, interviews, and written accounts which have to be content analysed. These methods make up 52% of the methods adopted or proposed for adoption.*

It is to meet this very need that the UsabilityNet project focussed on recommending methods and tools which are well attested in the commercial community. Over and over again, we were confronted with the question: which methods should we adopt first? The answer is not easy as it depends on the maturity level of the project with respect to usability processes. However, there is certainly a need for an elementary guide or road map for the project that wants to increase its expertise in this area.

## 3.7 Standards and Guidelines

The years since Baseline have shown a steady increase in the number of international standards to support usability engineering and there is now a good selection of standards to support the user centred design process. The activity of the International Standards Organisation in this area is useful as it serves to gather consensus among practitioners as to what is the best approach. EC projects increasingly turn to these standards for authoritative advice. These are summarised on the UsabilityNet web site (see UsabilityNet, 2003).

# 4 The UsabilityNet project assistance concept

UsabilityNet has drawn on the experience of previous support actions including Baseline to tailor its services to EU projects in the most strategic manner. Three basic principles of a support service could do well with being re-iterated here:

1. **Don't lecture; help**. Although general workshops and conferences are good for disseminating information about existence, in practice, real help for projects is so highly contextualised that one-to-one actions are the most effective way of conveying the message.

2. **Don't make it complex; make it straightforward**. Usability engineering is a complex subject and answers to questions are not always straightforward. However, simple first steps advice can and should be created.

3. **Don't confront; use guerrilla tactics**. Over and over again, in industrial consultancy, we have found a large benefit in identifying the critical issue facing a project and tacking that issue locally with the methods at our disposal.

The remainder of this section lists the ways in which we have been helping projects following this philosophy.

- **Guideline document:** The document "First steps for User-Centred Design" (Tscheligi et al., 2001) was produced for the target group of EU project participants. The document contains simple guidelines and a checklist that makes it possible for EU project participants to evaluate the status of their project work plan with regard to User Centred Design.

- **Analysis of potentially interested EU projects:** A list of projects that might need basic support with regard to Use Centred Design was produced (Tscheligi et al, 2001). Criteria for the selection of candidate projects were the degree to which user-centred activities were (not) included in the project description, the availability of UCD expertise, etc.

- **EU project support area on web site**: A new part of the UsabilityNet web site dedicated to EU project support issues, was set up (www.usabilitynet.org/eu.htm).

- **Contacting of EU projects**: The projects on the list were contacted systematically. A standard email was sent to the projects, announcing the usability web site services. Furthermore, the projects were encouraged to contact CURE to obtain the guideline document, and the possibility to receive a limited cost-free project-specific assistance in UCD was announced. In case of lacking response, telephone the projects were also contacted by telephone. A detailed contact log concerning the telephone calls and email communication was made. The projects who expressed interest in obtaining the guideline document, were asked to give feedback as to what extent the guideline document supported them in planning and conducting their project activities.

- **Forms of communication**: UsabilityNet developed a strategy to efficiently consult EU projects regarding UCD issues. A major principle for the assistance was to give punctual and specific support for each project, based on highly prioritized problems. Based on the specific conditions of each project, the forms of communication and assistance differed. In some projects, remote consultation was more appropriate, mainly because of travel costs. In these cases, project documents or online prototypes could be reviewed using E-Mail or telephone channels, and also collaborative workspaces. In other projects, workshops were conducted in order to discuss usability problems of a system and to give project participants a feeling for UCD fundamentals.

- **Form of assistance**: The assistance services offered by UsabilityNet can be divided into process-oriented and product-oriented consultation: The development *process* with regard to UCD principles was optimised and adapted to the specific project case. By means of using the checklist within the guideline document described above, project partners could make themselves a first impression by identifying tasks or activities that have not yet been included into the workplan. Furthermore, experts from UsabilityNet could detect improvement potentials with regard to UCD and enhance the development process to achieve user-centred systems. Concerning *product*-oriented consultation, recommendations for improvement of the user interface can be given.

## 5   Experience and Learning

The process of the project assistance itself was evaluated by obtaining feedback from the consulted projects.

Many projects were interested in obtaining the guideline document "First steps for user-centred design". The simple steps and clear references to in-depth descriptions, as well as guidelines applying specifically for EU projects are highly appreciated by the project participants. However, a guideline document was only seen as a starting point. This supports our concept, which includes, as a second step of providing project-specific guidance based on direct experience.

Concerning project-specific assistance: punctual, short and thus cost-effective interventions can be of high-value and acceptance for EU projects. Introductory workshops, help with test planning, advice on development solutions, an external perspective on user interface design issues are all valuable inputs.

It can be seen from the above that EC projects do want and are receptive to incorporating usability concepts. This should not be surprising, constituted as they are from leading-edge practitioners. As support actions in this area continue to expand their influence, we can only hope that a leverage effect will happen so that usability engineering knowledge is so well distributed among IST programmes as to make the achievement of the high goals of these programmes a certainty.

## 6   References

European Decision No 1513/2002/EC of the European Parliament and of the council of 27 June 2002, Official Journal of the European Communities, http://www.cordis.lu

ISO 9241-11 (1998) Ergonomic requirements for office work with display terminals. Part 11: Guidance on Usability. International Standards Organisation, Geneva.

Kirakowski J (1998): Baseline usability data and cost/benefit of the use of validation methods IE2013/D4.4.1 Human Factors Research Group, Cork. hfrg.ucc.ie/baseline/filearchive.html

Kirakowski, J. and Cierlik B., (1999): Context of Use: Introductory Notes. Human Factors Research Group, University College Cork. hfrg.ucc.ie/baseline/filearchive.html

Tscheligi, M., Fröhlich, P. and Giller, V. (2001). Draft Guide: First Steps to User Centred Design. Project IST-1999-29067 UsabilityNet, D3.1.

Tscheligi, M., Fröhlich, P. and Giller, V. (2001). Plan for Work with Projects. Project IST-1999-29067 UsabilityNet, D3.2.1.

UsabilityNet (2003): website home page at www.usabilitynet.org.

# Designing the UsabilityNet Web Site: A Case Study

*J Kirakowski*

Human Factors Research Group
University College Cork, Ireland.
jkz@ucc.ie

## Abstract

The user-centred activities involved in developing the UsabilityNet web site are outlined. There were two versions of the site, and the transition between versions is explained. The end user evaluation of the sites by means of the WAMMI questionnaire are presented. Two major topics of discussion are (1) the use of Flash technology in version 2, and (2) the relationship between the technical developer and the client who commissioned the site.

## 1    Support for Usability Professionals

The original objective of the UsabilityNet project was to provide services to support usability professionals. The primary vehicle for the delivery of such services was to be an accessible web site, linking a network of professionals in many different parts of Europe and beyond.

This paper is an account of the development of this web site over its currently two major iterations; focussing on the user-centred methods involved, the results obtained, and the lessons learnt.

## 2    Versions of the site

The primary activity in user centred design is the planning of the user centred process. In an European Community project, this is largely done when the technical annexe to the contract is being drawn up in accordance with the peer expert review of the original proposal. This plan had later to be revised on the basis of an EC expert review in month 10 of the project.

From the *context of use* analysis initial user groups for the UsabilityNet service were defined. Although the primary user group would be usability professionals, it was concluded that the service should also support other key groups: EU projects (Kirakowski et al, 2003), line and project managers (Claridge 2003), and IT procurers.

## 3    Product Definition and Version 1

Developing version 1 followed the ISO 13407 framework (ISO 1998), employing methods now documented by the UsabilityNet web site (Bevan 2003). A key point in the development of the requirements (Bevan et al, 2001) was a workshop attended by 24 experts from usability and allied professions from around the world. Each activity is outlined as much as possible in chronological order in the following account.

### 3.1 Comparative Analysis

Cross-classifying results from 14 popular search engines, a list was drawn up from the 10 top usability sites which featured most frequently when searching with the keyword 'usability'.

The successful sites exhibited the following characteristics:

- They offer a wide variety of information (usability and other non-usability but related information)
- They provide hard facts rather than opinions or ideas
- They contain references to published (non-web) work
- They point to useful things on the web but they also explain what they are pointing to and why this is a useful resource (rather than many 'blind' links)
- They have a pleasant non-commercial graphic presentation.

### 3.2 Analysis of topics appearing in usability discussion lists

A historical analysis of traffic on 6 discussion lists over the previous few months yielded 71 usability issues, which were summarised into 9 overall categories (eg, 'web design', 'how to do usability testing').

### 3.3 Focus groups in Sweden, UK, and Austria

Altogether four focus groups were held. The purpose was to get more detailed information with regard to users and their needs. The net result of the focus groups was to highlight a number of issues that usability professionals and managers thought the project ought to address; in all there emerged 20 issues when the reports were collated and assessed by UsabilityNet researchers, and this list fed naturally into a broader user survey.

### 3.4 Web questionnaire survey

The main focus was to administer a survey through the internet, and this was launched in March 2001. The bulk of the survey asked users to assess the importance of 22 different issues in usability engineering (eg 'Business cases and cost justifying usability', 'Usability case studies'.)

In addition to these 22 predefined topics, there was also a free-text field for the respondents' own suggestions. Examples of entries written by users were: 'User task analysis', 'Job bank', 'Book reviews', 'Contact database' and so on. The survey resulted in 680 responses in March - April 2001. Although usability professionals were well represented, there was also participation from managers, designers and developers.

In total some 200 items of required content were gathered from all activities up to and including this stage. These were reduced by collapsing similar worded items to 164.

### 3.5 Expert workshop

24 participants from all over the world were invited; they represented a wide range of skills in different applications domains. The important activities carried out during three intensive days were as follows:

#### 3.5.1 Personas

On the basis of preliminary analysis and context of use, five Personas were defined and a page of information about the activities of each were drawn up, as being possible end users of the

UsabilityNet services and web site. The group of experts was split according to which of the personas they preferred to represent.

### 3.5.2    Affinity diagramming

The 164 requirements items were arranged in alphabetical order and printed on 'stickits.' These were sorted on a wall by each persona group according to how their persona would expect to see them structured.  This process is better than card sorting in that a group consensual decision is made, decreasing the probability of idiosyncratic responses, but better than simply a 'concept wall' to which a potentially unlimited number of users may respond with different agendas

At the end of this exercise, the results were presented in the plenary. It was agreed that there was a substantial amount of overlap between the clusterings of the different groups. These results were remarkably well confirmed by a single-linkage cluster analysis of a similarity matrix created by putting the results of all five groups together. The clusters were: (1) *People and organisations*; (2) *Education and training*; (3) *Management and organisation*; (4) *Reference materials*; and (5) *Doing usability (methods)* .

### 3.5.3    Paper prototyping

The parallel design method was used: four groups worked in parallel to create paper prototypes of web sites that delivered information about the thematic areas established above. There was a plenary presentation, critique, and discussion.  The objective as always was not to pronounce a single 'winner' but to glean from each proposed solution what are the best aspects that will lead to success in a final version.

### 3.5.4    Conclusion from the workshop

The results of the workshop remain as a unique document outlining the aspirations of many hundreds of usability professionals and clearly went beyond the scope and budget of the project. The actual requirements document was much more modest in scope.

### 3.6    Formative co-evaluation by end users of near -final prototypes

A sub-contractor carried out the initial graphic design, but it became clear that the cost of implementing this design would bring us over-budget.  One of the partners took over this effort. The last few weeks of the development effort were therefore spent in that partner's usability laboratories, with development taking place in one room, and a stream of likely end users not otherwise associated with the project coming in to run through predefined tasks with the site. The feedback from these users was always used formatively, to inform the development team of what changes should be made to the design: visual appearance, wordings, and style were usually the main issues. Although we lost the input from a professional web developer, we gained by having very immediate feedback that could then be acted upon very quickly. The site was finally released on the 1st September, 2001.

### 3.7    Website evaluation by end users

The web site was evaluated with WAMMI (Kirakowski & Claridge, 2003). WAMMI is a 20-item questionnaire that measures end user satisfaction with a web site.  WAMMI results are scaled to percentiles, so that a score on the 75th percentile for instance means that the site being evaluated is better than 75% of the sites in the WAMMI standardisation base. WAMMI also provides the opportunity for a number of additional questions and free-form comments to be added to the survey. The WAMMI evaluation for version 1 ran from September to October 2001. In all, 40 users evaluated the site. Numeric data from the evaluation is shown in table 1 below (where it is compared to data from version 2).

It is not surprising, given the amount of attention paid to the iterative development of the contents of the web site, and the size of the user sample that participated in its development (more than 700 experts of various occupations assisted with the development effort) that **Efficiency** and **Helpfulness** were quite high. **Controlability** and **Helpfulness** were average but **Attractiveness,** was below average. Individual user comments and an analysis of the responses to each individual WAMMI question made a number of very specific points as to how the web site could be improved and a version two was obviously necessary, especially after the formal EC review of the project in September 2001 which extended the scope of the project.

# 4    Developing Version 2

The objective of the redevelopment work was to improve the graphic design and navigation taking account of feedback, and to modify and extend the structure in line with the recommendations of the 2 expert reviewers to include more information for managers and EU projects. Changes in the consortium structure required us to create the site afresh, using most of the material from the old site but re-arranging the structure and improving the graphics. The process followed picked up at the detailed design stage, and employed a new sub-contractor for the graphics and implementation.

## 4.1    Design workshops

Several design workshops were held in March 2002, involving a number of project partners working with paper prototypes. A new content structure to service the changed project objectives was also needed. The resulting specification was detailed and complete.  The graphics and navigation elements were created using Flash. The contract allowed for one major iteration, but after all the work was completed, yet more refinement was considered necessary and for this there was no budget.  One of the project partners therefore undertook to further refine and modify the site in line with co-evaluation feedback obtained informally after implementation.  It was this refined version that was eventually released and evaluated (Bevan et al 2002).

## 4.2    Website evaluation by end users

The site was evaluated using WAMMI in early January 2003.  33 users participated. To the question, can this web site support you in your work, 2 groups of responses were found with associated profiles.  Those that said *yes* gave the web site an overall better profile.  Table 1 shows the breakdown in WAMMI scores between the two versions.

|          |     | Attr  | Contr | Effic | Helpf | Learn | Global |
|----------|-----|-------|-------|-------|-------|-------|--------|
| **Ver 1** | Yes | 43.25 | 51.12 | 62.28 | 61.06 | 51.69 | 53.35 |
|          | No  | 8.20  | 18.80 | 22.80 | 14.80 | 23.60 | 17.20 |
| **Ver 2** | Yes | 42.31 | 53.89 | 68.63 | 57.79 | 52.58 | 54.63 |
|          | No  | 23.64 | 29.00 | 32.09 | 15.27 | 26.36 | 24.82 |

**Table 1:** WAMMI scores for version 1 and 2 of the web site broken down by the answers to the question: *can this web site support you in your work?*

For those users who feel the web site does support them in their work, **Controllability** is still average.  **Efficiency** remains high and has in fact improved; **Helpfulness** has slipped a little. Both **Attractiveness** and **Learnability** however do not show much discernible improvement despite the amount of effort put into **Attractiveness**. Interestingly, for those users who did *not* think the web site supported them in their work, large gains on most scales were in fact made, although the profiles for this group still do not reach a good industry average standard of 50. The poor scores on **Attractiveness** can largely be put down to the poor performance of Flash, as shown in the negative user comments in the second WAMMI evaluation.

The *Emerald Management Cool Sites* (2003) awarded the UsabilityNet web site 5 stars in January 2003: their comments focus on the value of the content, as do all the positive end user comments in the second WAMMI evaluation.

# 5    Concluding comments

The most difficult aspect of the development process for us has been finding a way to effectively integrate user-centred practices with the design and implementation activities of the sub-contractors.  Client-contractor relationships in this area are very much under-researched and in need of more work (Kirakowski, 2000).  Based on our experience, clients are well advised to look extremely carefully at the implementation practices of their developers before awarding a contract that may involve iteration of solutions, to see if these practices can stand the burden of repeated iterations and yet remain within budget.  The user comments from the WAMMI analyses, and the profiles themselves, show that for both versions there were important user satisfaction issues left unresolved at the time of release.

The use of Flash, while initially seeming to afford a short cut, a state of the art graphic treatment, and a solution that promised support for special needs users, has been a disappointment.  Flash, in the year we experienced it, was never able to mature sufficiently to meet special needs, and the available tools for creating sites with this technology have not grown out of the 'hand-craft' stage.

At present we envisage a third and final version of the site, keeping the same structure, but without using Flash, to further improve its user satisfaction for our targeted groups.

# References

Bevan, N. (2003): UsabilityNet methods for user-centred design.  Proceedings of HCI International 2003. Lawrence Erlbaum.

Bevan, N., Claridge, N., Frölich, P., Granlund, Å., Kirakowski, J., Tscheligi, M. (2002): UsabilityNet: Usability support network website.  www.usabilitynet.org

Bevan, N., Claridge, N., Granlund, Å., Kirakowski, J., Strasser, A. (2001):    UsabilityNet deliverable 1.1.  Methods workshop report and requirements specification.  Serco Usability Services, London.

Claridge, N (2003) Usability support for managers. Proceedings of HCI International 2003 Lawrence Erlbaum.

Emerald Management Cool Sites (2003): www.emeraldinsight.com/reviews/coolsites/im.htm

ISO 13407 (1998):   User centred design process for interactive systems. International Standards Organisation, Geneva.

Kirakowski J. and Claridge, N. (2003): WAMMI website.  www.wammi.com.

Kirakowski J., Tscheligi M., Giller V., and Frölich, P. (2003): Support for EU Projects. Proceedings of HCI International 2003 Lawrence Erlbaum.

Kirakowski J. (2000): Client developer relationships: a primary checklist. hfrg.ucc.ie/resources

# Usability Testing on All Products (UTAP): how it was incorporated into software development process

*Tadashi Kobayashi*

Quality Assurance Dept., Fujitsu Info Software Technologies Limited
Southspot Shizuoka Bldg., 18-1 Minami-cho, Shizuoka-shi, Shizuoka, 422-8572
JAPAN
kobayasi@ist.fujitsu.com

## Abstract

Fujitsu Info Software Technologies is a computer software developing company. After two years of preparation, the incorporation of usability testing into its software development process started from Apr. 2002. Our company is said to be the first IT company in Japan to define and to enforce a strict rule to apply usability testing on all of its products before going into product inspection work to be performed by quality assurance team. In this paper I will analyze why we succeeded in this challenge and discuss the testing results and what we are planning to do from now on.

## 1 Introduction

In Japan, usability testing has been so far tentatively applied to the important or representative products only. In a number of enterprises, a section or a department was established to enforce the enterprise level deployment of usability testing on their products. They succeeded in partial deployment without getting unanimous agreement on the application of usability testing on all the products. However, we succeeded in applying usability testing to all the products we release.

## 2 UTAP

Usually the benefit of usability is not fully understood by top executives because the effectiveness of usability testing is unclear and seems so much time and manpower are needed to perform usability testing. If top executives are quality-minded, they can welcome usability assets as differentiating points for their products.

Sometimes it is far better to directly appeal the benefit of usability testing to the top executives instead of middle class managements. As though this top-down approach seems so well working, this approach seems to be especially quite effective for a rather small-scale companies like us, because we are rather a homogeneous group of people and communication problem is nearly non-existent. One more advantage comes from the number of products to improve their usability. It is about 20 per year, which is appropriate for coordinating testing schedules among many products.

### 2.1 How UTAP began

Our usability lab was installed in April 2000. The layout of our lab is shown in Fig.1. Our lab's main feature is the communication room adjacent to tester's room and observation room. This room enables us to give testers instructions to the usability tests and to get hearing on fresh impressions just after the usability tests (IST Web design study group, 2000.)

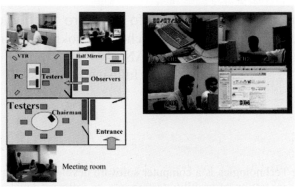

**Figure1:** Usability Test Laboratory

By utilizing this lab we started testing our own software products. Twenty-nine software products were tested between Apr. 2000 and Nov. 2001. Our software products include normal Windows applications, Web applications, Web sites, and mainframe products. Employed evaluation techniques were heuristic tests and user testing (Kurosu et al., 1999.)

In Dec. 2001, we submitted our plan to our president to employ UTAP into software development processes, based on the estimation that our basic skill and expertise are sufficient enough to perform UTAP within our company. Benefits of UTAP we listed in our employment plan are:
- Effective for differentiation with competitive products
- Reduction of so-called negative costs: man-power for field troubles or user-inquiries
- Reduction of developing cost in general
- Reduction of inspection cost

Our president understood the importance and effectiveness of usability and called for a board meeting to discuss the adoption of our UTAP plan. In Jan. 2001, UTAP was approved in the board meeting to make it a prerequisite process before going into product testing work.

## 2.2 UTAP products

Our software products are divided into two categories:
- General products for the public
- Customized Products with development contract (with government or corporations)

"Customized Products with development contract" were initially omitted from evaluation targets because they have many restrictions imposed by customer's intention or needs. So we decided to perform usability testing on the "General products for the public."

Enhance levels defined for our products are as follows:
- New products
- New products by porting from existing products
- Version-up products

- Level-up products
- OEM products

From these five types we eliminated "Level-up" type because the amount of interface changes and corrections are relatively small. Also we defined a new type of usability evaluation: "Simplified evaluation" for "New products by porting from existing products" and "OEM products" levels. Because there are possibilities to spoil usability in the new coding part and the modification part, we checked if they exist or not, and issued an official usability check report to the development section and the inspection section. Consequently, "New products" or "Version-up products" were taken as the terms of application of our usability test process for "company products" as mentioned in the above.

## 2.3 The evaluation method and timing

It was decided next which processes are to be targeted to do usability tests. Full-fledged usability testing has been enforced as mentioned in the previous section in our company since April 2000. The following three kinds of evaluations were defined and included into our usability test processes from the two years' expertise of usability tests.

- *DC (Design Check):* The experts of the usability and design evaluate screen layouts, screen links and so on from design viewpoints
- *HT (Heuristic Test):* The experts of usability evaluate the whole product from usability viewpoints
- *OT (Observation Test):* Scenarios and tasks are given to testers, and the usability of the product is evaluated by observing a series of works performed by testers

These three tests were performed in the earlier development phase and the program making phase, thus incorporated into the development process as our usability evaluation methods (see Figure 2.)

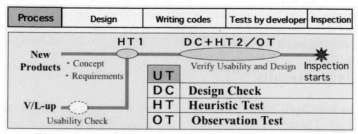

**Figure2:** Development process and usability test timing

In the earlier development phase, after the product concept and user requirements were consolidated, "HT" was scheduled to the function specifications and the prototype. In the program-making phase, "DC" and "OT" or "DC" and "HT" was carried out as a usability evaluation process to the product and the manual under development.

## 2.4 Test result report

A usability test report for the targeted product is made and presented to the development section in each usability test. It consisted of a comprehensive usability report, a problem list to inform detected problems, and problematic screen shots to explain the problem correctly.

## 2.5  Test result reporting and confirmation

When starting this usability evaluation process, development sections and inspection section show concern in what procedure to guarantee usability how. Our company's test result reporting and check procedure are as shown below (See Figure 3.)

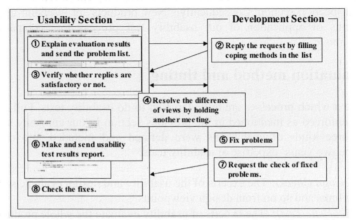

**Figure3:** Validation and confirmation process of usability problems

## 2.6  Acceptance report

If all the check of the usability test results by the usability section and problem fixing efforts by the development section are completed, the usability section will draw up a "usability evaluation report" and send it to the development section and an inspection section. This notifies that the usability test process in the usability section was completed. This usability evaluation report includes comments about the notice of "Acceptance" and the number of problems in each test, and the usability of a target product is assured.

## 3  Usability test results

During six months from April 2002, 24 usability tests for nine products and 11 simple evaluation tests were carried out. The coping rate was 90.9% that means the development section would cope with or intend to cope with in the future version for the problems found in the usability test.

## 4  The characteristics of our usability testing methods

This section states the feature of the usability testing in our company.

## 4.1  Problem list

Our problem list can be used not only to present problems and improvement proposals, but also to utilize it as a tool of communication between the usability section and the development section.

## 4.2 Third person evaluation

The third person evaluation is carried out in the HT. Since the evaluation target is the product of our company, a usability section person in charge has a high possibility of knowing the situation on the development of the product to some extent. Because there is a possibility that objective evaluation will be somehow affected, the outside usability experts are invited in evaluation tests.

## 4.3 Participation of inspection section to the usability process

In the usability test process, inspection section persons in charge are requested to participate in the evaluation. This has two-fold purposes to take in the opinion of many people for the product, and to cut down the inspection cost of functional tests in the following inspection process.

## 5 The effect of UTAP incorporation

The effects of incorporating the usability test into a development process are as follows:
- Detect serious usability problems
- Satisfy specifications with user's needs considered
- Improve problem coping rate
- Improve the usability awareness of developers
- Include the opinion of the inspection section
- Improve the understanding by the inspection section on the target product

## 6 Issues

Now we are facing following three challenges:
- More efficient usability testing
- Proof by quantitative evaluation
- Making of a check list or a guideline for developers

Because quantitative indices for usability improvements are difficult to establish, we decided to evaluate the effectiveness of our usability testing by the reduction of the number of Q/As (in the form of phone-calls and e-mails) from our customers on those usability improved products.

## 7 Conclusion

UTAP approach will inevitably lead our company to a useably engineered company full of usability-minded people. There we will evolve into the second stage usability level where we must recreate the entire usability procedures to suit our highly skilled "usable" people.

## References

Editorial-supervision Kobayashi and IST Web design study group (2001). Know-how of a successful website making. Tokyo: Kyoritsu shuppan Co., Ltd.

Kurosu, Ito, and Tokitsu (1999). An introduction to user engineering. Tokyo: Kyoritsu shuppan Co., Ltd.

# Usability in India – An Uneven Journey

*Apala Lahiri Chavan*

Human Factors International Pvt. Ltd.
Solaris Building No. 1, Saki Vihaar Road, Andheri (E), Mumbai 400072, India
apala@humanfactors.com

## Abstract

India has been reaping the benefits of an information technology boom for over a decade now. Indian systems integrators are to be seen and heard in the best and baddest of projects across the world. However, it is interesting that inspite of the presence of hundreds of high tech companies with the most sophisticated processes and tools in place, there was little or no interest in usability for a long time.

It has been an uneven and sometimes discouraging journey. It is only over the last 2 years that there has been a definite interest in usability, from various corporates, both at the tactical as well as strategic level. There have been some pioneering initiatives and these initiatives have had a major positive effect on the state of usability in India.

## 1    Is Usability a natural fit in India?

Based on the existence of the flourishing IT sector in India, one might expect India to be a center of intense usability activity.  Beyond the IT boom, however, there are some other factors, including inherent strengths, which could in principle contribute to making India a fertile ground for usability.  As Dr. Eric Schaffer of Human Factors International had envisioned three years ago, the core Indian values of hospitality, service, and craftsmanship are a good foundation for usability work.  After all, a usable web site is a hospitable one.  Another factor is that Indian society is very focused on family and community.  This cultural focus on others also seems like a good fit with the philosophy of user-centric design.  Next, the growing need for attention to designing for cross-cultural use seems to offer an area where Indian companies could excel. Indians are immersed in a multicultural society, and nothing in their context encourages them to think there is only ONE language in the world, ONE viewpoint, and ONE culture called Hollywood.  This presents India with another opportunity: become the acknowledged leader in developing globalized software.  Finally, if Indian IT is to move up the value chain, Usability Engineering must play a part.  The greatest value will come when the Indian IT industry is known not just for productivity in writing software, but also for producing software that is highly usable.

## 2    State of Usability in India

So what is in fact happening with usability in India?  Actually the situation is fairly complicated, and it will help if we consider it in terms of three different IT sectors.  First, one can divide the commercial IT industry into a component that focuses on developing software for internal Indian consumption whether for consumers or for business, and a component that focuses on developing software for customers in other (primarily European and North American) countries whose users are also not in India.  There is a third sector that is focused more specifically on increasing access to technology within India and on using technology to promote social and economic development. These initiatives often involve developing applications on an experimental or demonstration project basis.

Unfortunately, I have to report that in the first sector, commercial software developed for the Indian Market, there has so far been very little focus on usability.  A great deal of technical research is being done for local use, but with almost no investment in usability.  In contrast there are a growing number of corporate initiatives to usability in development work intended for overseas customers.  There have also been some very interesting research-oriented projects where usability issues have been addressed in relation to attempts to use technology to address social goals.  In the following sections, I will therefore focus on giving an overview of the usability work taking place in these two categories, corporate initiatives on the one hand and research/experimental initiatives on the other.

## 2.1   Corporate Initiatives

### 2.1.1   Satyam Computer Services

Satyam Computer Services Ltd., a diverse IT solutions provider, offers a range of expertise to its customers.  8,500 IT professionals at Satyam, its subsidiaries, and its joint ventures work onsite, offshore or offsite to provide customized IT solutions for companies in several industry sectors. Satyam has development centers in India, the USA, the UK, Europe, Japan, Singapore, and Australia, and has done projects for 260 global companies, including around 67 Fortune 500 corporations.

It was about 4 years ago that the top management at Satyam realized, that, apart from the quality of technical solutions, they also needed to provide an engaging and pleasant user experience to users of the systems they developed. As a result, Satyam now has a dedicated User Experience Management (UXM) Solutions group that provides usability and UI design expertise to Satyam projects worldwide.  UXM Solutions, headed by Anupam Kaulshi, focuses on interaction design, usability, online branding, and content architecture.  The World Bank has an ongoing project with this group, and, this year, the intranet that UXM has worked on for the World Bank was selected as among the ten best intranets of the year in the Nielsen Norman Group's "Intranet Design Annual 2002: Ten Best Intranets of the Year." Incidentally, the World Bank is an excellent example of a client with a global perspective turning to India for help with a system including its interface.

The Satyam Usability team in Bangalore has worked hard to integrate usability into their software development process. They have created a series of awareness building modules on issues related to usability.These modules are accessible to developers online,so they can act as a mechanism to build basic sensitivity to usability and convince developers of its value.  The usability team has also developed a set of checklists that are available to all project teams. These consist of the basic guidelines that developers need to follow in order to ensure that all projects adhere to basic usability principles.

### 2.1.2   Infosys Technologies Ltd.

Infosys Technologies Ltd. provides IT consulting and implementation services to organizations around the world.  Infosys works with clients, providing services such as business-technology consulting, application development, system integration, and management services.  The company has over 10,000 employees across locations in North America, Europe, and Asia.  Sridhar

Dhulipala,a product design graduate from the National Institute of Design, Ahmedabad, who has been with Infosys for the last 7 years, has been largely responsible for taking forward the management initiative of building usability capability within the organization.

The geographical separation of the usability staff in India from clients and users and even sometimes from the onsite development teams is an obvious challenge facing the usability groups within Indian software firms producing software for offshore customers. Most Indian-based usability groups use two basic strategies to deal with this, and Infosys' approach seems to be fairly representative. First, work (and travel) is organized to allow usability specialists to work closely with the client (and even the end-user at times) onsite during those phases where close interaction is required. During other phases, when frequent interaction is not required, the entire project team, including the usability specialist, work offsite. For example, when working in India, the usability specialists tend to focus on things like detailed design of the screens. In order to coordinate with parts of the project team that are on-site with the client, they of course make extensive use of Infosys's intranet system. The intranet is also used as an important means of disseminating usability related information to all developers, because, they keep tools and heuristics as well online learning modules on Usability available on their intranet.

The second strategy has to do with trying to organize the usability group's design input as efficiently as possible to have timely impact on the development process. For example, at Infosys, the usability group has worked to codify their usability experience in the form of tools and second level heuristics that can be used directly by the developers. In an attempt to make these as useful as possible for this audience, they have tried to make these tools more specific and targeted than familiar broad and general heuristics such as Nielsen's ten. Sridhar reports that the dissemination of these tools and heuristics for use by the development teams has greatly helped usability to be perceived as a 'doable' process by the developers. Further, the existence of detailed heuristics and usability checklists has also helped reduce the time taken for developers to integrate usability inputs when working on projects. Sridhar has seen a major shift in the developer community towards acceptance of usability inputs. The very fact that Sridhar is now no longer part of operational level usability, but has moved into Infosys' R&D organization, the Software Engineering Technology labs, is an indication of the growing recognition of usability in the company.

### 2.1.3 Human Factors International

Intrigued by the booming software industry in India, but also puzzled that this industry had largely ignored usability engineering, Dr. Eric Shafer, CEO of Human Factors International (HFI) USA visited India about 3 years ago. His visit eventually led to HFI setting up its Indian operations in January 2000, with a complement of 5 staff members. Today, HFI-India has a 22-strong team of specialists drawn from the National Institute of Design, Industrial Design Centre, Indian Institute of Technology (IIT), and various human factors programs in US universities. The Mumbai office is very different from HFI USA and has developed its own style of consulting, in keeping with its cultural gestalt.

Initially it took a great deal of evangelizing locally before any impact was noticeable. However, now, at the end of almost 3 years of operations in India, HFI has a portfolio that includes websites, hand held devices, GUIs and Interactive Voice Response Systems (IVRs). HFI India has also developed special expertise in the areas of internationalization and accessibility. Consistent with this, HFI India also works as an offshore center for several large projects undertaken by HFI USA, for clients in the US market. HFI offers a certification program in India, and has experienced increasing demand for it. It is interesting to note that there has been a marked change in the profile of participants who attend HFI's courses. While two years ago most course participants were still grappling with the "what" of usability, participants who attend the courses now are more aware of the concept of usability and want to move quickly into the "how."

One thing that distinguishes HFI India from other commercial usability groups is that it is interested in focusing on software designed for Indian users. It has pursued this, for example, in research oriented projects such as its collaboration with Media Labs Asia on the design of a micro credit application for rural illiterate/semi literate users.

## 2.2    Research/Experimental Projects

### 2.2.1    Hole in the Wall

National Institute of Information Technology (NIIT) is one of the largest computer training organisations in India, with over a million Indians trained in IT related skills and with over 3000 centres (many of which are franchised) in India and abroad.

The Center for Research in Cognitive Systems (CRCS) at NIIT, has, in fact, been conducting experiments on mass computer literacy for the last 20 years. One particularly interesting project that NIIT has undertaken is an initiative to investigate how groups of children can learn to use a computer with almost no adult intervention. The hypothesis of the project is that it is possible for groups of children to learn and use computers with minimal or no intervention from adults, if appropriate hardware and connectivity is provided.

The term that NIIT has adapted for this concept is "Minimally Invasive Education$^{TM}$," and the project itself is popularly known as the 'Hole in the Wall' project. The project's first computer kiosk was installed on January 26, 1999, in a slum bordering the NIIT's Headquarters at Kalkaji New Delhi. There are 52 such kiosks around the country today

The first outdoor kiosk was constructed outside the NIIT office in New Delhi.There was a well defined target group of children that NIIT expected would start using the kiosk. This target group consisted of children from the neighboring slum, most of whom were not school goers. There were a few who went to very poor quality government run schools with negligible resources and very de-motivated teaching staff. None of the children were familiar with English.

In order to record all activity at and around the kiosk, a video camera was placed on a tree near the kiosk. In addition, activity on the CPU was monitored from another PC on the network. NIIT project staff were always present to monitor activity from within the office and to take notes when necessary. This arrangement allowed the researchers to study the natural activity on at the kiosk, in order to evaluate whether and how the users would approach learning and using the kiosk, without any help from anyone, and whether it was even possible to operate such a kiosk with no supervision in an outdoor location.

## 2.2.1.1    Process of Learning

In the course of this project, NIIT has learned a great deal about the natural process of children's learning in group contexts with minimal adult intervention. The patterns observed during the experiments suggest a learning process in which children learn through experimentation and by observing and instructing each other in computer usage. Examples of several interesting observations have been recorded by the project team.

The CRCS at NIIT continues to work on verifying its hypotheses regarding the concept of Minimally Invasive Education with this ongoing and expanding project. In a country where a very large section of the population is illiterate, experiments of this kind could mean the difference between the darkness of ignorance and the light of knowledge and discovery.

## 2.2.2 *Media Labs Asia- The Hisaab Project*

As in many other developing nations, most of the poor in India, have re mained outside scope of the formal banking or financial system, particularly in rural and marginal areas. Opportunities for depositing their small savings and access to loans and micro-investments have been identified among their most urgent needs [Hans Seibel '01].

This is the motivation behind HISAAB (which means "accounting" in Hindi). HISAAB is an excellent example of a technology intervention aimed at social benefit and economic development. Specifically, it attempts to empower the rural poor to own, manage and operate the system of accessing credit in spite of the barrier of literacy. As described in a project summary available online, (http://sourceforge.net/docman/display_doc.php?docid=10428&group_id=50219)

> "HISAAB is group-level micro-finance management software, intended to document transactions at the lowest level of MFI operations….[This project explores]… the interface design space for illiterate and uneducated users, the typical users at the group-level of Micro Finance Institution operations." The plan is to "…explore a variety of interface design paradigms to test accessibility to such a user domain. One novel design paradigm is the idea of numeric interfaces, leveraging users' ability to remember, manipulate and enter numbers. (1)

The software is also being built with very little text and an emphasis on audio and graphics. Several interesting design iterations have happened and the final design solution is going to be prototyped very soon.

The project is funded by Media Labs Asia, Mumbai whose charter is one of digital inclusion.

## 3    Conclusions

Human Computer Interaction was a little known concept in India till even five years ago. However, fuelled by the requirements of the Information Technology industry which increasingly sees usability as a differentiator for their services, there has been a spurt in usability awareness.

There is also increased participation , in India, from usability related forums in the west. There has been a major initiative undertaken recently by the Computer Society of India(CSI) in partnership with the  British Computer Society (BCS) in relation to Software Usability and Human Computer Interaction (usability). The European commission has sanctioned sponsorship under this initiative known as the Indo European Systems Usability Partnership (IESUP).

With impetus such as the above, the Indian genre of usability seems ready to play a consistently major role in the global arena.

## References

http://sourceforge.net/docman/display_doc.php?docid=10428&group_id=50219

# An Account of Factors that Determine HCI Design Uptake in a Techno-Centered Country Like Singapore

*Kee Yong Lim*

Centre for Human Factors & Ergonomics
School of MPE, Nanyang Technological University
Nanyang Avenue, Singapore 639798

mkylim@ntu.edu.sg

## Abstract

When a country places a heavy emphasis on automation and advanced technology, one might expect a correspondingly greater concern in ensuring system effectiveness and acceptability. These concerns map nicely onto HCI concerns pertaining to system functionality and usability. Unfortunately, such an expectation may be too simplistic for many reasons. This paper examines various socio-economic factors and perspectives derived from a number of case-studies, in an account of how they may work to thwart or support the uptake of HCI contributions. It examines the implications for usability engineering as interpreted in the broadest sense; namely as encompassing the entire scope of HCI, rather than in the narrow sense as comprising usability evaluation in a laboratory.

## Position Statement

Taking a broad view of the scope of HCI, socio-economic factors that may determine due consideration or otherwise of HCI design contributions, may include the following:

1. whether HCI has been considered a mandatory part of the training of software engineers and computer scientists. In many developed countries where HCI is well established, the accreditation boards of computer science courses would require explicit inclusion of HCI in the curricula. This factor is important as it influences the attitude of software designers towards HCI at an early stage of their career path. An exposure to fundamental HCI knowledge and perspectives at this time, would not only empower these designers to address HCI concerns, but would also engender a more balanced view of user requirements and of trade-offs between technical and human aspects of

computer systems. This HCI training requirement, unfortunately, does not apply in Singapore. As a result, software designers graduate and enter into industry with little or no knowledge of HCI. Consequently, misconceptions about HCI prevail, for example:

- User friendliness is all about making screen displays and icons attractive and pretty.

- User interface design is all about screen display and layout design.

- A graphical user interface (GUI) is a panacea for usability. Providing a GUI guarantees a user friendly design.

- I follow the Microsoft style guide closely, so there can't be any usability problems. That's all I need to do for user interface design.

- User interface design is something I need to do at the end.

- User interface design is really not my job, the graphic artists should do that.

- HCI is something I do if I have time. It is really a luxury feature and not a necessity for system effectiveness.

- I can use this user interface so users should not have any problems.

- The design has been evaluated comprehensively by design team members and cleared.

- Acceptance testing has been done and there are no bugs. Why do we need users?

- The program is very effective. It gives you the result in 1 second, to 99% accuracy and 100% reliability.

- The system is good. It has been checked with the client who is paying for the project.

- To use this system, you must undergo intensive training. There is no way you could do it otherwise.

- The users are not well trained and shouldn't have done that. There is little we can do about these user errors as the system design is fine.

- I have checked the look and feel of the user interface design. The tasks the users perform previously is not relevant in this computerized system. They need to learn how to do it the new way.

2. whether there is a significant product/system design and development industry and/or demand for bespoke designs. If off-the-shelf procurement is the main stay, then HCI contributions may be confined to design evaluation. In Singapore, the design and development industry is in its infancy. HCI contribution to design specification is therefore still limited for this sector of the economy.

3. whether the pace of technological implementation and renewal is fast. If the pace is fast, then product/system failures due to poor design are quickly supplanted and forgotten, especially if the project scale and cost are small. Similarly, if competition is intense and the product life cycle is short (a characteristic typical of the IT industry), then poorly designed systems may be made obsolete more quickly and fade from the scene. The fast turnover of such products may be perceived as an unavoidable outcome of intense competition and a 'use and throw society', rather than 'premature death' due to the lack of appropriate design attention. Such cases have emerged in Singapore, where the product turnover rate is high. For instance, more than 10 post-implementation revisions had to be done within a short period of time, due to fundamental inadequacies in the design of a message display terminal for a GPS taxi booking system. The entire system is eventually replaced within a few short years.

4. whether systems are initiated by Government and commercial organizations, or paid for by end users. In the former case, development costs are not be borne directly by users. As such, they might be more tolerant of less optimal designs, or may choose to avoid using them if they could. In cases when system use is obligatory, users might be forced to learn to use the system and put up with poor designs. Lost productivity and hidden inefficiencies may remain undiscovered as a result. Alternatively, potential benefits of a hi-tech system may not be realized in full due to limited or partial use of the system. For instance, the reality for e- and m-commerce has so far fallen short of its promise of a major economic revolution. Similarly, the economic returns expected from the nationwide implementation of a digital information infrastructure in Singapore, have not been commensurate with the huge

investment made. A key reason for this dismal outcome is the neglect of user-centered concerns in the design of e-commerce services and user interfaces. As a result, users continue to be frustrated by poor functionality and by being lost in cyberspace. For consumer product companies, users may express their displeasure by voting with their feet. When confronted with such problems (usually discovered post implementation), these companies can only choose stopgap solutions such as offering incentives to coax customers to buy or use their systems. These sub-optimal fixes do not usually solve the problems. Sadly, such cases are rife in Singapore as evidenced by electronic banking, auto-teller machines, etc.

5. the educational level of the population. It may be anticipated that a population with a higher per capita education may be more resilient to less usable designs. However, user resilience does not equate with tolerance or an accommodating attitude. A reason for this is that users who are more educated also tend to be more sophisticated in their demands. They are also likely to be more self confident, vocal and defensive of their rights. Consequently, they are less likely to accept blame for errors and difficulties encountered in system use. However, these user responses may be moderated by cultural factors. In the case of Singapore, the population tends to be more accepting and tolerant, and less prone to speaking up. Thus, although poor designs are noticed, they can remain unaddressed.

6. the national perspective on automation and high technology. The Singapore Government has been very aggressive in exploiting automation and high technology as a way to increase productivity and to counter its scarce human resources. Although it is clear that such an emphasis constitutes a national imperative, there are instances in which the use of automation and high technology has been excessive or inappropriate. In particular, an imbalanced techno-centric perspective can lead to the undue predisposition of 'automatically' resorting to the replacement of human operators, rather than the use of technology to complement or extend human abilities.

7. the prevailing or predicted demographical trend. As with many countries worldwide, Singapore is experiencing a very fast ageing population. To aggravate matters, daily life is getting ever more complicated in Singapore. Work-wise, life-long learning is now considered the norm. However, more intensive user training should never replace efforts to reduce and manage system complexity through user-centered design. Thus, HCI design should emerge as a key concern to address the needs of this user group. The extent to which these users can continue to contribute to the socio-economic

development of the nation, will depend on how well the design of work systems and home applications has been tailored to meet their needs.

To conclude, the uptake of HCI is determined by the inter-play of many socio-economic factors prevailing in a country. HCI can only make substantial inroads if efforts are made to raise public awareness to influence government policies and shape national interests. Similarly, only with greater awareness can consumers be prompted to pressure commercial companies into paying greater attention on HCI design.

# Promoting Usability Engineering in China

*Zhengjie Liu  Haixin Zhang  Junliang Chen  Liping Zhang*

Chinese Center of EU UsabilityNet
Dalian Maritime University, Dalian, 116026 P.R.China
liuzhj@dlmu.edu.cn
http://usability.dlmu.edu.cn

## Abstract

For the rapid economic growth and the competition challenges faced, it is expected a great potential for Chinese industry to accept usability engineering in the near future. An assessment at some representative Chinese IT enterprises revealed the current status in the industry especially the obstacles that lie ahead, such as lack of multidisciplinary background, lack of expertise for UCD methods and convincing cases studies. Some feasible strategies are finally suggested to enable usability engineering gradually to be adopted in this country.

## 1  General Background

China is a country with an ancient civilization, a broad expanse of land and a population of 1.3 billion people. In the economic arena, China has experienced a consistent, rapid growth of 9.5% per year in the past two decades and now is becoming potentially the biggest market and an economic giant in the near future. In some of the most developed regions the GDP per person has reached the level of around 5000 USD. Therefore China has attracted attention from around the world in recent years for its rapidly increasing economic and political influence. The Chinese information technology (IT) sector has been a driving force in the economic growth under the government policy of promoting industrialization through digitalization. IT has already become the biggest sector in Chinese industry and foreign trading. Chinese IT sector is now the second largest in the world in comparison to being ninth in 1989. China is now also the second largest IT market in the world, has the largest user population for mobile phones and the second largest for Internet, and is expected to become the largest manufacturer for more and more IT products. Analysts predict that China will maintain a growth rate of some 22% in the IT sector and some 30% in the software and service area in the coming years. Chinese government recently launched the Policy No.47, which seeks to speed up development, enabling the software industry reach 30 billion USD in sales and 5 billion USD in exports by 2005.

The growth in economic scale and the evolution towards an increasingly mature market economy gradually brought Chinese enterprises into a situation in which they must face greater competition for their products and services. Especially after China entered the World Trade Organization (WTO) Chinese enterprises endured even greater competition internationally while at the same time more opportunities for development in a global market. Media reports and advertising clearly show that "customer orientation" has become a slogan for many Chinese enterprises. They are beginning to pay more attention to customers' needs and providing better services to win customers' loyalty. A shift from product economy to service economy is also taking place. This change is in sharp comparison to the circumstances under the planned economy of 20 years ago under which businesses cared little about users' needs and satisfaction. Businesses have gradually realized that they must find effective ways to strengthen their competitive edge in order to survive.

Usability, a key quality factor for an IT product, concerns whether users can do what they want with efficiency and satisfaction. Consequently, it is the determinant of the product's competitiveness. Usability engineering, or more specifically, user-centred design (UCD) is a proven, well recognized, and effective approach for developing usable products in Western countries since the 1980s. In light of the competition challenge that Chinese enterprises face, there is every reason to expect a great potential for China to accept usability engineering in the near future. To this end, the first step is to have a clear understanding of the current situation - in order to form a suitable strategy to promote the development of usability in China.

Before going further, a glance at the general HCI discipline in China may give us a broader understanding. From the 1990s professionals in China began to introduce concepts of HCI and to practice them. And later on in recent two or three years the concepts of usability, usability engineering as well as UCD were introduced. Now there are about twenty institutions working in HCI, including computer science, industrial engineering, and psychology departments in universities; research institutes; and industrial R+D departments. The majority of HCI people come from computer science rather than psychology or other disciplines. HCI research is concentrated on some limited HCI technologies like multimodal interfaces using speech and pen, emotion-based interaction, and HCI software architecture, while paying less attention to industry's needs. There are only a few universities teaching HCI to undergraduates or postgraduates in their curricula. There is still no dedicated journal or conference in the subject of HCI. However, a relatively small but active HCI community is forming, which has been working on setting up an HCI group. In November 2002 the 5th Asia-Pacific Computer-Human Interaction Conference (APCHI) was successfully held in Beijing attracting more than 120 participants. Although one might expect a promising future for HCI in China, the limited existence of this area inevitably has had some negative influence on the status of usability.

## 2  Obstacles: an Assessment

For better understanding the current situation of usability in China, we conducted an organizational human-centeredness assessment at Chinese IT enterprises. This assessment was based on the Human-Centeredness Scale (HCS) of the Usability Maturity Model (UMM), a model of usability processes (INUSE, 1998) (Bevan & Earthy, 2001). Human-centeredness assessment is an evaluation of an organisation's attitude to usability and is suitable for organisations that are new to usability. The evaluation focuses on how human-centred activities are managed in projects and can be used to diagnose the current situation of the organisation in planning a more detailed assessment. The HCS comprises five levels: recognized, considered, implemented, integrated, and institutionalised. Each level consists of one or more process attributes and each of them further includes several management practices.

We chose for the assessment three IT enterprises that are representative of leading enterprises in China. For the assessment at each company we interviewed one project manager and one developer. The interview, which was structured with note taking and audio recording, took about three hours altogether at each organization. After data collection and analysis, the assessment results showed that the usability maturity level for all three enterprises is the 'recognised' level in the five-point scale of HCS. This evaluation actually reflected the current maturity level of most leading enterprises in the sector. This means that these enterprises have already recognised the importance of usability and have had a kind of spontaneous, but not yet clear demand for usability engineering, but it is not yet deemed critical to their competitiveness and survival. Currently, there have been some practices in their process that collect information relevant to users' requirements and apply this information somewhere in development. This does provide a fundamental basis for promoting usability engineering in the IT sector of industry.

However, none of the enterprises studied reached beyond the 'recognised' level. Based on our understanding gained from this assessment, the following are obstacles that lie ahead:

- Meeting users' requirements, which is the core of usability, have been understood as very important by the enterprises while paying attention to the traditionally important quality factors like correctness, stability, and performance in operation. However, there is a lack of effective measures that could be used routinely to tackle usability problems. The enterprises tend to achieve this objective by emphasizing the rigid execution of software processes and strengthening requirements management following the principle of ISO9000 or CMM, which reflects the popularity of this approach nowadays in China. Although the importance of meeting users' requirements is stressed in the internal training for the employees, the focus of training is usually the regular development processes, programming skills, and tools.
- Usability engineering needs a multidisciplinary team, especially personnel with psychology and HCI design backgrounds. There is a severe lack of such personnel in the enterprises; almost all the employees in the development departments are trained in computer science or related technology oriented specialties. There is neither special departments nor special positions for UI design, just sometimes in a project a few developers are assigned by chance to do UI design. The specific skills for UI design and usability are not well understood, say nothing of gaining such expertises through training, recruiting or contracting.
- Communication with users has been assigned high priority. However the enterprises rely mainly on accumulation of experiences to enhance their communication without well-disciplined methods for this purpose. When analysing the reasons for not getting users' requirements correctly, companies used to assign to the users half of the responsibility, for example, because the users did not describe the requirements clearly and completely, changed their minds frequently, differed among themselves over requirements, did not know how to construct products technically, etc. They required users to adapt to the development process. Companies did not realise that the problem could be solved only by improving the process to make it more user-centred.
- There is some user participation at the stages of feasibility, requirements-gathering, as well as testing and maintenance. However there are almost no such activities at the design stage. Although users are requested to participate in reviews at the end of each stage, this procedure usually is not followed. Typical methods for communication include interviewing, prototype demonstration, and feedback collection. These methods are used in an 'at will' style, without well-established rules. Sometimes prototype demos are sent by email to the users for feedback to replace face-to-face contact.
- Although prototyping is used in product development, it is used only when the developers feel they need it to help communication with users, and merely as a demonstration. Rarely are users allowed to do real tasks with the prototype or product in order to find usability problems by observation, not to mention for measuring user performance and product usability quantitatively.

- Some enterprises have many contracts for exporting software for overseas market. This situation is quite representative for some coastal regions in China. In such projects, a foreign partner located in the target country usually takes charge of direct contact with the users at the feasibility, requirements-gathering, design, and maintenance stages, while the Chinese company takes charge only of detailed design, coding, and testing. Consequently, Chinese attention tends to be paid to quality measures like bug rate. This kind of project is not beneficial to cultivating an awareness of usability and the conception for user-centred design.

Although severe problems existed, we did find in this assessment some positive practices that are good for user-centered design and that deserve encouragement. For example, rules or guidelines about the scope and the process for requirements analysis are developed in some enterprises based on users' feedback as well as experiences gained by requirements-collecting personnel that strengthen communication with the users. Efforts have been made in some projects to get knowledge about users' working environments, to become knowledgeable in the application domains, to ask for domain experts for consultancy, or to make products conform to trade standards. User-interface style guides and templates are developed following popular user-interface styles. Users are categorised and different user interfaces are developed accordingly. Prototyping is used in the requirements-gathering stage to exchange ideas about the look and feel of user interfaces. In the test on delivery, users are sometimes asked to perform real tasks to find potential problems. Users are requested to provide a feedback report after delivery, and so on.

## 3 Challenges and Prospect

The assessment and the discussions above helped us to form the following picture of Chinese IT enterprises: Along with the increasingly greater competition pressure, they have started to realise the importance of usability and to try to find solutions for their problems. The major obstructs they are facing include lack of multidisciplinary background in the development teams and lack of expertise for UCD methods. Convincing cases of successful usability engineering application are needed to encourage them to really invest in this area. This summary suggests following measures to enable usability engineering gradually to be adopted in the Chinese IT sector of industry.

1. Discounted or cost-effective UCD methods (INUSE, 1997) (Nielsen, 1993) that are suitable for the development teams that is lack of psychology related backgrounds should be taken as the first choice when introducing usability engineering in the industry. Simplified formative user test with prototypes or finished products would be a good start to get developers realise how different users are from themselves. This would be very helpful to make them inclined to user-centered approaches and more willing to accept other UCD methods later on.

2. Usability has been perceived by most people as something hard to measure quantitatively. The quantitative evaluation methods for usability, like that recommended by CIF standard (ANSI, 2001), should be introduced to the enterprises to set up benchmarks for products. This approach would help make the effects of any UCD methods and the improvement in usability visible in order to convince the enterprises about the tangible benefits. It would also give people an impression that usability engineering is an established scientific discipline.

3. Training courses for user interface design and usability as well as corresponding certifications should be provided to show the value as the profession and to encourage enterprises to set up specific job positions for usability experts. This effort would foster the accumulation and development of related expertise in the industry.

4. Pilot projects need to be carefully selected and conducted to provide successful cases for convincing more enterprises to adopt UCD approaches. Emphasis of the pilot should be placed on the localization of the UCD methods (especially their suitability for technology people), cost-benefit analysis, as well as development of guidelines for the methods, training materials, case studies, and tools.

5. International or industrial standards, like ISO9000, CMM and product-oriented ones, have been well respected in the Chinese IT industry sector. Evaluations and certifications for product usability and process usability maturity might be a potential factor that can substantially drive the acceptance of usability engineering in industry.

6. Efforts should be made to make HCI and usability a component in the curriculum for computer and IT-related departments of universities and provide training to qualify the teachers.

7. Application-oriented research or joint research projects with industry in HCI should be especially encouraged to attract more industry involvement.

8. As user organizations are becoming more concerned about usage cost of the IT systems and products they use, efforts made on this side for promoting usability engineering might offer dramatic leverage.

The assessment consolidated our belief in the promising prospect for usability engineering development in China. At the same time, it provided a firmer basis for determining a feasible way to achieve our goals. However, further assessments and studies in depth must be done for a more comprehensive understanding of the circumstances. In the last few years some transnational companies such as Siemens, Microsoft, and Nokia set up their usability and HCI-related groups in China. We are even pleased to note that some leading Chinese IT enterprises like Legend, Neusoft, Kingsoft and Hier recognised the value of usability to their competitiveness. They began to pay attention, to learn about usability, set up usability groups, and to practise the discipline. The China ComputerWorld, the most influential IT bi-weekly, published a special column in issue No.36 of 2001 on usability engineering that was drafted by the author. A government-funded project is now working on introducing usability measures to the evaluation of highway toll-collecting systems. A top IT media company is considering adding usability and accessibility as part of its regular comparative evaluation of e-government websites. Although it is still far from mature for the current practice, these exemplary instances plus the demand for competitive products will definitely bring more people in China to understand the concept of usability and accept it as a necessity.

## References

ANSI. (2001). ANSI/NCITS 354 Common Industrial Format for Usability Test Reports.

Bevan, N. & Earthy, J. (2001). Usability Process Improvement and Maturity Assessment. *Proceedings of IHM-HCI'2001*. Cepadues-Editions.

INUSE. (1997). Handbook of User Centred Design. NPL project IE2016 INUSE Deliverable D6.2.1.

INUSE. (1998). Usability Maturity Model. Lloyd's Register of Shipping project IE2016 INUSE Deliverable D5.1.4.

Nielsen, J. (1993). Usability Engineering. Academic Press.

# Usability Metrics in Adaptive Agent-based Tutoring Systems

*Víctor López-Jaquero, Francisco Montero,*
*Antonio Fernández-Caballero & María D. Lozano*

Laboratory of User Interaction and Software Engineering
Computer Science Research Institute
University of Castilla-La Mancha, Albacete (Spain)
{ victor, fmontero, caballer, mlozano }@info-ab.uclm.es

## Abstract

Human-computer interaction in traditional application development is focused on the interaction between tasks and a single user interface designed for a single kind of user. A logical evolution should lead interaction to a development model where user skills and preferences are taken into account. In this paper, we introduce preference metrics and performance metrics as parameters that enable user interface adaptation in tutoring systems. The proposal includes a practical example for learning/teaching of an engineering course.

## 1    Introduction

The ultimate goal for Human-Computer Interaction (HCI) must be the creation of user interfaces based on each individual user preferences (López-Jaquero, Montero, Fernández-Caballero & Lozano, 2003). Those preferences can be captured initially, to a certain extent, in analysis development stages. Using those captured data user profiles can be created in concordance with the identified user stereotypes. However, the user advances in his knowledge, and his preferences change. We need to understand the way the user "uses" the application, and that is where usability metrics can do the job for us.

We introduce in this paper how usability metrics applied to intelligent tutoring systems can lead to the definition of high adaptive learning/teaching systems. These systems will be able to adjust to a specific student and even they will be able to recommend to teachers how to improve the course. Finally, to achieve our goal to create a highly adaptive teaching environment it will take some artificial intelligent techniques that will be modelled by means of multi-agent systems (MAS). An Intelligent Tutoring System is proposed, in which usability metrics are used to achieve intelligent behaviour of the system to improve learning.

## 2    Usability Metrics

Usability metrics are software quality metrics with a long history of successful application in software engineering (Card &Glass, 1990; Gilb, 1977; Henderson-Sellers, 1996). But, metrics also carry risks (Constantine & Lockwood, 1999). No simple number can completely represent anything as subtle and complex as the usability of a software system, but numbers can sometimes create the illusion of understanding.

Usability metrics have a number of uses, but mostly from the designer's point of view. Metrics for usability can be thought of as falling into three broad categories: preference metrics, which quantify the subjective evaluations and preferences of users, performance metrics, which measure the actual use of working software, and predictive metrics, or design metrics, which assess the quality of designs and prototypes. We shall focus on preference and performance metrics.

One of the most popular ways to assess usability is to use preference metrics. User satisfaction is a component of usability and also an important factor in success in the marketplace. One good example of a standardized set of preference metrics is the Software usability Measurement Inventory (SUMI) developed as part of the ESPRIT project (Porteous, Kirakowski & Corbett, 1994). SUMI is a 50-item questionnaire that includes five subscales measuring different subjective aspects of usability: affect, efficiency, helpfulness, control, and learn ability. Another approach is the Subjective Usability Scales for Software (SUSS) questionnaire, which measures six key elements of user interface designs affecting usability: valence, aesthetics, organization, interpretation, acquisition, and facility. Preference metrics are one of the pillars for user interface customization. However, because of their intrinsic characteristics, they are difficult to assess at run time. There are some preference metrics, such as the manipulation artefact used when commanding tasks (keyboard, menus, and toolbars) that can become especially useful for capturing user preferences.

Performance metrics are indices of various aspects of how users perform during actual or simulated work. Measurement studies form the basis of much traditional research on human factors. User performance is almost always measured by having a group of test users perform a predefined set of test tasks while collecting time and error data (Nielsen, 1993). Typical quantifiable usability measurements include: the time users take to complete a task; the number of tasks of various kinds that can be completed within a given time limit; the ratio between successful interactions and errors; the time spent recovering from errors; the number of user errors; and so on (Nielsen, 1993). Of course, only a subset of these measurements would be collected during any particular study. Performance metrics are especially useful for assessing overall usability. One important point for this kind of metrics is that most of them can be evaluated at run time in a simple manner. Performance metrics are one more input parameter to advance towards user interfaces adapted to the user. Our proposal is to leave behind user interfaces where the user must adapt to a given and fixed interface.

# 3    An Adaptive Agent-based Tutoring System

The architecture proposed so far is being tested in e-learning system as an Intelligent Tutoring System (ITS) for an Engineering course taught at the Polytechnic Superior School of Albacete, University of Castilla-La Mancha. One of the main goals is that the alumni learn more and better, that is to say, to be able to structure learning matter in such a way to facilitate the learning facilities. One characteristic to take into account in learning is the rhythm the student is able to learn. Thus, an ITS has to adapt rhythm it introduces the concepts to the learning rhythm of each student (for instance, to show more or less exercises, to show more or less tests, etc.). Another aspect widely considered in learning theory is reinforcement by rewarding a correct answer and penalizing the errors (by means of messages, sounds, etc.). Another goal in our environment is to enhance teaching as well as learning. One of the main problems a professor faces when teaching is that he does not know the skills of his alumni. Our proposal leads to conclusions that "teach how to teach".

In our teaching system (see figure 1) there are three multi-agent systems: (1) The *Interaction MAS*, which takes care of the user. It captures user preferences by means of usability metrics to build profiles. The contents shown to the user will be created according to the preferences and skills captured to improve learning experience. (2) The *E-Learning MAS* composes contents for the user. Contents are made of three different parts: theory, exercises and tests. All three parts are composed according to the information captured by the *Interaction MAS*. (3) The *E-Teaching MAS* is one of the most important parts according to our experience. It makes recommendations for improving our day-to-day classes for that course.

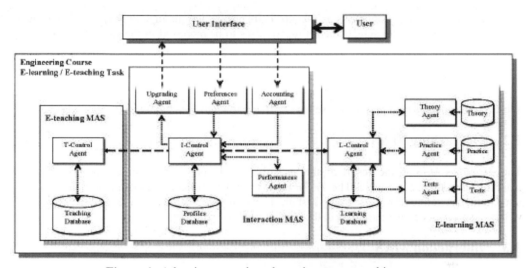

**Figure 1:** Adaptive agent-based tutoring system architecture.

The user (the student) is in front of the user interface. From the interaction of both entities, modelled by the *Interaction MAS*, different metrics that are stored in a *Profiles Knowledge Database (KDB)* are collected. This database contains the different profiles as a result of the use of the system by different students, with different aptitudes, motivations, etc. The multi-agent system for the learning (*E-Learning MAS*), gets data obtained from the profiles (analysis of the distinct metrics captured) and adequates the contents shown to the concrete student that accesses the Web site. On the other hand, the multi-agent system for teaching (*E-Teaching MAS*) obtains measures that permit to get recommendations to enhance the course. Finally, to offer a good learning of the course, the latter has been decomposed into theory, exercises and tests.

## 3.1 E-Learning MAS

The *Learning MAS* appears from the general goal to maximize the course learning. The *learning control agent* communicates bi-directionally (asks for and receives information) with the *theory agent*, the *exercises agent*, the *tests agent* and with the *interaction control agent* (*Interaction MAS*). This agent asks for/receives Theory Web Pages to/from the *theory agent*, asks for/receives Exercises Web Pages to/from the *exercises agent*, asks for/receives Tests Web Pages to/from the *tests agent* and communicates (through the *interaction control agent*) with the *performance agent* to record the performance of the student in order to decide if he needs a reinforcement. If the student needs some kind of reinforcement the *learning control agent* will elaborate a plan with the material that has to be shown to the student. In order to determine if the student needs reinforcement the *performance agent* will have access to a *KDB* where the minimum requisites for each subject are stored (quantity of exercises to be initially shown to the student, how many exercises the student has to answer correctly, and in how much time, maximum time to correctly answer an exercise, etc.).

The *theory agent* is constantly waiting for the *learning control agent* to ask for a Theory Web Page. When this occurs, it looks for the proper theory page and sends it to the *learning control agent*.

The *exercises agent* is autonomous as it controls its proper actions in some degree. The agent by its own means (pro-active) selects the set of exercises to be proposed in the subject studied by the student and adds to each exercise the links to the theory pages that explain the concepts related to the exercise. It sends to the *learning control agent* a Web page containing the exercises proposed. The *tests agent* is continuously listening to the *learning control agent* until it is asked for Tests Web Pages. The agent by its own means (pro-active) goes on designing a set of tests for the subject the student is engaged in. These tests will be shown to the student in form of a Web.

## 3.2 E-Teaching MAS

The *Teaching MAS* is the result of the second general goal fixed, namely, to maximize the teaching capacity of the course. The *Teaching MAS* will be collecting the goodness or badness of the parameters defined for the learning system. The *Teaching MAS* is pro-active in the sense that it will be providing recommendations to the teacher on those parameters.

## 3.3 Interaction MAS

The *Interaction MAS* has been conceived to facilitate the adaptive communication between the system and the user. The *interaction control agent* tells the *upgrading agent* what the user preferences are, as obtained by the *preference agent*, and which has to be the next Web page to be shown (*learning control agent* of *Learning MAS*). When speaking about the preferences of the student, we mean the type of letter, the colour, the icons, etc., the user prefers. The information collected is stored in the *Profiles KDB*. All information concerning time-related parameters and some of the user's behaviours are obtained through the *performance agent* and the *accounting agent*.

The *preference agent* perceives the interaction of the user with the user interface and acts when the user changes his tastes. The *preference agent* is continually running to know the student's preferences at any time.

The *performance agent* calculates the performance metrics when the student leaves the system (at the end of a working session) y goes evaluating everything the student does in order to know if he needs reinforcement. It is autonomous and pro-active; as it may calculate metrics at the same time the student performs other tasks. Some of the metrics the *performance agent* handles are: for each Theory Web Page, the mean time alumni spend there; for each exercise Web page, the mean punctuation obtained by the alumni, as well as the time spent to get the correct answer; for each Tests Web Page, the mean time spent to answer all questions, and the mean punctuation obtained in the tests.

The *accounting agent* perceives the interaction between the student and the user interface and acts (gets information) when the student changes to another Web page, scrolls up and/or down, performs an exercise or a test, etc.

Finally, the *upgrading agent* is constantly waiting for the *interaction control agent* to ask to update the user interface with the new information to be shown to the student (to show another Web page or to show the same Web page but changed to the new tastes of the student).

# 4    Conclusions

User interface generation has become a software engineering branch of increasing interest. This is probably due to the great amount of money, time and effort spent to develop user interfaces, and the increasing level of exigency of user requirements for usability and accessibility ("W3C", 2002) compliances. Besides it, users engaged in HCI are becoming more and more heterogeneous, and that is a fact we cannot ignore.

In this paper we have proposed an architecture that considers the high diversity of users' skills and preferences: a user-centred and adaptive interaction multi-agent system. This architecture is inspired in usability metrics.

## Acknowledgements

This work is supported in part by the Spanish CICYT TIC 2000-1673-C06-06 and CICYT TIC 2000-1106-C02-02 grants.

## References

Card, D., & Glass, R. (1990). Measuring Software Design Quality. Prentice-Hall.

Constantine.L.L., & Lockwood L.A.D. (1999). Software for Use: A Practical Guide to the Models and Methods of Usage-Centered Design. Addison-Wesley.

Gilb,T. (1977). Software Metrics. Winthrop Publishers, Inc., Cambridge MA.

Henderson-Sellers, B. (1996). Object-Oriented Metrics: Measures of Complexity. Prentice-Hall.

López-Jaquero, V., Montero, F., Fernández-Caballero, A., & Lozano, M.D. (2003). Towards adaptive user interfaces generation: One step closer to people. Proceedings of International Conference on Enterprise Information Systems 2003.

Nielsen, J. (1993). Usability Engineering. Academic Press.

Porteous, M., Kirakowski, J., & Corbett, M. (1993). SUMI User Handbook. University College Cork, Ireland.

W3C. (2002). http://www.w3.org/WAI/

# Procuring Usable Systems -
# An Analysis of a Commercial Procurement Project

*Erik Markensten*

Interaction and Presentation Lab and the Centre for User-Oriented Design
Numerical Analysis and Computer Science
Royal Institute of Technology
100 44 Stockholm, Sweden

## Abstract

This article presents a case study of how usability was dealt with in a procurement process of a content management system. The results indicate that the procurers found it difficult to define usability requirements, probably because they lacked tools and experience to do so. Difficulties also arose because the tools that they used were based on idealized models of how users worked. It is argued that proper usability activities at an early stage could have facilitated the procurement process and the discussion with suppliers, as well as integrated usability into the development process. A brief outline of how such an integration could be made is described.

## 1    Background

Methods for usability and user-centred design have mostly addressed suppliers' production models. Accordingly, positions as usability professionals have primarily been found in supplier organisations. Procurers, on the other hand, have relied on suppliers to create usable systems and have not focused explicitly on usability issues in their procurements (Holmlid & Artman, this volume). Unfortunately, although it might seem obvious to expect that the system you purchase should be usable and useful this is seldom the case. An important reason is that usability issues have not been dealt with consciously in organisations or projects and have been separated in development processes (Carlshamre, 2001). Usability professionals have, if at all, been involved in projects too late to have any impact on issues such as interaction design and utility. The separation has also brought with it a rather narrow view of the concept of usability. Usability is seldom discussed in relation to organisational change or business strategies, but rather as an isolated concept or as a property of the product or interface. The implicit assumption that suppliers should be responsible for usability may stem from this narrow view of usability - it is the responsibility of the supplier to develop the system.

In contract development (see Grudin, 1991 for a discussion of different development contexts) the development project and the work of the usability professional is often considered to start when the supplier signs the contract with the procurer. Although much work has often already been done in the procurement process this is not perceived of as usability activities. Nevertheless, the goal of the procurement is similar to the goal of the usability activities: to get from business goals to system requirements, while assuring that the system gets useful and usable. Thus, another reason why usability activities are often omitted in contract development is that the activities do not seem to add much to what has already been done in the procurement, even though this work may not have focused on actual use at all (Balic, Berndtsson, Ottersten, & Aldman, 2002).

This paper describes and analyses how a group of procurers attempt to define the requirements for a new content management system, and how requirements concerning usability were dealt with.

## 1.1 The Case Study

The study was carried out in a department within a large Swedish bank. The business goal of the department was to produce economical analyses. One of the key ideas behind the procurement project was to automate and streamline the analysis production process in order to cut costs and gain competitive edge. A project group consisting of three to four persons was put together to work with the procurement. Although the project leader was educated as a system analyst, none of the procurers had any formal training in either requirements engineering or user-centred design.

The bank had its own development process that should be used when purchasing and integrating products. Unlike many development processes found in the software engineering or HCI literature this one also included procurement. The first milestone, the Request for Information (RFI), is a tender that is sent out in order to find out which suppliers to involve in the forthcoming procurement process. It states the purpose of the new system and specifies important functions and business needs. The research project started after the RFI, when the work on the Request for Proposal (RFP) should begin. The RFP is the final requirement specification in the procurement process and is used to select a supplier to continue working with in a pre-study before the implementation project starts.

## 2  Method

Data was collected for six months, from March to September 2002, through participatory field studies including participatory observation, interviews, video and audio recordings. Data was also collected as documentation resulting from the ongoing work such as draft reports, mails, and meeting notes. The data was analysed qualitatively from an activity-theoretical perspective. Although the researcher role was seldom discussed my presence as a peripheral member of the team made it possible to participate in meetings and discussions. My role was at the outset to analyze usability issues in the procurement process but it soon changed to become more of a discussion partner or mentor for working with the requirements from a usability perspective. Thus, although I did not participate much in the activities, my presence had an effect on the RFP with respect to usability.

## 3  Results

When initiating the RFP project the motivation for procuring a usable system was high. Despite this interest in usability, it soon became apparent that the procurers found it difficult to define requirements. There was a frustration about not being able to get it right and they often exclaimed things like "This [specifying requirements] is so hard" or "This is the most difficult thing that I have ever done". Their goal was to define the requirements for a system that would not only fulfil the overall business requirements, but that would also fit the needs of the users.

At one meeting, there was a discussion about system requirements. The discussion was not grounded in knowledge about actual usage but more in on-the-fly inventions about what might constitute user requirements. For example, requirements and models gained bottom-up from the analysis of some users' work were applied top-down, without much consideration, to the work of some other users that had quite different duties. Incidents like this confused the participants and prevented progress. After a while, due to some interventions by me, the discussion in the meeting turned more towards actual usage and the user research results. This had immediate effect and moved issues forward. When finishing the meeting one of the procurers said *"Wow, what a relief! Now we really start with actual needs... this feels so good!"*

Thus, it appears that the procurers, with little experience of requirements engineering and user-centred design, could not find the right tools to reach their goal of defining requirements based on usage. Instead, they had to invent their own tools as they moved along. In a meeting with an internal expert on procurement and the bank's development process, the procurers complained about this: *"You know we have RFP templates, guidelines for managing the relation to the suppliers, supplier evaluation matrixes etc. but we don't have any support, internal course, or tools that guide you into how to define the system requirements."*

## 3.1 Understanding User Requirements

As is common (Rouncefield, Viller, Hughes, & Rodden, 1995) the procurers often based their reasoning on idealized models of work and use, influenced by a rational or technical perspective. This conflicted with their stated goal of actually understanding and starting from the actual use model. The following passage is taken from an early user interview and concerns whether the system should be structured according to a global standard (GIX) or a locally developed structure.

| | |
|---|---|
| Analyst | ... and now we have a common folder on a shared server where all our models are located (...) So it is not at all structured according to the GIX standard but more in a way that is practical for us so we know where each company belong. |
| Researcher | mmm... do you see any problem in using GROW [application using the GIX standard] together with this or do you know where to... |
| Procurer | ... to have the same structure on both your analysis and your shared folders and on your GROW data? I mean, that's what it's about, to structure information accordingly ... and structured... |
| Analyst | Well, I don't know if that'll pose any problems. I mean the GIX standard is a global standard and it is... well it's hard to apply to... certain sectors naturally. (...) |
| Procurer | And then... well, I think it feels natural to structure your data using the same structure as the data the companies are structured by. |
| Analyst | ... well (doubting)? If it wont cause any, any problems then...? |
| Procurer | It will be so messy otherwise because then you have one structure in one place and then you find the companies in another place... (...) But all work documents... all information folders... all INFORMATION should be structured in the same way...? |
| Analyst | ........... well, well I, I don't know.. but yes or I mean I don't know if it's so damn important cause then you could say that... if all information should be incorporated in all structure, I mean, the information that I keep in my head – that is not located in any structure at all. But it is still used and expressed in my written analysis sooner or later (...) maybe it doesn't matter how my own folder is structured, just as it doesn't matter how my brain is structured and the fact that I don't write everything down? |

In the passage above the procurer reasons about usage and tries to interpret and understand it from an idealized model of work. The user (analyst), on the other hand, bases his understanding on the actual use model. Since the procurer has little experiences of assessing use through activities with users, uses a different model for the user's work and has different objective than the user (to create a consistent requirement specification), she assumes that high quality must mean consistent structuring of information. The user, on the other hand, knows that they have defined their specific structure because the global GIX structure was too coarse and made work less effective. Although the aim of the interview is to get a better understanding of the use model and user requirements the operative procurer thinks more of the future setting and what would logically be most sound for the business, and tries to convince the user of this model.

## 3.2 What and How

Somewhat simplified, the user-centred design process can be seen as divided into two stages. The first stage, the "what" stage, aims at defining what kind of system and functionality that will solve identified user needs. The second stage, the "how" stage, addresses how the interaction with the identified functionality should be designed. Since activities with users in this project were not initiated until the start of the work on the RFP, there was only time to do a quick and dirty needs analysis, that is, completing the "what" part. Therefore, everyone had their own thoughts about the solution, which often confused discussions. Moreover, it made the choice of a supplier more difficult. Since the suppliers had only received "what" requirements their responses were also of this kind: "Yes, it is possible in our solution". All of the suppliers also said that they could fulfil more or less all of the requested requirements (although to vastly different prices). The decisive factor was therefore not if, but how the requirements would be solved. But since the group had different views of what would constitute a good system design and the suppliers had not been given any how-questions this was difficult to discuss. Consequently, the important discussion about choosing a supplier was largely based on the impressions of the supplier demonstrations.

# 4  Discussion and Conclusions

This article reports a case study where the responsible procurers worked with requirements specification and usability aspects already in the procurement process. The findings are interesting since it has often been taken for granted that usability is something that is dealt with only by suppliers, after a completed procurement. Even in the HCI community there has been a strong focus on suppliers and an absence of reports from procurements (Artman, 2002; Holmlid & Artman, this volume).

The case study presented here is based on the position that a user-centered methodology fits well into the procurement process; it would provide a way to bridge business and system requirements while still focusing on actual usage. It could also make it possible to integrate a user-centered perspective from the very beginning of a development project through the RFP - as requirements or wishes for the future system. However, in the procurement process that was studied this approach was not all successful and posed several problems. The conclusion is that most of these problems arose due to a lack of competence in user-centered design. The procurers involved were not trained in either requirements engineering or user-centered design. Lacking tools and experiences of working with usability, the procurers had difficulties accessing actual user needs. Moreover, the tools that were used often mediated an idealized view on usage. Internal discussions and communication with suppliers were hindered since the team did not penetrate interaction design issues, for example through an iterative prototype and evaluation process.

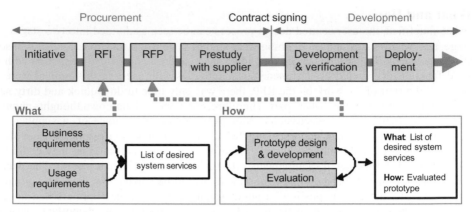

**Figure 1:** Integrating user-centered design in the development process.

Given this, it would be interesting to study a procurement where a trained usability professional would be responsible for writing user requirements in the RFP. Figure 1 describes how this could be realized in practice with the particular development process of the bank. Exploration of user needs and the definition of system services could be done in the RFI process. This initial "what"-requirement specification would form the basis for the first selection of suppliers. In the RFP process these requirements could be detailed into an information architecture and interaction design. When completed, the results could be presented both as textual requirements and as an evaluated prototype describing both which system services that are needed and how the interaction with these services should work. Together they could serve as a vision for the procurement and as a basis for concrete discussions with suppliers.

## Acknowledgements

I would like to thank my supervisors Henrik Artman and Stefan Holmlid for helping me throughout the research project and in writing this article. I would also like to thank all the people that I got an opportunity to work with at the bank and who made this project possible. This research project is supported by a research grant from Vinnova, the Swedish Agency for Innovation Systems.

## References

Allen, C. D. (1995). Succeeding as a clandestine change agent. *Communications of the ACM, 38*(5), 81-86.

Artman, H. (2002). *Procurer Usability Requirements: Negotiations in contract development.* Paper presented at the NORDICHI 02.

Balic, M., Berndtsson, J., Ottersten, I., & Aldman, M. (2002). *From Business to buttons.* Paper presented at the Design 2002.

Carlshamre, P. (2001). *A usability perspective on requirements engineering : from methodology to product development.* Linköping: Univ.

Grudin, J. (1991). The Development of Interactive Systems: Bridging the Gaps Between Developers and Users. *IEEE Computer, 24*(4), 59-69.

Norman, D. A. (1990). *The design of everyday things* (1st Doubleday/Currency ed.). New York: Doubleday.

Rouncefield, M., Viller, S., Hughes, J. A., & Rodden, T. (1995). Working with "Constant Interuption": CSCW and the Small Office. *The Information Society, 11*, 173-188.

# Usability of Software Online Documentation:
# A User Study

*Abbas Moallem*

PeopleSoft, Inc.
4460 Hacienda Drive
Pleasanton, CA 94588-8618
Abbas_moallem@PeopleSoft.Com

## Abstract

This paper discusses the usability of online documentation and summarizes the results of two usability surveys conducted to study the usability of online documentation for enterprise applications. Two groups of subjects were surveyed using paper questionnaires. The results show that users generally use documentation when they hit a roadblock. Even though the preference is for online documentation (49%), but the subjects like to use paper documentation 32% of the time. This paper analyzes user preferences and their current difficulties in using online documentation.

## 1    Introduction

Advances in the computer technology — more powerful computers, more complex applications and ultimately the Internet and Web browsers — have fundamentally changed the way people read, write and search for information. In the software industry the transition from the traditional way of delivering information by hard copy heavy manuals to soft copy, online documentation and online help appeared very fast. There are several advantages related to this method of delivering information. The ecological benefits include less paper thus less waste and the economical advantages include easy transfer, easy access to the information, easy maintenance and update.

Although most software companies now offer online documentation, much online documentation offers little more than paper documentation displayed on the computer screens. Paper documentations are not designed for viewing on a small computer display. Thus, putting documentation online by itself does not improve its use or its usefulness.

Online documentation uses the computer as a communication medium. It has two essential components. The first component is the content stored electronically. The second is the way for users to quickly and easily access that information. Thus there are two areas that usability is important in the online documentation. The usability of content: text picture, graphics and the usability of delivering techniques such as content, index, search engine and navigational tools. Several common types of error have been reported in writing user-friendly documentation. Weis (1991) summarizes these common errors in the following manner. First strategic, that includes searching several books, needing two books for one task, needing to ignore most pages. Second is structural, the problem such as jumping from front to back, never reading pages in sequence, searching for exhibits, tables. And the third type of common errors according to Weis is tactical. This includes: stopping to notice mechanical errors, getting stuck on inconsistent terminology, rereading difficult passages. To provide user-friendly design solutions in delivering and designing documentation several authors offer a variety of guidelines and methodology (JoAnn T. Hackos & al. 1996, W Horton, 1991&1994, Carroll J. M, 1998, Mulvany N. C. 1997). Microsoft Corporation in an extensive manual provides a complete online writing reference (Microsoft, 1998). In the research arena besides the traditional cognitive researches on human behavior, reading, writing and comprehension, and early human factors guidelines in computer documentation (Sullivan & Chapanis A. 1983, Solem A.1985) there is an extensive body of research published focusing on computer documentation and behavior of the users in using online software documentation.

In a study on behavior of the enterprise software documentation, we conducted surveys to better understand user behavior in using online documentation. Our goal was to understand on one hand the user's preference in navigational tools that deliver the information and on the other hand their behavior in reading the documentation. We performed two parallel surveys using paper questionnaires.

In this paper we will review the general findings of this investigation that can be generalized to all enterprise software documentation users and not only a specific product or company.

The first survey intended to study the preference of users regarding navigation/tools when documentation was offered through a browser.

The objective of the second study was to understand the behavior of the users when using online documentation.

The ultimate goal of this study was to provide input to all the professionals who are involved in creating, writing and distributing online documentation, that could help them enhance the usability of the documentation.

## 2    Method

Two paper questionnaires were prepared.

The first questionnaire included one question asking the subjects to assign a number from 1 to 10 (1, least important, 10 most important) to the following navigational/tools in delivery of online documentation. These objects include "Hide Reference Panel"(Panel(/frame that includes content and search) "Show Graphics" "Print" "Breadcrumbs", "Previous (backward)", "Next (Forward)", "History", "Exit", and "Clear"

## 2.1    Subjects

41 subjects from employees of a computer company were selected. This group includes 20% QA engineers, 24% technical consultant and 56% newly hired technical employees participating in a new-hire training program. The subjects were generally technical staff with a very mixed cultural and ethnic background.

The second questionnaire contained three groups of questions: demographics, computer usage and documentation usage and documentation usage for specific enterprise application (not discussed in this paper)

21 subjects completed the second questionnaire.75% of the subjects who completed the second questionnaire are male and 25% are females. Age ranges consist 45% 30-35, 14% 35-45, 23%, 45-55 and 18% under age 30. 59% of the subjects went to graduate school, 36% to college and 5 % high school level.

English is native language for 77% of responded and second language for 23%. They have all over 3 years experience with computer. They are all familiar with the Windows Operating system; some also used Mac OS or Unix in the past. They all use Internet Explorer as a browser but some use also Navigator. They use Internet on a daily base for a variety of tasks such as reading, shopping, banking, and researching. They have used Internet for over 3 years and are familiar with

Microsoft Office and Lotus Note; some regularly use other applications such as PageMaker, Quicken Photoshop.

## 2.2 Procedure

The questionnaires was sent by email or handed out after a training session. The returned completed questionnaires were kept anonymous. The participants for each survey were different.

# 3 Results

The results from the first survey show people (n=41) rating of navigation/tools item.

- 41% rate "Hide Reference Pan" Button between 7and 10.
  19 % rate "Show Graphics" Button between 7and 10.
- 41% rate "Print" Button between 7and 10.
- 65% rate "Breadcrumbs" Button between 7and 10.
- 49% rate "Next" Button between 7and 10.
- 59% rate "Previous" Button between 7and 10.
- 40% rate "History" Button between 7and 10.
- 34% rate "Exit" Button between 7and 10.
- 30% rate "Clear Search" Button between 7and 10.

The total score for each item stays almost equal for all items except for the "Hide reference pan". This shows that the even if one subject is rating an object lower, the other subjects are rating it higher. This seems to be related to the level of expertise of each subject. For example if a subject use less "Search" they rate "clear search" lower, but the users who are using more search rate "clear search" higher. (Chart 1)

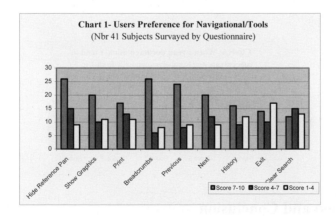

## 3.1 Behavior of the users in using documentation

The results of the second survey shows that most users who start working with a new computer system, start using the system without first consulting the documentation. Generally, people start using the documentation only when they hit a roadblock.

49% of people prefer "Online" documentation, and 31% "Paper Documentation" and only 12 % prefer on "CD-ROM Documentation". (Chart 2)

When the subjects use documentation. They tend between "Page through", "Indexes", and "Contents" use first content (41%), index (30%) and Page through 29%. Many users state that they use "Search" as an entry door to documentation. (Chart 3)

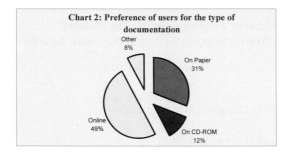

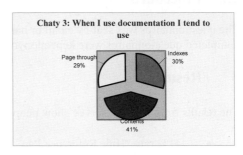

Surveyed People tend to concentrate on. "Text" then "Examples" and a last "Illustrations" when they use documentation. (Chart 4, 5 and 6)

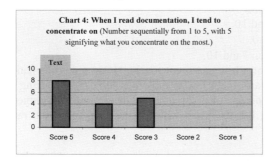

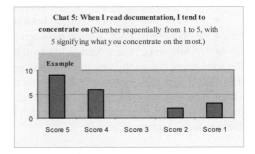

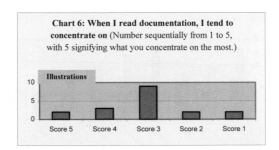

## 4    Discussion and Conclusion

The results of the first survey suggest that to improve the usability of the online documentation, it is important to provide all types of navigational tools and improve their effectiveness. The fact that the total score for each item stays equal shows that ultimately it would be useful to have all of these items in the site (Hide Reference Pan, Show Graphics, Print, Breadcrumbs, previous next, History, Exit and Clear search). The subjects also consider the tools such as "Hide reference pan", "Breadcrumbs", "Previous" and "Next" more important, thus it is highly recommended to not only make all of them available but also design them in a way that they act precisely and accurately. Some items such as "clear search" seem to be used more by advanced and moderate users who relay more on search.

The second study confirms already well-understood knowledge that users use documentation when they hit a roadblock. Even though the preference goes for online documentation (49%), but the subjects like to use paper documentation in 32% of the time. This study shows that software documentations need to be available on line but they should also be printable by users in order to give them opportunity to have a hard copy. The entry point to the documentation seems to be first through "Content" (41%) and equally, "Index" "Page through" or "Search". Users seem to prefer to look at the text first then, the examples and finally the illustrations.

This study also suggests that one major difficulty in using the software documentation is lack of examples and detailed descriptions.

The behavior of the enterprise software documentation users may differ from other communities of users such as: novice users, average Internet users or other categories of computer users. The subjects that we surveyed were expert users totally familiar with computer applications. Thus we should not generalize the findings of this study to all users. The study groups for both surveys were limited in number. Thus further study is needed to investigate more in depth behavior of online documentation in order to improve radically the usability of the computer manual and online documentation. It is generally understood that user-friendly documentation cannot improve an unfriendly application. A simple procedure, explained well is clearly simple. On the other hand a difficult procedure, explained well, is still difficult. Bad documentation makes procedures harder to follow. But clear documentation cannot improve an unfriendly system.

# 5    References

Barker T.T., (1998). Writing Software Documentation, A Task-oriented Approach, Neddham Heights: Allyn and Bacon

Carroll John M., (1998). Minimalism Beyond the Nurnberg Funnel (Technical Communication, Multimedia, and Information Systems) Systems Series, p.350, Cambridge: MIT Press

Coe M., Human Factors for Technical Communicators, (1996), New York: John Wiley & Sons,

Hackos J. & Stevens D. M, (1996), Standard for Online Communication, New York: John Wiley & Sons.

Horton W., (1994). Designing and writing Online Documentation, 2nd Edition, New York: John Wiley & Sons.

Horton W. K, (1991) Illustrating Computer Documentation: The Art of Presenting Information Graphically on Paper and Online, p.336 New York: John Wiley & Sons

Microsoft Inc., (1998), Microsoft Manual of Style for Technical Publications, Redmond: Microsoft Press

Mulvany N. C., (1997) What's Going on in Indexing? Journal of Computer Documentation, Vol. 21 (4), 10-15

Solem A., (1985) Designing Computer Documentation that Will be Used: Understanding Computer User Attitudes, ACM Fourth International Conference on Systems Documentation, p.55-56

Sullivan M. A & Chapanis A., (1983). Human Factoring a Text Editor Manual, Behavior and Information Technology, Vol. 2 (2) 113-125

Weiss Ed. H, (1991). How to Write Usable User Documentation, Phoenix: ORYX Press

# The Common Industry Format: A Way for Vendors and Customers to Talk about Software Usability

*Jean Scholtz, Emile Morse, Sharon Laskowski*
National Institute of Standards and
Technology
100 Bureau Drive, MS 8940
Gaithersburg, MD 20899
(jean.scholtz, emile.morse, sharon.laskowski) @nist.gov

*Anna M. Wichansky*
Oracle Corporation
500 Oracle Parkway, MS 2op2
Redwood Shores, CA 94065
Anna.Wichansky@oracle.com

*Keith Butler*
Boeing Phantom Works
POB 3703 M.S. 7L-40
Seattle, WA 98124
Keith.A.Butler@boeing.com

*Kent Sullivan*
Microsoft Corporation
One Microsoft Way
Redmond, WA 98033
kentsu@microsoft.com

## Abstract

One way to encourage software developers to integrate usability engineering into their development process is for purchasers to require evidence of product usability. Until recently this presented a difficulty because usability and "user friendly software" were vague, ambiguous terms. When large corporations purchase software, they use a number of quantitative measurements in their procurement decision-making process, such as the amount of memory needed, results from standard benchmark tests, performance measures, and measures of robustness. This paper describes our efforts to provide a standard method of quantifying usability and reporting on usability testing to include it in procurement decision-making.

## 1 Introduction

When large numbers of workers experience usability problems in their software, the result is an invisible source of uncontrolled overhead. Until recently, companies had no visibility of how a software product's usability would impact their overhead. Companies could not compare products or plan for the needed support. In 1997 the National Institute of Standards and Technology (NIST), a number of software vendors, and software consumer companies formed the Industry Usability Reporting Project (IUSR) (http://www.nist.gov/iusr). The goal of this project was to develop a methodology for ensuring that usability could be included as a criterion in software procurement decisions. The IUSR participants undertook two related efforts: the creation of a reporting format for objective, empirical usability tests, and methods for organizations to estimate the added value of usability, including it as a component in the total cost of ownership.

The first effort resulted in the Common Industry Format (CIF) for Usability Test Reports which is now ANSI/INCITS Standard 354-2001(http://www.ansi.org, http://www.incits.org). The Common Industry Format reports the results of summative usability evaluations. Summative evaluations are used to assess usability information on a product being released. Such an evaluation collects quantitative information on the usage of the product but does not attempt to capture usability issues and solutions for redesign. The CIF is designed to serve as a means to communicate these quantitative results from software vendors to consumer companies. The CIF is a codification of the current best industry practices. Having an agreed upon reporting format facilitates discussions as the software consumer can request the required information from vendors by requesting the delivery of the CIF report for a particular product. The vendor can also benefit from clarification from the consumer company about its users and primary tasks for the software.

## 2 The Common Industry Format

Figure 1 shows the outline of the various sections of the CIF. The CIF specifies how the summative tests and results should be reported. It does not specify content such as the tasks themselves. However, it does specify the measures to be used. The metrics of effectiveness, efficiency, and satisfaction are described and computational methods are specified by the CIF. These metrics can be used to support a business case for the product. The information requested by the CIF is sufficiently detailed for the evaluation to be replicated. A standard computation method for the metrics and reporting format allow vendors and consumers to easily exchange information.

---

**Common Industry Format**

    Executive Summary
    Introduction
        Full Product Description
        Test Objectives
    Method
        Participants
        Context of Product Use in the Test
            *Tasks*
            *Test Facility*
            *Participant's Computing Environment*
            *Test Administrator Tools*
        Experimental Design
            *Procedure*
            *Participant General Instructions*
            *Participant Task Instructions*
        Usability Metrics
            *Effectiveness*
            *Efficiency*
            *Satisfaction*
    Results
    Data analysis
            *Data Scoring*
            *Data Reduction*
            *Statistical Analysis*
        Presentation of the Results
            *Performance Results*
            *Satisfaction Results*

---

Figure 1: Outline of the Common Industry Format

## 2.1   Example scenario

This scenario explains how the CIF can be used for software product acquisition: Requests for Proposals (RFP) are a routine way that corporations initiate software acquisition. They typically issue RFPs to inform suppliers that they are looking for products in a given area (e.g., email, computer-aided design), and also to define the various categories of data that they will consider to make a selection. The RFP invites suppliers to respond with technical data for each category. In this scenario, the RFP manager for the consumer company, who leads the creation of the RFP, assures that the CIF is cited as a relevant standard, and that usability evidence is one of the requirements for the user interface section of the RFP. At Boeing, for example, the following language is used in RFPs:

> "The parts of the user interface that will be operated by end-users must be subjected to usability testing by the Supplier. Provide evidence of usability in the form of a report describing the most recent test and the results. For guidance on the requirements, refer to ANSI/NCITS 354-2001, the Common Industry Format. For further background information, see the NIST website on this standard:
> ( http://www.itl.nist.gov/iaui/vvrg/iusr/)."[1]

Suppliers who received an RFP have the opportunity to report an existing test, to conduct a new one and report it, or to risk that their response will get a low score from their customer. The citation of the CIF standard provides them with the needed guidance to respond. When the RFPs arrive, copies are distributed for scoring by the consumer's RFP team. Technical specialists on the RFP scoring team get the responses to sections that they wrote. Usability engineering skill is required to interpret the CIF reports. For the part of the RFP on user interface, the CIF report should carry a lot of weight because CIF metrics have a strong relation to the business case for a product.

## 2.2   A case study

Illustrating this scenario, Oracle and Boeing agreed to conduct a trial to communicate usability data concerning installation of the Oracle 8i database via the CIF. Representatives of both companies made numerous presentations to one another about usability processes and requirements over a one-year period. Oracle and Boeing teamed to define the key use-cases for the Installer product. The CIF was used as a communication vehicle for exchanging user requirements and usability data. Initial CIF results indicated that a key use-case could not be supported by the user interface. On the basis of the usability test results, Oracle refined the interface, and the task completion rate improved to excellent levels (the improvement was 91.75%). From the last CIF delivered by Oracle, Boeing was able to see that significant usability improvements had been made resulting in significant improvements in project productivity and major savings in database administrator overhead. Boeing applied this information in making the decision to renew the Oracle license for future versions of database software, training, and support.

## 3   Estimating The Usability Component of Total Cost of Ownership

A large, invisible source of uncontrolled overhead results when end-users cannot operate their tools effectively and efficiently. In addition, a significant portion of support costs is due to  users' difficulty in applying software products to their work.  In order to predict and evaluate overhead costs and include them in the total cost of ownership, corporations need to have a way of assessing

---

[1] Boeing Standard CW-BDSB-2002-166

the costs incurred when usability issues arise. Typically, workers, when confronted with a usability problem, will try to find some work-around. A worker might try several different methods to complete a task, look for on-line help, consult a co-worker, or call the company support desk. Regardless of which method is used, the result is that an additional cost is incurred, either due to lost time and lowered productivity or to increased support costs. Long-term costs might include additional training time for staff and even costs due to high turnover.

As an experiment in measuring overhead reduction, Microsoft and Boeing undertook a pilot study to compare the usability of beta releases of Windows 2000 and an earlier version of this operating system. The effort involved joint design of web-based survey tools for estimating the effects of improved product usability. The first part of the survey was the Software Usability Measurement Inventory (SUMI), a standardized subjective questionnaire based on industry norms (Kirakowski & Corbett, 1993). The second part contained questions that were specific to the features of Windows 2000. The third part gathered data on how often usability problems that caused users to stop work were encountered, and probed for the types of resources that were required to resolve them. The resources included: using on-line help, contacting in-house technical support, reading a hard copy manual, and asking a colleague for help.

Cost estimates were assigned to each of the resources. These data were used to estimate reduction in the overhead that results when end-users encounter usability problems. There were improvements in successive beta versions and an improvement over the usability of older versions of Windows.

Microsoft also provided Boeing with a CIF-compliant report, comparing the usability of Windows 98, NT4, and Windows 2000. Boeing used all the data provided to calculate the added value of the usability improvements in beta 3 over beta 2. The value of the reduction in overhead was $31.2m - $65m, over five years. The return on investment to Boeing (ratio of Boeing benefit to Boeing cost for collaborating on usability) was greater than 1000:1.

# 4 CIF Usage

## 4.1 Oracle

Oracle has standardized its usability evaluation methodology around the CIF and is currently running CIF summative tests on most major software applications and core technology products. Data from CIF tests are compiled into a usability scorecard, which is reported to the Oracle CEO and other managers by the User Interface Design Vice President at regular executive staff meetings. Based on the CIF data, more reliable comparisons between versions of products can be made. Some products are targeted for more user interface work, based upon internally set standards for key CIF metrics. Oracle uses task completion rates and SUMI global scores primarily for this purpose. More than 40 CIF tests on released software have been conducted and reported since 1999. In addition, CIF-style metrics are collected wherever possible for prototype and other pre-release usability tests. Oracle account teams also request CIF reports for products they are currently selling.

An interesting by-product of CIF usage has been increased communication between software vendors and their customers. Oracle started using the CIF to deliver usability results to Boeing. The two companies have been able to use this information to discuss the tasks that Boeing users need to accomplish and to better understand the needs of their users. As a result, the IUSR members are examining the CIF to determine what changes, if any, are needed to use the CIF to specify the usability requirements of a consumer organization.

## 4.2 Microsoft

Microsoft used the CIF to publish the results of summative comparison studies between Windows XP Pro and previous versions (Windows 2000 and Windows ME). These reports have been made available to prospective customers to help them understand the usability improvements in the XP release.

While Microsoft does not use the CIF for internal reports, its usability engineers are aware of the CIF and are comfortable using it for user study results that are published outside of the company.

## 4.3 European Union Studies

Bevan et al. (2002) completed four different studies involving usability requirements for a legal information website; usability requirements for a desktop travel expense reporting system; assessment of a web site for e-commerce; and assessment of travel management software. In one study a summative evaluation of a current system was used to generate requirements for a new system. These included at least equal if not better task completion rates. The requirements were delivered to the contractor for the web site and will be used as part of the acceptance criteria when the web site is delivered. Another study was also used to create a baseline and additionally to help the software supplier understand the consumer needs.

## 5   Status of the CIF

The CIF is currently available as ANSI/INCITS 354-2001 standard. The IUSR group is involved in three efforts at this time. The first priority of the IUSR group is to have the CIF become an ISO (International Organization for Standardization) standard. (See http://www.iso.ch for more information.) The IUSR group has an international membership, and, as such, would benefit from having an ISO standard. The current CIF was developed to report summative usability evaluation of software products. Many products consist of both hardware and software components, such as PDAs, printers, and other computer peripherals. A subgroup of IUSR participants is looking at the modifications of the CIF necessary to report evaluations that include hardware. As previously mentioned, there is an interest in using the CIF as a requirements document for usability and a third working group is examining changes that need to be made to the CIF to support this. These activities could result in several versions of the CIF in addition to the current standards. For those interested in obtaining and using the CIF, copies may be ordered from www.techstreet.com.

## 6   References

Bevan N, Claridge N, Maguire M, Athousaki M (2002) Specifying and evaluating usability requirements using the Common Industry Format: Four case studies. Proceedings of IFIP 17th World Computer Congress, Montreal, Canada, 25-30 August 2002, p133-148. Kluwer Academic Publishers.

IUSR. (1999) http://zing.ncsl.nist.gov/iusr/documents/WhitePaper.html

Kirakowski, J. and Corbett, M. (1993) .SUMI: the Software Usability Measurement Inventory. *British Journal of Education Technology*, 24 (3), 210-212.

# Ten Factors Affecting Adobe s Overseas Research

*Lynn Shade*

Adobe Systems Inc.
345 Park Avenue, San Jose, California 95110 U.S.A.
shade@adobe.com

## Abstract

Adobe Systems has a rather unique approach to Japan user research that is born out of the corporate organization and the employees involved. One of the challenges in doing overseas user research from a company based in the U.S. is navigating not only the cultural differences between the researcher and the customer, but also the cultural differences between the customer and the development teams, who are the ultimate recipients of the information. The researcher's challenge is to effectively translate the workflows, feedback, ideas, and essence of a customer into language and metaphors the people back home can relate to. In this paper I will describe the history behind our research in Japan, factors driving our research methods, and the challenges in conducting and communicating the results of overseas research.

## 1    Background

Adobe s overseas user research is focused on understanding the Japanese customer s needs. Although my group of two in the User Interface Team is called International Solutions, we currently focus on Japan for several reasons. Japan is the second largest market for Adobe Systems outside of North America. Japanese users also have the most distinct workflows and complex typography requirements. My personal and professional experience —growing up in Japan and working on Japanese products at Claris and Apple Computer before joining Adobe —have also affected my interest in seeing Adobe products truly meeting Japanese user s needs.

Many people are surprised to hear that Adobe does not have UI designers or researchers working in Japan. While we have plans to change this situation, this has something to do with the background of UI Design at Adobe. When I started working at Adobe five years ago, I was only the sixth employee of the Professional Products (Photoshop, Illustrator, Premiere, etc.) User Interface Team. While it was common to interact frequently with key users, it s safe to say that the role of the dedicated UI designer was relatively new to the company, and systematic usability testing or user research didn t exist. Overall, there was little understanding of the benefits of contextual user research. Paradoxically, I found myself in a very fortunate position: because there was little internal expertise on user research or Japan, I was free to experiment with methods and create a program of Japan user research.

Currently, the two of us in International Solutions play a hybrid role of both user research and UI design. We collect qualitative data on user tasks via in-depth interviews and observation. For example, I have studied the Japanese publishing workflow in detail, visiting all roles involved in creating a published work multiple times: editorial designers, publishers, printers, service bureaus,

DTP operators, typographers. Occasionally we have the opportunity to use different research methods, such as a diary study for a consumer product.

## 2 Ten Factors Affecting Current Overseas Research

At Adobe, I quickly found that my favorite field work model —slipping into a new subject s environment for an afternoon, observing workflows, having structured and unstructured conversations to discover new issues —was not going to work. Some of the reasons for this — logistical issues with conducting research, communicating the research results, and selling the research program —are described below and have resulted in solutions that work for us.

### 2.1 Conducting Research

#### 2.1.1 Complex Products

**Issue**: Many of Adobe s products are feature-rich tools for professional users.In my past research, it had been fine, even desirable, for the researcher not to be too wrapped up in product development. However, the sheer complexity of what we attempt —such as supporting full Japanese typographic traditions within an intricate application like InDesign —require direct in-depth interaction between development engineers, QE, product managers, and the users.
**Solution**: Site visit teams include a researcher, a product marketing manager or two, and often a programmer or Quality Engineer. More team members have direct knowledge of customer issues, and can represent those issues across different disciplines during the development cycle. The trip reports are also better when we are able to draft them together, as we can agree on wording that communicates across disciplines.

#### 2.1.2 Complex Research Topics

**Issue**: Complex products and research topics that span years also require specialized expertise on the part of the users, and more than one visit. Many of the top designers in Japan graciously give us their time, answering questions, patiently explaining issues, and occasionally preparing illustrated lists of requests.
**Solution**: We build and value long-term relationships with many of our customers and target users, even the ones that don t use and will never use our products. In Japan, those relationships allows even naturally reticent customers to give increasingly honest feedback. Essentially, what we give them in return is products that increasingly match their workflows. Conscious of the gift of time and energy they give us, I try to select gifts that will be useful or interesting to them.

#### 2.1.3 Busy Participants

**Issue**: In researching consumer products, it s fairly easy to pay participants or ask a friend of a friend to participate. Not so in researching professional products, where users and future users are often busy and often important. While we give these designers software and gifts, money is a poor substitute for the value of their time. The sheer busyness of most of these users means we either can t meet with them on every trip or are limited to a short amount of time.
**Solution**: This has caused us to widen our network of key users, resulting in exposure to a greater variety of workflows and approaches. For example, the InDesign team talks to Japanese

typographers who hold very different points of view about what beautiful Japanese typography is, some of whom bitterly disagree with each other. It s also very informative to talk to users who have chosen not to use our products, and find out why. All of this gives us more data points to consider when determining feature requirements.

## 2.1.4 Busy Stakeholders On the Road

**Issue**: People involved in product development are very busy. E-mail, meetings, and deliverables all pull at us when on overseas user research trips. I have observed co-workers that visit the target country, attend the user visits, and spend evenings on e-mail in the hotel, leaving without being exposed to local life.

**Solution**: I cannot over-emphasize the importance of truly experiencing life in the research country. Designers need to wander around, watching how ordinary people live: how they shop, use public transportation, leaf through popular-looking magazines, check out current fashions, etc. Curiosity about people and an interest in all kinds of life are important to being a good researcher and designer. One of my colleagues is an aficionado of tiny hidden Japanese design shops. He led a co-worker and I on a memorable hunt through narrow alleyways to signless shops, impressing the importance of rarity and exclusivity in aspects of Japanese design culture upon us. Another memorable visit was when colleagues who were both new to Japan and I decided to get inside the skin of our target users by trying their hobbies. Since we were studying Japanese enterprise workers, we tried salaryman hobbies of pachinko, karaoke, drinking, shopping for cell phone accessories, and train-platform umbrella golf swings. The experiences can be targeted or random, but without experiencing at least some of the everyday activities in the research country a true understanding of users is impossible.

## 2.1.5 No Usability Lab in Japan

**Issue**: While we have plans to create a User Research group in Japan that includes Usability Testing, we do not currently have a lab or researchers in Japan. Not only are we not able to provide regular and consistent feedback from usability testing, it is difficult to set up user research from overseas.

**Solution**: We are in the beginning stages of establishing research in Japan, laying the groundwork for hiring User Research staff and creating a Usability Lab in Japan. In the meantime, we continue to do validation on design concepts during the site visits. While at first I found it strange to ask users to evaluate A) My own designs with B) Team members there, some of whom disagreed with the design, it has resulted in some excellent discussions and more consensus amongst team members on the design solution.

## 2.1.6 Interpretation Necessary

**Issue**: On-the-spot interpretation is crucial to the success of overseas research. Most product team members don t speak Japanese, yet we want them to communicate face-to-face with Japanese customers.

**Solution**: My International Solutions colleague and I speak Japanese and English, and some product teams have bilingual team members. This situation is ideal; we can interpret for other participating co-workers, which should be done on the spot. When non-Japanese-speaking team members are not interpreted for until afterwards, it leads to boredom and ill feelings.

Despite the logistical factors affecting our overseas research methods, I still try my best to keep the focus on, as Liz Sanders says, the convergence of not just what users say, but what users do, and what users make.

## 2.2 Communicating Research Results

### 2.2.1 Hidden Cultural Differences

**Issue**: One of the challenges in doing overseas user research is cultural trends are often invisible to those immersed in the culture. It s very difficult to explain why users in another country value A over B when the audience doesn t understand that A and B are both relative cultural values.

**Solution**: We take several approaches. One approach is to draw analogies. For example, to understand why Adobe's Japanese customers respond so positively to Adobe's LE (Limited Edition) and Elements products, American stakeholders can recognize that most Americans equate 'more features' with 'more powerful' and would never choose to have less power unless they had to. Japanese customers, on the other hand, are interested in minimalism and modularity and the right product for me. It s also useful to think about the sort of shame in America with being a beginner or an intermediate that simply does not exist in Japan.

Because verbal analogies can only go so far, we put energy into explaining key cultural touchpoints with artefacts. In order to explain to Americans the purpose that 'cute' characters and objects serve in Japan and why it s important to have a friendly face on corporate entities, we take photos and bring back objects that are common or important in Japanese daily life and design. Sharing these artefacts is fun and communicates far better than words alone can.

### 2.2.2 Detailed Trip Reports

**Issue**: Trip reports need to be interesting, informative, and thorough enough to stand on their own. Posted on our UI Team site, they are often read years later by product teams I ve never interacted with.

**Solution**: The UI Team member is responsible for writing the trip reports, which include customer information, user requests, and workflows. Ideally all research participants sit down and jointly draft impressions after each visit, but as our time in Japan is short, we re often already on the run to the next visit. The key to creating interesting and helpful trip reports is in including background information; clearly conveying the reasoning behind the request. Without the information on why, recipients of the requirement wouldn t understand the logic behind the request. For Japanese user requests, adding background information requires research and dedication, as it often starts with describing a non-U.S. workflow or limitation. I often find myself adding not just why the request was made, but why a certain workflow or typographic standard is the way it is in Japan, anticipating questions. There s little doubt that this is a lot of work and complete trip reports can take weeks to write. But when a trip report informs product decisions in a scope larger than the people I interact with, the payoff is huge.

### 2.2.3 Translation Needed

**Issue**: We can t realistically expect Japanese users to use English-language products, or read and respond to surveys in English. Yet being far away both distance- and budget-wise from our users has resulted in us experimenting with eliciting feedback via e-mail. While e-mail and web-based feedback was pioneered by the UI Designers and User Researchers with American Alpha users, I was hesitant to try it with Japan because face-to-face communication is still the norm there. Also, while all team members can easily read American Alpha users responses if they wish to, our Japanese users feedback would have to be translated.

**Solution**: We are just starting to experiment with this form of user research, and here is still room for fine-tuning. While we have been communicating with the Localization team about developing earlier 80% translations of the product interfaces for user research, we still lack budget and resources for regular translation from Japanese to English. In addition to missing out on Japanese user s feedback during the development cycle, much of the available user feedback from sources like magazine reviews, online forums, and e-mails are never received by the product teams.

## 2.3 Selling Japan User Research

### 2.3.1 Creating a Solid Research Program with Sceptical Stakeholders

**Issue**: Because our UI Director is supportive of conducting user research in Japan, we are able to do research in Japan several times a year. However, to have the research results used, key decision makers must be convinced to buy in to investigating user needs in Japan.

**Solution**: To convince stakeholders that feedback from Japanese customers is worth getting, it s important to spend time selling the research program. Discussions with and presentations to upper management describing the benefits of early overseas research means sometimes sacrificing research and design time. I have also found it valuable to have key stakeholders interact directly with users, seeing and hearing their concerns first-hand. While the value of including additional people has to be balanced against the physical limitations of our customer's spaces, people in these positions can become powerful advocates. I ve also found it helpful to clarify for myself that the some visits are serving two purposes, research and politics.

## 3 Future

I have described the history and structure of our overseas user research as it exists at Adobe Systems today. Adobe as a company continues to evolve; for example, we are currently expanding our focus from Creative Professionals, such as Photoshop users, to Enterprise customers. Our User Interface Design Team continues to evolve as well, and future plans for International Solutions include increased collaboration with the experts on the User Research team, building a team of researchers and designers in Japan, and expanding research efforts to other countries. Our methods will doubtless continue to evolve as the products we research and design for require different approaches, and I am looking forward to experimenting with new research methods.

## Acknowledgements

Thanks to Liz Sanders, who is so inspirational in her research with people. I m sorry for continuing to use the word users all these years.

Thanks to our Japanese customers, who always amaze with their attention to detail, astute questions, and top-notch design sense. It s always a pleasure to be in their presence.

And thanks to my co-workers and managers at Adobe, who have been supportive of my work and so fun to work with.

# Usability and HCI in India: cultural and technological determinants

*Andy Smith*

Centre for Software
Internationalisation
University of Luton
andy.smith@luton.ac.uk

*Kaushik Ghosh*

MIT Media Lab Asia
Mumbai,
India
kaushik@medialabasia.org

*Aniruidha Joshi*

Industrial Design Centre,
Indian Institute
Technology, Mumbai
anirudha@idc.iitb.ac.in

## Abstract

This paper discusses a range of cultural and technological issues that are shaping the development and growth of usability and HCI within industry and academia in India. It outlines the role that the EU funded Indo European Systems Usability Partnership is playing in furthering usability.

## 1    Introduction

In comparison with developed nations India's present status as an information society is nascent. However the information revolution in India is clearly coming of age, encouraged by the declining cost of information and communication technologies world-wide, and the increased availability of high-quality and low-cost technology-enabled products and services in the country.

In May 1998, the Prime Minister of India formed a National Taskforce on Information Technology and Software Development in order to formulate a long term national IT policy. The main objective was to help India emerge as an *'IT software superpower'*. Although the industry is growing fast the penetration of IT within the whole Indian society is still very low.

According to NASSCOM (2002) - National Association of Software and Services Companies - a key strength of the Indian IT industry is a *'focus on a high value, software off-shoring model'* However, this strength is balanced by key weaknesses such as a low presence in the global packaged software market. We believe that India will not be able to claim IT superpower status without key developments, and that usability / HCI must play a key role. High levels of usability are critical to the quality of software products in a global market. Jakob Nielsen (2002) picked up these ideas in his recent Alertbox and, somewhat exaggeratingly, calculated that India will need to train 400,000 usability professionals in the next six years.

This paper will discuss some of the prominent characteristics of the information society in India with respect to the techno-economic paradigms that have evolved within the society, the research challenges already in place, and the requirements that need to be addressed by future work.

The observations will be documented to build an underlying rationale for the design for accessibility in terms of the hugely diverse user profile and the most appropriate technology interventions that organically support the social dynamics that is prevalent in the Indian society.

# 2 Technological determinants; a strange 'digital divide'

The 'skill base and proficiency' of the Indian IT industry provides a very interesting contrast compared to the information and communication needs of Indian society. On the one hand there are about 340,000 Indians employed by the Indian IT industry, including both hardware and software (NASSCOM, 1999). These are in addition another 180,000 people of Indian origin, now working in the US alone (March 2000). In 2002 the Economic Times reported that between years 2000 and 2002, 250,000 people migrated from India to US to work in the (then slowed down) US IT industry.

Indian IT industry is considered to be the third most significant foreign exchange earner in the country. On the other hand we have the extreme paucity of IT products and services in the Indian market (and perhaps a big hidden opportunity). There are 4.5 computers per 1000 people in India, 32 telephones and 3.5 cell phones (Bomsel and Ruet, 2000). The total annual revenue from IT exports to US is around US$ 2.7 billion and US$ 1.5 billion (2001/02) comes from IT enabled services, only US$ 5.65 billion in 2000/01 came from sales within India, whereas US$ 1.9 billion in 2000/01 came from sale of packaged software (NASSCOM, 2002).

While English is understood by less than 5% of the population, operating systems with Indian language support were launched only as late as 2000 and still have only a miniscule installed base. English continues to be the predominant language used even in products such as cell phones and ATMs. How can a country with such a large IT industry harbour such a strange digital divide? Two reasons can be attributed to this divide in India – the outsourcing based business model of the Indian IT industry and the lack of HCI education in the Indian IT education.

## 2.1 The outsourcing business model

In the past two decades, the Indian IT industry has relied on providing quality software services in a cost-effective manner. It has effectively leveraged the huge difference between the labour costs of equivalent skills in India and the developed part of the world. At the lowest end, this required Indian entrepreneurs to market 'skills' to international companies on 'costs plus time basis'. At the highest end, companies get outsourced projects – partly to be executed off-shore in India. The larger the off-shore component, higher is the profitability. Usually the Indian IT company deals with an IT group within the outsourcing organization. The highest end Indian IT companies have developed excellent software engineering processes to manage such projects effectively – the largest number of 'CMM level 5' companies are from India (NASSCOM, 1999).

From the perspective of designing human-computer interaction, this had terrible consequences. By this very nature of their business, a problem typically reached a group within an Indian IT company well after it had been identified and earmarked as 'one that needs solving' and clear enough for outsourcing. Usually (though not always) user requirements were already specified. At times, even the design requirements were ready before an Indian company got involved. This effectively transferred the responsibility of many HCI and usability issues in the first part of the project to the client.

At the end of the development life cycle, the product was developed and evaluated for quality against requirements by the Indian IT company and sent back for 'acceptance testing' to the client. Formal usability evaluations were rarely done until recently. Informal usability evaluations, if at all, were usually carried out as part of acceptance tests and were managed as 'upgrades' or 'change requests' as they were deviations from the original requirements. But can this continue for ever? Trends in the last three years already show changes to this pattern. Many of the top Indian IT companies have started incubating usability groups. Others have started hiring professional interaction designers on freelance basis for critical projects. There seem to be two reasons for this change.

Firstly, awareness about usability and HCI design has increased, not only among the Indian IT companies, but also among their international clients. Along with technical skills, clients now demand to see (though not routinely yet) proven capabilities towards usability and HCI design. Secondly, there is a growing need for Indian companies to move up the value chain – from proven provider of quality skills at highly competitive prices to providers of well-managed outsourced projects to providers of end-to-end solutions.

## 3 Educational determinants: HCI in Indian education

One strength of the Indian IT community as a whole is a strong educational orientation that produces high-class engineering graduates skilled in computer science. Currently however, very few universities address HCI in their curricula. An exception is the Industrial Design Centre (IDC) at the Indian Institute of Technology, Mumbai which offers formal courses in Human-Computer Interaction to students from both design and engineering backgrounds. Another exception is the Postgraduate Diploma in Design with a specialization in New Media offered by the National Institute of Design, Ahmedabad.

In fact these programs illustrate a common characteristic of the relatively few usability / HCI academics in India. It is also a reflection of the realities in the Indian industry. A large percentage of HCI / usability professionals working in the field in India originally come from the fields of design (particularly visual communication design or industrial design). A smaller percentage comes from a computer science background, and smaller still from cognitive psychology or usability.

Of course HCI is multi-disciplinary and involves individuals with backgrounds in a range of non-computing disciplines. However we believe that the lack of HCI within the computer science curriculum is a problem that needs addressing in order both to support the growth of the IT industry, and to facilitate India's integration into the global information society. We believe that by highlighting aspects of HCI and interaction design within the Indian computer science curriculum the next generation of systems developers will be more aware of the need for effective interaction design across the whole spectrum of interactive applications.

## 4 Cultural determinants: ICT of the people, by the people, for the people

In the very recent years, there has been a visible momentum towards building a national agenda for addressing the issues concerning user demographics, the digital divide and the emerging information society in India. India faces monumental social challenges and is hampered by bureaucracy, political strife and the legacies of a controlled economy. Yet the nation is endowed with so many highly trained, ICT savvy workers that many believe technology will launch India into the developed world.

In the emerging Information Society 'design for all' is the conscious and systematic effort to apply principles, methods and tools, in order to develop information technology and communications

(ITC) products and services which are accessible and usable by all citizens. This requires a democratised design framework to accommodate diversity of the possible users to avoid the need for a posteriori adaptation or specialised / segment specific design.

In spite of the bearings of a weak information infrastructure, high communication cost and poor service, the condition is improving everyday through private participation and state level initiatives. Break through enterprise initiatives are being taken in setting up a high bandwidth national backbone and a substantial amount of it has been made wireless, thus realising a huge potential of low-cost ICT products and services including rural mobile telephony. Other initiatives include ICT in education, health, digital trading platforms for local economic processes, localized software applications, software systems for illiterate users and Indian language related developments along with many more infrastructural as well as cultural issues that are critical to build an appropriate intervention of ICT in the Indian subcontinent.

As the majority of the Indian population live in villages that are completely beyond the reach of urban infrastructural facilities. Bridging the digital divide in India calls for radically innovative ideas that can be implemented in a sustainable manner at a low cost. Laying the ground work has already started. Rural Internet Access Centres are being set up in the villages of India that will serve as global information access points and e-commerce channels for about 100 villages. Another project, Computers on Wheels is a mobile ISP service that uses motorcycles to bring wireless Internet access to residents in remote villages that typically have no telecommunications infrastructure. Low cost handheld client devices based on smart cards are being developed for remote location computing.

NASSCOM recently conducted research on the domestic market for free software applications in India. The main highlights of the study are:
- A number of e-governance projects under implementation are either experiencing or intending free software usage.
- Virtually all leading IT vendors in India have developed products on the free software platform that they are actively marketing.
- Low cost applications can be used to the advantage by the fast growing SME sector.
- Lack of local language applications has been one of the major impediments to the growth of the Indian domestic market. The openness and customizability of free software platform is helping to take a major stride in this regard.

## 5 Indo European Systems Usability Partnership (IESUP)

Starting in October 2002 the Computer Society of India (CSI) joined in partnership with the British Computer Society (BCS) / British HCI Group and three European Universities in further developing usability and human-computer interaction in India. Sponsorship from the European Commission's ASIA IT&C Programme will fund a two-year project to support links between key experts in Usability and HCI throughout Europe and their counterparts in India.

Overall the aim is to support the integration of HCI and usability into both Indian IT education programmes, and software development projects, mirroring that which occurs in Europe and the USA. Activities will include seminars / workshops in India, visits from India to Europe, together with virtual communities and other methods of larger scale communication.

By facilitating discussions and debate in India and elsewhere, and establishing networks between groups of individuals in focused aspects of usability, IESUP will seek to promote HCI within academia and industry, develop methods for software localisation in the Indian context and foster greater awareness of the role of interaction design in the development of the next generation of online systems and interactive devices.

# 6    The scope of HCI in India

We believe that providing services to the developed part of world is only the 'tip' of the proverbial iceberg of the potential that usability professionals and HCI designers have in India. The market of the western or westernised, urban, office going, predominantly English-speaking, predominantly male (the market of the first billion, as it is sometimes referred to) will saturate out eventually, and as India emerges out of its status of a 'developing country', the attractiveness of the business model described above will decrease. On the other hand, India itself represents a huge, un-served market of ICT products and services. We believe that such products and services will go a long way in solving the problems and improving efficiencies in the resource starved developing countries of the world. There are many huge challenges of bringing about change in conditions here, but so are the opportunities.

The designers here will have every opportunity to exercise their skills. There will be no ready design specifications to follow, no specified user requirements to meet. There will be more freedom in the process – there will not be many legacy systems to take care of. The designers will have to bear the brunt of sceptics and pessimists. They will have to prove their concepts in the field several times and develop new techniques and processes to design. Every design idea will have to compete tooth and nail for limited resources, and only the best will survive. Those which do, will define the future.

# References

NASSCOM (2002). Indian IT Industry. Retrieved March 5th 2002 on the World Wide Web: http://www.nasscom.org

Nielsen, J. (2002). *Offshore usability.* Alertbox, September 2002 Retrieved from the World Wide Web on 10 October 2002: www.useit.com/altertbox/20020916.html

NASSCOM (1999). NASSCOM-McKinsey Report 2002. NASSCOM

Bomsel, O. and Ruet, J. (2000). *Digital India – Report on the Indian IT industry.* With the co-operation of Centre de Sciences Humaines, New Delhi. Details retrieved from the World Wide Web on February 3rd 2003: http://www.cerna.ensmp.fr/Documents/DigitalIndia-MainFindings.pdf

# Usability Challenges in Social Projects in Brazil: Lessons Learned about the Digital Divide

*Clarisse Sieckenius de Souza[1], Simone D.J. Barbosa[1], Raquel Oliveira Prates[2]*

[1]Informatics Department
Pontifical Catholic University of Rio de Janeiro
R. Marquês de São Vicente, 225, 4o. RDC
Rio de Janeiro, RJ, Brazil, 22453-900
{clarisse, simone}@inf.puc-rio.br

[2]Computer Science Department
State University of Rio de Janeiro
R. São Francisco Xavier, 524
Rio de Janeiro, RJ,Brazil, 20550-013
raquel@ime.uerj.br

## Abstract

In the scope of Brazil's Information Society program, we have carried out a user-centered software development project to support activities performed in organizations of social volunteers. The initial strategy was to adapt groupware applications that had been successfully developed for commercial companies to their needs. From the evaluation of a preliminary prototype, we realized we needed to change some of the basic project assumptions, due to the volunteers' diverse levels of computer literacy, as well as their strong feelings towards technology. Not only did we have to develop a unique set of applications for them instead of tailoring existing applications, but also great care had to be taken in order to introduce the new technology into their lives. In order to succeed in our endeavor, we had to shift from user-centered design to adopter-centered design.

## 1    Introduction

In late 1999, the Brazilian government launched its Information Society program, whose goals were twofold: to increase social and digital inclusion, and to encourage the growth of the IT industry in the country so as to compete in a globalized market. The main initiatives towards social and digital inclusion aimed at distributing low-priced computers to schools and community centers, and to connect them to the Internet. Such initiatives provide concrete results and a tangible impact that may be easily or objectively measured. However, hardware alone cannot improve the lives of users, their families and communities. We need to complement these initiatives with software-development endeavors for bridging the social and digital divide.

In Brazil, the challenge of using or developing computer applications to fight social inequalities is not fully understood. There is a widespread belief that giving people more information and education will improve their lives, but we don't exactly know what kinds of information and education are prioritary, and much less the format in which they should be conveyed. As a consequence, the Information Society program called for projects that would help the Government achieve its goals, but it did not set any requirements for software development or evaluation.

In this context, the Semiotic Engineering Research Group (SERG[1]), has proposed a project called ORÉ to support the activities performed in organizations of social volunteers (OSVs). Our primary goal was to take a user-centered design (UCD) approach to  provide OSVs with usable systems that could support their activities. The lessons-learned from this project point to the fact that UCD alone would not be sufficient to make the Information Society program successful, and that these projects must also raise relevant issues to be discussed before being able to fulfil goals of equality.

---

[1] SERG is situated in the Informatics Department of the Catholic University of Rio de Janeiro (DI/PUC-Rio).

## 2　The ORÉ Project

The ORÉ Project (ORÉ, 2002) aims at developing a suite of groupware applications to support the activities of Associação Saúde-Criança Renascer (ASCR), an OSV engaged in helping disenfranchised families whose children are hospitalized in one of the city's public hospitals (ASCR, 2002). ASCR is one of the most successful and respectable non-governmental organizations of its kind in Rio de Janeiro, operating with approximately 150 people, 80% of which are volunteers. The other 20% are full-time or part-time employees of the organization.

ORÉ's long-range target is to widen the participation of our civil society in social volunteering initiatives. It has been publicized that a vast amount of human resources ready to do volunteer work is wasted for lack of organizational infrastructure. The expected contribution of ORÉ lies in using IT to help volunteer organizations become functional and effective. From our experience with UCD of groupware applications for commercial companies, we initially believed that this kind of application would bring OSVs the most benefits. We hoped we would only need to adapt existing software to fit their needs.

We set up an interdisciplinary team including computer scientists, a linguist, a psychologist, and a graphics designer, and started off by interviewing some of the targeted users in ASCR. Our initial goals were to find out what kinds of groupware application should be tailored, and how the interviewees saw themselves using IT in their day-to-day tasks. We used the UDUM interview method, which allowed us to uncover deep personal feelings and attitudes towards computers in general (Nicolaci-da-Costa, 1989). We looked for ASCR members who already used computers to do their volunteer work. We wanted to explore values and attitudes within this group of people, with a special emphasis on the links between technology and subjective satisfaction. So, we started by interviewing 10 people, whose ages ranged from late 30's to early 60's. The interview script was structured around four major topics: the interviewees' profile, i.e., how they described themselves; their previous experience with computers; the tasks they performed at ASCR; and their dreams (or fears) about the introduction of IT in ASCR[2].

We have found that volunteers have strong emotional drives towards their activities in OSVs, such as motivation and pride in working for the organization, as well as equally strong negative feelings towards computer technology (Barbosa et al., 2002a). Volunteers are assigned to various activities based on their personal interest and availability, and usually have a wide margin of choice on how they want to perform them. These findings increased our awareness of how critical good HCI design had to be in OSVs, for imposing users to interact with poorly designed applications might drive them away and cause too much damage in this kind of organization. It also raised an issue we hadn't anticipated to its full extent: the need to carefully introduce technology in order to "break the ice" and gradually change users' negative feelings and attitudes towards computers.

We have also found that tasks that must be carried out daily are typically performed by different people in different ways. Thus, if IT is expected to support work in OSVs and to be adopted by all volunteers spontaneously, it must be flexible enough to accommodate extremely different work practices and subjective motivations. This variation lies at the opposite end of current standardization and work process quality control strategies.

From these findings, we learned that a successful strategy for introducing IT in a community of volunteers should be a prime goal of ORÉ, overriding our initial goals to adapt "well designed groupware" to this community. We believe that we first need to get volunteers involved and

---

[2] At a later stage we will be exploring these dimensions with a group of individuals who are not computer literate.

interested in IT, and only then we will be able to investigate the kinds of technology that will potentially improve their work practices, support OSVs in managing resources and opportunities *vis-à-vis* existing social challenges, and ultimately contribute to promoting this kind of volunteerism in Brazil. This change led us to take an adopter-centered approach (Seligman, 2000), in which steps are planned to get users to learn about technology and to have a favorable attitude towards it, before providing them with full support for their work practices.

## 2.1   Light-weight prototype for introducing technology

Due to the nature of their work, many volunteers don't know what happens during the days they aren't at the organization, or receive announcements about interesting events only too late. From the interviewees' remarks about the need for better communication mechanisms, we decided to start by a simple bulletin board application. We aimed at giving this community a chance to spontaneously get acquainted with IT in a situated context, using a tool that has been specifically tailored to ASCR, with familiar signs of their location, their practices the people who run the association, and so on.

Since a bulletin board is a technologically simple application, we assumed it could be easily introduced to this community, with minimal or no training. Our first prototype was customized for ASCR in the web environment. We gave special emphasis to including online help facilities in all ORÉ prototypes, given the purpose of introducing technology. To meet this end, we followed a model-based approach for both application design (Barbosa et al., 2002b) and help system design (Silveira et al., 2002). We also thought that using presentation patterns and functionality often found on the web would make it easier for infrequent users to learn and use the application, given the possibility to transfer knowledge from other applications to ours. As in typical bulletin board applications, announcements can be classified into sections, and a navigation bar allows users to browse through these sections. The help system was carefully designed to support the situated introduction of this technology and, as a result, multiple entry points for it were placed close to the contextual signs to which they applied. By providing pervasive help access, we hoped to allow users to clarify their doubts as soon as they emerged during interaction, answering such questions as "What's this?", "How do I use this?", and so on (Silveira et al., 2001).

Having developed the prototype, we set out to understand how users perceived it. When preparing the prototype for exploratory evaluation, we introduced announcements to give users the impression that the bulletin board was already in use. The announcements contained the same *kind* of information as real physical ASCR bulletin boards. For instance, a person's name would be a typical Brazilian name; an announcement would refer to an upcoming event or a donation request; and so on. We had both actual and fictional (although plausible) announcements in our database.

## 2.2   Prototype Evaluation

Six users participated in the test sessions, carried out according to the communicability evaluation method (Prates et al, 2000). Based on Semiotic Engineering, this method gathers both qualitative and quantitative data about interaction breakdowns. Our goals were to have indications of how well our solution would be understood and used by ASCR members; and to see which unanticipated challenges we were to face now.

Based on users' levels of previous experience, we expected they would have similar kinds of difficulties in using the prototype. However, it was not so. This made it clear to us that we were making a lot of strong assumptions about the knowledge that one can derive from previous experience with computers. For example, we were assuming that anyone who has used an Internet search engine or form understands *searches* and can use this knowledge in other contexts. We provided a structured search form in which, as is the case with most search forms, the content of each field is composed with that of others in an AND search expression. But instead of adding up words to *refine* the search, 2 participants tried to add up words to *increase the chances* of a term that was not found in a first trial being found in a second or third. It looked as though they thought that, by giving additional information on what they were looking for, the *system* would then figure out what it was that they were after, whereas it was just the opposite.

Another interesting finding had to do with the role of *familiar contexts* in test situations. One of the participants, while unable to find a specific announcement, realized that she had not provided its author's name. This information had not been defined in the test scenario she was given for the test, but she wanted to fill out the author's field in the search form with the name of the person in ASCR who she expected would have created it. Because that specific person was not in our fictional database, i.e., were not an option in the search form, this participant declared that she could not find the announcement. We took this as a sign of the difficulties novices may experience to understand technology in *simulated* environments such as ours. Two lessons were then learned by us. One was that ORÉ prototypes may be required to *make contextual sense* to ASCR users who agree to evaluate them, even in preliminary phases. In other words, when testing technology with users who have very little computer literacy, it may be necessary to use vertical prototypes (Preece et al., 2002), which are likely to require a lot of extra work when designing test cases. The other lesson is that we must be very careful with some ethical issues regarding the use of potentially sensitive information while testing technology. Test participants may be affected by test scenarios in ways that we cannot predict at this point, given the unexpected degree of involvement and realism that they seem to assume in trial situations.

Our prototype evaluation also showed us that, although online help systems are a particularly resourceful tool, we still have a long way to go. Although test participants were informed that the pervasive blue question marks on the screen provided access to help information, they tended to resort to help only as a last option, if at all. And when they did so, they either read only part of the information, or read it very quickly, overlooking the piece of information they needed. This suggested to us that help systems must perhaps be (re)introduced as a technology themselves, since the role and value of this technology was not perceived by participants of our experiments.

## 3    Concluding Remarks

We designed a second prototype to address some of the main problems uncovered during evaluation. Our first step in this direction was to build a set of mock-up web pages, in an attempt to simplify the bulletin board interface even further. For example, we decided to directly include contextualized tips and brief instructions within the working space, as well as verbose instructions in visually distinct areas, eliminating the need for users to explicitly request for help. Another example is related to the announcements search page. We moved from a structured search form to unstructured fields where users may type in words to appear in any of the database fields. In order to help users understand the effect of combining search terms, we now provide two search options, triggered by distinct buttons: to search for announcements in which *at least one* of the typed-in words appear, or to search for announcements in which *all* of the typed-in words appear.

Some of the changes in our prototype were not motivated by low usability alone, but mostly by the users' attitudes towards technology as a whole. During our project we broadened our view of the problem and focused on offering a solution that would solve a smaller range of the users' needs, but would help them overcome their negative feelings regarding technology. In other words, our project moved from the traditional user-centered to an adopter-centered design (Seligman, 2000). Adoption involves people becoming familiar with equipments and other physical artifacts and environment, but that alone is not sufficient. People also need to develop a positive attitude towards changing one's *modus operandi*, and accepting the mediation of new technology, especially in a context where spontaneous predisposition to collaborate is so vitally important to the organization.

Programs dealing with social and digital divide, such as the Information Society Program in Brazil, will certainly have greater challenges than those experienced in the ORÉ Project. In dealing with these challenges it seems to us that it is mister to involve disciplines which focus on attitudes, such as Psychology, Sociology, Anthropology, Arts and Design, and Linguistics, to name a few. However, at least in Brazil, they do not have the tradition of taking technology as a legitimate object of study in a design-and-engineering perspective. Instead, their contribution has always taken an analytic perspective on facts, telling us how things are and what *has been* the effect of technology here and there. But not how things are likely to be if this or that technology is introduced; much less what kind of intervention should be made if we want to reach this or that goal. Therefore, longer-term planning for a coordinated action involving Computer and Information Sciences with these disciplines is clearly called for. Among other things, we need to: raise awareness about the degree to which the multiple facets of technology affect individuals, groups and societies; revise academic curricula in colleges and universities to expose technologists to Social Sciences and vice-versa; and stimulate legitimate multidisciplinary research and development projects.

We believe that projects such as ORÉ can serve the dual purpose of the Information Society program. But only inasmuch as they generate the right research questions, and resonate in the academic, professional, and political spheres of this country.

## Acknowledgments

The authors thank PUC-Rio and CNPq for providing financial support to this work. Clarisse S. de Souza and Simone D.J. Barbosa thank CNPq and Raquel O. Prates thanks UERJ for their research grants. The authors also thank the members of ASCR and SERG for their valuable collaboration.

## References

ASCR (2002) http://www.saude-crianca.org.br

Barbosa, C.M.A., de Souza, C.S., Nicolaci-da-Costa, A.M., Prates, R. O. (2002a) "Using the Underlying Discourse Unveiling Method to Understand Organizations of Social Volunteers". *Proceedings of IHC'2002*. Brazil. October, 2002, pp 15-26.

Barbosa, S.D.J., de Souza, C.S., Paula, M.G., Silveira, M.S. (2002b) "Modelo de Interação como Ponte entre o Modelo de Tarefas e a Especificação da Interface" (in Portuguese). *Proceedings of IHC'2002*. Brazil. October, 2002, pp 27-39

Nicolaci-da-Costa, A. M. (1989) Análise de discurso e pesquisa qualitativa" (in Portuguese). In: Sociedade de Psicologia de Ribeirão Preto (Org). Anais da 18ª Reunião Anual de Psicologia, pp.501- 504.

ORÉ (2002) http://www.serg.inf.puc-rio.br/ore

Prates, R.O.; de Souza, C.S.; Barbosa, S.D.J. (2000) "A Method for Evaluating the Communicability of User Interfaces" *ACM interactions*. Jan-Feb 2000. pp.31–38.

Preece, J., Rogers, Y. & Sharp, H. (2002) *Interaction Design*. John Wiley & Sons. New York, NY.

Seligman, L. (2000) "Adoption as sensemaking: toward an adopter-centered process model of IT adoption". Proceedings of the 21st ICIS, p.361-370, December 2000, Australia.

Silveira, M.S.; Barbosa, S.D.J.; de Souza, C.S. (2001) "Semiotic Engineering contributions for designing online help systems". *Proceedings of SIGDOC 2001*, Novo México. October, 2001.

Silveira, M.S.; de Souza, C.S.; Barbosa, S.D.J. (2002) "*Design* de Sistemas de Ajuda Online baseado em Modelos" (in Portuguese). *Proceedings of IHC'2002*. Brazil, 2002, pp. 117-128.

# Example of a motorcycle manufacturer's approaches to usability

*Masamori Sugizaki*

YAMAHA MOTOR CO., LTD.
2500 Shingai, Iwata, Shizuoka 438-8501, Japan
sugizakima@yamaha-motor.co.jp

## Abstract

Unlike in the field of information devices, evaluations of usability are rarely conducted in the developmental stage for motorcycles. The reasons for this include that motorcycle structure is simple and once the basic operations are learned motorcycles can be used and enjoyed in accordance with the rider's level of skill, and that when learning to ride a motorcycle there is little of the sense of awkwardness that comes with entering a new field, since most people have experience riding a bicycle. However, if usability is reconsidered from the different viewpoints of functionality, safety, operability, cognitivity, and comfort, it is seen that the thinking behind usability is incorporated in the developmental stage for motorcycles.

## 1    Introduction

In recent years more and more companies are incorporating usability in the developmental process for their products. However, while some companies place great emphasis on usability, others feel it is not all that important. Motorcycle manufacturers are among the latter, and unlike in the information devices field, usability evaluations are rarely performed in the developmental process. There is also a low level of appreciation of or interest in usability within the company. Reasons that can be given for this include the fact that motorcycles have a simple structure; that since they are often used for pleasure the user does not mind if they are a little difficult to use as long as they are enjoyable; and that there are few parts of a motorcycle related to cognitivity.

The criteria for usability evaluations include functionality, safety, operability, cognitivity and comfort. If reconsidered from that viewpoint, we see that each element is taken in during the developmental process. The following will introduce usability activities as they have been adopted in motorcycle development.

## 2    Concepts of motorcycle usability

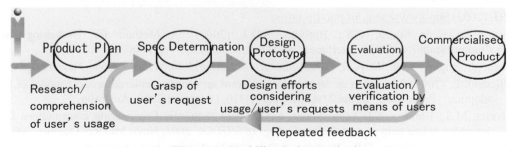

**Figure 1:** Usability design method

The kind of textbook usability evaluations that are done with information and home electronics devices are not generally performed with motorcycles. However, the developmental process does assume some of the techniques of usability design (Figure 1).

Kurosu presented the hierarchical structure shown in Figure 2 with regard to the criteria for the development of equipment. According to this, the criteria for equipment development are divided into the three categories of operability, cognitivity, and comfort. Added to these three criteria are the base categories of functionality and safety, with meaningfulness positioned at the top of the ultimate criterion.

Reviewing motorcycle development with respect to these categories, we see that each is incorporated somewhere within the steps of development. Thus, in this sense usability activities are also conducted in the development of motorcycles.

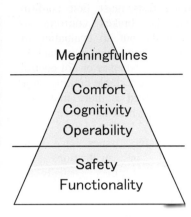

**Figure 2:** Hierarchical system of equipment development criteria

## 3    Usability activities for each criterion

Of the criteria in Figure 2, the ones of particular importance for motorcycles are functionality, safety, operability, and comfort.

### 3.1    Functionality

Functionality is the core aspect of motorcycles, but relatively few functionality items are directly related to the human user. Most functionality items are those such as durability or environment-related matters like fuel efficiency and exhaust gas. On the other hand, power output characteristics are extremely important in establishing the distinctive features of the motorcycle, and this is one of the items that receives the most attention during development. Since power output is something that users respond to most sensitively when a motorcycle is put on the market, it is important to realize output characteristics that match the product concept. This aspect of development is conducted by professional riders and the developers, but general riders are often also included in the evaluation phase.

### 3.2    Safety

Safety is extremely important for transportation machinery, and the standards are well established. Since crash tests cannot be done with real human riders, safety is investigated through computer simulation or experiments using crash dummies (Figure 3).

Motorcycles have a large degree of freedom movement at the moment of impact, and repeatability is poor in crash tests with real motorcycles. Therefore, computer simulation is used to supplement these tests.

Brake tests must first confirm that the brakes conform to standards, but an evaluation of brake feeling is also important. In particular, whether or not control is easy and a stable braking force is achieved is more important than the braking force itself (Figure 4).

Once safety issues have been cleared in evaluations by professional riders, virtually no safety problems will occur due to motorcycle function when used by the general rider. However, the possibility always remains of problems occurring due to the unrefined skills of the rider or rider negligence.

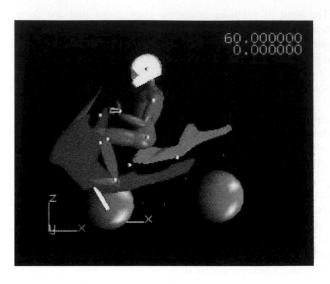

**Figure 3:** Crash simulation

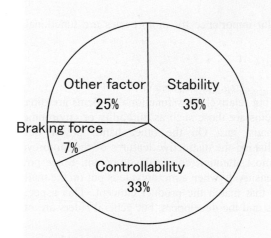

**Figure 4:** Factor of brake feel

## 3.3    Operability

When riding a motorcycle there are many instances when a number of operations must be performed simultaneously. The hands must work to steer the motorcycle at all times. The hands must also work the accelerator, light switch, turn indicators, brakes, and clutch; even an automatic transmission motorcycle has a shift switch. All these operations must be performed as needed while also handling the motorcycle. The foot brake pedal and shift pedal must also be operated in parallel with the hand operations. The handlebars, accelerator, brake, clutch, rear brake and gear shift must be operated simultaneously, especially when braking or speed control is necessary. In addition, since gloves are usually worn, particularly delicate operations are difficult. When such multiple operations are necessary, reliability and good operational feeling are demanded of the operations. To improve operability, compatibility between the human and machine is investigated from the viewpoint of ergonomics, and when a problem is found improvements are made. This process is done repeatedly. Specifically, evaluations are conducted from viewpoints such as the fit between human and motorcycle: whether the motorcycle fits the human shape and dimensions; whether the lever and pedal operating ranges are within the range possible by the human body, and whether they are positioned for most efficient operation.

Evaluations are also conducted from the viewpoints of compatibility between physiological and sensory characteristics, efficiency of operations, reduction of rider fatigue, and the like. Then, based on these evaluations, improvements are made. However, there are large differences in physique due to factors such as race, sex, and age, so that it is virtually impossible to satisfy all riders, and dissatisfaction from difficulty of use will arise in some.

**Figure 5:** Handlebar-mounted devices

## 3.4 Cognitivity

There are few aspects of motorcycles today that are related to cognitivity; instrument visibility is about the only concern. However, there are many switches concentrated in the area around the handlebars in motorcycles. Each one is simple and has just a single function, but still a certain amount of familiarization is needed by beginners. In addition, in terms of operability, as noted above, in many cases multiple operations must be performed at once, so that beginners may become confused when performing these simultaneous operations. For this reason, the level of cognitive engineering and ergonomic design incorporated is important. However, since there is not a hierarchical function structure, proficiency is quickly gained.

In addition, visibility in terms of the extent to which motorcycles can make their presence known to other vehicles or pedestrians is related to safety. Turning on the headlight during daytime is one such measure for this.

## 3.5 Comfort

Comfort is extremely important since motorcycles are often ridden for pleasure, and during development evaluations are conducted from the viewpoints of ergonomics and sensibility engineering. The riding feel is important among items of comfort. Riding a motorcycle for long periods will always be accompanied by pain in the buttocks. To produce a comfortable seat that reduces this pain, a wide range of evaluations are conducted during development, such as whether the pressure distribution when a rider sits on the seat is ergonomically appropriate, the pressure distribution and height of the backrest on motorcycles that have them, and whether there are adverse effects from the backrest when encountering sudden turbulence (Figure 6). In most cases seat evaluations are conducted

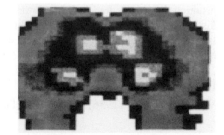

**Figure 6:** Pressure Distribution

not only by the developers but also by general evaluators.

Noise is another important item in terms of motorcycle comfort. Great effort is put into suppressing the noise level, but recently attention has also been shifting to the tone quality rather than simply reducing the noise level.

Vibration also affects comfort. The structure of motorcycle frames is simple, and the great design restrictions make it difficult to adopt a vibration absorbing structure. To reduce vibration, simulations are used in design and motorcycles with a comfortable vibration feeling are developed and assessed based on physical evaluations and sensory evaluations by people when riding. However, safety during tests must be considered for these motorcycle driving evaluations. A remaining issue, therefore, is the difficulty of conducting general user evaluations, since the evaluators are necessarily limited to the developers or professional riders for safety reasons.

Users have a great interest in design in terms of the aesthetic aspect of the motorcycle, and this strongly affects sales. A design is decided by those involved in development, mainly the designers, so that the motorcycle matches the developmental concept. In most cases the design is determined not from a survey of users, but is decided by those involved such as the designer and salespeople. However, some basic studies are also conducted as to what the user looks for in appraising a design, or the kind of design image users have.

## 3.6    Meaningfulness

Kurosu defined meaningfulness as "the ultimate criteria in machine design." Meaningfulness is determined by the ultimate goal in developing the machine. The developmental aim of motorcycles is not uniform. Some are developed simply as a means of transportation, in which case it is necessary to achieve the ultimate goals of superior operability and economy in terms of fuel efficiency and durability. On the other hand, leisure and sports motorcycles ridden for pleasure have greater emphasis placed on comfort, handling stability, and performance, since users demand a machine that takes them away from everyday life and stimulates them. In all cases, however, safety is essential.

## 4    Conclusion

Usability is thought to be only loosely related to motorcycles. However, from the standpoint of the user, incorporating the idea of usability is essential in development. A review based on the developmental criteria advocated by Kurosu for usability activities reveals that the idea of usability is incorporated in various instances during the developmental process for motorcycles. However, evaluation is largely made by the developers, with little input from users. Moreover, there has been virtually no evaluation by women at the time of development. The resulting products have thus had little opportunity for incorporation of the demands of women, as a result distancing themselves from women. In the future, it will be necessary to devise optimum methods from the standpoint of usability best suited to motorcycles. This may also lead to an increased number of elderly and female users.

## References

Masaaki Kurosu. (1996). Human Interface, *The Structure of the Usability Concept* (Vol.11 N&R 351-356)

# Scenarios, Models and the Design Process in Software Engineering and Interactive Systems Design

*Alistair Sutcliffe*

Centre for HCI Design
Department of Computation, UMIST
PO Box 88, Manchester M60 1QD, UK
a.g.sutcliffe@co.umist.ac.uk

## Abstract

Models are essential for engineering approaches to design, assumed in SE and HCI. However, models inevitably have limitations in their comprehensibility even with informal notations. Scenarios have received much attention as an effective means of user-designer communication; in SE, this is seen as a starting point for generating models, while in HCI scenarios are used as prostheses for design inspiration. The paper reviews different conceptions of scenarios and models with contribution they can make to the design process. The review explores the potential for constructive contrasts between scenarios as concrete, grounded examples and generalised, abstract models in an integrated view of systems development that encompasses both HCI and SE.

## 1    Introduction

Software engineering (SE) has tended to take a model-based approach to design either as informal graphical models, exemplified by the Unified Modelling Language, or as more formal models. However, more lightweight model-based approaches have been advocated for some time in RAD methods (DSDM, 1994), and this trend has been augmented by the appearance of agile methods and extreme programming. Similarly, HCI has had a model-based tradition of design-driven by task analysis, although this has never been as influential as conceptual modelling in SE. Both HCI and SE communities have been influenced more recently by use of scenarios, examples and use cases in the design process. Underlying these developments is a tension between model-driven and example-driven approaches to design, which this paper sets out to examine.

## 2    Contributions of Scenarios and Models

Scenarios have a host of definitions (see Rolland et al., 1998), ranging from narrative descriptions of system use expressed in natural language, frequent in HCI and the interactive systems design communities, to more formal representations of event sequences in SE (Kaindl, 1995). The range of definitions mirrors the process by which scenarios are transformed into models by a process of generalisation from specific examples. Use cases are usually considered to be models but they may be supported by narrative scenarios on which they are based. The tension between concrete detail and abstract model underlies design in both HCI and SE; however, little investigation has been directed towards understanding how scenarios and models can be profitably integrated in the design process. Scenarios have several roles in design. One role is to stimulate designers' creative imagination. Scenarios, examples and use cases can be used as lightweight instruments that guide thought and support reasoning in the design process (Carroll, 2000). A scenario, or even a set of scenarios, does not explicitly guide a designer towards a correct model of the required system. An extreme scenario might bias reasoning towards exceptional and rare events, or towards the viewpoint of an unrepresentative stakeholder. These biases are an acknowledged weakness of scenarios; however, we could trust designers as knowledgeable, responsible people who are capable of recognising such biases and dealing with them productively.

Scenarios arguably are the starting point for all modelling and design, and contribute to several parts of the design process (see Figure 1). Scenarios are gathered as examples of system use during requirements analysis and form the subject matter for creating models. The process of generalisation inevitably loses detail and the analyst has to make judgements about when unusual or exceptional behaviours are omitted. Models have to omit detail which may be vital, while scenarios can gather such detail but at the price of effort in capturing and analysing a "necessary and sufficient" set of scenarios.

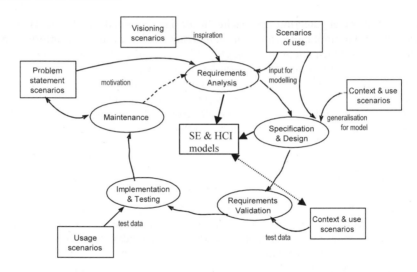

**Figure 1**: Use of scenarios and models in different phases of the design process

One productive juxtaposition of scenarios and models is to use scenarios as test data to validate design models. The Inquiry Cycle (Potts et al., 1994), uses scenarios as specific contexts to test the utility and acceptability of system output. By questioning the relevance of system output for a set of stakeholders and their tasks described in a scenario, the analyst can discover obstacles to achieving system requirements. Obstacle analysis has since been refined into a formal process for discovering the achievability of system goals with respect to a set of environmental states taken from scenarios (Van Lamsweerde & Letier, 2000). Scenarios, therefore, can fulfil useful roles either as test data, as a stimulant to reasoning in validating system requirements, or by providing data for formal model checking.

HCI uses scenarios in a similar manner in usability evaluation, although the role of scenarios is not articulated so clearly. Carroll has articulated several different roles for scenarios in the design process including as envisionment for design exploration, requirements elicitation and validation (Carroll, 1995). Other roles are usage scenarios that illustrate problems, and initiating or visioning scenarios that stimulate design of a new artefact, and projected use scenarios that describe future use of an artefact that has been designed (Sutcliffe & Carroll, 1999). Scenarios have been linked with generalised design principles by claims analysis in task artefact theory (Carroll & Rosson, 1992) which argues that scenarios of use with an example of the design artefact provide the context to understand a design principle. The results of evaluations are recorded in usage scenarios that describe the problem that motivated a general design principle, called a claim, with trade-offs expressed as upsides and downsides (Carroll, 2000; Sutcliffe & Carroll, 1999).

One criticism of conceptual models is that they do not capture the richness of interaction that occurs in the real world recorded in scenario narratives that concentrate on contextual description (e.g. Kyng, 1995). SE and HCI models all contain variations on a narrow range of semantic components, e.g. actions, events, states, objects, agents, goals, attributes and relationships. Models in both the HCI and SE traditions may be criticised for not representing the relationships between agents, activity and organisational structures, although these concepts are described in socio-technical system design frameworks such as ORDIT (Eason et al., 1996); and in requirements modelling languages such as the i* that analyses the dependencies between agents, tasks, goals and resources (Mylopoulos et al., 1999). Some extensions to describe the role of work artefacts in ecological interface design (Vicente, 2000) and communication between agents in collaborative tasks (Van der Veer et al., 1996; Sutcliffe, 2000b).

So model-based design might be accused of having a narrow scope of phenomena that it can represent. While narrative scenarios can represent a wider range of phenomena, they do so in an *ad hoc* manner and leave the responsibility of generalisation to the analyst.

## 3 Cognitive Affordances

To investigate the roles of scenarios and conceptual models, we enquire a little deeper into how scenarios and models function in cognitive terms. Scenarios use language and concepts that are readily accessible to users and domain experts, whereas tasks and SE conceptual models are expressed in a specialised language that users have to learn. Because scenarios invoke specific memory schema they help to recruit specific knowledge. This helps tune our critical faculties, since detail tends to provide more subject matter to detect inconsistencies and errors when we reason about models and specifications.

In contrast, models are harder to comprehend, because they represent abstract generalisations. While people form categorial abstractions naturally (Rosch et al., 1976), we are less efficient at forming categories of concepts and functions (Hampton, 1988). Unfortunately, formation of conceptual-functional categories is a necessary part of the generalisation process, so users can find reasoning about even simple conceptual models, such as data flow diagrams, difficult (Sutcliffe & Maiden, 1992). Once learned, models become memory schema that represent abstract concepts removed from everyday experience, so their effectiveness depends on how well connected they are to more specialised memory schema representing scenario-based knowledge. The importance of the connection becomes clear when we try to validate models. Without any connection to specialised knowledge I can accept the validity of the general concept simply because it has a wide scope of meaning. For example, I might accept the proposition that <all birds can fly> as a true type definition of the class <birds> in the absence of more specific knowledge of penguins, kiwis, rheas, ostriches and dodos. Models, therefore, need to be integrated examples and scenarios and, furthermore, cannot exist profitably without them; indeed, human categorial memory is probably an integration of abstract models and specific examples (Lakoff & Johnson, 1999).

While scenarios might be effective in grounding reasoning, their downsides lie in reasoning biases and partial mental model formation. Confirmation bias is a well known weakness of human reasoning (Johnson-Laird & Wason, 1983). We tend to seek only positive evidence to support hypotheses, so scenarios can be dangerous in supplying us with minimal evidence to confirm our beliefs. While problem statement scenarios and anti-use cases (Alexander, 2002) can counteract confirmation bias, we need to be wary of this downside. However, task models are not immune to this problem; we can read our own bias into a general model precisely because it does not contain detail, so the conjunction of scenarios and task model may present a cure. Another potential pathology is encysting, more usually described by the saying "can't see the forest for the trees". Since scenarios are detailed they can bias people away from the big picture of important design issues towards obsession with unnecessary detail. Models exist to counteract this pathology. Partial mental model formation is another weakness when we test hypotheses (Simon, 1973). Scenarios can encourage this pathology by reassuring us that we have covered all aspects of the problem with a small number of scenarios. This exposes the Achilles heel of scenario-based reasoning: it is difficult, if not impossible, to be confident that a necessary and sufficient set of scenarios have been gathered to escape from the partial mental model problem. Models can encourage cognitive dissonance, or conservative reasoning pathologies when we have invested in generating an idea (or model) and are reluctant to abandon it even in the face of contradictory evidence. Scenarios can provide such evidence, indicating a conclusion to this section, that a combination of task models and scenarios in critical review process can counteract many reasoning pathologies.

# 4    Integrating Contributions

So far I have argued that we might be able to have the best of both worlds by integrating specific grounded knowledge contained in scenarios, and more generalised knowledge in conceptual models. So is there a synthesis for model-analytic and creative scenario-exploration approaches to design? A partial answer is acknowledging the "horses for courses" argument. Model-analytic and creative exploratory design approaches will be necessary for safety-critical applications on the one hand, and those oriented to entertainment, education, and general commerce on the other. A more satisfactory answer is to examine the nature of methodological interventions in the design process. Methods, guidelines, principles and models are rarely used explicitly by expert designers (Guindon, 1987). Novices might use them *ab initio*, but design knowledge soon becomes internalised as the designer's skill. I argue that reusable knowledge in the form of generic models, claims and design rationale should become part of the skill-set of all system designers (Sutcliffe, 2002a). Models can contribute by providing representation for reusable knowledge via documentation. Scenarios can support the design process at run time as probes to test assumptions and stimulate creation. So both approaches have their contributions to make, albeit in different phases of design education and practice. Scenarios can stimulate thought, but knowledge can only be reused effectively in a generalised form as models, claims, principles and guidelines. The interesting research question is when explicit recording of generalised knowledge is necessary for design introspection and passing on for reuse.

# 5    References

Alexander I. (2002). Initial industrial experience of misuse cases in trade off analysis. In Proceedings of RE-02 IEEE Joint International Conference on Requirements Engineering, Essen, September 2002, (pp. 61-70). Los Alamitos, CA: IEEE Computer Society Press.
Annett, J. (1996). Recent developments in hierarchical task analysis. In S. A. Robertson (Ed.),

*Contemporary ergonomics 1996*. London: Taylor Francis.

Carroll, J. M. (Ed.) (1995). *Scenario-based design: Envisioning work and technology in system development*. New York: Wiley.

Carroll, J. M. (2000). *Making use: Scenario-based design of human-computer interactions*. Cambridge MA: MIT Press.

Carroll, J. M., & Rosson, M. B. (1992). Getting around the task-artifact framework: How to make claims and design by scenario. *ACM Transactions on Information Systems, 10*(2), 181-212.

DSDM (1995). *DSDM Consortium: Dynamic Systems Development Method*. Farnham Surrey: Tesseract Publishers.

Eason, K. D., Harker, S. D. P., & Olphert, C. W. (1996). Representing socio-technical systems options in the development of new forms of work organisation. *European Journal of Work and Organisational Psychology, 5*(3), 399-420.

Guindon, R. (1987). *A model of cognitive processes in software design: An analysis of breakdowns in early design activities by individuals. MCC Technical Report STP-283-87* Austin TX; Microelectronics and Computer Technology Corporation.

Hampton, J. A. (1988). Disjunction in natural categories. *Memory and Cognition, 16*, 579-591.

Johnson-Laird, P. N., & Wason, P. C. (1983). *Thinking: Readings in cognitive science*. Cambridge: Cambridge University Press.

Kaindl, H. (1995). An integration of scenarios with their purposes in task modelling. In G. M. Olson, & S. Schuon, (Eds). *Designing Interactive Systems: DIS 95 Conference Proceedings, Ann Arbor MI 23-25 August 1995*, (pp. 227-235). New York: ACM Press.

Kyng, M. (1995). Creating contexts for design. In J. M. Carroll (Ed.) *Scenario-based design: Envisioning work and technology in system development*, (pp. 85-108). New York: Wiley.

Lakoff, G., & Johnson, M. (1999). *Philosophy in the flesh: The embodied mind and its challenge to western thought*. New York: Basic Books.

Mylopoulos, J., Chung, L., & Yu, E. (1999). From object-oriented to goal-oriented requirements analysis. *Communications of the ACM, 42*(1), 31-37.

Potts, C., Takahashi, K., & Anton, A. I. (1994). Inquiry-based requirements analysis. *IEEE Software, 11*(2), 21-32.

Rolland, C., Achour, C. B., Cauvet, C., Ralyte, J., Sutcliffe, A. G., et al. (1998). A proposal for a scenario classification framework. *Requirements Engineering, 3*(1), 23-47.

Rosch, E., Mervis, C. B., Gray, W., Johnson, D., & Boyes-Braem, P. (1976). Basic objects in natural categories. *Cognitive Psychology, 7*, 573-605.

Simon, H. A. (1973). The structure of ill-structured problems. *Artificial Intelligence, 4*, 181-201.

Sutcliffe, A. G. (2002a). *The Domain Theory: Patterns for knowledge and software reuse*. Mahwah NJ: Lawrence Erlbaum Associates.

Sutcliffe A. G. (2002b). Requirements analysis for socio-technical system design. *Information Systems, 25*(3), 213-235.

Sutcliffe, A. G., & Carroll, J. M. (1999). Designing claims for reuse in interactive systems design. *International Journal of Human-Computer Studies, 50*(3), 213-241.

Sutcliffe, A. G., & Maiden, N. A. M. (1992). Analysing the novice analyst: Cognitive models in software engineering. *International Journal of Man-Machine Studies, 36*(5), 719-740.

Van der Veer, G. C., & Van Welie, M. (2000). Task-based groupware design: Putting theory into practice. In D. Boyarski, & W. A. Kellogg, (Eds). *Conference Proceedings: DIS-2000 Designing Interactive Systems: Processes, Practices Methods and Techniques, New York 17-19 August 2000*, (pp. 326-337). New York: ACM Press.

Van Lamsweerde, A., & Letier, E. (2000). Handling obstacles in goal-oriented requirements engineering. *IEEE Transactions on Software Engineering, 26*(10), 978-1005.

Vicente, K. J. (2000). HCI in the global knowledge-based economy: Designing to support worker adaptation. *ACM Transactions on Computer-Human Interaction, 7*(2), 263-280.

# A European Usability Forum
## Collaborating on Strategic Initiatives

*Manfred Tscheligi, Verena Giller and Peter Fröhlich*


CURE – Center for Usability Research and Engineering
Hauffgasse 3-5, 1110 Wien, Austria
www.cure.at
froehlich@cure.at

## Abstract

In this paper, it is argued that there is a strong need for concerted strategic activities of European organisations engaged in user-centred design, human-computer interaction, usability and related disciplines. The Usability Forum (working title), which has been initiated by the EU-funded project UsabilityNet, provides a global platform to coordinate comprehensive activities in the area of research innovation, lobbying, education, and publicity. The main benefits, organizational principles, as well as future plans are described.

# 1    Current situation and problems

In the research field comprising HCI, UCD, Usability and related areas, a lot of achievement has been made to communicate expertise and quality assurance to relevant target groups (managers, usability professionals, designers, etc.).

- Within the EU-funded IST-program, projects have been established that offer a large amount of information resources for usability and helped to promote usability-related issues (please see www.usabilitynet.org for an overview of these resources).

- Furthermore, a range of national professional organisations launches strategic initiatives and provides support for usability practitioners (e.g. the American-based UPA, or national organisations like the British HCI Group or the UK Usability Professionals Association).

- Research conferences and organisations (CHI, HCI, Interact, etc.) are assuring high quality standards and added value for the research community.

- Standards and accreditation associations are developing co-ordinated basic materials and guidelines to support the development of user centred design as a professional discipline.

Despite all these achievements, the fragmentation of the community into organisations, projects and networks has led to a situation that initiatives can only be made in a rather informal and bilateral way. Furthermore, the development of a strong and consistent voice towards the public, relevant funding organisations, and industry is aggravated. This fragmentation is rather unnecessary, since all actors in the field strongly agree in the fundamental concepts and benefits of user-centred thinking.

In several significant technological areas, strategic initiatives have emerged that provide a powerful means to foster the development of commonly shared visions and effective communication to relevant target groups. A prominent example for this kind of initiatives is the Wireless World Research Forum in the mobile research sector (see: www.wireless-world-research.org). Strategic networks that comprise all relevant major players in the field can provide a useful prioritisation of future research agendas. Furthermore, global strategic platforms provide the possibility to gain impact on influential institutions and research programmes.

In the field of usability and related areas, there is a strong need for such a strategic European platform. In order to give user-centred thinking the impact it deserves, strategic aggregations and concerted activities are indispensable.

## 2    The concept of the Usability Forum

In order to overcome these weaknesses, the EU-funded project UsabilityNet initiated the Usability Forum. The concept of the forum builds on the following basic principles:

Open and comprehensive organizational structure

The concept of the forum is very broad with regard to the covered application domains, organization types and technical paradigms. It is open to all interested parties in the field: research organisations, professional associations and commercial companies (from consultancies to respective departments of large industrial companies). Furthermore, the forum aims to cover all relevant sub-domains and paradigms of the research field. For instance, human factors research in domains like aeronautics is addressed as well as usability evaluation tool development.

This broad concept makes it possible for the forum partners to learn from other domains and to share knowledge, as well as to coordinate their opinions. Furthermore, it provides a clear external representation of the field, being a first contact point for the public, relevant institutions and economy.

High-level strategic and collaborative actions

The forum aims at combining the substantial efforts described above and to provide a powerful comprehensive means for lobbying and publicity. The target groups of the forum activities are – apart from the community itself – decision-makers within industry and the public sector. The main activities are:

- To communicate more effectively the business advantages and cost-benefit arguments to managers. This is done by collecting and sharpening the already existing publicity material.

- To provide a consistent and comprehensive contact point for public institutions (e.g. funding associations, strategic research initiatives, conference organisers).

- To gain impact in relevant institutions (such as the European Commission and national governments) by more powerful European-based lobbying activities.

Roadmapping of future research activities

Additionally to the strategic focus, the forum also aims at developing a long-term vision for future research activities in the field of usability engineering and user-centred design research. The definition of a road map will prepare the ground for research activities in the upcoming European Commission's 6th framework programme and future initiatives. In order to do this, the following steps are planned:

- Establishment of a think tank and conduction of expert workshops, together with focussed work in thematic working groups.

- Integration of visions from other or more general research areas (e.g. ISTAG scenarios, etc.).

- Preparation of a "Book of Visions for a USER FRIENDLY Information Society" containing the most promising and visionary interface solutions and methodologies.

- Dissemination of the results to relevant bodies, using all relevant communication channels.

European focus

Since there is a strong need to build up a common voice within the European research community, the Usability Forum is primarily focussed on Europe. However, to make overall strategic activities more efficient, it will also liase and coordinate with relevant American and other regional bodies. This certainly applies for the coordination of conference schedules, concerted publicity campaigns and strategic activities towards multinational industrial companies.

Sustainability

The initial set-up and beginning activities of the forum were supported by the EU-funded project UsabilityNet. The maintenance of the forum, in contrast, is provided by a voluntary contribution of the participants. The reason for the high commitment of the partners can be attributed to the shared opinion that it is worthwhile to invest in a more powerful European strategic representation. Furthermore, the development of roadmaps for future research provides an added value for key-players in the field. However, the sustainability of the forum as a whole does not mean that certain sub-parts will not strive for additional national or EU-funding (e.g., an innovative research initiative of a domain-specific working group).

Complementarity to other research and networking activities

A major goal of Usability Forum is to avoid doubling the work of other bodies. In contrast, since representatives of these organisations are also participating in the forum, significant content and initiatives can be reused. This especially applies for the following issues:

- **Information resources:** Several projects and organizations have established highly informative web sites about usability methods, quality principles, information for usability practitioners, etc. Examples for this are the UsabilityNet web site (www.usabilitynet.org) or the UsabilityNews web site (www.usabilitynews.com, provided by the British HCI group).

- **Usability accreditation:** In Europe, there have been many approaches towards usability certification as part of a professional development scheme. The EU-project UsabilityNet, which initiated the forum, has been very active in this area (see www.usabilitynet.org/professional/forum.htm and Bevan, 2003). The achievements and experiences from this project are fully integrated into the forum activities.

- **Education:** In most European countries, courses and curricula are already established (please see www.usabilitynet.org/usability/courses.htm for an overview). Thus, there already exists a high amount of learning materials and training expertise. However, approaches to a European curriculum for training and education are still missing. The

Usability Forum takes into account all singular training activities in Europe and provides a unified approach to foster quality and innovation.

- **Domain-specific expertise:** The Usability Forum offers a great chance for usability, human factors and HCI specialists to share their views concerning different application areas and research methods. The forum encourages domain-specific research initiatives to develop innovative concepts and to share knowledge with other domains represented in the forum.

- **Professional organizations:** The large range of Usability professional organizations offer many services for their target groups (information resources, publications, discounts on literature purchase and conferences, etc.). Furthermore, they have a substantial knowledge of national research and business conditions.

## 3  Set-up of organisational structure and procedures

The first step to set up the Usability Forum was to analyze and identify potential founding members who would be able to drive future activities and encourage further actors in the field to participate. This intensive inquiry covered the sector of industry (consultancies as well as specialist departments within multinational enterprises), research organizations and usability professional associations.

The official starting point was the Usability Forum kick-off meeting on 9 January in Vienna, Austria. The participants were the Usability Forum core-members - representatives of 19 organizations from 11 European countries, covering industry, research and professional networking organizations. The meeting had the objective to refine the detailed goals of the forum and to establish an organizational structure.

In order to streamline the work and to adapt the organizational structure to the main strategic issues, 6 working groups (WGs) were formed and responsibilities as well as timelines were specified.

- **WG1: Communication and Public Awareness:** This working group is mainly concerned with communication to interested parties (other professional bodies, public institutions, etc.). Additionally, internal communication is provided by the implementation of a working space and mailing lists. The UsabilityNet web site, being a powerful dissemination element, will be transferred to the forum after the end of the project time of the UsabilityNet project.

- **WG2: Innovation:** The definition of long-term visions for innovative user interfaces and roadmaps for research priorities is the goal of this working group. The results of this work are intended to have an impact on research priorities within relevant research bodies (e.g. the IST-programme of the European Commission). Furthermore, concrete projects can arise from the work being done in the innovation working group.

- **WG3: Policy:** The goal of the policy working group is to set-up efficient lobbying mechanisms directed to European and national governmental institutions. These activities have high priorities, since the lack of awareness and political representation is one of the most often complained weaknesses within the research community. Naturally, the work being done in the innovation and communication working groups is closely intertwined with these activities.

- **WG4: Business Case:** Relevant cost-benefit examples (i.e. recent and valid for a European context) and the demonstration of potentials arising from usability are the main focus of this group. These activities are important, because usability and related areas are not yet appropriately taken into account by business decision-makers. For this working group, valuable influence is given by the forum members coming from industry (e.g. departments of large companies).

- **WG5: Domains:** Specific application and sector-specific areas like eGovernment, High Integrity or aviation are addressed in this working group. Naturally, forum members with special interest in the respective areas are driving the activities.

- **WG6: Knowledge and Education**: The already existing course material and instructional know-how is taken into account to shape a European curriculum. Furthermore, a consistent concept for the different approaches for an accreditation scheme is formed.

## 4 Outlook

The activities done so far are promising and there has been a high interest in participation. The future challenges for the Usability Forum are to maintain sustainability and to amplify its impact. Further meetings are planned in the next 6 months. Both general meetings as well as specific working group meetings will be conducted. Representatives of the forum will attend and contribute to relevant conferences (e.g. CHI (5-10 April 2003), HCI International (22-28 June 2003) and Interact (1-5 September 2003). Although the forum has a clear European orientation, the activities to liase with American professional organisations (e.g. SIGCHI and UPA) will be intensified.

## References

Bevan, N. (2003) Accreditation of usability professionals. Proceedings of HCI International 2003. Lawrence Erlbaum

# A Usability Study of an Object-Based Undo Facility

*Iomar Vargas, José A. Borges*

Dept of Electrical & Computer Eng
University of Puerto Rico
Mayaguez, Puerto Rico
borges@ece.uprm.edu

*Manuel A. Pérez-Quiñones*

Dept of Computer Science
Virginia Tech
Blacksburg, VA
perez@cs.vt.edu

## Abstract

The undo facility is an essential feature of modern interactive systems. The predominant undo model used in commercial systems is the single level undo or the linear undo. In this paper we present a study that assessed the usability of selective undo facility that allowed the user to select any previously done action to be undo without undoing the intervening actions. The results showed that the object-based (selective) undo was preferred by the majority of the users. It was also the fastest way of undoing distant actions selectively. The linear undo mechanism proved to be faster for undoing the last action.

## 1    Introduction

The undo facility is an essential tool for graphical user interfaces (GUI). It is a mechanism that allows users to reverse the effects of their interactions enhances the usability of the system and encourages exploration. In this paper we present a selective undo mechanism for GUIs that makes it easy to select the action to undo, and the results of a usability evaluation its use.

Through the years, researchers have studied and proposed several types of recovery systems. These include non-linear undo systems [Berlage1994][Myers1996], hierarchical undo facilities [Myers1996] [Meng1998], timeline based recovery systems [Edwards2000], and graphical or visual facilities [Kurlander1988] [Derthick2000] among others. Undo has also been studied from a multi-user perspective [Sun2002]. In the selective undo model [Berlage 1994], the user selects the action to undo from the list of all actions performed without undoing any intermediate actions.

However, due to several factors such as the complexity, cost, and usability of many of these novel approaches, the single-level and/or the linear undo facility remain as the predominant recovery facility in today interactive systems. Many of the research papers that cover advanced undo facilities present usability evaluations as future work (e.g. [Myers1996] [Meng1998]). Wang [Wang1991] proposed the idea of attaching the undo history list from a linear model on an object-by-object basis, the approach taken here, but no usability evaluation was presented.

In this paper we present an object-based undo that allows selective undo to be implemented in a fairly straightforward fashion, avoiding some of the complexities of other approaches. The usability of this new model is assessed with a usability evaluation. We found that the object-undo was preferred over the traditional linear model. We also found that the object-based undo was faster for situations where a selective undo action was appropriate, but that the linear undo mechanism was faster in situations where undoing the last action was needed.

## 2 Object-Based Undo

This research studied, implemented, and evaluated an object-based undo facility. In this system the undo/redo options or actions are associated and applied directly to the objects themselves. The history of events representing the user interaction with is divided into two categories: object, and application histories. The object history records the events that change the attributes of an object. The application history records the events associated with the drawing canvas.

Through a right button mouse click over an object in the application, users can execute an undo or redo action related to that object. The undo options associated with each object are categorized according to the type of action or the object's attributes. Within each category, the commands are organized in reverse chronological order (i.e. most recent at the top). For example, for the specific application implemented, the undo/redo actions for one type of objects were categorized into actions related to the attributes *position*, *format*, and *text*. Of course, this classification could vary for different applications or different types of objects within a particular application. The user's actions concerning the application itself are similarly associated with an undo/redo facility for the application. Through a right mouse click on any place of the application's canvas or by selecting the undo on the application's toolbar, the user can undo/redo actions such as *add*, *copy*, or *delete*.

From a usability perspective, executing an undo in this model is more natural and direct since users interact directly with the object to be affected. This eliminates the confusion that can result by trying to identify which object is affected by an undo action, and the need to identify the objects in a global undo history list. In the linear undo model it is necessary to roll back the events done after the event desired to undo. Users must either redo all actions undone or, if possible, execute some action(s) that would restore an object to its desired state without using the undo facility. In the object-based undo model this is not necessary since only the object selected is affected by the undo action.

Figure 1 shows an example of using the object-based undo on a presentation designer application, called Jpresentation. The following list represents the actions executed to create the slide shown:
1. Type in Text1: "Human computer Interaction"
2. Type in Text2: "The effects in the children "
3. Change format of Text1
4. Type in Text3: "by Unknown person"
5. Type in Text1: "Object-based undo"
6. Type in Text2 "Nonlinear undo model"
7. Type in Text3: "by Iomar Vargas"
8. Change format of Text2
9. Type in Text1: "Human Computer Interaction"

In the sequence above, action 2 and 6 change object Text2. As illustrated in Figure 1 undoing action 2 on Text2 is accomplished by a right-click on the object, selecting the attribute category, and specifying undo. Only the actions associated with that object and category are displayed on the undo menu, and it only affects the object and category selected. Figure 1 shows in the pop-up menu the text that was typed by the actions (2 , 6) as possibilities to undo for this particular object.

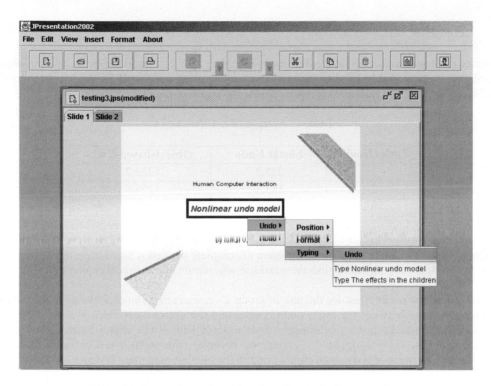

**Figure 1:** Example of the object-based undo in JPresentation

# 3 Usability Evaluation

To test the usability of the object-based undo facility we implemented two versions of JPresentation. One incorporated the undo model explored in this study and the other was implemented using the traditional linear undo. User tests were conducted by having 20 subjects create simple presentations on both systems. Half of them realized their tasks first in the linear undo system, and then in the other. The other half realized their task in the reverse order. After performing the tasks they were asked to fill out a subjective user satisfaction questionnaire.

The participants were asked to realize a total of 25 simple tasks on both versions of Jpresentation. The tasks for each of the two sessions were divided among three groups:
1. *Creating a slide* – typical tasks associated with the creation of a slide presentation.
2. *Consecutive changes* - tasks that asked the user to change an object's state to the state of the immediately preceding step. For example:
    task 9: Change the color of the text "Educacion Multicultural" to black
    task 10: Change the color of the text "Educacion Multicultural" to its previous color
3. *Unordered changes* – tasks that asked the user to change an object's state to a previous state done several steps before. For example:
    task 3: Change the text "por: Clyde Drexler" with "por: Reggie Miller"
    task 11: Change the text "por: Reggie Miller" to its original name

The tasks in group 2 should be better supported by a linear undo facility. The tasks in group 3 were design to take advantage of the object-based undo explored in this paper.

# 4    Results

A parity *t* test was used to compare the time to complete all the tasks in the linear, and the object-based undo system. Table 1 shows the average time measured in seconds for the twenty users that performed the test. The tasks are classified into the three groups mentioned above.

**Table 1:** Average time for all tasks

| Task Groups | Linear Undo | Object-based Undo |
|:---:|:---:|:---:|
| 1 | 225.3 | 223.2 |
| 2 | 183.4 | 228.9 |
| 3 | 215.0 | 165.3 |

The results of the parity *t* test for the tasks of group 1 - *creating a slide* did not reveal a significant difference on the average time it took the users to complete the tasks. Since users did not need the undo facility to be able to accomplish these tasks, it was reasonable to expect similar times.

The results of the parity *t* test for the task of group 2 - *consecutive changes* illustrate a significant difference in time between the linear undo system, and the object-based undo system. The users performed these tasks faster in the linear undo system than in the object-based undo system. However, for the tasks of group 3 - *unordered changes,* the parity *t* test revealed that users performed these types of tasks significantly faster in the object-based undo system.

Table 2 shows the average results from the questionnaire answered by the participants of the usability test. They were asked to rate a statement regarding the linear undo model and the object-based undo model. The scale ranged from 1- *Completely Disagree* to 5-*Completely Agree*. A Wilcoxon sign-ranked test revealed a significant difference in overall users' satisfaction. Users were significantly more satisfied with the object-based undo system.

**Table 2:** Average results of questionnaire

| Statement | Linear | Object-based |
|:---|:---:|:---:|
| 1. The undo was easy to use | 3.8 | 4.5 |
| 2. It was easy to recover from mistakes | 2.9 | 4.7 |
| 3. Selecting the task to undo was easy | 3.9 | 4.3 |
| 4. In the case I did a mistake, I preferred to undo the task, even if it was done several tasks before. | 2.0 | 4.5 |
| 5. The undo helped me to reduce the errors done in the presentation | 3.5 | 4.7 |
| 6. I felt comfortable and free to do any task without worrying of the mistakes. | 3.4 | 4.5 |
| 7. The use of the undo helped me finish the presentation in fewer steps. | 2.7 | 4.6 |
| 8. Overall, I was satisfied with the undo. | 3.0 | 4.7 |

The final part of the questionnaire asked participants to choose between the linear and object-based undo models to answer four questions. As shown in Table 3, 95% of the users answered that the object-based undo was most helpful and effective, while 75% answered that it was more usable. In addition, 95% of the users preferred the object-based undo system to the linear undo.

**Table 3:** Percentage of participant's preference responses

| Question | Linear | Object-based |
|---|---|---|
| 1. Which undo was most helpful | 5 | 95 |
| 2. Which undo was most effective | 5 | 95 |
| 3. Which undo was most usable | 25 | 75 |
| 4. Which undo do you prefer | 5 | 95 |

# 5 Conclusion

In this study we evaluated a model that provides a partial non-linear or selective undo facility. With this model, users can execute an undo/redo command selectively on any object or any object's undo/redo category. However, within the specific history list related to such object or category, the user is restricted to a linear selection. This approach is not as flexible as a selective undo facility but we feel that it reduces the complexities and dependencies associated with such systems and make this model easier to implement and understand.

The object-based undo takes advantage of "*deixis*", the ability to refer to an object and point to an object in the world of action during communication, minimizing the transition between action and communication. The user can refer directly to any object on the application for undoing an action on the specific object.

The users were very satisfied with the object-based undo model and preferred it over the linear model. The possibility of undoing distant actions selectively and the ability to refer to an object directly were two of the main factors that influenced the user's preference. This work was supported in part by the National Science Foundation under Grant No. IIS-0049075.

# References

[Berlage1994]. Berlage, T. "A Selective Undo Mechanism for Graphical User Interfaces Based on Command Objects." ACM Transactions on Computer-Human Interaction. **1**(No. 3): 269-294.

[Derthick2000]. Derthick, M., and Roth, S. "Data Exploration across Temporal Contexts." Carneggie Mellon University Robotics Institute. IUI 2000 New Orleans, LA, USA

[Edwards2000]. Edwards, W. K., Igarashi, T., LaMarca, A., Mynatt, E. D. "A Temporal Model for Multi-Level Undo and Redo." UIST 2000, San Diego, CA, USA

[Kurlander1988]. Kurlander, D., and Feiner, S. "Editable Graphical Histories." IEEE Workshop on Visual Languages, Pittsburgh, USA.

[Meng1998]. Meng, C., Yause, M., Imamiya, A. Visualizing Histories for Selective Undo and Redo. Third Asian Pacific Computer & Human Interaction, 15 - 17 July, 1998.

[Myers1996]. Myers, B., and Kosbie, D.S. "Reusable Hierarchical Command Objects." Proc. of ACM CHI 96 Conference on Human Factors in Computing Systems. pp 260-267.

[Sun2002]. Sun, Chengzheng, "Undo as a concurrent inverse in group editors," ACM Transactions on Computer-Human Interaction (TOCHI), Vol. 9, Issue 4, December 2002, Pages 309-361.

[Wang1991]. Wang, Haiying, and Green, Mark, "An Event-Object Recovery Model for Object-Oriented User Interfaces," UIST 1991.

# Usability Engineering in South Africa Today: Challenges and Opportunities

*Janet Wesson and Darelle van Greunen*

Department of Computer Science and Information Systems
University of Port Elizabeth, Port Elizabeth, South Africa
csajlw@upe.ac.za; csadvg@upe.ac.za

## Abstract

The goal of this paper is to review the status of usability engineering in South Africa today. The paper will also highlight the challenges facing South Africa and suggest some opportunities available for researchers, educators and developers in the Human-Computer Interaction (HCI) community in South Africa.

## 1    Introduction

Information and communication technology (ICT) is regarded as a vital catalyst for social change and economic development in developing countries. The South African government has recognized the potential benefits to be gained from harnessing the power of ICT and is working to create a technically literate workforce that can contribute to a dynamic economy and participate in the Information Society (Bridges.org, 2002). A national ICT strategy has been implemented that focuses particularly on the d isadvantaged segments of South African society. The current state of ICT policy in South Africa is that the Telecommunications Act, Act 103 of 1996, is under review (Bridges.org, 2002).

The necessary ICT infrastructure to provide technology access and help utilize technology is poorly linked and spread unevenly across the country. Telecommunications is, however, finally beginning to play a role in ICT development in South Africa. As a result, the growth in the number of mobile phone users in South Africa has been enormous. The current market size is 13 million users (32% of the population); this is expected to grow to 21 million (38%) by 2006. The coverage of mobile phone access in South Africa is also expected to be considerably higher than fixed line coverage (Figure 1). Internet connectivity in South Africa is by far the highest in Africa, with 2.8 million active users. This figure is growing at an estimated 20% per annum (Pater, Machanick, Booysen, & Hurst, 2002).

South Africa is facing a significant "brain drain" in technical and entrepreneurial ICT skills, with an estimated 200-300 ICT-skilled people leaving the country each month. This is caused by the rapid growth in demand for ICT skills worldwide, little opportunity (jobs, remuneration, and innovation) and social factors such as crime in South Africa. In response to this, various institutions in South Africa have called upon local and international experts to become involved in the revival and growth of African economies and industry (Hugo, 2002). Foremost among the areas singled out for urgent and immediate attention are the most powerful enablers of socio-economic growth in the new millennium: education, skills development, and the development of the ICT industry.

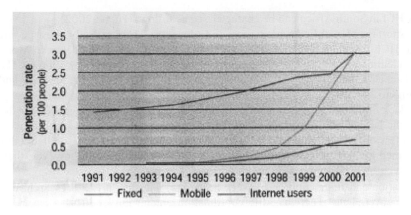

**Figure 1:** Historical growth of ICT services in South Africa (Pater et al., 2002)

The lack of developmental growth in the ICT industry in South Africa is aggravated by a lack of human factors and usability engineering training and expertise in the software industry. For the purposes of this paper, the term Human-Computer Interaction (HCI) will be used to describe both these concepts. A recent survey revealed that only 7 out of 25 universities (28%) in South Africa are actually teaching HCI principles as part of the typical Computer Science or IT undergraduate or postgraduate programme (Wesson, 2001). Consequently, very few software engineers in South Africa have any real knowledge or understanding of the implications of usability and user interface design. This situation is further complicated by the diversity in technological knowledge and expertise of the typical computer user in South Africa.

The goal of this paper is to review the status of usability engineering in South Africa today and to highlight the challenges and opportunities for researchers, educators and developers in the HCI community in South Africa.

## 2 Current Situation in South Africa

The South African society today finds itself in one of the most critical evolutionary phases in its history where it appears as if no aspect of everyday life and work is left untouched by the impact of ICT. For various reasons, computer-mediated communication technologies (such as cellular phones and the Internet) are becoming assimilated into certain sectors of South African culture more rapidly than anywhere else in the world (Hugo, 1998). People's attitudes toward new devices change as they discover new benefits. These devices often have considerable personal or social value, for example, cellular phones.

The South African user population has grown culturally, economically and educationally more diverse. Not only have users different expectations of technology, but these expectations are influenced by their frame of reference, educational level, cultural bias, career expectations, income, and many other variables that are often poorly understood by software developers. Productivity in South Africa is notoriously low, badly designed software can only make this situation worse (Van Greunen & Wesson, 2000). In a country with 11 official languages, it is clear that in order to understand user interface design and usability, we need to understand the complex interaction between culture, communication and technology (Hugo, 1998).

At present, HCI in South Africa can be characterised as follows (Hugo, 2002):

- A shortage of qualified practitioners and educators. There are only a few people involved in HCI teaching and research with the majority being academics. Most of these academics are situated at 7 universities, namely the University of Cape Town, Port Elizabeth, Natal (Durban), Natal (Pietermaritzburg), Free State, UNISA and Pretoria (Figure 2).
- A lack of awareness and implementation at industry level. The evidence for this is found in the lack of press coverage of HCI or usability matters and the lack of membership of professional HCI societies. A local chapter of ACM SIGCHI, called CHI-SA, was formed in 2001 to address this issue (Section 2.1). At present, however, there are only sixteen SIGCHI members in South Africa.
- Isolation and fragmentation between academia, industry, private research, development and government. There is little coordination among universities to ensure conformity with standards of HCI curricula. HCI in industry in South Africa is largely limited to usability testing.
- A lack of resources and inadequate training can result in inappropriate guidelines being adopted from the literature. Few structured methodologies for either software or usability engineering are followed.
- A lack of knowledge of standards for usability and user-centred design such as ISO 9241 (ISO, 1997) and ISO 13407 (ISO, 1999) exists in industry.

**Figure 2:** HCI Teaching and Research in South Africa

## 2.1   SIGCHI in South Africa

CHI-SA, a South African ACM SIGCHI Chapter, was established in 2001, comprising local HCI experts and usability professionals (http://www.chi-sa.org.za/). This special interest group focuses on promoting the field of HCI and usability in South Africa and provides a forum for CHI-SA members to share information and experiences. South Africa's first conference on HCI was held in Pretoria in May 2000 (http://www.chi-sa.org.za/CHI-SA2000/CHISA2000New.htm). For the first time, South African IT professionals, business people, students and academics were introduced to the field of HCI. The emphasis of this conference was on the unique needs of software users in a country characterised by diversity of culture, language, education, economic means and computer experience.

A second conference, CHI-SA 2001, was held in September 2001 (http://www.chi-sa.org.za/CHI-SA2001/chisa2001New.htm). Although smaller than the 2000 conference, there was a significant increase in the quality of presentations. This conference created a forum for the exchange of ideas and information about HCI in Africa.

A Development Consortium on HCI in South Africa was presented at CHI 2002 in Minneapolis in April 2002 (http://www.chi-sa.org.za/Devcon/DevConindex.htm). The theme of this Consortium was *"Changing the world, changing ourselves: in South Africa"*. Papers were presented on a number of different topics, including designing for multicultural web users, digital libraries and internationalization of user interfaces.

# 3    Challenges facing South Africa

The strategic issue for the South African ICT industry is to move beyond tradition and convention for user interface design. The software situation in South Africa is no different from that in the rest of the world (Van Greunen & Wesson, 2000). Most developers of interactive software and other systems want people to find their products effective, efficient and satisfying to use. People generally also want software to be easy to install, easy to learn and usable with an acceptable effort. Too many users in South Africa, however, still find it difficult to learn and use computers effectively (Shneiderman, 2000). The question is: *Why, and what can be done to address this issue?*

Well-designed user interfaces can contribute to the successful impact of ICT on society by making it usable by the wider population. In South Africa this population is characterized by a multiplicity of ethnic, cultural, language, education, economic, and other backgrounds. A concerted effort by all role players is required to draw together all those working within the fields of academia, research and industry to ensure the successful integration of HCI and its related principles and approaches.

To prepare ourselves for future challenges, we need to understand the role of the many cultural factors in the design of interactive software, and how these impact on organisations (Hugo, 1998):
- *Intra-cultural class differences* in the use of abstract and generalised information representations required to successfully use applications;
- *Cultural variants* based on differences in "locus of control" (i.e. internal or external);
- *Understanding of communication codes* (colours, symbols, metaphors, etc.), the role of ethnicity, class, gender, social factors and age in the design of interfaces.

Barriers to using supportive ICT (PDAs, fax machines, cellular phones, etc.) are both economic and educational. Access requires ownership or computer availability, technical knowledge and the financial resources to pay for access. Attitudes toward technology, exposure to technology and user acceptance, may also be significant factors. These human and social issues will lead to many new challenges and opportunities for HCI research in South Africa.

# 4    Opportunities

To encourage further HCI research in South Africa, it is important to leverage government, industry and university research funding. International experiences can provide valuable lessons for South Africa: an interdependent relationship between industry and academia is vitally important for future research and development. Universities should be encouraged to play a leading role in the development of new HCI principles, models and techniques that can be applied by industry in the South African context.

It is essential that South African Computer Science and IT students learn about HCI. Usability is likely to be one of the main value-added competitive advantages for the IT industry in South Africa in the future. If students do not know about HCI, they will not be in a position to serve the needs of industry. Without appropriate funding of academic research, there will be no HCI graduates to perform HCI research in industry, and fewer researchers in academic environments.

The HCI community in South Africa needs to be enlarged and extended. CHI-SA membership should be increased to include more academics and software developers. South African HCI practitioners should be encouraged to focus on promoting products that facilitate access, learning and empowerment. Active and collaborative links between academia and industry should be encouraged to provide an environment for HCI research and development in South Africa. Additional international links should also be created to assist the local HCI community.

## 5    Conclusions

This paper has reviewed the status of usability engineering in South Africa today. Several challenges facing South Africa were highlighted and opportunities suggested for HCI research and development. HCI is seen as a critical factor for the acceptance and adoption of ICT in South Africa. The wide diversity of the user population and the lack of adoption of HCI standards, however, present some unique challenges for the HCI community in South Africa. More collaboration between industry and academia is needed to encourage HCI research and development in this country.

## References

Bridges.org. (2002). *Progress towards ICT integration in South Africa: a survey of Government initiatives*. Retrieved, from the World Wide Web: www.bridges.org

Hugo, J. (1998). *HCI Research and Development in SA*. Retrieved, from the World Wide Web: http://www.chi-sa.org.za/research.htm

Hugo, J. (2002). *HCI and Multiculturalism in Southern Africa*. Paper presented at the CHI2002 Development Consortium.

ISO. (1997). *9241-1: Ergonomic requirements for office work with visual display terminals*.

ISO. (1999). *13407: Human-centred design processes for interactive systems*.

Pater, D., Machanick, I., Booysen, H., & Hurst, R. (2002). Communication Technologies: BMI TechKnowledge Group.

Shneiderman, B. (2000, September 2000). *The Future of Usability*. Paper presented at CHI-SA 2000, Pretoria.

Van Greunen, D., & Wesson, J. L. (2000, September 2000). *Usability Engineering for Interactive Multimedia Systems: A Formal Approach*. Paper presented at SATCAM'2000, Cape Town, South Africa.

Wesson, J. L. (2001). The Role of HCI Patterns in Software Development. *Journal of Research and Practice in Information Technology: Special Issue on HCI, 33*(1), 42 - 53.

# Usability of Usability Engineers:
## Usability Activities in Developing Office Products

*Makoto Yamasaki, Ryuichi Shimamura, Takako Inagaki*

Appliance Promotion Office, CSM Division, Ricoh Co., Ltd.
Yokohama, Japan
{makoto.yamasaki, rshima, takako.inagaki}@nts.ricoh.co.jp

## Abstract

This paper introduces usability activities being practiced at Ricoh and describes lessens learned through the activities. The role of Appliance Promotion Office of CSM Division where we belong to is to promote company-wide usability-related activities while performing direct supporting activities for product development projects and indirect back-up supporting activities.

## 1    Usability Activities at Ricoh

Ricoh provides worldwide various information products for office use, such as digital black-and-white and color copiers, printers, facsimiles, scanners, wide format copiers, digital cameras, etc. Ricoh also provides software products for document management and for specific industry solutions (for construction and real estate businesses, etc.). Usability activities at Ricoh are directed to those wide varieties of products. The functions of Appliance Promotion Office of CSM Division are introduced and usability activities we practice are described.

## 1.1    Appliance Promotion Office

The role of Appliance Promotion Office is to support and promote company-wide activities relating to usability. The group in charge of and dedicated to usability was first established in 1990 in Corporate Design Center. The main function of the group was usability evaluation. Thereafter, the function of the group was transferred to CSM (Customer Satisfaction Management) Division in charge of company-wide quality control, which is now the Appliance Promotion Office we are from.

Currently, the Appliance Promotion Office has 28 members, each assigned to a role, for example a context-of-use researcher or a usability evaluator. Assigning roles for respective members are not so strictly managed, and often the members collaborate with each other on their own judgments.

The activities of Appliance Promotion Office include, as described below, 1) direct supporting activities for product development projects and 2) in-direct back-up supporting activities.

[Direct Supporting Activities for Product Development Projects]
- Provision of references, such as usability design guidelines and a glossary of terms for user interfaces, etc.
- Research on context-of-use and identification of user's requirements
- Usability evaluation of products under development in predetermined developing phases
- Usability evaluation of competitive products

[In-direct Back-up Supporting Activities]

599

- Usability promotion through organizing case study seminars and distribution of pamphlets, etc.
- Assessment of product developing processes
- Human resource development relating to usability

These two kinds of activities are complementary to each other, and we are of the opinion that both activities are requisite for continuous achievement of usability of products.

## 1.2 Roles of Usability Engineers

For improving usability in wide-varieties of products, it is desirable that not only we, usability engineers, can understand the theory of a human-centered approach and practice necessary activities, but also product planners, designers and developers independently practice activities necessary from the viewpoints of usability, respectively.

At Ricoh, product planners conduct, as necessary, interviewing target customers and users. Further, product development and design sections respectively perform as needed field researches and usability evaluation necessary for examining design plans. The roles of usability engineers are therefore to provide higher level skills to support their activities, to assume activities in areas they cannot cover, to assemble references and effective methods, and to provide trainings of such methods.

We will describe the details of our usability activities below.

## 2    Direct Supporting Activities for Product Development Projects

With respect to product development in respective product development projects, we are responsible for activities of field researches and usability evaluations, and we provide to the respective product development projects information acquired through such activities and suggestions and proposals based upon such activities. For making such suggestions and proposals, it is indispensable that we, usability engineers, have knowledge about respective products, related products, and typical usability issues of the products.

Further, it is desired that usability engineers gain understanding of organizational cultures of respective sections involved in product development and diversities of product development processes, and individually approach the respective sections accordingly. Respective sections involved in product development may have different organizational cultures. For example, in one section, the product specification is prepared by a product planner, and in another section, the role of preparing the product specification may be born by product developers.

## 2.1    Providing Design Guidelines

In product development projects, normally, before product development starts, technological study starts for each element of the product development, such as electronic circuit design, mechanical design, and material selection (e.g. for toner). Often, the basic specification of a product, such as the performance, the size, the layout of the product, is almost determined at this phase of product development, which affects some aspects of usability.

Accordingly, at such a phase of product development developing a technological plan, it is necessary to provide design guidelines to product development engineers. We provide the following design guidelines;

- Design guidelines derived from successful and unsuccessful cases related to usability

- Standard terms (e.g., function names, etc.) for use in user interfaces and operational manuals
- Required specification based upon basic physical characteristics of a human being, such as sizes and forces necessary for operating a knob, etc., and sizes and brightness of displays
- Design guidelines required from the viewpoints of accessibility

## 2.2 Understanding User's Requirements

The activity of understanding user's requirements for a product is a work for understanding, through field researches, etc., characteristics and tasks of users and analyzing their requirements for the product so as to clarify functions and features that should be provided by the product. As research methods, we practice on-site interviews (see Figure 1), researches by questionnaires, and group interviews. We use a kind of scenario (McGraw & Harbison, 1997) in analysing their requirements.

Our Marketing Group collects information on businesses of users and requirements of the users through daily marketing and sales activities, and stores the information in a database accumulating sales cases as document information. We access the database to increase the efficiency of understanding businesses and requirements of users. Further, we interview experienced sales persons to understand outlines of respective markets. However, there is the possibility that information collected by Marketing Group through daily sales activities lacks details and/or objectivity, so that it is necessary to use the information with caution.

**Figure 1:** A Situation of an On-site Interview

## 2.3 Usability Evaluation

Usability evaluation is a work to verify if users can smoothly and comfortably use the functions of a product and to detect problems with respect to the product specification. Further, user's requirements for usability are made clear through usability evaluation. At Ricoh, usability evaluation is performed by usability engineers who have attended a predetermined in-house training course and have passed through a predetermined actual work experience.

Usability evaluation is performed in certain phases of product development using development prototypes and design prototypes. As techniques of usability evaluation, we use an inspection method (Nielsen, 1994) or usability testing, depending on the case. For usability testing, we often call on employees registered as in-house test participants. Thereby, we suppress the cost of usability evaluation and the risk that secret information on a product leaks.

We also evaluate usability of products from competing companies to learn from them.

# 3 In-direct Back-up Supporting Activities

## 3.1 Proposing and Assessing Development Processes

We also provide guidelines for enhancing product development processes to more ideal processes. We have developed the guidelines by incorporating a human-centered approach. Specifically, we have developed a checklist based upon the design process specified by ISO13407, and according to the checklist, we assess maturity of product development processes being actually practiced.

For practicing such assessing activities, it is important to gain understanding and cooperation of respective sections. Further, it is necessary to develop appropriate processes based upon the understanding of characteristics of respective product fields, organizational cultures of respective divisions, and the company's policy and culture as well.

## 3.2 Organizing Methods and Successful Cases

We are trying to share with our members and standardize processes and methods acquired through our practices in individual product development projects so that respective results are achieved more efficiently and effectively. For example, it is now popularly practiced to explain about business flows which have been understood through field researches, using icons that are shared by the members. By using such shared icons, a usability engineer analysing requirements of users can explain, without worrying about a way of expression, business flows of the users, and other usability engineers can smoothly understand the business flows.

Further, we assemble external successful cases relating to usability acquired through articles, etc., which we use to learn about a human-centered approaching method and to explain about effectiveness of the method. For assembling such external successful cases, we exchange information with firms in various industries, such as automobile and home-appliance industries, with respect to product planning and development businesses.

## 3.3 Developing Human Resources

We design programs of in-house trainings on usability and practice such trainings to product planners, developers and designers. Knowledge and skills about usability, such as, theory of a human-centered approach, techniques of field research and usability evaluation, etc. are included in the trainings. In designing the programs of in-house trainings, we have made use of know-how learned through our actual practices to adjust methods, etc. to the ones suitable for the situation of Ricoh.

Further, it is practiced that usability engineers informally convey to product developers and designers theory and methods of a human-centered approach while performing actual supporting activities, the effect of which is also important.

# 4 Conclusions

## 4.1 Fruitful Connections among Different Activities

When usability engineers belong to a product developing team to perform usability activities throughout the entire product development process, usability engineers can have a profound understanding of problem areas, so that a product superior in usability can be relatively easily realized (Rosson & Carroll, 2002). On the other hand, we belong to a usability-dedicated section that is independent from product development sections, so that the above-described advantages derived when usability engineers belong to a product development team cannot be obtained. However, we are advantageous in performing the following activities;

- Identify and approach sections appropriate for solving detected problems relating to usability
- Solve problems that should be cooperatively worked out by involved organizations, e.g., standardization of terms
- Learn effective and efficient methods for research, evaluation and analysis from activities in different product areas

## 4.2 Usability of Usability Engineers

It is needless to say that usability engineers must improve skills for performing major usability activities such as field researches and usability evaluations, but at the same time, for smoothly performing the activities, the usability engineers must have various skills for communicating with members of other sections. Usability engineers must first understand situations product planners and developers are in and analyze their requirements, and thereafter, assemble information necessary for enhancing usability and express the information in an easily-understandable manner. For enhancing usability of products, it is desired that usability engineers enhance their own usability. We, therefore, when we have performed direct supporting activities for a product development project, measure satisfaction levels of respective sections with respect to our activities, and based upon the result of the measurement, we review the details of our activities. Such satisfaction level measurement can be regarded as a usability evaluation of ourselves as usability engineers.

# 5 References

McGraw, K. L. & Harbison, K. (1997). User-Centered Requirements: The Scenario-Based Engineering Process. New Jersey: Lawrence Erlbaum Associates.
Nielsen, J. (1994). Heuristic Evaluation. In J. Nielsen & R. L. Mack (Eds.), Usability Inspection Methods (pp. 25-62). New York: John Wiely & Sons, Inc.
Rosson, M. B. & Carroll, J. M. (2002). Usability Engineering: Scenario-based development of human-computer interaction. San Francisco: Morgan Kaufmann Publishers.

# User Interface Design in Korea:
# Research Directions for a Digital Society

*Wan C. Yoon,  Seung-Hun Yoo and Dong-Seok Lee*
KAIST, IE Department
Yuseong-Gu, Guseong-Done 373-1,
Daejeon, Korea
wcyoon@kaist.ac.krl

## Abstract

As Korea has developed one of the strongest telecommunication infrastructures, Internet and mobile phones constitute a cyberspace over the country for shopping, banking, education, entertainment, friendship, and politics. Korea's electronics industry also pursues digital networks including digital TVs and home intelligent systems. As such environments are creating new cultures and trends in user-system interaction, the UI research is prompted to evolve so as to produce more than easy-to-use interfaces. This paper reviews the situation and trends in the areas of the Web, mobile devices, and electronics in Korea to identify the driving forces that demand the advance and evolution of UI design research. Then discussed are the prospective directions of future development of HCI in general to meet those observed trends of the digital culture.

## 1   Introduction

Since the introduction of Internet in 1996, Korea has experienced one of the most drastic changes in the social infrastructure and life style that any countries ever had. The country now boasts the best distributed high-speed Internet service in the world, the highest rate of regular Web surfers, and one of the highest rates of mobile phone users. It is widely agreed that the Internet even changed the outcome of the nation's president election in 2002. In this digital society, to design decent web pages is now one of the foremost missions of any organizations. Since the competition is already severe in any areas of Internet business in the country, the importance of good user interfaces for attracting customers and maintaining their loyalty is soaring steeply.

Another important source of motivation for UI research in Korea is the IT and electronics industry. Korea stepped into the information society as a major vendor. In 2002, Korea's IT product export amounted to about 30 percent of the country's total export value. The IT industry includes some of the world-largest suppliers of communication devices, digital electronics, and home appliances. As Korea's electronic products have to upgrade their images in order to expand the high-end international market, great emphasis begins to be placed on good user interface design.

The rest of this paper will review the environments of UI research in Korea in the three related areas: the Web, the mobile devices, and other electronics. For each product area, the industrial situation that may influence UI design will first be reviewed. Then the UI requirement trends and challenges will be discussed with improvisation of possible research strategies. Section 4 will describe the current status of UI engineering practice in Korea and the last section will summarize the desirable future UI research directions to pursue in the light of the emerging cultural phenomena.

## 2 The Web and UI

In a recent survey, the number of regular Internet users in Korea, who surfed the Web at least once a month, reached 23 million by the end of 2002. It was a 64.4 percent usage rate (Yang, 2002a). Almost 80 percent of the Internet users used the network primarily at home, and over 10 million, or 83.9 percent, of household Internet connections were broadband ADSL or VDSL, which was the highest figure in the world (Yang, 2002b). As for the tasks done on the Internet (multiple answers), 89.7 percent said searching information, 81.7 percent e-mail transmission, 62.0 percent playing games, and 58.6 percent listening to music or watching online movies (Yang, 2002a). The most active user age group in Korea appeared to be the one between 6 and 19, at 93.4 percent usage rate. (Yang, 2002a).

The above figures represent or lead to a few characteristic phenomena in Internet usage in Korea. Cyber-communities are replacing or strengthening off-line groups as young people are increasingly joining them to make friends and exchange information. Instant messaging and online chatting are prevalent so that even paid Avatars (i.e., personalized characters) are popular. The most culturally characteristic is the preference of multi-user computer games, where multiple users can log on and play at the same time without ever meeting, to home video games that are more popular in the United States and Japan. Considering the nature of those activities, mere convenience can no more be regarded the primary quality of UI design. The fun of interaction, cultural appeal, and emotive responsiveness are the other major axes that determine the performance of a user interface. The utility of conventional UI design guidelines stops here.

On the other hand, the Web is not only for entertainment or information mining. Based on the country's digital infrastructure, much of shopping, banking, education, and politics are now being done in the cyberspace. For example, online stock trading in Korea surpassed 70% of total trading in 2001 (Yang, 2002a). As Internet take up various social functions, the web pages are no more the interface between the computer and the user; they are now between the world of tasks and the human. Their design should become more of system development rather than shallow *design-and-evaluate* page design. The ease of use and convenience may no more be the dominant aspects of interface goodness. Instead, strong task support, appropriate and timely information display, supporting human reliability, and prevention of decision-making errors should be considered as the primary performance measures. Designing interfaces under such criteria needs system engineering approaches combined with domain technologies such as industrial engineering, manufacturing, financing, or management. When many people or organizations are involved, social science disciplines like organizational behavior, communication, and education may also have their roles in the interface design. This mission is clearly beyond that of conventional UI design practice that mainly dealt with ergonomic principles. (Yoon, 2001a)

## 3 The Mobile Devices and UI

At the end of October 2002 Korea had 32 million mobile phone subscribers, and 67.4 percent of the population owned a mobile phone, which is the world's 6th highest figure (Yang 2003). Samsung being ranked as the world's third largest mobile phone maker, mobile phones became the fifth largest single export item of the country after semiconductors, automobiles, computers and ships. The country is also leading the world's CDMA mobile phone market. As cdma2000 mobile technology is common within the country and full-color-display mobile handsets packed with multimedia features are readily available, Korean consumers are relatively less enthusiastic about PDA that is widely favored by American users.

Another reason for the surveyed preference of mobile phone to PDA in Korean market is the young generation's learned skills in using the phone interface. Korean teenagers tend almost to abuse mobile phones for SMS (sending message service) using the phone's small keypad as the only input device. For the agility of their input operation and the frequency at which they exchange messages with their peers, the teens of Korea are called the 'thumb generation'. To them, PDA's handwriting recognition is slower and less reliable. The young generation possesses about equal mastery of other operations of the tiny but function-packed hand sets. The skill and usage formed a reinforcing cycle so that the people, the culture and the device coevolved through an *artifact cycle* (Carroll, 1992). A representative observation of the mobile phone culture of Korea writes:

> *To try to understand this culture, I asked my friend, Jackie Kim, "Why do you need a mobile phone?" Her answer "I cannot imagine my life without my mobile phone. I get up in the morning with the alarm of my cell phone and keep in touch with all of my friends through this device. Even when I have nothing to do when I'm riding on the subway, I usually just play with it. It takes a big part of my every day life." (Ng, 2002)*

Thanks to this cultural development, the major mobile operators in Korea have been able to increasingly derive revenue from content and services. The competition to develop a value-added user experience is getting hotter. With various cutting-edge technologies such as multimedia messaging service and cell-broadcasting service, and upgraded software functions, the borders blurs between telecommunications and broadcasting, or even wired Internet.

The above observations reveal the following points that the researchers and practitioners have to consider for mobile device UI design. The interface design has to be differently exercised for different cultures; a designed interface can enforce user skills to the level of a social standard; the narrow interface problem is especially severe with the mobile phones; consistency is needed when multiple functionalities exist to support various life facets. In addition, the unique Korean character system calls for cognitive and physical ergonomics studies.

# 4    UI for Electronics and Home Appliances

Korea's major electronics manufacturers enjoy the largest world market share for a few electronics products besides semiconductors. The top-product list of Samsung in 2001 included color monitors, TFT-LCD panels, microwave ovens, VCR, and CDMA (code division multiple access) mobile handsets. Refrigerators and DVD players are gaining popularity in the global market. LG Electronics is leading the optical storage and air conditioning sectors. Its share in the LCD-TV market was the world's second in 2002 and in the PDP (plasma display panel) TV market, the third (Yang, 2002c). Analysts say that Korean manufacturers are leading the digital TV and set-top box sectors with better products and technologies.

The trend keywords in the above are *digital* and *networking*. Digital TV models are capable of handling more information and signals, and hence revolutionary interactivity, putting aside better color and sound quality. Home automation and networking is the next-generation technology that will harness a variety of electronic gears and gadgets under centralized control. The home appliances (e.g., air conditioners, refrigerators, washing machines, and microwave ovens) are hooked up to a communications network and to Internet, allowing users to operate each device through a wireless Web pad, PCs or mobile phones. As a cultural trend, the electronic environments are growingly *conversational*. This means future UI design in this area should be more than that of layout and shapes.

Other conspicuous trends in the recent evolution of electronic products that pose some challenges to UI design are the narrow interfaces, the functional variety, fusion of different devices, and culture-dependent acceptability. UI practitioners are asked to respond to these issues by putting more emphasis on the notions such as interactivity, integration, consistency, and life style. To expand electronic export markets abroad, the cultural differences in user behavior and preferences should also be considered.

# 5   UI Design Practice in Korea

As in other competing countries, most of Korea's electronic products including mobile phones are designed and manufactured by a few giant corporations. In 2002, those companies invariably heightened the level of UI design activities and some conducted a massive recruitment activity to establish new UI related teams or strengthen existing teams. Such concurrent alertness arose largely due to the recognition of the new UI challenges mentioned in this paper. As there exists an immediate shortage of UI experts, the electronics industry have to rely on inside education programs and on-job training. This pushes smaller electronic companies to a worse situation; most of them put UI design in non-experts' hands for lack of qualified people. UI enhancing activities in the electronics industry have been concentrated on interface evaluation of interim and final products. Suffering shortage of trained UI experts, user tests, in mostly observational forms, tend to be preferred to expert review or cognitive walkthrough. Some leading companies now begin to seek to establish their own style guides, UI-included product design process, and systematic UI assessment schemes, which movement will be soon followed by others as they are ready.

Most UI consultants in Korea are currently working in Web area, where countless shopping malls, banking and stock trading services, portal services, and data base suppliers strive to attract more customers via better site design. Those companies seek to improve their web sites through outsourcing since they do not possess design teams within organization while having great interest in the UI performance of their web sites. It is estimated that about 100 consulting firms are now operating, but it is apparent that most of them have grown up out of the design discipline than human engineering or HCI. The consulting market would be highly adversary to foreign consulting firms, however, due to the strong cultural factors in design, such as color, layout, accepted standard methods, common task knowledge, and the unique character fonts.

# 6   Required Research Directions

Introducing conventional techniques for UI design based on human-computer interaction and ergonomics is far from having been completed in Korea. Nevertheless, new required research directions are already strongly indicated. The industry cannot afford to ignore the new demands since the country currently plays the role of the most advanced test bed for innovative IT technologies and Web/mobile-based services due to its highly progressive market. Considering all the mentioned trends and phenomena, as the conclusions of this paper, five research directions to pay particular attention are summarized.

- *Application of more cognitive and systemic approaches*
  Multiple functionalities and task complexity of Web and devices as well as social dynamic systems behind the web pages (discussed in Section 2) require more systemic task analysis

and better engineering for interface optimization. Application of cognitive engineering technology to UI design process seems to be a promising direction.

- *Modeling and designing at interaction level*
  The discussed environments with functional multiplicity, device integration/fusion, and conversational interfaces demands more effort to explicitly describe, model, and define the interaction methods at an abstract level than the lower-level layout and sequences. User cognition and mental models should be in the center of UI research (Yoon, 2001b). At this level, design decisions may also be made in terms of cultural factors.
- *Introducing common UI schemas and standards*
  The complexity of interfaces in cognitive aspects increases inevitably. The most effective way to reduce the subjective complexity is introducing common schemas for functional operations and task methods into user knowledge. Companies may cooperate to build common standards for maximum benefit.
- *Establishing design process and knowledge management*
  Companies emphasize rapid development cycle to be responsive to the market. Without well-defined product design process including UI design, and without effective reuse of previous design knowledge, designing good UI in short time is impossible.
- *Extending UI to cultural technology*
  As the contexts of tasks are getting rich and highly cultural, interdisciplinary research and practice with the experts in games, VR, storytelling, and other areas become necessary. This synergetic effort may be called *Cultural Technology*. UIs are not to be designed only for adequate exchange of information and control but for fun and feel. Taking cultures into account is also important in designing export products.

## Acknowledgement

This study was carried on as a part of KOSEF Project 20015-315-02-2.

## References

Carroll, J.M. and Rosson, M.B. (1992). Getting around the task-artifact cycle: How to make claims and design by scenario. *ACM Transactions on Information Systems* 10, 181-212

Ng, M. (2002) Seoul Mobile Phone Culture and Its Effects. retrieved Jan.20, 2003 from Seoul Now, http://www.seoulnow.net/heartseoul/heartandseoul-028.php.

Yang, S. J. (2002a). 64.4 percent of Koreans surf Web regularly. *Korea Herald* Oct. 26.

Yang, S. J. (2002b). Korea boasts 10 mil. broadband users. *Korea Herald* Nov. 7.

Yang, S. J. (2002c). Electronics makers add to No. 1 item list. *Korea Herald* Dec. 9.

Yang, S. J. (2003). Korea`s videophone market growing fast. *Korea Herald* Jan. 22..

Yoon, W.C. (2001a). The emergence of large-scale logical systems and cognitive ergonomics. *Theoretical Issues in Ergonomics Science*, 2(3), 251-267.

Yoon, W.C. (2001b). Identifying, organizing and exploring problem space for interaction design. In G. Johannsen (Ed.), *8th IFAC/IFIP/IFORS/IEA Symposium on Analysis, Design, and Evaluation of Human-Machine Systems (pp.81-86)*. Kassel, Germany: International Federation of Automatic Control (IFAC).

# Section 4

Design and Evaluation Studies

# Establishing user requirements in HCI – a case-study in medical informatics

*Hans H. K. Andersen*

Risø National Laboratory
P.O. Box 49, DK-4000 Roskilde,
Denmark
hans.andersen@risoe.dk

*Verner Andersen*

Risø National Laboratory
P.O. Box 49, DK-4000 Roskilde,
Denmark
verner.andersen@risoe.dk

## Abstract

This paper reports from a user requirement, design and evaluation study on supporting educational activities in the medical autoimmune serology domain[1]. Establishing the user requirements has been based on the Cognitive Systems Engineering approach. The user requirements laid the foundation for designing the software system e-DOORS (extended Discrete Object Observation and Recognition System). This system can briefly be described as a tool that will assist the user in recording the classifications of (sets of) medical images. An evaluation of the system showed that the software satisfactorily assisted in quantifying improvements in the education and training process and the quality assurance process, by revealing quantitative changes in recognition skills and accuracy.

## 1 Introduction

The approach applied in the user requirement study discussed in this paper has mainly followed the principles and concepts offered by the Cognitive Systems Engineering (CSE) framework developed at Risø National Laboratory (Rasmussen, Pejtersen & Goodstein, 1994). It allows the work analyst to analyze a system of work in terms of means-ends relationships indicating the why, what and how relations among the layers in the hierarchy. Our role in the project was to elicit requirements from the users and communicate these to the designers. In this way we came to act as a sort of mediators trying to formulate expert medical knowledge into the language of the software designers and developers. In fact not a trivial task as it turned out. The paper focuses both on presenting the results of the user requirement study of medical domain and on discussing facilitating methodological refinements of the means-ends abstraction hierarchy in order to better communicate the user requirements. This way of working is in line with Bødker (1991) who argues that a distinct characteristic of a given method is that it has been created by a designer believing to have invented a good practice for design within a given domain. In addition, design methods or practices can be described as prescriptions for the application of a variety of design principles and guidelines in doing design (Andersen et al., 1990). The problem is that important experiences get lost and only certain aspects of the process are incorporated in the method. The consequence in applying a specific method (as for example the means-ends analysis) then is that the method should not be used as a recipe to be followed step by step but rather it should be perceived as a set of guidelines from which it should be possible to derive certain heuristics for doing design depending on the application domain as we have done in this case study.

---

[1] This work has in part been funded by the European Commission project CANTOR, Telematics Healthcare, HC4003.

## 2 Methods

The methodology was not so much to focus on what the participants do when classifying immunological patterns, but rather to approach the diagnosis or pre-diagnosis, respectively, at a semiological level. The methodics applied in the user requirement study all belong to the qualitative area of research:

- Interviews (qualitative, semi-structured, unstructured (Kvale, 1983)).
- Document inspection (worksheet reports, standards, quality assessment schemes, handbooks, laboratory manuals, classification lists, diagrams, drawings, etc.).
- Observations (activities at the microscope, use of existing image analysis software, presentation of labs).

The users came from six university hospitals and one national medical laboratory scattered across Europe. The designers were located in a small Danish software company. The elicitation of the user requirements was carried out during a period of 6 weeks visiting the seven sites. 15 persons have been interviewed (lengths of each interview 6-9 hours). All interviews have been tape-recorded. Notes were taken during the interviews. All tapes from the interviews have been transcribed.

## 3 Analysis of the work domain

The CSE means-ends abstraction hierarchy has been utilized as an analytic tool in formulating the user requirements (see Figure 1). The means-ends abstraction hierarchy provides a framework for identifying and integrating the set of goal relevant constraints that are operating in a given work domain. Each of the five levels in the hierarchy represents a different class of constraints. One way to think of the abstraction hierarchy is as a set of models of the system, each defining one level of the hierarchy. Higher levels represent relational information about system purpose, while the lower levels represent more elementary data about physical implementation (Vicente 1999).

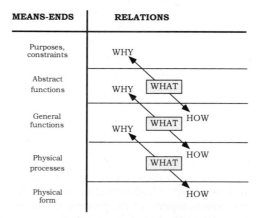

**Figure 1: A schematic representation of the means-ends relations' abstraction hierarchy**

When we first presented this way of categorizing our data at joint meeting with user representatives and software designers there was a concern among the participant on how exactly to interpret and understand the outcome of a means-ends analysis. There was also a strong opposition among the users against the designers' way of interpreting the requirements.

We then looked for alternative ways to organize the requirements that could satisfy both users and designers while maintaining the idea of a means-ends analysis. We found inspiration in the principles and concepts offered by the activity Theory (Leont'ev, 1978), and by the work analysis approach in (Schmidt & Carstensen, 1993).

From an activity theory perspective human activity mediates the relation between the organism and its environment (Ibid.). The structure of activity as a unit of analysis can be described as dimensions of activities, actions, and operations (see Table 1). At the topmost level of this structure, activity is motivated in relation to fulfillment of a need (why). Activities consist of goal directed actions (what) that are realized by operations triggered by specific conditions related to the goal (how). According to Leont'ev the entities constituting the structure is related to one another in a hierarchical manner. That is, operations are organized according to specific physical, instrumental and / or socio-cultural determined conceptual conditions in the environment that can serve to fulfill several goals while an activity is closely related to specific goals that can be met by several physical, instrumental, and / or socio-cultural determined conceptual conditions in the environment.

### Table 1 Model of the activity structure

| Activity dimensions | Psychological dimensions | Relations |
|---|---|---|
| Activity | Motives – needs | Why – social and personal meaning of activities |
| Action | Goals | What is needed to pursue motives |
| Operation | Conditions of actions | How to realize goals |

The means-ends model and the activity structure model share a number of qualities. In both approaches, goal-oriented behavior is fundamental, and both approaches classify behavior along a scale of increasing abstractness.

The hierarchy of means-ends conceptualizations can be conceived as a depiction of a work domain that mediates and models possible conflicts between human intentions and the constraints in the surroundings working against fulfillment of intentional activities. The activity structure can be seen as a characterization of dialectical subject-object relations that is mediated by artifacts rooted in socio-historical developed practices. In both cases the focus will be on characterizations of mediation artifacts that allows the actors immediate access to the work object at hand or in case of break down support actors to solve the problem at hand on an adequate level of knowledge-based activities or rule-based actions.

Within the work analysis approach for formulating user requirements three different levels of analysis has been identified: the strategic level, the functional level (or tactical level), and the operational level. At the strategic level focus is on identification of the requirements imposed by factors in the wider work environment (surroundings) and the problems, constraints and motives, etc, that are the core of a network of causal relations reflecting the work domain dynamics (Schmidt & Carstensen, 1993, pp .112). That is, in some sense this level is in accordance with the two top-levels of the means-ends hierarchy. With respect to the functional / tactical level focus is on identifying and characterizing the individual work functions, for example, the cognitive demands of tasks and how individuals respond or to the often contradictory task demands in terms of applying a series of different decision making strategies (Ibid, pp. 116). This level corresponds to third level of the means-ends hierarchy. At the operational level focus is on the actors, in identifying and characterizing components, facilities and processes that can support and mediate individual and cooperative work (Ibid, pp.118). In many ways this level corresponds to the two bottom levels in the means-ends hierarchy. Furthermore, inspired by approaches within the area of information management systems (Brender, 1997; Parker 1985) we found that the functional /

tactical level of the work analysis corresponds to what these theories has named the procedural level.

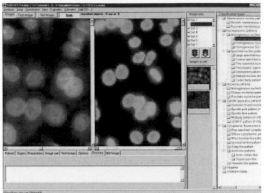

**Figure 2:** Snap shot of the "Matcher" tool of e-DOORS. The classification types (taxonomy) are indicated in the scrollable text at the right hand side

Therefore, based on this analysis we chose a structure having three levels: strategic, procedural, and operational, being the common nominator of the various approaches including the idea of means-end and the why-what-how relation, the psychological dimension, and having the lower level related directly to the system implementation specification. Our idea was to simplify our results, and going from five to three levels could be one solution. Next we also wanted to rename our categories into a language that where more familiar to the users and designers. This approach turned out to be a widely acceptable and also usable solution among the interested parties. The strategic requirements mirror the goals, purposes and constraints governing the interaction between the medical work system under consideration and its environment. Examples are: treatment planning of patients, consequences of mistakes, and improve the quality of autoimmune diagnosis. In addition, the strategic requirements represent concepts that are necessary to set priorities, such as quality of service and categories of diseases with respect to the diagnosis. The procedural requirements characterize the general functions and activities of classifying autoimmune sera based on pattern recognition of entire images. The operational requirements represent the physical activities, such as use of tools and equipment. Furthermore, the operational requirements signify the physical processes of equipment and the appearance and configuration of material objects, such as staining, clinical information and multi-head microscopes in the traditional set-up, etc. A screenshot of the final version of the software is illustrated in figure 2.

## 3.1 Evaluating the system

A validation of the final system was coordinated by one of the autoimmune serology sites involved (for more details on the test see Wiik & Lam, 2000). 12 people participated in the study (3 experts. 6 skilled persons and 3 novices). To be able to demonstrate any learning effects it was required that a participant classified both a baseline set of images and a certification set three times using the system. The threshold for expert ship was set at a kappa value of >0.95. One novice became an expert within 5 weeks because the kappa value went from 0.74 to 0.95. The two other novices did not improve their recognition skills significantly. As regards the trained staff, six persons went through all rounds whereas three did not. Two persons among those who completed all rounds turned out to be experts from the start, three persons reached an expert level and one person improved considerably (from k 0.84 to 0.90) during the exercise. These data are an illustration that the e-DOORS software adequately supported improvements in the education and training process among the participants. It was estimated that the present version could be perfected to become a novel tool for education, training, quality assurance, consensus formation and standardization in several areas of microscopic pattern recognition relating to autoimmune serologic reactions with cells and tissues. In a small experiment concerning the usability and user friendliness of the e-DOORS system for education and training we found that, except for minor suggestions related to the user interface, the general opinion of the participating physicians was very positive. They found the system not only valuable, but also inspiring due to the tools allowing direct feedback of their performance as compared to the expert opinion, and allowing objective indication of personal improvement by the Kappa value.

## 4  Conclusion

Whether these test results can be taken as an indication of the applicability of the means-ends analysis as a user requirement elicitation method in the form we chose is still an open question that needs further investigation. Looking back at the process there is no doubt that we as mediators between the users and the designers in some way contributed to the positive development in introducing the means-ends analysis as a common ground for all. The means-ends analysis also gave us a thorough understanding of the domain that we could draw upon in supporting the designers during the design process in an iterative manner and thereby leaving the experts to do their job.

## 5  References

Andersen, N. E., F. Kensing, J. Lundin, L. Mathiassen et al. (1990) Professional Systems Development — Experience, Ideas, and Action. Prentice-Hall, Englewood Cliffs, New Jersey.

Brender, Jytte. (1997). Methodology for Constructive Assessment of Medical IT-based Systems – In an Organisational Context. Amsterdam IOS Press, Studies in Health and Informatics, Vol. 42.

Bødker, Susanne (1991). Through The Interface: A Human Activity Approach to User Interface Design. Lawrence Erlbaum Associates, Inc., Hillsdale, New Jersey.

Kvale, Steinar. (1983). The Qualitative Research Interview. A Phenomenological and a Hermeneutical Mode of Understanding. In *Journal of Phenomenological Psychology*, Vol. 14, No. 2., pp. 9-33.

Leont'ev, A. N. (1978). Activity, Consciousness, and Personality. Prentice Hall. NJ.

Parker, B.R. (1985). A Multiple Goal Methodology for Evaluating Management Information Systems, Omega. Int. J. Mgmt. Sci., Vol 13, No 4, pp. 313-330.

Rasmussen, Jens, Annelise Mark Pejtersen, Len P. Goodstein. (1994). Cognitive Systems Engineering. Wiley series in System Engineering, ed. by A. P. Sage, John Wiley and Sons, New York.

Schmidt, Kjeld, and Peter Carstensen. (1993). Bridging the Gap: Conceptualizing Findings from Empirical Work Analysis Studies for CSCW Systems Design. Risø National Laboratory, 8 February.

Vicente, Kim. J. (1999). Cognitive Work Analysis: Toward Safe, Productive, and Healthy Computer-based Work. Mahwah, NJ: Erlbaum.

Wiik, Allan and King Lam: On The Usability Of Extended Doors For Education And Training, Quality Assurance And Consensus Formation. The Department of Autoimmunology, Statens Serum Institut, Copenhagen November, (2000)

# Too Many Hierarchies?
## The Daily Struggle for Control of the Workspace

*Richard Boardman*[†]        *Robert Spence*[†]        *M. Angela Sasse*[§]

Intelligent and Interactive Systems[†]
Electrical and Electronic Engineering
Imperial College London
London SW7 2BT, UK
*{rick, r.spence}@ic.ac.uk*

Department of Computer Science[§]
University College London
Gower Street
London WC1E 6BT, UK
*a.sasse@cs.ucl.ac.uk*

## Abstract

This paper reports research aimed at improving cross-tool support for personal information management (PIM). We present results from a study in which we investigated how users manage three collections of personal information: documents, email and web bookmarks. These findings have motivated the design of a software prototype that allows users to mirror folder structures between various PIM tools. We discuss the results of an initial evaluation that suggest that this approach may offers benefits to many users including improved control and consistency.

## 1    Introduction

Personal Information Management (PIM) describes the everyday process carried out by an individual as he or she gathers, handles and organizes information (Lansdale, 1988). In both the physical and digital domains, PIM is a pervasive ongoing activity. In the physical world, the accumulation of information resources, typically in the form of various types of paper-based documents, is a familiar feature of our personal environments, both at work and at home. In addition, the explosive growth in personal computing over the past two decades means that individuals are now able to maintain collections of *digital* information. Today's computers allow users to collect and manage a diverse range of information resource, such as document files, email messages, to-do items, contacts, and web bookmarks. However, there is much evidence that PIM is poorly supported by current technology, and that many users struggle to manage the information that they accumulate over time. Studies of PIM have highlighted the problems users encounter and the limitations of the organizational support offered by tools, typically based on the traditional hierarchy (Lansdale, 1988). Whittaker *et al.* (2000) highlight the general lack of progress within HCI towards improving tool support for PIM and other fundamental computer-based activities.

## 2    Our Cross-tool Perspective on Personal Information Management

Our research focuses on the problems that result from the *distribution* of digital PIM across a range of distinct tools, such as the file system and email. As a result of this distribution, those users who choose to manage multiple types of information must do so in parallel. We argue that many of the most pressing PIM-related problems encountered by users are not due to the design of particular tools, but instead can be attributed to this fragmentation of PIM across a range of poorly integrated and inconsistently-designed tools.

Several areas of related research have influenced our *cross-tool* perspective. Our work draws on the conceptualization of a computer as an *activity space*, populated by the tools and resources that facilitate user action, and the constraints that limit it (Kirsh, 2000). From this theoretical perspective both a user's production activities[1], as well as supporting activities such as PIM, are not confined to specific tools, but are distributed across a range tools throughout digital activity space. In order to provide effective support for such cross-tool activities, integration between tools is crucial. However there is evidence that this issue is not being given enough attention by designers. Bellotti and Smith (2000) note the *compartmentalization* of PIM activities due to poor integration between tools. For example, document collections are often divided between those stored in the file system and those stored as email attachments. In terms of the theoretical framework offered by Kirsh, compartmentalization may be considered as one set of constraints imposed on a user's activity space by poorly designed tools. Whilst we acknowledge the need to improve user interfaces to specific tools, such work often ignores user needs in the wider context. We suggest that some of the most pressing PIM-related issues faced by computer users can only be addressed through a cross-tool approach. Our work is based on this cross-tool perspective, and aims to provide more coherent, integrated support for PIM. We first discuss the findings from a cross-tool study of users' PIM practices.

# 3    Exploratory Study – Method and Results

We carried out a series of semi-structured interviews to investigate user practices in managing three collections of personal information: (1) documents, (2) email, and (3) web bookmarks. All twenty-five participants worked in an academic context and had at least five years of computing experience. Interviewees included users of Windows, Linux and MacOS. Interviews were carried out in each participant's workplace, and were centered on guided tours of the three collections on their primary desktop computer. We also enquired about the strategies they employed, and the problems they encountered in the three tools. Interview data consisted of our notes, and screenshots of the user's folder hierarchies. We carried out content analysis on the data to identify common user strategies and problems. Folder names were classified by type and compared between tools for each user.

As would be expected with such an individual activity, a wide range of behavior was observed, varying both between users, and between tools for individual users. In general documents and email tended to be collected much more extensively than bookmarks. Most users preferred to rely on other mechanisms such as search engines instead of devoting effort to managing bookmarks. In terms of structure, documents tended to be organized into folders most thoroughly, whilst there was a tendency for users to rely on message metadata for organizing email via sorting mechanisms. Any active bookmarks tended to be located in unstructured lists. The study highlighted the fact that individual users employ a variety of PIM strategies in different parts of the workspace to manage different types of information with varying degrees of success. Most users could not be described as being globally "messy" or "tidy".

---

[1] We find it helpful to differentiate between a user's production and supporting tasks. Production tasks are those that drive a user's computer usage (for a school teacher production tasks might include lesson preparation and administration). Supporting tasks such as PIM are carried out to enable their production activities, but are arguably not the prime motivator for using the computer.

We were often surprised at the vehemence expressed regarding PIM-related problems, and have coined the term *bugbear* for recurring problems that frequently or seriously affect users. Since PIM is an ongoing and often repetitive everyday activity, we found that even relatively minor bugbears can build up and have a negative impact on productivity and/or user experience. We were startled to find that a perceived failure to manage personal information can seriously dent users' self-image, e.g. they "feel bad" for "being untidy". All users emphasized the overheads of managing email, due to the higher (and uncontrolled) creation rate of messages compared to manually created files and bookmarks. However, subjects tended to be dissatisfied with the organizational state of *all* three collections, expressing feelings of guilt, stress, and lack of control.

A particular source of exasperation was the existence of old unfiled items, such as emails in the inbox, and icons on the desktop. Most users said that they did not have enough time to organize the collections, resulting in a lack of satisfaction regarding their tidiness. Twenty of the twenty-five users managed folders in two or more of the three tools (typically the file system and email tool). For many of these users, we noted a significant level of *folder overlap* – folder names that appeared in multiple PIM tools. Folder overlap was particularly evident between the document and email hierarchies (an average of 21% for the first seventeen users[2]). Overlapping folder names were generally based on participants' primary production activities, and were most commonly expressed in terms of role, project and interest. Folder overlap indicates that the study participants were devoting effort towards organizing resources relating to the same production activity in multiple tools. In other words, there are redundant aspects to user's information management activity when viewed from a cross-tool perspective.

We also observed a range of *cross-tool* problems that bridged multiple collections:
- Users complained that the management of certain types of information was compartmentalized between distinct tools (e.g. documents managed separately as files, desktop icons, and email attachments).
- Users complained about the need to coordinate production activities across multiple tools. Common scenarios included starting a new production activity (setting up folders in distinct tools), and finishing a production activity (archiving items in distinct tools).
- Annoyance was also caused by inconsistencies between different tools in terms of how they provided equivalent functionality such as "create new folder" or "mark this item as important". Users found this particularly irritating between tools from the same vendor!

## 4    Design of WorkspaceMirror

These findings, particularly those of folder overlap, lead us to question what benefits might be offered by sharing folder hierarchies between tools. In other words, do users really need the flexibility to develop distinct classification schemes for different types of personal information? In order to explore this idea we have designed a software prototype, WorkspaceMirror, which allows users to replicate changes to folder structures between their PIM tools. WorkspaceMirror has been implemented under MS Windows and synchronizes changes made to the folder hierarchies in three tools: (1) email folders in MS Outlook, (2) the user's document area in the file system, and (3) bookmark folders stored under Favorites. The tool works in one of two modes: automatic or prompted. In prompted mode the creation, deletion or renaming of any folder causes a dialog box to be displayed asking the user if they want to replicate the operation in the other two tools.

---

[2] Note that the quantitative results relating to folder overlap (folder names in common between tools) was carried out for seventeen users and is reported in more detail in Boardman (2001).

Our design can be considered as a step towards the full unification of personal information management that has been proposed in systems such as Lifestreams (Freeman & Gelernter, 1995). However such *revolutionary* technologies have been criticised for a lack of evaluation (Boardman, 2001). In contrast, a prime aim in our work was to facilitate evaluation by pursuing an *incremental* design based on relatively modest changes to standard software. This has the advantage of enabling evaluation in real user workspaces with minimal disruption to the users concerned.

# 5    Initial Evaluation

We have carried out an initial evaluation of WorkspaceMirror with a small number of users to determine whether our design is workable. A major challenge was the lack of any accepted evaluation methodology regarding PIM tools (Whittaker *et al.* 2000). The limitations of traditional performance-based measures of usability for complex, ongoing, interleaved activities such as PIM (Dillon 2001), lead us to steer away from a task-based experiment. Instead we based our evaluation on a longer-term field study. Four of our colleagues have been using WorkspaceMirror in their primary desktop workspaces over four months, and are providing feedback via diaries and weekly interviews.  We have also correlated this qualitative data with fortnightly logs of their evolving folder hierarchies to track their usage of any mirrored folders. We triangulated the data to build up a rich picture of the user's attitude to WorkspaceMirror, and investigate whether it influenced their PIM practices. Note that all four users had previously developed folders in the three tools. WorkspaceMirror was deployed in prompted mode, so as to give users more control over mirroring and allow them to retain the flexibility to organize each collection differently.

Three of the test users have provided highly positive feedback regarding the tool. They found the idea of sharing categories between tools both intuitive and compelling. In particular they welcomed the increase in consistency between the three folder hierarchies that resulted from mirroring. Although there was not always a direct one-to-one mapping between their folder requirements in each tool, they welcomed the chance to reflect on the relevance of the organizational decisions made in one tool, to other contexts. Occasionally mirrored folders were not always used for the storage of items in all tools, but the testers indicated that the improved consistency outweighed the side effect of increased clutter. The users also reported lower management overheads and easier retrieval of filed items, however we have not yet attempted to confirm these results objectively.  Two users mirrored mostly between the document and email collections, whilst the third mirrored between all three tools. In general mirroring was seen to be most useful for high-level folders, which tended to be based on cross-tool projects and roles. Feedback has also included a number of design requests that we are considering adding to future versions. These include support for cross-tool navigation (e.g. enabling traversal between mirrored folders via a context-menu option), better handling for email attachments (e.g. automatic saving of document/bookmark attachments in mirrored folders), and project management-like facilities (e.g. cross-tool high-level functionality such as "start project" and "archive project").

The fourth user provided a useful counter-example. He did not see any point in mirroring folders between the three tools, preferring the flexibility to each organize collection differently. He also found the prompting intrusive. However he has left the software running to test its robustness, and has indicated that he would find the tool more useful when setting up new project workspaces, or for users setting up a computer for the first time. Our initial evaluation indicates the potential benefits of our design, although the trade-off between consistency and reduced flexibility warrants further investigation.

# 6    Discussion and Future Work

One criticism that can be levelled at most PIM-related research to date, including our own, is that it has tended to focus on the needs of professional users – the so-called knowledge workers who manage information in a work context. We call for our field to devote increased attention to the needs of "social" users - people who use their computers for personal rather than work activities. Towards this end, we are currently extending both study and evaluation to users with less technical know-how. We envisage that these "social" users will find the simplification of workspace offered by cross-tool designs like WorkspaceMirror especially helpful.

We are also working towards extending our design in various ways. Firstly we are working towards including the feedback from our test users. In particular we aim to improve its configurability, so that those users who only want to mirror between certain tools can do so. We are also interested in how the scope of WorkspaceMirror can be widened beyond the desktop to encompass online tools such as web-based email and document management. A final intriguing, albeit longer-term, research direction is suggested by Chaffee and Gauch (2000) who have researched how user-defined sets of categories can be used to structure sets of search results. In the future, we hope to investigate whether mirrored folder categories could be used to in this way, and thus take a step towards the unification of information management and information retrieval

# 7    References

Bellotti, V. & Smith, I. (2000). Informing the design of an information management system with iterative fieldwork. *Proc. of DIS 2000 Conf. on Designing Information Systems,* 227-237.

Boardman, R. (2001). An investigation of unified personal information management. *Unpublished PhD. Transfer Report,* Dept. of Electrical and Electronic Engineering, Imperial College.

Chaffee, J. & Gauch, S. (2000). Personal Ontologies for Web Navigation. *Proc. of CIKM 2000 Conf. on Information and Knowledge Management*, 227-234.

Dillon, A. (2001). Beyond usability: process, outcome, and affect in HCI. *Canadian Journal of Information Science*, 26(4), 57-69.

Freeman, E. & Gelernter, D. (1996). Lifestreams: A Storage Model for Personal Data. *SIGMOD Record*, 25(1), 80-86.

Kirsh, D. (2000). A few thoughts on cognitive overload, *Intellectica,* 30.

Lansdale, M. (1988). The psychology of personal information management. *Applied Ergonomics*, 19(1), 55-66.

Whittaker, S., Terveen, L., & Nardi, B. (2000). Let's stop pushing the envelope and start addressing it: a reference task agenda for HCI. *Human Computer Interaction,* 15, 75-106.

# Do Interrupted Users Work Faster or Slower?
# The Micro-analysis of Computerized Text Editing Task

*Ivan Burmistrov*                    *Anna Leonova*

Moscow State University
Department of Psychology
8-5 Mokhovaya Ul.
Moscow 103009 Russia
ivan@psychology.ru, aleon@chair.cogsci.msu.su

## Abstract

Previous research on the effects of interruptions on the speed of performing the computerized tasks gave rather non-homogeneous results: many authors insist that interrupted users always complete tasks slower than when performing the same tasks without interruption, but others showed that interrupting a user during some categories of tasks caused that user to complete the tasks faster. The micro-analysis of concrete text editing operations conducted in our experimental study revealed the difference between the effects of interruptions on cognitively simple and cognitively complex tasks. While the performance of simple tasks was not influenced by interruptions, interruptions slowered complex task performance. The task re-orientation after the interruption was found to be responsible for performance degradation.

## 1    Introduction

Research on the effects of interruptions on the computerized work has extensively proliferated in the last five years (see McFarlane & Latorella (2002) for detailed review of the research findings and the existing user interface design literature relative to coordinating human interruption).

In many cases, this research gave rather non-homogeneous results. For example, Bailey, Konstan & Carlis (2000, 2001) and Cutrell, Czerwinski & Horvitz (2000, 2001) insist that interrupted users always complete tasks slower than when performing the same tasks without interruption. However, experiments of Speier, Valacich & Vessey (1997, 1999) and Zijlstra, Roe, Leonova & Krediet (1999) showed that interrupting a user during some types of tasks (e.g., the document-editing tasks and simple decision-making tasks) caused that user to complete the tasks faster.

Clearly, the conclusions derived from these independent studies are inconsistent and further investigation into the effects of interruptions on a user's task performance is required.

It must be noted, that in abovementioned experiments, researchers used "macro" measures of task performance such as "time on task" (TOT) and did not conduct a micro-analysis of concrete operations, of which the the whole task consists. The main idea of our experiment was to conduct this micro-analysis of concrete operations.

## 2 Experiment

### 2.1 Design

In our experiment, 30 subjects performed a computer-assisted text editing tasks. The experimental task was to make corrections in a computer file, based on a hard-copy version of a text containing hand-written corrections. No a priori time limits for completing the tasks were given, and subjects could work at their own pace. Experimental sessions took approximately 40 minutes each, depending on the individual speed.

During the experimental sessions, subject's work activity was disturbed by a number of interruptions – phone calls – when the subject was told to perform another task, referred to as secondary task. Interruptions affected three types of concrete editing operations: (a) typing in new text – *new*; (b) regular editing (making simple corrections) – *regular*; (c) moving a block of text to a new location – *move*.

The independent variable was the presence/absence of interruption. The dependent variable was editing latency for concrete operations.

### 2.2 Apparatus and Materials

The experiment took place in a simulated office environment. The 40 $m^2$ laboratory was divided into two rooms by a wall. One room has been equipped as an office workplace (with furniture, personal computer, intercom telephone), while the other was used as a control room. At the office location a tripod video camera was placed to monitor the subject. The video signals from the camera and from computer screen were routed to a video mixer in the adjacent control room. From this room the experimenter controlled the experiment and watched the mixed video signal (view of the subject plus contents of the subject's computer screen) via the video monitor. The mixed video signal supplied with 0.01 second precision timecode was also recorded on a VCR for further analysis.

### 2.3 Moments Measured and Time Intervals Calculated

The general scheme of moments measured in the experiment and time intervals calculated for data analysis is presented in Figure 1.

Moments measured:

- $T_{start}$ and $T_{end}$: start and end of operation;
- $t_{ring}$: moment of ringing of the phone;
- picking up the phone has been considered to be the start of the secondary task ($t_{start}$);
- the last visible operation of the secondary task has been considered to be the end of the secondary task ($t_{end}$);
- $T_{stop}$: stop of performing the main task, i. e. full switch to the secondary task;
- resumption of the main task ($T_{resumption}$): return of subject's attention to the main task after finishing the secondary task;
- point of continuation ($T_{continuation}$): first action in continuation of the main task after interruption.

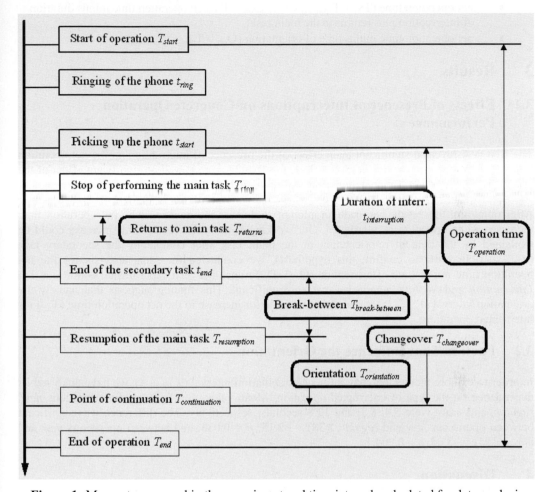

**Figure 1:** Moments measured in the experiment and time intervals calculated for data analysis

Time intervals calculated:

- operation time ($T_{operation} = T_{end} - T_{start}$): time spent to perform the operation including the secondary task;
- duration of interruption ($t_{interruption} = t_{end} - t_{start}$): time between picking up the phone and the last visible operation of the secondary task;
- returns to main task ($T_{returns}$): sum time of returns to the main task while working on the secondary task;
- break-between ($T_{break-between} = T_{resumption} - t_{end}$): time interval between the last visible operation of the secondary task and starting the resumption of the main task;
- orientation ($T_{orientation} = T_{continuation} - T_{resumption}$): time between starting the resumption of the main task and the first action in continuation of the main task;
- changeover ($T_{changeover} = T_{break-between} + T_{orientation}$): a sum of break-between and orientation;

- net operation time ($T_{net} = T_{operation} - t_{interruption} + T_{returns}$): operation time minus duration of interruption plus returns to the main task;
- net operation time minus time of orientation ($Q_{net} = T_{net} - T_{orientation}$).

# 3 Results

## 3.1 Effects of Presence of Interruptions on Concrete Operation Performance

The ANOVA revealed significant main effect of the presence of interruption on the net operation time ($T_{net}$) for operation *move*, $F(1,93) = 9.91$, $p = 0.0022$, but showed no significant effect of the presence of interruption for operations *new* and *regular*.

After obtaining this result, we made an attempt to answer the question why net operation time increases when operation is interrupted. Our working hypothesis was that this increase could be explained by the time of re-orientation in the main task after completing the secondary task ($T_{orientation}$). In order to confirm this hypothesis, we conducted the same analysis for the net operation time minus time of orientation ($Q_{net}$). Difference between experimental conditions *Yes Interruption* and *No Interruption* became nonsignificant. This finding suggests that namely the orientation interval ($T_{orientation}$) is mainly responsible for increase in the net operation time ($T_{net}$) for interrupted operations.

## 3.2 Factors That Influence the Orientation

In order to explore factors that may influence orientation interval ($T_{orientation}$), we have analysed its dependence on the type of interrupted operation. Mean values for $T_{orientation}$ for operations *new*, *regular*, and *move* were 5.4, 8.7, and 12.8 seconds, respectively. The difference was significant between operations *new* and *regular*, $t(38) = -2.15$, $p = 0.038$, and between operations *new* and *move*, $t(55) = -3.04$, $p = 0.004$.

# 4 Discussion

Statistical analyses revealed the significant effects of presence/absence of interruptions on the editing latencies for cognitively complex editing operations (such as moving a paragraph to a new location), while the performance for cognitively simple editing actions (e. g. typing in a new paragraph) were not affected by interruptions. A probable explanation to this fact may be that operation *new* is the simplest operation in text editing. It involve neither search and location of some point in the text (as for operation *regular*) nor include complex sequences of actions and additional mental load caused by the necessity to mentally track the contents of the clipboard (as for operation *move*). Operation *move* is an example of a "functional thread", i. e. a series of commands or actions, and effects of interruptions on this type of operations were more disruptive.

These results are consistent with findings of Speier, Valacich & Vessey (1997, 1999), where interruptions were found to facilitate performance on simple tasks, while inhibiting performance on more complex tasks, and also results of Bailey, Konstan & Carlis (2000, 2001), where interruptions slowered all categories of tasks except *registration* task, because the latter required the lowest memory load at the point of interruption and less effort to resume the main task after interruption.

Our results also suggest that the task re-orientation after the interruption is mainly responsible for increase in net operation time ($T_{net}$) for interrupted operations.

# 5    Acknowledgements

This work has been supported by the Russian Foundation for Basic Research grant # 02-06-80189.

# References

Bailey, B. P., Konstan, J. A., & Carlis, J. V. (2000). Measuring the effects of interruptions on task performance in the user interface. In *IEEE Conference on Systems, Man, and Cybernetics 2000 (SMC 2000)*, IEEE, 757-762.

Bailey, B. P., Konstan, J. A., & Carlis, J. V. (2001). The effects of interruptions on task performance, annoyance, and anxiety in the user interface. In *Human-Computer Interaction – INTERACT 2001 Conference Proceedings*. IOS Press, IFIP, 593-601.

Cutrell, E. B., Czerwinski, M., & Horvitz, E. (2000). Effects of instant messaging interruptions on computing tasks. In *Proceedings of the CHI 2000 conference on Human factors in computing systems, Extended Abstracts*. New York: ACM Press, 99-100.

Cutrell, E. B., Czerwinski, M., & Horvitz, E. (2001). Notification, disruption, and memory: Effects of messaging interruptions on memory and performance. In *Human-Computer Interaction – INTERACT 2001 Conference Proceedings*. IOS Press, IFIP, 263-269.

McFarlane, D. C., & Latorella, K. A. (2002). The scope and importance of human interruption in human-computer interaction design. *Human-Computer Interaction*, 17 (1), 1-61.

Speier, C., Valacich, J. S., & Vessey, I. (1997). The effects of task interruption and information presentation on individual decision making. In *Proceedings of the XVIII International Conference on Information Systems*. Atlanta: Association for Information Systems, 21-36.

Speier, C., Valacich, J. S., & Vessey, I. (1999). The influence of task interruption on individual decision making: An information overload perspective. *Decision Sciences*, 30 (2), 337-360.

Zijlstra, F. R. H., Roe, R. A., Leonova, A. B. & Krediet, I. (1999). Temporal factors in mental work: Effects of interrupted ativities. *Journal of Occupational and Organizational Psychology*, 72, 163-185.

# Experimental evaluation of the effectiveness
# of expert online help strategies

*Antonio Capobianco*

LORIA – INRIA LORRAINE
LORIA Campus Scientifique
F54506 Vandoeuvre-lès-Nancy
France
Antonio.capobianco@loria.fr

## Abstract

At present, most research on online help is focusing on the design of context sensitive help systems. The influence of contextual help on the performance of end users has not yet been analyzed extensively.

Using a bottom-up approach, we first analyzed a set of dialogues between novice and expert users of a standard application software, in order to elicit and model the experts' help strategies. We then performed a comparative experimental study, using the same software, in order to evaluate the effectiveness of the human experts' help strategies. These contextual strategies were implemented using the Wizard of Oz paradigm and compared with standard non-contextual online help. The wizard was assisted in the simulation of both simulated help systems by specific software tools.

The paper presents and discusses results on the influence of users' a priori task knowledge on the efficiency of contextual, versus non-contextual, help.

## 1    Introduction

In spite of recent improvements, current online help systems are still unsuited to the actual needs of novice users, and fail to help them effectively. After a few attempts, novices usually tend to ignore such systems. This can be due to the difficulties encountered by such users when consulting help material in electronic form. It has been shown that in spite of the numerous advantages of multimedia systems, lower user performance has been observed with online help support compared to paper help (Cohill and Williges, 1985). Thus, specific solutions must be implemented in order to provide users with appropriate access to usable online help information.

Several approaches have been experimented, stemming from research in various fields: databases (Borenstein, 1985 ; Roestler and McLellan, 1995), hypermedias (Edwards and Hardman, 1989), and multimodal information presentation (Palmiter and Elkerton, 1991 ; DeVries and Johnson, 1997). These approaches contribute to providing users with intuitive and easy search or navigation facilities in large help information repositories.

However, they only provide 'static' access to help information ; they neither take account of the evolution of users' skills and intentions during interaction, nor take advantage of available contextual information for helping them to carry out the tasks they have in mind. Recent research has brought out the usefulness of context sensitivity for implementing 'dynamic' help which is necessary for assisting novice users in overcoming the specific difficulties they are confronted with, such as error detection and correction, and for offering them strategic or didactic support and adaptive interaction (Coombs and Alty, 1980 ; Moriyon, Szelkely and al., 1994 ; Mallen, 1995). To be efficient, online help systems should be capable of:

- Adaptation to the current user's individual characteristics, that is their general knowledge and skills, motivations and objectives (Breuker, 1990), by implementing access methods or information selection support tailored to the actual needs of users, to their motivations and/or skills, using predefined 'static' knowledge (Bach, 1991 ; Roestler and McLellan, 1995).

- Sensitivity to the actual goals and needs of users (Coombs and Alty, 1980). Such 'dynamic' knowledge proves useful for improving error diagnosis and correction through comparisons between the actual state of the system and the expected one (Quast, 1993).
- Awareness to the current state of the system for generating help messages consistent with its current state and the actions that can modify it (Tattershall, 1990).

Many benefits may be anticipated from the implementation of such features in help systems. However, the efficiency of context-sensitive help systems and their impact on the end user have not yet been evaluated thoroughly. This paper presents and discusses empirical and experimental results stemming from the analysis of human experts' contextual help strategies and the ergonomic evaluation of a context-sensitive simulated help system implementing them.

## 2    Elicitation and implementation of human experts' help strategies

We defined the contextual help strategies that we meant to assess experimentally from the analysis of the behaviours of human experts. We analysed a set of 15 oral dialogues involving 2 human experts and 15 novice users. Experts were instructed to help the novices to carry out 19 text processing tasks using MS Word. Each speech act in the dialogues was characterised using, for the novices, criteria stemming from published taxonomies (Pilkington, 1992 ; Roestler and McLellan, 1995) and, for the experts, an ad hoc taxonomy established from a preliminary survey of the corpus. The analysis of the experts' speech acts yielded results that are summarised in (Capobianco and Carbonell, 2001), and fully detailed in (Capobianco, 2002) [1].

These results indicate that the type of contextual information the experts used most frequently is the progress of the current task execution (71% of their help speech acts). They used this information to provide subjects with opportunistic contextual help based on a hierarchical decomposition of the ongoing task. In particular, they delivered help information step-by-step, namely: for each step in the execution procedure of a complex task, the information needed to carry out this step was provided only after the previous step had been completed (see Figure 1). The help messages at the root of the tree point out the relationship between the user's goal and the function/procedure in the software that has to be executed to achieve it.

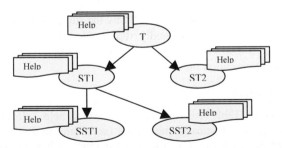

**Figure 1:** Task decomposition into sub-tasks with associated help messages. For instance, T = "Text centring", ST1 = "Text selection", ST2 = "<Centring> icon activation", etc.

---

[1] Capobianco, Antonio (2002). Stratégies d'aide en ligne contextuelles : Acquisition d'expertises, modélisation et évaluation expérimentale. PhD Thesis, Université Henri Poincaré, Nancy, France.

## 2.1 Implementation of the experts' help strategies

To evaluate the efficiency and usability of the experts' strategies, we performed a comparative experimental study based on the Wizard of Oz paradigm. Two help systems were compared, a contextual one implementing the experts' strategies and a standard non-contextual one.

We developed a software tool to assist the operator in charge of the simulation of both systems. The text processing tasks subjects had to carry out were decomposed into simple sub-tasks, using the GOMS model (John, 1990 ; Kieras, 1996). Appropriate, textual and graphical, HTML help messages were associated to each task and sub-task node (see Figure 1). Several messages were associated to each node, one per class of help requests in our taxonomy. The operator's main activity was to interpret the subjects' oral requests. To answer any request, he had only to select and activate the corresponding sub-task node in the displayed task tree, the appropriate message being selected, sent and displayed on the subject's screen by our software. This tool also saved the trace of the interactions between the subjects and both simulated help systems.

## 3 Experimental Set-up

Sixteen voluntary subjects performed eighteen word processing tasks using MS Word. Help messages appeared on an adjunct dedicated screen in answer to their oral requests which they could express without any linguistic or enunciation constraint. These requests were transmitted to two "wizards" through a microphone. One wizard translated them into written requests to a message database where expert knowledge of MS Word operation was represented using the GOMS model. Message selection was performed by our software according to request type and help strategy. Each subject was provided with contextual and non-contextual help successively, order being counterbalanced. The second wizard recorded subjects' interactions with the word processor manually, while the subjects' questions, task execution times and results were recorded automatically. All subjects were first year undergraduate students with the same background knowledge and a very superficial and limited familiarity with MS Word.

## 4 Analyses and results

To assess the effectiveness of each help strategy, we compared subjects' performances using the following three criteria:

- The percentage of tasks carried out successfully by the subjects (TPS).
- The percentage of tasks performed optimally, that is using the most efficient procedure, in particular without resorting to a trial and error strategy (TPO).
- For each task, average execution time (TE).

### 4.1 Users' performances

As regards TPS, results in Table 1 show that contextual help information leads to better performance, this result being statistically significant (according to the T-test). TPO results also seem to indicate higher performances (i.e. lower error rates) using contextual help. Therefore, context-sensitive help information seems easier to exploit by novice users than non-contextual information, thus reducing execution errors and improving task successful completion rates.

| | Contextual help | Non-contextual help | T-test |
|---|---|---|---|
| TPS | 92% | 81% | P = 0.041 |
| TPO | 65% | 59% | P = 0.217 |
| TE | 107 s | 99 s | P = 0.178 |

**Table 1:** Performance comparisons: contextual versus non contextual help.

However, contextual help does not reach statistical significance as regars TPO and TE indicators. Slower task execution in the contextual help situation (cf. TE results) may be explained by the fact that contextual messages described both the actions to be carried out and the pre- and post-conditions associated with their execution, whereas non-contextual messages only described the actions to be performed. Therefore, non-contextual help may be preferred over contextual help by some users, due to the higher "cost" of the use of context-sensitive help. Standard non-contextual help may also appear more appropriate in highly demanding task environments. However, further empirical and experimental evidence is needed in order to reach reliable conclusions on this issue.

## 4.2 Influence of task knowledge

To better assess experts' online help strategies, we classified the 18 prescribed tasks according to the subjects' average [2] performance for each task, assuming that average performance is indicative of task execution difficulty. For each task, we also defined another indicator (TPST) as the percentage of successful executions in each help situation.

Results reported in Figure 2 show that the gain in performance observed in the contextual help situation compared to the non-contextual help situation is strongly correlated with task difficulty. It increases from 0% for easy tasks to 60% for difficult tasks. Users' performances in the contextual help situation were even lower for a few tasks, indicating that contextual help can be less efficient than non-contextual help, when used unwisedly.

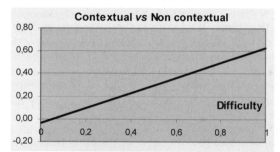

**Figure 2:** Correlation between gain in task performance, and task difficulty.
$$\text{Gain} = \text{TPST}_{CH} - \text{TPST}_{NCH}$$

Besides, as the difficulty of a task can be linked to the familiarity of the user with this task, the results in Figure 2 lead to the conclusion that the efficiency of contextual help compared to non contextual help is greater for the tasks users are less familiar with.

According to this finding, flexible help strategies are likely to prove most effective, namely strategies that provide users with either contextual or non contextual help, according to their familiarity with the current task. To implement such a strategy, help systems have to be able to initialise and update a dynamic model of the current user's knowledge of the software operation, and to infer, from the current state of this model, the type of message to generate in answer to the user's current request: several contextual messages (one for each step in the procedure), or a unique non-contextual one.

## 5 Conclusions

We described a study aimed at assessing the actual effectiveness and usability of a contextual online help system implementing the strategies of human experts. 18 potential users interacted with a text processor on the market, using two help systems successively, a contextual one versus

---

[2] computed over the results (TPS) obtained for each task by all subjects (i.e., with both help systems).

a non contextual one. Each system was simulated, within the framework of the Wizard of Oz experimental paradigm, by 2 human operators whose activity was supported by specific software. Results stemming from the analysis of task execution times, successful and optimal task execution, indicate that contextual help proved more effective than non contextual help as regards successful task execution, this result being statistically significant. In addition, contextual help proved most effective for the tasks users were less familiar with. However, comparisons between task execution times suggest that contextual help may increase the user cognitive workload compared to non contextual help. Further empirical and experimental studies should be realised to refine some of the results gained through this preliminary study.

Results on the relative effectiveness of the two help strategies according to the subjects' individual cognitive characteristics will be presented in a forthcoming publication.

# 6 References

Bach, C. (1991). A customizable direct manipulation user interface with automatic generation of help information. In *HCI International 91: Human aspects in computing*, Stuttgart, Lawrence Erlbaum Associates, September 1-6, pp. 920-924.

Borenstein, N. (1985). *The design and evaluation of online help systems*. PhD Thesis, Carnegie Mellon University ; Pittsburg.

Breuker, J. (1990). *EuroHelp: Developing intelligent help systems, Report on the P280 ESPRIT project EUROHELP*. Kopenhagen, Amsterdam, Manchester, Leeds.

Capobianco, A. and Carbonell, N. (2001). Contextual online help: elicitation of human experts' strategies. In *HCI International 2001*, New Orleans, USA, Lawrence Erlbaum Associates, pp. 824-828.

Cohill, A. M. and Williges, R. C. (1985). Retrieval of help information for novice users of interactive computer systems. *Journal of The Human Factors and Ergonomics Society,* **vol. 27 (3)**, pp. 335-343.

Coombs, M. J. and Alty, J. M. (1980). Face to face guidance of university computer users: Characterising advisory interactions. *International Journal of Man-Machine Studies,* **vol. 12**, pp. 389-405.

DeVries, G. and Johnson, G. I. (1997). Spoken help for a car stereo: an exploratory study. *Behaviour and Information Technology,* **vol. 16 (2)**, pp. 79-87.

Edwards, D. M. and Hardman, L. (1989). "Lost in hyperspace": Cognitive mapping and navigation in a hypertext environment. In *Hypertext: theory into practice*. R. McAleese (Ed), Intellect Limited, Oxford, pp. 105-125.

John, B. E. (1990). Extensions of GOMS analyses to expert performance requiring perception of dynamic visual and auditory information. In *CHI'90 Human Factors in Computing Systems*, ACM, pp. 107-115.

Kieras, D. E. (1996). Guide to GOMS model usability evaluation using NGOMSL. In *CHI'94The Handbook of Human-Computer Interaction, 2nd Ed.* T. Landauer (Ed), North Holland, Amsterdam.

Mallen, L. C. (1995). *Designing intelligent help within information processing system*. PhD Thesis, Leeds University ; Leeds.

Moriyon, R. ; Szelkely, P. and Neches, R. (1994). Automatic generation of help from interface design models. In *CHI'94 (International Conference on Human Factors in Computing Systems)*, Boston, MA, ACM Press & Addison Wesley, pp. 257-263.

Palmiter, S. and Elkerton, J. (1991). An evaluation of animated demonstrations for learning computer-based tasks. In *CHI'91 (International Conference on Human Factors in Computing Systems)*, New Orleans, LA, ACM Press & Addison Wesley, pp. 257-263.

Pilkington, R. M. (1992). Question-answering for intelligent on-line help: the process of intelligent responding. *Cognitive Science,* **vol. 16 (4)**, pp. 455-491.

Quast, K.-J. (1993). Plan recognition for context sensitive help. In *IWIUI'93 (International Workshop on Intelligent User Interfaces)*, Orlando, FL, ACM Press, Janvier, pp. 89-96.

Roestler, A. W. and McLellan, S. G. (1995). What help do users need? Taxonomies for on-line help information needs and access methods. In *CHI'95: International Conference on Human Factors in Computing Systems*, Denver, CO, ACM Press & Addison Wesley, pp. 437-441.

Tattershall, C. (1990). *Question-answering and explanation giving in on-line help systems: A knowledge based approach*. PhD Thesis, University of Leeds, School of Education ; Leeds, UK.

# Automatic vs. Intellectual Document Clustering: Evaluating 2D Topographic Maps

*Maximilian Eibl*

Social Science Information Centre
GESIS
Schiffbauerdamm 19
10117 Berlin – Germany
eibl@berlin.iz-soz.de

*Thomas Mandl*

Information Science
University of Hildesheim
Marienburger Platz 22
31141 Hildesheim – Germany
mandl@uni-hildesheim.de

## Abstract

Two dimensional maps have been constructed for a variety of purposes like e-mail management, music retrieval, web search or correlation analysis of medical symptoms. Nevertheless, those implementations lack thorough evaluation. The constructed systems appear to be a mere feasibility study than tools created for usefulness. It is necessary to establish a evaluation methodology for topographic maps. The paper is meant to be one step toward that endeavor. The main task of document maps lies in the reduction of the dimensions of a document set to merely two dimensions which can be displayed on a computer screen. This article reports on a thorough evaluation of different dimensionality reduction algorithms, ways of visualizing document maps and the differences between automatic and intellectual clustering.

## 1 Introduction: Space and Semantics

Evaluation methods in information retrieval and human-computer interaction differ significantly. In order to evaluate interfaces for information retrieval both evaluation strategies need to be combined in order to reach a deeper understanding of user interaction with retrieval systems. This is especially true for associative browsing systems which tightly integrate retrieval and user interaction.

Topographic document maps visualize large amounts of document sets. Their main task lies in the reduction of the dimensions of a document set to two or three dimensions which can be displayed. The resulting distribution of the documents expresses semantic similarity between the documents by spatial closeness. Several dimensionality reduction algorithms exist. The most popular are Latent Semantic Indexing (LSI) and Self Organizing Maps (SOM).

Two-dimensional displays like document maps show vague semantic relationships between knowledge objects by arranging them in an area. The spatial distances and arrangements express semantic knowledge about the objects. Recent systems like Kartoo[1] add value by explicitly displaying relations and allowing the analysis of social structures.

---

[1] http://www.kartoo.com

The basic idea of displaying semantics by space is usually well understood by users. However, the question whether the algorithm applied to arrange the objects expresses the relations as they are seen by the user and how an algorithm could be selected to better match between its features and the users cognitive concept remains unanswered. There exists even little knowledge about two-dimensional displays from empirical user experiments.

## 2 Reducing dimensions of data

In Mandl & Eibl (2001) we proved that different existing dimensionality reduction methods lead to completely different results. An extensive study in Eibl & Mandl (2002a) included user browsing strategies and confirmed these results. Furthermore, not only the mathematical formula chosen leads to different maps; the way of visualizing the documents can be quite different. Based on these results the next question for the current investigation is: Do dimensionality reduction methods result in document clusters which are semantically useful. The method we chose for our evaluation investigates whether these maps can be intellectually reconstructed or even predicted.

The dimensionality reduction methods evaluated differ greatly in their approach. Kohonen self-Organizing maps (SOM) are neural networks with a complex architecture including two layers of artificial neurons (Kohonen 1995). SOM are suited for clustering. In addition they form complex landscapes of similarity by placing similar clusters close to each other.

On the other hand, latent semantic indexing (LSI) is based on eigenvalue calculation in linear algebra (Berry et al. 1995). LSI carries out two tasks in one processing step. It combines several dimensions and ranks these artificial dimensions according to their mathematical importance for the data set.

LSI has been tested intensively in information retrieval, however, its capabilities for dimensionality reduction have not been applied for visualization. Evaluations need to investigate whether the two main dimensions found are perceived as to most important characteristics by the user as well.

## 3 Visualizing dimension reduced data

Not only the reduction algorithms lead to different maps. In addition, the reduced data can be displayed differently which highly affects usability. In Eibl & Mandl 2002b we demonstrated the differences in user interface design of document maps. Some examples are shown in Figure1. Figure 1b shows the simplest way of visualizing document maps. The documents are arranged on a two-dimensional flat pane. Accumulations of documents appear as dark clouds. Figure 1a shows the same map except that this time the accumulations are translated into the third dimension. The clouds from figure 1b appear as mountains. Though the display appears three-dimensional the document data was initially reduced to two dimensions.

Figure 1c shows the same data set this time reduced to three dimensions. The visualization uses three spatial dimensions to show the dimension reduced data. The resulting display resembles somewhat a star field. Eibl & Mandl 2002b show further examples and a pros and cons analysis.

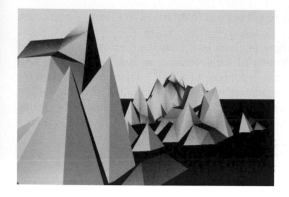

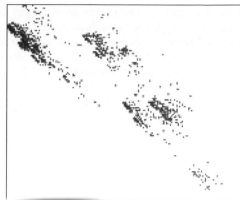

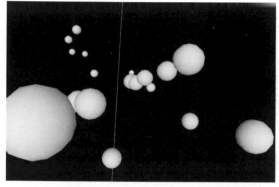

**Figure 1:** Examples of different map-based displays of the same data set

# 4 Automatic vs. intellectual clustering

As shown in section 2 and 3 the appearance of document maps highly depend both on the algorithms used for dimensionality reduction and display properties. But next to these rather mathematical factors there is another one which has been neglected: the human interpretation of the data. Do users interpret the data similar to the computer? In order to answer this question we are comparing the results of automatic and intellectual clustering.

The base of our experiment is the GIRT dataset. GIRT (German Indexing and Retrieval Test Database) is part of the American TREC (Text Retrieval Conference, cf. Voorhees & Harman 2001) initiative since 1992 and the European CLEF (Cross Language Evaluation Forum, cf. Peters et al. 2001) initiative since 1999. It contains some 80.000 documents and 90 topics expressing user needs. All documents found by retrieval systems during the campaign are evaluated as being relevant or irrelevant to these topics. In a test setting, users are provided with some of these topics and have to search the dataset for documents relevant to the topics. The traditional information retrieval measures recall and precision are generated in order to rate the tested information retrieval system.

Though GIRT is actually meant to evaluate information retrieval systems, it also can be used for evaluating automated clustering. Since the topics in GIRT describe sets of similar documents, these documents should also appear close to each other after applying an automatic clustering mechanism. In a visualization this could look like this: The topographic map visualizes clusters of documents as mountains or dense clusters. If every intellectually manifested topic can be reconstructed automatically, the visualization should show a mountain or a dense cluster for each topic. Coloring the documents according to the relevant topic should lead to effective interaction when browsing for relevant documents.

Figure 2 shows an example of how good matching between intellectual and automated clustering might look like. The gray dots show the distribution of the documents as it is generated by LSI. The black dots show the distribution of documents belonging to a GIRT topic. Initial tests show that some topic clusters resemble the automatic clustering by LSI some do not. The main problem here appears to find a regularity under which conditions intellectually and automatically created clusters overlap.

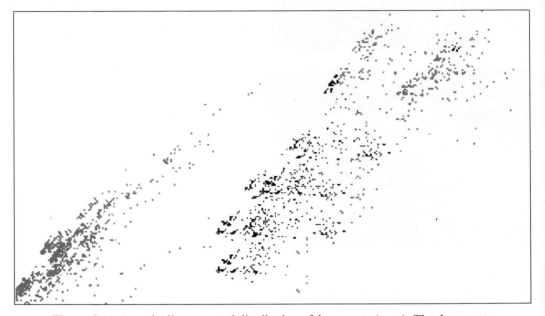

**Figure 2:** Automatically generated distribution of documents (gray). The documents belonging to a intellectually generated topic are highlighted (black).(Detail)

This methodology can be interpreted as a refinement of an evaluation by Rorvig & Fitzpatri 1998 and its adaptation to human-computer interaction.

## 5    Resume

This article discusses the issues involved in the evaluation of user interfaces for displaying similarity. It summarizes recent work and introduces a evaluation methodology for browsing systems which is built upon evaluation strategies for information retrieval.

## 6    References

Berry, M; Dumais, S, & Letsche, T. (1995). Computional Methods for Intelligent Information Access. In: Proceedings of ACM Supercomputing '95. San Diego, CA. pp. 1-38.

Eibl, M., & Mandl, T. (2002a). Including User Strategies in the Evaluation of Interfaces for Browsing Documents. Journal of WSCG. Special Issue 2002, Vol. 1, No.2, pp. 163-169.

Eibl, M.; Mandl, T. (2002b). Parameters for the Visualization of Document Sets. In: Proceedings of the 9th Conference on Electronic Imaging and the Visual Arts EVA, November 6-8 2002, Berlin, pp.96-99.

Lin X. Map Displays for Information Retrieval. In: Journal of the American Society for Information Science (JASIS), (1997)vol.48, no.1, pp.40-54.

Kohonen T. Self-organizing maps. (1995) Springer. Berlin.

Mandl, T., & Eibl, M. (2001). Evaluating Visualizations: A Method for Comparing 2D Maps. Proceedings of the 9th HCI Intl., New Orleans, August 5-10, 2001, Vol.1, p. 1145-1149.

Peters, C. (ed.) (2001). Cross-Language Information Retrieval and Evaluation: Proc of the CLEF 2000 Workshop. Lisbon, Portugal, Sept. 21-22, 2000 Berlin et al.: Springer [LNCS 2069]

Rorvig, M., & Fitzpatri, S. (1998). Visualization and Scaling of TREC Topic Document Sets, Information Processing and Management 31(2-3) 133-149

Voorhees, E., & Harman, D. (eds.)(2001). The Tenth Text Retrieval Conference (TREC-10). NIST Special Publication. National Institute of Standards and Technology. Gaithersburg, Maryland. http://trec.nist.gov/pubs/

# Evaluation of Story-Based Content Structure and Navigation for a Learning Module in SCORM

*Boris Gauss[1], Christopher Hausmanns[2], Rodolphe Zerry[2],*
*Günter Wozny[2], Leon Urbas[1]*

## Abstract

In this paper, we present the evaluation of a prototypic learning module. This module features a story-based content structure and a corresponding newly developed navigation tool, the *Process Control Navigation Display*. In a field study, we found a strong positive effect of the module on learning outcome. These results indicate that the prototype is a promising approach to design SCORM-compliant modules for the support of existing face-to-face courses.

## 1    Introduction

In the [my:PAT.org] project, partners from four German Universities are developing a web based educational system for the subject matter of process systems engineering. The purpose of the system is to support the face-to-face teaching at the University. Among other facilities, the system will contain several learning modules for existing courses and lectures. These modules supply an opportunity for the students to deepen their understanding of the topics by self-regulated learning. The content of the modules is based on existing course materials like lecture notes and slides. Additionally, new multimedia materials like animations, short films and interactive simulations are produced to be integrated into the modules.

## 2    The learning module

Our pedagogical vision for the educational system is a rich learning environment where the learners cooperatively solve real world engineering problems. This implies a didactical shift from frontal teaching to more constructivist methods with proactive learning. The first step towards our vision is to create learning modules with story-based scenarios.

### 2.1    SCORM

We develop the learning modules compliant to the Sharable Content Object Reference Model standard (SCORM, see ADL, 2002). Following SCORM, courses are built up modularly. Each course consists of various Sharable Content Objects (SCOs), which are in turn composed of assets, i.e. single media items. A SCO is the smallest logical unit of instruction and represents a single instructional objective. Thus, a SCO is conceived as a stand-alone lesson that can be integrated in different courses or learning modules without modification.

---

[1] Technische Universität Berlin, Zentrum Mensch-Maschine-Systeme, Jebensstr. 1, Sekr. J2-2, D-10623 Berlin; {gauss, urbas}@zmms.tu-berlin.de
[2] Technische Universität Berlin, Institut für Prozess- und Anlagentechnik, Str. d. 17. Juni 135, Sekr. KWT9, D-10623 Berlin; {christopher.hausmanns, rodolphe.zerry, guenter.wozny}@tu-berlin.de

The navigation between the SCOs of a learning module is controlled by the Learning Management System. The author of a course defines the structuring and sequencing of the SCOs in an XML-file, the *imsmanifest*. Since SCOs are considered as stand-alone lessons, direct linking is only possible within a SCO but not from one SCO to another SCO of a module. Therefore the linking in the SCORM-world is more restricted than in hypertext in general.

## 2.2 Content Structure

The prototypic learning module deals with the topic of stationary modelling of chemical plants, an essential chapter in the lecture of *Process Systems Dynamics*, which is held by Prof. G. Wozny at the Berlin University of Technology. The first step was to define the learning objectives of the module and to extract the according pieces of text, diagrams and images from the existing material. Additionally, new multimedia objects like films animations and simulations were produced to enrich the existing materials. The single media assets were combined in SCOs. The learning content of the chapter, which is structured according to the logic of the subject matter in the lecture, was reconfigured around the framework of a story from the engineering practice. This framework consisted in an industrial MeOH-H$_2$O distillation process. We structured the process into four hierarchical levels of different granularity: From the first level, where only the input and the output is considered (while the process is described as a black box), to the fourth level, where single components of process units are addressed.

## 2.3 The Process Control Navigation Display

According to the structural framework provided by the story, a new *Process Control Navigation Display* (PCND) for the navigation between the SCOs was developed and integrated into the Learning Management System.

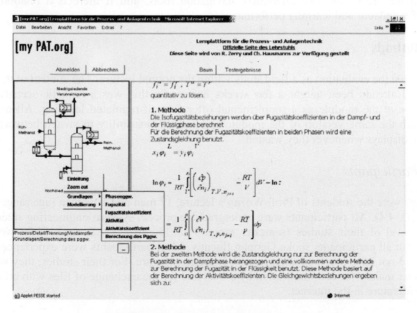

**Figure 1:** Screenshot of the module with the PCND (with context menu) and a text component.

The PCND resembles the display of a process control system of a chemical plant for MeOH-H2O distillation. It is placed in a frame in the left third of the screen while the other two thirds remain for the display of the learning content (see Figure 1). With the PCND, the students can zoom into the different levels of the content structure defined by the story. The learning content for each level is available through a context menu, which is opened by a right mouse click on an object of the PCND. Thus the students can always link the learning content to a certain level of detail and to the process unit it belongs to. Since the PCND does not provide a complete overview over the structure of all the SCOs of the module and this might cause orientation problems, the learners have also the possibility to switch from the PCND to an alternative navigation display by clicking on a button. This alternative display shows the table of contents of the module in an interactive tree component.

# 3 Evaluation

In an evolutionary approach of system development, we carried out repeated cycles of heuristic usability tests in the early stage of the design process. A highly-developed version of the prototypic learning module was subject to a controlled evaluation study, which we present in the following. Referring to Kirkpatrick's (1994) *Four Level Evaluation Model*, our purpose was to study the effects of the prototypic module on (1) reactions and (2) learning, the first two levels of the model. We expected positive effects of the story-based content structure and the process control navigation on both evaluation levels. The module presents the learning content from a different perspective compared to the lecture and creates a real world context. According to constructivist didactical theory (e.g., Murphy, 1997), multiple perspectives and real world contexts should enhance learning and motivation. On the other hand, the unfamiliar context and navigation could disturb or distract the students from learning. In particular, we were interested in how the students make use of the two alternative navigation tools, and if there is a relation between navigation behaviour and learning performance.

## 3.1 Methods

We conducted the evaluation as a field study within the normal lecture. The content of the learning module had already been taught a few weeks before. So this was a realistic scenario for the intended use of the module as a supplemental offer for self-regulated learning. All participants learned with the same version of the module and had the possibility to switch between the two navigation displays whenever they wanted.

### 3.1.1 Participants

Participants were the students of Prof. Wozny's lecture, 17 men and 1 woman (age range 22 to 41, $M=25.1$, $SD=4.4$). All participants were undergraduate process systems engineering students at an advanced level of their studies (semester range 4 to 15, $M=8.2$, $SD=2.3$). Eight were native Germans, but all participants spoke German fluently. Most participants were experienced internet users but did not have much experience with learning software. For their studies, they were using the computer mainly for communication by e-mail and for the exchange of files with others, or for searching literature in the internet.

### 3.1.2   Procedure

During the evaluation study, each student worked on a personal computer. At the beginning, the subjects completed a questionnaire about learning motivation and attitudes towards computer based instruction. Then the students logged in into the learning system and their prior knowledge was tested with 15 multiple choice questions, which had to be completed within 12 minutes. Subsequently, after a short introduction to the navigation tools by the experimenters, the subjects had around 40 minutes time to learn with the module. During this period, each student had access to a list of the topics in which they had made mistakes in the prior knowledge test. The log-files of the interaction with the module were recorded. After the learning, the subjects rated their reactions to the module in a detailed questionnaire with 42 items. Finally, the multiple choice knowledge test was applied for a second time. The whole session lasted about 100 minutes. In the following lecture, three days after, the knowledge test was applied once again in a paper and pencil version, without prior notice.

## 3.2   Results

As intended, the students spent about 40 minutes learning with the module (range 37 to 45, $M$=40.1, SD=3.6). Their initial attitudes towards computer based instruction were mostly neutral. In general, their motivation for participation in the lecture was high and rather extrinsic, nevertheless they were also moderately interested in the topic of the module and expected to improve their knowledge in the evaluation session.

### 3.2.1   Evaluation level 1: reactions

Altogether, the reactions of the learners to the system were slightly positive. On a 5-point-scale from 0=very negative to 4=very positive, the mean of the total score of the reactions questionnaire was $M$=2.5, with only little variation between subjects and no subject rating the system below the neutral point of the scale (SD=0.3, range 2.0 to 3.1). Furthermore, we asked the students what kind of additional functions or tools would be required to improve the system's effectiveness. By far the most wanted was the possibility to contact a tutor.

### 3.2.2   Evaluation level 2: learning

In the prior knowledge test, learning performance, measured on a 48-points-scale, was quite poor (range 5 to 20, $M$=11.0, SD=4.4). Since the data of the second knowledge test directly after the learning period was lost for 11 subjects (due to problems with the database), we consider only the data of the retest during the lecture, three days after (n=14). In this test, the students achieved better results (range from 8 to 36, $M$=19.0, SD=6.5). This increase in learning performance was statistically significant (Wilcoxon-test, Z=-3.2, p<.01) with a very strong effect size (d'=2.0).

### 3.2.3   Navigation display preferences

Overall, the participants navigated about half of their time in the module with the PCND (range 5 to 100%, $M$=50.6, $Md$=49.4, SD=34.4). Most of the students either switched only one time from the default PCND to the tree-navigation, or they did not switch at all and navigated with the PCND for the whole session. For further analysis, subjects were categorised into three groups of the same size (n=6), according to navigation display preferences: (1) PCND (> 75% of total time with PCND), (2) FLEXIBLE (25 to 75% with PCND), (3) TREE (<25% with PCND).

Navigation display preferences did not affect learning gain, measured as the difference between post test in the lecture and prior knowledge test (Kruskal-Wallis-test: $\chi^2$=0.2, df=2, p=.91). The navigation in the system was rated positive ($M$=2.8, SD=0.5, scale from 0 to 4) regardless of display preference and the students did not report any orientation problems.

## 3.3 Discussion

On both evaluation levels we considered, results were encouraging. First of all, the strong and stable effect on learning is remarkable. It should be noted that learning gain was measured three days after only 40 minutes of interaction with the module. The students had not even had time to go through the whole module and presumably did not prepare for the post test after the evaluation session. Concerning the use of the alternative navigation displays there was a wide range between subjects, though time was short and the sample was quite small. Since there was no influence of navigation preference on learning outcome, the learners seem to have found their individual best way to navigate through the module.

Nevertheless, this short term study is only the first step of evaluation. The good results for reactions and learning have to be confirmed when the module is online and learning is really self-regulated. The integration into the curriculum and the support by an online tutor will be crucial for the success of the module. Another point is that the questions in the knowledge test were mostly on lower levels of learning (knowledge and understanding, cf. Bloom et al, 1956). From the didactical approach of the module (real world context, multiple perspectives) we can expect even greater benefits on higher levels of learning like transfer.

## 4 Conclusions

The prototypic module with story-based content structure and process control navigation has shown to be a promising approach for designing learning modules in SCORM as a support of existing face-to-face courses in the subject matter of process systems engineering.

## Acknoledgements

The [my:PAT.org] project is sponsored by BMBF in the Programme "New Media in Education".

## References

ADL (2002). Advanced Distributed Learning initiative. SCORM Overview. Retrieved February 10, 2003, from http://www.adlnet.org

Bloom, C. S., Engelhart, M. B., Furst, E. J., Hill, W. H. & Krathwohl, D. R. (1956). Taxonomy of educational objectives. The classification of educational goals (Handbook I. Cognitive Domain). New York: Longman.

Kirkpatrick, D. L. (1994). Evaluating Training Programs. The Four Levels. San Francisco: Berett-Koehler.

Murphy, E. (1997). Constructivism. From Philosophy to Practice. Retrieved February 10, 2003, from http://www.stemnet.nf.ca/~elmurphy/emurphy/cle.html

# Assessment and Improvement of the Integrated Hazard Avoidance System for General Aviation Interface

*Sheue-Ling Hwang[1]*    *Wen- Ying Chen[2]*

*National Tsing Hua University*
*slhwang@ie.nthu.edu.tw*

*National Tsing Hua University*
*g893846@oz.nthu.edu.tw*

## Abstract

The paper aims to improve the human machine interface of Integrated Hazard Avoidance System for General Aviation (GA-IHAS), according to formal interface design principles. The GA-IHAS has been developed by the Institute of Information Industry (I I I) and been incorporated with the Global Position System, electrical data base, and the 3D image technology, in order to display complete flying condition and provide warning on time if necessary,

## 1    Introduction

Many evidences from literature review indicate that human factor principles should be taken into account in a new aeronautical assistant system design. In this system, the data that are shown on the computer monitor should be significantly meaningful so that aeronautical administrators or aviators can take correct procedures in time. According to the previous records, a plane under normal control may still crash because the flying vision is too vague or the pilot is not familiar with the changeable flying conditions. Based on the reasons above, the I I I developed an Integrated Hazard Avoidance System for General Aviation (GA-IHAS), incorporated the Global Positioning System, electrical data base, and the 3D image technology, in order to provide the warning of the crash earth and the complete display of the flying condition, and to make extra protection region under aviation safety system. However, no verification has been conducted on the user interface of this system.

The purpose of the present research is to improve the human machine interface of Integrated Hazard Avoidance System for General Aviation, and mainly emphasize related ergonomic principles to be referred to the HMI system. Considering the tasks to be executed under time pressure and preciseness, the concept of user-centered design as well as human factor principles are applied to meet the need of users and to improve system safety. As a result, in the process of system development, the principles of human factor in HMI should be taken into account to develop a user-oriented system to make users more comfortable, more efficient, more convenient, more precise and more effective.

Therefore, in order to achieve the goals of enhancing flight performance and increase the function of this system, this research refers to some design and methods of the human machine interface,

---

[1] Kuang Fu Rd., Hsinchu, Taiwan
[2] Kuang Fu Rd., Hsinchu, Taiwan

introduces the principles of the International Standard Organization 9241 on interface, and verifies the effectiveness of the proposed interface design.

## 2    Background information

There are many expected functions that GA-IHAS can provide, currently the system provide the warning of the crash earth and the complete display of the flying condition, and to make extra protection region under aviation safety system. The reason for this research is because that the establishment of the current interface has no guideline on considering the human factors principles. Considering the requirement of user when designing the interface is the most effective treatment. If the design of system interface is good, but user's ability does not mach it as much, the effect of the system will not be significant.

## 3    Experiment

The GA-IHAS include the warning of standard earth crash, the warning of premature drop, 3D relief map, 2D plane navigation map, the terrain sectional drawing, the aviation route plan, and these information are displaying in four different areas of the screen ( see Figure1).

At first, we evaluated a currently developed system, GA-IHAS, and some interface deficiencies had been detected. The color use and warning signal location were found to be critical design deficiencies of this system. Four variables of interest we focused on this study are:1)the background color of the status display 2)the size of characters 3)the color of characters, and  4) the location of the warning information.

Fig. 1:  The interface of GA-IHAS

### 3.1    Procedure

Two stages of experiments were designed to find out the optimal color arrangement and warning signal location. When completing the whole experimental task, every subject has to run 108 trials in the experiment. In order to reduce the experimental trials and save the time, we decide to separate the experiment into two stages.

The first stage was to find out the optimal color arrangement, and the independent variables were the background color of the status display, the size of characters, the color of characters. We design some interfaces with different background color(green and black),characters size(10,12,14) and characters color(blue, white, yellow). The interface setting is arranged according to the layout of GA-IHAS. The program was written by Flash5.0 to simulate the GA-IHAS interface Fifteen engineering graduate students at the National Tsing Hua University who have the potential to work as pilots are chosen as study subjects. And the fifteen volunteer subjects have normal vision. The main task that subjects had to do is to record the information appeared in the screen. Performance measurement index is the number of error of recorded information. The definition for this index was number of errors: number of wrong actions during the informing process. When analyzing the data, we find that the optimal color arrangement is the character size -14, the character color-yellow, the background color - black.

The second stage was to find out the optimal warning signal location. According the result of the first stage, we change the original interface of GA-IHAS (the character size -12, the character color - white, the background color - green) into the optimal color arrangement (the character size -14, the character color-yellow, the background color – black). And then we simulate a real environment to match the GA-IHAS interface, and three kinds of different virtual flight environment (high mountain, plains, and hill) were designed by 3D Studio. There were warning messages appearing on simulating GA-IHAS screen, and the warning signal appear in different location (the top of interface, the middle of interface and the below of interface) . He/she has to pay attention to the conditions appeared in very short time, and has to react immediately if warning signal appearing. After he/she has confidence to accept the test, each subject runs the testing program with three different environments. Performance measurement index include response time, and subjective rating. The definitions for these indexes are listed as follows:

- Response time: time from incident message appearing on display screen to trainee acknowledge its existence;
- Subjective rating: participants rated their subjective feelings about ease of finding warning signal, reading characters of different size and color, and different background color of the original interface and modified interface, comfort in original design and in modified design- - a scale of 1 = very difficult/uncomfortable to 5 = very easy or very comfortable was used for these items.

When analyzing the data, we find that warning signal appeared on the top of interface was the optimal warning signal location.

# 4    Results & discussions

Table1 indicated the difference between original interface and modified interface. Comparing the data of original design of interface and modified design interface, one could find that the reading accuracy of participants promoted from 77.9% up to 88.5%, and the reaction time of finding the condition appearance reduced from 1.29 seconds down to 0.96 seconds.. Table2 indicated the index value. And we can see the change of the interface (see Figure2).

Table 1: The difference of original design and modified design of interface

| Interface | original design | modified design |
|---|---|---|
| Background color | Light Green | Black |
| Characters size | 12 | 14 |
| Characters color | White | Yellow |
| Warning signal position | Middle of interface | Top of interface |

Table 2: The index value

| Interface | original design | modified design |
|---|---|---|
| Average subjects response time | 1.299 seconds | 967 seconds |
| Average subjects Number of errors | 22.1% | 11.5% |
| Subjective rating | 2.9 | 4.2 |

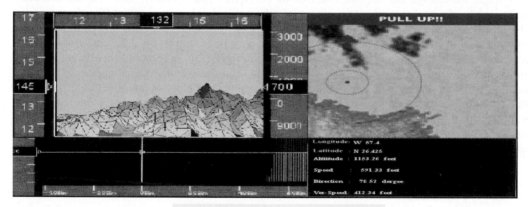

Fig. 2: The interface of modified GA-IHAS

We find that the modified design promoted the subjects' accuracy and speed of response. In addition, changing the warning signal location to the top of interface shortened reaction time of subjects. The experimental results indicate that rearrangement of color improved subjects' performance. As for the warning signal location, the improvement indeed shortens the reaction time of subjects and the subjects can easily perceive the signals.

## 5    Conclusions

The results of this experiment demonstrated that the modified interface satisfied the requirement of user, and allow users to possess more time to make a decision, and made the system more functional because of improving performance.

Finally, audio warning is suggested to be combined in the interface in order to help user to pay attention to the appearance of the warning faster, and have more time to deal with condition. The audio system is suitable for using as the warning and emergency signal.

## Acknowledgements

This study was conducted with exceptional support from the institutions and individuals: Institute of Information Industry (I I I). And the authors would like to express their gratitude to especially Yuang- Ming Gu, Kong-Pon Ha, and Yo-Chun Hwang for their kindly support for this research.

## References

Carey, M.S.(2000), " Human factor in the Design of Safety-related Systems" Computing & control engineering joural,11,pp28-32.

Filippi, G. & Saliou, G. & Palle,P.(1998), " Anacondas: data analysis to assess work activity in simulated control rooms. Setting up an Observatory of Nuclear power plant-operation at EDF" IFAC-MMS.pp16-18.

G. Johannes, (1995), " Knowledge-Based Design of Human-Machine Interface", Control Eng. Practive, Vol .3,No2, pp267-273.
http://www.taasa-web.org/hummanfactor.htm(in Chinese)

Deatherage,B.(1972), "Auditory and other sensory forms of information presentation", In H. VAN Cott and R. Kinkade, Human engineering guide to equipment design Washington:Government Printing Office

# Usability Evaluation for the Commercial Aircraft Cockpit

*David B. Kaber*

Department of Industrial Engineering,
North Carolina State University
328 Riddick Labs, Raleigh, NC 27695
dbkaber@eos.ncsu.edu

*Michael P. Clamann*

BOOZ Allen & Hamilton Inc.
Falls Church, VA 22041 USA
mpclaman@hotmail.com

## Abstract

The objective of this research was to identify usability evaluation techniques suitable for application in aviation systems design and development. This work is important because pilots remain a key element in aviation operations and major shortcomings exist in current design and certification processes leading to usability flaws in cockpit interfaces. We identify special requirements of the aviation domain that may affect usability evaluation, and characteristics of evaluation methods that may be effective in this domain. Formal usability testing and inspection methods are reviewed and compared in terms of potential to reveal system flaws, the accuracy of methods, and their relative costs. We present guidelines and recommendations for applying a combination of usability evaluation techniques to aviation systems design.

## 1    Introduction

Research in the field of Human-Computer Interaction (HCI) has shown that an early usability evaluation can reduce operator errors by optimizing functions for a specific population. This research has produced several methods for evaluating usability that have been proven effective in developing highly complex computer systems. Given the importance of the human pilot in the control loop of advanced commercial aircraft, it is likely that aviation systems may benefit from the application of usability research. The overall purpose of this work was to identify usability evaluation techniques, or some combination of techniques, suited for evaluation of contemporary commercial cockpit interfaces. We discuss usability issues in aviation systems and problems that currently exist in the development cycle of new systems. We also review formal usability evaluation techniques and identify those techniques that might best fit aviation applications.

## 2    The Aviation Systems Environment

Technology is driving rapid change in cockpit interfaces, which may only be limited by available cockpit space and the imagination of designers (Billings, 1997). Monitoring requirements associated with cockpit displays and the need for pilots to maintain up-to-the-minute situation awareness (SA) may cause excessive workload under certain flight circumstances. Unfortunately, there is little evidence that research on aspects of human information processing has been considered in cockpit design. Current interfaces have been found to provide inadequate feedback on behavior and intentions of aircraft system automation (Woods & Sarter, 1993). There is often a lack of visibility of functions, which inhibits accurate mental model development and compromises pilot SA.

The Flight Management System (FMS) in advanced commercial aircraft provides examples of flaws in aviation systems usability, including a lack of integration of command sequences with normal pilot operating procedures and long flight path programming times. The FMS is an excellent example of a system whose features were defined from an engineering perspective rather than a user's perspective. It was originally developed to optimize flight paths and its functions grew (over 20 years) to include calculations for wind, management of navigation data sources, remaining fuel calculations, error determination, presentation of performance data, processing of received transmissions, systems monitoring, etc. (Billings, 1997). Although all of these functions are available in the FMS, they can be difficult to execute. The general result of FMS usability problems is that pilots essentially don't understand why it works. To compound this, there is a lack of standardization of the design of interfaces to such systems in commercial cockpits.

With respect to the aviation systems design process, in general, there is little consideration of how new and existing interfaces integrate. To some extent the advent of "glass cockpit" technology has allowed for integration of cockpit sensor and indicator displays and controls, but often this has not been accomplished in a manner intuitive to pilots. With this in mind, recent work has advocated the use of pilots in new interface design, but they typically have no involvement early in the development process (Kaber et al., 2002).

Although aviation certification processes dictating airworthiness and operational requirements exist, the definition of the role of human factors in the Federal Aviation Administration and Joint Airworthiness Authority's processes are vague. Pilot certification requirements focus on fatigue and concentration in use of cockpit systems with little consideration of the high-level information processing that may be compromised by unusable interfaces (Singer, 1999). Furthermore, the procedures described as part of *Federal Aviation Requirements, Part 25* are dated and most applicable to single-sensor, single-indicator cockpit design and not contemporary "glass cockpits". In general, the overall certification process does little to contribute to cockpit interface usability.

## 3    Usability Evaluation Techniques

In light of the usability problems in aviation systems design and development, we reviewed and compared the more commonly used evaluation techniques in HCI towards identifying methods that might best "fit" aviation applications. The major classes of usability evaluation we considered were inspection and testing (or formal experimental evaluations) (Virzi, 1997). Here we briefly list the specific techniques, some of their characteristics, and advantages and disadvantages:

- Cognitive walkthrough - inspection method focused on evaluating learnability of systems. Requires use of experts to identify interface action errors users may make in task performance (Dix et al., 1998; Newman & Lamming, 1995).
- Direct observation (of user) – good for identifying inefficiencies in interface actions or errors in task procedures. Cooperative evaluation can also be conducted where expert evaluator asks user (domain expert) to describe actions to non-expert (Dix et al., 1998).
- GOMS (Goals, Operators, Methods and Selection Rules) - user model created based on detailed task analysis (Virzi, 1997). Best for evaluating and predicting expert performance of common tasks in an application (Dix et al., 1998).
- Heuristic evaluation - another inspection method where interfaces are judged based on select set of usability principles (e.g., visibility, modeless dialog, consistency, error prevention/recovery, flexibility, help) (Dix et al., 1998). Technique is very flexible, can be performed by non-experts at any point in system design, with or without actual interface ("warm" or "cold" evaluation) (Virzi, 1997). Inspections can be conducted by individual or group review. Number and expertise of evaluators directly effects cost.

- Usability testing - controlled experiments (in a lab or field setting) typically involving end users (Wixon & Wilson, 1997). Quantitative analysis of interface with goal of reducing errors. Goal must be matched to specific metrics for context (operationalized) and assigned quantitative levels for testing. One disadvantage is potential intrusiveness due to experimenters observing/recording user performance (Dix et al., 1998). Also requires involvement of one or more experts making method more costly. Testing is time consuming and can increase overall system development time.

In researching comparisons of techniques, we found that opinions are mixed on which evaluation approaches are most effective for finding problems and fostering usability solutions. Many studies have been conducted comparing testing and inspection in terms of implementation effectiveness, quality of results, cost, etc. (Jeffries et al., 1991; Nielsen & Phillips, 1993; Virzi, 1997). Here we very briefly summarize some of the results:

- Usability testing can identify more problems than some inspection methods, including the cognitive walkthrough (Jeffries et al., 1991; Virzi, 1997).
- Heuristic analysis with experts may identify even more problems than some testing methods (Virzi, 1997).
- Usability testing is most effective for identifying serious problems, but not low-priority problems, like heuristic analysis (Virzi, 1997).
- Usability testing is the most expensive of all methods, but it has also been found to be the most accurate, followed by model-based methods and heuristic evaluation (Nielsen & Phillips, 1993).

In summary, all the studies cited above agree that: (1) there is no single method that is superior in all aspects; (2) any evaluation is better than none; and (3) the best approach may be to use usability testing in conjunction with one other technique. It may be possible that techniques can be combined effectively to produce thorough results. In fact, most researchers say that inspection should either be a pre-cursor or enhancement to formal usability testing.

## 4    Putting Usability in the Aviation Context

Given the nature of existing aviation systems and aircraft design and certification processes, several criteria for application of usability techniques to the commercial cockpit include:

- Rapid execution – Technology is advancing faster than designers' capability to effectively integrate it into existing or new aircraft due to tedious certification processes. Any usability evaluation technique should not slow this process further.
- Cost effectiveness – Current certification requirements can increase the cost of designing new components by a factor of three in some cases (Abbott et al., 1999). A usability evaluation should not pose excessive additional costs for vendors.
- Integration in new systems development cycles – Many researchers have emphasized the need for usability to be built into the aviation systems development process (Abbott et al., 1999; Singer, 1999; Williges et al., 1988).
- Input from a variety of domain experts – As in designing any interactive system, it is crucial that aviation systems be designed for the intended user population. Pilots should be included early and continually in the development process.
- Transferability and scalability to cockpit design – Any usability evaluation technique must be adaptable to the context of cockpit interface design and should be scalable for application to a single new component or entire system.

On the basis of these criteria, we analytically evaluated the applicability of the various evaluation techniques reviewed. There has been some support for usability testing in aviation systems design because of the quality/accuracy of results, the flexibility of the technique, the preferences of usability experts, and the potential for wide acceptance and use (Williges et al., 1988). Even with these advantages, high costs have often been a deciding factor in applications. System designers must determine the importance of accuracy and cost of testing in their specific applications.

Heuristic evaluation fits all the unique requirements for usability evaluation in the cockpit. Its fast, inexpensive, can be performed at any phase of the development process, and usability experts are not required. Although non-experts can perform an evaluation, in the context of aviation systems, application of this technique would likely need to be overseen by an expert in both usability and the aviation domain of interest. It is also likely that heuristic evaluations are transferable to the aviation domain. Kaber et al. (2002) conducted a warm heuristic-based analysis of the Multi-Control Display Unit (MCDU) component of the FMS to identify usability issues. In their study, evaluators observed the use of an MCDU based on the design implemented in the McDonnell Douglas (MD)-11 passenger aircraft in a hypothetical flight task. Seven experts participated in a group evaluation of the MCDU based on principles under the major headings of learnability, flexibility, and robustness. The evaluators concluded that the MCDU violated several of the principles in these categories. However, there may be a need to adapt existing heuristics and develop new ones for application to specific cockpit interfaces. It may also be possible to use contemporary human factors design guidelines for aviation systems, which have recently been published, as a basis for this type of evaluation (Billings, 1997).

Direct observation of a user while they interact with a system, as part of cooperative evaluation, also fits with the requirements for evaluating aviation interfaces. The method is quick, inexpensive and utilizes the input of multiple experts by design. This method has also been used previously for development of aviation systems (Williges et al., 1988), so its transferability and scalability may not be in question.

Given the importance of cost in developing aviation systems, the completeness and accuracy of results should justify the specific approach. In addition, because of the high level of expert involvement that may be required, certain combinations of techniques may be very expensive. For example, combining labor-intensive methods like the cognitive walkthrough and usability testing should only be considered if results are expected to be far superior to rapid inspection methods (e.g., heuristic analysis). In general, if usability testing is used, it is not advantageous to consider walkthrough or model-based evaluation techniques because of the typical scope of testing. Testing may encompass both performance and learning assessments that are possible with GOMS and the cognitive walkthrough without the added involvement of experts. Furthermore, usability testing may produce more accurate results than GOMS, as a result of dealing with actual users and not a model (Nielsen & Phillips, 1993). In contrast, walkthroughs and GOMS models cannot be adapted to produce results equivalent to a usability test.

# 5    Recommendations

Based on our comparisons of the advantages and disadvantages of various evaluation techniques, and matching of the techniques to the needs of aviation systems design, usability testing, heuristic evaluation and cooperative usability evaluations involving experts may be most applicable to aviation systems and could greatly reduce potential usability flaws in production of systems. We recommend an approach involving a cold-heuristic evaluation (without a prototype) early in the design process during the task-function analysis step (Williges et al., 1988) in order to expose potential usability flaws. The use of a usability/domain expert can increase the quality of this evaluation. Another heuristic evaluation could be conducted after a prototype is developed to validate the design. Similar to Williges et al. (1988), we recommend conducting cooperative usability evaluations throughout the design cycle subsequent to the use of heuristic evaluation. These collaborative design reviews should involve experts with different backgrounds (human factors expert, computer scientists, pilot, etc.). The analyses should be conducted after the results of the heuristic evaluations have been applied to the system design, or prototypes, in order to limit the amount of time a group of experts must spend in evaluation meetings. Finally, usability testing should be conducted on a working model of the new component or system developed based on the results of the cooperative evaluations. In general, pilot performance and errors should be recorded as part of testing, but more specifically, usability metrics targeting the principles of concern to the design should be formulated in advance of testing (Wixon & Wilson, 1997). Lastly, any aviation system usability testing protocol should be designed to reflect the impact of new interfaces on the system as a whole (Billings, 1997; Woods & Sarter, 1993). For example, testing should compare pilot workload both with and without a new system integrated in the existing cockpit interface configuration. When the evaluation is complete, results should be preserved for benchmarking the next generation of systems. Several databases already exist including values for errors per flight hour, which could be used as baselines measures for initial testing (Singer, 1999).

## References

Abbott, D. W. et al. (1999). Underpinnings of system evaluation. In D. J. Garland, J. A. Wise, & V. D. Hopkin (Eds.), *Handbook of Aviation Human Factors* (51-66). Mahwah, NJ: Lawrence Erlbaum, Assoc.

Billings, C. E. (1997). *Aviation automation: The search for a human-centered approach.* Mahwah, NJ: Lawrence Erlbaum Assoc.

Dix, A. et al. (1998). *Human-computer interaction* (2nd ed.). Prentice Hall Europe.

Jeffries, R. et al. (1991). User interface evaluation in the real world: A comparison of four techniques. In *Proceedings ACM CHI `91 Conference on Human Factors in Computing Systems* (119-124). New York, NY: ACM.

Kaber, D. et al. (2002). Improved usability of aviation automation through direct manipulation and graphical user interface design. *The Int. J. of Avia. Psych.*, 12(2), 153-180.

Newman, W., & Lamming, M. (1995). *Interactive system design.* Boston, MA: Addison-Wesley.

Nielsen, E. & Phillips, V. (1993). Estimating the relative usability of two interfaces: Heuristic, formal, and empirical methods compared. In *Proceedings of ACM INTERCHI'93 Conference on Human Factors in Computing Systems*, 214-221, New York, NY: ACM

Singer, G. (1999). Filling the gaps in the human factors certification net. In S. Dekker & E. Hollnagel (Eds.), *Coping with Computers in the Cockpit* (87-107). Brookfield, VT: Ashgate.

Virzi, R. A. (1997). Usability inspection methods. In M. Helander, T. Landauer, & P. Prabhu (Eds.), *Handbook of Human-Computer Interaction* (705-715). New York, NY: Elsevier.

Williges, R. C. et al. (1988). Software interfaces for aviation systems. In E. Wiener, & D. Nagel (Eds.), *Human Factors in Aviation* (463-493). San Diego, CA: Academic Press.

Wixon, D., & Wilson, C. (1997). The usability framework for product design and evaluation. In M. Helander, T. Landauer, & P. Prabhu (Eds.), *Handbook of Human-Computer Interaction* (653-685). New York, NY: Elsevier.

Woods, D. & Sarter, N. (1993). Human interaction with intelligent systems in complex dynamic environments. In D. Garland & J. Wise (Eds.), *Human Factors and Advanced Aviation Technology* (107-110). Daytona Beach, FL: Embry-Riddle Aeronautical University Press.

# A «Combinatory Evaluation» Approach in the Case of a CBL Environment: The «Orestis» Experience.

*Karoulis Athanasis*       *Demetriadis Stavros*       *Pombortsis Andreas*

Dept. of Informatics – Aristotle University of Thessaloniki
PO Box 888 - 54124 Thessaloniki – Greece

karoulis@csd.auth.gr       sdemetri@csd.auth.gr       apombo@csd.auth.gr

## Abstract

In this paper several interface evaluation methods are correlated, applied in the case of a Computer Based Learning (CBL) environment, during a longitudinal study performed in three European countries, Greece, Germany, and Holland, and within the framework of an EU funded Leonardo da Vinci program. A «combinatory evaluation» approach is argued to provide in the case of complex environments, like educational ones, the best results. In the case studied, a combination of expert-based approaches with usability testing is dealt with, and the «ideal» combination is investigated, in order to consume the fewest possible resources by achieving maximum outcome for the assessment of the software. Both approaches, expert-based and empirical, have been found to perform adequately, although the empirical methodologies are always preferable. So, this study concludes by proposing a combination of an expert-based approach during the early stages of the design cycle with an observational evaluation, performed by the instructors and/or the designers, that proved to provide the best cost/performance factor.

## 1. Introduction

Computer Based Learning (CBL) environments have been used since the early days of utilisation of computer technology in education. A CBL environment is a piece of software that cognitively covers a particular domain and provides the student with all the means to gain knowledge on the domain. This definition implies two assumptions, that firstly the cognitive coverage and the presentation of the domain is sound (theory, exercises, simulations) and secondly the learner can interact with the piece (interactivity with the software, multimedia elements); in other words, there is a communication channel in order for the student to acquire the offered knowledge.

This study concerns the evaluation of the usability of a particular CBL piece, named "Orestis", which has been produced by the multimedia laboratory of the department of informatics of the Aristotle University of Thessaloniki. Three European countries participated in this assessment applying ten evaluation sessions utilizing six different methods belonging to two methodologies: expert-based and empirical (user-based). The scope of the study is to investigate the application of the combination of several evaluation methods applied to the same piece of software. It is argued that a «combinatory evaluation» approach provides the best results, in the case of complex environments such as educational ones. There seem to be agreement on this issue as regards the researchers in this field (Nielsen, 1994; Lewis & Rieman, 1994; Karoulis & Pombortsis, 2000). This study confirms this result and proposes a combination of an expert-based method with an observational evaluation.

# 2. Background

The main evaluation methodologies researched in this study are the expert-based and the empirical (user-based) evaluation. In an expert-based approach, a multidisciplinary team of experts (HCI, cognitive and domain experts) are asked to judge the interface and pinpoint problems that potential users will encounter during their work in it. In a user-based approach, users are observed during their work with the interface, or are given questionnaires and/or interviewed. Expert-based methodologies are said to be applicable early on and consume fewer resources, however they cannot predict the users' reactions precisely. On the other hand, empirical methodologies demand significant resources and can be applied only at a certain maturity level of the software, however they can elicit valuable qualitative results to improve the piece. So, this work attempts to investigate the «ideal» combination of both methodologies, in order to consume the fewest possible resources by achieving maximal outcome for the assessment of the software.

Four expert-based interface evaluations have been performed, the modified Cognitive Graphical Jogthrough (Karoulis et al., 2000), a phenomenographical approach (Marton, 1988), a formal expert review (Nielsen, 1993), a questionnaire-based expert survey using the QUIS questionnaire (Shneiderman, 1998), and six empirical: two observational, three survey (questionnaire-based; one is QUIS) and one in a usability laboratory.

# 3. Methodology

As all applied evaluation methods are very different in their nature and they provide qualitative as well as quantitative data, a minimalistic approach has been followed. Only quantitative data has been chosen for elaboration and the problems unveiled were categorized and grouped into more abstract problem categories, so that they could be compared and elaborated. Only the minimum common set of usability problems that all methods were able to unveil has been considered. After this categorization, a set of twenty questions belonging to the five main categories emerged: navigation and orientation, manipulation of the interface elements, multimedia information (quantity and quality), help, and use of exercises. The opinions of the evaluators are for every question represented by their mean values (MV). It is known that the mean value is sensitive to outliers, so it has been preferred over the median in order not to exclude any evaluators' opinion. As the main scope of this work was the comparative assessment of the used methods, the Pearson r correlation coefficient has been calculated between the various sets of data. (see Table 1)

**Table 1:** The 20-questions set, the mean values and the Pearson r correlation coefficient

| | Expert based | | | | Usability testing | | | | | | | | | Pearson | |
|---|---|---|---|---|---|---|---|---|---|---|---|---|---|---|---|
| | CGW | Phen | TZlexp | QUIS | TZlus | Rota-ob | Rota-qu | Holl-ob | Holl-qu | Erg-qu | MVexp | MVusab | MVgen | | |
| Entering the program | 3,25 | 3,67 | 2,00 | 4,00 | 3,00 | 3,67 | 3,67 | 4,00 | 4,00 | 3,91 | 3,23 | 3,71 | 3,52 | 0,60 | MVexp/MVusab |
| Start screen environment | 3,50 | 3,83 | 3,00 | 4,00 | 2,00 | 3,67 | 3,67 | 4,00 | 3,50 | 3,64 | 3,58 | 3,41 | 3,48 | | |
| Chosing a machine | 3,50 | 4,08 | 3,00 | 5,00 | 4,00 | 4,33 | 4,00 | 4,00 | 4,00 | 4,18 | 3,90 | 4,09 | 4,01 | 0,79 | MVexp/MVgen |
| Chosing a paragraph to study | 3,00 | 3,42 | 3,00 | 4,00 | 4,00 | 4,67 | 4,00 | 4,50 | 4,50 | 4,55 | 3,35 | 4,37 | 3,96 | 0,96 | MVusab/MVgen |
| Image manipulation | 2,50 | 3,58 | 2,00 | 3,00 | 3,00 | 3,67 | 3,67 | 2,00 | 4,00 | 3,64 | 2,77 | 3,33 | 3,11 | | |
| Video and sound manipulation | 2,00 | 4,17 | 2,00 | 2,00 | 2,00 | 3,67 | 2,67 | 2,00 | 3,80 | 3,36 | 2,54 | 2,92 | 2,77 | 0,56 | CGW/MVgen |
| Chapter navigation | 3,00 | 3,75 | 1,00 | 3,00 | 4,00 | 4,00 | 3,33 | 4,00 | 4,30 | 3,73 | 2,69 | 3,89 | 3,41 | 0,37 | Phen/MVgen |
| Navigation in general | 3,00 | 4,25 | 1,00 | 3,00 | 1,00 | 3,00 | 2,67 | 2,00 | 3,50 | 3,73 | 2,81 | 2,65 | 2,71 | 0,30 | TZlexp/MVgen |
| Entering the exersizes | 3,50 | 3,75 | 2,00 | 2,00 | 3,00 | 3,00 | 3,00 | 3,00 | 3,00 | 3,64 | 2,81 | 3,11 | 2,99 | 0,74 | QUIS/MVgen |
| Exersizes | 3,25 | 3,50 | 3,00 | 4,00 | 2,00 | 3,67 | 3,33 | 4,00 | 2,50 | 3,91 | 3,44 | 3,23 | 3,32 | | |
| Controling the answers | 3,75 | 3,50 | 2,00 | 3,00 | 1,00 | 3,67 | 3,67 | 3,00 | 3,00 | 3,82 | 3,06 | 3,03 | 3,04 | 0,81 | TZlusab/MVgen |
| Using the construction exers. | 2,00 | 2,92 | 2,00 | 2,00 | 2,00 | 2,67 | 2,67 | 3,00 | 2,00 | 3,73 | 2,23 | 2,68 | 2,50 | 0,87 | Rota-obs/MVger |
| Help in the constr. Exersizes | 3,75 | 3,00 | 3,00 | 2,00 | 3,00 | 3,33 | 3,67 | 3,80 | 4,00 | 3,82 | 2,94 | 3,60 | 3,34 | 0,84 | Rota-que/MVger |
| Choosing a simulation | 3,00 | 3,58 | 1,00 | 3,00 | 3,00 | 2,67 | 4,33 | 3,00 | 4,20 | 4,27 | 2,65 | 3,58 | 3,21 | 0,78 | Holl-obs/MVgen |
| Completing the exersize | 3,00 | 3,50 | 2,00 | 2,00 | 1,00 | 3,00 | 2,67 | 3,00 | 2,50 | 4,00 | 2,63 | 2,69 | 2,67 | 0,77 | Holl-que/MVgen |
| Proof of the comletion | 3,50 | 3,83 | 1,00 | 4,00 | 3,00 | 3,67 | 3,67 | 4,00 | 4,00 | 4,09 | 3,08 | 3,57 | 3,38 | 0,76 | Erg-que/MVgen |
| Calling help | 3,25 | 3,00 | 3,00 | 3,00 | 2,00 | 2,67 | 3,33 | 2,00 | 2,00 | 3,36 | 3,06 | 2,56 | 2,76 | | |
| Using help | 3,00 | 3,00 | 2,00 | 3,00 | 1,00 | 2,00 | 3,00 | 3,50 | 3,50 | 3,27 | 2,75 | 2,63 | 2,68 | | |
| Returning from the help | 2,50 | 3,42 | 2,00 | 3,00 | 1,00 | 1,67 | 2,67 | 2,00 | 2,00 | 3,27 | 2,73 | 2,10 | 2,35 | | |
| Efficiency of the help function | 2,50 | 3,08 | 2,00 | 2,00 | 1,00 | 1,67 | 2,33 | 2,00 | 2,00 | 3,09 | 2,40 | 2,02 | 2,17 | | |

To clarify the correlation between the different applied approaches, the following hypotheses have been stated. Null hypotheses:

$H_{01}$: No statistical correlation between the two methodologies (expert-based and empirical) exists.

$H_{02}$: Only the empirical methodologies provide a high statistical correlation to the total MV of all evaluations.

Alternative hypotheses:

$H_1$: The expert-based and the empirical methodologies provide a statistically significant correlation.

$H_2$: There is at least one expert-based method that provides a statistically significant correlation to the total MV of all evaluations.

## 4. Results and Discussion

The mean values (MVexp and MVusab) of the two methodologies (expert-based and empirical) were firstly considered. The scatter diagram (see Figure 1) of these values shows that they are in linear relationship, the samples of the study were related (the same software piece), and there is an interval scale of measurement. So the Pearson r criterion is applicable.

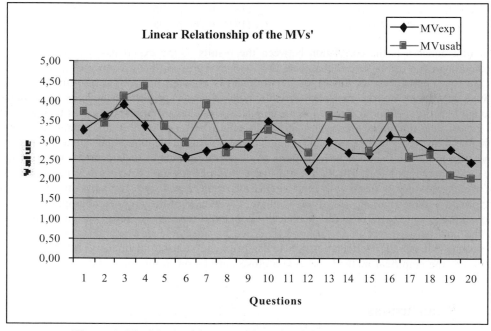

**Figure 1:** The MVs' of the expert-based and the empirical results

Secondly, the Pearson r for the above MVs' was calculated: for 18 degrees of freedom, for two-tailed hypothesis and for significance level of $p<,05$ the Pearson r has a critical value of 0,4438. The test showed that:

MVexp / MVusab                    $r_1 (18) = 0,60 , p<,05$                    (1)

Subsequently, the total MV (MVgen) of all evaluations has been calculated. From now on, all correlations are referred to this MV, in order to homogenize the results. The assumption is made that the total MV of all evaluations expresses the correct number of usability problems, concerning every question in the described context. So, more Pearson r correlation coefficients have been calculated, in different combinations. Below are only those presented that contribute to the stated hypotheses.

| | | |
|---|---|---|
| MVexp / MVgen | $r_2 (18) = 0,79$ , $p<,05$ | (2) |
| MVusab / MVgen | $r_3 (18) = 0,96$ , $p<,05$ | (3) |
| | | |
| CGW (Greece) / MVgen | $r_4 (18) = 0,56$ , $p<,05$ | (4) |
| Phen (Greece) / MVgen | $r_5 (18) = 0,37$ , n.s. | (5) |
| TZIexp (Germany) / MVgen | $r_6 (18) = 0,30$ , n.s. | (6) |
| QUIS (Greece) / MVgen | $r_7 (18) = 0,74$ , $p<,05$ | (7) |
| | | |
| TZIusab (Germany) / MVgen | $r_8 (18) = 0,81$ , $p<,05$ | (8) |
| Rota-obs (Greece) / MVgen | $r_9 (18) = 0,87$ , $p<,05$ | (9) |
| Rota-QUIS (Greece) / MVgen | $r_{10} (18) = 0,84$ , $p<,05$ | (10) |
| Holl-obs (Holland) / MVgen | $r_{11} (18) = 0,78$ , $p<,05$ | (11) |
| Holl-que (Holland) / MVgen | $r_{12} (18) = 0,77$ , $p<,05$ | (12) |
| Ergani-que (Greece) / MVgen | $r_{13} (18) = 0,76$ , $p<,05$ | (13) |

The first concern is the correlation between the results of the expert-based and the empirical methodologies. It can be argued that there is no longer a great difference between the performance of the two approaches, as Equations 2 and 3 show, although the empirical methodologies are always preferable. So, the first result of this study is that expert-based methods have matured enough to provide a fair alternative to the empirical ones, regarding the resources they consume. In general the empirical approaches are believed to be more reliable and qualitatively more valuable, however some biases in the opinions of the participants have been stated which experts have overcome. On the other hand, experts tend to overestimate or underestimate the severity of the encountered problems. This is obvious in the case of the expert review in Germany (Equation 6), which provided a statistically non-significant result, due to the formality in which it was performed and the rigorousness of the evaluators. The case of the phenomenographical approach (Equation 5), which also provided statistically non-significant results, relies, in our opinion, more on the qualitative nature of the method: it is hardly adaptable in a «combinatory evaluation» approach with statistical orientation, yet it has provided valuable qualitative results. Overall, as expected, the empirical methods provide a close correlation to the total MV, as Equations 8-13 show.

## 5.    Conclusions

Based on the aforementioned results, the null hypotheses must be rejected and the alternatives adopted. There is a statistically significant correlation between expert-based and empirical methods, which leads to the conclusion that the expert-based approaches provide a viable substitute to the empirical ones. In addition to this, two of the four applied expert-based methods performed in statistical significant correlation to the total mean value of the questions employed during the study. In generalizing these conclusions, it can be argued that the empirical methods have statistically proved to be of higher performance, yet some expert-based approaches could provide a valuable alternative, especially at the early stages of the design cycle of the software piece, where user testing is difficult to apply, as the corresponding literature reports (Nielsen, 1993; Nielsen & Mack, 1994; Shneiderman, 1998). The expert-based methodologies, as shown in this study, perform fairly and consume far fewer resources. However, user-based approaches are unavoidable in order to elicit valuable information of the users' performance. The results of this study show that the observational methods (Equations 9 and 11) and the usability testing (in a usability laboratory, Equation 8) provide the highest performance, while the questionnaire based methods (Equations 12 and 13) are less accurate. As reported in the literature (eg. Lewis & Rieman, 1994; Nielsen, 1993), this could be due to the biases of the test subjects.

Finally, there are some concerns that arise from this study, as well. The approach followed in the elaboration of the data provides many limitations. The most important of which is that it only takes the quantitative data collected during the study into consideration, in an attempt to compare the different evaluation methods. However, valuable qualitative data, eg. from the questionnaires and the phenomenographical approach, have been neglected. In addition to this, a more statistical elaboration could be performed, in order to investigate other issues as well, such as tendencies or differences. So, the outcome of this study provides only a first indication about the correlation of the methodologies used, yet this correlation is based on several sources over a long term evaluation. Therefore, it can be considered as a first step in statistically elaborating correlations of such issues.

# 6.    Acknowledgments

The authors want to acknowledge and thank all the participants in the three countries who have worked in the «Orestis»-project. There is not enough space in this paper to mention all their names, however a special Thank You must be stated to the Greek and Dutch faculties of law at the Aristotle University of Thessaloniki and the University of Groningen respectively, as well as the Technologiezentrum Informatik (TZI) of the University of Bremen. Last, but not least a great Thank You to all evaluators who participated and to the EU who funded this project.

# 7.    References

Karoulis, A., and Pombortsis, A. (2000 November). Evaluating the Usability of Multimedia Educational Software for Use in the Classroom Using a «Combinatory Evaluation» Approach. *Proc. of Eden 4th Open Classroom Conference,* 20-21 Nov 2000, Barcelona, Spain.

Karoulis, A., Demetriades, S., Pombortsis, A. (2000 February). The Cognitive Graphical Jogthrough – An Evaluation Method with Assessment Capabilities. *Applied Informatics 2000 Conference Proceedings,* 369-373. Innsbruck, Austria. Anaheim, CA: IASTED/ACTA.

Lewis, C. and Rieman, J. (1994). *Task-centered User Interface Design - A practical introduction,* Retrieved in October 2000 from ftp.cs.colorado.edu/pub/cs/distribs/HCI-Design-Book.

Nielsen, J. (1993). *Usability Engineering.* San Diego: Academic Press.

Nielsen, J. and Mack, R.L. (edts.) (1994). *Usability Inspection Methods.* New York, NY: John Wiley & Sons.

Marton, F. (1988). Phenomenography: Exploring Different Conceptions of Reality. In D.M. Fetterman (Edt). *Qualitative Approaches to Evaluation in Education: The Silent Scientific Revolution,* New York: Praeger, 176-205.

Shneiderman, Ben. (1998). Designing the User Interface, 3rd ed., Reading, Mass: Addison-Wesley

# A Comparative Study of Design Solutions for Industrial Process Control Systems

*T. Komischke*

*T. Govindaraj*

*K. Röse*

*M. Takahashi*

Siemens AG
Corporate
Technology, User
Interface Design
81730 Munich,
Germany
tobias.komischke@
siemens.com

Georgia Institute of
Technology
Industrial and
Systems Engineering
Atlanta, Georgia
30332-0205, USA
govindaraj@
isye.gatech.edu

University of
Kaiserslautern
Center for Human-
Machine-Interaction
67653 Kaiserslautern,
Germany
roese@mv.uni-kl.de

Tohoku University
Dept. of Quantum
Science and Energy
Engineering
Aoba-ku, Sendai
980-8579 Japan
makoto.takahashi@
qse.tohoku.ac.jp

## Abstract

Addressing intercultural considerations is increasingly important in the product development of globally active organizations. Since the topic, however, is new in the context of industrial process control, this project aims to investigate what happens when GUIs for process control systems are developed in different countries along the same abstract lines. A multi-dimensional methodology was developed and used to analyze and compare the GUIs. The results reveal similarities and differences in areas such as usability, intercultural dimensions and working methods.

## 1   Introduction

Today it is essential for companies that are internationally active to address intercultural considerations and to implement the findings in their product development processes. In this, they will profit from findings in industrial and university research projects. A look at the HCI conference proceedings from 1999 and 2001 shows that a number of globally active companies, including Siemens, Kodak, Honeywell and Nokia are actively publishing findings on this subject alongside those from universities. In the area of industrial process control the way in which the boundaries between automation technology and information technology have become more fluid since the end of the 1990s demonstrates that certain topics in software ergonomics are becoming increasingly important, although intercultural usability engineering in this area is still only in the initial stages. The objective of the project described here was to analyze the issues involved in intercultural usability engineering in the area of industrial process control. It aims to analyze the outcome when GUIs are developed in different countries on the basis of the same abstract and solution-independent definition.

## 2   Method

A scenario was defined that describes in a solution-independent way the work undertaken by a process operator when controlling a technical process. A distillation column was chosen for the technical process at the core of the scenario. When defining the contents of the scenario, care was taken to ensure that activities typical of process control were described (Komischke, 2001) and that the mapping of these to a GUI posed interesting problems from the industry. The scenario and an explanation of the technical processes were given to three universities in the USA, Japan and Germany. Design solutions were created in the form of prototypes that represent the demands on a process control system contained in the scenario. In addition, the relevant design decisions were documented in writing. The results were than analyzed and compared. Since the conclusion to be drawn from the design differences with respect to cultural tendencies and traditions is somewhat arbitrary due to the limited sample and unknown variance within each particular country, the German contribution represents a kind of control group. Since three student groups could be formed there who worked independently, three solutions were obtained, thereby allowing the variety of design concepts to be observed within a single culture. Where possible, an average was taken of the German results, in order to arrive at an overall German statement. The components upon which the analysis methodology is based are introduced below.

## 2.1 Functionality coverage

The extent to which the prototypes fulfilled the functional demands of the scenario (as defined by Siemens), and thus how far the requirements were met, was observed here. Categories included login, process visualization, navigation, messages & alarming, communication, process control, graphs, support system, logs and global system characteristics. The functional fit to the scenario was recorded on a "fulfilled / not fulfilled" checklist, and the individual questions were attributed weightings according to their relevance for the process. The data were totaled and standardized separately, according to functional areas.

## 2.2 Usability

In this analysis module, the extent to which the prototypes corresponded to the basic principles of dialog design according to ISO 9241-10 (1996) was heuristically evaluated. A list of qualities and system components was created which must be present in all prototypes that fulfill the functional demands of the scenario. The generally formulated criteria of the ISO standard were adapted to the special characteristics of process control and were re-formulated into specific Likert scale items. The result was a matrix of qualities and design principles, where around three items were formulated for each cell. The high detail resolution of the 174 Likert items should achieve the highest possible interrater reliability for qualitative evaluation.

## 2.3 Cultural dimensions

To describe and differentiate the individual participating cultures, the cultural factors of Hofstede (1991) were used. They can be described as separate instances in the dimensions Power Distance, Individualism/Collectivism, Masculinity/Feminity, Uncertainty Avoidance and Longterm-/Shortterm orientation. All countries studied by Hofstede have a particular pattern of distribution of scores in these dimensions. The analysis methodology developed for this project was also based on previous work by Röse (2002) who linked the Hofstedian cultural dimensions with factors in GUIs. Using design properties of the prototypes, a pattern of distribution of Hofstede scores was created. This was achieved by rating 35 bipolar scales. Relative frequencies were calculated from the ratings to obtain standardized values that are comparable to Hofstede's values. Since it is impossible to validate the precise values of the estimate, only a relative pattern comparison was performed.

## 2.4   Evaluation of the protocols

The universities were provided with templates for the documentation of the design decisions. Where possible, the documentation was enumerated, and relative frequencies were calculated. Since the question of the method of implementing functional requirements and real design solutions cannot be completely captured with an a priori tool, the project results were also qualitatively analyzed.

# 3   Results

## 3.1   Functionality coverage

The two analysis modules "Usability" and "Functionality coverage" are not independent of one another, since missing functions cannot be operated, leading to lower ratings values. For this reason, the specific distribution of the fulfillment of functional requirements can relativize the teams' usability results. This is evident in the Japanese solution: clear weaknesses are found for communication functions and can indeed be verified in the usability module's results. In the course of the analysis, it became clear that most of the usability weaknesses of the Japanese solution can be traced back to a failure to apply scenario requirements. As is clear from the use concept, those parts that were considered were well thought through. The general standard of the American solution is significantly higher; the prototype is clickable and most functions were implemented, although this was achieved at the expense of process visualization. The German teams on average achieve a similar high standard in fulfilling the functional requirements, although they do display some weaknesses. In some teams, for instance, the representation of graphs was relatively poor. Although the process of averaging evened out some of the differences, the conclusion can be drawn from the overall data that complete implementation of the scenario was one of the most important design maxims. All of the countries paid less attention to the functionalities "Messages & Alarming" and "Graphs" than they did to the other functionalities. One reason for this could be that these two points are thematically complex and demand comprehensive proposals, which are costly in terms of time and require a lot of practical experience.

## 3.2   Usability

The individual countries placed emphasis on different aspects. The Japanese prototype is impressive primarily as a result of excellent solutions in the message & alarming system, login/rights and the help system; the dialog principles "suitability for the task" and "conformity with user expectations" were best accomplished. The fact that some of the scenario requirements were not accounted for has had a corresponding effect on the pattern of results. Because of that, the team from the USA displayed a higher basic standard. There, communication and login/rights were the points that had been most successfully implemented. Weaknesses were observed in the navigation and the dialog principle "controllability". Both points are interrelated: the prototype allows several windows to be opened and displayed at the same time. This, however, can hinder navigation, overview and efficiency, which can in turn have a negative effect on controllability. The pattern of results averaged from the individual results of the German teams shows a consistently high standard. It should be remembered, however, that by averaging the results, some weaknesses were compensated by the better solutions from other teams, distorting the overall result.

## 3.3 Cultural dimensions

Although it cannot necessarily be assumed that the operationalization of the cultural dimensions was reliable and free from errors, many parallels can be observed with the intercultural differences discovered by Hofstede. Figure 1 shows the values in comparison with Hofstede's findings. In comparison to the American solution, the Japanese solution reveals heightened Power Distance and significantly lower Individualism, while in Germany slightly higher Masculinity and Uncertainty Avoidance values can be observed, whereas Long-Term Orientation is similar. The values of the German teams are quite similar, not only between the teams, but also in comparison with the Hofstedian values.

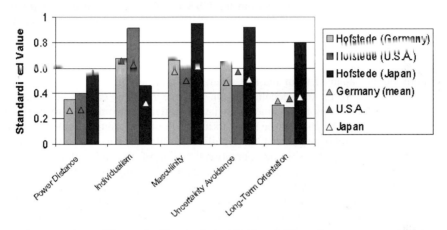

**Figure 1:** Comparison of cultural dimensions

It is interesting to observe that the Japanese prototype clearly deviates from the pattern of the Hofstede scores for the last three dimensions, namely Masculinity, Uncertainty Avoidance and Long-Term Orientation. It became evident that an Indonesian team member had been involved in the development process. According to Hofstede (1991), Japan and Indonesia display significant differences in precisely these three cultural dimensions. This suggests that in this investigation the three values were influenced by this team member.

## 3.4 Working methods

Ergonomic considerations exerted the greatest influence on the Japanese design, although existing solutions were also addressed. While no decision was taken on the basis of aesthetic considerations, despite the impressive appearance, intuition was allowed to come into play to some extent. When taking a decision, discussion was the main medium; moreover creativity techniques were employed more frequently here than in any other team. The team from the USA frequently referred to literature on usability and ergonomics and on existing software solutions. This is reflected in their very transparent and user-friendly prototype. The systematic alignment to ergonomics clearly took priority over aesthetics and intuition. In Germany, one team sought a detailed explanation of the distillation process at the neighboring control engineering institute at the university and profited from this knowledge in successful detailed solutions. On average, they worked most often on intuition. Furthermore, it was the only country that gave aesthetic considerations as the background to individual design solutions. Discussion was by far the most popular method for all teams in decision-making.

# 4 Discussion

The used evaluation methodology could answer the question about the similarities and differences demonstrated by the design solutions. One element that was common to all countries involved was the style of working, including the marked discussion culture. This style is not necessarily surprising, since it was heavily influenced by the external circumstances (working groups at universities). Differences in the design solutions occurred as a result of the different emphases placed by the teams. To generalize the results, the following statements could be made, based on the evaluations. The Japanese system boasts a most attractive appearance, but focuses more on structure rather on the given scenario, so that its usability cannot be completely assessed with the a priori tool developed here. The solution from the USA is easy to use, but this simplicity means that it is not capable of handling all situations. The German solution is function-oriented, without overlooking usability and aesthetics. Were the results at all typical of the countries involved? The question can be answered in the affirmative for the USA and Germany on the basis of the results for the cultural dimensions. The fact that an Indonesian made a significant contribution to the design solution on the Japanese team and that several cultural dimensions showed atypical results for Japan leaves much room for speculation. On the one hand, it cannot be denied that the Hofstedian cultural dimensions for Indonesia deviate significantly from those for Japan in some areas. On the other hand, the analysis of the working style of the group revealed a considerable amount of discussion, which means that the significance attributed to the Indonesian influence should be somewhat reduced. Also the operationalization of the cultural dimensions is not necessarily reliable. Since the influence of the Indonesian team member cannot be clearly determined in this case, the fact that the Japanese solutions produced atypical results for the cultural dimensions should lead to the conclusion that the solution is not typical for Japan. An explanation for the quite similar range of variation within and between cultures could be the very strict requirements from process control with regard to supervisory control systems, combined with the scenario. This may have left little room for creativity and thus concealed possible cultural differences. In future projects, this assumption should be more precisely conceived and investigated. Furthermore, the operationalization of the cultural dimensions must also be tested and, if necessary, optimized.

# 5 References

Hofstede, G. (1991). Cultures and Organizations: Software of the Mind: Intercultural Cooperation and its Importance for Survival. New York: McGraw-Hill.

ISO 9241 (1996). Ergonomic requirements for office work with visual display terminals (VDTs). Part 10: dialogue principles. Berlin: Beuth Verlag.

Komischke, T. (2001). Identifikation branchenübergreifender Kernarbeitsabläufe in der industriellen Prozessführung und Ableitung benutzer-orientierter Gestaltungslösungen. (The Identification of Cross-Industry Core Operating Procedures and the Derivation of User-Oriented Design Solutions). Dissertation. Kassel: University Press. ISBN 3-933146-55-0.

Röse, K. (2002). Methodik zur Gestaltung interkultureller Mensch-Maschine-Systeme in der Produktionstechnik. (Methodology for the design of intercultural human-machine systems in the production technology). Dissertation. Universität Kaiserslautern: Fortschritt-Bericht pak, Nr. 5.

### Acknowledgements

The authors want to thank the project sponsors Siemens Automation & Drives PT1 and Siemens Industrial Solutions & Services MP5 and especially all the students participating in this project.

# Multimodal Interfaces Evaluation
# with Virtual Reality Simulation

*Le Bodic L., De Loor P.*

Software Engineering Laboratory
Technopole Brest Iroise,
Parvis Blaise Pascal BP 30815
F-29608 Brest Cedex
Email : {lebodic,deloor}@enib.fr

*Kahn J.*

France Télécom R&D
38-40, rue de Général Leclerc
F-92130 Issy-Les-Moulineaux
Email:
julien.kahn@rd.francetelecom.com

## Abstract

Our proposal is to present the SIHMM, (Simulator of Interaction between Human and Multimodal Machine), a new approach of conception and evaluation of mobile interfaces. If it is proved that the paradigm of simulation can be relevant, notably for conception and evaluation applications of HCI, our matter is not to reduce the latter to this unique perspective. Indeed, we don't want to define a simulation relative to specific HCI criteria (Scapin, 1986). The aim of this study is to take into consideration of using's context in conception and evaluation of applications of HCI (Jones, 2002) and (Pascoe et al, 2000). These dimensions that widely were made obvious in literature (Dugduale et al, 2000), are doubtless determining but equally delicate to take into account notably for applications intended for usage in mobility. In collaboration with an ergonomists' team from France Telecom R&D, we explore an approach that consists in simulating; an user immersed in an environment and in interaction with a multimodality phone and mobile multimodal devices. This HCI will be endowed with new interaction technologies, several interaction modes, (Bellik, 1995) are possible between user and his machine (vocal, gestural, tangible …).

## 1    Introduction

We want to execute and evaluate, by simulation, the usability of interfaces inside of their use, hypotheses and models on role, and position and characteristics of user. Initially, the configuration and first parameter setting of the model are carried out starting from a review of literature in various disciplines (ethnomethodology, sociology, ergonomics, and software engineering ) on use (Licoppe & Relieu, 2002), utility and usability of multimodality (Coutaz & Nigay, 1995), or of the methods, etc. , with a preference for studies having milked to mobility (Calvet et al, 2001). This work pursues two goals:

- One scientific which consists in developing a multi-agents environment to model multimodal HCM in situation of mobility, by treating the questions relating to the realization of a model which takes into account contextual dimension related to the situation and an user in his global.
- The other industrialist: to have an environment allowing clarifying choices of ergonomics to be able to dialogue with users and developers and if it's possible to explore by simulation new uses, utilities and possible conditions of usability with new devices (a focus).

This evaluation is not more centered on users action but on user and on evaluation by scenario (Ackoff, 1979). Our aim is not to substitute end-users and ergonomists contributions to the design process by simulations. Our project should allow designers to have an overall investigation of some hypothesis of usage and usability during the design process of multimodal mobile communication services. Our work proposes conceptual propositions and some implementations in a software tool. It includes:

- A dynamic model of the HCI's user. This model is composed with:
  - Multimodal perceptive features which merge some informations coming from the environment (including HCI),
  - A dynamic cognitive model which can simulate output of the decision-making processes. This model allows to choose future action to do relative to the goal's users and routine skills,
  - Executive features (action and movement) allowing modification of the environment (including the HCI).

- A multi-agent based model of the virtual environment. Multi-agent technology allows modular and incremental simulation specification and semi-autonomous scenario generation.

This article is structured as following: the first part is talking about user's model, the second part presents multi-agent based environment and functionalities relative to the scenario specification and to finish, the fourth part introduces our simulating tools and first results it allow.

## 2    User's model

The user's model is based on the model of (Kang & Seong, 2001), which is relevant for interactions between human operators and control room equipment. This kind of machine is so complicated and its usability is so specific, that this model is generic, complete and adapatble for numbered HMI. This study show that several factors affect the information gathering and decision making processes of users: the contents of provided information, the way information is provided, and the knowledge of operators. This model allows us to provide outputs considering the complexity of the human cognitive process, but on no account, it could be considered as a plausible real human cognitive process architecture. We have applied this model to our domain as shown on Figure 1.

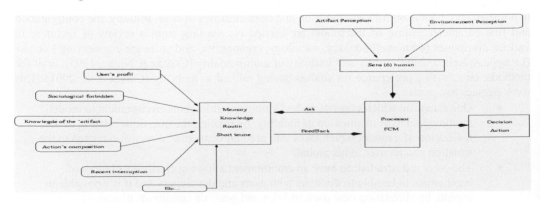

**Figure 1:** Global processes applied from (Kang & Seong, 2001)

The memory part of the model, permit to allow routine implication in the decision process, this implication is shown in (Noizet & Amalberti, 2000). In this study, we focalize explanation on the human processor; the user's processor is based on Fuzzy Cognitive Map (FCM) (Kosko et Dikerson, 1994). This is a descriptive and explanatory model based on concepts and influences relation between those concepts. Figure 2 represents a part of the first release of this model relative to one modality (Le Bodic et al, 2002). A numerical value is associated to each concept and mean its activation degrees. Each arrow traduces the influence between two concepts. It can be positive or negative (source concept value increases or decreases the target concept value).

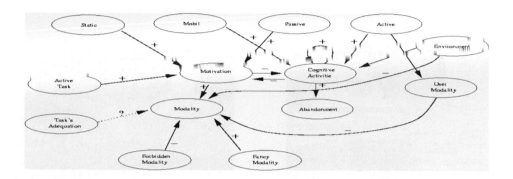

**Figure 2:** FCM

This map enables to have relations between elements of cognitive psychology, environment's elements and user's aspect (as his preference or social forbidden). It includes:

- Perceptive entrances. Indeed, in virtual environment different senses of perceptions (visual, auditive, tactile, smelling and tasting) and their unavailability due to context, action and user's considering elements, are representing by numerical values associated to each senses, this sensor values are entries of our FCM.
  - User environment,
  - User involvement in the task, (as active task and task's adequacy),
  - User physical states, (as mobile, static, passive, active and active modality),
  - User social acceptance of interactive modes, (as forbidden modality),
  - User preferential interactive modes, (as fancy modality).

- Internal concepts setting as follow: the psychological states "cognitive activities" which characterize tiredness of user in the meaning of (Theureau, 2001). Moreover the motivation is his antagonistic concept.
- Decisional exits (abandonment, modality). The most activated concept represents decision of the virtual user. He can give up a goal or choice to change the modality to accomplish it.

## 3    Environment as multi-agent system

The simulator is built with a multi-agent system to respect virtual reality paradigm as shown in (Tisseau, 2001). Using multi-agent technologies facilitates the specification and execution of

663

complex scenarios in complex environment with relevant and no-binary properties. Indeed, an agent can perceive its environment and decide alone to modifying it. Some classes of agents are pre-defined and can be introduced rapidly and easily in any environment at any time. The simulation scenario depends on different interactions between different agents and is not described explicitly. Therefore ergonomists can test different configurations by specifying behavioral constraints on theses agents which have disruptive force on each sense of user (for example, a car which makes noise).

## 4 The toolkit

The figure 3 is screenshots of our software.

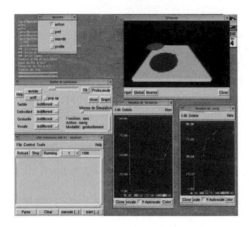 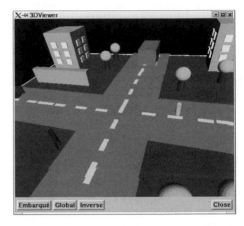

**Figure 3: the simulator SIHMM**

An ergonomist can visualize the simulation and make modification dynamically. He can make modification on physical states and observe decision's changing. For the first picture on the left, the two windows on the right of screenshot show evolution of use of each modality. The second picture on the right is a view of environment of simulation; this environment is built with multi-agent point of view. This tools permit to evaluate influence of environment on user's of multimodality HCI behavior.

## 5 Conclusion

We have proposed a simulation tool, SIHMM, based on multi-agent approach and a virtual user, whose comportment is specified with a fuzzy cognitive map (human processor), and this user interferes with a programmable environment. Now perspectives are to include user's knowledge and to complete our modeling of memory as developed in (Reason, 1990). Therefore, the goal of simulation is not to be proved anymore as shown in study (Dugdale et al, 2000) and allows foreseeing new perspectives for this study.

# 6 References

Ackoff, R. L., (1979). The future operations research is past. In journal of the Operations Research Society, 30, 93-104.

Bellik, Y., (1995). Interfaces multimodales : concepts, modèles et architectures, PhD thesis, Université Paris XI, Unpublished.

Calvet, G., Zouinar, M., Kahn J., (2001). Etude empirique de l'usage de la multimodalité sur un ordinateur de poche. *IHM-HCI2001*.

Coutaz, J., Nigay, L., (1995). Four Easy Pieces for Assessing the Usability of Multimodal Interaction: The CARE properties *INTERACT'95*, 115-120.

Dugdale J., Pavard, B., Soubie,J. L. (2000). A pragmatic development of a computer simulation of an emergency call center. *In Rose Dieng et al., Designing Cooperative Systems, Frontiers in Artifical Intelligence and Apllications*, IOS Press.

Jones, P. M., (2002). Model-based cognitive engineering in complex systems. *Transactions on systems,man and cybernetics*, 32(1).

Kang , H. G., Seong, P. H., (2001). Information therotic approach to man-machine interface complexity evaluation. *IEEE Transactions on systems, man and cybernetics,*31(A), 163-171.

Kosko, B., Dikerson J.A., (1994). Virtual worlds as fuzzy cognitive maps. *Presence*, 3, 173-189.

Le Bodic, L., Kahn, J., De Loor, P., (2002). Utilisation de la réalité virtuelle pour l'évaluation des interfaces multimodales. *Virtual Concept, Estia, 128-131.*

Licoppe, C., Relieu, M., (2002). Réseau, ISBN 2-7462-0516-5, 112-113.

Noizet, A., Amalberti, R., (2000). Le contrôle cognitif des activités routinières des agents de terrain en centrale nucléaire : un double système de contrôle. *RIA 14 PEC2000*, Eyrolles, 73-92.

Pascoe, J., Morse D., Ryan N., (2000). Using while moving : HCI issues in fieldwork environments. *ACM Transactions on Computer-Human Interaction,* 7, 417-437.

Strunk, W., Jr., & White, E. B. (1979). The elements of style (3rd ed.). New York: Macmillan.

Reason, J., E. B. (1990). Human error. Cambridge University Press.

Scapin, D. L., (1986). Guide ergonomique de conception des interfaces Homme Machine, une revue de la litterature.

Theureau, J., (2001). La notion de charge mentale est elle soluble dans l'analyse du travail et de la conception ergonomiques. *Introductive conference of Act'ing / Ergonomia, Cassis.*

Tisseau, J., (2001). Réalité Virtuelle – autonomie in virtuo, *HDR*, Université Rennes I.

# A Remote Camera Control Interface to Decrease the Influence of the Delay Time

*Kazuyoshi Murata, Yu Shibuya, Itaru Kuramoto, and Yoshihiro Tsujino*

Kyoto Institute of Technology
Matsugasaki, Sakyo-ku, Kyoto 606-8585 JAPAN
murata@hit.dj.kit.ac.jp, {shibuya, kuramoto, tsujino}@dj.kit.ac.jp

## Abstract

Nowadays, remote camera control systems on the Internet, such as a web camera, have become popular and widespread. However, there is an unavoidable problem in such network applications, that is, the influence of delay time. In order to decrease the influence, we introduce a remote camera control interface using a drag operation with pre-captured panorama image. We evaluated this interface experimentally and found out that the interface could decrease the influence of the delay time and was usable for the remote camera control.

## 1    Introduction

Nowadays, remote camera control systems on the Internet, such as a web camera, have become popular and widespread. However, there is an unavoidable problem in such network applications, that is, the influence of delay time. When the delay time is meaningfully long, the users cannot control the remote camera as he/she wants. However, it is difficult to eliminate the delay time completely. Therefore, we should take it into account when we construct the remote camera controlling system. In Virtual Dome (Hirose et al., 1991) system, the remote camera gathers the images surrounding it and the graphical workstation, in local site, generates spherical image using these images and displays it on the LCD of the head mounted display. Because these works independently, the influence of delay time can be decreased. However, these systems are not appropriate to a simple network application, like a web camera, because they often need some special input and output devices or workstations. In this paper, we focus on the delay time in the remote camera control and decrease it without using special equipments.

## 2    Remote camera control

A web camera is one of the most popular applications using the remote camera control. The user can pan or tilt the remote camera and change the zoom level of it. Following two interfaces have been commonly used for the web camera operation.

- Button interface:
  There are usually four direction buttons with arrow label and two buttons for zoom on the display. The user pushes down the buttons by a mouse to turn the remote camera and to change the zoom level.
- Scrollbar interface:

There are usually both horizontal and vertical scrollbars on the display. The user can control the remote camera with changing the position of the scroll box of scrollbar with a mouse. In addition, there is another scrollbar for zoom operation.

In the remote camera control, the delay time is caused by following three factors.

- Delay time with mechanical action of the remote camera
- Delay time with encoding and decoding video image data
- Delay time by transmission between the camera server and a client

However, it is difficult for the user to distinguish these delay time. In this paper, "delay time" is defined as "the time since the user begins to operate the camera, until the video captured image is displayed to him/her."

In our previous work, we explored the influence of delay time on the remote camera control with button interface (Takada et al., 1999). As the result, we found a typical operation manner of the user. When the delay time was perceivable, the user repeatedly pushed the button in short period. As the delay time became longer, each operation period became shorter, the number of operation became larger, and the completion time of a given task became longer. The user could not get visual feedback immediately when there was some amount of delay time. As a result, he/she failed to catch up the designated target because they would overreach or not reach it and repeat a little change of the camera direction. It was important to make up the lack of immediate visual feedback and to decrease the repetitious short operations. In this paper, we introduce a remote camera control interface which allows the user to operate the camera without taking care of the delay time.

# 3 Remote Camera Control Interface

## 3.1 Drag Operation

A drag operation, in this paper, makes that the user control the remote camera by dragging the video image on the local display window. A way of drag operation is shown in Figure 1. If the user intends to move the box on the top left corner (Figure 1(a)) to the center of the video image, he/she drags the image in the bottom right direction as shown in Figure 1(b). The user drops the image when the box comes in the center of the video image as shown in Figure 1(c). Then, remote camera begins to change the direction and captures the designated video image (Figure 1(d)). The user can also use the zoom function with a wheel button of the mouse. The remote camera zooms in or out with the rotation of the wheel instead of operation of neither zoom button nor scrollbar.

The remote camera does not work until the user drops the image. Therefore, the user can controls the remote camera without taking care of the delay time and easily grasps the state of the video image after an operation end. Furthermore, with the drag operation interface, the user can turn the remote camera to desired direction in single operation but he/she can not do that with button or scrollbar interface. However, when the target is out of the camera view, the drag operation does not work well because the user drags the current video image only and the user can not watch the out side of the image. So, as an improvement of drag operation, a panorama image is combined with the drag operation.

(a)          (b)

(c)          (d)

**Figure 1:** Drag operation

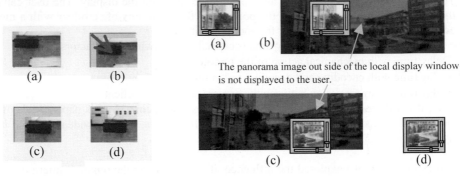

(a)     (b)

The panorama image out side of the local display window is not displayed to the user.

(c)                    (d)

**Figure 2:** Drag operation with panorama image

## 3.2 Panorama Image

The panorama image is the pre-captured static image covering the whole range of which the remote camera works. The panorama image is used instead of the online video image when the user begins to do drag operation. After the drag operation, the remote camera turns to the designated direction and the online video image is shown on the local display window. By using the panorama image, the user can find the target out of the local display window during the drag operation. An example of drag operation with panorama image is shown in Figure 2. Before the operation, an online video image is presented in the local display window as shown in Figure 2(a). When the drag operation begins, the panorama image is presented instead of online video image as shown in Figure 2(b). The user operates with seeing the panorama image as shown in Figure 2(c). After the operation, the remote camera turns and the online video image is presented instead of the panorama image as shown in Figure 2(d). The panorama image is also usable for changing the zoom level.

In order to use the panorama image, it is necessary to create it before the operation. From this reason, the panorama image is not a real time image but it does not matter. The user does not need a real time image so often because the user usually uses static landmarks to decide the camera view.

## 4 Experimental Evaluation

In order to evaluate the effect of the proposed interface, we conduct an experiment which is described below.

### 4.1 Conditions

The experiment was conducted in the environment as shown in figure 3. A client PC for the operation in a room and a camera server PC was set in another. The floor of the server room was divided into 7 x 6 domains. There were five boxes on the domains. An object was put on a high box and another one on a low box. Other three objects were put on the floor domains directly. An easy formula was written on the each object. Subjects were asked to search these objects by operating the remote camera and to write the formulas on the answer sheet. The delay time was 0.35, 0.7, 1.5, and 3 seconds. An experiment task was completed when the subject wrote down all formulas. We had 9 subjects and three kinds of interface, they were the button interface, the scrollbar interface, and our proposed drag operation interface. After the experiment was completed, subjects asked to answer the questionnaire about usability of the interface they used.

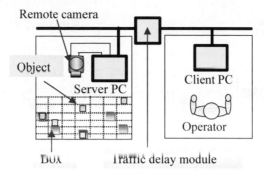

**Figure 3:** System configuration of experiment

## 4.2 Result

The result of the task completion time is shown in Figure 4. The task completion time of the button interface increased due to the delay time, and statistical significance was found between 0.35 and 3 seconds delay time (p<0.05). By contrast, the task completion time using the drag operation interface hardly increased due to delay time. However, when the delay time became 3.0 seconds, the task completion time increased significantly (p<0.05). The task completion time of the button interface and that of the drag operation interface was almost the same at 0.35 seconds delay. However, the difference of the task completion time between the button interface and the drag operation interface became larger as delay time increased. At the 3.0 seconds delay time, the task completion time of the drag operation interface was less than that of the button interface significantly (p<0.05). The task completion time of the scrollbar interface was always larger than other interfaces. These results indicate that drag operation interface is less influenced by delay time than other interfaces.

The result of the time duration per directional operation is shown in Figure 5 and the number of directional operation is shown in Figure 6. The time duration per directional operation of button interface was about 0.5 seconds at every delay time and that of scrollbar interface was less or equal to about 0.2 seconds. These were shorter than that of drag operation significantly (p<0.05). By contrast, the number of operation of the button interface and scrollbar interface was larger than that of the drag interface. These results indicate that the subjects needed to perform repetitious short operations to operate the remote camera under the influence of delay time when the button interface or the scrollbar interface was used. However, by using drag operation interface, it became unnecessary for them to perform repetitious short operations.

The result of the questionnaire is shown in Figure 7. Score-7 means "Very usable" and score-1 means "Not usable at all". The score of the button interface was decreased as delay time increased significantly (p<0.05). By contrast, the score of the drag interface hardly decreased. At 0.35 seconds delay time, no statistical difference was found between the score of the button interface and that of the drag operation interface. However, as delay time increased, the difference between the score of button interface and that of the drag operation interface also increased. Statistical significance was found between the score of the button interface and that of the drag operation interface at 0.7 seconds and above (p< 0.05). The score of the scrollbar interface was always very low. These results indicate that the usability of the drag operation interface is not influenced by delay time.

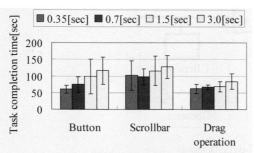

**Figure 4:** Delay time and task completion time

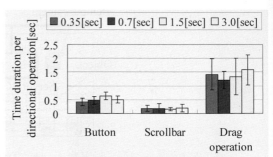

**Figure 5:** Delay time and time duration per directional operation

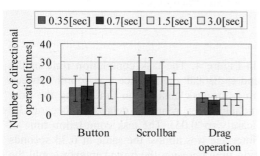

**Figure 6:** Delay time and number of directional operation

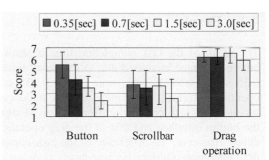

**Figure 7:** Questionnaire result

# 5    Conclusion

In this paper, the drag operation interface for the remote camera control was introduced to decrease the influence of the delay time. With the proposed interface, the user drags the pre-captured panorama image to control the remote camera. We compared the drag operation interface with the traditional interface experimentally. As the result, when the drag operation interface was used, the delay time did not increase the number of repetitious short operation and the task completion time. That is, the drag operation interface can decrease the influence of delay time on remote camera control.

## References

Hirose, M., Hirota, K., Kijima, R., Kanno, M., Hayakawa, K., & Yokoyama, K. (1991). A Study on Synthetic Visual Sensation through Artificial Reality. *Proceedings of the Seventh Symposium on Human Interface English Section*, 675-682

Takada, K., Tamura, H., & Shibuya, Y. (1999). Influence of Delay Time in Remote Camera Control. *Human-Computer Interaction Ergonomics and User Interface*, 421-425

# Mobile and Stationary User Interfaces
# – Differences and Similarities Based on Two Examples

*Erik Gøsta Nilsson*                    *Odd-Wiking Rahlff*

SINTEF Telecom and Informatics
P.O.Box 124 Blindern, N-0314 Oslo, Norway
egn@sintef.no                           owr@sintef.no

## Abstract

In this paper we present important similarities and differences between user interfaces on mobile and stationary equipment exemplified through two similar applications that let the user review and add annotations through a map-based user interface. First we present core differences between user interfaces on mobile and stationary equipment in general, with focus on exploitation of contextual information when designing mobile user interfaces. Then we present the two example applications, followed by an analysis and generalization of the core differences between the stationary and mobile variant. Finally, we give some topics for future research in this area.

## 1    Characteristics of Mobile and Stationary User Interfaces

One of the most evident differences between user interfaces on mobile and stationary equipment is the difference in screen size. This causes an actual dialog on a small screen to either be simplified or divided into smaller parts (using a tab folder type mechanism or a number of separate dialogs). The small screen size also usually makes it impossible to have more than one dialog available at a time. This influences both how a single application is designed, and how the user works with more than one application in parallel. These restrictions on mobile user interface design make it even more important to design user interfaces that are highly adapted to the users' tasks.

The limited repertoire of available interaction mechanisms is another evident difference between stationary and mobile equipment. The absence of keyboard and mouse on most equipment is most striking. Some Personal Digital Assistants (PDAs) have a miniature or foldable physical keyboard, but in these cases the user has to sit down to be able to type – which makes the usage situation less mobile. For equipment without physical keyboard there is a choice between an onscreen keyboard and a special area for writing single letters using special strokes. Either way it only makes it convenient to enter small amounts of text, which influences the design of mobile applications. The most common mouse equivalent is to use a pen. In actual use, the differences between pen based and mouse based interaction are larger than they seem. The traditional difference between select and open on a mouse based user interface has to be implemented differently on a pen based one because fewer event types are available. In mouse-based user interfaces, mouse position is often used to inform the user through cursor changes, tool tips etc. Movement is not recognized in a pen-based interface, so this type of information must be made available through other means. If we also consider mobile phones, the differences are even larger, both regarding screen size, interaction mechanisms and interface controls. Another difference is that mobile users usually use a computer for *secondary* task support, often combined with having the hands occupied elsewhere.

The context in which a mobile user operates changes much more rapidly than for a stationary user. An important challenge when designing mobile user interfaces is to exploit knowledge about these changes to enhance the user experience. The context changes are multidimensional – and sometimes rapid – comprising location, light, sound, task, network connectivity, and possibly biometrics (Herstad et al., 1998, Rahlff et al., 2000, Rahlff et al., 2001). Reflecting and utilising contextual information causes the need for new types of user interface designs for mobile users (Forman et al., 1994), covering the contextual aspects, and ways for the UI to adapt to changing contexts (Schmidt et al., 1999). This reveals a need for something more than just down-scaled versions of desktop-UIs, which is the case for many applications and services available on mobile computers today (Kuutti, 1999). It is also important to distinguish between augmenting existing applications with context information, and developing *new* types of applications and services that are enabled or driven by available contextual information. By this we mean applications and services that it usually does not make sense to offer to a stationary user, mainly because the context does not change very much (e.g. a map position moving as the user moves).

## 2    Examples: AnnI & Mocado – the Stationary & Mobile Annotator

**AnnI** (*Annotations through Internet*) is a prototype service that we developed to show how public users might exploit a map-based application on Internet. AnnI lets common arbitrary users browse and add information about events and arrangements that take place in a given area and period of time. The web application has a map-based user interface. It has two main modes: browsing and insertion. This figure shows and explains AnnI in browsing mode:

Icons representing various events

Selected event

Control panel

Basic information about chosen event

Target groups for events – used for filtering

Event categories – used for filtering

Time span in which events are shown – used for filtering

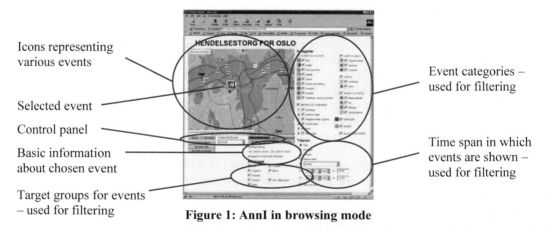

**Figure 1: AnnI in browsing mode**

In browsing mode, the user may focus on a number of events. This is partly done by *filtering* based on categories, time and target groups, and partly by *navigating* to the desired area in the map (using zoom and pan functions). Events are shown as icons in the map view. The design of the icons indicates the main category for the event. When the user selects an event, basic information (*short description*, *time* and *organizer*) is shown. If the user selects the details button in the control panel, all details about the selected event is shown in a separate forms-based view.

In event insertion mode, the cursor changes to a red X, which is used to mark the position of the new event. Once the location is given, the user must enter event details in a separate forms-based view. When the submit button is pressed, an icon for the new event is added to the map presentation. The information is transferred to the web server to make it available to other users.

**Mocado** (*MObile Contextual Annotation DemonstratOr*) is a prototype application that we developed to show how contextual information might be exploited to enhance the user experience in a mobile application. It lets a user with a PDA connected to a GPS-receiver (Global Positioning System) add and review multimedia (text, sound snippets and pictures) annotations connected to a physical location. Mocado has a map based user interface. "Map" meaning arbitrary two-dimensional projection – including traditional maps, orthophotos (aerial photos that show objects at their true positions, i.e. not skewered or stretched), architect drawings, etc.

When the user moves, Mocado shows the position and the user's track on the map. The position is shown as a small red circle, while the track is shown as a red line (tail) from the position indicator. Annotations in the area are indicated by yellow circles. Once the user is within the selection range of an annotation, it becomes selected. This changes the colour of the annotation to blue. Also, the text and picture for the annotation is shown, and the sound such as a tale told about the location is played back automatically through headphones. The next figure shows the map view in Mocado after the user has made a small stroll in a park, and an annotation has been selected (the position indicator and the selected annotation are the two adjacent circles).

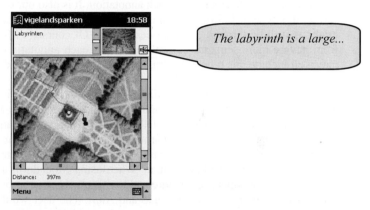

**Figure 2: Mocado map view with position, track and selected annotation**

At the top of the screen, a preview panel shows the text, a thumbnail of the picture for the annotation, and a speaker icon for replaying the annotation sound. The thumbnail picture may be selected to view a larger version of the picture. The user may add his/her own annotations at any location by using a menu choice. Detailed information is added in a dialog. Position, date and time are added automatically, while the user may add text, picture and/or sound details. The user may also record his/her tracks. These tracks may be replayed later at a chosen speed.

The main use of the Mocado is at the location of annotations, but it may also be used off-site. In "manual positioning" mode, all the functionality of the application is available (even position tracking) while the user simulates the changing position by tapping directly on the map. This functionality is essential for off-site entering or accessing remote annotations, and for debugging.

## 3 Discussion

One of the most striking differences between the two example applications is the user interface for finding information. Although the map interface is prominent in both applications, there is much focus on the filtering mechanisms in AnnI as well as much functionality for doing geographical filtering (zooming and panning) – functionality that is not present in Mocado. This difference – AnnI with much focus on filtering information geographically and thematically and Mocado with most focus on information close to the user's current position (and thus with limited or no need for filtering mechanisms) – reflects one of the most important differences between using these applications in a stationary and mobile context. Although it would not be easy to implement filtering mechanisms on the small screen on which the Mocado runs, it is our claim that the mechanisms are not needed because the user's position constitutes so much filtering that the thematic filtering becomes superfluous. That is to say that in the mobile case the user's position is the context, whereas the context has to be "created" in the stationary case.

The user interface for entering information in the two applications are different too, but in this case more on the amount and type of information than how it is handled. While AnnI requests quite detailed information, with much focus on thematic and semantic information, Mocado only requests a title and a short description in addition to the multimedia information. This difference has a strong connection to the different ways the two applications are used to browse information. For the flexible, thematic filtering mechanisms in AnnI to be effective, the relevant semantic information must be entered for each annotation. It is also our claim that this is the *main* reason for adding this type of information in a structural way. Once an annotation is identified, accessing it might as well be through a textual description. Of course, a Mocado user could generally benefit from having more semantic information about each annotation, and a user adding an annotation may well include this type of information as part of the textual description or the audio part.

The two applications also use media differently. In AnnI, all information is alphanumeric, while Mocado in addition uses pictures and sound. Experiences with Mocado show that the multimedia information is very valuable, especially the audible part of the annotations. In fact it has turned out that when using Mocado walking around on-site, sound is *the* most important part. The reason for this is partly that it is the least intrusive part for a mobile user, and partly because sound facilitates hands-free use when walking around. When using Mocado in a different location than where the annotations were added, the picture and text is much more important. It is probable that the same division is the case between Mocado and AnnI. The user would benefit from having pictures as part of the annotations in AnnI, but sound would in many cases likely be more annoying than useful. When sitting down, most users read faster than people talk, so having to wait for a person to say something that the user might as well read is often frustrating. For a mobile user the situation is quite different. To be able to read, the user has to stop, while listening is possible when s/he is walking anyway. This argumentation holds as long as the sound contains equivalent information to the alphanumeric. In AnnI (and of course in Mocado) it would anyway be valuable to use sound for other purposes. E.g. to add a sound sample to a concert annotation is very relevant. Also for pictures, the utilization is different depending on how the picture relates to the annotation. The considerations above about pictures in on-site and off-site use of the Mocado hold as long as the pictures show the position of the annotation. E.g. having a picture taken on the position of the annotation some time in the past may be very relevant also for a mobile user.

## 4    Conclusions and Future Work

In this paper we have showed that the differences between the user situation are just as important as the restrictions on screen size, interaction mechanisms etc. when comparing mobile and stationary user interfaces. To design usable mobile applications, exploiting context changes is very important. There is still much research to do to find design principles for how user interfaces on mobile devices should reflect context changes. We need schemes for characterising different types of context changes and guidelines for how different type of context changes may and should influence the user interface design (Gwizdka, 2000). There is also a need to find which meaningful higher-order context abstractions that can be deduced from first-order context information from sensors. The rapid context changes in a mobile setting cause the need for flexible user interfaces that are multitasking and exploiting multiple modalities, possibly also utilizing semi-transparent head-up displays, gesture recognition etc. Within these fields, a lot of unsolved problems remain. This is also the case regarding methods for evaluating contextual, mobile user interfaces (Følstad et al., 2002), and for cross-platform UI development (Nilsson, 2002a, Nilsson, 2002b).

In the EU project AmbieSense active tags are used as a means for enhancing both the type of context information to be available at the PDAs and the locations where this type of information may be available. This approach facilitates more advanced contextual services.

## 5 Acknowledgements

The work on which this paper is based is funded by the strategic projects AmbieLab and FAMOUS, and the EU IST project AmbieSense. The authors would like to thank Richard Moe Gustavsen from FFI, and Tor Neple from SINTEF who both have contributed to designing and implementing the Mocado application. Thanks also to our partners contributing to the design and implementation of the AnnI application.

## References

Forman, G. & Zahorjan, J. (1994). The Challenges of Mobile Computing. *IEEE Computer*, April 1994. (http://www.cs.washington.edu/research/mobicomp/ghf.html)

Følstad, A & Rahlff, O. W. (2002). Basic User Requirements for Mobile Work Support Systems - Three Easy Steps. *Proceedings of MobIMod'2002, ER/IFIP8.1 Workshop on Conceptual Modelling Approaches to Mobile Information Systems Development*

Gwizdka, J. (2000). What's in the Context? Position paper for workshop on The What, Who, Where, When, Why and How of Context-Awareness. *CHI'2000*

Herstad, J., Thanh, D. v., & Audestad, J. A. (1998). Human Human Communication in Context. *Interactive applications of mobile computing (IMC 1998)*.

Kuutti, K. (1999). Small interfaces – a blind spot of the academical HCI community? *Proc. of HCI International '99*, Lawrence Erlbaum Assoc.

Nilsson, E. G. (2002a). Combining compound conceptual user interface components with modelling patterns – a promising direction for model-based cross-platform user interface development. *Proceedings of DSV-IS 2002*

Nilsson, E. G. (2002b). User Interface Modelling and Mobile Applications – Are We Solving Real World Problems? *Proceedings of Tamodia'2002*

Rahlff, O. W., Rolfsen, R. K. & Herstad, J. (2000). Using Personal Traces in Context Space. Position paper for workshop on The What, Who, Where, When, Why and How of Context-Awareness. *CHI'2000* (http://www.informatics.sintef.no/~owr/Publications/chi2000/ConsTechCHI2000.htm)

Rahlff, O. W., Rolfsen, R. K. & Herstad, J. (2001) Using Personal Traces in Context Space: Towards Context Trace Technology, *Springer's Personal and Ubiquitous Computing, Special Issue on Situated Interaction and Context-Aware Computing*, Vol. 5, Number 1, 2001.

Schmidt, A. et al. (1999). Sensor-based Adaptive Mobile User Interfaces, *Proc. of HCI Interantional '99*, Lawrence Erlbaum Assoc.

# Understanding the Tradeoffs of Interface Evaluation Methods

*Jose Luiz Nogueira and Ana Cristina Bicharra Garcia*

Universidade Federal Fluminense
Rua Passos da Patria, 156 sl 326, Niteroi, RJ, Brazil
bicharra@dcc.ic.uff.br; jltnogueira@hotmail.com

## Abstract

This paper discusses current usability evaluation techniques emphasizing the costs and benefits of each method. Costs consider required time and expertise for applying the method. Benefits include method efficiency and effectiveness to reveal usability problems. The comparison is based on an ideal scenario with available usability experts. Given this expertise is still a scarce resource, we conducted an experiment using software engineers and domain experts as evaluators after a short-period training.

## 1    Introduction

Interface is an embedded part of any software and responsible for the user experience with it. The effort users spend to accomplish their goals may compromise the software successful use. It is through the interface users perceive the functionalities of a software. Consequently, evaluating a software is intrinsically connected to evaluating its interface. When evaluating an interface, we are evaluating the dialog between two different entities: the user and the computational system.

Interface usability is the metric for evaluating interfaces. It addresses not only the adequacy of the software functionalities to the users' needs, but also the ease of use, learn, memorize, adjust to individual needs, task accomplishment and overall users' satisfaction when interacting with a software interface.

Usability methods can be divided into two categories:

- Direct methods: users evaluate the interface; and
- Indirect methods (Inspection methods): there are no real users involved in the testing process. Evaluators simulate the user's behaviour when interacting with an interface.

There are tradeoffs that should be considered when selecting a method (or set of methods) to evaluate an interface. This paper presents a framework for letting costs and benefits explicitly stated to help interface designers consciously decide and also justify the costs to owners in a software development environment.

Furthermore, we discuss the effectiveness of each method considering a scenario without usability expert (UE). We observe a team of non- experts evaluating the usability of web-based software to gather data for the Brazilian Institute of Geography and Statistics (IBGE) in deployment phase.

## 2    Evaluation Methods

### 2.1    Indirect Methods

Indirect methods consist of letting evaluators simulate users' behaviour when interacting with a software interface to predict probable usability problems. This paper addresses two very popular indirect methods: Cognitive Walkthrough (CW) and Heuristic Evaluation (HE).

CW (Wharton, 1994) has its root in exploratory learning theory. Evaluators, after a careful study on software functionalities and users' profile, create task descriptions to be followed during evaluation. They walk through the interface, following their task descriptions, annotating positive and negative interactions. It requires a great deal of time and is biased towards easy of use and easy of learn usability criteria. The method can be applied during any design stage, but it is more suitable for the early design stages.

HE (Nielsen, 1994) consists of inspecting an interface as a whole looking for desirable and undesirable behaviours according to a list of heuristic rules or their own experience. The method suggests using at least two experts. The evaluators separately navigate the interface and annotate possible problems they have noticed. Afterwards they share their notes and generate a list to be shown to the interface designer. It takes generally two to four weeks to complete the job for moderate complexity software. Generally two evaluators identify 50% of interface main problems, three identify 60% and fifteen around 90% (Nilsen, 1994). The method can be applied at any design stage, even over an interface sketch. It is a very simple and fast method to use. However there are some drawbacks related to the subjectivity of the method and the lack of actual users in the process. Additionally, the list of mistakes may sound to interface designers just as a set of different opinion.

## 2.2 Direct Usability Methods

Direct methods utilise a set of actual users to evaluate the usability of the interaction. This paper includes two fundamental methods, such as Think Aloud (TA) and Questionnaire (Q).

TA (Nilsen, 1992) is the most popular usability testing method due to its simplicity and intuitive application. It has its roots in experimental psychology (Duncker, 1945). The method aims to let users' cognitive model emerge. Evaluators ask users to verbalize their thinking while interacting with a software interface. Evaluators work as facilitators. They observe users interacting with an interface, making comments to instigate users to talk. Users work as co-evaluators, verbalizing their thinking during interactions. Although each session is very time consuming, just a few subjects are needed. Nielsen claims five subjects are enough to evaluate an interface using think aloud method. The method imposes a high cognitive load to users because they have to decide, do and verbalize their actions at the same time [Preece, 1994]. In addition, users may verbalize their rationalization of actions instead of their rationale.

Questionnaire (Q) is the traditional method of evaluating anything. Questionnaires are sent to a significant sample of the population (people that will be using the software). Questions should be carefully elaborated to allow interaction problems to be reported. After collecting the questionnaire, data gets a statistical treatment. Results are generally easy to interpret. Users are likely to find problems software engineers do not foresee.

## 2.3 Evaluation Methods' Tradeoffs

The selection of an evaluation method to a specific software development scenario must consider the pros and cons of each method. Making these tradeoffs explicit helps interface designers rationally decide and also justify costs to their client. Table 1 presents a summary of the features to be considered when selecting an evaluation method.

Table 1: Evaluation method analysis framework.

| | Cognitive Walkthrough | Heuristic Evaluation | Think Aloud | Questionnaire |
|---|---|---|---|---|
| Required People | 1-4 UEs<br>No users | 3-5 UEs<br>No users | 1 UE<br>3-5 users | 1 UE;<br>Sample of users<br>(at least 30) |
| Evaluation Duration | 1 week | 2-3 weeks | 1-2 weeks | Variable |
| Milestone | Task description planning | Results acceptance | User verbalization | Questionnaire Elaboration |
| Best suitable for | Predicting easy of use and ease of learn issues | Predicting general problems | Verifying easy of use, ease of learn | Verifying user satisfaction and productivity |
| Best applied during | Early design phase | Any development phase | Any development phase | Deployment phase |
| Method efficiency* | Average-High | Average-High | High | Average |
| Effort during test | Annotate successful & unsuccessful interactions | Instantiate general expertise in a specific case | Make thinking explicit | Answer questions |
| Post evaluation work | Merge evaluations | Merge evaluations | Merge evaluators results | Statistical analysis |

*Note: CW and HE point potential problems that may or may not occur while TA and Q point to users' detected problems.

# 3    Case Study

IBGE was implanting a new web-based system to collect data to a national research on income distribution. The system was developed in the company to be deployed throughout the country. The system runs in the company's intranet through the MSExplorer Browser.

We used a set of forty volunteer employees: eleven of them were software engineers (SE) and the rest were domain experts (DE). Only four of them did not have a college degree. We gave a one-day workshop explaining all techniques to each participant. Even though the tasks were not complex, they all have problems understanding CW because they were not familiar with GOMS (Card, Moran, & Newell, 1983). Since the company's expertise was in questionnaire, elaborating one for our experiment was not a challenge.

Participants have to evaluate the software using the four methods. In addition to the fast training, a written material with each method's guidelines was available. Their task was to annotate usability problems. They all applied each method in different order. We noticed a slightly better performance when they start with heuristic evaluation.

## 3.1.1    Insights on Test Duration

We separated method duration, as presented in Table 2, in three different components:

- Preparation time: time spent preparing for method application.
  - For HA, this is the time used for reviewing the heuristic guidelines.
  - For WC, this is the time for developing the task descriptions;
  - For TA, this is the time for creating the scenarios;
  - For Q this is the time for elaborating the questionnaire.
- Application time: time spent actually applying the method.
- Data analysis time: time spent transcribing (in the case of TA) and analysing the results.

678

**Table 2:** Average time spent per evaluation session considering the two types of evaluators.

| Method | Preparation | | Application | | Data Analysis | |
|---|---|---|---|---|---|---|
| | SE | DE | SE | DE | SE | DE |
| HA | 110 | 290 | 271 | 518 | 79 | 137 |
| WC | 243 | - | 238 | 491 | 36 | 87 |
| TA | 243 | - | 99 | 290 | 492 | 1363 |
| Q | 371 | - | 144 | 430 | - | 140 |

### 3.1.2 Method Effectiveness

We counted the number of usability problems encountered using each of the method. We also considered the number of mistakes as a function of the number of SE (1 to 11 SEs), as well as domain expert people (1 to 29 DEs). As illustrated In Figure 1, in the absence of experts TA was the method with the best performance. SE and DE, working as co-evaluators presented similar results since their role did not require expertise as described by the method. The results also suggested, evaluators must have at least computer science background (SE) to obtain a cost effective result when using HE and CW. It also suggests the unsuccessful use of DEs as evaluators. We used a single evaluator when applying TA varying the number of co-evaluators (users) "thinking aloud". Questionnaire was elaborated by a single evaluator (SE) and applied to all 40 people.

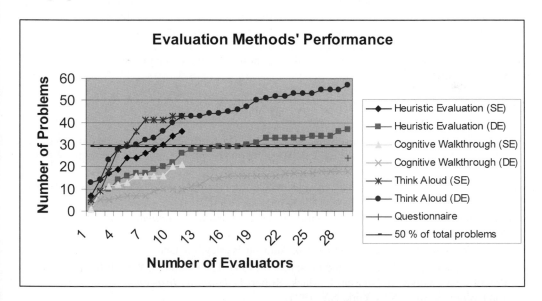

**Figure 1:** Evaluation Method performance.

Although questionnaire accounted to a small number of problems discovery, it is the only method that accounts for user satisfaction in a broad sense; i.e., looking at the users' community instead of a couple of users as in TA.

# 4    Final Remarks

This paper presented a comparison among four usability evaluation methods pointing out the costs and benefits of each one. The requirements for applying each method were drawn considering the availability of usability experts. Costs were discussed in terms of time and effort, while benefits were discussed in terms of the effectiveness of the method, i.e. number of problems each method reveals.

We also presented a case study applying the four different methods to evaluate an intranet-based software to be used by the Brazilian census organization.  The software is in the deployment phase. There was no usability expert available. We trained software engineers and other company people to perform the tests acting as evaluators.

Test results suggested a potential advantage of HA over the other evaluation tests, but using a bigger number of evaluators (five at least). However, since the results of HA may sound more like an opinion to the software developers, there may have a political problem accepting the list of problems provided by non-usability experts.

No matter the method usability tests pay.  When usability experts are not available, software engineers with basic training may conduct evaluation tests.  The number of evaluators increases to get average evaluation coverage. There is almost no use to apply any of the method using lawman (DE) as evaluators.  The number of problems found is much smaller and less relevant than the ones found by SEs. Heuristic evaluation seems the best cost-effective method in this scenario.

The results from WC may be low due to the need of a better training. All evaluators complaining Our observations should not be faced as a conclusive.  They are indications of benefits of developing usability test even with software engineers. The results also helped justify to clients the cost effectiveness in using UEs.  They would save money for the company and provide better and trustworthy results.

# References

Duncker, Karl(1945) On problem-solving, in Dashiell, John F.: Psychological Monographs, The American Psychological Association, Inc., Washington DC, vol. 58, pp. 1-114. (reprint, Westport, Conn. Greenwood, 1976).

Nielsen, J. (1992). Evaluating the thinking aloud technique for use by computer scientists. In Hartson, H. R. and Hix, D. (Eds.): Advances in Human

Nielsen, J. and Marck, R. L. (1994). *Usability Inspection Methods Computer*. John Wiley & Sons, New York, NY, 1994.

Preece, J. (1994), *Human-Computer Interaction,* Addison-Wesley, England.

Wharton, C., Rieman, J., Lewis, C. and Polson, P. (1994). The Cognitive Walkthrough Method: A Practitioner's Guide. In Nielsen, J.; Mack, Robert L. *Usability Inspection Methods Computer*. John Wiley & Sons, New York, NY, 1994.

# Live the Vision
## character- and plot-driven scenarios in case-based material

*Rikke N. Orngreen, Ph.D.*

HCI Research Group, Department of Informatics
Copenhagen Business School, Denmark
orngreen@cbs.dk

## Abstract

The use of scenarios as discussed in this paper is adapted to the area of education by inspiration from the movie industry's use of two tools: *treatments* and *storyboards*. It is the work with plots and character in treatments for movie-scripts, which is in focus here. The presented research looks at experiences from use of character- and plot-driven scenarios in multimedia applications. The applications are built on a case-based learning paradigm, more specifically teaching cases in business education. In case-based learning a firm impression of the situation presented and its implicated characters are very important. Using knowledge from use of treatments regarding character and narrative plots supported the analysis and design of these elements as well as the content design. The presented research shows that the use of these scenarios sustained motivating factors (i.e. motivating use of the application), by having visionary descriptions of the plot and characters in the case-situation. However, it was very difficult for the project team to adapt and keep such visionary descriptions alive with respect to the characters using the case – the students.

## 1 Research Design and Problem Area

The paper presents the research design and the theoretical problem area, outlining briefly the empirical basis of the research (a multinational EU project) and the theoretical foundation of both teaching cases and scenarios. The treatment tool and experiences with character- and plot-driven scenarios in design of case-based educational material are then discussed and concluded upon.

### 1.1 Teaching Cases and BUSINES-LINC E-Case Series,

Teaching cases in business education present companies, and the problems/opportunities they are facing. (Maufette-Leenders, Erskine & Leenders, 1997, Heath, 1998) The primary characteristic of teaching cases is that the cases are based on real events taking place in an existing company. Very often the case description follows a decision-making situation. (Mauffette-Leenders et al., 1997) A case can be open ended, i.e. a given problem is presented, but no solution provided, or a closed case, which means the problem and the solution, which the company chose are presented. A case is often prepared by students individually, then in smaller groups and finally discussed in class. The discussion in class is seen as a vital part of the case-based learning process. The objective of the class discussion is to analyse the company's situation, evaluate possible solutions and come up with viable strategies for the future (Mauffette-Leenders et al., 1997). Traditionally teaching cases have been written descriptions, but as everything else in the digital era, multimedia teaching cases began to emerge in late 90′ies.

In 1998 a two year EU project named BUSINES-LINC (Business Innovation Networks – Learning with Interactive Cases) commenced. In this project 18 multimedia teaching cases were developed in six different countries. The cases run in an Internet browser environment (basic and dynamic HTML combined with embedded sound, animations and video recordings on many pages). The cases primarily focus on the different ways e-commerce has been implemented in 18 very different types of companies around Europe. Six business schools participated in the project.[1] Each partner was responsible for the development of three of the 18 cases, which became a mix of open ended and closed cases, with very different designs (i.e. some cases focused on providing the users a kind of journalistic or objective viewpoint of the company story, others took on a more role-playing perspective, some cases provided menu-structures, others sequential navigation etc.). The cases are published through the European Case Clearing House under the name: E-Case Series, as they are *e*lectronic (the media), *e*-commerce (the subject) and *E*uropean (the origin) cases.

The author followed this project, and the research was carried out as a participatory action research project (Kemmis & McTaggart, 2000), where grounded theory informed and improved the theory generation part (Baskerville & Pries-Heje, 1999).

## 1.2    Problems with Scenarios in Case-Based Educational Design

A scenario is a simple, but efficient tool, which consists of a narrative sequence. "*Scenarios are stories - stories about people and their activities.*" (Carroll, 2000, p. 46). The objective of a scenario is to create a common ground, a common vision about the system being built (Preece, Rogers & Sharp, 2002).The scenario can be used very early in the development process, both in the analysis and the design phases. This may lead to two dominating (but not excluding) ways of using scenarios. The scenario on one side is a way of forcing the development team to begin analysing the environment/context and its influence on the final system, and also consider design issues, such as functionality plus interface and interaction design. The scenario on the other hand also provides a mean to make detailed design assumptions / explicit requirements, and considering the appropriateness of the discussed design. (Carroll, 1995, Dobsen & Riesbeck, 1998)

In Carroll's books, which contain very well documented and rigorously researched work on scenarios, there are typologies for scenarios that contain or evolve around plots and character, as well as goal- and/or task-driven scenarios (Carroll, 1995 & 2000). However, the language in the examples seems quite distant and without the emotions, daily actions and sense of character, which a case-based scenario for education should contain, thus involving people (students) relating to situations and people (managers and employees) in companies. When compared to the theories or models behind these examples it seems to be due to the focus on user tasks. Even when a scenario is written from a perspective of goals, this is translated to tasks, like wanting to operate a system, rather than on people (students) learning goals and peoples (managers/employees) objectives and needs, which would be interesting here.

For example, the scenario in Table 1 was written for an application in instructional design for programmers, but is oriented towards the tasks carried out, not the users. This may be adequate in this example, but not for teaching cases. E.g. Carroll's seven methods for creating scenarios give thorough and useful experiences on analysing and interpreting scenario descriptions (Carroll,

---

[1] The six business schools were: University of Cologne (Co-ordinator), Copenhagen Business School, Norwegian School of Economics and Business Administration (Bergen), Rotterdam School of Management, SDA - The Business School of Bocconi University (M ilan) and Stockholm School of Economics.

2000, chapter 10), but does not provide ways for users to get an empathetic insight or for creating motivating experiences through (case-) stories that would allow further questioning, reflecting and relating to other experiences (as the case paradigm would require of a student).

**Table 1** - Excerpt from the "Redesigned scenario", (Carroll, 2000, p. 146 – Box 6.11)

| |
|---|
| The programmer is using a workspace and the Bittitalk Browser to work with a function-level view of the blackjack model (that is, exploring the functionality of the game without its user interface and working at the level of application objects and messages, not their implementations) ….<br>….., The programmer follows an instruction to explore methods defined for the four other major model objects: BJPlayer, BJHand, Card and CardDeck. |

## 2 Character- and Plot-Driven Scenarios based on Treatments

Motivating students to read a written teaching case is done by creating some sort of tension or plot (Heath, 1998). In a multimedia teaching case there are particularly three factors that support this, namely the emotions and reasoning of the people in the company, through narrative and/or through multiple pathways in the system (Orngreen, 2002). The "traditional" scenarios, as presented above, do not seem to have room for this.

### 2.1 Treatments

A treatment is typically a couple of pages (from 3 to 25 pages on average) describing the story-line in a movie (Atchity & Wong, 1997). A treatment has two functions. One is as a tool for working on the plot, the characters and their actions while a movie script is being written (whether for cinema or television, and whether it is an animated movie, drama or comedy, fiction or a "true-story"). In this stage the treatment is a kind of work-in-progress tool (Frensham, 1996). The other function is as marketing material, which should "sell" the idea of the movie to the producers. It is this description, which is forwarded to possible producers, and it thus makes up the first basis on which the producer / script reader decide, whether this is a film-project they would like to consider pouring their money into or not. (Atchity & Wong, 1997, Frensham, 1996.)

Characters are displayed in the treatment as individual persons. So even if a person represents an archetype, as the programmer in the Table 1 scenario, it is a believable personality and not an unidentifiable "John Doe", which makes actions in a case-environment unnatural, and therefore difficult to extract any design principles and requirements from. The character of an individual is the quality, which was needed in the multimedia teaching case design, as opposed to anonymous persons, whose actions are difficult to perceive or artificial in their description.

According to Frensham, the treatment should be a chronological description, which uses "*active verbs and descriptive nouns to capture the action, verve and pace of your script. Pace is important,*" (Frensham, 1996, p. 177). The treatment thus also focuses on the use of the language to capture our interest. In other words, the treatment has to be written in a language, which seems true to the context it presents and the persons involved - the elements, which were lacking in scenarios, when dealing with the multimedia cases.

### 2.2 Working with Scenarios in Design of Teaching Cases

Scenarios in BUSINES-LINC were written early in the design phase, and were not used as an analytical tool. They were based on information derived from work in the analysis and content collection phase, i.e. after the first visits to the case company and after some managers/employees had been interviewed. This was because in order to work with the character- and plot-driven scenario, it was necessary to have a quite well established sense of the content, the case company. From this point it was possible to generate interesting case stories, from which learning objectives / lessons learned and possible interaction and navigation forms could be derived.

The scenario used in BUSINES-LINC covered two types of characters, the users / the students and company employees/managers (in other case-based learning environment this could be compared to teachers/students in education, doctors/patients in medicine etc.). The scenarios depicted a use situation, including descriptions of who the user is (his/hers objective for using the case, and how the user prepares for the class discussion), the case company actors and actions, as well as the context, problems and opportunities of the company's situation.

**Table 2** - Excerpt from a BUSINES-LINC case extended scenario (The LEGO case)

| |
|---|
| .... The student is now in the role as an external consultant. He enters the LEGO headquarters through the glass doors – [he is looking out through the video camera as if everything is seen from the eyes of the consultant]. At first his eyes pan-around the reception of the LEGO headquarter which is decorated with LEGO bricks and toys ...... <br><br> ..... Clicking on the first door activates the opening of the door and the consultant enters the office of the manager of the electronic commerce department. Tau Steffensen is sitting behind his desk...... |

As an example of the high priority given to the narrative elements and thoughts about interaction and navigation, compare Table 2 to the scenario in Table 1. The plot or story and the character of the company are more visible here. For example instead of writing that "the multimedia system will include 3 screens looking like offices. First office illustrates the manager of electronic commerce office, including folders with information about....", an attempt to provide a situational feeling of what the users will experience when using the system was provided, by for example stating that the consultant enters the office and meets Tau Steffensen. The character- and plot-driven approach thus worked well with respect to creating a motivating case story that called for analytical skills and reflection of the user.

However, in BUSINES-LINC it was not at all easy to maintain or "keep alive" the different user- or student-characters identified. There was a lot of reluctance in the consortium to such a "game play", which was not seen as serious development work. For example, it was observed that some scenarios started out with the personal character, but became more and more "a user" as the scenario work proceeded. The scenarios used as final versions for design requirements did not at all refer to an individual character with respect to the user. This is seen in for example how the text uses the anonymous term "the student" instead of Michael, Susan etc (see table 2). The consequence was that it became difficult to distinguish and derive different learning situations possible with one teaching case design as opposed to other designs, for example based on different types of students (bachelor/graduate/MBA).

It should be noted that recently new initiatives within scenario design also contemplate the issues of characters, with respect to the users. Alan Cooper's use of *personas* (Cooper, 1999) & Lene Nielsen's use of *rounded user* or *rounded character* (Nielsen, 2002), both of which challenge the archetypal use of the concept "the user". A scenario according to Cooper thus becomes "*a concise description of a persona using a software based product to achieve a goal*, (Coopers, 1999, p.179). Such initiatives could provide the necessary focus on the importance of having a development team engage in visionary user descriptions, though it may be that the lack of commitment is rather a result of the background of the development team than the tools available. The development team that Coopers and Nielsen are involved with are experienced designers, whereas the development team from BUSINES-LINC included many university students and professors with a business viewpoint rather than design.

# 3    Conclusion

Treatments opened for discussions on people and stories, which should seem real. Both the users of the teaching case and the employees at the case company as well as the case story described in the teaching case are very real, the fiction characters and stories of treatments, however, are not. Nonetheless the treatment was able to provide character- and plot-driven scenarios, which took on an objective or goal perspective, which made it possible to visualise, analyse and derive design assumptions about content design, interaction and navigation, as opposed to the more task-oriented scenario approach.

The major task seems to be to adapt and maintain for longer periods the use of such visionary descriptions in a project team, with respect to users, because many within the environment of business schools find it a rather unconventional or a "too creative" approach to educational design. However, the character- and plot-driven scenarios did provide visionary descriptions of narrative aspect and actions, as well as the individual character, culture and context of the case company in the teaching case, which was the other reason for using the treatments.

# References

Atchity, K. & Wong, C. (1997). *Writing treatments that sell*, 1st edition, Henry Holt, USA.
Baskerville, R. & Pries-Heje, J. (1999). Grounded action research: a method for understanding IT in practice, *Accounting Management and Information Technologies,* Vol. 9, 1999, pp.1-23
Carroll, J. (editor) (1995). *Scenario based design – envisioning work and technology in system development*, John Wiley & Sons, Inc., New York, New York, USA:
Carroll, J. (editor) (2000). Making Use – scenario-based design of HCI, The MIT Press, USA
Cooper, A. (1999). *The inmates are running the asylum – why high-tech products drive us crazy and how to restore the sanity*, SAMS, Macmillan Computer Publishing, Indianapolis, USA
Dobsen, W. & Riesbeck, C. (1998). *Tools for incremental Development of Educational Software Interfaces*, in proceedings of CHI98 Human Factors in Computing Systems, p. 384-391
Frensham, R. (1996). *Screenwriting*, Hodder & Stoughton Educational, London, UK.
Heath, J. (1998). *Teaching and writing Case Studies - a practical guide*, The European Case Clearing House, Cranfield University, Bedford, UK.
Kemmis, S. & McTaggart, R. (2000). Participatory Action Research, in Denzin, Norman & Lincoln, Yvonna (editors), *Handbook of Qualitative Research*, second edition, Sage publications, Thousand Oaks, California, pp 567-606.
Mauffette-Leenders, L, Erskine, J & Leenders, M. (1997). *Learning with cases,* Richard Ivey School of Business, the University of Western Ontario, Canada.
Nielsen, L. (2002). From user to character - an investigation into user-descriptions in scenarios, The HCI research group, *Department of Informatics working paper-series*, Jan. 2002, no. 1
Orngreen, R. (2002). *Multimedia Teaching Cases*, Ph.D.-series 27.2002, Samdunslitteratur, Frederiksberg, Denmark.
Preece, J.; Rogers, Y. & Sharp, H. (2002). *Interaction Design – beyond human computer interaction*, John Wiley & Sons, Inc., USA

# A Comparison of Four New Communication Technologies

*Ruth Rettie*

Kingston University
Kingston Hill, Kingston, KT2 7LB, U.K.
R.Rettie@Kingston.ac.uk

## Abstract

This paper describes a study of four new communication channels: Instant Messenger, email, text messages and mobile phones. The research develops a new model of communication channel choice in which media richness and social presence are important factors, but the core concept is channel-connectedness. The channels studied facilitate different levels of connectedness and this helps to explain usage. The degree of connection desired varies by both participant and occasion, and channels are chosen accordingly. The four channels play different communication roles, consequently, despite convergence of their technologies, the different formats will persist.

## 1    Introduction

Venkatesh (1998, p 670) writes, "the recent convergence of communication and information technologies has created possibilities unthinkable only a few years ago". Mobile phones, email, SMS (Short Message Service) messages and IM (Instant Messenger) are new communication technologies, which all contribute to the 'death of distance' (Cairncross, 2001). This research explores and compares consumer usage and attitudes to the four technologies, developing a new communication choice model. Although the research focuses on leisure use among young people, the model should be more generally applicable.

## 2    Literature Review

### 2.1    Media Communication Theory

Social presence (Short, Williams and Christie, 1976) and media richness (Daft and Lengel, 1986) help to explain media choice. Social presence is the extent to which a medium conveys the actual presence of participants. The 'richness' of a medium is measured by its capacity for multiple cues and immediate feedback. Computer-mediated communication (CMC) is low in social presence (Rice and Love, 1987) and lean in media richness (Walther, 1992). Flaherty, Pearce and Rubin (1998) found that face-to-face communication was rated higher than CMC for all motives, including social ones, such as, inclusion and affection. Clark and Brennan (1990) identify eight factors that constrain media choice: co-presence, visibility, audibility, co-temporality, simultaneity, sequentiality, review-ability and revisability.

### 2.2    The Four Communication Channels

### 2.2.1 Instant Messenger

Instant Messenger (IM) is a proprietary, simplified version of Internet Relay Chat, which allows two or more people to carry on a conversation, in real-time, using text based messages with context awareness. In the U.S. 40% of Internet users use messenger (Nielsen NetRatings, 2002). IM is used to avoid boredom, to socialise (Schiano et al. (2002), Leung 2001), and to maintain contact with casual acquaintances (Lenhart et. al. 2001). Leung (2001) found seven motives for messenger use among college students: affection, inclusion, sociability, entertainment, relaxation, escape and fashion. Nardi, Whittaker and Bradner (2000), found that in the inactive state IM participants sometimes monitor the presence of others, and use the medium to sustain a sense of connection.

### 2.2.2 Text Messages

Short Message Service (SMS) or text messages were introduced in 1992. GSM (Global System for Mobile communication) estimates that 250 billion SMS messages were sent through their networks in 2002. (http://www.gsmworld.com/news/statistics/index.shtml). SMS messages are quick, cheap, convenient and discrete (Eldridge and Grinter, 2001), less formal and more private than email (Clarke and Strong, 2000), and used socially for networking, co-ordination, and managing relationships (Döring, 2002). Grinter and Eldridge (2001) found that 63% of UK messages are sent from home; they identify the 'goodnight' text as a new type of message content.

### 2.2.3 Mobile Phone Calls

Globally, the number of mobile subscribers is estimated at 1 billion (Gibney, Swain and Hooper, 2002). Research on mobiles has found they are useful for hyper-coordination, security, socializing, relieving boredom and as a vehicle for parental control (Baursch et al., 2001; Ling and Helmersen, 1999) and to express identity (Alexander, 2000). For some, the mobile becomes almost a body part, an extension of the hand (Hulme and Peters, 2001).

### 2.2.4 Email

The number of email messages sent daily, worldwide, is expected to increase from 31 billion in 2002 to 60 billion in 2006 (Levitt & Mahowald, 2002). Lee (1996) described email as a hybrid medium combining elements of the phone and letter, i.e. conversational informality in text format. Research on email has found that people are more uninhibited, non-conformist and conflictual when using email, and that email broadens communication circles (Ducheneaut, 2002). Although the primary use is communication, use includes socializing and developing relationships (Finholt and Sproull, 1990). Schiano et al. (2002) found that teenagers mainly use email for non-personal communication.

## 3    Methodology

The objective of the research was to understand and explore communication channel choice and therefore, qualitative research was used. Six (three male/three female) 1½-hour focus groups were held, with four groups of university students and two groups of teenagers. Respondents were all users of mobile, email, text messages and IM. The groups were analyzed using grounded theory.

# 4 Results

Respondents frequently had all four technologies available and so channel choice was often pertinent. Respondents were aware of the advantage of near-synchronicity afforded by SMS and IM. These technologies provided 'thinking time' without the disruption and discontinuity of asynchronous communication such as email. Less socially confident, or time-pressured participants, sometimes chose leaner media with low social presence to avoid social embarrassment, or to save time through quicker communication or multi-tasking. Channels are not exclusive and can be complementary; simultaneous and sequential use were both common.

Media choice depended on functional factors, communication motives, relationship between the participants, personal preference and 'connection need'. Functional aspects include: cost, availability, time, typing proficiency, and message-specific characteristics such as sensitivity, confidentiality, quantity and urgency. Email, IM and text are less appropriate for personally sensitive communication: respondents were also conscious of the ease with which email and IM can be forwarded. Preference or predisposition for visual, auditory or kinaesthetic cognitive style (Sarasin, 1998) will also influence choice of communication channel.

Communication motives were intrinsic or instrumental, and included relaxation, entertainment, social, and affection. The main motivation for mobile use was affection, for email it was social, and for IM it was entertainment. Text message motives were usually either social or affection.

The most important, and least obvious choice factor, was the need for connection, a concept which emerged from the research. The need for connection varied, sometimes respondents just wanted connection without conversation, which IM could provide, at other times there was no desire for connection, and they would just send an email. Although mobiles had higher connectedness they could generate anxiety and feelings of social inadequacy, therefore less connected channels were sometimes preferred. Respondents agreed that they generally felt most connected when using mobiles, followed by IM and text, with email providing least connection.

The perceived connectedness of a medium appeared to be a function of media richness, social presence, interactivity, duration, and information processing mode. Media richness affects the quantity and quality of cues, e.g. voice tone; social presence creates awareness of the other party in the connection. Interactivity creates the experience of connection through two-way communication, and is facilitated by synchronicity and near-synchronicity. A longer duration of communication increases the experience of connectedness. Audio information processing has to be cotemporaneous with audio source, which prevents scanning, discourages multi-tasking, increases focus and generates a greater sense of connectedness.

# 5 Discussion

To ensure experience with all four technologies all respondents were under 25 and therefore the findings of this research may be specific to this age group. For older people social confidence could be less relevant and time pressure may be more important. Although IM and text are increasingly used in the work place, lack of typing skill inhibits the adoption of text-based communication, especially among older, non-working, women.

Channel connectedness can be defined as the extent to which a channel enables the participants to feel connected. The concept of channel-connectedness is similar to social presence but it is not equivalent. Social presence relates to the perception of the other participant while connectedness is an emotional experience. The difference between the social presence and connectedness of a channel is illustrated by IM and text messaging; there is virtually no social presence, but used interactively in a 'conversation' IM conveys connectedness, as does an exchange of 'goodnight' text messages. On the other hand an Internet web-cam conveys social presence but not connectedness.

Previous research has suggested the superiority of richer media with more social presence for gratifying communication needs (Flaherty, Pearce and Rubin, 1998); however, the young people in this study often preferred less rich media. Each of the four technologies researched has its own inherent advantages and different degrees of connection; these create specific roles and gratify different communication needs. These roles derive from communication norms (such as use of abbreviation, absence of social niceties, length, etc) as well as from the intrinsic characteristics of the original technology (e.g. word limit, synchronicity, sensory type). Contrary to predictions, as technologies converge, with email and messaging being available on mobiles, and SMS and VOIP available on PCs, the different formats are likely to be retained because of their specific roles.

There is scope for interfaces that extend the advantages of the different channels, for instance, the use of context awareness technology for telephony, and the development of email-style storage, organisation and subject notification on SMS. User penetration could be increased by voice to text conversion for non-typists, and text to voice for those intimidated by text interfaces. To increase user choice, design for convergent personal communication devices should focus on enabling multiple formats, for instance enabling SMS messages from email interfaces. Designers could also develop new formats, for example, mobiles with simultaneous text conversion, or context awareness channels designed to provide connection rather than communication.

# References

Alexander, P.S. (2000). Teens and mobile phones growing- up together: understanding the reciprocal influence on the development of identity, *Wireless World Workshop*, University of Surrey, April 7th.

Baursch, H., Granger, J., Karnjate, T., Khan, F., Leveston, Z., Niehus, G., & Ward T. (2001). An investigation of mobile phone use: a socio-technical approach. *IE449 Socio-technical Systems in Industry*, Summer Session 2001.

Cairncross, F. (2001). The death of distance: how the communications revolution will change our lives. Boston, MA: Harvard Business School Publishing.

Clark, H. H., & Brennan, S. E. (1990). Grounding in communication. In R. M Baecker (Ed.), *Readings in groupware and computer-supported cooperative work, assisting human-human collaboration,* (pp. 222-233). San Francisco, CA: Morgan Kauffmann.

Clarke, B., & Strong C. (2000). Kids Net Wave 5. London: NOP Research.

Daft, R.L., & Lengel, R.H. (1986). Organizational information requirements, media richness and structural design. *Management Science*, 32(5), 554-571.

Döring, N. (2002)."1x Brot, Wurst, 5Sack Äpfel I.L.D." – Kommunikative funktionen von kurzmitteilungen (SMS). *Zeitschrift für Medienpsychologie*. Retrieved February 3, 2003, from http://www.nicola-doering.de/publications/sms-funktionen-doering-2002.pdf

Ducheneaut, N. (2002). The social impacts of electronic mail in organizations: A case study of electronic power games using communication genres. *Information Communication and Society*, 5 (2), 153-188.

Gibney, O., Swain, K., & Hooper, G. (2002). Global mobile forecasts to 2010, 3rd edition. Colchester, U.K.: Baskerville Telecoms

Eldridge, M. & Grinter B. (2001). Studying text messaging in teenagers. CHI 2001 Workshop: M Communications: Understanding Users, Adoption and Design.

Finholt, T., Sproull, L., & Kiesler, S. (1990). Communication and performance in ad-hoc task groups. In J. Galegher, R. Kraut & C. Egido (Eds.) *Intellectual Teamwork*. (pp. 295 -325). Hillsdale, NJ: Lawrence Erlbaum Press.

Flaherty, L., Pearce, K. & Rubin, R. (1998). Internet and face-to-face communication: Not functional alternatives. *Communication Quarterly*, 46(3), 250-268.

Grinter R.E. & Eldridge, M.A. (2001). y do tngrs luv 2 txt msg? Proceedings of the European Conference on Computer-Supported Cooperative Work, 16-20 September 2001, Germany.

Helmersen, P. (1999). It must be necessary, it has to cover need: The adoption of mobile telephony among pre-adolescents and adolescents. Telenor Report FoU R 9/2000, Telenor.

Hulme, M. & Peters, S. (2002). Me my phone and I: The role of the mobile phone. CHI 2002 Workshop on Mobile Communications. April 1 -2.

Levitt, M. & Mahowald, R.P. (2002). Worldwide email usage forecast, 2002-2006: Know what's coming your way. IDC #27975, September 2002.

Lee, J. Y. (1996). Charting the codes of cyberspace: A rhetoric of electronic mail. In L. Starte, R. Jacobson, & S. B. Gibson (Eds.), *Communication and Cyberspace: Social Interaction in an Electronic Environment* (pp.175-296). Cresskill NJ:Hampton Press.

Lenhart, A., Rainie L. & Lewis, O. (2001) Teenage life online: The rise of the instant-messenger generation and the Internet's impact on friendships and family relationships. PIP. Retrieved February 3, 2003, from http://www.pewinternet.org/reports/pdfs/PIP_Teens_Report.pdf

Leung, L. (2001). College student motives for chatting on ICQ. *New Media and Society*. 3(4), 483 -500.

Nardi, B., Whittaker, S. & Bradner, E. (2000) Interaction and outeraction: Instant messaging in action. Proceedings of CSCW 2000, 79-88.

Nielsen NetRatings (2002). Instant messaging used by more than 41 million home Internet surfers. Retrieved February 1, 2003, from http://www.nielsen-netratings.com/pr/pr_020617.pdf

Sarasin, L.C. (1998). Learning style perspectives: Impact in the classroom. Madison, WI: Atwood Publishing.

Schiano, D.J., Chen, C.P., Ginsberg, J. Gretarsdottir U., Huddleston M., & Isaacs, E. (2002). Teen use of messaging media. Human Factors in Computing Systems, CHI April 2002.

Short, J., Williams, E., & Christie, B. (1976). The social psychology of telecommunications. London: John Wiley.

Rice, R.E. & Love, G. (1987). Electronic emotion: Socio-emotional content in a computer-mediated communication network. *Communication Research*, 14, 85-108.

Walther, J.B. (1992). Interpersonal effects in computer-mediated interaction: a relational perspective. *Communication Research*, 19, 52-90.

Venkatesh, A. (1998). Cybermarketscapes and consumer freedoms and identities. *European Journal of Marketing*, .32(7/8), 664 -675.

# Design Process for Product Families – a case study of a software application package for hearing acousticians

*Nina Sandweg[*], Heinz Bergmeier[*], Sonja Pedell[**],*
*Benno Knapp[***], Eduard Kaiser[***]*

[*] Siemens AG, Corporate Technology
81730 Munich, Germany
Nina.Sandweg@siemens.com
Heinz.Bergmeier@siemens.com

[**] University of Melbourne, Department of Information Systems
3010 Victoria, Australia
Pedell@acm.org

[***] Siemens Audiologische Technik
91050 Erlangen, Germany
Benno.Knapp@siemens.com
Eduard.Kaiser@siemens.com

## Abstract

The concept of the user interfaces of the Siemens Audiologic Technology (SAT) products for hearing acousticians are described. Usable and user accepted solutions of a hearing instrument fitting software were reused for all products establishing an entire product family. Results were an enhanced usability of the products due to a seamless workflow across the applications.

## 1    Significance

For economic and quality reasons software reuse becomes more and more important in the software development process (eg. Sherill et al., 2001; Savolainen & Kuusela, 2001). One possibility to reuse software is the concept of product families. According to recent publications in this area (eg. Sherill et al., 2001; Clements & Weiderman, 1998), product families are meant to have the following benefits:

- Shorten the time to market
- Minimize the costs of developing new products
- Increase reliability of new products within the family

Depending on the kind and similarity of products and the main intention of the people involved they might have the same look & feel, a shared software architecture or the same interaction elements. Beside time and cost savings, the development can benefit from the insights that were gained from the first usage. Furthermore the user will accept gladly a familiar appearance of a subsequent product.

The main challenge is that on the one hand the above mentioned benefits should be met and on the other hand the structure has to be open to support the diversity of future products in the family and to support the target user.

## 2    Objective

In this paper the usability of the user interfaces of the Siemens Audiologic Technology (SAT) product family for hearing acousticians are emphasised. This means that proven usable and user accepted solutions for one product should be reused for all products of the product family. During the re-design process of the hearing instrument fitting application the analysis of the acoustician's workflow made clear: one of the major usability challenges lays in the continuous switching between different applications such as office management system for client administration, measurement application for diagnosing the client's hearing loss, electronic forms and fitting applications of different hearing instrument manufactures. The competitive advantage of a manufacturer who can cover the complete workflow with applications in the same look & feel and the same generic user interface elements is obvious.

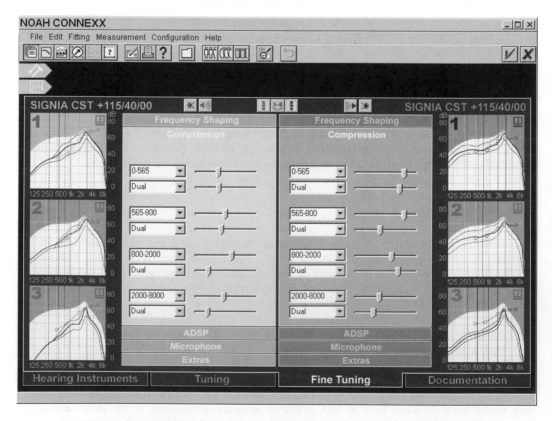

**Figure 1: The re-designed hearing instrument fitting software CONNEXX**

# 3    Procedure

In the following, we describe our approach for the SAT software product family, starting from the re-designed fitting application CONNEXX.

The hearing instrument fitting software CONNEXX (see figure 1) was functionally and visually revised in an iterative usability engineering process (Sandweg et al., 2002). The interaction and design concept has proven its usability in systematic user evaluations and customers feedback. Its layout, structure and generic user interface elements are described in a detailed UI specification that lead into a widget set library for further use.

Some of the main concept rationales are introduced in the following: Task Cards are based on a tab card metaphor. Each contains all functions and objects necessary for the completion of a self-contained task. The Task Cards structure the program visually according to the daily routine process of work and therefore provide guidance for the user. A consistent and symmetric screen layout optimizes the binaural fitting procedure. Colour coding refers consequently to their audiometric meaning. The colours red and blue stand for the monaural representation of the right and the left ear, grey stands for the binaural representation.

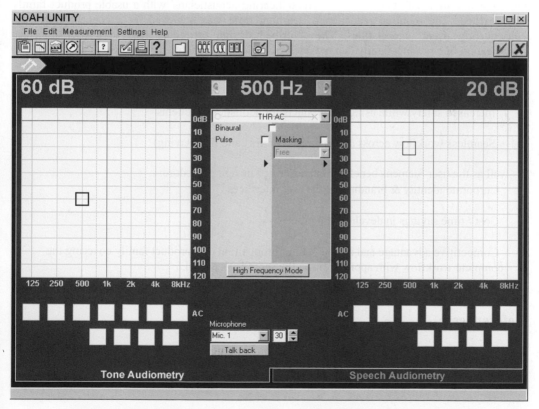

**Figure 2: The design of the hearing loss measurement software UNITY**

Depending on the degree of congruence, the elements of this concept were taken over without change or adopted as necessary to the other applications. Unmodified transferable parts are e.g. symmetric screen layout and the colour coding as well as interaction solutions for identical features. Similar features or similar behaviour (such as browsing or parameter setting) have to be modified to their different purposes based on previously used generic elements. This supports a common look and feel, the recognition of the elements and so the conformity with users' expectations. For new features and use cases new solutions were developed that are derived from existing elements and patterns.

The results of applying this procedure can be seen in figure 2 which shows the re-designed hearing loss measurement software UNITY with the same look & feel as the fitting software CONNEXX. Currently, the new order creation module SIFORM for ordering electronically Siemens customs hearing instruments is being designed in the same style, following the above described procedure as well.

# 4    Conclusion

Our objective to satisfy the target user group 'hearing acousticians' with a usable product family was reached. The required reduction of learning time was achieved by a concept following consequently users' expectations. The described design procedure lead to a user interface specification of a product family that provides the following benefits for the users:

- Higher satisfaction of the acousticians

- Increased safety while using the products for acousticians' clients

- Seamless workflow across applications (comfort and time saving)

As well as the user Siemens benefits from described design procedure:
- Corporate identity & branding by the user interfaces

- Cost / time savings in development process

- Competitive advantage and unique selling proposition of a product family covering the complete workflow with applications in the same look & feel.

# References

Clements, P.C. & Weiderman, N. (1998). Notes on the second international workshop on development and evolution of software architectures for product families. ACM SIGSOFT Software Engineering Notes, Volume 23 Issue 3

Sandweg, N., Pedell, S., Platz, A., Schneider, K.P., Honold, P., Hermann, D., Kaiser, E. (2002). Re-Design of CONNEXX Hearing Instrument Fitting Software - a Case Study. Proceedings of the 6th international Scientific Conference on Work with Display Units WWDU 2002. Berchtesgaden, Germany, May 22-25 2002

Savolainen, J. & Kuusela, J. (2001). Volatility Analysis Framework for Product Lines. ACM SIGSOFT Software Engineering Notes, Proceedings of the 2001 symposium on Software reusability: putting software reuse in context. May 2001, Volume 26 Issue 3

Sherrill, J., Averett, J., Humphrey, G. (2001). Implementing a product line-based architecture in Ada. ACM SIGAda Ada Letters, Proceedings of the annual conference on ACM SIGAda annual international conference (SIGAda 2001), Volume XXI Issue 4

# Impact of Cognitive Style upon Sense of Presence

*Corina Sas and Gregory O'Hare*

Department of Computer Science
University College Dublin
Belfield, Dublin 4, Ireland
{corina.sas; gregory.ohare}@ucd.ie

## Abstract

The role played by cognitive style upon sense of presence has been addressed in the presence literature. However, no experimental study was carried out in order to investigate this hypothesized relationship. This paper highlights the relationship between each of four bi-polar dimensions of cognitive style, such as extraversion–introversion, sensing–intuition, thinking– feeling and judging –perceiving, and the experienced level of sense of presence. Implications of these individual differences for understanding sense of presence and for designing virtual environments to address these differences are discussed.

## 1 Introduction

One of the psychological phenomena experienced by users while (and not only) they interact with virtual reality systems, is a *sense of presence*. It allows them to *be there* (Schloerb & Sheridan, 1995), to feel themselves immersed and moreover to perceive the virtual world as another world where they really exist. In our previous work, we defined presence as a psychological phenomenon, through which one's cognitive processes are oriented toward another world, either technologically mediated or imaginary, to such an extent that he or she experiences mentally the state of being (there), similar to one in the physical reality, together with an imperceptible shifting of focus of consciousness to the proximal stimulus located in that other world (Sas & O'Hare, 2001). Sense of presence is particularly experienced when the task being carried out requires a high involvement of both cognitive and affective resources. The experience within the remote world is a complete one, encompassing cognitive, emotional and behavioural aspects. In other words, the more the users think, feel and act in the remote world and the more collateral activities are inhibited within the real worlds, the more sense of presence they will experience (Sas & O'Hare, 2002).

Understanding users' preferred manner of processing information opens a door towards their perception of world, either physical or virtual. The term of cognitive style was coined by Allport (1937) and is rooted in Jung's theory of psychological types (1971). Despite the large number of meanings attributed to it, cognitive style refers to enduring patterns of cognitive behaviour (Grigorenko, 2000). It describes the unique manner in which the unconscious mental processes are used in approaching and/or accomplishing cognitive tasks.

Curry's Onion Model (1983), presented by Riding (1991) proposes a hierarchical structure of cognitive styles, with the outmost layer referring to the individual's choice of learning environment, with the middle layer referring to the information processing style and with the innermost layer consisting of cognitive personality style. Defined as the individual's tendency to assimilate information, cognitive personality style is an enduring and context-independent feature. Therefore, it should make little difference if the context of providing cognitive stimulation is technologically mediated or not, as long as the given task involves information processing.

Cognitive style was referenced in presence literature as a possible significant issue affecting presence (Lombard, Ditton, 1997, Heeter, 1992). However, to the best of our knowledge no experimental study has been carried out to investigate this relationship.

## 2 Methodology
### 2.1 Procedure
The Virtual Environment (VE) utilised was that of the ECHOES system (O'Hare, Sewell, Murphy, Delahunty, 2000) a non-immersive training environment, which addresses the maintenance of complex industrial artefacts. Adopting a physical world metaphor, the ECHOES environment comprises a virtual multi-story building, each one of the levels containing several rooms: conference room, library, lobby etc. Subjects can navigate from level to level using a virtual elevator. The rooms are furbished and associated with each room there is a cohesive set of functions provided for the user.

After users gained familiarity with the environment and particularly learned movement control, they were asked to perform an exploration task. The exploration task lasts for approximately 25 minutes. In order to induce motivation for an active exploration, users were asked to find a valuable painting hidden within the virtual building. The sample consisted of 30 undergraduate and postgraduate students from the Computer Science Department, 18 males and 12 females, within the age range 20-38. The study hypothesis states that different dimensions of cognitive style have an impact upon presence.

### 2.2 Methods
#### 2.2.1 Presence Questionnaire
Presence was measured using a questionnaire devised by the authors, which was shown to lead to measurements which were both reliable and valid (Sas, O'Hare, 2002). It contained initially 34 items, typical for tapping the presence concept (Lombard, 2000), measured on a 7-point Likert scale, ranging from 1 (not at all) to 7 (completely). The presence score was computed as the averaged score of the items composing the questionnaire, with the minimum value of $Min = 1.74$, the maximum value of $Max = 5.17$ and the mean of $Mean = 3.38$.

#### 2.2.2 Myers-Briggs Type Indicator
Myers-Briggs Type Indicator (MBTI) (Myers & McCaulley, 1998) measures the strength of preference for the manner in which one processes information. Its development is grounded on Jung's theory of personality types (Jung, 1971), and the four basic dimensions of which are: Extraversion (E)–Introversion (I); Sensing (S)–Intuition (N); Thinking (T)–Feeling (F) and Judging (J)–Perceiving (P). The (E)–(I) continuum explains the orientation of attentional focus as a source of energy. While (E) are energized by interacting with others, (I) are energized by their inner world of reflections and thoughts.

The (S)–(N) continuum suggests the manner of perceiving and acquiring information. (S) people are usually realistic, organized and well structured, relying heavily on their five senses to perceive information. Quite contrarily, (N) individuals are creative and innovative looking at the overall picture rather that its details and acting on their hunches.

The (T)–(F) continuum refers to how one filters and organizes information in order to elaborate decisions. While analysis and logics are fundamentals for (T) people, leading them to make decisions, which are strongly coherent with their principles, (F) individuals value more feelings, kindness and harmony, which drive them to decide.

The (J)–(P) continuum describes the preferred life–style and work habits. (J) individuals are those which try to order and control their world, well-organized, good planners and potentially not very

697

open-minded. On the other hand (P) people are spontaneous, flexible, multiplex, but with a risk of not accomplishing the multiple approached tasks.

## 3 Discussions

In order to test the impact of cognitive style upon presence we conducted t-tests, comparing the level of sense of presence experienced by groups of users, identified on the basis of their scores for cognitive style dimensions. Thus we considered two independent groups for each dimension, with the cutt-off point of the second quartile. The findings suggest significant differences between groups of users formed along the feeling-thinking. With respect to the introversion-extroversion and the intuitive-sensitive dimensions, the cutt-off point had to be moved to the first quartile in order to reveal differences between level of sense of presence experienced by the two groups. As shown in Table 2, the differences were noticeable, with two significant at the level .05, indicating that persons who are more sensitive and feeling type experience a higher level of presence. Without being statistically significant, findings suggest that individuals who are more introvert or more judging type are more prone to experience presence.

**Table2:** *T-Tests* Comparing Presence Experienced by Users Grouped along Cognitive Style Dimensions

| Variables   (N = 30) | Group 1 | | Group 2 | | Presence |
|---|---|---|---|---|---|
| | *Mean* | *SD* | *Mean* | *SD* | *t*-Test |
| (I) – (E) | Introvert | | Extrovert | | |
| | 3.74 | 0.65 | 3.29 | 0.82 | 1.49 |
| (N) - (S) | Intuitive | | Sensitive | | |
| | 3.24 | 0.76 | 3.90 | 0.92 | 1.95* |
| (T) - (F) | Thinking | | Feeling | | |
| | 3.03 | 0.62 | 3.70 | 0.89 | 1.97* |
| (J) - (P) | Perceiving | | Judging | | |
| | 3.16 | 0.44 | 3.46 | 0.44 | 1.51 |

* $p< .05$.

Furthermore we present an interpretation of these findings in the light of study hypothesis, which states that different dimensions of cognitive style have an impact upon presence. Breaking down this general hypothesis, we summarize the following results. Along the (E)–(I) dimension, the contemplative nature of (I) individuals allow them to construct the mental model of the virtual world, providing also the energy needed to explore, understand and eventually become immersed within it. The level of presence experienced by (I) individuals is significantly greater than that experienced by (E) individuals (Fig. 1).

In the special case of the present study, where the participants have undertaken solitary tasks, our finding seems appropriate. However, it is expected that in Collaborative Virtual Environments, (E) individuals will experience a greater level of *social presence*.

In the context of our research, the results indicate that (S) individuals experienced a greater level of sense of presence (Fig. 2). However, this finding should be considered in relation with task characteristics. The main task of our experimental design consisted in wandering for 25 minutes within the virtual building and searching for a hidden painting. It was a highly perceptual task. Probably during learning curve, intuitive people were highly stimulated by learning new skills (i.e. navigating), while after this stage is met, the routine involved in practising it could lead to less involvement of cognitive and affective resources.

On the contrary, the more time (S) individuals spent within the environment, carrying the same task which requires attention and precision, the more focused they become. It seems that sensitive people are better anchored in the concrete, tangible reality (even when it is a virtual one), fact which enables them to achieve a superior level of spatial orientation.

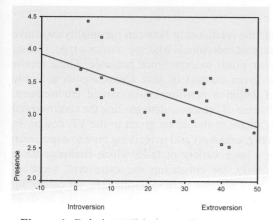

**Figure 1:** Relationship between Presence
and (I)-(E) dimension

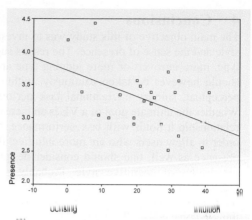

**Figure 2:** Relationship between Presence
and (N)-(S) dimension

They easily become absorbed within the activity they get engaged with, while the remote world offers the context for *here and now*. Heeter (2001) posed a very interesting question: "is presence for an intuitive more conceptual, while presence for a sensate is more perceptual?" The answer seems to be affirmative. Probably in order to feel presence, (I) individuals need to be stimulated with novel, symbolic information which challenges their abilities of grasping ideas.

Along the (T)–(F) continuum, (F) type is the empathic one. Empathy was already discussed as a quality which increases the experienced degree of presence (Sas, O'Hare, 2001). Since (F) people can potentially experience a greater level of empathy, they experience also a greater sense of presence (Fig. 3). This result can be better understood by analysing the relationship between willingness to be transported within the remote world and (T)-(F) dimension. The results indicate that (F) individuals are significantly more willing to be transported, than the (T) individuals ($t(28)$ = -2.43, $p < .05$) (Fig. 4).

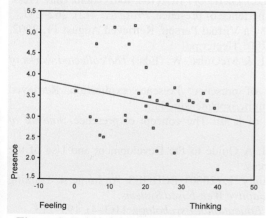

**Figure 3:** Relationship between Presence
and (T)-(F) dimension

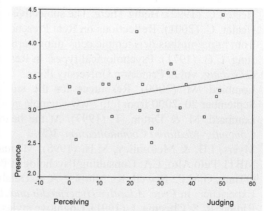

**Figure 4:** Relationship between Presence
and (J)-(P) dimension

# 4 Conclusions

The main objective of this study was to investigate the relationship between personality cognitive style and the sense of presence. The results suggest that individuals who are sensitive type, feeling type, more introvert or more judging type are more prone to experience presence. These results should however be taken cautiously, within the given context of task characteristic: a highly perceptual, solitary navigational task performed within a non-immersive virtual environment. Whether the ultimate goal of a VE is an increased sense of presence, disregarding the controversial relationship it holds with task performance, then attention should be given to the VE design. In order to allow users who are more intuitive, thinking, extrovert and perceiving type to experience presence as well, one should consider designing a large variety of tasks which challenge these dimensions of cognitive style. Probably social tasks for enhancing the extroverts' sense of presence, or more abstract, strategic tasks consisting of manipulating symbolic data, for enhancing the intuitive users' sense of presence, tasks which require more reasoning to address the needs of thinking type could be solutions in this direction.

It is more likely that the cognitive style dimensions do not in and of themselves carry a great impact upon presence. Everything else being equal, they would nonetheless manifest themselves in particular ways, giving a distinct *flavour* to the sense of presence experienced by users. However, these cognitive style dimensions should be considered in the broader context of other personality traits, of willingness to suspend disbelief, of activity being undertaken, of media and so forth.

Future work will focus on the relationship between the dimensions of personality cognitive style and the others cognitive factors such as creative imagination, absorption, empathy and willingness to be transported within the virtual worlds, which has been proven to impact upon sense of presence. Another future direction will assess the controversial relationship between presence and task performance, and in this light will try to formulate suggestions for a better design the VE in order to accommodate these individual differences impacting upon sense of presence.

# 5 References

Allport, G.W. (1937). Personality–A psychological interpretation. New York: Henrt Holt & Cmp.

Curry, L. (1983). An organization of learning styles theory and constructs. *ERIC Document*, 235–185.

Grigorenko, E. (2000). Cognitive style. In Kazdin, A. (Ed.), *Encycl. of Psych.*(163–166) Oxford: Univ. Press.

Heeter, C. (1992). Being There: The subjective experience of presence. *Presence*. 1(2), 262–271.

Heeter, C. (2001). Reflections on Real Presence by a Virtual Person. Retrieved August 11, 2002, from http://nimbus.ocis.temple.edu/~mlombard/P2001/Heeter.pdf

Jung, C.G. (1971). Psychological types. In Ress, L & McGuire, W. (Eds.) *The collected works of C.G. Jung*, vol 6. Princeton University Press.

Lombard, M. (2000). Resources for the study of presence: Presence explication. Retrieved September 20, 2000 from http://www.temple.edu/mmc/ispr/measure.htm

Lombard, M. & Ditton, T. (1997). At the heart of it all: The concept of presence. *Journal of Computer-Mediated Communication,* 3(2).

Myers, I.B. & McCaulley, M.H. (1998). Manual: A Guide to the Development and Use of the MBTI. Palo Alto, CA: Consulting Psychologist Press.

O'Hare, G.M.P., Sewell, K., Murphy, A. & Delahunty, T. (2000). ECHOES: An Immerse Training Experience. In *Proc. Adaptive Hypermedia and Adaptive Web-based Systems*.

Riding, R. & Cheema, I. (1991). Cognitive styles. *Educational Psychology*. 11(3–4), 193–213.

Sas, C. & O'Hare, G.M.P. (2001). Presence Equation: An Investigation Into Cognitive Factors Underlying Presence Within Non-Immersive Virtual Environments. *Presence Workshop 2001*.

Sas, C. & O'Hare, G.M.P. (2002). Presence Equation: An Investigation Into Cognitive Factors Underlying Presence. *Presence: Teleoperators and Virtual Environments*. (in press)

Sheridan, T.B. (1992). Musings on telepresence and virtual presence. *Presence*. 1(1), 120–126.

# 3D Modelling Is Not for WIMPs

*Silvia Scali[1], Dr. Mark Wright[2], Ann Marie Shillito[3]*

[1,3]Edinburgh College of Art,
79 Grassmarket,
Edinburgh EH1 2HJ
e-mail: s_scali@yahoo.it
a.m.shillito@eca.ac.uk

[2]Edinburgh Virtual Environment Centre,
James Clerk Maxwell Building,
The King's Buildings, Mayfield Road,
Edinburgh, EH9 3JZ.
e-mail: mark.wright@ed.ac.uk

## Abstract

This study compares a traditional 3D WIMP (Window Icon Menu Pointer) modeller to a prototype of a novel system with a 6DOF haptic feedback device, stereovision and a co-located display, both in quantitative and qualitative terms. The novel system was conceived to overcome limitations of traditional interaction techniques and devices when dealing with three-dimensions. Experimental results confirm the fundamental role of spatial input for 3D modelling and the significant contribution of haptics and stereovision to qualitative and quantitative performance. A review of relevant research and motivations for the study is presented along with a discussion of main outcomes.

## 1 Introduction

Although research highlights shortcomings of WIMP systems for 3D tasks, popular 3D modellers based on those systems, such as 3D studio MAX,™[1] are widespread in Design practice. However commercially available 3D modellers tend to overcome constraints imposed by the physical set up of the workstation by using advanced software functions. This introduces the disadvantage of a steep learning curve, but also promises efficient performance once the software is mastered; thus expert users might find such systems satisfactory and efficient.

However, 6DOF haptic interfaces may be greatly more efficient and usable.

## 2 Related Research

It has been argued that traditional WIMP systems are inadequate for effective 3D modelling due to a mismatch between DOF (Degrees of Freedom) required for the task and afforded by the input device. Manipulation of 3D objects exhibits a parallel structure of translation and orientation that, if transformed into a serial structure, may significantly increase the total task completion time (Wang Y. and MacKenzie C L., Summers V. A. and Booth K. S., 1998). Interacting with a virtual model for a positioning task means defining separate parameters, expressing orientation and position components for X, Y and Z axes (Butterworth J., Davidson A., Hench S. and Marc T., 1992), which must be input separately with a mouse device. Lack of depth cues, typical of traditional 2D-displays, can also hinder 3D interaction: using a 6DOF Polhemus device with a 2D display makes it difficult to select a target (Badler, N. I., Manoochehri, K. H., Baraff, D., 1986). Other research shows that a stereoscopic display in a positioning task with a 6 D.O.F. input device reduced error rate by 60% (Beaten R.J., DeHoff R.J., Weiman, N. and Hildebrandt, W., 1987). Overall, Jacobs' statement that the structure of the perceptual space of an interaction task should

---

[1]3ds max is a registered trademark. Discreet is a division of Autodesk, Inc.

701

mirror that of the control space of its input device (Jacob, R.J.K. & Sibert L.E., 1992) summarizes the above issues well, since the effectiveness of an input device is relative to the task rather than being absolute. In turn, different devices "suggest" specific cognitive strategies, influencing how users "think" about a task (Hinckley K., Paush R., Proffitt D., 1997). Research in psychology (Parsons L., 1995) and in HCI shows that experiencing 3D space helps the user to understand it (Hinckley, K., 1996): this should be taken into account in implementing novel systems for 3D, since cognitive activities are fundamental in HCI models. A system providing coherent depth cues, spatial input and a more "natural" environment for spatial interaction could be regarded as desirable for 3D interaction. However practical implementation presents many drawbacks, arising from limitations in available technologies. Exploitation of physical devices for interaction with computers such as tangible interfaces and physical props works towards restoring a more natural interaction. However this can in turn limit the flexibility offered by the digital medium. If "natural" interaction was to be achieved via software, this could be preferable, although it might impose a cognitive load on the user, as software widgets must be "understood" (Hinckley, K., 1996). Haptic displays, such as force feedback devices with 6DOF, could represent a valuable compromise, being both "programmable" and affording the use of our natural skills in manipulation. It has been shown that haptic force-feedback improved performance in a peg-in-hole task (Massimino, M. J., & Sheridan, T. B., 1994) and that use of haptic widgets, such as "gravity wells" improved precision in a 3D-targeting task. (Wall S. A., Paynter K., Shillito A.M., Wright M., Scali S., 2002). Force feedback devices could be further enhanced through co-located displays particularly in situations requiring a 3D rotation (Ware C. & Rose J., 1999).

These considerations led us to devise a study to test if a novel system with the above specifications would outperform a WIMP system running 3D studio MAX™[2].

## 3   Hypothesis

Independent variables to be tested on a macro-level are the above two systems and the provision of haptics, stereovision and spatial input in related subsystems. Dependant variables are performance and user's perception of workload and of system's usability in completing a 3D Combining task. Two systems were compared: (a) a PC with 2D-display, 2DOF input device, with standard 3D modelling software and (b) a workstation with 6DOF spatial input device which selectively affords stereo vision and haptic feedback, running a prototype 3D modelling software, with an optional snap alignment tool. Subsystems of (b) were tested against (a), and against each other in order to clarify contribution of single elements of (b) to the overall performance. This produced a total of seven conditions:

    I.  WIMP system running 3Dstudio MAX
    II.  6DOF input device, stereovision, haptic feedback, with snap tool
    III. 6DOF input device, stereovision, haptic feedback, with no snap tool
    IV. 6DOF input device, no stereovision, haptic feedback, with snap tool
    V.  6DOF input device, no stereovision, haptic feedback, with no snap tool
    VI. 6DOF input device, stereovision, no haptic feedback, with no snap tool
    VII.  6DOF input device, no stereovision, no haptic feedback, with no snap tool

In the context of related work we hypothesized the following:
   i.    System (b) and subsets will lead to a better performance compared to (a)
  ii.   System (b) and subsets will lead to lower workload scores compared to (a)
 iii.   System (b) and subsets will lead to a higher usability compared to (a)
  iv.   A correlation will be found between performance, usability and workload perception.

---

[2] According to on-line pools, 3D studio Max™ is one of the most used 3D modeling software. (e.g. http://www.renderology.com/polls/default.asp). It also offers a solid scripting utility that has been exploited.

v.  The spatial input device will be the most important contributor to the increase in performance (due to affordance of 3D rotation / translation and better cognitive fit to 3D).

vi.  Haptics and stereovision will decrease completion times and perception of workload.

## 4  Experimental Design and Procedure

The experimental method is a "within subjects" design. A total of 12 subjects were recruited among expert users of 3Dstudio Max™. A WIMP (Window Icon Menu Pointer) system running 3D Studio Max™[3] (a) and a system equipped with a Reachin Developer Display[4], a PHANToM™[5] haptic force feedback stylus device, a co-located display and a non-dominant hand input device, Magellan Space Mouse ®[6], (b) were utilized.

An equivalent task was performed on both workstations. Task entailed placing 4 different geometric elements, randomly scattered and orientated in space, against non-movable "target" surfaces, recognisable by matching dimensions and colors, also placed in various positions and random orientations.

As users were asked to repeat the given task three times under the seven different experimental conditions (I to VII)[7], random orientation and positioning of target surfaces and shapes were predetermined for each repetition. The presentation order of the different conditions and repetitions was randomised to neutralise any learning effect or other relevant interactions. Subjects were allowed to adopt any preferred strategy within the system's limitations, and time limits were not imposed, although accuracy and completion times were measured to evaluate performance[8]. Qualitative data was gathered using a computer version of the TLX [9] Task Load Index test (NASA) and in a SUS (System Usability Scale, Brook, J. 1996) questionnaire. Qualitative data was gathered after each subject had completed the three trials of each condition (I-VII). Instructions for completing the questionnaires and operating the Reachin's system were given before each experimental session. Users were given time to familiarise themselves with the task under the various conditions.

## 5  Results and Discussion

Data analysis confirmed the main hypothesized outcome as stated in (i). Figure 1 summarizes data for completion times, used as a measure of performance: a striking difference was found between condition I, which tested system (a), and all other conditions. (e.g. condition II against I: $F_{(1,70)}$ = 106.7671, $p<0.01$). Significant discrepancies in performance held true with no regard as to whether or not haptics and stereovision were used. (VII against I: $F_{(1,70)}$ = 87.95473, $p<0.01$). This is in agreement with hypothesis (v), as the 6DOF device affording spatial input produced the most significant decrease in completion times. However, performance increased significantly when graphic stereo cues and haptic feedback were provided in addition to the spatial input, in agreement with hypothesis (vi) (VII against III: $F_{(1,70)}$ = 4.833404, $p<0.05$). Times dropped further under condition II (stereo, haptics, and snap tool) and this was found to be statistically

---

[3] Wide variability in data obtained from 3D studio MAX probably indicates a variation in subjects' expertise.

[4] http://www.reachin.se/products/reachindisplay/ (Retrieved February 21, 2003)

[5] http://www.sensable.com/products/phantom_ghost/phantom.asp (Retrieved February 21, 2003)

[6] The Space Mouse affords 3D rotation. This function was not enabled in this study, in order to maintain a coherent equivalence between systems (a) and (b) and to avoid introducing additional variables.

[7] Under all conditions which included haptic feedback (I, II, III) this was provided in the form of a force feedback activated by the point of the stylus hitting a target surface.

[8] Each user clicked a button to start and end the task. Precision results will be omitted in this paper.

[9] The NASA Human Performance Research group defines six different factors for workload (mental demand, physical demand, time pressure, effort expended, performance and experienced frustration), measured through the TLX index tool. Elements are rated, weighted with pair-wise comparisons, and then combined.

significant (condition II against III: F (1,70) = 55.58223, p<0.01). Although the sole use of haptics or stereovision failed to reach significance in the increase of performance over spatial input alone,[10] haptics provision did reach significance when ANOVA tests where repeated excluding one of the subjects, whose measures were extreme outliers. Other results did not vary.

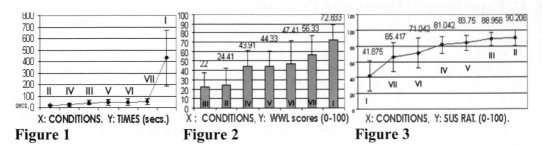

X: CONDITIONS. Y: TIMES (secs.)  X : CONDITIONS, Y: WWL scores (0-100)  X: CONDITIONS, Y: SUS RAT. (0-100).

**Figure 1**          **Figure 2**                    **Figure 3**

## 5.1    Perceived Workload

Paired T tests were applied to WWL (weighted workload) scores,[11] which are summarized in fig. 2. Lowest perceived workload (mean: 22/100) was obtained when stereo and haptics without the snap tool were used to complete the task (condition  III). Highest WWL score was attributed to the WIMP system running 3D MAX (condition I, mean: 72.83333/100). Overall results also confirm hypothesis (ii), with all WWL scores attributed to system (b) and subsystems significantly lower than whose attributed to system (a)

WWL scores for all conditions where haptics was provided exhibited lower workload rates than all other conditions; this suggests further investigations to better understand effect of haptics in lowering specific elements of workload.

Introduction of the snap tool failed to decrease WWL scores significantly. Perceived workload under condition V (haptics and spatial input) and in condition VI (stereo and spatial input) was lower than in condition VII (spatial input only), although results failed to reach significance in this latter case (condition VI against VII). These findings partially confirm hypothesis (vi).

## 5.2    System Usability Ratings

Paired T tests were also carried out on SUS ratings, whose mean values and standard deviations are shown in fig. 3. Results strengthen findings obtained from previous analysis on Workload Scores.  Condition I (3D studio Max) was rated as the less usable system (41.875/100), followed by condition VII (spatial input). Condition II was perceived as the most usable (90.208/100), whereas the greatest significant gap between ratings was found between condition I and VII ($T_{11}$= - 3.61654, p= 0.002025). Hypothesis (iii) is thus confirmed, since the less usable of the (b) subsystems is significantly more usable than system (a).  Again, a significant increase in usability ($T_{11}$= 4.147522, p= 0.000812) was found between the haptics and spatial input condition (V), compared to spatial input alone (VII). Stereovision (V) failed to determine a significant increase in the perceived usability of the system compared to condition VII (spatial input), although it was rated as more usable. All conditions exhibiting haptic feedback were significantly more usable than other conditions (e.g. condition VI against IV: $T_{11}$= -2.48504, p=0.015151)

The latter is in agreement with WWL results, which exhibit a strong negative correlation to the SUS ratings, accordingly to the Spearman's ranked correlation coefficient (s =-0.76122084,

---

[10] (condition V: F (1,70) = 2.834078, p>0.05; condition VI:F(1,70) = 0.295738, p>0.05)

[11] III, II : $T_{11}$= -0.50769, P= 0.310846612209752 - II, IV: $T_{11}$ =3.03782, P= 0.005645 -IV,V: $T_{11}$= -0.12101, P= 0.452932 - V,VI: $T_{11}$= -0.38631, P= 0.353317 - VI ,VII: $T_{11}$= -1.23036, P= 0.122113 - VII, I: $T_{11}$= - 1.94813 P= 0.038682 - V, VII: $T_{11}$= -2.2532 P= 0.022816

p=4.31264E-17). Significant negative correlation was also found between SUS ratings and time measures (s = -0.2205, p=0.043851), thus validating hypothesis (iv). However, the positive correlation found between time measures and perceived workload failed to reach significance (s=0.153915, p=0.162154).

## 6 Concluding Remarks

The study highlights the fundamental role of spatial input for 3D tasks, suggesting that its use could greatly improve 3D modelling systems. Stereovision and Haptics seem also to ease operating within three dimensions. Haptics seems beneficial in lowering perception of workload and increasing usability, while interface widgets such as the tested snap tool could contribute to lower completion times. Further investigations should clarify related issues in greater depth.
Cautious interpretation of these clear results rests mainly with the issue of systems' equivalence.

## References:

Dudlu N. I., K. H. Manoochehri K. H., Baraff, D (October 1986). Multi-Dimensional Input Techniques and Articulated Figure Positioning by Multiple Constraints. *Proc. 1986 ACM Workshop on Interactive 3D Graphics* (pp. 151-170)

Beaten R.J., DeHoff R.J., Weiman, N and Hildebrandt, W. (1987). An Evaluation of Input Devices for 3-D Computer Display Workstations, in *Proceedings of SPIE, vol. 761,*The International Society for Optical Engineering, (pp. 94-101).

Brook, J., (1996). SUS: A 'Quick and Dirty' Usability Scale. *Usability evaluation in industry*. Digital Equipment Corporation. Retrieved 21.02.2003 from http://www.cee.hw.ac.uk/~ph/sus.html

Butterworth J., Davidson A, Hench S. and Marc T. Olano (1992). 3DM: A three Dimensional Modeler Using a Head Mounted Display, *Proceedings of the 1992 Symposium on Interactive 3D graphics* (pp.135-138). New York: ACM Press.

Hinckley K. (1996). Haptic Issues for Virtual Manipulation. Faculty of the School of Engineering and Applied Science, University of Virginia, Ph.D. dissertation, retrieved: 05.02.2003 from http://research.microsoft.com/Users/kenh/thesis/front.htm

Hinckley K., Paush R., Proffitt D. (1997). Attention and visual feedback: the bimanual frame of reference, *Proceedings of the 1997 symposium on Interactive 3D graphics* (pp.121-126). New York: ACM Press

Jacob, R.J.K. & Sibert L.E. (1992). The Perceptual Structure of Multidimensional Input Device Selection. *Proceedings CHI '92* (pp.211-218). ACM

Massimino, M. J., & Sheridan, T. B., (1994). Teleoperator performance with varying force and visual feedback. *Human Factors*, 36 (1), (pp. 145-157).

NASA Human Performance Research Group. *Task Load Index (NASA-TLX) v1.0 computerized version*. Ames Research Center Moffett Field, California (415) 694-6072

Parsons L. (1995). L., Inability to reason about an object's orientation using an axis and angle of rotation, *Journal of Experimental Psychology: Human Perception and Performance*, Vol.21, No.6, (pp. 1259-1277)

Wall S., Paynter K., Shillito A.M., Wright M., Scali S., (2002). The Effect of Haptic Feedback & Stereo Graphics in a 3D Target Acquisition Task*., Proceedings of Eurohaptics 2002* (pp. 23-29).

Wang Y. and MacKenzie C L., Summers V. A., & Booth K. S. (1998). The Structure of Object Transportation and Orientation in Human-Computer Interaction, *CHI '98 Proceedings*. (pp. 312 - 319). New York: ACM Press.

Ware C. & Rose J., (1999). Rotating virtual objects with real handles. *ACM Transactions on Computer -Human Interaction (TOCHI),* volume 6, Issue 2, June 1999 (pp. 162 - 180). New York: ACM Press.

# PICK – A Scenario-based Approach to Sensor Selection for Interactive Applications

*Jennifer G. Sheridan, Jen Allanson*

Lancaster University
Lancaster, UK LA1 4YR
{sheridaj, allanson}@comp.lancs.ac.uk

## Abstract

An explosion of new interaction paradigms makes the design of contemporary interactive systems particularly difficult. System designers are swamped by often conflicting models of interaction which can obscure the fundamental issues that need to be addressed in order to build an application for given users in prescribable situations. This paper describes on-going work to create tools to assist designers of sensor-rich systems. Our approach is paradigm-light and is concerned with working methodically from a scenario of usage through to potential technical solutions.

## 1 Introduction

More than a decade ago Nielsen proposed that the "fifth generation user interface paradigm [will be] centered around non-command-based dialogues" [5]. Three years later in their *Agenda for Human-Computer Interaction Research* Jacob et al. acknowledged that support for the new paradigm would come in the form of new input devices, interaction techniques and software approaches [4]. Saffo identified what the input devices would be [6] and the first questions about the new sensing-based paradigm were asked by Bellotti et al. [1].

However, unlike traditional mouse and pointer interaction, the possible ranges and number of sensors to be employed in any one interactive system is both diverse and potentially variable. While an abundance of HCI resources exist, very few of these resources focus on tools and techniques specifically for the design of sensor systems.

## 2 PICK Framework

Our approach centres around a model we call PICK (Phenomena, Intimacy, Constraints and Knowledge). The PICK model borrows ideas from HCI theory to help designers to methodically deconstruct a narrative describing the circumstances in which a system may be deployed. The PICK approach is both scenario-based and context-sensitive. It encourages system designers to consider not only interactive phenomena themselves, but also the affects of those phenomena in specific circumstances and environments. Use of the model involves the following steps:

1) Preliminary investigation of the design space.
2) Application of the PICK Model.

Preliminary investigation of the design space involves definition of the situation in which a technical solution may be appropriate. A simple way to develop a problem statement which describes the design space is by simply writing a narrative describing the situation. This gives us a context-sensitive scenario, which we can methodically deconstruct into a formal specification of the problem space. In its current form the PICK approach requires system designers to extract appropriate words from the narrative scenario (Table 1). The rules for breaking the narrative scenario are as follows:

- *User/Effect (Nouns):* Define a typical user or effect on the system using nouns.
- *Action (Verbs):* Using verbs and adverbs define the action that is taking place.
- *Location (Clause/Phrases):* Describe the area in which the sensing will be taking place, using phrases.
- *Context (Clause/Phrases):* Describe occasional situations that may occur when in this particular situation.
- *Objects/Hazards (Any Descriptor):* Describe potential hazards that may affect the system.

Terms defined in the problem statement are then applied to the scenario

Table 1: Charting a scenario to determine interaction issues to be addressed

| | Problem Statement | | | | |
|---|---|---|---|---|---|
| **Colloquial** | I am | dancing jumping | in a club | indoor, lights and music | while drinking, smoking |
| **Grammatical** | Subject | Verb | Phrase/ Clause | Phrase/ Clause | Phrase/ Clause |
| **Purpose** | User + Artefact | Active | Environment/ Conditions | Description/ Desired Condition | Description/ Negative Condition |
| **Scenario** | User/ Effect | Action | Location | Context | Hazards |

The next step in the process is to apply the PICK Model.

## 2.1 Phenomena

Using our scenario, we can then consider which actions are candidates for augmentation. We can consider phenomena in terms of things that can be sensed – things that we might look at technically in terms of physical and chemical transduction principles for example. However, understanding of these principles rather obscures the issues we want to consider at this stage. Therefore, in PICK we deal with phenomena using familiar language. Actions identified from the scenario should be categorised in terms of desired and undesired outcomes (Table 2). Desired outcomes are those actions we wish to avoid and undesired outcomes are those we wish to achieve. Once the actions have been defined, they can be used to determine both the human sense that is being affected and what that affect produces.

**Table 2:** Plotting actions to determine phenomena

| | Condition | Desired? | Affects on subject/object | Phenomena |
|---|---|---|---|---|
| **Action** | 1. dancing | Yes | Increased heart rate, perspiration | Heart rate, body temperature, skin conductance |
| | 2. sitting | No | Low heart rate, low movement | Heart rate, body temperature |

Without using complex table of principles, a designer can determine the transduction principles that apply to their scenario (the phenomena). This in turn provides them with potential input and outputs.

## 2.2 Intention

Intention, the second PICK principle, considers the particularities of both active and passive systems. Passive sensing requires that a system become engaged, at some definite point, due either to an environmental trigger or through a user-initiated action. Either way, users have a measure of control over when passive sensors receive input. Consequently, passive sensors require some form of interface through which users may manipulate variables such as threshold levels or trigger points. Passive sensors can be seen to have discrete parameters. In contrast active sensors are continuously engaged with their user/environment and thus receive a constant stream of data. Active sensors, then, will have pre-defined thresholds and thus users will have less active roles in how and when the sensor receives input.

Ubiquitous communication between users and non-users provide a way of looking at mediated communication in terms of *Intimate Computing*. We define and develop intimacy in four separate stages, each of which are linked through users' growing certitude:

- *Appeal:* How is interaction initiated?
- *Engagement:* How do we prolong interaction?
- *Experiential Indulgence:* How do we ensure continuous motivation?
- *Experiential Performance:* How do we encourage repeated engagement?

## 2.3 Constraints

Designers are constantly required to resolve conflicting design demands whilst working with a variety of constraints. Such constraints include budget, time constraints, and availability of components. Another possible constraint is appropriate safety and health practices etc. As in any engineering discipline, trade-offs must be continuously identified, evaluated, and decided on the basis of the best information available. Tradeoffs such as cost, weight and size will also help to determine what sensors are available.

In the Sense and Sensability framework model, Benford et al. provide a systematic comparison of what is sensible, sensable and useful [2]. While this framework proves useful for revealing the potential constraints of an interface, the PICK framework considers that constraints may take the form of design standards and guidelines that fluctuate in generality and authority [3].

Constraints determine how users can stabilize the design of a sensor-rich system. Following these constraints, users can draw attention to potential hazards, detail sensor performance and determine design trade-offs for sensor systems.

The PICK framework includes facilities for the ranking and sorting of design requirements (Table 3).

**Table 3.** Constraints determine the authority, tradeoffs and importance of design variables.

| | | Constraint | Authority | Tradeoff | Importance |
|---|---|---|---|---|---|
| VARIABLE | **User/ Effect** | age | guideline | diversity | critical |
| | **Location** | bar | guideline | range | bon-critical |
| | **Context** | indoors | standard | range | critical |
| | **Objects/ Hazards** | body temp | standard | threshold | critical |
| | **Appeal** | tempt | standard | number of users | non-critical |
| | **Engagement** | visual | guideline | non-sighted users | non-critical |
| | **Indulgence** | patience | guideline | complexity | non-critical |
| | **Performance** | stimulation | standard | familiarity | critical |

## 2.4 Knowledge

The final step in the PICK model involves the consideration of what we may know by looking at the data. This part of the model is concerned with data gathering/ analysis. Specifically we consider:

- Input: How does the system begin to stream data (user, system or admin input)?
- Values: What values are required?
- Effects: What effect will this have on the system/user?

**Table 4.** Consideration of data collection

| | | DATA | | | |
|---|---|---|---|---|---|
| | | **Constraint** | **Input** | **Value** | **Effect** |
| VARIABLE | **User/Effect** | user age | user | min/max | stop/start |
| | **Action** | dancing | system | threshold | maintain beat |
| | **Location** | bar | admin | parameters | in/out of range |
| | **Context** | indoors | admin | parameters | in/out of range |
| | **Obj/Hazards** | body temp. | system | min/max | stop/start |
| | **Appeal** | tempt | user | on/off | stop/start |
| | **Engagement** | visual | user | timer | record time |
| | **Indulgence** | patience | system | max level | record end level |
| | **Performance** | stimulation | system | times used | count number of times used |

### 3   Conclusions and Future Work

The result of running the PICK model on an initial scenario should be a fairly clear set of system requirements. In this way the *actions* of users are augmented through the use of sensors. This is the approach advised by Tolmie et al. [7] and is in contrast to existing approaches where artifacts are augmented because new sensors become available.

The next step for us is to ascertain the utility of the PICK model. This can only be achieved by engaging the interactive systems design community on order to test the robustness of the model in designing real-world sensor-rich interactive systems.

## 4   Acknowledgements

This work is funded by the EPSRC Grant Number GR/R45253/01.

## 5   References

[1] Bellotti, V., Back, M., Edwards, W. K., Grinter, R. E., Henderson, A., Lopes, C. "Making Sense of Sensing Systems: Five Questions for Designers and Researchers" in proceedings of CHI2002, pp 415-422, 2002.

[2] Benford S., Schnadelback H, Koleva B., Paxton M., Anastasi R., Greenhalgh C., Gaver B. (2002). Sense and Sensability: a framework for designing physical interfaces. Technical Report Equator-02-009, Equator, September.

[3] Dix, A., Finlay, J., Abowd, G., Beale, R. (1998). Human-Computer Interaction (2nd ed.). Essex: Prentice Hall.

[4] Jacob R.J.K., Leggett, J.J., Myers, B.A., Pausch, R. (1993). "Interaction Styles and Input/Output Devices" Behaviour and Information Technology, Vol. 12, no. 2, pp. 69-79.

[5] Neilson J. (1990). "Trip Report: CHI'90," SIGCHI Bulletin, Vol. 22, no.2, 20-25.

[6] Saffo P. (1997). "Sensors: The Next Wave of Innovation" in Communications of the ACM, Vol. 40, No. 2.

[7] Tolmie, P., Pycock, J., Diggins, T., MacLean, A., Karsenty, A. (2002). Unremarkable Computing. In Proceeding of CHI2002, 99-106.

# A Proposal of Guideline for Colour Arrangement on Screen Design Used in VDT Work

*Masanori TAKEMOTO, Yusaku OKADA*

KEIO University, Graduate School of Science & Technology,
Department of Science for Open & Environmental Systems,
Y.OKADA Laboratory (Human Factors & Ergonomics)
Hiyoshi 3-14-1, Kohoku-ku, Yokohama city, Japan, 223-8522
takemoto@ae.keio.ac.jp, okada@ae.keio.ac.jp

## Abstract

This study intends to examine the way of changing background colour on computer display. In particular, gradual change, which is called 'the Gradation Type', was focused on.

To begin with, images of some colour combinations used for the Gradation Pattern were evaluated using subjective methods. Next, the Gradation Type was applied to indicating that a time limit is approaching in order to examine its effect on emotion through an experiment.

As a result, it is obtained that the Gradation Type could control users' psychological stress given by time pressure. Furthermore, the effective colour combinations in the Gradation Type were found out. This study would suggest more effective use of colour when users monitor computer display passively.

## 1 Introduction

Today, primary work in industrial plants becomes the supervisory control with display units, and so operators are required to detect the symptoms of the accident immediately. However, it is difficult for operator to notice the unusual signs before the alarm sounds and the colour of annunciators changes. If the operator finds out the unusual signals early, it will become possible to prevent an accident. For that purpose, the indication that operator recognize unconsciously should be introduced into VDT work in industrial plants.

In this paper, we intend to discuss the information from the viewpoint of operator's consciousness, that is, the information is divided into two following types.

(1) Information of User-Active Type: Information that users get actively (with consciousness).

(2) Information of User-Passive Type: Information that users get passively (without consciousness).

In this study, we examine the indication of information which users can get uncomsciously not depending on their ability. And this study proposes the gradual changing colours (it is called 'the Gradation Type') which inform users of elapsed time, and we discuss its psychological effects and their influences on performance.

## 2 Colour Image and Changing Colour Type

### 2.1 Evaluation of Colour Image

711

First, we evaluated colour image on computer display for four categories: 'Prohibition', 'Warning', 'Intention' and 'Safety'. As a result, Red(R=255, G=0, B=0), Yellow(R=255, G=255, B=0), Blue(R=0, G=0, B=255) and Green(R=0, G=192, B=0) gained the highest value in each category. And it was confirmed that colour image was influenced by not only hue but also tone of colour.

## 2.2 Proposal of Changing Colour Type

This study proposes the gradual changing background colour as information of User-Passive Type.
(1) Hue-Gradation Type: A starting colour gradually changes to an ending colour for the decided time. It is the Gradation Type that hue of colour changes.
(2) Tone-Gradation Type: It is the Gradation Type that hue and tone of colour changes, based on the result that pale colours give safe image. A starting colour gradually changes to an ending colour via a pale middle colour.
In order to consider the characteristics of the Gradation Type, following type of displaying information are used.
(3) Sudden Changing Type: A background colour changes to another colour at half of the decided time.
(4) Progress Bar: The lengths of bar changes at a uniform pace, and become max at ending time.
(5) Digital Time: The rest of time is displayed using numerical value.
In changing colours, four colours which had the strongest image in each category (Prohibition, Warning, Intention and Safety) are used for a starting colour and an ending colour.

## 3 Psychological Influence on Users

### 3.1 Method

The subjects observed changing colours on computer display and evaluate the each type subjectively by use of questionnaires. 12 kinds of Hue-Gradation Type, 16 kinds of Tone-Gradation Type and 12 kinds of Sudden Changing Type were used in this experiment. The questionnaire had 5 items on feeling of colour, 3 items on visibility of colour and 5 items on elapsed time and changing colours. The number of subjects in this experiment was ten.

### 3.2 Value on Influence of Psychology by Subjective Evaluation

As the result of the subjective evaluation, application of principal component analysis revealed characteristics of the each type. Sudden Changing Type would discompose user's mind and Hue-Gradation Type would compose users' mind at base. However, some of them cause discomfort because their middle colours are not comfortable. Tone-Gradation Type would bring users comfort due to pale colours.

### 3.3 Chromaticity Diagram Model

We made a model which expresses psychological influences of changing colours of Hue-Gradation Type by use of the chromaticity diagram (Figure 1: Chromaticity Diagram Model).
Four basic colours could be divided into two groups from the meaning of category. That is to say, the first group includes red meaning 'Prohibition' and yellow meaning 'Warning' which give negative image. The second group includes blue meaning 'Intention' and green 'Safety' which give positive image. If a starting colour and an ending colour of Hue-Gradation Type are different

groups of image, middle colour could have different image from the image of the two colours. We called this case 'Confusion of the Meaning of Colour Image', and the combinations of red and blue, red and green were applicable to this from the result of value on psychological influence. If the middle colours of the Gradation Type cause discomfort, we called this 'Feeling Bad with the Way of Changing Colours'. The combination of red and green was applicable to this.

Theoretically, the combination of yellow and blue is applicable to 'Confusion of the Meaning of Colour Image' and 'Feeling Bad with the Way of Changing Colours'. However, this combination wouldn't discompose users' mind because the middle colour of yellow and blue is achromatic colour which have not image of colour.

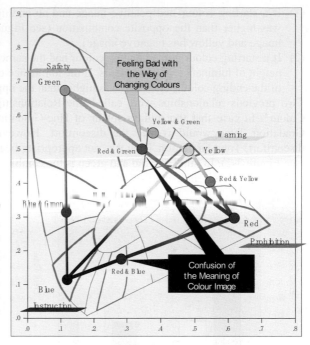

**Figure 1:** Chromaticity Diagram Model

# 4 Value on User's Performance in VDT Work

## 4.1 Method

We evaluated user's performance in VDT work by use of five types of information explained in section 2.2. The screen used in this experiment is shown in Figure 2. Forty-five characters which mean nothing was arranged in a quadrangle placed in the left side (see Figure 2). These characters were composed of one capital letter and seven small letters. Subjects selected one of the characters and a button which corresponded to the capital letter in the right side as many times as possible within thirty seconds. Eight subjects performed this experiment.

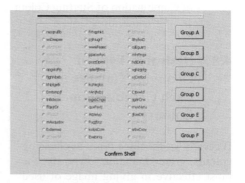

**Figure 2:** Experimental Program

## 4.2 Result

We considered the effects of changing colours based on error rate. As the result, in more case than half of Sudden Changing Type and Hue-Gradation Type and Tone-Gradation Type, error rate was lower than that of Progress Bar and Digital Time which are conventional display of time. In other words, changing colours could work effectively on user's performance.

In the case of Hue-Gradation Type, the relationship between a starting colour and an ending colour of changing colour led to the following remarks.

713

(1) If a starting colour had positive image and an ending colour had negative image, error rate was higher than the opposite combination (see Figure 3-1. In this case, green has positive image and yellow has negative image).

(2) If a starting colour and an ending colour had the same image, error rate was influenced by the height of luminance of two colours. If the luminance of the starting colour was lower than one of the ending colour, error rate was higher than the opposite combination.

Two previous relationships were called 'the Relationship between Starting Colour and Ending Colour'. In case that the middle colour of Hue-Gradation Type made users discomfort, Tone-Gradation Type would reduce the discomfort. However, if the middle colour didn't make discomfort, Tone-Gradation Type was not appropriate because pale colours reduced colour image (see Figure 3-2). In this case, red and green is the combinations which make users discomfort, and yellow and green is the combinations which make users comfort.)

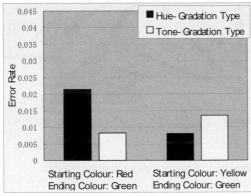

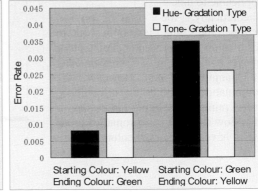

**Figure 3-1:**

Influence on Error Rate
of Combination of Starting Colour
and Ending Colour

**Figure 3-2:**

Influence on Error Rate
of Comfort of Middle Colour

## 4.3 Desirable Type of Changing Colours to Each Combination of Colours

Basically, of the three types of changing colours, Hue-Gradation Type is the most effective in giving image of used colours. However, in some cases, Hue-Gradation Type makes error rate higher by 'the Relationship between Starting Colour and Ending Colour' and so on. And we made a flow chart. This chart reveals a procedure for deciding the type of changing colours in order to reduce error rate (see Figure 4).

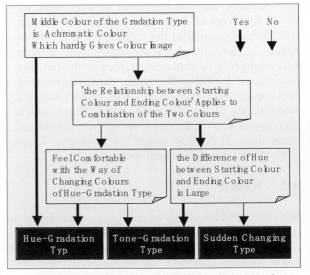

**Figure 4:** Procedure for deciding the type of changing colours

# 5　Conclusion

Five types of displaying information are characterized as Table. Their important characteristics are shown as followings; (see Table 1)

(1) Ease of recognition of elapsed time:

- Using Sudden Changing Type, users cannot recognize elapsed time at all.
- Two kinds of the Gradation Type help users recognize elapsed time, and change users' motivation for work in closing time limit. But they have an influence of time pressure on users.

(2) Psychological influences:

- Users would be disturbed by changing background colour, using Sudden Changing Type.
- Two types of the Gradation Type wouldn't have a great influence on users by changing colour because background colour change gradually.

This study revealed the basic effects and characteristics of the Gradation Type from the viewpoint of their psychological influence and their effects on user's performance. In future, we would suggest more effective way of displaying information in getting information passively in consideration of users' character.

**Table 1:** Characteristics of Five Type of Displaying Information

| | Sudden Changing Type | | Hue-Gradation Type | | Tone-Gradation Type | | Progress Bar | | Digital Time | |
|---|---|---|---|---|---|---|---|---|---|---|
| the object informing time | background colour | | background colour | | background colour | | indicator on the screen | | numerical value on the screen | |
| the way of expression of elapsed time | background colour changes at set time | | a starting colour gradually changes to an ending colour | | a starting colour gradually changes to an ending colour via a pale middle colour | | the length of bar changes at a uniform pace, and become maximum at ending time | | the rest of time is displayed using numerical value | |
| ease of recognition of elapsed time | very difficult | × | can feel passing time, but be difficult to recognize ending time accurately | △ | can feel passing time, but be difficult to recognize ending time accurately | △ | easy | ○ | very easy | ◎ |
| awareness of coming of the time limit | cannnot aware | × | be easy to aware | ○ | be easy to aware | ○ | be difficult to aware if the object is not large | △ | be difficult to aware if the object is not large | △ |
| influence of time pressure | nothing | ◎ | strong | × | strong | × | depends on the size of the object | △ | depends on the size of the object | △ |
| change of motivation for work in closing time limit | nothing | × | be enhanced | ◎ | be enhanced | ◎ | depends on the size of the object | △ | depends on the size of the object | △ |
| psychological influence — before the time at which display changes | not be influenced | ◎ | be influenced | △ | be influenced | △ | be influenced | △ | be influenced | △ |
| psychological influence — at the time at which display changes | be much influenced | × | be influenced as much as before | △ | be influenced as much as before | △ | be more influenced than before | ▲ | be more influenced than before | ▲ |
| psychological influence — after the time at which display changes | be much influenced | × | nothing | ○ | nothing | ○ | could be influenced | △ | could be influenced | △ |
| effects of used colours | strong | ○ | very strong | ◎ | strong | ○ | not so strong | △ | a little | ▲ |
| total value for information of User-Passive Type | not appropriate (because users couldn't recognize elapsed time and are disturbed by changing colours) | × | effective (used colours are limited) | ○ | effective (used colours are not so limited) | ◎ | not appropriate (this type would be effective in information of User-Active Type) | △ | not appropriate (this type would be effective in information of User-Active Type) | △ |

# Development and Validation of a Tool for Measuring Online Trust

*Christy Thomas*, Meridian Incorporated, Omaha, NE 68117,
cthomas@meridianmap.com
*Cynthia L. Corritore*, College of Business, Creighton University, Omaha, NE
68178, cindy@creighton.edu
*Beverly Kracher*, College of Business, Creighton University, Omaha, NE 68178,
bkracher@creighton.edu
*Susan Wiedenbeck*, College of IST, Drexel University, Philadelphia, PA 19104,
susan.wiedenbeck@cis.drexel.edu

## Abstract

Trust is necessary for successful online interactions. In this paper, we present a tool to measure online trust and its antecedents. We report on construction of the tool, its testing, and the analysis of its reliability and validity. Future plans for refinement of the tool are discussed.

## 1 Overview

As the World Wide Web matures into its second decade, it has become host to a wide range of interactive activities that have gone beyond the visions of its original creators. Its users, including academics, scientists, entrepreneurs, large and small businesses, online hobbyists, and a variety of non-technical people, have shaped the World Wide Web to meet their needs. Yet underlying the web's exponential growth and eclectic usages, there remains a central truth: the World Wide Web is a key communication technology used by humans today. As in many other human activities, trust is necessary to facilitate successful interactions on the web [2]. It is easy to believe that without trust, the robust online environment of the web would not be possible. However, while trust and trust relationships in the off-line world have been a topic of research in many fields since the 1950s, little research has been done on the role of trust in an online context.

We, along with other early trust researchers, are exploring the concept of online trust. We identify two major problems facing online trust researchers today. First is the issue of synthesizing previous research on trust in the off-line world. Trust has been studied for decades in a wide variety of fields. Work in each of these fields has produced its own definitions and findings. In fact, even within a given field, there is often a lack of agreement and focus of effort [12]. This is not surprising in the study of such a core concept as trust, where researchers typically work against the background of their own discipline-specific paradigms, which are often contradictory and inconsistent [11]. However, it does make it difficult to build upon previous work. Nevertheless, this large and dispersed body of research is currently being examined in order to tease out the conditions of off-line trust that are essential for online trust [5,10]. Special issues in trust and technology are being published [18,21]. This effort of synthesizing previous research on trust is on-going and leading toward better understanding in the interdisciplinary online trust community. At this point in time, online trust work is advancing to the stage of proposing models of online trust, based on the body of trust literature [4,13].

The second problem facing online trust researchers is more serious. Presently there is a lack of tools for empirically measuring trust in an online context.. This is a serious problem as it holds back any empirical work that could study, test, and revise proposed online trust models. Some have tried to address this problem by looking to reliable and valid trust measurement tools previously developed in the general trust literature. However, this is problematic. While such tools do exist, they primarily focus on interpersonal or organizational trust [eg, 8,17], making their use in a typical online context questionable. Similarly, there are also existing tools that measure concepts related to trust, such as credibility or relationship commitment [9,14]. However, once again, these tools do not directly address online trust.

What is needed to advance research on online trust is a trust tool that is theory-based, general enough to be useful in the study of trust in the wide variety of online contexts that exist, and thoroughly tested for its psychometric properties. A useful analogy is Fred Davis's [7] work on the Technology Acceptance Model (TAM). He developed and validated measurement scales that were broadly applicable to measuring the perceptions of ease of use and usefulness of computers in organizations. Today dozens if not hundreds of researchers have used his scales. Such a meta-model, along with a reliable and valid tool for measurement, is needed for the next step in online trust research.

This paper presents our work to date in developing and testing an online trust scale. We begin by describing our model of online trust. Then we discuss the development of a tool for measuring online trust. Next, we describe the process by which we tested the tool. We conclude with our plans for finalizing the tool in order to use it in future empirical research.

## 2 The Model

Our Online Trust tool was designed to measure online trust as it is proposed in our Web Trust Model (see Figure 1). This meta-model of online trust is firmly based on the rich body of trust literature from the fields of HCI, human factors, psychology, sociology, business, and philosophy. An extensive discussion and justification for it can be found in Corritore, Kracher, & Wiedenbeck [4]. Note that all of the trust factors in the model are perceptions of a given user about a specific website, and so are accessible through questions posed to a user.

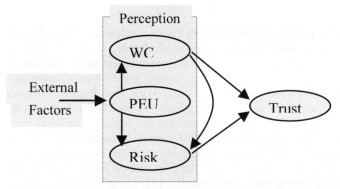

Figure 1: Depiction of the Web Trust Model

## 3 Development of the Online Trust Measurement Tool

We developed multiple item scales to measure online trust and its factors. The tool is comprised of a total of 29 statements. The distribution of the questions was: Web Credibility: Honesty (5), Expertise (4), Predictability (4), Reputation (4); Ease of Use (4); Risk (4); and Trust (4). The statements about Ease of Use were taken directly from Davis's Perceived Ease of Use scale [7]. We developed the remaining 25 statements based on the previous trust research. Each statement is rated using a five-point Likert-type scale running from Strongly Disagree to Strongly

Agree. Three experts evaluated the form and content of the tool. As a result of their evaluation, several items were modified, deleted, and added.

# 4 Testing the Online Trust Measurement Tool

## 4.1 Participants

One hundred and eighty-four students in a United States mid-western university performed tasks on predetermined websites, then completed the trust measurement tool to describe their experiences. They were undergraduate students in the College of Business Administration and graduate students in the School of Pharmacy and Allied Health. Participants were seated in computer classrooms in groups of 30 to 50. Some received extra credit points for their participation in the test.

## 4.2 Running the test

Participants completed one of two possible online scenarios, interacting either with an existing transactional or informational website. Each scenario consisted of three tasks. Participants were asked to read their scenario, perform each task, then complete the trust measurement tool which was presented online. All twenty-nine statements of the tool were presented on the same page, one per line with a space between every five statements to facilitate easy reading. Next to each statement were five radio buttons, one for each Likert choice. The statements were numbered and presented in the same order for all participants. When participants were done responding to the statements, they clicked on a button labeled 'Finished' to submit their answers. Answers were sent to an email address. All responses were anonymous.

## 4.3 Results

We analyzed the data using Confirmatory Factor Analysis (CFA) and Cronbach's alpha. CFA is a standard way to evaluate construct reliability, confirming the underlying structure of a tool [1]. CFA allowed us to identify which items on our measurement tool correlated with the underlying factors of our tool [19]. The underlying factors were Web Credibility (which has four components), Perceived Ease of Use, Risk, and Trust (see Figure 1). Specifically, Principle-components analysis was used for factor extraction. We used Kaiser's greater-than-one eigenvalue criterion. Varimax rotation was used to transform the initial factors into more meaningful and interpretable factors. According to Stevens [20], a cases-to-variables ratio of 5:1 assures a reliable factor analysis procedure. Our test had 29 variables and 180 cases, resulting in a case-to-variable ratio of 6:1, well over the suggested ratio. Varimax rotated factor loadings were evaluated using the threshold of 0.35 recommended by Churchill [3]. All of our items loaded much higher than this recommendation. We ran three factor analyses: first, with seven factors over all 29 items (this included all four components of Web Credibility); second, with three factors over 25 items (four risk items were removed from consideration); and third, with four factors over the 17 items related to Web Credibility.

The results of first analysis with seven factors over 29 items were mixed. Some of the items we expected to load on the Web Credibility factor, namely, honesty, expertise, reputation, as well as all of the Trust items loaded strongly and as expected. However, the items related to the Web Credibility Predictability component, Perceived Ease of Use (PEU), and Risk did not load as expected. The Web Credibility Predictability component and PEU loaded together as a single factor. The other unexpected finding was that three of the items we expected to load together for a

Risk factor instead loaded on other components of Web Credibility, most notably with Honesty and Trust. The remaining risk item that asked directly if the user thought the website was 'risky' loaded separately from all of the other items.

To obtain a clearer picture of the data, we removed the four risk items from the analysis and ran a second factor analysis using the three factors of Web Credibility, PEU, and Trust over 25 items. This time all of the items loaded cleanly onto their postulated factors except that items related to the Web Credibility Predictability component and PEU again loaded together as a single factor. Lastly, we ran a third factor analysis over the 17 items for Web Credibility. These results showed four clear factors for Web Credibility that we identified as honesty, expertise, reputation, and predictability.

Reliability of the items was estimated using the Cronbach-Alpha test. Reliability refers to stability over a variety of conditions and time, and reflects the homogeneity or internal consistency of a set of items [6,15]. We omitted the risk items for this analysis due to the factor analysis findings. We calculated Cronbach's alpha for the factors of Web Credibility (.94), Predictability (.91), Predictability + PEU (.95), and Trust (.89). We also calculated Cronbach's alpha for the Web Components of Honesty (.89), Expertise (.88), and Reputation (.90). All of the alpha values are above the .80 level recommended for basic research [16].

## 5 Discussion and conclusion

Analysis of our tool items yielded mixed results. We found that our items measuring Trust had high reliability and validity. Also, the Web Credibility components of honesty, expertise, and reputation had high reliability and validity. However, our analysis indicated that the items were not able to separate PEU and Predictability. This suggests a possible, strong interrelation between Predictability and Ease of Use that is not discussed by Davis. Davis created the PEU to explain the adoption of computer technology, but not explicitly to measure the ease of use of a website in an online environment. When PEU is extended to the online environment, there may be a link between predictability and ease of use that does not show up in the offline environment of adopting computer technology. Based on our results, we will have to reconsider the interplay between our Predictability items and the PEU items. The question that must be answered is whether items can be developed to measure solely Predictability as an element of Web Credibility, or if Predictability is best understood in relation to PEU and should be detached from Web Credibility.

Our most surprising results had to do with risk. We found that our risk items did not strongly identify a single factor. This can be explained in one of two ways. Either risk is not a single factor, or our items did not correctly identify risk. Since the general trust literature identifies risk as an important antecedent of trust, our next step will be to develop new items to replace the three that loaded on other factors. Then we plan to retest the new items and repeat the overall analysis of our web trust tool. The creation of a valid and reliable tool for measuring online trust is essential for future research in the area of online trust. A copy of our tool may be obtained upon request.

## References

1. Byrne, B. M. (2001). *Structural Equation Modeling with AMOS.* Mahwah, NJ: Lawrence Erlbaum Associates.
2. Cheskin Research and Studio Archetype/Sapient. (1999). Ecommerce trust study. http://www.sapient.com/cheskin, accessed 5/9/2000.

3. Churchill, G.A. (1979). A paradigm for developing better measures of marketing constructs. *Journal of Marketing Research*, 16(1), 64-73.

4. Corritore, C.L., Kracher, B., & Wiedenbeck S. Online trust: concepts, evolving themes, model. *International Journal of Human Computer Studies*. Forthcoming.

5. Corritore, C.L., Kracher, B., & Wiedenbeck S. (2001). Trust in the online environment. In M.J. Smith, G. Salvendy, D. Harris, and R.J. Koubek (Eds.), *Usability Evaluation and Interface Design: Cognitive Engineering, Intelligent Agents and Virtual Reality*. Mahway, NJ: Erlbaum, 1548-1552.

6. David, D. & Cosenza, R. (1993). *Business Research for Decision Making.* Belmont, California: Duxbury Press.

7. Davis, F. (1989). Perceived usefulness, perceived ease of use and user acceptance of information technology. *MIS Quarterly*, 13(3), 319-340.

8. Driscoll, James W. (1978). Trust and participation in organizational decision making as predictors of satisfaction. *Academy of Management Journal*, 21(1), 44-56.

9. Fogg, B.J., Marshall, J., Laraki, O., Osipovich, A., Varma, C., Fang, N., Paul, J., Rangnekar, A., Shon, J., Swani, P., & Treinen, M. (2001). What makes web sites credible? A report on a large quantitative study. *Proceedings of CHI 2001,* NY: ACM, 61-68.

10. Gefen, D. (2000). E-commerce: the role of familiarity and trust. *International Journal of Management Science*, 725-737.

11. Kuhn, T. S. (1970). *The Structure of Scientific Revolutions,* second edition. Chicago: University of Chicago Press.

12. Lewicki, R.J. & Bunker, B.B. (1995). Trust in relationships: a model of development and decline. In B.B. Bunker & J.Z. Rubin (Eds.), *Conflict, Cooperation, and Justice: Essays Inspired by the Work of Morton Deutsch*, San Fransicso: Jossey-Bass, 133-173.

13. McKnight, D.H., and Chervany, N.L. (2002). What trust means in e-commerce customer relationships: An interdisciplinary conceptual typology. *International Journal of Electronic Commerce*, 6(2), 35-59.

14. Morgan, R.M. and Hunt, S.D. (1994). The commitment-trust theory of relationship marketing. *Journal of Marketing*, 58, 20-38.

15. Nunnally, J.C. (1967). *Psychometric theory*. New York: McGraw-Hill.

16. Nunnally, J.C. (1978). *Psychomatric theory* (2nd Ed). New York: McGraw-Hill.

17. Rotter, J.B. (1971). Generalized expectancies for interpersonal trust. *American Psychologist*, 26, 443-452.

18. Sambamurthy, V., and S. Jarvenpaa, S. (2002). Eds. Special Issue on Trust in the Digital Economy in the *Journal of Strategic Information Systems,* vol. 11, issues 3 and 4.

19. Schumacker, R. and Lomax, R. (1996). *A beginners guide to structural equation modeling*. Mahwah, NJ: Lawrence Erlbaum Associates.

20. Stevens, J. (1986). *Applied multivariate statistics for the social sciences*. Hillsdale, NJ: Lawrence Erlbaum.

21. Wiedenbeck, S., Corritore, C.L., and Kracher B. Eds. Special Issue on Trust and Technology in the *International Journal of Human Computer Studies*. Forthcoming.

# User Interface Evaluation Methods for Internet Banking Web Sites: A Review, Evaluation and Case Study

*David Wenham*

School of Informatics
City University, London, EC1V 0HB
d.w@totalise.co.uk

*Panayiotis Zaphiris*

Centre for HCI Design
City University, London, EC1V 0HB
zaphiri@soi.city.ac.uk

## Abstract

The project reviewed twenty-seven available user interface evaluation methods and selected a shortlist of methods, which are appropriate for evaluating mature, post-implementation Internet Banking Web sites. The selected methods were applied to two Internet banking Web sites. Based on the experience and the results of these evaluation exercises, the methods were evaluated for their usefulness. Finally a set of heuristics was developed that can be used when evaluating internet banking web sites.

## 1 Introduction

There are numerous methods for evaluating user interfaces. Hom (1998) lists over thirty usability methods. More recently, various theorists/researchers have developed or adapted methods to suit the specific characteristics of Web applications (for example, Nielsen's (1999) Web design specific guidelines). In addition, specific classes of application engender specific design issues, which may make some design principles invalid or irrelevant, whilst other areas are neglected. For example, privacy and security issues are key to Internet banking sites, but relatively irrelevant to information browsing portals.

Consequently, the job of selecting relevant methods for evaluating a particular type of interface can be daunting. This project will review the available methods, selecting a subset of the most suitable ones for evaluating Internet banking Web sites. They will then be tried out and evaluated.

A further problem that this project will seek to solve relates to the Heuristic Evaluation method. This is a form of expert review based on generally accepted guidelines for interface design, which tend to be well supported by theory and research. This method is quite popular because it is relatively cheap and easy to apply. Nielsen's (1994) guidelines tend to be presented as the definitive list (for example, Nielsen, 1994 and Brinck et al, 2002), but this is not necessarily the case – there are lots of relevant guidelines available. Nielsen (1999) has created a list of 10 usability guidelines relevant to Web design, although few of these are relevant to Internet banking.

### 1.1 Literature Review

In the early days of Internet banking, many organisations rushed to provide Internet based services in order to gain competitive advantage. The Internet only online bank Egg was one of the first

721

success stories, whose perceived threat spurred the larger high street banks on to create their own Internet banking services (Goldfinger, 2002). Now, with so many high street retail banks having an online presence – not to mention the online only banks - just providing an Internet banking service will not offer any real advantage over competitors.

Virtual Surveys (2002) note that the satisfaction of users with their Internet banking services is improving. The number of UK customers describing their online bank as 'excellent' rose from 17% to 24% in the year to 1st Quarter 2002. In addition, Petry (2001) noted that the frequency of sign-ons by existing customers is increasing. These figures may indicate that the banks are working to improve usability, thus making their online services more agreeable – or perhaps the existing customer base are just learning their way around any problems. This more pessimistic view would not justify neglecting usability though, because the number of online banking customers is increasing (Petry, 2001), with the greatest Internet banking penetration being achieved in Europe (Goldfinger, 2002). These new users will have to like a site from the start if they are to be retained.

# 2 Methods and Results

## 2.1 Introduction

This section describes how each method was applied in this study and the resulting insights into the usefulness of the method provided by it's application. These methods were applied to two UK Internet banking websites (namely: LTSB and HSBC). Descriptions of the methods can be found elsewhere (e.g. Hom, 1998).

## 2.2 Task Analysis

This method was applied using Heirarchical Task Analysis diagrams, which helped the evaluator to focus on the structure of each task when performed on the system, rather than physical features of the interface. In this way it offered a different view of the system to screen based methods such as Feature Inspection and Heuristic Evaluation, and therefore complements those methods. However, the use of Task Scenarios means that the method tends to focus on common pathways through the system and could miss problems in alternative scenarios, so should really be used in conjunction with a non-scenario based method, such as Heuristic Evaluation.

## 2.3 Interviews

This study used a small sample – 2 users of each Web site. Presumably as a result of this, few problems were identified in comparison with the other methods. Difficulties were experienced in finding users willing to participate and arranging and performing the interviews. Transcribing and analysing the data was onerous and time consuming. The key advantages of the method are that problems identified by users seem more valid and may differ to those identified through methods that do not involve users. With additional subjects, agreement between users on areas of poor validity would imply some reliability. Other methods could then be used to investigate areas that users report as having low usability. The costs and benefits of this method should be carefully considered before choosing to use this method.

## 2.4    Cognitive Walkthrough

This method requires less knowledge of usability guidelines and best practice, because the evaluator is guided by the three questions (does the user know what to do? can they see how to do it? and can they determine if the action they took was right?) towards potential usability problems. Feature Inspection might be more thorough for a usability expert, because the 3 questions could restrict the evaluator's focus – for example, you could miss aspects of the site such as the aesthetic appeal of the interface and how easy it is to escape from places you did not intend to go. Having said that, the Feature Inspection and the Cognitive Walkthrough are restricted to the scenarios you use, so you could easily miss important scenarios using either method.

## 2.5    Feature Inspection

This study applied the method by documenting each task scenario as a procedure, based on Hom's (1998) theory that features that are troublesome to describe are probably troublesome to use. It was found that this method is not particularly useful if the evaluator does not know what to look for – that is to say, if they are not an expert in interface design usability.  There seems little advantage in using this method, over a Heuristic Evaluation, which benefits from taking user interface guidelines into account, and thus points evaluators at the kind of thing they should be looking for.  Similarly the Cognitive Walkthrough prompts evaluators to consider the users thought processes as they step through the task, which seems more likely to give rise to the identification of problems.

## 2.6    Heuristic Evaluation (Incorporating Guideline Checklists)

This study involved developing an Internet Banking specific set of guidelines for use in the Heuristic Evaluation (see Table 1).  These were based on Nielsen's (1994) 10 heuristics, some of which were adapted to make them relevant to Internet Banking Web sites.   The list was supplemented and modified with reference to complementary heuristics/guidelines by other researchers and through application of the other methods.

The guidelines tell the evaluator what to look for, so this method is ideal for evaluators who are familiar with the type of system, but who are not usability experts.  However, also good for experts, as the guidelines help ensure that most potential types of usability problem are considered.  Provides a broad but detailed view of the system, due to the range of guidelines and by not being restricted to specific scenarios. Non-scenario based methods like Heuristic Evaluation might identify some of the problems that could effect the less common scenarios, but on the downside, they might not focus strongly enough on areas needed in the main task scenarios.

## 2.7    Comparative Analysis

In this study, Comparitive Analysis was used with each method except for the interviews, where the sample would not have allowed meaningful comparisons.  Comparitive Analysis did give rise to additional insights, benefitting from the alternative perspectives of the different methods.   It seemed most useful for identifying alternative solutions where usability problems were identified. The method is least compatible with Feature Analysis, because the features on one site do not necessarily have a parallel on the other.  This means that a higher level view of features or functional areas must be taken, which limits the potential for a detailed methodical review.

**Table 1:** Internet Banking Specific Guidelines

| No. | Guideline |
|-----|-----------|
| 1. | Make users feel secure<br><br>Users need to feel secure when doing Internet banking. Sites need to be secure, make security measures visible and explain to users how to use sites in the most secure manner, providing appropriate warnings where necessary. |
| 2. | Easy navigation<br><br>Are there adequate site maps, navigation bars, menus and so on, to help users find their way around the site? (Shneiderman, 1998) Are menus broad and shallow? Avoid deep, narrow and hierarchical menu structures that force users to immerse themselves into the depths of the structure (Zaphiris and Mtei, 1997; Larson and Czerwinski, 1998), and thus cannot be easily navigated without practice and route memorisation. |
| 3. | Visibility of system status<br><br>The system should always keep users informed about what is going on, through appropriate feedback within reasonable time (Nielsen, 2002, page 1). The feedback however, must not detract from the perceived or actual security of the Web site. |
| 4. | Match between system and the real world<br><br>The system should speak the user's language, with words, phrases and concepts familiar to the user, rather than system-oriented terms. Follow real-world conventions, making information appear in a natural and logical order (Nielsen, 2002, page 1). |
| 5. | "User control and freedom<br><br>Users often choose system functions by mistake and will need a clearly marked "emergency exit" to leave the unwanted state without having to go through an extended dialogue. Support undo and redo". (Nielsen, 2002, page 1). |
| 6. | "Consistency and standards<br><br>Users should not have to wonder whether different words, situations, or actions mean the same thing. Follow platform conventions" (Nielsen, 2002, page 1) – that is to say, do not just make the site internally consistent, but consistent with the majority of other sites (Nielsen, 1999). |
| 7. | "Error prevention<br><br>Even better than good error messages is a careful design which prevents a problem from occurring in the first place" (Nielsen, 2002, page 1). |
| 8. | "Recognition rather than recall<br><br>Make objects, actions, and options visible. The user should not have to remember information from one part of the dialogue to another. Instructions for use of the system should be visible or easily retrievable whenever appropriate." (Nielsen, 2002, page 1) For example, provide mouse-over text to explain further where each menu item / link will take you (Nielsen, 1999). |
| 9. | "Flexibility and efficiency of use" (Nielsen, 2002, page 1).<br><br>The interface should be suitable for novices as well as experienced users (Keith Cogdill, 1999). Avoid unnecessary steps towards a user goal, making the process as simple and logical as possible. Convoluted and complex navigation should be avoided, making all parts of the site available from the homepage. |
| 10. | "Aesthetic and minimalist design<br><br>Dialogues should not contain information which is irrelevant or rarely needed. Every extra unit of information in a dialogue competes with the relevant units of information and diminishes their relative visibility. " (Nielsen, 2002, page 1). Textual information should be structured by breaking it into separate meaningful chunks to help users scan and locate the information they are seeking (Lynch & Horton, 1999; Nielsen, 1999). |
| 11. | "Help users recognize, diagnose, and recover from errors<br><br>Error messages should be expressed in plain language (no codes), precisely indicate the problem, and constructively suggest a solution." (Nielsen, 2002, page 1) |
| 12. | "Help<br><br>Even though it is better if the system can be used without", it may be necessary to provide help. "Any such information should be easy to search, focused on the user's task, list concrete steps to be carried out, and not be too large." (Nielsen, 2002, page 1) |

# 3   Conclusions

All the selected methods were suitable and valuable in assessing the usability of Internet banking Web sites.   If you were selecting just one method, then we would recommend Heuristic Evaluation, as it seems likely to identify more problems than the other methods, if conducted with care and with the Internet banking specific Guideline Checklist developed /sourced for this study.

Further to the methods applied in this study, fifteen methods were identified which could be usefully applied to Internet Banking Web sites, where the goals and resources of the evaluation exercise are different to those of this project.   These were Focus Groups, Questionnaires, Journalled Sessions, Self Reporting Logs, Screen Snapshots, Formal Usability Inspection, Pluralistic Walkthrough, Consistency Inspection, Standards Inspection, Thinking Aloud Protocol, Question Asking Protocol, Competetive Analysis, Affinity Diagrams Blind Sorting and Card Sorting (See the full version of this report along with its appendices and Hom, 1998 for descriptions of these methods).   Further studies could be done with the aim of using and evaluating these methods on Internet Banking Web sites.

# 4   References

Brinck, T., Gergle, D & Wood, Scott D. (2002). *Usability for the Web: Designing Web Sites that Work*. San Francisco: Morgan Kaufmann.

Cogdill, K. (1999). MEDLINEplus Interface Evaluation: Final Report. College Park, MD: College of Information Studies, University of Maryland. Cited in Preece et al, 2002.

Goldfinger, C. (2002). Internet Banking Update. Available: http://www.fininter.net/retail%20banking/internet_banking_update.htm

Hom, J.T. (1998). The Usability Methods Toolbox. Available: http://jthom.best.vwh.net/usability/usable.htm [10 Aug 2002]

Larson, K. & Czerwinski, M. (1998). Web page design: Implications of memory, structure and scent for information retrieval. In *Proceedings of CHI*, 1998, 25-32.

Lynch, P.J. & Horton, S. (1999). *Web Style Guide.* London: Yale University Press.

Nielsen, J. (1994a). Heuristic Evaluation. In J. Nielsen and R.L. Mack (eds.) Usability Inspection Methods. New York: John Wiley & Sons.

Nielsen, J. (1999). Top Ten Good Deeds in Interface Design. Available: http://www.useit.com/alertbox/991003.html [10 Aug 2002]

Nielsen, J. (2002). Ten Usability Heuristics. Available: http://www.useit.com/papers/heuristic/heuristic_list.html [25 Nov 2002]

Petry, M. (2001). eBanking: Is it Becoming More Popular? Available: http://www.bankersonline.com/ebanking/gurus_eb1203b.html [3 Nov 2002]

Shneiderman, B. (1998). Relate-Create-Donate: A teaching philosophy for the cyber-generation. *Computers in Education*, 31(1), 25-39.

Virtual Surveys Online Banking Awards (2002). Available: http://www.virtualsurveys.com/news/press_release_5.asp

Zaphiris, P. & Mtei, L. (1997). *Depth Versus Breadth in the Arrangement of Web Links*. Available: http://otal.umd.edu/SHORE/bs04/ [15 Nov 2002]

# Analysis of interaction for shape modification during conceptual design

*Tjamme Wiegers*    *Raluca Dumitrescu*        *Joris S.M. Vergeest*   *Chensheng Wang*

Faculty of Industrial Design Engineering
Delft University of Technology, Delft, The Netherlands
http://www.dynash.tudelft.nl
t.wiegers@io.tudelft.nl

## Abstract

During shape ideation, designers use mainly traditional methods, in spite of the availability of modern CAD (Computer-Aided Design) methods. Apparently, in some contexts interacting with clay, foam or paper is more appropriate than modeling with CAD. We studied designers' interaction methods while generating product models in clay. We developed a method to identify and classify the activities of subjects working on clay modeling assignments. We have applied this method and identified parameters that played a role within the different contexts of the interaction process. Further research can apply these data to generate guidelines for more appropriate shape interaction with computer tools for shape ideation.

## 1    Introduction

When there is only a vague notion of shape instead of precise data, designers use often traditional means like pencil, clay and foam. Obviously, these traditional means have their advantages. But CAD also has advantages, for example easy modification, storage and sharing of concepts. How to benefit from both? One approach can be to adapt CAD to let it support shape ideation in the same intuitive and flexible way as physical means like clay, foam, etc. Another approach could be to integrate CAD and physical means, and let the choice to the designer to apply in each context the tool that is most appropriate. Both approaches require extensive knowledge on designers' preferences and abilities to modify shape in different contexts. Knowledge of traditional methods is required, and knowledge of CAD. This study concerns a traditional method, namely clay modelling. A study on a CAD method can be found in (Dumitrescu, Vergeest and Wiegers, 2002).

We did not find much literature on clay modelling for industrial design. Most clay modelling books for artists or amateurs. Several authors stress the importance of clay modelling for the automotive (Stein, 2001) and there are a few books that describe the clay modelling process for stylists (Hoadly, 1999 and Yamada, 1993). However, to our best knowledge, there are no studies about which activities can be intuitively or effectively be performed in clay modelling. We developed a method to identify clay modellers' activities, to determine the relevant parameters and to indicate the effectiveness in a specific shape context. This paper describes the method and the data it delivers. Some results are shown and it is discussed how these data can be used to identify what parameters a designer varies to control a shape modification.

## 2    Method

The first part of the method is an experiment in which test persons have to perform clay-modelling assignments. The test persons should perform the following assignments (see Figure 1):

1. Modelling an existing soap box in clay
2. Enlarging the box by 20%, for a larger bar of soap.
3. Rounding the top of the box, to make it suitable for holding a larger, rounder bar of soap.

**Figure 1:** The original soapbox, the enlarged one and the rounded one

The experiments must be video recorded. From the video, individual frames must be extracted to be able to analyse the process from moment to moment. A rate of 10 frames per second appeared to be appropriate. The analysis has an exploratory character. The data is analysed frame by frame to identify activities. *A priori* inventory of possible activities could be done, for generating a tick list for activities to be identified. Such a list could speed up the analysis. However, it would also increase the chance of overlooking activities that are not foreseen. For this reason, a tick list is not used. For each activity that is identified, the relevant data should be recorded.

## 3    Recorded data

The method is developed to analyse interaction for shape modification. Therefore, data should be gathered to record how the shape evolves and with what type of interaction the shape modification was achieved. For each *activity*, the object that is modified should be recorded. It will be mentioned *operand*. Sometimes, only a specific part of the operand is affected. This part should be recorded as the *region of interest* (ROI). Also, the means of interaction should be recorded. They will be called *tools*. In many cases, these *'tools'* are the hands or fingers of the test subject. If necessary, details on the execution of the interaction can be stored; they are named *execution details*. Additionally, for each activity the *start time* should be recorded. If necessary, a *comment* may be added. Table 1 summarises the data to be recorded.

**Table 1:** The data to be recorded

| Start time (seconds) | Activity | Operand | Region of interest | Tool | Execution details | Comment |
|---|---|---|---|---|---|---|
| 755.6 | impressing | top face | lengthwise axis | stick | with both hands | Just for illustration |

Analysing ten frames for each second of the video recordings appeared to be an elaborate task. Moreover, the test persons started with kneading the clay for at least half a minute, before giving any relevant shape to it. We adapted the method to be able to focus on the relevant activity sequences, those where shape was modified. We defined multiple detail levels for the recording of the data. Using the enhanced method, data is first recorded at a coarse level of detail. At this level, activity sequences are identified, not separate activities. From each activity sequence, only its start time is recorded. This enables quick selection of relevant activity sequences, prior to elaborating all details. The selected activity sequences can then be analysed in more detail. Four different levels of detail are defined. Each higher level adds more details, as follows:

- Level 1 - Activity sequences and their start times
- Level 2 - Individual activities and the operands on which they are performed
- Level 3 - Region of interest (ROI) and tools used
- Level 4 - Details on the execution of the activity and additional comments

## 4    Results

The above experiment was performed 16 times, each time with a different test person. The described method has been applied to one of these sessions. This section presents some of the first results. We start with a summary of the test person's activities.

**Figure 1:** Hitting and pressing to make flat faces, smoothening, and making a disk, respectively

For assignment 1, the subject cut off an appropriate amount of clay and kneaded it. Flat faces were made, e.g. by hitting the work piece with a hand and by pressing it between both hands (Figure 1). Next, some smoothening was done, for example, by gently rubbing with one or more fingers. Then the subject took another piece of clay, rolled it over the table and flattened into a thin ribbon. The ribbon was stuck on the box shape and again some touching up was done. After that, grooves were impressed with a stick. The assignment ended with some final touch up.

For the second assignment, the subject had to increase the soapbox by 20%. The subject made carves into the modeled box, to divide it into 5 parts. He added an equal part of extra clay started kneading again. The next activities were similar to those during assignment 1.

For assignment 3, a rounded top had to be modeled. The subject took extra clay, kneaded it into a ball and rocked the ball between his hands (Figure 1). The result was an 'M&M-shaped' disk. The disk was cut into two halves and one halve was stuck on top of the modeled soapbox. Next, grooves were made in the added top and some final touching up was done.

The duration of the whole experiment was about 31 minutes (1578.9 seconds). The data table contains 460 rows, with different detail level. Level 1 consists of 39 activity sequences. They are summarized in Table 2. This table is simplified, e.g. by leaving out sequences that did not involve shape. The table is enhanced with a categorization.

**Table 2:** Identified activity sequences

| Assignment 1 | Assignment 2 | Assignment 3 | Categories |
|---|---|---|---|
| Modelling into a box<br>Making flat faces<br>Touching up | Modelling into a box<br>Making flat faces | | Basic shape |
| Rolling into a tube<br>Flattening into ribbon | Rolling into a tube<br>Flattening into ribbon | Rolling into a ball<br>Rocking into disk<br>Touching up | Separate part |
| Sticking to box<br>Touching up | Sticking to box | Cutting into half disk<br>Sticking to box<br>Touching up | Sticking |
| Making grooves<br>Touching up | Making grooves<br>Touching up | Making grooves<br>Touching up | Grooves |

The time spent on the different types of activity sequences is depicted in Figure 2. Generating the global shape was done in assignments 1 and 2 in about the same time. The ribbon was made as an add-on and then stuck to the basic shape. Sticking the ribbon to the box was done about 4 times quicker during the second assignment, and impressing the grooves even about 7 times quicker. In assignments 1 & 2, most time was spent on refining activities (shape level *fine* in the graphs). In assignment 3, the global shape could be re-used. The rounded top was made as an add-on and stuck to the basic shape. Much time was spent on touching up. The subject gradually modified the rounding of the top, until he was satisfied about it. In this shape context, clay modeling has the advantage of immediate feedback. Changing double curved surfaces in CAD often requires the change of multiple parameters before the total effect can be seen. This requires the designer to be more focussed on the shaping method and less on the result. In clay modeling, the designer can already see and feel the result *while* he is making the modification. This process can be seen as a metamorphosis of one shape into another. The designer can stop this morphing as soon as he is satisfied. The morphing process can be described as a gradual change of an (often complex) geometric parameter. Implementing similar methods in CAD requires to make this parameter explicit (Vergeest, Horváth & Spanjaard 2001, Podehl 2002).

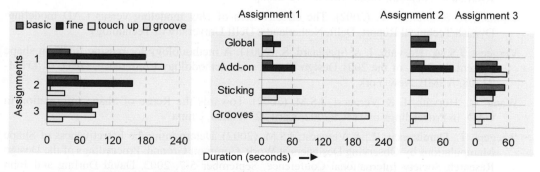

**Figure 2:** Time spent on different types of activity sequences

729

Four parts of the experiment were recorded in detail. Together they consist of 414 activities with a cumulative duration of 347.6 seconds. These parts are:

- Measuring the required clay to increase the soapbox. (Assignment 2, 80.0 seconds).
- Impressing grooves (assignment 2, 38.8 seconds).
- Making an 'M&M'-shaped disk (assignment 3, 177.0 seconds).
- Touching up the soap box and impressing grooves (end of assignment 3, 51.8 seconds)

The type of activity that occurred most (106 times) was positioning, mostly done to prepare the next shape manipulation. Other frequent activities were pressing (52 times) and rubbing (37 times). Some activity sequences occurred repeatedly, up to 15 times. These sequences could be speeded up if re-use of earlier designed shape were possible (Wang, Horváth & Vergeest, 2002). Because of page limitations of this paper, for more information about these and other results we refer to descriptions by (Timmer & Brands, 2002) and (Wiegers, Dumitrescu & Vergeest, 2002).

## 5    Discussion

We developed a method to analyse the way designers apply clay modelling for shape ideation. The developed method enabled us to discard irrelevant activity sequences and to analyse shape relevant activities in detail. We found many repeating cycles of activity sequences. We identified parameters that were used by the test subject to control shape modification. We identified gradually performed shape modifications with immediate feedback. If this morphing can be explicitly described as a gradually varying parameter in a geometric function, it will be possible to include similar shape modification methods in CAD. In some cases, the effectiveness could be improved if the subject could apply re-use of earlier defined shape aspects.

## References

Dumitrescu, R., Vergeest, J.S.M., Wiegers, T. & Wang C. (2002). Towards context-sensitive modeling tools. In Proceedings of International Design Conference 2002, Vol. I, pp. 471-476, Design 2002, Dubrovnik, Croatia

Hoadley, F. E. (1999). Automobile Design Techniques and Design Modeling. TAH Productions

Podehl, G. (2002). Terms and measures for styling properties. In Proceedings of International Design Conference 2002, Vol. I, pp. 471-476, Design 2002, Dubrovnik, Croatia

Stein, J. A. (2001). Clay modelers survive into the digital age. *Automotive News*, June 4, 2001, www.autonews.com, Crain Automotive Group

Timmer, R. & Brands, V. (2002). The effectiveness of clay modeling versus CAD modeling. Dynash Technical Report, Delft, Netherlands: Delft University of Technology

Vergeest, J.S.M., Horváth, I. & Spanjaard, S. (2001). A methodology for reusing freeform shape content. In Proc. of the 2001 Design Theory and Methodology Conference, DETC'01/DTM-21708, ASME, New York

Wang, C., Horváth, I. & Vergeest, J.S.M. (2002). Towards the Reuse of Shape Information in CAD. In Proceedings of TMCE2002, April 22, Wuhan, China

Wiegers, T., Dumitrescu, R., & Vergeest, J.S.M. (2002). Determining the Effectiveness of Shape Manipulations by Observing Designers at Work. Common Ground. Proceedings of the Design Research Society International Conference, September 5-7, 2002, David Durling and John Shackleton (eds.), Staffordshire University Press, Stoke on Trent, UK, pp. 1200-1209

Yamada, Y. (1993). Techniques for giving three-dimensional forms to idea. www.mp-artware.de

# Section 5

## Web Design and Usability

# A Fuzzy Model to Measure Colour Contrast as an aspect of Web Usability

*Maysoon Abulkhair*

Department of Computer Science
University of Sheffield
Regent Court, 211 Portobello Street,
Sheffield S1 4DP
m.abulkhair@dcs.shef.ac.uk

*Siobhán North*

Department of Computer Science
University of Sheffield
Regent Court, 211 Portobello Street,
Sheffield S1 4DP
s.north@dcs.shef.ac.uk

## Abstract

Colour is not only an aesthetic feature of web design; it can be employed to emphasize specific items and to send signals to the user. There are many aspects of effective colour use; the most obvious are the consistency of the colours, their number and the relationship between the text and background colours. This last is often the most important but all of them affect web usability to some extent.

This paper describes a fuzzy colouring model to measure objectively the relationship between the text and background and to produce a colour usability rating based on both web experts recommendation, to have the maximum contrast between the text and the background colours, and experimental results. This model is built in two phases, the intensity measuring phase and the colouring usability rate phase. The fuzzy model described here gives better usability rates than the colouring difference model suggested by other authors.

## 1    Introduction

Many web usability recommendations exist to guide web authors in the use of the colour. Most of these guidelines are based on the contrast between the text and the background colours, however almost all of them use very imprecise measures like "sufficient contrast" (Kerr, 2001; Sklar, 2000; Badre, 2002; Lynch and Horton, 2001), "high contrast" (Nielsen, 2000; Nielsen and Tahir, 2002; Rigden, 1999) etc.. These guidelines are inappropriate for automated usability measuring and there are too many pages on the web now to evaluate them consistently without some sort of automated tool. A more useful mechanism for measuring colour contrast in an appropriate way needs to be evolved. We have developed a scheme derived from the common web usability guidelines on colour but one which is quantifiable and testable.

There is one existing quantifiable procedure, suggested by (Manley, 2001), which consists of adding the three values of the red, green and blue (RGB) colour components together for both the text and the background and then finding the difference. According to Manley the colour difference should be at least 255 (decimal) to provide sufficient contrast between the text and the background colour. This colour difference measure does provide a starting point but is too crude to be useful.

Our model involves measuring colour usability using fuzzy logic to incorporate the usual guidelines. It is constructed in two phases; the colour intensity phase and the colour usability rate phase. Both are described below in sections 2 and 3 and the results are discussed in sections 4 and 5.

## 2 Colour Intensity Phase

It is necessary to determine the colour intensity of both the text and the background colours prior to the colour contrast assessment, which in turn is necessary to measure the usability rate. This phase uses the fuzzy rules to infer the colour intensity from its RGB components and their variations. Each of these components can be split into five fuzzy sets: Dark, MidDark, Medium, MidLight and Light, according to the amount used by the web designer and then combined to give an overall colour intensity measure. To determine the overall colour intensity various fuzzy rules are applied to combine the RGB components in order to give a specific colour intensity value. The colour intensity is quantified as: Darkest, VeryDark, Dark, MildDark, Darkish, Medium, Lightish, MidLight, Light, VeryLight and Lightest. The ranges of each element in the colour intensity fuzzy set are presented in figure 1. The fuzzy rules involve combining the RGB intensities determined earlier in the fuzzy sets to produce a fuzzy colour intensity. This process can be illustrated by the following example:

**IF (Red IS Dark) AND (Green IS Dark) AND (Blue IS Dark) THEN**
**(ColourIntensity IS Darkest)** (1)

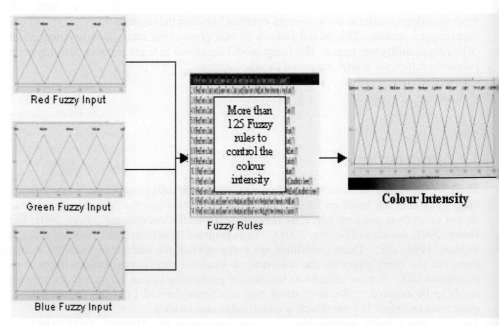

**Figure 1:** The Intensity of the Colour

734

Having established a general rule for colour intensity measuring, we also considered the special case of red/green colour blindness which is fairly common. The red/green colour blindness case is dealt with using the RGB colour intensities to produce an intensity of either Red or Green. This fuzzy rule is of the form:

**IF (Red IS Dark) AND (Green IS Medium) AND (Blue IS Dark) THEN**
**(ColourIntensity IS Dark) AND (ColourBlind IS Green)** (2)

There are around 125 fuzzy rules to derive the colour intensity measure, and all of them followed the form of either example (1) or (2). Both forms of fuzzy rule are based on experimental results. The colour intensity crisp value, computed from the deffuzzification process, varies from (0-100) where 0 is the darkest and 100 is the lightest.

# 3    Colour Usability Rate Phase

Once the colour intensities have been established they are employed in the colour usability rate phase. This is a crucial part of the model, because it integrates the two colours intensities with the most common colour blindness case to produce an appropriate colour usability rate. Thus the specification of the colour intensity fuzzy set is inherited from the previous phase to infer the colour usability rate from the colour contrast measure. The colour usability rate can be categorised as VeryHigh, High, Highish, Medium, Lowish, Low and VeryLow. The VeryHigh colour usability rate is achieved by the most widely recommended usability guidelines for colour. Most of the fuzzy rules that determine the colour usability rate are derived from previous studies together with experimental results. For instance, the combination of white background colour together with black text (but not the other way around) will be given the highest colour usability rate (Nielsen, 2000; Nielsen & Tahir, 2002; Lynch & Horton, 2001; Shneiderman, 1998) whereas a low contrast between the text and background colour indicates low colour usability evaluation. This guideline is applied in the following form of the fuzzy rules:

**IF ( TextIntensity IS Darkest) AND (BackGroundIntensity IS Lightest) THEN**
**(ColourUsabilityRate IS VeryHigh)**

Even where there is sufficient contrast between a dark background and light text, some usability experts give this a lower usability rate than the reverse (Nielsen, 2000; Nielsen & Tahir, 2002; Lynch & Horton, 2001; Shneiderman, 1998). However, there are other usability guidelines which strongly recommend the use of the dark background, light text combination (Preece et al 1994; Rivlin, Lewis & Davies-Cooper 1990). After some experiment, it was decided to adopt the more recent recommendations and the colour contrast fuzzy model will give this a lower usability rate than the dark background/light text combination. The corresponding fuzzy rule used to assess this is in the form:

**IF ( TextIntensity IS Lightest) AND (BackGroundIntensity IS Darkest) THEN**
**(ColourUsabilityRate IS High)**

Whereas, low contrast between the text and background colour results in a low colour usability evaluation. This gives us a group of rules as follows:

**IF ( TextIntensity IS Dark) AND (BackGroundIntensity IS Dark) THEN**
**(ColourUsabilityRate IS VeryLow)**
**IF ( TextIntensity IS Light) AND (BackGroundIntensity IS Lightest) THEN**
**(ColourUsabilityRate IS VeryLow)**
**IF ( TextIntensity IS Medium) AND (BackGroundIntensity IS Darkish) THEN**
**(ColourUsabilityRate IS VeryLow)**

There are more than 140 fuzzy rules in the model to cover different aspects of colour contrast. These fuzzy rules are associated with each other to give a reliable colour usability rate, and all the fuzzy rules are illustrated in figure 2. In this figure different intensities values from the text and background colours can give rise to different usability rates. In other words, the right hand section of the surface in figure 2 represent the light text over dark background colour combination, which is rated lower than the opposite colour combination presented on the left surface section. The crisp value, produced from the deffuzzification process, again varies from (5 – 95) where 5 is the lowest and 95 is the highest colour usability rate.

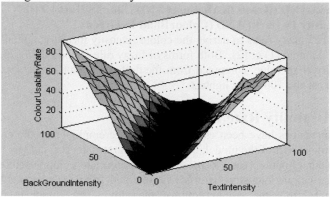

**Figure 2:** Colour Usability Rate [the dark colour shaded point to the lower colour usability rate, where the two upper edges to the right and left indicate high usability rate]

## 4 Evaluation

The fuzzy model has been used to produce a reliable colour usability rate. For example, the combination (white text on green background) gives a usability rate of 12.2% (Low) but the calculated intensity of the background is 77 (Light) and the text intensity is 100 (Lightest). So, in the colour difference model, it would be acceptable at 33.3%. The other example, involving the most common form of colour blindness, with the combination of green shade (# 99 ff 99) text and red (# ff 00 00) as a background is acceptable in the colour difference model and rated at 40% whereas, the fuzzy model produces 23.7% colour usability rate.

Unlike the previous models for calculating the contrast between the text and the background colours, this fuzzy model deals with the combination differently. It is not necessary for the same combination to have the same usability rate when exchanging the text with the background colour and vice versa. As discussed earlier, black text on a white background is not the same as white text over a black background, so, the highest usability rate, 95%, is given to the black text over white colour combination by our fuzzy model whereas reversing this combination to black text on a white background is only rated as 83.3%.

There are some other combinations that commonly annoy users even with normal vision, and, as result they might face difficulties when reading them. For example, light pink text (# ffccff) over green background (#00ff00) where both colours picked from the safe web colours, is given a rating of 34%, but even though it provides a reasonable degree of contrast. With the difference model mentioned earlier this would be given 60%.

## 5 Conclusion and Future Work

Our fuzzy model considers the most recent and common usability recommendations and transforms them into a consistent and quantifiable form suitable for automated evaluation of web pages. This is, in itself, useful but an even more useful feature of our model is that it can easily be adapted to take into account other aspects of colour as it affects usability. We intend to extend the model to include further experimental results on different users' and cultural groups' preferences in terms of colour. It can of course also be adapted to take into account other researchers results in the same area as they emerge in the future.

# References

Badre A. (2002).  Shaping Web Usability Interaction Design in Context.  Boston: Addison Wesley.

Kerr, M. (2001).  Tips and Tricks for Website Mangers. London: Aslib-IMI.

Lynch, P., Horton, S. (2001). Web Style Guide: Basic Design Principles for Creating Web Sites (2nd ed.) New Haven: Yale University Press.

Manley B. (2001), quoted in Tips and Tricks for Website Mangers.  Edited by Mark Kerr London: Aslib-IMI

Nielsen J. (2000).  Designing Web Usability: The Practice of Simplicity, Indianapolis: New Rider publishing.

Preece J., Rogers Y., Sharp H., Benyon D., Holland S., Carey T. (1994).  Human-Computer Interaction. Harlow, England: Addison-Wesley.

Rigden, C. (1999).  'The Eye of the Beholder'- Designing for Colour-Blind Users. *British Telecommunications Engineering*, Vol.17.

Rivlin, C., Lewis R., Davies-Cooper R. (1990). Guidelines for Screen Design.  London: Blackwell Scientific publications.

Shneiderman B. (1998).  Designing the User Interface Strategics for Effective Human-Computer Interaction (3rd ed).  Reading, MA: Addison-Wesley.

Sklar, J. (2000). Principles of Web Design.  Cambridge, MA: Course Technology.

Tahir M., Nielsen J. (2002).  Homepage Usability 50 Websites Deconstructed.  Salem, Virginia: New Rider publishing.

# Brazil: Corporate Web Sites and Human-Computer Interaction, a Case Study

*AGNER, Luiz C. and MORAES, Anamaria*
Pontifícia Universidade Católica do Rio de Janeiro, PUC-Rio
Rua Marquês de São Vicente, Gávea, Rio de Janeiro, RJ - Brasil
moraergo@rdc.puc-rio.br

## Abstract

The problem to be researched is that users of a specific web site stop navigation and evade it when they reach the home page. The hypothesis is that the home page is not adequate to support the tasks users would like to complete. Our research methods included log analysis, content analysis and online questionnaires.

## 1 Introduction

The Brazil National Service is a professional training facility. The Institute has almost 650 professional schools and is established in 2.000 cities all over the country. It is administered by a confederation of corporations. Statistics compiled on access to National Service's web portal in the internet show that users stop navigation at the very beginning of its hypertext. In a large percentage of consultations, navigation is interrupted on the first page. A survey was conducted to provide a basis for our Master's degree thesis at Pontifícia Universidade Católica (PUC), in Rio de Janeiro (AGNER, 2002). Our study describes an ergonomic framework to guide redesign of user interface.

Our major objective is to discuss methods for the performance improvement of organizations in the World Wide Web considering the users' goals, needs, opinions and tasks. It is a descriptive research - a case study. We aim to establish guidelines to redesign the user interface - based on user requirements for information and easy navigation. Therefore, we aim to make a contribution to web design process.

The hypothesis for this theme is that the home page for this specific group of users is not adequate to support the tasks they would like to complete. Evidence and facts collected through research techniques contributed to prove this hypothesis is correct.

As explained by NIELSEN and TAHIR (2001), home pages may present a number of usability problems which will cause difficulties in the localization of the information. Accumulated effects of many usability problems in the home page may confuse the users, even though these usability problems are not "catastrophic" and don't prohibit the site utilization.

As stated by IBM (2001), geographic and cultural differences should always be considered. For instance, novice and occasional users in some countries, outside the USA, might be less familiar with computers than North-american ones. Designers should get to understand their community of users and take advantage of that knowledge to get the best results. Therefore, we organized a survey to understand our specific community of users.

## 2 Research method 1 – log analysis

Access statistics of the analyzed portal have been increasing and reached 40.000 users monthly. However, such web portal has a problem which is a challenge to the HCI point of view, as explained below:

- Thousands of users currently access the home page of this portal;
- Only 33% of those users access the links to the web sub-sites, were the information about professional courses is. This should be the main task for users to perform.

Although, the application of the new communication media in the case's Brazilian organization seems to be a relative success, these figures show that a great percentage of the internet users stops navigation at the beginning of the hypertext.

Consequently, matters for investigation are: Why navigators get lost inside the web portal? Why don't they get to the exit page (sub-site links)? Is the design of the home page too bad? Is it important to make a redesign of the web site?

## 3    Research method 2: content analysis

In order to overcome the limitations of log analysis, we applied a qualitative technique – the content analysis. According to RICHARDSON (1999), content analysis is specifically applied to qualitative materials – mainly to get a first picture, organize ideas and analyse the elements and its rules. It has a scientific nature and must be rigorous, precise and efficient. Our content analysis was based on the framework of Ergonomics and HCI and aims at understanding the system usability as well as the Web design effects.

We collected spontaneous messages containing questions, suggestions and intentions, sent by users from almost all of the Brazilian States. We believe that a systematic content analysis of these messages could bring us closer to a picture of the type of audience involved, in order to establish guidelines for the redesign of the user interface.

As we know the usability design process should not end when the interfaces are installed. The systematic *feedback* collection is important to justify the future development of the web projects, as well as its costs and organizational structures.

## 4    Research method 3: questionnaires

As we know, the web needs *"zero learning time"* (NIELSEN, 2000). Usually, nobody uses a web site sufficiently to become an expert. Even when some users return to the site frequently enough to become experts, it is necessary that the system gives attention to the novice. Nobody will enter into a site if it is not absolutely obvious, within a few seconds, how to use it.

According to MANDEL (1997), everything the user says represents his/her personal perception or individual preferences about the product which is being designed or evaluated. So, the importance of collecting data from a large sample of users is to determine shared suggestions and not individual opinions.

We created an online questionnaire, inspired by the one presented by SHNEIDERMAN (1998). We took care to modify and adapt it to the specific conditions we had encountered. The questionnaire form was designed in HyperText Markup Language (HTML) and presented to 1591 random users. A semantic-opposition scale of five alternatives was presented with each one of the proposed questions. Open questions were also included. Some results will be presented in this case study.

### 4.1    The application of questionnaires

The questionnaire consisted of nine parts, and an introduction to get the general profile of the user. 155 questionnaires were answered and returned via email. The application of the online questionnaire method was successful. Research as to the geographic origin of users showed that they came from all over the country, particularly from the states of São Paulo (27,1%), Rio de

Janeiro (23,2%), and Minas Gerais (7,1%). 24 States of Brazil responded this questionnaire (of a total of 27 states).

## 5 Content analysis – results

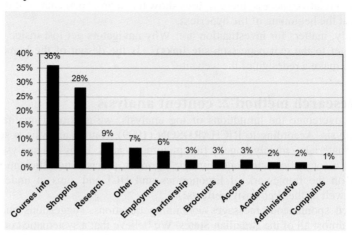

**FIGURE 1** – *Content analysis shows the information users want to find in the site.*

## 6 User education profile – survey results

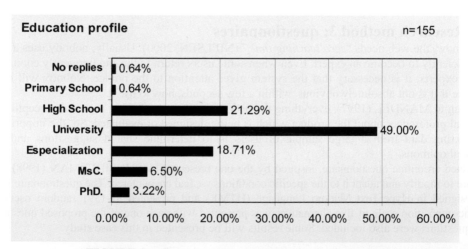

**FIGURE 2** – *Questionnaire results: Formal education user profile.*

The people who visit the analyzed site form a sofisticate and elevated level in terms of a formal education profile. Almost 80% of the users are attending University or are University graduates - some of them have Master or Doctor degrees. This high level of education profile seems to represent the majority of the users.

This situation is basically different from the dominant public who attend the professional courses of the Institute in the "real world". Courses are commonly directed to the poor middle-class and teenagers and are focused on the very operational side of professional skills. Often, these courses do not require formal academic qualifications. Contrary to commonsense, the figures suggest that "virtual users" of the National Service are completely different from their clients of the "real world".

# 7    User experience in Internet – survey results

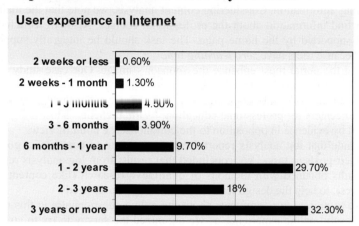

**FIGURE 3** – *Questionnaire results: The user experience in Internet shows an advanced profile.*

# 8    User opinion about the site – survey results

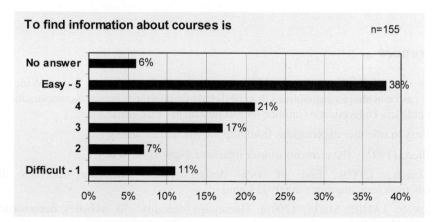

**FIGURE 4** – *Web users evaluated if it is easy is to get information about the offered courses.*

As we know, the support provided to the user for achieving his/her goals and completing his/her tasks is the main objective of the home pages. 35% of those interviewed do not seem to agree that the site helped them to complete their first task – procuring information about courses. As an academic research, our hypothesis was proven to be correct.

# 9    Conclusion

Home pages of some organizations can present a number of usability problems which cause difficulties in finding information. Considering content analysis, we note that the main task of the internet user is to find information about the professional courses. Also, we concluded that this task was not well supported by the home page. The task should be integrally supported by the portal home page because users have *zero learning time*.

The home page and the portal must enhance the sponsor's image. Our case study indicates that National Service should redesign its home page as well as its whole site in order to optimize the dialogue with "virtual clients", since they are very different from the "real clients" who attend training schools and centers for professional education. "Virtual clients" show a big different user profile, as indicated by evidence in opposition to the commonsense point of view.

We can also conclude that log analysis reports were incomplete to construct a good picture of virtual users and their goals or tasks. We concluded that results from log analysis reports must be compared with results obtained with methods of qualitative research (like content analysis and online questionnaires), to help the design team to draw a map of user needs.

We would also like to present our comments about the online questionnaire in this research. This method was a successful online experience as we organized the survey to try to comprehend our specific community of users. Considering the feedback obtained from this questionnaire and some complaints about its size, we tend to believe that SHNEIDERMAN's questionnaire is a good way to obtain usability data. However, it should be applied in a compact and brief form to address the Brazilian internet users' cognitive style.

The major goal of our academic research was to discuss methods for improving the presence of local organizations in the Web considering the users' goals, needs, opinions and tasks. We feel we were sucessful in contributing to the usability design process for web sites within the scope of Brazilian reality.

# 10    References

AGNER, Luiz C., (2002). Otimização do diálogo usuários-organizações na World Wide Web: estudo de caso e avaliação ergonômica de usabilidade de interfaces humano-computador. Ms. Thesis. Pontifícia Universidade Católica do Rio de Janeiro, PUC-Rio.

IBM (2001) Easy to use: user expectations from http://www.ibm.com/easy/.

MANDEL, Theo., (1997). The elements of user interface. New York: Wiley.

NIELSEN, Jakob., (2000). End of web design. Jakob Nielsen's Alertbox from http://www.useit.com/ alertbox / 20000723.html

NIELSEN, Jakob; TAHIR, Marie., (2001). Homepage usability: 50 websites deconstructed. Indianapolis: News Riders.

RICHARDSON, Robert J. et al., (1999). Pesquisa social: métodos e técnicas. São Paulo: Atlas.

SHNEIDERMAN, Ben., (1998). Designing the user interface; strategies for effective human-computer interaction. (3rd ed.). Chicago: Addison Wesley.

# Usability Evaluation of Architectural Web Sites

*Canan Akoglu*

Research Assistant
Yildiz Technical University
Faculty of Art and Design
Department of Communication Design
Istanbul, Turkey
akoglu@yildiz.edu.tr

*Oguzhan Ozcan*

Assoc. Prof. Dr.
Yildiz Technical University
Faculty of Art and Design
Department of Communication Design
Istanbul, Turkey
oozcan@yildiz.edu.tr

## Abstract

The medium, internet, provides different user groups a wide range of possibilities in many platforms from research fields to entertainment. When we evaluate the situation from this point of view, the interaction between the user and the product-web site- becomes more and more. The variety of user groups makes experts create idealist common solutions for them who are at different levels when interacting with web sites.

The aims of this paper are to bring out whether it is possible to define design criterions according to users' reactions in architecture based web sites; if possible, to fix out what these criterions can be during the design period. In other words it is aimed to put out a design guideline for architecture based web sites according to the results of usability tests. It is also aimed to find out what type of a usability test should be conducted for such a special topic: whether a general usability evaluation is efficient; if not, what type of a usability test should be applied.

## 1 Introduction

With technology beginning to take an important role in our daily lives, internet has become a tool which can not be given up in many fields such as from communication to having information, from education to entertainment. As mentioned in the abstract, many people having different qualifications from many different fields use internet intensively. Consequently, there comes out many different user demographics. As these demographics change, it becomes inevitable for interfaces and structures to be developed in order to target different user groups.

The situation mentioned above is an important issue for web sites of which content is design such architecture. Although architecture based web sites are designed due to interaction design criteria like all other web sites, according to the changes in social life and cultural values, arising of new needs and in order to increase the number of users and supply the continuity of usage easiness, they should be updated. In this paper, the type of usability testing method which will be the most suitable so as to get user reactions in architecture based web sites, is interrogated. As mentioned, it is researched that whether a general usability testing method is sufficient or not, if not what kind of a method should be conducted. It is also discussed that whether it is possible or not to put out basic design criteria in architecture based web sites, if so then what can these criteria be.

In the first part of the paper, it is discussed on the determination of the test environment we will use and on the development of a usability evaluation method.

Secondly, we had a research on web sites in terms of their contents in order to find out the ones on which we would conduct usability tests (Fleming, 1998).

In the third part of the paper, the method used in constitution of user groups, who would evaluate architecture based web sites, and in preparation of questions which would be asked to these users is brought out.

In the fourth part, the results of usability tests are analyzed under titles such as user liking, navigation, access to information, satisfaction of knowledge given to users, and attention of users one by one.

In the last part, according to the data gained from these tests a guideline when designing architecture based web sites is proposed under 3 subtitles such as information design, content design and application, and graphics design.

## 2    Determination of the Test Environment to be Conducted

As known, internet which is used in many fields as a tool causes network communication become wider and wider. It is seen that 'this tool' is also used in usability evaluations. Test environments used in usability evaluations are classified into 2 main groups (Castillo, Hartson, Hix,1997):

- Traditional Lab Environment (Traditional usability evaluation)
- Internet (Remote usability evaluation)

In the first one of these 2 environments, usability tests are conducted in a traditional laboratory with tools such as video capture and camera which users can be interrupted psychologically. Because this situation can prevent users to interact with the test product freely, trustful data may not be taken. In the second environment, user can participate in usability evaluations from work, home, in short from his/her natural environment. In this type, usability tests are held through internet (Castillo, 1999).

In a traditional lab environment, the users and the observers can be in the same room or observers can watch users when interacting with the product synchronously from a one-way mirrored room. They also can watch users reactions from a video record (Osterbauer, 2000). On the other hand, in remote usability evaluation, because the user and the observer/evaluator are different places, the evaluation is usually taken asynchronously (Rubin, 1994). It is important to observe what users do; at which part of the product do they spend much time etc. As this observation is hard to achieve in remote usability evaluation, specially prepared software can be used. To get technical support for preparing such a special tool for our remote usability evaluation, we had a collaborative work with Prof. Dr.Lale Akarun and Mehmet Ali Orhan from Department of Computing Engineering, Bogazici University, Istanbul. This tool worked through internet. When users connect to the URL of this tool, it makes a record for every new user. If users want, they can register and sign up. So it will be possible to recognize users next times. The tool takes IP addresses and puts cookies into users computers. Even if users don't want to register, with the help of these cookies and IP addresses. This tool is prepared to record mouse clicks of users, the navigation of users in web sites, and the duration of spending time in each web page.

The tool has 2 different parts: the first one is the admin part where questions of usability tests are updated online and results are taken. The second part is the users part where users click URLs of the pilot web sites and answer questions about these sites. As mentioned above, every result is recorded into the related user's database.

# 3    Fixation of the Architectural Web Sites to be Conducted Usability Evaluation

To find out which type of architectural web sites consist, it is useful to overview web sites in terms of their contents: web sites are classified into 6 main groups by means of their contents (Fleming, 1998): Shopping based web sites, learning based web sites, entertainment based web sites, community web sites, information web sites, corporate identity based web sites

Within the classification above, we found out that architectural web sites can be classified into 4 groups in terms of content:
- Shopping Aimed Architectural Web Sites
- Learning Aimed Architectural Web Sites
- Information Aimed Architectural Web Sites
- Corporate Identity Aimed Architectural Web Sites

When compared with other web sites, it is seen that there are many interface  and navigation design alternatives in corporate identity aimed architectural web sites.  When determining about the below web sites we put forward 3 factors such as "being well-known internationally by means of professionalism", "information structure", and "graphic and typographic design" .
- Bernard Tschumi Architects
- Richard Rogers Partnership

# 4    Constitution of User Groups and Questionnaires

Target user groups of architectural web sites are naturally architects and students from departments of architecture. It is possible to talk about  other users except from target user groups such as designers of web sites and another group of which members visit architectural web sites with or without any reason. Starting out from this point of view, target users were determined to be students from Istanbul, Yildiz Technical University(YTU), Faculty of Architecture, Department of Architecture and architects working as professionals. It was also determined to test the ones as the designers of web sites from Istanbul, Yildiz Technical University, Faculty of Art and Design, Department of Communication Design. In addition to these two groups a general user group was constituted. Usability tests were conducted to 13 architects and 16 students from YTU Department of Architecture as being the target user group, 17 students from YTU Department of Communication Design as being designers of web sites, and 6 people representing the general user groups having a total number of 52.

# 5    Evaluation of Usability Test Results

Firstly, it should be indicated that the questions asked to 3 user groups are classified into 5 groups such as discrimination, navigation, access to information, satisfaction level of given information.

## 5.1 Discrimination

When users are asked which of the 2 architectural web sites they felt themselves the closest, target users consisting of architects and architecture department students preferred Bernard Tschumi Architects' web site. On the other hand, other users including communication design department students and general ones preferred Richard Rogers Partnership's web site. We prepared another question asking users the reason for their choices. Architects indicated that the reason is not the manner of the architect, but visual design of the web site. Students from architecture department stated that the reasons are both the manner of architect and the visual design of the web site.

## 5.2 Navigation

Users were asked to make judgments about navigation structures of the pilot architectural web sites. Some of the judgments we put forward are as follows:
- It is difficult for me to understand where I am at this web site.
- I feel myself in comfort when I am navigating through the pages of this web site.

According to reactions of users, we can say that target user groups might not have taken care of the details about the navigation structure if they were not disturbed very much. On the other hand, students from department of communication design where information organization design plays an important role might have more careful when making judgments about navigation structures.

## 5.3 Access to Information

Users were also asked to make judgments about access to information at the pilot architectural web sites. One of the judgments we put forward are as follows:
- I can easily find whatever I want at this web site.

According to the results of the above judgment, we can say that users can understand and use a mixed information organization more easily. Moreover, it can be said that linear information organization does not provide users sufficient interactivity.

## 5.4 Satisfaction Level of Given Information

According to the answers of questions about this topic, target user groups-architects and architecture department students- want more detailed visual information about the projects such as sketches, QTVRs, more photos etc. It is obvious that because other users naturally have no exact ideas, they chose the answer 'no comment'.

## 6 Conclusion

Gathering results from usability evaluation, a design guideline for corporate identity aimed architectural web sites can be proposed. By means of information organization, a mixed structure can be used. This type of information structure provides users more interactivity than other models do. For example, if linear information structure is used in such kind of a promotional web site, users can be bored to go through the pages one by one until the structure allows. In a mixed information structure users have freedom to select the path they would like to go through; they can access the information they want to have fast.

In terms of content design, it is obvious that target users-architects and students from architecture department- want to have more visual information about projects from sketch steps to detailed design and construction levels. They also want to analyze some special projects by 'virtually walking in buildings' which they mean QTVRs technically.

By means of graphic and typographic design, the corporate identity aimed architectural web sites should have the architectural manner. For example, if an architect has 'deconstructive' designs, then the web site he/she owns should not have the quality of 'bauhaus modernism'. Otherwise, a complete corporate identity can not be carried out

In such type of an interactive design project, interactive and interface designers should work together with an architect. In this team, the most important part of architect's role will be during the content design and organization. Especially gathering visual information and relating 2D images and drawings with 3D images are the works of an architect rather than an interface designer.

# References

Castillo, J. (1999), Motivation for conducting remote evaluation. Retrieved March 8, 2000 from http://www.miso.cs.vt.edu/~usab/ remote/motivation.html,

Castillo, J., Hartson, R., Hix, D. (1997). Remote usability evaluation at a glance. Retrieved March 10, 2000, from http://www. miso.cs.vt.edu/~usab/remote/docs/TR_remote_evaluation.pdf

Fleming, J. (1998). Web navigation, California:O'ReillyL&Associates Inc.

Osterbauer, C., Köhle, M., Grechenig, T., Tscheligi, M., (2000),Web usability testing- A case study of usability testing of chosen sites. Retrieved January 18, 2002 from http//www.alpha.swt.tuwien.ac.at/publications/papers/ausweb2k/paper.html

Rubin, J. (1994). Handbook of usability testing: how to plan, design and conduct effective tests. New York:John Wiley&Sons Inc.

# MiLE: a reuse-oriented usability evaluation method for the web

*Davide Bolchini*
TEC-lab
University of Lugano
Switzerland
davide.bolchini@lu.unisi.ch

*Luca Triacca*
TEC-lab
University of Lugano
Switzerland
luca.triacca@lu.unisi.ch

*Marco Speroni*
TEC-lab
University of Lugano
Switzerland
marco.speroni@lu.unisi.ch

## Abstract

MiLE (Milano-Lugano Evaluation Method) is a usability framework that offers a practical toolset to carry out a user-centered validation of complex web applications. MiLE combines assessed evaluation approaches (user testing and inspection) with widely used usability techniques (heuristics and task-based evaluation) to gain analytical and reasoned results. In order to support the widespread reuse of usability knowledge, MiLE offers a usability kit (U-KIT) to project teams that comprises user profiles, scenarios, usability factors and a process to employ during the usability evaluation in a given domain.

## 1    Introduction and Related Works

The main goal of web usability evaluation is to detect usability breakdowns of a web application and provide analytic feedback to redesign.

The most commonly adopted approaches to web usability are *user-based methods* (or user-testing methods) and *usability inspection methods* (or expert reviews) (Matera et al., 2002).

*User-based methods* mainly consist of user testing, in which usability properties are assessed by observing how the system is actually used by some representatives of real users (Whiteside Bennet & Holtzblatt, 1988) (Dix A. et. al., 1998). User-testing evaluation provides the trustiest evaluation, because it assesses usability through samples of real users. However, it has a number of drawbacks, such as the difficulty to properly select correct user samples and to adequately train them to manage advanced functions of a web site (Matera et al., 2002). Furthermore, it is difficult, in a limited amount of time, to reproduce the actual usage situation. This condition is called the "Hawthorne effect" (Roethlisberger & Dickson, 1939): observed groups can be affected by observation alone. Failures in creating real-life situations may lead to "artificial" conclusions rather then realistic results (Lim, Benbasat & Todd, 1996). User testing is considerable in terms of time, effort and cost. However, it is effective to evaluate quickly the look and feel of the interface, as it is possible to verify at "real-time" the reactions of the users.

*Usability Inspections methods* is the generic name for a set of methods based on having expert evaluators analytically examine usability-related aspects of a user interface (Nielsen & Mack, 1994). With respect to user-testing evaluation, usability inspection is more subjective, having heavy dependence upon the inspector's skills (Matera et al., 2002). The main advantage of the inspection methods is the relationships between costs and benefits. As a matter of fact, performing an inspection can "save users" (Nielsen & Mack, 1994), (Jeffries, Miller, Wharton & Uyeda, 1991) and does not require any special equipment. The inspector alone can detect a wide range of usability problems and possible faults of a complex system in a limited amount of time (Matera et al., 2002). For these reasons, inspection methods have achieved widespread use in the last years, especially in industrial environments (Madsen, 1999).

However, current usability inspection methods have two main drawbacks. Firstly, they focus on "surface-oriented" features of the graphical interface (mainly at page level) (Green & Benyon, 1996). Only few of them address the usability of the application structure, e.g., the overall information architecture, organization of content or navigation patterns. Secondly, the reliability of the results is often entirely dependent on the individual know-how, expertise and skills of the inspectors. Under these circumstances, effective inspection is almost impossible to reproduce without a usability expert.

Both user-based methods and inspections methods are alternatively based on two techniques: *heuristic-driven evaluation* and *task-driven evaluation*.

Essentially, *heuristic-driven evaluation* provides checklists and usability principles (Nielsen, 1999). The quality of the web site is assessed against these principles (e.g. consistency, reliability, status visibility, etc.). In user testing, heuristics are used to ask users to comment in a structured way their experience with the web site (e.g. heuristic questionnaires or structured interviews). During inspection, heuristics guide the expert to explore the site and check compliance with usability principles.

*Task-driven evaluation* assumes that usability is assessed by trying to complete actions with the web site. Tasks are provided which describe potential goals or sequences of actions that users might want accomplish with the application. In user testing, tasks are defined and given to the users. Tasks are also employed in walkthrough and other structured inspection techniques (Rosson & Carroll, 2002) (Brinck, Gergle & Wood, 2002).

Heuristics and task-driven techniques are usually adopted in alternative and separately, thus loosing the opportunity to obtain a more comprehensive evaluation.

Moreover, one of the main disadvantages shared both by heuristics and task-based techniques is that they are not *reuse-oriented*, i.e. they have not been defined to be effectively reused. Most of usability techniques are proprietary methods or expert-dependent techniques; in other words, they can be difficult for less-experienced evaluators who do not have the necessary conceptual tools to gain appreciable results. The problem of reuse is strongly connected to the difficulty of teaching and communicating the essence of a method in a way that other people can successfully apply it. Projects teams are acknowledging the importance of usability evaluation but are still reluctant to make considerable investment in consultancy for an "ad-hoc" evaluation, especially if the web project is at the end and the remaining budget is very limited. Effective reuse of usability knowledge and practices would enhance the adoption of usability techniques by designers and project teams.

## 2   MiLE: pushing reuse in web usability

MiLE (MIlano-Lugano Evaluation method) is an experience-based usability evaluation framework for web applications that strikes a healthy balance between heuristic evaluation and task-driven techniques. MiLE offers reusable tools and procedures to carry out both expert reviews and user testing within budget and time constraints. Several inspections of complex web sites in the field of e-banking, e-commerce and cultural heritage (e.g. museums and digital libraries) (Di Blas et al., 2002) allowed the MiLE method to grow, refine and strengthen, thus consolidating a framework that is being extended to other domains.

In extreme synthesis, MiLE employs in combination the following heuristic concepts.

*Concrete Tasks* are specific actions (specific in that they are defined for a concrete application) that users might want to perform. Concrete tasks lead the inspection and are then given to users while exploring the application during the empirical testing. *Abstract Tasks* are generic actions (generic in that they can be applied to a wide range of applications) capable of leading the inspector through the maze of the different parts and levels a web application is made of. Abstract tasks can be used only for inspection (see Matera et al., 2002). *User Profiles* are potential target types of the web application. *Scenarios* (Carroll, 2002) are essentially considered as a combination of a user profile and a task. *Usability Attributes* are usability properties of the web application.

749

Usability attributes are used to detail the evaluation of the tasks and provide an analytical assessment of different application judgement on several concerns (navigation, information structure, accessibility, layout, labelling, etc.).

MiLE provides the inspector with a reusable set of evaluation tools (U-KIT, the usability evaluation kit) tailored on the specific domain. The U-KIT is a library of specific evaluation tools. Notably, a U-KIT comprises a set of tasks, user profiles, scenarios and usability attributes to be used on a given family of application (e.g. banking web sites, museum web sites, etc.).

# 3 MiLE in action

In this section, the process of the MiLE method is illustrated through an example. The example is taken from the results of the usability evaluation of a famous museum web site (www.louvre.fr).

A U-KIT for museum web sites has previously been defined. An extract of the U-KIT is shown in Table 1. Lack of space prevents us to describe in detail how the U-KIT has been created. In general, user tasks and profiles are defined on the basis of the user requirements analysis and previous evaluation experience with several web applications in a given domain.

**Table 1:** Excerpt from U-KIT for museum websites

| MiLE: U-KIT for Museum Web sites | | |
|---|---|---|
| *TASKS LIBRARY* | | |
| **TASK NUMBER** | **DESCRIPTION** | |
| 1 | Find information about physical address of museum | |
| 2 | Find the city's map and/or area where the museum is located | |
| 3 | Find the charge of the ticket | |
| 4 | Find information about guided tour and/or special guided tours (special events) | |
| 5 | Find information about organisation of events (shows, concerts, etc.) within the "real-museum" | |
| 6 | Find information about history of museum collections | |
| 7 | Find information about didactic activities organised by museum | |
| *USER PROFILES* | | |
| **U. PROFILE NUMBER** | **PROFILE** | **DESCRIPTION** |
| 1 | High-school teacher | He usually surf the site to find content, which can help him for preparing visits to the real museum. The educational material is particularly relevant for his/her. |
| 2 | Art-lover | He usually looks for accurate and in-depth information current exhibitions events update. Technical content about the state and conservation of the works of art might be relevant. |
| *ATTRIBUTES* | | |
| **ATTRIBUTE NUMBER** | **NAME** | **DESCRIPTION** |
| 1 | Accessibility | The information is easily and intuitively accessible |
| 2 | Completeness | The user can find all the information required |
| 3 | Currency | The time scope of the content's validity is clearly stated. The info is updated. |

Given the relevant U-KIT, the MiLE process for evaluating web sites comprises of six phases:
1. Modelling the application under inspection
2. Performing some selected tasks
3. Evaluating the tasks through usability attributes
4. Weighting the results according to user profiles, communication goals/requirements
5. Empirical testing (user testing)
6. Reporting the evaluation results

1. *Modelling the application under inspection.* The inspector draws a high-level mental model – either informally or by adopting a semi-formal model– of the application under inspection. The expected output is represented by a general schema of the most relevant features of the level under inspection; for example, the content structure, the navigational capabilities offered, or the interface elements.

2. *Performing the selected tasks.* According to salient user scenarios, the reviewer selects relevant tasks from the U-KIT and tries systematically to perform them on the site. For each task, the reviewer assesses whether or not it can be properly accomplished. Let us consider *Task number 6* and the corresponding scenario information (see Table 2).

**Table 2:** Example of user scenario

| SCENARIO | |
|---|---|
| USER PROFILE | Art-lover. |
| TASK | Find information about the history of museum collection |
| SCENARIO DESCRIPTION | Joe is an art-lover. He would like to find some information about the history of a particular collection of the museum (e.g. paintings). He wants to know how and when the museum has acquired some works of art. |

The tool used for accomplishing the next phases is the evaluation matrix shown in Table 3.

3. *Evaluating the tasks through usability attributes.* Inspectors score each usability attribute (see Table 3) for each task. In this way, tasks are not only evaluated as feasible or infeasible. Tasks are assessed taking into account the different aspect of the application that might have an impact on the user experience. Attributes augment the accuracy of inspection because they decompose the evaluation of a task in different usability concerns.

4. *Weighting the results according to user profiles and communication goals.* Inspectors weight the score given according to the user profile and the goals of the applications. Low weight means low relevance for the user profile of the scenario showed in table 2; high weight means high relevance. Weights limit the subjectivity of inspection because they balance the general score of the attribute with the needs and expectations of a user profile.

**Table 3:** Inspection matrix for task number 6.

| | ATTRIBUTES | | | | |
|---|---|---|---|---|---|
| **TASK 6:** Find information about history of museum collection | **Accessibility** | **Orientation** | **Richness** | **Clarity** | Global score for this Task |
| **Scores** | 8.0 | 8.0 | 5.0 | 6.0 | 6.7 (just average score) |
| **Weights** | 0.1 | 0.1 | 0.5 | 0.3 | |
| **Weighted scores** | 0.8 | 0.8 | 2.5 | 1.8 | **5.9** ("weighted" average) |

The museum web site in the example obtained a pass mark 5.9/10 for task number 6. In this scale, 6/10 is considered "pass" value. Analysing carefully the partial results, it is evident that both the *richness* and *clarity* of the information regarding the collection's history should be improved.

5. *Empirical testing (user testing).* To empirically validate the most critical tasks identified during inspection a user testing is carried out in a usability lab. The user accomplishes several critical tasks and reports the results obtained. An inspector ensures that the user testing is carried out correctly and gathers the impressions, satisfaction and problems of the users. The expected output is a final usability report that shows the results obtained during user testing.

6. *Reporting the evaluation results.* In the final phase the inspector should draw a report, which highlights the problems of the application for each level of analysis; notably issues on the usability attributes and problems in the tasks. This document should summarize both inspection and user testing results.

# 4 Conclusions

As shown in the example, MiLE provides the inspectors with a usability toolset that records the evaluation experience and allows reuse across different web sites within a domain.

The distinctive features introduced by MiLE are diverse. An *efficient combination of inspection and empirical testing* allows reducing the main drawbacks of these techniques. The *use of tasks as guidelines for inspection* leads the evaluators during the complex analysis of the application; the *use of attributes* is as a way to define "unit of measurement" of the task and provide more analytic usability results. The *use of weights* is a way to translate scores into evaluation; through weights it is possible to take into account the user profiles and the specific objectives for the (portion of the) application during inspection. The *U-KIT (Usability Kit)* allows reusability and in-depth evaluation within a domain. In fact, every application domain has distinctive features that inspectors should take into account for a more reasoned and accurate usability evaluation.

U-KITs for e-banking web sites and digital libraries are under validation at Tec-Lab, University of Lugano (www.tec-lab.ch).

# 5    References

Brinck, T., Gergle, D., Wood, S.D., Usability for the web, Morgan Kaufmann, 2002.

Carroll, J., Making Use – Scenario-based design of Human-Computer Interactions, MIT Press, 2002.

Cato, J., User-Centred Web Design, Addison Wesley, 2001.

Di Blas et al., Evaluating The Features of Museum Websites, Museum & Web Conference, Boston, April 2002.

Green T.R.G., and D.R. Benyon, The skull beneath the skin; Entity-relationship modeling of information artifacts, *int. J. Hum.-Comput. Stud.*, vol.44, no.6, pp.801-828, June, 1996.

Jeffries R., Miller J., Wharton C., and Uyeda K.M., User interface evaluation in the real world: A comparaison of four techniques, *in Proc. ACM CHI*, 1991, pp.119-124.

Lim K.H., Benbasat I. and Todd P.A., An experimental investigation of the interactive effects of interface style, instructions, and task familiarity on user performance, *ACM Trans.Comput.-Hum. Interact.*, vol.3, no1, pp.1-37, January 1996.

K.H. Madsen, Special issue on the diversity of usability practices, *Communication ACM*, vol.42, no.5, 1999.

Matera, M. et al., SUE Inspection: An Effective Method for Systematic Usability Evaluation of Hypermedia, *IEEE Transaction*, Vol.32, No. 1, January 2002.

Nielsen, J., Mack R., Usability Inspection Methods, Wiley 1994.

Nielsen, J., Designing Web Usability, New Riders, 1999.

Rosson, M.B., Carroll, J., Usability Engineering, Morgan Kaufmann, 2002.

Roethlisberger, F.J. and Dickson,W.J., Management and the Worker, Boston, Mass.: Harvard University Press, 1939.

Whiteside J., Bennet J., and Holtzblatt K., Usability engineering: Our experience and evolution, in *Handbook of Human-Computer Interaction*, M.Helander, Ed. Amsterdam, The Netherlands, North-Holland, 1988, pp.791-817.

# Evaluation of Tourism Website Effectiveness: Methodological Issues and Survey Results

*Adriana Corfu*

Universidade de Aveiro
DEGEI, Campus de
Santiago, 3810-193
Aveiro, Portugal
acorfu@egi.ua.pt

*Manuel Laranja*

Universidade Técnica de
Lisboa – ISEG
Rua do Quelhas, n.º 6, 1200-
781 Lisboa, Portugal
mlaranja@iseg.utl.pt

*Carlos Costa*

Universidade de Aveiro
DEGEI, Campus de
Santiago, 3810-193 Aveiro,
Portugal
ccosta@egi.ua.pt

## Abstract

Web site analysis is an absolute no-brainer to embark on, but it also creates a hurdle: without relevant, understandable and up to date information, web site data is like all the books in a library, dumped in one big pile on the floor. All the information is there – you just can't find or use it. This is true all the more in an environment where tourism is a global industry (with some part of the value chain in the country of origin of the tourist and the remaining in the countries that are destinations) and extremely dependent on information and communication industries. According to (Kierzkovski, McQuade, Waitman, & Zeisser, 1996, p. 11) travel is one of the products "natural fits" with interactive media. Understanding what to analyze is going beyond hits and page viewings; understanding what actions to take as a result is a difficult job, moreover as information and communication services should be made usable for every citizen. "In a fair society, all individuals would have equal opportunity to participate in, or benefit from, the use of computer resources regardless of race, sex, religion, age, disability, national origin or other such similar factors" (ACM Code of Ethics cit. in Shneiderman, 2000, p. 85). However, there is appealing evidence that to get such goals the campaigns to be taken should be designed for being measured, analyzed and iteratively improved. This paper aims to: (1) establish a methodological framework for tourism related websites effectiveness' evaluation; (2) find out the comparative perceptions of Portuguese people on 2 national travel related websites features' effectiveness. Based on these results, some of employable findings will be discussed. To dig deeper into those business implications, further research is necessary. However, it seems reasonable to conclude that society is becoming more familiar with technology, in particular with the PCs and the Internet. This means that the travel industry has now the opportunity to deploy the Web in a more accurate, agile and innovative way.

## 1    E-Tourism: A Brief Overview

It is more than a decade since hyped stories about the impact of Internet Technology in the tourism industry has caused much of a stir in the minds of academics, practitioners and public policy makers, among others, and most of the discourses figured throughout around the theme painted pictures of a myriad shades and hues of the days to come. Internet based marketing and e-commerce are two major, but interrelated areas of development impacting the sector. The former reduces information asymmetry between tourists and ultimate service providers in destinations. The latter was predicted to eliminate intermediaries who are either information brokers or those guaranteeing services at the destination. Intermediaries such as tour operators and travel agencies were demanding economies of scale and this led to production of more standardized packages. Dis-intermediation should ideally lead to differentiated tourism products matching the demands of post-modern tourist at lower cost capitalizing the benefit of economies of scope. This then would lead to price advantages for the tourist and higher returns for the host society. So, with the Web now an essential part of business, is the gospel of the 4Ps[1] being rewritten? In fact, all 4 of the Ps are being transformed due to a major shift in the power structure of commerce with the adoption of the Internet. With this fundamental power shift of producers to consumers (today, the consumer has more information to make decisions and to access sellers globally), the way in which marketers think needs to adapt or run the risk of faltering in the face of more dynamic competitors. Philip Kotler, the legendary father of modern marketing, says that intelligent management of information and the use of technology-supported costumers' interactions are among the e-marketing rules of the 'New Economy' (Kotler, 2000). However, in general, most of these 'predictions' remained armchair musings, having not much research to substantiate their claims.

## 2    Methodological Issues and Survey Results

The research methodology reflects those two aims stated behind. First, a methodological structure has been proposed. Given that, a synopsis of variables was established based, on one hand, on concepts that existing literature amply covers and, on the other hand, on successful web applications. With the use of exploratory research, a structured-disguised questionnaire developed according with 4 broad categories (see Figure 1): (i) the web as an information source; (ii) the web as a booking system; (iii) the web as a customer retention platform; and (iv) the generic evaluation of the websites effectiveness, according with usability criteria, was spread between March and June of 2002 by e-mail, among a convenience

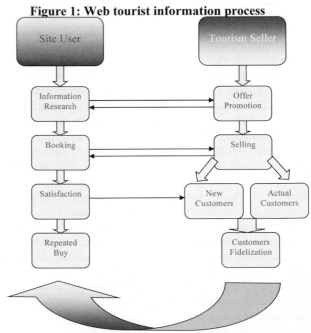

**Figure 1: Web tourist information process**

sample in a post-graduated community. Although only 73 persons responded to our questionnaire, response rate was considered high. Based on those responses, non-parametric tests (Mann-Whitney) were performed. Conclusive evidence about website effectiveness has been reached. Firstly, the overview we made on both websites (www.pousadas.pt and www.turihab.pt) suggests that not all the features we have considered in the model are offered. Therefore, the website www.pousadas.pt was the one that better responded to our approach. Secondly, according to the websites effectiveness evaluation's results (5 points Likert scale), was the same site that better satisfied the inquiries' needs.

---

[1] Product, price, promotion, place – set of marketing elements defined by MacCarthy in 1964 as "Marketing Mix".

## 2.1 The Web as an Information Source

The effectiveness of the Web as an information source was evaluated according with 17 specific features (see Table 1). The performance of Mann-Whitney test points in almost all the analyzed items to statistically significant differences (sig.<0.05) at the 95% significance level, being the website www.pousadas.pt better classified than www.turihab.pt.

**Table 1: Evaluation of the Web as an Information source**

|  | Sig. | www.turihab.pt | | www.pousadas.pt | |
|---|---|---|---|---|---|
|  |  | N | Mean | Mean | N |
| Electronic brochure | na |  | (*) | (*) |  |
| Company's history | 0,000 | 68 | 3,41 | 3,93 | 70 |
| Generic presentation of products/services | 0,000 | 72 | 3,74 | 4,36 | 69 |
| Press publications about company's activities | na |  | (*) | 3,54 | 61 |
| Online directory | 0,024 | 71 | 4,10 | 4,32 | 68 |
| Offered services & facilities | 0,000 | 72 | 3,74 | 4,27 | 60 |
| Maps, how to get there? | 0,009 | 72 | 3,75 | 4,11 | 70 |
| Prices | 0,094 | 68 | 3,54 | 3,86 | 70 |
| Address & useful contacts | 0,003 | 71 | 3,52 | 4,03 | 69 |
| What is "new" on site | na |  | (*) | (*) |  |
| Virtual tour | na |  | (*) | 4,02 | 69 |
| Last minute discounts/special promotions | na |  | (*) | 4,16 | 67 |
| Opportunity to choose the language | 0,002 | 70 | 3,67 | 4,06 | 67 |
| Currency conversion | na | 28 | 3,67 | (*) |  |
| Weather | na | 26 | 2,85 | (*) | na |
| News, press release, awards | na | (*) | 3,62 | 61 | na |
| Links to complementary sites | na | (*) | 3,65 | 68 | na |

na – not available (Mann-Whitney test wasn't performed); (*) – the site doesn't exhibit this option

The Web can give access to a greater store of information than other traditional communication media, and provide users with the means to select and retrieve only that which appeals to them. In this 'Age of Individual' that means that customized brochures, *à la carte* itineraries and guides, *interalia*, could be produced 'at the touch of a key'. So, *"electronic brochures"* and *"virtual tours"* are features that should be included on a site, to positively influence buying intentions (especially when trying to appeal to potential clients). The challenge for the managers is to monitor the interests and to be flexible in responding to them.

## 2.2 The Web as a Booking System

The Web as a booking system was evaluated through 6 specific features (see Table 2). An important finding is that 44% of the inquiries didn't engage in that simulation process, being the main reasons the time needed to spend on it and the concerns with data confidentiality. Those concerns reflect that having to pay by credit card might be a barrier to shop online, being no statistically discernible difference between the two companies. However, buying through the Internet depends upon not only the connectivity, but also the fundamental matter of social acceptance and patterns of buying. This should be taken into account when creating an e-commerce site. The website planners should be aware that it is very important for travel related sites to have a secure site where customers are confident that will receive the product/service that they purchased quickly and in the right time. However, to induce new internet-shoppers to buy, a payment method other than credit card hence seems to be profitable. On average, there is a clear direction that the site www.pousadas.pt was better classified than www.turihab.pt.

**Table 2: Evaluation of the Web as a booking system**

| | Sig. | N | www.turihab.pt Mean | www.pousadas.pt Mean | N |
|---|---|---|---|---|---|
| Web form booking | **0,004** | 34 | 3,47 | 4,13 | 47 |
| E-mail booking | **0,050** | 29 | 3,55 | 4,10 | 41 |
| Online booking notification | **0,050** | 27 | 3,44 | 4,06 | 33 |
| Opportunity to cancel online the reservation | 0,501 | 22 | 3,73 | 3,57 | 30 |
| Real time booking | **0,047** | 28 | 2,75 | 3,35 | 34 |
| Online payment / credit card | 0,858 | 24 | 3,00 | 3,07 | 31 |

## 2.3    The Web as a Customer Retention Platform

The performance of the Web as a customer retention platform was analyzed through 12 specific features: *"special services for frequent clients"; "visualization of own account"; "create/modify the own profile"; "special offers"; "newsletters"; "customized services"* (Welcome Paul Smith, for example)*; "guestbook"; "feed-back"* (comments or suggestions through web form or e-mail)*; "frequent asked questions"* (FAQ); *"special facilities for associates and members"; "employees space"* (Intranet); *"own company' statistic information about the customer".* An important finding is that no one of them is offered by the website www.turihab.pt. Only 4 features are offered by www.pousadas.pt, being all of them positively evaluated (*"guestbook"* – 3,7; *"feed-back"* – 3,6; *"FAQ"* – 3,4; *"own company' statistic information about the customer"* – 3,4). Overall, the level of performance suggests that much attention should be given by web designers on create such loyalty programmes; this is true all the more that great part of inquiries didn't identify these options in the website (*"guestbook"* – 65,8%; *"feed-back"* – 30,1%; *"FAQ"* – 60,3%; *"own company statistic information about the customer"* – 57,5%). This aspect is particularly important the more the target of such loyalty programmes is the existing, or even frequent, guests. It is quite assumed that "it is between five and ten times as expensive to win a new customer as it is to retain an existing one" (Rosenberg & Czepiel; Barnes & Cumby; Liswood cit. in Gilbert, Powell-Perry &Widijoso, 1999, p. 25). There is no doubt that the Internet is nowadays the most comprehensive and universal medium, able in providing managers with a set of useful tools for, on time, generate synergies between clients and companies, in general and, for increase the profitability of existing client relations and a extension of the period that such relationships last through the creation of an atmosphere of trust, satisfaction and commitment, in particular (Bauer, Grether & Leach, 1999).

## 2.4    Web Effectiveness Evaluation through Usability Criteria

This section's results have as main references (Shneiderman, 1998) and (Nielsen, 1994) interface evaluation research specifically guidance concerning usability criteria. From this point of view, the website www.pousadas.pt was better classified by the inquiries, being all differences statistically significant. Aspects like information's structure/organization, access' easiness; information's updateness; intuitiveness of navigational process; interface' attractiveness; have been tested (see Table 3). Comparing the means, we can conclude that the site www.pousadas.pt was on average the one that better satisfied the users' needs, being all the differences statistically significant at a 95% confidence level. The challenge for the managers is to make the information and communication services through the Web more usable for every citizen. However, this is a tough task. According to (Shneiderman, 2000, p. 85), designing for experienced frequent users is difficult enough, but designing for a broad audience of unskilled users is a far greater challenge".

**Table 3: Evaluation of the Web effectiveness through usability criteria**

| | Sig. | N | Mean | Mean | N |
|---|---|---|---|---|---|
| | | | www.turihab.pt | www.pousadas.pt | |
| Sufficient information | 0,000 | 73 | 3,30 | 4,06 | 69 |
| Structured/organized information | 0,000 | 73 | 3,44 | 3,99 | 70 |
| Relevant information | 0,000 | 71 | 3,23 | 4,09 | 70 |
| Up to date information | 0,000 | 72 | 3,50 | 4,01 | 69 |
| Attractive interface | 0,000 | 73 | 3,14 | 3,79 | 68 |
| Easy access | 0,049 | 73 | 3,68 | 3,97 | 68 |
| Navigation process as intuitive | 0,001 | 71 | 3,32 | 3,89 | 66 |

# Conclusion

Choosing the collection form of data of a convenience sample led to some degrees of bias, which could be reduced by replacing the convenience sample by probability samples, extend the sample to other nationalities than Portuguese people, as well as to non-academic communities, is another issue that deserves attention. Also, in developing an analysis of a travel related site, many different choices should be considered. In fact, the methodology proposed was, on time, one of many other possible approaches. However, we believe that a successful Web presence depends upon more than just the technology used, the design and the layout i.e. the 'look & feel' of the site. Butler Group suggests that effective Web applications in the context of the hotel industry are those that demonstrate an understanding of network limitations, demographics and culture. Moreover, what the lodging industry needs "is a framework to bridge the gap between simply connecting to the Web and harnessing its power for competitive advantage" (Gilbert, Powell-Perry &Widijoso, 1999, p. 21).

# References

Bauer, H. H., Grether, M. & Leach, M. (2002). Building customer relations over the Internet. *Industrial Marketing Management*, 31 (2), 155-163.

Gilbert, D. C., Powell-Perry, J. & Widijoso, S. (1999). Approaches by hotels to the use of the Internet as a relationship marketing tool. *Journal of Marketing Practice: Applied Marketing Science*, 5, 21-38.

Kierzkovski, A., McQuade, S., Waitman, R. & Zeisser, M. (1996). Marketing to the digital consumer. *The McKinsey Quarterly*, 3, 5-21.

Kotler, P. (2000). Administração de marketing – A edição de novo milénio (10ª ed.). São Paulo: Prentice Hall.

Shneiderman, B. (1998). Designing the user interface: Strategies for effective human-computer interaction (3rd ed.). Massachusetts: Addison Wesley Longman.

Shneiderman, B. (2000). Universal usability: Pushing human-computer interaction research to empower every citizen. *Communications of the ACM*, 43 (5), 85-91.

Nielsen, J. (1994). Usability inspection methods (1st ed.). John Wiley & Sons.

# At the Right Time:
# when to sort web history and bookmarks

*Alan Dix*

Lancaster University
Computing Department
Lancaster, LA1 4YR
UK

alan@hcibook.com

*Jason Marshall*

Loughborough University
Sports Technology Research Group
Loughborough, LE11 3TU
UK

J.M.Marshall@lboro.ac.uk

## Abstract

In open-ended interviews web users preferred text interfaces for sorting bookmarks to more complex graphical 2D and 3D interfaces. They also expressed a desire to sort web pages as they bookmarked them. However, in an experimental study we found that recall performance of web pages sorted during browsing was significantly poorer than performance when they sorted bookmarked pages at the end of a browsing session. This effect appeared to decay in a retest a week later. This work shows that users are able to articulate meta-knowledge strategy, but it questions whether users' expressed preferences are a good guide for design.

## 1  Introduction

This paper describes an empirical study of the effects of different sorting times for web bookmarks and history: incrementally 'during' browsing or all together 'after' browsing.

Studies by Tauscher and Greenberg (1997) have shown considerable revisiting of the same web page during browsing. Some of this is clearly due to backing up after following mistaken links or hub-and-spoke behaviour; indeed Catledge and Pitkow (1995) found that 30% of all navigation is the use of the 'back' button. However, there are a considerable residual number of pages that are 'really' revisited because the user wants to see the content again. Browsers support this behaviour both for short-term revisitation (back button and visit stack) and for the long term (history, bookmarks, favourites). In a formal analysis of several hypertext and web browsers Dix and Mancini (1997) found that the history and back mechanisms were subtly different in them all. This emphasises results found in other studies that users find back and history confusing and this is reflected in behaviour with comparatively little use of history or multi-step back. Bookmarks are more heavily used, but still are known to have many problems. There have been some more radical interfaces proposed and used at an experimental level including the data mountain (Roberston et al., 1998), which allows users to arrange thumbnails of bookmarked pages in a 2D landscape, and the WebView interface unifying history and bookmarks (Cockburn et al., 2002).

In order to investigate these issues we performed a short study in two phases. The first phase consisted of a series of exploratory interviews aimed at understanding users views on different proposed visualisation techniques for history and bookmark mechanisms. This was followed by a more formal experimental phase. Our original intention was to perform the more detailed work on

2D and 3D interfaces, but our initial study showed that the most critical issue for users was not how history was displayed, but *when* classification was performed. So, the more formal experiments during the second phase was focused instead on comparing outcomes when users sorted during or after browsing.

## 2 Phase I – what the users want

Our initial study was focused on a small number of open-ended exploratory interviews to discover typical browsing patterns, understanding and use of bookmarks and history. In addition, interviewees were presented with a number of 2D and 3D visualisation techniques proposed in the literature including data mountain (Roberston et al., 1998),, WebView (Cockburn et al., 2002). and WebBook (Card Robertson & York, 1996). It should be noted that in contrast to Tauscher and Greenberg (1997) and Catledge and Pitkow (1995) numerical studies of web data, this was about users self-reported behaviour and attitudes. There were six interviewees, two female and four male, with ages from 19 to 31 years. All were experienced web users.

The users reported hub-and-spoke browsing behaviour using either search engines or directly typed URLs to start their browsing. Surprisingly when asked what features were most useful for remembering a site four out of the six said that the URL or some derivative of it was most useful. None mentioned actual page content and only one mentioned keywords. This may be a reflection of the web expertise of the interviewees, or may be a reflection of existing history interviews, however, it does call into question interface suggestions that rely heavily on recall of page images and thumbnails. In fact, the users here may be being more knowledgeable than those designing novel interfaces as empirical studies by Kaasten, Greenberg and Edwards (2002) found that thumbnails have to be quite large before they become useful, especially if one is interested in identifying pages rather than simply sites.

When offered images (not live use) of the various proposed history interfaces it was the text based 2D interfaces (Apple Hot List and WebView) that were most popular. Again this should be interpreted with care as they are more familiar looking and also included explicit annotation which most interviewees felt was more important than the particular visual aspects of the history mechanism. This latter point also supports Amento et al. (1999) study where all the participants wanted to record comments about sites as they visited and collected them.

The interviewees were given the opportunity to comment freely on web history and favourites mechanisms and to suggest what they would like to see including drawing sketches etc. Although the main reason for the interviews was to investigate history visualisation, the issue that repeatedly arose was the interviewees' desire to be able to classify bookmarks/favourites at the moment they were 'remembered' rather than as a secondary exercise. Currently all common browsers force a bookmark-now, sort-later mode of working. The strength of the interviewees' reactions led to a refocusing of our empirical studies towards understanding this 'when to classify' issue in more depth.

## 3 Phase II – what the users need?

Based on the results of phase I study, we decided to focus our more detailed experiments on the issue of sorting during browsing compared with sorting afterwards and its effect on later (online) recall.

Although the interviewees had expressed a desire for 'during' sorting, we postulated that this would in fact lead to less clear classification. This is because when items are sorted during the process of browsing it is not clear the full range of future pages that will require classification, whereas when classifying pages after navigation it is easier to produce a balance and sensible classification. For example, if the first few pages seen are about aspects of football should one classify them as 'football' pages or 'sport' pages. If the following pages include one further page on golf and many on completely different topics then 'sport' would have been the best classification. However, if the rest of the pages were about golf, rugby, cricket etc., then 'football' would have been best. Classification 'after' is able to take this into account.

This therefore lead to a hypothesis that sorting during browsing would be less 'good' than sorting afterwards.

The main condition was the *during* vs. *after* sorting. In both conditions the participants used the Internet Explorer favourites mechanism to classify 50 web pages. In the *during* condition the participants were asked to look at each page and then immediately save it as a favourite and classify it. In the *after* condition the participant would look at all the pages saving each as a favourite and then after all had been seen open the sorter and classify all 50 pages.

For the experiment ten paid participants were recruited, five male and five female, aged between 18 and 36 years. All were experienced users of Internet Explorer and of its favourites mechanism. Two datasets of 50 web pages were selected from 100hot.com's 2002 list of top web sites. The experiment consisted of a within subjects design where each participant performed both a *during* and *after* sorting. The experiment was balanced for order of presentation and for the site set. So half the participants had set 1 with the *during* condition and set 2 with the *after* condition and half the other way round. The quality of participants' classification was judged by asking them to use the classified bookmarks to answer a series of questions. In each case this was immediately after the relevant sorting. Because different sets of pages were used there was no learning effect for content and in fact the results showed no measurable order effect.

Our hypothesis was indeed borne out by the results, which did show significantly ($P<0.04$) better recall for the *after* condition compared with *during* sorting (see table 1). Also in a post-test questionnaire the participants preferred the *after* sorting, in direct contrast to the interviewees' imagined preference. Other results included a correlation between time spent sorting and performance.

**Table 1:** Mean retrieval times for sorting method and retrieval cue (in seconds)

|                 | Mean  | Std. Deviation |
| --------------- | ----- | -------------- |
| During browsing | 584.2 | 232.73         |
| After browsing  | 421.8 | 106.77         |

A small number of participants were retested a week later. The number was very small (only four participants) and so any results are merely suggestive. However it did appear that the advantage of *during* vs. *after* sorting disappeared almost completely. If the quality of the *during* classification were indeed worse then one would have expected to have had even worse results on retesting after the immediate memory of the classification process had faded.

As the numbers were too small for statistical testing these counter-intuitive results may well be just a random effect. However, they have made us question whether the strength of the *after* sorting may be partly explained by the fact that the sorting process occurs closer to the post-test. More sophisticated experiments may be required to separate all the potential causes.

# 4 Reflection

There are several lessons from this study beyond the raw results of the experiments.

First the interviewees were able to articulate desires relating to meta-knowledge issues – the timing of bookmark classification. This ability to reflect on as well as engage in knowledge discovery has been investigated further in Dix, Howes & Payne (2003).

Second, the actual running of the experiment also showed how complex these issues are, especially because we are looking at relatively long-term effects that are hard to capture fully within a laboratory setting. We deliberately chose an experimental set-up that was at least partially ecologically valid rather than a more controlled and specific pure psychological experiment. This allowed us to find some real and strong effects, but by its nature admits multiple interpretations.

Thinking about the differences between the experimental condition and 'real life' several things are obvious. In a real situation the pages visited would be ones that held some personal interest for the users and hence fit within existing mental structures. In contrast our participants were faced with pre-selected pages. However, the chosen pages were of broad interest, which we hope reduced this effect. Harder to control was the interface itself. Using an off the shelf and familiar interface rather than a do-it-yourself one means that we avoid the frequent situation of testing a ropey prototype The downside is that the IE favourite mechanism is designed for periodic *after* sorting, which could have biased against the *during* condition. Finally, real use would not fall neatly into one of these camps. Instead on-the-fly sorting would happen when the user had and existing set of categories and so some of the problems of on-the-fly sorting with a blank slate would only occur during the earliest browsing. Also even if on-the-fly sorting were the norm it is likely that heavy users would perform some periodic tidying up of categories.

Finally, the two phases of this study show, what everyone in HCI knows, but we are often reluctant to admit: users don't always know best. Although the initial interviewees were heavily in favour of on-the-fly sorting, this preference was reversed when faced with doing it in practice. As noted, this latter preference may have been due to a poor interface, but this would be equally worrying if bad experiences influence user requirements so heavily. This was an experimental study, not an exercise in participatory design, but does emphasise that good participatory design should be user focused and may be user led, but always requires strong expert guidance and aid.

# 5 Acknowledgements

This work was supported by EPSRC under Masters Training Package GR/N63277 and the EQUATOR Interdisciplinary Research Collaboration (www.equator.ac.uk). A complete description of the results in this paper and further results from the same experiments can be found in Marshall (2002).

# 6 References

Amento, B., Hill, W., Terveen, L., Hix, D. & Ju, P. (1999). An Empirical Evaluation of User Interfaces for Topic Management of Web Sites. In *Proceedings of CHI 99*, 552-559.

Card, S.K., Robertson, G.G., and York, W. (1996). "The WebBook and the WebForager: An Information Workspace for the World-Wide Web" in *Proceedings of ACM SIGCHI '96*, Vancouver, Canada, 111-117, April 1996.

Catledge, L., & Pitkow, J. (1995). Characterising Browsing Strategies in the World-Wide Web. *In Proceedings of the 3rd International World Wide Web Conference, Darmstadt, Germany*, Published in *Computer Networks and ISDN*, 27, 1065–1073, Elsevier Science. http://www.igd.fhg.de/www/www95/papers/

Cockburn, A., Greenberg, S., McKenzie, B., Jasonsmith, M., and Kaasten, S. (1999). WebView: A Graphical Aid for Revisiting Web Pages. *OzCHI'99: Australian Conference on Computer-Human Interaction*, Wagga Wagga, November 28–30, 1999. pages 15-22.

Dix, A., & Mancini, R. (1997). Specifying history and backtracking mechanisms. In P. Palanque and F. Paternó (Eds.),.*Formal Methods in Human-Computer Interaction* (pp. 1–24). London, Springer-Verlag. http://www.hcibook.com/alan/papers/histchap97/

Dix, A., Howes, A., & Payne, S. (2003). Post-web cognition: evolving knowledge strategies for global information environments *International Journal of Web Engineering Technology*, 1 ( 1) (in press).

Kaasten, S. and Greenberg, S., & Edwards, C. (2002). How People Recognize Previously Seen WWW Pages from Titles, URLs and Thumbnails. In X. Faulkner, J. Finlay, F. Detienne (Eds) *People and Computers Volume XXVI* (pp. 247–265), Springer Verlag

Marshall, J. (2002). An Exploratory Study in World Wide Web Navigation History. MRes Dissertation. Lancaster University, UK
available from this paper's web page: www.hcibook.com/alan/papers/HCII2003-history/

Robertson, G., Czerwinski, M., Larson, K., Robbins, D., Thiel, D., & Dantzich, M. (1998). Data Mountain: Using spatial memory for document management. In *Proceedings of UIST'98* (pp. 153–162). San Francisco, California, November 1998, ACM Press..

Tauscher, L., & Greenberg, S. (1997). How people revisit web pages: empirical findings and implications for the design of history systems. *International Journal of Human Computer Studies*, 47 (1), 97–138.

# ANTS: An Automatic Navigability Testing System for Web Sites

*Marcos González Gallego; María del Puerto Paule Ruíz; Juan Ramón Pérez Pérez; Martín González Rodríguez*

University of Oviedo
Department of Computer Science, Calvo Sotelo, s/n, 33007 OVIEDO – Spain
marcos@petra.euitio.uniovi.es, paule@pinon.ccu.uniovi.es,
jrpp@pinon.ccu.uniovi.es, martin@lsi.uniovi.es;

## Abstract

We are going to introduce a new system to help the designer to build up a web site more usable. Before that, we made a simple introduction of what is the usability and how can it be measure. Then we evaluated the different existing tool to measure the usability and we considered the advantage and disadvantage of those systems to make our system more powerful. We show in the following pages how our system works and the improvements that it has.

## 1 Usability Lab and its Disadvantages

The success of any web site depends on the users who are going to use it in the future. Those users should find the site satisfactory, effective and efficient (ISO Ergonomic Requirements, ISO 9241 part 11: Guidance on usability specifications and measures). If the user feels uncomfortable while surfing the web, it is very simple to close the browser and look for another web site. In E-Business this is equal to waste a very amount of money and let the rival enterprise to gain those clients.

If we want to know how is our user's navigation, we have to use a usability lab. The usability lab offers powerful and precise information that helps the designers to analyze the weak points of their site: difficulty of use, incapacity of understanding the interface, incapacity of understanding the classification of contents and more questions referred, of course, to user's interaction. However, the usability labs have the following objections:

- The evaluation of the navigation requires a high precision and constant observation of the volunteers. Besides, the time of observation should be long.
- Due to the high prices of the evaluation and to the examples have to be small, the quality of the evaluation decrease.
- The testers choose volunteers among people with the same cultural level believing to see in them the medium user, so it is difficult to discover new types of user no yet identified.
- The volunteers know that they are being observed. This fact adds external factors to the evaluation process. We speak of nervousness, confusion, timidity and similar reactions.
- Know certain details of the experiment also influences in user's navigation. Volunteers know what is expected of them so they can adapt their own techniques of navigation to the model of navigation that are supplied by testers.

- The usability evaluation is make in computers that match with some specific technical specifications needed to run the application. Probably those computers and the user's computer are very different.
- As the evaluation of the usability takes place in a laboratory, the testers forget that the environment affects the behavior of the user.

# 2    Remote Usability Evaluation

To avoid the usability lab disadvantages, the usability test should be done in the user's computer, in his own environment and without the physical presence of the testers. This is call Remote Usability Evaluation. Using this method, the advantages obtain are:
- Users ignore the nature of the evaluation, as well as the role they are playing in it. They only interact with the tested system.
- Users are not under pressure. They feel comfortable to explore the navigational map provide. External factors do not affect the test results.
- It is possible to obtain information about how the characteristics of the computer influence in the form that users explore the site.

When the remote evaluation is carried out, the prototypes can be freely distributed as beta versions to be evaluated by all users interested in the application.

# 3    Enterprises and Software for Remote Usability Evaluation

We are going to comment the characteristics of some enterprises and available products to analyze the web usability using remote evaluation.

## 3.1    Ergolight

Ergolight offers the Ergolight Interactive Site Usability Reports to evaluate a web site. That product allows the designer to obtain reports about the design problems that affect the user's experience. The report includes a list of pages sorted by cost of effectively. For each page, the reports also includes a sorting list of problematic links. To use this service, the designer needs an archive with the visits done to the site. Also the domain is needed to allow direct references from the report to the analyzed pages. That register archive should be created by the server automatically. The analysis is composed of:
- Site ranking. It will be evaluated the access facility, the use facility and the information quality.
- Site diagnostic. It will be evaluated if the pages are going to be hard to find, if they are going to irritate the users, if they are going to be displayed slowly, and so on.
- Statistics. It will show statistics regarding the number of visits, the number of leavings of each page, the percentage of inputs and outputs, the size of the downloads realized, the number of access to each page sorting by the size of them, and so on.

### 3.1.1    Advantages

The number of variables analyzed regarding the user interaction with the analized web site. To see the results, we only have to access the result page. This page is always updated. It brings suggestions to improve those things that have made worse the site evaluation. Each page is important. Concrete actions like downloads are important. The results include links to concrete pages.

### 3.1.2 Disadvantages

To use this service we have to know how the web server works. This program doesn't obtain information directly from the users. It obtains it and classifies it from the information included into the log archive of the web server. We don't know what we will have to do if we want to reset the statistics. Because the results are statistical, if we want to obtain a satisfactory result we will depend on the number of visits to our web site.

## 3.2 WebMetrics NIST

WebMetrics NIST offers a group of products to facilitate the task of determinate if the web site is usable. This group is composed by:

- WebSAT (Web Static Analysed Tool). It indicates if the HTML code carries out with the IEEE specifications or with the WebSAT usability guides.
- WebCAT (Web Category Analysis Tool). We can create different kinds of categories and different kinds of objects. Each object belongs to a category. During the test the users have to determinate which is the better category for a concrete object.
- WebVIP (Web Variable Instrumenter Program). This product allows to insert code into the pages to make automatic advice of how the user can do some tasks. For it, we have to edit a config file. This file is stored using the FLUD format.

### 3.2.1 Advantages

The tools are orientated to make easier the web design, not to evaluate the web after it was developed. When the user is required to do something, what he has to do is very simple and he doesn't need to learn new especial things. The division of the options into different tools becomes a great advantage if we will only need one of them.

### 3.2.2 Disadvantages

The division of the options into different tools becomes a disadvantage when we will need more than one tool, because we will no be able to obtain them together. The FLUD format is a WebMetrics special format that need a special application for its management. On this way the collaboration between different applications becomes more complex.

## 4 ANTS

We have developed a new remote evaluation system: Automatic Navigability Testing System - ANTS. This system improves the disadvantages previously cited. It analyses multiple user's factors, and the installation is very simple and it takes advantage of the network possibilities. We can group the ANTS objectives by consider the technical used:

- Automatic Data Gathering. ANTS obtains information about the user navigation through the analyzed web site.

- Remote Collaborative Evaluation. ANTS obtains information about the user's behaviour when he is trying to complete concrete tasks proposed by the tester.

## 4.1 ANTS Architecture

Based on client-to-server technologies, the ANTS project simulates the behaviour of a community of ants. The ants are agents that come out of the hill in search of food. When they find it, they carry them back to the anthill for its storage. In our case, we have a central server (anthill) that contains data-bases storing data about users (warehouses of food). We obtain that data after the insertion of agents (ants), applets in the pages of the analysed web site.

Therefore, when a user visits the web site, whatever is the page he choose, he discharges the agent. The agent will register the user's behaviour automatically. It will send back the data to the server, and the server will store it. With the data stored properly to distinct different navigational sessions, different users an so on, the experts in web design and in usability of interactive systems will find very important information to solve problems of usability and use of the web site.

## 4.2 Automatic Data Gathering

With the data obtained in each page by the agents, the following information is delivering to the designer: list of the sessions opened, list of the session pages, list of users ordered by the number of sessions opened, path followed by the average user through the web, list of user's countries registered, list of user's languages registered, list of pages ordered by number of visits, list of pages never visited, list of links never used and list of links more used
 All these results are shown to owner of the web site analised by a web site.

## 4.3 Remote Collaborative Evaluation

### 4.3.1 Completing a Proposed Task

The user has to complete a proposed task in the analysed web site. The information given to do the test has to be simple and evident. In other war, few users will not realise the test because if the text is too long, they will not read it and they will cancel the test. When the user accepts the test, it will be opened the page where the tests starts. Using the automatic data gathering previously cited, we are going to obtain if the user is able to finish the test in an efficient way.

### 4.3.2 Distributing Elements into Categories

In essence the web sites are content stores. Usually, those contents are very numerous so it is very important the way they are organized. We have to group them to let final users who visit our site control those contents. The web site often use a content grouping by categories. It can be for example: bibliography, similar themes, and so on. The problem arises when we think that everybody has got the same sorting schema and that content sorting proposed during the web site design is going to be understood by the average user. The reality proves us that the content organisation is different for each user, so if we sort the contents making them similar to the average user's sorting we will obtain the ideal case.

For it, we show a list of elements (contents) to the user, and he has to distribute each of them into a determinate category. After computing those results, it will be obtained true data about what is the way of sort the elements more similar to the one that understand the average user. Thus we are making visible the user's form of sorting the information of our web. The use of this technique is applicable so much to the diffusion of beta versions of our web site as to a continuous control of the behaviour of the final user of our definitive web site. The users aren't statical entities, they are clearly dynamic. So we can never assure that after some months of perhaps after a few days, our users organise the contents in another way. We must remember that our site is not the only site over the net and that others sites influence in the behaviour and in the requirement are going to follow our users.

### 4.3.3  Visual Interpretation

When we are designing a web site, specially the composition that will have the page, we use to forget that the way of organize this information on the screen is very important. We talk about the buttons arrangement, of the index, of the access by icons to another nodes. We can think that this information is seen clearly, when in reality it can be totally inadvertent by the user. We have a good example with banners. The banners are a tool very powerful to advice user using them by an advertisement panel. But if we put that banner on the top an central zone of our pages, and if we use it with the classical extended form, because this is the form to show publicity, the user will not look to the banner and he will think that is only publicity.

To solve this kind of problems, we propose to the user to make a visual interpretation of our web site. We show him a page during 30 seg. When the time finish, we will show a blank window with a simple group of tools for drawing. We will ask him to draw exactly what he has seen. When the server receives that graphic we will observe the things that the user saw and those things unnoticed by him. If we consider their results, we will obtain pages more intuitive to the user, because he will see the important information where he wants to find it.

## 5    Conclusion

ANTS allows to obtain real data about the user's behaviour relative meanwhile he is surfing the web and to his understanding of the analyzed web site. The existing utilities in the market are centred in a specific analysis, as can be Ergolight and its navigation analysis. ANTS is the only utility that considers necessary the analysis of the major number of characteristics of the user.

## 6    References

González Rodríguez, M; Álvarez Gutierrez, D: Proceedings of the EuroMedia 2000 Conference: Fifth Annual Scientific Conference on Web Technology, New Media, Communications and Telematics Theory, Methods, Tools and Applications, Antwerp, Belgium, ISBN 1-56555-202-4.

González Rodríguez, M: Proceedings of the International Software Quality Week   Europe: Lessons Learned, Brussels, Belgium, 1999.

Nielsen, J: Guerrilla HCI: Using Discount Usability Engineering to Penetrate the Intimidation Barrier  http://www.useit.com/papers/guerrilla_hci.html  (1994).

Nielsen, J: Discount Usability for the Web  www.useit.com/papers/web_discount_usability.html  (1997)

Nielsen, J: Usabilidad, diseño de sitios web: Prentice Hall (2000)

Preece, J: Human Computer Interaction: Addison and Wesley (1994)

# Presenting Results of a Search Engine for Recorded Lectures in order to Support Relevance Decisions by the User

*Wolfgang Hürst*

Institut für Informatik, Universität Freiburg
Georges-Köhler-Allee, D-79110 Freiburg, Germany
huerst@informatik.uni-freiburg.de

**Abstract**: In addition to correct results, search engines often return documents with less or minor relevance. Therefore, one of the main challenges for search engine interface design is to present the results in a way that supports the users in selecting the documents that are really relevant for their particular information needs. In addition, different search tasks make varying demands on the interface design and different application domains allow to present different information to the users. In this paper, we discuss some issues evolving when designing a search engine for recorded lectures. An evaluation with different user interfaces for such a search engine is presented in order to identify the problems and issues that arise for this particular task.

## 1    Introduction

Often, the result of a search engine contains many documents that have no or very low relevance in a particular situation. The reason for this is generally limited precision of the retrieval algorithms as well as vague and ambiguous user queries. Therefore, it is very important for a search engine interface designer to present the retrieved documents in a way that supports the users in quickly identifying the ones that are really relevant for them. In this paper, we address the question of how the results of a search engine for recorded lectures should be presented. Since the end of the 90s, research projects such as *Classroom2000/eClass (Abowd, 1999)* or *Authoring on the Fly* (AOF) *(Müller, 2000)* have gained a lot of experience in automatic presentation capturing and in putting the resulting multimedia files (each containing one recorded lecture) on the Web for further access and usage by the students. Meanwhile, many universities and institutions as well as industrial companies have adopted these approaches and started to capture and record single talks as well as whole courses. In addition, commercial tools are now available what makes automatic presentation capturing an easy and commonly used task. The intensive usage of such tools results in a tremendous amount of data making it necessary to offer some sort of search functionality for these documents. Therefore, we started to develop a specialized search engine for recorded lectures (see *aofSE, 2002* and *Hürst, 2002*). The aim of this search engine is to study and evaluate the problems that appear in retrieval of recorded lectures and to develop and investigate new, alternative approaches for this task. In the following, we present an evaluation we did in order to identify how the results of such a search engine should be presented.

## 2    Different UIs for a Search Engine for Recorded Lectures

When designing the interface of a search engine for recorded lectures, various options and parameters exist. First of all, there is a lot of information available in the documents that might be useful for the users and therefore should be presented at the result page. In addition, there are numerous ways to format and show this information. The answer to the question about the best possible way to do this generally depends on the intention of the users. Earlier studies we did with the usage of our search engine for recorded lectures indicated that users typically follow two types of search patterns that can roughly be classified as (a) searching for whole documents, i.e., one complete presentation about a particular topic, and (b) searching for particular parts of a document, i.e., one specific information within a larger presentation (e.g., the definition of a technical term). Hence, one of the most critical questions in designing an appropriate result presentation is at what granularity level the retrieved documents should be presented. Is it more appropriate to show them on a detailed level, such as a list of relevant slides (independent of the documents they appear in) or should they be presented at a higher level, e.g., as an ordered list of presentations? To answer this question, we set up three user interfaces for our search engine that present the retrieval results at different granularity levels. Each of them provides exactly the same information to the users but in a different way.

The first interface – in the following called "*Slides*"-Interface (see *Fig. 1a*) – presents the results on slide-level, that is, each entry in the result list represents one slide, independent of the lecture it appeared in. Retrieved slides are represented through their titles (or first line of the slides-text if there isn't any title) and are ordered according to their relevance for the user query, as it is common in general search engines. The interface shown in *Fig. 1b* – in the following called "*Lectures-or-Slides*"-Interface – leaves the decision about the granularity level to the users. By using one out of two search buttons, the users can select between different result presentations. In the first case, the relevant lectures are shown, without any detailed information about particular slides. In the second case, again, a sorted list of relevant slides is presented (*Fig. 1b* shows the result that is presented after pressing the "Search for Lectures"- button. The result presented after pressing the "Search for Slides"-button is identical to the one shown in *Fig. 1a*). The third interface – which is shown in *Fig. 1c* and subsequently called "*Lectures-and-Slides*"-Interface – again presents the relevant slides to the users, but this time they are sorted according to the lectures they belong to. For each retrieved lecture, the titles of all of its relevant slides are shown. Lectures are sorted according to their relevance. An evaluation comparing these three interfaces with each other is presented in the next section.

The documents with the recorded lectures usually contain much more information than just the lecture and slide titles that are indicated in all user interfaces introduced so far. Providing some of this information to the users might help them during the search process. In our search engine (and most other search engines for recorded lectures) the text from the slides is the main information source on which the decision of the retrieval engine is based. In addition, slides offer a very good initial way to look at the content of the files. Because of this, it seems very promising to present slide-based information in the search engine result.

*Fig. 1d* shows an interface – in the following called "*Thumbnails*" – that is very similar to the "*Lectures-and-Slides*"-Interface. The only difference is that instead of the slide title a thumbnail is shown. This presentation has the advantage that the user gets more information more quickly. Its main disadvantage is that it takes up more space and thus more scrolling is required. *Dziadosz, 2002* reports that the presentation of thumbnails of Web pages has proven successful in case of Web search engines, what suggests a high usability of this interface, as well. However, showing thumbnails of slides can sometimes be a problem because the text can become to small and therefore unreadable. For this reason, we set up another interface – in the following called "*Slides-Text*"-Interface (see Fig. *1e*) – where the text from the slides is automatically extracted and shown instead of the thumbnails. Here, the advantage is that the text is always easy to read, independent of the font size used on the original slide, and that it allows us to easily highlight the search terms that appear on a slide. However, the disadvantage compared to the thumbnail representation is that the layout information is lost and the automatic extraction of text from the original slides sometimes results in grammatically incorrect sentences.

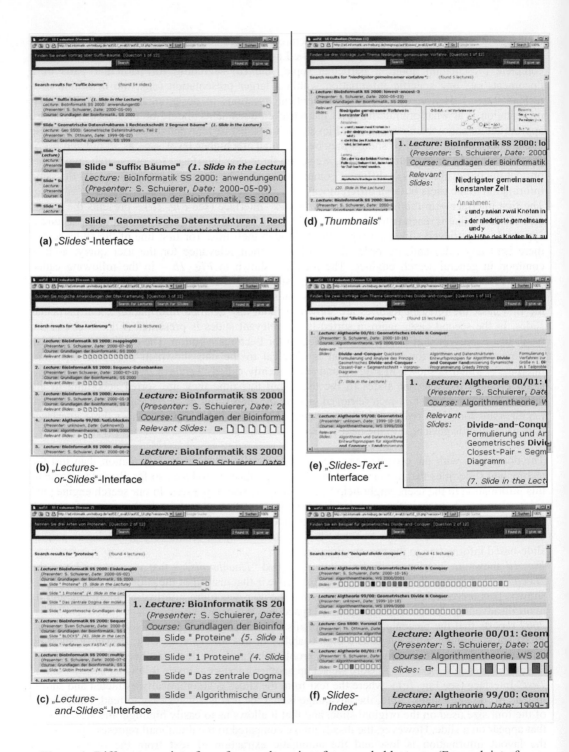

**Figure 1:** Different user interfaces for search engines for recorded lectures. (For each interface enlargements of the first result entry are given for better reading.)

*Fig. 1f* shows a user interface – in the following called "*Slides-Index*" – where every slide that was used in a relevant lecture is presented through a little icon independent if the slide itself is relevant or not. Relevance of a slide is indicated through a shade of grey: A slide is more relevant the darker its corresponding icon is, slides with no relevance are represented through a white icon. This interface has the advantage that all slides from a relevant lecture can be accessed very easily, what is especially important if a relevant one was accidentally misclassified as not being relevant. In addition, it is very easy for a user to identify the overall position of a slide in the whole lecture.

## 3   Usability Study and Evaluation

To evaluate the different interfaces presented in the last section we set up a comparative user study. Two tests were done to evaluate the interfaces presented in *Fig. 1a-c* and *d-f*, respectively, with each other. The main goal of the first evaluation (*Fig. 1a-c*) was to gain information about the best granularity level for result presentation. While in this test all interfaces presented exactly the same information (but in different ways), the purpose of the second evaluation (*Fig. 1d-f*) was to identify how the presentation of different information can improve the overall search experience.

**Evaluation setup**. For each of the six interfaces, users had to answer twelve questions. Three of those questions were more general, such as "*Find a lecture about topic xy*" while the others were rather specific, such as "*Find the definition of xy*" or "*Find an algorithm to solve the xy problem*". The data base contained about 120 lectures from six different courses, all from the area of computer science (three about the theory of algorithms, two about geometrical algorithms, and one about bioinformatics). The first query was given by default. The slide or lecture containing the correct answer to each question was always retrieved with this first query, however, in some cases it was not within the top most results and therefore difficult to find. All questions could be answered by just looking at the slides (For the evaluation we disabled audio replay of the corresponding lectures. Users only had access to the slides). After clicking on the slide-shaped icon in a result entry, a separate window opened containing the corresponding slide. Ten users, all members of our faculty and experienced search engine users, participated in the evaluation. In both tests, the three interfaces were presented to them in a counterbalanced order. The first two questions were considered as test questions where the participants could get familiar with each interface. After that, they were asked to solve each of the remaining ten questions as soon as possible. The time to solve them as well as any interaction with the interface was logged. The size of the search engine window was 800x700 pixels with a screen resolution of 1280x1024. After answering the questions for each interface the users had to fill out a questionnaire. At the end of evaluating each group of interfaces (*Fig. 1a-c* and *d-f*, respectively), the users were asked to fill out another questionnaire with questions that compared the interfaces with each other. The whole evaluation process took about two hours for each user.

**Evaluation 1 (Interfaces in *Fig. 1a-c*)**: In most of the cases the participants did not use any query reformulation. The only exception was the "*Lectures-or-Slides*"-Interface (*Fig. 1b*), what is no surprise since query reformulation in this case also allows to change the result presentation style. On average, the "*Lectures-and-Slides*"-Interface (*Fig. 1c*) required much less access to the original slides in order to solve the questions, what indicates that it is easier for the users to separate relevant from irrelevant results with this kind of interface design. However, the average time needed to answer the questions was the longest (six and nine seconds longer than with the other two interfaces). Nevertheless, most of the users considered this interface to be the best solution, especially compared to the "*Slides*"-Interface (*Fig. 1a*), which was generally seen as the worst of them. Although most users first said that they liked the possibility to have the choice between different presentations ("*Lectures-or-Slides*"-Interface), some of them mentioned after the evaluation that it was hard to make the decision about which presentation to choose when they initially started their search.

**Evaluation 2 (Interfaces in *Fig. 1d-f*)**: The "*Slides-Index*" (*Fig. 1f*) was seen as the best interface of all by most of the users. Beside the clear and compact presentation they mainly preferred that the original order of the slides was kept. This was useful in many situations, for example, when searching for definitions (which are more likely located at the beginning). Obviously, this interface required more accesses to the original slides, because less information about the slide content is presented here. However, to our surprise, this increase in interaction did not lead to a significant increase of search time. The average time the users needed to answer the questions was in about the same range as with all the other interfaces (except if compared to the "*Slides-Text*"-Interface, see below). The results for "*Thumbnails*" (*Fig. 1d*) and "*Slides-Text*" (*Fig. 1e*) are somehow inconsistent. Most of the users liked the idea of having the thumbnails of the slides available, but only few liked the "*Slides-Text*"-Interface. However, although the subjective user judgments of this interface were often very negative, it was by far the fastest of all (including the ones from the first evaluation) with an average search time for all users of about 25 seconds (compared to the 31 till 40 seconds needed with the other five interfaces). One possible reason for this might be that the search terms were highlighted in the retrieved results. Most of the users remarked this as positive and very helpful. The average number of accesses to the original slides was less for both interfaces. One problem with the thumbnails is that their usability depends very much on the quality of the original slide, e.g., if the presenter used a large font size as well as reasonable colors and highlighting on the slide. The main problem with both interfaces was the necessity to scroll into two directions, what was considered as very disturbing and uncomfortable by most users. Some also complained about being overwhelmed by too much information.

**Final Discussion and Conclusion**. The evaluation showed that with the given interfaces most users preferred to have a compact, clearly arranged result representation, although presenting more and detailed information about the documents was helpful in many situations. This general trend could not only be identified from the (subjective) user comments but also by the objective time measurements and evaluation of the logging information. For the final interface design we therefore propose to start the search with an easy and clear interface, such as the "*Slides-Index*", and then offer the possibility to the user to interactively extend the initial result to present more and more detailed information, depending on the individual needs.

# References

Abowd, G.D. (1999). "Classroom 2000: An Experiment with the Instrumentation of a Living Educational Environment." *IBM Systems Journal*, Volume 38/ 4, pp. 508-530.

"aofSE - the AOF Search Engine" (2002). Web site: http://ad.informatik.uni-freiburg.de/aofSE

Dziadosz, S. Chandrasekar, R.. (2002). "Do Thumbnail Previews Help Users Make Better Relevance Decisions about Web Search Results?" Proceedings of ACM SIGIR Conference on Research and Development in Information Retrieval, Tampere, Finland

Hürst, W. (2002). "Indexing, Searching, and Retrieving of Recorded Live Presentations with the AOF (Authoring on the Fly) Search Engine." Proceedings of ACM SIGIR Conference on Research and Development in Information Retrieval, Tampere, Finland

Müller, R., Ottmann, T. (2000) "The Authoring on the Fly system for automated recording and replay of (tele-)presentations." ACM/Springer Multimedia Systems Journal, Volume 8/3.

**Acknowledgments**: This work is supported by the German Research Foundation (Deutsche Forschungsgemeinschaft DFG) as part of the strategic research initiative "V3D2".

# Characteristics of Web Site Designs: Reality vs. Recommendations

*Melody Y. Ivory*

The Information School
University of Washington
myivory@u.washington.edu

## 1 Introduction

Although most prominent web sites are created by professional design firms, an enormous number of smaller sites are built by people, who, despite having little design experience or training, need to make information available online. As a consequence, the usability and the accessibility of sites with local reach, such as sites for non-profits, academic courses, or small businesses, are often substandard. Much has been said about the way to design usable and accessible web sites, yet there is often a wide gap between the recommendations, such as ``make the site consistent,'' and their application. Furthermore, guidelines tend to conflict with one another and consequently require careful study and practice.

As one way to address the limitations of prescriptive web design guidelines, we have developed a methodology to derive statistical models of effective design practices from empirical data [1]. The methodology entails computing 157 highly accurate, quantitative page-level and site-level measures. The measures assess many aspects of web interfaces, including the amount of text on a page, color usage, and consistency. We use these measures along with ratings from Internet professionals to build statistical models or profiles of high quality web interfaces. Our models have been able to predict, with over 90% accuracy in most cases, whether or not the measures and relationships among them on individual pages or the site overall are consistent with design patterns that are used on highly or poorly rated pages and sites. We use these models in the automated analysis of web sites.

Another application of these quantitative measures and statistical models is to revisit web design recommendations, specifically for recommendations that are contradictory, vague, or not empirically validated. The statistical models elicit quantitative ranges or thresholds that can then be used to provide better recommendations to occasional or novice web designers.

## 2 Analysis of Web Design Profiles

### 2.1 Objective

Our objective is to highlight key characteristics of highly rated web site designs and to furthermore contrast these characteristics to recommendations in the literature. Ideally, quantifying effective design patterns and revealing concrete thresholds (i.e., numerical ranges for measures) will provide the extra guidance that novice or occasional web designers need to build better sites.

### 2.2 Procedure

Our methodology entails analyzing quantitative measures and expert ratings for a large collection of web interfaces. We analyzed over 4400 web pages from 570 sites. The sites were submitted for the initial review stage of the 2002 Webby Awards;[1] anyone could submit a site for review. At least three expert judges (described as Internet professionals) evaluated each site. Our assumption is that site ratings also apply to its individual pages, because the judges explore pages on a site before evaluating it. Judges rated the sites on six criteria–content, structure and navigation, visual design, functionality, interactivity, and overall experience; we used the average overall experience ratings in our analysis, because it was significantly correlated with the other criteria.

Using the quantitative measures and expert ratings, we developed a number of statistical models or profiles for predicting the *quality* of web pages and sites [1]. (We use the term *quality* to refer to the rating categories that we derived from the average overall experience ratings; sites that were rated in the top 33% form the *good* category, in the middle 34% form the *average* category, and in the bottom 33% form the *poor* category.) Each model encompasses key relationships among the measures that have a significant role in the predictions. These models enable us to determine how a web interface's measures are similar to or different from common design patterns that we found for high quality interfaces (i.e., interfaces rated in the top 33% of sites). For example, we can use the mean and standard deviation for the number of words on highly rated pages to determine a concrete range that can inform other designs.

In this paper we contrast thresholds derived from our 2000 and 2002 models and show that fundamental design patterns have not changed radically over recent years. We developed three cluster models--small-page, large-page, and formatted-page–by analyzing pages from the highly rated sites in 2000. Pages in the small- and large-page clusters are distinguished by the amount of content on them, and pages in the formatted-page cluster have slightly more content than pages in the small-page cluster and use tables considerably to control page layouts. We also developed models for predicting the quality of individual pages and the site overall from our 2002 data. The cluster models provide more context than the latter quality models, because they elicit thresholds based on the type of page being designed (e.g., a small web page). Ivory [1, Chapter 10] contains a detailed discussion of the cluster and other predictive models, our prior analysis of the 2000 data, relevant design recommendations, and literature references. We do not provide complete literature references in this paper.

# 3    Results

We examine guidelines for seven aspects of web interfaces, including the amount of text, fonts, colors, and consistency, and provide quantitative thresholds for these aspects; Ivory [1] discusses additional aspects like accessibility and download speed. Our intent is not to suggest that these are the only important aspects of web interfaces, rather we highlight aspects that are relevant to many sites and can be assessed with our tools. In addition to providing a quantitative range for each aspect, we discuss design patterns that we discovered by inspecting correlations between measures for the aspect and other complementary measures. For example, we show that headings are introduced in proportion to the amount of text on a page. The highlighted design patterns demonstrate how occasional or novice web designers might apply the derived thresholds to improve the usability and accessibility of their designs. Finally, we discuss how the derived thresholds validate and in some cases invalidate guidelines.

## 3.1    Amount of Text on a Page

The literature contains contradictory heuristics about the ideal amount of text for a web page. Some texts say that users prefer all relevant content to appear on one page, while other texts

---

[1]  Information about the Webby Awards, including the judging criteria, can be found at http://www.webbyawards.com/.

suggest that content should be broken up into smaller units across multiple pages. Furthermore, there is no concrete guidance on how much text is enough or too much. Table 1 depicts ranges for the total number of words on a page (word count). Contrary to the design heuristics, ranges suggest that pages with both a small and a large amount of text are acceptable. However, our profile analysis revealed that text formatting needs to be proportional to the amount of text on a page [1]. Table 1 shows that headings (display word count), text clustering (text cluster count), as well as the number of columns where text starts on pages (text column count) varies for the three clusters. The ranges are all significantly different across clusters as determined by analyses of variance. The ranges for the overall page quality (i.e., across page types) are consistent with the cluster models and illustrate the use of proportional text formatting as well.

**Table 1:** Word count and other text element and formatting ranges for good pages.

| Measure | Good Page Cluster | | | Overall |
|---|---|---|---|---|
| | Small-Page | Large-Page | Formatted-Page | Page |
| Word Count | 72.8–371.1 | 710.4–1008.7 | 218.6–517.0 | 61.4–627.2 |
| Display Word Count | 1.1–18.7 | 20.8–38.4 | 14.7–32.2 | 0.0–28.9 |
| Text Column Count | 0.6–4.4 | 1.9–5.7 | 6.4–10.1 | 0.5–7.3 |
| Text Cluster Count | 0.0–2.3 | 1.9–4.2 | 2.6–4.9 | 0.0–3.8 |

## 3.2    Length and Quality of Link Text

Nielsen [4] suggests that web designers use 2–4 words in text links; however, Sawyer and Schroeder [5] suggest that they use links with 7–12 "useful" words (i.e., words that provide hints about the content on a destination page). Our average link words measure suggests that text links on good pages contain from two to three words. Furthermore, the average good link words measure suggests that one to three of these words are not common words or the word 'click,' thus they are potentially useful. Ranges for each of the three good page clusters are very similar. Hence, the data suggests that link text on good pages is consistent with Nielsen's heuristic. There is one caveat to this finding: Our tool does not distinguish between links within the body text or outside of the body text (e.g., in a navigation bar) and there could be differences for the two types of links.

## 3.3    Number and Type of Links on a Page

There is an ongoing debate in the literature about the appropriate number and types of links to use on web pages. Table 2 provides ranges for the link, text link, redundant link, and link graphic counts on pages. The table also depicts ranges for link text cluster counts (i.e., regions on web pages wherein links are grouped together in colored or bordered regions) and shows that the numbers of link text clusters are somewhat proportional to the number of links on pages. For example, pages in the formatted-page cluster appear to have the most links as well as the most link text clusters, most likely because they contain one or more navigation bar on them.

The table shows that redundant or repeated links are used on good pages and graphical links are not avoided as suggested in the literature; it is possible that the graphical links have text in them, which is not currently detected by our metrics computation tool. Given the ranges for redundant links along with the ranges for text and graphical links, it appears that links are repeated in both text and image formats. However, the measures do not currently reveal how often redundant links correspond to those that are also graphical links. Table 2 shows that good pages tend not to use within-page links (page link count), which is suggested in the literature.

**Table 2:** Link element and text formatting ranges for good pages. The page link count is not used in the good cluster models.

| Measure | Good Page Cluster | | | Overall |
| --- | --- | --- | --- | --- |
| | Small-Page | Large-Page | Formatted-Page | Page |
| Link Count | 12.4–41.2 | 35.8–64.6 | 54.0–82.9 | 10.4–56.7 |
| Text Link Count | 5.0–28.0 | 24.3–47.3 | 35.8–58.8 | 2.4–37.7 |
| Link Graphic Count | 5.3–14.3 | 5.8–14.8 | 12.8–21.9 | 2.4–21.3 |
| Redundant Link Count | 1.4–9.1 | 4.6–12.3 | 9.0–16.7 | 0.0–12.1 |
| Page Link Count | – | – | – | 0–1.2 |
| Link Text Cluster Count | 0.0–1.4 | 0.5–1.9 | 1.7–3.1 | 0.0–3.8 |

## 3.4 Use of Non-Animated and Animated Graphical Ads

Most texts suggest that graphical ads should be minimized or avoided, but there has been one study to suggest that ads increase credibility [2]. Our measures of the number of graphical ads (animated and non-animated) suggest that good pages are likely to contain no more than two graphical ads. Pages in the 2002 dataset used graphical ads to a lesser degree than pages in the 2000 dataset; the latter pages contained one graphical ad on average. Good pages in both datasets tend to have zero or one animated graphical ad, which suggests that animation is used sparingly.

## 3.5 Font Styles and Sizes

The use of serif and sans serif fonts as well as appropriate font sizes has been examined in a number of empirical studies. Our analysis revealed that sans serif is the predominant font style used on good pages; studies have shown these fonts to be more readable than serif fonts online. The 2002 dataset suggests that serif fonts and font style variations (i.e., combinations of font face, size, bolding, and italics) in general, are being used to a lesser degree than in the past (2.3–7.0 combinations on a page). Similarly, slightly larger font sizes are used (9–12 pts on average, with a minimum font size of 8–10 pts).

## 3.6 Unique Colors and Color Combinations

The literature offers various recommendations on using a small number of colors, browser-safe colors, default link colors, color combinations with adequate contrast, and so on. We have developed over twelve measures related to color usage, which we do not present in detail due to space constraints. Our analysis revealed that good pages tend to use from one to three colors for body text as well as one to two colors for headings; it appears that the number of heading colors is proportional to the amount of text or formatting on pages. Although it is not clear whether different colors are used for body and heading text on these pages, unique color counts (5.4–11.2) suggest that this may be the case. The 2002 dataset reflects a small decrease in the number of different colors used for text (one to five), which is more consistent with guidance in the literature. We also found that good pages tend to use good text color combinations more so than neutral and bad text color combinations; we used results from Murch's study of color combinations [3] to determine the quality of color combinations (i.e., good, neutral, or bad).

Two to four colors are used for links, and the default browser colors (red, blue, and purple) are not always used; these ranges suggest that good pages do not closely follow the guidance in the literature. Ranges for the total number of unique colors (5.4–11.2) show that good pages do not adhere to the guidance of using no more than six discriminable colors [3]. Furthermore, they do not strictly use browser-safe colors (3.4–6.8) as recommended by the literature.

## 3.7    Consistency Across Pages

Some texts advocate consistent use of design elements across web pages, while other texts claim that such elements become invisible. The literature also suggests that page titles (i.e., title tag) should be varied. We derived twelve measures to reflect the percentage of variation for groups of related page-level measures [1]; a large variation percentage reflects less consistency and vice versa. Our analysis suggests that page layouts vary more so in the 2002 dataset (5–55%) than in the 2000 dataset (0–22%); thus, it appears that the trend is to introduce more variation into page layouts as opposed to keeping them consistent. This trend is somewhat disturbing, because some studies show that performance improves with consistent interfaces. The link element and formatting variation measures suggest that navigation elements are fairly consistent across pages on good sites. Finally, the page title variation suggests that titles vary considerably on good sites (10–130%).

## 4    Discussion and Future Work

We summarized design patterns derived from our analysis of page- and site-level quantitative measures for sites rated highly in the 2000 and 2002 Webby Awards. The design patterns extend beyond individual measures to show, for instance, that the amount of text formatting is proportional to the amount of text on a page (i.e., a change in one aspect may necessitate a change in another). We also demonstrated that the guidance varies slightly depending on the design context (e.g., page style). The derived patterns should provide some concrete guidance for improving web site designs. In some cases, such as the number and types of links and graphical ads, what is done in practice contradicts the literature; perhaps these areas need to be examined with further empirical studies. For the most part, design patterns were fairly consistent for the two years. We have derived similar thresholds based on the type of page (e.g., home or form page) and the page or site genre (e.g., community or education) [1]. We have embedded these design patterns into an automated evaluation prototype that is available for public use at webtango.ischool.washington.edu. Future work entails validating the new prediction models, providing designers with more guidance on interpreting model predictions and implementing design changes, and building an integrated web site design and evaluation environment.

## 5    Acknowledgments

We thank Maya Draisin and Tiffany Shlain at the International Academy of Digital Arts and Sciences for making the Webby Awards data available for analysis. We thank Tina Marie for preparing the 2002 data for analysis, Mary Deaton for constructive feedback, and the anonymous reviewers.

## 6    Bibliography

[1]    M. Y. Ivory. *An Empirical Foundation for Automated Web Interface Evaluation*. PhD thesis, University of California, Berkeley, Computer Science Division, 2001.
[2]    N. Kim and B. J. Fogg. World wide web credibility: What effects do advertisements and typos have on the perceived credibility of web page information? Unpublished thesis, Stanford University, 1999.
[3]    G. M. Murch. Colour graphics -- blessing or ballyhoo? *Computer Graphics Forum*, 4(2):127-135, June 1985.
[4]    J. Nielsen. *Designing Web Usability: The Practice of Simplicity*. Indianapolis, IN: New Riders Publishing, 2000.
[5]    P. Sawyer and W. Schroeder. Report 4: Links that give off scent. In *Designing Information-Rich Web Sites*. Bradford, MA: User Interface Engineering, 2000.

# The Effects of Expertise in Web Searching

*Christine Jenkins & Cynthia L. Corritore*

Creighton University
Omaha, NE 68178 USA
christine_jenkin@hotmail.com
cindy@creighton.edu

*Susan Wiedenbeck*

Drexel University
Philadelphia, PA 19104 USA
susan.wiedenbeck@cis.drexel.edu

## Abstract

This verbal protocol study investigated the patterns of web searching among groups of professional nurses differing in domain expertise and Web expertise. The group with high domain and high Web expertise carried out depth-first searches following deep trails of information. The groups with low Web expertise searched breadth-first and often were confused or lost. The results suggest distinct patterns in searching which may be exploited in development of searching tools.

## 1 Introduction

Information access via the Web has radically changed the way people seek information. Instead of using libraries and commercial databases, people turn to the Web for many of their information needs. They search for information largely on their own, self-taught and without the aid of a professional search intermediary. Thus, while the Web makes full-text information available in abundance, it places the heavy burden of crafting the search squarely on the user. In this study we examine the effects of domain and Web expertise on information seeking on the Web. We ask the question, "What are the effects of domain expertise and Web expertise on Web searching?"

## 2 Related Research

Much research on Web searching involves using transaction log data to analyze the queries of users of search engines. A comparison of three such studies (Jansen & Pooch, 2001) showed that users pose an average of two queries per session, and the queries are short, consisting on average of two terms. Queries are simple with very little use of Boolean operators and low use of other modifiers. The number of results typically viewed by users is 10 or less per session. Research on Web browsing also has shown that the browser Back button is used very frequently in navigation, while the history list mechanism is not (Catledge and Pitkow, 1995). During browsing only a small number of pages are visited frequently (e.g. personal home pages, search engine pages); however, 58 percent of pages visited by individuals are revisits (Tauscher & Greenberg, 1997). Revisited pages usually come from the last 6 pages visited, and extreme recency is the best predictor of revisitation.

Research has also addressed characteristics of novice and expert behavior in information seeking in small hypertext databases (Marchionini, Dwiggins, and Lin, 1990). This work compared retrieval by domain experts without search training, professional search intermediaries without domain expertise, and novices with neither domain nor search expertise. The two expert groups differed in their approaches. Domain experts were content-driven, using their greater conceptual knowledge in disambiguating the search problems, choosing the appropriate query terms, and evaluating the relevance of retrieved information. Search experts were search-driven, using their expertise in searching procedures to form more sophisticated queries and using all the search features of the system. Compared to domain experts, they had more difficulty understanding the questions asked in the search tasks, carried out more distinct searches, and spent less time reading the material retrieved. Unfortunately, Web studies have only occasionally distinguished searchers by expertise. The contribution of this study is that it yields an in-depth picture of different patterns of user behavior on the Web from direct observation.

## 3    Methodology

Sixteen practicing nurses, ages 25 to 61, took part in the study. The domain expertise for the study was the medical condition of osteoporosis, a common condition generally familiar to all nurses. Half the nurses had special training in osteoporosis and half did not. Participants carried out two Web searching tasks. Task 1 was to search for information for a family member who had recently been diagnosed with osteoporosis. Task 2 was to find material on the Web to use in preparing a short professional presentation on steroid use and its effects on osteoporosis. The goal was to find five appropriate websites for each task. Participants were run in individual two-hour sessions. They had a computer with Web access. An experimenter was present at all times. First, two short practice tasks were done. Then the participant did Task 1, followed by Task 2. Participants were videotaped and talked aloud as they worked. At the end of each task, the participant and experimenter conducted a retrospective review of the task by jointly viewing the videotape.

## 4    Results

The data are qualitative and come from the verbal protocols and retrospective reviews/interviews. The protocols were transcribed, segmented, and encoded using four types of statements proposed by Ericsson and Simon (1993): cognitions, intentions, planning, and evaluation. Here we report on cognitions, intentions, and planning.

Both groups of Web novices, *Domain Novice Web Novice* (DNWN) and *Domain Expert Web Novice (DEWN)*, behaved very similarly. They lacked knowledge about how to search the Web and did not understand clearly the concept of a search engine. When they began their first task, they were uncertain how to use the search engine to begin searching. Questions were posed such as, "What is a search engine?" Once they began searching, DNWNs had great difficulties with the mechanics of searching and navigation. They used few browser features and tended to abandon features if they were not immediately successful with them, for example, adding websites to the Favorites list. Participants became lost during navigation, and were not able to get back to an earlier location, such as the search engine results page. As one participant said, "I go back, forward, back, forward, where am I?" The overload experienced by new Web searchers is evident in the fact that they frequently forgot navigation options that they had previously used. For example, some participants dropped the Back button from their navigation repertoire even though they had used it successfully several times earlier in the session. Web page design contributed to the confusion. In particular, images, color, animation, and advertisements distracted these users and made searching more difficult. Overall, the focus of the Web novice groups was on the mechanics of searching, rather than on the osteoporosis tasks. In their searching the Web novices used only single concept terms such as 'osteoporosis,' 'osteoporosis treatment,' and 'steroids.'

After their initial experiences, Web novices became very cautious to avoid getting lost on the Web. To deal with the lostness problem, they adopted a classic hub-and-spoke, or breadth-first, pattern of searching. All searching started from a central hub, the search engine results page. Web novices only went one level from that starting point, returning to the hub each time after they investigated a Web page. They did not browse further by clicking links in a website that would take them either deeper into the same site or to another website (Figure 1).

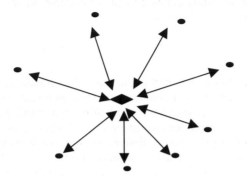

**Figure 1:** DNWN search pattern (dots represent websites)

*Domain Novice Web Expert* (DNWE) participants were at ease with the technology. They showed no signs of confusion or becoming lost on the Web. The DNWE participants tended to be very focused in their searches. They did not lose sight of their goals while using the technology. These participants wanted to get the job done quickly. They worked rapidly, using a wide repertoire of browser features for navigation. They expressed strong preferences about search engines and websites. For example, they disliked search engines that required many steps in the search process or that returned results in an unfamiliar format, and they were critical of websites that did not provide an internal search function.

On the other hand, the DNWEs' lack of subject knowledge led them to do simple, broad searches with little planning. The pattern of their searching resembled a hybrid breadth-first/depth-first search. They began at the hub, followed a link from the hub to a site, then followed links within the site, exploring deeper along an information trail. The high degree of browsing within the sites was the result of their undifferentiated, single term searches. Generally, they searched about two links down a trail before returning to the hub. They sometimes searched even deeper within a site if they thought they were on the right track. The DNWE participants sometimes moved from one site to another by following links in the sites. However, most often they returned to the hub.

*Domain Expert Web Expert* (DEWE) participants worked quickly and comfortably. They were familiar with browser and search engine conventions. They used a variety of search engines and took advantage of browser features, for examples, quickly saving a website in their Favorites list, rather than writing the URL on paper. Their high confidence is reflected by a participant who said, "I am not intimidated by searching." At the same time, the DEWEs were highly focused on the osteoporosis task goals. Unlike the other groups, they questioned the intent of the tasks until they felt they fully understood them. Their queries were more complex, for example, they entered names of specific drugs and classes of steroid drugs that affect osteoporosis. They formed queries using Boolean operators and modifiers. They displayed their domain expertise by sometimes typing URLs of known osteoporosis websites directly into the browser address box.

The DEWEs searched deeply, following long trails of information. Their search pattern can be described as depth-first. They began at a central hub, then followed links within the site and from one site to another, until they found an appropriate website (Figure 2). They then returned to a central hub. During a search, they created many lower-level hubs and followed paths radiating from those hubs. DEWEs were confident about moving very deep along a trail beyond the main hub in their search, and they did not have difficulty returning to their main or lower-level hubs.

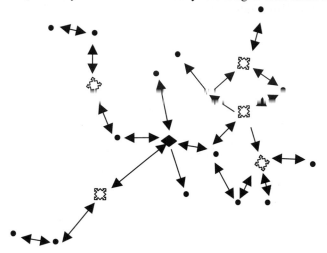

**Figure 2:** DEWE Search Pattern (dots represent websites; snowflakes represent following links within a website)

# 5    Discussion

In general, being a novice in either the domain or the Web led to problems. Search times were longer and less planning was seen. Among Web novices, confusion and frustration were high. All novices used simplistic search terms despite their professional background. In addition, Web novices experienced problems with navigation and orientation on the Web. These difficulties may be linked to lack of a good mental model of the Web and searching, as well as cognitive overload.

One would expect the lack of a good mental model to lead to an inability to predict the effects of one's actions, a lack of planning, and disorientation in the search space. Web novices reduced their cognitive load by restricting their behaviors to a small virtual space. This is reflected in their strong hub-and-spoke searching pattern. Additionally, the Web novices were all very focused on the mechanics of searching the Web with little mention of the task domain, even among the DEWNs. This was evident in their simple search terms and minimal use of domain language in the DEWNs' protocols. The demands of the search task may have suppressed the DEWNs' ability to use their expertise. By contrast, Web experts, both domain novice and expert, kept both their end-user and the domain of the task in mind during the search process. For example, they expressed concern about the suitability of the material to the target user, as well as the format for presentation of the material.

With respect to the number of searches, all the novice groups (DNWN, DNWE, and DEWN) conducted more searches than the DEWE group. This is probably related to a common problem:

not knowing when to stop searching. However, the problem stemmed from different sources for the three groups. The DNWEs possessed the skill to conduct a search, but not the expertise to evaluate their results. Thus, they often did not accept sites they found that were, in fact, relevant and continued to search further. The DEWNs, while able to evaluate obtained information, continued to find relevant information in search after search, and lacked the experience to lead them to a decision to stop. They were caught in the trap that many Internet novices experience – the information appears to go on and on, and so the point at which to stop is unclear. The DNWNs were subject to both of these difficulties.

# 6    Conclusion

The findings of this study cannot be broadly generalized because of the small sample size and specific domain. However, we can conjecture about the practical implications of these findings. Systems suggested by Chi, Pirolli, Chen, and Pitkow (2001) that infer the expertise of a user could be used to adapt the search engine interface to the user. A search engine results page for Web novices could display results in a hub-and-spoke. In this format, clicking a link on the diagram would display the website in the lower part of the screen with a small image of the diagram appearing in the upper part of the screen. Novices would always have their hub in sight, and could return with a click, or they could explore the 'new' site in the lower portion of the screen. Their current path would be highlighted on the graphic. This kind of representation could facilitate orientation in their search space, reducing their cognitive load and helping them to acquire a more accurate mental model.

## References

Catledge, L. & Pitkow, J. (1995). Characterizing browsing strategies in the World Wide Web. In *Proceedings of the 3rd International World Wide Web Conference*, Darmstadt, Germany. Available:http://www.igd.fhg.de/archive/1995_www95/papers/80/userpatterns/ UserPatterns.Paper4.formatted.html. Accessed February 11, 2002.

Chi, E.H., Pirolli, P., Chen, K. & Pitkow, J. (2001). Using information scent to model user information needs and actions on the Web. In *Proceedings of the Conference on Human Factors in Computing, CHI 2001* (pp. 490-497), NY: ACM Press.

Ericsson, K.A. & Simon, H.A. (1993). *Protocol analysis: Verbal reports as data.* Cambridge, MA: MIT Press.

Jansen, B.J. & Pooch, U. (2001). A review of Web searching studies and a framework for future research. *Journal of the American Society for Information Science and Technology*, 52(3), 235-246.

Marchionini, G., Dwiggins, S. & Lin, X. (1990). Effects of search and subject expertise on information seeking in a hypertext environment. *Proceedings of the 53rd Annual Meeting of the American Society for Information Science* (pp.129-142), Washington, D.C.: American Society for Information Science.

Tauscher, L. & Greenberg, S. (1997b). Revisitation patterns in World Wide Web navigation. In *Proceedings of the Conference on Human Factors in Computing, CHI'97* (pp. 399-406), NY: ACM.

# Web-site quality evaluation,
# a case study on European cultural web-sites.

*Sofia Z. Karagiorgoudi*

*Emmanouil G. Karatzas*

*Dimitrios K. Tsolis*

*Theodore S. Papatheodorou*

Building B, University Campus, Rion 26500, Patras, Greece
Karagior, ekaratzas, dkt, tsp@hpclab.ceid.upatras.gr

## Abstract

In this paper the issue of a web-site quality definition and evaluation is addressed. Quality and usability of web-sites are two very closely related issues, and more specifically quality could be considered as a superset of usability. In this sense, quality and usability share some common criteria. Herein we present a set of quality criteria for web-sites of cultural content and an on-line evaluation system. Finally we present a first use of these criteria and the corresponding results.

## 1    Introduction

In general, quality is a very broad, generic and subjective concept. To make quality more objective, applicable and measurable, it is essential to adopt a set of quality criteria and a set of evaluation methods. Although, there are many general approaches to quality in general, there are scarce results which investigate the notion of quality in the specific scope of cultural web sites. Based on this observation, we have provided a methodological tool which cultural institutions can use both during the evaluation of cultural web applications and during the design and development process. The main constituents of our approach are a set of quality criteria, a questionnaire for the assessment of the abovementioned criteria and an on-line system implementing the questionnaire. Our approach is based on the Brussels Quality Framework [10] that appeared in the conclusions of the experts' meeting "The digitization of European cultural heritage on the web". However it has been enriched with more detailed information, after a complete and deep study about the quality [5], [6], [9] and usability [1] on web.

## 2    Quality Criteria

The goal of defining a set of quality criteria is to decompose the broad and partially subjective concept of quality into a set of more objective and measurable attributes which explicitly take into account the specificity of cultural web sites. For this reason, we have first defined a quality space. The idea is that we can analyze a cultural web site from the viewpoint of its content, its presentation, its design, its interactivity, its adopted policies and we can identify a number of quality attributes for each viewpoint, or dimension. The assessment of quality results from the systematic evaluation of each set of attributes, dimension by dimension. The following table (Table 1) presents a summary of the criteria (presentation, content, policy, design, interactivity) used in our approach. A complete list of the evaluating criteria may be found in the reference [2].

**Table 1:** Quality Criteria

| CRITERIA | SUBCRITERIA |
|---|---|
| 1. Presentation of the site | 1.1 Scope Statement<br>1.2 Authority of the site |
| 2. Content | 2.1 Coverage/Completeness<br>2.2 Accuracy/Objectivity/Validity<br>2.3 Usefulness<br>2.4 Logical Organization of Information/Comprehensiveness<br>2.5 Authority of the Content<br>2.6 Currency |
| 3. Policy | 3.1 Legal Policy<br>3.2 Maintenance Policy |
| 4. Design and Usability | 4.1 Accessibility [11]<br>4.2 Navigability<br>4.3 Quality of Links<br>4.4 Aesthetic Design |
| 5. Interactivity | |

# 3 Quality Evaluation Methods

Quality criteria must be complemented by evaluation methods, i.e., guidelines and procedure that define how the quality of a cultural web application can be measured. Following, we investigate the state of art in quality evaluation methods, mainly focusing on those addressing quality from the user perspective as they can found in the HCI literature on usability.

## 3.1 The state of art in quality evaluation methods

A general acknowledged classification of usability evaluation methods distinguishes among two broad categories of methods: User Testing and Inspection [4]. User testing deals with real behaviours, observed from the representative of real users, performing real task and operating physical artifacts, being them prototypes or final systems. In inspection methods, evaluators - usability specialists inspect or examine usability related aspects of a product, trying to detect violations of established usability principles without involving end users during the evaluation process. The most commonly method used for inspecting an application is heuristic evaluation, in which usability specialist judge whether the system properties conform to established usability principles. The main advantage of this method, with respect to user testing, is that it involves fewer (experienced) people. The main disadvantages of inspection methods are the great subjectivity of the evaluation - different inspectors may produce incomparable outcomes - and the heavy dependence upon the inspector skills.

## 3.2 Description of our evaluation method

The quality criteria identified provide an excellent basis for evaluating quality by heuristic inspection: these quality criteria can be regarded as quality heuristics. However, the criteria per se are not enough to guarantee the quality and correctness of inspection based evaluation. The quality measurement process needs to be structured and organized, by defining what the evaluator should do with the application when trying to assess the various criteria. There are some approaches to the problem of defining a quality evaluation method [7], [8]. In our work we have followed a questionnaire-based method. Particularly, we have developed a model questionnaire, intended to be used by quality inspectors to analyze the quality of cultural web sites. The questionnaire contains detailed questions to collect the maximum available information for each quality criterion. The various questions are organized into groups, each group addressing a different quality criterion. Apart from the questions themselves, we present the weight of each question / group of questions to the result of the questionnaire. Some questions require a binary value: yes/no; others adopt finer grained metrics (e.g., on a scale 0-5). In order to gather measurements in an automated way, the above questionnaire has been implemented as an online system, which is described in brief below.

# 4    Description of the system

The on-line system for the evaluation of cultural web-sites was implemented using open, web oriented technologies (Apache, PHP, MySQL, Linux). The description of the system is separated in two areas: the functional part of the system and the technical part of it. A more detailed description of the system may be found in the reference [3].

## 4.1    Description of the functional part of the system

The system supports three main functions: user authentication, completion of a questionnaire and presentation of the results. Before the interaction between users and the system, the user must log on or register (if he/she is not) to the system, as only authorized users can access the questionnaire and view the results. After the user can either to complete a new questionnaire or view the results table. The user must complete all the required questions, which are stored in the database. If the user selects to view the results, then the results table, the graphical representation of them and some horizontal results of the most important questions are presented to him/her. From the result table, the user can have access to the web site of a particular evaluated site or to a read-only version of the specific questionnaire.

## 4.2    Description of the technical part of the system

From the technical point of view, the on-line system for the evaluation of cultural web sites is web-oriented. This means that the system is easily accessible on the web and supports user authentication. The on-line system is now available at http://www.benchmarking.gr. Also, the system is database driven. The questions that form the questionnaire and are the satisfaction indicators of the abovementioned criteria, as well as the answers are stored in the database. In consequence, the definition of the questions and the changes made by the users (quality experts) are conducted in an intuitive and fast way. Also, the way of defining the answers is dynamic. Therefore, the definition of a new answer type or the change of an existing type is conducted easily and requires no changes in already answered and stored questionnaires. Furthermore, the results are based on a weighted questions system. Each question is assigned a different weight depending on its importance in the evaluation of the cultural web site. These weights are also stored in the database. Therefore, the system supports easy result processing and has the possibility for re-evaluation with new weights. Finally, the system provides automatic report generation; each time a new questionnaire is filled the results are updated. The results are visually presented using custom-made graphic representations.

# 5 Results of the first iteration of quality evaluation

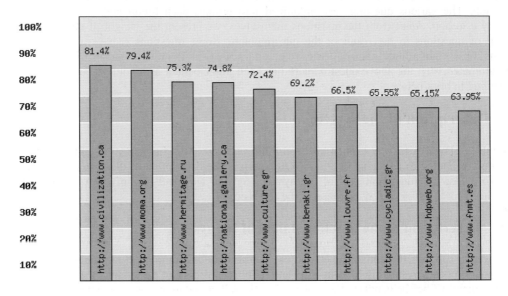

**Figure 1:** Graphical Representation of the Questionnaire Results

In order to test the reliability and precision of our evaluation method, we have evaluated ten cultural web sites as a test. In this attempt we have selected well-known sites, such as the Louvre museum, the· Hermitage museum, the Canadian Museum of Civilization etc. The graphical representation of the results of the first attempt of using the questionnaire is presented in **Figure 1**. Most of the evaluated sites have reached a score of 60%-70%, that certifies that an important part of the information accessible on the web is not of quality matching the importance of the content. Taking into account that the sample of the web sites evaluated has not been randomly selected, this percentage may not reflect the real world situation (it is expected to be lower).

# 6 Future actions

The next step is the design and implementation of the second iteration of the process. The second iteration is aiming at approaching the issue from a different perspective. The criteria will be divided in two main categories. A set of criteria based on the end user view and a set of criteria based on the view of the technically expert. In this framework three questionnaires will be produced. The first is the general, brief questionnaire (1 page maximum) for the end-users (cultural organizations and quality experts), aiming mainly at the quick and easy questionnaire completion. The second is a complete, wide-range questionnaire for the end-users (cultural organizations and quality experts). Finally, the third is a complete technical questionnaire for the technically experts, web site creators and administrators, aiming mainly at evaluating the technical part of the Cultural Web Sites.

In addition, the results so far will be assessed extensively leading to the further refinement of the questionnaires. Judging from the results of the questionnaires we aim to publish guidelines pinpointing common mistakes in the creation of a cultural web site, in order to help developers create "better" sites.

## 7 Conclusions

The technical implementation of the system was a very good chance to test our ideas on creating generic and easily manageable web-based systems. Based on this implementation we are now finishing the implementation of a new system, where except from having a dynamic definition of questions and answers we have implemented the dynamic and abstract definition of any kind of questionnaire. The questionnaire testing and evaluation has leaded us to the following qualitative conclusions regarding the creation of cultural web sites. First of all, we have observed that all the evaluated sites provide diversity, several navigation mechanisms and have a common design through the pages, avoiding the use of broken links and advertising. Also, they provide alternatives to auditory and visual content that enable the use of the site by physically challenged people. Unfortunately, the majority of them doesn't avoid the use of frames and doesn't provide information like the publication date, the last update date, etc. Also, only a small percentage of the sites use video, sound and animation to attract the interest of the user and to add vividness to the content. Therefore, the design of sites offering cultural content should be made very carefully so as to reflect to the importance of their content.

## 8 Acknowledgement

Part of the work presented here has been presented to and accepted by the Quality Workgroup of the National Representatives Group of the European Commission promoting the Lund Principles on Culture and Digitization.

## 9 References

[1] Jakob Nielsen. (2000). Designing Web Usability: The Practice of Simplicity. New Riders Publishing, Indianapolis, USA.

[2] Karagiorgoudi S., Karatzas E., Tsolis D. and Papatheodorou Th. (2002). Quality Criteria for Cultural Sites Accessible Over the Web. IADIS WWW/Internet 2002 conference (pp 636-641). Lisbon, Portugal.

[3] Karagiorgoudi S., Karatzas E., Tsolis D. and Papatheodorou Th. (2003). A system for the evaluation of cultural web-sites. IADIS e-Society 2003 conference. Lisbon, Portugal.

[4] Jakob Nielsen. Usability issues. Retrieved from http://www.useit.com.

[5] Evaluating and choosing WWW sites. Retrieved from www.goethe.de/os/tok/bib/gerin5.htm.

[6] Evaluating Information Found on the Internet, www.library.jhu.edu/elp/useit/evaluate/

[7] Evaluating WebSites: Criteria & Tools, www.library.cornell.edu/okuref/research/webeval.html

[8] Evaluating Websites. Retrieved from http://trochim.human.cornell.edu/webeval/webeval.htm

[9] Quality Criteria for Website Excellence, www.worldbestwebsites.com/criteria.htm

[10] Quality standards for cultural sites accessible on the Web: The Brussels Quality Framework, Retrieved from http://www.cfwb.be/qualite-bruxelles/gb/notecrit_FR38170EN.doc

[11] Web Accessibility Initiative (WAI). Retrieved from www.w3c.org/WAI

# Improving web site usability through a clustering approach

*Martha Koutri[1], Sophia Daskalaki[2]*

[1]Electrical & Comp. Eng. Dept, HCI Group, [2]Engineering Sciences Dept.
University of Patras, 26500 Rio Patras, Greece
mkoutri@ee.upatras.gr,  sdask@upatras.gr

## Abstract

One of the main parameters of web usability is related to the smooth navigation of the user and easy access to the information sought. In this paper, we propose a process for improvement of web site structure design by interpreting discovered patterns of web usage logs, with the objective to improve the usability parameters of a given web site. We discuss a cluster-mining algorithm for grouping together related web pages of the site. The clusters found in the process are to be used for redesigning the site and improving connectivity of related pages, thus, facilitating navigation.

## 1    Introduction

The poor structure of a web site along with the complexity of the information provided quite often result in disorientating its users. On the contrary, user-centered design efforts for web sites almost always lead to the construction of usable web sites, which potentially offer an efficient and agreeable navigational experience. *Usability* is one of the most important features of any software tool and concerns the effectiveness, efficiency, and satisfaction that it gives to the user in a given context of use and task (Tselios, et al., 2001). It is further concerned with making systems safe, easy to learn and easy to use (Preece, 1994). While these facts also hold for web sites, the need for improving the interaction between a site and its visitors, through personalization, is an additional necessity in this domain. Moreover, web personalization requires the development of new methodologies, including web mining techniques. *Web mining* has emerged as a special field during the last few years and refers to the application of knowledge discovery techniques specifically to web data. In addition, the usage patterns, which result by applying web- mining techniques to a particular web site log files, can also be used for improving the design of the site. As described in this paper, clustering techniques applied to a web site usage data are especially suitable for discovering those subsets of web documents in the site that need to be connected and those that even if they are connected, their connection should be more visible. Thus clustering results can be used for improving design –and therefore augment the usability– of the web site.
The next section associates the usability of web sites with the browsing behavior of the visitors. Section 3 presents some of the existing approaches for clustering along with our proposed algorithm. Finally, section 4 presents the application of the algorithm to a particular web site in order to demonstrate ways of improving the design of the web site.

## 2    Exploitation of navigational patterns for improving usability

The application of design guidelines should always direct the design of web interfaces in order for them to be usable. Web design is described using specific terms referring to different areas of concern within the web design space (Newman & Landay, 2000). In particular, the term *information design* refers to the identification of groups of related content and the structuring of information into a coherent whole. *Navigation design* concerns the design of methods for helping users find their way around the information structure. *Graphic design* is the visual communication of information using elements such as color, images, typography, and layout. Many researchers propose supporting web site design from the early phases of the design process. Site maps, storyboards, schematics, etc. are among the tools suggested for representing a web site throughout the design process (Newman and Landay, 2000). It is also very likely that a combination of templates and design conventions will make it easier to design usable web sites (Nielsen, 1999).

Supporting web site design aims at obtaining a desired level of usability. Likewise, in an effort to continuously improve the web site design, even after its implementation, we propose a reconstruction phase where usage patterns revealed from a web usage mining procedure are interpreted to better structures and more usable interfaces. *Web usage mining* concerns the discovery of users' access patterns by analyzing web usage data (Han & Kamber, 2001). *Usage data* is stored into web server access logs and contain information regarding visitors IP address, time and date of access, files or directories accessed, etc. As mentioned earlier, designing the structure of a web site belongs to navigation design. Thus, improvement of the structure of a web site results to more efficient browsing activity, which in fact is related to its usability. Thereby, increased usability is achieved by analyzing navigational behavior of users during their interaction with a given web site.

Besides, even if the design of a web site is user-centered, the mental model of the web site designer differs from users mental models. The concept of mental model describes the cognitive layout that a person uses to organize information on his/her memory (Lokuge, Gilbert & Richards, 1996). Each visitor of a web site has his/her own mental model, which is modulated by his/her needs, desires, cultural and educational level. Thus, each user browses a particular web site by following a specific individual sequence of hyperlinks. The application of web usage mining techniques aims at discovering different navigational patterns, so to facilitate them in a future re-structuring effort of the site.

## 3    Existing approaches for clustering

*Clustering* is a knowledge discovery technique based on the idea that similar objects are grouped together and clusters of them are then created. Clustering –unlike classification– is an *unsupervised technique* and this means that there are not any predefined classes and examples of previously observed cases. Standard clustering algorithms partition the set of objects into non-overlapping clusters. Fu, Sandhu & Shih (1999) have presented a system that groups web documents into clusters using the BIRCH algorithm (Zhang et al., 1996). The incremental construction of a CF (Clustering Feature) tree of web documents corresponds with the tree-like graphical representation of web sites. BIRCH is sensitive to the order of entering data records and classifies each document into exactly one cluster. However, the classification of a document into more than one clusters is an important requirement in web mining. Since a web document reserves its own structure, content semantics, and presentation uniformity, different groups of visitors of the site could mentally relate a web document with several different clusters of documents. In (Perkowitz and Etzioni, 2000) a new *cluster mining algorithm*, specifically designed to satisfy the requirements of the web domain, was presented. By learning from visitors' access patterns, they developed PageGather, a cluster mining methodology, for identifying a small set of possibly overlapping clusters, which in fact are collections of mentally related - but currently unlinked - documents in a web site.

The intention of using the results of a cluster mining algorithm to improve navigation design in fact raises the requirement for finding mentally related web documents regardless of their inter-linking with hyperlinks. Suppose, for example, that a cluster consisting of two web documents is the result of a clustering algorithm. This means that a reasonably large number of users tend to visit these two documents in the same session, even if these are not directly linked. The web designer could then improve navigation design of the underlined web site by adding a new bidirectional hyperlink between these two web documents. In another example, consider a cluster consisting of some web documents. Two of these are physically connected via a hyperlink. The classification of both two documents in the same cluster shows their strong mental relation. The designer could therefore highlight the specific hyperlink, so as to help visitors to easily "find their way". In (Avouris et al., 2003) a set of modifications concerning the linkage of web documents and the formatting of hyperlinks are presented. Thus, we ascertain that the application of the appropriate clustering algorithm reveals useful access patterns, which could be used by the web designer in order to improve the usability of a given web site.

## 3.1    Description of a cluster mining algorithm

The cluster mining algorithm presented here aims at clustering the documents of a web site using information about the presence or absence of each one document during the interaction of different users with the web site. The algorithm takes as input the preprocessed web access logs and generates first all possible singleton clusters from the related web documents. Next, the algorithm successively inserts a second document into the existing clusters to create clusters of two documents. The construction of these two-document clusters is based on the value of a properly defined similarity measure. This measure is a function, which determines a degree of correlation between two or more documents. The process continues by entering more documents to the existing clusters based on the value of the similarity measure, until a desired size for the clusters is reached. The output of the algorithm consists of a set of possibly overlapping clusters of documents that users tend to visit together during their interaction with a web site. We denote that the algorithm has the main characteristics of *agglomerative clustering* approaches (Jain et al., 1999). It begins with each document in a discrete class and proceeds by iteratively inserting documents into the already formed clusters. Moreover, it is insensitive to the order of input data.

In particular, the first step concerns the generation of *user visits* by pre-processing the web access logs. User visits are next used for mining useful information about the usage statistics of web sites. A user visit (or *user session*) is a sequence of page transitions for the same IP address, where each transition is done at a specific time interval (Pierrakos et al., 2001). Thus, each user visit may be represented using a $1 \times n$ vector with "0" and "1", where $n$ is the total number of documents in a given web site. A "0" implies that the user has not visited a particular web page, while the value "1" implies that the user has visited it. We next build a $v \times n$ matrix $V$, where $v$ denotes the number of users' visits available and $n$ the total number of documents in a given web site. The algorithm begins with the formation of singleton clusters by inserting each document into a distinct cluster. Web documents are denoted with the variable $i$, where $i \in \{1, 2, \dots , n\}$. The algorithm continues by inserting iteratively one document at a time into an existing cluster from previous step according to the value of the *similarity measure* $F_{[i,j,\dots,k]}$, defined as:

$$F_{[i,j,\dots,k]} = \frac{(\# v_i\text{'s in } V, \text{ where } p_i = 1, p_j = 1,\dots, p_k = 1)}{(\text{total } \# v_i\text{'s in } V)} \quad (1),$$

where $[i,j,\dots,k]$ is a subset of documents, potentially a cluster, depending on the value of $F_{[i,j,\dots,k]}$, and $p_i$ is a binary variable that denotes the presence ($p_i=1$) or the absence ($p_i=0$) of document $i$ in a certain visit, with $i,j,\dots,k \in \{1,2,\dots,n\}$. If, for example, a singleton cluster from the first step consists of document $i$, we compute the similarity distance $F_{[i,j]}$, so as to decide if another web document, supposing $j$, may form a cluster with $i$ in step 2 . The algorithm proceeds by iteratively inserting new web documents into clusters formed in the previous step, until there is not any other insertion to do. The insertion of a web document into an already formed cluster is feasible, if this particular document has not been included yet into the given cluster, and the similarity distance exceeds a predefined threshold $t$. The threshold is an empirically defined parameter by the web miner. Therefore, a group $[i,j,...,k]$ of web documents belong to the same cluster $C$, if:

$$(C \ni i) \wedge (C \ni j) \wedge ... \wedge (C \ni k) \wedge F_{[i,j,...,k]} \succ t \quad (2)$$

A generalized description of this clustering approach in a Pascal-like pseudocode follows:

*Input:*
*Set of n web documents, set of v visits in V, threshold t, size s of each cluster.*
*Procedure:*
    **Step 1.0:** *Form singleton clusters, by inserting each document to a single page clusters.*
    **Step 2.0:** *Compute $F_{[i,j]}$, for all $i,j \in \{1,2,...,n\}$ and $i \neq j$.*
    **Step 2.1:** *For all $i,j \in \{1,2,...,n\}$ if $F_{[i,j]} > t$ then form the 2-page cluster $[i,j]$.*
    **Step 3.0:** *While $|\{i, j, ..., k\}| = s' \leq s$ compute $F_{[i,j,...,k]}$, where $i, j, ..., k \in \{1, 2, ...n\}$.*
        **Step 3.1:** *For all $i,j, ..., k \in \{1,2,...,n\}$ if $F_{[i,j,...,k]} \geq t$ then form the $s'$-page cluster $\{i, j, ..., k\}$.*
*Output:*
*Set C with overlapping clusters of size s.*

# 4   An illustrative example

As an example of the algorithm's applicability, we considered an experimental web site, depicted in Figure 1. After analyzing the web access logs collected from this web site with the cluster algorithm presented above, we received some interesting results regarding its structure.

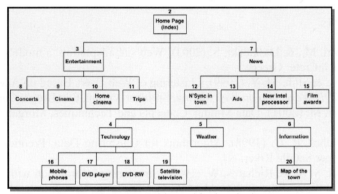

**Figure 1.** A hierarchical web site

In particular, for threshold $t=0.30$ and clusters size $s=2$, the algorithm returned the following 2-page clusters: $\{7,3\}$, $\{9,3\}$, and $\{11, 20\}$. The web documents "7" and "3" stand for "News" and "Entertainment", respectively. Cluster $\{7,3\}$ thus indicates that a significant number of users who visit the page with the latest news, also visit the page concerning entertainment. These two documents are not physically linked, because in the designer's mental model were not relate. However, they are related in the users mental model, so, the web designer could add a bidirectional hyperlink conducting directly from the one document to the other. We now consider the cluster $\{9,3\}$ concerning web pages "Cinema" and "Entertainment", which are physically connected by a hyperlink. Cluster $\{9,3\}$ indicates that a large number of users who visited page "9", also visited page "3" during the same visit. Regardless of the existence of a hyperlink between the particular documents, an improvement in navigation design would be to make this

791

related information easily accessible. In such a case, adaptation tasks include highlighting, using different colours or fonts, inserting a small icon, etc. Finally, we consider the cluster formed by "11" and "20" representing the web pages "Trips" and "Map of the town". A direct hyperlink could be added, in order to shorten the users navigational paths.

Conclusively, one may say that the clusters of web documents revealed patterns of the users browsing activity. The discovered patterns may then guide the navigation redesign process, in order to improve the usability of a given web site. We note that for different values of threshold we receive different number of clusters. In particular, a medium to large threshold implies a very small number of clusters. In such a case, the web documents, which constitute the corresponding clusters, have large degree of correlation.

# 5 Conclusions

Clustering the documents in a web site may reveal useful patterns in the browsing activity. The presented innovative cluster-mining algorithm aims at finding groups of related documents of a web site, regardless of their existing links. The presented algorithm is claimed to be more efficient and simpler than other similar approaches. As shown, the clusters found may subsequently be used by the designer for improving navigation design of the site, thus, improving its usability.

# 6 Acknowledgments

Funding by the UoP Research Committee under the basic research program K. Karatheodoris, Project "Development of probabilistic models of web use" is gratefully acknowledged.

# 7 References

Avouris, N., Koutri, M., & Daskalaki, S. (2003). Web site adaptation: a model-based approach. In Proc. of *HCII2003*, Crete: Greece.

Fu, Y., Sandhu, K., & Shih, M. Y. (1999). Clustering of Web Users Based on Access Patterns. In Proc. of the *1999 KDD Workshop on Web Mining*, Springer-Verlag, San Diego: Canada.

Han, J., & Kamber, M. (2001). Data Mining: Concepts and Techniques, Morgan Kaufmann, San Francisco.

Jain, A. K., & Dubes, R. C. (1998). Algorithms for Clustering Data. Prentice Hall advanced reference series, Upper Saddle River: NJ.

Lokuge, I., Gilbert, S. A., & Richards, W. (1996). Structuring Information with Mental Models: A Tour of Boston. In Proc. of the *CHI96*, ACM Press, Vancouver, British Columbia.

Newman, M. W., & Landay, J. A. (2000). Sitemaps, Storyboards, and Specifications: A Sketch of Web Site Design Practice. In Proc. of the *DIS2000*, ACM Press, New York, 263-274.

Nielsen, J. (1999). User Interface Directions for the Web. *Com. of the ACM*, 42 (1), 65-72.

Perkowitz, M., & Etzioni, O. (2000). Towards adaptive Web sites: Conceptual framework and case study. *Artificial Intelligence 2000* (118), 245-275.

Pierrakos, D., Paliouras, G., Papatheodorou, C., & Spyropoulos, C. D. (2001). KOINOTITES: A Web Usage Mining Tool for Personalization. In Proc. of *PC HCI 2001*, Patras: Greece, 231-236.

Preece, J. (1994). Human-Computer Interaction. Addison Wesley.

Tselios, N., Avouris, N., Dimitracopoulou, A., Daskalaki, S. (2001). Evaluation of Distance-Learning Environments: Impact of Usability on Student Performance. *Int. J. of Educational Telecommunications*, 7 (4), 355-378.

Zhang, T., Ramakrishnan, R., & Livny, M. (1996). An Efficient Data Clustering Method for Very Large Databases. In Proc. of the *SIGMOD*, Montreal: Canada, 103-114.

# Web Usability: Its Impact on Human Factors and Consumer Search Behaviour

*Bernie Lydon, Tom Fennell*

Dublin Institute of Technology
bernienlydon@eircom.net, tom.fennell@dit.ie

**Abstract:** Usability testing methods and phenomenology based qualitative techniques were applied in naturalistic settings in consumers' homes to establish factors which are perceived as hindering and facilitating consumers in finding product/service information, and making e-commerce purchases. A facility to see an overview of site structure in order to make quick evaluations about content and navigation schemes emerged as prominent user concerns with regard to human interface design factors. The placement of a search engine on the homepage of a website, so that users can easily establish the starting point for specific search tasks, was found to be critical. The absence of such search functionality on homepages led to users navigating many irrelevant pages and in some instances failing to find sought product/service information. In such cases, consumers sometimes opted to abandon the site altogether and access alternative sites to complete the tasks (which can impinge directly and negatively on e-commerce sales). Search engines that returned inaccurate results within a site also led to dissatisfied customer experiences. Layout of price information was found to be essential to aid readability and interpretability. The results suggest that more positive user attitudes are associated with the vertical layout of such content as opposed to a horizontal style layout.

## 1.0    Introduction

In the literature, web usability is highlighted as a key factor likely to affect web search and subsequent buying behaviour. In fact it has been claimed that "usability is a prerequisite for e-commerce success" (Nielsen et al, 2001: 1). The main elements of usability may be categorised as the quality of interaction between primarily personal factors and primarily human interface factors (Turk 2000: 12). Usability refers to the interaction between the personal factors and site attributes in relation to the specific tasks that users need to perform (Mcleod, 1994). The main personal factors concerned with usability are constructs such as consumer attitudes, expectations, satisfaction and consumer commitment. The main site attributes are encapsulated in the human interface design, which incorporates elements such as web content, web structure, speed of accessing webpages, and search functionality. These personal factors and human interface design factors interact to form the total web experience for consumers.

### 1.1    Personal Factors

In relation to personal factors, it has been claimed that in order to develop, evaluate and improve the usability of virtual environments, a comprehension and appraisal of human performance and satisfaction is necessary (Gabbard & Hix 1997). A large number of web sites are poorly designed, because user requirements are often not incorporated into the web design process (Vora, 1998; Nielsen 1999; 1996; 2001). Online purchasers' attitudes and behaviours towards the site is affected by a mixture of web design evaluation, brand loyalty (Supphellen & Nysveen 2001), and the outcome of their first online purchase experience (The Boston Consulting Group, 2000). If a consumer encounters a positive experience on a web site, it is likely that it will increase their time spent at the site (Hoffman & Novak 1995). Furthermore, if consumers are satisfied with the site it is plausible that they may revisit.

### 1.2    Human Interface Design

In regard to human interface design factors, taking the content aspect of design first, Nielsen (1999) suggests three main guidelines that he feels should be obeyed when writing content for any web site. These include creating a web site that is concise, scannable, and objective. In addition, many sites lack fundamental up-to-date information, and appear more interested in content quantity than content quality (Jerrams-Smith, 2000). As regards the structural aspect of design, web content should be presented and structured in a scannable format that supports navigation (McGovern, 2000:56). Many users express frustration at getting lost within sites (Badii & Murphy, 2000). Modjeska & Marsh (1997) found that site structure significantly affects user navigation. Therefore, web structure needs to be carefully considered in web design guidelines to assist in user navigation, and to deter users feeling disorientation and getting "lost in hyperspace"(Modjeska & Marsh, 1997). Speed of accessing or downloading pages another key site attribute involved in human interface design. GVU's 10th User Survey reports that one of the top pressing issues facing the Internet is speed of downloading a page to access content on particular web sites (www.cc.gatech.edu 1998, Jerrams-Smith, 2000). Research conducted on this problem has indicated that if web designers indicate a downloading time to consumers', it may reduce consumers' negative evaluations of the website (Dellaert & Kahn 1997). A major concern, therefore, for any e-commerce site, is how easily users can find the information they need and how accessible it is. In relation to search functionality, the reality for many users is that finding their desired product is neither easy nor quick (Reda 1997: 60). The search tools available online influence consumers perceived efficiency of search (Jiang 2002: 184). As research conducted by Lauren Freedman, a retailing consultant at Chicago-based E-tailing Group notes: 'research shows that 70 per cent of shoppers know what they want to buy. If you can't offer them a faster or better alternative, you've lost them' (Reda 1997: 60).

## 1.3 Research Objectives

Accordingly, the following research objectives have been set:
To determine the attitudinal reactions, search behaviour and levels of commitment of Irish consumers towards e-commerce web sites on the following usability dimensions:

- searching and navigating mechanisms
- sequencing, and linking mechanisms
- ease and speed of accessing information
- quantity, quality, and presentation of information content
- overall usability

To establish whether and to what extent the following factors are associated with variations in consumer attitudes, search behaviour and commitment:

- web expertise of consumers
- age
- social class
- gender
- education levelVarious methodologies and web usability guidelines in the literature are drawn on in order to research the above objectives.

## 2.0   Research Approach

Usability methods such as task based scenarios (Dunliffe 2000; Nielsen et al 2001) and the think aloud protocol (Nielsen 2001) were used in conjunction with the phenomenology method (Thompson et al 1989) in an attempt to get as close as possible to informants' web experiences. Rather than examine a range of product sectors, with concomitant etiolation of the data, it was decided to concentrate on one sector where richer data might emerge. The travel industry was chosen because it represents a substantial proportion of online sales. According to BizRate total online sales amounted to $35.87 billion in 2001 (Greenspan, 2003). A consumer survey conducted by Jupiter claimed that online travel sales for 2001 were $24 billion with airline tickets accounting for $16 billion of online travel transactions (Greenspan, 2002). Airline businesses are increasingly migrating significant portions of their business to the web. Low-cost carrier Ryanair (the largest low-cost airline in Europe (Guardian 2003)) receive 100 per cent of bookings via their website (Pastore, 2001) and they boast that their profits are up by 50 per cent for the last three months of 2002 (www.ryanair.com 2003). In comparison Aer Lingus, a more traditional full service airline currently has 45 per cent of the company's customers booking via the web. Given that air tickets are among the biggest online purchases for Irish Internet users (Amarach, 2002), the Irish airline industry has been chosen for the current study.

## 3.0   Research Process

Consumers living in Dublin were recruited, using the following selection criteria: web expertise, age, social class and gender. Unlike a typical usability study where the test is carried out in a lab based environment (Nielsen 2001, frontend.com 2001), the facilitator's intent here was to carry out the interviews in the homes of the chosen respondents on a one to one basis. Informants were asked to carry out specific scenario based tasks on both the Ryanair and Aer Lingus websites using their own computer equipment. This was critical to the study in order to elicit the true perceived speed and ease of carrying out the tasks as experienced by participants in their naturalistic setting. Prior to carrying out the tasks users were encouraged to talk out loud and describe their thoughts and feelings towards different aspects of the sites during the execution of each task. Tasks were modified in accordance to problems highlighted from the pilot study. Accordingly, the following tasks were set:

- **Task 1(a):** Imagine you are searching for flight and fare information for a return trip from Dublin to Paris. Departing Dublin on the morning of Sep 13th 2002, and departing Paris on the evening of Sep 15th[th], 2002.
- **Task 1 (b):** Imagine you wish to compare the flight and fare information from task **1(a)** to another airline.
- **Task 2(a):** Imagine you are booking a connecting flight from Dublin to Rome. Departing Dublin at any time on Sept 13th 2002, and departing Rome at any time on Sept 15th 2002. Please organise your trip using connecting flight information. Find the total cost of flight in Euros, and also give your chosen flight times. Proceed with the booking until the site prompts you for your credit card details.
- **Task 2 (b):** Imagine you are booking a direct flight from Dublin to Rome. Departing Dublin on the morning of Sept 13th 2002, and departing Rome on the evening of Sept 15th 2002.

An interactive style of facilitation was used during the tasks. The ensuing dialogue was emergent, led by informants' own focus and comments. In this way informants went into depth describing their experience in lived terms often either commenting on specific aspects of the site or pointing towards them. Whenever consumers appeared stuck, frustrated, or indeed satisfied, neutral questions were asked such as 'Can you describe what are you thinking now?' 'Can you describe how you are feeling now'? The format of the in-depth interview aided interpretation as illustrative comments reflecting informants' lived web experience were captured so that the true meaning of the phenomena could be established. Behavioural, affective and cognitive responses were assessed in depth. Facilitator reports of informants' reactions and body language were used in order to convey user attitudes and motives and to provide an overall picture of responses. Twelve usable interviews were completed.

## 4.0 Discussion of Results

Many complex and inter-related issues concerning usability emerged. The following summary points regarding consumers' experiences of human interface design factors are presented here as illustrative of the findings:

### 4.1 Site Content

Ryanair's 'summary of flights selected' page presented flight fares in a vertical fashion and therefore was seen as easier to read than the Aer Lingus page, which presented the prices horizontally. This was seen as an issue because participants associated the presentation of total fare price information with the vertical method that is generally used to display accounts. It was considered that *'flight times are easy to see'* on the Ryanair 'summary of flights selected' page whereas on the Aer Lingus page it was felt that the flight times were difficult to read *'you wouldn't know is it the flight number or the flight times'*. Some attributed this to the size of the font used and the colour of the background and text. As product/service information, particularly price information, is becoming increasingly important to the online consumer (Shanker et al, 1999), it is important that this information is laid out in a manner that facilitates legibility. Turban & Gehrke (2000) support this view and claim that simple background, textures, colours, and clear text should be used.

### 4.2 Site Structure

Participants felt that the task map on the Aer Lingus site acted as a useful navigational aid that allowed them to see where they were at each stage of the booking process. *'I always liked that... They tell you what part of the process you are at. At any stage before you get to the final booking stage you know that you can stop.'* (See Figure 1). The Ryanair site also had a task map located on the top of the screen, but some did not recognise that it actually represented a visual guide to where the consumers were in the booking process. *'I do see itinerary up here though. I am wondering what that is so I might go in there and see'* (See Figure 2). Other consumers clicked on it thinking that it represented a search engine. Such confusion can be eliminated if the steps of the shopping process are presented in a clear manner to users while they are executing this important online task (Nielsen, 2001).

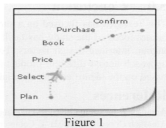

Figure 1

Participants claimed that the hyperlinks (see Figure 3) on the Ryanair homepage were not well labelled because they did not accurately describe the content accessed when clicking on them. This led to informants clicking on many links before reaching their sought information.

The following text describes some experiences that people encountered when trying to search for flight and fare information from the Ryanair homepage. *'I had the problem of getting to the page that I wanted to get to, I had to go through all this confusion, you know, which link do I click on, am I doing the right thing'*. As a result, respondents were confused and thus their mental model of the site's overall structure was affected. The linking mechanisms on the Aer Lingus homepage were seen as more comprehensive and thus navigation was more straightforward.

Figure 3

Participants experiences confirm that effective screen layout and linkage structure should be incorporated into the design process in order to reduce the number of steps taken to locate sought information, and facilitate users' search and navigation patterns (Schneiderman, 1997).

### 4.3    Search Functionality

Participants were primarily concerned with the ease of establishing from the sites, how to find flight information. It was generally felt that it was '*much easier*' to navigate the Aer Lingus site to retrieve the required information than on the Ryanair site. Participants seemed more in tune with where to go and attributed this mainly to the position of the search engine on the Aer Lingus homepage. *'I mean to me a screen with just that (search engine) on it is all that I need'*. Whereas the absence of a search engine on the Ryanair homepage meant it was '*..vague to know where to go from the first page*. Another informant suggested abandoning the site altogether '*I would look at another site, I suppose Aer Lingus would be the next one'*. Consequently, in line with guidelines already established by Nielsen & Tahir (2002) it is vital to provide users with a search facility on the homepage. When consumers eventually found the search engine on the Ryanair site, the search results proved more accurate in terms of fulfilling the search criteria specified by users. In contrast, the Aer Lingus search engine was seen as providing users with flight information that they did not select. Providing users with information that they specify is seen to be critical, and designers should ensure that search functionality on the site will produce accurate results.

### 4.4    Accessibility

Pages on both sites were generally considered '*Quick*' to download in order to access the information required. The Aer Lingus site was perceived by some respondents to be a '*bit slower*' in comparison to the Ryanair site. Some attributed the large company logo presented on the pages as being the cause for slower web page response time. According to a report by Zona Research, consumers expect to be able to download web pages in less than eight seconds; if this does not occur it could lead to vast losses in sales (Lake, 1999). The speed at which webpages download is key to success in the online marketplace.

## 5.0    Conclusion

This research has focused on a substantial sector of the online market (airline travel). The naturalistic (in-home) locus of the study has facilitated an in-depth analysis and understanding of consumer experiences of human interface design factors. To improve the generalisability of this work, it is intended to extend this approach across different product and market sectors. It is also planned to undertake quantitative surveys, based on the identified consumer language utilisation and response patterns.

## References

Amarach Consulting (2000) 'iMarketing Insight 3: eGuarantees' *iMarketing The Future of Direct Marketing: pg 1-65*
Amarach Consulting (2002) 'E-Commerce' Retrieved 11 January, 2003, from http://www.etcnewmedia.com/

Badii, A. & Murphy, A, (2000) 'Point-of-Click: Managed Mix of Explicit & Implicit Usability Evaluation with PopEval_MB & WebEval_MB' *Encompass 2000*

Burke, Raymond, R (2002) 'Technology and the Customer Interface: What Consumers Want in the Physical and Virtual World' *Journal of the Academy of Marketing Science*, Vol. 30, No. 4, 411-432

Boston Consulting Group (2000) 'Bridging the Gap Between the Online Promise And The Current Consumer Experience' *Winning the online consumer: Insights Into Online Consumer Behaviour,* March, pp1-32. Boston: Houghton Mifflin Company

Dunliffe, Daniel (2000) 'Developing usable Web sites – a review and model' *Internet Research: Electronic Networking Applications and Policy*, Vol 10, No.4, pp 295-307

Dellaert, Benedict G.C. and Kahn, Barbara, E (1999) 'How tolerable is Delay? Consumers' Evaluations of Internet Web Sites After Waiting' *Journal of Interactive Marketing*, Vol 13, No 1, 41-54

Frontend.com 'A Brief User Test: Aer Lingus and Ryanair' Retrieved January 22, 2003, from http://www.frontend.com/

Gabbard, J.L. & Hix, D (1997) 'A Taxonomy of Usability Characteristics in Virtual Environments' Department of Computer Science Virginia Polytechnic Institute and State University Blacksburg, VA 24061. Pg 1 –182.

GVU's WWW Surveying Team (1998) 'GVU's 10th WWW User Survey' Retrieved August 28, 2001, from http://www.cc.gatech.edu/gvu/user_surveys/survey-1998-10/

Oehlke, B & Turban, E (2000) 'Determinants of E-commerce website *Human Systems Management,* Vol 19, Part 2, 111-120

Greenspan, Robyn (2002) 'Online Travel Expected to Fly High' Retrieved 11 February, 2003 from http://cyberatlas.internet.com/markets/travel/article/0,,6071_1002561,00.html,

Greenspan, Robyn (2003) '2002 E-Commerce Holiday Wrap Up'. Retrieved 11 February, 2003 from http://cyberatlas.internet.com/markets/retailing/article/0,,6061_1563551,00.html#table1.

Guardian (2003) 'Ryanair flies a risky route' Retrieved January 31, 2003, from http://www.guardian.co.uk/airlines/story/0,1371,886448,00.html

Hoffman, D.L., and Novak T.P., (1995) 'Marketing in Hypermedia Computer-Mediated Environments: Conceptual Foundations. *Journal of Marketing*, Vol 60, 50-68

Jeramms-Smith, Jenny (2000) 'Helping e-customers satisfy their information needs' *Encompass 2000,* pg

Macleod, Miles (1994) 'Usability: Practical Methods for Testing and Improvement' in *Proceedings of the Norwegian Computer Society Software '94 Conference* (Sandvika, Norway, 1-4 Feb).

Lake, David. (1999) 'Slow Sites Cost Vendors Billions' Retrieved on 25 September 2002, from http://www.thestandard.com/article/0%2C1902%2C5374%2C00.html

Modjeska, D. and Marsh, A. (1997) Structure and Memorability of Web Sites. *Technical Report of the Computer Science Research Institute of the University of Toronto* , Toronto: University of Toronto

Nielsen, J. (1999) 'User Interface Directions for the Web' *Communications of the ACM*, Vol. 42, No.1.

Nielsen, J., Molick, R., Snyder, C., Farrell, S (2001) *E-Commerce User Experience:* USA

Nielsen, Jacob & Tahir Marie (2002) Homepage Usability: New Riders Publishing

Pastore, Micheal (2001) 'Europeans Increasingly Turn to the Net for Travel' Retrieved 13 February, 2003 from http://cyberatlas.internet.com/markets/travel/article/0,,6071_554271,00.html

Rowley, Jennifer (2000)'Product Search in e-shopping: a review and research propositions' *Journal of Consumer Marketing* , Vol. 17 No. 1, pp20-35

Schanker, Venkatesh, Rangaswamy, Arvind & Pusateri, Michael (1999) 'Customer Price Sensitivity and the Online Medium' *Working Paper*, University of Maryland, College Park, MD 20742

Schiffman Leon, G. & Kanuk, Leslie L (2000): Consumer Behaviour (7th Ed): Prentice Hall

Schneiderman, Ben (1997) 'Designing information-abundant web sites: issues and recommendations' *International Journal of Human-Computer Studies*, Vol 47, 5-29.

Schneiderman, B. (1998) 'Designing the User Interface: Strategies for Effective Human Computer Interaction. Third Edition: Addison-Wesley.

Supphellen, M & Nysveen, H (2001) 'Drivers of intention to revisit the websites of well-known companies' *International Journal of Market Research*, Vol 3, Quarter 3, 341-352

Thompson, Craig J, Locander, William B., Pollio, Howard R. (1989) 'Putting Consumer Experience Back into Consumer Research: The Philosophy and method of Existential-Phenomenology' *Journal of Consumer Research*, Vol 16, September

Turk, A. (2000) 'A Contingency Approach to Designing Usability Evaluation Procedures for WWW Sites' *Encompass 2000,* pg 12

Rowley, Jennifer (2000)'Product Search in e-shopping: a review and research propositions' *Journal of Consumer Marketing,* Vol. 17 No. 1, 20-35

ryanair.com (2002) 'Ryanair delivers record Q3 Profits' Retrieved 13 February, 2003 from http://www.ryanair.com/

Vora, Pawan (1998) 'Human Factors Methodology for Designing Web Sites' in: Forsythe, C., Grose, E. and Ratner, J. (1998) *Human Factors and Web Development.*

Yoon, Sung-Joon. (2002) 'The Antecedents and Consequences of Trust in Online-Purchase Decisions' *Journal of Interactive Marketing*, Vol 16, Number 2, 47-63

# Deconstructing Web Pages

*A. Maidou*

Department of Architecture
Aristotle University of Thessaloniki
54124 Thessaloniki, Greece
anthoula@limnos.physics.auth.gr

*H. M. Polatoglou*

Physics Department
Aristotle University of Thessaloniki
54124 Thessaloniki, Greece
hariton@auth.gr

## Abstract

The many internet users and their daily growing number necessitate the critique of the human computer interface (HCI). The HCI is far more important than its name actually states, namely a means to interact with the computer, since through the web, it is mainly used as a means of communication using the computer. In order to understand how this communication functions and why it has become so important, we examine web pages in terms of a spatial and a deconstructive analysis.

## 1    Introduction

Originally, computers were designed by engineers for engineers – and little attention had to be paid to the interface. Later on, the use of computers by a broader, non-specialized user group, necessitated the use of interfaces to enable them ease of use. The way we understand HCI design nowadays has become more complicated, moving from efficient ways to design understandable electronic messages (Norman, 1988, Shneiderman 1998) to include also cross-cultural aspects (Marcus 2001, Marcus & Gould 2001) – a fact, which reflects the broadening of the users to the whole globe.

People of very diverse interests find excitement using the Web either as researchers, explorers, browsers, or just as connectors (Dillon, 2000), and their number rises daily. In order to understand how and why it has become so important, we will examine web pages in terms of a spatial and a deconstructive analysis.

The space we produce though the computer is virtual, it doesn't exist anywhere, it is not made of material, it doesn't obey to physical laws, unless we program it to. Neither do the restrictions we have as human beings, such as our dimensions and abilities apply necessarily to virtual space - we can "see" a large building form any height, walk through walls, jump from one place to another. Humankind has constructed a new kind of space. The experience of a new kind of space isn't something new. Since the implementation of the telegraph and later on the telephone and television, humankind is experiencing a new kind of perception, the "perception at a distance", or telesthesia (McKenzie, 1994). This experience is perceived as real, like the real world experience - it differs only in the fact that things are not bounded by the rules of proximity. Virtual space is also experienced as a real space - we use virtual space to get information on any subject, read the news, buy, visit libraries, museums, listen to music, etc. Furthermore, the terms we use to refer to virtual space has a close analogy to the physical world: we talk about "virtual communities", "homepages" or "sites" that have "addresses", etc.

Virtual geographers study the geographies of the virtual space (Dodge & Kitchin, 2001) using geographical metaphors. Additionally, we talk about the law of virtual space, protection of privacy, etc. Virtual space is perceived as a notional mechanism beyond the real world. Spatiality takes a new dimension; it can be electronically constructed and experienced. Through our memory we transform these experiences into possibly experienced realities. Virtual space is an extension of real space and can thus be analyzed in spatial terms.

## 2 Method

If we would like to ask questions about the meaning and content of something, in our case web pages, we would have to apply an analytical tool, a critique method. Deconstruction is such an analytical tool, although not quite a method (nor a well-defined analytical tool) - it explores the content and meaning of something, in connection with the society, which produced it.

Deconstruction is currently widely used to understand underlying cultural, social, historical, political, etc. aspects in many fields. It started as a philosophical theory and was directed on the (re)reading of philosophical writings and found very soon application in literature critics, arts, advertisement, architecture, psychology and psychoanalysis, etc.

In a deconstructive analysis everything is considered as a text. "There is nothing outside the text" (Derrida 1976, p. 158) and texts are suitable for deconstruction. Through the deconstructive analysis all texts (and this word has here a very extended meaning, containing everything, from a text to a building facade, to a look of a face, to the way of dressing up, etc.) are seen as complex historical and cultural processes rooted in the relation of texts to each other and in the institutions and conventions of writing. Furthermore, our knowledge works by differences and through codes, not by identification. In a deconstructive approximation the basic structuralist principle of difference is located ontologically and semiotically. Something can only be through difference, because it is not something else instead (Norris, 1991). As difference comes before being, in the same way a trace comes before the presence of a thing. Something is itself by virtue of not being something else, by differing and that from which it differs remains as a trace. It is it's absence that is necessary for it to be. All texts are constituted by difference with other texts, which makes them similar to them. A text includes that which it excludes - opposites compose a unity, otherwise they cannot be opposites.

Derrida argues (Derrida, 1982) that in a traditional philosophical contradiction two opposite concepts cannot coexist peacefully, but are in a violent hierarchy - for example good/bad, presence/absence, life/death, etc. Deconstruction of the contradiction is the reversion of the hierarchy. The interplay of two opposite concepts like real space/virtual space will allow their understanding.

## 3 Analysis

In this study we apply the above briefly described tools to analyze web pages. We restrict our analysis to the study of the main page (structure and semantics) and to the way the links are displayed. Since architects are the most appropriate group to design space, our sample consists of three groups: web pages of architectural university departments, web pages of architects, and web pages of "others".

## 4 Results

### 4.1 Web pages of architectural university departments

From the analysis of our samples we find that all web page creators use spatial tools to display the content of the web pages. This includes also abstracts forms of geometric space, where in the simplest case the text is used as a shape put on the background of the page. An example for this case is the web page of the Department of Architecture of the Helsinki University of Technology (http://www.hut.fi/Units/Departments/A/). In the center of the web page the word "ARKKITEHTIOSASTO" and the words "DEPARTMENT OF ARCHITECTURE" under the first one, separated by a line, together with the logo of the University in front of them, form a small shape, a parallelogram – small in relation to the size of the page. These two lines are also the links – one to the Finnish pages and the other to the English pages, and the logo is the link to the University. In a similar way the does the Department of Architecture of the Cornell University (http://www.architecture.cornell.edu) present itself to the visitor: on the top right of a white page appears the word CORNELL in black letters, followed by the word ARCHITECTURE in smaller red letters, initially with the same width as the first word, and afterwards animating to about the half width. The visitor is somehow surprised expecting and waiting for the links, which do not appear – one has to put the mouse over the letters of the word CORNELL to see the links. During this process each letter transforms to an animated grid and turns to an abstract shape, and at the same time the links appear under those two words. Although in both cases there is no evident architectural space the sites are interesting and refer well to architecture, especially the Cornell site, which uses the animated grid, to refer to architectural design, and the creation of shape, through this animation.

In all cases we found that the homepage was handled as a two-dimensional (2D) surface, without perspective, or common scale. E. Tufle (Tufle, 1997) developed the term "confection" to describe such divided patterned surfaces, without perspective or a single space, and noted that confection is the characteristic employment of images on the Web. Text, colored surfaces or lines, are usually used to organize these 2D surfaces, which are often enriched with pictures, drawings, or animated material. The arrangement of these patterns is in the most cases in rows-and-columns style table, reflecting a preference to depict a balanced rather then a dynamical web page.

From our sample of 200 university departments, only in one case, the homepage of the Department of Architecture of the Aristotle University of Thessaloniki (http://www.arch.auth.gr) a three-dimensional (3D) graphic was used. Over the background, which looks like an ochre sheet of paper with gray-brown architectural drawings, and close above it, a 3D granite-like tile with engraved letterings of the faculty, casting shadow on the background, is displayed. The links to the Greek and English pages are embossed casting shadows, like the plummet, which serves together with the plumb-line and the logo of the university as a symmetry axis and dividing line among the Greek and English side of the site.

Examining the home web pages for cross-cultural aspects, as defined by Marcus (Marcus & Gould 2001), namely power distance, individualism vs. collectivism, masculinity vs. femininity, uncertainty avoidance, and long term orientation, no characteristic particularities could be observed referring to a special group, but all kind of patterns were found distributed all over the globe.

## 4.2 Web pages of architects

Architects use also spatial tools to present their work on the net. This can be achieved using text as a shape on a surface, or it can have a more complicate synthesis, with texts, pictures and animations, or videos.

In many cases the home page contained elements referring to the drawing process, as for example a grid (mostly as a background element), drawings (either as background elements, or as an organizing element of the web page composition), or building elements, as for example a door, a pillar, or a capital were used instead of links' bullets.

As for cross-cultural aspects, from the sample of 200 web pages only one (http://www.adler.co.il) exhibited a right to left writing composition, with the links also positioned in a right to left direction. This fact reflects only the different way of writing of this ethnic group, while the content of the image, which is an architectural drawing, is readable for all architects.

In addition women architect' s web pages did not show differences in general, neither in the use of colors, patterns or content.

## 4.3 "Others'" web pages

Under the group of "other's web pages" we examined web pages of private businesses, organizations, shops, museums, net communities, etc. The picture we got from 200 home web pages was about the same as mentioned above, with the backgrounds, colors, etc reflection the either the logo of the firm, or being designed according to the content.

## 5 Conclusions

From our analysis of the web pages, we found that in all cases of our sample, the pages were considered as 2D spaces, which were organized using only text, or text and images, and/or animations. Web pages had unifying elements characterizing their content, while no differentiations concerning cultural aspects were observed, at least not in relation with specific ethnic groups. Furthermore, concerning the gender, female and male architects could not be distinguished according to their web sites. It seems that representation in virtual space is the global pattern despite languages and cultural differences, although a lot is still to be done in three-dimensional representation on the web.

## 6 References

Adler Architectural Office. Retrieved at February 15, 2003 from http://www.adler.co.il

Department of Architecture of the Aristotle University of Thessaloniki. Retrieved at October 10, 2002, from: http://www.arch.auth.gr.

Department of Architecture of the Cornell University. Retrieved October, 10, 2002 from: http://www.architecture.cornell.edu.

Department of Architecture of the Helsinki University of Technology. Retrieved February, 2, 2002 from: http://www.hut.fi/Units/Departments/A/.

Derrida, J. (1976). Of Grammatology, trans. Spivak , G.C. Chicago: Chicago University Press.

Derrida, J. (1982). Différance (trans Alan Bass), Margins of Philosophy. Chicago: Chicago University Press.

Dillon, G. (2000). Dada Photomontage and net.art Sitemaps. *Postmodern Culture*. Retrieved April 12, 2000, from http://jefferson.village.virginia.edu/pmc/current.issue/10.2dillon.html

Dodge, M., & Kitchin R. (2001) An Atlas of Cyberspace, Addison Wesley.

Marcus, A. (2001). Cross-Cultural User-Interface Design. In Smith, M.J. and Salvendy G (Eds.) *Proceedings of the Human-Computer Interface International Conference, 5-10 Aug., 2001, New Orleans, USA,* Vol. 2, pp. 502-505, Mahwah: Lawrence Erlbaum Associates.

Marcus, A. & Gould, E.W. (2001). Cultural Dimentions and Global Web Design: What? So What? Now What? Retrieved October 12, 2002, from http://www.AmandA.com.

McKenzie, W. (1994). Virtual Geography, Indiana University Press.

Norman, D. (1988). The design of everyday things. New York: Doubleday.

Norris, C. (1991). Deconstruction Theory and Practice. London: Routledge.

Shneiderman, B. (1998). Designing the user interface: Strategies for effective human-computer interaction (3rd ed.). Reading: Addison-Wesley Publishing.

Tufle E. (1997). Visual Explanations: Images and Quantities, Evidence and Narrative. Cheshire: Graphics Press.

# A Quality Model For Testing the Usability of Web Sites

*Francisco Montero, Víctor López-Jaquero, María Lozano, Pascual González*

Grupo de Investigación LoUISE. Escuela Politécnica Superior de Albacete
Universidad de Castilla–La Mancha. Avda. de España s/n.
02071 – Albacete – Spain
{fmontero, victor, mlozano, pgonzalez}@info-ab.uclm.es

## Abstract

As the Internet expands, and the quantity of the information that we can find on the web grows along with it, the usability of the pages gets more important (Shneiderman, 1998). Many of the sites still got quite low evaluations from experiments when it came to certain aspects of usability. This paper proposes a set of quantitative and qualitative metrics under an usability-centred quality model and an usability testing experiment where this model could be validated.

## 1    Introduction

Research by User Interface Engineering, Inc.(http://world.std.com/~uieweb), shows that people cannot find the information they seek on Web sites about 60% of the time. This can lead to wasted time, reduced productivity, increased frustration, and loss of repeat visits and money.

Nielsen (2000) reports Studies of user behaviour on the Web find a low tolerance for difficult designs or slow sites. People don't want to wait. And they don't want to learn how to use a home page. There's no such thing as a training class or a manual for a Web site. People have to be able to grasp the functioning of the site immediately after scanning the home page — for a few seconds at most.

Traditional software design and information appliances could afford to neglect usability considerations, since users only experienced the usability of a product after making their purchase. In contrast, on the web, users experience a site's usability *before* making any purchases – on the web, the user interface becomes the marketing materials, store front, store interior, sales staff, and post-sales support all rolled into one.

This paper deals with the problem of usability of web sites and proposes an usability-centred quality model-based on quantitative and qualitative metrics. A quality model can be used to understand, control, and improve a product. Nowadays we have several quality models (Olsina, 1999, Ivory, 2001, ISO 9126-1, 2000) but these proposal are quality-centred but they are not usability-centred. We think that in Web environments a usability-centred quality model is more interesting and meaningful in function of Web features than other software products and new proposal of quality model should be proposed. Our proposal is an usability-centred quality model where accessibility factors, for example, are considered.

This paper is organised as follows. In Section 2 we give basic definitions usability-related; usability, understandability, learnability, operability, satisfaction, quality model. In Section 3 several kinds of evaluation methods will be presented. In Section 4 an usability quality model will be proposed and in Section 5 a user testing design and related considerations will be introduced.

## 2 Basic Definitions

Usability is the measure of the quality of a user's experience when interacting with a product or system — whether a Web site, a software application, mobile technology, or any user-operated device.

ISO/IEC 9126 (1991) has recently been replaced by a new four part standard that has reconciled the two approaches to usability. ISO/IEC 9126-1 (2001) describes the same six categories of software quality that are relevant during product development: functionality, reliability, usability, efficiency, maintainability and portability The definition of usability is similar:

> *Usability*: *the capability of the software product to be understood, learned, used and attractive to the user, when used under specified conditions.*

The phrase "when used under specified conditions", equivalent to "context of use" in ISO 9241-11 (1998) was added to make it clear that a product has no intrinsic usability, only a capability to be used in a particular context. In Figure 1, definitions of usability are provided by international standard can be seen. Both usability definitions are compatibles.

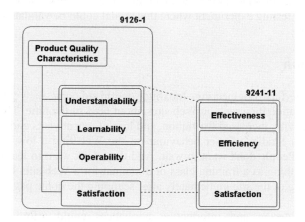

**Figure 1:** Usability and International standards

Usability is a combination of factors that affect the user's experience with the product or system, including:

- easy of understanding: How fast can a user who has never seen the user interface before learn it sufficiently well to accomplish basic tasks?
- easy of learning and memorability, if a user has used the system before, can he or she remember enough to use it effectively the next time or does the user have to start over again learning everything?
- operability of use how fast can he or she accomplish tasks?
- subjective satisfaction. How much does the user like using the system?

The objective of designing and evaluating for usability is to enable users to achieve goals and meet needs in a particular context of use. A quality model defined for supporting a given kind of analysis, is a description of which attributes are important for the analysis, which one more important than others, and which measurement methods have to be used to assess the attributes values (Brajnik, 2001). In web site design, a quality model is essentially a set of criteria that are used to determine if a website reaches certain levels of quality. Quality models require also ways to assess if such criteria hold for a website.

# 3 Usability Evaluation

ISO 9241-11 (1998) explains how to identify the information that it is necessary to take into account when specifying or evaluating usability in terms of measures of user performance and satisfaction. Guidance is given on how to describe the context of use of the product and the measures of usability in an explicit way. It includes an explanation of how the usability of a product can be specified and evaluated as part of a quality system.

In the software engineering community the term usability has been more narrowly associated with user interface design. ISO/IEC 9126, (2000), developed separately as a software engineering standard, defined usability as one relatively independent contribution to software quality associated with the design and evaluation of the user interface and interaction.

The term usability is sometimes used with a similar meaning to quality in use but excluding safety. So, effectiveness defined as the accuracy and completeness with which users achieve specified goals can be expressed in terms of the quality and quantity of output. ISO/IEC DTR 9126-4 provides effectiveness metrics. And efficiency is measured by relating the level of effectiveness achieved to the resources used. ISO/IEC DTR 9126-4 (2001) provides efficiency metrics.

A suite of user interface-related metrics was proposed by Constantine (1998) and Noble.

There are generally three types of usability evaluation methods: Inspection, Testing and Inquiry. In Usability Inspection approach, usability specialists examine usability-related aspects of a user interface. Inspection methods include: cognitive walkthroughs, feature inspection, heuristic evaluation, pluralistic walkthrough (Hoyler, 1993) (Hom, 1998) (Mayhew, 1999) (Ivory, 2001). In Usability Testing approach, representative users work on typical tasks using the system (or the prototype) and the evaluators use the results to see how the user interface supports the users to do their tasks. In Usability Inquiry, usability evaluators obtain information about users' likes, dislikes, needs, and understanding of the system by talking to them, observing them using the system in real work (not for the purpose of usability testing), or letting them answer questions verbally or in written form. Examples o inquiry methods are field observation, focus groups, interviews or questionnaires.

# 4 An Usability-Centered Quality Model

With proposed metrics in the previous sections we can begin the construction of an usability-centered quality model. Metrics have been chosen in order to determine, in a reliable and accurate way, the value for sub characteristics of usability such as, understandability, learnability, operability, attractiveness and compliance.

Proposed metrics are based in ISO/IEC 9126-4 (2001) document and Constantine's metrics (1998). Each metric is has associated a kind and can be assessed by using a different method, for instance by using questionnaires or automatic tools. The following nomenclature is used to highlight these considerations: **ct**: quantitative metric, **cl**: qualitative metric, **cu**: questionnaire, **a**: automatic assessment, **ISO**: metric proponed by ISO 9126-4 **Cn**: metric proponed by Constantine.

## Usability quality model

### Understandability
Visual coherence (ct, Cn)

Indication of the current location (cl)
A site map is available (cl)
Site structure is simple, with no unnecessary levels (cl)
Functionality is clearly labeled (cl)

All graphics links are also available as text links (cl, a)

The language used is simple (cl)

It is always clear what is happening on the site (cl)

Users are informed if a plug-in or browser version is required (cl)

Standard colors are used for links and visited links (cl)

## Learnability

Layout Uniformity (ct, Cn)

Task visibility (ct, Cn)

Only one word or term is used to describe any item (cl)

Links match titles of the pages to which they refer (cl, a)

If is necessary, online help is available (cl)

Terminology is consistent with general web usage (cl)

## Operability

Task effectiveness (ct, ISO)

Task completion (ct, ISO)

Error frequency (ct, ISO)

Task time (ct, ISO)

Task efficiency (ct, ISO)

Economic productivity (ct, ISO)

Productive proportion (ct, ISO)

Relative user efficiency (ct, Cn)

Task Concordance (ct, Cn)

No unnecessary plugins are used (cl)

All appropriate browsers are supported (cl, a)

Users can give feedback via email or a feedback form (cl)

Page size is less than 50Kb/page (cl, a)

Unnecessary animation is avoided (cl)

## Navigation facilities

There is a link to the homepage (cl, a)

All major parts of the site are accessible from the homepage (cl, a)

If necessary, an easy-to-use search function is available (cl)

## Control

The user can cancel all operations (cl)

There is a clear exit point on every page (cl)

## Attractiveness

Satisfaction scale (cu, ct, ISO)

Satisfaction questionnaires (cu, ct, ISO)

Discretionary usage (cu, ct, ISO)

## Compliance
## Accessibility

WAI de la W3C (cn, a)

Section 508 (cn, a)

---

# 5    Usability evaluation testing design

In the previous section, we have proposed a usability-centred quality model. And now, we want to establish a framework for usability evaluation testing using that quality model. We will follow the recommendations presented in the annex G and F of (ISO/IEC 9126-4, 2001).

The framework describes the components of usability and the relationship between them. In order to specify or measure usability it is necessary to decompose understandability, learnability, operability and satisfaction and the components of the context of use into subcomponents with measurable and verifiable attributes.

We must specify a description of the components of the context of use including users, equipment environments, and tasks. The relevant aspects of the context and the level of detail required will depend on the scope of the issues being addressed. The description of the context needs to be sufficiently detailed so that those aspects of the context which may have a significant influence on usability could be reproduced. In order to specify or measure usability is necessary to identify each component of the context of use: the users, their goals, and the environment of use. The Common

Industry Format (Annex F of ISO/IEC 9126-4, 2001) provides a good structure for reporting usability and will be used to provide a comprehensive report.

# 6    Conclusions

This paper has proposed a usability-centered quality model where several quantitative and qualitative metrics were compiled. Many of these metrics were introduced by international standards such as (ISO/IEC 9126-4, 2001) and others are user interface metrics (Constantinc, 1998). These quality model not have been validated.
We propose an usability testing design to validate the quality model and his associated metrics. Quality models can enable development and maintenance processes that consistently achieve high quality standards based on standardized data acquisition and measurement methods.

## Acknowledgments

This work is supported by two grants CICYT TIC 2000-1673-C06-06 and CICYT TIC 2000-1106-C02-02.

## References

Brajnik, G. (2001). Towards valid quality models for websites. Human Factors and the Web, 7th Conference, Madison, Wisconsin.
Constantine, L., Lockwood, L. (1999). Software for Use: A Practical Guide to the Models and Methods of Usage-Centered Design. Addison-Wesley.
Hom, J. (1998). The Usability Methods Toolbox. http://jthom.best.vwh.net/usability/usable.htm. (Last upadate.
Holyer, A. (1993). Methods for Evaluating User Interfaces. Technical report CSRP 301, School of Cognitive and Computing Sciences, University of Sussex.
Ivory, M. (2001). An Empirical Foundation for Automated Web Interface Evaluation. University of California at Berkeley.
ISO/IEC 9126-1. (2000). Software Engineering. Product quality. Part 1: Quality model.
ISO/IEC 9126-4. (2001). Software Engineering. Product quality. Part 1: Quality in use metrics.
ISO 9241-11. (1998). Guidance on usability.
Mayhew, D. (1999). The Usability Engineering Lifecycle: A practitioner's Handbook for User Interface Design. Morgan Kaufmann Publishers.
Nielsen, J. (2000). Designing Web Usability: The Practice of Simplicity. New Riders Publishing
Nielsen, J. (2001). Alertbox: Usability Metrics. http://www.useit.com/alertbox/20010121.html
Olsina, L. (1999). Metodología Cuantitativa para la Evaluación y Comparación de la Calidad de Sitios Web. PhD. Universidad de la Plata.
Section 508: The Road to Accessibility: http://www.section508.gov
Shneiderman, B. (1998). Designing the User Interface: Strategies for Effective Human-Computer Interaction. Addison-Wesley Publishers.
The World Wide Web Consortium: http://www.w3.org/

# Usability Evaluation of a Web-based Authoring Tool for Building Intelligent Tutoring Systems

*Maria Moundridou*

University of Piraeus - Department of Technology Education and Digital Systems
80 Karaoli & Dimitriou str., Piraeus
18534 Greece
mariam@unipi.gr

*Maria Virvou*

University of Piraeus - Department of Informatics
80 Karaoli & Dimitriou str., Piraeus
18534 Greece
mvirvou@unipi.gr

## Abstract

Authoring tools for Intelligent Tutoring Systems (ITSs) provide environments that can be used by authors-instructors who are not necessarily computer experts to easily develop cost-effective ITSs. However, the quality of the ITSs to be generated depends heavily on these authors-instructors who will provide the content for the courses. This means that authoring tools should be carefully designed, developed and evaluated in order to ensure that they are usable, friendly and effective. This paper reports on a study we conducted in order to evaluate the usability of WEAR, which is a Web-based ITS authoring tool. The results of the study were very encouraging. Authors were quite satisfied with the functionality of the system, they found it very friendly and easy to use and most importantly they stated that it could be really useful.

## 1   Introduction

The main goal of Intelligent Tutoring Systems (ITSs) is to reproduce the behaviour of a human tutor who can adapt his/her teaching to the needs and knowledge of the individual learner. As a number of successful evaluations of ITSs have shown, these systems can be educationally effective compared to traditional instruction either by reducing the amount of time it takes students to reach a particular level of achievement or by improving the achievement levels given the same time on task (Du Boulay, 2000). A common criticism for ITSs concerns the complex and time-consuming task of their construction: more than 200 hours are required to develop an ITS to cover just one hour of instruction (Woolf & Cunningham, 1987).

Authoring tools for ITSs are meant to simplify the ITS construction by providing environments that can be used by a wide range of people to easily develop cost-effective ITSs. However, it would be extremely optimistic to think that authoring tools are simply the perfect solution for the creation of ITSs. In particular, the fact that authoring tools depend heavily on the instructors for the quality of the ITSs to be generated, links these systems with various problems. For example, instructors may face several difficulties during the design process and they may become frustrated or they may provide inconsistent information to the tool that may lead to the generation of ITSs with problematic behaviour. This means that authoring tools should be carefully designed, developed and evaluated in order to ensure that they are usable, friendly and effective. The usability of authoring systems will promote the production of ITSs and improve their quality because it will increase the instructors' degree of acceptance towards them and decrease the possibility of errors that may lead to the generation of non-effective ITSs.

This paper reports on a study we conducted in order to evaluate the usability of a Web-based ITS authoring tool that is called WEAR.

## 2   The system

WEAR is a Web-based authoring tool for ITSs focusing on problem construction and problem solving in Algebra-related domains (Virvou & Moundridou, 2000). However, WEAR also offers authors the facility of building adaptive textbooks in every domain even if it is not Algebra-related (Moundridou & Virvou, 2001). These textbooks offer navigation support to students, adapted to their individual needs and knowledge. In this section we will briefly describe only the way that the authoring of the adaptive textbooks is performed and not the rest of the system's operation, since the usability evaluation study that will be reported concerns this part of WEAR's functionality.

Although most of the existing authoring tools for adaptive educational textbooks approach the adaptivity issue in quite similar ways, they differ a lot in the authoring process they impose to their users (authors). For example, in Interbook (Brusilovsky, Eklund & Schwarz, 1998) the author should provide a specially structured, annotated MS-Word file. In AHA (De Bra & Calvi, 1998) the author should write annotated HTML files. MetaLinks (Murray, Shen, Piemonte, Condit & Thibedeau, 2000) on the other hand, provides a GUI interface for authoring all aspects of the electronic textbook.

In WEAR, we address authoring in a way that in its first steps resembles the simple one adopted by commercial tools like WebCT (Goldberg, Salari & Swoboda, 1996). It should be noted that this similarity between WEAR and WebCT concerns only the authoring procedure and not the resulting courses, which in the case of WebCT are not adaptive. In particular, the authoring procedure is the following: The author should prepare HTML files for the topics that would be contained in the electronic textbook. The next step is to use WEAR's facilities for uploading these files to the WEAR server. For each uploaded file the author must also specify a title, a difficulty level and the position that it should have in the topics hierarchy. Finally, the author must edit the is_prerequisite_of relationships between topics. To perform this, the author is presented with the hierarchy of topics and s/he should write next to each topic the section numbers of its prerequisite topics.

The author may also create multiple choice tests or problems and associate them with the appropriate topics. The procedure is again quite simple: the author should fill in forms stating the question, the possible answers and the right answer. Then s/he should specify the difficulty level of the exercise as well as the topic that this exercise refers to.

## 3   The study

Five instructors-authors were asked to work with the system to build part of an adaptive textbook for a Software Engineering course. Before working with the system the authoring and learning environments of WEAR were introduced to the participants. Authors were provided with 7 HTML files. Each of these files contained a section from the Software Engineering book chapter "Requirements specification".

The authors were requested to use WEAR's facilities to:
- upload these files to the server,
- create the topics' hierarchy,
- edit the prerequisite relationships between topics,
- insert students to their virtual class, and
- construct some multiple-choice questions.

This book chapter as well as the multiple-choice questions were also given to the participants in a hard copy so that they could easily explore the available teaching material before working with the system. The time spent by the authors to perform the requested tasks was recorded. Finally, each author was interviewed in order to gather information concerning the usability of the system.

# 4 The results

Concerning the time that was needed by the authors to build their course, the result was very impressive: all the authors finished their tasks in 15-20 minutes. This is considered as a very good result since the teaching material covered corresponds to more than 2 hours of classroom instruction. Furthermore, all of the authors stated that they found extremely useful to use a tool like WEAR to produce adaptive Web-based textbooks.

The answers that the five authors gave when they were interviewed were measured in the scale from 1 to 5. The mean ratings are summarised in Table 1 and indicate a high degree of usability for the system.

In particular, concerning the participants' general impression of the system, this was recorded as very good. This was also the case for the satisfaction that users felt when working with the system. The participants also stated that the system was rather motivating and easy to use.

As for the system's characteristics the authors were very pleased with the fact that the system was able to recover from errors without losing the already entered data. They were also very satisfied with the system's response time, as well as with its reliability. Users stated that the system does not inform them adequately about their errors (mean rating: 2.8) but the validity of this finding should be investigated since most of the participants did not make many errors.

Authors were quite satisfied with the amount of explanation that the system provided as well as with the clarity of the system's instruction messages. The system was also considered consistent as for the terminology it uses and the feedback messages it provides. Users also stated that the results of their actions were rather predictable.

Concerning how easy will be for a user to learn the system, the authors stated that this will be very easy and it will take minimum time. Finally, the number of actions that should be carried out in order to accomplish a task was considered by the participants as adequate and their sequence as logical.

From the participants' general comments on the system we came up with a useful finding: most of them even if they liked the system a lot, they would prefer to perform some tasks in a more visual way. For example, when authors arrange the topics in the topics' hierarchy they would prefer to do this by dragging and dropping rather than by altering the topic's position number in a text box.

**Table 1:** Mean ratings in questions measuring the system's usability

| Questions | Mean rating (n=5) |
|---|---|
| What is your general impression of the system? (1: I disliked it... 5: I was very impressed) | 4.2 |
| How satisfied you felt when working with the system? (1: it was annoying... 5: I was very satisfied) | 4.0 |
| Was the system motivating? (1: it was boring... 5: it was very motivating) | 4.0 |
| Was the system easy to use? (1: very difficult... 5: very easy) | 4.6 |
| Was the system able to recover from errors without losing the already entered data? (1: unable... 5: very able) | 0.0 |
| How fast was the system in its responses? (1: very slow... 5: adequately fast) | 4.2 |
| Was the system reliable? (1: unreliable... 5: very reliable) | 4.0 |
| Were the system's error messages informative? (1: never... 5: very often) | 2.8 |
| Was the amount of explanation that the system provided adequate? (1: inadequate... 5: adequate) | 3.8 |
| Were the system's instruction messages clear enough? (1: they were confusing... 5: they were very clear) | 4.2 |
| How consistent did you find the terminology that the system was using? (1: inconsistent... 5: consistent) | 5.0 |
| How consistent did you find the system's feedback messages? (1: inconsistent... 5: consistent) | 4.6 |
| Could you predict the results of your actions? (1: never... 5: always) | 4.4 |
| How easy can it be to learn using the system? (1: very difficult... 5: very easy) | 4.8 |
| How long does it take to learn using the system? (1: a long time... 5: minimum time) | 4.2 |
| How did you find the number of actions that were needed to accomplish a task? (1: large... 5: adequate) | 4.0 |
| Was the sequence of actions to accomplish a task logical? (1: never... 5: always) | 4.2 |

# 5   Conclusions/Discussion

The study that was conducted to evaluate the usability of the authoring environment of WEAR provided very encouraging results. Authors were quite satisfied with the functionality of the system, they found it very friendly and easy to use and most importantly they stated that it could be really useful.

However, it should be taken into account that the subjects participating in the study were computer scientists. On the one hand, this strengthens the positive results because it means that these subjects would probably have very high expectations as compared to instructors of other domains.

On the other hand, it was easier for these authors to work with the software and so the opinion of the novice computer user is not reflected in the results of the study.

It is within our future plans to run more experiments in order to confirm the positive results of the discussed study. In these subsequent evaluations, we plan to involve instructors of domains that are not computer-related. In this way, the possible problems that the non-competent computer user may face when working with WEAR will be revealed.

Finally, based on the results of this evaluation study and those to follow, we plan to redesign part of WEAR's interface in the direction that the users' opinions point. However, the results of the current study were very much in favour of the system and the only implication on redesigning the authoring environment of WEAR concerned the participants' preference to perform some of the tasks in a more visual way. Until the forthcoming evaluations, this issue is what we will be working on.

# References

Brusilovsky, P., Eklund, J., & Schwarz, E. (1998). Web-based education for all: A tool for developing adaptive courseware. *Computer Networks and ISDN Systems*, 30, 291-300.

De Bra, P., & Calvi, L. (1998). AHA: a Generic Adaptive Hypermedia System. In P. Brusilovsky, & P. De Bra (Eds.), *Proceedings of 2nd Adaptive Hypertext and Hypermedia Workshop at the 9th ACM International Hypertext Conference - Hypertext'98*, Computing Science Report No. 98-12, (pp. 5-11). Eindhoven: Eindhoven University of Technology

Du Boulay, B. (2000). Can We Learn from ITSs? In G. Gauthier, C. Frasson, & K. VanLehn (Eds.), *Intelligent Tutoring Systems, Proceedings of the 5th International Conference on Intelligent Tutoring Systems*, Lecture Notes in Computer Science No. 1839, (pp. 9-17). Berlin: Springer

Goldberg, M. W., Salari, S., & Swoboda, P. (1996). World Wide Web - Course Tool: An environment for building www-based courses. *Computer Networks and ISDN Systems*, 28, 1219-1231.

Moundridou, M., & Virvou, M. (2001). Authoring and delivering adaptive Web-based textbooks using WEAR. In T. Okamoto, R. Hartley, Kinshuk, & J. P. Klus (Eds.), *IEEE International Conference on Advanced Learning Technologies; Issues, Achievements, and Challenges - ICALT 2001*, (pp. 185-188). Los Alamitos, California: IEEE Computer Society

Murray, T., Shen, T., Piemonte, J., Condit, C., & Thibedeau, J. (2000). Adaptivity in the MetaLinks Hyper-Book Authoring Framework. In C. Peylo (Ed.), *Proceedings of the International Workshop on Adaptive and Intelligent Web-based Educational Systems* (held in Conjunction with ITS 2000), Technical Report, (pp. 61-72). Osnabrück: Institute for Semantic Information Processing

Virvou, M., & Moundridou, M. (2000). A Web-Based Authoring Tool for Algebra-Related Intelligent Tutoring Systems. *Educational Technology & Society*, 3 (2), 61-70.

Woolf, B. P., & Cunningham, P. A. (1987). Multiple knowledge sources in intelligent teaching systems. *IEEE Expert*, 2, 41-54.

# WebTracer: Evaluating Web Usability with Browsing History and Eye Movement

*Noboru Nakamichi[†], Makoto Sakai[‡], Jian Hu[†], Kazuyuki Shima[†], Masahide Nakamura[†], Ken'ichi Matsumoto[†]*

[†] Graduate School of Information Science, Nara Institute of Science and Technology

8916-5 Takayama Ikoma, Nara, 630-0101, JAPAN

{noboru-n, jian-hu, shima, masa-n, matumoto} @is.aist-nara.ac.jp

[‡] SRA Key Technology Laboratory, Inc.

Marusho Bldg, 3F, 3-12, Yotsuya, Shinjuku-ku, Tokyo, 160-0004, JAPAN

sakai@sra.co.jp

## Abstract

This paper describes a WWW site evaluation tool, WebTracer, which can record user's gazing points, a user's operational data, and the screen image of browsed pages. In addition, the WebTracer can replay a user's browsing operations. In an evaluation experiment, we record without interruption, a user's browsing operations using WebTracer. The average execution time per task in the experiment was 2 minutes and 48 seconds. We interviewed users by applying the usability-testing-support function based on a replay and summary of the WebTracer. The average time required for an interview was 19 minutes. 16 comments on the average were obtained for the execution time. The experimental results show that the various summarized user-operational data were helpful for getting more comments about web usability from the subjects.

## 1    Introduction

Designing attractive Web sites is a crucial problem in business, since Web sites directly reflect the images and sales of companies (Goto & Cotler, 2002). Therefore, usability evaluation for web pages is now an important concern in finding flaws and shortcomings in the pages with respect to usability (Jakob, 1993).

Web usability testing is a popular way to conduct usability evaluation. Web usability testing requires subjects (users) to browse a target web site, and then evaluators get feedback from the users based on an interview. Usability testing has been widely studied and various methods have been proposed to date. However, most conventional methods must occasionally (or periodically) interrupt the user's browsing operations, to get opinions within a certain period of time on the web pages browsed. This discontinuous browsing creates difficulties in evaluating the usability of a "whole" Web site, which consists of a number of pages.

To achieve effective continuous testing, we need to first "record" how users browse the entire site, and then we need to perform the interview by "replaying" the recorded data. Finally, we need to justify the feedback by "analysis" of the recorded data.

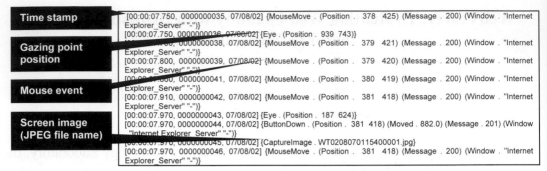

**Figure 1:** Example of data collected by the WebTracer

However, no tool currently exists to allow the cooperation of the above tasks effectively, although each task requires sophisticated methodologies. Therefore, conventional methods had to utilize much less data than was actually performed by the users.

In this research, we have developed an integrated usability evaluation tool, called WebTracer, which can record user operational data without discontinuation of user operational data and can be replayed. In addition, we demonstrated WebTracer's effectiveness through an experimental evaluation.

## 2 WebTracer

WebTracer is an integrated environment for web usability testing. It can record a user's browsing operations, replay the user's recorded browsing history, and provide analysis tools which can depict graphs and calculate statistical equations. WebTracer is optimized especially in the following two features.

### 2.1 Recording web operation

WebTracer records the various user operational data needed for replay and analysis. Specifically, WebTracer records user's gazing points via the camera eye, mouse movements and clicks, keyboard inputs, and the screen image of the browsed pages. An example of data collected by WebTracer is shown in Figure 1. Unless the appearance of the browsed page changes, WebTracer does not record browsed screen image. The image is captured only when a transition of the browsed page is triggered by a user's events (e.g., mouse click to follow the next links). Thus, the size of the recorded image can be significantly reduced to 1/10 to 1/20 of the size of recorded data when compared with data recorded in an Mpeg-2/4 format.

### 2.2 Usability testing support based on replay and summary

WebTracer can support usability testing by using a replay of the user's operations, summarized data, and graphs derived from the recorded data. By using the summarized data, we can capture the characteristics and statistics of each page, which helps with the analysis of a Web site. Recorded data are summarized in the form of a table for every page, as is shown in Figure 2. The data can also be shown in graph form. An example of an eye movement statistics graph is shown in Figure 3. In addition, an example of the replay screen with the eyemark of the user (the user's gazing point) is shown in Figure 4. The replay feature reproduces operations, such as the eyemark

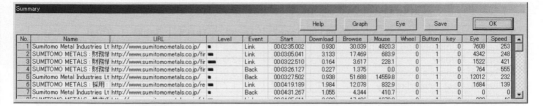

| No. | Name | URL | Level | Event | Start | Download | Browse | Mouse | Wheel | Button | key | Eye | Speed |
|---|---|---|---|---|---|---|---|---|---|---|---|---|---|
| 1 | Sumitomo Metal Industries Lt | http://www.sumitomometals.co.jp/ | ■ | Link | 00:02:35.002 | 0.930 | 30.039 | 4920.3 | 0 | 1 | 0 | 7608 | 253 |
| 2 | SUMITOMO METALS : 財務情 | http://www.sumitomometals.co.jp/fir | ■ | Link | 00:03:05.041 | 3.133 | 17.469 | 683.9 | 0 | 1 | 0 | 4342 | 248 |
| 3 | SUMITOMO METALS : 財務情 | http://www.sumitomometals.co.jp/fir | ■■ | Link | 00:03:22.510 | 0.164 | 3.617 | 228.1 | 0 | 1 | 0 | 1522 | 421 |
| 4 | SUMITOMO METALS : 財務情 | http://www.sumitomometals.co.jp/fir | ■ | Back | 00:03:26.127 | 0.227 | 1.375 | 0.0 | 0 | 1 | 0 | 764 | 555 |
| 5 | Sumitomo Metal Industries Lt | http://www.sumitomometals.co.jp/ | ■ | Back | 00:03:27.502 | 0.938 | 51.688 | 14559.8 | 0 | 1 | 0 | 12012 | 232 |
| 6 | SUMITOMO METALS : 採用 : | http://www.sumitomometals.co.jp/re | ■ | Link | 00:04:19.189 | 1.984 | 12.078 | 832.9 | 0 | 1 | 0 | 1684 | 139 |
| 7 | Sumitomo Metal Industries Lt | http://www.sumitomometals.co.jp/ | ■ | Back | 00:04:31.267 | 1.055 | 4.344 | 410.7 | 0 | 1 | 0 | 0 | 0 |

**Figure 2:** Example of a summary (summarized browsing history)

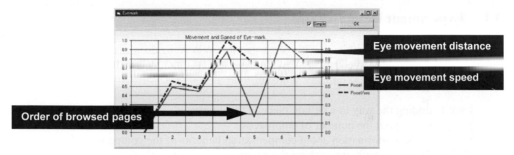

**Figure 3:** Example of eye movement statistics graph

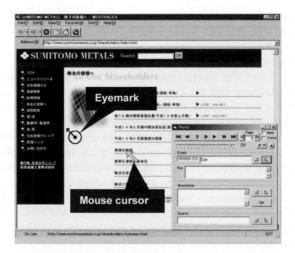

**Figure 4:** Example of replay screen with eyemark

and mouse cursors, operations performed when the page is being browsed. In another window, WebTracer can display other events, such as a keystroke. Moreover, at any time during the recording, we can insert annotations and replay these annotations later.

## 3 Experimental Evaluations

We have conducted an experiment, to evaluate the effectiveness of WebTracer in a Web usability evaluation. In the experiment, we asked the three subjects to find objective information within a

| | Task1 | Task2 | Task3 | Task4 | Task5 | Total | (%) |
|---|---|---|---|---|---|---|---|
| Phase1: Summary and Graph | 0.7 | 2.3 | 2.3 | 1.7 | 2.3 | 9.3 | 11.6 |
| Phase2: Replay | 7.7 | 13.7 | 12.0 | 8.0 | 18.7 | 60.0 | 74.4 |
| Phase3: Subject's memory | 2.3 | 0.0 | 0.0 | 1.3 | 0.3 | 4.0 | 4.9 |
| Phase4: Fast forward replay | 0.7 | 0.7 | 1.7 | 1.0 | 3.3 | 7.3 | 9.1 |

company Website. Then, with the support of WebTracer, the subjects were interviewed and their comments were recorded.

## 3.1 Experiment Design

First, we asked the three subjects who often use the WWW to do a task, and we recorded the operational data using the WebTracer. Tasks for the subjects included gathering the following five pieces of information from a company Website.

Task 1: finding the way to a certain place in the company
Task 2: finding out the number of employees
Task 3: finding information about compensation benefits of the company
Task 4: finding specific news about the company
Task 5: finding a technical method of construction

Secondly, we interviewed each subject about the ease of using an object Website after the end of a task. The interview is divided into four phases, and the outline of each phase is as follows.

Phase 1: An interview based on statistics obtained from recorded data.
WebTracer shows the summarized browsing history. We interviewed each subject based on the summary and the eye movement graph as shown in Fig. 2. We had the subject point out the difficulty of use in focusing on a web page with both the large moving distance and moving speed of the subject's gazing point.
Phase 2: The interview based on the replay screen in which the eyemark was placed.
We had the subject point out the difficulty of use with the replay screen as shown in Fig. 3 regarding the web page from which the comment was obtained by phase 1.
Phase 3: The interview based on the memory of the subject.
We conducted an interview based on the memory of the subject for whom the whole task was performed. Comments which were obtained and which overlapped by phase 2 are not included in the number of comments obtained by this phase.
Phase 4: The interview based on the fast forward replay screen.
We had the subject point out the difficulty of use with the fast-forwarding replay screen. Comments which were obtained and which overlapped by phase 3 are not included in the number of comments obtained by this phase.

## 3.2 Results and Discussion

The average execution time per task in the experiment was 2 minutes and 48 seconds. Also, the average time of the interview was 19 minutes. Therefore, the total time spent for entire process was 21 minutes and 48 seconds.
The average number of comments given by the three subjects is summarized in Table 1. For execution time, we obtained 16 comments on average. The comments include usability problems,

thinking during operation. We assume that the unit of a comment is every sentence uttered by a subject. The number of comments becomes material data of the usability problem for evaluator judges.

As shown in Table 1, about 95% of the entire comments are obtained from the graph and replay (Phase 1, 2 and 4) using the WebTracer. Moreover, about 74% of the entire comments are obtained from Phase2 by replaying of WebTracer. We can see that the replay function of the user operation which added the eyemark in WebTracer is effective from the experiment results. Subjects can remember their operations, can point out the points, which is hard to use, and a evaluator can ask an interview more easily by replaying of WebTracer.

We believe that the WebTracer made the phases of this evaluation experiment possible. For example, we consider the case where "Usability laboratory" carries out the interview phases. The "Usability laboratory" can be applied to Phase2 and Phase4, if the evaluator records the subjects' operations with a video camera. However, by replay of the video camera, we still cannot know a user's gazing point. Therefore, we assume that the number of comments obtained from subjects decreases compared with the WebTracer. Phase3 is also made possible by performing a questionnaire-based evaluation at the end of the task. However, "Usability laboratory" cannot be used for Phase1, since no quantitative result is available. "Usability laboratory" cannot record a detailed user's operational data and show the summarized data. Consequently, 9.3 comments in Table 1 would not be available. As a result, the "Usability laboratory" would miss 11.6% of the entire comments shown in Table 1.

# 4    Conclusions

In this research, we have presented an integrated usability evaluation tool, WebTracer, and have also conducted an experimental evaluation. WebTracer allows evaluators to perform an efficient Web usability evaluation with optimized features, i.e., with recording operations, replay, summaries and graphs. Web Tracer records the user's gazing point as well as the user's operations, and replays the user's operation in the same screen. From the user's operations performed in each page, we were able to capture the characteristics of the page. As a result of the evaluation experiment, we have obtained many useful comments with respect to usability.

Our future work includes refining the evaluation procedure, as well as comparing the proposed method with other evaluation methods.

If WebTracer spreads widely, research of Web usability will become less difficult. Moreover, a usability evaluation in an actual development also becomes easier. Furthermore, we expect that software which is easier to use will increases.

# References

Etgan, M. & Cantor, J. (1999). What does getting WET (Web Event-logging Tool) mean for web usability?, HFWEB99.

Goto, K. & Cotler, E. (2002). Web ReDesign. Peason Education.

Nielsen, J. (1993). Usability Engineering. Academic Press.

Okada, H., & Asahi, T. (1999). GUI TESTER:  log-based usability testing tool for graphical user interfaces. IEICE Trans. on Information And Systems, Vol.E82-D, No.6, pp.1030-1041.

Paganelli, L. (2002). Intelligent analysis of user interactions with web applications. IUI'02.

Torii, K., Matsumoto, K., Nakakoji, K., Takada, Y., Takada, S., & Shima, K. (1999). Ginger2: An environment for computer-aided empirical software engineering. IEEE Trans. on Soft. Eng., Vol.25, No.4, pp.472-492.

# IOWA
## Intuitive-use Oriented Webtool for the creation of Adapted contents

*Sergio Ocio, Mª del Puerto Paule*

HCI Group- Labtoo. Dpto. Computer
Science of University of Oviedo-Spain
Facultad de Ciencias
C/Calvo Sotelo S/N 33007 Oviedo
paule@pinon.ccu.uniovi.es,
djrekcv@terra.es

*Martín González , Juan Ramón Pérez*

HCI Group- Labtoo. Dpto. Computer
Science of University of Oviedo-Spain
Facultad de Ciencias
C/Calvo Sotelo S/N 33007 Oviedo
martin@lsi.uniovi.es,
jrpp@pinon.ccu.uniovi.es

## Abstract

Nowadays, e-learning is a very interesting option to the teaching community. However, actual systems force teachers and students to adapt their ways to work to what the system demands. There are several e-learning systems in the market today: the learning environments. These systems provide a whole set of tools to set up courses to be studied through a web interface. However, they are only a first step: they provide online courses, but students must adapt their way to learn to what the system offers them. Also, teachers must create new contents just how the system forces them to. In this paper, we present a system called IOWA which automates the adaptation process to teachers, so, IOWA guides teacher when inserting contents. Contents are being inserted in an adapted way, so when students visualize them they will see them adapted to their learning style. Finally, we include an evaluation of our system.

## 1  Introduction

Usually, studying something is a hard task, but there are some tricks each one uses to make it easier. As we are studying and studying, we are shaping our learning style, this set of tricks and paths we follow to accomplish the task the best way we can.

A step to know how students process  information is to get a classification of them. A classification widely known is CHAEA's one (Alonso, Gallego & Honey, 1999) . It offers an acceptable reliability and validity proved to be applied to Spanish Universities. This classification determinates the learning style with a test.. This test has 80 questions. Perhaps, 80 questions are a lot of questions, but it is thought to be done in 15 minutes, and the answers aren't difficult .

The test returns four learning styles: Theoretician, Active, Reflexive and Pragmatic.

Actually, we are doing adaptation to the  Theoretician and Active learning styles. This is because the adaptation to these two learning styles is the base for Reflexive and  Pragmatic ones.

### 1.1  How can Learning Styles be applied to e-Learning?

Actual distance learning systems have some contents and then show them to students, independently of who is the user.

We can have different contents prepared for each learning style and show the appropriate content depending on the type the user belongs to.

Once the style is identified, the system will choose which contents are appropriated to this user and they will be shown according to what the teacher has said in creation-time.

The process to get the adaptation working is divided in three steps: first, the system administrator adds some courses to the system; then, teachers create the adapted contents; finally, these contents are shown to students.

## 2 Creating new contents

Learning environments always provide a tool to add new contents to their courses. Some of these tools are very powerful. However, as these systems do not do adaptation, they are mere "editors". IOWA is a full environment which not only has the functionality of these editors, but it establishes the foundations to the entire adaptation framework provided by Feijoo.net.

### 2.1 Types of content

In Feijoo.net, adapted information is composed by two types of contents: exercises, examples, etc. and textual contents

Exercises are the main way to provide "practical" contents to users, which can serve to help users to fix new knowledge better in their memory, while text is the "theory".

#### 2.1.1 Textual contents

Basically they are composed by written contents. When we designed Feijoo.net's knowledge base, we decided that there could be contents that were common to the four learning styles and that there would be style-specific contents.

When a teacher wants to create new textual contents, he/she insert them telling which of them are common and which are specific; this last case will be also have another component: the learning style which must be specified.

This kind of contents is where all the power of adaptation resides. The results obtained from our tests point that providing different contents depending on the learning style works. (See Section Evaluation)

#### 2.1.2 Exercises

In a prior time, we worked with adapted exercises, following the same philosophy we used with textual contents: we had different exercises depending on the learning style.

However, we found that, although adapted written contents works very well, exercises do not. Results pointed that when something is learned well, this knowledge can be demonstrated always, independently of which learning style the question was focused to, i.e. a student can answer any question belonging to other styles as well as he/she answers questions proposed to his own learning style. (See Evaluation Section)

## 3 System's Architecture

We offer teachers a complete webtool for create and manage contents for their online courses. Although they don't have to worry about how adaptation works, the tool is very powerful, letting them to personalize almost anything they want: they can insert new contents, update or delete existing ones, describe the way the want their contents to be shown, define the navigation model that will be used when contents are shown, i.e. how to go through the tree of contents, and even define related topics for each content they create.

All of these operations are made through a simple web interface in which usability is considered as the most important factor, because teacher's work must be simple, we cannot demand more effort at the time of creating these contents, because then all the system would be somehow useless.

We use the pattern of the Model-View-Controller as a base for our system.

In our case, the model is formed by the system's core, and an error manager. It is the main part of our solution. It is composed by five modules: the Learning Style Selector, the Creation of Contents module, the Organization of Contents module, the Templates of Visualization module and the Help Subsystem.

The view is a set of JSP and the controller is a servlet which dispatches requests to the appropriate handler from a set of them that are used by IOWA.

We can see our system's architecture in figure 1.

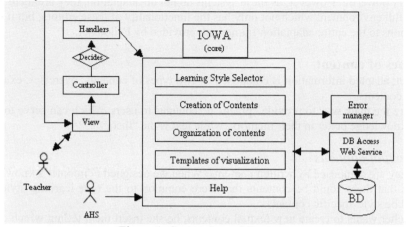

Figure 1: **System's architecture**

Two important figures in the system are the AHS and the DB Access Web Service.

The AHS is a system actor (thus, it is represented with that icon). It controls the adaptation inside IOWA. This adaptation is, at least by the moment, not related to learning styles, but to make system's use more comfortable. Depending on some parameters, as the learning style selected to work with, the number of times a page have been visited... the AHS decides what should be shown, guides the teacher through each task

The DB Access Web Service is what we use to access our database. Built on the .NET Framework, it could process any SQL statement we wanted and it returns the data requested in a XML format. This way, all the data we use from the database can be easily processed and, at the same time, we have a centralized access point to the knowledge base.

The name of each module belonging to the system's core is self-explicative so we are not going to explain what the functionality of each part is.

However, the organization of contents is almost the most important module in all the systems, so it will be explained a little bit deepen now.

## 3.1 Organization of contents Module

This module is subdivided in two different parts. First one allows teachers to see the tree of contents of their didactic units. This tree shows links between nodes, exercises related to a node...

However it is the second part of the module, the context subsystem what is really important.

Context is a concept that appeared in the OOHDM methodology (Scwhabe, Rossi, 2003). This concept allows teachers to define which path should their students follow to go through the tree of contents. It is possible to define different paths for the same three, one path per learning style. To build these trees, we use a graph editor. Teachers simply drag nodes from the tree of contents and drop them into the graph editor. Then, these nodes are linked using arrows to define the graph. Through this simple procedure, we get one of the adaptation techniques proposed by Paul de Bra and Peter Brusilovsky, which is the "conditional inclusion of fragments" (De Bra, Brusilovsy, Houben, 1999).

When contents are going to be visualized, the web page showing them will be prepared "on the fly". Looking for these "context" information, some links will be added to each web page: they could be only two links (forward and back), or they could be a set of links, when the current node has more than one edge departing from it.

Defining paths, teachers could have real freedom to configure what is exactly the way the wanted their course to be followed, just as they would do if it was a traditional course and not an e-learning one.

## 4 Evaluation

In this experiment a Course about HTML was designed .This course is very basic and it had five lessons: First Page with HTML, Headings, Paragraphs, Design with Style, List and Links.

The experiment has to prove the following hypothesis:

1.  The learning with Adaptation Model is more effective then the learning without Adaptation Model.
2.  The evaluation of the knowledge does not depend on the learning style.

The population was students of the subject Information Systems of the Business School of University of Oviedo.

54 students formed the sample. 27 of them were theoretician and 27 of them were active. All of them were the first time that they were registered in the subject and they don' t know anything about HTML.

We formed three groups: A Control group, experimental group and the third group called Not_Adapted_Test Group. These group were homogeneous in their composition, so, there were nine theoretician and nine active students in each group. The distribution of student was random in each group.

We designed two types of course:

1.  An adapted course with the proposed Adaptation Model
2.  A course without adaptation.

Too, we designed two types of tests:

1.  Test with Adapted questions to learning styles
2.  Test with questions without specific adaptation, so, there were questions of both learning styles.

The experimental sessions were developed in the Business School of University of Oviedo. Each session was 110 minutes length. These sessions had four parts:

1.  Attitude Test[1].
2.  Chaea's Test.
3.  Navigating on the Course
4.  Acquired knowledge test.

Parts 1 and 2 were equal for all groups, but parts 3 and 4 were different depending on the group. Control group students had to navigate on the course without adaptation and they had to do a test adapted to the learning style. Experimental group students had to navigate on the adapted course and they had to do a test adapted to the learning style. Not_Adapted_test group had to navigate on the adapted course and they had to do a test without adaptation.

The students of theses groups were different. The test was formed by 15 questions and it was the multiple answer tests. At the end of it, the students knew the score of it.

The test had 3 questions of each lesson of the course. Each correct answer was 1 point and the incorrect answer was –1 point and the questions without answer was 0 point.

---

[1] This test was made in base of Likert Scale with numeric answers where 1 represents the most disagreement and 5 represents the most agreement.

821

At the beginning of the session the student was told that the session was part of teaching quality evaluation of the University of Oviedo, so the student was not conditioned by the test. With this experiment, we obtained information about the attitude and what knowledge the student acquired after the navigation on the course. With this information, we can determine if the adaptation has influence on the learning and if the way the evaluation depends on learning style.

We used the statistical software SPSS to get this analysis.

Due space problems, the results of this analysis will be shown in the presentation of this work. However, we are going to discuss these results now.

First, we have to determine if the groups were similar on attitude. We made a TStudent to prove this. Second, we checked if the results of control group were better than the results of the experimental group. Third, we had to compare the results of the test of experimental group (adapted test) and the Not_Adapted_test group (test with a merge or theoretician and active questions).

In the first part of this analysis, we can see that all groups (Control Group, Experimental Group and Not_Adapted_Test group) have a normal distribution for each learning style. Too, the attitude has a normal distribution in Control Group and Experimental Group.

In the second part of this analysis, we can see that the control Group and the Experimental Group has a homogeneous variance and too, the Experimental Group and the Not_Adapted_Group has a homogeneous variance. The Attitude variance is homogeneous in the Control Group and Experimental Group. So, we can apply the TStudent to get the differences and the improvements.

The application of TStudent in the Control Group and Experimental Group returned an improvement, so, the scores obtained in the Experimental Group are lightly better than the scores obtained in the Control Group. This result proves the first hypothesis: There is an increment of learning with the adaptation.

The application of Tstudent in the Experimental Group and Not_Adapted_test returned that the different is not significant, so, the evaluation is not important, the most important is the learning. When a student learns, she/he can answer any type of answer. The evaluation is independent of learning style.

## 5   Conclusions and new features

Nowadays, although there are very good e-learning systems available, there is no a real adaptive system. This is a great disadvantage comparing these systems to our Feijoo.net. Using adaptation, in our case, to the learning styles using CHAEA as our cognitive model, we offer a tool with far more possibilities than what was offered till today.

Using adaptation is really worth, as our studies have shown, because there is learning and this learning is easier to students, because they see want they want to. This makes the use of e-learning a real alternative to traditional teaching and not only as a strengthening to that kind of teaching. We can now offer real solutions to students which want to study when they want and where they want but, at the same time, they seek something personalized and not only an e-book but a place to learn.

## 6   References

Alonso, C.M.; Gallego, D.J.; Honey, J.: *Cuestionario Honey-Alonso de Estilos de Aprendizaje (CHAEA)*. 2001, http://www.ice.deusto.es/guia/test0.htm.

Rossi, G.; Schwabe, D.: *"The Object-Oriented Hypermedia Design Model (OOHDM)"*. http://www-lifia.info.unlp.edu.ar/~gustavo/oohdm.html.

Brusilovsly, P, De Bra P., Houben, G. "Adaptive Hypermedia: From Systems to Framework". ACM Computing Surveys, Vol.31, Number 4es, December 1999

# From Web Usability to Web Comfortability: A Paradigm Shift

*Roberto Okada and Yuri Watanabe*

School of Project Design – Miyagi University
1 Gakuen Taiwa-cho Kurokawa-gun Miyagi 981-3298, Japan
okir@myu.ac.jp

## Abstract

Information Technology is becoming ubiquitous. Our lifestyle is changing due to this digital revolution. Many researchers had been working on exploring the meaning and ways to realize the comfortability in physical sites. With the appearance of huge amount of Web sites, which provide a variety of services like electronic commerce, network game, search engines and so on, there are many researchers working on the definition of Web Usability. In this research, we go one step forward, by exploring the meaning of Web Comfortability, based on Kansei Engineering methods.

## 1 Introduction

With the appearance of huge amount of Web sites, which provide a variety of services like electronic commerce, network game, search engines and so on, there are many researchers working on the definition of Web Usability. As a result, there are rule books and suggestions about building "Usable Web Sites" (Nielsen, 2000).

Many researchers had been working on exploring the meaning and ways to realize the comfortability in physical sites, in planes as well as in spaces. We believe that it is important to study the meaning of comfortability in Web sites as well, where this new paradigm of Web Comfortability goes beyond the concept of Web Usability. We define Web Comfortability as "the means to provide not only useful functions, but also the feeling of pleasantness in the virtual cyberspace", and Kansei Engineering takes into account human's feelings in order to create products. In the same way that we visit more frequently and stay longer in physical places where we feel "comfortable", we believe that applying the same feeling in virtual Web sites will be a key factor to become winner in the battle of Web sites to retain users.

The way of interacting with Web sites is based on two senses: audio and visual. Thus, it is very important to pay attention to the visual aspect, i.e. the design. In the case of design elements like layouts and combination of colors, the user can get at first sight the feeling of pleasant or unpleasant. If he feels unpleasant or uncomfortable, he might go immediately to other sites.

## 2 Kansei Engineering Approach

Kansei Engineering is a consumer-oriented technology process used to develop products. It uses the consumer's "feelings" or Kansei as a guideline in creating the product. In order to do Kansei Engineering, one must first determine the Kansei words suitable for the product to be designed. Designers would then create different concepts out of these words. After creating the concepts, they would then be presented to the consumers and rated with the same Kansei words gathered before to determine if the product has matched the Kansei. The rating test contains scales of 1 to 5 with antonym Kansei words on both ends. The process may continue to cycle until the people involved in the development are satisfied with the results (Quality Portal, 2002).

The Kansei Engineering steps in order to explore the meaning of comfortable in Web sites are: (1) Collection and arrangement of target Web sites, (2) Collection and arrangement of Kansei words (as adjectives) which are able to express the image of target Web sites, (3) Sensitive evaluation of the target Web sites, based on Semantic Differential method with the selected adjectives, (4) Analysis, such as categorical regression analysis, correlation analysis, etc.

Semantic Differential (Osgood, Suci & Tanenbaum, 1957) -or SD for short- uses a rating scale, e.g. a 5-scale with -2,-1,0,+1 and +2. Users rate bipolar "adjective pairs", e.g. good-bad, heavy-light, according to this scale. This is a technique for obtaining meaning space of man's feelings, by giving numerical values to bipolar or opposite adjectives.

# 3 Investigation Process

## 3.1 Extending the meaning of comfortable from physical sites to web sites

In the same way that it is important the visual aspects of color and structure in planes and spaces, the same is valid for virtual spaces. Thus, we will investigate with focus on the influence of Web sites layouts, based on Kansei Engineering approach, i.e. sensibility measurements, such as the Semantic Differential.

Most of the adjectives were obtained from brain-storming sessions about comfortability in Web sites, with words expressing sensibility and emotions. Some of the adjectives selected to express the Comfortability are: Uniform, Simple, Calm, Cute and Beautiful.

## 3.2 Elements for the study

For our investigation, we have used the following elements:

**(1) Layout patterns:** 8 common layouts patterns (Uchida, 2001) shown from Fig. 1 to Fig. 8 were used. The corresponding abbreviations are shown in parenthesis. The detailed explanations of each of the layouts were omitted due to the space constraints.

**Fig. 1:** Left-Right (R/L)      **Fig. 2:** Panorama (Pan)      **Fig. 3:** Satellite (Sat)

**Fig. 4:** Photo (Ph)      **Fig. 5**: Enclosed (Enc)      **Fig. 6:** Dispersion (Disp)

**Fig. 7:** One Slope (1Slp)        **Fig. 8:** Simmetry (Sim)

**(2) Kansei words:** The Kansei words related to comfortability used in the evaluation were extracted from brainstorming session as well as from Kansei Engineering literatures. We have got a set of 9 bipolar or opposite adjectives to be evaluated on a 5-scale with the SD method: Comfortable-Uncomfortable, Calm-Dynamic, Simple-Complex, Beautiful-Awkard, Familiar-Unfamiliar, Uniform-Not uniform, Like-Dislike, Reliable-Unreliable, Safe-Unsafe.

# 4    Performed Analysis

## 4.1    Simple Statistical Analysis

50 male and female students from 18 to 24 years have participated in the experiments of evaluating the 8 web layouts with the SD method. Based on such evaluations, Simple Statistical Analysis, such as calculation of average were performed. The results are summarized in Table 1. We can see that the bigger the value of the average is, the better that particular Web layout corresponds to that Kansei word.

**Table 1:** Average of Simple Statistical Analysis of SD Method applied to the 8 Layouts

|  | R/L | Pan | Sat | Ph | Enc | Disp | 1Slp | Sim |
|---|---|---|---|---|---|---|---|---|
| **Comfortable** | 0.30 | 0.44 | 0.44 | 0.30 | 0.02 | 0.58 | 0.78 | 0.18 |
| **Calm** | 0.12 | 0.38 | 0.04 | -0.04 | 0.38 | 0.50 | 0.82 | 0.62 |
| **Simple** | 0.92 | 1.46 | 0.64 | -0.1 | 1.16 | 0.86 | 1.08 | 1.38 |
| **Beautiful** | 0.46 | 0.30 | 0.44 | 0.78 | -0.02 | 0.50 | 0.84 | 0.32 |
| **Simple** | 0.54 | 0.68 | 0.66 | 0.58 | 0.04 | 0.46 | 0.44 | 0.26 |
| **Like** | 0.46 | 0.24 | 0.44 | 0.54 | -0.28 | 0.10 | 0.48 | 0.04 |
| **Uniform** | 0.68 | 0.58 | 0.16 | 0.08 | 0.34 | 0.14 | 0.38 | 0.32 |
| **Reliable** | 0.34 | 0.24 | 0.30 | 0.38 | 0.01 | 0.38 | 0.18 | 0.16 |
| **Safe** | 0.42 | 0.26 | 0.24 | 0.36 | 0.10 | 0.28 | 0.26 | 0.24 |
| **OVERALL RATE** | 0.47 | 0.51 | 0.37 | 0.32 | 0.19 | 0.42 | 0.59 | 0.39 |

First, we have asked to the participants in a straight way the image of "Comfortable" for the 8 layouts, and the average are shown in the first row of Table 1.  From the results -regarding "Comfortable"- we can see that 1Slp (av. 0.78) gives the best image, while Enc (av. 0.02) gives the worst image.  In the same way, in Table 1, it is possible to see the influence of each factor or Kansei words Calm, Simple, Beautiful, Like, Uniform, Reliable and Safe on the different Web layouts.

825

In the last row of Table 1, we have calculated the Overall rate as the average of the individual factors previously mentioned. We can observe from the values that 1Slp (av. 0.59) gives the best overall value. On the other hand, it is important to note that Enc (av. 0.19) gives the worst value for the overall calculation, as well as for most of the individual factors.

## 4.2 Correlation Analysis

In order to check the correlation among the Kansei words used, we have performed a Correlation Analysis, using SPSS. The results of the correlation factors are shown in Table 2.

**Table 2:** Results of Correlation Analysis for each of the Kansei words used.

| | Comfort. | Calm | Simple | Beauty | Familiar | Like | Uniform | Reliable | Safe |
|---|---|---|---|---|---|---|---|---|---|
| Comfort. | 1.000 | 0.645 | 0.272 | 0.553 | 0.588 | 0.651 | 0.414 | 0.482 | 0.53 |
| Calm | 0.645 | 1.000 | 0.416 | 0.495 | 0.399 | 0.547 | 0.479 | 0.433 | 0.54 |
| Simple | 0.272 | 0.416 | 1.000 | 0.165 | 0.200 | 0.190 | 0.353 | 0.080 | 0.24 |
| Beauty | 0.553 | 0.495 | 0.165 | 1.000 | 0.460 | 0.669 | 0.459 | 0.506 | 0.53 |
| Familiar | 0.588 | 0.399 | 0.200 | 0.450 | 1.000 | 0.701 | 0.363 | 0.417 | 0.47 |
| Like | 0.651 | 0.547 | 0.190 | 0.669 | 0.701 | 1.000 | 0.444 | 0.564 | 0.60 |
| Uniform | 0.414 | 0.479 | 0.353 | 0.459 | 0.363 | 0.444 | 1.000 | 0.480 | 0.56 |
| Reliable | 0.482 | 0.433 | 0.080 | 0.506 | 0.417 | 0.564 | 0.480 | 1.000 | 0.65 |
| Safe | 0.536 | 0.548 | 0.249 | 0.534 | 0.473 | 0.602 | 0.569 | 0.657 | 1.00 |

Correlation factors bigger than 0.6 mean that there are some correlations, while factors less than 0.3 mean no correlation. We can observe from the results that there are some correlations between Comfortable-Calm (0.645) and Comfortable-Like (0.651). The other correlated factors are Beautiful-Like (0.669), Safe-Reliable (0.657), Like-Familiar (0.701) and Like-Reliable (0.602).

## 4.3 Categorical Regression Analysis

In order to check the influence of the Kansei words on the comfortability of web sites, we have performed Categorical Regression Analysis, using SPSS. As a result, we have obtained a "comfortability function" which expresses the comfortability as a weighted sum of Kansei words, as shown below.

Y ( comfortability ) = 0.373×Calm + 0.04871×Simple + 0.108×Beautiful + 0.239×Familiar + 0.150×Like - 0.05251×Uniform + 0.05249×Reliable + 0.138×Safe

For this case, we have the values R=0.787, $R^2 = 0.619$, where R > 0 indicates that the obtained function is good to represent the comfortability in terms of the given Kansei words.
On the other hand, the value of $H_0$ is 0.00 < 0.005, which indicates this function's ability to predict values of comfortability.
By replacing the average of each of the factors or Kansei words (obtained in the Simple Statistical Analysis), we have calculated the comfortabity for each of the 8 Web layouts.
From the results of calculation with the comfortability function –summarized in Table 3- we can see that the layouts 1Slp (0.648), Disp (0.457) and Pan (0.456) give the higher values for comfortability, while the layouts Sat (0.356), Ph (0.349) and Enc (0.156) gives the lower values. Note the poor value of Enc (0.156) in comparison with the other layouts.

**Table 3:** Layouts ordered by results of comfortability function

| ORDER | 1 | 2 | 3 | 4 | 5 | 6 | 7 | 8 |
|---|---|---|---|---|---|---|---|---|
| LAYOUT | 1Slp | Disp | Pan | Sym | R/L | Sat | Ph | Enc |
| SCORE | 0.648 | 0.457 | 0.456 | 0.423 | 0.371 | 0.356 | 0.349 | 0.156 |

What we can say in common from the Simple Statistical Analysis and the Categorical Relation Analysis is that in both analyses, the best corresponds to 1Slp, while the worst corresponds to Enc. The differences in order between both analyses for the other layouts may be due to the low influence of some of the Kansei words on the comfortability function.

# 5    Results

From the results of the analysis we give the following recommendations.
(1) The layouts giving positive images of comfortability are those with the"sense of relief" and "open world". For example, 1Slp with plenty of free blank spaces at the right side, Disp and Pan which give the "sense of relief", where the user's glance can come from any of the directions.
(2) The layouts giving negative images of comfortability are those with the "sense of closed world", and "center-stable". For example, Ph and Sat, with pictures concentrated in the center of the page, and Enc, in which the central picture is surrounded by colors, giving the image of "closed world".
(3) Also, layouts giving negative images where those with too much text, like R/L and Sat.

# 6    Conclusions and Future Works

In this research, we have explored what is the proper layout of a comfortable web site, based on the Kansei Engineering approach. To do so, we first have investigated the layout of several Web sites in order to extract Kansei words which are able to express the image of comfortability for such sites. Next, based on Categorical Regression Analysis, we have found a function which expresses the comfortability as a weighted sum of Kansei words. According to our findings, the weights of the Kansei words Calm, Familiar and Like were the higher in this comfortability expression. We have applied this function to each of the design layouts, in order check the layouts that give the image of comfortability in Web sites.
According to this calculation, the more comfortable layouts are those giving the sense of "relief" and open, like 1Slp, Disp and Pan layouts.
On the other hand, the "uncomfortable" layouts are those providing the sense of "closed and limited world", such as Enc, Ph and Sat layouts.
Our future works include applying the method we have proposed, to study the influence of other design factors such as combination of colors on the comfortablility. We will extend our research to more dimensions, as the current study is mainly on two-dimensions or planes.

# References

Osgood, C., Suci, G. and Tannenbaum, P.: The Measurement of Meanings, Univ. Illinois Press, 1957

The Quality Portal, Kansei Engineering definition in http://thequalityportal.com/glossary/k.htm

Nielsen, J: Designing Web Usability – The Practice of Simplicity, New Readers Publishing, 2000.

Uchida, H.: Basics of Web Design (Web Design Kiso Kouza, in Japanese), 2001

# Tools for Remote Web Usability Evaluation

*Fabio Paternò*

ISTI-CNR
Via G.Moruzzi, 1 – 56100 Pisa - Italy
f.paterno@cnuce.cnr.it

## Abstract

The dissemination of Web applications is enormous and still growing. This raises a number of challenges for usability evaluators. Video-based analysis can be rather expensive and may provide limited results. This paper presents a discussion of what information can be provided by automatic tools for remote Web usability evaluation as well as an analysis of existing approaches, including our proposal.

## 1   Introduction

There are many motivations for automatic tools able to support the evaluation process (Ivory & Hearst, 2001). Tools that implement the design criteria incorporating usability principles as well as the total or partial automation of usability evaluation can reduce the time and costs involved and release evaluators from repetitive and tedious tasks. Examples of automated usability tools include code parsers and evaluators, image analysis tools, usage measurement tools (such as hit log analysis and instrumented browsers), semi-automated tools to aid the human evaluator, design evaluation and advice in web development tools, and automated online surveys. The goal of applying automatic tools is not to provide designers with an overall, definitive evaluation. Rather, a more meaningful approach is to provide a number of pieces of information that can be helpful to evaluators and developers in order to improve their applications.

Automatic evaluation is going to be an extraordinarily common technique, though it is currently in a germinal state. A good number of organizations are seriously investigating and beginning to consider automatic evaluation tools. Thus, we are at an excellent point to obtain engineered tools for widespread application. For example, with over 30 million of web sites in existence, web sites are the most prevalent and varied form of computer-human interface. At the same time, with these many web sites being designed and maintained, there will never be a sufficient number of professionals to adequately address usability issues without automation as a critical component of their approach.

With the advent of the Web and the refinement of instrumentation and monitoring tools, user interactions are being captured on a much larger scale than ever before. Automated support for the capture, representation, and empirical analysis of user behaviour is leading to new ways to evaluate usability and validate theories of human-computer interaction. It enables remote testing, allows testing with larger numbers of subjects, and motivates the development of tools for in-depth analysis. The data capture can take place in a formal experimental setting or on a deployed system. Particular attention must be paid to remote usability evaluation, where users and evaluators are distant in time and/or space. This type of approach can overcome some limitations of usability laboratories: it is often difficult to bring a considerable number of users to such laboratories, and then they have to interact with the application in an environment different from their daily working environment. In practice, lab-based tests are mostly performed with a small, local sample of the user population, which renders them inadequate for the evaluation of products to be used by people across a wide geographical area and demographic groups.

## 2   Approaches to Web Remote Evaluation

In empirical testing the actual user behaviour is analysed during a work session. This type of evaluation requires the evaluator to observe and record user actions in order to perform usability evaluation. Manual recording of user interactions requires a lot of effort thus automatic tools have been considered for this purpose. Some tools support video registration but also video analysis requires time and effort (usually it takes five times the duration of the session recorded) and some aspects in the user interaction can still be missed by the evaluator (such as rapid mouse selections). In model-based evaluation, evaluators apply user or task models to predict interaction performance and identify possible critical aspects. For example GOMS (Goals, Operators, Methods and Selection rules) (John & Kieras, 1996) has been used to describe an ideal error-free behaviour. Model-based approaches have proven to be useful but the lack of consideration for actual user behaviour can generate results that can be contradicted by the real user behaviour.

It becomes important to identify a method that allows evaluators to apply models in evaluation still considering information empirically derived. To this end the main goals of our work are:

- To support remote usability evaluation where users and evaluators are separated in time and/or space;
- To analyse possible mismatches between actual user behaviour and the design of the Web site represented by its task model in order to identify user errors and possible usability problems;
- To provide a set of quantitative measures (such as execution task time or page downloading time), regarding also group of users, useful for highlighting some usability problems.

As we are considering remote evaluation without direct observation of the user interactions, it is important that testing furnishes logs with detailed information. An ideal log should provide information such as who has visited the site, the paths that they have followed during the visit, how long users stay on a page, where users leave the site, and the success of the users' activities. Logs can be considered at three different levels: browser, proxy server, and web server. The most straightforward approach is to look at Web server logs. However, a number of problems arise with such an approach. It is difficult to identify the users: even the same host can receive different IP addresses in the case of modem connections, or proxy servers can hide the real original address. It is also difficult to identify the actual pages visited because the browser cache memory may hide the accesses subsequent to the first one. It is possible to determine when a page download starts, but not how long the page is actually in front of the user. It is difficult to identify the end of a user session or whether or not users have achieved their goals successfully. In the case of proxy servers, logging is done through the proxy, which captures user accesses without having to modify the participants' software or accessing the web server (Hong, Heer, Waterson, and Landay, 2001). However, there are some drawbacks also in this case because it is not possible to detect the local interactions with the browser. Thus, the use of browser logging seems to be a more promising solution as it allows detecting all the user interactions at any level of granularity, as well as all the interactions between the browser and the server. The latter information is useful, for example to distinguish the page download time from the display time. An example of tool supporting client-logging is WebVip (Web Visual Instrumenter Program) (Scholtz, Laskowsi & Downey, 1998) that has been developed at NIST. This tool allows logging of user interactions and the resulting log files can be analysed through a graphical tool that visualises the paths followed by the users during the site visit. The logging tool proposed requires a number of modifications in the HTML pages that must be evaluated because each tag representing a user interface component calls for adding

Javascript code to record the interaction. Because of the many modifications required, WebVip needs a copy of the entire site. Unfortunately copying the entire site can generate many problems. Also WET (Etgen & Cantor, 1999) considers client-side logs but they are obtained more efficiently without requiring copy of the entire site. In this case it is sufficient to include the javascript file in the heading of the page. This javascript file includes the specification of the events that can be detected and the handling functions able to capture them. In WET only the some types of events are recorded. This limitation is due also to the lack of automatic tools able to analyse the data. Since the analysis is performed manually it is important to have readable log files with content useful for the evaluator.

# 3   WebRemUSINE

At the HCI group of ISTI-CNR, we have developed a method and an associated tool to detect usability problems in Web interfaces through a remote evaluation (Paganelli & Paternò, 2002). Our approach combines two techniques that usually are applied separately: empirical testing and model-based evaluation. The reason for this integration is that models can be useful to detect usability problems but their use can be much more effective if they can be related to the actual use of a system. Our tool is able to analyse the possible inconsistency between the actual user interactions and the task model of the Web site that describes how its concrete design assumes that activities should be performed. To support remote evaluation, we have developed a technique that allows recording user actions during a site visit. The analysis of the logged data is based on the comparison of the traces of actions performed with the structure of the task model. This analysis provides evaluators with a number of results that are related to the tasks that users intend to perform, the Web pages and their mutual relationships.

The method is composed of three phases: *Preparation*, which consists of creating the task model of the Web site, collecting the logged data and defining the association between logged actions and basic tasks; *Automatic analysis*, where WebRemUSINE examines the logged data with the support of the task model and provides a number of results concerning the performed tasks, errors, loading time, .. *Evaluation*, the information generated is analysed by the evaluators to identify usability problems and possible improvements in the interface design.

The environment is mainly composed of three modules: the ConcurTaskTrees (Paternò, 1999) editor (publicly available at http://giove.cnuce.cnr.it/ctte.html); the logging tool that has been implemented by a combination of Javascript and Java applet to record user interactions; WebRemUSINE, a java tool able to perform an analysis of the files generated by the logging tool using the task model created with the CTTE tool.

The WebRemUSINE analysis can point out usability problems such as tasks with long performance or tasks not performed according to the task model corresponding to the Web site design. These elements are useful to identify the pages that create problems to the user. Thus the evaluation performed provides information concerning both tasks and Web pages. These results allow the evaluator to analyse the usability of the Web site from both viewpoints, for example comparing the time to perform a task with that for loading the pages involved in such a performance. WebRemUSINE also identifies the sequences of tasks performed and pages visited and is able to identify patterns of use, to evaluate if the user has performed the correct sequence of tasks according to the current goal and to count the useless actions performed. In addition, it is also able to indicate what tasks have been completed, those started but not completed and those never tried. This information is also useful for Web pages: never accessed web pages can indicate that either such pages are not interesting or that are difficult to reach. All these results can be provided for both a single user session and a group of sessions. The latter case is useful to understand if a certain problem occurs often or is limited to specific users in particular circumstances.

During the test phase all the user actions are automatically recorded, including those associated to the goals achievement. The evaluation performed by WebRemUsine mainly consists in analysing such sequences of actions to determine whether the user has correctly performed the tasks complying the temporal relationships defined in the task model or some errors occurred. In addition, the tool evaluates whether the user is able to reach the goals and if the actions performed are actually useful to reach the predefined goals. In order to determine whether the sequence of tasks performed is complying with the temporal relations defined in the task model we used an internal simulator. For each action in the log, first the corresponding basic task is identified and next there is a check to see whether the performance of that task was logically enabled. If not then a precondition error is identified. If yes, then the list of the enabled tasks after its performance is provided. In addition, also the list of accomplished high level tasks after its performance is provided and it is used to check whether the target task has been completed

In the report analysing the user session, for each action there is an indication whether it was correctly performed or a precondition error occurred. The analysis of the user actions allows the detection of problems generated from the execution task order. The precondition errors highlight what task performance did not respect the temporal relations defined in the model describing the system design and consequently mismatch between the user and the system task model occurred.

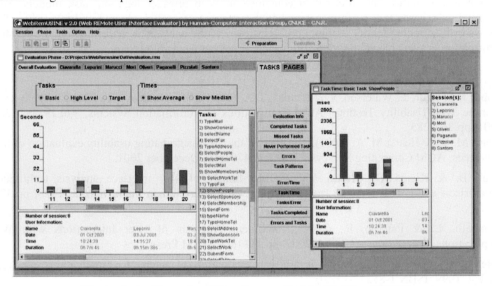

**Figure 1:** An Example of Results provided by WebRemUSINE.

The presence of events not associated to any task can indicate parts of the interface that create problems to the user. For example, if in the log there are events associated to images that are not associated to any link then evaluators can understand that the image confuse the user. In this case designers can change the page in such a way that it is clear that has no associated function or decide to associate a link to it.

In addition to the detailed analysis of the sequence of tasks performed by the user, evaluators are provided with several results (an example in Figure 1) that provide an overall view of the entire session considered, for example:

- The basic tasks that are performed correctly and how many times they have been performed correctly.
- The basic tasks that the user tried to perform when they were disabled, thus generating a precondition error, and the number of times the error occurred.
- The list of tasks never performed either because never tried or because of precondition errors.
- The patterns (sequences of repeated tasks) occurred during the session ad their frequency.

Such information allows the evaluator to identify what tasks are easily performed and what tasks create problems to the user. Moreover, the identification of tasks never performed can be useful to identify parts of the application that are difficult to comprehend or reach. On the basis of such information the evaluator can decide to redesign the site trying to diminish the number of activities to perform and make the task performance required of the user simpler and easier.

# 4 Conclusions

This paper provides a discussion of how approaches for remote usability evaluation of web sites. After a brief discussion of the motivations for this type of automatic evaluation, we analyse different solutions to logging user interactions.

Lastly, we discuss the results that can be obtained through a tool (WebRemUSINE) able to automatically analyse the information contained in Web browser logs and task models. Such information regards task performance and Web page accesses of single and multiple users. This allows evaluators to identify usability problems even if the analysis is performed remotely. More information is available at http://giove.cnuce.cnr.it/webremusine.html

# 5 References

Hong, J. I., Heer, J., Waterson, S. & Landay, J. A. (2001), WebQuilt: A Proxy-based Approach to Remote Web Usability Testing, ACM Transactions on Information Systems, Vol.19, N.3 July 2001, pp.263-285.

Ivory, M. Y. & Hearst M. A. (2001), The state of the art in automating usability evaluation of user interfaces. ACM Computing Surveys, 33(4), pp. 470-516, December 2001.

John, B. & Kieras, D., (1996), The GOMS family of user interface analysis techniques: comparison and contrast, ACM Transactions on Computer-Human Interaction, 3, 1996, pp.320-351.

Paganelli, L. & Paternò, F., (2002), Intelligent Analysis of User Interactions with Web Applications. Proceedings ACM IUI 2002, pp.111-118, ACM Press.

Paternò, F., (1999) Model-based design and evaluation of interactive applications, Springer Verlag, 1999. ISBN 1-85233-155-0.

Scholtz, J., & Laskowski, S., Downey L., (1998), Developing usability tools and techniques for designing and testing web sites. Proceedings HFWeb'98 (Basking Ridge, NJ, June 1998). http://www.research.att.com/conf/hfweb/ proceedings/scholtz/index.html

Etgen, M. & Cantor J., (1999), What does getting WET (WebEvent-logging Tool) mean for web usability?. Proceedings of HFWeb'99 (Gaithers-burg, Maryland, June 1999). http://zing.ncsl.nist.gov/hfweb/proceedings/etgen-cantor/index.html.

# Feijoo.net: An Approach to Adapted Learning Using Learning Styles

*Mª del Puerto Paule , Juan Ramón Pérez*
HCI Group- Labtoo. Dpto. Computer
Science of University of Oviedo-Spain
Facultad de Ciencias
C/Calvo Sotelo S/N 33007 Oviedo
{paule, jrpp}@pinon.ccu.uniovi.es

*Martín González Rodríguez, Sergio Ocio*
HCI Group- Labtoo. Dpto. Computer
Science of University of Oviedo-Spain
Facultad de Ciencias
C/Calvo Sotelo S/N 33007 Oviedo
martin@lsi.uniovi.es, djrekcv@terra.es

## Abstract

This paper proposes an adaptation model called Feijoo.net. Feijoo.net is a system, which adapts contents and the presentation of these contents to the learning style of each student. In this model, contents are separated from their presentation on the Web. Feijoo.net is focused on university students. The main goal is to get an adaptation model of the learning of each student following the directives proposed by the cognitive psychology and pedagogy. The article includes, too a study in which the model is evaluated.

## 1    Introduction

One of the major obstacles in Educational Web Sites is the imitation of the traditional way of teaching based in the master class of a teacher due overcrowding in universities. In these circumstances, the teacher cannot do a full adaptation to the learning style of each student because there are a lot of students in the classroom and it is impossible to do. So, Internet can be an adequate resource to do this adaptation because it is a mass media and the content of individual dynamic pages can be adapted to each learning style.

In the design of Educational Web Sites it is very important to have prepared the objectives of learning, the motivations, the previous knowledge and the preferences of each student. It is evidence the first step towards the design of high quality courses is the user classification.

Feijoo.net uses the CHAEA's test (CHAEA's test, 2003) (Catalina Alonso, Domingo Gallego and Peter Honey) to classify the user. This test offers an acceptable reliability and validity proved to be applied to Spanish Universities.This test returns the learning style of the student. The learning styles are the way of thinking, processing information, and learning of each individual student. This test returns the preferences of the student at the time of learning.

There are four styles in this classification: Theoretician, Active, Reflexive and Pragmatic. The test returns a value between 0 and 20 for each style. With these values the cognitive style of the user is known. For example, if the student gets 20 in the Active style, he/she is Active and she/he is going to learn like an Active.

The CHAEA's test is a test with 80 items. Perhaps, 80 items are a lot of items, but it is thought to be done in 15 minutes, and the answers aren't difficult. Each style has its own characteristics and particularities.

Actually, Feijoo.net is adapting the learning styles Theoretician and Active. The selection of these learning styles is because the Theoretician and Active Model Adaptation is the base for Pragmatic and Reflexive learning styles.

The goal is to adapt the content and the presentation of these contents in a course to students belonging to the cognitive style Active or Theoretician. This adaptation will increment the quality and the usability of Educational Web Sites. The usability is a subjective factor and it is difficult to

833

measure it, but when a user is comfortable with the site, the usability will increment (Nielsen, 2003).

Feijoo.net has three main agents: the University Student, the teacher and the Adaptation Core, that is how to link and organize the adapted contents and it is the part most important. Too, Feijoo.net automates the adaptation process for the teacher.

## 2 Adaptation Core

This part is how to link and how to organize the information.

In this part, FEIJOO.NET has three parts:

- POL: PrOgrammed Learning management (Web Platform)
- IOWA: Intuitive-use Oriented Webtool for the creation of Adapted contents (in a e-learning environment).
- VIC: Visualization of Contents.

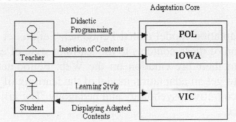

**Figure 1:** Feijoo.net

### 2.1 POL (PrOgrammed Learning Management - Web Platform)

POL will help the teacher to elaborate and organize his work on a Web platform, adjusting contents to a flexible and recognizable structure, showing the way he should carry out the educational management.

Normally, begging to distribute a subject, the teachers make a planning about how they are going to carry out that work. POL tries to automate this work, so it is made of a more systematic way, rational and reflective.

The teacher is going to do the didactic programming with POL; so, he/she teacher has to decide the general and specific learning objectives of each didactic unit. The general learning objectives will be reached when the specific objectives did. The system will warn the teacher when general objectives aren't fulfilled or when specific objectives from each didactic unit do not fit into general one.

POL has three types of users: Administrator, Teacher, Student. Each kind of user has different privileges, can make different functions and has access to the information in different ways. Each user has a login and a password to access the system. POL shows different contents according to the type of the user.

### 2.2 IOWA (Intuitive-use Oriented Webtool for the creation of Adapted contents - in an e-learning environment).

When the teacher finishes inserting the didactic programming, then he will add the contents of the didactic unit or lesson. IOWA is the part of the system that allows the teacher to add contents adapted to each student learning style ( Del Moral, Alvarez, Cabrero & Esbec, 2002).

It is designed with a very intuitive interface. It is thought to use by a teacher who doesn't know anything about computers or he /she has user-level knowledge.

IOWA has two main functions:

- Helping the teacher: IOWA automates the process of adaptation. IOWA guides the teacher in the task of creating adapted contents to the learning style. IOWA offers the teacher a set of steps to follow when he inserts the theory, examples o exercises depending on learning style. So, the teacher introduces the appropriate content for each learning style.

- Contents will be shown according to the type of the individual student. To get this goal, IOWA will have a template editor. We will use templates to define how we want contents to be displayed. The editor will be a wizard-like one, so we can guide the teacher easily through the process. Each course will have an associated template.

The information of each didactic unit is organized as tree. Feijoo.net uses the concept of "context" proposed in the methodology OOHDM (Rossi, 2003). The traversing of the tree depends on the learning style. We are going to traverse the tree using different paths according to the learning style.

## 2.3 VIC (Visualization Of Contents)

The main functions of VIC are:

- Knowing the learning style of the student with the CHAEA's test and
- Showing the user the adapted content according to her/his cognitive style.

The student has to do CHAEA's test. VIC uses a subsystem that uses fuzzy logic to know the learning style. With the second function, VIC gets the learning style of the student from the database and the core of the VIC shows the adapted content.

How does VIC work for the user? When the student enters the login and password, he/she is into the system. If he/she is registered, VIC loads the learning style of him/her. If the student has selected the subject for the first time, then the course starts from its earliest stage, but if the student had started previously it, he/she can continue at the stage he left it last time.

If VIC doesn't recognize the student, he/she has to do the test.

# 3 Architecture

The Application Model bases on three tiered model.

The first tier consists of the presentation layer, which is responsible for assembling the data into a presentable format. The second tier is the application layer. Finally, the third tier provides the second tier with data that it needs.

In this architecture, the presentation layer is composed by Web Pages, which are encoding with Java Applets, or XML with XSLT StyleSheet. The XSLT Style Sheet comes from templates, which are making by the wizard explained in the module IOWA (See 5.2 IOWA). The application layer is the Adaptation Core and it is going to be explained next because it is the most important part in the system.

The third tier is composed by a Web Service, which is made by. Net platform. This Web Service provides a great flexibility to the system because it is independent. The user introduces the database name, the login and the password, and the SQL query that he wants obtain and the Service returns the result of this query in XML . On this XML code, it applies a XSLT StyleSheet, which has been designed by the teacher with the wizard of the module IOWA.

Adaptation Core (see figure below) has the explained components: POL, IOWA, VIC. First, POL allows introducing the didactic programming (See 5.1 POL), so there is a didactic unit inserted into database. At this moment, the teacher introduces the unit didactic content step by step with IOWA for each learning style (this process is possible because IOWA automates it for the teacher, so the teacher knows when he must to introduce the theory, exercises, examples, etc.). If the teacher wants, he can design a presentation with the template wizard. This template is stored into database and it is a XSLT StyleSheet. Too, the teacher can do pre- visualization of the presentation of the content and the content with its template for each learning style.

When a student enters in the systems, VIC is activated. VIC shows the CHAEA's Test to the student, who has to do it if he did not do it. With the test, VIC knows the student learning style with a fuzzy logic subsystem. When it knows the learning style, it obtains from database the appropriate template and the contents of the lessons of the subject.

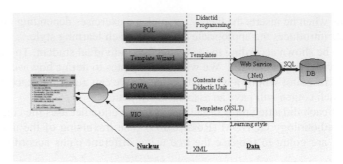

**Figure 2:** Architecture

# 4 Evaluation

When we started to develop FEIJOO.NET, we wanted to prove: Is there an increment of learning with the adaptation, really? We know various parameters that affect the learning of a student as: intelligence, motivations, environment, etc. These parameters depend on the individual student, and we cannot "manage" them. However, the learning style is a parameter we can "manage".

To prove this question we make one experiment called Priority Queue.

## 4.1 Experiment: Priority Queue

We designed a prototype. This prototype was about a lesson of subject Data Structures. This subject is a subject of the second course of studies of Computer in University of Oviedo. The lesson was about Priority Queues.

We designed a Web Site[1] where the theoretical contents are showed in a Web Page. This Web Page explains the concept of Priority Queue and the basics operations of this. The theory is complemented with examples. The Web Site has an applet. The students can interact with the applet, adding dates and watching the created structure with the operations as: To Adding, deleting, changing, etc.

The goals of this experiment were:

- If a question is the same than other but it is formulated with different way, what is the most appropriate way for each style in this subject?
- How each student's style answers the same questions in the subject Data Structures?

Our sample was six students of a practice class of the subject[2]. These students did not know about Priority Queues, they knew that it was a data structure, but they did not anything else and all of them were registered in the subject for first time. [3]

There were three Theoretician and three Active students.

The experiment was done in one session and it was composed by three phases:

- Doing the CHAEA's Test (15 minutes)
- Surfing the Web Site. (1 hour)
- Doing the knowledge test. (30 minutes).

The test had ten questions and had the same proportion of Theoretical and Active questions. We wanted to evaluate the abilities (following Bloom's Taxonomy): Knowledge, Comprehension, Application and Evaluation.

The result was: The Theoretician students spend more time with the theory, and some did not open the applet because they think "it isn't necessary, because, the theory is explained good".

---

[1] The prototype was implemented with Java and XML.
[2] These students are volunteers and all of them were very interested in subject.
[3] The subject of Data Structures is a difficult subject in the studies of Computer of University of Oviedo and there are a lot of repeated students.

The Active students were spending more time with the applet and less time with the theory and they think the applet is very important to understand the priority queues.

The result of this evaluation is: The students pass the test. They get a score greater than 7. However, if we are analyze each individual answer, we have obtained that questions that which do not belong to the learning style are less complete in their response than the ones which do.

For example: The question 1 was the same question as 2, but, the question 1 was more appropriate for Theoretician and the question 2 was more appropriate for Active.

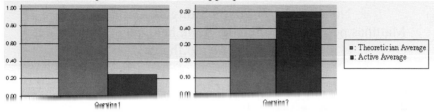

**Figure 3:** Question 1 (Theoretician) and Question 2 (Active)

The figure 4 shows the Theoreticians answered Question 1 better and that Question 2 was answered better by Actives.

The test was questions with free answerable-questions, so, the students can answer these questions with contents and presentation that they thought it was better.

The result obtained from these free-answerable questions was: Theoretical students choose textual representation of the contents, while Active users write an example and then, a little written explanation.

Question number 8 is a free-answerable question and the result is:

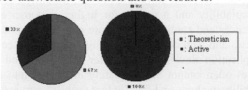

**Figure 4:** Question 8 (Theoretician) and Question 8 (Active)

Conclusions of the Priority Queue Experiment
The obtained conclusions in the Data Structured subject were:
Theoretical students choose textual representation of the contents, while Active users write an example and then, a little written explanation to answer a question. The theoretician students answer better the theoretician questions, and the active students answer better the active questions, but this claim does not mean that the theoretician students can answer only theoretician questions and the active students can answer only active questions.

# 5   References

Brusilovsly, P, De Bra P., Houben, G. "Adaptive Hypermedia: From Systems to Framework". ACM Computing Surveys, Vol.31, Number 4es, December 1999

CHAEA's test. Deusto University. http://www.ice.deusto.es/guia/test0.htm.

Del Moral Pérez, M. E.; Álvarez Fernández, S; Cabrero Álvarez, J.C. y Esbec Montes,I. (2002):"Rur@lnet: Un espacio para la teleformación y el trabajo colaborativo". En 2nd European Congress on Information Technologies in Education and Citizenship: a critical insight.

Nielsen, Jakob. "Designing Web Usability". http://www.useit.com.

Rossi, Gustavo. Rossi's Home Page. http://www-lifia.info.unlp.edu.ar/~gustavo/oohdm.html.

# WebSCORE Expert Screening – a low-budget method for optimizing web applications

*Matthias Peissner, Frank Heidmann, Inga Wagner*

Fraunhofer Institute for Industrial Engineering (IAO)
Nobelstr. 12, D-70569 Stuttgart, Germany
Matthias.peissner@iao.fhg.de, frank.heidmann@iao.fhg.de,
inga.wagner@iao.fhg.de

## Abstract

The WebSCORE Expert Screening is designed in order to address the specific requirements of web application evaluation. It takes an holistic perspective on web application quality that goes beyond traditional usability issues. On the other hand, it is a low budget service as it follows a structured and standardized evaluation scheme.

## 1    Introduction

User-friendly conception and design are key factors for successful web applications. Especially in the areas of e-commerce, e-business and corporate portals, usability becomes a strategic aim as it is a precondition for acceptability and a means for higher customer satisfaction and customer loyalty. Our recent experiences and studies demonstrate a very high and even growing awareness for usability in web agencies and web content providing companies (see Peissner & Röse, 2002). However, many web application developments suffer from low and continuously decreasing budgets. Usability activities often cannot be adequately integrated into the development process. The need for discount usability methods and low-budget usability services is greater than ever.

Moreover, the world wide web demands a broader perspective on human-computer interaction than just considering the concepts of software-ergonomics. Some important supplementary aspects are the suitability for the goals of the content provider and subjective and emotional factors of web site use such as trust and credibility. Currently leading guidelines and textbooks (e.g. Nielsen, 1999; Spool et al. 1999) hardly take an integrative view that comprises all these issues.

This paper presents the WebSCORE Expert Screening as a structured method for expert evaluations of web applications. Besides the "classical" usability issues, this method covers a wide range of web-specific quality issues beyond usability. Due to its detailed structure its application is quick and the results are reliable. First, we introduce the WebSCORE reference model. Then, we describe the basic ideas of the method and the evaluation procedure. Finally, we present results from customer feedback questionnaires and empirical studies that validate the value of the method.

## 2    WebSCORE – a web usability reference model

The WebSCORE reference model provides an holistic framework for the conception, design and evaluation of web applications (cf. Bullinger, Heidmann, & Ziegler, 2002). The model comprises a set of design domains and a set of evaluation domains (see figure 1). The design domains cover all components that are conceived and designed within the development process: "strategy", "content & functionality", "navigation & interaction" and "media design & information presentation".

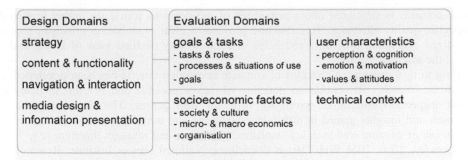

| Design Domains | Evaluation Domains | |
|---|---|---|
| strategy | **goals & tasks**<br>- tasks & roles<br>- processes & situations of use<br>- goals | **user characteristics**<br>- perception & cognition<br>- emotion & motivation<br>- values & attitudes |
| content & functionality | | |
| navigation & interaction | | |
| media design &<br>information presentation | **socioeconomic factors**<br>- society & culture<br>- micro- & macro economics<br>- organisation | **technical context** |

**Figure 1:** The WebSCORE reference model for the design and evaluation of web applications

Designing a web strategy includes reflections about how to reach the superordinate goals of the website, e.g. economic goals, PR goals, goals of organisation development, etc. As a result, target user groups and typical user tasks will be prioritized, a certain style of communication and a strategy for the positioning against competitors will be outlined.

The domain "content & functionality" pays attention to the mere supply of content and functionalities. This includes aspects that are specific to the genre of the web application (e.g. the suitability for the user groups and their tasks), and aspects that are basic for any web site such as internationalization issues and editorial information. Further criteria are the quality and credibility of the content and the supply of added value, e.g. by providing free downloads, external links to interesting sources of information or the opportunity for user feedback and engagement.

In the domain of "navigation & interaction" the structure and the interactive parts of a web application are addressed. Does the information architecture represent the mental models of the user groups? Does the navigation concept allow for easy and secure navigation and orientation? Do the interactive elements and controls accord with relevant standards and user expectations?

Finally, "media design & information presentation" is about the visuals and the use of multimedia. Are the texts easily readable and do they support scanning? Is the visual design made intuitive by the use of clear and consistent layout templates and by applying the Gestalt Laws? Is the use of different media suitable and effective? What about aesthetics and the emotional effect?

Each of these design domains is considered with regard to the set of the evaluation domains "individual user characteristics", "tasks & goals", "socioeconomic factors" and "technical context conditions". WebSCORE is a framework for a context- and domain-specific evaluation. As the ergonomic quality of each design domain is defined in terms of the suitability for the given target users, their main tasks, the socioeconomic context and the specific technical constrains, the evaluation results can be compared even across different genres and different types of websites. As the label WebSCORE implies, one of the evaluation results is a score between 0 and 1 for each of the four design domains. For different evaluation methods the scoring procedures differ. The following section will describe the scoring procedure in an expert review.

## 3    WebSCORE Expert Screening

The WebSCORE Expert Screening has been developed as a low-budget service for an efficient WebSCORE evaluation. It is based upon the WebSCORE reference model and refers to its design domains. As the web strategy cannot be evaluated by mere inspection, it has not yet been addressed in the present screenings. In the future, we will use a specific web strategy questionnaire in order to investigate the strategic aims and considerations of the web site provider.

Until now, the Expert Screening relies upon the three central design domains "Content & Functionality", "Navigation & Interaction" and "Media Design & Information Presentation". Each of these domains is structured into a certain set of sub-domains which are further broken down into specific quality attributes. This structure provides a clear organisation principle, supports a standardized evaluation procedure and helps to attain a very detailed view of strong and weak points of the assessed website.

According to this structure, a checklist of concrete inspection criteria has been developed which refers to the WebSCORE evaluation domains whenever possible and helpful. In many cases, different inspection criteria are given for different web genres. The checklist is based on experiences and insights gained in manifold evaluations and usability tests, as well as on an in-depth review of existing web usability standards, guidelines and research literature (e.g. Johnson, 2002; Donelly, 2000; IBM Web Design Guidelines; National Cancer Institute: Research-Based Web Design & Usability Guidelines; etc.)

The evaluation proceeds as follows:

- *Short exploration and scenario elaboration*:
  The evaluator gets an overview of the website and defines 2-3 scenarios of use and target user groups as a basis for the cognitive walkthrough.
- *Cognitive walkthrough:*
  A cognitive walkthrough is conducted. Positive and negative findings are described and allocated to the appropriate quality attributes.
- *Screening:*
  Several central pages are inspected in detail regarding the WebSCORE quality attributes and the corresponding checklists.
- *Completion of the evaluation scheme*:
  The evaluator revises the listed findings and completes missing aspects.
- *Severity rating and calculating the WebSCORE*:
  The evaluator assigns a score to each quality attribute (0: "Usability catastrophe" to 4: "Positive finding – strong points of the website"). Based on these scores an over-all score is derived for each sub-domain. For an integrated view on the usability of the evaluated web site, each design domain is characterized by the ratio of the summed sub-domain scores to the best possible score sum. The resulting percentage represents the quality of use and the presence and severity of usability problems within a domain.
- *Prioritizing problems and recommendations for optimization:*
  The scores allow a rapid identification of the most important and severe fields of problems. For these, the evaluator develops recommendations for optimization.
- *Summarizing the results*
  A figure illustrates the evaluation results regarding the assessed design domains. This figure allows for a quick overview of the strong and weak design areas and for a comparison with other web sites. Moreover, each design domain is briefly characterized and described by the main evaluation findings.

The Expert Screening requires an experienced evaluator who is familiar with the WebSCORE reference model. According to the size and complexity of the web application, an evaluation including a standardized report can be done within one or two person days.

## 4    Evaluation of the WebSCORE Expert Screening

In order to evaluate the reliability and effectiveness of the WebSCORE Expert Screening, we used customer feedback questionnaires and conducted a study that compared Expert Screening results with the results of usability tests of the same web applications.

## 4.1 Usability Tests

Two web sites that were already evaluated by the Expert Screening were usability tested with six users each. The positive and negative findings of the test were structured into the WebSCORE design domains and sub-domains to allow for a comparison to the screening results. Some design sub-domains could not be evaluated with both methods: emotional aspects can hardly be assessed without user participation, whereas most of the content-related sub-domains cannot be addressed in scenario-based usability tests. The findings of this comparison can be summarized as follows:

- The correspondence regarding the sub-domain scoring is very high. For the first (second) website, six (six) of the eleven sub-domains were scored identically, in two (one) sub-domains the scoring differed by 1, in no (one) sub-domain the scoring differed by 2, three (three) sub-domains could be scored by only one method.
- The expert screening uncovers more usability problems than usability tests. This holds especially for problems regarding content & functionality. Figure 2 illustrates the results of one study. The results for the other website are very similar.
- The structured approach of the WebSCORE Expert Screening increases the percentage of actual usability problems that can be anticipated by expert reviews.

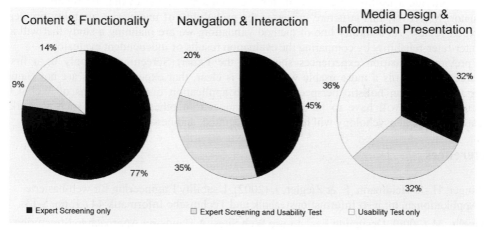

**Figure 2:** WebSCORE Expert Screening and usability tests: uncovered usability problems

## 4.2 Customer Feedback Questionnaires

The customer feedback questionnaire covers the over-all satisfaction with the evaluation report, the question if the structuring into the WebSCORE design domains and sub-domains is understandable and helpful, a usefulness assessment of the report sections (management summary, detailed screening results, recommendations for optimization), a cost-value ratio assessment and the opportunity to post other comments. Together with the evaluation report, we always send this questionnaire as an e-mail attachment. By the time this paper is written, we have received only three filled-in questionnaires. Six other reports and questionnaires have been sent very recently. So we are still waiting for the feedback. The results can be summarized as follows:

- *Over-all quality:* All three customers were "satisfied" or "very satisfied".
- *Structuring into WebSCORE design domains* was assessed as understandable and very helpful. Once this structuring was stated to be one of the strengths of the report.

Another customer criticized that some problems were stated twice in our structure. He would prefer a structuring regarding to the navigation structure of the website.

- *Usefulness of the report sections:* All sections were judged as "useful" or "very useful". Additional comment: "the recommendations for optimization could be more helpful if the evaluator knew more about the background and internal decisions. Nevertheless, good and helpful ideas."
- *Cost-value ratio:* All three customers were "satisfied" or "very satisfied".
- *Other comments:* Many of the identified problems were stated to be already known by the customers. The report was said to be very helpful as a third-party's qualified opinion.

# 5 Conclusions

The WebSCORE Expert Screening is an efficient and low-budget web usability service that meets the web-specific requirements. It broadens the usability perspective by integrating "soft" human factors and contextual factors such as socio-economic, organisational and technical aspects. Customers recognize the Expert Screening as very helpful and valuable. The WebSCORE design domains provide a useful structure both for the recipients of the evaluation report and for the evaluator. Moreover, this structure increases the probability of uncovering usability problems in an expert review. As a further line of method validation, we are planning a study that will assess the inter-rater-reliability by comparing the evaluation results of independent evaluators.

Our previous evaluation experiences show that the Expert Screening can only be a first but valuable step towards a more usable web site. It is clear, that expert methods are not enough in order to achieve an holistic perspective on web application quality in terms of WebSCORE. Existing methods will have to be adopted, methods from other disciplines such as marketing research and social psychology will have to be integrated, and new methods will be needed.

# References

Bullinger, H.-J., Heidmann, F. & Ziegler, J. (2002). Usability Engineering für web-basierte Applikationen. In: it+ti Informationstechnik und Technische Informatik 44 / 1, pp. 5-13.

Donnelly, V. (2000). Designing Easy-to-use Web Sites: A Hands-on Approach to Structuring Successful Websites. Boston: Addison-Wesley Professional.

IBM Ease of Use Web Design Guidelines from http://www-3.ibm.com/ibm/easy/eou_ext.nsf/publish/572

Johnson, J. (2002). GUI Bloopers. Don'ts and Do's for Software Developers and Web Designers. San Francisco: Morgan Kaufmann.

National Cancer Institute: Research-Based Web Design & Usability Guidelines from http://usability.gov/guidelines/

Nielsen, J. (1999). Designing Web Usability: The Practice of Simplicity. Indianapolis: New Riders Publishing

Peissner, M. & Röse, K. (2002). Usability Engineering in Germany: Situation, Current Practice and Networking Strategies. In: Proceedings of the 1st European Usability Professionals Association Conference, London, 2002.

Spool, J.M., Scanlon, T., Schroeder, W., Snyder, C. and DeAngelo, T. (1999). Web Site Usability: A Designer's Guide. San Francisco: Morgan Kaufmann Publishers.

# Interactive Design Elements to Improve Information Presentation on Web Pages

*Christian Rathke and Valerie Schreiweis*

Hochschule der Medien
Wolframstr. 32, D-70191 Stuttgart
rathke@hdm-stuttgart.de, vschreiweis@web.de

## Abstract

In order to find evidence for our hypothesis that interaction may improve the design of information on web pages we conducted a survey of how interactive elements are used with the presentations of digital cameras on the net.
A catalogue of interactive design elements has been identified and is presented along with how these design elements are used. Finally, it is suggested, that interaction could be used in more contexts to improve the understanding of information.

## 1    Problem Statement

In our research, we are concerned with how well information is presented and received on web pages. Web-based information differs from information accessible on other media in that the users are able to satisfy their information needs by interactively directing themselves to the information sources. Web-based information is information "on demand". By using the web, a huge amount of information sources are at the users' fingertips. Concerning the variety and accessibility of information, browsing the web has gone far beyond browsing books even in a very large library. There are many dimensions for comparing media and how they are used. One significant property of using the web is the self directedness and interactivity when searching and gathering information. Users interact with a web-browser when they follow a link or specify search options.

In order to find relevant information, interaction is used for navigation purposes almost all of the time. Following a link, going back and forth between web-pages, and issuing a search command focus on navigating. Once the "right" web page or web pages are found, interaction stops and users turn to reading, occasionally supplemented by clicking in the scroll bar. With the exception of the elements used for navigation, information on a web page is not very different from information on a book or catalogue page. Indeed, many companies use the same data sources for both their printed and their electronic representations.

This means, that information on a web page is "passive". It cannot be interacted with. Even if a web page contains movies and animations, they are just being looked at. Users may start and stop them but they are not able to investigate them any further. Designers focus on how well they can structure and present information on a web page, but they often restrict themselves to what can be displayed on a printed page.

We argue that information presentation und reception can be enhanced by allowing the users to interactively explore them, thereby looking for additional perspectives and hidden properties.

Being able to interact is seen as a natural extension to well designed information. As much as graphics enhances textual information, interaction may enhance graphical information representation. The key issue here is how users can comprehend.

In order to find evidence to our hypothesis we conducted a survey on a potentially suitable part of the web, namely web pages containing information about digital cameras. The properties of the cameras make graphical preparation of information quite suitable and at the same time exhibit opportunities for interaction. From this analysis it became quite clear that interaction is still very sparsely used.

## 2 Catalogue of Interactive Design Elements

Information about digital cameras and how it is typically represented on web pages can be roughly categorized according to Table 1:

**Table1:** Information about Digital Cameras and How It Is Typically Represented

| Type of Information | Typical Representation |
| --- | --- |
| material properties | data sheet |
| performance figures | data sheet, graphics |
| looks, proportions, controls | picture, 3D-motion |
| operation | handbooks, movies, simulations, interaction |
| picture quality | picture |
| style, impression | text, associative pictures |

Obviously not all kinds of information are subject to interaction. Most commonly, operation demonstrations and operation instructions and – to a lesser extend – looks and visual properties are made accessible by interaction elements.

With Direct Manipulation (Shneidermann, 1983) as an interaction style, which causes immediate alteration in the system status as a consequence of the user's input, mouse actions are at the core of almost all interactive design elements. They comprise most of the user's possibilities of how to interact with the information on the page.

The basic mouse actions are:
- Mouse Over: Displays context-dependent information like system status/feedback, aid or functional dependence.
- Mouse Click: Initiates some action on the object.
- Mouse Move and Drag: Moves or turns objects in a 360 degree angle or adjusts parameters.

We identified several interactive design elements which in connection with graphical representations make use of the basic mouse actions. Table 2 relates these design elements to the context in which they are used with cameras. It also mentions generalizations for use with arbitrary information objects and provides examples found on different web pages.

As interactive elements on web pages become more common, users will become more familiar with this dimension of browsing the web. Sometimes designers don't trust the capabilities of their

**Table 2:** Interactive Design Elements

| Interactive Design Element | Description | Example use for camera related functions | General use for information objects | Example |
|---|---|---|---|---|
| **Hotspot** | An area as a part of an information object which reacts to mouse clicks. | Operation of camera controls. | Initiate a function associated with a specific location on the displayed object. | Clicking on the controls of the Fuji FinePix virtual camera causes simulated effect on the virtual camera (Fujifilm 2003). |
| **Slider** | Movable button for clicking and dragging along one dimension. | Turn the view of a camera. | Adjust parameters. | The slider below the Canon virtual camera turns it horizontally (Canon, 2003). |
| **Mouse-Gesture** | Click-and-drag the mouse in a two-dimensional space | Turn 2,5-view of the camera. 2 dimensions. | Navigate in a 2-3 dimensional space. | Moving the mouse in the lower right area around the red spot causes the camera to turn horizontally or vertically (Leica, 2003) |
| **Timed-Button** | Click, press and hold key or mouse button. | Show more detail; enlarge a specific part of the camera. | Zooming in and out of an information object. | Pressing the shift and control keys zooms in and out of the Canon Powershot S200 camera (Steves Digicams, 2003) |
| **Tool Tip** | Pop-up containing text and/or graphics specific to mouse location | Explain specific parts of a camera. | Provide context dependent help in information. | Moving the mouse to one of the designated areas yields a description of the corresponding camera control (Canon, 2003). |
| **Push Button** | Separate area on the page which reacts to mouse clicks. | Change color of camera. Take a picture. Switch views. | Initiate some action on the object. Change some property of the object. | Pushing the button at the upper right corner switches between front and rear view (Canon 2003). |

users and provide special supportive elements to explain the interactions. A short animation is shown of what can be done interactively or contextual information is presented to the user which guides him/her during all the stages of the interaction.

Introduced by Hassenzahl et al. (2002), the "hedonic quality" of a product describes the individual feeling of social prestige, exclusivity, excitement and degree of innovation which can be associated with the product. It is the idea that some information may not only be useful but dealing with it may also lead to an enjoying experience. "Playing" with the objects on a web page, making interesting discoveries, and in general combining fun with usefulness will make web pages more likely to be remembered and visited again.

In the domain of digital cameras, interaction is most adequately used for illustrating how to operate the camera. This has the same advantages as "playing around" with a physical or software system for gaining a better understanding of the system and its functions. This does not replace a manual but it is well suited to give a first impression of how the camera works. Compared to operating the physical device directly there are even some advantage to the virtual surrogate:

- Users receive constant and reliable system feedback. Errors become apparent immediately and need not be traced in written material.
- The "table of contents" becomes contextualized and can be shown on demand.
- Additional information and related hints may be provided at the right moment and the right location.
- The virtual object may be linked with additional material such as availability of the product or merchants.
- The digital camera's specification data can also be prepared. The memory size, for instance, could be represented by the number of pictures in a certain resolution.
- Searching and comparing different camera types is comfortable and easy to do.

# 3 Extending Interaction

Our analysis has shown that mostly certain properties of the camera are qualified for interactive representation, generalized in operation manual and capability characteristics. But it is also possible to adopt the potentials of interactive information representation to other kinds of information. For digital cameras it makes sense to explain the terms and definitions not only by a glossary but also in an interactive way.

As an example, we have developed an application with Macromedia Flash MX to demonstrate the user what the resolution of pictures means. An interactive slider varies the resolution from the minimum of 320x440 pixels to the maximum of 2048x1536 pixels thereby adjusting the picture in terms of detail and sharpness (Figure 1).

For the future it is desirable that more interactively represented information enter the internet, with the addition that they are used reasonable. Designing the GUI still depends on the personal taste of the designer himself. Direct Manipulation at the information object remains the most effective way to represent networked information and complex data stocks.

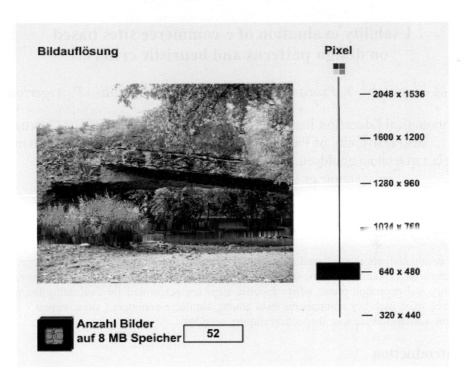

Figure 1: Illustrating resolutions

# References

Canon (2003), Virtual Camera of the Powershot A40 Digital Camera Model, Retrieved February 14, 2003, from http://www.powershot.de/facts/a40/content.html

Hassenzahl, M., Burmester, M. & Beu, A. (2001). Engineering Joy. IEEE Software, 1&2, p. 70-76.

Leica (2003), 3D-View of the Leica Digilux, Retrieved February 14, 2003, from http://www.leica-camera.com/digitalekameras/digilux1/3dansicht/index.html

Fujifilm (2003). Simulator of the Finepix F601 Zoom. Retrieved February 14, 2003, from http://www.finepix.de/simulator/601_applet.php?sprache=de

Shneiderman, B. (1983). Direct manipulation: A step beyond programming languages, IEEE Computer, 16, 8, 57-69

Steves Digicams (2003), Quicktime VR Movie, Retrieved February 14, 2003, from http://www.steves-digicams.com/2002_reviews/S200/S200_qtvr.mov

# Usability evaluation of e-commerce sites based on design patterns and heuristic criteria

*M. Sartzetaki[1], Y. Psaromiligkos*

Technological Education Inst. of Piraeus
[1]also University of Paisley
maria.sartzetaki@gouldseu.ittind.com,
jpsa@teipir.gr

*S. Retalis, P. Avgeriou*

University of Cyprus
{retal, pavger}@softlab.ntua.gr

## Abstract

DEPTH (evaluation approach based on DEsign PaTterns & Heuristic criteria) is an approach for performing scenario-based heuristic usability evaluation for e-commerce sites. It is comprised of a preparatory and execution phase where specific steps are performed for evaluating the usability along three axis: usability comparison tests among similar e-commerce sites, expert / heuristic evaluation, scenario-based user inspection/enquiry sessions.

## 1   Introduction

The e-commerce market has exploded over the last three years and is projected to continue its rapid expansion. Although in the early years of e-commerce sites, companies mainly aimed in designing a system the quickest possible in order to achieve an early presence in the market, nowadays usability matters.  Systems' user interfaces are now redesigned, taking into account the user needs. The benefits they anticipate are as follows: increased sales, customer satisfaction, customer retention, reduced support, stronger brand equity.

In contrast to mainstream business applications relatively little emphasis has been given on formal usability within the e-commerce sector (optimum.web: http://www.optimum-web.co.uk/ improvewebsite.htm). Most of the researchers and practitioners agree that E-commerce is indeed task based and these sites are therefore relatively easy to evaluate in a quantitative way. In addition, they are special types of websites on a bit more explorative in nature that offer flexibility and personalisation in navigation, thus typical evaluation heuristics can be applied.   In the literature one can find three dimensions particularly applicable to e-commerce usability evaluation:
- Usability comparison tests among similar e-commerce sites.
- Expert / heuristic evaluation.
- Scenario-based user inspection/enquiry sessions.

Scenarios provide a versatile and reproducible means of evaluating a system. A scenario is an instantiation of one or more representative work tasks and transitions linking those tasks (Rossi, Lyardet & Schwabe, 2000). The granularity of the scenario is not fixed; a scenario can be highly scripted or loosely defined. One of the main difficulties is how to create such scenarios.

In this paper we present the DEPTH approach, an innovative approach for usability evaluation of e-commerce sites based on design patterns and heuristics criteria.  This approach tackles all three of the aforementioned dimensions of usability evaluation. Moreover, DEPTH prescribes how to perform the steps of the evaluation giving emphasis on how to compare the e-commerce sites as well as how to easily create scenarios for user inspection. The main aid in this prescription is the usage of design patterns. Design patterns describe a problem and a solution for this problem in a particular context, together with the rationale for using that solution and the consequences (pros and cons) of using it (Gamma, Helm, Johnson, & Vlissides 1994). "The pattern is, in short, at the same time a thing, which happens in the world, and the rule which tells us how to create that thing, and when we must create it. It is both a process and a thing; both a description of a thing which is alive, and a description of the process which will generate that thing" (Alexander et al, 1977). The structure of the paper is the following: Section 2 gives an overview of the DEPTH approach while section 3 illustrates an example of its application. The paper concludes with few ideas about the effectiveness of the approach as well as our future plans.

## 2    The DEPTH approach

As in other evaluation methods (e.g. SUE method (Garzotto, Matera, & Paolini, 1998)) we require two operational phases of evaluation: the *preparatory phase* and the *execution phase*. The preparatory phase aims to define the conceptual framework that allows the evaluation to be carried out in a systematic and effective way. We propose four different activities:
- Gathering all design patterns related to domain under evaluation
- Generation of a list of user-centred tasks for each design pattern according to the underlying functionality of the pattern
- Generation of the scenarios for user inspection
- Selection of usability criteria that apply to each scenario (or task of the scenario) and generation of the toolkit (e.g. questionnaire) for quantitative and qualitative evaluation

The execution phase occurs each time a specific site is evaluated. We propose the following activities:
- Execution of the inspection utilising the toolkit
- Gathering of results
- Calculation and comparison of partial and global quality preferences of sites using the LSP mathematical approach
- Analysis of data gathered and presentation of results

The ultimate aim of both phases is to support the measurement of the usability of an e-commerce site by examining three dimensions: i) usability comparison tests among similar e-commerce sites, ii) expert / heuristic evaluation and iii) scenario-based user inspection/enquiry sessions.

### 2.1    The preparatory phase

Most of the times, the preparatory phase is entirely dependent on the experience of the usability expert/engineer. In order to systematize the whole process and provide an effective way for identifying a) the comparative matrix, and b) the main scenarios for user inspection, we propose the utilisation of design patterns. According to (Alexander et al, 1977): "each pattern describes a problem which occurs over and over again in our environment, and then describes the core of the solution to that problem, in such a way that you can use this solution a million times over". A "problem" is normally a specific functional or non-functional requirement from the system. Because we are interested in solutions that improve the usability of the system in use, we focus on customer-oriented problems and not developer-oriented. Aid in the 'pattern mining' approach is the work on hypermedia patterns (Rossi, Lyardet & Schwabe, 2000), as well as the HCI patterns (Hypermedia Design Patterns Repository, 2003). Before we proceed to the next step, each design pattern is categorized as in (Stefani & Xenos, 2001) in high, middle, and low category. High level comprises of those characteristics of e-commerce systems under evaluation that are most important. Middle level consists of those characteristics that are related to the services provided, but are not as important as those of the high level. Finally, low level includes the least important characteristics.

After gathering the design patterns related to domain under evaluation e.g. "Shopping basket", "Landmark", "Opportunistic Linking", etc., a list of user-centred tasks/subtasks can be derived according to the required functionality proposed by the underlying design pattern. These tasks describe (or can be derived from) "good practices" on how to implement the underlying functionality, which then will be used a) for comparing a specific e-commerce site to an "ideal" one, and b) for evaluation during the inspection process. However, the existence of required functionality is not enough to make the e-commerce site usable since how each design pattern is implemented will make a difference in usability terms. Thus, appropriate scenarios should be easily generated to measure the usability of the site. These scenarios will guide the user inspection. For each scenario (and sometimes for a group of tasks) we identify the main heuristic evaluation criteria that should be measured.

The next step of the preparatory phase is the generation of a toolkit (e.g. a questionnaire) that will provide data for quantitative and qualitative evaluation. In the case of a questionnaire, it should consist of two sections: a specific and a general section. The specific section is comprised of questions that measure the existence as well as the user satisfaction, easy of use, and/or usefulness of all the functionality points identified by the list of tasks included in the previous step. The questions are grouped according to the task they belong while an appropriate (small) scenario is preceding in order to guide the user to answer the questions. The general section consists of questions that measure general aspects of each design pattern according to Nielsen's heuristic evaluation criteria.

## 2.2 The execution phase

At the execution phase we perform the user inspection of e-commerce sites. The underlying scenarios incorporated into the toolkits developed at the preparatory phase guide user inspection. With the execution phase we compare the completeness of the functionality of the e-commerce site under examination against an "ideal" site which contains the full range of functionality of a site as described by a design pattern. Moreover, we measure the usability of the functionality offered. Thus, we do not only care about what the site offers but how well it does offer it.

After gathering the results we propose the Logic Scoring of Preference (LSP) model and continuous preference logic as mathematical background for the calculation of partial and global quality preferences of sites (Jozo, 1996). The same approach has been proposed by Web-QEM [Santos, 2002] for the evaluation of academic web sites.

## 3 An example

In this section we give a small example to demonstrate the steps of our approach. Suppose we want to evaluate some e-commerce sites with main emphasis on checking the usability of the online purchase of a product. Starting the preparatory phase we should firstly gather all relevant design patterns and categorize them in high, middle and low level category. We can gather a series of design patterns such as:
  - Shopping Cart/Basket: Allows users to gather all products first and pay for them all at once and whenever they want.

- _Advising_: Help the user find a product in the store, assist him according to his wishes.
- _Opportunistic Linking_: Keep the user interested in the site. Seduce him to navigate in the site even when he has already found what he was looking for.
- _Explicit Process_: Help the user understand the buying process when it is not atomic.

We identify the "Shopping Cart/Basket" as the most essential capability of an e-commerce site according to our objective. The description of the design pattern "Shopping Cart/Basket" can be found at (Welie, 2003).

As we can see from the solution element of the design pattern there are a number of "good practices" describing how to implement the underlying functionality. The total number of user-centred tasks identified was 19 as shown in table 1.

**Table 1.** The list of user-centred tasks

| 1.1 | Shopping cart/basket |
|---|---|
| 1.1.1. | Appropriate name |
| 1.1.2. | Ability to add items from anywhere |
| 1.1.3. | Any type of item can be included |
| 1.1.4. | Contents viewable at any time |
| 1.1.5. | Properties "Description", "Qty", "Price", Availability", "Category" defined for each line item |
| 1.1.6. | Additional properties defined describe each line item appropriately |
| 1.1.7. | Delete Line Item |
| 1.1.8. | Modify quantity |
| 1.1.9. | Link to detailed description |
| 1.1.10. | Total costs calculated according to changes performed |
| 1.1.11. | Help customers proceed with order |
| 1.1.12. | Provision of label next to the shopping basket image |
| 1.1.13. | Links related with shipping and handling costs and their calculation |
| 1.1.14. | Links for applicable taxes |
| 1.1.15. | Link for return policy |
| 1.1.16. | Validation within shopping basket contents |
| 1.1.17. | Shopping carts saved period |
| 1.1.18. | Crossing selling |
| 1.1.19. | Up selling |

Now, we can generate for each task appropriate scenarios to guide the user inspection. These scenarios are integrated into questionnaire along with the necessary questions that measure the existence as well as the user satisfaction/easy of use/usefulness of each criterion/characteristic defined in table 1. Table 2 shows an example of how a small scenario is integrated into questionnaire.

**Table 2.** Tasks/Scenarios generation.

| 1.1.7 Delete **item** |
|---|
| **Task**: By now more than one item are located in the basket. Delete one item from the shopping cart |
| **Question & Evaluation mark scale definition** |
| **Question 1.1.7.1:** Did you find easy to perform this task? Mark with any value between 0-100% (Too difficult 0% - Very Easy 100%) **Question 1.1.7.2:** Are you satisfied with the task implementation? Mark with any value between 0-100% (Not satisfied at all 0% - Too much 100% |

# 4    Conclusions

E-commerce sites are a particular kind of Web applications with similar requirements, as for example, good navigational structures, usable interfaces and so on. They present new challenges to the designer: we not only need to help the user find what he wants (a product he will buy) but also ease the shopping process. The DEPTH approach is valuable for examining the completeness of the functionality of the site under evaluation, illustrates the use of scenarios in performing user inspection, and identifies easy-to-measure correlates of more important, but complex, behaviours. Third parties have not extensively applied this approach as yet. Thus we cannot provide any evaluation data of the efficiency and the effectiveness of DEPTH. We plan to organise systematic user trials of this approach as well as to start experimenting with applying DEPTH for evaluating the usability of other types of hypermedia systems like Learning Management Systems.

## References

Alexander, C., Ishikawa, S., Silverstein, M., Jacobson, M., Fiksdahl-King, I. & Angel,S. (1977). *A Pattern Language: Towns, buildings, constructions*, Oxford University Press,New York. .

Dujmovic J Jozo (1996). A method for evaluation and selection of complex hardware and software systems. *The 22nd Int'l Conference for the Resource Management and Performance Evaluation of Enterprise CS.CMG 96 Proceedings, Vol. 1*, pp. 368-378

Douglas K. Van Duyne, James A. Landay, Jason I. Hong (2002). *The design of sites*, Addison – Wesley.

E.Gamma, R.Helm, R.Johnson, John Vlissides (1994). *Design Patterns – Elements of reusable object oriented software*. Addison –Wesley.

F.Garzotto, M. Matera, P. Paolini (1998). To Use or Not to Use? Evaluation Usability of Museaum web sites, *Museums and the Web, An International Conference*, April 1998

*Hypermedia Design Patterns Repository*. Accessed 11 February 2003, from http://www.designpattern.lu.unisi.ch/PatternsRepository.htm

J. Nielsen (2002). *Designing Web Usability*, New Riders Publishing.

G.Rossi, F.Lyardet, D.Schwabe (2000). Patterns for e-commerce applications", *Proceedings of EuroPLop 2000*.

M. Beth Rosson, J. M. Carroll, D. D. Cerra (2001). *Usability Engineering: Scenario-Based Development of Human Computer Interaction*, Morgan Kaufmann Publishers, 1st edition, ISBN: 1558607129

L.Olsina Santos (2002). Web-site Quality evaluation method: A case study of Museums. *ICSE 99-2ND Workshop on software engineering over the internet*

A.Stefani, M. Xenos (2001). A model for assessing the quality of e-commerce systems. *PC-HCI 2001 Conference*.

Martijn van Welie (2003), *Design patterns for E-commerce sites*. Accessed 11 February 2003, http://www.welie.com/patterns/index.html

# A Web Agent for Automatic Extraction of Language Resources from Hypermedia Environments

*Kyriakos N. Sgarbas, George E. Londos,*
*Nikos D. Fakotakis and George K. Kokkinakis*

Wire Communications Laboratory, Electrical & Computer Engineering Dept.
University of Patras, 265 00 Patras, Greece
sgarbas@wcl.ee.upatras.gr

## Abstract

The WATCHER project aims to automate the extraction of language resources from the Internet via an intelligent agent able to actively search and collect subject-specific and language-specific texts and build corpora and lexicons based on them. The agent is still under development. This paper presents an overview of its architecture and functionality and reports recent progress.

## 1 Introduction

The development of language engineering applications usually involves the creation of specific corpora and/or lexicons. In order to be efficient, these language resources have to be gathered from a large amount of data. Thus, their collection is a necessary but tedious task. Apart from the inherent difficulties of collecting these data, there are a few additional issues that create problems to their use. An important one is that the language resources gathered for a certain application may not be suitable for a different one. Moreover, the resources, once built, usually remain static. Although we would like to update them from time to time, such a task increases dramatically the cost of the application.

On the other hand, there exists a huge multilingual corpus that is continually growing and updated and it is available to everyone for free: the Internet. Indeed, the information contained in the Internet can be seen as a huge, constantly evolving text corpus. Moreover, the language used in this corpus is not the strictly-syntaxed formal language that parsers are used to work on. It is a free and highly expressive language of human interaction requiring robust processing tools.

The above considerations indicate the need for a tool for automatic extraction of language resources from the Internet. In order to face the problems of building large resources from scratch and the problems arising from the aforementioned issues, we proposed in 1998 the implementation of such a tool and described its function and use during an ELSNET workshop (Sgarbas, Fakotakis & Kokkinakis, 1998). We called this tool the "WATCHER" ( = Web Agent for Text Collection from Hypermedia Environments and Resources), since it is an intelligent agent able to constantly watch the Internet, providing up-to-date language resources upon request.

This tool can be useful in a few more ways than the direct exploitation of the corpora gathered. Since the corpus collection operation will be automatic, it can be initiated in regular time intervals, thus tracking the evolution of language over time. For example, a general corpus can be collected annually which will be analyzed statistically to produce an annual report. This report may contain new words, the frequency of foreign words within the corpus, etc. The same procedure can be repeated for every language and/or every thematic subject in question. If we consider this procedure in long-term, the WATCHER project will provide a permanent eye in the Internet, tracking the evolution of languages over time.

## 2    Functionality of the WATCHER

The WATCHER agent is accessing the Internet via a non-dedicated server, and users can have access to the resulted language resources. An Operator is responsible for any manual tasks during the resource collection process. The agent is able to read publicly-available documents from the Usenet, e-mail lists and World Wide Web (WWW), preprocess these texts and gather the resources locally. Measurement of statistical data such as occurrence frequencies can be performed afterwards.

According to the requested resources, the WATCHER is able to work in two *modes* (*passive* and *active*), two types of *sessions* (*constant* and *temporary*) and two types of *operations* (*on-line* and *off-line*). An Administrator (in practice the Administrator and the Operator can be the same person) will be able to set the precise function parameters along with additional language and subject-specific ones during the initialization of the session. Following are explained the functional parameters by category.

Modes: The WATCHER can function in two modes: passive and active, according to the level of initiative taken by the agent in order to collect the information. The two modes can also be seen as development stages. In the passive mode the agent receives electronic messages from e-mail lists or other similar services. It is able to analyze these messages and use them to built language resources. Thematic and multilingual resources can be developed by subscribing Watcher to the appropriate lists. In the active mode the WATCHER is able to actively search for language resources in the Internet. It will be intelligent enough to find new relevant lists and subscribe to them and to recognize used resources if it meets them again, to avoid multiple inclusions of the same data.

Sessions: The agent can collect texts from the Internet and build thematic and language-specific corpora and lexicons automatically. It can establish multiple constant and temporary search sessions. The constant sessions are used to build continually updated language resources. The temporary sessions are used for the collection of static resources. For each target language, there

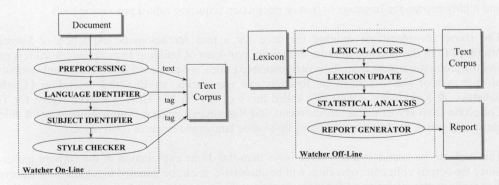

**Figure 1:** Modules (on-line operation)        **Figure 2:** Modules (off-line operation)

should be at least one permanent session producing national lexicons and corpora. The objective is to create general-purpose national resources, as complete as possible, that will be constantly updated. The precise specifications will have to be defined so that the words will be tagged properly.

Types of Operation: The WATCHER operates on-line when it is connected to the Internet searching and collecting the documents and off-line when it processes the corpus to obtain statistical data. The agent is composed of several modules, some of them being activated during the on-line operation and some activated during the off-line operation.

Figure 1 displays the modules for the on-line operation: Each of the collected documents is first processed by a preprocessing module. This is necessary because the collected documents can be gathered from many different sources. First, a checksum test is performed to each document, combined with size and source information, in order to ensure that the current document has not been already included in the corpus. Any non-text related items (such as pictures, figures, etc.) are filtered out and finally the document is normalized into plain ASCII text. A language identification component and a subject identifier determine the language-specific and subject-specific sites and texts. Thus the agent can concentrate in specific subjects or specific languages, in order to build special purpose resources. Finally, a style-checking component (Michos, Stamatatos, Fakotakis & Kokkinakis, 1996) can determine the text style. Some keywords and statistical data of a text indicate general attributes concerning its written style. For instance, a newspaper text will have a different adjective-to-noun ratio than a legal document or a literature text. This way, the texts can be categorized not only according to their subject but according to their written style as well. The normalized text is then added to the corpus tagged according to its language, subject and style. An additional tag is inserted to the text, indicating that this is a new text and no statistical analysis has been performed on it yet.

Figure 2 shows the off-line operation: The lexical access module inputs words from the new texts of the corpus and crosschecks each of them through the lexicon. The lexicon update module updates the frequencies of the words and stores any words not previously appeared in the lexicon as new words with a special tag so that the human Operator will eventually distribute them to the other categories. Then the statistical analysis module is responsible for the further processing of words and/or groups of words (Boguraev & Pustejovsky, 1996) according to the objective measurements, while a report generator module collects all the statistical data from the analysis module into an easy-readable report. The new texts that have been processed are finally re-tagged as processed like the rest texts in the corpus.

Interfaces: There are also two interfaces accompanying the WATCHER. The *Operator Interface* is used by the human Operator who tags the new words added to the lexicon. The interface enables quick link to the text from which the new words have emerged, in order to provide the context of the word, thus easing the task of the Operator. The *Administrator Interface* provides an environment from where the various sessions will be initiated.

# 3   Uses of the WATCHER

Corpus and lexicon creation is the main task of the agent. It first gathers corpora from the Internet. Their analysis is performed in a second stage. The WATCHER uses a word database with nine categories (Common, Idiom, Jargon, Proper, Acronym, Foreign, Special, Erratic, and New Word) to classify the encountered words (Sgarbas et al., 1998). Despite the human Operator involved, the described process can be considered automatic for two reasons: (a) the interference of the Operator is occasional, not crucial for the effectiveness of the process, and predetermined: he will check once every new word encountered, (b) the interference of the human Operator becomes progressively less frequent as the size of the databases grows. Of course, it is logical to assume that the process will start with a large amount of words already enlisted in the databases, although their frequencies can be set to zero.

In the same way, grammatical and morphological information can be incorporated in the lexicons. Already developed morphological lexicons can be easily incorporated in the initial word database, whereas new morphological information can be added manually for new words by the Operator.

The WATCHER is also able to determine the subject and the style of the input text (Stamatatos, Fakotakis & Kokkinakis, 2000). A semantic database is needed, containing keywords that determine specific subjects, and data that indicate certain text styles. The semantic database combined with statistical methods, can categorize the texts according to their style.

An intriguing long-term use of the WATCHER will be the tracking of language evolution over time through the generation of annual reports containing certain statistical aspects of the languages examined and all statistical data will be evaluated from these sources. The agent will establish a permanent session in the Web and it will update its corpus in a daily basis, being able to determine if an occurring text has already been considered. At the end of each year it will perform statistical measurements on the corpus collected the whole year and it will provide an annual report determining several aspects of the language in question. For instance, the word frequencies will determine how many and what new words are introduced each year, what is the percentage of the foreign words in the language, which proper names are the most popular, any new acronyms occurred, as well as more complex phenomena, like foreign loans, words that have significant changes in their frequency over time, etc. Successive annual reports for a language will give an indication to its evolution over time.

# 4   Implementation Issues and Recent Progress

The WATCHER project is still in progress. A prototype has been developed and tested, able to follow web-links (Aggarwal, Al-Garawi & Yu, 2001) and download web-pages in a breadth-first manner (Najork & Wiener, 2001). It uses *crawling* (Cho, Garcia-Molina & Page, 1988), a multi-threaded process that although it has been used extensively for the hypertext transfer protocol (http) it can be extended to other protocols as well (e.g. ftp, gopher, etc.).

Nearly all string-related-structures (e.g. URL lists, keyword sets, etc.) have been implemented as acyclic Finite-State Automata (FSA), incrementally built and updated (Sgarbas, Fakotakis & Kokkinakis, 1995). This way, the string or substring search can be performed in various flexible ways (including content-addressable and approximate matching) and the WATCHER is able to store effectively (and very quickly) all string-related information that it handles, i.e. the URLs it visits, the words it encounters, even the morphological features of the words in the lexicons. The prototype integrates a language identifier and a module providing grammatical and morphological information for Greek (Sgarbas, Fakotakis & Kokkinakis, 2000), also built as an FSA.

The performance of the WATCHER prototype was measured in a series of lengthy tests. The prototype was tested on a Pentium III 733MHz with 256 MB RAM, using a LAN connection to the Internet. The agent collected Greek texts for 5,352 minutes during a week of operation. It retrieved 52,314 files in total. The Rabin filter was active and filtered-out 27,179 of them. Thus the agent finally gathered 25,135 files with total volume 94.3 MB. The average storage rate was 4.73 files per minute with a peak of 75 files per minute.

# 5   Conclusion

In this paper we have presented an overview and recent progress of the WATCHER project, which concerns the construction of an intelligent agent watching the Internet, able to provide large-scale up-to-date general-purpose language resources with the minimum of human effort. Continually updated multilingual language resources can be extracted from the Internet, which can be seen as a huge, dynamically evolving text corpus. Specific or thematic resources will be easy to build, whereas functions such as grammatical/morphological checking and subject/style classification of texts can also be incorporated to the agent. Furthermore, the agent will be able to track down the evolution of languages over time, performing statistical processing on the pieces of corpora gathered every year and producing annual reports. Aspects of the current working prototype have also been mentioned concerning architecture, string representation and on-line update, grammatical / morphological representation of collected resources and actual performance results.

# Acknowledgements

The authors would like to thank Dr. Philippos Sakellaropoulos for optimizing the code of the WATCHER agent.

# References

Aggarwal, C., Al-Garawi, F., & Yu, P. (2001). Intelligent Crawling On the World Wide Web with Arbitrary Predicates. In *Proc. 10th International World Wide Web Conference*. Hong Kong, May 1-5, 2001.

Boguraev, B., & Pustejovsky, J. (Ed.). (1996). Corpus Processing for Lexical Acquisition. MIT Press.

Cho, J., Garcia-Molina, H., & Page, L. (1998). Efficient Crawling through URL Ordering. In *Proc. of WWW7 Consortium*. Brisbane, Australia, April 14-18, 1998.

Michos, S., Stamatatos, E., Fakotakis, N., & Kokkinakis, G. (1996). An Empirical Text Categorizing Computational Model Based on Stylistic Aspects. In *Proc ICIAI '96, 8th IEEE International Conference on Tools with Artificial Intelligence* (pp. 71-77). Toulouse, France, November 16-19, 1996.

Najork, M., & Wiener, J. (2001). Breadth-first search crawling yields high-quality pages. In *Proc. 10th International World Wide Web Conference*. Hong Kong, May 1-5, 2001.

Sgarbas, K., Fakotakis, N., & Kokkinakis, G. (1995). Two Algorithms for Incremental Construction of Directed Acyclic Word Graphs. *International Journal on Artificial Intelligence Tools*, 4 (3), 369-381.

Sgarbas, K., Fakotakis, N., & Kokkinakis, G. (1998). WATCHER: An Intelligent Agent for Automatic Extraction of Language Resources from the Internet. In *Proc. ELSNET in Wonderland* (pp. 100-104). Soesterberg, The Netherlands, March 25-27, 1998.

Sgarbas, K., Fakotakis, N., & Kokkinakis, G. (2000). A Straightforward Approach to Morphological Analysis and Synthesis. In *Proc. COMLEX 2000, Workshop on Computational Lexicography & Multimedia Dictionaries* (pp. 31-34). Kato Achaia, Greece, Sept. 22-23, 2000.

Stamatatos, E., Fakotakis, N., & Kokkinakis, G. (2000). Automatic Text Categorization in Terms of Genre and Author. *Computational Linguistics*, 26 (4), 471-495.

# Supporting Novices in Detecting Web Site Usability Problems: A Comparison of the Think-Aloud and Questionnaire Methods

*Mikael B. Skov*

*Jan Stage*

Department of Computer Science
Aalborg University, Denmark
dubois@cs.auc.dk

Department of Computer Science
Aalborg University, Denmark
jans@cs.auc.dk

## Abstract

This paper reports from an empirical study of web site usability testing. The study compares two usability testing methods which we denote the think-aloud method and the questionnaire method. The empirical study involved 36 teams of four to eight first-semester university students who were taught the two methods and applied one of them for conducting a usability test of a commercial web-site. The results of the study show that the teams performed differently in terms of their ability to detect usability problems. We conclude that the think-aloud methods provides support to weak teams by enabling them to produce acceptable results. Several of the teams that used the questionnaire method lacked this support as they were only able to detect very few problems.

## 1    Introduction

The discipline of usability engineering is characterized by numerous methods for evaluating the usability of software products. Key textbooks on usability engineering and human-computer interaction describe a broad variety of usability testing methods and techniques, including expert evaluations, usability audits, usability testing, thinking aloud, participatory design, focus groups, surveys based on questionnaires, open interviews, cognitive walkthroughs, heuristic inspections, field studies, logging, user feedback, observation (Nielsen, 1993, Rubin, 1994). Some of these textbooks include advice about the situations in which specific usability testing methods and techniques are relevant, e.g. tests with and without users (Karat, Campbell, and Fiegel, 1992), test monitor affection (Jacobsen, Hertzum, and John, 1998), or user-based test methods (Henderson et al., 1995). The WWW challenges these established experiences. It is usually argued that the web-sites are qualitatively different from conventional software systems, e.g. software products for web-sites have considerably shorter life times both in terms of development and use (Anderson 2000). Furthermore, many web-site development projects involve people with no or little formal training in usability or human-computer interaction issues (Braiterman, Verhage, and Choo, 2000).

This paper aims to improve our understanding of web site usability testing. We report from an empirical study that examines two classical usability-testing methods in supporting novices in detecting web site usability problems. These two usability-testing methods include the think-aloud protocol, cf. (Rubin, 1994, Molich, 2000) and a questionnaire-based approach, cf. (Spool, 1999).

## 2    Method

The experiment involved a large number of first-semester university students and hence, we focus on novice usability tester. We decided to rely on two different methods that are fully described in

858

the literature. We denote these two methods the think-aloud method and the questionnaire method. The think-aloud method was specified with the description by Molich (2000). This method embodies the fundamental approach as well as many guidelines from classical usability testing literature (Nielsen, 1993, Rubin, 1994). The questionnaire method was specified with the description by Spool et al. (1999). With this method, test subjects are also working on a set of predefined tasks. After completing each task, a test subject fills in a questionnaire. Another questionnaire is filled in after all tasks are completed.

## 2.1 Experimental Design

We designed and conducted an empirical study to compare the relative strengths and weaknesses of the two usability-testing methods. The study was conducted in connection with a course that is part of the first semester at a university. The course is mandatory for students in architecture and design, informatics, planning and environment, and chartered surveyor. The authors of this paper were instructors for each of the two classes that attended the course in the fall semester of 2000. Within the boundaries defined by the overall purpose of the course, we had considerable freedom to design the content of the course. We chose the two methods that were described above and combined them with general techniques for test planning, interviewing, and questionnaire design.

The course comprised ten class meetings. One class meeting lasts four hours that are roughly divided into a two-hour lecture for the whole class and two hours of exercises that are carried out in smaller groups. Thus, the total time spent for each individual student on the course and the experiment should be around 40 hours. The content of the course was designed to cover the required general issues and give a detailed account of the two methods. This was done in the lectures. The exercise time was allocated to the experiment. In the first exercises they should try out the two methods on small examples, but as soon as possible that should start to work in the exercise time on the experiment. The experiment was limited by the exercise time in order not to conflict with other activities on the semester. We chose the Hotmail web site (www.hotmail.com) as the subject for the evaluation. From the functional point of view, we chose this site because it involves a high degree of interaction, and this interaction involves both data entry and display.

## 2.2 Participants

The two classes were attended by 234 students. Independently of the course, the students were already divided into 36 groups in which they worked on a project assignment and did all class exercises. The average team included six or seven members. It was suggested that they divided the team by appointing one test monitor, a number of loggers (they were recommended to use two loggers), and three or four test subjects (they were encouraged to get as many test subjects as possible). Thus the team filled all the main roles pertaining to the usability test.

## 2.3 Procedure

Each student team was required to apply one of the two methods described above. In addition, they were allowed to supplement this method with other techniques according to their own choice. The distribution of teams on the two methods was made randomly by the instructors under the condition that each method should be used by approximately the same number of teams within each of the four categories of. The result was 19 teams working with the think-aloud method and 17 teams working with the questionnaire method. The test monitors and loggers received an email stating what method they should apply and a two-page scenario that described the web site and included a list of the web site features that they were supposed to evaluate. The test monitor and loggers were to examine the system, design task assignments for the test subjects, and generally prepare the test. The usability test sessions were planned to last approximately one hour for each test subject. Due to the pedagogical approach of the university, each team has their own office that was used for the test. The office was equipped with a personal computer and Internet access. After the test, the entire team worked together on the analysis and identification of usability problems and produced a usability report. The purpose of the report is to describe and characterize identified usability problems. the usability report should consist of a summary, description of the approach applied, results of the evaluation, and a conclusion and discussion of methodology.

The usability reports were the primary source of data for our empirical study. All reports were evaluated and marked by both authors of this paper. Through a number of iterations, we designed an evaluation sheet with 17 factors categorised within quality of test, results, and report. When the list and description of factors was complete, both authors marked each of the 36 reports independently of each other. For each report, we marked each of the 17 factors on a scale of 1 (worst) to 5 (best). The overall grade was based on a scale of 1 (worst) to 10 (best). Then we compared all the markings and negotiated each of them in order to reach an agreed evaluation of each usability report. One factor was the number of usability problems that each group found. This factor was treated differently from the others. We went through the reports and noted all problems that were found by the team, first independently and then together. This produced an absolute number of problems found by each team. In this paper, we focus specifically on that factor.

## 3    Results

The 19 think-aloud teams detected between 5 and 16 problems. Eight of the teams (42%) detected at least 10 problems. This is a remarkable number for a group of novice usability testers. At the other end of the scale no team detected less than 5 problems. The average number of problems detected by these teams is 8.6 problems. The 17 teams that used the questionnaire method detected between zero and 14 problems. Eight of these teams (47%) detected at least 10 problems. Again, this is a remarkable number for a group of novice usability testers. At the other end of the scale, five teams (29%) detected four or less problems. Even for novice usability testers, this is a disappointing result. The average number of problems detected by these teams is 7.6 problems. Figure 1 illustrates that for the think-aloud teams, there are three numbers of problems that were detected by more than one group (3, 10, and 11). The remaining ten numbers were achieved by one group. For the questionnaire teams, there are four numbers of problems that were detected by more than one group (5, 8, 11, and 12). The remaining five numbers were achieved by one group.

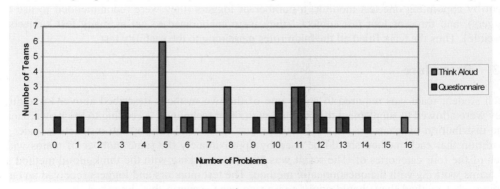

**Figure 1.** Number of teams detecting specific numbers of problems

The average number of problems detected by the think-aloud teams was 8.6 compared to 7.6 problems detected by the questionnaire teams. This difference originates primarily from the fact the five questionnaire teams that only detected between zero and four problems. This supports the immediate conclusion that the think-aloud teams did better than the questionnaire teams. This conclusion is, however, challenged if we focus exclusively on the teams that detected many problems. With both methods there are eight teams that detect at least 10 problems. Because the total number of questionnaire teams is lower than the total number of think-aloud teams, this result is slightly in favor of the questionnaire method.

The 36 teams detected 123 different usability problems. This can be considered a rather high number of usability problems for a web site. 78 of the 123 usability problems (63%) were detected by only one student team. This means that only approximately 1/3 of all detected problems were identified by two of more teams of the 36 teams. This may indicate that the student teams identify and detect a significant number of usability problems that might not be problems to other users. In fact, it is probably questionable whether all of these problems are reasonable usability problems at all. Only seven usability problems of the total of 123 (6%) was detected by at least 10 teams.

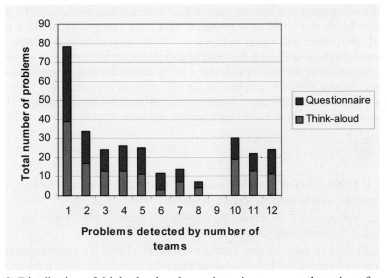

**Figure 2.** Distribution of think-aloud and questionnaire teams on detection of problems

Out of the 78 usability problems that were detected by only one student team, 39 of these usability problems were detected by think-aloud teams and other 39 usability problems detected by questionnaire teams (figure 2). It seems that there are no significant differences in the distribution of problem detection between the two methods. The problems detected by 10 or more teams are fairly even distributed between questionnaires and think-aloud teams. The slight difference in favor of the think-aloud method may be explained by the fact that the study involved 19 think-aloud teams against 17 questionnaire teams. Two usability problems were detected by 12 student teams and both problems relate the registration procedure. One problem relates the registration where more test subjects had problems in identifying the link for signing up. Six think-aloud teams and six questionnaire teams detected this problem. The other problem relates the required number of letters for the password during the registration procedure. More test subjects did not see this information and tried to enter an invalid password. Seven questionnaire teams detected this problem compared five think-aloud teams.

# 4    Conclusion

The teams that used the think-aloud method performed on average better than the teams that used the questionnaire method. The think-aloud teams detected on average 8.6 usability problems whereas the questionnaire teams detected 7.6 problems. Nearly one third of the questionnaire teams detected between zero and four usability problems. No think-aloud team did that bad in fact the lowest number of problems detected by these teams were five. At the opposite end of the scale, the teams performed more equally. With both methods, more than 40% of the teams detected at least 10 usability problems. This result is quite remarkable as they were achieved by usability testing novices who had only received around 40 hours of training in this subject. The results indicate that the think-aloud method provides basic support to weak teams in the sense that they do not completely miss the task of usability testing. That was not the case with the questionnaire method. This deficiency originates from weak methodological support to the process of identifying usability problems from a questionnaire. This process is not supported very well.

The specific conditions of the experiment limit the validity of our conclusions in a number of ways. First, the environment in which the tests were conducted was not optimal for a usability test session. In some cases, the students were faced with slow Internet access that influenced the results. Second, motivation and stress factors could prove important in this study. None of the teams volunteered for the course (and the experiment) and none of them received any payment or other kind of compensation; all teams participated in the course because it was a mandatory part of their curriculum. Finally, the demographics of the test subjects are not varied with respect to age and education. Most test subjects are a young female or male of approximately 21 years of age with approximately the same high school background and have recently started on an education.

## References

Anderson, R. I. (2000). Making an E-Business Conceptualization and Design Process More "User"-Centered. *interactions*, 7, 4 (July–August), 27-30.

Braiterman, J., Verhage, S., & Choo, R. (2000). Designing with Users in Internet Time. *interactions*, 7, 5 (September–October), 23-27.

Henderson, R., Podd, J., Smith, M. and Varela-Alvarez, H. (1995) An Examination of Four User-Based Software Evaluation Methods. *Interacting with Computers*, Vol. 7, No. 4, pp. 412-432.

Jacobsen, N. E., Hertzum, M. and John, B. E. (1998) The Evaluator Effect in Usability Studies: Problem Detection and Severity Judgments. In *Proceedings of the Human Factors and Ergonomics Society 42nd annual meeting*, HFES, Santa Monica, pp. 1336-1340.

Karat, C. M., Campbell, R. and Fiegel, T. (1992) Comparison of Empirical Testing and Walkthrough Methods in User Interface Evaluation. In *Proceedings of CHI'92*, pp. 397-404

Molich, R. (2000) *User-Friendly Web Design* (in Danish). Teknisk Forlag, Copenhagen

Nielsen, J. (1993) *Usability Engineering*. Morgan Kaufmann Pub., Inc., San Francisco, California

Rubin, J. (1994) *Handbook of Usability Testing – How to Plan, Design, and Conduct Effective Tests*. John Wiley & Sons, Inc., New York

Spool, J. M., Scanlon, T., Schroeder, W., Snyder, C., & DeAngelo T. (1999) *Web Site Usability – A Designer's Guide*. Morgan Kaufmann Publishers, Inc., San Francisco, California

# Web Browsing Activity Visualization System
# For Administrator Assistance using Browsing Information

*Satoshi Togawa\*   Kazuhide Kanenishi\*\*  and  Yoneo Yano\**

**\*Dept. of Information Science and Intelligent Systems**
Faculty of Engineering, Tokushima University
2-1 Minami Josanjima Tokushima 770-8506, Japan
{togawa,yano}@is.tokushima-u.ac.jp

**\*\*Center for Advanced Information Technology, Tokushima University**
2-1 Minami-Josanjima Tokushima 770-8506, Japan
marukin@cue.tokushima-u.ac.jp

## Abstract

In this research, we have built a system that visualizes the web browsing activities of network users for assistance of network administrator. This system visualizes the Web browsing activities of an organization by making the processing object not an individual user but a user group. The network administrator can comprehend the Web browsing activities of the organization by referring to the map. This system extracts a keyword from an HTML file that the users browsed, and this system gets the concepts from the extracted keyword and thesaurus. In addition, this system creates a browsing model that makes use of a Vector Space Model. The features of the browsing model are emphasized by weighting and after that, the model is visualized by Kohonen's Self-Organizing Maps.

## 1    Introduction

Today, because of rapid popularization of the Web, Internet users can easily get various kinds of information. This information ranges from news to bomb making activities. Network providers such as companies and universities provide rules of use, which establish a variety of limitations on Web browsing activities. However, not all users necessarily observe these rules.

Network administrators must also prevent hacking and the attacking of external sites by internal users. As well, system administrators should employ preventative measures against users placing slanderous remarks on bulletin boards and mailing lists on the Internet. The network administrator should not rely on users' morality only. If the network administrator can grasp the users' behaviour, the administrator can usually ascertain problems at an early stage. As a result, a filter that adapts to the realities of the organization can be created. At the present time, if an administrator wants to understand a user's web browsing activities, protocol analysis can be used. However, this method is very labour intensive, and this method only provides basic information like domain classification. The administrator really wants a result that shows "What is the concept of the Web that the users browsed".

So, We have developed a Web-browsing-activity visualization system for administrator assistance. This system presents a feature map of concept made up from what the internal users browsed. This system assists the monitoring operation by the administrator by showing the feature map that this system presents. As a result, we think that we can assist the monitoring operation by the administrator.

# 2 Assistance of Web Browsing Monitoring

Figure 1 shows a framework of administrator assistance. We assist the monitoring operation of the administrator by showing the type of information that the internal users browsed. We paid attention to log information preserved in the HTTP proxy server. The users browsing history is recorded in this log. This system extracts keywords from the HTML files that the users browsed. In addition, this system gets the concepts from the keywords and a thesaurus. Consequently, a browsing model is generated from the concepts and URL history that the users browsed. The method of generating the model is the Vector Space Model, the similarity problem between concepts is replaced with the cosine measure between vectors. Weighting is added to the obtained browsing model to emphasize feature quantity. Afterwards, a feature map is generated by using Self-Organizing Maps (SOM). SOM is an algorithm to map multi-dimensional vectors on a two-dimensional plane. As a result, this map expresses the typical concepts that the user browsed. The administrator gets the bird's-eye view of the organizations browsing activities by referring to the map. Therefore, the administrator is assisted in grasping user behaviour by this feature map.

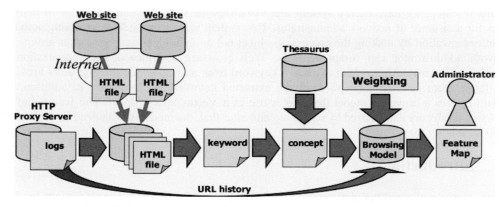

**Figure 1 : Framework of Administrator Assistance**

## 2.1 Browsing Model

The Vector Space Model (VSM) has features expressible by a mathematical model. It is simpler than other models, such as Semantic networks and Frame, that describe logical structure knowledge. Therefore, the similarity problem between concepts is replaced with the similarity problem between vectors like cosine. In addition, VSM is tied closely to SOM input, which we tried to use in this research.

## 2.2 Reinforcing of Feature Amount

Weighting is added to each element of the vector that composes the browsing model. As a result, the features of the models become clear. TF-IDF is a popular weighting method in the IR (Information Retrieval) field. The method that TF-IDF adopts is to add the products of TF (Term Frequency) and IDF (Inverse Document Frequency) to each element of the vector as weight.

TF is the number of times that the specific term appears in the document. IDF shows which documents the term appears in. Our concern was how much a large number of concepts could be extracted from user browsed Web pages. That is, we were concerned that IDF did not become too important a weighting factor. For the model used this time, it is thought that concepts that appear in the URL at high frequency express the features of the URL.

## 2.3 Visualization

The generated browsing model is constructed as a multi-dimensional vector set. Generally, humans can intuitively understand space up to three dimensions. However, grasping four-dimensional space is not easy. We believe that map information presented to the administrator should be simple. Therefore, compound operations should not be necessary to understand the entire image of the model. A Self-Organizing Map (SOM) is an algorithm to map multi-directional vectors on a two-dimensional plane. Consequently, the features of the model are expressed on a two-dimensional map. This facilitates the administrator's understanding.

## 3 System Configuration

We show the configuration of the proposed system in Figure 2. This system has a Log Analysis Module, HTML Analysis Module, Conceptualization Module, Modelling Module and a Visualization Module. A detailed description of each module is provided below.

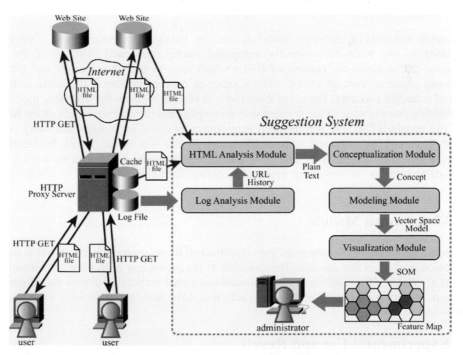

**Figure 2 : System Configuration**

## 3.1 Log Analysis Module

The Web browsing record is extracted from the log of the proxy server. It contains the URL that the user actually browsed. Extracted URL information is passed to the HTML Analysis Module.

## 3.2 HTML Analysis Module

This module acquires the HTML files that the user actually browsed, using extracted URL information. On the other hand, the HTML file to which the user referred is stored as cache data in the proxy server. If the local cache data can be used for acquiring the HTML files, it is used. When the proxy server's cache data is not available, this module attempts to acquire the HTML file from the original Web server pointed to by the URLs. The obtained HTML file is pre-processed, which includes the HTML tag removal and character code conversion. As a result, the HTML file is converted into plain text.

## 3.3 Conceptualization Module

This module extracts keywords from the plain text obtained by the HTML Analysis Module. It then executes a morphological analysis of the plain text and breaks the plain text down into morphemes. After the morphological analysis, the system searches for general and proper nouns. And, this module searches for the thesaurus by using the keyword, and acquires the concept of corresponding. The concept value of the HTML file is decided by collecting all concepts.

## 3.4 Modelling Module

This module generates a browsing model defined by the Vector Space Model. One concept corresponds to one multi-dimensionally composed vector, and each element of the multi-dimensional vector stores the number of URLs which refer to that concept. We call the multi-dimensional vector a "concept vector". The number of concept vectors is the same as the total number of extracted concepts. The set of these concept vectors becomes the browsing model.

The weighting process done to the concept vectors emphasizes the characteristics of the browsing model. If a certain HTML file is referenced many times by users, we believe that many concepts included in that HTML file will appear. Therefore, if the module discovers frequently used concepts, it is possible to search the HTML files that were frequently referred to by users.

This module executes its weighting process according to the frequency with which concepts appear.

## 3.5 Visualization Module

This module visualizes the feature map from the obtained browsing model. The SOM is used as a visualization method in this module. The concepts of the processing object are self-organized by the SOM algorithm. This results in a well-consolidated visual output that allows the administrator to get an overall view of users' Web browsing activities. (The SOM program used was SOM_PAK developed by Kohonen.)

## 4 Experimental Use and Result

The system was tested to confirm its effectiveness. We collected a log from users belonging to the Faculty of Management and Information Science of one university from July 8th to July 12th, 2002.

We show the feature maps of experimental result in Figure 3. Each feature map that the system presented has 320 elements, and each element corresponds to summarized concept. The number of concepts before clustering is 1,752 that the system extracted. Therefore, the concepts appearing in the map are browsed many times by users. In addition, the administrator can grasp a users' behaviour by recognizing the concept of appearing. For example, the oval of each map shows the concept "Game".

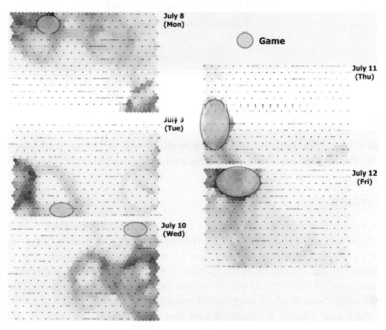

**Figure 3 : Feature Maps**

# 5    Conclusion

In this paper, we proposed a Web-browsing-activity visualization system for administrator assistance. This system extracts records of users' browsing activities from the proxy server log, and presents the administrator with a feature map. We developed a prototype system and experimented to confirm its effectiveness. It was shown that an administrator could inspect the results of the feature map.

## References

Takeda, T., Koike, H. (2000). MieLog: Log Information Browser with Information Visualization and Text Mining, *IPSJ Journal*, 41-12, 3265-3275.

WEBSOM (1999). WEBSOM research group, NNRC, Helsinki University of Technology, from http://websom.hut.fi/

Kohonen, T. (2001). Self-Organizing Maps (3rd ed.). Springer-Verlag Belrin Heidelberg New York.

Marc M. Van Hulle (2001). Self-Organizing Maps: Theory, design, and application, Tokyo, Kaibundo.

# Creating sophisticated web sites using well-known interfaces

*Fabio Vitali*

Dept. of Computer Science
University of Bologna, Italy
fabio@cs.unibo.it

## Abstract

ISA, the Immediate Site Activator, is a novel approach to web page production developed at the University of Bologna. It does not require authors to learn the features and quirks of the interfaces of specialized tools, or even less to master the number of technologies that are needed nowadays to create professional looking web pages (HTML, CSS, Javascript, not to mention XML, XSLT, etc.); rather, ISA lets them keep on using their well-known authoring tool (such as Microsoft Word or Adobe Photoshop) and builds sophisticated, professional looking, clean HTML pages behind the scene.

## 1 Introduction

The distinction between authors and readers of hypertext artifacts has always been a source for heated discussions among researchers and professionals in the field: many early hypertext prototypes had a very blurred distinction of roles. In many cases, the hypertext system was meant as a idea-collector for individuals or small groups of people (Hypercard (Goodman, 1987) is the most famous example of such systems), in others (such as Xanadu (Nelson, 1993)) the main philosophical principle was to allow everyone the right to access and tailor all published material for one's own purposes: clearly a hypertext tool that had little in the way of editing hypertext nodes was not considered appropriate. Only some publishing tool (for instance Hyperties (Shneiderman, 1987), meant for the creation of on-site museum displays) clearly distinguished between authoring and browsing activities and role types.

Things changed with the World Wide Web. Although the first WWW client written for NeXT computers by Tim Berners-Lee (Connolly, 2000) was in fact an editor, the number of plain browsers very soon largely exceeded editors (Marc Andreessen, it is said, found too difficult to create an editor for the Solaris operating system, and simply dropped off the feature in his early Mosaic prototype), and for almost everybody in the world the activity of creating content for the web has always been separated from the activity of browsing and reading existing content.

This separation has continued ever since. Browsers have been given away for free, and have rapidly become extremely intuitive tools readily usable by the majority of technically unaware Internet users. On the contrary, web editors have concentrated in providing more and more support for complex content and presentation effects, much to the detriment of simplicity, intuitiveness and ease of learning. These tools cater mostly for a wide range of web professionals, highly sophisticated and technically proficient, that need tools to increasing productivity and impact in porting to the web all sorts of content that have been created by others.

Current professional tools for creating web sites range from page-oriented HTML authoring tools that let users create single finely-tuned HTML pages (e.g., Macromedia DreamWeaver, 2003), to very sophisticated multi-user content management systems that support complex authoring processes for large scale dynamic web sites (e.g. Vignette, 2003). At the other end of the spectrum, many editing tools also allow to output content in HTML, but the result is usually simple, ugly, and visibly naive: the HTML created by graphic applications (Adobe Photoshop or

Macromedia Flash) is just a wrapper for the inclusions of images or multimedia animations, while what is created by word processors such as Ms Word is ugly, esthetically plain, and even simply and clearly incorrect.

The choice when authoring web resources is therefore restricted to either a complex and expensive tool that needs to be learnt and mastered to be used proficiently, but creates sophisticated, state-of-the-art web pages, or a simple and well-known word processor or graphic application that can create HTML code (not natively), but creates highly unsatisfactory and not professional-looking pages. It is hardly surprising that the former is bought by web professionals that already know the technologies, are not afraid to use them, and strive for good-looking results, while the latter is preferred by non-professionals that need a quick and easy output on the web, regardless of the final esthetical effect.

A separate discussion is needed for weblogs and wikis. Weblogs (Blood, 2000) are tools for fast web publishing of personal diaries, mostly free-form outburst of individuals and small communities. Weblog software is usually open source, requires some sweat in the installation and configuration, but once properly working it allows easy addition of content through web forms. Wikis (Leuf & Cunningham, 2001) are collaborative tools for the shared writing and browsing, allowing every reader (or, in some case, every authorized user) to access and edit any page of the site, with simple to use web forms and a very intuitive and reasonable text-based syntax for special typographical effects. Both may allow HTML templates to be automatically added to the content to improve the esthetics of the final result. At the moment, both weblogs and wikis seem to address the needs of special communities and marginal interest groups.

Clearly, an additional category of tools is appropriate: tools that may lack the sophistication and control that are provided by professional applications, but provide better results than those obtained by exporting HTML from a word processor. The purpose of this paper is to provide some details about one specific experiment in this direction: ISA, the Immediate Site Activator.

ISA is a server-side application that can create automatically both static and dynamic web sites by receiving in input layouts from well-known graphical packages, and content from well-known word processors. The main purpose of ISA is **not** to provide easy-to-learn interfaces for a new tool, but to allow users to exploit their previous knowledge of specific tools for the new task of creating web sites. Complete adherence to the existing interface of the selected tools is in fact at the core of the ISA philosophy, and it is in our opinion an interesting and innovative approach in the creation of new computer application.

## 2 Providing content for Web sites

If we exclude these approaches and home-made amateurish efforts, "real" web sites are still mostly created with traditional tools. Depending on the size of the site and the economical investment, these tools may range between some hundred dollars to several hundred of thousands dollars. Web site production happens in three or four usually well-differentiated stages:

- **Architecture design**: the overall structure of the site is designed, the nature and organization of the content is decided, the overall look and feel of the site is thought of. This task can be done with a variety of tools, or even on paper, or it could simply reside within the designer's head.
- **Graphic design**: one or many layout templates are created that provide the overall look of the final pages. The layouts are composed of a variety of content boxes, graphic decorations and any other textual or graphic part that is repeated across a number of similarly looking web pages (navigation menus, copyright notices, logos, etc). This job often requires a specialized graphic application used by a professional graphic designer. No professional looking web site can do without this step.
- **Back-end application design**: whenever the final pages are created dynamically, out

of queries to a server-side information system, a database has to be designed, implemented and integrated with the web server. This task can be performed by hand or with a wide range of extremely sophisticated tools, but it requires a computer professional, often a programmer, that can deal with the complexity and the technicalities of such task.

- **Content generation**: the generation of the actual information content of the site, usually composed of text blocks (from small text paragraphs to multi-paged styled documents), images, multimedia objects and any content which is not repeated across the site but belongs to a single web page.

There are many cases (e.g., simple sites composed of some dozens or hundred pages) where the architecture design can be done without specialized tools, and no real back-end database is necessary. But always content and layout need to be blended together to build the final web page.

This is done in a variety of ways, depending on the complexity of the project: a web designer can be asked to use an HTML editor to insert the content within the page layout, therefore creating by hand each single page of the site, or a complex DBMS-backed solution can be employed, that provides the authors with a web interface for the insertion of the content, which is then dynamically used for the creation of the final web pages.

In the first case, a web designer proficient with HTML and (say) Macromedia Dreamweaver is necessary every time any content of the site needs to be added or changed. In the second case, authors are required to learn the specifics of the web interface to the database, which often lack common word processing features such as styles, spelling checker, temporary saves, etc; it is common practice in such situations that the actual writing is performed locally on a desktop word processor and the final result is then copied and pasted in the appropriate boxes of the web interface. Packages of an intermediate complexity (such as Macromedia Contribute, 2003) provide support for both direct modification and copy and paste. In all cases, a proficient web professional is required to set up and maintain the overall structure of the site, establish and control the publication steps and (in simpler cases) actually create and place the final pages.

## 3    ISA for web page creation

ISA (the Immediate Site Activator) provides a completely different approach to web content creation. The main idea of ISA is to exploit standard desktop tools for the creation of content and layout, and employ a server-side application for the delivery of the final web pages. This allows both types of users to keep on using well-known interfaces for doing web-related tasks, and eliminates the need for a web professional in any part of the production and maintenance of the site.

In the scenario of producing a web site with ISA, a graphic designer (having little or no awareness of the technicalities of the web technologies that will be employed) creates the overall graphical aspect of the page using a desktop tool such as Fireworks or Photoshop. She will then use the slicing tool of the application (many graphical packages implement it) to draw the active areas (the zones where the ISA package will act), and specify their properties.

ISA allows three kinds of areas to be created:

- **Text areas** In a text area (*fvblock*) one can name the parts of the content document that must be displayed or ignored. When the content document use styles (MS Word styles work fine, as well as CSS styles or HTML tag names), ISA bases the selection on them. The "body" style refers to the main section of the document. It is thus possible to separately specify what to display and to ignore as in "Include all of the body except for the "relatedlink" paragraphs".
- **Inclusion areas** In an inclusion area (*fvinclude*) sub-layouts or additional content documents can be placed. Navigation buttons that depend on the actual content, or

content fragments that are in external files (e.g. to create a *News* section, etc.) can be specified with the inclusion areas.

- **Stretch and tile areas** A layout is liquid when it dilates to accommodate larger or longer content than expected or also to fill larger or longer browser windows than designed. A layout is usually composed of a layout table whose cells contain a slice of the overall graphics. Cells in the same column or row of the text content will therefore get taller or wider according to the dimension of the text. Stretch and tile areas (*fvstretch* and *fvtile*) allow these cell to contain not just the graphic as originally sliced by the table, but a stretched or tiled version of it. The overall result is of a homogeneous, liquid layout that correctly surrounds the text content of the page regardless of the dimension of the content. Of course, the decision between tiling or stretching the slice heavily depends on the type of graphics.

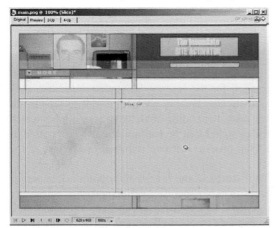

**Figure 1:** Drawing and slicing a layout   **Figure 2:** The final Web page created by ISA

The type of areas and the additional parameters are specified without any ad-hoc modification of the graphic package. Many graphic applications (including Macromedia Fireworks, Adobe Photoshop and ImageReady, Jasc Paint Shop Pro, Corel CorelDraw, Deneba Canvas, and countless others) allow slicing of the overall image in a table of cells; all of them include a mechanism to specify linking properties to the area. ISA uses the forms for specifying the link properties for the specification of its parameters. This requires no modification in the application package and a very low learning effort for using all of ISA features.

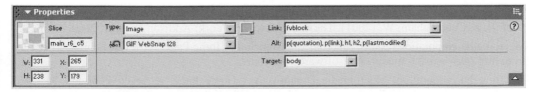

**Figure 3:** Specifying ISA parameters as link properties

After the layout has been created, the properties of the text are specified via CSS styles. ISA allows to specify a cascade of up to four stylesheets:

- A fallback sheet for basic default properties
- A browser-determined sheet for properties that are different in different browsers

- A layout-determined sheet for the properties that are specific of a layout
- A document-determined sheet for the properties specific to a single content document.

Each of them can be either created by hand (requiring the author to know CSS and deal with its syntax), or via MS Word: by saving a Word file with as many styles as foreseen in the real content documents, and the appropriate typographical properties, ISA will be able to deduce the CSS styles automatically and apply them to all the real content documents.

After the layout has been created, the content producer (having no knowledge of HTML or any other markup language) can proceed to write the content documents. He will use either an HTML editor, or, more frequently, a word processor. ISA knows and handles MS Word files with ease. The content producer will thus create any number of Word files, using styles as instructed by the layout designer, and saving them on the site as Web documents.

By providing the corresponding URLs in a browser window, ISA will merge the layout and the content document to form a complete web page. In usual web scenarios there would be the additional step of the web technologist joining the layout and content into a final artifact. ISA provides the services usually supplied by this intermediary.

## 4 Conclusions

While several tools already exist that remove the need for the intermediary web specialist, they still require layouts to be created in advance by sophisticated technologists, and indeed the task of creating such layouts can often be more complex and demanding than with less sophisticated applications. In our scenario, on the other hand, the designer can be technologically naive, just as the content author, and still be able to deliver the final product with no required awareness of the actual web technologies used.

ISA is a batch tool completely controlled by the data stored in the layout and content documents. No complex setup of the publication process is necessary. No specific web expertise is necessary in any part of the production process. The main goal of ISA is for users to keep on using their well-known tools, and to learn no new interfaces in order to create professional-looking web sites.

ISA can be tested on line at http://130.136.2.226:6969. We consider it the first example of the leveraging of existing tools with no additional interface elements for a task that is somewhat outside of the main purpose of the tool itself.

## References

Blood, R. (2000). Weblogs: a history and perspective, September 2000, retrieved February 14, 2003 from http:// www.rebeccablood.net/essays/weblog_history.html

Connolly, D. (2000). A Little History of the World Wide Web, retrieved February 14, 2003 from http://www.w3.org/History.html.

Goodman, D. (1987). The Complete HyperCard Handbook, New York: Bantam Books.

Leuf, B. & Cunningham, W. (2001). The Wiki way. New York: Addison-Wesley, 2001.

Macromedia corporation (2003). Macromedia Contribute Home Page, retrieved February 14, 2003 from http://www.macromedia.com/software/contribute/

Macromedia corporation, Macromedia Dreamweaver Home Page, retrieved February 14, 2003 from http://www.macromedia.com/software/dreamweaver/

Nelson, T.H. (1993). Literary machines, Ed. 93.1. Sausalito, CA: Mindful Press.

Shneiderman, B. (1987). User interface design for the Hyperties electronic encyclopedia, 1st ACM Conference on Hypertext, November 1987, ACM Press, pp 189-194.

Vignette corporation (2003). Vignette Content management Server home page, http://www.vignette.com/

# The Effectiveness of the Common Industry Format for Reporting Usability Testing: A Case Study on an Online Shopping Website

*Chui Yin Wong\* and Martin Maguire\*\**

Ergonomics and Safety Research Institute (ESRI)
Loughborough University
Holywell Building, Holywell Way,
Loughborough, Leicestershire, LE11 3UZ, U.K.
wchuiyin@hotmail.com\*, m.c.maguire@lboro.ac.uk\*\*

**Abstract**

The Common Industry Format (CIF) was used to help bridge the procurement decision for a software system between a consumer organization and a supplier. This paper describes the application of CIF to an online shopping website in the UK. It aims to examine the effectiveness of the CIF format in the case study. The results show that the CIF is generally a good framework for validating usability measures to report the usability testing to the client (consumer organisation) and the Web developer (supplier). However, there is room for improvement, for example the explanation of efficiency rate, inclusion of user comments and evaluator's observations, and the listing of the severity of usability problems found. This paper goes on to suggest possible issues to be included in the content of the CIF.

## 1 Introduction

In 1997, the National Institute of Standards and Technology (NIST) developed a new industry-based reporting format for usability evaluation study, namely the Common Industry Format (CIF) to help bridge the procurement decision for a software system between a consumer organisation and a supplier. The main reason behind the initiative was the need to establish a standard criterion for a minimal indication of usability for companies or organisations in making large procurement decisions for software products. As such, this has prompted a difficult situation for the companies in comparing products, planning for training or support tailored to users' needs, or estimating total cost of ownership. Hence, NIST recognised the importance of usability in the procurement process and initiated an effort to increase the visibility of software usability.

A common format to report usability test results is crucial because there are various ways in reporting such results among usability practitioners and the human factors community. The CIF reporting format will give a standard format to usability professionals highlighting the minimum level of reporting, and to identify the minimum format of shared usability information to allow consumer organizations to evaluate test results or replicate the tests if desired. Then again, the procurement team from consumer organizations may evaluate the validity and relevance of any test that it uses to support its decision-making. A common format for reporting usability tests will facilitate an evaluation and standardize the results, which will reduce misinterpretation among the supplier and consumer organizations. As such, CIF will bring benefit to both parties with the aim of reporting usability results by usability professionals within supplier organizations to be used by consumer organizations in the procurement decision-making process for interactive software products.

The content of CIF is consistent with existing standards such as ISO9241 Part 11 (Guidance on Usability), and ISO13407 (Human-Centred Design Process for Interactive Systems), and recently received the ANSI/NCITS (American National Standards Institute) – 354-2001 certification. To date, NIST is seeking international standards recognition.

## 2 PRUE (Providing Report for Usability Evaluation) Project: An Online Shopping Case Study

From 2000 to 2001, an EU-funded PRUE project (Providing Report for Usability Evaluation) was carried out by four partners from European Usability Support Centres to specify, test and report usability requirements by employing CIF as part of a contractual relationship between a supplier and a consumer organization (Bevan, 2002).

A case study conforming to CIF format was applied to an online shopping website in the UK. The online shopping website is fundamentally a mail-order company based in the UK Midlands, supplying premium quality gifts to the public that are not commonly available in the ships. The client (consumer organization) commissioned a Web development company (the supplier) to develop its Web-based online store. This paper describes the effectiveness of applying the CIF format in reporting two usability evaluation studies from 2001-2002.

### 2.1 The Design of Usability Testing Conforming to the CIF Format

Two usability studies were conducted from 2001 to 2002 on two versions of Prezzybox website to justify the effectiveness of CIF and to identify improvements made from the recommendations reported in the first trial. According to the CIF, a carefully designed usability test needs at least 8 representative users to carry out realistic tasks to evaluate the usability requirements (Bevan, 2002). As such, two user trials with 12 participants each were recruited in a usability laboratory to evaluate the user performance and satisfaction scores for the online shopping website.

The first user trial showed that 2 (16%) out of 12 (100%) subjects failed to 'purchase' online, thus highlighted a potential loss in the number of sales. A few factors contributed to the potential loss which included the lack of user confidence on the issue of "trust" whilst conducting an online purchase, speed of response, slow downloading time, slightly cluttered page layout, and poor navigational 'back' button. As such, the second user trial was conducted to examine whether the improved Web site could achieve the desired 100% purchase rate.

### 2.1.1 Task Scenarios

A task was designed in an open manner rather than following pre-determined scenarios. The subjects were given a budget to find something they were interested to purchase for themselves or for others. They were allowed to leave the site at any point during the trial. The open task was designed in such a manner to be closer to the real context to testify whether the participants would successfully complete an online purchase transaction.

## 2.1.2 Usability Metrics

The usability metrics used to examine the website are as follows:
- Effectiveness:
    - Task Completion Rate: the percentage of success for each user in completing the task. It is measured either 0% (failure to purchase) or 100% (successfully completed the purchase) at the end of each user trial session.
    - Errors: errors are instances where users did not complete the task successfully, or had to attempt portions of the task more than once.
    - Assists: when the users could not proceed with the task, the tester needed to prompt with assistance to ease the situation.
- Efficiency:
    - Task Duration: the mean time taken to complete each task, together with the range and standard deviation of times for all users. During the user trials the assisted and unassisted completion rate were not much difference, thus only the total task time was recorded.
    - Completion Rate Efficiency: task completion rate (%) / task duration (time)
- Satisfaction Scores: user satisfaction rate was measured in the Web Acceptance Questionnaire in terms of usefulness, clarity, efficiency, supportiveness, and satisfaction.

## 2.1.3 Summary of Two Usability Testing Results:
A table below shows a summary of both usability evaluation studies:

**Table 1:** A Summary of Two Usability Evaluation Studies:

| Average Scores | First Trial | Second Trial | Changes |
|---|---|---|---|
| Mean Effectiveness (Completion Rate %) | 83% | 75% | -8% |
| Mean Time to complete task | 45.25 | 46.5 | -1.25 |
| Mean Efficiency (% Effectiveness Completion/Minute) | 2.28 | 1.65 | -0.52 |
| Mean Errors | 0.9 | 1.42 | -0.52 |
| Mean Number of 'Assist' | 0.2 | 1 | -0.8 |
| Overall Satisfaction with site Scores (1 = poor, 7 = good) | 4.2 | 4.6 | +0.4 |
| Usefulness: | 4.5 | 4.2 | -0.3 |
| Clarity: | 4.6 | 4.7 | +0.1 |
| Efficiency: | 4.5 | 4.7 | +0.2 |
| Support/Help: Satisfaction: | 3.8 | 4.4 | +0.6 |

The results show that there is a 8% drop of mean effectiveness in the second trial (75%) compared to the first trial (83%). This may be due to the cluttered page layout, dissatisfaction in finding something interesting to purchase, finding difficulty in completing the online registration form during the checkout process, faulty 'back' navigational button, and lack of trust. On the contrary, the satisfaction score increased to 4.4. The reason behind these findings is possibly due to a wider range of products compared to the previous trial as expressed by the three users with participated in both trials.

# 3 The Effectiveness of CIF on an Online Shopping Website: a Discussion

Generally, Table 1 shows that the CIF provides a good framework for validated usability measures to report the usability testing to the client (consumer organisation) and the Web developer (supplier). The CIF is good for the supplier to provide measurable usability scores to gain a contract acceptance with a consumer organisation. Thus, the CIF for usability testing is set as a meaningful benchmark for the supplier if the usability scores are acceptable by the consumer organisation.

In the first user trial of 2001, the benefits of CIF to the consumer organisation were (Bevan, 2002):
- To find out how successful consumers will be in making a purchase from their site, i.e. what percentage can actually make a purchase?
- To provide a benchmark for user performance and attitude, which can be used for comparison with the shopping site when it is revised;
- To obtain insights into any problems that users face when using the site (to complement) the summative results and to receive suggestions for improving the site.

The benefit of CIF to the supplier organisation were:
- To obtain objective feedback on the success of the design they produced;
- To identify the most important usability issues that will enable the shopping site to support more successful purchases – and therefore improve the profitability of the site for the consumer organisation;
- A new contract for the supplier to improve the site, based on the test results and the comments and suggestions for improving the site.

CIF is also set as a benchmark to follow the prior requirement in the previous usability testing for revised design, especially to ensure the consistency of task carried out in the same study. The second trial was conducted according to the first trial procedure and follows the same usability benchmark. Having said this, it also indicates that setting the right usability requirements at the beginning are very important to produce a validated test result. To do a comparative study, 12 subjects recruited in each trial (minimum 8 users as requirement stated in CIF) have shown no significance differences in the usability metrics. This notion aims for the supplier to produce an industry-based report of usability evaluation studies for the customer organisation, and not targeting for rigorous statistical result. As such, the CIF showed some useful trends for the comparative usability evaluation, rather than significant differences.

In the CIF, there is no section mentioned to include user comments and evaluator's observation. An assessment of the effectiveness of first CIF survey (Maguire, 2001) indicated that the client (consumer organization) and the Web developer (supplier) are interested to receive priority ratings from the comments. For instance, how important and severe the changes should be made from the user comments and evaluator sessions are not taken seriously into account for Web re-design. It was thought the first report did not give much impact to the supplier to resolve matters from the recommendations in the revised Web site. It can be seen from the two sets of results that there is a percentage drop of effectiveness of the revised trial conducted after one year of the first trial. It was found that only partial modification of the design Web site had taken place as a result of the first evaluation. Two possible reasons behind the 8% drop of mean effectiveness in the second trial compared to the first trial are:

- No weighted or severe ratings given on the problems in the first report which meant the supplier was not sure which problems to focus on;
- The target users are not well defined and so it is not clear whether the users in the two trials were natural purchasers of goods from the site. However, all were enthusiastic to take part.

Some possible suggestions are made to be included in the content of CIF:

- Screen shots of an online shopping Web site as self-explanatory visual descriptions highlighting features or problems user experienced during the evaluation;
- Inclusion of a special section on design recommendation or evaluator's comments on particular issue(s) if any arise in context-sensitive approach, especially on online shopping behaviour, navigation path, and so forth. The report stated that responses received from the suppliers and consumer organizations as appreciating feedback provided on any design recommendations are provided, apart from the usability problems highlighted.

The CIF was found to be good at reporting traditional usability metrics (effectiveness, efficiency, and satisfaction scores) particularly on software system. In terms of efficiency, the definition has to be refined to be applied on an online shopping website where a good site that encourages browsing may lead to long completion times. It is difficult to justify whether the efficiency rate was due to the satisfactory reason for spending longer time to explore the Web site. Different shoppers have different shopping behaviours. For instance, users spending a longer time may indicate their pleasure in shopping online. However, users who spend less time searching or browsing may be preoccupied with other matters. For this reason, there are limitations to measure user experience goals (such as enjoyment, engagement, fun, entertainment, motivation and so forth) in emerging technologies such as infotainment (i.e. online games), hand-held communication devices, and interactive TV.

## 4    Summary

Generally, CIF is good at reporting a summative evaluation result for an existing system. There is room to develop an appropriate CIF tailored to formative evaluation studies, and other systems such as online shopping Web site in this instance. In addition, though CIF is good at reporting usability test results to consumer organizations and suppliers in a standard industry format, it was not found completely to be an effective communication tool to report diagnostic usability issues across to the organizations and ensure that they are addressed. No matter how well written or documented the CIF report, it does not guarantee that the client/supplier will take account of the results contained within it. As such, it is vital to have a presentation or meeting to the customer organization and suppliers highlighting the severity of core usability problems for future development.

## References

Bevan, N. (2002) Final Report to Specifying and Evaluating Usability as Part of a Contract, version 0.2. In Providing Reports of Usability Evaluation (PRUE) Project, 16 January 2002.

Maguire, M. (2001) Usability Evaluation of the Prezzybox Website, version 2.02. For Common Industry Format for Usability Test Report (PRUE) Project, HUSAT/RSEHF, Loughborough, UK, 15 June 2001.

# An Ergonomical Analysis of the Information Architecture of Websites: Developers vs Users; a Case Study of Brazilian University's Websites

*Renata Zilse*

Leui – Laboratory of Ergonomics
and Interface Usability – PUC/RJ
Av. Marquês de São Vicente, 225 -
Rio de Janeiro, RJ, Brazil
rzilse@infolink.com.br

*Anamaria de Moraes*

Leui – Laboratory of Ergonomics
and Interface Usability – PUC/RJ
Av. Marquês de São Vicente, 225 -
Rio de Janeiro, RJ, Brazil
moraergo@rdc.puc-rio.br

## Abstract

Internet is still a young media, and, mainly for the fact that it is developing and disseminating very fast, many studies are still needed in order to make it simple to use and truly democratic.

Under the hypothesis that website developing is done according to the viewpoint of the programmers, and therefore the solutions encounter a deficient usability. The present work evolves a research in order to know the mental models of both the website developers and the users, regarding the Information Architecture of websites. The study outlines a panorama of the present situation of Internet use as a truly effective tool, by analyzing articles and books so far published, about issues selected as a theoretical basis. Two semi-structured surveys were made with developers and users, comparing their respective mental models. The purpose was to draw a profile of how a website is made and how its usability is considered. On the other hand, users have been questioned in order to identify their difficulties in the execution of the navigation task on university websites, and define what they usually expect of a website.

## 1   Introduction

The creation of solutions focused on the technology has been increasingly questioned for their true usefulness. In the same way, the technologist, who dives into the process of an extremely complex development, forgets (or ignores) the user. This can be mainly noticed in the Internet, which, being a media still in its period of adolescence, as said by Donald Norman (Norman, 1998), tries to affirm itself through technological excesses and extravagances.

This premise in the development gives the product an extremely technical caracteristic feature, very often utilizing technological solutions of new generation, but inefficient from the point of view of the user. This obscure interaction, developed by and for these people, managed to create what Alan Cooper call a "software apartheid" (Cooper, 1999), where regular people are usually forbidden to enter the job market, and even to join the community because they cannot use this tool efficiently.

## 2    Problem

Besides the problem pointed out in the research development, other factors contributes to the bad usability scenery in websites:

- The computer itself, which began as a tool used by a specific category of users is used today by every kind of professionals.
- In the same way the internet, which started as a restricted use system for a specific user population, today has spread out to the most diverse necessities and to all kind of users.
- When trying to "modernize" and to be ahead in terms of technology, the firms ignore their consumers.

## 3.    Concepts

### HCI – Human-Computer Interaction

According to the Handbook of Human-Computer Interaction, the interest in HCI grows in the same proportions as the number of people who use computers to accomplish the most diverse tasks. "First, we believe research can have its most significant effect on future design (…) by revealing the aspects of human tasks and activities most in need of argumentation and discovering effective ways to provide it. We believe that such research can be fundamental, and foundational to the synthesis and invention of computer applications that simply would not occur by intuition or trial and error" (Helander, Landauer and Prabhu, 1997, p. 11).

### Usability

According to Anamaria de Moraes and Cláudia Mont´Alvão, "usability implies the system to offer it's functionality in a way that the proposed user can be capable to control it and use it without great constrains over it's capacities and abilities."(...) (Moraes and Mont´Alvão, 2000). According to Nielsen (Nielsen, 1993) and Shneiderman (Shneiderman , 1998), it is important to consider individual differences and the user categories. In Nielsen's opinion, it is important to consider each user not only in the commonly situation experienced Vs beginner, but in a 3 dimension point along the difference of users experience: with the system, with the computers in general and with the task control.

### Mental Model

According to Preece, when we interact with anything, like the environment, someone else or technological artifacts, we create internal mental models of ourselves interacting with them.

When this mental models are executed or repeated  from the beginning to the end, they offer the basis where we may predict or explain our interactions (Preece, 1997, apud Caldas, 2000). In this research we are analyzing and compensating the mental models of developers and users, knowing that it puts together their system model, through experiences and interactions with it and the image that each one has of this system.

### Navigation

According to Shneiderman (Shneiderman , 1998), navigation may be understood as the way to make progress in hypertexts, or hypermedia, associated to nets (also known as articles, documents, archives, pages, frames and screens), that contains information ( text, graphics, videos, sounds, etc) which are connected by links. According to Pierre Lévy (Lévy apud Caldas, 2002) and Rosenfeld and Morville (Rosenfeld and Morville, 1998), there are basically two kinds of navigation: hunt or the search for a known item, when the users know exactly what they are looking for; stacking or casual navigation, when users don't know what they are looking for and sometimes may not know the correct name of what they desire or don't know if it exists.

## Information Architecture

Rosenfeld e Morville think that the architectural analogy is a powerful tool to introduce a complex nature and multidimensional informational spaces. As buildings, website's architecture takes us to the most diverse reactions. And adds: the users don't realize the information architecture in a website, unless it is not working (Rosenfeld and Morville, 1998).

# 4.    Research's methods and techniques

A semi-structured research was made together with developers and users, from specific basic matters, based on theories and hypothesis interesting to the research. This way, the informant, following spontaneously it's line of thought and it's experiences on the main focus specified by the investigator, starts to participate and work on the research content (Triviños, 1987).

A.) Developers: Eleven web systems developers were interviewed, to understand their work process when elaborating a fake university website. They were asked to narrate the process and make a list of it's main preoccupations when building the website's hierarchical information structure.

B.) Users: Eleven university students (also graduated students), from 22 to 37 years old that had already searched for an article published on the internet. The interviewed students should list their main difficulties during the search process.

# 5.    Results

## 5.1.    Mentioned items on the development work process

*About users:*

Five people stated to try to understand who their users were, and three of them mentioned that an interview with the target group should be done and only one person stated that usually do it in the beginning of the development process. Two people mentioned that the user should participate on the final tests. Two others emphasized the "information divisions" according to each website's user profile, and one person affirmed that "watching with the user's eyes" is very important.

*About information's architecture*

Although this term hasn't been mentioned, six people showed concern when grouping the information by affinity, by a direct relation, per concept or function. One person stated that is important to "position the user by the context to help the navigation not causing cognitive overweight". Five people stated the use of software to help on the website structure and to visualize the relations the system should predict to facilitate the programming and built the hierarchical information structure.

*Client's opinion focus*

Five people stated it should be done as the client demands, "according to the focus and the business model established by them. Four people stated that is important to understand the client's structure (in this case, the university), how the departments are divided before the website development, and one person said this criterion should be transferred to the website itself.

## The design

The design was mentioned by six people as a finishing ("aesthetic") in the end of the project's definition made by the developers, being considered as a parallel stage, without associated functionality.

## About the work process

Two people stated that after the research, an internal work analysis is done (with the team) to study the system viability. Five people affirmed that the focus is the database model and it's interaction through the system

## 5.2.   User's navigation process

### Time of download

Ten users observed that the main problem to execute the task is the webpage download speed. This observation became obvious when it's necessary to transit through various links on the hypertext. Nine people stated that most of the time it's necessary to come back to the first stage many times until they feel satisfied or give up (seven people established a maximum period of time for this process, sometimes because they feel tired or because of the telephone line).

### Learning

Three users stated that the learning process of a website is a task that doesn't concern the media. Even if this process has to be repeated in every researched new website.

### Navigation's bad habits or website's problems?

Six users stated preference to navigate on university websites using a search tool (Google, Yahoo etc). On this cases, the results page work as a "home" to the navigation. The user clicks on the offered link, and after reading the two presented phrases, opens the desired link, take a quickly look, observes if it is what they are looking for otherwise, returns to the search results page. Only one person stated to navigate "two or three levels"on the researched university website. One user also considers the navigation through the links on the menu faster. These links (hypertexts or buttons) were considered confusing to four users, and received a dubious label, making it difficult to understand.

### Apparent excess of information

Six users stated that there's an excess of low interest and sometimes useless information, which makes the research more complicated and stressing. A user stated to feel bothered with flicking information, that pollutes the page and makes it more difficult to execute the task. Six people said that usually the research target is hidden under the website's layers, multiplying the difficulty when executing the task.  At last, all users stated to give up the research at least once and two of them felt unable by not finding the desired article.

## 6.    Conclusion

### About the interview with the developers:

- 63% gives the client the total responsibility of the work direction and the hierarchy of the information contained on the website.
-  91% priories the system.
- 73% leave the user on a third  level.

The absence of tests with real users on the development process is clearly a factor that compromises the usability navigation of the generated product. The lack of a practical usability criterion to build the website's semantic, addresses the work to a strictly technical point of view, with the information architecture turned to the system performance

## About user's interviews:

The problem of structure on websites where information is not easily found, is clear. It's also clear that the excess of resources (images, animations, etc) contributes to the delay of the page loading and add confusion to the navigation. Links with dubious titles are a serious problem. Even the ones who found difficulties to reach their aim and that already experienced failure when executing the task, considered that the website learning process doesn't concern the media. The items in a context and the grouping of information mentioned by developers as a development concern, wasn't mentioned by users as a facilitating factor. It was clear that the desired information being hidden under diverse layers is a problem.

This research confirms that the interests doesn't align and the analysts intended results wasn't necessarily desired by users. The solutions proposed doesn't make easier the navigation of the user who wants to accomplish a research in universities websites and can't execute it . It is worth to think about these aspects and the "software apartheid" proposed by Cooper (Cooper, 1999) as a thought on how to make this media a useful tool for every human being.

# References

CALDAS, Luiz Carlos Agner. Otimização do Diálogo Usuários-Organizações na World Wide Web: Estudo de Caso e Avaliação Ergonômica de Usabilidade de Interfaces Humano-Computador. PUC, 2002. 2 v.

COOPER, Alan. "The Inmates are Running the Asylum". 1999. Sams, Indianopolis, Indiana. 261p.

HELANDER, Martin G.; LANDAUER, Thomas K.; PRABHU, Prasad V. Handbook of Human – Computer Interaction. Elsevier, North-Holland. 1997.

LÉVY, Pierre. Cibercultura. São Paulo. 2000. 2a Edição, 260 p.

MORAES, Anamaria e MONT'ALVÃO, Cláudia. Ergonomia, conceitos e aplicações. Rio de Janeiro, 2AB, 2000. 2a Ed. 132p.

NIELSEN, Jakob. Usability Engeneering. San Francisco (California), Morgan Kaufmann, 1993. 362p.

NORMAN, Donald A. The Invisible Computer: Why Good Products Can Fail, the Personal Computer Is So Complex, and Information Appliances are the Solution. London (England), The MIT Press, 1998. 302 p.

ROSENFELD, Louis; MORVILLE, Peter. Information Architecture for the World Wide Web. Beijing, O'Reilly. 1998.

SCHNEIDERMAN, Bem. Designing the User Interface. Reading (Massachusetts), Addison-Wesley.

TRIVIÑOS, Augusto N. S. Introdução à pesquisa em ciências sociais. Editora Atlas S.A. São Paulo. 1987, 175p.

# Automatic Web Resource Discovery
# for Subject Gateways

*Konstantinos Zygogiannis[1], Christos Papatheodorou[2],*
*Konstantinos Chandrinos[3], Konstantinos Makropoulos[4]*

[1]Dept. of Communication Systems, Univ. of Lancaster, Lancaster, UK
[2]Dept. of Archive & Library Sciences, Ionian University, Corfu, Hellas
[3]Institute of Informatics & Telecom., NCSR Demokritos, Athens, Hellas
[4]Division of Applied Technologies, NCSR Demokritos, Athens, Hellas
[1]k.zygogiannis@lancaster.ac.uk, [2]papatheodor@ionio.gr,
[3]kostel@iit.demokritos.gr, [4]cmakr@nh.gr

## Abstract

Subject gateways have been heralded as the qualitative answer to information overload and the meagre ability of current-generation search engines to provide context-sensitive query answering. However, human maintenance proves problematic as soon as the domain broadens slightly. In this paper we report on an attempt to leverage the maintenance problem of a successful subject gateway for chemistry, by providing automatic tools. The tools collect and rank quality Web resources, utilizing the existing ontology of the gateway. Evaluation of such a Web search mechanism against classic web queries to popular and topic-specific search engines gives promising results.

## 1    Statement of problem

Recent years have seen a number of efforts attempting to leverage information overload. Although the most obvious aspect of information overload is quantity, the problem of information quality has started to receive special attention, particularly when treating the Internet as a source of educational or professional information. Information quality attempts to answer the question: how can one select from "everything" only those information items that meet one's information needs and at the same time carry a certain validity or authority?

The immediate reaction to this was the design and implementation of hierarchical directories that attempted to capture the real-world subject hierarchy and assist the user in narrowing down the number of resources he/she had to consult. Two critical virtues of such efforts, soundness (the fact that all links pointing to the intended resources are valid) and completeness (nothing equally or more interesting exists outside these links) were vanishing on a daily basis. Soundness, although not guaranteed at 100%, is often supported by automated link-checkers, while completeness or near completeness has become an issue of commercial competition. The larger the army of human indexers, the higher the completeness of the directory is.

To address content quality, long-standing traditions and techniques coupled with novel technology have given rise to a new type of information gateways, the so-called "quality-controlled subject gateways". These are "Internet services which support systematic resource discovery" (Koch, 2000). Important aspects of a quality-controlled subject gateway are that it provides access to resources that are mostly on the Internet along with descriptions of these resources and the ability to browse through the resources via a subject structure. In the following paragraphs the term "subject gateway" is to be interpreted as a quality-controlled Internet resource catalogue, with descriptive documentation and a hierarchical browsing facility.

Subject gateways employ one or more quality criteria for selecting resources. The employed criteria manifest themselves in the resource selection, in the hierarchy decisions and/or the descriptions that accompany every link to a resource. Subject gateways are hard to maintain, primarily because the criteria for content inclusion are not necessarily compatible with the indexing mechanisms provided by Web search engines. Although it is relatively simple for the administrator of a subject gateway to construct automated Web queries and augment the gateway with the links returned, it is often the case that the results contain marginally relevant or even irrelevant answers. This is not alleviated either by the so called "relevance ranking" performed lately by the search engines, although a number of algorithms have been proposed (Yuwono & Lee 1996; Carrière & Kazman; Kleinberg 1998; Page et al., 1998). Relevance ranking within the result set is often used because there is no hierarchical structure of concepts to signify the context when issuing a query. Relevance metrics are tuned to popularity and cross-reference, rather than quality, which is of course a subjective judgement and hence cannot be automatically computed by a search engine.

Although the difficulty of maintaining a subject gateway, there are continuing efforts to provide automatic "intelligent" software components that will attempt to employ "quality" criteria while they are crawling the Web for resource discovery. Personalization techniques learn, model and utilize the subject gateway users behavior to discover interesting information resources and customize personal portals (Anderson & Horvitz, 2002). The Focused Crawler system (Chakrabarti et al., 1999) bounds its crawl to find links that are relevant to web pages indicated by a user and tries to identify the pages that are great access points to many relevant pages. Ontologies address the semantics problem of the query words issued to the search engines. An ontology (Noy & McGuiness, 2001) is a formal and explicit description of concepts in a domain of discourse as well as their properties and features and the restrictions on these properties. Concepts are usually organized hierarchically. Ontologies include machine-interpretable definitions of basic concepts in the domain and relations among them and define a common vocabulary for sharing information in a domain. Many intelligent search engines use ontologies to organize web pages according to their topical hierarchical structure (Tiun et al., 2001; Tandudjaja & Mui 2002).

In this paper we report on work contributing to the automation of maintenance of a quality-controlled subject gateway. Our work aims to provide a subject gateway administrator with a system, which aids him in discovering new resources using the controlled vocabulary offered by a domain specific taxonomy. In doing so, we expect that the quality criteria employed by humans in crafting taxonomical hierarchies will be maintained in the newly discovered resources. Our experiments were based on a well-established and popular subject gateway on Chemistry, but the design and implementation of the mechanisms for automatic resource discovery is domain independent.

## 2    The Chemistry Information Retrieval

"Information Retrieval in Chemistry" subject gateway (http://macedonia.chem.demokritos.gr) has developed a 3 level taxonomy followed by a controlled vocabulary for representing the chemical knowledge and related domains. For each level there is a set of codes.

- 1st level: Chemistry or Chemistry Related scientific domains (two-digit codes for example "Chemistry" code=01 or "Energy" code=06)
- 2nd level: Chemistry or Chemistry Related sub-domains (Biochemistry, Organic Chemistry, Food Chemistry, etc., three-digit codes, for example "Biochemistry" code=050 or for the Chemistry related "Biotechnology" code=200)
- 3rd level: Information Resources (Books, Journals, Databases, Conferences, Mailing lists, Academic Servers, News/newspapers/magazines, etc., two-digit codes for example "Database" code=15)

In view of the above, the combination of codes defines a query to the subject gateway (i.e. 01-050-15 means "Biochemistry databases").

The administrator maintains the subject gateway by performing a search into popular search engines, extracting the results and finally selecting the best. Given the current size of the gateway a full cycle through the taxonomy can be achieved twice a year on a part time basis or up to four times a year on a full time basis. We designed and implemented a set of automatic mechanisms that allow the administrator to go through this cycle much more rapidly, by utilizing the implicit ontology to construct queries to popular search engines. We re-rank their results minimising the number of links requiring visual inspection for quality assurance.

# 3    The proposed architecture

We designed and implemented a client/server architecture using Java for content collection and processing and HTML for the Web interface. The proposed system consists of three modules. These are the searching module, the ranking module and finally the end-user interface.

## 3.1    Searching

The administrator issues a query through a web form, using the controlled vocabulary terms provided by the taxonomy. The query is re-formatted appropriately and propagated to a list of predefined popular search engines including a large number of Chemistry related search engines and portals. Moreover he/she can choose the number of the results that will be analysed from each response set, including "all results". In this version, we have used hard-coded wrappers to extract the links pointing to content from the returned results. The searching procedure removes duplicates and invalid URLs and stores the requested number of links in a database. Since the search engines rank their results, the stored links are the top ranked from each response set.

## 3.2    Ranking

The Ranking servlet extracts the search results from the database where they were stored from the Search procedure, processes them and finally stores the ranked results to a different database. During its operation the algorithm methodically examines all the various links provided and uses regular expressions, substring matching and the edit distance for errors and misspellings since the domain contains a lot of technical terminology.

In particular, if the exact taxonomy term has been found in a specific link, then the algorithm marks this link with a score accordingly to (i) the frequency of the taxonomy term as well as (ii) its position in the retrieved page (e.g. if the taxonomy term is found in the meta tags instead of the main text, for example in the title of the retrieved page, then the page takes a bonus). Moreover if a taxonomy term consists of more than one word, then the algorithm combines the abovementioned criteria with the proximity of the taxonomy term words in the retrieved page. If the proximity of two words or phrases is greater than a maximum allowed distance in the retrieved page, then the algorithm considers that the taxonomy term does not exist in the retrieved page. For instance, for a given example of a taxonomy term consisted of two words, it could be stated that these words should occur within four words. In the case of a relevant matching, the algorithm ranks the retrieved page with respect to the (i) edit distance that the taxonomy term has with the words in the retrieved page (e.g. if taxonomy term is 'journal' and a word in the retrieve page is 'journalism' then edit distance equals to 3) and (ii) above-mentioned three criteria (position, frequency and proximity) of the similar words in the retrieved page.

## 3.3    User Search algorithm and user interface

The ranked results can be browsed by the users through a web form. The users can select at a query one or more Chemistry related sub-domains (Educational, Databases, Glossaries, Journals), but only one domain   (Organic, Inorganic, Physical, Analytical, and Environmental Chemistry). Furthermore they can select a sub-set of the search engines used by the searching module of our system to collect links. The query is performed offline, on the previously aggregated, ranked and potentially approved links. The results are returned in the form of clickable HTML links and their ranking is mentioned.

Given the positive results of the work so far, we intend to experiment by allowing the user, after appropriate warning, to send queries to the selected search engines at real time, when the existing results do not satisfy the original query. This will allow us to define a trade-off between the delays of answering with the topicality of the real time results.

## 4    Evaluation

An early evaluation test was performed with the first prototype. Queries were posed to four search engines, including two Chemistry related, Yahoo and Google, as well as to the proposed automation mechanisms.  During this evaluation, each search engine was queried ten times for ten different subjects. The queries issued were: Organic chemistry journal, Chemical suppliers, Chemical magazines, Chemical portals, Chemical DNA structure, Spectroscopy, Nuclear Magnetic Resonance, Apolipoprotein Antigen, Acetyloleolinesterace, Ion chromatic analysis system. The subject gateway administrator browsed the results and selected the relevant answers summarised in the following table, after removal of duplicates.

**Table 1:** Relevant Answers for a Sample of 100 Queries

| Query | 1 | 2 | 3 | 4 | 5 | 6 | 7 | 8 | 9 | 10 |
|---|---|---|---|---|---|---|---|---|---|---|
| ChemIndustry. | 18 | 19 | 23 | 20 | 32 | 42 | 16 | 32 | 29 | 09 |
| Chemie.de | 34 | 15 | 12 | 29 | 28 | 30 | 19 | 41 | 16 | 13 |
| Google | 23 | 28 | 22 | 11 | 13 | 19 | 6 | 9 | 0 | 1 |
| Yahoo | 19 | 24 | 14 | 12 | 10 | 6 | 2 | 17 | 3 | 0 |
| Total | 56 | 53 | 43 | 42 | 41 | 23 | 29 | 51 | 14 | 16 |
| Our system | 69 | 73 | 78 | 69 | 71 | 49 | 53 | 74 | 32 | 43 |

The *Total* row of Table 1 accumulates the overall number of correct results.  This was produced by adding the number of correct results returned by each search engine excluding duplicate findings than may occur during this addition. Table 1 shows that the designed gateway maintenance method produces in total, more relevant results than the rest of the engines do together. Also, our system performs best for highly specific queries (e.g. Acetyloleolinesterace or Ion chromatic analysis system). It that cases it returns almost three times more correct results than all the engines. This is because our system utilizes all available results, not only the first one hundred.

# 5    Conclusions and future work

In this work we have tried to provide automated tools for the assistance of a subject gateway administrator in maintaining its content. We did so, by exploiting the implied ontology from the taxonomic representation of the domain. However, our tools are at no point domain specific, so they could be easily re-used for different domains and gateways. The evaluation results are a strong indication that our system could easily become the automated assistant of the subject gateway administrator, collecting appropriate material in minimal time and reducing it to a tractable quantity that would allow the administrator to keep the gateway up-to-date

A number of implementation choices will be revised in a second version. For example, the use of open queries with term matching against the ontology via thesauri or the hard-coded vs. automatic wrapper induction for the various search engines. In addition, we intend to explore learning techniques from the content of links already approved and included in the directory structure so as to achieve a more thorough modelling of each concept and utilize query expansion for ranking.

# References

Anderson, C.R. & Horvitz E. 2002. Web Montage: A dynamic personalized start page. In *Proc. 11th Intl. World Wide Web Conference*. ACM Press.

Carrière, J. & Kazman, R.  WebQuery: Searching and visualizing the web through connectivity. Retrieved February 7 2003, from http://www.cgl.uwaterloo.ca/Projects/Vanish/webquery-1.html

Chakrabarti, S., van der Berg, M. & Dom, B. 1999. Focused crawling: A new approach to topic-specific web resource discovery. In *Proc. 8th Intl. World Wide Web Conference*, 545-562.

Kleinberg, J. 1998. Authoritative sources in a hyperlinked environment. In *Proc. 9th ACM-SIAM Symposium on Discrete Algorithms*.

Koch, T. (2000). Quality-controlled subject gateways: Definitions, typologies, empirical overview. *Online Information Review*, 24(1), 24-34.

Noy N.F. & McGuiness, D.L. 2001. Ontology Development 101: A guide to creating your first ontology. Stanford Knowledge Systems Laboratory Technical Report KSL-01-05.

Page, L., Brin, S., Motwani, R. & Winograd, T. (1998). The PageRank citation ranking: Bringing order to the web. Stanford Digital Libraries Working Paper.

Tandudjaja, F. & Mui, L. 2002. Persona: A contextualized and personalized web search. In *Proc. 35th Hawaii Intl. Conference on System Sciences*. IEEE Press.

Tiun, S., Abdullah, R. & Kong, T.E. 2001. Automatic topic identification using ontology hierarchy. In *Proc. 2nd Intl. Conference on Intelligent Text Processing and Computational Linguistics*, LNCS 2004, (pp. 444-453). Springer-Verlag.

Yuwono, B. & Lee, D.L. 1996. Search and Ranking Algorithms for Locating Resources on the World Wide Web. In *Proc. 12th International Conference on Data Engineering*, 164-171.

# Section 6

Learning and Edutainment

# Social mechanisms for content quality control in web-based learning: An agent approach

*Nikolaos Avouris, Konstantinos Solomos*

Electrical & Computer Engineering Dept, HCI Group,
University of Patras, 26500 Rio Patras, Greece
{ N.Avouris, solomos }@ ee.upatras.gr

## Abstract

This paper investigates interaction of users with multiple agents in the context of web-based implicit communities of learners. In particular a user-centred quality-of-content control mechanism is presented, inspired by social protocols of interaction, which is used for building and maintaining cohesion of such communities. This takes the form of an innovative algorithm for ranking offers of educational material from heterogeneous educational sites, modelled as autonomous information agents. This algorithm takes into consideration the social acceptability of each sites' contents. The described user interaction with this multi-agent system allows for adaptive participation in the educational process, while maintaining the user as the principal locus of control in the user-system interaction.

## 1 Introduction

The web has been considered, since its early days, a medium suitable for learning. However the role of the student in most current web-based educational systems is that of a passive receiver of educational content. This is despite the fact that modern pedagogical views suggest active participation of the student in the learning process. In addition, learning is considered a social process (Bruckman, 2002, Dillenbourg, 1996). So collaborative learning has gained prime importance in most realms of contemporary learning sciences. Since the web has recently started playing the role of effective enabling technology for collaboration and community building besides that of the more traditional information gathering (Bruckman, 2002, Preece 2000), the use of the web as means for building communities of active learners is very interesting perspective.

In this context, one can distinguish two types of learners' communities: Those that necessitate explicit membership and the ones in which participation is implicit, e.g. collaboration emerges as the result of sharing peer members' views on the common resources. The latter case, i.e. that of implicit learners communities, is particularly interesting and is the subject of research reported in this paper.

In particular, we refer to the design of a prototype to facilitate online and web-based study. The main problem in this context is to provide easy and flexible access to various data residing in multiple, unstructured, heterogeneous information resources, and to integrate these data into semantically coherent information. An objective is also to support social cohesion through meta-information sharing about the educational resources. The proposed solution is the introduction of autonomous, goal-driven agents that permit collaboration and facilitate access to heterogeneous educational resources through an open-system approach. The users of this environment can direct their attention to various complementary sources of educational material. An important mechanism developed in the frame of this architecture is the quality control of the content by the users themselves, thus creating a social network of individual and community-wide preferences. This is the main mechanism used for maintaining the implicit community of the users of these shared resources.

In this paper we present and discuss the design of this multi-agent distance learning system. Issues of control, transparency, heterogeneity and protection of user privacy are discussed in relation to existing system architectures. Subsequently special focus is given to the user interaction with this distance-learning platform. It is argued that the proposed approach can tackle effectively some of the key issues of user interaction with multi-agent systems and community support and is relevant to other application areas with similar characteristics.

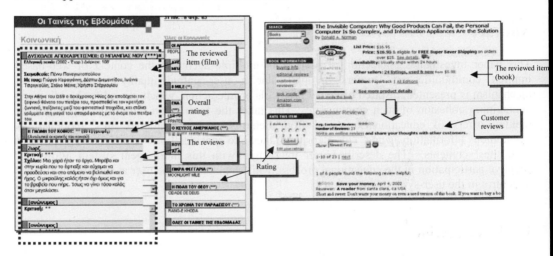

Figure 1. Implicit community of cinemagoers in the area of Athens, Greece (www.athinorama.gr) and of the amazon bookshop customers (www.amazon.com).


## 2   On implicit communities

Web-based implicit communities are formed by users who share resources and maintain meta-information relating to them without getting engaged in direct interaction. Examples of such communities are often found in the web. Two typical examples are discussed in this section. The first example is the community of performing-arts spectators of the Athens region in Greece. This community is built around the website *www.athinorama.gr*, which provides information on cultural events in the Greek capital. The designers of the site, owners of a popular weekly on city life, allow for the public to provide their views on films and theatre plays watched. Long chains of viewers' comments are included in the site. See figure 1(a). A rating on a 1 to 5 star scale is provided as a subjective quantitative measure of the expressed views. The most active reviewers provide usually a nickname, many remain anonymous, while some reviewers provide just a rating with no additional comments. It is common for writers of comments to read earlier reviews and refer to them, however no threading of this exchange of views is allowed, since the same person cannot provide a second opinion on the same item. The site provides an overall rating of the reviewed item as a mean value of the reviews' ratings. In the screenshot of figure 1(a), the film has been reviewed by 55 users and has received a 3 out of 5 rating. The reviews are presented in inverse chronological order, the most recent on top. While items with associated average ratings are shown on the screen, no ranking of items is provided according to spectators' views. Users of the website may identify frequent reviewers who provide comments on many of the films and plays that are on show. So a real community of people who share interests and exchange views, without getting engaged in direct communication is formed through this website.

A second example of implicit community is that maintained by the popular online bookshop amazon (*www.amazon.com*), see fig. 1(b). The items to be reviewed in this case are books, sold by the bookshop. The comments are relatively longer and the reviewers can come from any geographical location in the world. Fewer people provide anonymous reviews in this case. The overall rating of a book, is provided by the site, as in the previous case. It should be noticed that in this case the website provides this reviewing facility not as its main mechanism for influencing prospective buyers, since there are additional ways for making suggestions on books, e.g. thematic grouping, books bought by other shoppers, comments by publishers, authors etc. However in this case the customer reviews facility is supporting building of an implicit community, even if this might not have been the prime objective of the website builders.

In these two discussed examples the website makes reference to items that are locally stored in the database, there is an invested interest in the relationship between the web site visitors and the website owner, while the items rating is performed in both cases in addition to verbal comments. The discussed implicit communities have many differences from usual on-line communities, the most notable of which is the lack of direct communication of the community members. In the following we discuss a new framework for web-based learning in an open environment that uses some of the presented ideas of implicit communities.


# 3    The learners community-building architecture

The proposed architecture (Solomos & Avouris 1999, Avouris & Solomos 2001) contains an open number of educational systems that can register and withdraw from the society temporarily or permanently without affecting other agents' tasks. Each educational source can interact with other peer sources through an agent who acts as representative of the educational source. The content can be an electronic multimedia book, an intelligent tutoring system or any other web-based educational system. The agent contains a meta-level description of the content, according to a commonly used educational ontology, hierarchically structured in a four-level model. A specifically designed communication language has been defined (EACL, see Solomos & Avouris 1999), supporting agent interaction, based on the widely used KQML language performatives (Finin, Labrou, Mayfield, 1997), implementing an adaptation of the contract net collaboration protocol for the educational domain.

The typical scenario of user interaction with this environment, involves the following steps:

(a) The user searches the available resources and sends a request ( r ) through the user agent, see figure 3 (a).

(b) The user agent seeks suitable agents to satisfy the request ( r ) through the server. The server searches the meta-description of registered agents and selects the most suitable ones to which it forwards the request ( r )

(c) The agents send their bids to the user. The bids are ranked according to their content and previous history of interaction with the bidding agents (social acceptability factors) and are presented to the user, see figure 3(b).

(d) The user selects one of the offers and the corresponding agent is requested to provide the content through a new interaction window, figure 3(c).

893

(e) After completion of interaction with the selected source, the user is asked to evaluate the provided content, see figure 3(d). Only quantitative measures are requested, in contrary to other implicit communities, discussed earlier. This evaluation is stored in the relevant databases in order to influence future user-agents interaction.

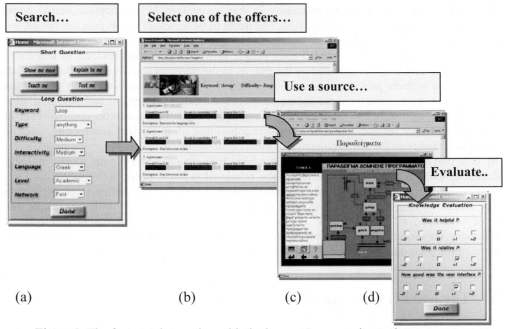

(a)  (b)  (c)  (d)

Figure 2. The four step interaction with the learners' community environment.

This peer user evaluation is stored both locally in the User Database and in the Server, where a record of users' evaluation of provided services is stored in the Social Acceptability Database. These evaluations are used for ranking the offers presented in future users. The ranking of offers is important, as with web search engines. The ranking criteria used are based on an adapted version of the contract net protocol modified by a social acceptability factor. In the traditional contract net protocol (Smith, 1988, Sandholm & Lesser, 1998) the offers of the bidders relate the bid request to their capabilities, and performance characteristics. In the proposed architecture the bids express the capability of the bidder to satisfy the requirements of the requesting agent, the relevance of their content, the similarity in educational level, the style of presentation etc, called the bid factor (BF). In the proposed architecture however the bid-ranking algorithm is not based on just the offering agents' bids. BF is considered a subjective evaluation factor, since all bidders do not use similar standards for describing the quality and the characteristics of their content. A more objective criterion is to adjust the offers by a measure of credibility of the offering bidders. This is based on two factors: the previous history of the particular user interaction with the offering agent (the *Trusted Agents Factor, TAF*) and the social consensus on the quality of the agent (the *Social Acceptability Factor, SAF*). TAF is stored in the local user database, while SAF is provided by the server database. These factors model a social network of evaluation of services by the human agents built within the agents' society.

Agent Ranking (AR) is thus estimated according to the following formula:

$$AR(EA_j) = (w_1*TAF_i+w_2*SAFi+w_3*BF_i)/S\ w_j$$

Where: w1 and w2 are weighting factors calculated as functions of N, i.e. the number of evaluators that participated in the estimation of the corresponding factors, according to the following formulas:

$$w_1 = (N1/NTmax)*w_T \quad and \quad w_2 = (N2/NSmax)*w_S$$

N1 and N2 are the number of evaluations that are contained in the corresponding databases, NTmax , NSmax the maximum number of recent evaluations that can be taken in consideration and $w_T$ , $w_S$ are constant weight factors. In case that N>= Nmax, then the Nmax more recent evaluations are taken into account. In our example the weights used are $w_T = w_S = 0.3$ and $w_3 = 0.4$.

## 4  Conclusions

The architecture of an open system that can integrate heterogeneous educational resources has been presented in this paper. An essential part of the architecture is the algorithm for ranking offers of material through the search engine of the user. This algorithm takes into account the evaluation of the peer users of the available resources. This way the human society's rules that establish evolution of trust and commitment, based on the history of social interaction are modelled. According to these rules, the selection of a service provider is based on a combination of previous positive experience with the candidate and other people's positive experiences. An implicit community of peer learners is build through this mechanism. In contrary to other communities, in this case, the views of peers influence the search engine and is interweaved in the system. During the evaluation of the described prototype, the users did not object to the imposed short evaluation questionnaire, since it was task-related and it provided them with the opportunity to express their view on the service. However longer experimentation with wider communities of users of this architecture are needed in order to confirm the first positive findings.

### Acknowledgements

The architecture has been developed in the frame of the Greek national research project PENED 99ED234 "Intelligent Distance learning Tutoring Systems", funded by GSRT.

## 5  References

Avouris N., Solomos K., (2001), User Interaction with Web-based Agents for Distance Learning, J. of Research and Practice in Information Technology, 33 (1), pp. 16-29, February

Bruckman A. (2002), The future of e-learning communities, Communications of the ACM, 45 (5), pp. 60-63.

Dillenbourg, P., Some technical implications of distributed cognition on the design of interactive learning environments, J. of Artificial Intelligence in Education, 7 (2), pp. 161-180, 1996.

Finin T., Labrou Y., Mayfield J., (1997). KQML as an Agent-Communication Language, Chapter 14 in Bradshaw J.M. (ed.) Software Agents, pp. 291-316, AAAI Press.

Preece J. (2000), Online Communities, Designing Usability, Supporting Sociability. J. Wiley, Chichester, UK.

Sandholm T., Lesser, V. (1998), Issues in Automated Negotiation and Electonic Commerce: Extending the Contract Net Framework, in Huhns M.N., Singh M.P. (eds.) Readings in Agents, pp. 66-73, Morgan Kaufmann, San Fransisco.

Smith R.G (1988). The contract net protocol: High-level communication and distributed problem solving. In Readings in Distributed Artificial Intelligence, Morgan Kauffman.

Solomos K., Avouris N. (1999), Learning From Multiple Collaborating Intelligent Tutors: An Agent-Based Approach, J. of Interactive Learning Research, 10 : 3/4, pp. 243-262.

# Reinventing the Lecture: Webcasting Made Interactive[1]

*Ron Baecker[2], Gale Moore, and Anita Zijdemans[3]*

Knowledge Media Design Institute (KMDI), University of Toronto
40 St. George Street Room 7228, Toronto Ontario M5S 2E4 Canada
rmb@kmdi.utoronto.ca

## Abstract

We present a novel system (called ePresence) for highly interactive webcasting with structured archives. We discuss how various kinds of interactivity among both local and remote participants in lectures are enabled by this technology. We suggest how ePresence may be an effective tool for creating a knowledge base and enhancing learning. Finally, we illustrate these concepts with an example of using the system to encourage and facilitate the formation of a learning community.

## 1    Background

The lecture has long been a dominant method for transmission of information from instructors to students, particularly in undergraduate education. In an age of increasing use of technology-based distributed education, lectures are still cost-effective presentations and performances of carefully designed sequences of material that can be attended and viewed concurrently by many students. Yet this type of delivery model also has disadvantages, particularly within the context of current research on how people learn (Donovan et al, 1999). Perhaps most serious is that it is typically a one-way broadcast medium from instructor to student, with little back-channel communications and little dialogue between instructor and student and among the students themselves.

Despite its prevalence, this transmission model is not conducive to the learner-centred model presently considered to be the more effective approach to pedagogy (Olson & Bruner, 1996). Furthermore, it does not take into consideration the social nature of learning or the need to accommodate different learning objectives with a variety of teaching tools and techniques (Branson, 1998). Thus lectures are increasingly under challenge (Kerns, 2002), as we understand more about learning and the kinds of environments we should be creating to support learning

## 2    The ePresence system

The technology of webcasting gives us the opportunity to reinvent the lecture. *Webcasting* is the Internet broadcasting of streaming audio possibly accompanied by streaming video so that it can be viewed via a Web browser on a personal computer. Webcasting use grows as Internet broadband communications becomes more available and more affordable, as people seek to avoid travel, and as teachers and learners see demonstrations of its effective use.

To allow scaleable visual communications at a distance, our research seeks to make webcasting:
- Interactive, engaging, flexible, scalable, and robust
- Accessible in real-time and via archives
- Useful for communications, information sharing, knowledge building, and learning.

---

[1] We thank Bell Univ. Labs for research support, and Videotelephony Inc. for advice & production support.
[2] Also with the Department of Computer Science.
[3] Also with the Ontario Institute for Studies in Education at the University of Toronto.

To date, we have created a viable and innovative webcasting infrastructure called ePresence (Baecker, 2002). This currently supports video, audio, slide, and screen broadcasting; slide review; integrated moderated chat; question submission; the automated creation of structured, navigable, searchable event archives, and automated data collection for evaluation. Capabilities for audience polling and for linking in wireless mobile devices are under development.

The ePresence system architecture, functionality, and user interface has been developed in response to a set of design requirements which have been classified into 5 categories: **P**(articipants), **M**(edia), **I**(nteractivity), **A**(rchives), and **S**(ystem) We "reinvent the lecture" by seeking to support the needs of a variety of participants, by transmitting rich media to remote participants, and by enabling interactivity among participants. Unlike traditional lectures, the primary content is available in useable archives after the event. Primary goals are to increase the potential for collaborative learning and to support the development of a learning community

The requirements are.

P1: Design keeping in mind the needs of various classes of participants: speaker, moderator, local attendees, real-time remote attendees, and retrospective remote attendees.
P2: Support scalability to hundreds of viewers.
P3: Support both local and remote audiences.
P4: Do not severely impact the local audience experience to support the remote audience.
P5: Work hard on room design issues, which are more critical than one might expect.
P6: Do not allow slide display to depend upon receiving digital versions in advance, but exploit this to advantage when it does happen.
P7: Plan for a significant role for a moderator.

M8: Ensure quality sound even at the expense of sacrificing quality video.
M9: Do not force speakers to use only Powerpoint, but support a variety of rich media.
M10: Transmit video of the speaker, but emphasize delivery of content media even more.
M11: Enhance the sense of presence with high-quality cinematography.

I12: Support interactivity.
I13: Allow slides to be controllable by the viewer independently of the speaker.
I14: Afford viewers easy Web access to relevant material.

A15: Make events available retrospectively through video archives.
A16: Allow video archives to be randomly accessible via navigation and searching mechanisms.
A17: Allow archives to be viewable interactively with the capability for annotations.
A18: Support archive construction that is as automatic as possible.

S19: Strive for elegance and simplicity.
S20: Ensure robustness.
S21: Ensure malleability and extensibility.
S22: Provide for logging and data collection in forms usable for social science research.

Most of these requirements have been met to some extent, although much remains to be done. Our work can be compared and contrasted to Abowd (1999), Cadiz et al. (2000), He, Grudin, & Gupta (2002), Isaacs and Tang (1997), Jancke, Grudin, & Gupta (2000), and Rowe, et al. (2001). The approach may be distinguished relative to this body of work in that:
- We do not force speakers to use Powerpoint, and support the Internet transmission of a variety of rich media presentation formats

- We encourage dialogue among remote viewers and questions to the speaker using an integrated chat facility and (currently) a moderator as an intermediary
- We produce automatically structured, navigable, searchable video archives
- We evolve our technology via an iterative design process, introducing new system prototypes as often as weekly and evaluating them in real use.

## 3   Enabling and Exploiting Interactivity in Webcast Lectures

Of great importance is supporting and enhancing the experience of 'presence' through interactivity among all actors in a distributed presentation room, including the lecturer, local attendees, and remote attendees. We do this by enabling various kinds of interactivity:

*1. Interaction between lecturer and remote audience*

A distinguishing feature of our system is a mechanism for remote attendees to ask questions of the speaker. This is currently done via a moderator, although we will be investigating techniques for communicating directly with the speaker. This poses a design challenge because this must be done in such a way as not to distract the speaker or make her lose concentration. On the other hand, there are opportunities for lecturers to use techniques that will help engage the remote (as well as local) audience. For example, the lecturer will also soon be able to "ask questions" of the remote audience via a polling mechanism.

*2. Interaction among members of the remote audience*

Members of the remote audience can communicate via an integrated chat subsystem and can also send private messages to one another. Unlike attendees at a traditional lecture, they can do this without disturbing other attendees and without distracting the speaker. An example of the use of this capability is discussed below.

*3. Interaction between lecturer and local audience*

Speakers will soon be able to augment traditional verbal and nonverbal communication with the local audiences by allowing those who have mobile wireless devices to ask questions of the speaker and be polled by her, either directly or via the moderator.

*4. Interaction among members of the local audience*

Local attendees who have mobile wireless devices will also soon be able to participate in the chat and private messaging. An open question is the effect of this capability on the concentration and understanding of the local audience, on the ambience in the lecture hall, and on the concentration and effectiveness of the speaker.

*5. Interaction between local and remote audiences*

Local and remote participants will then also be able to communicate with one another via the chat and private messaging capabilities.

*6. Interaction among the retrospective audience viewing the archives*

Finally, interactions can continue (or start) anytime after the event while viewing the archives. We are developing mechanisms to allow chat and structured dialogue to occur as the archives are viewed and reviewed.

There are many other open questions. One is the extent to which these kinds of interactivity are useful and useable. Another is the extent to which successive viewers of lectures will add layers of annotations and engage in conversations, and the extent to which this will support and enhance collaborative learning and foster the development of a "learning community."

# 4    The Millennium Dialogue on Early Child Development

An interesting case study of ePresence has been its use by the Millennium Dialogue on Early Child Development (MDECD) project. MDECD is the Atkinson Centre for Society and Child Development's [see http://www.acscd.ca] first step towards establishing a learning community for child development based on an iterative theoretical model for developing a learning society network (Keating & Hertzman, 1999; Matthews & Zijdemans, 2001; Zijdemans, 2000; see also http://www.webforum2001.net/]. The research as envisioned involves several stages:

* Pulling together a cross-disciplinary knowledge base from diverse and traditionally separate fields in child development;
* Forming a research team of researchers and practitioners to examine existing pedagogical practice and develop curricula grounded in the knowledge base;
* Developing effective tools for communicating the research knowledge to a geographically distributed and broad audience;
* Extending the knowledge base by developing and testing new educational materials,
* Continuing collaborative partnerships to translate and communicate the knowledge to a variety of audiences in ways that affect practice.

To this end, the first phase of the model involved bringing together eight internationally renowned experts from different areas in child development, and preparing the context for cultivating the learning community through the creation of an inclusive technological infrastructure designed to support a variety of technology-based distributed learning initiatives. These included a cross-Canada collaborative curriculum development project, a face-to-face and online graduate level course, and a public webcast and face-to-face conference featuring the work of the eight scientists.

The two-day conference (held in November, 2001) was attended by roughly 200 local participants and was webcast using ePresence to 20 remote North American groups. Over 600 public and private chat messages among the remote groups were exchanged. Table 1 shows how the composition of the chat messages changed over the two days. Of particular interest is the increase in the percentage of messages related to the content of the sessions, from an average of 4% on day 1 to 13% on day 2, and in the percentage of social messages, from 8% on day 1 to 26% on day 2.[4]

**Table 1**: Categorizing chat messages over the four half-days of WebForum 2001

|                   | a.m. Day1 | p.m. Day1 | a.m. Day2 | p.m. Day2 |
|-------------------|-----------|-----------|-----------|-----------|
| Content-related   | 11        | 5         | 13        | 16        |
| Technology-related| 116       | 112       | 44        | 41        |
| Administrative    | 38        | 21        | 13        | 10        |
| Social            | 30        | 1         | 28        | 30        |
| Other             | 18        | 19        | 13        | 14        |

The post conference phase of the project has established the ePresence multimedia archive of the scientist presentations as the knowledge base and continues to nurture the learning community through several ongoing activities. Subsequent to the conference, the knowledge base has been incorporated into three separate courses for graduate students and professional development. One extension of the knowledge has led to the creation of *Conversations on Society & Child Development* [see http://www.cscd.ca]. *CSCD* is an interactive ePublication that uses CD and Web

---

[4] White, et al. (2000) report that text exchanges went from 27%:62%:11% content:technology:social messages to 60%:14%:26% over the last 3 sessions of their course.

technologies to create an environment for accessing the knowledge and supporting exchange among those who generate research and those who want to apply the findings. Ongoing curriculum development is now taking place in the area of early child education in the form of a collaborative project involving colleges across Canada. The knowledge is being used as base for developing an interactive multimedia resource, and is slated to be implemented in Early Childhood programs nationally and internationally in the fall of 2003. Finally, discussion is in progress with a national media outlet to translate the knowledge for public consumption for use by parents, educators, and policy makers.

The conference oral presentation will discuss this case study and others in progress in more detail.

# References

Abowd, G.D. (1999). Classroom 2000: An Experiment with the Instrumentation of a Living Educational Environment. *IBM Systems Journal* 38(4), 508-530.

Baecker, R. (2002a). Highly Interactive Webcasting with Structured Archives. Poster presentation at CSCW'2002, New Orleans, LA., November.

Branson, R.K. (1998) Teaching –Centered Schooling Has Reached Its Upper Limit: It Doesn't Get Any Better Than This. *Current Directions in Psychological Science* V7, N4, 126-135.

Cadiz, JJ., Balachandran, A., Sanocki, E., Gupta, A., Grudin, J., and Jancke, G. (2000). Distance Learning Through Distributed Collaborative Video Viewing. *Proc. CSCW2000,* 135-144.

Donovan, M. S., Bransford, J. D., Pellegrino, J. W. (Eds.) (1999). How People Learn: Bridging Research and Practice. National Research Council. Washington: National Academy Press.

He, L., Grudin, J., and Gupta, A. (2002). Designing Presentations for On-Demand Viewing. *Proc. CSCW2000,* 127-134.

Isaacs, E.A. and Tang, J.C. (1997). Studying Video-based Collaboration in Context: From Small Workgroups to Large Organizations. In Finn, K., Sellen, A.J., and Wilbur, S. (Eds.) (1977), *Video-Mediated Communication.* Erlbaum, 173-197.

Jancke, G., Grudin, J., Gupta, A. (2000). Presenting to Local and Remote Audiences: Design and Use of the TELEP System. *Proc. CHI 2000,* 384-391.

Keating, D.P. & Hertzman, C. (Eds.) (1999). Developmental health and the Wealth of nations: Social, biological, and educational dynamics. New York: Guilford Press.

Kerns, C. (2002). Constellations for Learning. *Educause Review*, May-June 2002, 20-28.

Matthews, D., & Zijdemans, A. S. (2001) Toward a Learning Society Network: How being one's brother's keeper is in everyone's self interest. *Orbit V31*, N4, 50-54.

Olson, D.R., & Bruner, J.S. (1996) Folk Psychology and Folk Pedagogy. In Olson & Torrence (Eds.) The Handbook of Education and Human Development. UK: Blackwell Publishers.

Rowe, L.A., Harley, D., Pletcher, P., and Lawrence, S. (2001). BIBS: A Lecture Webcasting System. BMRC Technical Report June 2001, University of California, Berkeley.

White, S.A., Gupta, A., Grudin, J., Chesley, H., Kimberly, G., Sanocki, E. (2000). Evolving Use of a System for Education at a Distance. *Proc. HICSS-33.*

Zijdemans, A. S. (2000). Creating Cultures for Learning: Toward a New Paradigm for Education. Masters Thesis. Ontario Institute for Studies in Education of the University of Toronto.

# Development of an Instructional Training Model and Diagnostics in Support of e-Learning

*Philip Callahan*

University of Arizona South
Sierra Vista, AZ 85635 USA
pec@u.arizona.edu

## Abstract

RapID is a pragmatic product-oriented instructional model based on the premise that educators have little time and minimal design experience for the development of instruction augmented with technology. The model, developed as browser-based training, encourages integration of instructional technology and learner centered education techniques to produce learning object oriented instruction. Closely associated test diagnostic and "gradebook" software, allowing comprehensive data aggregation for multiple-measures learner assessment, provides insight into traditional classroom and non-traditional learning environments. A formative evaluation of the browser-based instruction by a diverse group of instructors from elementary, secondary, community college, and university levels suggests the model provides adequate training for most instructional needs. While designed as a minimalist approach, espousing relevancy and ease, for the development of technology-augmented instruction, participants in this study desired additional content and support. Content changes include a more personable interface, in the form of embedded talking-video, as an orientation and introduction to key content areas. Other revisions address evaluation related templates and examples with a more rigorous integration with cross-curriculum development.

## 1    Introduction

As education transitions to technology augmented instructional delivery, educators face the daunting task of converting their traditional instructional practices into these newer methods of delivery. Although these newer methods provide greater educational opportunity, they also demand greater attention to the design of instruction and subsequent evaluation of student learning. Thus, the intent of this ongoing University of Arizona study is to develop an e-learning system designed to realize three objectives concentrating on training educators, facilitating instructional delivery, and evaluating outcomes.

The first objective is to train educators in the use of learner-centered and learning-object oriented procedures for use in traditional classroom and e-learning environments. Learner-centered education (LCE) emphasizes a more active participatory role for the learner through greater control of planning, implementation, and evaluation of learning. LCE methods, such as problem solving, discovery, simulations, and cooperative group learning, can be readily facilitated through technology. Technology, and more specifically the products and the process of technology as they are applied to learning, may actually provide for learning opportunities well beyond the limitations imposed by the traditional classroom. Thus, a goal of this training is integration of relevant technologies for developing LCE instruction with emphasis placed on the use of learning objects.

Wiley (2000) describes learning objects as comparable to object-oriented programming (OOP) methods whereby components (objects) can be reused in multiple contexts. Paralleling OOP, small instructional components can be reused a number of times in different learning contexts as learning objects. These learning objects are often digital entities, deliverable over the Internet, allowing any number of people to simultaneously access them.

The use of learning objects presents the opportunity to more efficiently design instruction through reusable chunks of instructional media whereby instructors may reassemble materials that support individual learning targets and LCE processes. To facilitate a realistic implementation of LCE and learning objects within the instructional design process, a pragmatic model emerges whereby the instructor is essentially orchestrating three key instructional elements: a learning object; an evaluation object; and a teaching object.

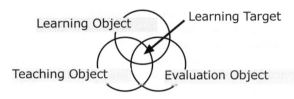

Figure 1. RapID Model describing the relationship between objects and learning target.

Within the context of this model, the learning object, in the simplest sense, contains the learning content and corresponds to the learning session. An evaluation object contains the method by which the learning outcome is measured. A teaching object contains the rule set or methods for coordinating the instruction and corresponds to the lesson plan. A point of commonality, the learning target, or objective, provides the conditions, degree, and behavior that form the intent of the learning. Thus, the instructor is able to orchestrate instruction and implement both LCE and the learning object using the traditional concept of the lesson plan.

One of the more difficult issues concerning learning objects is that of "granularity" or the size of the learning object. Pragmatic factors including the scope and sequencing of the instruction, as well as the need to eventually catalog the learning object for reuse, limit the size of the learning object to one of functional reusability (Wiley, 2000). While RapID's orientation to the development of learning objects may seem to challenge granularity, the resulting learning objects are consistent with the instructor's design of the teaching objects (instructional rule sets) and evaluation objects that control the operations and evaluate the learning objects.

The first objective, training educators, has resulted in the development of instructional training and software tools in support of the RapID model. The training is delivered in a browser-based format using the same software and methods the educators are expected to use to develop their instruction. The intent of this browser-based instruction is to provide an expeditious yet rigorous product-based approach to instructional design for use in traditional and browser-based settings. A tiered development approach (Table 1) permits progressive development of instruction by building from matter developed in earlier tiers. While the development of instructional materials is presented in a pragmatic and sequential top-down manner roughly paralleling the Dick and Carey (2001) systems approach model, the instructor can elect to work from any tier.

Table 1: Tiered Approach to Instructional Design

| Tier | Process | Outcome |
|------|---------|---------|

| A | Basic course syllabus | Identify the learner, learning context, and learning targets |
|---|---|---|
| B | Calendar of events (learning targets) | Organize and sequence learning targets |
| C | Evaluation milestones (evaluation object) | Evaluate learner success |
| D | Lesson plan (teaching object) | Organize learning sessions |
| E | Learning session (learning object) | Facilitation of session activities through technology |
| F | Distance learning | Facilitation of course through technology |

Tiers A, B, and C comprise development of a progressively more refined course syllabus based on learning objectives or learning targets. This process of systematic refinement of broad key learning targets yields progressively more detailed learning targets and their associated evaluation objects. Using the learning targets and evaluation objects, tiers D and E develop the concept of "lesson plan" and the session elements and learning objects. The lesson plan forms the teaching object, a rule set for directing the learning sessions. The learning session may involve the augmentative use of technology and can emerge as a learning object.

The second objective, assisting educators in the delivery of instruction, addresses the development of support tools. These tools include, but are not limited to, classroom diagnostic software for analyses of performance-based (rubric-oriented) and traditional tests, a "gradebook" allowing comprehensive data aggregation for multiple-measures learner assessment, and highly modularized course management software. Educators are not always in visual contact with their learners and the diagnostic software provides insight into classroom confusion and the identification of potentially "at-risk" learners using traditional objective evaluative measures, such as classroom tests (Callahan, 2001). Browser-based diagnostic software with a strong graphical orientation has been developed to better address accessing remote instructional servers, as well as educators lacking the rigor of a test and measurement background particularly when dealing with rubric-based tests.

A further consideration is orienting the diagnostic software to the learning targets to better facilitate LCE, and the use of learning objects, by producing instructor and learner test outcomes with a direct visual association of test item and learning target. Diagnostics extend to performance-based testing, as well as providing multiple measures capabilities. Multiple measures improves assessment and comes as a result of educators aggregating a broader range of learner outcomes, such as discussion board activity, occurring in e-learning environments.

The third objective, study the effects of applying these instructional principles and practices, focuses on: educator success in mastering and applying the materials; the integration of the new learning into traditional teaching activities; the development of e-learning courses; and student success in courses making use of this new approach. Feedback is intended to drive revisions to the training model as well as supporting software. Due to the longitudinal nature of this data, this continues as a work in progress. Educator participation is described in this study.

## 2 Methodology

During the fall of 2002, twelve instructors from elementary, secondary, community college, and university levels participated in a formative evaluation of the RapID system. The intent of a formative evaluation is to obtain data from participants for the purposes of revising instruction. Participants in the formative evaluation are intended reflect the diversity of the final audience. Thus, the participants were selected based on interest in the program and their diversity in curriculum and technical abilities. They were charged with using the RapID system for one month to sufficiently develop a course so as to include at least one technology-augmented lesson.

A WebCT-based discussion board was maintained for posting instructors' works and for interaction with other instructors and RapID researchers. Instructors were expected to implement their instruction within one semester of development.

Consistent with formative software evaluations (Apple Computer, 1995), participants were not prompted. Rather, participants were requested to post their queries and concerns to either the discussion board or to an identified RapID researcher. Following completion of their instructional development, the participants were interviewed and questioned on the usefulness and facility of the system, recommendations for change, and their resultant instructional materials. The interviews were conducted individually and recorded with the permission of the participant.

# 3    Results

Although still a work-in-progress, these results reflect qualitative interviews with instructors participating in a formative evaluation of the RapID system. Participant completion of their respective instructional materials was closely associated with their immediate need to implement the resultant instructional materials. That is, those participants not expecting to immediately implement their instruction were least likely to fully complete their materials within the one month period. Results varied from no completion to full completion within the one month period. Two of the twelve participants developed an entire one semester college-level course using the RapID process.

Usefulness and Facility of RapID. None of the participants challenged the RapID concept and, in general, found the browser-based format to be acceptably intuitive and workable regardless of their computer background. But, those participants with minimal computer background and exposure to browser-based instruction desired a more "personal" interface to RapID. One participant characterized this need as an orientation to RapID possibly in the form of an integrated multimedia-based instructor. This integrated instructor would introduce RapID and potentially each tier of training.

Regarding the RapID content, several comments emerged concerning a more rigorous development of evaluation materials with more emphasis on linking learning target and associated evaluation. Additionally, there was a desire to include additional examples and templates focusing on rubric-based testing and grading in growth/progress learning situations.

The RapID development process tends to be linear and progressive with focus on a single-subject curriculum. This approach was problematic within the elementary school domain where a learning target is developed across-curricula. Thus, the elementary grade instructor typically needs to develop instruction both vertically (chronologically) and horizontally (across curricula). The calendar of events approach of tier B focuses on vertical development and thus provides some degree of usefulness to elementary grade instructors in its current form.

Separate from the RapID was the supporting WebCT course management system. Several participants had difficulty using the discussion tool and requested technical support for using the tool and posting their instructional materials. Additionally, access to technology resources, such as scanners, were difficult to locate and access in some educational institutions.

Recommendations for Change. Participant comments regarding modification to RapID focused largely on the inclusion of content and examples; specifically, more examples addressing writing learning targets, additional templates and guidance for rubric writing particularly in the area of evaluation of progress, examples of good and bad instructional practices, and information regarding copyright issues relating to browser-based learning.

Where specific hardware, software, or related resources are required to facilitate an aspect of RapID training, inclusion of a succinct list of resources should be identified in the RapID training. Inclusion of human introductions through embedded video was recommended for an orientation to RapID, as introductions to each tier, and to introduce complicated topics. Additionally, an orientation to WebCT or any course management system augmenting the RapID was requested.

Finally, more advanced participants asked for middleware support and additional references for optimization procedures and software, e.g. Impatica to reduce PowerPoint file size. Middleware included the desire for more fully automated processes to facilitate sharing of information, e.g. learning targets, between tiers to minimize copy and paste operations.

Instructional Materials. Participants completing the RapID training were able to develop instruction that was fully adaptable to their particular learning environment. The exception was related to use in elementary school settings where limitations exist.

# 4    Conclusion

RapID is a pragmatic product-oriented instructional model based on the premise that educators have little time and minimal design experience for the development of instruction augmented with technology. The model, developed as browser-based training, encourages integration of instructional technology and learner centered education techniques to produce learning object oriented instruction. Closely associated test diagnostic and "gradebook" software, allowing comprehensive data aggregation for multiple-measures learner assessment, provides insight into traditional classroom and non-traditional learning environments.

A formative evaluation of the browser-based instruction by a diverse group of instructors from elementary, secondary, community college, and university levels suggests the model provides adequate training for most instructional needs. While designed as a minimalist approach, espousing relevancy and ease, for the development of technology-augmented instruction, participants in this study desired additional content and support. Content changes include a more personable interface, in the form of embedded talking-video, as an orientation and introduction to key content areas. Additional evaluation related templates and examples with a more rigorous integration with cross-curriculum development are anticipated. Consistent with the outcomes of other technology augmented instruction studies, an assurance of technical support, and readily available hardware and software resources is necessary to assure that participants can fulfill development of their instructional materials (Callahan & Shaver, 2001).

# References

Apple Computer. (1995). Involving users in the design process. *Macintosh Human Interface Guidelines*. Retrieved July, 30, 2002, from: http://developer.apple.com

Callahan, P. (2001). Diagnostic methods in distance education and non-traditional learning environments. In M. J. Smith & G. Salvendy (Eds.), *Systems, social and internationalization design aspects of human-computer interaction* (pp. 697-700). Mahwah, NJ: Lawrence Erlbaum Associates.

Callahan, P. & Shaver, P. (2001). Formative considerations using integrative CALL. *Applied Language Learning*, 12(2), 147-160.

Callahan, P. (2002). *Rapid instructional design (RapID)*. Retrieved February 7, 2003 from: http://150.135.84.221/design/design.htm

Dick, W., Carey, L., & Carey, J. O. (2001). *The systematic design of instruction* (5th ed.). New York: Longman.

Wiley, D. A. (2000). Connecting learning objects to instructional design theory: A definition, a metaphor, and a taxonomy. In D. A. Wiley (Ed.), *The Instructional Use of Learning Objects*. Retrieved February 7, 2003 from: http://reusability.org

# Rapid Development of IMS compliant
# E-Learning Modules

*Jörg Caumanns, Hatice Elmasgünes*

Fraunhofer-Institute for Software and Systems-Engineering
Mollstr. 1, 10178 Berlin, Germany
{joerg.caumanns, hatice.elmasguenes}@isst.fhg.de

## Abstract

Content authoring is still the blood, sweat, and tears part of E-Learning. Authors are faced with either very technical and complicated or very monolithic and inflexible tools. They usually spend more time with reading manuals than with writing content assets. It is often more challenging to get a piece of text into an authoring workbench than to write this text.
In this paper a modular authoring workflow is sketched that allows for prototyping E-Learning courses using well-known tools like Microsoft Word and Powerpoint. The workflow can easily be enhanced by automated metadata retrieval and nearly arbitrary filters and converters to produce any target format desired.

## 1   Introduction

Almost every publication about E-Learning starts with an enumeration of E-Learning's advantages over courses and textbooks. But while some years ago media enrichment  was most highlighted, today a stronger emphasis is put on adaptability, on-demand access and maintainability.
The main approach to reaching these goals is on-demand and individualised recontextualisation of modular learning fragments (assets). Recent specifications like SCORM ("SCORM", 2002) and IMS ("IMS", 2002) adopt this approach by defining different granularities of learning objects and taking care of storage considerations. A lot of research is done in domains like peer-2-peer content infrastructures (e. g. edutella (Nejdl et al., 2002) or P-Grid (Aberer, 2002)), query routing and semi-automated content aggregation (e. g. Teachware on Demand (Caumanns & Hollfelder, 2001)). The problem with most of these research activities is that they remain theoretical as there is no relevant amount of content available to proof the concepts and to establish running networks of such highly modular and distributed infrastructures.
The main reason for this unsatisfying situation is that nearly no one cares about reusable tools and processes for content and metadata creation that are suitable for not just creating single, stand-alone WBTs. Especially the annotation of content with standardised metadata is crucial for intelligent services and queries on top of storage infrastructures. With the exception of some simple LOM editors, nearly no tool support is available for metadata creation and maintainance.

## 2   Authoring Workflow

The creation of E-Learning modules is usually an interdisciplinary task. Domain experts, pedagogues, and multimedia developers have to work hand in hand. Experience has shown that especially domain experts are not familiar with creating content in a page-oriented way with a very loose context between single assets. For this reason, prototyping is crucial: the earlier domain experts (and pedagogues) take a look at the whole module as it will be seen by the learner, the fewer development cycles will be needed. This is very important as the later changes have to be done the more expensive and time consuming they get.

For this reason we propose an authoring workflow that supports the following, consecutive steps:

1.  domain experts and pedagogues agree on a set of content structures based on didactical ontologies, etc.
2.  domain experts write texts, sketch images and animation storyboards using tools they are familiar with
3.  pedagogues and multimedia experts check and improve texts and storyboards
4.  a prototype is generated to give domain experts and pedagogues a view of the final module
5.  storyboards, texts and images are improved until domain experts and pedagogues agree on them
6.  if single assets of the module should be stored in a repository infrastructure (e.g. edutella), metadata annotation and classification of the appropriate assets is done next
7.  multimedia developers implement the specified assets, e. g. by rewriting them using Macromedia Flash

This workflow can be used for turning existing textbooks into WBT prototypes as well as for authoring new E-Learning assets and modules.

# 3    Tool Support

Figure 1 (below) sketches an implementation of the authoring workflow described above.

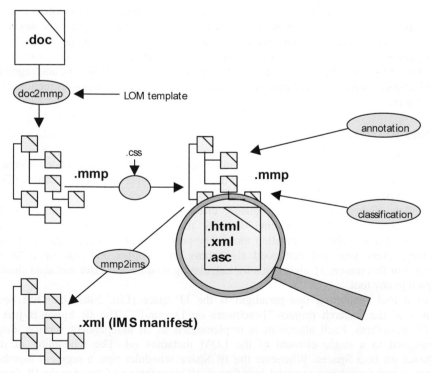

**Fig. 1: Tool-Supported authoring workflow**

The structure of the content and all texts and images are encoded within a Microsoft Word document. In a first step the document is segmented into its sections and chapters to create a tree of smaller Word documents. Each segment is encapsulated by a temporary file, which is no more then an aggregation of the segment, its metadata, and an ASCII representation of the segment's text (mmp: media, metadata, and plain text). Basic metadata is assigned to each asset by using templates.

In a following step, the segmented content is converted into HTML. By using style sheets, the layout of the assets can be brought close to the final look and feel. The final step is encoding the content structure (table of contents) as an IMS manifest. Now the prototype view of the complete module can be displayed by using tools like Microsoft's LRN viewer (part of IE6) or any of the existing tools for converting IMS manifests into HTML framesets. As all these steps are passed through automatically, changes and improvements can be made within the original source document.

As soon as the author agrees on the result, each asset is annotated and classified for further reuse and passed to multimedia experts for implementation.

## 3.1 Automated Segmentation

The Source of the whole workflow is a Word representation of the module. The hierarchical structure of the module is described by using different heading levels. Sections and chapters may contain nearly any kind of content representation Word provides: paragraphs of text, tables, diagrams, images, etc.

Segmentation of the document into its sections and chapters is done by a tool named "doc2mmp". All this tool does is to split the document on chapter and section boundaries to create a tree of smaller Word documents. Metadata as segment title (heading of the section/chapter) and segment file size are extracted automatically. Each segment is encapsulated by an mmp-file, which is no more then an aggregation of the segment, its LOM metadata, and an ASCII representation of the segment's text. Basic LOM metadata is generated by filling segment size, MIME type and segment title into a LOM template which initially may contain generic information as author, creation date and intended end user.

## 3.2 Metadata Retrieval

Metadata that is valid for the whole module (e. g. author, creation date) can be encoded within a template for each segments' metadata. Metadata such as title, media type, and file size can be filled in during segmentation.

Many other metadata elements - such as language, difficulty, semantic density etc. - can be automatically calculated using IR algorithms. Usually these algorithms are integrated into metadata editors or other monolithic annotation tools. The problem with this approach is that IR algorithms improve every year and that good algorithms are complex and take up a lot of computing power. For this reason, IT algorithms for calculating module and asset metadata should be loosely coupled to any tool.

One example of a tool supporting this paradigm is the IT space (Liu, 2002) that has been developed as part of the research project "Teachware on Demand". The IR Space is just a scheduler for IR algorithms. Each algorithm is implemented as an independent service. Each algorithm is assigned to a single element of the LOM metadata set. The scheduling of the algorithms is based on Java Spaces. Whenever the IR Space scheduler puts a segment together with its current metadata record into a special Java Space, all algorithms registered at the IR Space check whether they can provide an metadata element that has not been recently set. If so, the segment is locked until the algorithm has calculated the element and updated the metadata record. Part of the IR Space is a library for reading and writing metadata records that can be used for wrapping existing IR algorithms.

## 3.3 Workflow Support

The automated segmentation of source documents, the retrieval of metadata, and the final encoding and sequencing of the segments into courses are performed by different tools. This modular separation of functionality allows for very individual authoring workflows that can be adapted to different source and target encoding. E. g. if Powerpoint would be used instead of Word as the primary authoring tool only the segmentation and the HTML encoding tool would have to be replaced by appropriate Powerpoint based tools. All tool concerned with metadata retrieval and IMS encoding could remain unchanged. Another advantage of this disintegration of tools is the flexibility to add, replace and remove tools for metadata retrieval based on the needs of the target platform. Even functionality that is not part of the generic workflow (e. g. creation of SCOs out of the final IMS packages) can be added at will.

Within the last two years nearly ten tools have been developed that can be stuck together to set up individual authoring workflows. These workflows support Word and Powerpoint as source formats, IMS and PHP as target formats, and any combination or mixture of these source and target formats. Some more tools were developed to support check-in and check-out of metadata into a web-based repository.

To support authors in selecting and sequencing the appropriate tools a graphical scheduler was implemented (see screenshot above). Predefined workflows can be loaded into the scheduler. The scheduler is then responsible for calling segmentation, retrieval, and conversion tools in the approriate order and for passing segments and metadata from one tool to another.

## 4    Conclusion

The major benefits of the development cycle described in this paper are:

- Authors create their content using tools they are familiar with (in this case Microsoft Word). Especially existing textbooks can easily be turned into online courses this way.
- A separation of concerns is supported. Authors create the table of contents, texts, and images using Word. Media designers can afterwards enhance and redesign each single segment.

- Much metadata is generated automatically. Even if all frames are rewritten in Flash, the metadata is still valid.
- Setting up courses from single modules is implicitly supported as IMS content packages can be aggregated and nested.

We are sure that tools based on existing editors and using a modular authoring workflow can speed up the development of E-Learning modules and may lead to some kind of mass production as it is already common with scientific papers and books.

## References

Aberer, K. (2002). P-Grid: A Self-Organizing Access Structure for P2P Information Systems. In *Proc. Sixth International Conference on Cooperative Information Systems* (CoopIS 2001), Trento, Italy. Lecture Notes in Computer Science 2172, Heidelberg: Springer Verlag

Advanced Distributed Learning (2002). SCORM Overview. Retrieved December 2002, from http://www.adlnet.org/

Caumanns, J.& Hollfelder, S. (2001). Web-Basierte Repositories zur Speicherung, Verwaltung und Wiederverwendung multimedialer Lernfragmente. In R. Schmidt (Ed.), *Information Research and Content Management* (pp. 130-140). Frankfurt/Main: DGI

IMS Global Learning Consortium (2001). IMS Content Packaging Information Model. Retrieved December 2002, from http://www.imsglobal.org/content/packaging/cpinfo10.html

Liu, N. (2002). Entwicklung eines IR-Space für Metadaten. Diploma Thesis. Berlin: Technical University.

Nejdl, W., Wolf, B., Qu, C., Decker, S., Sintek, M., Naeve, A., Nilsson, M., Palmer, M. & Risch, T. (2002). Edutella: A P2P Networking Infrastructure Based on RDF. In *Proc. 11th Iternational World Wide Web Conference* (WWW2002), Hawaii, USA, May 2002

# Context-Based Autonomous Monitoring Module for Web Educational Environments

*Despotakis Theofanis, Palaigeorgiou George, Siozos Panagiotis*

Computer Science Department, Aristotle University of Thessaloniki
Thessaloniki Greece
{tdespota, gpalegeo, psiozos}@csd.auth.gr

## Abstract

Monitoring modules are internal or external components of Web applications that collect information about user activities. Various modules that monitor and advance the interactivity protocol have been proposed but most of them are embedded to systems in ad hoc ways. In this paper, we propose WebInspector, an autonomous, programmable, multipurpose activity monitoring and analysis system. WebInspector differs in its ability to take into account the design hypothesis of the application's usage and offers more hermeneutic capabilities. This framework (a) provides a formalized language (API) for the manipulation of the recorded data stream, (b) is oriented on event-handlers tracing, (c) provides conditional (pattern-based) updates and insertions in the stream and (d) tries to maintain a standardized recorded data stream that can be utilized as data provider for additional independent services. Exemplary additional services that we have developed are "ESDA Explorer", "Activity Recorder", "Reaction Creator" and "History Objects Manager".

## 1    Introduction

Most modern web applications try to transcribe the interactions' development with their users and exploit them in many ways. Usually, the recorded data are used in order to evaluate the usability of the web application's hypertext structure (Scholtz, Laskowski & Downey, 1998), to obtain interest indicators (Claypool, Phong, Wased & Brown, 2001), to recognize patterns in users' behaviors (Card, Pirolli & Wege, 2001) etc. Typically the recorded data streams are linear, they include events as elementary interaction units and describe the event object instances and navigation selections among web pages. The most common recording and analyzing mechanisms do not use the semantic assumptions of application's design, although user behaviors are indirectly confined and enhanced by the application's interaction space. Moreover, the monitoring modules are instantiated in an exclusive way depending on their specific utility and therefore are far from autonomous and reusable. The lack of autonomy prohibits them to utilize the recorded stream as a data provider for additional services.

In response to these prospects, we have developed WebInspector, an autonomous, programmable, multipurpose activity monitoring and analysis system that takes into account the design hypothesis of the prospective user behavior. WebInspector is easily incorporable and removable to existing educational applications and easily configurable in order to support multiple mechanisms of additive value in the educational application.

## 2 WebInspector Functionality

### 2.1 WebInspector Presumptions

User interface events constitute the human-computer communication mechanism in windows environments and are regarded as a fruitful source of information. Users fire events in order to commence system processes corresponding to their needs. Events are associated with function-handlers, code sections that are executed as reactions to their appearance. In web applications, every event generation corresponds to an event object that includes information about the firing action (the mouse position, the object that generated the event, the type of event etc). So, the first alternative in monitoring a web application is the recording of event object instances.

However, many times events fired by the user are not adequate enough to disclose the hermeneutic attributes of user actions.

User interface event sequences have often a grammatical structure and higher level user activities can be specified in terms of combinations of lower level events (Hilbert & Redmiles, 1998). Also, the interaction context of each event (parameters-conditions at the time the actions revealed, the set of possible interaction activities from which the selected action was originated etc.), is disseminated among many events (Hilbert & Redmiles, 2000). Tracing event object instances supply evidences only for their functioning and not for user-computer interactions development.

Therefore, firstly WebInspector is oriented to monitor the functions that implement the event's effect (event-handlers), instead of event objects. We consider important the enrichment of the monitoring stream with contextual information while user is acting. In this way, we can have access to data that are irrelevant to the event object and contribute substantially in the realization of the specific functionality such as function call parameters, global variables etc. Additionally, we avoid the exhaustive tracing of all events that can activate an application service. WebInspector provides a formalized language (API) that enables web developers to precisely define what they wish to monitor. Naturally, the monitoring process is not autonomous but ensures the successful and efficient attachment of the tool to the substantial different approaches and architectures of web educational environments.

Secondly, a set of pattern recognition rules that takes into account the previous considerations can capture tacit segments of the interaction's reality and produce a more informed alternative recorded data stream. In educational applications, where the interaction space is usually task-oriented and users behave with clearly defined goals, this set of rules can be more effective in the production of hermeneutic attributes since the distance between educational applications' design assumptions and user's intensive behavior, is minimized. This set of rules comprises a description of the educational design and the user's behavior assumptions.

### 2.2 WebInspector Operation

The incorporation of WebInspector is implemented with convenience following the next two steps:

Initially, the web developers of the educational application have to update all web pages by adding an include call to an external file that carries out three operations: (a) the insertion of a hidden iframe element that its source page resides in WebInspector's web server. (b) the insertion of a JavaScript library that includes all functions that the developer will use in order to add records in the monitoring data stream and (c) the automated overriding of browser's default behaviors. Additionally, they have to add a call to the function "Monitor" in each event-handler they wish to monitor.

Afterwards, web developers have to adapt WebInspector's functionality, to their specific needs following a set of configuration-description actions that further determine the content and the formation of the recorded stream. As we have mentioned before, the developers have to describe the semantic and operational characteristics of the educational application in terms of possible user activities. We distinguish two possible user activities: "actual" and "intellectual". The concept of "actual" activities corresponds to the recorded event-handlers while "intellectual" activities are descriptions of implicit user actions that derive as patterns of other user activities.

Every activity is a vector consisting of metadata properties and requires two levels of description. In the *operational description level*, the developers declare the activity's basic attributes that indicate its identity (name, type, parameters etc.). In the case of "actual" activity, function's requirements and results may also be specified. If the activity is intellectual, the developers fill out only the attributes that can be ascribed to them. In the *hermeneutics description level*, the developers can form derivation rules that dictate WebInspector to insert a new activity to the data stream when predefined patterns of "actual" and/or previously derived "intellectual" activities are detected. Developers can also modify the default coding of each activity in the stream, asserting different metadata properties to their appearances in different contexts. Sample rules that can be defined are:

- **Default Metadata properties insertion (MPI)**: If a student searches for a keyword in the educational material (e-book), then insert into the appropriate metadata property the phrase "The student searched the e-book".
- **Parameterized MPI**: If a student searches for the keyword "crawling" in the educational material, then insert into the appropriate metadata property the phrase "The student searched the e-book for the keyword $par1[='crawling']$".
- **Pattern based MPI**: If the results of a previous search with the keyword "crawling" have not satisfied student's expectations and the student searches *again* then insert into the appropriate metadata property the phrase "The student searched again the e-book".
- **Pattern based activity generation**: If a student stays for at least three minutes in an assessments' question then generate the activity "The student delayed in answering the question".

The developers can use temporal, sequential, logical or conditional operators for the formation of the rules.

# 3    Additional Services

From the above descriptions, it appears that the operation of the WebInspector requires a standardized data stream that is easily parameterized, updated and adapted to different situations. Additionally, we have extended WebInspector's API for manipulating the recorded data stream in a way that can support the development of new independent services exploiting the data recorded. The nature of these services must be independent of the particularity of each application and its internal logic and must exclusively use the recorded data stream as data provider for their functionality. Each service has to be accompanied by a distinct description level that configures its operation.

Exemplary independent services that we have developed are:

- **ESDA Explorer:** "Explorative Sequential Data Analysis (ESDA) is any empirical undertaking seeking to analyze any … observational (usually recorded) data in which the sequential integrity of events has been preserved" (Sanderson, Fisher 1994). The developer can utilize the tools provided by ESDA Explorer in order to explore and query the sequential recorded data and answer quests for their meaning (e.g. Fischer's Cycle, Maximal Repeating Pattern, Lag Sequential Analysis etc.).

- **Activity Playback:** Activity Playback service represents and replays user actions. Its administrative features offer opportunities for selecting and watching past user activity sequences. This functionality, which can be considered as a form of learning history, is very important for multiple reasons (Plaisant, Rose, Rubloff, Salter, Shneiderman, 1999).

- **Reaction Creator:** Many times, developers wish to author micro-scenarios of interactions that are activated when specific patterns of usage are detected. For usability, cognitive apprenticeship or help reasons, we want to interrupt the normal flow of user's interactions and inform or question him. For example, in case the user always selects to navigate to a search page from a menu choice (although an easily accessible toolbar exists) then an interface evaluator would like to ask exactly at that time if the user knows about the toolbar and its functionality (Hartson, Castillo, Kelso, Neale, 1996). Reaction Creator includes the interactive presentation authoring tool ACT (Palaigeorgiou, Korbetis, Siozos, Tsoukalas, 2002) which uses MS Agent and dynamic HTML behaviors in order to support structured dialogues with the users.

- **History Object Manager:** Activities can be considered as actions with or toward interface objects. In opposition to digital objects, physical objects maintain their interactions' history with humans and time, either with immediate ways (notes in a writing book remain in the same place, with the same writing style and maybe with a drop of coffee favoring the recalling of logical and emotional attributes of a previous circumstance), or with indirect ways such as their natural wear. History Object Manager enables the definition of "History Objects" that preserve their interaction history and use it in order to enhance their functionality and appearance. (Hill, Hollan, 1993), (Wexelblat, Maes, 1999). Prescribed types of history interactivity embedment are offered for specific HTML objects.

# 4    Conclusions

In web educational applications, developers, administrators and teachers intensively pursue the exploration of student actions in order to better understand their learning needs and selections. We detected a need for an overall effective reengineering of the usage of the monitoring process and proposed an activity recording framework that requires a formalization of possible user actions. The monitoring stream is standardized and enriched with supplementary semantic data and also can be used for the development of independent additional services.

Now, the efficiency of the monitoring module depends on the developers' creativity in providing the different description levels. The process is characterized as semi-autonomous and produces different results under different description approaches.

# References

Card S. K., Pirolli P., Wege M. (2001), Information Scent as a Driver of Web Behavior Graphs: Results of a protocol Analysis Method for Web Usability, *Proceedings of SIGCHI'01,* ACM Press.

Claypool M., Phong L., Wased M., Brown D. (2001), Implicit Interest Indicators, *Proceedings of IUI'02*, ACM Press.

Hartson H., Castillo J., Kelso J., Neale W. (1996), Remote Evaluation: The network as an Extension of the Usability Laboratory, *Proceedings of CHI'96*, ACM Press.

Hilbert D., Redmiles D. (1998) Agents for Collecting Application Usage Data Over the Internet, *Proceedings of Autonomous Agents 1998.*

Hilbert D., Redmiles D. (2000) Extracting Usability Information from User Interface Events, *ACM Computing Surveys*, 32(4), 384-42.

Hill W., Hollan J., (1993), History-Enriched Digital Objects, *Third Conference on Computers Freedom and Privacy, Computer Professionals for Social Responsibility – CPSR* (URL: http://www.cpsr.org/conferences/cfp93/hill-hollan.html).

Palaigeorgiou G., Korbetis A., Siozos P., Tsoukalas I. (2002), ACT: Acting Cartoons for Trainers' Presentations, *Proceedings of EDMEDIA 2002*, AACE.

Plaisant C., Rose A., Rubloff G., Salter R., Shneiderman B. (1999), The design of history mechanisms and their use in Collaborative Educational Simulations, *HCIL Technical Report No. 99* (URL:http://www.cs.umd.edu/hcil).

Sanderson P, M. Fisher C. (1994) Exploratory Sequential Data Analysis: Foundations, *Human Computer Interaction Special Issue on ESDA*, 9(3), 251-317.

Scholtz J., Laskowski S., Downey L. (1998) Developing Usability tools and techniques for designing and testing web sites, *Proceedings of HFWeb'98* (URL: http://www.research.att.com/conf/hfweb/proceedings/scholtz/index.html).

Wexelblat A., Maes P. (1999), Footprints: History-Rich Tools for Information Foraging, *Proceedings of CHI'99*, ACM Press.

# Shared 3D Internet environments for education: usability, educational, psychological and cognitive issues

*Nicoletta Di Blas*

*Paolo Paolini*

*Caterina Poggi*

Hoc – Politecnico di
Milano
Via Ponzio 34/5,
20133 Milano (Italy)
diblas@elet.polimi.it

Hoc – Politecnico di
Milano
Via Ponzio 34/5,
20133 Milano (Italy)
paolini@elet.polimi.it

University of Italian
Switzerland
Via G. Buffi 13, 6900
Lugano (Switzerland)
poggic@lu.unisi.ch

## Abstract

Today's cooperative virtual environments are mainly built for social and entertaining purposes: MUDs, chat-rooms, online massive multi-player games are the most common examples. However, they also possess a high educational potential: students may learn much from a virtual experience. There are some difficulties, however, to overcome. Creating a 3D virtual setting, and allowing people to connect, does not necessarily make a successful cooperative environment. Users neither cooperate with each other spontaneously, nor have meaningful interactions with the virtual world on their own initiative. They rather "have a look around" and quickly loose interest.

Users need a specific *purpose* to act and interact in a virtual world. Users need to be provided with precise guidelines about what to do and how; they also must be given specific objectives. A virtual 3D environment for educational purposes is no exception. Users are supposed to learn something besides enjoying themselves: making their experience effective is even more complicated.

Let us examine more in detail how online virtual worlds can be successfully used for educational purposes, using as an example SEE - Shrine Educational Experience, developed jointly by the Israel Museum and Politecnico di Milano (Di Blas, Hazan & Paolini, 2003). SEE brings together in a shared virtual space students from around the world, to learn about the Dead Sea Scrolls[1], and the two thousands year old culture that produced them (Roitman, 1997).

During the experience, educationally relevant content is delivered, and interaction among remote participants is encouraged. Users may belong to different countries, and learn much from cross-cultural exchange. Moreover, they become familiar with innovative state-of-the-art technologies.

The very complex design and implementation of SEE has raised a number of crucial issues, pertaining to different fields of research (see the works of Barbieri et al.) and thus reflecting the interdisciplinary nature of the team that works to build it:

- How can students be engaged in a rewarding experience, stimulating their reaction?
- How can we make the environment and the overall experience cognitively acceptable?
- How can it be ensured that the SEE experience has an effective educatinal value?
- How can we measure the usability of 3D educational cooperative environments?

These issues, and similar ones, will be discussed in the following of this paper.

---

[1] The "Dead Sea Scrolls" were found in caves near the archeological site of Qumran, probably occupied (from approx. 150 B.C. to 70 A.D.) by a religious Essene community. They contain the earliest version of books from the Bible, a description of the community life and insights on Israeli culture in the same years.

# 1  The Shrine Educational Experience

The Shrine Experience includes 4 online cooperative sessions, distributed through a couple of months, and a set of off-line activities taking place in schools in the intervals between a session and the following (see Table 1).

**Table 1:** SEE structure

| INTRODUCTION | QUMRAN SCROLLS | TOPIC | HOMEWORK |
|---|---|---|---|
| Israel Museum and Jerusalem | and the Essene Community | *Rituals* or *The Bible* or *Roman Empire*... | by students on the chosen topic |
| Virtual training for games Reading material on Qumran | Reading material on chosen topic | Preparing homework | Reading others' homework |
| Week 1 | Week 2 | Week 3 | Week 4 | Week 5 | Week 6 | Week 7 | Week 8 |

The choice of planning several cooperative sessions was essentially motivated by the attempt to take into account all the valuable goals that a cooperative online experience may achieve:

- Arousing the students' interest on a cultural theme presented in an intriguing way.
- Allowing students from different countries to meet together and discuss with an expert.
- Stimulating students' critical thinking, discussion, and interpretation of themes.
- Encouraging students to share and discuss the result of their elaboration.

In a typical cooperative experience, four classes of students between 12 and 19 years of age, from different geographical areas (e.g. Israel, Italy, U.S., Australia), join a Museum guide in a set of 3D virtual environments. Some of them reproduce the Israel Museum's Shrine of the Book, where the Scrolls are preserved (see Figure 1); the others are artificial settings, favouring users interaction.

**Figure 1:** The "virtual" Shrine of the Book

Participants meet together and with a museum guide in the shared online 3D environment, where only 9 *avatars* (i.e. graphical human-like representations of users) will be visible: 8 students plus the guide. Two students per class log in, move their avatar in the shared space, manipulate objects, and chat with remote users. The rest of the class support them by performing various activities.
In order to avoid the typical "idleness" of 3D worlds, where users end up with hanging around aimlessly, the Shrine Experience has been structured in detail, through a sort of storyboard: every single slot of time in a cooperative session is dedicated to a very precise activity. The actions allowed (or forbidden) to users in any situation are well defined in advance.
At the beginning of a cooperative session, after a short "welcome", the guide introduces he cultural theme, inviting then students to explore the environment. Cultural Games are the core of the experience: "quiz", "treasure hunts", "Olympic games", offer a highly engaging experience, at the same time requiring previous knowledge about Qumran, the Scrolls and related issues.

917

Cooperative sessions last about 45 minutes. During this time, students are never left "idle": the guide coordinates them, invites them to perform activities, to actively collaborate with each other and to interact with the virtual environment. The guide is provided with extended powers, which may be used to maintain discipline, or to assist avatars that encounter technical problems.

## 2 "Edutainment" in a 3D world

How can we make students enjoy themselves dealing at the same time with serious cultural content? How can we promote their interaction and cross cultural exchange? How can we provoke interest towards a culture remote both in space and time?

A rich set of introductory material is offered to students and teachers, so that they may prepare before the experience and exploit at best the time shared online. This material is composed of:

- Interviews to leading experts about Qumran and the Scrolls
- Editorial insets describing in detail events or characters mentioned in the interviews
- Anthologies collecting all the excerpts from the Scrolls, the Bible, historical sources or other texts, mentioned or quoted in the interviews
- Auxiliary educational resources, providing background information on historical and geographical issues that may be obvious to part of the audience, and obscure to others.

Key concepts from the introductory material are then recalled during the online experience with the help of "boards" (pop-up browser windows, see Figure 2).

**Figure 2:** An avatar activates a board

If the discussion with the guide is essential for stimulating critical thinking about the subject (Chang, 2002), interaction with peers offers an effective opportunity to revise and fix the most significant concepts in the students' memory. This is the objective of the Cultural Games.

The rationale behind SEE, confirmed by initial trials in schools (Di Blas, Paolini & Poggi, 2003), is that teachers appreciate this innovative way of learning and students are motivated to study and revise what they have learnt by the pleasure of the game and of the competition.

Collaboration among participants extends also to off-line activities; students are requested to work with their remote colleagues, researching on what they learnt (e.g. relate Qumran rituals to their own local culture and traditions). The virtual museum is thus not only a space where an antique culture is reflected upon (Falk & Dierking, 2000): it becomes a lively setting where people from different cultures are confronted with each other and with bi-millenary - yet still topical - issues.

## 3 Is the 3D world "usable"?

The problem of usability of "traditional" web sites is nowadays widely discussed in literature and certainly one of the issues to which designers pay the greatest attention. Although of course we cannot borrow directly existing usability guidelines, we can nonetheless inherit the basic "ontology" (concepts and terms) of "classic usability".

Usability is "the effectiveness, efficiency and satisfaction with which specified users can achieve specified goals in particular environments" (ISO 9241-11). Usability is therefore a combination of factors that affects the user's experience with the product.

Concerning the SEE experience, the following were specific items to which we paid attention:

- *A carefully designed experience.* The overall organization into four sessions and the detailed storyboard of each session aim at "pressing" users to interact and cooperate. We wanted to avoid avatars hanging around aimlessly. The storyboard "discipline", constantly enforced by the guide, builds a very precise frame of activities, alternating games and lectures, and none of them lasts more than 15 minutes.
- *Delivery and recollection of robust cultural content.* The background material (interviews, anthology etc.) is constantly recalled during the cooperative experiences, by means of the boards and games. Classes are also asked to prepare a homework based on the contents learnt. Therefore the experience holds a highly educational value.
- *The pleasure of competition and interaction.* Games are designed so that it is impossible to win "playing alone": students must intensely cooperate with each other, mainly via chat. For example, they need to compare the objects found during the treasure hunt, possibly using advanced features (e.g. looking through another avatar's eyes). First trials in schools have shown that students are excited by the cooperating "atmosphere" and feel dismayed when the session is over: they want more.

When the project will be fully deployed we will test usability features, by using questionnaires and direct systematic observation of users in action.

# 4 Conclusions

A while ago 3D virtual environments were considered a new frontier, with a kind of "gold rush" going on. In videogames, the engagement of high quality graphics, high performances, fast-paced interaction and play ensured success. For educational or cultural-heritage applications, however, we have more failures than success stories; an exception can be considered the field of historical reconstruction (e.g. the Virtual Olympia Project, Kenderdine, 2001), since there 3D can help in visualizing "lost worlds". In many cases 3D environments create an immediate excitement for the new technology, followed by the user's terrible question: "what should I do now?" Shared 3D worlds over Internet allow meeting other users in the cyberspace, but have the disadvantage of low graphic quality, slow performances, etc. In addition, the excitement of social interactions can quickly vanish unless each user has a good reason for relating with the others.

Our approach, to make 3D world educationally relevant, can be synthesized in a small checklist:

- Users must cooperate with each other, rather than interacting only with the 3D world.
- We conceive the overall environment as a competitive game.
- A few elements are required in order to win the games:
  - Knowledge about the content (educational value)
  - Interaction ability (interactive skills)
  - Cooperation (cooperative skills)
  - Ability to relate with other cultures (intercultural value)
- Users are always kept in action, and never idle.

Initial field tests have shown enthusiastic acceptance by the market (schools and teachers love the idea) and by our potential users (i.e. the students). We are aware that field test is very different from field deployment: we are ready to find remedies to whatever may go wrong (technology, organization, educational value, cultural clashes). We are confident, however, that positive indications will prevail and that actual deployment will be successful.

# 5 Acknowledgements

We wish to thank all the people who passionately contribute to make the Shrine project a successful experience: the editorial staff and the development team of SEE, including the invaluable contribution of Sophie s.r.l, the scientific committee, and the curatorial staff of the Israel Museum. Due to space constraints we cannot mention them all by names: you can find them listed in the Web site (www.seequmran.it).

# References

Barbieri, T. (2000). Networked Virtual Environmentsfor the Web: The WebTalk-I and WebTalk-II Architectures. *IEEE for Computer Multimedia & Expo 2000 (ICME), 2000*, New York, NY.

Barbieri, T., Paolini, P. (2001). Cooperation Metaphors for Virtual Museums. *Museums and the Web, 2001*, Seattle, WA.

Barbieri, T., Paolini, P. (2000). CooperativeVisits to WWW Museum Sites a Year Later: Evaluating the Effect. *Museums and the Web, 2000,* Minneapolis, MN.

Barbieri, T., Paolini, P., et al. (1999). Visiting a Museum Together: how to share a visit to a virtual world. *Museums and the Web, 1999,* New Orleans, LA. 27-32.

Chang, N. (2002). The Roles of the Instructor in an Electronic Classroom. *Educational Multimedia and Hypermedia*, 2002, Association for the Advancement of Computing in Education, Denver, CO.

Di Blas, N., Hazan, S., Paolini, P. (2003). The SEE experience. Edutainment in 3D virtual worlds. *Museums and the Web 2003,* Charlotte, SC.

Di Blas, N., Paolini, P., Poggi, C. (2003). SEE (Shrine Educational Experience): an Online Cooperative 3D Environment Supporting Innovative Educational Activities, *ED-Media 2003*, Honolulu, Hawaii.

Falk, J.H., & Dierking, L.D., (2000). *Learning from Museums. Visitor experiences and the Making of Meaning*. Walnut Creek, CA: Altamira Press.

Kenderdine, S. (2001). 1000 Years of the Olympic Games: Digital reconstruction at the home of the gods. *Museums and the Web 2001,* Seattle, WA.

Roitman, A. (1997). *A Day at Qumran: The Dead Sea Sect and Its Scrolls*. Jerusalem: The Israel Museum.

# Main Features of a CBIR Prototype supporting Cartoon Production[1]

*Tania Di Mascio*

Dipartimento di Informatica e
Sistemistica
Università di Roma La Sapienza
Via Salaria 113, I-00198 Roma Italy
tania@ing.univaq.it

*Laura Tarantino*

Dipartimento di Ingegneria Elettrica
Università degli Studi dell'Aquila
I-67040 Monteluco di Roio
L'Aquila - Italy
laura@ing.univaq.it

## Abstract

We present the main features of a CBIR system prototype for vector images, which is being validated in the framework of the production of 2D animation. For retrieval purposes a vector image is discretized to be viewed as an inertial system in which material points are associated with basic elements obtained by discretization. This allows us to obtain a representation invariant to translation, rotation, and scaling. The result of the searching process is a ranking of database images based on a metric obtained as weighted combination of the first seven moments of the inertial system.

## 1   Introduction

The paper describes the main features of a content based retrieval system for vector based images, applied to the management of cartoon images. There are three fundamental bases for Content-Based Image Retrieval (CBIR), namely Visual Feature Extraction, Multi Dimensional Indexing, and Retrieval System Design. Our work is presently focused on aspects related to Visual Feature Extraction. In broad sense, features may be either text-based (keywords, annotations, etc.) or visual (color, texture, shape, faces, etc.). We are interested here in *visual features*, which can be classified as domain-specific (e.g., human figure or finger prints) and general (color, texture, and shape). The former case is better covered in Pattern Recognition, while the latter is shortly reviewed here. More specifically, at this stage of our project, without loss of generality, we deal with shape extraction, which, in our application domain, well identify objects. In our approach the inclusion of color and texture can be easily treated as a direct generalization of the shape case. Hence though we did not face this problem yet, for the sake of completeness we briefly overview here also color and texture extraction (for a general discussion on CBIR systems see, e.g., [Smeulders et al, 2001]).

The *shape representation* is generally required to be invariant to translation, rotation, and scaling. In general, shape representation techniques can be classified into two categories, boundary-based and region-based. The former uses only the outer boundary of the shape while the latter uses the entire shape region. The most successful representatives for these two categories are Fourier Descriptor and Moment Invariants. The main idea of Fourier Descriptor is to use the Fourier

---

[1] Partially supported by the European Union under the IST project *SEWASIE*: SEmantic Webs and AgentS in Integrated Economies (IST-2001-34825), by the Italian Ministry of Education and by the University of L'Aquila under the project Representation and Interaction Techniques for Spatial Data.

transformed boundary as the shape feature. For example, Rui proposes a modified Fourier Descriptor that is both robust to noise and invariant to geometric transformation [Rui et al., 1996]. The main idea of Moment Invariants is to use region-based moments, which are invariant to transformations, as shape feature are. Based on [Hu, 1962] many improved versions appeared in the literature during the course of the years. Yang and Albregtsen [Yang et al., 1994] propose a fast method of computing moments in binary images. In [Kapur et al., 1995] algorithms are discussed to systematically generate and search for a given geometry's invariants. In some recent review papers [Babu et al., 1997], authors compare the performance of boundary based representations, region based representations, and combined representations. Their experiments show that combined representations outperform the simple representations.

Diverse *color* representations have been applied in Image Retrieval, notably, Color Histograms (see, e.g., [Swain et al., 1991] and [Niblack et al., 1998]), Color Sets (see, e.g., [Smith et al., 1995]), and Color Moments (see, e.g., [Stricker et al., 1995]). In particular, in the moment-based approach, any color distribution can be characterized by its moments, with most of the information concentrated in the first, second, and third central moments (mean, variance, and skewness). As in the case of moment-based shape extraction, weighted Euclidean distance can be used to calculate the color similarity.

Finally, *texture* contains important information about the structural arrangement of surfaces and their relationship to the surrounding environment. Currently, most texture representation techniques are based on the Wavelet transform, originally proposed in [Chang et al., 1993], and used, e.g., in [Smith et al., 1994], where statistics (mean and variance) extracted from the Wavelet sub-bands are used as texture representation, and in [Ma et al., 1995] where diverse kinds of Wavelet transform are used to evaluate the texture image annotation.

Differently from most of the techniques proposed in the literature, focused on raster images, we deal with vector images. The work presented in this paper is part of the Paperless system[2], an advanced high quality 2D animation environment based on the combination of user-friendly software tools integrated with innovative interaction devices [Vennarini et al., 2001]. More specifically, the CBIR module supports cartoon episode management. It is very common to reuse animation scenes and frames from previous episodes into new episodes. Possibilities for scene reuse usually stem from the memory of the animators, with little or no computational aid. Efficient archival and searching of animation material is hence appropriate.

## 2    The Application Domain

The creation of cartoons is based on two fundamental aspects: realization and animation. While the former aspect is related with the authors' skill, the latter, nowadays, is related to a specific technology. In general, authors prefer to use traditional paper for realization, because the new technologies do not assure them the artistic signature, thus not guaranteeing high quality artistic drawing. To solve this problem and to offer an environment as close as possible to the natural one, Paperless supports an interactive slate. This input device, shown in Figure 1, is composed by a digital display and an electronic, pressure sensitive, pen. This technology, based on new *intelligent pen* systems, protects every natural element of a paint realized in the traditional environment.

Generally, to create an episode is necessary to animate a large number of scenes, in turn composed by a number of frames. The traditional realization of a scene utilizes an appropriate device, called *rostrum* (Figure 2). A single frame is shot by a camera perpendicular to a series of parallel *transparent trays*, each carrying a slide with one or more objects of the frame (e.g., background, characters, etc).

---

[2] developed within the EU IST project Tools for Paperless animation (IST-1999-12329)

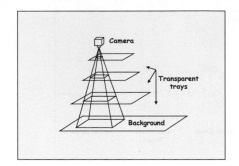

**Figure 1**: Interactive slate                    **Figure 2**: Schema of the rostrum

Therefore a single scene is composed by one fixed background and one or more animated characters. Each of these animated characters is formed by a fixed part (the profile of the character) and a moving part (the face of the character) to simulate, for example, talk actions.

Hence images in our domain often possess few characteristics, and often belong to well-defined categories (background, characters, faces, etc). It is therefore appropriate that the retrieval system be able to answer queries of two level of abstractions: while the level 1 comprises retrieval by *primitive* features (such as shape), in response to a query formulated by an example drawing, the level 2 comprises retrieval by *logical* features, aimed at extracting images of a given category (in other words the retrieval system should discriminate among categories).

# 3  Our approach

The purpose of image processing in image retrieval is to enhance aspects in the image data relevant to the query, and to reduce the remaining aspects. It is customary to include in a CBIR system a feature extraction module that associates an image with a vector of descriptors representing visual features (e.g., color, texture, and shape).

Without loss of generality, we focus on shape, which, in our application domain, adequately identify and classify an object (it has to be noticed that, due to the characteristics of the vector image format used in Paperless, treatment of other visual features is a direct generalization of the shape case). In our application the shape representation has to be invariant to translation, rotation, and scaling. These affine transformations are to be regarded as applied to a selected point belonging to the image and representative for the image. Our approach is to consider the image like an inertial system and to use the center of mass as selected point. The inertial system is obtained by discretizing the vector image, and associating material points with basic elements obtained by the discretization process. The origin of the inertial system is then moved to the center of mass, to which transformation can be applied (see Figure 3). Similarly to Hu [Hu, 1962], we use the first seven moments of the inertial system (computed from 2nd and 3rd order central moments), which constitute our vector of shape descriptors.

The similarity between any two images is computed as the similarity between the two corresponding descriptor vectors. The similarity measure is based on the Euclidean distance (an optimal combination of weights for the seven moments was empirically determined). A gross dataflow scheme illustrating the treatment of queries of level 1 is depicted in Figure 4. Given a query image, provided by authors through the interactive slate, database images are ranked based on the similarity with the input image [Di Mascio, 2002]. The retrieval efficiency of moment invariants for queries of level 1 is documented in the literature for raster CBIR systems; our experimental results for vector images indeed met the expectations.

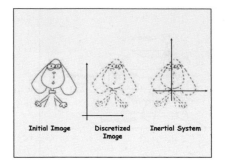

Initial Image     Discretized     Inertial System
                Image

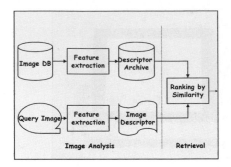

Image Analysis          Retrieval

**Figure 3**: Creation of Inertial System        **Figure 4**: Architecture of the system

Differently from CBIR systems that operate at the primitive feature level, our application domain requires also that the discriminating power of the retrieval system be adequate for the logical feature level. We therefore carried on a series of experiments to single out the role of individual moments in the treatment of logical level queries. For example the Figure 5 shows how the background category is easily identified by using the first moment (an example image of the background category is depicted in Figure 6).

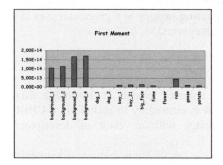

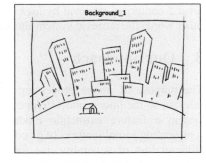

**Figure 5**: Results of the first Moment        **Figure 6**: Example background

Actually, the seven moments are computed from terms representing distribution, skew, and kurtosis of the inertial system, helpful in determining the image category. We experimentally determined combinations of weights to be used in the similarity measure appropriate to process three different types of queries: given a query image, find database images similar to it, given a query image, find database images of the same category of the given image, and given a category, find database images belonging to the required category (appropriate weight vectors are associated with categories).

## 4    Conclusions and future work

Besides more thoroughly evaluating the prototype, we plan to continue our research along new work lines: additional visual features (e.g., color and thickness) will be taken into consideration, and different application domains will be considered to test the retrieval and discriminating power of the prototype (initial work is being done in cooperation with researchers working in satellite images, generally stored as vector-based images).

# References

Babu M. M., Kankanhalli M., and Lee W. F. (1997). Shape measures for content based image retrieval: a comparison. Information Processing & Management, 33(3), 319-337.

Chang T. and C.-C. Jay Kuo (1993). Texture analysis and classification with tree-structured Wavelet transform. IEEE Trans. Image Proc., 2(4), 429-441.

Di Mascio T., (2002). Un motore di ricerca visuale per database di immagini vettoriali. Master Thesis, Università degli Studi di L'Aquila.

Hu M. K. (1962).Visual pattern recognition by moments invariants. IRE Trans. on Information Theory, 8, 179-187.

Kapur D,. Lakshman Y. N, and Saxena T. (1995). Computing invariants using elimination methods. In Proc. IEEE Int. Conf. on Image Proc.

Ma W. Y. and Manjunath B. S. (1995). A comparison of Wavelet transform features for texture image annotation. In Proc. IEEE Int. Conf. on Image Proc.

Niblack W., Barber R., and et al. (1994). The QBIC project: Querying images by content using color, texture and shape. SPIE Storage and Retrieval for Image and Video Databases.

Rui Y., She A. C., and. Huang T. S. (1996). Modied Fourier descriptors for shape representation: a practical approach. In Proc. of First International Workshop on Image Databases and Multi Media Search.

Smith J.R. and Shih-Fu C. (1995). Single color extraction and image query. In Proc. IEEE Int. Conf. on Image Proc.

Smith J.R. and Shih-Fu Chang. (1994). Transform features for texture classication and discrimination in large image databases. In Proc. IEEE Int. Conf. on Image Proc

Smeulders A.W.M., Warring M., Santini S., Gupta A., Jain R. (2001). Content-based image retrieval at the end of the early years. IEEE Trans. on Pattern Analysis and Machine Intelligence, 22(12).

Stricker M. and Orengo M. (1995). Similarity of color images. SPIE Storage and Retrieval for Image and Video Databases.

Swain M. and Ballard D. (1991). Color Indexing. Int. J. Comput. Vis., 7(1), 11-32

Vennarini V. and Todesco G. (2001). Tools for paperless animation. IST project fact sheet. Retrived August 22, from http://inf2.pira.co.uk/mmctprojects/paperless.htm

Yang L. and Algregtsen F. (1994). Fast computation of invariant geometric moments: a new method giving correct results. In Proc. IEEE Int. Conf. on Image Proc.

# Instructional Use of Engineering Visualization: Interaction-Design in e-Learning for Civil Engineering

*Martin Ebner*

*Andreas Holzinger*

Institute of Structural Concrete (IBB)
Graz University of Technology,
Lessingstraße 25, A-8010 Graz
martin.ebner@tugraz.at

Institute of Medical Informatics,
Statistics and Documentation (IMI)
Graz University, Engelgasse 13, A-
8010 Graz
andreas.holzinger@uni-graz.at

## Abstract

The main course at the Institute of Structural Concrete (IBB) of Graz University of Technology has been supported by the e-Learning project iVISiCE (Interactive Visualizations in Civil Engineering) using a web-based course management system since the year 2000. Within this project a large number of animations, simulations and visualizations have been created that are used as Learning Objects (LO). The most interesting part, however, was the creation of Interactive Learning Objects (ILO). These require the students to operate the visualizations interactively by themselves. During the design and development of these ILOs we considered aspects of Human-Computer Interaction (HCI) and User Centered Design (UCD).

## 1    Introduction

The lecture Structural Concrete (Sparowitz (1995, 2001)) is a required subject for the study of civil engineering. It is one of the largest lecture elements of the whole course. Every year about a hundred students attend the lecture to learn more about structural concrete buildings. The basic content is the design and construction of reinforced or pre-stressed concrete structures using the European Standard Norm (EC2 (1992)). Because of the non-linear behavior of concrete it is very difficult to describe engineering models and to understand the complex connections between, for example stress and strain. All too often it is not possible to explain such problems with a few words or only a drawing on the blackboard (see section 2). To show the coherences between load and required reinforcements many complex images are required. As a course management system we rely on the "Hyperwave e-Learning Suite" (eLS, http://www.hyperwave.com).

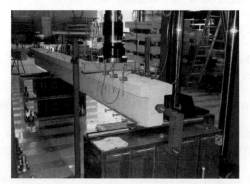

Fig. 1    Reinforced Concrete-Beam, the practical problem within the laboratory

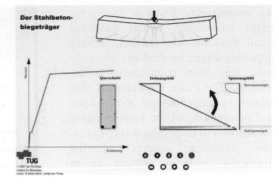

Fig. 2     The Project iVIiCE

Fig. 3     Animation of a reinforced concrete beam

## 2     Animations and Visualizations

Our considerations were that the most common mode of instruction is verbal. In explaining how the reinforcement of a concrete beam works, for example, an instructor is most likely to rely on printed or spoken words. There are several findings which show that lectures supported by animations are advantageous for a positive effect on learning (Mayer & Moreno, 2002), (Tversky, Morrison, & Betrancourt, 2002). Although the effect on learning is disputed, the motivation of the student can be enhanced and thus an indirect positive effect on learning occurs (Holzinger, 1997, 2001a). Thus it is important that the animations point out the basic matter without showing all the details. The goal is to improve student understanding of engineering problems by animating essential topics. Furthermore, animations save valuable lecture time because it is not necessary to draw all the explanations on the blackboard. The simulations can be shown via laptop and projector, followed by a discussion about the main topic. Consequently and having in mind that learning processes can be activated by doing, we decided to take further steps towards Interaction.

## 3     Interactive Learning Objects (ILO)

Our guiding principle for the development was "Learning By Doing", where  (Dewey, 1916) argued that one learns through direct experience which means by engaging in authentic tasks. Learning is thus not a process of transmitting information from someone who knows to someone who doesn't ~ rather, learning is an active process on the part of the learner, where knowledge and understanding is constructed by the learner (Holzinger, 2000). Moreover, we consider learning also as a social process: learning proceeds by and through conversations.

### 3.1     Development

During the development of the first interactive learning objects the main idea was to create visualizations that require the students to act independently. It is important that the exercise be defined very precisely for the students because a learning process can only be activated when the student operates the objects themselves. Therefore the animations are designed so that the students can interact with them, thus aiding learning in real-time. The interactions were designed and programmed with Macromedia Flash (http://www.macromedia.com) by using the programming language Action Script, which is an object-oriented language comparable with a very small version of java but it is possible to make the output dependant on user interaction.

An advantage of Flash is the vector based technology, which generates small files for the internet, a very important aspect for students who only have very slow connections to the internet. The Flash environment also allows visualizations of experiments and descriptions of the corresponding calculations to be designed very easily. The didactical concept of the interactive learning objects is in accordance with the principles of instructional design of Gagne (1992).

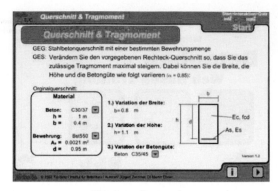

Fig. 4    Start screen

## 3.2    Major Parts of the ILO

### 3.2.1    Information and Learning-material

There are 3 elements that must be presented to the student for a structural learning process: the information to be learned (advanced organizer), the core material and a carefully selected problem. The first screen of the ILO is a start screen comparable with a homework assignment. Only a few words sketches explain the problem. The explanation must be understandable for the target group. It is particularly important to gain the attention of learners at the beginning of a lesson (Gagne, 1985; Wilson 1993) so the students can focus on the main instructional points of the lesson. With the aid of a noticeable button, the learner can navigate to the next screen. Here, the tool is explained. First the learning target of the tool is pointed out and then an overview of the screens is given. The necessary previous knowledge and the estimated learning time are also provided. The main part of the ILO consists of the main screen, where the exercise must be solved by the students. As a result of the user based input, the tool calculates and animates the engineering model. Thus learning in real-time is made possible. It is very important, that the exercise be randomly generated, allowing the students to practice the lesson repeatedly. The next screen is the help-screen. Here the main principles of the interactive example are explained. The students are provided with information about the basics of the engineering model, but are not given the solution. The content of this screen shows the theoretical background in sketches and formulas.

### 3.2.2    Communication

The second part of our tool consists of communication and co-operation. The great possibilities of the internet have been used to give appropriate assistance during the learning process. With the aid of the learning-management system, discussion forums on the several topics dealt with in the interactive objects were opened. Chat and e-mail have also been used extensively to discuss problems that occurred. Only after numerous discussions in the newsgroups were the real problems in understanding the content of the course material properly identified. Due to this intensive conversation, the quality of learning has been noticeably increased. Communication via the Web and the learning-management system became the main focus for the lecture course because these tools also enabled the tutors to address misunderstandings during the learning process.

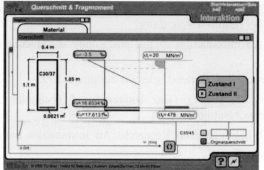

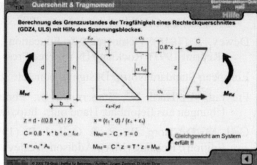

Fig. 5    Interactive Learning Object        Fig. 6    Help-screen

### 3.2.3   Assessment

The last part of the ILO is a multiple-choice test where the major points of the example are tested. The questions are carefully worded and require the students to understand the entire coherences of the tool. This allows the students to monitor their progress and determine where they need more practice on their own.

## 4    Methods

Observation and testing our work with real end-users proved to be the best way to understand the effectiveness and suitability of our ideas. During the development of our ILOs we committed ourselves to User-Centered Design (Carroll, 1987), (Vredenburg, Isensee, & Righi, 2002), including thinking aloud, cognitive walkthrough and video analysis and incorporated knowledge from the area of interaction design (Preece, Sharp, & Rogers, 2002), (Shneiderman, 1997) following the recommendations of (Stephanidis et al., 1999). Concerning the research in motivation and learning we relied on qualitative methods including interviews (Gall, Borg, & Gall, 1995). We are currently carrying out a quantitative research using the pretest/posttest experimental control group design with questionnaires, but the results were not yet available at the time this paper was printed.

## 5    Results and Discussion

Generally it was interesting to notice the similarity between Civil Engineering and Medicine - learning in both fields are supported well by using ILOs. Especially the mix of (real) lectures together with the online material proved to be successful. The basic findings were that through the use of ILOs the students were able to understand the content more in-depth. Some students and teachers argued that "it is more play than serious and hard studying". Most of the students and teachers, however, urged us to carry on. A further perspective is that our ILOs are internationally reusable by applying the concept of metadata (Holzinger, 2001b).

### References

Carroll, J. M. (1987). Interfacing Thought: Cognitive Aspects of Human-Computer Interaction. Boston (MA): MIT.

Ebner, M; Holzinger A. (2002): e-Learning in Civil Engineering. Journal on Applied Information Technology (JAPIT), Vol. 1, Iss.1, 2002, S.1-9 (http:www.japit.org)

Dewey, J. (1916). Democracy & Education. An introduction to the philosophy of education (Reprint 1997). Rockland (NY): Free Press.

European Standard Norm: Design of concrete structures – General rules for buildings: 1992-1

Freytag, B.; Hartl, H.; Stebernjak, B.; Ebner, M., (2001), Graz University of Technology: Übungen aus Betonbau, Hand out, http://www.bau.tugraz.at/ibb

Gall, M. D.; Borg, W. R. & Gall, J. P. (1995). Educational Research: An Introduction (6th Edition). Reading (MA): Addison Wesley.

Gagné, R. M. (1992): Principles of Instructional Design. New York: Holt, Rinehart & Winston

Holzinger, A (2002): Multimedia Basics. Volume 2: Cognitive Fundamentals of multimedial Information Systems. New Delhi: Laxmi-Publications. Available in German by Vogel-Publishing, http://www.basiswissen-multimedia.at

Holzinger, A (2002): Multimedia Basics. Volume 3: Design. Developmental Fundamentals of multimedial Information Systems. New Delhi: Laxmi-Publications (also in German).

Holzinger, A. (1997): Computer-aided Mathematics Instruction with Mathematica 3.0. Mathematica in Education and Research. Vol. 6, No. 4, Santa Clara, CA: Telos-Springer, 37-40.

Holzinger, A.; Kleinberger, T. & Müller, P. (2001b): Multimedia Learning Systems based on IEEE Learning Objects Metadata (LOM). Educational Multimedia, Hypermedia and Telecommunication, 2001, AACE, Charlottesville, VA, 772 - 777.

Holzinger, A.; Pichler, A.; Almer, W.; Maurer, H. (2001a): TRIANGLE: A Multi-Media test-bed for examining incidental learning, motivation and the Tamagotchi-Effect within a Game-Show like Computer Based Learning Module. Educational Multimedia, Hypermedia and Telecommunication, 2001, AACE, VA, 766 - 771.

Mayer, R. E., & Moreno, R. (2002). Aids to computer-based multimedia learning. Learning and Instruction, 12(1), 107-119.

Preece, J., Sharp, H., & Rogers, Y. (2002). Interaction Design: Beyond Human-Computer Interaction. New York: Wiley.

Shneiderman, B. (1997). Designing the User Interface, 3rd Ed. Reading (MA): Addison-Wesley.

Sparowitz, L. (1995), Graz University of Technology: Konstruktionsbeton, Hand out of the lecture Structural Concrete , http://www.bau.tugraz.at/ibb

Sparowitz, L. (2001), Graz University of Technology: Betonbau, Supporting documents for the lecture Structural Concrete, http://www.bau.tugraz.at/ibb

Stephanidis, C., Salvendy, G., Akoumianakis, D., Arnold, A., Bevan, N., Dardailler, D., Emiliani, P. L., Iakovidis, I., Jenkins, P., Karshmer, A., Korn, P., Marcus, A., Murphy, H., Oppermann, C., Stary, C., Tamura, H., Tscheligi, M., Ueda, H., Weber, G., & Ziegler, J. (1999). Toward an Information Society for All: HCI challenges and R&D recommendations. International Journal of Human-Computer Interaction, 11(1), 1-28.

Tversky, B., Morrison, J. B., & Betrancourt, M. (2002). Animation: can it facilitate? International Journal of Human-Computer Studies, 57(4), 247-262.

Vredenburg, K., Isensee, S., & Righi, C. (2002). User Centered Design: an integrated approach. Upper Saddle River (NJ): Prentice Hall.

# Multimedia in Education: Myths and Realities

*Andreas Evangelatos,*

Athens University of Economics and Business,
12 Derigny Street, Athens 104 34
aevan@aueb.gr

*Maria Constantopoulou*

Athens University of Economics and Business,
12 Derigny Street, Athens 104 34
marconst@aueb.gr

## Abstract

In this paper we are trying to investigate theoretically the contribution of multimedia in education. We start with a short survey of the academic views and of the relevant empirical studies that appeared in the past decade. We then go on to propose a framework in which the opposing views are shown to uncover different aspects of the issue in question. We finally put forward a theoretical model integrating the different aspects of the effect of multimedia in education. The next step would be to test empirically the relevance of this model.

## 1. Introduction

The introduction of interactive multimedia in the decade of the 90's has brought back with renewed vigor the discussion concerning (multi)media effects on learning. At the same time a myth was born, or more precisely re born, concerning the alleged superiority of multimedia in learning compared with traditional instruction. Thanks to this myth multimedia has enjoyed a prestige and a hype while a lot of research effort was devoted to the investigation of the alleged unlimited potential of multimedia. The advent of the current decade has revealed that the results of the endeavor were not proportionate to the expectations and effort spent. And the question is: what went wrong? What was badly accounted for from the start?

## 2. Views about the influence of media on learning

The discussion of the effect of media on learning started with Edgar Dale (1969) who propagated the potential of audiovisual technology to improve education in his famous book "Audiovisual Methods in Teaching" published for the first time in 1946. Dale introduced the Experience Cone (diagram 1 below) in an attempt to classify means of conveying knowledge according to the degree of sensory involvement of the student. The cone shows that the information conveyed verbally requires smaller sensory involvement compared, for instance, with animation or movies (video or television).

The mechanical connection of Dale's views with the ability of multimedia applications to incorporate educational material in audio and video form and through planned interactivity to offer the possibility of 'devised experience' for the student through simulation, formed the basis for the development (Hasebrook J, 1997) of what came to be called "naïve theories of multimedia learning".

**Diagram 1:** Dale's experience cone

These naïve theories propagated that the combination of many media (text, graphics, animation, etc.) in one sole source of transmission would produce a rich sensory environment thus permitting a higher rate of understanding and retention of the information transmitted[1] (Yanerbaum G. et al, 1997).

---

[1] It is maintained that "people remember 20% of what they hear, 40% of what the see and hear simultaneously, and 75% of what the see, hear and perform at the same time.

Among the earlier exponents of the view that multimedia affect learning positively are Levin (1989) και Kozma (1991). Levin considers that multimedia enhance motivation and consequently understanding. Kozma maintains that media differ in their influence on the receiver according to the manner in which his mind represents and processes information; thus, the use of multiple means succeeds in the recall of cognitive material.

Opposing the views outlined above Richard Clark continues to uphold his earlier position, maintaining that media form the vehicle through which knowledge is transmitted and as such they do not contribute to the success of learning (Clark 1991). In his words "media are mere vehicles that deliver instruction but do not influence student achievement any more than the truck that delivers our groceries causes changes in our nutrition" (Clark 1983).

Thus an academic controversy started concerning the effect of media on learning, which has its empirical extension, since some empirical studies show that multimedia have a positive effect on learning (Leidner 1994, Mayfield-Stewart, Morre & Sharp 1994, Dillon & Gabbard 1998) while others show the opposite (Clark R. & Graig T., 1992/ Yaverbaum & Nadaragan 1996). The existence of these two tendencies is clearly outlined in the meta-analyses performed by Liao (1999), who found a limited positive effect of multimedia on learning, depending on the type of instruction with which it was compared. In the empirical studies he analysed a number of studies revealing a superiority of traditional teaching methods which were not negligible (10 out of 35), thus offering a clue as to the relative validity of opposing views.

### 3. Definition of the theoretical framework

In order to examine the contribution of multimedia on learning, which remains still a controversial issue, we must define clearly a framework in which we shall investigate the possible effect of multimedia on learning. The model chosen is the so called "teaching triangle" (diagram 2), which describes the three main functions developed in every learning case, except self-learning[2].

**Diagram 2:** Instruction triagle

1. The first function deployed in each case of guided learning is that of the student's mentoring from the teacher. Through this function the teacher tries to guide the student in his attempt to acquire knowledge. This function is characterised by the instructional method that the teacher will pursue and all researchers consider it as the most important parameter among those influencing learning. As instructional method, according to Salomon (1979) we mean the particular transformation of the information that the student must acquire and which facilitates or obstructs the required cognitive process. The instructional method comprises lectures in large or small audiences, small groups, mentoring for the acquisition of experience in devised or experimental or professional environments, distance guidance, etc. The instructional method, in its turn is influenced by the organizational environment in which it is developed (school, university, enterprise, etc.), by the existence or not of an institutional setting (analytical curriculum, program of studies, in-house training, etc.) and by the dynamics of the team constituted (school class, academic course, training seminar, etc.)

2. The second function developed is that of the arrangement of the cognitive material by the teacher. The content of this function consists in the teacher's attempt to collect material, to update it, to restructure it, etc. Depending on the case it is possible for the teacher to be himself the producer of the cognitive material or he may use material that pre-exists and which has been produced by

---

[2] Please note that distance learning is not classified as self-learning, since in this case we do have a teacher, but teacher-student communication is asynchronous.

himself or by others. Publishers or content providers, who create material to be used in education do contribute heavily in this function. This is a function that generates costs for the teacher. Capper and Fletcher (1996) in their analysis of distance learning studies determined five factors that are responsible for the costs incurred by the teacher (a) the number of courses offered (b) the rate of revision of the course (g) the type of teaching support used (d) the number of students supported and (e) the rate of course depreciation

3. The third function is that of the assimilation of cognitive material by the student. The result of this function is the main objective of the whole process, the acquisition of knowledge by the student. The acquisition can be defined and therefore tested in various levels (known as target classifications) according to the goals that the teacher has set at the beginning of the learning process.

## 4. Construction of the Framework

The framework put forward above for the investigation of the effect of multimedia on education is independent of the use of multimedia. The introduction of multimedia in this framework affects directly all three functions, mainly because of the parameter of technology use implied. Among all the theories proposed up to date for the explanation of users' behavior with regard to the acceptance and use of information technology, one of the most widely accepted among researchers is TAM (the Technology Acceptance Model). TAM was put forward by Davis (1989) in order to explain the adoption of new applications by the users given the additional effort needed on their part. The model is based on the psychological theory of Reasoned Action developed by Aizen and Fishbein (1980) according to which the beliefs of an individual influence his attitudes, which in their turn generate intentions which finally shape his behavior, as shown in diagram 3.

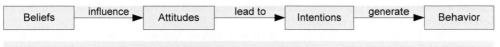

**Diagram 3:** Reasoned Action Theory

Davis constructed his TAM model as an application of the theory of Reasoned Action in order to predict the acceptance of an IT application from its potential users. In order to do this one must investigate the views of the users about the IT application in two main axes (a) the axis of usefulness - the degree in which the application will improve the user's productivity and (b) the axis of ease of use - the degree to which it will be used with more or less additional effort. The adaptation of the TAM model so that it can be used to predict the adoption of multimedia in education is shown in diagram 4.

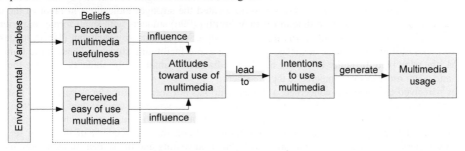

**Diagram 4:** Multimedia Acceptance Model or MAM

This model suggests that the fundamental data which will enable use to predict the adoption of multimedia is (a) perceived usefulness and their (b) perceived ease of their use.

## 5. Application of the model

The question we set out to answer is whether multimedia influence learning. The framework put forward (the teaching triangle) combined with the MAM will be used to predict the adoption of multimedia investigating its potential influence on each function (cost – benefit).

Guidance function: The question is whether a teacher would develop the instructional method he uses so as to comprise multimedia. On the basis of MAM he will do it if he believes that multimedia will help him to achieve better his goals - the guidance of the student - and in particular if the perceived cost of this action is low. Although we are not aware of any research on teachers' perceptions, there is an indirect indication on their views, from relevant research of teachers' views concerning the contribution of technology in distance education. Their views on this subject, as presented by a number of researchers (Johnstone & Krauth 1996, Frost & Fukami 1997, Mangan 1999) are negative. If we use these studies to predict teacher's views concerning the contribution of multimedia in education in general, we can predict that teachers are unlikely to adopt multimedia in order to achieve an improvement in the guidance of their students. Of course perceived usefulness of multimedia is not the only factor in MAM. Ease of use by the teacher is equally important: the training of teachers themselves in information technology is a crucial although very difficult matter.

The logical consequence of the above mentioned thoughts is that it is improbable for a teacher to include multimedia in his instructional method in order to improve the guidance of his students. The more satisfied the teacher with his performance and his ability to achieve his goals with traditional methods, the less probable it will be to adopt multimedia. Furthermore, the better educated in the use of multimedia a teacher is, the more probable it will be to comprise multimedia in his instructional method. We can therefore develop testable hypotheses concerning this part of the model proposed.

The function of arranging cognitive material is directly associated with the use of multimedia. Yet multimedia content is not available in sufficient width (theme coverage) nor in sufficient depth (number of multimedia titles in every subject). Actually, multimedia content is minimal in comparison with the needs of current education for appropriate cognitive material. Publishers have only started to produce multimedia material suitable for education. The task of developing multimedia material is currently undertaken by governmental or community institutions. And yet, inspite of the fact that the relevant initiatives[3] concentrate a large amount of effort in the direction of the development of multimedia material, the material already existing and that which will be produced soon in the European Union and North America is insufficient for the support of the educational tasks conducted.

Multimedia material is produced by a team of experts with various specialities. The teacher alone cannot produce multimedia material of quality. The question, however, is whether he/she is willing to contribute to the production of multimedia material. By applying the MAM we see that this is rather unlikely: Experience shows that perceived cost of the production of multimedia material is especially high and is connected with (a) the type of media used, (b) the frequency of revision of the material required and (c) with the rate of depreciation of the material. As concerns the perceived usefulness on the part of the teacher, his/her participation does not seem to be connected with any benefit except for the use which the teacher himself/herself will do, contrary to writing and publishing a book, which can be sold.

The support and enhancement of the third function of acquisition of cognitive material by the student, is that which has been mainly investigated, and in this were located the stronger indications in favor of the use of multimedia in education. Thus, according to Relan & Smith (1996) students with small former experience in computers could easily take to multimedia instruction, while Yaverbaum & Nadarajan (1996) suggest that instruction with multimedia has the advantage of satisfying and facilitating students. Thus, in the framework of MAM, the perceived difficulty of the use of multimedia seems small for the students, because they learn quickly and adapt more easily compared to their teachers.

As far as perceived usefulness is concerned, a student who has not seen multimedia as part of his instruction is very unlikely to deem their contribution useful. This means that when the teacher does not impose the use of multimedia and resorts instead to traditional teaching methods, the student has no incentive to use multimedia. On the contrary, their use may make more difficult the achievement of the goals set by the teacher. One should not forget that exams are usually taken in writing and the student finds it more useful to have the educational material in the form that he will be asked to reproduce in his exams. On the contrary, the use of the world wide web can crucially contribute to the support of students in their effort to carry out projects, a fact that impels them to use the www, irrespective of whether it has been incorporated in the instructional method. In the case of www what impels student to use it is not its multimedia character, but the huge amount of data and the possibility of its almost costless acquisition[4].

---

[3]These efforts are financed through initiatives of support of multimedia content i.e. initiative MEDIA in the past and the initiative eContent nowadays.

[4] This was manifested by the Napster case which concentrated the fire of the whole of music industry. Napster's users were mostly students who cared for the costless acquisition of material, compatible with the technological infrastructure they possessed.

## 6. Conclusions

To round up the paper, we must admit that we do not believe any more that multimedia will bring the 'revolution in learning ways' propagated ten years ago. Yet we do not agree with Clark (1994) either "media will never influence learning". Multimedia is difficult to produce, scarce to find and costly to adopt. Thus the likelihood that it will be included in teaching methods, and if included that the teacher will insist that students refer to it, is small. On the other hand, verbal (mostly printed) material comes in the form that best suits students in their preparation for exams and this turns in favor of traditional teaching methods. The characteristic of passivity in the absorption of information that some means possess compared to others, despite the fact that they engage many more senses of the receiver, does not seem to confer an advantage to the increasingly rising need for learning. On the contrary, the ability of selective retrieval of information that some means offer in a society submerged by loads of information constitute an increasingly valuable – albeit overlooked - advantage.

## References

1. Ajzen I., Fishbein M., 1980, «Understanding Attitudes and Predicting Social Behavior», Prentice-Hall: Englewood Cliffs, New Jersey.
2. Capper J., Fletcher D., 1996, «Effectiveness and cost-effectiveness of print-based correspondence study» Paper prepared for the Institute for Defense Analyses, Alexandria V A
3. Clark R., 1983, «Reconsidering research on learning from media», Review of Educational Research, 53 (4), 445-459,
4. Clark R., 1991, «Reconsidering research on learning from media», Review of Educational Research, 61(4), 179-211.
5. Clark R., Graig T., 1992, «Research and theory on multi-media effects», in Giardina M. (Ed.), Interactive multimedia learning environments. Human factors and technical considerations on design issues, Heidelberg: Springer, 19-30.
6. Clark R., 1994, «Media will never influence learning», Educational Technology Research and Development, 42(2), 21-29.
7. Dale E., 1969, Audiovisual methods in teaching, 3d ed., New York: Holt, Rinehart, Winston
8. Davis F., 1989, «Perceived usefulness, perceived ease of use, and user acceptance of information technology», MIS Quarterly, 13(3), 319-340.
9. Dillon A., Gabbard R., 1998, «Hypermedia as an educational technology: A review of the quantitative research literature on learner comprehension, control and style», Review of Educational Research, 68(3), 322-349.
10. Frost P., Fukami C., 1997, «Teaching effectiveness in the organizational sciences: Recognizing and enhancing the scholarship of teaching», Academy of Management Journal, 40(6), 1271-1281.
11. Hasebrook J., 1997, «Learning with multimedia and hypermedia: Promise and Pitfalls», Fifth European Congress of Psychology, Dublin.
12. Johnstone S., Krauth B., 1996, «Balancing quality and access: Some principles of good practice for the virtual university», Change, 28(2), 38-41.
13. Kozma B., 1991, «Learning with media», Review of Educational Research, 61(2), 179-211.
14. Leidner D., 1994, «An examination of learning outcomes in an advanced presentation electronic classroom», working paper, Information Systems Department, Baylor University, Waco, TX.
15. Levin R., 1989, «A transfer-appropriate-processing perspective: Pictures in prose, In H. Mandel & J.R. Stevens (Eds.), Knowledge acquisition from text and pictures», North-Holland: Elsevier.
16. Liao Y., 1999, «Effects on hypermedia vs. traditional instruction on students' achievement: A meta-analysis», Journal of Research on Computing in Education, 30(4), 341-360.
17. Mangan K., 1999, «Top business school seek to ride a bull market in online MBA's», Chronicle of Higher Education, S: Information Technology, 45 (19).
18. Mayfield-Stewart C., Morre P., Sharp D. et al., 1994, «Evaluation of multimedia instruction on learning and transfer», paper presented at the Annual Conference of the American Education Research, New Orleans.
19. Oz E., White D., 1993, «Multimedia for better training», Journal of Systems Management, 44(5), 34-38.
20. Relan A., Smith W., 1996, «Learning from hypermedia: A study of situated versus endemic learning strategies», Journal of Educational Multimedia and Hypermedia, 5(1), 3-22.
21. Salomon G., 1979, «Interaction of media, cognition and learning», San Francisco: Joy Bass.
22. Yaverbaum G., Nadarajan U., 1996, «Learning basic concepts of telecommunications: An experiment in multimedia and learning», Computers and Education, 26(4), 215-224.
23. Yanerbaum G., Kulkarni M., Wood C., 1997, «Multimedia Projection: An Exploratory Study of Student Perception Regarding Interest, Organization, and Clarity», Journal of Educational Multimedia and Hypermedia, 6(2), 139-153.

# Innovating Web Based Collaborative Learning by Applying the Case Method

*Christine Frank*

DS&OR Lab University of Paderborn
Warburger Strasse 100
33098 Paderborn
Germany
frank@dsor.de

*Leena Suhl*

DS&OR Lab University of Paderborn
Warburger Strasse 100
33098 Paderborn
Germany
suhl@dsor.de

## Abstract

Learning on the web offers new possibilities to teachers and learners. However, this technology is still very new, and at the time being it makes sense to apply it in a blended learning approach. Blended learning combines virtual and traditional approaches and thus enables the learners to profit from the advantages of both approaches, without having to face the specific disadvantages. When creating web based lessons, it is important to consider the methodological and didactical prerequisites of the chosen teaching approach. In dependence on traditional learning settings the level of interactivity has to be considered. While certain traditional learning settings obviate collaborative learning, e.g. the lecture, other traditional settings encourage cooperative learning, e.g. seminars or tutorials. Collaborative and cooperative learning have always been essential components of the case method. Translating the components of the conventional case method into a web based environment is reasonable, because it facilitates diverse learning scenarios which encourage collaborative work. When using the web as a medium for teaching and communication, it is very important to stay within the limits of what seems natural communication to the users. Evaluation of what natural communication means to them is as indispensable as formative evaluation of the prototype.

## 1    Introduction

In the 1970s and 80s the term E-learning was seen as a generic term to describe all activities that had to do with electronically enhanced learning. In the middle of the 1980s learning software spread rapidly. This is when learning methodologies became medial. In the 1990s these programs then became multi-medial. This enabled computers to combine interaction of different media, e.g. films, pictures, diagrams, text, language and music – for the first time active and interactive handling of learning with media was made possible. For a long time the main emphasis was placed on the "E" (electronic) part as a success guarantee of innovative technologies. The "learning" part of the term was often seen as a secondary matter. Didactical concepts were frequently missing, which limited the possibilities of E-Learning (Uesbeck, 2001). Today, a crucial role is attributed to the actual learning part within E-Learning.

When integrating E-learning into existing settings a question which is often asked, is to which extent E-learning can extend, or can it even replace traditional teaching and learning methods in the future? It is certain that not all areas of training offer the same favourable conditions to be supported by learning through multimedia. While some areas are suitable for virtual learning, other fields are dependent on traditional education methodologies. Every subject has learning processes that require traditional learning sequences – but these sequences can be designed to be supported by multimedia. The main reason to implement media into learning processes is not to simply replace the teacher. No medium can do this. Media can, however, be used to increase the quality of the learning process. This means that E-learning means to extend traditional teaching, not to replace it. E-learning most commonly includes communication between learners and learner and teacher. The trend within E-learning is developing towards a blended learning approach, which combines traditional and virtual phases, thus combining the best of both worlds. This paper describes the attempt to transfer a teaching method, called the case method, to the web using a blended learning approach. This is done within the project VORMS which is funded by the BMBF (Federal Ministry of Education and Research) of the German Government and focuses on the subject Operations Research/Management Science.

## 2    The case method

The case method originated at the Graduate School of Business Administration of the Harvard University in Boston, Massachusetts. The aim was to involve students into learning processes, rather than letting them be passive participants as their role is conceived within many conventional teaching methods, e.g. the lecture. The traditional case method facilitates many prerequisites concerning interaction which are not utilized within traditional lectures due to - at least in most cases - missing phases of interaction between learners and learner and teacher. For a detailed definition of the traditional case method see (Lynn, 1999) and (Erskine, Leenders & Mauffette-Leenders, 1998).

Working through a case within the case method can be divided into three phases. The first phase is a single working phase, this means, that the learner is to individually prepare the case, and also prepare a way of solving the given problems. After accomplishing this the learner meets with a small study group. This is the environment, where individual opinions are stated and discussed. At the end of this phase, the small study group has to come up with and agree upon one answer. This answer is then taken into the plenum, where the entire course, made up of several small groups, meet together with the teacher. The teacher moderates and facilitates the discussion amongst the small groups. Small group and plenum phases are regarded as collaborative work, and are very important within this educational setting.

### 2.1    Collaborative learning/cooperative learning

The need for self-directed teams has increased in the business environment as global competition and other changes in today's business requirements are increasing. This shift in organizations must also be reflected within Universities (Bobbitt, Inks, Kemp & Mayo, 2000). Universities need to teach students effective teamwork and communication skills, in order to prepare them to work in such an environment.

The underlying philosophy and methodology of the terms collaborative and cooperative are completely different. Flannery describes that cooperative learning uses "student learning groups to support an instructional system that maintains the traditional lines of classroom knowledge and authority" (Flannery, 1994, p. 17). The definition of collaboration is different, it stresses that the synergy develops something that has not been there before as en essential component of collaborative work. Collaborative learning not only challenges and changes students, it also challenges teachers, who in turn need to facilitate groups and group work.

## 2.2 Collaborative learning via the web

Within Operations Research (OR) learners use quantitative techniques - e.g., mathematical optimization - to solve given problems. If they solve these problems on their own or through lectures, social competences are not being learned. The case method is a complex setting which realizes collaborative learning and thus enforces learning and developing social competences. Because the traditional case method facilitates these features, we are designing a web based version which also integrates aspects of collaborative work. We are implementing a blended learning approach, which partially relocates social interaction to the web, and partially encourages live and personal interaction. This way we are trying to transfer all of the typical advantages of the traditional blended approach. Typical advantages of the case method are that learners don't simply memorize facts and that they actually experience situations. Learning social competences is another advantage of the case method opposed to many other teaching styles.

# 3 Implementing a web based version of the case method

Learners receive the case study via the web. In a self study process, they begin working on the problem. If they need to acquire skills, they can use guided tours or free exploration to search for the information within the VLE. Learners can make use of asynchronous communication possibilities e.g. e-mailing questions to tutors and/or other learners, or by posting questions on moderated and/or unmoderated discussion boards. Synchronous communication and whiteboard features will also be implemented in the future, if evaluation shows, that learners favor these functions to other features. The same applies for a whiteboard feature. Once the learner has an overview of the problem, he is to independently build up or integrate himself into a small discussion group of 3-6 learners. Groups can be built by making contact to each other over the web, or by making acquaintance in lectures. Meeting processes are facilitated by teachers and/or by offering technological support such as mailing lists or blackboards for students searching for a group, etc.. Groups can meet and discuss virtually within the VLE, or meet in real life – this decision is up to the group. Attendance to certain lectures within the web based case method is obligatory. During the obligatory lectures, the small groups must present their solutions to the plenum and discuss/defend why they chose their way of solving the problem. Cases usually do not have only one correct way of solving the given problem so it is likely that another group will have different results or at least different procedures towards the results. Discussing these different results builds up the social skills mentioned earlier. Discussion skills like these cannot be learned virtually through the web, they need to be embedded into face to face interaction. Due to the direct dialogue between the learners and between learner and teacher, traditional learning and teaching methods are suitable to teach subjects that are complex and connected to social interaction processes. All participants can integrate themselves into the learning process through communication and open discussion. The professor moderates the discussion, rather than giving correct answers.

Many preparatory steps towards these discussions can be facilitated through the web. A homogenously prepared class with independently working learners is one possible result when preparation of cases is done via the web, not having to neglect the advantages of presence phases.

In addition to teaching and learning, traditional teaching scenarios fulfil the elementary function to establish a structure within the group of learners. Group-finding and group-building processes cannot take place sufficiently within virtual settings. If a phase of "getting to know each other" is missing, groups cannot work together constructively and efficiently.

# 4 Evaluation

Web based communication possibilities were assessed before implementing them into the prototype. This gave the developing team the priorities which communication processes to integrate into the prototype. After implementing the prototype of the web based case method, which did not yet have all the above mentioned features for it was still in development, the prototype was qualitatively evaluated by interviewing ten participants who had worked with it. Within the project VORMS we believe in the importance of a combination of formative and summative evaluation. Formative evaluation takes place during the process of the development, opposed to summative evaluation which takes place at the end of development. All results gathered in the phase of formative evaluation are integrated into further developments.

## 4.1 Evaluation of communication

Communication is an important part of the traditional case method. Therefore it can be concluded, that it is also a very important part within the online version of the case method.

Many of our communication settings are changing at a very fast pace. Communicating face to face is still done by almost all of us, but the use of e-mail, chat, discussion boards, white boards and the internet in general are expanding rapidly. Before implementing electronic communication into our prototype of the web based case method we assessed the readiness of our target group towards these new communication possibilities. Detailed results are described by (Brunn & Frank, 2002). The characteristics students strive for when they use electronic communication, which are therefore also our success factors are:

- Natural communication
  Students need to be aquainted with the communication possibility and accept it as "natural communication". One example for accepted "natural electronic communication" is the telephone. Within the target group of VORMS e-mail is also considered to be natural electronic communication.
- Mobility
  Students want to be mobile. Accessing communication possibilities from different sources at varying times is important to them.
- Communication rules
  Communication rules need to be set, in order to facilitate successful communication.
- Goals
  A common goal in mind of the users (e.g. preparation for an exam) supports successful electronic communication.
- Personnel contact
  Personally knowing who one communicates with electronically makes communication through the web easier. This could be facilitated e.g. by integrating a kick-off meeting.

## 4.2 Evaluation of prototype

The results of the interviews were very rich, and can only be introduced in the following. For a complete overview of the results see (Frank, Reiners & Suhl, 2003). The results can be divided into the categories: discussion, feedback, communication, motivation and overall improvement. The following paragraphs give a short overview of the categories feedback and motivation.

Within the category feedback the students mentioned that receiving feedback was a very important feature. Students criticized that they did not always receive feedback within the discussion board. This lead to less productive work on the case, for they needed the feedback to continue their work. When implementing an online version of the case method it is very important to ensure that feedback is given to all participants. This can be done, e.g. by reducing the number of participants, by increasing the number of teachers and tutors, or by encouraging the students to give each other more feedback.

Concerning the motivation within the virtual case method the students stated, that the common goal they had in mind, which was the preparation for an exam, motivated them to participate regularly within the web based case method. It was also said, that the authentic context of the cases encouraged the students to participate. Seeing actual application fields of their theoretical knowledge was another motivating factor for the students. Some students were de-motivated by the amount of work they had to invest into the case method. They described it as being too much work. Others stated that because this specific case was not relevant for the exam they were not really interested in working through it.

## 5 Conclusion

New forms of information technology are opening a wide range of innovative E-learning possibilities. Implementing new technologies into traditional learning scenarios calls for integration strategies in order to successfully improve these. One of these strategies is to introduce virtual learning opportunities in a blended learning approach. At the time being this is especially advisable, because web based learning opportunities are still in development and lack e.g. motivational and personal components. This can be compensated by integrating traditional phases into the web based approach. Another main aspect of integrating new technologies into traditional settings is the acceptance of the users. User acceptance is a main success factor. User acceptance can be increased by involving the users in the formative evaluation of the product. Choosing methodologies that facilitate motivation, group work, and offer authentic learning contexts support the constructivist learning approach, which is the approach best suited for web based applications.

## References

Bobbitt, L. M., & Inks, S. A., & Kemp, K. J., & Mayo, D. T. (2000). Integrating marketing courses to enhance team-based experiential learning. *Journal of Marketing Education*, 22, 15-24.

Brunn H., & Frank C. (2002). Online Communication: A Success Factor for Blended Learning. Published in: Proceedings of Elearn, World Conference on E-learning in Corporate, Government, Healthcare, & Higher Education, Montreal, Canada.

Erskine, J., & Leenders M., & Mauffette-Leenders, L. (1998). Teaching with Cases. London, Ontario: Richard Ivey School of Business.

Flannery, J. L. (1994). Teacher as co-conspirator: Knowledge and authority in collaborative learning. In K. Bosworth & S. J. Hamilton (Eds.), *Collaborative learning: Underlying processes and effective techniques* (pp. 15-23). In R. J. Menges & M. D. Svinicki (Series Eds.), *New Directions for Teaching and Learning #59*. San Francisco: Jossey-Bass.

Frank, C., & Reiners, T., & Suhl, L. (2003). Implementing an Online Version of the Case Method: A Qualitative Evaluation. To be published in: ICDE World Conference on Open Learning and Distance Education, Hong Kong, China.

Lynn, L. (1999). Teaching and Learning with Cases: A Guidebook. New York: Chatham House Publishers of Seven Bridges Press.

Uesbeck, M. (2001). Effiziente Strategien und Werkzeuge zur Generierung und Verwaltung von e-Learning-Systemen (dissertation). Tuebingen, Germany.

# A prototype application for helping to teach how to read numbers

*Diamantino Freitas, Helder Ferreira, Vítor Carvalho, Dárida Fernandes \*, Fernando Pedrosa\**

Faculty of Engineering of the University of Porto    \*Polytechnic Institute of Porto
LPF-ESI / DEEC / FEUP / Portugal      Calculus / IPP / Portugal
dfreitas@fe.up.pt, hfilipe@fe.up.pt, vitor@fe.up.pt, darida@mail.telepac.pt, fpedrosa@sc.ipp.pt

## Abstract

Nowadays we are several multimedia math applications [1,2] in support of teaching the basic operations, geometric analysis, algebra and other complex operations; however there's no math application for the children at starting school age that helps to teach how to read a number in Portuguese. The *Laboratory for Speech Processing, Electro acoustics, Signals and Instrumentation* (LPF-ESI) at the *Faculty of Engineering of the University of Porto* (FEUP), started a project using a text-to-speech (TTS) engine and a Visual Basic multimedia application in order to fulfil that gap.

A prototype has been produced, called "*A Quinta dos Números*" (AQN) (*The farm of numbers*), that intends to create a bridge between engineering, computer science and pedagogy, using psycho-pedagogic concepts inherent to the teaching of maths, artificial intelligence, speech/text processing and multimedia systems, as scientific basis [3]. Developed to help the teacher to consolidate the student's knowledge and to support a progressive evaluation of her/his performance, AQN offers a multimedia interaction with speech synthesis and text boxes. It allows a greater accessibility at the same time it will offer a wide choice of didactic games, with learning and evaluation aims.

## 1    The role of ICT in learning Mathematics

School is changing. Programs, methodologies and attitudes are being renewed. The Information and Communication Technologies (ICT) are said to be tools, which facilitate the development of cognitive skills and enhance the power of communication. Specifically, the computer is accepted as an excellent instrument of education. Papert believes that the computer will provide the learners with the "materials" they need to build their knowledge in a constructive perspective, enhancing their autonomy and enriching the context where such construction evolves [4][5][6][7][8]. The National Council of Teachers of Mathematics [9] also points out as a pedagogic purpose the need of the curriculum from the 1st to the 4th school year to allow the appropriate and progressive use of technologic materials, namely the computers. The thoughtful and creative use of technologies can improve both the quality of the curriculum and the quality of the learning. That organisation also mentions that simulations of mathematic ideas in the computer can be an important help for children to identify the fundamental issues of mathematics. In fact, this association as well as many researchers [4][6] among others strongly believe that a significant way of improving the students' involvement in learning Mathematics is to use the computer as a mean of encouraging them to do research, explore mathematic ideas and to make use of mathematic relations in different contexts.

Fernandes [10] also supports that the learning context of the first cycle of basic school is favourable to the use of the computer as a work instrument, therefore, through suitable software, it is possible to promote the acquisition and the consolidation of concepts, as well as the development of the skill to solve problems. The mathematic reform implemented in the United States was significantly influenced by technology, which changed the way of "doing" elementary mathematics. All shows that an approach of Mathematics based on the computer leads to a better understanding of mathematic ideas and releases time for the development of new concepts.

## 2    Application features and requirements

The main concept explored in AQN is the simultaneous integration of the three media: text, graphics/animation and sound/speech. This allows the student to try and practice the in-full reading of numbers, and to play in free activities while performing evaluation.

The reading of numbers develops in a bi-directional structure of numeric categories and their readings, associated with graphic representation, animation and speech synthesis, with a clear correlation between text and numeric representation.  A wide range of number types is available, for instance, cardinal, ordinal, roman, scientific, dates, phone numbers, and many other types of numbers. For the moment only cardinal and ordinal types are implemented into activities/games.

### 2.1    Visual Design and Interfaces

AQN is designed with a maximum resolution of 640x480 pixels, using only 256 colours (Windows palette) and Windows default fonts, allowing a good portability between different systems and equipments.

**Figure 1:** Screenshots from AQN. The first image shows the welcome screen, the second image illustrates the scenario where the children choose their school grade, and the third image shows where the children choose their animal companion that will guide them through the application.

In a design thought also in environmental terms, the student can travel through several farm-like scenarios, made of pictures with dynamic icons. These elements were created with the purpose of approaching children's universe (animals, fairy tales and objects from the classroom). Several animations and hidden elements were placed along the scenarios, providing a few surprises. The navigation is provided by menus and icons, the input devices are the keyboard and mouse. The practical and simple interface of AQN, based on pictographic animated icons and various sound events, allows this application to be used either as a didactic product integrated in school classes or as a self-learning product elsewhere. Figure 1 and 2 shows some scenarios from AQN.

## 2.2 Accessibility features

AQN is a disguised interface based on Visual Basic, with soft animations, not aggressive colours and intuitive sounds. All the buttons/icons have tool tip text. Messages to the user are of two types: static and dynamic. Static messages are reproduced by playing pre-recorded human voice (*wave* files). Dynamic messages are reproduced with the help of a Text-to-Speech (TTS) engine with a european portuguese grammar. All the numbers categories are marked up with LPF-ESI's mark-up language called *TPML* that identifies the type of number and directs the TTS's text converter in the conversion's decision.

**Figure 2:** Screenshots from AQN. The first image shows AQN's Talking Calculator, the second image illustrates an example game before start playing it, and the third image shows the resulting messages and icons after playing a game.

The combination of text, graphics and speech, enables a much stronger interaction and comprehensive teaching for the student, since she/he watches and listens to the number in question, relating the object and characters to the specific sounds. The capability of a TTS engine, also allows an extraordinary interactivity and comfort. When the student input's her or his name, the application will speak a greeting sentence including the entered name. That's found a warmer reception and is gratifying for the student.

## 2.3 Application requirements

AQN has the following requirements: 256 colours minimum, screen resolution of 640x480, Windows 95 to XP, nearly 25MB on hard disk, at least 32MB RAM, soundcard, speakers, keyboard and mouse. Visual Basic runtime files and Microsoft Speech API runtime files are provide at the AQN's installation.

## 3 Usability/Accessibility Evaluation and Results

A few preliminary tests were already made. AQN was tested with a group of students with ages between 5 and 7 years old enrolling the 1st and 2nd year of primary school. The tests were performed individually as well as in group. Some of the students had there the first contact with a computer. Each test consisted in a free experimentation phase followed by a questionnaire and a task. The duration of each test was about 15 minutes. The application was evaluated in 3 different aspects: design and interface, text-to-speech evaluation and subjective opinion and evaluation.

## 3.1 Design and Interface

A few questions were made with the intention to evaluate visual perception and memory, navigation, animations, pictographic icons and spatial organization. The majority of the users collected the main visual information of the application (menus, titles and animations). A few problems were found with navigation between screens, mainly because the use of a few non intuitive icons. The selective learning options menu was found very confusing by the users. The "choosing favourite animal" screen was found very simple to understand and very intuitive to choose an animal and continue to the next screen. The main theme of a farm was enjoyable to the children. Some visual and audio elements were concealed in the scenarios. A few users found this disturbing at the first impression.

## 3.2 Text-to-Speech evaluation

The use of a TTS engine proved to be a valuable tool to improve the accessibility features and comprehension of the application, especially if the student has reading difficulties. However, we have found that good students with a good reading capability tend to almost ignore the TTS messages.
A few tests were made to evaluate the perception of the TTS voice. Most of the students were able to understand the text-to-speech engine; however the adaptation to the voice was fully visible only at the second or third experience during the application games. One of the features of the TTS is speaking the user's chosen name. This was found very rewarding to the children.

## 3.3 Subjective opinion and evaluation

In a scale of 1 to 4 (bad, normal, good, very good), AQN was evaluated in average by the children as a 3.
All users enjoyed very much the application's calculator and the animals present on the scenarios. The game was found very intuitive and the TTS was found a helpful tool. However the TTS engine was found a little difficult to understand at the first time it is heard.

## 4 Conclusions and future trends of work

The accessibility of information was one of our prime aims, and the use of a TTS engine was a great tool to accomplish this requirement.
This application, if using a different interface but the same approach, could be used in other situations such as: Portuguese number's teaching for blind students (with the help of a Braille-output), for foreign persons or teaching older people (night-school). It could even be used as a kind of speech dictionary for numbers.
The immediate goal is to overcome some of the criticised aspects and test again the application with real users and also to evaluate its added value in the teaching activity. The extension to other types of numbers and operations is also envisaged. User customization also needs to be incorporated in the application. AQN will also provide a basic expert system that will try to understand the student's progress on the proposed exercises, detecting difficult levels and advising. Also, the teacher or pedagogic responsible, will be able to access the application's records for control, supervision and evaluation of the student's progress.

The text-to-speech engine and the basic expert system that guides and evaluates the student knowledge, combined with the inherent pedagogic aspects, make AQN an interesting application in its area, and a tool to consider.

# 5    Acknowledgments

The authors wish to thank all the teachers involved in testing and the following students: Nuno Santos, Clara Sousa, Filipa Torrão, José Loureiro and Alexandra Grandão, enrolling at the " Agrupamento de Escolas de Valongo: EB1/JI nº 4 ".

## References and litterature

[1] Porto Editora Multimédia (2002) from http://www.portoeditora.nt

[2] Tooobing Ideas    Numeracy Numbers   http://www.teachingideas.co.uk/maths/contents2.htm.

[3] Jenny Preece. (1994). Human Computer Interaction. Addison-Wesley.

[4] Papert, Seymor. (1980). Teaching Children Thinking. The Computer in the School: Tutor, Tool, Tutee. (pp.161-176).

[5] Freitas, João. (1992). As Novas Tecnologias de Informação e Comunicação. Desenvolvimento dos Sistemas Educativos. Educação e Computadores ("The New Technologies of Information and Communication. Development of educational systems. Education and Computers."). GEP. Ministry of Education. II, (pp.27-84).

[6] Moreira, Maria L. (1989). A Folha de Cálculo na Educação Matemática ("A calculus sheet in mathematic education"). Department of Education. University of Lisbon. Portugal.

[7] Matos, João. (1991). Logo na Educação Matemática: um estudo sobre as concepções e atitudes dos alunos ("Logo in Mathematic Education: a study."). Department of Education. University of Lisbon. Portugal.

[8] Estrela, Maria T. (1992). Relação Pedagógica. Disciplina e Indisciplina na Aula ("Pedagogic Relation. Discipline and Indiscipline in the classroom"). Porto Editora.

[9] National Council of Teachers of Mathematics. (1992). Normas para o Currículo e a Avaliação em Matemática Escolar Quinto Ano ("Norms to evaluate mathematics in the 5th school year."). Institute for the Innovation in Education. Portugal.

[10] Fernandes, D. (1994). A utilização da folha de cálculo no 4º ano de escolaridade, estudo de uma turma ("The use of calculus sheet on the 4th year of primary school, study of a classroom"). Masters Degree. University of Minho. Portugal.

Ben Shneiderman. (1998). Designing the User Interface: Strategies for Effective Human-Computer Interaction. (3rd Ed). Addison-Wesley.

Microsoft Windows Guidelines for Accessible Software Design: Creating Applications That Are Usable by People with Disabilities - http://www.cs.bgsu.edu/maner/uiguides/msaccess.htm.

Program Nonium. Estratégias para a acção. As TIC na educação. ("Strategies for action. ICT in education"). Portuguese Ministry of Education.

Roe, P. (Editor). (2001). Bridging the Gap? COST, European Commission.

Text Processing Markup Language (TPML) was developed to mark-up text used in TTS. Contact the authors for more information.

# Usability Engineering in Computer Aided Learning Contexts
## Results from usability tests and questionnaires

*Ronald Hartwig, Inga Schön, Michael Herczeg*
Institute for Multimedia and Interactive Systems, University of Luebeck
Willy-Brandt-Allee 31a, D-23554 Luebeck
Hartwig|Schoen|Herczeg@informatik.uni-luebeck.de

## Abstract

This paper describes a process to consolidate observations and findings from usability evaluation phases into a comprehensible list of usability deficiencies as a basis for quality assurance methods. A so called "evidence test" process is transferred from the world of software usability engineering into the context of computer aided learning (CAL) and computer supported cooperative learning (CSCL). It is based on experiences from quality management and evaluation efforts in the German flagship project "VFH" (Virtual University of Applied Sciences in Engineering, Computer Science and Economics) and the project "medin". The paper is intended as a basis for continued research in the area of usability of computer aided learning, as well as a founded best practice description for practitioners who work in this context.

## 1 Introduction

The work and results we describe are taken from the context of two projects called "Virtual University of Applied Sciences" (Virtuelle Fachhochschule - VFH) and "medin", which intend to offer location independent learning. Both are sponsored by the German Federal Ministry of Education and Research (BMBF). The focus of interest lies in computer-supported multimedia-based teaching and the production of learning material which offers distance learning to students dispersed all over Germany. Quality management, especially in projects of this size, needs a well defined process as a guideline for all participating parties. It has been one major task to implement a procedure which is suitable to enhance or even enforce the usability of the products, namely multimedia learning modules for the use in a virtual university. The main interest is to closely connect usability tests and reviews to the original context of use by defining a comprehensible and reliable method on how to extract usability requirements from use scenarios and to test the compliance of the produced material with these requirements. This paper describes how results from the usability testing within the installed iterative process were analyzed and classified in order to validate their relevance for further improvement of the modules.

## 2 Didactical models

Unlike the working tasks known from usability engineering, the task "learning" is not a home homogenous process. Along with the external context of learning (e.g. the environment) the internal conditions (learning prerequisites and admission conditions) must be considered. The premises for learning change during the learning process; this causes the sequencing of the learning material to play an important role (Niegemann 1995). In traditional instructional design it is an important rationale to have processes that take care of learning premises and goals for the learning content. For example (Gagne, Briggs & Wager 1992) differentiate five learning goals categories (knowledge represented by words, cognitive abilities, cognitive strategies, settings, motoric abilities). Instructional design is mainly based on the assumption that at the beginning of planning and developing of teaching instructions, the desired abilities should be analyzed on the basis of these categories. One example is the ARCS-Model (motivation as a task of instructional design). The model was extended with concrete recommendations for multimedia learning environments by (Keller & Suzuki 1988) and (Niegemann 2001). Different instructional design theories try to specialize on this theory to the field of computer supported learning. Examples are "Principles of Instructional Design" from Gagné, Briggs and Wager as mentioned above and, the more complete and less analytical, Instructional Transaction Theory (ITT) from (Merrill 1999). All theories have in common that students, through work with the computer program, should be stimulated to study the learning material intensively. The goal is not only to present the learning material in an attractive way, but also to support explorative and individualized learning. The connection to the usability engineering is that the above mentioned didactical theories imply certain learner activities. These activities correspond to tasks and sub-tasks known from usability engineering. Please note that an activity ("making an annotation") is not the task ("learning algebra") but a vital part of it. In reverse analyzing activities helps to identify tasks and subtasks.

Usability, as a central software quality, is seen as a prerequisite for giving the students the opportunity to learn efficiently. Typical requirements from learning contents are that users must always be able to keep an overview of learning paths in order to plan their proceeding. Students must be offered the possibility to handle parts of the learning material in the learning process by marking, excerpting and annotating. Explorative use of the acquired knowledge must be supported by communication possibilities (chat, whiteboard, newsgroups). In order to leave a maximum of mental resources for managing the complexity of the learning contents only an absolute minimum of elaborateness may be allowed for the handling of the tools. It is important to let the learner learn the contents and not the system usage. Therefore, the ergonomic design is a necessary but not sufficient precondition for the success of didactical models. Of course it must still be the major concern to plan and implement the appropriate didactical model in time.

It is difficult to engineer and evaluate a complex interactive system with respect to a complex task like "learning", so we use the "divide et impera" method to cope with this complexity. In order to apply usability engineering methods it is essential to identify the subtasks from the didactical model and their desired concrete results. Then potential usage problems are identified and their severity is consolidated in an evidence testing process (as described in (Dzida&Freitag 2001)). Of course it is a necessary but not sufficient precondition to optimize the subtasks but it is a starting point at all. Still the whole task should be considered as well. Another important point while transferring usability engineering methods into the world of CAL is, that the shortest way to a solution is not always the preferable one in terms of learning support. One has to distinguish between desired slow-downs that support learning, because they force the learner to intensify his work with the contents and usability problems which make the user think about the system instead of the content. Efficiency has to be measured with this in mind and if there is a planned detour in the learning path it must be justified by the didactical model but not by technical insufficiency. Another aspect is that the task "learning" ideally has to be evaluated over a longer time frame (e.g. a whole study program) which is often hard to achieve in practice.

# 3 Transferring the usability quality concept to the world of CAL

The approach used in the project is based on well established usability evaluation methods (see (Herczeg 1994)) and engineering approaches ((Nielsen 1993), (Mayhew 1999), (ISO13407 2000)).

First, common usability issues were covered by expert walkthroughs (Reeves et al. 2002) mainly concerning the presentation of information and some consistency issues. As (Dimitrova 2001) showed, relying on such expert reviews is not sufficient, because experts will overlook surprises. Second, empirical data from questionnaires showed which usability issues were considered by users to be most important and how those issues were managed in the offered learning material. Questionnaires were analyzed to give an importance rating as an indicator of what users expect from computer aided learning support. The questionnaire items refer to issues which already had been identified as potential problems during the iterative process. The questionnaire yields how grave these problems are from the user's point of view. In addition, the questionnaire reveals how content the user are with the current offer (module). As the third and major step, usability tests were conducted on site with real students in their living rooms or studies (See (Hartwig, Triebe & Herczeg 2002) for a more detailed description). In usability engineering it is assumed that this so called "Triangulation" of three usability evaluation methods delivers the most promising and reliable results. Each method has its strengths and limitations but put together they cover the potential problem fields of interactive systems. The result of this evaluation is a list of potential problem items: objections from the expert walkthroughs, observations of users struggling with the system from the user testing and questionnaire items with low satisfaction values or a negative impact. All potential problems are analyzed using an evidence test described below. This process also yields information on how severe a problem is and how it can be overcome.

## 3.1   Evidence test

Usability engineering is defined, according to the international standard (ISO9241 1996-2000), as a product of the following three criteria:
1. Effectiveness: It is seen as the degree of completeness and correctness regarding the goal.
2. Efficiency: The effort that has to be made in order to reach the goal. Effort can be rated absolutely as time consumption or (mental) work load or relatively compared to alternative methods/tools.
3. Satisfaction: The degree of user contentment when using the system. It is normally calculated on an empirical basis.

In a first step it is analyzed if the potential problem deters the problem solving at all or leads to faulty task results (effectiveness "fail"). In this case the potential problem is a severe failure and no further analyzing has to take place. Presuming that the effectiveness of the observed part of the system did not fail, in a second step the efficiency is considered: the user effort for completing the subtask is compared to the available amount of time for this subtask. Finally, if the potential problem did pass the effectiveness and the efficiency testing, a traditional empirical enquiry is done. Using a scale from 1 (very good) to 5 (poor) and 1 (very important) to 4 (not important) and considering a slight positive bias, low satisfaction (<2.5) together with higher importance ratings (<3) are taken as evidence for user stress and make the satisfaction criteria fail (see (Hartwig et al. 2002) for a more detailed description of this questionnaire).

The support for "using individual learning strategies and working with the material" complex is taken as an example to clarify the process of consolidating an observed potential problem analytically into a deficiency: The students had problems in working with the offered contents instead of just reading them. Their subtasks included all kinds of reusing the contents such as summarizing, transcribing, annotating, linking them into a personal hypermedia network or discussing them with fellow students. It could be observed that many of the students referred to paper as their working media so that they could work in the manner they were used to instead of using the possibilities of computer supported techniques. The following example should help to understand, why students did this and how the decision was made that this "fallback" is considered to be a usability problem. "**Annotations**": If students used the offered annotation tools they only were able to annotate web pages as a whole, but not to annotate specific text pieces. Strictly speaking it did not affect the effectiveness, because at least theoretically it was possible to complete the desired subtask of annotating. They theoretically could write line numbers as references into their annotations in order to link to specific paragraph. But if you compare this to the efficiency of writing some short notes (e.g. containing some mathematical formulas or drawings) directly into a printed version, it gets obvious that the computer supported method implies a lot more effort for the user and therefore the efficiency is rated as "fail". Having said that, the new question is: Is it bad to have the students make their annotations on paper instead of the computer? The missing of a computer supported tool is rated as a failure as well, because the use of paper based annotations implied many new sources of error, problems with re-referencing to the online-contents and it implicitly lead to higher efforts when students try to put it back into the computer based media again if they wanted to communicate their personal annotations to fellow students. This lack of efficiency and the probability of new effectiveness problems lead to the consolidated decision that this complex really is a severe usability problem, because both alternatives (using the computer for the annotations vs. using pen and paper) imply severe usability problems.

**Tabular 1:** How typical problems affected effectiveness, efficiency and satisfaction

| | Usability problems while... | Effectiveness | Efficiency | Satisfaction |
|---|---|---|---|---|
| 1 | using individual learning strategies and working with the material | OK | Fail | Fail |
| 2 | being informed about news and changes | Fail | Fail | - |
| 3 | orienting and navigating within the presented material | OK | Fail | Fail |
| 4 | using a chat | OK | Fail | - |
| 5 | trying to work parallel with different information sources on one screen | OK | Fail | - |
| 6 | detecting and recovering interactive possibilities of the media | Fail | Fail | - |

# 4    Results

In order to come to general results the faults are categorized in more general classes. These classes reflect problem complexes encountered during the project. The results are taken from testing 12 different modules in 15 on-site user tests and 91 questionnaires from about 50 users (from about 150 students). They are specific for the described project but we expect similar findings for other modules and contexts. Therefore table 1 is intended as a warning, how bad problems in those areas *can* affect the usability of the product with respect to the learning task.

Many of problems already had been foreseen or did already appear in earlier iterations of the development process. The results of the questionnaire, which mainly contains questions related to these already anticipated problem areas, were predominantly positively rated by the students. This shows that the use of a project guideline was effective at least to avoid well known problems. But the fact that in spite of this still severe faults persisted shows that relying only on questionnaires and expert reviews would be careless. This would be a systematical mistake because grave problems could be overseen this way.

# 5    Summary

While the three evaluation methods contribute different views on the usability of computer aided (cooperative) learning, the combination of these three approaches is a promising starting point on what a general guideline for usability and a user- and task-centered process should cover. The results offer clues where, in our opinion, usability experts should be present in the design and development process in order to prevent technical constraints from interfering with the success of implementing the didactical models. Currently a semantic web with specific usability knowledge data (observations, guidelines, test results) as well as learning content and development instructions is developed in order to support the knowledge transfer throughout the whole project and the participating developers. They should be enabled to comprehend all decisions and corroboration processes in order to improve the overall usability of the product.

# References

Dimitrova, M. (2001): "Are Experts Able to Predict Learner Problems During Usability Evaluations?" In: Proceedings of ED-MEDIA 2001. AACE: Finland, pp.1023 -1028

Dzida, W. and Freitag, R (2001); Usability Testing - The DATech Standard. In: Wieczorek, Meyerhoff (Editor): Software Quality - State of the Art in Management, Testing And Tools. Springer. pp. 160-177. ISBN 3-540-41441-X

Gagné, R. M., Briggs, L. J., Wager, W. W. (1992). Principles of Instructional Design. 4. Ed., Fort Worth: Harcourt Brace Jovanovich.

Hartwig, R.; Triebe, J.K.; Herczeg, M. (2002): Usability Engineering as an Important Part of Quality Management for a Virtual University. In: Proceedings of Networked Learning 2002 - Challenges and Solutions for Virtual Education. Technical University of Berlin, Germany, ICSC-NAISO Academic Press, Canada/The Netherlands.

Herczeg, M. (1994): Software-Ergonomie – Grundlagen der Mensch-Computer-Kommunikation, Addison-Wesley, Bonn;Paris;Reading (Mass.)

International Organization for Standardization (1996-2000): ISO 9241 - Ergonomic requirements for office work with visual display terminals, Parts 1-17. International Standard

International Organization for Standardization (1999): ISO 13407 - Human-centred design processes for interactive systems. International Standard

Keller, J.M., & Suzuki, K. (1988): Use of the ARCS motivation model in courseware design. In D. H. Jonassen (Ed.), Instructional designs for microcomputer courseware (pp. 401-434). Hillsdale, NJ: Erlbaum

Mayhew, D. (1999): The Usability Engineering Lifecycle. Morgan Kaufmann Publishers. (1999)

Merrill, M.D. (1999): Instructional transaction theory (ITT): Instructional design based on knowledge objects. In C.C. Reigeluth (Ed.), Instructional Design – Theories and models. A new paradigm of instructional theory (pp. 397-424) Mahwah, NF: Erlbaum

Niegemann, H. M. (1995): Computergestützte Instruktion in Schule, Aus- und Weiterbildung. Theoretische Grundlagen, empirische Befunde und Probleme der Entwicklung von Lehrprogrammen. Frankfurt am Main: Peter Lang

Nielsen, J. (1993): Usability Engineering, AP Professional, Boston

Reeves, T.C.; Benson, L.; Elliott, D.; Grant, M.; Holshuh, D.; Kim, B.; Kim, H.; Lauber, E.; Loh, S.: "Usability and Instructional Design Heursitics for E-Learning-Evaluation"; in Proceedings of ED-Media 2002, AACE, p. 1615-1622

# Developing Context- and User Groups Sensitive Learning Scenarios with XML Configuration

*Michael Hellenschmidt*

*Norbert Braun*

Fraunhofer Institute for Computer
Graphics, Dept. e-Learning
& Knowledge Management
Fraunhoferstraße 5
64283 Darmstadt
Germany
Michael.Hellenschmidt@igd.fhg.de

GRIS, FB Informatik, TU Darmstadt

Fraunhoferstraße 5
64283 Darmstadt,
Germany
NBraun@gris.informatik.tu-darmstadt.de

## Abstract

We present an approach that separates the presentation of a simulation from the processing logic. This approach fulfils the requirement to generate simulation learning scenarios with tutorial help functionalities. Furthermore, the architecture provides the possibilities for analysing user interactions, classifying learners by the help of the inverse coefficient of orientation and giving context- and user group sensitive helps. With this approach learners could examine different uses of complex technical devices. Developers of simulations only have to prepare HTML pages representing different status within the learning scenarios. Once created, the HTML pages can easily be reused by building new XML documents for the configuration of the simulation system.

## 1    Introduction

Nowadays users of highly complex technical devices have to learn multiple ways of menu-driven handlings in more and more decreasing time. As an example of menu-driven handling one can think of the usage of an automated teller machine or the operating of a machine by touch pad. Often the control panel of such technical devices are used to accomplish several tasks (e.g. to withdraw money or to check the current account balance). Authors wanting to teach the usage of such functionalities therefore need the possibility to produce simulations representing special workflows. Furthermore, the production of new learning scenarios and the modification of existent scenarios should be possible in best time. To face this requirement the reusability of each part of a simulation is necessary (cf. Rada, 2000). Learners facing the requirement to train on the job need the support of methods of intelligent tutoring to learn the operation of machines by interacting with the simulation (Farin et al., 2000 gives a discussion of the problems of intelligent tutoring system in the industrial environment). They need instructions – in case they need help - within every single step according to the given learning context and according to their expertise. That means that the simulation system should be able to classify the learner by labelling (e.g. beginner, intermediate, professional). Thus the author responsible to develop help functionalities needs a convenient instrument to describe this functionalities of the simulation system.

951

# 2    The Simulation Technology

The simulation technology we developed consists of an internet-based system where a number of HTML pages represent the single user steps  (e.g. screenshots of the real user interfaces). Each single page only transmits the interaction of the user to the server. This means that as a basic principle each user interaction is free of any workflow context. The whole processing logic and thus the context of the learning scenario and therefore the analysis of the user interactions is located on the server. This approach provides the author with reusable HTML page and eases the way of developing different learning scenarios with the same technical device in a faster way. Having on hand HTML pages that represent every possible occurrence of the user interface of the device which is to be simulated the system could be configured. Within an XML document the learning scenario will be described (see the possibility to solve reusable problems using XML in Hiddink, 2000). The description follows the metaphor of status.

```
<status>
        <initialPosition> Z4 </initialPosition>
        <truePosition> Z5 </truePosition>
        <errorPosition> Z4 </errorPosition>
        <trueInputs>
                <trueInput type = "Button" name = "button"> L2 </trueInput>
        </trueInputs>
</status>
```

Each status has an identifier (<initialPosition>) and uses two other identifiers of other status. The element <truePosition> indicates the designator of the status which follows in case the user interacts with the simulation according to the given workflow. The element <errorPosition> indicates the following status in every other case. A status is further described by the possible successful workflow steps. A workflow step during operating a machine could be e.g. pressing a certain button of the control panel of the simulated user interface or typing certain strings or numerical values (e.g. a password interface or an interface to adjust technical values). The element <trueInput> describes  the list of possible required user interactions within this status. The XML example above shows the description of a step from an initial status with the identifier Z4 to the status Z5 in case the control panel button with the internal marker L2 is pressed. In general the identifier of a status means a certain HTML page representation the view of the simulated graphical user interface. The next example of a status shows how two input fields for an integer value and for a float value can be described:

```
<status>
        <initialPosition> Z6 </initialPosition>
        <truePosition> Z7 </truePosition>
        <errorPosition> Z6 </errorPosition>
        <trueInputs>
                <trueInput type = "Integer" name = "E1">
                        <range type = "include"> 0 </range>
                        <to type = "include"> 10 </to></trueInput>
                <trueInput type = "Float" name = "E2">
                        <range type = "include"> 0 </range>
                        <to type = "exclude"> 100 </to></trueInput>
        </trueInputs>
</status>
```

To define ranges of values the elements <range> and <to> within the element <trueInput> can be used. It is also possible to include or exclude the beginning and / or the end of the co-domains. In the same way the specification of text inputs or the multiple declaration of true inputs is possible. Thus certainly all interaction possibilities given by the usage of a menu-driven touch pad and the adjustment of technical values are describable. Consequently an XML document including status can describe step by step a workflow and thus it represents the whole learning scenario. The description now can be analysed by the server and the simulation system is adapted to the learning scenario. There is no limitation to the numbers of different scenarios running within the simulation system at the same time.

## 3  Adaptive Tutoring Technology

Another XML document describes the activities of the tutoring assistance. During the whole simulation the learner has access to a help functionality. First the author can define the number of user groups and their denominations (e g. the beginner group, the intermediate group and the professional group) and with this their range within the inverse coefficient of orientation (see below). Then for each status of the simulation and for each defined user group a source of help assistance could be given. Thus each learner belonging to a certain user group is provided with learning support which matches best within the current learning status. The inverse coefficient of orientation is calculated by the simulation system at every time the user interacts with the system. Each interaction (e.g. doing errors or activating the help functionality) is recorded and used to calculate the coefficient. The range of values of the inverse coefficient of orientation is normalized. Consequently it is possible to define different user groups within this range. The possible numbers of user groups is unlimited.

### 3.1  The inverse coefficient of orientation

Many ways of observing user interactions and drawing conclusions for tutorial help functionalities are presented (e.g. (Encarnação & Stoev, 1999) or (Murano, 2001)). We decided to record every user interaction with the simulation system (doing a simulation step or activating help functions for instance) and calculate them into the formula:

$$Q_I = \text{Quotient of Helps} + \text{Quotient of Errors} = Q_H + Q_F = (0.5) * (n_H / n_{HM} + n_F / n_A)$$

We call this formula the inverse coefficient of orientation. Definitions:
- $n_H$ represents the number of help calls of the user
- $n_{HM}$ stands for the number of help possibilities, that means the number how often the learner can activate the help function
- $n_F$ represents the number of errors the learner made
- $n_A$ represents the number of interactions the learner made within the simulation

Obviously we have a strict differentiation between the interaction with the simulation (calculated by the Quotient of Errors) and the interaction with the tutoring function (calculated by the Quotient of Helps). This can be used to distinguish learners attending the simulation without the usage of any helps (learning by try and error) from learners using the help function very often. It is obvious that a professional completing the simulation gets an inverse coefficient of orientation calculated to zero. A learner using the help function at every step of the simulation and each interaction with the system does not follow the given workflow will get a value of one.

## 3.2 The predefinition of user groups

Based on the inverse coefficient of orientation the author can define an unlimited number of user groups, e.g.:

```
<arrangement>
        <group range="0" to="0.40"> professional </group>
        <group range="0.40" to="0.72"> intermediate </group>
        <group range="0.72" to="1"> beginner </group>
</arrangement>
```

If the system configured with the example given above calculates an inverse coefficient of orientation within the range of 0 to 0.4, the learner belongs to the group of professionals and thus will get the help for professionals the author of the simulation provides. It is evident that the learner switches between different user groups during the process of a simulation, because at each step the learner does the coefficient responsible for the learners´ classification is recalculated.

## 3.3 The sources of help assistance

For each status within the simulation and each learner group the author can give a different source of help. The following listing shows an example for the status with the designator Z4 and the three user groups defined above:

```
<status identifier = "Z4">
        <help group="professional"> professional_help_Z4 </help>
        <help group="intermediate"> intermediate_help_Z4 </help>
        <help group="beginner"> beginner_help_Z4 </help>
</status>
```

Representation of help can either be text or a link to an HTML resource for example. It should be a basic principle that different user groups get different types of help according to their knowledge.

## 4    Applications

After the implementation of the simulation system two scenarios had been tested. The first scenario covers the topic of the "Setting-up Operation of a Servo Drive" consisting of 33 status. Within this scenario a learner uses a system which simulates the user interface of a servo drive. The system is meant to provide the possibility of doing its initial operations. The second scenario (see Figure 1.) deals with the application of a laser from its initiation till its availability. This simulation consists of 18 status. Both scenarios have been configured with three user groups assisting each learner with context and user group based help.

## 5    Perspectives

With the presented simulation system two potencies are shown. Single simulation status are reusable and learning scenarios can be developed only by configuration with XML documents analysed by the server. Separation of simulations into single steps makes it also possible to define context- and user group specific help functionalities. Analysing of user interactions by the help of the inverse coefficient of orientation by the server makes intelligent tutoring within simulations possible. In the future we want to expand this technology into larger scenarios and assist the authors simulations with suitable graphical user interface tools.

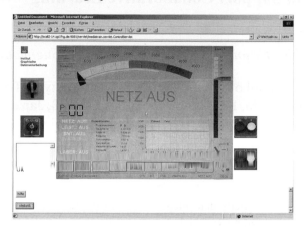

**Figure 1:** One status of the scenario of the application of a laser operating system.
On the bottom buttons to access tutorial help and statistics of the
group classification at each status of the simulation are shown.

# 6    Acknowledgements

These results are developed within the project "Medi@Train" which is funded by the German Bundesministerium für Bildung und Forschung (BMBF). The consortium of this project consists of academic staff from different Fraunhofer Institutes and former GMD-Institutes.

# 7    References

Murano, P. (2001). A new software agent ´learning´ algorithm. *Human Interfaces in Control Rooms, Cockpits and Command Centres. The Second International Conference on People in Control*, IEEE Conf. Publ. No. 481, 297 – 301

Hiddink, G. (2000). *Using XML to Solve Reusability Problems of Online Learning Materials*. International ICSC Cogress, Intelligent Systems & Applications, ISA´2000, December 11 - 15

Encarnação, L. M. & Stoev S. (1999). An Application-Independent Intelligent User Support-System Exploiting Action-Sequence Based User Modelling. *Proceedings of the 7th International Conference of User Modeling*, 245 – 254

Rada R. (2000). *Levels of Reuse in Educational Information Systems*. International ICSC Cogress, Intelligent Systems & Applications, ISA´2000, December 11 - 15

Faria, L., Valez, Z., Ramos, C., Marques, A. (2000). An ITS for control centre operators training : issues concerning knowledge representation and training scenarios generation. *Fourth International Conference on Knowledge-Based Intelligent Engineering Systems and Allied Technologies*, Brighton, 129 -132

# Integrating Shared and Personal Spaces to Support Collaborative Learning

*Kazuhiro Hosoi*        *Masanori Sugimoto*

Graduate School of Frontier Sciences, University of Tokyo
7-3-1 Hongo, Bunkyo-ku, Tokyo, 113-0033, Japan
hosoi@r.dl.itc.u-tokyo.ac.jp        sugi@k.u-tokyo.ac.jp

## Abstract

In this paper, a system called Caretta that integrates personal and shared spaces for supporting face-to-face collaborative learning is described. A personal space of Carretta is used for supporting individual user's activities, while its shared space is for sharing users' activities and enhancing discussions with other users. We use PDAs and a multiple-input sensing board for personal and shared spaces, respectively. Caretta allows users to participate in collaborative learning by interchangeably using both personal and shared spaces. The first version of Caretta has been developed to support learning about urban planning and environmental problems. Preliminary user studies to evaluate Caretta are described.

## 1    Introduction

We have so far developed several systems for supporting collaborative learning (Kusunoki, Sugimoto & Hashizume, 1999; Sugimoto, Kusunoki & Hashizume, 2002; Sugimoto et al, 2002). These systems were constructed with sensing and augmented reality technologies: we devised a multiple input sensing board that can rapidly identify types and locations of multiple objects, and created an immersive environment by overlaying a computer-generated virtual world onto the physical board (Sugimoto, Kusunoki & Hashizume, 2001). The systems have been used to support elementary school children or university students in learning about urban planning and environmental issues.

Throughout three-year evaluations of the systems, we have confirmed that they could support learners in participating in their learning situations and activate their discussions. We have also received various comments from users. One of the strongest requests was related to a support for learners' reflection in collaborative learning processes. Each learner could utilize the sensing board as a shared space among learners and externalize his/her own ideas by manipulating physical pieces on the board. However, when he/she wanted to individually deliberate over new ideas, there was no support given by the systems. If collaborative learning is a process that comprises reflection by individual learners and shared understanding through discussions among a group of learners (Koschmann, 1996), a computational support for their own reflection processes must be provided by the systems. In this paper, we propose a new system called Caretta that integrates personal spaces and shared space for supporting learners' individual reflection and promoting their shared understanding.

Each learner of Caretta sits around the sensing board with a PDA. When a learner discusses with other learners, he/she could manipulate physical pieces (elements for constructing a town, such as a house, a factory, a store, and so on) on the board. Then, the system starts environmental and financial simulations based on the arrangement of pieces on the board, and projected the visualized simulation results on the board. When a learner uses a PDA, he/she chooses one of the two modes: the personal mode and shared mode. In the personal mode, each learner can use his/her PDA for taking notes, or accessing to web sites to search necessary information. He/she can also use his/her PDA to test their own ideas through simulations: a learner can virtually add, move or remove virtual pieces on his/her PDA. These simulations are not visualized in the shared space (in this case, on the sensing board), but on his/her PDA. In the shared mode, a learner can manipulate virtual pieces on his/her PDA and immediately visualize them on the board (therefore, the pieces are shared with other learners). The reason for using a PDA is that its recent models have sufficient computational power, and a PDA is portable. While having it, learners can move around the board and manipulate pieces.

The merits of our proposed system are summarized as follows:

1. By allowing learners to construct towns in a physical world and share simulation results in an immersive environment, the system can enhance learners' experiences and raise the level of their engagement in collaborative activities.

2. By using a PDA in the personal mode, a learner can confirm his/her ideas through personal simulations. Based on the simulation results, the learner can evaluate and examine the ideas before showing them to other learners.

3. By allowing learners to input virtual objects that appear on the board through PDA's, more realistic simulations become possible. For example, it is difficult to input a railroad or a highway as physical objects. It is, however, easy to place a railroad object in any shape, by drawing it on a surface of a PDA's display with a stylus pen.

## 2    Related Works

In this section, systems related to issues of integrating personal and shared spaces are described. Research on Single Display Groupware (SDG) (Stewart, Bederson & Druin, 1998) has so far investigated how to best support group collaborating in face-to-face or co-present situations. Several SDG applications accept simultaneous inputs by multiple users through their graphical user interface (GUI). One critical issue in SDG research is related to a technical problem associated with supporting interactions among multiple users working in a shared space. A user of SDG applications usually has no way to conceal his/her activities from the other users, because any information or activities on a shared space are visible to all users. Single Display Privacyware (Shoemaker & Inkpen, 2001) proposes one solution by making private information visible to only the corresponding user through his/her head-mounted displays (HMDs). Another approach is to give each user a mobile device. Pick-and-drop (Rekimoto, 1998) allows users to use a special stylus pen with which they can "pick-up" and "drop" information between the shared space (whiteboard) and a personal space (PDA). Also in Caretta, PDAs are used for a personal space. However, it is different from these previous systems from the following points:

1. Caretta is used for simulation-based applications, such as urban planning or entertainment games. Users can try their own idea privately and find best one for collaborating with or outwitting others. Therefore, Caretta's personal space by PDA is regarded not only as an

input method to the shared space, but also as a space that gives users the functionally same workplace as the shared space.

2.  Intelligent support modules are implemented in Caretta. Based on manipulation histories, Caretta can make inferences on each user's intention and automatically suggest several alternatives for individual or a group of users. Such modules can be effectively utilized, when a group of users have to solve complicate problems concerning different stakeholders such as urban planning and environmental problems.

# 3 System Configuration
## 3.1 Overview

Figure 1 shows an overview of Caretta. It comprises a multiple input sensing board and an LCD projector for a shared space, PDAs for personal spaces, and a simulation server. The sensing board was devised by arranging RFID (Radio Frequency Identification) readers like a checkerboard (Sugimoto, Kusunoki & Hashizume, 2001). It can rapidly recognize locations and types of multiple objects, each of which an RFID tag is attached. By projecting visualized results of simulations generated by the server on the board, an immersive and shared space is created. The data transmission between the simulation server and the sensing board is through RS-232C interfaces, and between the server and PDAs is through wireless network.

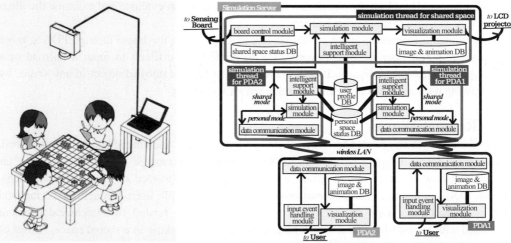

Figure 1. An overview of Caretta        Figure 2. System configuration of Caretta

## 3.2 Integration of Personal and Shared Spaces

In Caretta, a user uses his/her PDA to work privately and manipulates pieces on the board to work collaboratively. Figure 2 shows the software architecture of Caretta. It provides developers with a substratum to construct applications for collaborative activities by integrating personal and shared spaces. As shown in Figure 3, a user's PDA visualizes the same information as that projected on the sensing board. A user can switch between the "personal mode" and "shared mode" on his/her own PDA. In the "personal mode", users' manipulations on PDAs, such as placing, deleting or moving an icon with a stylus pen, change the simulations results and its visualization on their own PDAs, but does not change the visualization on the shared space (we call it a "personal simulation"). In the "shared mode", users' manipulations on the PDAs change the simulation results and its visualization of both PDAs and the shared space (we call it a "shared simulation"). Manipulations of pieces on the shared space change the simulation results and visualization of users' PDAs in the shared mode, but do not affect those in the personal mode.

The current version of Caretta (version1.0) is used for learning about urban planning and environmental problems. In urban planning in the real world, different stakeholders have to discuss with each other and find acceptable solutions through negotiations. In this version, the personal spaces allow each user to find individually most desirable solutions. The shared space supports a group of users in negotiating in a face-to-face setting by manipulating physically sharable pieces. Users can also access to the shared space through their own personal spaces. For example, a user can put a highway on the shared space, when he/she places it on his/her PDA in the shared mode. In using Caretta, users firstly decide which role they plays, for example, residents, public officials, building constructors, and so on. They then use their own PDAs, and manipulate physical pieces to design a town on the board.

### 3.3 Intelligent Support Module

In Caretta, several knowledge bases to support users are implemented. When a user is in the personal mode, Carretta analyzes the user's manipulation histories, detects the users intention and gives him/her a suggestion. When users are in the shared mode, Caretta infers from each user's intention and generates several alternatives satisfactory and acceptable to as many users as possible. Currently we have revised several existing machine learning algorithms (such as support vector machine or k-nearest neighbor) and are testing them through users studies.

In our future work, Caretta will be evaluated and improved through more intensive user studies. We have several plans to develop different types of applications such as entertainment games.

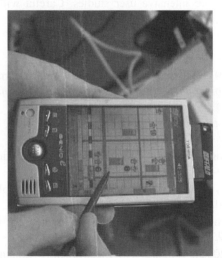 

Figure 3. The personal space (left) and public space (right) in Caretta

## 4 User Studies

Preliminary user studies of Caretta were conducted. 12 university students formed three groups of four, and participated in the experiments. Although the analyses of the user studies have not been completed, following issues were confirmed:

1. Users could interchangeably use their personal and shared spaces without any trouble. They spent longer time in the personal mode than in the shard mode. Users were absorbed in

959

finding their own best idea by testing as many ideas as they could imagine, before showing the idea in the shared space.

2. When two users in a group shared interests, they often discussed privately by using their PDAs. A PDA worked not only as an individual private space, but also as a private space for a subgroup (a pair of users).

3. The shared space worked as a trigger for discussions. When a change of the shared space by one user was serious to the other users, they started negotiations to improve their situations. It did happen frequently, because any user was allowed to manipulate the physically shared space anytime he/she liked.

4. Users answered that suggestions given by Caretta were not always satisfactory, but they could be hints or clues for their own further thinking or discussions.

The post-experimental inquiries showed that the participants' comments were positive. We are now investigating how the integration of personal and shared spaces by Caretta affected users' collaborative learning processes.

## 5 Conclusions and Future Works

In this paper, a system called Caretta, which is used for supporting collaborative learning is described. The important feature of Caretta is to integrate personal and shared spaces, with mobile and augmented reality technologies, respectively. Although the preliminary user studies might not be enough to prove how effectively Caretta could support collaborative learning, they depicted that Caretta was acceptable to users as a new type of learning environment. In our future works, Caretta must be evaluated and improved through more intensive user studies. Caretta will be useful for developing different types of applications such as entertainment games.

## References

Kusunoki, F., Sugimoto, M., & Hashizume, H. (1999). A System for Supporting Group Learning that Enhances Interactions. In *Proceedings of Computer Support for Collaborative Learning*, 323-327.

Koschmann, T. (1996). CSCL: Theory and Practice of an Emerging Paradigm. Mahwah, NJ: Lawrence Erlbaum.

Rekimoto, J. (1998). A Multiple Device Approach for Supporting Whiteboard-based Interactions, In *Proceedings of CHI'98,* Los Angeles, CA, 344-351.

Shoemaker, G. & Inkpen, K. (2001). Single Display Privacyware: Augmenting Public Displays with Private Information, *Proceedings of CHI2002,* Seattle, WA, 522 – 529.

Stewart, B., Bederson, B., & Druin, A. (1998). Single Display Groupware: A Model for Co-present Collaboration. In *Proceedings of CHI' 99*, Pittsburgh, PA, 286–293.

Sugimoto, M., Kusunoki, F., & Hashizume, H. (2001). E$^2$board: An Electronically Enhanced Board for Games and Group Activity Support," In *Proceedings of Affective Human Factors Design*, Singapore, 227-234.

Sugimoto, M., Kusunoki, F., & Hashizume, H. (2002). Design of an Interactive System for Group Learning Support. In *Proceedings of DIS2002*, London, UK, 50-55.

Sugimoto, M., Kusunoki, F., Inagaki, S., Takatoki, K., &Yoshikawa, A. (2002). A System for Supporting Group Learning in Face-to-face and Networked Environments, In *Proceedings of CSCW2002*, New Orleans, LA, (video submission).

# Exploring Medium-Tool Relations:
# Field Trials in Construction of Hypermedia in Schools

*Anders Kluge*

Department of Informatics, University of Oslo,
Pb 1080, Blindern, 0316 Oslo, Norway
anders.kluge@ifi.uio.no

## Abstract

Digital appliances are increasingly used both as a medium and a tool. This makes it relevant to study use situations of consumption and production of content. How do users approach these possibilities? These issues are explored by studying pupils aged 12-15 creating hypermedia presentations in groups. They vary a lot in how they approach the task and in what they develop, although the pupils have similarities in how they link information together. To distinguish between production and consumption of content as separate activities did not seem as a fruitful analysis of the pupils' work.

## 1   Introduction

We are to an increasing extent using the computer as a media machine. Now a standard, high-end computer can work as a TV set, a video player, stereo equipment or radio, and for on-demand content of various media types from the web. Computer models have also began to come with media-player buttons in the hardware similar to e.g. VCR-players, in addition to the regular keyboard (see e.g. Z-note2002 http://www.zepto.no). However, while having possibilities as a general media consumption machine, the computer still maintains its role as a tool in most every domain of professional and leisure life.

To consume information and to use a tool may be viewed as two very different processes and may call for divergent design strategies. To consume can be understood to be passive and in opposition to a tool that requires operation for it to work. However, activity can also be seen as an important part of consuming information, in order to grasp the content of it. This is in accordance with the constructivist view on learning (Fosnot, 1996).

This paper explores how use unfolds as pupils employ information technology both as a medium and a tool, both to conceive and produce content. It is based on a field trial where 14 groups of pupils at 4 schools in Norway use a special-purpose tool named Syncrolink (see http://www.nr.no/lava/) over a period of 3-5 days intensive project work. Through participative observation the trial gave an opportunity to study how the pupils worked with the software as a medium and a tool, how they handled the prepared material and content in general, and ultimately to study the resulting product.

## 2   The Syncrolink software

Syncrolink consists of an editor and a player as separate applications. The player runs in a browser as a regular plug-in. It consists of two frames, one for video and one for html-pages (see figure 1). Links may be added to the video by use of the editor. The links can mainly be of two types: (1) Time-triggered links where the html-pages will appear in the neighbouring frame when the video reaches a particular point in time. The video proceeds as normal. (2) User triggered links, where they appear as texts or coloured areas in the video frame (see figure 1). The user will have to click on it to make the html-page appear, the video will pause, and must be started again by the user clicking the play-button. The user-triggered links must be assigned with its' time of appearance and disappearance in the video frame. Adding 'time-jumps' in the video timeline does the simple video editing. The video area of the Syncrolink player includes a 'play' and a 'pause' button, available for the end-user studying the presentation. The timeline of the video is visible and includes a sliding bar, movable for the end-user. The links will also be visible as red lines in the timeline.

## 3   The field trial

The general theme for the pupils project work was the multi-ethnic society. They had to choose one video among six documentary- / coverage-style TV-programs. This was to be the video they would continue to work with throughout the project. The videos were of 15-40 minutes duration, but a 2-3 minutes clip of each video were available for the students as triggers, and were their basis for selecting a video. The students had three types of content to base their work on: (1) the video they selected, (2) Web-pages, (3) text books at school or from the library. The students were

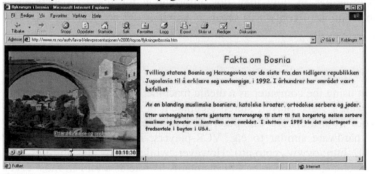

also encouraged to interview people who knew something about the subject, either from their experiences or professionally.

The technical tools at their disposal were the Syncrolink editor and player, an html-editor, digital still-cameras and a scanner, in addition to a standard PC.

**Figure 1:** A snapshot of a Syncrolink presentation.

### 3.1   The initial project work

The project had a well-defined starting point for the pupils. However, when they sat down in front of the computer they obviously expected to be active, – they seemed unable to handle a passive position of watching. The pupils in control of the computer jumped back and forth in the material in most cases, making it difficult for the other pupils in the group to keep track of the story, and they soon lost interest. In general, the pupils had little attention to devote to the video material at this stage. Also the ones who did manoeuvre in the video seemed to play more than they actually looked for anything. At the last school of the trial some simple structure were put on the videos to try to engage the pupils. The already short video-snippets were split in two parts, and after the first half a request appeared for the pupils to take some notes of what they had seen. They did not take notice and just continued as nothing had happened when this appeared. What seemed to be the most successful approach was at the school were they had problems with the equipment and had to show the videos on a large screen for all the pupils. This session resembles their regular way to watch video material, and they did not expect to be active as they did in front of the computer. One illustration of how they felt when they were confronted with the video material were when one pupil asked: "Aren't we supposed to do some work?" when he was told that they should watch video on the computer. A video session, according to the teachers would normally bring cheers in the class, but were experienced as a nuisance in front of the computer.

## 3.2 Construction of links

The pupils devoted the majority of their time to work with the links. Of the 55 links in the 14 presentations, 24 are associated to a dramatic event. In the video dealing with Bosnian refugees in Norway, the bombing of the old bridge in Mostar is shown. The two groups using this video as their point of departure both have links at the very point where the bridge falls down, and both the links leads to material concerning reconstruction of buildings in Bosnia. These two groups worked in different schools and did not have possibilities to influence each other.

Two of the tree groups selecting a video dealing with racial harassment had a link appearing at the point were a window is smashed. The third group started their presentation at that very point. Three of the four groups working with a video from Bangladesh, made a link were a young boy takes of his bandage and a serious wound is shown. The other two categories of link-points put in the video stream, were related either to the introduction of a new object or the start of a new scene in the video. It was obvious that this kind of information 'bursts', a dramatic event, a new object or a new scene in the video, prompted the pupils to add links. All the 55 links could be categorised in these three groups.

One group wanted to supply content in opposition to the existing material when they constructed the presentations, what is referred to as 'mirrorworlds' by Bernstein (Bernstein, 1998). These groups discussed contrasts in the information as a way of extending the understanding and to underline issues brought up in the video material, as a quotation indicates: "When she says [in the video]:'Mostar used to be a beautiful city', we can show how it is today as a contrast to what she is saying, with a picture on the side [of the video frame]".

There were also discussions in the groups about making an alternative view to the existing content, and also on some occasions to try to weaken the points made in the video. They were also quite aware of the possibility to change the impression from the material. One group noted that when they had a picture of themselves in the html-frame while a violent attack was shown in the video, the total impression of the presentation was that the persons on the picture (the group themselves!) were responsible for the attack.

## 3.3 Four approaches

A Syncrolink presentation is inherently video centric. When a user initially starts a presentation, a video frame appears with an html-frame next to it. The further advance of the presentation is dependent on the video. Still, within this tight grammar, the pupils had very different approaches on how to utilise the application and the content, and made quite diverse presentations, a result tat confirms with (Pohl and Purgathofer, 2000). Four identifiable approaches emerged in this trial and are presented below.

*Video extension:* One way of relating to the video as existing content, is to make the accompanying web-pages relate very close to the video material, as a visual and verbal deepening of the video presented. This can be viewed as the most obvious way to create and relate content to the existing video material with Syncrolink. The tool encourages this line of thinking by having video material as the starting point for a presentation, and the visual pointers physically located in the video frame. In spite of these properties in the tool, only half of the presentations (7 of 14) are dominated by a video-centric approach. What came to be considered as most advanced of these presentations had extensive visual representations of the facts presented in the video material. For instance, when the name of a city in Bosnia was mentioned in the video, a text with the city name appears in the video frame and activation of the link opens a map of Bosnia in the html-frame with the city name emphasised. The pupils had found a map on the web and altered it by emphasising certain elements to make it more relevant to the video. A similar illustration technique was used when facts and numbers of refugees and unemployment were mentioned in the video. The same group created graphical illustrations and made the links visible when facts were mentioned in the video. These were illustrations made partly out of information found in the video, and partly of additional information from web. The information was put together graphically to e.g. by comparing unemployment figures in Norway and Bosnia with a powerful visual effect, and to illustrate levels of refugees in different Norwegian municipalities. The pupils synthesised information from several sources and added this to the video content.

Another way to apply the same approach were to more or less mechanically create links as various topics appeared in the video. E.g. a girl was interviewed when she attended to a horse. The pupils then searched for horse-pages and made links to these pages without reflecting on the relevance and coherence of the total presentation, the horse were never actually an issue in the video. Two presentations applied this mare arbitrary procedure in their work.

*Parallel strands:* Another approach the users applied was to create material as a parallel strand in the presentation. Three of the 14 presentations had an inclination to this way of working. The most representative example was a group who spent most of their time interviewing people about the issue of racial harassment dealt with in a video. They took pictures of the people interviewed and made html-pages combining the textual interviews with images. The group selected a part of the video material with quite dramatic and violent events from the video, and made links to the interviews with regular time intervals. This resulted in a presentation with two separate stories: a strand of the presentation conveying the edited video material, and the other strand consisting of two interviews from fellow pupils at school and two interviews with people passing by the school. These three groups only occasionally used material that they had not created themselves.

*Text-pages to guide the presentation:* Three of the groups can be viewed as trying to escape the video-centric approach in the Syncrolink tool. They framed their selected parts of the video by using html-pages to guide the presentation, by having the pages appearing by time-triggered links before the actual scenes appear. One starts with: "You will now be able to see a video about racism". They continued to frame the video material throughout the presentation e.g. by introducing a dramatic event: "Six racists attack two young boys". These groups spent a lot of time on the video material; selecting, editing and making the text they found suitable for the particular parts of the video. Here the combination of consumption (looking at the video), creating (making text to frame the video material) and constructing (selecting parts of the video which they consider most relevant) seem impossible to distinguish from each other in the observations of the groups working.

*Material as inspiration:* One presentation have a free associative relation to the video material, and use the video mainly as an inspiration to investigate a related topic. The video material worked as a starting point for the group to discuss related issues. From there on the group more or less left the video and followed their own path to interview immigrants and search the web. The links to the web and textual material written by the pupils where quite coherent in itself, but the original video worked as a rather loose context and detached background material. The links in the presentation were distributed freely over the video timeline, and seemed to be put there to make a suitable spread in order for an end-user to get some distance in time between the readings of different textual material.

# 4    Discussion and concluding remarks

The educational setting is a well-suited test-bed for the combination of consumption and creation and to study applications as media and tool, in particular in learning viewed as a constructive activity where active pupils are an essential element (Fosnot 1996). From the field trial it is obvious that a sequential process of consuming content, constructing relational structures to prepare for own production, and then creating new material in the structure, is not in accordance with reality. They engaged in a continuously oscillating process between constructing structures, creating material, and studying content. In parts of the trial, these processed were impossible to identify as separate activities by observation of the action going on. To try to design tools and media for a pre-defined set of phases and activities (Shneiderman 2000) is not in accordance with these results. Also to try to impose structure to the process were useless in this field trial. Even if the pupils were strongly encouraged to start up with exploring the content in this trial, they seemed uninterested to do so. The main reason for this appeared to be their expectations as they sit down in front of a computer. The computer signalled activity to the pupils. They entered an active mode in front of it to the extent that even looking at video-material, which usually is a popular part of school, became boring and made them restless. They wanted to be in touch with the material, control it and play with it, and not only to watch it (Arias, et al., 1999).

Information bursts, as it occurs in a dramatic event (Laurel 1993), triggered the pupil's to create relational links. All the 55 links in the 11 presentations can be attributed to a sudden appearance of information in the video: – the beginning of a new scene, a new object or subject enters the frame, or a dramatic event occurs. The pupils vary in their presentations and activity in all the other variables under investigation, but this linking strategy becomes the common trait for the pupils' work processes and production.

The groups varied in their emphasis on content creation. The video-centric presentations were mainly relational. They created little new content, more attention were devoted towards linking external web-pages and adding information to them. The groups constructing a strand in parallel with the prepared video material created their own content in order to present an alternative story. The pupils applying the parallel strands' approach did very little linking to other material than what they have made themselves and their emphasis was clearly on content creation.

The groups trying to escape the video centric approach do not fit into the dichotomy between links and content, between relations and creation. Their presentations were directed towards making a context for the video material. Their time-triggered links were on a meta-level, describing the course of the events, and they were less inclined to make links to contrasting or diverging material. They accepted the video material as the main content, and used their efforts on understanding what was happening in the video, by framing the material their way.

Within its limited capabilities of expression, Syncrolink was able to serve for construction of these diverging presentation styles and link logic. Even constrained by a rigorous grammar, the possibilities of the hypermedia format itself seems to ensure a large degree of freedom in presentation genres.

# References

Arias, E., H. Eden, et al. (1999). Beyond Access: Informed Participation and Empowerment. *Computer Supported Cooperative Learning*, Stanford.

Bernstein, M. (1998). Patterns of Hypertext. *ACM Hypertext*, Pittsburgh, ACM.

Fosnot, C. T. e. (1996). Constructivism : theory, perspectives, and practice, New York : Teachers College Press, 1996.

Laurel, B. (1993). Computers as Theatre, Addison-Wesley.

Pohl, M. and P. Purgathofer (2000). Hypertext authoring and visualisation. *International Journal of Human-Computer Studies* **53**.

Shneiderman, B. (2000). Creating Creativity: User Interfaces for Supporting Innovations. *ACM Transactions on Computer-Human Interaction* **7** (1).

# Learning to Dance via Multiple Representations

*Lefteris A. Kolleros*

Evias 29
Alimos, Athens, GR-174 56
Greece
lkolleros@hotmail.com

*Alan P. Parkes*

Lancaster University
Computing Department
Lancaster, LA1 4YR, UK
app@comp.lancs.ac.uk

## Abstract

Multiple representations have been used successfully in many educational applications, and their strengths and weaknesses have been widely discussed. However, there are unanswered questions about what multiple representations can contribute to the training of predominantly physical activities, such as dance. This paper describes a project that prototyped and evaluated an interface intended to provide a learning environment for a specific dance. The interface features multiple representations of the dancing steps.

## 1 The potential of multiple representations in dance teaching

Dance is a multifaceted expressive form. We cannot completely describe a dance using one representation medium (e.g. video). Video can obviously *support* the learning process. However, the greatest drawback of video is the viewer's inability to see, in sufficient detail, the various stages of the dance. Guest (1984) asserts that the viewer receives mainly an impression, rather than detailed information. There is also the problem of point of view. For example, if we show only a rear view of a dancer, the viewer obtains an incomplete representation of the dance. Finally, we can verbally explain dance movements, but words are not ideal for describing movement. However, again, they can support the understanding and mastering of movement.

Multiple representations need to complement each other. We argue that the interaction and integration of representations can lead to significant developments in dance teaching.

## 2 The Objectives of the Project

The objectives of the dance tutor project can be summarized as follows:
- *To develop a prototype learning environment for a specific dance.* As an experimental domain, we chose the dance Kalamatianos, due to its popularity (in Greece) and relative simplicity.
- *To use multiple co-ordinated representations of the dancing steps.*
- *To evaluate the prototype using actual learners of the dance.*

# 3 The dance tutor system

We now describe the dance tutor system that was developed to meet the objectives outlined above. We describe the various representations used in the tutor. We then describe some of the features, such as quizzes, that the system supports.

## 3.1 Multiple Representations

The representations of the dancing steps used in the interface are:

- Videos of actual dancers in action
- Drawings and animated characters
- Thumbnails (of the drawings and the video frames)
- Footprints to represent the dance steps
- Textual descriptions of dance movements and contextual information

When the user switches to a new step in the dance, all open representations are modified to match the current step (see Figure 1). The user can view the pictorial representations of all preceding steps in order to integrate the steps up to the current step. The pictorial representations (drawings, thumbnails, and video) are available in various views (back, front, side) so that the user can form a complete image of the dance. The top view is actually provided by the *footprints* representation.

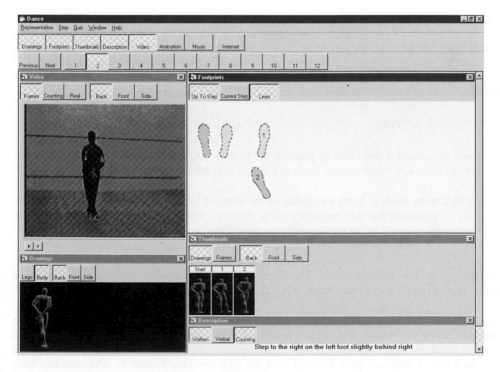

**Figure 1:** The interface at step two

The video representation shows the still frame corresponding to the current step. The user is able to switch to various views (back, front, side) and play the video clip of the currently active view up to the current step. Video clips are also used to show the dance accompanied by a count of the steps, so the steps can be linked with the rhythm, and to show experienced dancers performing the dance. The learner can thus appreciate many important aspects of the dance (Raftis, 1987).

Drawings represent various orientations of the dancer's body. Posture is rendered more faithfully in the video, but the drawings can isolate salient elements.

Thumbnails are small versions of original images and thus less demanding in terms of screen space. They display all steps up to the current step, in the form of drawings or video frames.

The footprints representation provides an option for displaying only the steps. The user can add connecting arrows between successive footprints to indicate the orbit of feet.

There are currently three options for verbal descriptions: displaying a description, replaying the verbal description of the current step, and playing a verbal count of the steps.

## 3.2    Additional Functionality

### 3.2.1    Animation

In the prototype, animation involves the simultaneous and coordinated presentation of the various representations (even of the non-pictorial ones).   It is based on showing and hiding open representations (thus creating the illusion of movement) up to the current step. The user is able to apply animation with or without music. In the first case, all open representations are animated in the correct rhythm, so that the user can associate the steps with the music. In the second case, the time interval between successive steps is the same, but the user can vary this interval, as desired.

### 3.2.2    Problem Solving

The interface also enables the users to practice the material. Quizzes cover the actual steps, the timing between them and other aspects, such as the way that the dancers hold hands.

The *timing quiz* helps users to learn the timing of the steps. A voice counts the first set of steps in time in order to introduce the user smoothly to the rhythm.  The user then indicates the points in time where the steps should occur by clicking a button on the screen.

In the *drag & drop* quiz, images of the steps are presented in random order and the user has to re-arrange them into the correct order. The user can switch between various representations (drawings, frames) and views (back, front, side), as required.

In a further quiz, a collection of footprints is displayed. The user must select the correct footprints for each step, using the mouse. To make it possible for the user to actually perform the steps, a specially prepared mat was used, in which switches were secured in positions corresponding to the steps of the dance. Footprints were fastened over the switches (see Figure 2). The switches were connected to the keys of the keyboard. Hence, stepping on a switch was equivalent to a key press. The quiz can be invoked with or without music. When the quiz is invoked without music, the user has only to step on the correct footprint for each step, with no regard to the timing of the steps. If the quiz is invoked with music, the user must perform the steps with the correct timing. Of course, this quiz focuses entirely on foot movement, at the expense of all other aspects of the dance.

**Figure 2:** The interactive carpet

# 4   The Evaluation

This section discusses some results obtained from the evaluation, which examined the overall effectiveness and usability of the interface, and of each one of the representations used. We used pre- and post-questionnaires, observation (video recording and note-taking), and semi-structured interviews (to expand on the answers provided in the post-questionnaire). The set of tasks consisted of three parts. The first was the interaction with the interface. The user was expected to acquire the knowledge and skills required in the next parts. The second part was the execution of the quizzes. The tasks were designed to be of progressive difficulty. In the third part, the users performed the steps first without music, and then with music.

## 4.1   Multiple Representations

Users readily understood each representation and its relationships with the domain. The understanding of relations between representations was not problematic, possibly due to their co-presence, the high level of *redundancy* and the common (pictorial) modality of most representations (Ainsworth, 1999a, 1999b). Users occasionally encountered difficulties in associating the textual descriptions with the pictorial representations. The former could not compete with the latter, due primarily to the physical nature of dance. The users actually had to translate the words into mental images of steps, whereas with pictorial representations the visual information did not have to be pre-processed in this way. This relates to the so-called *picture superiority effect* (Nelson, Reed, & Walling, 1976; Paivio, Rogers & Smythe, 1968).

Each representation exists in a multi-representational environment and therefore cannot be simply considered in isolation. Their effectiveness could be strongly related to their implementation in the current interface. Care should therefore be taken in drawing general conclusions.

The footprints were the most effective representation, followed by video frames (static) and video clips. Animations were also effective because of their dynamism. No representation was found to be ineffective, so they all had something to offer. In fact, each representation was considered effective by certain subjects. The most effective combination was considered to be the footprints and video, mainly because the former taught the steps and the latter illustrated the dance. The most effective view was the back view, and the least effective the front view. Most of the subjects did not find the co-ordinated use of multiple representations confusing as they could (consecutively) focus on individual representations. All subjects preferred a multi-representational environment.

## 4.2 Quizzes

The timing quiz was considered of neutral difficulty. The drag & drop and carpet quizzes (when the latter was performed on the carpet with music), were considered to be the most difficult to accomplish. No quiz was found to be very easy or too difficult. The carpet quiz was considered the most stimulating. It also proved effective for learning the steps.

## 5   Conclusions

Our evaluation demonstrated the effectiveness of the system, given the limited time of interaction and the limited prior dancing experience of the subjects. It is encouraging that subjects did not need to be assisted by a human tutor during interaction. However, it is difficult for the machine to achieve the reactivity to the learner of the human dance tutor. All subjects confirmed that the present interface can only complement human-centered dance teaching. The dancer can fully learn the dance only if he or she performs it with experienced dancers. Learning a dance via the computer, even interactively, lacks the characteristics of public performance.

## References

Ainsworth, S. E. (1999a). A functional taxonomy of multiple representations. *Computers and Education*, 33(2/3), 131-152.

Ainsworth, S. E. (1999b). Designing effective multi-representational learning environments, Technical report No 58, ESRC Centre for Research in Development, Instruction and Training, University of Nottingham.

Guest A. H. (1984). Dance Notation: The process of recording movement on paper, *Dance Books*, 8-14, 163-173.

Nelson, D. L., Reed, V. S., & Walling, J. R. (1976). Pictorial superiority effect. *Journal of Experimental Psychology*: Human Learning and Memory , 2 , 523-528.

Paivio, A., Rogers, T. B., & Smythe P. C. (1968). Why are pictures easier to recall than words? *Psychonomic Science* , 11, 137-138.

Raftis A., (1987). The World of Greek Dance, FineDawn Publishers.

# Educational Software Interfaces and Teacher's Use

*Walquíria Castelo-Branco Lins*

Centre of Education - UFPE
Rua Alcides Codeceira, 320/203 –
Iputinga –Recife-PE 50800-090 - Brazil
wcblins@ufpe.br

*Alex Sandro Gomes*

Centre of Informatics - UFPE
Rua Prof. Luiz Freire,s/n, Recife-PE
Po. Box: 7851, 50732-970, Brazil
asg@cin.ufpe.br

## Abstract

Designing educational software is normally approached as a pure creative or interdisciplinary activity. Teachers' activities are rarely considered in initial requirements elicitation phase. The aim of our research was to propose a qualitative approach to analyse teachers' classroom technology mediated activity as a source of information to educational software design. Our main results show the need to consider aspects correlated to how flexible the interface is to allow teacher to make changes according to his/her didactical choices.

## 1   Introduction

In the educational context, there is a gap between expectations generated from the potentiality of interactive and digital technologies and the way they are used mediating pedagogic activities (Resnick, 2001; CastroFilho e Confrey, 2000; Dugdale, 1999). In our viewpoint, many of these problems are related to usability: methods and operations, that these tools materialize, are inadequate (Leont'ev, 1975) to pedagogic culture concretised on professional day-life (Tardif, 2002). The *National Center for Education Statistics* (2000) indicates that half of the teachers of United States of America who have access to computer and to the web at classroom do not use them in class. The majority of them use these tools to search models in class, to plan their activities, to elaborate teaching materials and to communicate, but they do not feel confident to introduce them in pedagogic activities with pupils in classroom.

Researches in Mathematics Education points to the same issue: teachers use technologies in their teaching activities, in a limited way. Some researchers point to the need for improvement in their teacher training (Kennedy, 1990; Ball, 1991 apud CastroFilho e Confrey, 2000). So, the point is how to train these teachers? According to Handler & Strudler (1997), Thomas (1999) and Wang & Holthaus (1999) quoted in Pope *et al.* (2002), in general, teacher's undergraduate degrees (*colleges of Education*) include courses to introduce teacher on computer use, however, the methodological and educational courses do not use computer as a tools to discuss teaching and learning process on different content. Teachers learn how to use technology, but do not learn how to teach with them (Pope, Hare & Howard, 2002). Literature on school use of computer, normally, focuses researches on students' learning. There are a small quantity of researches on the relation between technology and teaching activities. Therefore, it is necessary to investigate teachers practice in the context of their pedagogic activities (CastroFilho & Confrey, 2000; NCES, 2000).

Another historic source of problem is raised by considerations of educational software design quality (Frye & Soloway, 1987; Hinostroza & Mellar, 2001). The literature points to the limited quality of these materials as a factor that leads to a small percentage of use in pedagogical practice. According to Hinostroza & Mellar (2001), educational software designers give priority to questions related to learning, and this would be a cause of teachers' difficulties to incorporate these materials in their class activities.

Our purpose is to investigate teaching activities using computational interfaces in classroom. In the second section of this paper, we discuss elicitation of requirements to educational software. In the third section, we describe details of method used in this research. The fourth section, some data illustrate the way we identified requirements related to teaching practice. Finally, we conclude discussing the results of this research.

# 2   Eliciting requirements for Educational Software

Researches on Human-Machine interactions have been pointed how e how much user-computer interactions are affected by technology and by social context, in which user is engaged. Tools quality should be related to the conditions that allow users to effectively and satisfactorily reach the aims of their activities. Then, to discover the real users' needs and in which conditions they act, it is necessary to investigate their use context (Nardi, 1996). In our research, the users are mathematic teachers and our aim is to elicit their requirements when the activities are mediated by computational interfaces to teach fractions.

Our theoretical framework is the activity theory – developed by Leont'ev – based on researches of Vigotsky, Rubinstein and others, started in 1920. This theory is used as a tool to investigate the development of human activities mediated by context and socio-cultural artefacts. According to their theoretical principles is through human activities, mediated by cultural artefacts and people, that consciousness are developed. Artefacts are, among other things, laws, rules, procedure, division of labour, tools, language and concepts (Kuutti, 1996). They are created and modified during an activity and they can be used to control our behaviour (Bellamy, 1996). The artefacts come from the context in which it is inserted and are internalised and externalised by subjects through their acts, speech, been able to be evaluated and corrected.   Tool is something that is produced to overcome some needs of the activity, to teaching aims, which shape ways and operations of use, but do not shape actions and aims (Leont'ev, 1975). This transformation of objects into tools are not arbitrarily produced, it is necessary an appropriation of the object in itself, appropriation of "natural logic" of the object to be inserted into the logic of human activity. They are oriented by reasons directed to objects (material or ideal). For example: the reason that structure teacher's activity to teach fractions is students' learning of concepts involved in understanding fractions. The object is to learn fractions. According to the aims to which human procedures are oriented, we can relate hierarchic and differentiated levels of procedures: operations, actions and activities. The activity is oriented by one reason; the actions by aims and the operations are related to conditions to do the actions. The subject of the activity, in turn, is inserted in the community where people relationships are mediated by artefacts, rules and division of labour (Bellamy, 1996).

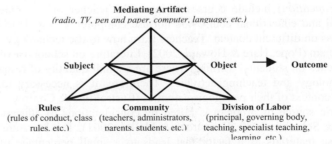

**Figure 1:** Model of Activity Analysis of Cole and Engerström (1991) to K-12 Education

These issues raised in this research will support the development of software to teach fractions. On the one hand, in Brazilian education, this is a concept pointed as to be difficult for a large percentage of our students to cope with. On the other hand, there are few educational software programs to teach fractions, and the majority of them repeat a teaching approach, which works associating iconic representations of area (pizzas and rectangles) with numeric representations. In our research, we reach the problem of developing the interface of this material by identifying requirements of two kinds. The first is related to the conceptual field (fractions on multiplicative structures) and the second is related to the teaching practice. In this paper, we present results of a study to elicit requirements related to practice to design a software program to teach fractions.

## 3  Method

Aiming to reach some elements of teaching practice mediated by computational interfaces, a research method was designed. The research comprised some phases undertaken by ten mathematic teachers during an in-service training, which is part of AMaDeuS[1] project. The first phase of the study was to characterize teacher practice using a semi-structured interview and an observation of teacher class while teaching fraction in their teaching context. During the course, the teachers were required to choose an educational software and were oriented to build an activity to approach fractions with the chosen software, experience it with a par of students while video-recording the session, analyse the session, then, plan a class for the whole class. So, our data are: transcriptions of the interviews, video-records of teacher participation during the course, email communication of a e-group created for the course, with participation of the teachers and the instructors, a first activity planning, video-recording of the classes with and without the software with teachers' pupils, a report of the session undertaken by the teacher. The analyses presented here is the results of two of the teachers who focuses their activities on the concept of fraction. The other teachers chose to work with other mathematic concepts. The research is still analysing the data to refine categories to elicit requirements of the mathematic teachers.

### 3.1  The analysis

In other to understand a phenomenon is necessary to know its development process, the changes made on it, up to its actual form. It is in the totality of this principle that the use of the theory of activity lies as a tools to elicit and to analyse the development of different human practices, in particular, the teaching practice. The subject of the activity is inserted in a community (teachers, parents, principals, students, etc). In this community, the relations between people are mediated by: artefacts (tv, radio, pen and paper, chalkboard, computer, etc), rules (conduct, classroom rules, etc) and division of labour (orientation, administration, teaching, learning, specialized learning, etc) (Engeström, 1993 apud Bellamy, 1996).
Our analysis focus the way teachers plan, execute and assess the impact of their own teaching with the use of the educational software.

## 4  Analysis and Discussion

The *teaching practice* using the software are structured from the choice of the content and of a software to teach the chosen content. A diversity of educational software was let available to the teachers and they were taught to use them. However, our teachers chose the software that the scenarios were close to students' day-life. The teachers did not analyse the quality of

---

[1] Partially financed by CNPq.

representations and properties of the focused concept, neither the meaning of this concept in the software (fraction). They chose to work with a hypermedia software program that simulate a supermarket, and despite of the mathematical content be defined, tracing as actions in classroom the students exploited the software following the levels proposed in the software. Both teachers justified their choice saying that it was to be close to students' reality. They claim for the inclusion of a clock in the software to observe how long students take to collect goods, to go to pay, and to pay. They argued that, in day-life, everyone has time constrains to act. Terms such as context, day-life were frequent in their justifications that approximate their decision to educational movements in Brazil, that claims for mathematic teaching more close to students reality (Brasil, 1998). However, the teachers had difficulties to identify, in the contextual representations, properties of the concepts they have as teaching aims. The difficulty allows one to delegate to the software the aims and the actions, which were defined, without the software, in their planning. The interface seems to be something to be followed by student, following the levels defined in the software, as it happens in the textbooks. In the case of our two teachers, they felt inhibited to interfere in students' motivation to exploit the software. This data lead us to propose as requirement related to the teacher practice 'how the software allow a adaptation of its interface to teacher practice', because there is a clear tension between what the teacher can do without the material and what s/he can do with the software. Teacher Anne, one of our subjects, is a typical example, her class was planned to work a mathematical concept and ends working with another concept, when students decided to change the level of the software. Her first aim was to exploit additive structures with natural number is change by the aim of the students to reach the highest level of the software. The professional tries to use a new type of material, but perceive, even that implicitly, its limited flexibility when compared to touchable and manipulative materials. The software does not allow teachers to define tools, context, time and action to reach the object of its activity, as they did when using pen and paper and manipulative material to teach fractions. The flux of activities proposed by the educational software interferes directly on the flux of activities that are normally undertaken by teacher in their rhetoric to teach something. In this sense, we identify the need of the software to be more flexible to allow teacher to define the teaching variable. A criterion to consider in designing a interface is how it allows teacher to choose their didactical variables and to adapt the platform to it; actions, operations and teaching activities to reach the learning of the concepts, the development of abilities, procedures, among others, which involve institutions, curriculum, students, parents, etc. It deals with intangible technologies as didactical transpositions, administrations and knowledge of the content and of the available materials (Tardif, 2002). It involves coordination of curricular time. Regarding this point, one of the main teacher's worry, while planning and executing their activities, is adapting this activities to class time, and the time they have to do all these activities required by the curriculum, which is claimed by parents, administrators, principals, etc. Since the beginning of the activities, the teachers select materials adapting them to the time the have to appropriate the material to teaching, difficulties the material offers use and what can contribute to improve quality of teaching conditions.

## 5   Conclusion

The results show the need to consider aspects correlated to how flexible the interface is to allow teacher to make changes according to his/her didactical choices, integrating teacher's decision within the interface. The results also points to the need to consider practice and curricular limitations as requirement in software design.

# References

Bellamy, R.K.E. (1996). Designing Educational Technology: Computer-Mediated Change. In B.A. Nardi (Ed.), *Activity Theory and Human-Computer Interaction* (pp. 123-146). London: The MIT Press.

Brasil, Ministério da Educação e do Desporto, Secretaria de Educação Fundamental (1998) *PCNs Parâmetros Curriculares Nacionais – Matemática*. Brasília: MEC/SEF.

CastroFilho, J.A. & Confrey, J. (2000). Discussing technology, classroom practice, curriculum, and content knowledge with teachers. *Anais da RIBIE 2000*.

Cole, M. & Engeström, Y. A cultural-historical approach to distributed cognition. In G.Salomon (ed.) *Distributed Cognition* (pp. 1-47). Cambridge: Cambridge University Press.

Dugdale, O. (1999). Establishing Computer as an Optional Problem Solving Tool in a Nontechnological Mathematics Context. *International Journal of Computer for Mathematical Learning*, 4, 151-167.

Frye, D. & Soloway, E. (1987). Interface Design: A Neglected Issue in Educational Software. Anals of CHI+GI, pp. 93-97.

Hinostroza, J.E. & Mellar, H. (2001). Pedagogy embedded in educational software design: report of a case study. *Computer & Education*, 37, 27-40.

Kuutti, K. (1996). Activity Theory as a Potential Framework for Human-Computer Interaction Research. In B.A. Nardi (Ed.), *Activity Theory and Human-Computer Interaction* (pp. 17-44). London: The MIT Press.

Leont'ev, A.N. (1975). *Actividad, conciencia, personalidad*. Habana: Editorial Pueblo y Educación, edition in Spanish 1981, edited from original in Russian 1975.

Nardi, B.A. (1996). Activity Theory and Human-Computer Interaction. In B.A. Nardi (Ed.), *Activity Theory and Human-Computer Interaction* (pp. 7-16). London: The MIT Press.

National Center for Education Statistics (NCES) (2000). Teacher's tools form the 21st Century: A report on teacher's use of technology. Retrieved February 01, 2003 from http://www.nces.gov/pubsearch/pubsinfo.asp?pubid=1999080.

Pope, M., Hare, D.& Howard, E. (2002). Technology integration: Closing the gap between what preservice teachers are taught to do and what they can do. *Journal of Technology and Teacher Education*, 10 (2), 191-204.

Resnick, M. (2001). Rethinking Learning in the Digital Age. Retrieved November 17, 2002, from http://www.cid.havard.edu/cr/pdf/gitrr2002_ch03.pdf.

Tardif, M. (2002). *Saberes Docentes e Formação Profissional*. Petrópolis: Editora Vozes.

# Acknowledge

We thanks to the volunteer teachers for their participations and the financial support from CNPq (CNPq/ProTeM-CC Proc. n. 680210/01-6 and n. 477645/2001-1).

# Mindshifts - An adventure journey into the land of learning

*Christine M. Merkel*

Head of Division for Culture and Communication/Information,
German Commission for UNESCO
Colmantstr. 15
D-53115 Bonn-Germany
merkel@unesco.de

## Abstract

The recent world-wide explosion of interest in online learning communities and modes of e-learning coincides with a deep crisis in the new economy. The emerging public debate about the necessary reconstruction of the IT market offers a window of opportunity. It can help to identify promising avenues to foster the right to read, to write and to communicate. The paper argues that the prefix 'e' is quite irrelevant: the heart of the matter is about learning. What do we know today about learning and what is not understood (yet) ? How can new technologies in e-learning and edutainment help develop a critical inquisitive human mind, the 'embodied mind' (Varela 1992) ? How do they contribute to the development of equitable knowledge societies ?

## 1 Getting to the airport

A hundred-fifty years ago, correspondence courses turned up in the education landscape of old Europe and North America, subsequently spreading as part of human capacity building in the age of empires. In the 1920ies, BBC broadcast its first radio classes. When the second World War hadn't even ended, South Africa pioneered its first open university , UNISA. Just one year later, Latin America went into distance education, establishing Radio Sutatenza by Acción Communitaria Popular, with educational broadcasting spread throughout most countries of the continent in a couple of years. The sixties of the last century saw the international emergence of educational TV, serving major student populations, teachers and parents (UNESCO 2002).

The coming into being of multimedia systems in the last decade coincided with the second wave of international policy talk about shifting towards a culture of life-long-learning. This reflects growing demands on the workforce and self-organised aspirations of civil society. The landscape of learning started changing impressively. School and university buildings, vocational training centres were complemented by network structures of self-study groups. Multimedia systems would usually combine text, language, audio, video, photography, and computer-based material with some face-to-face learner support delivered to both individuals and groups of learners.

This starting shift in the place of learning saw a quantum leap with the advent and establishment of internet-based learning systems. Roughly it can be said that those systems have been standing on their own over the last five years. At least three developments converged (Heydenrych 2002): the use of computer mediated communication and the Word Wide Web through Internet connectivity combined with the experience of open and distance learning. This produced the delivery of courses in radically new formats. Internet-based systems enable multiple interactions, as teacher-student and student-student, one-to-one, one-to-many and many-to-many interactions, place independence (any place) and time independence (synchronously or asynchronously) through e-mail and computer conferences or bulletin boards (computer mediated messaging). Multiple digitised multimedia resources can be tapped, along with access to databases and electronic libraries. If they use voice rather than text, they open new avenues even for illiterate people.

## 2    Checking in: 'newness' revisited

The spread of broadband internet communication is stimulating new types of networked learning communities as well as new types of educational organisations, just as each previous generation of distance learning technology. At the same time, re-thinking about the effectiveness of those older generations and their possibly new potential in the web-connected world has started on a world-wide scale.

The most important characteristics of the current and future stages may well be their potential for true media *integration*, making all forms of broadcast media and interpersonal communication accessible via a unique interface (Servaes,2002). At the same time, "the digital revolution has disrupted and will continue to disrupt what we mean by learning and how we organize our disciplines" (Burnett.,2002, p. 21).

While the new technologies have become ever so more powerful in their capacity to foster communication and facilitate learning communities, many learning offers have become ever more limited in scope, serving narrowed goals of knowledge transmission by enterprises and public bodies (Visser, 2002). How can this seeming erosion of depth and width of learning in times of increasing bandwidth be explained ?

## 3    Take off: delete the 'e' - it is all about learning

The heart of the matter is about learning. The prefix 'e' is rather irrelevant. Where principles for effective learning are respected, ICTs can enhance the learning process. Several principles for effective learning have been identified by studies in cognitive psychology and brain science (Caine & Caine, 1994; Haddad, 2003, p.5):
- learning engages the entire physiology
- learning is influenced and organized by emotions
- memory is organized both spatially and through a set of systems for rote learning
- humans possess a need to make sense of the environment
- the brain downshifts under perceived threats
- concepts are learned best when they arise in a variety of contexts
- learning to do well involves practice in doing
- effective learning requires feedback

## 4    Travel height: what we don't know about learning

Given the widespread policy discourse of emerging knowledge societies, the manifold e-learning initiatives and the recognised need for media literacy, the international programs for life-long as well as life-wide learning, one may get the impression that the job is already done: It seems sufficient to shift the paradigm from education and traditional delivery thinking to learner's communities, a knowledge infrastructure and a need based catering for learning needs. True, it is only quite recently that the close linkage between instruction and learning is starting to disappear (Visser,2001).

As important as those transformations are to overcome recognised deadlocks of educational systems developed in the age of the first modernity, they bear the risk of missing the point of the complexity of the world of learning ahead. Overlooking a necessary research agenda today may well created unexpected deadlocks for tomorrow's learning communities. It is in this spirit, that an international research community started organising around a web-enabled 'Book of problems' about what is not known about learning has taken shape over the last two years (Visser,2001).[1]

# 5 Landing: meeting the spirit of inquiry

In the constructivist learning experience, the learner has to construct an internal representation for knowledge, constantly open to change (Heydenrych, 2002). It is through reflective comparison and interaction with others that knowledge structures become solid over time. In line with this, (web-based) learning environments need to be open and problem-centred. This fosters an active, self-directed and explorative type of knowledge construction. Visser pleads to use the awesome technology capability to finding ways how to encourage people to develop mind, an spirit of inquiry and critical web-preparedness rather than to continue age-old patterns of knowledge transmission by some and knowledge acquisition by others(Visser,2002, p.47).

In the spirit of inquiry, it is crucial to link the work on new technologies for e-learning with the question of alternative energy provision and telephone-line independent technologies. In his breath-taking montage of satellite photos of the earth at night, Serres speaks about the new" City of Light" (Serres,1995, p.60). This 'City of Light' leaves large parts of the world in darkness, while denying the calm of darkness to those who are continuously exposed to electricity in all its forms. If one cares about all those worlds, the over-serviced as well as the countries in the shadow, the vision of surfing the net via satellite radio is both bold and feasible.[2]

## Where do you want to go today ?

New developments in e-learning and edutainment are interesting only to the degree that they help to midwife these critical scientific minds for the XXI century and nurture the spirit of responsible inquiry. After all, e-learning and open and distance education are "not about technology, not about information, not about the web and not about interaction with computers" (Lim,2001, 11-12). In a summary of UNESCO's input to the preparatory process to the World Summit in the

---

[1] ). Inspired by the history of mathematics in the first half of the 20th century ('The Scottish Book', 1957), Visser initiated an international club of scholars who are ready for continuous collaborative reflection about what one does not know and understand, in short, "to ask the right questions and to ask them the right way" (Visser,2001, p.1). This trans-disciplinary research community has since started to identify key problems in the development of the science of learning.

[2] The WorldSpace satellite-radio broadcaster will enable Internet data-downloads in India at 128 Kbps directly to a computer (t approx. 1 MB of data per second) without having even a telephone line (Noronha,2001). WorldSpace aims to deliver radio programming to three-quarters of the world populations that lacks today adequate radio reception and quality news, knowledge and entertainment at affordable cost.

Information Society (UNESCO 2003) so far, four principles are considered essential for the development of equitable knowledge societies:

Cultural diversity
Equal access to education
Universal access to information in the public domain
Freedom of expression

Enhancing information flows alone is not sufficient to grasp the opportunities for development that are offered by knowledge. Therefore, a more complex, holistic and comprehensive vision and a clearly developmental perspective are needed. The reality test for new technologies in e-learning and edutainment are their contributions to human needs, decent work and the conversation across cultures.

The willingness and audacity of scholars and technicians is invited to keep challenging each other at the frontier of what is known, looking out over the vast unknown. The distance we travel matters little – if the horizon is vast !

---

Any opinions expressed in this paper are entirely those of the author and do not necessarily reflect official policy of either UNESCO or the German Commission for UNESCO

## References

Barnett, R., (2002) Shifting the ground for our conversations. Contribution to The book of Problems . Retrieved April 4, 2003, from http://www.learndev.org/BOP-AECT2002.html

Beck, U. & Sommer, W. (Eds.) (2002). LearnTec2002. 10th European congress and trade fair for education and information technology. Proceedings of UNESCO activities organised in the framework of LEARNTEC 2002. Karlsruhe: KMK

Caine, G., Caine, R. N., & Crowell, S. (1994). Mindshifts: A brain-based process for restructuring schools and renewing education. Tuson: Zephyr Press

Dreyfus, H.,(2001). On the Internet: Thinking in Action. London/New York: :Routledge

Haddad, W. D., (2001). The challenge of access to and quality of distance learning. In: U.Beck & Sommer, W. (Eds.). (2001). LearnTec2001. 9th European congress and trade fair for education and information technology. (pp. 11-19). Karlsruhe: KMK

Haddad, W. D.,(2003). Is Instructional Technology a Must for Learning ? *TechKnowLogia, January-March 2003,* 5-6. Retrieved April 4, 2003 from http://www.TechKnowLogia.org

Heydenrych, J., (2002). Global change and the online learning community. *TechKnowLogia, April-June 2002*, 14-17. Retrieved April 4, 2003, from http.//www.TechKnowLogia.org

Lim, C. P.,(2001). What isn't E-learning ? *TechKnowLogia May-June 2001,* 11-12. Retrieved April 4, 2003, from http://www.TechKnowLogia.org

Merkel, C., (2002). Wege zur Lerngesellschaft. Wissen, Information und menschliche Entwicklung. In C. Wulf & Merkel, C. (Ed.), Globalisierung als Herausforderung der Erziehung (pp. 127 –176). Münster/New York: Waxmann

Noronha, F., (2001). Surf the net, via your satellite radio. Bytesforall_readers. 31.10.2001. Retrieved November 19, 2001, from http://www.bytesforall.org

Noronha, F. (2002). Computers....battling illiteracy. BytesForAll: Updates [Bytesforall_readers], April 2002. Retrieved April 10, 2002, from http://www.BytesForAll.org

Ouane, A. & Merkel, C., (2001) Global dialogue between UNESCO and World Bank at Expo2000. In U. Beck & Sommer, W. (Eds.). (2001) LearnTec2001. 9th European congress and trade fair for educational and information technology. Proceedings of UNESCO activities (pp. 21-26). Karlsruhe: KMK

Serres, M., (1995). Die Legende der Engel. Frankfurt a.M./Leipzig: Insel

Servaes, J., (2002) Constraints to 'learning' in the knowledge society. Contribution to The book of problems. Retrieved April 4, 2003, from http://www.learndev.org/BOP-AECT2002.html

The book of problems (or what we don't know about learning) (2002). Introduction – Abstract of AERA 2002 Alternative Session Proposal . Retrieved August 9, 2001, from http://www.learndev.org/BookOfProblems.html

The "Book of Problems" community of scholars – inputs into a collaborative dialogue (2002 onwards) Retrieved April 4, 2003, from http://www.learndev.org/BOP-AECT2002.html

UNESCO (2001). Distance education in the E-9 countries. The development and future of distance education programs in the nine high-population countries (Bangladesh, Brazil, China, Egypt, India, Indonesia, Mexico, Nigeria, Pakistan) Paris

UNESCO (2002). Open and distance learning. Trends, policy and strategy considerations. Paris

UNESCO(2003). UNESCO and the World Summit on the Information Society. Briefing document for permanent delegations. Paris, February 2003

Varela, F., (1992) The embodied mind: Cognitive Science and Human Experience. Massachusetts: MIT Press

Visser, J., (2002). Technology, learning, corruption and the development of mind. Retrieved April 4, 2003, from http://www.learndev.org

Wulf, C, & Merkel, C. (eds.) (2002). Globalisierung als Herausforderung der Erziehung. Theorien, Grundlagen, Fallstudien. European Studies in Education. Münster/New York: Waxmann

# Designing Appropriate Technology for Group Learning

*Ingrid Mulder and Janine Swaak*

Telematica Instituut
Enschede, The Netherlands
Ingrid.Mulder@telin.nl
Janine.Swaak@telin.nl

*Joseph Kessels*

University of Twente
Enschede, The Netherlands
kessels@edte.utwente.nl

## Abstract

Why do virtual team members learn and reflect less than co-located teams? One explanation is that current technology does not invite them to collaborate and does not stimulate them to reflect. In the current work, we attempt to design appropriate technology for group learning. Hereto, we view reflection as an important part of (group) learning. More specific, we expect that stimulating questioning behaviour results in more reflection, and thus in more learning. First, we report on a collaborative design workshop, in which we explored how we can design technology that stimulates people's questioning behaviour. Then, we describe a user pilot in which we evaluated initial prototypes yielded by the workshop. This study helped us to select a tool that supports questioning behaviour. Finally, we give details of the validation of the selected tool in a large-scale experimental study.

## 1    Introduction

Insights from empirical studies showed us that virtual team members hardly reflected and hardly asked each other questions (Mulder, Swaak & Kessels, 2002; Veerman, 2000). Also, we found that current technologies are designed to support collaboration, e.g., sharing documents, however, mostly they fail to support learning, e.g., understanding and evaluating on the documents (Agostini, De Michelis & Susani, 2000; Mulder & Slagter, 2002). If this is true, the importance of proper technology support for group learning increases. Therefore, our aim is to design technology in such a way that it not only enables people to collaborate, but even stimulates them to learn.

## 2    Design workshop

We organised a design workshop to explore technology that stimulates people to learn. The workshop's goal was to come up with three ideas for prototypes that could be evaluated in a user pilot. Whereas this goal has a technological annotation, we explicitly tried to design out of people's interaction. We invited 'designers' with diverged expertises and backgrounds; only one designer had an engineering background. Workshop participants started brainstorming on social and cognitive functions of questioning behaviour. To put it differently, they brainstormed on what kind of interaction is involved when group members ask each other questions. Note that by thinking out of the interaction, thinking in terms of features and technologies was avoided. This brainstorm on functions of questioning behaviour resulted in a list of identified functions, which were discussed plenary, and clustered. Main functions that designers came up with were social and cognitive functions such as indicate the need for questioning, structure the process of questioning, collect questions, and explicating a question. After this plenary discussion on what could and or should be supported, duos were grouped and worked on their prototypes. Figure 1 shows one of the ideas that were yielded at the end of the session, the system control timer. This system control timer indicates a moment for raising questions. A window with 'time for questions' appears in the middle of the videoconferencing screen. The system control timer asks then if there are any questions, and if there are more questions. A more advanced version can also indicate for whom (team or individual person) the question is. Other ideas include a 'question token' that structures questioning and controls one person speaking at a time, a 'question box' and a 'question bucket' that both collect questions, and a 'question mark button' that indicates the need for questions.

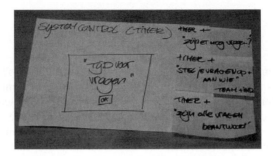

**Figure 1:** Example of a workshop result, i.e., the system control timer

## 2.1 Towards a prototype

Next to a plenary discussion at the end of the workshop, we asked the participants which of the final ideas they preferred. Individually, they indicated the prototype or function they liked most, and why. Based on the workshop results and the individual reflections, we concluded that we should support the following functions: indicating the need for questioning, and explicating a question. Other conditions, we had also taken into account were that the questioning tool should not interfere with the task of the virtual team, and that the tool would not seduce group members to chat. In sum, we came up with three ideas that are to be developed into prototypes: a question mark popup; a system timer; and a question mark popup extended with the formulation of the question. These prototypes are explained below.

### 2.1.1 Prototype 1: A question mark popup
The main motivation for the question mark popup is that it is really simple, and probable therefore less intrusive. The question mark popup structures a conversation; first a question should be answered, before another question can be raised. At the same time it stimulates related questions (see Figure 2).

**Figure 2:** Initial prototype: the question mark popup

If a person wants to raise a question, he has to press the button 'question', and a question mark appears on the screen of the remote group. The idea is that the question mark icon is enough to let the remote group realise that they have to listen to what kind of question the others have. When the group or person who raises the question feels satisfied with the answers, he or she presses the button 'answered'. The question mark disappears, and another person can raise a question. The tool prevents asking more questions at a time, while stimulating asking related questions to get the question more focused.

### 2.1.2 Prototype 2: the system timer

The second prototype has been based on the previous one, however the main difference is that the system indicates a moment for raising questions. Therefore, this prototype contributes to a culture on questioning. The system makes group members aware of questioning by explicit indicating a moment to raise questions, and thus stimulates group members to raise questions. Every x minutes the question mark prompts on the screen, and 'asks' if there are any questions

### 2.1.3 Prototype 3: the extended question mark popup

Also, the third prototype has been based on the first one, the question mark popup. The difference is that this one has been extended with the possibility to formulate the question more carefully. First, a person who likes to raise a question presses the 'question' button, which results in a question mark icon on the videoconferencing screen of the remote group. This icon alerts the remote group members that the others have a question. Below the question mark icon presence information of the questioner's activity is shown: 'other party is typing the question'. By showing the activity status we try to avoid situations in which a person formulates a question very carefully, and when this formulated question appears on the remote video screen after some while, the other group members react as follows: 'where have you been, we just discussed this topic!'

## 3    Pilot study

We employed a user pilot to get insight in these three distinct functions of questioning behaviour. Two teams, each of 6 persons that worked together in two subgroups, were invited to design collaboratively a university portal. Each team was divided at two different locations, and had videoconferencing and visualisation tools such as a shared whiteboard, pen and paper facilities at their disposal. Teams worked together for an hour and a half. After each half hour we interrupted shortly and installed another prototype on their videoconferencing system. The main aim of this pilot was to find evidence which function (and which prototype) was most important to support questioning behaviour while designing together by means of videoconferencing. To identify which prototype supported questioning behaviour in videoconferencing teams best, we observed questioning behaviour and the group process. Log-files were made to count how often the distinct question tools were used. At the end of the pilot we asked the respondents which tool they preferred and why. Moreover, we asked them to describe positive and negative aspects of the three tools they used in their design task. After they returned the questionnaire we evaluated the whole pilot plenary. The idea behind this plenary evaluation was to be able to ask more questions on specific topics, i.e., topics such as each of use, fun, and obtrusiveness they experienced.

### 3.1    Tool selection

The pilot participants preferred the question mark popup, as it was easy to use, less intrusive, and there was no need to type. However, they lacked feedback on whether a question mark was still at the others' screen. They also noticed that a visual alert was not enough, they preferred to have also a sound alert to get the remote party focused to the screen. Both the sound alert and the awareness of an awaiting question were added to the final questioning tool (Figure 3). Both sub-teams have the button 'question' next to the video screen. By clicking on this button one expresses the desire for questioning, and a red question mark appears on the video screen of the remote team. Only the sub-team that presses the button can remove the question mark by ticking 'we've got an answer'.

**Figure 3:** The final questioning tool

# 4 Experimental study

In an experimental study we investigated whether this questioning tool stimulated reflective behaviour and therefore stimulated learning and understanding. We compared twenty teams (N=20; n=110) that performed a complex design task (i.e., the same as in the pilot study); ten of these teams had next to audio and video support a questioning tool available. We tried to make the experimental setting as realistic as possible: students were working on a complex design task, in two sub-teams, using collaborative technology. The unrealistic part was that teams were not really geographically dispersed, but were working in two different rooms in the same building. The experimental conditions, in which we measure reflection and learning, are videoconferencing with and without the questioning tool. We used several instruments to collect data to assess shared understanding and learning in the design teams (for more analyses see Mulder, Graner, Swaak & Kessels, submitted). We describe the main results related to the validation of the tool. We measured the *perceived shared understanding* in teams. After each half hour team members indication their perception of shared understanding on a self-score rating scale. The numbers in Table 1 increase from T0 to T3; as expected the perception of shared understanding increased. According to our hypothesis we noticed that teams with the questioning tool had a better perception of shared understanding than teams without.

**Table 1:** Perception of shared understanding at start (T0) and after each half hour (T1, T2, T3) (mean and sd)

|  | T0: mean (sd) | T1: mean (sd) | T2: mean (sd) | T3: mean (sd) |
|---|---|---|---|---|
| Questioning tool + (N=10) | 3,48 (0,52) | 4,02 (0,33) | 4,37 (0,34) | 4,71 (0,36) |
| Questioning tool – (N=10) | 3,40 (0,59) | 3,98 (0,48) | 4,05 (0,42) | 4,39 (0,40) |

*Log-files* were made to monitor teams' tool usage. Table 2 shows how often the questioning tool has been used in the teams. Interestingly, the team that used the tool most (73 times) wrote that they experienced the questioning tool as a very nice way to get attention of their remote team members (*report of participants experiences*). A consequent use seemed to yield positive experiences. On the other hand, the team that used the tool 51 times indicated that they used the tool primarily for fun. To conclude, it is not straightforward how to interpret these frequencies.

**Table 2:** Tool usage in teams

Tool usage in teams: 20  12  16  19  53  18  73  51  28  19

# 5 Conclusions and discussion

In the current work, we designed appropriate technology for group learning. More specific, we organised a design workshop to design technology that stimulates people's questioning behaviour. This design workshop resulted in three initial prototypes, which were evaluated in a user pilot. Insights from this pilot lead to the development of a questioning tool, which has been validated in an experimental study. We hypothesed that teams with a questioning tool learn and understand each other better than teams without. We assessed the perception of shared understanding, and found that teams with a questioning tool had a better perception of shared understanding than teams without; which is in line with our hypothesis. Our informal observations indicated that social influences appeared to be stronger than technical ones. Professional behaviour, including facilitator behaviour, influenced virtual communication in a positive way. Also teams that knew each other beforehand indicated that they would have performed well with sub-optimal technology support. At the same time, teams without a tool that faced negative social influences such as dominance and not listening to others, indicated that they had no means to get into the remote conversation. Another interesting observation was that there seemed to be thresholds to use the tool, unless our main goal was to make the questioning tool as intuitive and simple as possible. Activating the questioning tool by touching the screen or pressing a button instead of clicking with the mouse seems to be a solution to make the tool more intuitive. Low thresholds are crucial, however, as we found that a consequent use yields positive experiences. In conclusion, there seemed to be evidence that the questioning tool enhanced perceived shared understanding. Moreover, there seemed to be more fun in teams with a questioning tool.

# References

Agostini, A., De Michelis, G., & Susani, M. (2000). From user participation to user seduction in the design of innovative user-centered systems. In R. Dieng & et al. (Eds.), Designing Cooperative Systems - The use of theories and models (pp. 225-240). Amsterdam: IOS Press.

Mulder, I., & Slagter, R. (2002). Collaborative Design, Collaborative Technology: Enhancing Virtual Team Collaboration. In: Callaos, N., Leng, T., Sanchez, B. (Eds.), *Proceedings of the 6th World Multi-Conference on Systemics, Cybernetics and Informatics (SCI2002)*, Vol. V, pp. 74-79, Orlando, Florida, July 14-18, 2002.

Mulder, I., Graner, M., Swaak, J., & Kessels, J. (submitted). Stimulating questioning behaviour in video-mediated design teams: an experimental study on learning and understanding. *Paper submitted to CSCL 2003 conference.*

Mulder, I., Swaak, J., & Kessels, J. (2002). Assessing group learning and shared understanding in technology-mediated interaction. *Educational Technology & Society, 5,* 35-47.

Veerman, A. (2000). *Computer-supported collaborative learning through argumentation.* Ph.D. Thesis, Universiteit Utrecht, Utrecht.

# Visual Knowledge Construction Algorithms for Supporting Learner-Instructor Interaction

*Shoichi Nakamura*

The University of Aizu
Ikki, Aizuwakamatsu,
Fukushima, Japan
d8021202@u-aizu.ac.jp

*Kazuhiko Sato*

Muroran Institute of
Technology, Mizumoto,
Muroran, Hokkaido, Japan
kazu@csse.muroran-it.ac.jp

*Youzou Miyadera*

Tokyo Gakugei University
4-1-1, Nukui kita,
Koganei, Tokyo, Japan
miyadera@u-gakugei.ac.jp

*Akio Koyama*

Yamagata University
4-3-16, Jonan, Yonezawa,
Yamagata, Japan
akoyama@yz.yamagata-u.ac.jp

*Zixue Cheng*

The University of Aizu
Ikki, Aizuwakamatsu,
Fukushima, Japan
z-cheng@u-aizu.ac.jp

## Abstract

This paper mainly describes algorithms for supporting instructors to grasp learning situation, which is quite important for effective guidance. Specifically, algorithms for constructing support knowledge based on visualization of learning histories are developed. Finally, the effectiveness of developed support algorithms has been shown by experiments.

## 1 Introduction

In recent decades, rapid development of network technologies has enabled people to perform several intelligent activities on the network. Particularly, learning and research activities on the Internet have become prosperous more and more (Brusilovsky, 1996). These popularisations of network environment and activities on it brought about remarkable changes in needs of learning. In other words, learners' demands are diversifying increasingly with pointing out of the problems in previous uniform education styles. Therefore, there are pressing needs for realization of flexible learning / education styles in which several important factors, e.g., learners' interests, purposes of learning, can be satisfied. To come true such a flexible learning style, novel characteristics should be supported. At first, learning targets should be decided unrestrictedly by learners. Next, learning materials should be selected freely by learners. Furthermore, learning with pliable speed according to each learner's situation should also be supported. In flexible learning style with these characteristics, much more difficulties take place than uniform styles. Specifically, huge numbers of learning materials are required to satisfy learners' various needs. Effective supports for selecting learning materials, which are very important but difficult tasks for learners, are also required. Moreover, learning situation becomes really complex compared to the uniform styles. On the other hands, proper guidance is still essential in flexible styles since unrestricted learning is one thing and self-learning is another thing. Therefore, it is quite difficult but important for instructors to

986

grasp learning situation successfully. Smooth communication of conditions in learning between instructors and learners is also important.

Many researches to support e-learning and distance education have been reported (Timothy, 2003). Some related researches, e.g. (Latchman, 1999), aim to support the communication between instructors and learners utilizing video or audio technologies. On the other hand, there are some approaches for supporting the grasping learning situations (Kanenishi, 2000) (Kuwabara, 2000) and for providing navigation to learners (Kayama, 2000) (Hasegawa, 2002). Although these approaches have some merits for conventional learning styles, they could not solve the problems enough in novel unrestricted learning styles mainly because support systems in these related researches could deal with only the exclusive or so limited learning materials. Some related researches also have another week point that special preparation or knowledge are required to use the systems. Moreover, almost no related researches have provided the synthetic supports.

These backgrounds arise the motivation to develop the synthetic supports to solve problems in novel unrestricted learning styles. Therefore, this research has aimed to develop the synthetic support environment that consists of three supports, 1) Supports for managing materials and conditions in learning, 2) Supports for selecting learning materials, and 3) Supports for grasping learning situations (Nakamura, 2002). This research supposes that web documents are utilized as learning material. Although some related researches adopted web documents as learning material, it was the only way in almost all cases that instructors produce all material by themselves and line up them in the order to be learned before the class. This kind of approach is quite difficult in unrestricted learning styles since learners find out learning targets and select materials based on their own interests, understanding state, knowledge and so on. Therefore, in this research, it is supposed that not only materials produced by instructors but also any web documents on the Internet are utilized. To realize these assumptions, at first, the novel methods for catching referring histories to web documents precisely, which are not at all inferior to the related methods, e.g. the use of proxy server, have been developed as the software. This method demands no special knowledge for instructors. Next, supports for managing learning materials and learning conditions have been developed. Navigation to support selecting learning materials has also been developed. But, it is still difficult for instructors to utilize raw learning histories to grasp learning situations.

Therefore, this paper mainly focuses on support algorithms for grasping learning situation based on visualization of learning information. These supports enable instructors to grasp learning situations properly and to make effective guidance based on it.

## 2    Supports for Grasping Learning Situations

In this research, learning histories (learner ID, URL of referred web document, material ID of referred web document, date of referring, and host name from which learner referring) can be caught skillfully with proposed methods even if a learner refers to any web documents on the Internet (Nakamura, 2002). However, it is quite difficult for instructors to utilize the raw histories for grasping learning situation. To solve this problem, algorithms for constructing visualized support knowledge that helps instructors to grasp learning situation are developed. In this section, the functions of support knowledge are described.

**Learners' distribution** expresses all learners' distribution to each learning material (Fig. 1). This knowledge is presented in bar graph style and is updated automatically when instructors request it. This knowledge enables instructors to grasp the learning situation that how many learners are trying to which material easily.

**Referring times** expresses each learner's referring times to a learning material (Fig. 2). This knowledge is also presented in bar graph style. This knowledge enables instructors to grasp the learning situations that how many times each learner referred to a material.

**Figure 1:** Learners' distribution

**Figure 3:** Personal learning history

**Figure 5:** Progress state

**Figure 2:** Referring times

**Figure 4:** Current learners

**Figure 6:** Neighbor sequence state

**Personal learning history** expresses each learner's detailed history (Fig. 3). This knowledge is presented in map style and is updated automatically. Each referred material on the Internet is expressed as a node with hyper link. This knowledge enables instructors to grasp each learner's detailed learning history; a set of materials he referred and their order. Instructors can also refer to any materials in a map actually just by clicking nodes.

**Current learners** expresses a list of learners who are referring to a material currently (Fig. 4). This knowledge enables instructors to grasp the learning situation that who are referring to a material. Instructors can utilize other detailed knowledge of each learner with learner ID in a list.

**Progress state** expresses the detailed progress situation of each learner in all materials (Fig. 5). This knowledge enables instructors to grasp the learning situation that each learner finished or not each material easily.

**Neighbor sequence state** expresses the detailed referring histories of learners, who are trying to a specified material, focusing on that material (Fig. 6). This knowledge enables instructors to grasp the sequences of the referred neighbor materials for learners who referred to specified learning materials easily.

**Previous neighbors rate** expresses the rate of previous referred neighbors to a specified learning material (Fig. 7). This knowledge enables instructors to grasp the situation that relatively many learners referred to which material before the specified one.

**Behind neighbors rate** expresses the rate of behind referred neighbors to a specified material (Fig. 8). This knowledge enables instructors to grasp the situation that relatively many learners referred to which material after the specified one.

**Similar state** expresses learners who referred to similar set of materials and their learner ID (Fig. 9). This knowledge enables instructors to grasp learners with similar learning situation and their learner ID easily by inputting a set of learning materials.

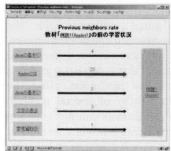

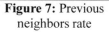

| **Figure 7:** Previous neighbors rate | **Figure 8:** Behind neighbors rate | **Figure 9:** Similar state |

Although support knowledge can help instructors' grasping learning situation, it is still difficult for them to select proper knowledge and to utilize it for actual guidance. Therefore, guidance navigation, which manages each instructor's guidance styles and assists their selecting support knowledge, has also been developed.

## 3    Experiments and Discussion

Three experiments with prototype system implemented mainly by Servlet and CGI ware performed to evaluate the effectiveness of the proposed supports. Support knowledge is available in experiment 1 and 2. Guidance navigation is available only in experiment 3. The rough explanation about this system was performed for examinees of instructors and learners side before the experiments. After the experiments, the questionnaire for experimental subject of instructor side and oral interview for learner side were performed.. Questionnaire consists of Q1 "Could you grasp learning situation successfully?" Q2 "Could you perform effective guidance?" Q3 "Could you select support knowledge easily?" Q4 "Did you feel difficult to use this system?" and Q5 "What kinds of supports are needed, do you think?" (Q3 and 4 are only for experiment 2 and 3) The answers for Q1 - Q4 were performed by putting grades from 1 (poor) to 5 (good). The answers for Q5 were performed by writing free comments.

The results of Q1 - Q4 in these experiments are shown in Table 1. The result of Q1 shows the effectiveness of visualized support knowledge for grasping learning situations. The relations between the results of Q1 and Q2 show that skillful grasping learning situations has advantageous effect on guidance. The result of Q3 shows the effectiveness of guidance navigation for selecting proper support knowledge. As a result, it can be said that supports provided by this system can support instructors' grasping learning situations and actual guidance successfully. Furthermore, there ware mainly three comments for Q5, "Supports to produce and improve viewpoints for guidance navigation are needed." "A help function is needed. " and "Several visualization styles of knowledge are required." On the other hands, three comments, (1) "It is seemed that guidance in experiment 3 is more carefully than the case in experiment 1 and 2." (2)"Learner-oriented education style in which learners can select learning materials freely is important and welcome." and (3) "The function to support communication among learners is needed." were given by experimental subjects of learner side as answers for an interview. The comments (1) also show the effectiveness of this system for instructors' grasping learning situations. Furthermore, the comment (2) shows the importance of the learning style proposed in this research. It is seemed that the comment (3) is an important hint for future improvement.

Consequently, the effectiveness of support knowledge for instructors' grasping learning situations is proved. The effectiveness of guidance navigation for instructors' selecting proper

**Table 1:** Results of questionnaire

| | Average in Experiment 1 | Average in Experiment 2 | Average in Experiment 3 |
|---|---|---|---|
| Q1 | 2.0 | 4.2 | 5.0 |
| Q2 | 2.0 | 3.5 | 4.2 |
| Q3 | - | 2.7 | 4.0 |
| Q4 | - | 3.5 | 4.5 |

support knowledge based on their guidance styles and actual guidance with the knowledge is also certificated. Furthermore, the results of experiments show the important fact that precise grasping learning situations has good effect on guidance. The worthy of special mention is that proposed system can support the novel learning styles in which learners can discover and select learning contents and materials based on their interests, understanding states and so on, that has not been supported by other researches.

# 4    Conclusion

Recently, rapid developments of network environment have made learning activities on it adopting web documents as learning materials prosperous. This change has brought about diversification of needs in learning. Therefore, development of novel learning styles, which can satisfy various needs based on each learner's interests, knowledge and so on, is strongly demanded. Therefore, novel unrestricted learning style in which learners can select and discover learning targets and materials freely was proposed. In such a learning style, it becomes quite difficult for instructors to grasp learning situation. To solve this problem, algorithms for constructing knowledge to support grasping learning situations based on visualization of learning information were developed. Guidance navigation, which manages each instructor's guidance styles and assists effective guidance based on the styles, were also developed. These supports have enabled instructors to grasp learning situations skillfully and perform effective guidance based on it. They have also enabled instructors to perform flexible guidance based on their guidance styles.

   Experiments with developed system have shown the effectiveness of these supports. As a future works, further experiments in actual class should be performed to improve support algorithms.

## References

Brusilovsky, P. (1996). Methods and Techniques of Adaptive hypermedia. *User Modeling and User-Adapted Interaction.* 6, 87-129.

Hasegawa, S., et al. (2002). Reorganizing Learning Resources on WWW and Its Application to an Adaptive Recommendation for Navigational Support, *Trans IEICE*, J83-D-1 (6), 671-681.

Kanenishi, K., et al. (2000). LOGEMON: The Teacher Support System in Class Used Web-Based Materials -Visualization of Learning History for Teacher Aid. *Trans. IEICE*, J83-D-I (6), 658-670.

Kayama, M., et al. (2000). A Navigation System Based on Self Organizing Feature Map for Exploratory Learning in Hyperspace, *Trans. IEICE*, J83-D-I (6), 561-568.

Kuwabara, T., et al. (2000). Support Functions for Stalled Students and Their Effect in a Multi-Media Assisted Education System with Individual Advance (MESIA). *Trans IEICE*, J83-D-I (9), 1013-1024.

Latchman, H.A., et al. (1999). Information technology enhanced learning in distance and conventional education. *IEEE Trans. on Education*, 42 (4), 247-254.

Nakamura, S., et al. (2002). A Support System for Teacher-Learner Interaction in Learner-oriented Education, *IPSJ Journal*, 43 (2), 671-682.

Timothy K.S., et al. (2003). A Survey of Distance Education Challenges and Technologies, *International Journal of Distance Education Technologies*, 1 (1), 1-21.

# Bringing History Online:
# Plimoth Plantation's Online Learning Center

*Lisa Neal*

EDS and eLearn Magazine
3 Valley Road, Lexington, MA 02421 USA
lisa@acm.org

## Abstract

Plimoth Plantation, a recreated 1627 village located in Plymouth, Massachusetts, sought to expand its online presence and to provide online standards-based education for children and teachers. Plimoth Plantation's Online Learning Center (OLC) uses multimedia in an interactive environment to teach about the lives of the Wampanoag Indians and the early Colonists who settled in the US. The OLC challenges children to become historians and teaches them what an historian is, how primary source and oral histories are used, and how historical events are interpreted and reinterpreted. The OLC focuses on the first harvest feast, and through exploration and activities, children learn what is known and guessed about this celebration and how it evolved into the modern Thanksgiving celebration.

## 1  Introduction

Museums have been well established for centuries as physical places that people go to on planned and purposeful visits. Museums are increasingly leveraging online technologies to help people plan a visit, view collections, or take a virtual tour (Tedeschi, 2003). The unlimited virtual gallery space has the potential of allowing museums to show stored collections. Museums are also helping educators and children to better prepare for a visit, so that the visit is the culmination of a thematic study (Carliner, 2001). The online presence of museums is important for the many people who are interested in the collections but are unable to visit. But beyond the actual collections, museums house a wealth of expertise and understanding that is used to plan exhibits, write publications, and guide tours. As standards-based education is becoming more common, museums are bringing their considerable expertise to helping teachers develop curriculum and provide online educational experiences for their students. According to the Institute of Museum and Library Services, 71% of US museums work with curriculum specialists to tailor educational programming to support school curriculum standards (Maxwell and Bittner, 2003).

## 2    Plimoth Plantation and the Online Learning Center

Plimoth Plantation is a "living history museum" that provides an immersive experience where visitors walk through a recreated 1627 village and meet and talk with "interpreters" who are dressed in period costumes, speak in dialect, and talk about their lives. The challenge was how to bring the richest educational parts of the visitor experience online to educate children and teachers about the lives of the early Colonists and the Wampanoag Indians.

The Online Learning Center (OLC), entitled "You Are the Historian: Investigating the First Thanksgiving", is designed to meet the social studies standards for elementary school children in the Commonwealth of Massachusetts and eventually the US. The goal of the OLC is to educate children and teachers about the origins and evolution of the Thanksgiving holiday using new research taken from the multiple perspectives of English and Native American sources. The OLC enables teachers and students to see the abiding impact of 17th-century events on modern culture, as well as develop an understanding of what an historian is.

## 3    Design of the Online Learning Center

The primary challenges in designing the OLC were how to capture the richness of the physical Plimoth Plantation experience online and how to provide online activities that were both engaging and educational. Many museum web sites depict collections, and the few that venture to provide a more experiential visit do so through a virtual tour. The museums that provide online activities, as well as other educational sites, often have trouble finding a balance between fun and education, so that children focus more on the gaming and little learning results.

### 3.1    "You Are the Historian"

A visitor to the OLC is challenged to become an historian and to develop an understanding of people and events, rather than just read text or perform isolated activities. This perspective breaks down the distinction between the producer and the consumer of the media, thus personalizing each child's experience (Miyagawa, 2002). While children learn about historical events, they also learn what a historian is, how primary source and oral histories are used, and how historical events are interpreted and reinterpreted over time. They even learn that Plimoth Plantation has a different understanding of the events of 1621 now than they did only a few years ago.

## 3.2 Use of multimedia

The hallmark of the site is its visual and auditory richness, incorporating graphics, audio, and video. The OLC in part replicates the immersive experience of Plimoth Plantation by using people as characters, experts, and guides where the visual impact of the people or the information they provide aids learning and understanding. The use of people also makes the pages more engaging and authentic. The home page introduces two children, a Wampanoag boy and Colonial girl, who act as guides and interpreters. These children offer 1621 and modern perspectives, dressed in appropriate attire as they voice their views. In addition to the children, Plimoth Plantation's expertise is embedded in the site, offering optional in depth auditory, video, and text explanations (see Figure 1). Figure 2 shows another example of a page. It provides information about the Wampanoag tradition of giving thanks and uses audio and video to engage children and add realism, since what they are learning about is not just historical information, but perspectives and practices that have endured for centuries.

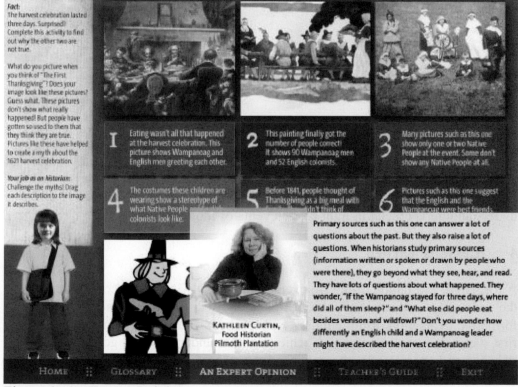

Figure 1: An Expert Opinion

A circle!

Wampanoag means "People of the First Light." We, the Wampanoag, have lived here on our homeland for over 12,000 years.

Today, we live in modern houses and have jobs in mainstream life. My People continue to give thanks for each season as we always have. We still practice our ceremonies. We also hunt and fish, and we tell our children the oral history of our People.

In 1621, Massasoit and 90 men attended the harvest celebration with the colonists.

Your job as a historian: See what life was like for my People around 1621. How did we give thanks? Click on each stone in the circle to find out..

Before planting in the spring, my family gives thankful prayers to Mother Earth who grows all our food. My daughter and I dig holes and some years use herring fish for fertilizer. We cover the fish with soft earth and make mounds. After a few weeks, we plant four or five corn kernels in each mound. Later, beans and squashes are planted alongside the corn.

| HOME | GLOSSARY | AN EXPERT OPINION | TEACHER'S GUIDE | EXIT |

Figure 2: Wampanoag traditions of giving thanks

## 3.3 Understanding history and culture

While it is important that children learn facts, it is perhaps more important that they develop an understanding of how historians learn about and interpret history. In addition, they learn about how two cultures met and lived together, and this understanding is crucial in light of the events of modern times.

## 4 Lessons learned

A formative evaluation of the first version of the OLC, done in collaboration with the Harvard Graduate School of Education, led to significant redesign. The formative evaluation, using children and teachers, found that many aspects of the first version were effective at achieving the understanding goals. The visual richness and the use of audio enhanced the appeal of the site. However, we also found out that children and teachers wanted more activities, more audio, and a stronger role for the children as guides. Children who took part in the formative evaluation identified with the guides and were curious to find out more about them. They liked the juxtaposition of the 1621 and modern children. The subsequent redesign incorporated this data, enhancing the existing features that were especially appealing to children. The final version is currently being tested.

Through good design, online learning can incorporate fun to meet learning objectives for children while increasing motivation and engagement. We struggled with how to add the additional activities children requested. Some of their suggestions were skewed toward increasing interactivity rather than learning. For instance, one child wanted "a runaway turkey game...you have to press forward and backward to run away from the bullets." Another requested "a game that teaches you to sail a ship or a game where you guide the ship through obstacles." While these and the other ideas were loosely related to Plimoth Plantation or Thanksgiving, they did not further the OLC's education goals. The final version added new activities that bring in the desired interactivity but do so in keeping with the theme and understanding goals of the site. Other enhancements, to further engage children, include increasing the sense that the child is investigating a mystery and drawing children in using an intriguing question. The final testing and the subsequent use of the OLC will allow us to see if we were successfully able to bridge the gap between learning and fun, while achieving our educational objectives.

## References

Carliner, S. 2001. Reflections on Learning in Museums: The Inspirational Power of Museums, http://saulcarliner.home.att.net/museums/inspirearticle.htm.
Kearsley, G. and Shneiderman, B(1999) Engagement Theory, http://home.sprynet.com/~gkearsley/engage.htm.
Maxwell, E. and Bittner, M. 2003. Museums Spend Over a Billion, Commit Over 18 Million Hours To K-12 Education Programs: Study Finds 72 Percent Use Web Sites To Teach, Institute of Museum and Library Services, http://www.imls.gov/whatsnew/current/012903.htm.
Miyagawa, S. 2002. Personal Media and the Human Community, Technos Quarterly Summer 2002 Vol. 11 No. 2, http://www.technos.net/tq_11/2miyagawa.htm.
Tedeschi, B. 2003. CYBER SCOUT; Pixels At an Exhibition. NY Times, March 16, 2003, http://query.nytimes.com/search/article-page.html?res=9D0CE4DD143EF935A25750C0A9659C8B63.

# User-Centered Design of Workflows in E-Learning

*Genésio Gomes da Cruz Neto*

*Alex Sandro Gomes*

FIR- Faculdade Integrada do Recife
R. Abdias de Carvalho, 1678,
Recife/Brasil
genesio@fir.br

Centre of Informatics - UFPE
R. Prof. Prof. Luiz Freire, s/n
Recife/Brasil
asg@cin.ufpe.br

## Abstract

In this article, we describe an user-centered design process for an educational workflow in a virtual environment for the teaching of subjects related to Computer Science. We adopted a socio-cultural theoretical framework to model user activities for the learning based in projects. Our results point to a series of user requirements obtained from experimental design, qualitative analysis, and modelling techniques based on activity theory.

## 1 Introduction

The socio-cultural learning approach (Vygotsky,1998) suggests that the learner must have the initiative to question, discover, and understand the world through his interaction with the other elements of the historical context of which he is part. Having this in mind, we singled out the use of technology in cooperative learning based in projects (Honebein, Duffy & Fishman 1993), whose objective is not only to incorporate up-to-date access to information but, mainly, to promote a new learning culture through the creation of environments that foster the production and exchange of knowledge.

We believe the most natural way to promote the teaching based in projects in a web-based learning environment is to use workflow technology as its foundation (Georgakopolous, Honick & Sheth, 1995). However, the majority of Virtual Learning Environments (VLE) do not offer management and automation tools for educational workflows. Special instances are VLE with Workflow Flex-eL ([www.flex-el.com]) (Lin, Ho, Sadiq & Orlowska, 2001), and VLE based on Zebu Projects (Tiessen & Ward, 1999).

Our problem is to identify and incorporate the user needs and their practices to guide the process of developing an educational workflow system to be used in the context of computer science higher education. This article is organized in the following way: the second section describes the teaching and learning environment. Then, the process of generating requirements through user-centered elicitation is described. Finally, some results, and implications to the system design are presented.

## 2 An Educational Workflow

Our proposal aims at creating a virtual project-based learning environment, whose cooperation process is promoted by the integration of communication functionalities with an educational workflow. This environment, named Amadeus (Tedesco & Gomes, 2002), is a generic framework offering adaptative teaching tools, called "user software components" (USC), centered in group work and monitored by intelligent agents (Weiss, 1999). There are communication mechanisms between the project management module, the evaluation module, the teaching-learning process, and the multi-agent system.

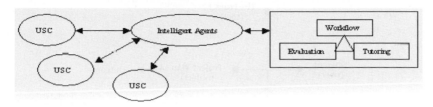

Figure 1: Amadeus

# 3 The User-Centered Design Process

The user-centered requirement analysis employed in the construction of the environment is divided in stages as shown below:

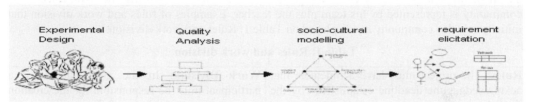

Figure 2: Requirement process

**Experimental Design**

We performed an ethnography study of the practice of the students and software engineering teachers of an undergraduate computer science course. We adopted a prototyping approach advocated by (Bødker, 2000), where is attempted to build an estimate scenario for the final product using a similar system as an artifact. In our context, we used the yahoogroups environment [www.yahoogrupos.com.br], Microsoft Messager, and MS-Project. The students are analyzed via taped observations and interviews.

**Qualitative Analysis**

After the initial data gathering, we applied a qualitative research paradigm (Denzin & Lincoln, 1998). The information is processed by a computer (Rourke, Anderson, Garrison & Archer, 2001) to obtain clues for the structural characteristics of the activities. The outcome of this process is a categorization of the hierarchically collected data.

**Socio-Cultural Modelling**

We adopted activity theory (Leont'ev, 1978) as a model for social analysis because of its high degree of abstraction. According to this theory, an activity is the way a subject acts directed to an object in order to achieve an outcome. The mutual relationship between the subject and his object is always mediated through one or more tools. The existing systemic relationships between the subject and his environment is represented by the concepts of community, rules, and work division (Martins & Daltroni, 1999). See (Figure 3: Systemic model ) for an example of an activity modelled according to a socio-cultural approach.

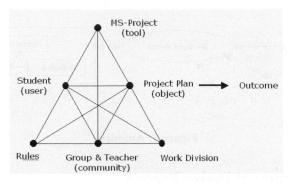

**Figure 3: Systemic model**

This activity involves a negotiation of responsabilities during the creation of the workflow. The action of the student is directed to his project plan and measured using the MS-Project. The community is represented by his team plus the teacher. Examples of rules and work division that initially rule this community are displayed on Table 1: Rules and work division.

**Table 1: Rules and work division**

| **Rules**: The project must have a timetable that acknowledges the deadline established by the discipline. | **Work Division:** In the group activities, a participant must be responsible for the creation and management of the workflow. |
| --- | --- |

**Requirement Elicitation**

Researches demonstrate that it is easier to use qualitative information to describe the activities when they are connected to the basic tasks described in the use cases (Kujala, Kauppinen & Recola, 2001). Therefore, the tasks performed by the system are identified first and then associated to tables of user needs and use cases. (seeFigure 4: Requirement elicitation)

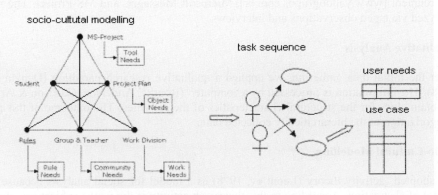

**Figure 4: Requirement elicitation**

In the table of needs (see Table 2: User needs related to activity dimensins ) we imbed the requirements associated to each one of the activity dimensions. We identify, thus, users' needs related to the object (*object needs*), the tools (*tool needs*), the rules (*rules needs*), the community (*community needs*), and work division (*work needs*). From the analysis of the diverse needs we can list the requirements of the user and of the context. Below, there is a table of needs generated from the activity of responsability negotiation.

**Table 2: User needs related to activity dimensins**

| Task | Tool Needs | Rule Needs | Community Needs | Work Needs |
|------|-----------|-----------|-----------------|------------|
| Project creation | There must be a guide for the creation of projects | The time schedule for the project must be respected. | The projects must be saved as an html file in the group environment | Only the project manager must have authority to make changes in the workflow |

To each task, use cases are generated from the respective table of needs. The description of each use case follows the usual pattern: with information about the task, the actors, the preconditions for the task to take place, the exceptions, and the post-conditions. A great number of activities can be identified from qualitative studies performed through the observation of a group work. Each activity modelled can originate an unspecified amount of tasks to which tables of needs and tables of use cases are associated.

# 4 Implications to Software Development

The process of requirement analysis presented above, resulted in the specification of a series of activity diagrams, tables of users' needs, and use cases documents. From this process we could produce requirements for the development of the system, such as:

- **Negotiation and Follow up**: a great care should be dispensed to the process of negotiation of responsabilities, and to the following up of activities. In many situations, only part of the team collaborate, leaving some individuals idle. The requirements were generated in a way that the intelligent agents follow up the activities of each individual.
- **Conflicts**: Conflicts between the students usually take place. Requirements were identified in such a way as to give us a tutorial of how to solve conflicts in the project and mechanisms to activate quickly the teacher for mediation. .

# 5 Conclusions

In this article, we present an user-centered requirement analysis process of an educational workflow system to be used in Computer Science higher education. Methods of qualitative analysis and techniques based on the activity theory were used to model the practice of Software Engineering students and teachers of an undergraduate Computer Science course. The process of analysis described worked satisfactorily for the generation of the workflow system requirements.

The tutoring system, a previous version of the AMADEUS system, was developed using only qualitative analysis techniques starting from interviews with distance-course teachers. Our research shows that the process presented in this article resulted in a definition of requirements more intuitive to the developers and analysts, and also produces information about the context of activities.

# 6 Acknowledgements

We wish to Express our gratitude to the FIR for the encouragement, the students that volunteered to help in the research for their invaluable help, and to the CNPq for the financial support (CNPq/ProTeM-CC Proc. n. 680210/01-6 e n. 477645/2001-1).

# References

Bodker, S. (2000) Scenarios in user-centred design-setting the stage for reflection and action. *Interacting with Computers* 13 61-75.

Bustard, D. W. and Wilkie, F.G. (2000) Linking soft systems and use-case through scenarios. *Interacting with Computers* 13(2000) 97-110.

Denzin, N.K. & Lincoln, Y.S (1998) Strategies of qualitative inquir. *Thousand Oaks* CA:SAGE Publications.

Georgakopolous,D. ,Honick,M. and Sheth,A. (1995). An Overview of Workflow Management: From Pocess Modeling to Workflow Automation Infrastructure. *Distributed and Parallel Databases*, 3, 119-153.

Honebein, P., Duffy, T.M. and Fishman,B.J. (1993) Constructivism and the design of learning environments: Context and authentic activities for learning. In T. M. Duffy, J. Lowyck & D. H. Jonassen (Eds). *Designing environments for constructive learning*, Vol 105, pp 87-108. Berlim. Springer-Verlag. 1993.

Kujala, S., Kauppinen, M. and Rekola, S. (2001) Bridging the Gap between User Needs and User Requirements. In Avouris, N. and Fakotakis, N. (Eds.) *Advances in Human-Computer Interaction* I (Proceedings of the Panhellenic Conference with International Participation in Human-Computer Interaction PC-HCI 2001), Typorama Publications, pp. 45-50.

Leont'ev, A. N. (1978). Activity, Consciousness, and Personalit. Prentice-Hall

Lin, J. , Ho, C., Sadiq, W. and Orlowska, M. E. (2001) On Workflow Enabled e-Learning Services. In Proceedings of the *IEEE International Conference on Advanced Learning Technologies*, ICELT'2001, 6-8, Madison, USA.

Martins, L. E. G. e Daltrini, B. M (1999) .An Approach to Software Requirements Elicitation Using the Precepts from Activity Theory. *ASE'1999* 15-23

Rourke, L., Anderson, T., Garrison, D. R. and Archer, W. (2001) Methodological Issues in the Content Analysis of Computer Conference Transcripts. *International Journal of Artificial Intelligence in Education* (2001), 12,8-22.

Tedesco, P. and Gomes, Alex Sandro (2002). Amadeus: a Framework to Support Multi-Dimensional Learner Evaluation. *ICWL'2002*. China.

Tiessen, E. L. & Ward., D. R.(1999) Developing a Technology of Use for Collaborative Project-Based Learning. In *Proceedings of the Computer Support for Collaborative Learning (CSCL) 1999 Conference*, C. Hoadley & J. Roschelle (Eds.) Dec. 12-15, Stanford University, Palo Alto, California. Mahwah, NJ: Lawrence Erlbaum Associates.

Vygotsky S. L.(1998), A formação social da mente: o desenvolvimento dos processos psicológicos superiores. 6ª ed. São Paulo: Martins Fontes, 1998.

Weiss, G. (1999) Multiagents Systems - A modern Approach to Distributed Artificial Intelligence. MIT Press.

# A case-study application of tour & time travel metaphors to structure an e-learning software

*W.L. Ng and K. Y. Lim*

Centre for Human Factors & Ergonomics,
Nanyang Technological University, School of MPE,
50, Nanyang Avenue, Singapore 639798
Email:  mkylim@ntu.edu.sg, weilieh@pacific.net.sg

## Abstract
This paper describes the use of metaphors to enhance the effectiveness of an e-learning software. Specifically, tour and time travel metaphors are applied to empower learners to creatively organise their navigation of the software, and to facilitate the exploration and assimilation of knowledge. By offering learners some customisation in the form of implicitly structured interaction paths that may be selected, learning may thus be made more engaging, interesting and exciting. A case study involving a history and current affairs subject taught in Singapore schools, is used to exemplify the application of these metaphors.

## 1      Nature and application of metaphors

With the advent of the Internet, there has been a tremendous surge in computing power available to the common household.  In support of this development is better software usability, for example, the shift from command line interfaces to graphical user interfaces (GUI). For the latter, human computer interaction is made even more intuitive by the use of metaphors.  An excellent example is the 'desk-top' metaphor that is pervasive in all Windows and Macintosh operating systems.  By porting over to the computer system domain objects commonly found in the physical office (such as files, folders and wastebasket), the 'desk-top' metaphor exploits existing knowledge of users to help them access and interpret a new application context (Smith *et al*, 1982; Neale and Carroll, 1997). In software, metaphors work by providing the user with a means of mapping system functions (which the user may be unfamiliar) with metaphor functions (which the user is likely to be familiar), to help them navigate and perform desired tasks. According to Hammond and Allinson (1987), this process comprises two steps, namely a mapping abstraction (where the user attempts to understand information about the system that is conveyed by the metaphor), and a mapping invocation (where the user calls upon the metaphor to interpret and achieve specific goals).  During mapping abstraction, known primary entities are matched. The abstractions are later matched with secondary entities during mapping invocation. The same solution scheme has been applied to other metaphors that have found their way into the computer, e.g. rooms, windows, menus, spreadsheet-ledger sheet, word-processing-typewriter, etc. The efficacy of using metaphors to support software navigation is thus well established. Bearing this background in mind, two metaphors of interest may now be introduced.

## 2      Conceptualising tour and time travel metaphors

In many software applications, the 'tour' metaphor has been used to introduce new features of the software. Typically, a 'wizard' would appear after software installation, to take and guide the user on a 'tour' of new and noteworthy features of the software. However, if the users are familiar with the software, they can opt instead for a 'free and easy itinerary' that allows them the freedom to visit only software features of interest. Finally, all users may return to 'Home', a main menu where they can commence using the software. We can clearly see the application of the two step mapping process here.  When the 'wizard' appears to offer to take user on a 'tour', the mapping of primary entities will allow the user to immediately recognize that perhaps, he/she will be taken on a predetermined sequence of features. This is because the knowledge of what it means to go on a tour is now practically universal, and thus the exploitation of this metaphor is only logical or second nature. The mapping of secondary entities occurs during mapping invocation, when the user identifies a lack of knowledge in the system domain to perform a task and thus draws upon metaphorical parallels provided by the 'wizard' to assemble a solution. For example, when the user wishes to replay a particular segment, he/she 'asks the tour guide again'. The 'wizard' then proceeds to repeat the segment of interest. In the case of education software, the user interface could also be designed to map the 'digital tour' concept directly onto a physical tour. For instance, this could be done for software documentaries targeted at specific countries or historical events and sites. In summary, a 'tour' metaphor can thus contribute to both software and knowledge navigation.

Time travel is a fictitious concept popularised by modern fiction. Time travel purports to allow people to 'warp' back to a specific point in history. Its use by Hollywood and the silver screen (e.g. Back to the Future, Time Machine) is so pervasive that it is fast becoming well understood and independent of a user's culture and age. However, unlike the 'tour' metaphor, the 'time travel' metaphor has not been as widely explored, let alone used. Perhaps the closest example of its use might be the historical trace of Internet sites and pages visited in the Uniform Resource Locator (URL) bar of web browsers. Alternatively, one could cite time sensitive features found in some software such as the repeated 'undo' function, 'last saved' function and more complex macros for recording an interaction sequence for repeated activation. In these cases, however, the trace is characterised more by sequence rather than by time. With reference to the two-step mapping process proposed by Hammond and Allinson (1987), the mapping abstraction of 'time travel' invokes the notion that a person could either follow a time line from the past to the present and to the future, or 'jump' back and forth to specific time periods. At an abstract level, mapping invocation may be applicable to tasks and events for which time is a dominant dimension. An example of such an application may be the "identikit" software with subject ageing capabilities used by some police forces. Another example may be version control in multi-authoring applications, where the main user may remember that it is only after lunch yesterday that his/her colleague made specific changes to a particular document. In this case, both the identity of the user and the time of the event are key determinants of the version of a document. The main user may then instruct the software to revert back to the document that is saved before lunch yesterday. The same would apply if the main user remembers a specific event that is unique rather than a specific point in time. This concept may be considered analogous to the modelling of a user's memory for organising personal files as suggested by Jones (1986).

By exploiting dominant features of interactive movies and 'tour' and 'time travel' metaphors, one can design user interaction with education software more creatively, to provide learners with an engaging learning experience involving entities, events and accounts encountered through space and time navigation under their control. With the addition of scenario manipulation and analysis functions within the software (to enable exploration of various outcomes through scenario or event manipulation, followed by projection through time), one would even be able to offer learners the ability to navigate context. An example of such an interaction design may include role playing individually or with peers, linked up digitally on a stage set in a particular location and at a specific point in time. In this way, the interactive power of computers may be exploited to the full.

## 3        A case study for metaphor exemplification

To exemplify the application of these metaphors and ideas, a project concerning e-learning is selected to instantiate the application of these metaphors and ideas. e-learning is targeted because through the SingaporeOne initiative, Singapore is fully connected with fibre optics for broadband communication of digital information nation wide. Further, e-learning would provide an interesting case to examine the dynamics and effectiveness of peer and group-based learning mediated digitally beyond the classroom. The subject matter targeted by this case-study is another recent national initiative named National Education (NE). Initiated by the Ministry of Education in May 1997, the objective of NE is to create a common culture that will give Singaporeans a shared perception of the future ahead, and to draw them closer together as one people if confronted with serious problems. This was to be achieved by developing in them an awareness of facts, circumstances and opportunities facing Singapore, and a sense of emotional belonging and commitment so that they will be able to make decisions for their future with conviction. Although NE involves national history, its delivery has to be very different for its objectives to be met. It has to expose students to a rational account of historical events that underlie and explain present day situations. It also may be conceptualised to comprise an analytical exposition of national events and current affairs. This is very different from the teaching of history, where there is a heavy emphasis on facts and events. Specifically, the goal for NE demands deeper learning and internalisation, in that students are required to understand how past events may have shaped present day situations, and to think for themselves the future consequences of various present day decisions. In summary, the objectives of the project described in this paper, may thus be conceptualised as follows:

- To exemplify the 'tour' and 'time travel' metaphors and investigate their effectiveness for enhancing the ease of navigation and the organisation of subject content.
- To develop a prototype e-learning solution for NE that applies basic pedagogical principles in its design.

The effectiveness of the proposed solution will be tested with learners.

## 4    Requirements and conception of an e-learning software

The advantages for delivering the NE content via the SingaporeOne national digital information infrastructure are obvious, namely:

- Learners will have access to NE content when and where they want it, as they will no longer be limited to the classroom. This freedom to choose when and what content to access may encourage them to view NE with a more positive attitude, as opposed to the view of NE as an additional lesson taught during class.
- Learners can enjoy 'personalised' education and learn at their own pace by choosing to replay particular segments of a lesson or review concepts that they are unclear about (Stantoni et al 2001).
- By using rich media, NE content may be delivered to learners in an interesting manner with a pedagogically sound and meaningful use of text, graphics, narration, animation and video. The multi-media content may be augmented further with a high level of interactivity, in the form of challenging quizzes to motivate students to learn.
- The e-learning software can exploit the broadband infrastructure to incorporate email and other network functions to enable students to interact with their teachers and peers. Thus, an active exchange of information and opinions may be promoted, as opposed to a 'traditional' style of learning characterized by the passive reception of information.

To take advantage of the above benefits, the basic requirements for an e-learning software should include the following:

- A transparent software navigation and engaging user interface design developed using a user centered approach. The need to ensure transparent navigation is pressing in education software because inter-related visual, textual and auditory information have to be packaged in a meaningful manner to support comprehension and retention (Mousavi et al, 1995). As the NE e-learning software has a large amount of information, the 'tour' metaphor may be used as a framework to organise the presentation of subject content. In particular, the 'tour' metaphor can be used to structure and present thematic information in the e-learning software (e.g. sites of massacres in the Second World War, ethnic enclaves during colonnial times) according to the domain implied by the metaphor. For instance, the user interface may be designed to present various tour buses corresponding to thematic tours for the learner to select. During the 'learning journey', the learner may be shown either a geographical or concept 'map' updating the current location on a physical route or navigation path as appropriate. Further, if a learner so chooses, he/she may hop onto another 'tour bus' at specific 'bus interchanges' to join another thematic tour. Thus, by capitalizing on a learner's recognition of primary entities in the physical tour domain (tour buses and interchanges) as opposed to knowledge of secondary entities (e.g. mouse clicks and menus of historic events), a metaphor based user interface design can make the software navigation more transparent. Further, by configuring software navigation in this manner, the learners may be encouraged to explore whatever captures his/her attention during a 'learning journey'. Thus, the 'tour' metaphor may be used as a framework to organise user computer interaction and the presentation of subject content.

Similarly, the 'time travel' metaphor may be used to enable learners to trace events along a time or sequence line, e.g. the path of the Japanese invasion of Singapore or the growth of Singapore as a harbour. In addition, for learners at higher level education, the software may support exploration or postulation of particular scenarios. This software function would engage and encourage a learner to examine and think about alternative outcomes in two cases; namely how a change in key events in the past might alter present day situations, and how certain present day decisions may be extrapolated to predict future developments.

To ascertain the efficacy of such a user interface design, the software will be tested with learner subjects.

- Subject content tailored appropriately to the learner. This requirement is usually addressed well by domain experts since curriculum and learning objectives and outcomes are normally well defined. For instance, the Ministry of Education has established particular emphases in the NE curriculum, according to the intellectual and emotional maturity of the learners at each level of education. The curriculum begins with an exposure to basic historical facts leading up to nationhood and a review of national achievements. From the pre-university education level onwards, the emphasis of the curriculum shifts to current affairs, national challenges and leadership development.

- Application of basic pedagogical principles to deliver the subject content. Mayer and Moreno (2000) reported that much education software is not only useless in aiding knowledge assimilation, but in certain cases, may even hamper the user's learning process. This is because many education software attempt to compensate for the lack of pedagogical principles in their design, by 'padding' up with irrelevant text, graphics and animations (Stantoni et al 2001). With the proposed e-learning software, learners can customise their learning preferences by choosing a particular learning style, presentation style and emphasis of the NE content. They will be given full control of their learning pace. Concerning learning style, the learner can choose to go on a 'learning journey' with a companion, or to explore the NE content on his or her own. Network capabilities of the e-learning software (including email, internet chat, netmeeting or video-conferencing) would be exploited to enable learners to interact in a community comprising their peers and teachers, promoting discussions and active exchanges of views and opinions, as opposed to passive uni-directional knowledge dissemination. Context sensitive web-links to connect to relevant SingaporeOne websites such as the Ministry of Education NE website and the Picture Archives of Singapore (PICAS), will be included to prompt learners to explore links to specific information sources where further detail on a topic may be gathered. In such a versatile learning environment, teachers may assume a number of roles, including mentor, facilitator, consultant and umpire. Role playing, play acting, quizzes, 'what if' scenario exploration and creative/competitive games, would also be included to make learning an interesting, creative, challenging and engaging experience. By using appropriately text, graphics, narration (including originator or first person speeches and commentaries), animation and video, the NE content could be delivered in a much more amenable and interesting manner. The rich multimedia experience, coupled with a high level of interactivity with both the software and their peer group, would motivate learners to move away from passive learning to assume a more active and independent learning style.

The NE e-learning prototype will be developed using Macromedia Flash MX and Dreamweaver MX. The applications are selected for their capability to support the development of an e-learning software with high interactivity and comprehensive data tracking functions. The prototype would then be used to support user testing with learners to determine the effectiveness of the software design concept and metaphors

# 6    Concluding summary

To produce effective e-Learning software, it is necessary to ensure ease of use and navigation, the provision of appropriate content and the application of sound pedagogical principles. Sadly, these crucial considerations have frequently been side-lined by pushing for 'edutainment' as a marketing pitch. The result is a detrimental pre-occupation with the provision of computer graphics, animations and other effects. Consequently, effective teaching and learning has not always been achieved by existing e-learning software.  This paper reports an attempt to enhance the effectiveness of an e-learning software targeted at NE. A tour metaphor is used to make the navigation of the software transparent to the user.

A second metaphor involving time travel, is applied together with the tour metaphor to enable learners to navigate both time and space.  Together, the metaphors support a versatile means of content presentation to enable learners to visualise better the context of the events addressed by the NE content. A deeper understanding and better assimilation of the subject may thus be derived by the learners through vicarious experiences afforded by the e-learning software.

As development of a prototype of the e-learning software is currently ongoing, a complete account of the software and subject tests will be provided at the time of paper presentation.

# References

Hammond, N. and Allinson, L. (1987). The Travel Metaphor as Design Principle and Training Aid for Navigating around Complex Systems, *Proceedings of the Third Conference of the British Computer Society Human-Computer Interaction Specialist Group University of Exeter 7-11 September 1987*, (Cambridge University Press: Cambridge).

Jones, R.J.K. (1986).  On the Applied Use of Human Memory Models: The Memory Extended Personal Filing System. *Communications of the ACM*, **26**, 259-394.

Mayer, R.E. and Moreno, R. (2000). Aids to Computer-Based Multimedia Learning. *Learning and Instruction*, **12**, 107-119.

Mousavi, S.Y.; Low, R. and Sweller, J. (1995). Reducing Cognitive Load by Mixing Auditory and Visual Presentation Modes. *Journal of Education Psychology,* **87**, 319-334. Neale, D.C. and Carroll, J.M. (1997). The Role of Metaphors in User Interface Design. In M. Helander (Editor), *Handbook of Human-Computer Interaction. Second Edition.* (Elsevier Science B.V.: The Netherlands, Amsterdam).

Smith, D.C.; Irby, C.; Kimball, R.; Verplank, B. and Harslem, E. (1982). Designing the Star User Interface. *Byte*, **7, 4**, 242-282.

Stanton, N.A.; Stammers, R.B. (1990). Learning Styles in a Non-Linear Training Environment. *Hypertext: State of the Art.* 114 –120 (Oxford, England: Intellect)

Stantoni, N.A.; Porter, L.J. and Stroud, R. (2001). Bored with Point and Click? Theoretical Perspectives on Designing Learning Environments. *Innovations in Education and Teaching International*, **38, 2**, 175-182.

Wickens, C. and Hollands, J. Engineering Psychology and Human Performance, Prentice Hall.

# Making the Network Visible to the User in Virtual Environments and Online Games

*Manuel Oliveira   Mel Slater*

University College London
WC1E 6BT London
{m.oliveira, m.slater}@cs.ucl.ac.uk

*Jon Crowcroft*

Cambridge University
CB3 OFD Cambridge
jon.crowcroft@cl.cam.ac.uk

## Abstract

The trend in Distributed Virtual Environments and online games has been to try to hide the state of the network from users. This paper presents a novel approach to the impact of degraded network performance on people's experience of Virtual Environments and online games. Instead of trying to hide the impact of the network from the participants, the state of the network is used to determine a set of global and local properties that are used to determine visual aspects of the displayed scene. These are called Perceptual Network Metaphors (PNM).

With PNMs representing the behavior of the network, the idea is that it may be possible to influence people's expectations and behavior, in similar way that in everyday life people naturally change their behavior according to the weather. The paper describes the functionality and operation of the approach, along with descriptive examples.

## 1   Introduction

'Cyberspace', a term first coined in science fiction literature, describes an immersive three-dimensional environment where people meet to engage in work or entertainment. It is an alternative reality, parallel to the physical world of everyday reality with its boundaries only limited by the imagination. However, independently of the supporting technology or the content available, the essence of cyberspace remains the same: it is all about people.

When two or more people share the same social space within a Virtual Environment (VE), they may in fact be geographically dispersed across the globe. This implies complex problems that need to be handled by the supporting VE systems, which must maintain the desired illusion of real-time interactivity amongst users, notwithstanding the network in-between along with the associated fallacies. The current trend of all solutions has been to isolate the user from the network by either adopting network compensation techniques or assuring particular Quality of Service (QoS) from the network.

The design principle of network isolation seriously hampers the utility of the application for the users, who associate most disruptions to network problems. An alternative approach is the Perceptual Network Metaphors (PNM) framework, which is based on the design principle of exposing the state of the network to the user. The additional information is seamlessly integrated into the application feedback loop to the user by means of metaphors that are context sensitive thus minimizing the possibility of disruption. In similar fashion to when a person takes an umbrella whenever the skies are threatening to rain, so will the user modify their behavior according to the sensorial cues provided by the metaphors which reflect the state of the network.

The remainder of the paper is structured into four additional sections. The next section will describe previous work in assessing the importance of network awareness to the users. This is followed by a description of the types of network relationships perceived by a local host. Based on these concepts, section four provides an overview of the operational framework. Finally, section five presents some concluding remarks.

## 2   Network Awareness

In (Bouch, Kuchinsky & Bhatti, 2000), a study carried out tried to assess the users' perception of Quality of Service (QoS) when interacting with commercial websites. Although the target application was online shopping, the results may be applicable to other application domains, including VEs. The users demonstrated a more favorable response when a better understanding of the network state was conveyed explicitly in some form.

In VE, a preliminary study (Vaghi, Greenhalgh & Benford, 1999) was carried out to evaluate the behavior of users when playing a simple 2- player ball game in the presence of various degrees of network delay. The users not only perceived the network delay beyond a given threshold, but also developed and adopted behaviors aimed at compensating the problems encountered. However, this adaptation was only possible because the delay was constant with no variation. The results were used to implement a visual widget consisting of a wire-framed volume that encompassed an avatar experiencing network delays (Fraser et al., 2000). The assessment of the current delay was visually displayed as a sliding bar on top of the volume. Although the interface paradigm probably improves utility by providing some feedback similar to progress indicators, the mechanism is tightly coupled to the network property (Genter & Grudin, 1990). A more adequate interface design was adopted in (Conner & Holden, 1997), which investigated the use of visual cues to provide information regarding the impact of network latency on lock ownership acquisition of objects in the VE. The approach consisted of allowing people to manipulate replicas of the original object until ownership resolution was achieved. Once the owner was known, the respective object would become solid and all other instances would fade. However, the mechanism did not reflect the network state, but the delay in acquiring the ownership lock.

## 3   Perceptual Network Metaphors

The Perceptual Network Metaphors (PNM) framework identifies two types of perceived relationships between users and the network:

- **Global Perceptual Network Metaphor (GPNM).** These PNMs embody the relationship between the local host and the network.

- **Entity Perceptual Network Metaphor (EPNM).** These metaphors represent how the local host perceives the remote hosts' network connectivity.

One could dispute the need for EPNMs, since it may be sufficient to have a single global metaphor shared amongst all the users within close spatial proximity of each other. However, the nature of the network forestalls any attempts at triangulation during the duration of a session, since the Internet path properties are not associative, commutative or transitive in nature. Also there is rarely symmetry in paths between two hosts, so even then it would be difficult to have a metaphor mechanism that portrayed the state of the network. Both classes will be described in further details in the subsequent subsections.

## 3.1 Entity Perceptual Network Metaphor

For every remote entity that is a source of data traffic, there should be at least one associated Entity Perceptual Network Metaphor (EPNM). A particular EPNM is triggered whenever the local host perceives that the associated remote entity is experiencing network problems. The particular response of the metaphor depends upon the nature of its implementation. The design of an EPNM should take into account the particular context of the VE, thus enabling the user to react accordingly. Consider the following EPNM as an viable example:

- **Probabilistic EPNM**. The metaphor is based on a probability model that produces small number of mirror images of the remote entity constrained to an area centered on the last known position. Associated to each image is a density function that reflects the likelihood of where the next position of the entity will be. This will allow the local user to make the necessary assumptions and act accordingly.

  The typical application of the EPNM is any VE where it is imperative to know the exact location of entities, such as First Player Shooters (e.g.: Half-Life™).

## 3.2 Global Perceptual Network Metaphor

The other relationship to consider is between the actual local host and the network. This metaphor affects the user's interaction with the VE itself, not just a set of remote entities. Taking into account its encompassing nature, this class of metaphors is designated Global Perceptual Network Metaphors (GPNM).

A GPNM is triggered whenever the local host detects that it is experiencing network problems. There may be either a single GPNM, which incorporates the state of the network, or several GPNMs with each associated to a single characteristic of the network. The aim of GPNM is not merely to inform the user of the network state, but also to constrain their area of interaction within the VE. The consistency requirements beyond the sphere of influence of the user may be relaxed, considering that the user may not interact. This allows for the application to easily accommodate the lower QoS provided by the network without degrading the immersive experience of the user. As an example, consider the weather as a GPNM:

- **Weather GPNM**. In ancient civilizations, the weather was always seen as the medium that the gods would use to disclose their feelings, ranging from contentment to anger. Although this perception no longer holds true, the various connotations of the weather still remain to this day. Thunderstorms and rain are foreboding, while sunshine with clear skies is auspicious.

  Taking advantage of this common association, the Weather GPNM associates bad weather with poor network conditions and good weather with optimal state. The more severe the network problems, the worse the weather becomes, reducing the area of interaction of the user. This GPNM works best in VE that are open spaces rather than enclosed environments.

# 4 Operational Model

The overall operational model of the metaphors is depicted in the block diagram of Figure 1 with three main types of building blocks (from the perspective of a particular local host), along with their relationships. In an application, there is no restriction on the number of PNM that may be operational. As illustrated, each remote entity may have associated to it one or more EPNMs. In similar fashion, there may be one or more GPNMs active for the local host.

## 4.1 Network Model

As illustrated, each entity and the local host have an instance of a Network Model, which may all be different from one another in the extreme case. The network model is a simplified model of the underlying network, mapping its current state with a tuple of variables. This model may either represent the local network state or the perceived network state experienced by a remote host.

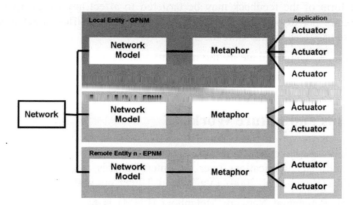

**Figure 1 - Overview of the metaphor operational model**

Each variable represents a network property and their number varies depending on how detailed the particular model is. Although it is possible to have a different network model per entity, in most cases the set of variables will remain the same even if independent instances of the model may exist. Normally, most network models will be a variant of the following tuple (Delay, Loss, Throughput, Jitter). This information is normally available within most network protocols, thus overhead may be avoided if necessary.

## 4.2 Metaphor

The Metaphor block exists for every PNM, independently of its class being either Entity or Global. There are three sub components:

- **Pre-conditions**. The network model will contain a current assessment of the network state, however this does not imply that the metaphors should always produce responses. The volatile nature of the network, where the pattern of data traffic is mostly in bursts, requires some filtering and smoothing.

    The role of the Pre-Conditions block is to only trigger the metaphor when a specific criterion is matched. This permits for metaphors to be associated to a common Network Model, where customization may be achieved by applying combinatorial or filtering operations to the associated tuple.

- **Logic**. The Logic corresponds to the actual model of the metaphor, determining its operation. The actual model may vary and depends on the application, ranging from probability to keeping track of past states. Considering the highly customized nature of the logic, it is most likely that it will be tightly coupled with the Response block.

- **Response.** The response triggers all the associated Actuators, passing the appropriate parameters as necessary. This design decouples the metaphor from the Actuator, thereby

1009

permitting Actuators to be shared across several metaphors. Another advantage of this approach is the possibility of having complex responses by the response of a metaphor activating several Actuators rather than a single one.

## 4.3  Actuator

The actuator block is responsible for providing the actual feedback of the metaphor to the end user. The actual form of the feedback may be targeted to affect any of the five human senses. There is one actuator per response, but it is possible to combine different actuators to achieve a concerted response triggered by the same metaphor.

The activation of an actuator is done by the mere invocation of its interface, which may contain zero or more parameters. It is convenient to make the interface as generic as possible, to facilitate reusability of the actuator with different metaphors.

## 5  Conclusions and Future Work

Although not widely recognized, the end-users of today are increasingly more aware of the role of the network in multi user applications over the Internet. This result has been found in small experimental studies (Bouch et al, 2000), (Vaghi et al, 1999), but with different social settings.

The approach with Perceptual Network Metaphors (PNM) is to convey to the user additional digital information, namely the state of the network, to increase their satisfaction. The exposure is not done in the periphery of the user, but rather in an integrated fashion with the user feedback system (visual display, haptic device, sound device, etc). An early prototype of a distributed game has been implemented, and the next stage consists of conducting user trials to evaluate the effectiveness of the chosen EPNM and GPNM.

Although PNM have been presented as the means to convey the state of the network, the metaphors could convey other digital information such as hardware and software limitations of the hosts.

## Acknowledgements

This research work has been partially funded by Alfamicro.

## References

Bouch, A., Kuchinsky, A. and Bhatti, N., "Quality is in the Eye of the Beholder: Meeting Users' Requirements for Internet Quality of Service", Proc. ACM CHI'2000, Netherlands, 2000

Conner, B. and Holden, L., "Providing a Low Latency Experience in a High Latency Application", Proc. ACM Interactive 3D Graphics'97, Providence, 1997

Fraser, M., Glover, T., Vaghi, I., Benford, S., Greenhalgh, C., Hindmarsh, J. and Heath, C., "Revealing the Realities of Collaborative Virtual Reality", Proc. ACM CVE'00, San Francisco, 2000

Gentner, D. and Grudin, J., "Why Good Engineers (Sometimes) Create Bad Interfaces", Proc. ACM CHI'90, April 1990

Vaghi, I., Greenhalgh, C. and Benford, S., "Coping with Inconsistency due to Network Delays in Collaborative Virtual Environments", Proc. ACM VRST'99, London, December, 1999

# Dream3D: Design and Implementation of an Online 3D Game Engine

*Tae-Joon Park, Soon Hyoung Pyo, Chang Woo Chu, Seong Won Ryu, Dohyung Kim, Kwang Hyun Shim, and Byoung Tae Choi*

Virtual Reality Research and Development Department, Electronics and Telecommunication Research Institute
161 Gajeong-Dong, Yuseong-Gu, Daejeon, 305-350, Republic of Korea
{ttjjpark|shpyo|cwchu|ryusw|kdh99|shimkh|btchoi}@etri.re.kr

## Abstract

In this paper, we propose the Dream3D system, an online 3D game engine for 3D MMORPGs. We analyzed requirements to build 3D MMORPGs together with the techniques to satisfy them. And then, the techniques are classified into four categories: 3D rendering and animation techniques to build 3D game client systems and server and network techniques to implement server-client systems. We design the Dream3D system to provide all the required functionalities. Related with the technique classification, the Dream3D system consists of four subsystems: server engine, network engine, rendering engine, and animation engine. For each of the subsystems, we propose an implementation model to satisfy the functionality requirements.

## 1    Introduction

Due to the improvement of graphic hardware technologies, it becomes possible to generate high-quality images even on conventional personal computers in real-time. Without any expensive hardware devices, 3D virtual environments such as virtual shopping malls and virtual museums can be constructed on personal computers. Recently, 3D computer games on conventional personal computers are also developed and announced.

Moreover, due to the advances of network technologies, it becomes possible to build a large distributed virtual environment shared by several client systems. The most typical example of these kinds of applications is on-line game services, especially massive multi-user online role playing games (MMORPGs). There have been released several MMORPG services around the world, and the number of game players enjoying such kind of games have also been increased.

In this paper, we proposed an online 3D game engine called Dream3D for 3D MMORPGs. Software techniques required to build 3D MMORPGs are analysed and used to make an implementation model of the engine. We subdivided the engine into four subsystems: server engine, network engine, rendering engine, and animation engine. And then the functional requirements for each of the subsystems and software techniques to satisfy them are analysed.

## 2    Advantages of Employing Game Engines to Develop MMORPG

Server and network techniques and computer graphics and virtual reality techniques that are used once to develop a commercial MMORPG service are usually re-used to produce another game services. Therefore, it is common for a game provider to maintain the developed software techniques as a software library or a set of software modules. We call the set of libraries and modules that is re-used during the game development procedure as a game engine (Park, Pyo, Chu & Choi, 2002a, Park, Pyo, Chu & Choi, 2002b, Ryu, Park, Kim & Ko, 2002). Recently, some game providers even sell their engines.

Usually, typical MMORPG services form very large virtual environments shared by more then ten thousands of client systems. Therefore, game server systems must be stable, fault tolerant, and consistent to every client and the game server systems must provide functions to control the amount of network traffics between servers and clients. It is not easy to develop all the components of MMORPG services from the beginning. Almost game developers usually employ more than one game engines to reduce the time and resources required to accomplish the development project.

# 3    Designing the Game Server Engine

Typically, there are thousands of client systems simultaneously connecting to a certain server system, therefore it is necessary to develop a server engine structure to share a game world by several server and reduce the amount of network packet transferred to the server system. In Dream3D server system, the following game world subdivision scheme is used to accomplish the system requirements.

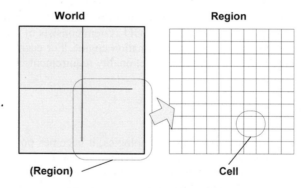

**Figure 1:** Game world subdivision

A game world is subdivided by several regions and each region is also subdivided by regular sized cells (See Figure 1). The region is a area controlled by a server system. If there is only a server system controlling a given game world, the game world and the region assigned to the server become identical. We made it possible to control the size of each region according to the amount of the network traffic. If there occur lots of traffic around a certain area, more server systems take control around this area and thus the size of regions assigned those server systems become small.

The cell is a fixed sized unit area. The size of a region is controlled by the number of cells included in the region. To accomplish the seamless multi server environment, the cells at the boundary area of each region transfer the occurring game events to the server controlling the region together with those of adjacent regions.

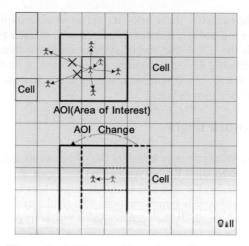

**Figure 2:** AOI control based on cell structure

To control the amount of network traffics and maintain consistent game environments, the Dream3D server system provides area of interest (AOI) control based on the cell structure. For a user character controlled by a client system, only the game events occurred in the cell including the character and its neighboring cells are transferred to reduce the network traffic. If the user character moves, the AOI is changed according to the movement.

# 4 Designing the Network Engine

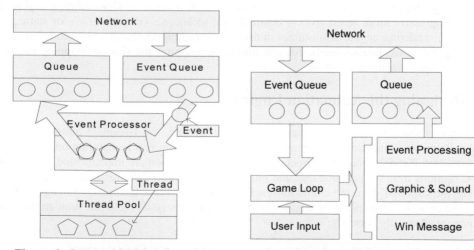

**Figure 3:** Server side network engine          **Figure 4:** Client side network engine

## 4.1 Server Side Network Engine

We built the Dream3D server system on MS Windows NT platforms and thus the server side network engine can utilize the input and output completion port (IOCP) provided by the Windows system. The network engine sends and receives game events as network packets transmitted through the IOCP and then transfer them into the server engine implemented on the network system. The server engine maintains a number of pre-created threads as a thread pool. For each event transmitted through the IOCP, the server system assigns a thread from the thread pool, processes the event by activating the corresponding event processor on the thread, and then sends the processing results through the IOCP again as resulting event packets. The basic structure of the server side network engine is shown in Figure 3.

## 4.2　Client Side Network Engine

At the client side, the network engine only transfers the game events from server to client and vice versa. Game events transferred from the server engine are input to the game main loop of the client system, and analyzed by the predefined game logic together with other game events occurred by the user input and certain game situation. The results of the analysis change the game situation by modifying the position of the game objects and activating several game effects. If necessary, the analysis results are transferred back to the server engine as forms of resulting game events (See Figure 4).

## 5　Designing the 3D Rendering Engine

To generate impressive game situation on computer screens, a 3D rendering engine must draw high quality game scenes in real-time. The Dream3D rendering engine is designed to satisfy such requirements. By adapting geometric culling algorithms that cull out a part of polygons that are not visible from the camera among the whole polygons, the rendering engine can generates game scenes rapidly. To perform such kind of culling operation, additional geometric data structures such as the BSP-PVS structure (Moeller & Haines, 1999, Watt & Policarpo, 2001) used to choose polygons to be culled out must be constructed at the game design phase. The rendering engine also provides progressive mesh based level of detail (LOD) techniques (Hoppe, 1996) for static game objects. The rendering engine also utilizes multi-texture techniques that are common features of advanced graphic acceleration hardwares and thus light mapping and shadow mapping functions can be provided using those features.

## 6　Designing the Animation Engine

To produce realistic motions of human-like game characters, the Dream3D animation engine provides functions to handle captured motion data obtained from real human motion. Techniques such as motion blending, motion retargeting, and motion transition are implemented to generated realistic motions from the captured data and interpolate them. Furthermore, to get more flexibility on using the captured data, the animation engine provides inverse kinematics rules to edit and modify the data. And the animation engine also has the physics system that can simulate the game environment physically correctly.

## 7　Experimental Results: Implementation of a MMORPG

We implemented a prototype MMORPG system to verify the soundness of the proposed game engine. The server system is built on a Windows platform with four Xeon processors. The client system is implemented on a Pentium3 with an NVidia GeForce3 acceleration board. We also implemented a special purpose plug-in software that converts 3D Studio MAX model data into the implemented game system. It is possible to use all image formats including tga, bmp, and jpg if

Direct3D supports them. In Figure 5, we show screen shots of the implemented MMORPG system as the implementation results.

**Figure 5:** Screenshots from the implemented MMORPG service

# 8    Conclusion

In this paper, we proposed the Dream3D game engine for developing 3D MMORPG services. We analyzed requirements for developing 3D MMORPGs and then design an implementation model of a game engine that satisfies the requirements.

As future works, we plan to adjust the proposed engine so that it can run on several platforms including game consoles such as PlayStation2 and X-Box. We are also supporting development of a commercial 3D MMORPG service using the proposed engine.

## References

Hoppe, H. (1996). Progressive Mesh. *Proceedings of ACM SIGGRAPH 1996*, 99-108.

Moeller, T. & Haines, E. (1999). Real-Time Rendering. A K Peters.

Park, T.-J., Pyo, S. H., Chu, C. W, & Choi, B. T. (2002a). Design and Implementation of a Rendering Engine for Developing 3D Computer Games. *Proceedings of Japan Korea Computer Graphics Workshop 2002*.

Park, T.-J., Pyo, S. H., Chu, C. W, & Choi, B. T. (2002b). Implementation of a Rendering Engine for 3D Computer Games. *Proceedings of Annual HCI conference 2002* (in Korean).

Ryu, S. W., Park, T.-J., Kim, D., & Ko, D. (2002). Design and Structure of an Online 3D Game Engine Dream3D. *Proceedings of Annual HCI conference 2002* (in Korean).

Watt, A. & Policarpo, F. (2001). 3D Games: Real-Time Rendering and Software Technology. Addision-Wesley.

# Added-Value Functionality for Learning Management Systems: the selection process within a German insurance company

*Nadja Reckmann, Paula M. C. Swatman*
University of Koblenz-Landau
Institute for Management, Universitätsstraße 1, 56070 Koblenz, Germany
reckmann@uni-koblenz.de, paula.swatman@uni-koblenz.de

## Abstract

The authors will provide an exemplary insight into the relevant processes required for the selection and development of an effective Learning Management System. The paper, therefore, will describe the actual procedures which took place inside a German insurance company to arrive at the choice of appropriate learning-management-tools. This involves analysing the approaches used which were characteristic of this particular company, and describing the major activities and goals. Of special interest is the relation between the individual approach, the current eLearning market offer, and the acknowledged state-of-the-art eLearning theory.

## 1    Universal eLearning theory meets the individual organisational context

The driving forces for the implementation of eLearning in the commercial world are now widely recognised and accepted across virtually all industry sectors. Many even speak of the Merrill Lynch eLearning "megatrends" (Learnframe, 2001), such as the changing demographics which have resulted from the aging of the baby boomers and the reduced "knowledge half-life", which has meant that many books are out-of-date even before they are printed, together with the rapid expansion of technology force firms and their employees into ongoing, lifelong learning and training activities. At the same time, organisations are witnessing a number of the benefits of eLearning, such as cost savings, and increased flexibility and productivity (Hall, 2001a).

In parallel with the rapid growth of eLearning, the need to reengineer Human Resources (HR) development is becoming an increasingly well-accepted concept (ProSci, 2002). The same technology growth which has spurred eLearning has also meant an increased demand for skilled and capable people able to work in today's more agile and flexible organisations – and the hiring and retention of human capital is more than ever a priority for companies of all sizes. "In the 21st century, the education and skills of the workforce will be the dominant competitive weapon" (Rosenberg, 2001, p. 6). These two trends are clearly inter-linked as technology makes it possible for this skilling and education to be integrated into the workday, making flexibility and lifelong learning ever more possible.

More recently, however, the euphoria of the early years of the eLearning and Business Process Reengineering phenomena has given way to a more sober realisation that the process of integrating new learning technologies is not easily or quickly achieved. KPMG Consulting recently published a survey of the German market, in which they interviewed 604 companies concerning the state of their training development and their eLearning experiences (KPMG *et al.*, 2001). The results showed that although very optimistic expectations of eLearning growth and uptake were common, German HR departments have tended to stick to the "traditional" classroom mode for their training seminars. Fewer than half of the surveyed companies were actually using eLearning, and another 25% had no plans to introduce it in the near future. And of those actually using eLearning, conceptual and implementation problems frequently led to planning periods of longer than six months, prior to the actual roll-out of the system.

The most likely explanation for this dilemma lies in the fact that eLearning markets across Europe are still volatile and lacking in transparency (Payome, 2002) – partly because research has not yet managed to provide universally-applicable guidelines for a state-of-the-art eLearning strategy. We all are familiar with recommendations for successful eLearning such as: "focus on building a learning culture" (Rosenberg, 2001, p. xvii); foster a "culture of open access to information and knowledge" (Rosenberg, 2001, p. 15); or "elevate learning to the highest levels of the firm" (Rosenberg, 2001, p. 13). Abstract directives such as these, which target the existing corporate culture, are a familiar form of advice for eLearning initiatives. At the opposite extreme to this approach are handbooks supplying: catalogues of questions to answer (Hall, 2001 a); "quick and dirty" step-by-step alternatives for selecting the most appropriate technology (Lguide, 2001); or techniques for finding one's way through the eLearning "jungle" (Baumgartner, Häfele & Maier-Häfele, 2002).

None of these approaches, however, no matter how correct, properly -funded, or well thought-through they may be, is guaranteed to be implementable – or even desirable – for every company, because each company is unique and each needs an eLearning solution which meets its own unique mix of technical, managerial and cultural needs. For example, while a company may be attempting to change its HR development practices by means of eLearning, and may approve of the fact that "... an effective and durable e-learning strategy is not just about technology or instruction / information design. It must also be about the culture..." (Rosenberg, 2001, p. 304), a change in the company *culture* is another, and major, issue. A company's ability to change its learning culture is often overestimated (Koller, 2002). Further, a strategic management decision that "the time has now come" for eLearning will take precedence over the fact that the company in question may not yet be ready for eLearning – in this case the adoption process may well take longer than initially anticipated. This was the case with the research project currently underway between the Faculty of Informatics at the University of Koblenz-Landau and a large and fairly traditional German insurance company, which we will call CivilCo.

In this paper we initially describe the CivilCo case study – the background to the case, and the stages through which the project has already passed. We discuss the insights into cultural context which our growing awareness of CivilCo's goals and philosophy brought us and, finally, consider the benefits which ethnographic research approaches have to offer those engaged in developing eLearning solutions in a corporate environment.

# 2 The eLearning research project inside CivilCo

## 2.1 Project background, goals, players

In the context of the restructuring of their Information Technology and within the scope of their long-term eBusiness strategy, CivilCo is designing an Enterprise Network (the "Intranet Project") and its connection to the global Internet – effectively an extranet extension. One of the major goals of the "Intranet" project is to support and complement continuing training for the company's staff, paying special attention to the needs of the external insurance agents, with the help of Internet technology. Taking advantage of an existing strategic alliance with the University, CivilCo created an "eLearning" sub-project with the eBusiness Workgroup of the Faculty's Institute for Management, which will design and implement a Learning Management Platform tailored to the needs of this particular company.

Goals for the eLeaning project are:

- Engineering of a Blended Learning concept to improve the staff's education and training as well as their information supply.
- Provision of up-to-date, relevant and uniformly high quality content on a user-friendly learning platform which is easily accessible by all target groups.
- No *reduction* of classroom training, but rather *reorganisation* of the learning approach
- Processing of training organisation on an electronic platform.

Target groups are office as well as field staff, including both trainers and trainees in a variety of apprentice training programmes.

The project team consists of two part-time researchers within the eBusiness workgroup, working exclusively on this project, as well as the Professor as director of the project (working only part-time on the project), and three to four CivilCo staff members, all being exempted part-time from their normal jobs for the project. The director of the eLearning project from CivilCo's side is also the overall director of the company's Intranet project.

Space limitation for this paper prevent us from giving a detailed description of CivilCo. What is relevant to our discussion of the eLearning project, however, is that CivilCo is one of the ten largest German insurance organisations with more than 12,000 staff members across Germany, but with virtually no international business The company originated as an insurance provider for civil servants, and thus has a fairly mechanistic, hierarchic, and bureaucratic organisational style. CivilCo's philosophy is based on its legal structure as a Mutual Association, i. e. clients are not customers but members. CivilCo's main goal, to offer members the highest possible benefits in return for their contributions, is the basis for all business and management decisions.

## 2.2    Selecting a suitable LMS

The term "eLearning" was originally used to denote any kind of learning supported by electronics, such as television, audio/video tapes, or CD-ROMs etc. In the research literature today, however, it is generally used synonymously with Internet-based learning, meaning any distribution of learning material possible via browser (Rosenberg, 2001, Köllinger, 2001, Hall, 2001 b). Other researchers apply a more pragmatic definition of eLearning as being learning supported by a learning system, i. e. a software system with applications for learning like a CBT, WBTs with collaborative tools, or even an integrated Learning Management System (Back, Bendel & Stoller-Schai, 2001). It is this latter definition which is used by CivilCo.

What is a LMS? An LMS "uses Internet technologies to manage the interaction between users and learning resources." (Rosenberg, 2001, p. 161). It is "software that automates the administration of training events." (Hall, 2001 b). Whatever the individual nuances of understanding eLearning, it is essentially based upon technology. Ultimately, CivilCo must tackle the question of how to select an appropriate LMS for the organisation and its training problem.[1] The creative task behind this is to design the interaction between learner and trainer – effectively, the human-computer interface, since the trainer is "in the machine" – in order to efficiently address the company's educational context.

While it is clear that learning depends on the type and quality of interaction the technology is able to support, it is equally clear that new technologies do not of themselves lead to improved training quality (Kerres, 2001, Rosenberg, 2001). The design of a suitable LMS is, then, a complex procedure where pedagogic, technical and organisational variables have to be considered. The executive team must agree on a bundle of functionalities promising a real and traceable added-value for all stakeholders, in order to satisfy needs on a long-term basis (Kerres, 2003).

Starting out on a project like this one, the first move was to understand the current status of the company's HR department. Questionnaires on the HR development concept and the training plan, interviews with stakeholders about the prevalent learning culture, and analyses of relevant documents, e. g. the current training material, all added to the detailed analysis. Accompanying this was an investigation of the technological situation inside the company, the technical equipment and experience of staff, the information structure and, above all, the technological vision for the Intranet project. Simultaneously, information was needed on what the market had in store, i. e., which segments, suppliers and products could be identified within the German eLearning market. We made a special effort here to identify a short-list of 30 LMS suppliers, as well as elicitating the availability of relevant eContent or, alternatively, the costs for internal / external production.

This process was nurtured by theoretical input from eLearning research and empirical studies. An analysis of the *status quo,* like the one described here, can be done by either an internal or external team, as long as the analysts can draw on insider knowledge from local informants. That is what made a difference for the precise evaluation of the situation in the company's relevant environment. Only then was the project team able to work out a plan which could serve as a framework for the LMS selection procedure.

## 2.3    The selection procedure

### 2.3.1    The didactic scenario

Having a detailed picture of what *is*, the next task for the company was to find out what *should be*. During the process of selecting an LMS, a company is largely dependent upon describing the  future education environment and the wanted processes for teaching and learning by means of "scenarios". Primarily, these scenarios illustrate in a down-to-earth way how the technological support the LMS enables would change the wider learning environment, (eg. the HR development practices, the learner administration processes, the

---

[1] Although we focus here on the selection of the LMS we are, of course, aware that there are essentially *two* major parts to any eLearning program: the LMS *and* the electronic content.

management of all types of resources (human and infrastructural), the related information and communication procedures, and also the organisation and management of training courses. At a more concrete level, the scenario also includes the changes affecting the design of courses in general, eg. the learning organisation (place and time), the hybrid (or blended) course arrangement (i.e. both traditional classroom phases and computer-interaction phases), and the kinds of content – both electronic and traditional – which will be needed. In the case of CivilCo, information relating to the necessary technology was also included, as long as it helped to describe the target situation. The frame for the document is provided by the goals of the overall didactic concept, a SWOT[2] analysis of the concept, and an estimation of likely costs and benefits. The scenario additionally included the needs and expectations of all stakeholder representatives, who were consulted during the associated meetings and serves as a *factual basis* for the technical requirements specification.

At this point, insider knowledge was *the* crucial input to reach at a feasible and viable scenario, because only members of the firm can appreciate if an aspect of the necessary organisational or cultural change can actually be implemented.

### 2.3.2 The interim test installation and evaluation

The next planned step was to come up with the "technical specification" which would identify the required eLearning functionalities in detail; and which forms the heart of the selection process. Once the company has agreed on a scenario, the requirement specification is also part-way to being settled – which is why the importance of a carefully-crafted scenario cannot be overestimated. CivilCo had to acknowledge a lack of practical experience with eLearning, which led to a need to reinforce, as well as protect, its scenario. It was decided to have a test installation of an LMS and an evaluation of the still-uncertain aspects of the scenario. The evaluation procedure included theoretical work such as agreeing on the general evaluation objectives, as well as the pedagogic, organisational, and technical goals and criteria (Kromrey, 2001) before designing an appropriate course and implementing and interpreting the test in practice. The LMS for the test was selected at random, as it was not the objective to test a platform, but rather to test aspects of the didactic scenario such as: the learning organisation, the value of communication components, or the course design at an engineering level. The test installation and the evaluation serve as a *practical and experiential basis* for the requirements specification.

### 2.3.3 The technical requirement specification and the make-or-buy decision

This is the project's current status. With the experiences and lessons learned from the test evaluation, the project team will go back into the scenario to refine it. From the basic scenario, the company will have to identify the technical functionalities needed and generate the catalogue of requirements. The decision about whether to make or buy the agreed-on functionalities will be made in cooperation with CivilCo's Intranet project and the company's IS department to reveal possible internal synergies. This process will reveal, finally, whether purchase of a commercial LMS is still an option. If so, vendors will be contacted and a request for proposal issued. The evaluation of the proposals should result in a small number of vendor presentations and, again, evaluations of test installations.

## 3  Importance of the CivilCo case for research and theory

After spending more than a year with this project, what we as researchers and members of the executive team have gained so far is not merely practical expertise in handling an eLearning implementation. The really striking experience has been learning to understand an individual organisational culture. Initially, we were tempted to characterise it as "technology resistant". Despite an abundance of technologically refined learning solutions available on the market and various resources more than willing to be of intellectual service – including IS doctoral students in the project team, who were always at hand with funded arguments in favour of change and technological support – the company insisted on *doing it their way*. Over time we realised that, firstly, the company *did not want our change*, and, secondly, that CivilCo was actually trying to find a way to integrate eLearning strategic theory and practical advice into its own cultural reality. Theory and market practice, no matter how appealing they might seem, were only helpful or useful to CivilCo if they provided support for the company's philosophy and strategic goals.

---

[2] SWOT = strengths, weaknesses, opportunities, threats

While some researchers have already found the relation between eLearning theory and practice to be a stress ratio, and managed to identify the problematic issues which need to be addressed (Habermann & Kraemer, 2001), neither theory nor the eLearning market is able to supply a recipe. The key to a successful learning solution, promising added value for all users of the system within a particular company, is an instructional conception professionally tailored to the precise problem situation within the specific cultural context of that company.

It is, therefore, essential for the successful introduction of eLearning to be familiar with the specific cultural aspects of the company – and not seek to transform it quickly into a learning organisation, which is a venture doomed to fail. We argue in favour of making a cultural analysis of a company, taking advantage of methods drawn from the field of ethnographic research. Ethnography as a research methodology allows a deep understanding of the people, the organisation, and the broader context within which they work. It is thus well suited to providing rich insights into the human, social, and organisational aspects of information systems (Myers, 1999). The aim of ethnography is not the generation of universalistic knowledge or the understanding of human action through a set of universally applicable lenses which, for the ethnographer, conceal far more than they reveal (Prasad, 1997). Key elements of a method which is founded on ethnography are the predominance of local knowledge and a focus on the cultural context with its shared values, rituals, and myths and, consequently, the individual preconditions for learning.

# References

Back, A., Bendel, O., & Stoller-Schai, D. (2001). E-Learning im Unternehmen: Grundlagen, Strategien, Methoden, Technologien. Zürich: Orell Füssli.

Baumgartner, P., Häfele, H., & Maier-Häfele, K. (2002). E-Learning Praxishandbuch. Auswahl von Lernplattformen. Marktübersicht – Funktionen – Fachbegriffe. Innsbruck: StudienVerlag.

Habermann, F., & Kraemer, W. (2001). Envision E-Learning – Von der Strategie zum detaillierten Projektplan. In Kraemer, W., & Mueller, M. (Eds.), *Corporate Universities und E-Learning. Personalentwicklung und lebenslanges Lernen* (pp. 233– 258). Wiesbaden: Gabler.

Hall, B. (2001 a). E-Learning Guidebook. Six Steps to Implementing E-Learning. Retrieved January 16, 2002, from http://www.brandonhall.com/public/forms/sixstepdb/

Hall, B. (2001 b). New Technology Definitions. Retrieved February 4, 2003, from http://www.brandonhall.com/public/glossary/glossary.html

Kerres, M. (2001). Multimediale und telemediale Lernumgebungen. Konzeption und Entwicklung. München: Oldenbourg.

Kerres, M., de Witt, C., & Stratmann, J. (2002). E-Learning. Didaktische Konzepte für erfolgreiches Lernen. Retrieved February 17, 2003, from http://online-campus.net/edumedia/

Koller, W., Flum, T., Müller, M., & Tockenbürger, L. (2002): Kulturelle und personelle Bedingungen für E-Learning vor Ort klären. In Hohenstein, A., & Wilbers, K. (Eds.), *Handbuch E-Learning. Expertenwissen aus Wissenschaft und Praxis.* Köln: Deutscher Wirtschaftsdienst.

Köllinger, P. (2001). E-Learning. Eine Marktanalyse für Deutschland. Düsseldorf: Symposion Publishing.

KPMG Consulting; MMB Michel Medienforschung und Beratung, & Psephos Institut für Wahlforschung und Sozialwissenschaft (2001). E-Learning zwischen Euphorie und Ernüchterung. Eine Bestandsaufnahme zum E-Learning in deutschen Großunternehmen. Retrieved January 16, 2002, from http://www.mmb-michel.de/New_Learning_Zusammenfassung.pdf

Kromrey, H. (2001). Evaluation – ein vielschichtiges Konzept. Begriff und Methodik von Evaluierung und Evaluationsforschung. Empfehlungen für die Praxis. *Sozialwissenschaften und Berufspraxis,* 24 (2), 105-131.

Lguide (2001). Ten Steps to Successfully Selecting a Learning Management System. Retrieved on January 30, 2002, from http://www.docent.com/elearning/tessteps.html

Learnframe Inc. (2001). Driving Forces Behind e-Learning. Retrieved January 15, 2003, from http://www.learnframe.com/aboutelearning/page8.asp

Myers, M. (1999). Investigating Information Systems with Ethnographic Research. *Communications of the Association for Information Systems,* Vol. 2, Article 23.

Payome, T. (2002). E-Learning im europäischen Vergleich. *wissensmanagement online.* Retrieved on January 10, 2003 from http://www.wissensmanagement.net/online/archiv/2002/09_1002/e-learning.shtml

Prasad, P. (1997). Systems of Meaning: Ethnography as a Methodology for the Study of Information Technologies. In Lee, A., Liebenau, J., & DeGross, J. (Eds.), *Information Systems and Qualitative Research* (pp. 101 – 118). London: Chapman & Hall.

ProSci Research (2002). Innovative Practices in Human Resources: Benchmarking Results from 67 Companies. Loveland, Connecticut: ProSci Research,.

Rosenberg, M. J. (2001). E-Learning: Strategies for Delivering Knowledge in the Digital Age. New York: McGraw-Hill.

# Effects of WWW Cooperative Learning on Children Education

*Teresa Roselli , Eleonora Faggiano, Antonella Grasso, Paola Plantamura*

Università Degli Studi di Bari – Dipartimento di Informatica
Via Orabona, 4 – 70125 Bari - Italy
{roselli, faggiano, grasso, plantamura}@di.uniba.it

## Abstract

Cooperative learning is one of the most widespread and fruitful areas of theory, research and practice in education. The use of the Internet and the World Wide Web to achieve cooperative type activities is attracting increasing interest. In this work we introduce Geometriamo, a WWW system for cooperative learning about geometric figures, where each student communicates online with the rest of the group. We also propose a study aiming to evaluate the web-based cooperative learning effectiveness using Geometriamo, by comparing it with the traditional in-class cooperative learning.

## 1    Introduction

The use of the Internet and the World Wide Web to achieve cooperative type activities and, in particular, cooperative learning, based on the principles of constructivism, is attracting increasing interest. There are many studies on cooperative learning on the pedagogical field, where this interest originated before it was extended to the computer field. Comparisons of the results obtained with cooperative groups versus both competitive groups and groups of individuals showed that the former made the greatest progress because, as reported by Johnson & Johnson (Johnson & Johnson, 1991), the more conceptual the learning process, the more active the cognitive analysis and problem solving techniques needed, and therefore the greater the advantage for cooperative learning over competitive or individual methods. Moreover, Johnson & Johnson carried out comparative studies on cooperative learning with the aid of the computer and concluded that the principles established in the classroom were valid for computer learning groups, too, and that the interaction involved in asking for help by formulating specific answers contributes fundamentally to the improvement shown.

The value of using WWW for educational purpose can thus be greatly enhanced by the addition of cooperation to the classic dimensions of computer-aided learning. In this context, Internet enables students to become the inventors, designers and implementers of their learning system using hypertextual techniques and environments that extend their contacts beyond the bounds of their own class and school (Ewing, Dowling & Coutts, 1999).

However, for this to be successful, Internet must be seen not only as a new communications technology but also as a cognitive tool to be integrated in the scholastic environment. The most critical element in designing system is the cooperative model by means of which learners should act and interact to achieve the common goal. Various models have been suggested in the literature for cooperative learning (Kagan, 1989),(Johnson & Johnson, 1994), (Johnson, Johnson, Holubec & Roy, 1994a), (Johnson, Johnson, Holubec & Roy, 1994b), (Johnson, Johnson, Holubec & Roy, 1994c) and some essential factors have been identified that must be present in any cooperative didactic situation involving the use of telematics. These include: active, visible collaboration, sharing of the decision process when formulating group projects, highly structured work groups, reciprocal commitment among students and teachers (Bagley & Hunter, 1992), well-defined tasks, a sense of responsibility toward the group and the task undertaken, strong leadership within the group, and a final assessment mechanism (Riel & Levin, 1990).

In this work we introduce Geometriamo, a WWW system for cooperative learning about geometric figures, where each student communicates online with the rest of the group. We also propose a study aiming to evaluate the web-based cooperative learning effectiveness using Geometriamo, by comparing it with the traditional in-class cooperative learning.

## 2    The Cooperative System Geometriamo

In designing the system, we referred essentially to Slavin's model and, in particular, to the cooperative technique known as Student Teams Achievement Divisions (STAD), that is best suited to learning mathematics. Its strong points are those of enabling interaction among small groups while emphasizing individual responsibilities that have an important effect on the group's final assessment.

In the implementation of the domain, each topic has been organized into Theory pages, Examples and Exercises. Solution of the exercises is the goal of the group work. The system keeps track of everything that happens during a working session and thanks to a tutorial component, implemented through AI techniques, it can intervene and stimulate group activities.

Each member of each group must register. Groups consist of four members that use a notice-board for exchanging messages, which are an essential part of cooperative learning. A Leader is nominated in each group, who also has the task of dialoging with the tutorial component and supervising the group work of his/her companions. Geometriamo also features an area where students carry out assessment tests enabling grading of the individual learning gain, and that of the whole group. Students learn to explore the data available online and, through dialogue with their peers, to derive knowledge rather than memorizing it passively, to gain answers to their questions and give and receive help in the shared task of problem-solving. Thus, cooperative learning occurs in a constructivist context where two or more people can attain goals that they would be unable to aspire to alone. This is one of the strong points of Geometriamo.

The fundamental components embedded in our system are:

- the **Knowledge Base** represents the domain to be communicated to the learner by means of links and nodes;
- the **Tutor Module** is intended to emulate the teacher, according to Slavin's STAD technique;
- the **Student Model** contains the data identifying each student registered at any time during the work session;
- the **Group Model** keeps track of the activities carried out by the group, the exercises done and results achieved and the interaction among its members and any tutorial intervention required;
- the **Student and Leader Interfaces** present the course topics and exercises, allowing access to each single logical node of the hypermedia;
- interpersonal relationships are fostered by the **Messages Module** specifically set up to enable messages, both one-to-one and one-to-many, to be sent in natural language to the other members of the same group and from the group leader to the tutor.

## 3    The Experiment

The study was aimed to evaluate the effectiveness of Geometriamo as an instrument for cooperative learning and to answer two fundamental questions by comparing the use of our system with traditional in-class cooperative learning:

- Can the children actually improve their knowledge of geometric concepts by using the system?
- Can the use of the cooperative learning system be as effective as cooperative in classroom mediated by a teacher?

As a basic experimental hypothesis, we have predicted that Geometriamo had pedagogical potentialities that could be equal than cooperative learning in classroom with a human teacher.

The following sections describe the method adopted for testing the learning effectiveness and report the results of the experiment.

## 3.1 Participant

Participants were 152 students attending the fifth class at the primary schools: "XX Circolo Didattico E. De Amicis" and "XX Circolo Didattico C. del Prete" of Bari. Nobody had previous experiences with learning hypermedia but some students were familiar with the use of PC's primarily for playing games and for navigating on Internet to find information.

Initially, children were administered a pre-test, consisted of some exercises referring to standard concepts, to evaluate their knowledge about geometry. The test was developed in collaboration with the teachers of the school. Final scores ranged from 0 to 10. In addition, children were required to fill out a questionnaire aiming to infer the student social ability.

## 3.2 Design

The experiment is based on a mixed design, with pre-test post-test as the within-subjects factor and the between-subjects factor. Children were divided in the Experimental Group (EG) and the Control Group (CG). According to the pre-test evaluation and the estimated student social abilities, each group was divided into subgroups of four elements to realize cooperative learning. Moreover, EG Group was divided into EGC Subgroup, with students belonging to the same class, and EGM Subgroup, with students belonging to different classes. The same procedure was made for the CG Group (CGC and CGM). Great attention has been devoted to counterbalance gender and previous knowledge on geometry among the groups.

## 3.3 Procedure

The actual experiment was composed of three sessions. Each session lasted an hour and a half, with a 2-day interval during the sessions, children were required to do a revision about the Geometriamo's contents, to examine related example and to do related exercises. Students assigned to the EG subgroups revised geometric concepts, discussed and solved complex problems proposed by the system. Children assigned to the CG subgroups revised geometric concepts in class attending lessons and, according to the STAD method, they were required to solve some complex problems proposed by the teacher. One week after the experiment, all the students were given a post-test. In the attempt of controlling confound variables related to the participants' history, no other geometric lessons were taught during the period of the experiment and students were not given any homework on the topic.

## 3.4 Results

Figure 1 illustrates the average value of the pre-test and post-test scores in the two cooperative learning conditions. The learning improvement is evident: the children have increased their knowledge about geometry.

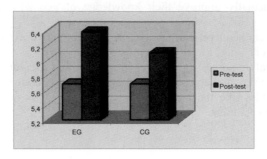

**Figure 1:** Average pre-test and post-test score

Table1 reports the t-test analysis, which was applied to the results of the post-test for testing the difference between the performances of the experimental and control groups. The predetermined alpha level adopted for hypothesis testing was .05. No significant difference due to the mode of instruction was found between the two groups (T(150) = 1.007, t =1.67), and therefore our null hypothesis of the equality of the means was not rejected.

**Table 1:** Control vs. Experimental Group results

|  | N | Mean | Standard Deviation |
|---|---|---|---|
| **EG** | 75 | 0.66 | 2.30 |
| **CG** | 75 | 0.21 | 3.08 |

| | |
|---|---|
| Total Standard Deviation = | 2.722 |
| T(150) = | **1.007** |
| $t =$ | **1.67** |

Results confirmed the experimental hypothesis claiming an equal improvement over time in the two experimental conditions. This implies that all the children have learnt during the experiment and that this learning is almost identical for both groups.

## 4    Conclusion and Discussion

Maximum use of the Net's potential may lead to the development of learning environments that will enable the creation of new knowledge and knowledge links, on the basis of the proposed one, and a valid group cooperation, working with the aim of improving the level of knowledge and ability of each single member.

This work presented a study aimed to evaluate two different modes for cooperative learning. In one treatment, students discussed and solved complex problems proposed by the cooperative learning system, in the second they were required to discuss and solve complex problems assigned by the teacher according to the STAD method. The results indicate that, on average, the learning is almost identical for both groups. We noted that in the EG Group a higher learning gain has been obtained (as shown in Figure 2) by subgroup with students belonging to the same class (EGC). A factor which may explain the different achievement of the experimental students could be that the use of computer allowed a better and freer communication among group component, above all if they already knew each other.

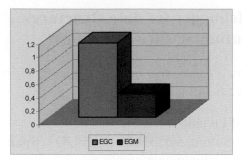

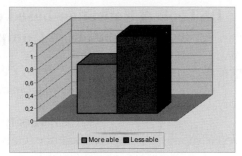

**Figure 2:** Learning gain in EG Group

**Figure 3:** Learning gain in subgroups with children belonging to the same class

We can say that all the children significantly enhanced their knowledge during the experiment independently of the cooperative learning condition they were assigned to. Furthermore, analysing the learning gain in subgroups with children belonging to the same class, we noted that (as shown in Figure 3), the cooperative learning mediated by computer has strong potential for improving the performance of less gifted learners. Our results prompt us to continue along these lines, conducting further investigations aiming to evaluate the effectiveness of WWW cooperative systems as valid support in the learning process.

# References

Bagley, C., & Hunter, B. (1992). Restructuring, constructivism and technology: Forging a new relationship. *Educational Technology*, 32 (7), 22-27.

Ewing, J. M., Dowling, J. D., & Coutts, N. (1999). Learning Using the World Wide Web: A Collaborative Learning Event. *Journal of Educational Multimedia and Hypermedia*, 8 (1), 3-22.

Johnson, D. W., & Johnson, R. T. (1991). Cooperative Learning and classroom and school climate. In B. J. Phrase, & H. G. Walberg (Eds.), *Education Environments-Evolution, antecedents, and consequences* (pp. 55-74). New York: Pergamon Press.

Johnson, D. W., & Johnson, R. T. (1994). Learning together. In S. Sharan (Ed.), *Handbook of Cooperative Learning methods* (pp. 51-65). Westport, CT: Greenwood Press.

Johnson, D. W., Johnson, R. T., Holubec & Roy (1994a). Cooperative Learning in the classroom. Alexandria, VA: Association for Supervision and Curriculum Development.

Johnson, D. W., Johnson, R. T., Holubec & Roy (1994b). The new circles of learning: Cooperation in the classroom and school. Alexandria, VA: Association for Supervision and Curriculum Development.

Johnson, D. W., Johnson, R. T., Holubec & Roy (1994c). The nuts and bolts of Cooperative Learning. Edina, MN: Interaction Book Company.

Kagan, S. (1989). Cooperative Learning: Resources for teachers. Riverside, CA: University of California at Riverside.

Riel, M., & Levin, J. (1990). Building electronic communities: success and failure in computer working. *Instructional Science*, 19, 145-169.

# DSTool: A Reflection-based debugger for data structures comprehension in Computing Science Learning

*Sergio Sama Villanueva, Juan Ramón Pérez Pérez, Sergio Ocio Barriales, Martín González R..*

University of Oviedo

Department of Computer Science, Calvo Sotelo, s/n, 33007 OVIEDO – Spain

sergio_samav@yahoo.es, jrpp@pinon.ccu.uniovi.es,
i1652800@petra.euitio.uniovi.es, martin@lsi.uniovi.es;

## Abstract

Data structures debugging is a complex task. This complexity is mainly caused by the lack of debuggers which show the data structures in an intuitive way. DSTool is a tool that tries to solve the mental models conflict that is produced when these debuggers are used. To get this , DSTool will show the data structures in a graphic way, letting perform any kind of operation through the same graphic representation, achieving a user mental model as close as possible to the design model.

## 1    Introduction

Debugging and bug fixing are two of the most important tasks that must be done when any kind of computer application is being developed, the quality degree of the final product depends mainly on these tasks. In spite of this fact, debugging is an activity almost handmade yet, where factors like experience, intuition, or even luck, become very importants. All these factors make debugging a slow, complex and inefficient process in most cases.

At debugging time, progammers usually resort to the tools given by the IDE which they are working, generally watches, tracing systems and breakpoints. These tools are not very sophisticated, and they rarely satisfy every necessity of the programmers. This situation gets worse when the object to debug is a complex data structure, because the difficulty of the task grows, but we can not use more powerful tools, situation where the programmer could feel lost in his own data structure.

One of the main problems that users have when they are working with these debuggers, is to find the necessary data among the big amount of information that is shown by the tool, information which could not be useful in that moment. This situation can be explained with a simple example.

An user wants to see the value of a variable stored in a node of a list. To reach the value, it is necessary to spread every node until the one searched is found. When the user has localized the node, he can see a great amount of data about the variable, where he has to look for the value he wants, because it is very possible that he had not localized the value at a first view. The remainder data is information that have not special importance to the user  (in most cases), so that values could be omitted in order to make a clearer interface.

Another problem that users have when they are working with these debuggers is the difficulty to fix through the same tool the bugs found. There are some debuggers where the modification of the variables values is very complex, and there are very little debuggers than let the user insert or remove nodes from the data structures.

This is a common problem for every programmer, but it is specially dangerous for data structures students. These students have difficulties to make abstractions, for example, they do not identify easyly raw data shown by the debugger with the leaves of a tree. This way, their capacity to interpret the information shown by the debugger is limited, and the efficacy of the tool goes down.

This problem is caused by a mental models conflict (Norman, 1988). In an ideal situation the design model, model that shows data structures the way programmers have it in mind, that is, a tree data structure as a set of leaves, must agree with the user's model, the image the user obtains from the interaction with the system (see Figure 1). This way, it could be said that user is interpretting the information the way designer wanted, the right way.

To reach this target is essential the system image, because the user obtains his mental model from it.

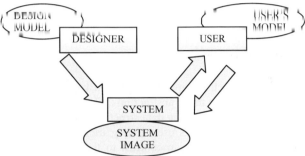

**Figure 1**: Mental models.

In classic debuggers, the system image does not reflect the design model in an effective way, it is too abstract and complex, especially for students. This way, those students can get a wrong user model, causing a slower and more complex debugging process.

The system image of the debuggers included in IDE's are very similar . The next examples show the proccess of some debuggers to debug a simple five elements queue.

The debugger included in Forte for Java IDE displays the information the way it is implemented (see Figure 2).

First of all, it shows only the initial node, and to see the next one, the first node has to be spread. This debugger offers a disproportionate amount of information in comparision with the simplicity of the data  stored in the data structure (strings, characters, integers and boolean values). This situation causes that the user has to look for the value that he wants among a great quantity of data, possibly useless in that moment.

The debugger included in JBuilder IDE is very similar to the last one, although JBuilder debugger offers less facilities. This debugger presents the same troubles than the last one, the information is shown the way it is implemented and the data which is being searched is inside a great amount of additional information.

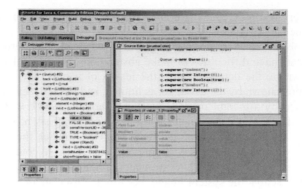

**Figure 2**: Forte for Java debugger.

## 2    A multimodal approach

DSTool tries to avoid that abstraction capacity become a key factor. Therefore, this tool tries to achieve that users with limitations to make abstractions, like the students previously referenced, do not find more difficulties than the debugging data structure task causes.

To get this target, DSTool debugger lets users debug their data structures in a logic way, with graphic representations of the structures, in contrast to the physical way that is imposed by the classic debuggers.

For the user not to need to resort to the abstraction, it is necessary to understand immediately the way the debugger works and to recognize clearly the data structures the tool displays.

These targets are absolutely necessary to get a correct behaviour of the debugger, and to reach those targets DSTool has to produce in the user a mental model the closest possible to the design model. To get this, the system design is essential, because the user's model is defined by its image. The solution used by DSTool to achieve the user obtains a correct mental model from the system is the graphic representation of the data structures. This representation shows the information stored in the structures according to the typical layout of each data structure (see Figure 3).

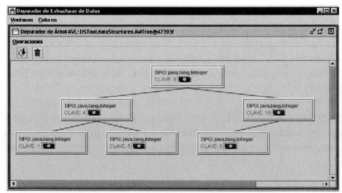

**Figure 3**: AVL tree representation in DSTool.

DSTool debugger only shows the essential information, this tool avoids to show useless data (useless to the user in that moment). Therefore, it will be shown the object type and value, and every property of the object that can be modified by user (see Figure 4).

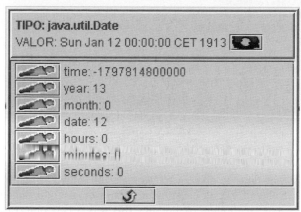

**Figure 4**: Data shown in every node of the data structure.

Obviously, it is essential the information to be right all the time. To get this, it has been used the structural reflection (Holser, 2001) of Java, in order to get the necessary information. The Model-View-Controller (Gamma, 1995) paradigm is the other technology, used to keep the consistency between real data and graphic representations.

By means of structural reflection are got every piece of data that is shown by the graphic representation. Thanks to this technology, restrictions will not be imposed on users at programming time. This way any class, included in the Java library, or user made, can be explored by DSTool debugger without need of modifications or additional programming.

Model-View-Controller has been used to keep the consistence permanently between the information stored in data structures  and the graphic grepresentation, achieving this way the absolutely essential reliability.

In the first section it was criticized the lack of flexibility in the classic debuggers to modify values, to insert new elements, or to remove nodes from the data structures. These limitations disappear in DSTool, because of this tool allows to perform all those operations through a graphic representation of the data structure, getting a simpler and more intuitive procedure.

Every value shown by the DSTool debugger can be modified through this tool by means of simples dialogs. This characteristic is offered by classic debuggers, but in those tools is very difficult to insert or to remove nodes of the data structures.

DSTool lets insert and remove objects into/from the structure. The biggest complexity it is found in the insertion of new elements, because it is necessary that can be inserted any kind of object. To make this operation as easy as possible, DSTool develops a wizard to build objects, wizard that guides the user through three steps.

In the first step, user must introduce the class name of the object to insert. In that moment, the wizard will be show all the constructor that the class has with its parameters (see Figure 5).

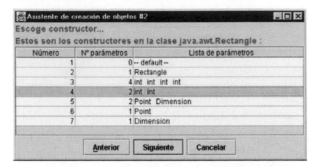

**Figure 5**: Constructor choosing wizard screen.

This way, the user is choosing the constructor to make his object. Finally, in the third step, is necessary to introduce values for each constructor parameter, using a dialog where the user can introduce the data he wants (see Figure 6), building this way a new object that can be inserted in the data structure.

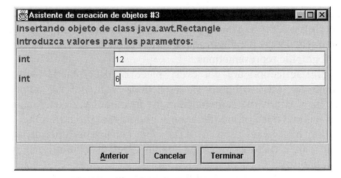

**Figure 6**: Data introducing wizard screen.

# 3    Conclusion

DSTool debugger helps students and more experienced users in debugging process, thanks to the graphic representations of the data structures, a more intuitive method than the one classic debuggers offer, avoiding this way, the abstraction problem of the students.

# References

Gamma, E. (1995). Design patterns. Elements of reusable object-oriented software. Ed. Addison-Wesley.

Holser, P. (2001). Limitations of reflective method lookup. *JavaReport*, August.

Norman, D. (1988). The psychology of everyday things. Basic Books, Inc.

# Artificial Ant Colonies and E-Learning: An Optimisation of Pedagogical Paths

*Y. Semet[1], Y. Jamont, R. Biojout, E. Lutton, P. Collet*

Projet Fractales - INRIA Rocquencourt, B.P. 105
FR-78153 Le Chesnay Cedex, France
Yann.Semet@tremplin-utc.net
[Evelyne.Lutton;Pierre.Collet]@inria.fr
[Yannick.Jamont;Raphael.Biojout]@paraschool.com

## Abstract

This paper describes current research on the optimisation of the pedagogical path of a student in an existing e-learning software. This optimisation is performed following the models given by a fairly recent field of Artificial Intelligence: Ant Colony Optimisation (ACO) [1,2,4]. The underlying structure of the E-learning material is represented by a graph with valued arcs whose weights are optimised by virtual ants that release virtual pheromones along their paths. This gradual modification of the graph's structure improves its pedagogic pertinence in order to increase pedagogic success. The system is developed for Paraschool, the leading French E-learning company. Tests will be conducted on a pool of more than 10,000 users.

## 1 Introduction

The e-learning software of the French Paraschool company offers a complement to high-school teaching. The software is used in schools, over a LAN, with a supervising teacher, or from home over the Internet. It contains small tutorials, exercises and multiple-choice questions that allow students to practice on their own. The original software provided deterministic HTML links and Paraschool was looking for a system that would enhance the navigation by making it adaptive and user-specific, so that both individual profiles and collective characteristics could be taken into account in an automatic and dynamic fashion. For instance, some successions of lessons may prove particularly successful in helping students understand a particular notion and those successions, leading to high success rates in subsequent exercises, should be automatically detected and highlighted.

### 1.1 Evolutionary Computation and Ant Colony Optimisation

Numerous difficulties pertaining to this problem (multiple contradictory objectives, fuzziness, complexity) immediately ring the bell of evolutionary techniques (a set of AI engineering tools of bio-mimetic inspiration), among which Ant Colony Optimisation (ACO) seems particularly well suited. This subfield of Evolutionary Computation comes from the observation of actual ant colonies and social insects in general such as bees or termites, and of their extraordinary abilities

---

[1] Université de Technologie de Compiègne, BP 60319, 60203 Compiègne Cedex, France.

to co-operate at the individual level to trigger complex and intelligent behaviour at the global level, an emerging phenomenon also known as "Swarm Intelligence" [1,4]. The ants' ability to come up with optimal paths to fetch food, for example, through the release of chemicals along their way is very remarkable and modelling this simple idea yielded exciting results in the field of combinatorial optimisation [5] (efficient heuristic solving of the Travelling Salesman Problem (TSP), routing problems, etc.).

Applying ACO techniques in Paraschool's context seems relatively straightforward when one sees the E-learning software as a graph with valued arcs through which students navigate, following suggestions made by the system and where:

- nodes are pedagogical items (exercises, lessons, quizzes, etc.);
- arcs are hypertext links between those items;
- weights on the arcs reflect the probabilities to suggest subsequent nodes to students;
- original weights are determined by the pedagogical team of Paraschool.

The task of the ACO is to optimise the weights on the arcs in order to maximise student success.

Besides their efficiency to quickly reach near-optimal solutions, ACO algorithms are also especially appreciated for their robustness and adaptability: just as natural ant colonies quickly find a new source of food when one disappears, ACO algorithms quickly find new optimal paths when the underlying graph suddenly changes. In Paraschool's case, optimising a graph with respect to some dynamic cognitive behaviour at both an individual (with multiple instances) and a collective level therefore seems like a perfect job for ACO algorithms. The transposition, in particular, from the work successfully carried out with ants on TSPs is again straightforward: each student going through the graph is represented by a virtual ant that releases virtual pheromones (concretely by incrementing floating point values carried by the arc) proportionally to its amount of successes and failures. Out of this information, stored in the "environment" and called "stigmergic" information [1], emerges a representation of the interaction between the students and the pedagogic material. This representation is used to derive probabilities that dictate the forthcoming behaviour of the software. The key advantage of this system is that this representation is both reactive and robust. And this is so, firstly because pheromones evaporate with time -which prevents the system from freezing or converging towards a particular state- and secondly because students, by browsing the graph, continually update the representation, thereby reflecting the dynamics of their needs.

## 2 Features and specifications

### 2.1 Pedagogic weights

The pedagogical team gives a weight $W$ to each arc, reflecting its importance with respect to other arcs coming out of the same node. This describes the pedagogic structure of the site: after a given lesson, the user can follow several possible arcs; the relevance of which is indicated by $W$. The higher $W$, the more adequate it is for students to follow the corresponding arc.

### 2.2 Pheromone release and evaporation

Following the validation of a node, an ant (i.e. a student) releases pheromones along the way that led it to that node. There are two kinds of pheromones: one for successes ($S$), one for failures ($F$). Pheromones are released backwards in time along the ant's path starting from the last validated node with decreasing amplitude. This is meant to reflect the fact that all the nodes a student went through previously have an influence on its ability to succeed in validating its current node. Of course, this influence should decrease with time: the more ancient a visit to a node, the less influence it has. This "back propagation" of pheromone release is limited in scope for obvious reasons (from both the algorithmic and pedagogic standpoints) and a number of nodes is thus set by the pedagogical team after which the back propagation stops. A typical value of 4 nodes has been agreed upon. In addition, pheromones released as stated above evaporate with time: their values tend to go back to 0 if the corresponding arc is unused for a long time. This is meant to make the system adaptive and to prevent it from being trapped in a particular state.

## 2.3 *H-node*s: historic weight computation and evaporation

In order to adapt the system to each student, track is kept i.e. stored in a database- of each node visited by a student, not only in the present sessions but in all of his/her previous sessions as well. For each node and for each student a historic weight $H$ is stored in a *H-node* with a default value of 1.0, meaning that the node has not been visited yet. When the node is visited, the value is multiplied either by *h1* (if it is a success) or by *h2* (if it is a failure). Values for *h1* and *h2* can be tuned, but typically *h1*=0.9 and *h2*=0.95. This $H$ value is going to be used (see below for details) on how) to discourage a student from visiting a node he/she has already seen, this discouraging being weaker when the node was failed. To reflect the fact that a student has limited memory, this $H$ value tends to go back to 1.0 with time, along the following equation where $x$ is the time elapsed since the last consultation:

$$(1) H_t = H_{t-1}\left(1 + \frac{1 - H_{t-1}}{H_{t-1}} \cdot \frac{1 - e^{-\alpha}}{1 + e^{-\alpha}}\right)$$

In the next equation, $\tau$ is a time constant that sets the speed of the phenomenon. It should be calibrated to correspond to the volatility of the students' memory:

$$(1) \Leftrightarrow \tau = \frac{1}{x} \cdot \ln\left(\frac{1 + \alpha}{1 - \alpha}\right) \text{ with } \alpha = \frac{H_t - H_{t-1}}{1 - H_{t-1}}$$

This latter equation can be used by the pedagogical team to tune the value of $\tau$ in a convenient way: provided one defines what "forgetting an exercise" means, for instance if its weight, starting from $H_{t-1} = 0.5$ (1 visit with success), grows back to $H_t = 0.9$, this gives $\alpha \approx 2.2$ and the pedagogic team then only has to estimate the time it takes to "forget an exercise": 1 week for example ($x$=604800 sec.) gives $\tau \approx 3.6E - 6$.

## 2.4 Fitness calculation

Using all the information described above, each arc $a$ is given a *fitness value*:

$$f(a) = H \cdot (\omega_1 W + \omega_2 S - \omega_3 F)$$

This value unifies in a weighted average all the factors that make an arc "desirable" or not: $f$ is high when:
- The arc's ending node was last visited a long time ago ($H$ is close to 1)
- The arc is encouraged by professors (high $W$)
- People have succeeded a lot around that node (high $S$)
- People have failed a little around that node (low $F$)

## 2.5 Arc selection and subsequent suggestion

After a node has been validated, the outgoing arcs are sorted according to this computed fitness value. One arc is randomly selected among the whole list, with a probability that is proportional to its fitness. It is suggested as an adequate follow-up to the student who pressed the ACO-powered NEXT button. A variety of selection procedures has been implemented, taken from the field of genetic and evolutionary computation, among which: roulette-wheel selection, ranking based methods and stochastic tournament selection. Choosing one method or another gives more or less control on the phenomenon by allowing to tune more or less precisely the amount of randomness this selection procedure is going to have. (cf. [3] for details).

The three approaches have been implemented, but tests have not been thorough enough to determine which method was the best.

## 3 First Results

Numerous tests have been conducted. First, a simulation procedure has been defined to allow for stabilization and calibration of the various parameters. The algorithm was then applied to the actual Paraschool software.

### 3.1 Simulations

#### 3.1.1 Modelling the population

A model of user population has been derived to conduct automatic simulation tests:

Each virtual student (i.e. ant), is given a certain *level* represented by a floating point value between 0.0 and 1.0. This value is normally distributed over the population of students with mean 0.5 and standard deviation 1/3. Each exercise is assigned a *difficulty* value, also between 0.0 and 1.0. When an ant arrives at a given node, if its level allows it to validate the node (*level>difficulty*), it succeeds, otherwise, it fails. Pheromones are released accordingly. General calibration of the algorithm was performed on a "real" graph, i.e. corresponding to an actual part of the Paraschool website (the "Vectors" chapter of a mathematics course for high school students around age 14). Arcs between nodes and corresponding weights have been assigned by the Paraschool pedagogical team. The sample case is therefore realistic (20 nodes, 47 arcs) and constitutes a meaningful structure with real size.

#### 3.1.2 A particular test case

Several features are expected from the ant colony. In particular, it should be able to correct inappropriate arc weight values. To investigate this properly, after a rough calibration and observation process conducted on the real sized graph mentioned above, experiments are conducted on a reduced graph that exhibits such a situation:

After solving exercise 1, the student can either go to exercise 2 or to exercise 3 before he ends, in both cases, with exercise 4. Exercise 3 is encouraged by the pedagogical team as the arc leading to it is assigned a weight of 5 versus 1 for the arc leading to exercise 2. The problem is that the success rate of exercise 4 is much higher when the student comes from exercise 2 than when he/she comes from exercise 3. What is expected from the system in such a case is to detect the situation and to reverse the two probabilities so that students are encouraged to follow the right path. This should be achieved naturally, i.e. without any human intervention, thanks to the release of virtual pheromones along the arcs. The arc leading to exercise 2 will hold a large amount of success pheromones and a low amount of failure pheromones. The arc leading to exercise 3, on the contrary is going to be in the opposite situation and this double discrepancy is going to be reflected in the arcs' fitness, thereby modifying their probabilities to be followed. Progressively, the arc leading to exercise 2 takes over the arc leading to exercise 3 and experiments show that a reasonable situation is promptly re-established

In the real-world version, such a discrepancy between weights given by the pedagogic team and evolved weights will issue a warning so that measures can be taken to solve the problem.

## 3.2 Real world application

The application to the real Paraschool system is only in its early stage. The ant colony algorithm has been integrated to the entire website in a downgraded mode where pheromones are only used to gather information and do not yet influence arc probabilities. Ten days after the integration, 566 "ants" have browsed the site, 2419 arcs have been visited and 3021 *H-nodes* have been created. First observations tend to show that singular nodes (i.e. too easy or too difficult exercises) see corresponding amounts of pheromones cumulate around them (e.g. high $S$ and low $F$), giving its first credits to the pheromone representation of the pedagogic structure. From an algorithmic point of view, these initial observations also show that the system is stable and able to handle all the additional computations without any noticeable overhead.

## 4 Conclusions and outline for future work

Time has now come to analyse the algorithm behaviour while in passive mode and tune the different parameters. When results show that the system is really stable, it will be switched to active mode wherefrom it is hoped that the ACO heuristic will provide:
- a seemingly intelligent system that improves the behaviour of the web site from the student's viewpoint,
- a refined auditing tool to help the pedagogical team identify the strengths and weaknesses of their software and pedagogic material.

From a theoretical perspective, this work brings encouraging elements of answers as to whether the emerging properties of social insect-based models can:
- make proper tools to enhance e-learning systems,
- adequately describe the cognitive behaviour of a social system (students and teachers),
- scale well from small experimental environments to a real-world application involving several thousands of individuals.

As this technique is original in the field of E-Learning and as the observation phase is only beginning, the present study should be seen as pointing out a potentially interesting research direction while great expectations are put in the observation of the forthcoming behaviour of the system.

## References

[1] E. Bonabeau, M. Dorigo, and G. Theraulaz, Swarm Intelligence: From Natural to Artificial Systems, Oxford University Press 1999. ISBN 0-19-513159-2.

[2] E. Bonabeau, M. Dorigo, and G. Theraulaz . Inspiration for optimization from social insect behaviour. Nature, vol. 406, 6 July 2000.

[3] D. E. Goldberg, K. Deb, A comparative analysis of selection schemes used in genetic algorithms (1991),in FOGA91, vol. 1, pp 69-93.

[4] M. Resnick, Turtles, Termites and Traffic Jams. Explorations in Massively Parallel Microworlds. Complex Adaptive Systems series MIT Press, 1994 .

[5] T. Stützle and M. Dorigo. ACO Algorithms for the Travelling Salesman Problem. In Proceedings of the EUROGEN conference, M Makela, K Miettinen, P Neittaanmaki, J Periaux (Eds), John Wiley & Sons, 1999, ISBN: 0471999024.

# The electronic bulletin board system "IS-Board" which supports the information education

*Yoshihisa Shinozawa*

Faculty of Science and
Technology, Keio
University
3-14-1 Hiyoshi Kouhoku-ku
Yokomaha 223-8522, Japan
shino@ae.keio.ac.jp

*Tomofumi Uetake*

School of Business
Administration, Senshu
University
2-1-1 Higashimita Tama-ku
Kawasaki 214-8580, Japan
uetake@isc.senshu-u.ac.jp

*Shinji Takao*

NTT Advanced Technology
Corporation, Hit Center
Higashi-Totsuka-West
Bldg., 9F, 90-6 Kawakami-
cho Totsuka-ku Yokohama
244-0850, Japan
takao@hit.ntt-at.co.jp

## Abstract

In the information education, instructor's load is very large, because he/she needs to teach to the students who have various questions. So, a support system which reduces the instructor's load and supports the information education has been required. Moreover the system which does not obstruct natural communications has been needed. In this situation, we propose the electronic bulletin board system "IS-Board". IS-Board extracts useful information from the electronic bulletin board without requiring users' extra efforts and provides them to the students effectively. IS-Board also has the function of accumulating reusable information as FAQ to reduce instructor's load. Using this system in the classes of the information literacy education during three months, it is confirmed that our system is effective to support the information education.

## 1 Introduction

In the information education, especially the information literacy education, all students must understand the wide-ranging lecture contents (i.e. operating methods of the computer, information ethics, etc). Moreover, the student's understanding level is different, and he/she often has various questions. So, instructor's teaching load is very large. In addition, the support methods in the self-study are not sufficiently offered. To solve these problems, TA (teaching assistant) is arranged, and Q&A by E-mail is done. However, the instructor's load is still large, and it is not sufficiently answering for the student's questions.

Recently, the various systems which support the learning activity by using the Internet have been proposed (ALIC (ed.), 2001)(Satou, 2001). But there are few systems which the student can easily use. Because, many of these systems are inconvenient, and there are constraints in the content of the input. In this situation, we have constructed an electronic bulletin board system for the purpose of solving above problems (Takao and Mihira, 2000). By managing this system, we confirmed that the electronic bulletin board system could effectively support the information education. But we found following problems (Shinozawa , Uetake and Takao, 2002).

- If the utterance content is restricted, there are few questions and useful information. Because students hesitate to utter their opinions.
- If students are allowed to utter freely, it is difficult for them to find useful information from all utterances on the bulletin board system. Because there are a variety of utterances concerning the lecture, operating methods of the computer, and campus life.
- The instructor's load is still large, because he/she provides most of useful information.

So we propose the electronic bulletin board system "**IS-Board** (Information Sharing Board)" to solve above problems without requiring users' extra efforts. IS-Board has following functions.

- IS-Board has no constraints to utterances and does not obstruct natural communications.
- To prevent students from overlooking useful information, IS-Board extracts them from all utterances automatically and provides them to students effectively.
- To reduce the instructor's load, IS-Board extracts reusable information from useful information and accumulates them as FAQ.

## 2 Useful information and Reusable information

In the information education, useful information means the information which contains the further explanations of the lecture, the question and answer about examinations or assignments the operating methods of computer. By analysing the log of the normal electric bulletin board system, we found that useful information is classified into following two types.

- *Provide type*   : Useful information which is voluntarily provided.
- *Q&A type*   : Useful information which is provided through Q&A.

We also found the following features of useful information.

A) Most of instructor's utterances are related to the lecture and are usually useful.
B) Useful information often includes URL, or keywords related to the lecture.
C) Useful information includes relevant nouns included both in the title and the article.

Moreover, we found that we can use part of useful information as reusable information. Reusable information contains the information about the operation methods of the computer and application software. The articles which contain examinations and assignments for a particular class are not reusable information, because students in other classes cannot use them. We also found that the articles of reusable information do not include keywords related to time restrictions, examinations and assignments.

## 3 IS-Board

An outline of IS-Board is shown in Figure 1. IS-Board has no special constraints. IS-Board can extract useful information automatically and provide them to the students effectively. IS-Board can accumulate reusable information as FAQ to reduce the instructor's load.

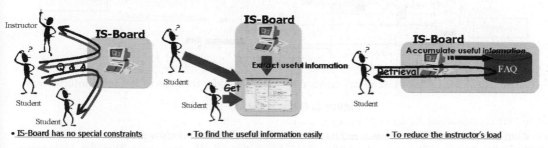

**Figure 1:** Outline of IS-Board

## 3.1 Rules

We constructed extract-rules which can extract useful information automatically by using the features of the useful information. IS-Board extracts useful information from all utterances by the following extract-rules.

### Extract-Rules
1. In case of the provided type, if an instructor writes new article or an article includes keywords related to the lecture, our system extracts it as useful information.
2. In case of the Q&A type, if an instructor answers student's question or the article includes keywords related to the lecture, our system extracts it as useful information.
3. Our system accumulates the nouns included both in the title and in the article of useful information extracted by extract-rule-1 and extract-rule-2 as keywords. If an article includes these keywords, our system extracts it as useful information.

After that, our system extracts reusable information from useful information. Using the feature of the reusable information, we constructed following FAQ-rule which can automatically extract reusable information from useful information.

### FAQ-Rule
1. If an article of useful information does not include keywords which indicate time restrictions, examinations and assignments, our system extracts it as reusable information. Moreover, IS-Board extracts nouns in the title and the words related to a lecture in the article as keywords for indexing. IS-Board accumulates reusable information with these keywords as FAQ.

## 3.2 Overview of IS-Board

IS-Board has some improvements upon the electronic bulletin board system on WWW. A student can utter about his/her questions and opinions on IS-Board with the same operation as normal electronic bulletin board system.

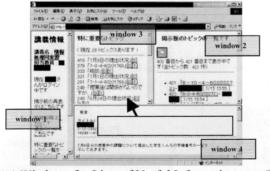 

(A) Windows for Lists of Useful Information     (B) Window for Lists of Reusable Information

**Figure 2:** Overview of IS-Board

Our system uses HTML frames and has five windows. The first window provides the information about the lecture. The second window provides a list of all utterances. This list includes a contributor, date and title of utterance. The third window provides a list of useful information extracted automatically. By using this window, the students can get useful information easily. The fourth window provides the content of the article. The fifth window provides a list of reusable information. By using this window, students can get reusable information. IS-Board also has a function to correct the lists manually. So instructors can correct mistakes in the lists easily.

# 4   Evaluation

To evaluate the effectiveness of IS-Board, we used our system in two classes of the information literacy education during three months. Each class has about 75 students. In the last lecture, we inquired questionnaires on IS-Board to students.

As a result of the analysis, there were 443 articles and 422 topics on average. There were 27 topics on average as useful information. Students used IS-Board 1.4 times per week on average.

## 1.1   Evaluation of the rules

To evaluate the extract-rules, we compared with the topics which are judged as useful information by the instructor and the topics which are extracted by IS-Board automatically. The FAQ-rule was also evaluated by the similar method. The results are shown in Table 1.

**Table 1:** Evaluation of the Rules

|  | Correct answer rate |
| --- | --- |
| Extract-rules | 78 % |
| FAQ-rule | 70 % |

As a result of this analysis, it is confirmed that our rules were sufficiently practicable. Moreover instructors can correct mistakes easily by using IS-Board. So, students can get useful information effectively. But there is a future work to improve these rules.

## 4.2   Effectiveness of IS-Board

Is-Board provides the list of useful information and the list of all utterances on each window separately. To evaluate the effectiveness of information which was extracted by our system, we did a questionnaire about the methods of our system use to the students. The result of the questionnaire is shown in Table 2.

**Table 2:**  Methods of IS-Board Use

|  | Answer rate |
| --- | --- |
| 1. He/She read the articles on the list of useful information. | 48% |
| 2. He/She read the only articles provided by instructors. | 28% |
| 3. He/She read all articles. | 5% |
| 4. He/She hardly read articles. | 19% |

This result shows that 80% of the students often used IS-board and a half of students read the articles from the list of useful information by priority. By this result, it is confirmed that useful information provided by our system is useful for the students and our system is effective to support the information education.

Finally, we evaluated the educative effectiveness of IS-Board. First we classified the student into a beginner (who can use only e-mail and web browsing), middle class person (who can use word processor, excel, etc.), and upper grade person (who can do programming). Next, we analyzed the relation between the frequency of IS-Board use and the result of this subject (See Figure 3).

These results show that the more student (except upper grade person) uses IS-Board, the more effectively he/she can understand the lecture contents. These results also lead that IS-Board is more effective to the beginner who especially needed useful information.

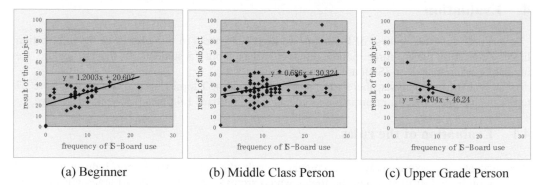

(a) Beginner       (b) Middle Class Person       (c) Upper Grade Person

**Figure 3:** Educative Effectiveness of IS-Board

## 5   Conclusions

We proposed the electronic bulletin board system "IS-Board" which supports the information education. IS-Board has no constraints to utterances and has following functions without requiring users' extra efforts, (1) extracting useful information from utterances and providing them to the students easily, (2) accumulating reusable information as FAQ. We used our system in the classes of the information literacy education to evaluate the effectiveness of IS-Board. The results found by the experiments are as follows.

- There were a lot of utterances which include useful information, because our system has no constraint in input method and its content.
- Our system extracted useful information from utterances successfully and provided them to the students effectively. So students do not overlook useful information, and can have better understanding to the lecture.
- Our system extracted the reusable information successfully and was able to accumulate reusable information as FAQ. So, it is expected to reduce the instructor's load.

These results show that IS-Board is effective to support the information education. To support the information education further, we need to improve the rules, and the interface of IS-Board.

## References

ALIC (ed.). (2001). e-learning white paper 2001/2002, Ohmsha,.(In Japanese)

O. Satou. (2001). Net learning, Chuo keizai sha. (In Japanese)

Y. Shinozawa, T. Uetake, S. Takao. (2002). Evaluation of the electronic bulletin board system "IS-Board" which assists information education, IPSJ SIGNotes Groupware and Network services Abstract No.045 – 011, pp.59-64. (In Japanese)

S. Takao, Y. Mihira. (2000) Participation Promoting Functions on a BBS Service for Lectures, IPSJ SIGNotes GroupWare Abstract No.037 – 008, pp.43-48. (In Japanese)

# MARILYN: A Novel Platform
# For Intelligent Interactive TV (IITV)

*Soha Maad*

Fraunhofer
Institute For Media Communication (IMK)
Germany
Soha.Maad@imk.fhg.de

Abstract

This paper discusses the needs and requirements for Intelligent Interactive TV (IITV) and proposes MARILYN as a platform for providing it. MARILYN "Multimodal Avatar Responsive Live News Caster" is a prototype for business television (BTV) that reveals the prospects of the use of virtual human avatars for intelligent interaction with digital TV. MARILYN motivates an analysis of the requirement engineering for IITV and suggests a framework deploying natural language understanding, speech recognition, and an experience-based approach to decision support for establishing an IITV platform.

## 1   Introduction

Interactive TV (ITV) aims at combining the traditional style of TV watching with the interactivity of the personal computer and the internet. The development of ITV depends on several factors including:

*First, the development of the ITV production-delivery chain.* This chain involves producing the media content of a TV program, broadcasting it on a particular TV channel and receiving it by the TV viewer system.

*Second, the development of the medium and paradigm of interactivity with the ITV.* So far, the interaction with various types of ITV content is still limited, in general, to click-able interactivity as the sole paradigm of interaction and the use of remote control by ITV viewers as the sole medium of interaction.

*Third, the development of AI techniques that serves an IITV platform.* Speech recognition, natural language understanding, and decision support systems should be deployed in the development of an IITV platform.

This paper proposes a MARILYN as a platform for IITV. In section 2, a general model of the ITV production delivery chain is proposed. With reference to this general model, section 3 discusses the advantages of using 3D virtual human avatars for the presentation of TV programs. Section 4 depicts MARILYN "Multimodal Avatar Responsive Live News Caster". The requirements engineering for developing IITV is motivated in light of the analysis of various features of MARILYN.

## 2 The ITV production-delivery chain: the general model

The author proposes a general model for the ITV production- delivery chain (depicted in Figure 1). This model consists of three parts: i. content authoring and encoding at the production end; ii. transmission which refers to the broadcast of the digitally encoded content; and iii. delivery, decoding, and play-out at the ITV viewer end.

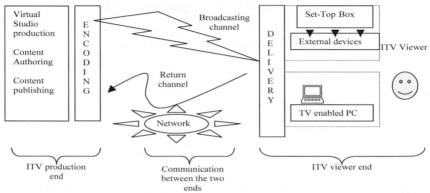

**Figure 1:** The general model of ITV production – delivery chain

At the ***ITV production end***, content authoring, virtual studio production, and content publishing takes places. The ITV content is digitally encoded for ***digital broadcasting*** through a particular broadcasting medium. At the ***ITV viewer end,*** the digitally broadcasted signal are received by the ITV viewer hardware (Set-Top-Box, or TV enabled PC) where they are decoded and played out. The return channel allows interactivity from the TV viewer to feed the content production, this might be in real time. A network enables this return channel.

## 3 Virtual Human Avatars for an IITV Production Delivery Chain

This section discusses the advantages of using 3D virtual human avatars for the presentation of ITV programs with reference to the general model introduced in section 2.

***At the ITV production end***, *3D virtual human avatars helps in bypassing the rigidity of traditional styles of TV production:* By taking off the role of the human TV presenter, the virtual human avatar helps in bypassing the rigidity introduced by the need to resort to digital video recording of a human TV presenter. Through appropriate modeling of the virtual TV presenter various edutainment needs of ITV viewers may be attained. Voice replies and facial gestures may be additional medium of interaction with the virtual human TV presenter.

***At the ITV viewer end,*** *3D Virtual human avatars facilitates on-demand continuous availability and engaged interactivity with the ITV viewer:* This may be possible with an appropriate technology set up. An intelligent virtual human avatar supports the development of interactive programs that responds intelligently to the ITV viewer ***on demand***. Video recording of a human TV presenter delays the response to the ITV viewer and makes 24 hours availability of the ITV programs hardly possible.

***At the ITV viewer end,*** *3D Virtual human avatars can potentially provide greater accessibility to the handicapped ITV viewer:* Physically handicapped ITV viewers may resort to other medium of interaction beyond clickable interaction. Speech interaction for the blind and hand gestures for the deaf are more easily produced in TV programs presented by virtual human avatars.

*The communication medium* between the ITV production end and the ITV viewer end is re-enforced through the use of 3D Virtual human avatars: Authoring content for ITV should now rely on intelligent paradigms of interaction that involves the user in the ITV content production to suit his/her edutainment needs. The ITV viewer may be engaged in the ITV production by having an on-demand choice of virtual human characters, of the mode of interaction, and of the speaking language used in the presented TV program. Abstract virtual characters may be more popular to ITV viewers and might help in better directing the attention of the ITV viewer to the domain specific TV content being presented. Moreover, the multilingual feature of ITV programs widens the scope of ITV audience.

With reference to the proposed general model of ITV production – delivery chain, there is evidence of intelligent interactivity and promised universal access[1] when resorting to virtual human avatars for TV presentation. The use of virtual human avatars provides a support for a completely automated ITV production – delivery chain.


# 4    MARILYN

This section describes a platform for ITV presentation of financial news by a virtual human avatar to create an innovative prototype for BTV[2]. MARILYN, developed at Fraunhofer Institute of Media Communication – ITV competence center, features a "Multimodal Avatar Responsive Live News Caster". Depicted in Figure 2, MARILYN gives insight on the use of virtual human character for developing the IITV platform. Five stages are adopted in the development of MARILYN:

**Stage 1.** This stage involves the development of a template for an animated virtual human avatar (3D humanoid) enriched with facial expressions, voice replies, and hand gestures. The template would also give the option to choose a desired virtual human character.

**Stage2.** This stage involves developing several multilingual templates for live financial news content (tickers, and news) taken from various international stock exchanges.

**Stage 3.** This stage involves establishing a bi-directional communication medium between the virtual human avatar (TV presenter) and the financial content. The virtual human avatar is conceived as a mediator between the ITV viewer and the financial content, thus providing the mean for interactivity with this content. The virtual human avatar has two responsibilities: (1) to serve the ITV viewer intelligently by channeling the ITV viewer's demand to the appropriate ITV content; and (2) to edutain (educate and entertain) the ITV viewer through voice replies, hand gestures, and facial expressions.

**Stage 4.** This stage involves establishing an interactive communication medium between the animated virtual human avatar and the ITV viewer. In this respect various levels of interaction can be identified: the constrained, the semi-constrained, and the open-ended levels of interaction.

– The constrained interaction reduces the interaction between the ITV viewer and the virtual human TV presenter to preconceived modes of interaction that follows a menu driven style of interaction with audio and visual multimedia enhancement. In this case, the embedded

---

[1] The concept of universal access is supported, among others, by research in the area of User Interface for All (UI4ALL). This research is described in (Stephandis, 2000) as "*being rooted in the concept of Design for All in HCI, and aiming at efficiently and effectively addressing the numerous and diverse accessibility problems in human interaction with software applications and telematic services*".

[2] This case study was also considered in (Maad, 2002) to support ongoing research on universal access to multimodal ITV content.

intelligence of the virtual human avatar is restrained to the reply to specific queries from the ITV viewer and to the serving of pre-published ITV content that fulfils the viewer demand. All voice replies as well as the facial expressions and hand gestures of the virtual human avatar are fully preconceived.

– The semi-constrained interaction relies on conventional artificial intelligence techniques for retrieving and serving the ITV financial content on demand. The boundary of the financial content and the boundary of the intelligent interaction are pre-defined in this case. Spontaneous responses of the virtual human avatar to a wider space of ITV viewers' interaction are aspired, however, the ITV financial content might still be restrained to pre-published content.

– The open-ended medium of interaction suggests access to on-the-fly and on-demand authored ITV financial content. As such the virtual human avatar gives the ITV viewer an indirect mean for the access to the ITV content and the potential to intervene in authoring the ITV content to suit his/her edutainment needs.

**Stage 5.** This stage involves checking the compliance of the used technology with prevailing standards for ITV technology.

**Figure 2:** MARILYN: "Multimodal Avatar Responsive Live News Caster"

# 5 Requirements Engineering For IITV

MARILYN suggests novel foundations for artificial intelligence serving the development of IITV. By studying the various aspects of MARILYN the following requirements engineering of IITV emerge:

*Providing intelligent open-ended interaction with the ITV content:* Novel paradigms for artificial intelligence are needed to cope with open-ended interaction[3] with an ITV viewer querying a domain specific content for edutainment purposes. Authoring ITV content with non-preconceived modes of interaction is essential to serve a wider audience need and edutain the viewer in a less constrained and dictated way.

*Coping with various categories of ITV viewers:* Studying various categories of ITV viewers is essential to establish various modes of interactions with various categories of viewers. This would involve the classification of the ITV viewers in pre-defined categories based on the initial interaction of the virtual human TV presenter with the ITV viewers.

*Identifying various medium of interaction beyond clickable interactivity:* This necessitates the deployment of voice recognition and natural language understanding to produce synthetic voice

---

[3] Open-ended interaction designates interaction evolving with users' needs and requirements, involving various medium of interaction, and relying on well-established paradigms for user-developer collaboration in the corresponding Software System Development activity (Beynon, Rungrattanaubol, & Wright, 1998).

replies of the virtual TV presenter. Pattern recognition is also needed to capture the ITV viewer video recorded gestures in the course of his/her communication with the virtual TV presenter. *Supporting the decision making of the virtual human avatar:* An experience-based approach[4] is needed to support the virtual human avatar decision in serving the needs of the ITV viewer. New foundation for artificial intelligence that favour an experience based approach to decision support were proposed in (Beynon, Rasmequan, & Russ, 2000)

# 6    Conclusion

This paper discussed the needs and requirements of IITV with reference to MARILYN, a newly proposed platform for IITV. MARILYN involves virtual human avatars interacting with the ITV content and the ITV viewer. By exposing MARILYN to a wider audience during a CEBIT2003 exhibition held in Germany, viewers expressed their interest in seeing the future deployment of MARILYN for TV programs including: personalized TV news presentation, EPG (Electronic Program Guides) where interaction is through virtual human avatars rather than a desktop style of navigation, and documentary TV programs. Besides its popularity for an ITV platform, MARILYN raised interest in its deployment on other pervasive platforms such as mobile and home devices. Conservative TV viewers (old generation and traditionalists) showed preference for a human TV presenter and simple modes of interaction with the TV. Nonetheless, there is a wide acknowledgement of the importance of 3D virtual human avatars in enhancing availability and interaction with TV programs provided that 3D technology gets more mature to serve various platforms including the ITV one.

# 7    Acknowledgement

The author is indebted to many members of the Fraunhofer-Institute For Media Communication (IMK) group for relevant ideas, in particular to Sepideh Chakaveh for proposing the project of virtual TV presenter and supervising its development. The author acknowledges the efforts of Helmut Ziegler, Tamer Messih, Oloaf Geuer, Stephan Putz, Sorina Borggrefe, and Ralf Haeger in developing MARILYN. The author would like to thank ERCIM for sponsoring her fellowship.

# References

Beynon, W. M., Rasmequan, S., and Russ, S., A New Paradigm for Computer-Based Decision Support, *the Journal of Decision Support Systems*, Special Issue, Autumn 2000.
Beynon, W.M., Rungrattanaubol, J., Sun, P. H., Wright, A. E. M., Explanatory Models for Open-ended Human-Computer Interaction, Research report CSRR-346, Warwick Univ., 1998
Maad, S., Universal Access For Multimodal ITV Content: Challenges and Prospects, *SPRINGER LNCS proceedings of the* 7th ERCIM Workshop "User Interfaces for All", Paris, France 2002.
Stephandis, C., From User interfaces for all to an information society for all: Recent achievements and future challenges, Proceedings of the 6th ERCIM Workshop UI4AL October 2000, Italy

---

[4] An experience based approach to decision support would aim at bypassing the rigidity of formal approaches to decision support that circumscribe this activity in problem formulation, solution exploration, and selection of the best possible solution.

# Speech Interaction for Networked Video Games

*Eleni Spyridou, Ian Palmer*
School of Informatics
University of Bradford

Bradford BD7 1DP, United Kingdom
{e.spyridou, i.j.palmer) @Bradford.ac.uk

*Elric Williams*
Digital Media
Trinity and All Saints
College
Leeds, United Kingdom
e_williams@TASC.ac.uk

## Abstract

In this paper we are presenting our ongoing research on multimodal human computer interaction in video games. We give an explanation of multimodality and identify its role in video games interfaces. A summary of the preparatory phase is given before describing the concept and methodology of the acquisition phase, leading to the final experimental phase, where again concept and methodology are presented.

## 1   Introduction

Adding a speech interface to a virtual environment changes the relationship between the user and system. With direct manipulation, the system is relatively transparent; the user is directly embodied as an actor alone in the virtual world. Speech, however, requires a dialogue partner; somebody or something the user will talk to.

Combining a speech interface with a direct manipulation interface results in a multimodal interface where the user can act upon the world by issuing physical or speech commands and, conversely, the system can respond by speaking and/or by making changes in the virtual world. Examples exist of such systems (McGlashan 1995), (Karlgren et al. 1995), (Everett et al. 1998), (Cavazza and Palmer 1999), (Cavazza, Bandi & Palmer 1999), (Everett 1999).

In the following paper, we will give an overview of what multimodality is and what can offer in video games. We will then discuss our research on speech interfaces in video games and after describing the implementation of the speech interface, draw conclusions for future work.

## 2   Multimodality in video games

We refer to a multimodal interface when a computer system can receive human directed instruction using different modes, such as vocal manipulation (speech) and direct manipulation (mouse, keyboard, joystick). Multimodal interfaces can also feedback information using different modes, such as visual display or sound. A multimodal interface combining speech and direct manipulation can provide more efficient interaction than a single modality interface, and give the user benefits of both modalities (McGlashan, 1995). It allows one modality to compensate for the limitations of the other; a direct manipulation interface can compensate for limitations of speech by making immediately visible the effects of actions upon objects, and indicating through the display which objects are currently most important for the system. In addition, the user is free to decide which modality to use for the actions; the user may use direct manipulation for transportation within the virtual world, but the speech modality for manipulating objects.

It has been suggested (Bolter, Grusin, 1999) that in computer games, the player is acting both as a spectator and a director. Natural language input would certainly reinforce the latter role by providing the user with more abstract control. Using natural language and speech to control a third-person video game also gives rise to certain design decisions. Natural language essentially enables high-level, abstract commands and its processing is likely to be longer that using the mouse or keyboard. However, this means that natural language control would shift the game experience towards tactics rather than immediate action. It is also likely to affect the nature of the embodiment relation between the player and the avatar. These new interfaces could substantially alter the balance between acting and viewing, i.e. between the director and spectator role of the user.

# 3  Previous Research

Prototype systems were built to test the effects achieved when introducing speech into an interface (Spyridou, Palmer 2002). Computer games were created using the Unreal Tournament® (first person shooter video game) and Age of Empires®, The Age of Kings™ (real time strategy video game) engines. The experiments aimed to find how speech influenced the different types of game play and whether it enhanced the interaction of either.

The results of the experiment have shown that Automated Speech Recognition (ASR) interfaces are practical and desirable but only in certain type of games. Participants were conscious of the speech interface in both situations, when playing Unreal Tournament® and Age of Empires®, the Age of Kings™ (Spyridou, Palmer 2002). Delays in recognition in Unreal Tournament made the game slow, boring, awkward and uncomfortable to play. Speech was not the right interface for this particular game since it gave another level of interface for them to deal with. In Age of Empires®, the Age of Kings™, speech changed the level of interface of the game from being in control when using the mouse and keyboard to being in command when using speech. Speech recognition appeared faster and more accurate in this game and on more than one occasion convenient in saving mouse clicks and time to choose from the icons in the menu.

At this stage of the research we focused on two video game genres; action and strategy. For our present work we concentrate only on First Person Shooter (FPS) action games. This time instead of single-player mode we will be working on a team multi-player mode and make a transition from low-level commands to high-level commands using Natural Language Processing (NLP).

# 4  Current Research

## 4.1  Introduction

We are preparing to conduct experiments using a team multi-player FPS video game using Natural Speech (NS), not ASR, to test what are the most frequent vocal commands used in the game by the users. The subjects will use Natural Speech, and therefore the commands issued are expected to be lengthy and of high complexity.

Jurafsky and Martin (2002), state that high frequency spoken words are accessed faster or with less information than low frequency words. They are successfully recognised in noisier environments than low frequency words, or when only parts of the words are presented.

## 4.2   Methodology

We are using Unreal®, a commercially available 3D graphical game engine and Microsoft's Sidewinder for speech communication with the players. SideWinder provides four communication channels for the team members to communicate individually with, along with another channel to communicate with all of the team members. Mini DV cameras record the dialogues and reactions from the participants.

The aim is to collate a dataset of vocal instructions and feedback, which will be used as part of a larger lexicon in the experimental phase.

There are two teams of two players each, playing against each other. In order to win they have to work as a team. To achieve this they need to communicate. Each individual will be in a separate room and given full instructions of the use of the game and hardware. The players will be given 15 minutes unrecorded playing time to familiarise themselves with the control systems.

The team will compete in a 'Capture the Flag' mod (http://www.captured.com) for thirty minutes. The most common multiplayer FPS is 'death-match'. In a 'death-match' game the players have essentially one goal: to kill all the enemy players. This invariably leads to players using the 'lose cannon' playing technique; (they run around and shoot everything on an individual basis). It is rare to find evidence of teamwork and communication.

In 'Capture the Flag', each team has a flag at their base point and their aims are twofold: to capture and return with the enemy's flag and to prevent the enemy from capturing their flag. In order to be successful teamwork is essential, as the players must know what their teammates are doing. Otherwise both players could be defending or worse attacking leaving their flag undefended.

The results of this experiment are pending.

## 5   Future Research

An experiment will be conducted, which will be the same as the experiment performed during the acquisition phase, except that although the player will be told he has a teammate in another room, he will in fact be playing alongside a bot. The player will be told that since he has a good understanding of the software and hardware, he will not be allowed a familiarisation segment.

In order to achieve wide variance of recognised commands, and produce a better grammar for the speech interface, we will use as a working corpus, a compilation of Unreal® spoilers available from the Internet, of Unreal Tournament® official strategy book and the results from the previous experiment. The recommended actions are described as specific instructions, such as "Go left and jump into the water", "Go down the ramp into the library", "Jump down to the next floor".

For speech control we will use Natural Language Processing (NLP), which is based on a parser integrating syntax and semantics. The process of applying a grammar to an input sentence to determine its grammaticality and appropriate syntactic structure is known as parsing. However, in order to apply this grammar to one of the sentences it is necessary to provide a lexicon, which tells the parser which lexical syntactic categories, parts of speech, are associated with the words in the input sentence.

The example (see Figure 1), states that a sentence (S) can be divided into two sub-sentences (S1) and (S2). These two sub-sentences can be joined together with the conjoined word (and). The first sentence (S1) consists of a verb (V) and an adjective (Adj.). The second sentence (S2) consists of a verb (V) and a prepositional phrase (PP). The latter can break down to the preposition (Prep.) and a noun phrase (NP), which consists of the determiner (the) and the noun (N).

Although it has been suggested (Spyridou, 2002) that First Person Shooter (FPS) games are incompatible with simple low level speech recognition, NLP allows recognition of complex structured sentences. Such commands are more common in 3Dimensional gaming environments. Therefore it is considered that NLP should offer greater and more advanced control for the user in an FPS game.

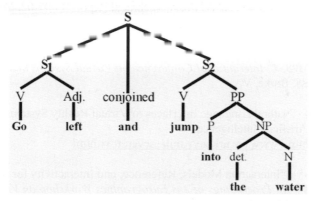

**Figure1:** The spoken command "Go left and jump into the water" parsed.

# 6   Discussion

Human beings are essentially social animals. We have the need to communicate in the real world with speech; posing questions and giving answers. Transferring this social activity to the virtual world in a video game, is another matter. When we are talking to a machine we feel uncomfortable, embarrassed, like talking to an empty room. That is because we expect a dialogue from the environment, a verbal feedback that will give us a confirmation that we became understood or that an action has been completed. This kind of interaction makes the video game more realistic and adds depth to the characters of the game.

# 7   Conclusion

In this paper we presented the three stages of our research as they stand at the present time.

What this research aims to prove is that a high-level multimodal interface utilising vocal and direct manipulation, when introduced in a video game changes the immersion, the interaction and the perception of the game from the users perspective.

The more communication the user has to employ, and the more varied the sensorial feedback, the more immersed the user will become with the virtual environment. This increases interaction and will change the user's perception of the game from 'playing a game' to 'having an experience'. In effect, the user's role within the game has become neither a spectator nor a director, but instead the player is now a participant.

# References

Bolter, J. D. & Grusin, R. (1999), *Remediation: Understanding New Media*, MIT Press, Cambridge.

Cavazza, M. & Palmer, I. J. (1999), "Natural Language Control of Interactive 3D Animation and Computer Games", *Virtual Reality*, Vol.3, pp.1-18

Cavazza, M., Banti, S. & Palmer, I. (1999), *""Situated AI" in Video Games: Integrating NLP, Path Planning and 3D Animation"*, *Proceedings of the AAAI Spring Symposium on Artificial Intelligence and Computer Games*, AAAI Press, Technical Report SS-99-02, March 1999.

Everett et al. (1998), "Creating Natural Language Interfaces to VR Systems: Experiences, Observations and Lessons Learned", *Future Fusion: Application realities for the virtual age Proceedings of VSMM98, 4ᵗʰ International Conference on Virtual Systems and Multimedia*, Vol. 2, pp. 469-474. IOS Press, Burke, VA.

Everette S. S., (1999), "Natural Language Interfaces to Virtual Reality Systems", Navy Center for Applied Research in Artificial Intelligence.
[Online], Available: http://www.aic.nrl.navy.mil/~severett/vr.html

Klargen, J. et al. (1995), "Interaction Models, Reference, and Interactivity for Speech Interfaces to Virtual Environments". *In: Proceedings of 2ns Eurographics Workshop on Virtual Environments, Realism and Real Time*, Monte Carlo.

McGlashan, S. (1995), "Speech Interfaces to Virtual Reality", *International Workshop on Military Applications of Synthetic Environments and Virtual Reality.*
[Online], Available: http://www.sics.se/~scott/papers/masevr96/masevr96.html [1996, July 9]

Spyridou, E. and Palmer I. J. (2002), "Investigation into Speech based Interaction for Video Games", *Proceedings of the 3ʳᵈ International Conference of Intelligent Games and Simulation, Game-On 2002*, ISBN: 90-77039-10-4, pp.65-72. A Publication of SCS Europe Bvba, 29-30, November 2002.

Speech and Language Processing, D. Jurafsky, J. H. Martin, Prentice Hall, 2000, New Jersey, 0130950696

Investigation into Speech based Interaction for Virtual Environments, Transfer Report, E. Spyridou, Dept. of Electronic Imaging and Media Communications, School of Informatics, University of Bradford, Bradford, U.K.

Noname CTF v0.6.0, A CTF Mod for Unreal 1, [Online], Available: http://www.captured.com/nonamectf

# Is every kid having fun? A gender crossover approach to interactive toy design

*M.A. Stienstra*

*H.C.M. Hoonhout*

University of Twente
Postbox 217, 7500 AE Enschede
The Netherlands
marcelle_stienstra@hotmail.com

Philips Research
Prof.Holstlaan 4, 5656 AA Eindhoven
The Netherlands
jettie.hoonhout@philips.com

## Abstract

In the development of interaction toys for children aged 8 to10 yrs, two design strategies were used: a gender specific strategy, resulting in one masculine and one feminine toy, and a gender crossover strategy, resulting in a toy that should appeal to both genders by combining feminine and masculine elements. A test was conducted to determine which toy was seen as the most fun to play with. The results show that, though the toys were not easy to use, most boys and girls experienced this as a highly enjoyable challenge. The toys supporting a physically active way of interaction were enjoyed the most by both genders. The results partly corroborated our hypotheses.

## 1    Introduction

While children used to play outside in backyards and playgrounds, nowadays many children can be found behind computers and game consoles. Especially with children in the midst of their development process these devices are feared to offer not enough physical and cognitive stimulation. Another issue is that many existing electronic toys and computers in general are more appealing to boys than to girls. Recent studies have indicated that while boys use computers more for fun and games, girls use them mainly for schoolwork (Jenkins, 1998). This implies that boys and girls learn to associate computers with different task-settings, and probably develop different attitudes towards computers and related technologies. Our aim in this study is to investigate the possibility to develop electronic toys *equally* appealing to boys and girls.

## 2    Design for boys and girls: the gender crossover strategy

In order to try to bridge the differences between boys and girls in their perceptions of technology, we adopted the so-called 'gender crossover' strategy as opposed to a 'gender specific' strategy. Jenkins (1998) laid the foundation for this 'gender crossover' strategy, and elaborating on his notions we state that it is possible to create products that appeal to both boys and girls if masculine *as well as* feminine product characteristics are designed into products. These are more likely to appeal to all users irrespective of their gender. The general idea is that if users are offered a product that contains elements that will appeal to the one and elements that will appeal to the other gender, there is something appealing and familiar to be found in the product for everyone, but at the same time the product offers the possibility to get acquainted with elements that the user would not explore so easily otherwise. Whereas electronic toys such as fighting or shooting games

1051

or Barbie® software for the computer, are seen as highly gender-specific, a 'gender crossover' product cannot be so easily labeled as 'for boys' or 'for girls'.

## 3 Make it fun to use: design heuristics

A requirement was that the interaction toys could be used in connection with different computer games. Another requirement was that the toys had to be fun to use – since enjoyment is seen as a means to change attitudes towards technology. To this end we adopted a set of design heuristics outlined by Malone and Lepper (1987). Although primarily directed at educational software design, we considered these heuristics to be usable starting points for the design requirements of our toys. Following the heuristics the toys should:

- Provide a cognitive and physical *challenge* to the children, i.e., provide a difficult enough task that will put the child's skills and abilities to the test, without becoming too challenging.
- Raise the children's *curiosity* (by providing novel and surprising but not completely incomprehensible interface elements) and stimulate *exploration* to find out how it works.
- Contain *co-operative* and *competitive* elements, as they were designed to be played by two children. Both elements contribute to a challenging and motivating experience.
- Invite the children to take an *active* attitude, cognitively as well as physically (N.B. this requirement is not in Malone & Lepper's set of heuristics).

The children were to be cognitively challenged by the fact that they had to co-operate and co-ordinate their actions and by the fact that it was not obvious how the devices should be used. Research on child development and gender differences (e.g., Geary, 1998; Jenkins, 1998) provided information on appropriate physical challenges that could be designed into the toys. Significant differences appear to exist between the motor capabilities of boys and girls aged 8 to 10 years: boys appear to have better developed gross motor skills, while girls have better developed fine motor skills. (Geary (1998) provides a detailed account on this topic).

## 4 Design of the toys

In the design of the toys (see Figure 1), all the aforementioned requirements were incorporated. The differences between boys' and girls' motor capabilities formed the basis for the differences between the interaction toys and how to operate them. The first toy would require fine motor skills, the second toy gross motor skills, and the last toy would require both fine and gross motor skills. Thus, gender differences were designed into the manipulation and operation of the interaction toys rather than in their appearance. In addition, a game was designed, to provide a setting in which the toys could be used as interaction devices; this will be described in section 5.

**Figure 1:** The toys; from left to right Stickysticks, Twistyertouch, and Tunemein

The first toy, **Stickysticks**, is a magnetic table that can be operated by two magnetic sticks, for each child one, challenging their fine motor skills. The top of the table is like a chessboard, with 4 by 4 fields. Under each field a magnet is positioned. The interaction is realized by positioning the magnetic sticks just above specific fields, however without touching the game board. Feedback is given through sound and magnetic force. In the version of Stickysticks that was used in our study, the fields on the board contained colored blobs or grey geometric figures. Each child manipulated one stick that could either be used for the color or the figure fields. They had to create particular combinations of colors and figures as input for the game ('making a particular move') by 'activating' a color field and a figure field, such as a green square or a blue star, at the same time.

The second toy, **Twistyertouch**, is a large soft mat. On the base, which measures 160 by 160 cm, 4 cubes are placed, each measuring 40 by 40 by 40 cm. Each visible surface of Twistyertouch functions as a control button. Children can activate a surface by hitting, jumping or pushing on it, requiring gross motor skills. When a surface is activated, feedback is given through sound. To realize a certain action, children have to activate a number of specific surfaces. This, in combination with the dimensions of Twistyertouch, stimulates co-operation: to realize a certain outcome as quickly as possible, the players need to divide the surfaces that have to be activated, to save time. And, the quicker they are, the more points they can gain. In this study, the Twistyertouch was implemented as a color keyboard. There were seven differently colored groups of surfaces. Particular game actions could be realized by activating all surfaces of the same color. Feedback was given on the number of surfaces that were already activated, but not on which surfaces were activated, to add to the challenge.

The third toy, **Tunemein**, is based on the theremin, a music instrument developed in 1919 by a Russian physicist named Lev Termen. Tunemein is an instrument that consists of a small base with a long aerial sticking out of it. Total height of the Tunemein is 160 cm. Tones can be produced by moving a hand towards the aerial. The closer a player moves his/her hand towards the aerial, the louder the tone. It requires the fine motor skills of children to make a very loud tone, while not touching the aerial. Different tones can be produced by moving a hand vertically along the aerial, calling upon the children's gross motor skills. Feedback and feedforward information is provided through sounds and leds. In this study two Tunemeins were used, one for each child, and each with 3 unique tones. Each tone was linked to an action in the game (e.g. forward, left, or right), which would only be completed if a certain volume was produced. Each Tunemein could be used to initiate three different actions, two of which could also be initiated by the other Tunemein, and one action unique to the specific Tunemein. The children were not allowed to touch the aerial. To add extra challenge the mapping of the tones on the aerial would change every two minutes.

## 5    Testing the toys

An experiment was set up to determine which of the three toys children would find the most fun to play with, and whether the toys were appreciated differently by boys and girls. Our hypothesis was that children would indicate Tunemein as the most fun to play with; as for boys it calls upon their gross motor skills and for girls it calls upon their fine motor skills. This would provide them with a strong basis to play the game to a satisfactory level, and challenge them to try activities requiring skills not as well developed yet. It was expected that this would appeal to children's curiosity and exploratory inclination. Furthermore, boys were expected to prefer Twistyertouch, which calls upon their better developed gross motor skills, to Stickysticks that calls upon their less-well developed fine motor skills. Girls were expected to prefer Stickysticks, which calls upon their better developed fine motor skills, to Twistyertouch, which relies upon gross motor skills.

In order to compare the children's response to the new toys with their appreciation of more common devices, we used a keyboard and a mouse connected to a laptop as a baseline. Nowadays these input devices are familiar to most children – which we verified: all participants had indeed experience with using a keyboard and mouse. One child had to operate the keyboard, which could only be used for actions involving moving up and down in the game, the other child would use the mouse for actions involving moving left and right – essential actions in the game used in this test.

For this study a game, the **Rabbitmaze**, was designed. The aim of the game was to earn points by guiding a rabbit through a maze towards a carrot. When the rabbit would have reached the carrot, a new one would appear in a different position in the maze. The toys were used to steer the rabbit through the maze. The game finished after a fixed period of time, showing a screen with the score. For Stickysticks, Tunemein, and Twistyertouch the game was projected on the wall with a beamer, for the keyboard/mouse combination it was presented on the screen of the laptop. Depending on the specifics of each toy, there were slight variations in the visual presentation.

In total 66 children (aged between 8-10 years) participated in the study. These children came either from a primary school in the neighborhood of the Philips Research Laboratories in Eindhoven or their parents, employees of Philips, had reacted to a call on an internal message board. The children participated in same-sex and same age pairs that they themselves had formed beforehand, resulting in 33 pairs (18 girl- and 15 boy-pairs). Participation was voluntarily.

The study was conducted in a room that was decorated and furnished as a children's play/bed room. The sessions took place after school hours, and lasted approximately one hour. All teams started their session with the familiar keyboard/mouse combination. In this way the children could get used to the type of task (playing the Rabbitmaze game), and get at ease with the setting. The order in which the other toys were presented was balanced over gender and age to prevent bias. The 'Rabbitmaze' was always played two times per input device: the first time to practice, the second time 'for real'.

After finishing playing with the last toy, the children were asked to individually complete a paired comparison test. In this test, each child was presented with pictures of the input devices two at a time (resulting in 6 different pairs). Each time they were asked to indicate which of the two they found the most fun to play with. They also had to describe why they liked it best. In addition to the paired-comparison test, we collected also other data (for example, game scores, behavioral observation data), which will not be further discussed here. In order to check for individual differences, one or two weeks after the toy test session, the children's cognitive and motor capabilities were assessed, and via questionnaires the children's gender identity and attitudes towards novel stimuli were determined. The aim was to be able to deduce the challenge the different toys would pose for each child and thus clarify the choices made by the participants. For a detailed account of all tests and other data collection methods employed, and the respective outcomes, the reader is referred to Stienstra & Hoonhout (2002), and Stienstra (forthcoming).

## 6  Results

A detailed account of all results will be presented in Stienstra (forthcoming). In this paper we will focus on the paired-comparison data. The toys appear to be more difficult to use than the keyboard/mouse combination. For example, the children on average get higher scores playing the Rabbitmaze game with the keyboard/mouse, than with the other toys. The children's verbal comments also indicated that the new toys were seen as more difficult to use. However, 60% of the children indicated that they enjoyed Twistyertouch most (56% of the girls, and 63% of the boys). A small number of boys liked the keyboard/mouse combination best (7%); none of the girls had this opinion.

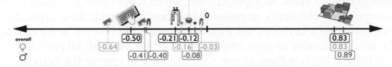

**Figure 2:** Preferences for the different toys based on the paired-comparison test.

Thurstone's *Law of comparative judgment* (1927) allows transforming the children's preferences for the different toys into a one-dimensional scale. The resulting scale values are depicted in Figure 2. The shown values represent relative, not absolute values.

# 7 Discussion and conclusions

Even though the toys appear to be more difficult to use, judging by the obtained scores and the children's reports, this does not have a negative influence on their appreciation. On the contrary, the highest score was obtained with the keyboard/mouse combination, but this device was appreciated the least (and as was to be expected, more so by girls than by boys). One of the reasons why the children prefer the new toys is the challenge offered by them: according to the children "playing with these toys is really difficult, and that is what makes it fun", and "it is different from other things and much more difficult, and that is why I like it".

Contrary to what was expected, the Tunemoin, designed according to the gender oriented strategy, was generally well liked but did not end first, but second in appreciation. As it turned out, both boys and girls highly enjoyed the physical activities required by Twistyertouch. We aimed for devices that would promote an active attitude in the children, but we did not expect that being physically active would be valued so much by boys and girls alike. An interesting finding with respect to future electronic game and toy design. We realize that also the appearance of the toys might have played a role in their appreciation – an aspect certainly warranting further study.

It was a challenge to us, to translate the design heuristics of Malone & Lepper (1987) into design specifications, since few concrete examples do exist. Challenge, curiosity, and novelty, definitely play a role, as well as playing together in a team. And also being physically active. Further studies should help to clarify the respective contributions of each factor to the overall enjoyment of toys.

# 8 References

Geary, D.C. (1998). Male, Female: The Evolution of Human Sex Differences. Washington: American Psychological Association.

Jenkins, H. (1998). "Complete Freedom of Movement": Video Games as Gendered Play Spaces. In J. Cassell & H. Jenkins (Eds), *From Barbie® to Mortal Combat. Gender and Computer Games*. Cambridge, MA: The MIT Press.

Malone, T.W., & Lepper, M.R. (1987). Making learning fun: A taxonomy of intrinsic motivations for learning. In R.E. Snow & M.J. Farr (Eds.), *Aptitude, learning and instruction*. Hillsdale, NJ: Erlbaum.

Stienstra, M.A. (forthcoming). Is every kid having fun? A gender approach to interactive toy design (PhD Thesis). Enschede: University Twente, The Netherlands.

Stienstra, M.A., & Hoonhout, H.C.M. (2002). TOONS Toys. Interaction toys as means to create a fun experience. In M.M. Bekker, P. Markopoulos, & M. Kersten-Tsikalkina (Eds.), *Interaction Design and Children*. Maastricht: Shaker Publications.

Thurstone, L.L. (1927). A law of comparative judgment. *Psychological Review*, 34, 273-286.

# Acknowledgements

We would like to thank professor dr. Oudshoorn of the University of Twente, dr. Overbeeke and dr. Djajadiningrat of the Technical University Eindhoven, G. Hollemans of Philips Research Laboratories Eindhoven, and D. van Duijn.

# Investigating the Role of User Cognitive Style in an Adaptive Educational System

Evangelos Triantafillou, Athanasis Karoulis, Andreas Pombortsis

Computer Science Department
Aristotle University of Thessaloniki
P.O. Box 888, 540 06 Thessaloniki, Greece
{ vtrianta, akarouli, apombo} @csd.auth.gr

## Abstract

Many experimental studies have identified a number of relationships between cognitive style and learning. The objective of our research was to comply and synthesize findings in current literature in an effort to develop an Adaptive Educational System based on Cognitive Styles (AES-CS), a prototype system that includes accommodations for cognitive styles in order to improve students' interactions and learning outcomes. This paper discuss the design issues that were considered for the design of the system and the way they were implemented in the development of AES-CS.

## 1    Theoretical Background

Hypermedia environment is considered to be a flexible instructional environment in which all the learning needs can be addressed (Ayersman & Minden, 1995). Many researchers have been working to construct sophisticated hypermedia systems, which can identify the user's interests, preferences and needs and give some appropriate advice to the user throughout the learning process. Adaptive Hypermedia (AH)  was introduced as one possible solution. Adaptive Hypermedia (AH) is an alternative to "one-type-suits-all" approach.  AH build a user model of the goals, preferences and knowledge of the individual user and use this model to adapt the content of pages and the links between them to the needs of that user. Since the user's goals, preferences and needs may change over time, AH observe these changes in order to update the user's model (Brusilovsky, 1996). There are two major technologies in adaptive hypermedia: a) adaptation of the content of pages, referred as adaptive presentation, b) the adaptation of hypertext links which mainly affects navigation within a hypertext system, referred as adaptive navigation.

Adaptive Educational Systems (AES) can be developed to accommodate various learner needs; is the ideal way to accommodate a variety of individual differences, including learning style and cognitive style. Although, cognitive styles are one of the several important factors to be considered from designers and instructors of hypermedia-based courseware, little research has been done regarding the adaptation of hypermedia system to students' cognitive styles (Liu & Ginther, 1999) and this is the focus of the research presented here.

Cognitive style is usually described as a personality dimension, which influences attitudes, values, and social interaction. It refers to the preferred way an individual processes information. There are many different definitions of cognitive styles as different researchers emphasize on different aspects. However, Field dependence/independence (FD/FI) is probably the most well known division of cognitive styles (Witkin, Moore, Goodenough & Cox, 1977).

FD/FI dimension refers to a tendency to approach the environment in an analytical, as opposed to global, way. Studies have identified a number of relationships between this cognitive style and learning, including the ability to learn from social environments, types of educational reinforcement needed to enhance learning, amount of structure preferred in an educational environment (Summerville, 1999).

Field independent (FI) learners generally are analytical in their approach while Field Dependent (FD) learners are more global in their perceptions. Furthermore, FD learners have difficulty separating the part from the complex organization of the whole. In other words, FD individuals see things in the entire perceptual field (the forest than the trees). Additionally, FI individuals tend to be intrinsically motivated and enjoy individualized learning, while FD ones tend to be extrinsically motivated and enjoy cooperative learning. Specifically, FD individuals are more likely to require externally defined goals and reinforcements while the FI ones tend to develop self-defined goals and reinforcements (Witkin, Moore, Goodenough & Cox, 1977).

## 2    Adaptive Educational System based on Cognitive Styles (AES-CS)

The current study is an attempt to examine some of the critical variables, which may be important in the design of an adaptive hypermedia system based on student's cognitive style. As a case study a Higher Education module was developed, called AES-CS (Adaptive Educational System based on Cognitive Styles) that includes accommodations for cognitive styles in order to improve student interactions and learning outcomes. Next we will discuss the design issues that were considered for the development of the system and the way they were implemented in the design of AES-CS.

*Program control versus learner control:* The amount of learner control seems to be a central variable when integrating adaptive methods in educational settings. There are several arguments in the literature for and against learner control. On the one hand, learners' motivation is increased when they control the navigation of a hypermedia environment. On the other hand, research seems to indicate that the amount of learner control depends on the pre-skills and the knowledge state of a learner (Williams, 1993). Furthermore, many studies have demonstrated student preference and improved performance using a linear structure. With regards to cognitive styles, there is evidence that FD individuals perform better using program control while FI ones prefer more learner control (Yoon, 1993). Since these findings are consistent with theoretical assumptions in FD/FI dimension, AES-CS provides both program and learner control option. In the case of learner control option, AES-CS provides a menu from which learner can choose to proceed the course in any order. In the program control option there is no menu, but the system guides the user through the learning material via adaptive navigation support.

*Instructions and Feedback:* Studies have shown that FD are holistic and require external help while FI people are serialistic and possess internal cues to help them solve problems. FD learners are more likely to require externally defined goals and reinforcements while FI tend to develop self-defined goals and reinforcements (Witkin, Moore, Goodenough & Cox, 1977). Jonassen and Grabowski (1993) in their study summarized the research on the implications of the individual differences based on FD/FI dimension. We consider these implications of style characteristics in order to design the instructional support and the instructional environment of AES-CS. As a result, the system provides clear, explicit directions and the maximum amount of guidance to FD learner, while it provides minimal guidance and direction to FI learner. Moreover, it provides extensive feedback to FD learner, while it provides minimal feedback to FI learner.

*Structure:* Several problems of learning in a hypermedia environment arise from the structure of the environment itself. In an ideal web site, the structure is evident to the user and the information is organized coherently and meaningfully. Navigational tools are essential in order to assist learners to organize the structure of the web site as well as the connections of the various components. A coherent resource collection will allow the user to construct an accurate mental model of the topic. Research has indicated that FD learners are less likely to impose a meaningful organization on a field that lacks structure and are less able to learn conceptual material when cues are not available (Witkin, Moore, Goodenough & Cox, 1977). Furthermore, Jonassen and Wang (1993) argue that the FI learners generally prefer to impose their own structure on information rather than accommodate the structure that is implicit in the learning materials. In our approach, AES-CS provides two navigational tools in order to help learners organize the structure of the knowledge domain: concept map and graphic path indicator [Figure 1].

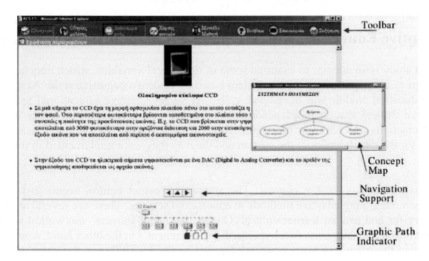

**Figure 1.** System screen with the initial adaptation for FD learners.

*The use of contextual organizer:* Another feature that is embedded in AES-CS is the use of contextual organizers according to FD/FI dimension. Field Dependent learners appeared to benefit most from illustrative advance organizers, while Field Independent learners preferred illustrative post organizers (Meng & Patty, 1991). An advance organizer is a bridging strategy that provides a connection between one unit and another. It also acts as a schema for the learner to make sense out of the new concept. A post organizer serves as a synopsis and supports the reconstruction of knowledge. Usually, it is available after the presentation of new information.

# 3   Evaluation

Throughout the development of the AES-CS, formative evaluation was an integral part of the design methodology. Formative evaluation is the judgments of the strengths and weakness of instruction in its developing stages, for the purpose of revising the instruction (Tessmer, 1993). In our research, three types of formative evaluation were used to evaluate the AES-CS: expert review, one-to-one evaluation and small group evaluation. Recommendations from expert reviews, suggestions from the students and results of the small group evaluation are summarized below:

- *Design:* The design and the development of an AES that includes accommodations for cognitive styles was effective and efficient.
- *Initial adaptation:* The initial adaptation based on research results and theoretical assumptions in FD/FI dimension was consider successful. However, the adaptive systems need to be controllable by the user because they cannot be intelligent enough to appropriately adapt in all possible cases.
- *Instructions:* Students should be able to access the maximum amount of guidance and instructions whenever they needed.
- *Structure:* The system should provide tools such as the concept map and the graphic path indicator, in order to help learners organize the structure of the knowledge domain. Furthermore, these tools should be active so to be used as an extra navigation tool.

After the final revision, summative evaluation took place in order to assess the effectiveness of the system with reference to other educational material used for the instruction of the particular module. The course "Multimedia Technology Systems" was used as the example learning material in this phase. Typically the students follow the lectures and they study through a hypermedia-based environment. The experiment took place over a four-week period and during that period AES-CS was an integral part of the course together with the lectures and the hypermedia-based environment. Fourth year undergraduate students volunteered to take part in the study. They were allocated in two groups: experimental (36 students) and control (30 students). The students of the experimental group studied through the adaptive educational system AES-CS, while the control group studied through the traditional hypermedia based environment.

The main research direction was the evaluation of the educational effectiveness of system's adaptation. The results from the summative evaluation of the prototype system, support the evidence that the adaptivity based on cognitive styles can ensure that all students can learn effectively in a hypermedia environment. Statistical analysis indicated that students in experimental group (AES-CS) performed significantly better than students in control group (HTML). These findings indicate that the adaptive educational system AES-CS, which was designed to be adapted to individual cognitive styles, can be an effective tool to support and promote learning.

A secondary purpose of the study was to see if the adaptivity based on cognitive style influence the performance of FD students in order to reach the same level of performance as FI. The statistically significant findings confirmed the evidence that the adaptivity based on student's cognitive style could be beneficial for the observed learning outcomes, especially for FD students. Finally, an analysis of the data collected showed that the subjects were satisfied with the initial adaptation based on their cognitive style. In addition, the qualitative data analysis indicated that the adaptation granularity contributed to the overall user satisfaction.

Summarizing, this paper showed the important role of cognitive style with regards to student success in hypermedia learning environments. Instructors, therefore, might well be encouraged to consider cognitive style as a valuable factor throughout the development of a hypermedia environment. Further research is needed in order to expand the adaptive features of the system focusing on a dynamic detection of student's cognitive style throughout his/her interaction with the system. With these features, the adaptation of the system would be constant and more effective throughout the educational experience.

## References

Ayersman, D.J. & Minden, A.V. (1995). Individual differences, computers, and instruction. *Computers in Human Behavior*, 11(3-4), 371-390.

Brusilovsky, P.(1996). Methods and Techniques of Adaptive Hypermedia. *User Modeling and User-adapted Interaction*, 6, 87-129.

Jonassen, D.H & Grabowski, B.L (1993). Handbook of individual differences, learning & Instruction. Hiilsdale, NJ. Lawrence Erlbaum Associates.

Jonassen, D. & Wang, S. (1993). Acquiring structural knowledge from semantically structured hypertext. *Journal of Computer-based Instruction*, 20(1), 1-8.

Liu, Y., Ginther, D. (1999). Cognitive Styles and Distance Education. *On-line Journal of Distance Learning Administration*, 2,3.

Meng, K. and Patty, D.(1991). Field-dependence and contextual organizers. *Journal of Educational Research*, 84(3), 183-189.

Summerville, J.(1999). Role of awareness of cognitive style in hypermedia. *International Journal of Educational Technology*, 1(1).

Tessmer M. (1993). Panning and Conducting Formative Evaluations. Kogan Page Limited.

Williams, M.D. (1993). A comprehensive review of learner-control: The role of learner characteristics. Paper presented at the annual meeting of the Association for Educational Communications and Technology, New Orleans, LA.

Witkin, H.A, Moore, C.A., Goodenough, D.R., Cox, P.W.(1977). Field-dependent and field-independent cognitive styles and their educational implications. *Review of Educational Research*, 47(1), 1-64.

Yoon, G.S.(1993). The effects of instructional control, cognitive style and prior knowledge on learning of computer-assisted instruction. *Journal of Educational Technology Systems*, 22(4), 357-370.

# A system for e-learning via annotated audio/video clips and asynchronous collaboration

*Nikolaos Tsoutsias*

University of Cyprus - Department of
Computer Science
75 Kallipoleos Str., P.O. Box 20537
1678 Nicosia, Cyprus
cs98nt1@ucy.ac.cy

*S. Retalis*

University of Cyprus-Department of
Computer Science
75 Kallipoleos Str., P.O. Box 20537
1678 Nicosia, Cyprus
retal@softlab.ntua.gr

## Abstract

This paper presents a system that supports asynchronous multimedia communications (AMC). AMC is a combination of multimedia conferencing with asynchronous conferencing philosophy. AMC can support the collaborative 'discussion' and 'critique' of sharable representations of video, audio streaming, over time and anywhere in space, among learners, and teachers. In addition to presenting representations in textual as well as audio and video format, voice and textual annotations to these representations can be added. The use of AMC offers new ways to encourage learners to make practice of their declarative and academic knowledge.

## 1 Introduction

The widespread use of information and communications technologies (ICT) has enabled many new forms of collaborative distance learning activities, which are based on the integration of communication/collaboration space with information access and organisation, within a commonly accessible hyperlinked environment (Khan, 1997). There is an increasing amount of research supporting the critical influence that the choice of communication technology can have on both the process and the product of collaborative distance learning (Collis & Smith 1997). It is possible to think of two communications-based learning "worlds" which have had surprisingly little to do with each other. On the one hand, we have the world of asynchronous text-based communications - E-mail, computer-mediated conferencing – threaded discussion fora, etc (e.g. McConnell, 1994). On the other hand, we have the world of synchronous multimedia communications - live audio-conferencing, live video-conferencing, etc. Synchronous multimedia communications make it possible for people at different sites to partake in the same conference at the same time through the "magic" of two-way audio and two-way compressed video. The advantages of synchronous multimedia communications are: a) live connectability, b) availability to share data and c) synchronous and vivid interactions. Its disadvantages: a) time and space constrains, b) requires specialized equipment, c) availability of bandwidth and telecommunication costs.

Asynchronous computer mediated communications (ACMC) can effectively and efficiently support the collaborative distance learning process, due to the fact that they offer flexibility in the use of time as well as space. Taking all the above into consideration we decided to perform research in the field of asynchronous multimedia communication (AMC).

The AMC supports the creation of vivid representations of real world cases (e.g. concise digitised video demonstrations and explanations by practitioners). In addition to capturing representations in videoclips, we are also interested in voice and textual annotations to video clips. The use of voice annotations will offer new and effective ways to encourage learners to make practice of their declarative and academic knowledge (Steeples, 1995). A videoclip provides a direct and demonstrable example of a real case, while making an annotation allows the subject to focus visual attention on actions in the clip. Annotations can be used to encourage learners not only to pose questions but also to explain issues as well as to reflect upon the real world case. Thus AMC can support the collaborative 'discussion' and 'critique' of sharable representations, over time and anywhere in space, among learners, teachers and other practitioners, using audio, video and/or textual 'annotations' on a digitised video resource.

Our team has started developing an AMC system in order to serve as an integrated environment that supports the incorporation of audio-visual representations, which then serve as base material for asynchronous, multimedia discussions within a community of learners. Furthermore, the system provides a means for exchange and review of the base material and the capture and hyper-linking of multimedia annotations to this material.

This paper gives an overview of the AMC system developed emphasizing on use cases for learning. The structure of this paper is as follows: Section 2 illustrates the nature of AMC, presents some usage scenarios of AMC and provides a more concrete example for e-learning. Section 3 presents the architecture of a system and Section 4 contains some concluding remarks

## 2   AMC in e-learning process

AMC could be said to combine richness of multimedia communication and the possibility of quickly creating vivid accounts and demonstrations of practice with flexibility in the use of time (participants in the communication do not have to be available at the same instant; there is opportunity for reflection on what is seen and heard). The price of temporal flexibility is that asynchronous communication cannot benefit from the rapid turn-taking and negotiation which is characteristic of, and highly valued in, synchronous communication - especially face-to-face synchronous communication (Boden & Molotch, 1994).

### 2.1   Usage scenarios in AMC for e-learning

AMC is based on an audio-visual representation of a real world cases, the exchange and reviewing of the base material and the capture and hyperlinking of multimedia annotations to this material. Three main sets of activities need to be understood:
- Producing digitized audiovisual representations of real world cases (i.e. video clips)
- 'commenting' or 'discussion' on the video-clips. Focus should be given on the processes of annotation and linking that create a 'web' of multimedia objects thus creating hypermedia structures in an AMC environment.
- 'looking through' the materials in an AMC 'web' in order to find what is relevant; The focus here is on searching and browsing in a hypermedia web or database.

Imagine the following usage scenarios for learning. A video clip is created or just posted by a teacher for triggering a learning activity. This videoclip is accompanied with metadata for facilitating its retrieval. This videoclip can be accompanied by a series of questions that intend to make the learners describe what they see. These questions can be comments referring to the whole videoclip or annotated comments referring to a specific part of the base material. Such comments can be in a textual (with or without attachments), audio or video format. Imagine how such a scenario in elearning, for example in language learning, for listening comprehension. The learner taking part in this learning activity must answer to the posed question either with a textual comment or using an audio message. The asynchronous mode of communication enables reflection before answering. The teacher can assess the answer and provide her/his feedback as a treaded comment

Moreover, AMC participants (learners and teacher) can initiate and/or elaborate in a threaded discussion. Apart from encouraging learners to participate to an asynchronous discussion, a teacher should moderate the discussion for keeping it at a certain quality level. AMC participants can make textual and/or audiovisual annotations upon an animation clip, offering other perspectives. Such exchanges of viewpoints in the target language offer the learners unique opportunity for practice. The teacher can intervene during the discussion in order to correct learners misconceptions. Over time a web of linked multimedia objects evolves as the "discussion" proceeds, and further contributions can be added, including as "topics" and "responses".

Apart from the above features of the AMC system, more specialized ones have been implemented such as:
- Keeping a portfolio of messages where the user can store for future process some of the messages appearing in the forum
- Presentation of statistics where the user can see how many messages have been posted by how many users as well as the number of the different types of messages (text, videos, audios, still images)

Of course basic functionality for the administration of the forum has been constructed (i.e. deletion of a forum, creation of new forum, user management, etc.)

# 3 The architecture of the AMC system

One way for describing the architecture of the AMC system is to follow the Rational Unified Process (Jacobson, Booch, & Rumbaugh 1999). According to its logical view, the first-level decomposition of the system is performed by specifying the very coarse-grained discrete subsystems in the design model, as they have derived from the use case and analysis model. The decomposition is combined with the enforcement of the "Layered Systems" architecture pattern, which helps organize the subsystems hierarchically into layers, in the sense that subsystems in one layer can only reference subsystems on the same level or below. The communication between subsystems that reside in different layers is achieved through clearly defined software interfaces.

The proposed layered architecture for the AMC system that identifies all first-level subsystems, organizes them into the following layers:
The application-specific sub-systems are:
- Login Subsystem: The subsystem that is responsible for accepting or rejecting a user.
- User Management: Adding, deleting, activating or deactivating a user or sending mail to a specific user.
- Conference Management: Creating a new conference or deleting an existing conference.
- Message Management: Deleting, adding or viewing a message.
- Statistics Viewer: Viewing statistics about the system or users.

The application-general sub-systems of the layered architecture are:
- Charts Creator.
- Upload File.
- Video/Audio Presentation: Presents an audio or video file.
- Raw Data Management: Performs raw queries to the database.

The middleware-layer sub-systems of the layered architecture are:
- Java APIs (JMF, jfreechart, Oreilly utility classes, e.t.c)

The protocol-layer sub-systems contains the software for the computing and networking infrastructure, are the TCP/IP, HTTP, RTTP protocols. The software system-layer sub-systems are:
- MS ACCESS as RDBMS
- Helix Real streaming server
- Java Virtual Machine.

The following two figures (Figures 1, 2) illustrate the user interface of system. The reader should try correlating these figures with the usage scenarios as presented in section 3.

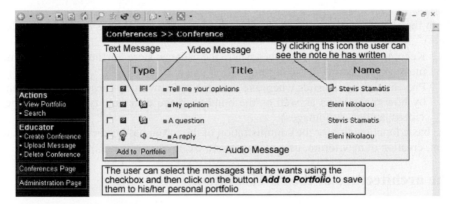

**Figure 1.** Screen shot with the list of messages at a forum of the AMC system

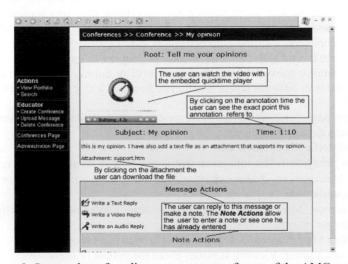

**Figure 2.** Screen shot of reading a message at a forum of the AMC system

# 4   Concluding remarks

AMC is a new and very promising technique for transferring knowledge and augmenting learners skills. E-learning can take advantage of it. Our philosophy as educators, researchers and developers in AMC is that, in addition to capturing representations of practice in videoclips, we are also interested in voice and textual annotations to video clips. We have developed an AMC whose architecture is component-based and quite open. Up to our knowledge not any similar system exists today.

This system is still in a prototype phase but fully functional in two course (undergraduate and post graduate) at the University of Cyprus. Additional improvements of the AMC system include the integration of user notification mechanisms (via e-mail and SMS) when new messages have been posted as well as for building a business model for a virtual community using such an AMC system. Very soon this system will be an off-the-shelf AMC tool and its full-scale evaluation in subject domains where it can add significant value such as language learning and vocational training on working practices.

# 5   References

Boden, D and Molotch, H (1994). The compulsion of proximity, In Friedland, R and Boden. D, (eds), *NowHere: space, time and modernity*. Berkeley: University of California Press, 257-286

Collis, B. and Smith, C. (1997). Desktop multimedia environments to support collaborative distance learning. *Instructional Science*, 25, 433–462.

Goodyear, P. (1996) Asynchronous peer interaction in distance education: the evolution of goals, practices and technology. *Training Research Journal*, 1, 71–102.

Goodyear, P., & Steeples, C. (1999). Asynchronous multimedia conferencing in continuing professional development: issues in the representation of practice through user-created videoclips. *Distance Education* (20, 1) 31-48

Khan, B. (1997). *Web-Based Learning. Englewood Cliffs*, NJ:Educational Technology Publications.

McConnell, D (1994). *Implementing computer-supported cooperative learning*, London: Kogan Page.

Röscheisen, M., Mogensen, C. and Winograd, T. (1997) Shared Web Annotations As A Platform for Value-Added Information Providers: Architecture, Protocols, and Usage Examples. *Technical Report [STAN-CS-TR-97-1582]* from http://www-diglib.stanford.edu/~roscheis/TR/TR.html

Sgouropoulou, C (2000) Web-Orama: a Web-based system for ordered asynchronous multimedia annotations, *PhD thesis*, National Technical University of Athens , Greece, December 2000

Steeples, C. (1995)  Computer-mediated collaborative writing in higher education: enriched communication support using voice annotations.  In J. D. Tinsley & T. J. van Weert *World Conference on Computers in Education VI: WCCE '95 Liberating the Learner*.  London: Chapman & Hall, 337-347

I. Jacobson, G. Booch, J. Rumbaugh (1999). *The Unified Software Development Process*. Addison-Wesley.

# Lecture Enhancement by Community Portal

*Hiroshi Tsuji*

Osaka Prefecture University
1-1 Gakuen-cho, Sakai, Osaka, Japan 599-8531
tsuji@ie.osakafu-u.ac.jp

## Abstract

This paper describes our experimental study on using a community portal. The portal is provided for teacher and students for communication enhancement. However, students have not always become active for "student to teacher" communication in the virtual class yet. We classified the obstacles into four categories for e-learning promotion and propose how to clear them.

## 1    Introduction

It is well-known that most students in Japan are not good to make questions and to debate on a topic in their class. In fact, most Japanese lectures are one way. Our assumption is that some students hesitate to ask teachers even if they have opinion because others do not like to be interrupted the class. Further, most Japanese students are apt to be afraid that their question may be stupid even if it is interesting and valuable.

To overcome this problem and enhance the lectures, we suppose the Web-based community portal (sometimes, it is called "community computing" (Ishida, 1998)) is useful. This paper describes our experimental study on using the system developed by a US company (Jenzabar, 2002). Our purpose is to evaluate not only the system function but also to analyze participants' literacy. We also would like to evaluate whether the e-learning system developed for US students fit to Japanese students or not from the viewpoint of cultural and cognitive aspects.

## 2    Experimentation

### 2.1   Lecture Model

There are varieties of e-learning systems taxonomy: for universities versus for companies, by synchronous communication versus by asynchronous communication, in the same place versus among the distance places, contents intensive (without teacher) versus communication intensive (with teacher), and process managerial versus context free.

Our model for lecture enhancement is illustrated in Figure 1. The lecture is done in the university, and each course consists of fourteen lessons and one examination day. In the real class, the teacher uses the traditional black board to teach students. The teacher and the students exchange their electronic materials and opinions asynchronously via communication portal after school.

The teacher should describe the course overview at first, specify syllabus items next, and upload handouts for each syllabus item. On the other hand, the students download the course works on each syllabus item and upload the reports. Then, the teacher assigns grade to the students. Thus, our model is process managerial for the teacher while there are forum databases for context free discussion.

**Figure 1**: Lecture Model in Experimentation and Contents in Community Portal

## 2.2 Overview of Community Portal

There are four kinds of participant roles in the portal: system administrator, management officer, teachers, and students. The functions and the assessable information for one are different from those for others. There is a security mechanism among them. In the experimentation, the administrator and the teacher is the author of this paper and there is not a management officer.

Both teachers and students should sign up to get user ID. Then the teacher can open the courses and the students can apply registration to the courses. The relationship between teacher and students is originally N: M while there is one teacher in the experimentation.

The communication functions among students, teacher, and administrator include contents uploading and its downloading. Figure 2 is an example of screen. Because the system does not provide any educational contents, the teacher should prepare contents for the class. He should also manage forums in order to exclude the inappropriate messages.

## 2.3 Students and their Internet Literacy

The experimentation started in October 2002. We selected four classes for our experimentation in the industrial engineering of technology department:
- Case 1: twenty eight undergraduate students for a lecture (Business Process Restructuring),
- Case 2: fifteen graduate students for a lecture (Data base & Knowledge base),
- Case 3: four undergraduate students for group discussion (Reading on UML),
- Case 4: three undergraduate students for computer operation (Web page data analysis).

The undergraduate students in Case 1 and Case 4 do not have the office in the university. They use the shared PC at school and can not access to the system at home unless they have their own PC. The other students have office at school where they can access to the community portal.

The students were asked on their IT literacy at the beginning of the October in the first class. The result is summarized in Table 1. It seems that most students prefer KEITAI (mobile phone) mail to e-mail. Note that their literacy is upgraded day by day.

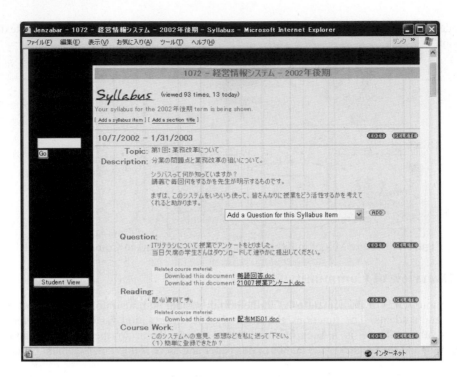

**Figure 2:** Example Screen in Experimentation

# 3 Lesson Learned

## 3.1 Sign-up and Early Exit

At first, teacher asked students to sign up to the community portal by themselves in two weeks. The procedure is not special. The students are requested to fill name, grade, ID, password, and e-mail address in the Web form. Then, some students might hesitate to sign up. They may afraid their load in the class. Sixteen percent of students dropped out from the class at the early stage.

Case 3 and Case 4 were small classes. There are active face-to-face discussions between the teacher and the students in the real classes. It is time consuming for the teacher to use communication portal in such classes. Then teacher had not updated the course contents in the system. For the small class, e-mail is powerful enough to enhance face-to-face discussion.

## 3.2 Material Delivery and Communication in Forum

Let us show the access statistics on "login", "course view", "handouts view", and "forum main" in Figure 3. At first, the students were interested in course description. In November, the counts of handouts view became increased. It might be because the students had accustomed to use the portal. However, the forum was not active.

Teacher became busy to prepare materials for the virtual class in November. He had to manage to the virtual class after the real class. Then, the access frequency decreased in December. Although there is a forum database in the virtual class, the debate "student to teacher" communication is still inactive unless the teacher does not assign the coursework for the discussion.

In January, the teacher asked students to submit their comments on class to the forum database. Then, the access frequency suddenly increased. Thus, in the case that the teacher can not update his contents, the theme selection for debate is the most important issue for student access.

**Table 1**: IT Literacy of Students in October 2002

| Items | Case 1 | Case 2 |
|---|---|---|
| Grade | Undergraduate | Graduate |
| Initial Registration | 28 | 15 |
| Personal PC | Yes (23) No(5) | Yes(14) No(1) |
| I3T Connection | Broadband (15) Dialup (8) No (5) | Broadband (6) Dialup (0) No (0) |
| Web Access | Daily(11) Weekly(15) Rare(2) | Daily(10) Weekly (5) Rare(0) |
| e-Mail | Daily(5) Weekly(12) Rare(11) | Daily(8) Weekly (5) Rare(2) |
| Forum | R&W (7) Read only (12) No (5) | R&W (6) Read only (6) No (3) |
| Shopping | Yes (11) No (17) | Yes (4) No (11) |
| File Download | Program (14) Game(7) Others (6) No(10) | Program (10) Game (8) Others (5) No (1) |
| Sign up to System | 25 | 11 |
| Completed | Yes(24) No(1) | Yes(8) No(3) |

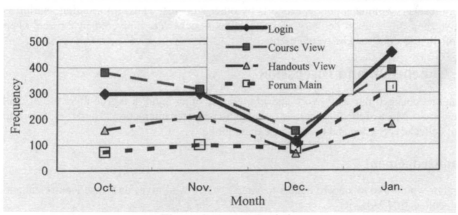

**Figure 3**: Access Statistics

# 4 Consideration on Obstacles

According to the student feedback, the community portal is attractive. However, the students have not always become active for "student to teacher" communication in the virtual class yet.

## 4.1 Infrastructure of University:

Although there are many shared PCs in the special area of the university, the students in Case 1 do not use them in daytime. They claim the location of the shared PCs because they like to check the portal at the interval time between the classes. The PCs area and the classrooms should be located in the same building. Thus, the location of shared PCs is critical for undergraduate students.

## 4.2 Internet Literacy and Motivation:

They have submitted neither their opinions nor questions to the forum unless the teacher assigned the issues. Incentives for the students are indispensable. Currently, the forum in the portal does not accept anonymous submission. Japanese students may hesitate to be identified in the forum.

## 4.3 System Functions

Many students are interested in the security policy on the personal information. One student claimed her e-mail address is referred to the other students in the participants list. She intended to let only teacher know her address. This is reasonable. Other students are also nervous who can access to the forum although they do not refuse that the third party access to it. Thus, the security level and operation guideline should be announced to the students.

It is difficult for students to know if there are new contents from the teacher or not and if the teacher reads their opinions and questions. Therefore, the system is required to express the status on the communication. The web usability (Pearrow, 2000) is critical problem for communication intensive system.

## 4.4 Contents from Teacher:

Without community portal, it has been difficult for the teacher to feedback the homework and examination. The feedback was welcome by the students. And students are strongly interested in not only teacher comments but also reports by other students. Thus, Information sharing is more attractive than information delivery while the files which were delivered in the real class were welcome by the students who were absent.

## 5 Conclusion and Discussion

The internet changed our daily work and life. On the other hand, it is true there are obstacles for this new style. Most of them are issues on human-computer interaction. Therefore, it is important tasks to collect experience and to exchange them globally.

## Acknowledgement

The author would like to sincere thanks to Mr. S. Iwami who gives us to use lecture enhancement system, called JENZABAR.

## References

Ishida, T. (Ed.) (1998). *Community Computing and Support Systems*, Lecture Notes in Computer Science 1519, Springer-Verlag.

Jenzabar, Inc. (2002) Smart Solutions for Higher Education, from http://www.jenzabar.com/

Pearrow, M. (2000). *Web Site Usability Handbook*, Charles River Media, Inc.

# SpyCam and RoboCam: An Application of the Future Technology Workshop Method to the Design of New Technology for Children

*Giasemi N Vavoula, Mike Sharples, James Cross, Chris Baber*
University of Birmingham, UK
g.vavoula@bham.ac.uk, m.sharples@bham.ac.uk,
crossjw@eee-fs7.bham.ac.uk, c.baber@bham.ac.uk

## Abstract

The Future Technology Workshop is a method for the design of future socio-technical systems. It builds upon existing participatory design techniques and provides a way of exploring the interactions between technologies and activities, now and in the future. The method has been applied to the design of future imaging technologies for children. Prototypes of such technologies have been produced and tested with children. The paper describes the method and its application in this design exercise.

## 1 Introduction

There are various methods for encouraging participation in design, many stemming from Scandinavian research. In general, these approaches encourage brainstorming, storyboarding of concepts based on current experiences, design templates, and role-play. (Druin, 1999a, 1999b; Druin et al., 1999) has developed methods for including children in the design process as legitimate design partners. In outline, the methods are based around participatory envisioning, contextual inquiry (Beyer & Holtzblatt, 1997), and lab-like observations. The children are observed in their normal environments, or in specially constructed labs that are technology rich, or are given various materials to construct a product and asked to explain how it is used. (Inkpen, 1999) has explored the design of handheld devices for children. Her methods have included (a) the administration of a questionnaire, from which she gathered information on children's general requirements for handheld computers, (b) a participatory design exercise where children worked in groups of 2-4 to produce lowtech prototypes of the handheld computers they would like to have, and (c) a diary exercise where children had to report the places where they would want to use a handheld computer and the activities for which they would like to use it. (Scaife & Rogers, 1999) worked with children as informants at various stages of the process of designing an educational software system for the teaching of basic ecology concepts. The children's role was primarily concerned with refining the user interface, whereas the decision of what technology to design was suggested by the curriculum.

The methods and techniques described above focus on children constructing and acting with objects, and encourage them to lead design. However, design is grounded in current experiences, which may constrain the children's designs, with the danger of missing more radical design opportunities. As (Inkpen, 1999) notes, "while this activity was successful in identifying important issues for the design of handheld computers, many of the children were constrained by their preconceived notions about what constitutes a computer and the functionality it can provide". Therefore, a methodology that encourages users to postulate future uses of technology and provides a transition from current to future thinking could be beneficial for the creative design of new technologies.

The Future Technology Workshop (FTW) method is being developed to address these shortcomings. The method is intended to serve both as a process of "discount sociocognitive engineering" (Sharples et al., forthcoming) and a means of envisioning future technologies and technology-mediated activities. It is a method of "imagineering" – envisaging and designing human interactions with future technology.

A series of the FTW have been undertaken with children aged 10 to 13. The workshops led to design ideas that were subsequently transformed into working prototypes and fed into further iterations of the workshop. The remainder of this paper presents the results of this design exercise. Section 2 briefly presents the FTW method; section 3 describes the applications of the method and the prototypes; and section 4 concludes with a discussion of our experiences and plans for future research.

## 2 The Future Technology Workshop Method

The FTW consists of a creative exploration of the interaction between activities and technology, now and in the future. The interactions can be shown as a grid (see Table 1).

The four items in the grid are explored in turn during the workshop. The aim is to reach an informed understanding of item 4 – how, in the future, we might perform new activities enabled by new technologies.

Unlike conventional systems design, the starting point is not item 1 of Table 1 above – current technology and activity – as this might anchor the process to the present and miss more innovative design opportunities. In contrast, the starting point is item 4, and the workshop then circles through 3, 1, 2, and back to 4. This is done through a series of seven workshop sessions. The sessions and their correspondence to the four items of Table 1 are shown in Table 2 below.

|  | CURRENT TECHNOLOGY | FUTURE TECHNOLOGY |
| --- | --- | --- |
| **CURRENT ACTIVITY** | 1. Everyday technology-mediated activity | 2. Familiar activities supported by new technology |
| **FUTURE ACTIVITY** | 3. New activities that current technology might support | 4. New activities with new technologies |

**Table 1. Interactions between activities and technology, now and in the future**

The seven workshop sessions are described in (Vavoula, Sharples, & Rudman, 2002). In brief, during the first three sessions the participants envisage and create designs for future technologies and activities. In particular, in the first session the facilitator primes the group with a question regarding the design task at hand, in a way that it limits the scope of the ideas while at the same time does not pre-empt them to think about current technologies, and a brainstorming follows. The purpose of the session is to set the scene and to get the participants to think in terms of the future, with respect to both the technology and the needs satisfied by it. In the second session (see fig. 1) the participants are divided into 2 or 3 groups and are provided with a set of low-tech prototyping materials, such as PlayDoh and collage kits to build a model based on ideas produced earlier. The purpose of the session is to set the participants to imagine the future and produce models of useful and meaningful technology.

| Interactions / Sessions | 1. Everyday technology-mediated activity | 2. Familiar activities supported by new technology | 3. New activities that current technology might support | 4. New activities with new technologies |
| --- | --- | --- | --- | --- |
| 1. Imagineering |  |  |  | ☑ |
| 2. Modelling |  |  |  | ☑ |
| 3. Role-play |  |  |  | ☑ |
| 4. Retrofit |  |  | ☑ |  |
| 5. Everyday | ☑ |  |  |  |
| 6. Futurefit |  | ☑ |  |  |
| 7. Requirements |  | ☑ |  | ☑ |

**Table 2. Correspondence of FTW sessions with technology-activity interactions**

**Figure 1. FTW participants at work**

In the third session the groups exchange models and are asked to plot a scenario demonstrating how the model might be used, and enact it. The purpose of this session is to bring the future into the present, by getting participants to "act" as if future technologies were already there to support new activities, and also to have them engaged in the future activities and make their ideas of them more tangible.

Session 4 explores new activities that current technology might support. The groups are asked to modify their scenarios so that they only use existing technologies to enact them. The purpose of the session is to bring the future into everyday life, getting the participants to think how the futuristic activities they have imagined so far might be adapted into their current lives. Session 5 explores everyday, technology-mediated activities. The whole team is shown photos of current technologies and existing gadgets relevant to the design task, and is asked to discuss their current activities in relation to this design task and to identify relevant problems and shortcomings of the existing technologies. The purpose of this session is to remind the participants of the things they currently do and the technologies they currently use, as well as to set them thinking about how existing practices could be improved. Session 6 explores how familiar activities might be supported by new technologies. The group is asked to look at the outcomes of session 5 and discuss how they think those activities will be performed in the future, in relation to the models they had built in the earlier sessions, and to modify their initial models to accommodate as many of the current activities as possible. The purpose of this session is to set the participants thinking about what sorts of future technologies will be used to support the activities they currently perform. The final session relates to how both current and future activities will be supported by future technologies. The whole workshop team is asked to produce a set of requirements for each model, and are then asked to rate the requirements based on their importance/appeal.

This process can be repeated over a series of half-day workshops (preferably at least two), so that when the participants revisit item 4 they have an enhanced insight into 'new activity' and 'new technology'. Ideally, the process leads to a 'spiral' of design ideas, with each revisit to item 4 building on and pushing forward earlier conceptions and prototypes.

The method has been found to be appropriate for use both with adults and children. When used with children, however, minor modifications might be necessary, mainly in adapting the terminology and language used in a way appropriate to the specific age group. For example, in the final session, rather than asking the children to list requirements for future technologies, we asked them to produce a list of "instructions for the engineers who are going to build the models". Otherwise, all the activities involved were easily attainable by children.

# 3 Applying FTW to the design of new imaging technologies for children

We have undertaken a series of four FTWs with children, aged 10-13, focusing on the design of new digital imaging technologies and activities. After the first two workshops, prototypes were built based on the common themes that appeared to be emerging. The final two workshops built upon these prototypes, taking the designs further. The prototypes were also tested independently of the workshops, with 32 children at a residential children's "education and adventure" centre. The remainder of this section presents the four workshops, the prototypes, and the results of the tests.

## 3.1 Initial FTWs, Prototypes, and Testing

The initial research that led to the conception of the method took place at a children's holiday camp. At that point, only the first two sessions were carried out. Six boys and six girls, aged approximately 11 years, participated. The concept of a 'spy camera' was a primary theme for the boys – to be able to send the camera off by remote control to capture images without being noticed, while being able to see the images on the screen of the remote controller. For the girls, a robot-type camera capable of a friend-type relationship was important. Reflections on the conduct of this exercise, and further piloting with adults, led to the full development of the method as it was applied in subsequent workshops.

The second children's workshop was carried out with six different children, three boys and three girls aged 11-12. The design task related to "capturing and sharing visual events". Again, spy cameras and independently mobile cameras were the main theme. The girls produced two models, one consisting of a spy camera operating through glasses that allow the user to secretly take photographs of anything they see, by pressing a disguised button on the glasses' frame. The second consisted of an independently mobile camera, that is intelligent enough to remember objects and people and knows where to find them, and can therefore be instructed to "go off and take a picture of my dog's basket". The boys produced a model of an amphibious camera that has the ability to propel itself safely in deep sea and take pictures of the bottom of the sea on behalf of its owner.

(a)          (b)          (c)

**Figure 2. (a) the SpyCam, (b) the RoboCam, (c) children around the RoboCam control table**

The common themes that emerged from the workshops are 'spy' cameras (miniature cameras hidden on the body that can capture everyday events, or relay the images to another person) and 'robot' cameras, where a camera is attached to another person, an animal or an object such as a remote control submarine, with the images viewed at a distance. As a consequence of these findings, we built a "SpyCam" device that combined elements of both the spy and remote camera. The prototype consisted of an inexpensive wireless mini colour camera, mounted on a pair of sunglasses, which transmits a colour composite video signal to a Panasonic Toughbook computer that has a separate wireless-connected handheld screen. Thus, the camera and the view screen can be carried separately, each with a range of about 100 metres from the base-station. The quality of the video signal was variable, depending on range and interference, ranging from near-VHS video quality to flickering between colour and black and white. We used the SpyCam in two configurations. In the first, one child wore the glasses and another held the Toughbook screen. The child with the screen had a continuous transmission of what the other child was seeing, and could tap the screen at any time to capture a still frame. The children were able to communicate by voice through "walkie-talkie" handheld radios. This configuration exemplified the remotely controlled camera idea. In the second configuration, a child wore the glasses with the attached camera and also had a wireless remote control key-fob. A press of the key-fob's button sent a command to the Toughbook to take a picture. This configuration exemplified the spyglasses camera idea (see fig. 2a).

Different scenarios were tested for both configurations, with 32 children at a residential children's "education and activity" centre. We also carried out tests using two blindfolded treasure hunt games, designed to exploit the features of the SpyCam prototype. All sessions, including the treasure hunts and the subsequent interviews, were recorded on video.

The trials revealed that the children found no difficulty in operating the equipment. All the children gave very positive responses to the question of whether they enjoyed it:

"It was cool!" (Griffin, age 10)

"I had the power to do anything to him!" (Mark, age 12)

"Yes! It was really fun!" (Yasmin, age 11)

"It was just like, you didn't have to worry about holding it or nothing it was just, always on you and easy to use, because if you wanted to take a photo just tell them, just push the button" (Scott, age 13)

"It was just cool, being able to see what someone else saw" (Jamie, age 12)

"I liked that you could see what they were seeing instead of just seeing what you could... instead of just looking at them and trying to imagine what they could see" (Annie, age 11)

Their responses were also positive to the question of whether they would like to own a similar device:

"Yes, definitely!" (Mark, age 12)

"Yes, it's excellent!" (Nick, age 10)

"I would like to have the sunglasses and the hat, I think they are cool!" (Annie, age 11)

The children did not appear to be concerned by the poor quality of the captured images in this prototype or with the transmission problems. Although they did mention it as a problem during the interviews, while doing the task they did not seem frustrated. If they moved out of transmission distance, they patiently came back into range.

The children very rarely asked to see the pictures they had captured; they appeared more interested in the live transmission of video. This was verified during the interviews, where what seemed to impress them the most was the ability to see what someone else was seeing in real time, the ability to control where someone else goes, the need to trust someone's instructions on how to move and what to do, and the choice of whether to take a photograph or not.

## 3.2 Follow-up FTWs, Prototypes, and Testing

The third children's workshop was carried out with a different group of six children, again three boys and three girls. This workshop built on the outcomes of the previous ones, with the children being shown the prototypes and asked how they could be altered or improved. The design task therefore related to "capturing visual events remotely, without being noticed". Both the boys and girls group worked on spy-camera ideas such as cameras hidden in glasses and wrist watches, with the boys also producing the further idea of a 'remote landmark camera', a camera situated on a lamppost at some remote landmark and being able to transmit images to people who cannot visit the place.

The children in the third FTW seemed to converge on a design that includes a camera, which can attach to a number of different objects – sunglasses, wrist watches, etc. Our initial prototypes were therefore expanded to include, besides the SpyCam on the sunglasses, a RoboCam – a model car operated by remote control with the miniature camera mounted on top (see fig. 2b). The camera can easily be removed from the glasses frame and attached to the roof of the car, which the children can then operate by the remote control and receive the image from the camera on to a computer monitor or on a small-screen TV.

The new prototypes were tested with the same six children who participated in the second children's workshop, in the following two scenarios: The first scenario was, again, a blindfolded treasure hunt, where one child was blindfolded and wore the spyglasses and the other children were instructing her to wander around another room and collect clues. Six pieces of evidence were scattered around the room, one for each child to discover. The second scenario involved a collaborative mystery solving and made use of the RoboCam. Six pieces of paper containing key information about the mystery were put around a room different to the one where the children were, at floor level. One child was navigating the RoboCam through the remote control, and the other five children were watching the image sent by the RoboCam and instructing the navigator-child where to send the car (see fig. 2c).

The children enjoyed both scenarios and had no trouble in operating and making use of either the SpyCam or the RoboCam. As with the previous trials, shortcomings of the technology in terms of performance were not off-putting and the children were more interested in the live image than the captured still frames.

The same children then participated in the final children's workshop. They were asked to go through the workshop refining the designs and expanding the ideas of the RoboCam and the SpyCam. The imagineering session mainly produced a list of things on which the SpyCam could be mounted: briefcases, wristwatches, sunglasses, dog collars, model boats and many more.

## 4    Conclusions and Future Work

The SpyCam and RoboCam prototypes are now being developed further: We are looking at producing a base for the camera that can be fixed on different objects. For the task of designing new imaging technologies for children, the FTW method has proved a reasonable success.

The field studies thus far have shown that the Future Technology Workshop method can be successful as a means of rapidly generating and refining designs for new interactions between technology and activity. The FTW method has benefits over more traditional methods of product design and engineering, which include: the combination of creativity with practical design; the focus on interaction between activity and technology; the ease with which a FTW session can be run, at low cost, in one half-day; a clearly-defined set of activities within an easily-understood framework, the method is readily taught to other designers and engineers.

As part of the process of refining the method, we are planning to apply it to different design tasks. More specifically, it will be used in the design of new mobile learning technologies under the framework of the Mobilearn project (http://www.mobilearn.org/). Further work is needed in refining and documenting the "design, implement, deploy, evaluate, and redesign" cycle so that the FTW can become a practical tool for product and interaction designers.

## References

Beyer, H., & Holtzblatt, K. (1997). *Contextual Design: A Customer-Centered Approach to Systems Designs*: Morgan Kaufman Publishers.

Druin, A. (1999a). *Cooperative Inquiry: Developing New Technologies for Children with Children.* In Proceedings of CHI 99, Pittsburgh, PA, May 15-20.

Druin, A. (1999b). *The Role of Children in the Design of New Technology* (Technical Report HCIL Technical Report No. 99-23): HCIL.

Druin, A., Bederson, B., Boltman, A., Miura, A., Knotts-Callahan, D., & Platt, M. (1999). Children as Our Technology Design Partners. In A. Druin (Ed.), *The Design of Children's Technology* (pp. 51-72): Morgan Kafumann Publishers.

Inkpen, K. (1999). Designing handheld technologies for kids. *Personal Technologies, 31*(1&2), 81-89.

Scaife, M., & Rogers, Y. (1999). Kids as Informants: Telling Us What We Didn't Know or Confirming What We Already Knew? In A. Druin (Ed.), *The Design of Children's Technology* (pp. 27-50): Morgan Kaufmann Publishers.

Sharples, M., Jeffery, N., du Boulay, J. B. H., Teather, D., Teather, B., & du Boulay, G. H. (forthcoming). Socio-cognitive engineering: a methodology for the design of human-centred technology. *European Journal of Operational Research*.

Vavoula, G. N., Sharples, M., & Rudman, P. (2002). *Developing the 'Future Technology Workshop' method.* In M. M. Bekker & P. Markopoulos & M. Kersten-Tsikalkina, Proceedings of Interaction Design and Children, Eindhoven, The Netherlands, Aug 28-29.

# Enriching the Pedagogical Value of an Asynchronous HCI Course: Adding Value Through Synchronous Collaborative Knowledge Building

*Rita M. Vick[1], Brent Auernheimer[2], Martha E. Crosby[3], Marie K. Iding[4]*

University of Hawaii, USA[1, 3, 4], California State University, Fresno, USA[2]
{vick, crosby, miding}@hawaii.edu[1, 3, 4], brent@CSUFresno.edu[2]

## Abstract

Actively engaging students in distance-learning courses is a challenge that can be met by promoting interaction among students and with instructors through "mixed" synchronous-asynchronous course design. A cross-institutional, cross-cultural, distance-learning course in Principles of Human-Computer Interaction (HCI) Design was offered separately, asynchronously, and in parallel to 36 globally-distributed graduate level computer science students at two geographically and temporally separate universities. The course was conducted asynchronously until the two classes joined at a common Blackboard (Bb) site for a synchronous Final Project. The synchronous phase enriched what had been a purely asynchronous course by introducing synchronous same-university teamwork followed by matched inter-university tandems for project presentations. This design added to the pedagogical value of the course by facilitating deeper understanding of HCI principles through collaboration with peers.

## 1   Motivation for the Study and Theoretical Foundation

Online distance learning has gained in value as classroom education costs have risen. Driven by changing personal schedules, the need for lifelong learning, and the desire to access distributed domain expertise, educators and students have turned to e-learning for delivery of course value that is not or cannot be provided by in-classroom curricula. Having already taught an online Introductory Human-Computer Interaction (HCI) course successfully in asynchronous mode, we wanted to discover whether adding a synchronous element in the form of project-based team learning would facilitate increased student involvement in the course. We also wanted to find out how students working collaboratively might share and build knowledge to effectively bridge the gap between theory and application. To answer these questions, and to allay some of the unevenness of interaction that tends to characterize asynchronous course activity, we joined asynchronous and synchronous elements in a unique cross-institutional, cross-cultural, version of the same course. We hoped to increase student involvement and so enhance student motivation to engage in persistent interaction.

The Final Project task for the course was designed to invoke the benefits of problem-based learning where, rather than dividing the project into parts and working outside the group process, project work was to be done collaboratively so that knowledge building and sharing would take place within the group interaction process (Stahl, 2002). Because of the chat-based distributed teamwork nature of the project, students were called upon to perform linguistic, cognitive, and task coordination, which enabled them to develop sufficient mutual understanding, or grounding, to achieve the learning goal (Mäkitalo, Häkkinen, Salo, & Järvelä, 2002). The first exercise, critiquing of their own and others' work, was designed to engage students in social-constructivist generation of explanations, questions, and debate about their common task. This type of focused, sustained discourse would enable creation of communal inquiry, enabling mutual motivation among team members (Lally, & De Laat, 2002). At the same time, possible frustration with learning a difficult task would be reduced to encourage persistence in completing the task (Barron, Martin, Roberts, Osipovich, & Ross, 2002). Taken together, these facets of synchronous distributed collaboration were expected to facilitate deeper understanding of HCI principles and engage students in actively working to understand the relationship between theory and application.

## 2 Method and Results

The course was offered separately, asynchronously, and in parallel to thirty-six globally distributed graduate level computer science students at two geographically and temporally separate universities. It was conducted asynchronously until the two classes joined at a common Blackboard (Bb) site for a synchronous Final Project. Students worked synchronously online in same-university teams to review their initial individual attempts at performing the task. Subsequently, teams were paired with other-university teams to form tandems for a synchronous online Final Project presentation session. The classes consisted of culturally diverse students who were from the People's Republic of China (PRC), Taiwan, Northern Europe, South East Asia, the Continental United States, Hawaii, the Indian Subcontinent, and South America. During online sessions, students were geographically and temporally dispersed. Students were in Hawaii and California, one student was in the PRC, another in Chicago, and another in Boston. Students successfully met online for synchronous pre-presentation discussions and final presentations. Figure 1 indicates substantial interactivity despite differences in time zones.

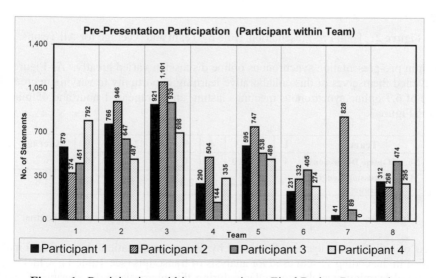

**Figure 1:** Participation within teams prior to Final Project Presentation.

To develop students' understanding of HCI "principles-in-use," we introduced a synchronous four-phase Final Project that required students to perform a cognitive walkthrough (CW) of a Web site: (1) Individual students performed CWs. (2) Four-member same-university student teams carried out synchronous online review of their individual work, combining results into a single

PowerPoint presentation. (3) Teams from each university paired with other-university teams to reciprocally present and discuss results. During this course capstone, two teams and three instructors (eleven participants) were present simultaneously in a Bb Virtual Classroom. Communication was via text chat, the support system proved robust, and sessions went smoothly. (4) Same-university teams discussed the presentations and critiqued the course as a whole. Given voluminous data archived by the common Bb system (pre-presentation team meetings, tandem presentation sessions, and post-presentation debriefing sessions) and the complexity of the course design, the focus here is on pre-presentation synchronous online team conversations (Final Project Phase 2). During Phase 2 students discovered they did not understand the rationale behind the CW, which seemed the same as the usability study done earlier in the course. To assist discussion, study questions and relevant Web sites[1] were posted on the common Bb Web site.

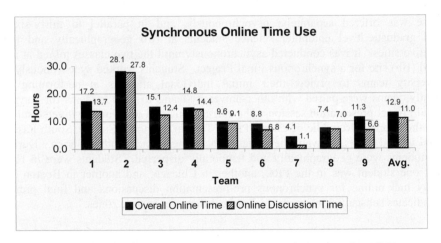

**Figure 2:** Total Pre-Presentation Time by Team and Average for All Teams.

Time spent in pre-presentation synchronous online discussion varied greatly. As Figure 2 shows, students availed themselves of this collaborative learning opportunity to varying degrees, holding an average of 6.7 online synchronous meetings lasting an average of 1 hour and 45 minutes each as shown in Figure 3.

| Team | 1 | 2 | 3 | 4 | 5 | 6 | 7 | 8 | 9 | Average |
|---|---|---|---|---|---|---|---|---|---|---|
| **No. of Online Discussions** | 4 | 16 | 10 | 9 | 4 | 7 | 1 | 3 | 6 | 6.7 |
| **Avg. Discussion Length (Hours)** | 3.4 | 1.7 | 1.2 | 1.6 | 2.3 | 1.0 | 1.1 | 2.3 | 1.1 | 1.8 |

**Figure 3:** Pre-Presentation Time Use by Team and Average for All Teams.

Key aspects of collaborative learning facilitated by the pre-presentation phase of the Final Project are embodied in the coding system used for analysis of transcripts (see Table 1). Chief elements of sociocultural learning available to students during this phase were the sharing of knowledge-containing artifacts (SKA), collaborative knowledge capture (CKC) in the form of student-created

[1] http://www.cs.umd.edu/~zzj/CognWalk.htm
http://www.acm.org/sigchi/chi95/Electronic/documnts/tutors/jr_bdy.htm
http://www.cc.gatech.edu/computing/classes/cs6751_96_winter/handouts/cognitive-walkthrough.html

| Speaker | Comment | SKA | CKC | PKR | PKS | CKB |
|---------|---------|-----|-----|-----|-----|-----|
| TH12 | starting by putting some of the questions that rita put on the assignment page. | ■ | | | | |
| TH12 | the ones that I think are relevant. | ■ | | | | |
| TH12 | Then maybe we can discuss our results pertaining to the questions. | | | | | ■+ |
| TH11 | That is a good idea, I'm going to take a look at you walkthrough.... | ■ | | | | |
| TH11 | I started assuming that they would already have the "insight browser" on the screen, | ■ | | | | |
| TH12 | Can I bring up files without losing chat? | | | ■ | | |
| TH11 | this was probably a bad assumption. | ■ | | | | |
| TH11 | probably right, just discuss the results, | | | | | ■+ |
| TH12 | Well, I can start the ppt project and then upload it and anyone can change/add to it. | | ■ | | | |
| TH11 | I have the questions in front of me now. | ■ | | | | |
| TH12 | should we email when we are working on it so changes are synchronized. | | ■ | | | |
| TH12 | Can I open up a new browser with out losing this session? | | | ■ | | |
| TH11 | It will screw it up, you should probably get a new browser from scratch. | | | | ■ | |
| TH11 | OK, in reference to Rita's guiding question number one. | ■ | | | | |
| TH12 | number one should be answered last. | ■ | | | | |
| TH12 | After we discuss more. | ■ | | | | |
| TH11 | OK, I have something to say about #2. | ■ | | | | |
| TH12 | I think that 4 and 5 are good to discuss. | ■ | | | | |
| TH11 | OK, number 2 ... | ■ | | | | |
| TH11 | 2.a) While empirical user testing with the "real" users is probably the ultimate way to test a design, | ■ | | | | |
| TH12 | Number 2 maybe second to last, since it is kinda summing things up. | ■ | | | | |
| TH11 | ok, let me finish this thought... | | | | | ■! |
| TH11 | a cognitive walkthrough, by an expert can spot many problems of the interface... | | | | | ■+ |
| TH11 | Yes, I agree, and since 5 users is enough to catch a majority of errors that shouldn't be that expensive, but if we have an "expert" that can spot the major problems, why not do that first? | | | | | ■- |
| TH12 | But I don't know if the cog. walkthrough is needed to find obvious problems. | | | | | ■- |
| TH11 | isn't the cog walkthrough just formalized way of "checking out" the user interface? | | | | | ■- |
| TH11 | the other day when I had a smart computer person use an app I had written. | | | | | ■+ |
| TH11 | ... even if you think you are an expert at usability you probably shouldn't be doing a cog walkthrough of you own system. | | | | | ■+ |
| TH12 | ... you would have a biased pt of view. | | | | | ■+ |

**Table 1:** Pre-presentation chat excerpt. (Minor editing performed to conserve space.)

artifacts (e.g., the PowerPoint presentation), and helping one another by providing feedback on others' work as well as dissemination of information through personal knowledge requests (PKR) and personal knowledge sharing (PKS). In this way, reciprocal production and distribution of information resulted in collaborative knowledge building (CKB) through discourse and enabled collaborative knowledge emergence (CKE) as students learned from one another. New knowledge gained in this way is "an emergent" since you "can see the conceptual object taking shape but you cannot find it in the bits and pieces making up the discourse" (Bereiter 2002). Because emergence of shared understanding through discourse in our study resulted from synthesis of contributions from many individuals, instances of collaborative knowledge building in chat excerpts are marked by a "-" (minus sign) to indicate a knowledge building process (KBP) loss, a "+" to indicate a KBP gain, or an "!" (exclamation point) to indicate a request to stop for reflection.

## 3    Discussion

Analysis of pre-presentation conversations revealed the extent to which each team achieved common ground, becoming conversant with the CW method through discussion with peers. Our analysis focused on how use of self-created and instructor-supplied cognitive, physical, and digital artifacts facilitated understanding the rationale behind CW as an evaluation method. This case study represents a significant contribution to management of diverse pedagogical tools in a complementary way to advance the quality of what diverse student groups learn beyond the context of course material. It demonstrates maximal leveraging of collaborative web-based interaction. Our analysis will benefit educators and trainers who will increasingly use information and communication technologies as connectivity improves and reliance on e-learning grows. Connectivity and courseware (e.g., WebCT, Blackboard) enable persistent interaction among students and educators along with tools instructors can use to tailor content to the knowledge requirements of specific learning domains. The combination of e-learning and pedagogical tools (asynchronous WebCT and Bb, synchronous Bb, and teamwork) we have described is an example of deployment of minimal resources to productively join culturally and geographically diverse students and educators in construction of understanding.

## 4    Acknowledgements

This research was supported in part by DARPA grant NBCH1020004 and ONR grant no. N00014970578 awarded to Martha E. Crosby.

## 5    References

Barron, B., Martin, C., Roberts, E., Osipovich, A., & Ross, M. (2002). Assisting and assessing the development of technological fluencies: Insights from a project-based approach to teaching computer science. Paper presented at CSCL '02 (Jan. 7 - 11), Boulder, CO, USA.

Bereiter, C. (2002). Education and Mind in the Knowledge Age, Lawrence Erlbaum Associates, Hillsdale, NJ.

Lally, V., & De Laat, M. (2002). Cracking the code: learning to collaborate and collaborating to learn in a networked environment. Paper presented at CSCL '02 (Jan. 7 - 11), Boulder, CO, USA.

Mäkitalo, K., Häkkinen, P., Salo, P., & Järvelä, S. (2002). Building and maintaining common ground in Web-based Interaction. Paper presented at CSCL '02 (Jan. 7 - 11), Boulder, CO, USA.

Stahl, G. (2002). Contributions to a theoretical framework for CSCL. Paper presented at CSCL '02 (Jan. 7 - 11), Boulder, CO, USA.

# Learning to Learn:
# HCI-Methods for personalised eLearning

*Christian Voigt*

Paula M.C. Swatman

Faculty of Informatics, University of Koblenz-Landau
Universitätsstr. 1, 56070 Koblenz, Germany
Tel: +49-261-287 2850  Fax: +49-261-287 2851
Email: cvoigt@uni-koblenz.de

## Abstract

One of the major problems in eLearning systems is the very high rate of user drop-out. While there are many explanations for this phenomenon, one issue relates to the lack of attention to users' real needs paid by many of today's Learning Management Systems. In this paper we introduce the concept of personalising the LMS User Interface (UI) – not merely in terms of changing colours, but in terms of mental models and the application of relevant learning theory (behaviorist, cognitive or constructivist). Rather than endeavouring to learn each system anew, the paper provides an argument for "learning to learn", ie. the investment of time initially spent to understand the UI which will be saved many times over by the learner who can then personalise each eLearning system s/he uses to suit his/her own mental models.

## 1.     Introduction

The unfortunately high rate of learner drop-out in many of today's eLearning systems is, to a certain extent, caused by lack of attention to the learner's actual needs within those environments. Significant research has been undertaken concerning the integration of new technologies, rich multimedia and collaborative learning tools – but these components alone are not sufficient to ensure that eLearning is used in an instructionally meaningful way. Learning is a highly complex process which depends on individual cognition as well as on the surrounding organisational factors (Martinez, 2000). Drifting learners are therefore often a result of heavily technology-oriented eLearning scenarios where either little pedagogic theory is applied (Hannum, 2001) or where the integrated instructional intentions cannot be communicated clearly enough to the learners (Kerres, 2001).

Instruction theory suggests that one of the major problems for didactics-driven eLearning is the matchmaking between learning objectives, learning situation (e.g. motivation, capacities, environmental support) and the form of instruction most likely to succeed (Jonassen & Grabowski, 1993). Solutions vary along a continuum at one of which the machine tends to do all the personalisation (eg. intelligent tutoring systems), and at the other end of which learners have total freedom to evoke instructional events (examples, glossaries, exercises or feedback) when they want and for as long as they want (Kinshuk, Patel & Russel, 2002). How to determine an optimal balance between guidance and control within eLearning environments is still an open question (Moonen, 2002). Despite the difficulties of striking exactly the right balance between total machine control and total human control, the UI is clearly the entry point to eLearning. Here the learner discovers how to interact with the learning environment.

In this paper we discuss the appropriate role of personalised User Interfaces in the creation of a more theoretically-based, and potentially more successful, approach to eLearning. Rather than hoping for an algorithm-based artificial intelligence solution which can completely control the process of personalising an eLearning environment, we propose encouraging the learner him/herself to take the initiative, in an approach to the Learning Management System (LMS) interface which supports a variety of instructional strategies and which encourages self-directed learning. The UI can then offer pre-designed instruction models which can be chosen by the learner to suit his/her specific needs and skills. Such an approach allows learners to become the co-designers of their own courses (Lowyck, 2002); and leads to a more active approach to learning – what we describe as "Learning to learn".

In the remainder of the paper we discuss the issues associated with personalisation, introducing the concept of mental models and their relationship to eLearning and identifying the complexities which learning theories raise for personalisation. We then apply this theoretical approach to UI personalisation to a real-world case study of eLearning and finally draw some conclusions about the likely effectiveness of our suggested approach.

## 2.    Personalisation: Learners at stake

In this section we want to explain what is meant by personalised eLearning and show how personalised eLearning is best supported by personalised User Interfaces. Our argument is based on the concept of mental models used by learners when interacting with the learning environment. The aim is that learners are motivated to reflect on their own models of learning and understand learning models embedded in the learning management system (LMS). User Interfaces thus need to present learning models in such a way that users can easily access and use them.

### 2.1 Individual differences and mental models

Before we can define personalisation we need to discuss the role of mental models in comprehending and learning about complex systems. Mental models are conceptual beliefs a user holds about a system's purposes and how these might be achieved (Carroll & Olson, 1988). These models are based on the user's previous knowledge and expectations and are unstable, being inclined to change over time as a result of self-enquiry, training, observation and accidental encounters (Fischer, 1991).

Learners approach eLearning in a very similar way – and we can expect them to have mental models about learning and how learning should be supported by the Learning Management System (LMS). When learning starts, the learner's model is set against the designer's eLearning model, implemented in the system's model of eLearning. If we take mental models as basis for personalisation, we assume that the system is most effectively used when the system's model and the learner's models can be mapped. If the UI reflects an appropriate model then the learner needs less guidance and may perform more accurately (Carroll & Olson, 1988).

If eLearning is to successfully transmit information about the subject matter, it must also modify underlying mental models. Modification is needed because learners do not always have appropriate mental models – and consequently usage of the LMS may also be: irregular (not based on concept knowledge), erroneous (the expected concept differs from the one actually used) or incomplete (concepts are not expected to be part of the system) (Fischer, 1991).

Mental models enable system usage from a meta-learning perspective, ie. learners improve their learning competencies in terms of different forms of learning activities supported by the LMS (Kerres, 2001). A forum does not, by itself, automatically improve learning, it has to be understood as a structured way to discuss open issues, where different views relate to one another. Personalised User Interfaces therefore have a twofold mission: to support the learner's current mental model as far as possible; and to enable learners to change their mental models, if necessary, to make appropriate use of the system's functionalities.

### 2.2 Personalise the User Interface

Differences in learning are easy to observe, individuals perform differently in the same task because they differ in traits, thinking processes, aptitudes, abilities and preferences (Jonassen & Grabowski, 1993). The quest for an optimal learning condition has led instructional designers to use a variety of learning theories. These theories are not so much to be interpreted as exclusive explanations of the learning phenomena, but rather as complementary views focusing on different aspects of the same process (Kerres, 2001).

Personalised learning aims to integrate these theories into various instruction strategies which affect learning outcomes. It is beyond the scope of this article to explain the topic of learning theories in depth, but in Table 1 we provide a brief overview of three major learning theories and their implications for User Interface design (Lowyck, 2002; Kerres 2001; Petraglia, 1998).

**Table 1:** Behaviorist, cognitive and constructivist learning theories

| theory | learning is | implications for User Interface design |
|---|---|---|
| behaviorist | ▪ an overt reaction to external stimuli<br>▪ based on repetitive task execution | ▪ external- control learning<br>▪ decomposition of complexity<br>▪ focus on content |
| cognitive | ▪ the construction of meaning as a reaction to cognitive stimuli<br>▪ based on information processing by the learner | ▪ self-controlled learning<br>▪ complexity is reduced by cognitive schemes<br>▪ focus on presentation |
| constructivist | ▪ situated in social activities where meaning is negotiated<br>▪ based on group activities | ▪ group-controlled learning<br>▪ complexity is accepted (multi-dimensional problems)<br>focus on collaboration and contextualisation |

Just as many proponents of eLearning now prefer the idea of blended learning, because not all learning activities can or should be executed online, we propose the analogical concept of "blended personalisation". Many existing personalisation solutions are superficial or lack instructional background. Simple strategies which rely on next- and forward buttons (the behaviorists' external control) or the integration of chat and mail (the constructivists' claim for interaction) are not adequate solutions to the eLearning problem – learning becomes either too stereotyped or too demanding. Making allowances for individual differences in learning styles and abilities should be done when the LMS is introduced. Current explanation mostly refer to technical aspects of the learning platform – we propose the investment of a little time into the introduction of learning theories, which would give the learner a real chance to participate in the personalisation process. A further benefit concerns the vocabulary level, where the User Interface could just name the concept and thereby indicate the UI's purpose to the learner. Learners could choose between behaviorist, cognitive or constructivist learning rooms independently of the material they are studying.

Another reason why learners should be consciously aware of learning models is that the learner is the most appropriate person to analyse the complexity of current constraints and decide which instruction method fits best at this moment. Existing eLearning applications seem to assume that all e-learners are highly motivated people, eager to exploit all the functionalities of the system and try to find a way around even tricky procedures to complete the course. But what if the learner lacks motivation and commitment or has less time available? if they don't share the same aspirations as their instructors, the bursting interest in new insights? What if s/he is someone who says: "Just tell me what I have to know and I will learn it" (Petraglia, 1998). Then personalisation goes beyond the learners' cognitive abilities and needs a more holistic perspective which also includes emotions, intentions or social factors (Martinez, 1999).

Recognising different levels of motivation and ability sheds a new light on what personalisation can mean for us. The design of personalised eLearning goes beyond a mere change of colours, to provide adaptation on a functional level, supporting a variety of instructional strategies to achieve a particular learning objective.

We therefore propose keeping the control of the matchmaking process between the learning situation itself and the strategy to be applied by the individual learner. This does not mean that we simply pass the problem to the learner but, given the lack of proven rules for automated personalisation, this is how we preserve instructional quality. For this approach to work, learners must reflect on their own learning situation – and this requires some previous learning about how to learn. Questions such as: What does a certain terminology mean?, What can this functionality be used for?, What is my preferred learning style? What are the underlying concepts supported by the learning environment? have to be clarified in advance. If for any reason, cognitive or social, learners need to change their learning conditions, they can amend the personalisation and avoid both the boredom and the distraction of many existing eLearning solutions.

## 3. Practice: Learner-centered design

Within our current eLearning project with a major insurance company, our current task is the specification and prototypical use of a Learning Management System. The overriding evaluation goal is pedagogical effectiveness, that is, technology is relevant only if its usage adds real value to the learning process. Although technically the configuration, using the highly customisable XML/XSLT architecture, is not a major problem, pedagogically there is still the challenge of creating a convincing learning environment. Therefore the early involvement of learners is a basic principle of our design process, for which we use mixed teams from the Human Resources Development and the IT Departments, contributing both organisational and technological knowledge from within the company. These user "guinea pigs" are asked to specify the functionalities the LMS should offer once it is implemented and to describe the expected benefits.

Our underlying concern is that simply putting technology at the learners' disposal would not provide sufficiently realistic feedback about the pedagogical relevance of the various system functionalities. This concern led us to pay more attention to the role of mental models, initially within the design team and then with the learners themselves.

The fit (or misfit) between learners and learning system is only observable at the physical level. Although most of the system resources are already in place (LMS trial version, integration with existing IT-infrastructure, organisational specifications for a 6 week test run) the final hurdle is the interoperability between externally-produced e-learning courses and the LMS itself. This principally concerns the functionalities which manage learner activities and report test results. Although content and LMS vendors naturally all claim to adhere to communication standards such as the AICC Standard (Aviation Industry CBT Committee), the extent of interoperability only becomes clear when it is actually tested with the given LMS. We hope to obtain suitable learning material within the next few weeks, which will enable us to run the prototype and thus evaluate the design approach described in section 2 of this paper – fostering the learners' awareness of learning models and improving their learning competencies.

To ensure that environmental or technological changes are immediately taken into account, we use a scenario-based, iterative design framework. Scenarios are narrative task-descriptions which are: flexible (avoiding premature commitment) and concrete (applying to real world stories). This is a good solution for situations where many external constraints (diverse stakeholders, copyrights, network infrastructure, authoring systems) have to be managed (Carroll, 2000). We have already gained an advantage from this relatively simple approach, when a discussion about the learners' on site environment revealed that bandwidth could not be guaranteed at the same level across all the company's subsidiaries. In this case we had to change some of the test participants, including only those subsidiaries which already had access to stable bandwidth.

At the end of the pilot phase we will measure the usability of each scenario in terms of Shneiderman's five human factors: ease of learning, speed of performance, user error rate, user retention over time and subjective satisfaction (Shneiderman, 2002). Data will be collected by means of: observation, interviews, surveys and automatically generated reports. The form(s) of data gathering selected are determined by the accessibility of the information holder and the need for further inquiries. We are also taking the political dimension of questions into account. This certainly implies some trade-off management, but we hope that our scenario-driven approach to implementation will offer sufficient benefits to make this worthwhile. Contradictory design evaluation criteria can be analysed in a specific context where it is easier to decide which human factor should be prioritised.

# 4.    Conclusion

Personalisation does not provide a recipe for successful eLearning system implementation, nor is there a hierarchal order of learning theories which indicates the most successful strategies under all circumstances. This suggests that participatory personalisation, i.e. where the locus of control is kept with the learner, is a viable approach. We stress learner participation as a form of acknowledging learning as an individual process with very few rules that can be imposed from outside.

One might say that personalisation is not the primary goal if a company is struggling to get eLearning running. Although the interaction between the learner and the learning environment for the purpose of personalisation is only a small part of the whole learning process it is nonetheless an important one, where a well-designed interface may have a decisive impact on the overall acceptance of eLearning. Learners see themselves accepted as individuals and can be included in the development process at an early stage. This saves time for developers and avoids creating artifacts which will never be used despite sound underlying theories. Clearly, further research is needed to actually develop suitable content, and it is equally clear that personalisation has economic limits. Not every course will support a variety of cognitive learning styles or can be easily integrated into different instruction strategies.

One approach which offers promise in this context is the concept of "learning objects" – digital entities which are deliverable over the internet and which can be reassembled into courses on the base of metadata description (Hodgins, 2002). Which set of metadata to use and an ontology of the terms used are some of the issues to be solved if learning objects are going to be used on a large scale (Simon, 2002). Future research in this project will look at whether and how learning objects might be applied to the user-directed personalisation of LMS.

# 5.    References

Carroll, J.M. (2000). Making use: Scenario-based design of human-computer interactions. Cambridge: MIT Press.

Carroll, J.M., & Olson, J.R. (1988). Mental models in Human-Computer Interaction. In M. Helander (Ed.), *Handbook of Human-Computer Interaction* (pp. 45-65). Amsterdam: Elsevier.

Fischer, G, (1991). The importance of models in making complex systems comprehensible. In M.J. Tauber, & D. Ackermann (Eds.), Mental Models and Human Computer Interaction 2 (pp. 3-36). Amsterdam: Elsevier

Hannum, W. (2001). Web-based training: Advantages and Limitations. In B.H. Khan (Ed.). *Web-based training* (pp. 13-20). New Jersey: Englewood Cliffs

Hodgins, H.W. (2002). The future of learning objects, A United Engineering Foundation Conference 11-16 August 2002 Davos, Switzerland, Retrieved December 7, 2002, from http://www.coe.gatech.edu/e-TEE/pdfs/Hodgins.pdf

Jonassen, D.H., & Grabowski, B.L. (1993). Handbook of individual differences, learning, and instruction. Hillsdale: Lawrence Erlbaum.

Kerres, M. (2001). Multimediale und telemediale Lernumgebungen: Konzeption und Entwicklung. (2nd ed.). München: Oldenbourg.

Kinshuk, Patel, A., & Russel, D. (2002). Intelligent and adaptive Systems. In H.H. Adelsberger, B. Collis, & J.M. Pawlowski (Eds.), *Handbook on Information Technologies for Education and Training* (pp. 79-92). Berlin: Springer

Lowyck, J. (2002). Pedagogical design. In H.H. Adelsberger, B. Collis, & J.M. Pawlowski (Eds.), *Handbook on Information Technologies for Education and Training* (pp. 199-218). Berlin: Springer

Martinez, M. (2000). Designing Learning Objects to personalize learning. In D.A. Wiley (Ed.), *The Instructional Use of Learning Objects*, Online Version. Retrieved November 11, 2002, from http://reusability.org/read/chapters/wiley.doc

Moonen, J. (2002). Design Methodology. In H.H. Adelsberger, B. Collis, & J.M. Pawlowski (Eds.), *Handbook on Information Technologies for Education and Training* (pp. 153-179). Berlin: Springer

Petraglia, J. (1998). The real world on a short leash: The (mis)application of constructivism to the design of educational technology. *Educational Technology Research and Development*, 46 (3), 53-65.

Sasse, M.-A. (1991). How to t(r)ap users' mental models. In M.J. Tauber, & D. Ackermann (Eds.), *Mental Models and Human-Computer Interaction 2* (pp. 59-79). Amsterdam: Elsevier

Shneiderman, B. (2002). User Interface Design (3rd ed.). Bonn: Mitp.

Simon, B. (2002). Do eLearning standards meet their challenges? Retrieved November 11, 2002, from http://www.rz.uni-frankfurt.de/neue_medien/standardisierung/simon_text.pdf

# Saccadic Processes in Listening-Comprehension Processing as Cognitive Interactions between Listeners and Texts in a Computer-Based Learning Environment

*Setsuko Wakabayashi*　　　*Koichiro Kurahashi*
Himeji-Dokkyo University
7-2-1, Kamiohno,Himeji,670-8524,Japan
setsuko@himeji-du.ac.jp　　kurahasi@himeji-du.ac.jp

## Abstract

In the unique environment where free relistening is allowed, listeners' attempts  to reach an interpretation exhibit saccadic processing.  In these processes a feature of listening-comprehension processing is well demonstrated as in reading- comprehension processing.  The result is also considered to provide a good basis for a theoretical foundation for interface design:  that is, an environment in which learners are able to relisten to any part and any length of text in order for them to deliver the necessary input and to make their adjustment to earlier interpretations plausible in a very short time.

## 1　Introduction

In order to investigate (1) how listening comprehension processing takes place, and (2) a basis for theoretical foundation for interface design of listening comprehension practice, the sizes of listening processing units and their dynamic patterns of combinations are discussed.

Our previous studies with various types of materials, modes and tasks show, firstly, that listeners' selected listening units at 'their will' have four sizes, and the sizes of these units seem to relate to the levels of information used in processing (Wakabayashi & Kurahashi 2002 (a); Wakabayashi 2003 (b)).  Secondly, combinations of these different sized units show static patterns of combinations of processing units, from which various levels of information are assumed to have been employed for different types of processing (Wakabayashi & Kurahashi 2002 (b); Wakabayashi 2003 (a)).

Following these findings successive usages of processing units exhibiting dynamic patterns of processing are discussed in order to investigate how listeners reach their interpretations. Directions along which listeners feed their intakes (parts of texts) and levels of information processed/listened are the focus of our investigation.  The characteristic usage of various sizes of processing units and the multi-directions of relistening are introduced as saccadic processing, and the implications are discussed.

## 2　How sizes of processing units and levels of information were determined

### 2.1　Listening Practice Environments

#### *2.1.1　Listening Environment*

Digitized visual-audio text was indexed by frame numbers (interval of 1/30 sec).  One click of any play button allows listeners to instantly receive whatever they have chosen.  The chosen part of the text which begins at the frame specified by clicked play button and continues until the next subsequent action is defined as 'the listening processing unit'.

Five experienced researchers were acting as speakers of each text and segmented texts. The concept of unit size from the smallest units of words to complete topic level units (groups of paragraphs) (Baddeley 1992; Roach 1991; Miller 1962; Chafe 1980; Ebbinghaus 1885) as well as specifically speakers' supra-segmental information such as intonation and pausing were highly taken into consideration for segmentations (Altenberg 1987; Crystal 1969 & 1975; Roach 1991). In order to provide listeners with flexible control over playing or replaying parts of texts, different sizes of segmentations at four levels of information unit: from 'Clause', 'Sentence'(*), Paragraph, and Topic were introduced.

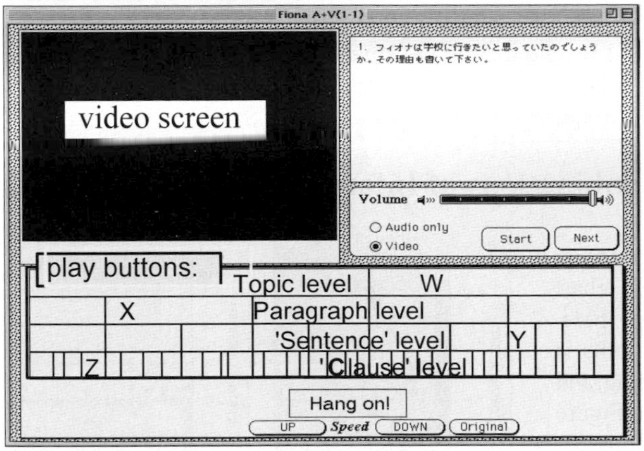

Figure 1: Listener control screen with Play Buttons giving immediate access to different parts of the oral text.

Figure 1 shows the basic listener control screen. When there is accompanying video it is shown in the upper left screen. The panel in the upper right quadrant displays comprehension questions in Japanese. The grid-like display is a 'Segmental map' of the text. Clicking on the box (that is 'paly button') labeled W in Figure 1 would immediately start playing from the beginning of the second topic; clicking in the box X would start from the beginning of the second paragraph of the first topic, and so forth. It only stops playing when listeners take the next subsequent action at any parts of texts.

### 2.1.2 Conditions of Tasks and Subjects

Various types of materials (e.g. 'Fiona's Story': A UNESCO video, made and distributed by Vanuatu Community Health Education-[F], and 'Kouyou': A clip from an NHK documentary on autumn colours and of life (in Japanese)-[K] in this study); modes of presentation: audio-only <A> and audio plus visual <AV>); and types of tasks: discrete point questions (DPQ) about vocabulary meanings and integrative questions (IQ) for text understanding (Davis 1990)) were given in order to study how different aspects of input influence listeners' intakes.

A cumulative total of 760 lower-intermediate (TOEFL scores of about 400) learners of EFL at a Japanese university were examined over four years, and 4,560 study histories of those learners were collected and analysed.

## 2.2 Sizes of Listening Units and Combinations of Sizes

### 2.2.1 Sizes of Listening Units

The Expectation Maximizing (EM) Algorithm (Dempster et al. 1977) was employed to obtain 'the most likely' estimation of sizes (mean) of comporment-units from observed length of listening units. The listening units have four components: small sized unit (designated as S), medium

---

* Single quote on two smaller units are there to draw attention to a certain imprecisions here, aside from the fact that it is difficult to demarcate the sentence as a unit is in spoken discourse.

sized(designated as M), large sized (designated as L) and extra large (designated as LL). Each unit has similar value for different conditions , texts and modes.

The unit-sizes are highly correlated ($p >> .01$) to the sizes of information units shown in Figure 1; LL to Topic, L to Paragraph, M to 'Sentence' and S to 'Clause', therefore the sizes of listening units are considered to be levels of information units. (Wakabayashi 2003 (b))

### 2.2.2 Combinations of the Units

The cluster observations of listening units were carried out to find the features of combinations of the component-units of individual subjects. The result shows that listeners are clustered into three or four groups in all conditions. Four groups of subjects with their characteristic features are designated as: C0, C1, C2 and C3 and the features are summarized using component-units as follows: (Wakabayashi & Kurahashi 2001 (b); Wakabayashi 2003 (b))

C0,C1 : large proportion of LL and L , and medium to small proportions of M and S,
C2     : small proportion of LL and mixed use of L, M and S ,
C3     : small proportions of LL, L and M, and large proportion of S.

## 3   Dynamic Features of  Combinations of Information Units

### 3.1   method of Analysis

In order to analyse how individual listeners construct their interpretation, the time chart showing dynamic pattern is introduced here. Each of accessed listening processing units is shown by a line indicating the beginning and the end of each access in sequence (See Figure 2).    The Y-axis shows the positions of accessed units of texts in frame numbers, and the X-axis shows listening steps.

From the patterns of intake chunks the directions of listening and the combinations of various sized units are explained.

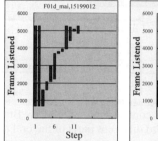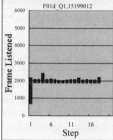

Figure 2: Foward move   Figure 3: Repetitive

### 3.2   Directions along which Listeners Feed their Intake Units

The basic dynamic pattern is formed by the directions of listening: (1) forward moves -- listening along the stream  of text (Figure 2), repetitive listening to the same and/or neighbouring units (Figure 3),  (3) backward moves -- against the stream of text (Figure 4) and the mixture of these moves (Figure 5).

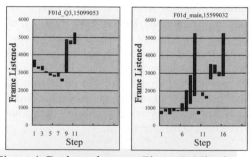

Figure 4: Backward move   Figure 5: Mixed

The saccadic movements, abrupt changes in accessed position and the sizes of intakes, are also observed (see 4.1 & 4.3).

## 4   Discussion

### 4.1   Intakes for processing --information levels in processing

Characteristic patterns of the usage of  information levels for constructing listeners' interpretations in the task of  **'main ideas'** and subject groups are:

**(i) S units:** ( 'Clause', 'Sentence')  (Fig.6)
Listeners in C3 group show S units usages which implies they used smaller levels of information such as 'Clauses' and/or 'Sentence'.
**(ii) L units:** (Paragraph, Topic)  (Figure7)
Listeners in C0 and/or C1 groups show the usage of LL and/or L units, which implies they utilised larger levels of information such as topic and paragraph.
**(iii) S, M and L units:** (from 'Clause' to Topic) (Fig.5)
Listeners in C2 group show M units usages with L and/or S units which implies they used mixed levels of information from 'Clauses' to paragraph.

In the **DPQ tasks** listeners use S and/or M units in a specific Paragraph and/or a 'Sentence'.
**(i)** An example of a pattern in extracting a meaning in small context (Figure 8)
**(ii)** A typical patterns in getting sounds of words (Figure 3)  S units by C3 listener.

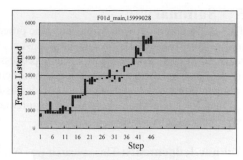

Figure 6: Main ideas, by C3

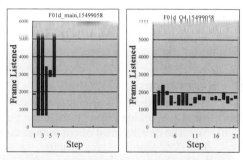

Figure 7: Main ideas    Figure 8: DPQ
      by C0                  by  C2

## 4.2   Why saccadic processing takes place

Levels of information processed / listened in relation to multi directions can be observed.  There seem to be typically three types of saccadic movements in processing:  (i) horizontal moves where listeners repeat listening to similar parts of a text, especially in small sizes of units for extracting lexical information (See Figures 3,7 and 8), (ii) forward moves where listeners listen to the stream of a text, in various sizes of units for constructing and/or confirming an interpretation (Figures 2,5,6 and 7), and (iii) backward moves where listeners listen to parts of a text against the stream of a text, in smallish sizes of units for seaching the coming information (Figure 4).  Such processing are considered as saccadic processing where listeners are assumed to confirm information which they have heard just before in an attempt to achieve a precise interpretation.

## 4.3   Cognitive Interaction

The virtual interactive setting is a kind of simulation in which the program facilitates interaction between computer and listeners (Crookall, Coleman and Versluis 1990).  Situations where listeners can interact have the potential of creating a supportive, low-anxiety environment which fosters 'positive affect' since learners have a chance to try out new language activity in simulation (Scarcella and Crooall, 1990).  Computer-assisted language learning should be based on sound theoretical studies based on cognitive considerations (Chapelle 2001).

The results obtain with the listening environment where free relistening is allowed indicate that usages of various sizes of units show not only which information was used in processing, but also how these pieces of information were used to construct an interpretation.  The results also show that flexiblility in accessing texts as well as sponteneous accessing provides cognitive interactions between listeners and the text with informational feedback in comprehension processing.

# 5 Summary

This study empirically describes saccadic processes in listening-comprehension processing. In the unique environment we have developed where free relistening is allowed, listeners' attempts to reach an interpretation exhibit saccadic processing. It is in these processes that a feature of listening-comprehension processing is well demonstrated. The analyses also demonstrate (1) that spontaneous access to audio input provides listeners with cognitive interactions, and also indicates (2) that flexible interface is a crucial requirement for making a computer-based learning environment interactive.

**Acknowledgements:** This research was supported by Research Project, Grant-in-Aid for Scientific Research (C) (No. 13680260, Chief researcher Setsuko Wakabayashi) and in part with special research funds by Himeji Dokkyo University. To the University of the South Pacific (USP), Fiji, we owe thanks for kind permission to use the video 'An Introduction to Physical Geography' by Professor Patrick Nunn and the Media Centre, and the video 'Fiona's Story' by UNESCO.

# 6 References

Altenberg, B.: Predicting text segmentation into tone units. *Meijs,* 49-60 (1987).

Baddeley, A.: Working memory. *Science*, vol. *255*: 556-559 (1992).

Chafe, W.: The development of consciousness in the production of a narrative. In Chafe, W. (ed.) *The Pear Stories*, Ablex (1980).

Chapelle C.: *Computer Applications in Second Language Acquisition.* Cambridge University Press (2001).

Crookall, D., Coleman, D.W. and Versluis, F.B.: Computerized language learning simulation. In Crystal, D.: *Prosodic Systems and Intonation in English*. Cambridge University Press (1969).

Crystal, D.: *Prosodic Systems and Intonation in English*. Cambridge University Press (1969).

Crystal, D.: Prosodic features and linguistic theory. In D. Crystal, *The English Tone Voice*. Edward Arnold (1975).

Dempster, A.P., Laird, N.M and Rubin,D.B.: Maximum-likelihood from incomplete data via EM algorithm, *Journal of the Royal Statistics*, Society B(methodological), 39, pp.1-38,(1977).

Ebbinghaus, H.: *Uber das Gedachtnis* (1885). [Reprinted as *Memory* (H. A. Ruger & C.E. Busenius, trans) New York: Teachers College (1913).]

Miller, G. A.: Decision units in perception of speech. *IT, 8* (1962).

Roach, P.: *English Phonetics and Phonology.* Cambridge University Press (1991).

Scarcella, R. and Crookall, D.: Simulation/gaming and language acquisition. In D. Crookall and R.L. Oxford *Simulation, Gaming, Language Learning*, New York: Newbury House Publisher (1990).

Wakabayashi, S. and Kurahashi, Text Segmentations of EFL Authentic Materials Based on Cognitive Processing., Proceeding of the 2001 conference of Japan Society for Educational Technology (JET). pp.481-482 (2001 (a)).

Wakabayashi, S. and Kurahashi, K.: FL Listening Comprehension Processing. In *Human Interface 2001*, pp.225-228 (2001 (b)).

Wakabayashi, S. and Kurahashi, K.: Dynamic patterns of processing units in listening comprehension, In *Human Interface 2002*, pp.557-58 (2002).

Wakabayashi, S. and Kurahashi, K.: Sizes of Listening-comprehension Processing Units. In *Journal of the Japan Society for Speech Science,* vol.4 (2003(b)).

# Designing A Tool for Taking Class Notes

*Nigel Ward and Hajime Tatsukawa[1]*

University of Texas at El Paso
El Paso Texas, 79968-0518
nigelward@acm.org

## Abstract

Although digital devices are replacing paper and pencil in ever more domains, class notes have so far resisted this trend. There are both hardware and software reasons: today there appears to be no software designed specifically to support note-taking in class. Based on an analysis of class notes and the note-taking process, we identify the special needs of students and propose software features required to meet these needs. Preliminary experience with classroom use of a prototype supports the basic soundness of our design and shows that, for some students, taking notes with the computer is feasible and preferred to pencil and paper.

## 1    Introduction

Many students own a notebook computer, some carry it around with them all the time, but it is rare to see anyone in lecture taking notes with one. Naively this seems to be a missed opportunity, as digital documents are superior to hand-written documents in many ways: being searchable, editable, easily sharable, and, most of all, more legible.

This paper addresses two questions: First, what software features are required to support note taking in class? Second, how feasible is this today?

## 2    About Class Notes

The use of technology to assist student note-taking has been considered before, but the present study is unique in its focus on three issues.

First, many discussions of classroom note-taking treat it only tangentially, as part of a larger agenda, such as enabling collaboration, providing access to multimedia and audio content, allowing better organization, and exploiting wireless and digital classrooms (Landey 1999, Truong et al. 1999). This paper discusses of technology to support note-taking itself, independently of other classroom improvements or reforms. The goal is merely to let students produce notes that are very similar to what they produce today with pen and paper, but different in being digital documents, and thus better than handwritten notes for purposes of review.

---

[1] Most of the work reported here was performed at the University of Tokyo. Tatsukawa is currently with NTT Communications.

Second, it is sometimes claimed that lecturing and note-taking is just an archaic method for conveying information. Experiments in educational psychology have shown, however, that note-taking has value also as a technique for engaging the student's mind, thereby facilitating learning --- taking notes itself promotes learning, whether or not the student ever looks at the notes again (Kiewra 1985). While there is great potential for technology to improve the lecture hall in many ways, there is no reason to believe that such advances will render note-taking unnecessary or alter its basic nature.

Third, student note-taking is sometimes considered to be a mere special case of note-taking in general. However student note-taking has special properties. Compared to meeting notes, class notes tend to be large in volume. They are often less sloppy than meeting notes, since they must be understandable at the end of the semester, not just a few days later. They are often highly two-dimensional, reflecting the instructor's use of the blackboard. Text generally appears in short chunks, with full sentences rare. Text and graphic elements tend to be tightly mixed, not only in labelled diagrams but also where conceptual relations are represented with arrows, importance is indicated with underlining and circles, etc. Class notes are usually produced under time pressure. Finally, for class notes, the process, not just the product, is important.

## 3    Features Needed in a Tool

Based on the above considerations, on observations of a small corpus of class notes gathered mostly from engineering students at the University of Tokyo, and on experience with early versions of the system, both in the lab and with one early-adopting user in the classroom, we believe that a tool for taking class notes should have the following properties.

Since notes often include diagrams and other graphic elements, a system should of course **provide a pen for drawing**.

A pen computer by itself, however, is not enough for taking class notes, because most class notes consist mostly of text, and here the keyboard is of course at an advantage. Typing produces text that is more legible than handwritten notes, no matter how good digital ink capture becomes. While handwriting recognition seems to offer an alternative, today's recognizers demand that the user either write carefully or be able to take the time to correct errors, and so are not adequate for classroom use. Also, for many users, keyboard input is faster and less tiring than writing even on paper. Thus a note-taking system should also **provide a keyboard for text input**

Going back and forth from pen to keyboard being time-consuming, a system should **provide a mouse** or equivalent for pointing, such as input positioning and menu selection. Since the time to move the hand between keyboard and mouse is generally greater than that to move between keyboard and trackball/trackpoint/touchpad the latter may be better.

For class notes, where text is the bulk of the content, and the text appears in little chunks scattered over the screen, the overhead of creating text boxes or clicking before typing can be significant. To avoid this a system should be modeless and **always accept text input**. This can be done simply by having text input always appear at the current cursor location, unless the cursor is over an existing chunk of text, in which case text is appended or inserted.

Similarly a system should **always accept pen input**, rather than having a separate mode for pen input. Since the pen should be available both as an input (drawing) device and as a pointing (selection and cursor positioning) device, this can lead to ambiguity. For example, in a mode-free system, user attempts to use the pen to select objects or to position the cursor may, when the pen slips, be misinterpreted as input, resulting in unwanted short lines. Various clever fixes are possible (Jarrett and Su, 2003), however we found it helpful to simply accept this, while providing the user a ``remove flak" function to let him recover by deleting all very short lines which have not been grouped into larger objects.

Class notes tend to be rich in text-related graphic objects such as underlining, brackets, ovals, and boxes around text, and arrows connecting text chunks. If such ``text decorations" are produced using menus and/or the pen, there is tool-switching overhead and the extra time cost involved in positioning such graphics in close alignment with the keyboard-entered text. Thus a system should **provide text-decorations invokable from the keyboard**.

# 4 Implementation

We developed a prototype, dubbed "NoteTaker", based on the above design principles, and built, for the sake of rapid development, in Java. The source code is about 4000 lines. Both Japanese and English versions can be downloaded from http://www.nigelward.com/notetaker/. The system was tested on the most suitable hardware available at the time, namely the Panasonic CF-02 and Fujitsu FMV-BIBLO MC3/45 Notebook PCs with passive touch-sensitive displays.

# 5 Preliminary User Studies

We recruited three students to try out the final, full-featured version of NoteTaker in class. Two kept using it beyond the ten class sessions we had requested: one for a semester and one for two. Altogether they have produced the equivalent of 100 A4 sheets of notes, for classes including HCI, Artificial Intelligence Programming, Software Engineering, Control Theory, and The Essence of Humanity. Figure 1 shows a screenshot including notes. In addition we had 10 students take notes in the laboratory from videotaped lectures with both pencil and NoteTaker.

Asked what they considered the advantages of NoteTaker over pencil and paper, all users said that the notes produced were more legible. Other factors mentioned included the ability to take notes faster, the ease of adding, correcting, and moving objects, the ability to search, and the ease of back-up.

Asked about the disadvantages of NoteTaker, all users mentioned that they couldn't draw complicated graphics neatly; this was expected given the hardware and software used. There were other comments about hardware issues, notably battery life. Some users felt self-conscious using a computer in the classroom. One user found it hard to keep up in equation-heavy lectures. Another pointed out that if the instructor was providing lecture notes on paper it was easier to just annotate these.

In general, notes taken with NoteTaker were very similar in content and layout to notes taken with pencil and paper, suggesting that students were able to keep up with the lecture while using it, and that it was not altering their note-taking style. In the laboratory study one of the subjects remarked that he was able to pay more attention to the blackboard while typing than while handwriting, but one reported the opposite. In the literature it has been said, regarding meeting notes, that ``people preferred handwriting to typing, saying that it was easier to listen while writing than while typing"

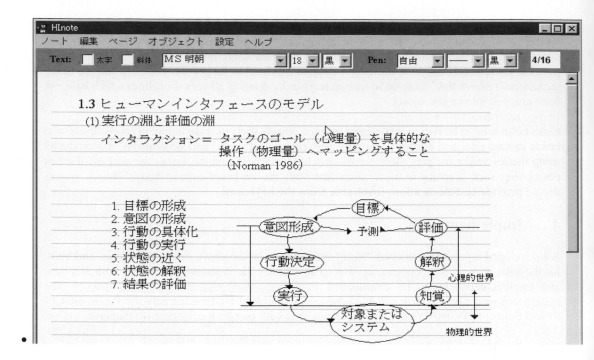

Figure 1: NoteTaker Screenshot. The main pane shows class notes from a lecture on cognitive models of interface use.

(Wilcox 1997). Our long-term users didn't mention this as an issue at all. Probably this depends on individual preferences.

Although the laboratory subjects generally rated NoteTaker highly overall, one subject rated it negatively (3 on a scale from 1 to 7). This subject remarked that she preferred pencil and paper because it let her write exactly what she wanted more easily. As this subject was one of those who had let us make copies of their old lecture notes, we examined them to see what she meant. These were unusually well organized and written, suggesting that NoteTaker is currently useful only for students who tolerate a certain amount of sloppiness.

More details on the user studies appear elsewhere (Tatsukawa 2002, Tatsukawa and Ward 2002).

## 6    Conclusion and Prospects

Regarding the software question, we identified the needs of students taking lecture notes, and showed how they are not satisfied by existing drawing tools or editors. We proposed design features which, barring the appearance of new technologies that magically change the nature of text entry, note-taking, or human learning, any application for taking lecture notes should probably have.

Regarding the hardware question, we found that digitizer quality was the main issue. However even hardware with a low-quality digitizer was felt usable by some, and now, with the availability of Tablet PCs, which use active digitizers and have operating system support for pen event handling and ink rendering, the situation is much better. Although designed for office workers, this hardware should be well suited for use also by students in the classroom

Finally, regarding the overall question of the feasibility of taking class notes with the computer, we found that, at least for some students, this is possible today and can be preferred to paper. We predict that within a few years, as the general student population continues to improve in computer literacy and typing speed, as hardware improves and becomes cheaper, and as class hand-outs become digital, taking notes with computers in class will become common. We do not yet know whether this will be merely a seductive distraction, or whether it will actually improve learning outcomes. This is a question of great interest.

# References

(Jarrett and Su 2003)   *Building Tablet PC Applications*. Rob Jarrett and Phillip Su. Microsoft Press, 2003.

(Kiewra 1985)   Investigating Notetaking and Review: A Depth of Processing Alternative. Kenneth A. Kiewra. *Educational Psychologist*, 20, pp 23-32, 1985.

(Landay 1999)   Using Note-Taking Appliances for Student to Student Collaboration. James A. Landay. 29th ASEE/IEEE Frontiers in Education Conference, pp 12c4-15--20, 1999.

(Tatsukawa and Ward 2002)   A Tool for Taking Lecture Notes (in Japanese). Hajime Tatsukawa and Nigel Ward.  Interaction 2002, pp 209-216. Information Processing Society of Japan, 2002.

(Tatsukawa 2002)   A Note-Editor Combining Pen and Keyboard (in Japanese). Hajime Tatsukawa. M.E. Thesis, Mech-Info Engineering, University of Tokyo, 2002.

(Truong et al. 1999)   Personalizing the Capture of Public Experiences. Khai N. Truong, Gregory D. Abowd and Jason A. Brotherton. UIST (12th Annual Symposium on User Interface Software and Technology), pp 121-130, ACM, 1999.

(Wilcox et al. 1997)   Dynomite: A Dynamically Organized Ink and Audio Notebook.   In *Proceedings of CHI'97*, pp 186-193, 1997.

# Satisfaction and Learnability in Edutainment: A usability study of the knowledge game 'Laser Challenge' at the Nobel e-museum

*Charlotte Wiberg*

Department of informatics,
Umeå University,
901 87 Umeå, Sweden
cwiberg@informatik.umu.se

*Kalle Jegers*

Department of informatics,
Umeå University,
901 87 Umeå, Sweden
kjegers@informatik.umu.se

## Abstract

This paper is a report on the initial findings of a study conducted in the project FunTain with the main purpose to find general guidelines for edutainment games, in order to guide designers of such games. Usability evaluations, with users and experts, were conducted on the edutainment game in order to find usability problems. These findings were then analyzed and used as input in focus group meetings, held with joint teams consisting of game designers and HCI experts. The result was a proposal of a list of design guidelines. In this paper they are grouped in three general categories; (1) game experience, (2) balance between entertainment and education, and (3) general understanding. Findings indicate that users had problems in understanding the underlying model for the game as well as finding the knowledge related content. Experts, further, gave comments about feedback problems and different types of inconsistencies. Some of the implications from the findings, as discussed in the focus group, were guidelines for earning and loosing points, scoring and performance feedback and game object characteristics.

## 1    Introduction

Entertainment is a factor that recently has become important for a number of different areas. One of the areas where entertainment is applied with purposes beyond just creating an amusing experience is the area of edutainment, where entertainment is used in combination with education in order to create a motivating and successful environment for learning. An example of how edutainment could be defined is:

*"...the marriage of education and entertainment in a work or presentation such as a television program or a Web site."*(Jones et. al., 1999)

Considering the definition of the edutainment concept we might conclude that design of edutainment includes the design of both entertainment and educational aspects in a design artifact. This may cause some difficulties. The pedagogical aspects that are of importance for the educational part of the artifact may in some cases be in opposition to the aspects of importance for the entertainment part of the artifact. There seem to be a need for some kind of trade offs to be made, in order to achieve a good result in the design of both the entertainment and the education in the artifact. Furthermore, existing guidelines are developed to cover more general usability aspects or with the intention to regard only entertainment aspects. The purpose of this paper is to report on initial usability evaluations on an edutainment game performed in order to provide design implications for design of edutainment games, for future research to refine and revise.

## 2    Usability and entertainment

Previous findings in the related area of interactive entertainment evaluation reveals that evaluation of entertainment web sites based on methods from the usability discipline, and user testing in particular, tend to provide findings that are focused on basic usability problems concerning navigation, design of menu buttons, etc. This implies that more subtle factors such as immersion, absorption and engagement, all potentially important to both entertainment and education, are difficult to grasp with the user testing method (Wiberg, 2001). Other related work include, for instance, Malone (1982) where the researcher spot four (4) characteristics of games. However, these characteristics do not consider any educational aspects.

## 3    The edutainment game

The game evaluated in the study is called "Laser Challenge" and was designed in order to educate the player/user about appliances of the laser technique. No specific knowledge about the laser technique was required for playing the game, but the user was supposed to be inspired by the game to learn more about laser. The game followed a linear, platform metaphor, and consisted of four episodes with increasing difficulty in the interactive parts. The main theme was supposed to be non-violent and the basis was that the user should collect CD's to give a party. Further, the user got points when answering questions.

## 4    Evaluation method

Methodologically, several studies reveal that usability inspection methods, such as Design Walkthrough (Karat, 1997) and Heuristic evaluation [c.f Nielsen, 1993) in many cases identifies problems overlooked by user testing, but also that user testing may identify problems overlooked in an inspection (Nielsen, 1993). In this study, we therefore use a combination of evaluation methods including both user testing and inspection methods; (1) An empirical usability evaluation, (2) Evaluations using inspection methods, in this case *Desigh Walkthrough* (DW) and (3) *Focus Groups* (FG).

### 4.1    User testing

The subjects performed the test individually, and each test took about 30 minutes in all. The user tests consisted of three parts;(1) 10 minutes of free surf (Wiberg, 2001) with Think Aloud, (2) 10 minutes of Walkthrough, performed by the test subject in collaboration with the test leader (collaborate evaluation), and finally (3) 10 minutes of post-interaction interview. Below, the subjects are described:

| Sub ject | Age | Sex | Computer literacy (1=Novice,5=Expert) | Computer gaming literacy (1=Novice, 5=Expert) | Comment |
|---|---|---|---|---|---|
| 1 | 25-30 | Female | 3 | 1 | IS researcher |
| 2 | 25-30 | Female | 5 | 5 | IS researcher |
| 3 | 50-60 | Male | 3 | 1 | Engineer |
| 4 | 20-25 | Male | 4 | 4 | IS lecturer |
| 5 | 20-25 | Male | 3 | 3 | IS lecturer |

In the first part of the session, the subjects played the game without any specific task to solve or instructions to be carried out. They were asked to verbalise their thoughts throughout the interaction, and they finished the session when they wished to do so. In the second part, the subjects performed a Walkthrough of the whole game prototype in collaboration with the test leader. Different aspects of the game were discussed, and the subjects were asked to give their opinions about specific features and parts of the design. They were also able to express any thoughts and comments they wanted to share. The post-interaction interview gave the subjects an opportunity to give comments and thoughts on general aspects of the game, the interaction and the performed test procedure. Here, the subjects could develop or refine their opinions and ideas from the previous parts of the test, and the test leader could follow up on issues that needed to be clarified.

## 4.2 Design Walkthrough

| Exp ert | Age | Sex | Computer literacy (1=Novice, 5=Expert) | Computer gaming literacy (1=Novice, 5=Expert) | Comment |
|---|---|---|---|---|---|
| 1 | 20-25 | Male | 5 | 4 | HCI expert |
| 2 | 30-35 | Female | 5 | 3 | HCI expert |
| 3 | 25-30 | Female | 5 | 3 | HCI expert |
| 4 | 30-35 | Female | 5 | 3 | Interaction designer |

The evaluators investigated the game prototype and made comments on possible problems or design improvements. The comments were written down and discussed in the last part of the evaluation, the focus group. The instructions were very brief, and the experts had a large degree of freedom in the evaluation procedure. In a large extent they relied on their personal experience and opinions in their evaluations.

## 4.3 Focus group

When the User tests and Design Walkthrough parts were finished, the test leaders and the expert evaluators (which in some cases were the same persons) performed a focus group meeting. In the focus group, the results from the previous parts of the study were discussed and reported. This was done in order to conduct design implications or guidelines based on found problems. From the results, a more general picture of the reported problems in the prototype was constructed. This picture was then used to generate a number of implications for the next step in the overall design process; design implications. Since the study was performed as a collaborative part of the process of designing the edutainment game, implications were kept at a level that was considered to be meaningful for the overall design process in terms of redesign of this specific game. From a research point of view, these findings could be considered as input for further revisions and refinement in future studies of other types of edutainment games.

# 5    Usability problems found

In order to highlight the research process, examples of the usability problems found are stated below. These are kept short, with the only purpose to pinpoint the overall picture of what occurred. In the expert walkthroughs, three main findings were found. (1) It was unclear how to gain points. Strange question marks and other moving objects confused and search of "hidden", point giving objects was fruitless. (2) Not obvious what to look out for. What is really dangerous in the game? (3) The skateboard kid somewhat seemed dangerous, however not clear at all how he could harm you. Further, in the empirical usability evaluations, the above usability problems were also found, and also three more. (1) A lack of interest in reading initial instructions results in frustration later in game was noted. (2) A loss of only some points was confused with a total loss of points. (3) The music is not connected to the actions in the game which confuses player and do not highlight level of danger

# 6    Design implications

The above stated usability problems are examples of some of the occurred issues from evaluation of the game. In the focus group session, a thorough discussion of all sessions was conducted and the general guideline list below was created. The initial list of guidelines includes ten (10) guidelines. Below, these guidelines are divided into three general groups; (1) Game experience, (2) balance between entertainment and education, and finally (3) general understanding.

## 6.1    Game experience
(1) *Task performance and feedback*: In order to achieve good game experience and competition, a failure to achieve a certain task that successfully performed will result in a large amount of points scored, should lead to the disappearance of the opportunity to score that particular set of points. (2) *Scoring and performance feedback*: The points should be summarized in a visible and easily interpreted counter, placed at a location in the environment according to conventions in the game genre.

## 6.2    Balance between entertainment and education
(1) *Promoting exploration*: There should be "hidden points" in the game environment, to reward the user when exploration of the environment is performed and to provide variation and discrimination in the overall performance of users considering points scored. (2) *Earning and loosing points:* The overall scoring system should be clear, unambiguous and provide distinct feedback to the user. Here the balance between entertainment and education seems critical. The points system seems to be one of the most important triggers for the user to enter the parts of the game connected to learning, i.e. if the scoring for the knowledge parts is to low, the users hesitate to enter these parts or objects.

## 6.3    General understanding

(1) *Game objects characteristics*: The difference between objects that affects the gaming procedure and objects that constitutes the background surroundings of the environment should be clear and unambiguous. (2) *Real world inheritance*: When designing objects in the game environment, it is important to be aware of the conventions considering the specific object generated by other similar types of games, but also conventions and affordances provided by real

world connections. (3) *Understandable menus*: Menu buttons and choices should be clear, descriptive and context sensitive. (4) *Supporting tools and their layout*: Pop up menus and additional tools for problem solving (i.e. information databases or dictionaries) should never occur on top of the main element (i.e. a particular question) which they are supposed to support, but should occur beside that particular element. (5) *Differences in valuable objects*: There should be intuitive, easily understood representations of objects and actions that result in scoring points when performed. (6) Game instructions: Instructions dealing with basic movements and actions in the game environment should be visually presented and explained in a short and compact fashion.

## 7    Conclusions

In this paper we have presented an initial study with the main purpose to find design guidelines for edutainment games, as it seems to exist a lack of guidelines both in research as well as practice. After the evaluation process, where expert walkthroughs as well as empirical usability evaluations were conducted, focus group sessions with HCI experts and game designers were performed. This resulted in the above-described guidelines. Briefly described, the guidelines could be grouped into (1) game experience, (2) balance between entertainment and education, and (3) general understanding. These findings could work as input in future research with purpose of further revision and refinement of the guidelines, in order to strengthen the generalizability of the guidelines.

## References

Jones, A., Scanlon, E., Tosunoglu, C., Morris, E., Ross, S., Butcher, P., Greensberg, J. (1999) Contexts for evaluating educational software. In Interacting with Computers 11 (1999) 499-516, Elsevier.

Karat, J. (1997) User-Centered Software Evaluation Methodologies. In Handbook of Human-Computer Interaction, 2nd edition, Helander, M., Landauer, T. K., Prabhu, P. (eds), Elsevier.

Malone, T. W. (1982). Heuristics for Designing Enjoyable User Interfaces: Lessons from Computer Games. Eight Short Papers in User Psychology. T. P. Moran. Palo Alto, Palo Alto Research Centers: 37-42.

Nielsen, J. (1993) Usability Engineering. Academic Press, San Diego

Wiberg, C. (2001). From ease of use to fun of use : Usability evaluation guidelines for testing entertainment web sites. In Proceedings of Conference on Affective Human Factors Design, CAHD, Singapore

# Section 7

Virtual, Mixed and Augmented Environments

# Improving Interaction in an Augmented Reality System Using Multiple Cameras

*Ingmar D. Baetge, Gregory Baratoff, and Holger T. Regenbrecht*

DaimlerChrysler AG, Research and Technology
RIC/EV, Virtual and Augmented Environments
P.O.Box 2360, 89013 Ulm / Germany
{Ingmar.Baetge|Gregory.Baratoff|Holger.Regenbrecht}@DaimlerChrysler.Com

## Abstract

In this paper we focus on an Augmented Reality (AR) approach to enhance and enrich a meeting scenario. In this scenario several people join together to hold a meeting, during which they discuss 3D-models, for example engineering parts in the automotive industry. The aim of this system is to give the users the possibility to actually see the virtual objects and interact with them in natural ways, not in the way a VR system would enable users to view objects in a purely virtual environment, but to put the objects out in the real world, on the meeting table. One of the most challenging issues to be addressed in building such a system is to provide adequate support for effective interaction between users and virtual objects and for collaborative interaction amongst the users. Our system relies on marker tracking as source of information about the position and orientation of users and objects they interact with. Based on our experience with the system we identified several problems mainly related to occlusion which arise from this tracking approach. As the virtual objects to be overlaid on the view of the real world can only be shown when specific markers associated with them are tracked, the key to a continuos and stable visualization is a stable marker tracking system which is robust in the presence of marker occlusion. The main cause being the line-of-sight problems due to occlusion of the markers by the users' hands while interacting, we decided to implement a solution to this problem using a multi-camera tracking system. Our quantitative and qualitative evaluations showed this approach to be significantly more robust against occlusions and hence more effective in supporting interaction.

## 1    Introduction

Augmented Reality (AR) is used to enrich a user's real world environment by virtual parts [1], which helps him to perform better in his real world tasks. AR technologies have been applied to a large variety of fields, for example medical visualization, annotation and explanation of the real world, or applications in aircraft navigation.

**Figure 1: Users discuss virtual objects in MagicMeeting**

The main focus of our research are collaborative AR-Systems which enrich the user's visual perception of the world using head-mounted displays (HMDs). Examples of such systems are the Studierstube[3] and the EMMIE[4] system. A specific application we focus on is our demonstration application MagicMeeting[2]. The goal of the MagicMeeting setup is to support a group of users during a meeting, where the design of engineering parts in the automotive industry is discussed. The system provides the necessary hardware and software tools to enable users to view and manipulate virtual 3D models which are placed on the meeting table in the real world.

The users can view the models by wearing video-see-through HMDs (Figure 1). Each of these HMDs has been equipped with a camera which captures the field of view of the user. The virtual objects are digitally added to the image of the video camera, and the user sees the enriched video image through the HMD.

The virtual objects have to be displayed at the correct position and orientation, and to determine the position and orientation of the user's viewpoint, a marker tracking approach is used. Markers are printed symbols of a given size which can be detected and identified using simple real-time image processing techniques. For example, one marker is placed on a wooden tablet (the 'cake platter') at the center of the meeting table. This is the designated spot for virtual models under discussion, and by moving or turning the cake platter, the position and orientation of the model can be controlled. Other markers are attached to small paper cards or office utensils. They can be used to track the pose of a variety of virtual tools. These tools can be used to perform interactions like pointing at parts of the virtual model or clipping parts of the object[6].

## 2 Motivation

Interaction with virtual models using a marker based tracking approach is susceptible to a variety of disturbances that limit the utility of an AR application. These can be categorized into the following cases: (1) The fidelity of the marker tracking itself can be very low, mainly due to the facts that (a) only one camera (mounted on the HMD) is used to estimate the pose between each user camera and the markers, (b) this camera is constantly moving, causing motion blur and comb effects if interlaced mode has to be used, (c) the resolution of the camera is limited (from 320x240 for USB cameras to 720x576 PAL resolution for analog cameras). (2) The fidelity of the design and printout of the markers. The markers have to be planar and have to have very sharp and orthogonal edges. Their patterns should be sufficiently different to prevent mismatches in the pattern matching stage. (3) Errors in the marker detection stage can strongly affect the pose calculation. (4) The environmental (esp. lighting) conditions can (and will) vary in a real setup. Finally (5) there are many user interactions which cannot be foreseen in a predictive manner. Firstly, the user continuously moves his head to different positions and changes the direction of his gaze. Secondly, he interacts with the real and virtual environment using his hands or virtual tools. The most common cases of interaction that heavily disturb the marker tracking are illustrated in figure 2. (A) Markers can only be identified if they are fully within the field of view of the camera. Since markers move out of the camera's field of view when the user changes his gaze, this happens quite often. (B) Interaction of users with the virtual models or with real world objects often causes the user to occlude markers. This leads to an interruption of the display of virtual objects or tools. (C) Markers can only be found in the image when they are not tilted away or too far from the user, as their projection in the video image must have a certain minimum size.

A                                                          B                                                          C

**Figure 2: Occlusion of markers disrupts the visualisation of virtual objects. (A) marker partly out of view, (B) marker covered by user hand/finger, (C) marker too distant or tilted away too much to be detected.**

Since we implemented the demonstration setup many people have used MagicMeeting, and although users quickly adapt to the limitations of the marker tracking once we explain under what circumstances virtual objects attached to markers disappear, we wanted to improve the interaction and visualization stability of the system. Our approach to improve the stability of the tracking is to (1) incorporate additional static cameras observing the setup from different viewpoints to reduce the effects of marker occlusion on interaction stability, and (2) to place additional markers in the setup to increase redundancy.

There has been some previous work on multi-camera setups. In the Computer Vision field, the focus has mainly been on 3D reconstruction of the environment [11,9], which requires a careful calibration of the cameras with respect to each other. In surveillance applications, the focus is usually on determining in which camera's field of view some action is happening and switching to that camera. Most similar to ours is the "dynamically shared optical tracking" approach [10]. Our work differs from theirs in the way information is fused from the cameras. A more detailed account can be found in [13].

## 3    Prerequisites and Concept

### 3.1    Implementation of the MagicMeeting system

Users in the MagicMeeting system wear a combination of HMDs and a camera, and these components are connected to a PC dedicated just for that user, and where the image processing as well as the augmentation of the video images is performed. All these machines, as well as a central tracking server, are connected over a LAN to enable distributed interaction. For example, modifications of the 3D model like highlighting are distributed over the network, so that other users see these modifications, too.

The marker tracking implementation is based on the popular AR-Toolkit library[5]. We added improvements to better deal with varying illumination and reduce marker mismatches, incorporated a pose estimation algorithm that is more accurate, and that can fuse the information from several markers [12].

### 3.2    Implementation of the multi-camera tracking

To address the limitations of the marker tracking approach mentioned in the previous section, we introduced several modifications to enhance the reliability of the tracking. First, additional static cameras were placed around the meeting table. Since static cameras do not suffer from motion blur and interlaced images, they produce more reliable pose information about the markers. Also, they can be placed in such a way as to minimize possible occlusion of the markers by the users' hands and tool markers. Second, we place additional markers on the table near the "cake platter". This makes it more likely that at least one of the markers will be seen, which - together with the fact that the static cameras usually see all the of the markers - allows the pose of occluded markers to be determined.

We chose a client-server software architecture to implement our multi-camera tracking approach. With each camera – and, thus, each user - we associate one client, which computes the marker poses relative to that camera and sends them to the server. The server collects all the relative marker poses, combines them to a consistent model and sends the results back to the clients. The enhanced model is then used by each client to obtain information about those markers which are occluded in its user's view.

One of our goals was to keep the additional configuration and calibration overhead as small as possible. Consequently, our system does not have a central reference system with respect to which all poses are calculated. Each client only sends relative pose information about markers to the server. Cameras and markers can be moved freely if necessary without requiring any recalibration. Although this makes it a little bit more complicated to integrate the different tracking data into the server - as the cameras all have different coordinate systems – the flexibility of the setup far outweighs the extra complication.

The integration of the collected data by the server can be seen as a graph problem, where cameras and markers are nodes, and where there is an edge between a marker and a camera node whenever the camera has (relative) pose information about the marker. The next section explains the pose integration in detail.

## 4    An Algorithm to Combine Relative Pose Information

We first want to take a look at an example situation to explain our algorithm in detail. In figure 3, two cameras observe a scene containing four markers, with each camera having information about only three of the four markers. For example, the relative pose between camera 1 and marker 2 is known, as well as the relative poses between camera 2 and both marker 2 and marker 3. Since there is a path in the graph from camera 1 to marker 3, a relative pose between camera 1 and marker 3 can be calculated by chaining the relative pose transformations. Alternatively, a path passing through marker 4 can be generated. Our combination algorithm searches paths from each camera to those markers which are not seen by that camera to find out the markers' relative poses with respect to the camera. There is no single global world model, but rather a model for each camera, which is enhanced by (relative pose) information from other cameras.

There are two phases of the algorithm: In the first step we generate a list of paths from each camera to markers not seen by it, and in the second step – since there can be several such paths between a camera and a marker – we either select a path (the best one according to some measure), or we fuse the paths by combining the relative poses associated with the paths.

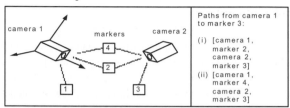

Figure 3: graph interpretation of a multi-marker multi-camera ssituation.

Path-generation is performed in three steps: First, all trivial paths from the selected camera to all seen markers are added to a list. Second, all these simple paths are then expanded by edges that lead to static cameras. Third, the resulting paths are expanded by edges that lead to markers which are not seen by the selected camera. The resulting path list contains all paths in the graph of length 3 which lead from the original camera to markers which the camera does not have any information about.

Once the paths are determined, we compute the pose $p_{injm}$ for the path from camera i via marker n and static camera j to marker m by chaining the transformations corresponding to the relative poses :

$$p_{injm} = p_{in} * (p_{jn})^{-1} * p_{jm},$$

where $p_{in}$ is the relative pose of marker n with respect to camera i. Next, we have to fuse the pose estimates from the multiple paths. In order to do this we need a measure of the reliability of each individual pose estimate. In our system we use the covariance matrix of the pose estimates as a measure of their uncertainty. The covariance matrix is computed by our pose estimation algorithm for the relative marker to camera poses. Using covariance propagation techniques [7] we then compute the covariances of the poses corresponding to the individual paths from the covariances of the relative poses along the path.

We implemented two methods for fusing the paths. In the selection method the "best" path is chosen, where the best path is the one with the lowest uncertainty. A suitable measure is the trace of the covariance matrix. In the fusion method the poses are combined using a weighted averaging scheme based on each pose's covariance matrix[8]. Both methods produced good results during test runs. Their behavior differs in the case of faulty pose estimations. In the case of the selection method, faulty marker poses tend to disturb the results of the combination process strongly, but not very often, as faulty pose estimations usually have a higher uncertainty and are therefore not selected very often. The other method of fusing the produced paths always incorporates all markers poses, but the impact of misinterpretations of video images is not as strong as in the case of selection as the pose is an average of the estimates.

## 5    Results and Future Work

We conducted an informal usability study with 18 subjects to get an impression of the qualitative and quantitative improvements. The subjects were asked to use the system with both implemented algorithms and the original system without improvements for ten minutes each. During the test phase, the users were asked to perform a set of increasingly complex interactions, from simple observation of the virtual object to collaborative interactions using various interaction tools (e.g. turning the virtual object on the platter, highlighting parts of it). After the test phase, the subjects' impressions of the system performance concerning interaction stability were evaluated using a questionnaire.

In the test runs the multi-camera approach (using either of the two fusion algorithms) showed significantly higher stability, as measured both quantitatively by a much reduced failure rate of the tracking and qualitatively by an increase in user satisfaction concerning the effectiveness of the interaction with the system. Figure 4 illustrates various situations where the multi-camera system still correctly shows the virtual object despite occlusion or non-detection. This should be compared to Figure 2, where in corresponding situations tracking fails, thereby preventing the virtual object to be shown and resulting in a breakdown of the user interaction with the system.

The new multi-camera tracking approach is now under integration into our VR/AR in-house software. This will make it possible to improve interaction for a wider range of applications than just MagicMeeting. Currently, the system performs a fusion based on the current pose estimates available from all cameras at a given time instant. In the future, we are planning to further improve stability by extending the fusion to the time domain, e.g. by means of Kalman filtering. Further formal user tests at our customer sites will allows us to better evaluate the benefits and limitations of the system as a whole and of individual interaction techniques in particular.

**Figure 4: Multi-camera tracking stabilizes the visualization of virtual objects**

## 6    REFERENCES

[1]     Azuma, R. (1997), A Survey of Augmented Reality. *Presence:Teleoperators and Virtual Environments*, 1997. 6(4): pp. 355-385.

[2]     Regenbrecht, H., Wagner, M., & Baratoff, G. (2002). MagicMeeting - a Collaborative Tangible Augmented Reality System. *Virtual Reality - Systems, Development and Applications*, Vol. 6, No. 3, Springer Verlag.

[3]     Schmalstieg, D., Fuhrmann, A., Szalavari Zs., Gervautz, M. Studierstube – Collaborative Augmented Reality. In. *Proc. Collaborative Virtual Environments '96*, Nottingham, UK, 1996.

[4]     Butz, A.T., Höllerer, T., Feiner, B., MacIntyre, C., Beshers. Envelooping Computers and Users in a Collaborative 3D Augmented Reality. In *Proc. IWAR'99*.

[5]     Kato, H., Billinghurst, M. (1999), Marker Tracking and HMD Calibration for a Video-based Augmented Reality Conferencing System, In *Proc. 2nd Int. Workshop on Augmented Reality*, 1999, pp. 85-94.

[6]     Regenbrecht, H. and Wagner, M. Prop-based interaction in mixed reality applications. In Smith, Salvendy, Harris, & Koubek (eds.): *Proc. HCI International 2001*, Lawrence Erlbaum Associates, Publishers, Mahwah/NJ, Vol. 1, pp. 782-786, 2001.

[7]     Haralick, R., Shapiro, L. *Computer and Robot Vision*, Addison Wesley, 1997.

[8]     Hoff, W., Vincent, T. Analysis of Head Pose Accuracy in Augmented Reality. *IEEE Trans. Visualization and Computer Graphics*, Vol. 6, No. 4, 2000.

[9]     Tomas Svoboda, Hanspeter Hug, Luc Van Gool. ViRoom --- Low Cost Synchronized Multicamera System and its Self-Calibration. In *Proc. DAGM'2002*, Zuerich, Switzerland, Van Gool (ed.), Springer LNCS Vol. 2449, Sept. 2002.

[10]    Ledermann, F., Reitmayr, G., Schmalstieg, D. Dynamically Shared Optical Tracking. *In Proc. 1st IEEE Int. Augmented Reality Toolkit Workshop*, Darmstadt, Germany, 2002.

[11]    Narajanan, P.J., Rander, P.W., Kanade, T. Constructing Virtual Worlds using Dense Stereo. In Proc. Int. Conf. Computer Vision, Bombay, India, pp. 3-10, 1998.

[12]    Baratoff, G., Neubeck, A., Regenbrecht, H. Interactive Multi-Marker Calibration for Augmented Reality Applications. In *Proc Int. Symp. On Mixed and Augmented Reality*, September 30 - October 1, 2002, Darmstadt, Germany.

[13]    Baetge, I. Automatic Calibration of a Multi-Camera Marker Tracking System for Augmented Reality Applications. (in german). Diploma theses, University of Ulm, Germany, July 2002.

# Non-Zero-Sum Gaze in Immersive Virtual Environments

*Andrew C. Beall, Jeremy N. Bailenson, Jack Loomis, Jim Blascovich &*
*Christopher S. Rex*

Research Center for Virtual Environments and Behavior
Department of Psychology, University of California
Santa Barbara, CA 93106
beall@psych.ucsb.edu

## Abstract

We discuss the theoretical notion of augmenting social interaction during computer-mediated communication. When people communicate using immersive virtual reality technology (IVET), the behaviors of each interactant (i.e., speech, head movements, posture, etc.) are tracked in real time and then rendered into a collaborative virtual environment (CVE). However, it is possible to change those behaviors online, and to render these changed behaviors for strategic purposes. In the current paper we discuss one such augmentation: non-zero-sum gaze (NZSG). An interactant utilizing NZSG can make direct eye contact with more than one other interactant at a time. In other words, regardless of that interactant's physical behavior, IVET enables him to maintain simultaneous eye contact with any number of other interactants, who each in turn may perceive that he or she is the sole recipient of this gaze. We discuss a study in which an experimenter attempted to persuade two participants in an CVE, and manipulated whether gaze was natural (i.e., rendered without transformation), augmented (i.e., each participant received direct gaze for 100% of the time) or reduced (i.e., neither participant received any gaze). We measured participants' head movements, subjective perceptions of the experimenter's gaze, the attitude change for the persuasion topic, and recall of information. Results indicated that participants were unaware of augmented and reduced gaze behaviors despite the fact that the participants' own gaze behavior changed in reaction to those conditions. We discuss these results in terms of understanding mediated communication and nonverbal behavior.

## 1    Overview and Rationale

Real-time augmentation of one's social behavior during interaction is an appealing, albeit Orwellian, prospect that is made possible by recent advances in immersive virtual environment technology (IVET). Using this technology, which tracks and renders a person's nonverbal behavior, one can use intelligent social algorithms to enhance the manner in which an individual's nonverbal behaviors are conveyed to others.

IVET also allows us to examine complex patterns of visual nonverbal behaviors within realistic social contexts with nearly perfect experimental control and high precision. In the current study, we augment gaze in Collaborative Virtual Environments (CVEs). Mutual gaze occurs when individuals look at one another's eyes during discourse. In face-to-face conversation, gaze is zero-sum. If interactant A maintains eye contact with interactant B for 60 percent of the time, it is not possible for A to maintain eye contact with interactant C for more than 40 percent of the time. However, CVEs are not bound by this constraint; in a virtual interaction with *avatars* (virtual human representations), A can be made to appear to maintain mutual gaze to both B and C for a majority of the conversation. In the following, we describe a paradigm that allows interactants to achieve non-zero sum gaze (NZSG). Figure 1 demonstrates the concept of NZSG.

Gaze in general is one of the most thoroughly studied nonverbal gestures in psychology (Gibson and Pick, 1963; Anstis, Mayhew, and Morley, 1969; Rutter, 1984; Kleinke, 1986). According to Kendon (1977), speakers use gaze to regulate the conversation. Gaze can provide cues for intimacy, agreement, and interest (Argyle, 1988). Consequently, a CVE that augments interactants' capacity to transmit gaze can provide an excellent tool to study social interaction.

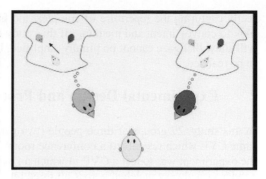

Figure 1: A conceptualization of Non-Zero-Sum Gaze. The balloons above each person represents his or her belief state concerning the experimenter's gaze.

Gaze can be expressed by both head and eye movements. In previous work, we argued that both cues are important sources of information (Bailenson, Beall, & Blascovich, 2002) in social interaction. Head and eye direction are highly correlated and therefore, with caution, head pose can be used to estimate focus of attention. Head pose also conveys unique symbolic information, such as indications of agreement or disagreement.

In real face-to-face interaction, gaze has been shown to significantly enhance performance of information recall. This positive effect has been shown for both children (Ottenson & Ottenson, 1979) and adults (Fry and Smith, 1975; Sherwood, 1987) using a simple fact-recall task. The authors of these studies generally attribute the enhanced performance to there being an increased sense of intimacy between interactants, which in turn better captures attention.

Realizing accurate gaze in CVEs is challenging. Video Teleconferencing often fails to convey effective gaze information because the camera's lens and the monitor's image of interactant's eyes are not optically aligned. To overcome this, various ingenious techniques either optically align camera and monitor (Buxton & Moran, 1990; Ishii, Kobayashi, & Grudin, 1993) or alter the display to "correct" the gaze (Vertegaal, 1999; Gemmell, Zitnick, Kang, & Toyama, 2000).

In IVEs, however, assessment of performance as a result of gaze is still very much work in progress. Recently, Gale and Monk (2002) devised a two person CVE after the ClearBoard demonstration of Ishii et al (1993). They found that gaze behavior traded off with other communication channels, reducing the number of turns and speech required to complete the collaborative task. In other experiments, subjective ratings made by the interactants indicated significant enhancements to the social communication when gaze information was conveyed (Müller, Troitzsch, and Kempf, 2002; Bailenson, Beall, and Blascovich, 2002).

Thus far, computer science and behavioral research has focused on the difference between having gaze cues available and not. What is novel about the work here is a possibility that emerges during n-way interactions with more than two persons, namely NZSG. Consider the fact-recall task. Normally, the constraint of zero-sum gaze imposes a hard limit on a speaker's ability to capture the attention of individual listeners via gaze. As such, we speculate that average fact recall after a group presentation would be worse than the same average recall had the speaker presented the material to each person dyadically. We believe CVEs offer an intermediate possibility, namely that even in simultaneous n-way interaction, each interactant can be led to believe that she is being gazed upon more than in reality. Specifically, we hypothesize that this form of augmented gaze can serve to enhance performance as compared to either natural gaze (zero sum) or gaze absent conditions. While this hypothesis provided the motivation for the current study, we recognize that augmenting interaction with such a simple social algorithms may in fact fail as a result of not correctly capturing the repertoire of complex and linked head and eye motions that individuals employ to convey intent and meaning. If this study does in fact find that our social algorithm fails, it will show that gaze cannot be blindly amplified, but likely requires a more sophisticated algorithm to be realized.

## 2    Experimental Design and Procedure

In this study, 27 groups of three people (two participants and one experimenter) interacted in the same CVE which resembled a conference room. The 54 participants were told that the purpose of the experiment was to test a CVE in which an experimenter was going to lead a discussion. Group gender was always matched across all three interactants. We employed two 2 male experimenters and 2 female experimenters. Figure 2 shows images of the conference room.

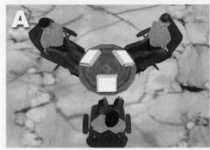
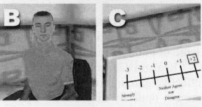

All three interactants were placed in physically different rooms with the door closed and remained seated throughout the study. Each participant's perspectively correct view of the virtual environment was rendered stereoscopically and updated at 60 Hz. Head orientation and position were tracked by a hybrid inertial/optical tracking system with low latencies (less than 5 and 20 ms, respectively). A full duplex intercom system provided natural audio communications among all participants. Mouth movements were tracked via a microphone that sensed sound amplitude, which in turn controlled simple mouth animations of each person's avatar. Figure 3 shows a participant in his own room wearing the head-mounted display (HMD). We chose for both scientific and technical reasons (i.e., the challenge of accurately tracking eye movements in IVEs) to use avatars in which head and eye directions are always locked together.

**Figure 2: Scenes from the CVE. Panel A - bird's eye view. Panel B - avatar close-up. Panel C - Likert response screen on each computer monitor.**

We manipulated interactants' perception of gaze in three conditions. The first was *natural interaction* (head movements of all interactants were veridically rendered). The second was *augmented gaze* (each participant saw the experimenter's avatar making direct gaze for 100% of the time). The third condition was *reduced gaze* (neither participant received any gaze from the experimenter's avatar). There were 9 groups (i.e., 18 experimental participants) in each condition. Participants were never told of the gaze manipulation and the experimenters themselves were kept blind to condition to ensure that experimenters behaved similarly. We encouraged our experimenters to be as persuasive as possible and to use as much eye contact as possible. To implement the augmented and reduced conditions, our software scaled the experimenter's actual head motions by a factor of 20 and re-centered the effective straight-

**Figure 3: A participant uses an HMD, intercom, and gamepad.**

ahead position to point either at the participant's head or the experimenter's screen, respectively.

Participants went into their own physical room without meeting the experimenter. We demonstrated how to use the equipment and respond to the questionnaires. Once the three were immersed and online, the experimenter read two passages to the participants. We measured as dependent variables: 1) head movements, 2) subjective estimation of experimenter's gaze direction, 3) information recall, and 4) persuasion for the passages.

## 3    Experimental Results and Conclusions

One of the most striking findings of this study is that participants did not detect either the augmentation or the reduction of gaze. Despite the fact that, from a given participant's point of view, the other participant received absolutely no gaze from the experimenter in the augmented condition, participants did not notice. After the study, we asked each of the participants to estimate the percentage of time that the experimenter looked at each participant. Figure 4 demonstrates those differences by condition and participant. Estimation in every condition and participant was statistically different from zero. Consequently, participants did not notice the lack of gaze given to their counterparts in the augmented condition.

Next, we analyzed the head movements of our participants. If they accepted the augmented gaze as real gaze, then we would predict that participants would return the gaze (i.e., look towards the experimenter) most often in the augmented condition. Figure 5 demonstrates the percentage of time that participants looked (oriented the head) toward the experimenter or the other interactant.

To test the significance of this difference in looking we ran a two factor ANOVA: gaze condition (natural, augmented, and reduced) and head orientation (towards experimenter or other participant). The predicted interaction was significant $(F(2,49)=3.70, p<.05)$, demonstrating that the difference in looking percentage between experimenter and other interactant was greatest in the augmented condition.

Our other dependent measures such as persuasion and recall did not show a discernible pattern across gaze conditions. We have no compelling explanation for

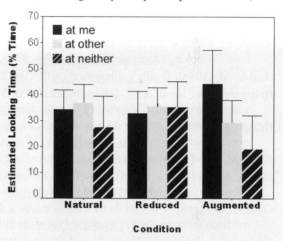

**Figure 4: Participants' estimation of where the experimenter was looking by condition**

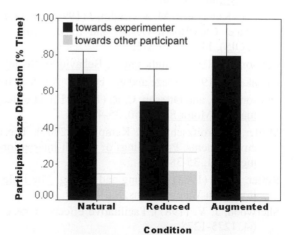

**Figure 5: Participant's gaze direction during passage presentation**

this. It is possible that this study lacked the power to find differences that have been found in previous real face-to-face and IVE interactions. At worst, social augmentation algorithms such as ours may be too simple and the lack of real connectedness between experimenter and participant may have undermined its potential effectiveness.

However, we feel our data suggests such algorithms may succeed. Participants were not aware in the augmented condition that their partner was being entirely ignored. Equally important, the augmented was as successful if not better at capturing their attention compared to the natural condition. In sum, this study points out the possibility for augmenting social interaction within computer mediated environments and shows that technology available today can usefully investigate this phenomenon.

# References

Anstis, S., Mayhew, J., & Morley, T (1969). The Perception of where a Face or Television 'Portrait' is Looking. American Journal of Psychology, p. 474-489.

Argyle, M. (1988). Bodily communication (2nd ed.). London, England UK: Methuen.

Bailenson, J.N., Beall. A.C., & Blascovich, J. (2002). Mutual gaze and task performance in shared virtual environments. Journal of Visualization and Computer Animation, 13, 1 8.

Buxton, B. & Moran, T. (1990). EuroPARC's integrated interactive intermedia facility (IIIF): early experiences, Proceedings of the IFIP WG 8.4, p.11-34, Heraklion, Crete, Greece.

Cline, M. (1967). Perception of Where a Person is Looking. American J. of Psychology, p. 41-50.

Fry, R. & Smith, G. F. (1975). The effects of feedback and eye contact on performance of a digit-encoding task. Journal of Social Psychology, 96, 145-146.

Gale, C. & Monk, A.F. (2002) 'A look is worth a thousand words: full gaze awareness in video-mediated conversation'. Discourse Processes 2002, Volume 33, Number 1.

Gemmell, J., Zitnick, L., Kang, T., and Toyama, K., (2000). Software-enabled Gaze-aware Videoconferencing, IEEE Multimedia, 7(4), 26-35.

Gibson, J., & Pick, A.(1963). Perception of Another Person's Looking Behavior. American Journal of Psychology, 76, 386-394.

Ishii, H., Kobayashi, M. & Grudin, J.(1993). Integration of Interpersonal Space and Shared Workspace: ClearBoard Design and Experiments, ACM Transactions on Information Systems (TOIS), 11(4), 349-375.

Kendon, A. (1977). Studies in the Behavior of Social Interaction. IU: Bloomington.

Kleinke, L. (1986). Gaze and eye contact: A research review. Psychological Bulletin, 100, 78-100.

Otteson, J. P., and Otteson, C. R. (1979). Effect of teacher's gaze on children's story recall. Perceptual and Motor Skills, 50, 35-42.

Müller, K. Troitzsch, H. & Kempf, F. (2002). The role of nonverbal behavior in 3D-multiuser-environments. Proceedings of the 4th International Conference on New Educational Environments 2.1, 35-38.

Rutter, D. R. (1984). Looking and Seeing: The Role of Visual Communication in Social Interaction. John Wiley & Sons: Suffolk.

Sherwood, J. V. (1987). Facilitative effects of gaze upon learning. Perceptual and Motor Skills, 64, 1275-1278.

Vertegaal, R. (1999). The GAZE groupware system: mediating joint attention in multiparty communication and collaboration, Proceeding of the CHI 99 conference on Human factors in computing systems: the CHI is the limit, 294-301.

# Augmented Reality on Handheld Computers to Flight Displays

*Reinhold Behringer, Jose Molineros, Venkataraman Sundareswaran, Marius Vassiliou*

Rockwell Scientific
1049 Camino Dos Rios, Thousand Oaks, CA 91360
{rbehringer, jmolineros, sundar, msvassiliou}@rwsc.com

**Abstract**

In recent years, Augmented Reality (AR) research and applications have gained strong momentum. Rockwell Scientific (RSC) has pursued activities in this domain, ranging from AR applications on hand-held computers to displays for commercial airplanes. The main technical topic addressed in our work is "Registration," namely the alignment of the synthetic information overlay with a view of the real world. This paper presents an overview of the recent work by RSC on this topic.

## 1    Augmented Reality

Since the inception of Augmented Reality in the early 1990s, this initially small and "exotic" topic turned into a major research field (Azuma et al., 2001). The concept of presenting (graphical) information that is spatially registered with the surrounding environment (Behringer, 2001) is a compelling tool for intuitive human-computer interface, and has been demonstrated in various applications (e.g., Sundareswaran, Behringer, Chen, & Wang, 1999). Registration for AR applications, i.e, the alignment of information with the environment, is the topic that still draws significant attention of researchers in the AR community: at the 2002 International Symposium on Augmented Reality, about a third of the presentations dealt with the technical issues of sensing and tracking, the core issues in Registration. The overall consensus among AR researchers is that hybrid methods that include a Computer Vision approach are best suited for achieving the required precision. However, Computer Vision approaches must deal with a large number of technical hurdles. Real-time performance is crucial to AR applications. RSC has continued to develop and refine its real-time model-based Computer Vision tracking approach, presented in the following sections.

## 2    RSC's Vision-Based Registration

In order to achieve real-time performance for tracking, RSC has focused on developing and refining its model-based Computer Vision tracking approach, implemented on several hardware platforms. The core of this approach is the use of a technique known as "Visual Servoing" (Espiau, Chaumette, & Rives, 1992). Visual Servoing is suited to track changes of camera position and orientation (6 DOF) by tracking the 2-D motion of visual features on the image of the observed object. The Visual Servoing approach dynamically determines the 3-D motion of the camera from the 2-D motion of these features (point or line features). The 3-D model of the object, along with the location of the tracked features, must be known. The approach is an iterative error (between measured and predicted feature location) minimization performed in closed-loop. The resulting registration is easily verified by an overlay of the 3-D model rendering onto the video imagery (Sundareswaran & Behringer, 1998). Finally, this approach requires that the correspondence between the 2-D features in the image and model features be established.

## 2.1   Initialization – Establishing Correspondence

Aligning a 3-D model with an image of the real object without prior knowledge of user location and orientation is necessary for initialization. We have implemented two methods for achieving an initial match of a 3-D model with the image of the real object: one method is based on testing the validity of six different initial viewpoints by projecting the 3-D model from these viewpoints through an application of the Visual Servoing algorithm. The viewpoint that results in least residual error is the one chosen as the correct initial viewpoint. The advantage of this method is the direct integration with the Visual Servoing algorithm – no separate initialisation algorithm is needed. Its shortcoming, however, is that when the feature model is complex, six viewpoints may be inadequate – the iterative process may converge to a local minimum that does not correspond to the correct relative pose. The following, second method is more applicable to this situation, and when salient feature extraction is unreliable. It consists of a multi-resolution hierarchical search strategy, combined with a probabilistic template matching approach (Olson, 1998). The 3-D model of the known object is projected onto the image plane from a set of viewpoints corresponding to an exhaustive set of translations and rotations. The set is subdivided hierarchically into non-overlapping *cells* at multiple levels. This subdivision enables the search to be executed rapidly in a multi-resolution fashion, by eliminating large cells early in the process. Each cell is assigned a score that is a measure of the probability of the model being present in the video image under the subset of transformations corresponding to that cell. Cells with a low score are eliminated first. This step requires the use of appropriate goodness-of-fit metrics, as well as probabilistic approaches that do not produce false negatives, to avoid elimination of a cell that might contain a potential solution. A metric that has been shown to be robust for detecting point set patterns in complex scenes is the Hausdorff measure (Rucklidge, 1996). The directed Hausdorff measure from point set A to point set B is defined as:

$$(1)$$

$$h(A, B) = max \ min \ \| \, a - b \, \| \qquad a \in A \quad b \in B$$

This represents the maximum distance from a point in set *A* from it nearest point in set *B*. In our case, the directed Hausdorff metric would be used to measure the distance between the worst matching object point to its closest image point.

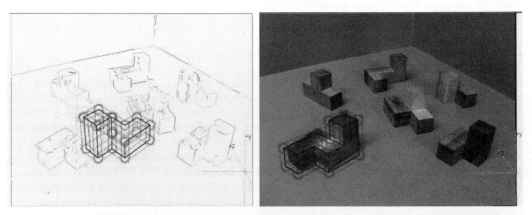

**Figure 1:** <u>Left:</u> Model on ground plane and edge-processed image. <u>Right:</u> Model match after initial pose estimation.

The initial pose space can be reduced by constraining the model to lie flat on the ground plane. We demonstrated this approach on parts of a 3-D puzzle (Figure 1).

## 2.2 Visual Servoing Registration in Urban Environment

The Visual Servoing approach has been used in the tracking module of RSC indoor AR applications (Sundareswaran et al., 1999). For outdoor applications, as for example in the TINMITH project (Piekarsky, Gunther, & Thomas, 1999), visual Registration is more difficult due to the relatively less constrained environment. Instead of fiducial markers (that are typically applied to objects in indoor applications), one is limited to the features in the scene. In an urban environment, there are numerous man-made features suitable for visual tracking (e.g., windows). Using efficient edge detection techniques we implemented an AR registration system on a laptop (750 MHz), using 320x240 image capture at 15 fps. Figure 2 shows snapshots of the resulting overlay of the 3-D model of a building on live video. Outdoor AR applications such as BARS (Julier, Baillot, Lanzagorta, Brown, & Rosenblum, 2000) are expected to benefit from the improved precision of this registration method.

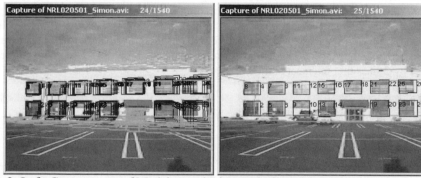

**Figure 2:** <u>Left:</u> Convergence of Model onto the image during Visual Servoing iterations. <u>Right:</u> Converged overlay.

## 2.3 Registration for AR on a Pocket PC

**Figure 3:** PocketPC screen with live video and overlay.

While the algorithms described above have been implemented on desktop PCs and laptops, a subset of the real-time registration approach has been ported to a PocketPC. The application captures and displays live video on the PocketPC display (using an Compaq Ipaq® with a LiveView® jacket). We have implemented image processing to track a 5-point fiducial set. This in turn is combined with Visual Servoing to overlay a 3-D model that appears to be aligned with the 5-point pattern (Figure 3). The upper half of the display shows the live video with the overlay. The lower half shows the segmented point pattern determine in the image processing step. PocketPC AR could be used to annotate live video without the need for a head-worn display.

## 2.4 Runway Detection during Approach

Another application of AR is in the airplane cockpit. See-through heads-up displays (HUD) can provide conformal overlay information for pilot guidance. This is useful during approach and taxiing, where for example taxiing instructions (e.g., taxiway closures) can be shown in an intuitive manner. In this application, the AR system needs to know precisely the attitude of the airplane, to provide exact registration with the outside view. Onboard sensors can provide attitude information with limited accuracy. We employed Computer Vision-based approaches to improve the registration for a HUD overlay. These methods rely on taxiway and runway markings for visual registration.

During the approach towards an airport, the runway is first detected in the video image. To determine the attitude of the video camera mounted in the cockpit, the horizon can be a useful cue. In case of a mountainous terrain, the horizon silhouette could be determined from a rendition of 3-D terrain data. Evaluation of the horizon under the assumption of a known azimuth angle provides information about the pitch and roll angle (we obtained precision of about 0.1°). This can be used to search for the runway as the plane is approaching. White runway border markings are useful to this process. Our algorithm searches for white marking segments below the horizon in a set of search paths. If sufficient markings aligned on a straight line are found, one marking is determined to have been found. If the other marking is also found and if constraints on the marking angle and location are verified, the runway is considered to be detected (Figure 4). It is then marked in the video image by 2 lines that intersect at the horizon.

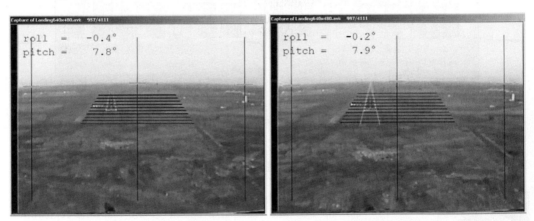

**Figure 4:** Runway detection during approach (applied on a recorded video scene from an approach at Moses Lake airport, Washington).

The current implementation of runway detection is video-based. By using mapped airport data, for example runway threshold location, the relation of the video to the real world can be established by applying model-based methods such as the Visual Servoing described in previous sections. Also, tracking the moving ground plane texture by tracking point features (Tomasi & Kanade, 1991) the airplane heading and speed could be estimated. The application of the improved overlay generation in airplane cockpits can be both for see-through heads-up displays (HUD) or for console mounted Enhanced Vision displays, merging synthetic rendering with real imagery.

# 3    Summary and Conclusion

We apply Computer Vision based Registration techniques in a variety of Augmented Reality applications that range from PocketPC displays to displays in airplane cockpits. The core real-time dynamic registration method– Visual Servoing – has proved to be a valuable tool a range of AR applications and has a footprint small enough to be implemented on computing platforms with limited capabilities, such as PocketPCs or wearable computers. More work is needed on the initialization issue for establishing the correspondence, since this problem requires relatively higher computing power. We are also investigating interactive methods to assist the system in the initialization where applicable.

## References

Azuma, R., Baillot, Y., Behringer, R., Feiner, S., Julier, S., MacIntyre, B. (2001). Recent advances in Augmented Reality. *IEEE Computer Graphics and Applications*, 22(6), 34-47.

Behringer, R. (2001). Augmented Reality. In Kent & Williams (Eds.), Encyclopedia of Computer Science and Technology, 45(30), 45-57, Marcel Dekker.

Behringer, R., Park, J., Sundareswaran, V. (2002). Model-Based Visual Tracking for Outdoor AR Applications. *Proceedings of ISMAR 2002*, 277-278.

Espiau, B., Chaumette, F., Rives, P. (1992). A new approach to Visual Servoing in Robotics. *IEEE Transactions on Robotics and Automation*, 8(3), 313-326.

Julier, S., Baillot, Y., Lanzagorta, M., Brown, D., Rosenblum, L. (2000). BARS: Battlefield AR System. *NATO Symposium on Information Processing Techniques for Military Systems*.

ISMAR (2002). Proceedings of IEEE and ACM Int. Symposium on Augmented Reality 2002. Darmstadt, Sept.30-Oct.1, 2002. IEEE Computer Society.

Olson, C.F. (1998). A probabilistic formulation for Hausdorff matching. *IEEE Computer Society Conference on Computer Vision and Pattern Recognition*, 150-156.

Piekarski, W., Gunther, B., Thomas, B. (1999). Integrating Virtual and Augmented Realities in an Outdoor Application. *Proceedings of 2nd IEEE and ACM IWAR '99* (pp 45-54).

Rucklidge W. (1996). Efficient Visual Recognition Using the Hausdorff Distance. *Springer, Berlin.*

Sundareswaran, V., Behringer, R. (1998). Visual Servoing-based Augmented Reality. *In Augmented Reality – Placing Artificial Objects in Real Scenes Proceedings of IWAR '98* (pp 193-200). A.K.Peters, Natick.

Sundareswaran, V., Behringer, R., Chen, S.L., Wang, K. (1999). A distributed system for device diagnostics utilizing AR, 3D audio, and speech recognition. *Proceedings Human Computer Interface (HCI) '99*, Munich.

Tomasi, C., Kanade, T. (1991). Detection and tracking of point features. Carnegie Mellon University Technical Report CMU-CS-91-132, April 1991.

# Interacting with Hierarchical Information Structures in Immersive Environments

*Roland Blach, Hilko Hoffmann, Oliver Stefani, Manfred Dangelmaier*

CC Virtual Environments - Fraunhofer Institute for Industrial Engineering
Nobelstr. 12,  70569 Stuttgart, Germany
Roland.Blach@iao.fhg.de

## Abstract

The ongoing process of collection and storage of knowledge with computer technology leads to highly complex information environments. The efficient access of information and the structure itself gets more and more complicated. Techniques for handling information have been developed mainly for desktop environments. The presented work investigates the usage of immersive environments for visualization and manipulation of hierarchically structured data and describes an application prototype.

## 1    Introduction

Today complex tools for the access and administration of knowledge information spaces have been developed. Techniques for handling information have been researched mainly for desktop environments (Card, Mackinlay & Shneiderman, 1999), but the framed and two dimensional computer screen is often a restricting factor for the layout of and the navigation in data spaces.
Immersive environments might have the potential to overcome some of these shortcomings. These Environments differ from conventional desktop in the sense that they embed the user in a computer generated data environment. These systems have the following key properties i.e. as described in (Kalawsky, 1993):

- Stereoscopic 3D image generation in real-time (approx. 10-30 frames per second)
- First person viewpoint
- Spatial interaction in real-time
- Perceived 1:1 Representation

Well known examples are systems which drives head mounted displays or 3D stereoscopic multiwall environments as i..e. CAVE-systems (Cruz-Neira, Sandin & DeFanti, 1993).

## 2    System Concept

The presented research investigates the usage of immersive environments for visualisation and manipulation of data structures found in enterprise resource planning (ERP) and product data management (PDM) systems. A main concept in these databases are tree-like hierarchies.  The immediate access of product component lists of a SAP R/3 database from within an immersive CAD Evaluation session defines the application context (Figure 1).

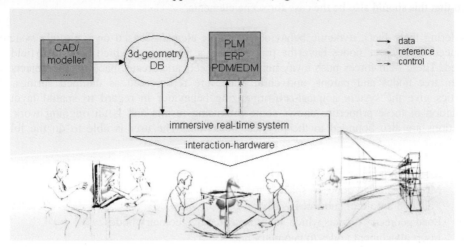

**Figure 1:** Overview on application context

The following functionality is implemented:

- Tree-like product component data structure is visualized and navigable
- 3D CAD parts can be loaded and visualised.
- Results of the evaluation session as f. e. annotations or snapshots can be written back to the database

# 3   Interaction concept

The focus of research was the evaluation of existing and the development of new representations of hierarchical tree-like data structures, which are equally suited for navigation and manipulation in immersive environments. Basic approach was the improvement of navigation through and interaction with the tree-like data structures. Classical Representations have been mixed with new derivations of known information visualisation paradigms. The Cone-Tree (Robertson, Card & Mackinlay, 1991) was used as a base representation for our concept (Figure 2).

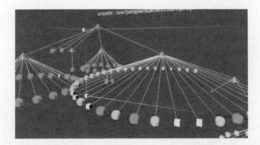

**Figure 2:** Cone-Tree representation

One of the major problems in early prototypes was the flexible spatial layout of the data structures, Especially if new elements were inserted (i. e. fold and unfold operations of subgraphs) the structure has to be spatially extended. User expectations for coherent translocation of elements are mostly known in the navigational area (Bowman, Koller & Hodges, 1997). Our observation have shown that this should also be the case for scene elements.

Considering these facts, dynamic behaviour of single elements based on a particle system was introduced. Single tree nodes have the properties of a point mass which have also field forces attached. These field forces push away neighbour objects and create space for the objects. Links between tree nodes and parent- and child nodes are represented as damped springs. These properties give the system a quasi-self-organizing behaviour in regard to spatial layout. The exploitation of these properties shows some promising results and is an ongoing work . User Interactions are also bounded to the physical behaviour. The user is able to do the following operations:

- Selection of tree-nodes, selection is done by ray-intersection
- Moving selected nodes in space
- Fold and unfold of sub-trees
- Load geometry objects which are represented by geometry nodes
- Insert attachment nodes to geometry nodes

Because of the large variety of objects context is important for the available functions. This is addressed with a context-based menu based on the selected element.

# 4    Implementation

The application prototype was developed with the Virtual Reality application development system Lightning (Bues et al., 2001). The specific information visualisation modules have been integrated with the plug-in API for external program modules. The underlying compute hardware can be SGI graphic-workstations or a PC-based cluster which will be connected to the projection environment. The target projection environment is a two wall L-shaped system which can easily be extended to more walls.

**Figure 3**: Information Visualization in a CAVE

The general approach follows the classical model-view-controller paradigm (Krasner & Pope, 1988) where the representation is simply one of several views onto the data. This is an important abstraction not only for well-known software engineering issues but especially for the prototyping in research of new representations and interactions.

# 5    Related Work

Navigation and interaction with data structures in immersive environments has been done in the past 10 years by several researcher. Unfortunately none of these experiments resulted in a convincing amelioration of representation of  or interaction with abstract data, compared to classical desktop systems.

The first implementations have been basically pure representational approaches, where the new immersive technology was used to fly through these structures in search of new perspectives. As i.e. (Ebert, Shaw, Zwa, Miller & Roberts, 1996) or (Risch et al, 1997).

Because of the immaturity of the virtual environment technology in the mid-nineties, information visualization has not used this technology as intense as it could have. So not much is known about the actual benefit of immersive technology in the area of information visualization.

Using physical based modelling in user interaction has been done by other researchers as well. Two general directions can be seen:

- Physical system behaviour
- Physical interface behaviour

Physical based system modelling is a general approach to  realistic representation of real world behaviour. For our research the usage for  abstract representations with no real world experience is more interesting. For the 2D visualization of graphs spring and mass approaches have been used with good results (Fruchterman, Reingold, 1991).  An approach which is very close to our is the work of Osawa (Osawa, 1999), which incorporates thermodynamics to represent content proximity. He has designed the interface also in an immersive environment. However Osawa does not use the physical behaviour for the interface itself.

Physical interface behaviour has also been introduced in the research area of immersive environments as pure interaction metaphors very often applied to assembly scenarios as i.e. (Koutek Hess, Post, Bakker, 2002) or  (Fröhlich, Tramberend, Beers & Barraff, 2000).

Our approach tries to combine both domains in a coherent over all system behaviour.

# 6    Conclusion and Future Work

A working prototype for the visualization of and interaction with hierarchical data-structures was realized. To obtain greater layout flexibility and coherent translocation behavior a dynamic particle system which is based on physical based modeling was introduced.  As mentioned already the particle system for user interaction show some promising properties.

The prototype has shown the general feasibility of the approach. Although promising, the usage of immersive environments as interface paradigm for information visualization has not been proven being superior to classical systems. The future work has to concentrate on general user studies and comparisons with established approaches.

## Acknowledgments

The development of the information visualization application is funded by the BMBF-Leitprojekt INVITE.

## References

Bowman, D.A., Koller, D. & Hodges, D.F. (1997) Travel in immersive virtual environments: an evaluation of viewpoint motion control techniques, *Proceedings of VRAIS 1996*, pp 45 – 52

Bues et al. (2001), Towards a Scalable High Performance Application Platform for Immersive Virtual Environments. In: *Proceedings of IPT/EGVE 2001*, pp 44-52

Card, S.K., Mackinlay, J.D. & Shneiderman, B., (1999). Readings in Information Visualization – Using Vision to Think. San Francisco: Morgan Kaufmann

Cruz-Neira C., Sandin, D.J. & DeFanti, T. (1993). A. Surround-screen projection-based virtual reality: The design and implementation of the CAVE. In: *Proceedings of SIGGRAPH 93*, Annual Conference Series, pp 135-142

Ebert, D., Shaw, C.D., Zwa, A., Miller, E.L. & Roberts, D.A. (1996). Minimally-immersive Interactive Volumetric Information Visualization. *Proceedings of IEEE Symposium on Information Visualization*, InfoVis'96.

Fröhlich, B, Tramberend, H., Beers, A. Barraff D. (2000). Physical-Based Manipulation on the Responsive Workbench. In: *Proceedings IEEE Virtual Reality 2000*, pp 5-11

Fruchterman, T. M. J. & Reingold E.M. (1991). Graph Drawing by Force-directed Placement, In: *Software – Practice and Experience*, Vol. 21, No. 11, pp. 1129-1164, 1991.

Kalawsky, R.S (1993). The science of Virtual Reality and Virtual Environments, Addison-Wesley Publishing Company, New York.

Koutek, M., van Hees, J, Post, F.H. & Bakker, A.F. (2002). Virtual Spring Manipulators for Particle Steering in Molecular Dynamics on the Responsive Workbench, *Proc. Eurographics Virtual Environments 2002*, Barcelona, pp 53-62.

Krasner, G.E. & Pope, S.T. (1998). A Cookbook for using the Model-View-Controller User Interface Paradigm in Smalltalk-80. In: *Journal of Object-Oriented Programming*, volume 1 number 3, pp 26-49,

Osawa, N. (1999) , A multiple-focus browser for network structures using heat models," *The 59th National Convention of Information Processing Society of Japan*, 5K-09

Risch, J.S et al. (1999). The Starlight Information Visualization System. In: Readings in Information Visualization – Using Vision to Think. Eds. Card, S.K., Mackinlay J.D. & Shneiderman, B. pp 551-560, San Francisco: Morgan Kaufmann

Robertson, C.G., Card, S.K. & Mackinlay, J.D. (1991). Cone Trees: Animated 3D visualizations of hierarchical information. In: *Proceedings of SIGCHI'91*, pp. 189-194

# Virtual Reality - Ergonomic Solutions for Overcoming the Complexity Trap

*Alex H. Bullinger\*, Marcus F. Kuntze, Franz Mueller-Spahn,*
*Angelos Amditis and Ralph Mager*

\*Center of Applied Technologies in Neuroscience - Basel (COAT-Basel),
University of Basel, Wilhelm Klein Str. 27, 4025 Basel, Switzerland
ahb@coat-basel.com

## Abstract

The still ongoing rapid technological development leads to Virtual Reality – Systems becoming complex ever more. This growing complexity makes it more difficult for the users of such sophisticated systems, to carry out the respective tasks effectively and efficient. In order to facilitate an anthropocentric design and development process of both, Virtual Reality – Systems and Virtual Environments, objective usability criteria together with a suitable assessment approach is needed.

## Introduction

VR technology is still a rapidly evolving and diversifying field. For example, new display technologies and interaction devices are continuously being developed and modified, having an impact on the way in which users view, navigate and interact with Virtual Worlds. The technology has penetrated a wide spectrum of application domains, reaching from highly specialized, custom tailored single purpose solutions over adaptable and collaborative industrial applications to embedded home based enter- or infotainment installations. Where technological terms and concepts have shaped the discussions of the past, more and more concepts like cost-effectiveness, user centred health considerations, social and performance impacts on the actual working environment or consequences for organizational structures seem to dominate the discussions today.

All these different developments added to an already high level of system complexity that came with the introduction of a third dimension into the human computer interaction concept. This tendency seems to be far from over.

Before we go into more detail here, we would like to define a small set of terms as they are used in this paper:

- *Virtual World (VW)* describes worlds existing with the support of computers or computer networks only. Virtual worlds can be completely fictive as well as representations of real worlds. While some authors consider even text based environments (e.g. chatrooms, etc.) as virtual worlds, in the context of this paper a Virtual World is composed of three dimensional objects, light sources, grids, etc. and three dimensional user interfaces.
- *Virtual Reality (VR)* is used to describe the technology to display and interact with virtual worlds in general. It addresses the required hardware, basic software and applications, which are necessary to give a user access to a virtual world.
- *Virtual Environment (VE)* in this context is defined to consist of a VR-installation and at least one virtual world. A VE provides the user(s) with the possibilities to fully immerse into and interact with virtual worlds.

## The Complexity Trap

An interesting an well known observation was made in 1965 by Gordon Moore, co-founder of Intel, that the number of transistors per square inch on integrated circuits had doubled every year since the integrated circuit was invented. He predicted that this trend would continue for the foreseeable future. In subsequent years, the pace slowed down a bit, but data density has doubled approximately every 18 months, and this is the current definition of Moore's Law[1]. Most experts, including Moore himself, expect Moore's Law to hold for at least another two decades from now[2]. With regard to Virtual Reality this massive increase in computing power triggered an astonishing development process leading from VR-Systems that were merely able to visualize complex data sets over passive, later active interactivity concepts to systems aiming at autoevolutionary capabilities.

This development not only left the R&D departments with a ever growing challenge but also the users of these technologies. More and more specific knowledge, training and expertise is needed on their side in order to operate these complex systems, interact efficiently and perform the respective task(s) effectively.

This leads us to another problem, sounding that trivial that it is taken into account quite seldom: While the VR-systems clearly are adaptable to an enormous degree and – within the given technological boundaries - can be even upgraded massively when needed, their human users are not. This was not of greater importance for years. The problem back then was to improve the VR-systems capabilities to match the respective user's need. This of course still holds true but more and more in terms of content only. Instead of this technology oriented adaptation process nowadays another adaptation is needed: The adaptation of the VR-systems to their respective users capabilities and limitations. While this might be a somewhat polemic remark it still is an interesting observation, that modern fighter pilots only seem to be able to keep up with the massive inflow of multisensory information coming along with the absolute necessity to react on this information within fractions of a second by consuming amphetamines, that is psychomotor stimulating drugs. Their sensory and cognitive system simply seems not to be able to cope with the technology otherwise.

It can be stated that there is an ongoing shift from designing merely technology driven over content driven towards user requirements driven development processes of VR-systems and VEs. These user requirements consist of content based components as well as of physiological, cognitive, emotional and social requirements.

## Anthropocentric Evaluation of Complex VR-Systems

Human Computer Interface and Human Computer Interaction evaluation is an accepted and grown concept, understood as a possibility to measure the usability of a given computer-based system: "… the ease with which a user can learn to operate, prepare inputs for, and interpret outputs of a system or component"[3].
This definition is still valid. What we have to keep in mind is that interfacing technology in relation to VEs is quite different from the graphical interfaces that led to and where used in the early years of the science of usability. Within Virtual Worlds the usability assessment needs reach clearly beyond traditional measurement approaches:

- The in- and output devices are multimodal, addressing a whole set of sensory channels and cognitive entities on the user side
- The third dimension has impact on a variety of interfacing domains:
  - Navigation
  - Object manipulation
  - Sensory perception

- Presence and immersiveness
- Comfort, side- and aftereffects
- [emotional and affective attitudes] – put in brackets because it is not a direct usability criteria. Nevertheless it can not be neglected as it has massive impact on subjective user evaluation, awareness and acceptance of virtual worlds.

This bullet list clearly is far from being complete[4]. To address all of these entities is not an easy task to undertake. But trying to do so by using mainly subjective instruments like questionnaires, rating and impression scales is not only "not easy", it is quite hopeless. This is mainly due to the impossibility of filtering and correcting all possible influencing disturbances and biases for such a huge number of independent and dependent variables. What is needed instead is clear in theory, at least: A set of objective parameters, not to be influenced by the user or the external evaluator/examiner, thus leading to an objective assessment of usability issues with respect to virtual environments. This is easier said than done by far as it requires real time physiological data acquisition on the user side. Why is that required? Because physiological parameters derived from the human autonomous nervous system provide information on the users level of comfort, strain, stress, relaxation, anxiety, etc. - by and large without the possibility on the user side to influence these parameters voluntarily. By adding neurophysiological parameters to such a set of measurements (for example EEG), even concepts such as mental work load, cognition, etc. can be addressed.

To our best knowledge, there is no integrated evaluation methodology for usability testing and evaluation within VEs so far. For our own development efforts we defined a minimal set of two conditions to be met:
- In order to achieve the highest degree of objectivity possible, only entities that cannot be addressed otherwise will be investigated, rated and evaluated by questionnaires and (self)assessment scales
- In order to achieve the highest degree of reliability and validity possible, measurement technologies for such an approach should be embedded within the respective VR-system. Where this is not possible, the respective hardware should be miniaturized to a fully and easily portable degree

An application study of this approach is presented within HCII 2003 elsewhere[5].

Topics under current investigation (amongst others) are influences of
- Age (user)
- Gender (user)
- Level of education (user)

- Level of knowledge (user)
- Level of pre-experience (user)
- Colour and texture (VE)
- Frame rate (VR-System)

on performance effectiveness and efficiency within a given VE.

## Future Work

This work is aimed at the possibility of defining a set of parameters that enables objective assessment of usability criteria with respect to VEs. Taking this development another step forward would mean to aim at a predictive set of variables that can be taken into account even in the early stages of VE-design and development in order to reach a certain performance level with a pre-defined and specified user population.

## References

1. G.E. Moore, 1965. "Cramming more components onto integrated circuits," *Electronics,* volume 38, number 8 (19 April), at http://www.intel.com/research/silicon/moorespaper.pdf
2. I. Tuomi, 2002. "The Lives and Death of Moore's Law," *First Monday,* volume 7, number 11 (November 2002), at http://www.firstmonday.dk/issues/issue7_11/tuomi
3. [IEEE 90]: Institute of Electrical and Electronics Engineers. *IEEE Standard Computer Dictionary: A Compilation of IEEE Standard Computer Glossaries.* New York, NY: 1990
4. D. Hix & J.L. Gabbard, 2002: "Usability engineering of virtual environments". *In K.M. Stanney (Ed.), Handbook of virtual environments: Design, implementation and applications* (pp. 681-699). Mahwah, NJ: Lawrence Erlbaum Associates, Inc.
5. R. Mager et al., 2003: "Cognitive Ergonomics in the Development of Virtual Reality: A Neurophysiological approach". HCII 2003 Proceedings.

# Crossing from Physical Workplace to Virtual Workspace: be AWARE!

*Nina Christiansen*

Copenhagen Business School
Howitzvej 60
DK-2000 Frederiksberg
nin@cbs.dk

*Kelly Maglaughlin*

University of North Carolina
CB 3360, 100 Manning Hall
Chapel Hill, NC 27599-3360
maglk@ils.unc.edu

## Abstract

This conceptual paper synthesizes the existing body of studies on awareness in four overall categories: (w)orkspace, (a)vailability, (g)roup, and (c)ontext. The four categories (WAGC) define coordination mechanisms that bridge the traditions physical work*place* and virtual work*space*.
The proposed WAGC-framework suggests on consolidating the overwhelming variations of awareness definitions can aid building a solid base for ongoing and future studies of awareness. Although the impressive body of HCI research has demonstrated substantive insight in the general awareness concept, more work is needed on the cumulative, epistemological and methodological knowledge gathering for the awareness challenges related to bridging the physical workplace and the virtual workspace.

## 1 Introduction

The use of IT have made it possible to use the workspace to collaborate across boarders and thereby invite the whole world into your workplace. Work*space* denotes the system designed to support collaborative work, rather than the physical location or work*place* where that system is used. As the traditional physical work*place* has expanded into the virtual work*space*, the interaction between people has to relies and use other coordination mechanisms known by heart by each individual in the traditional face-to-face setting. Collaboration in groupware workspace is often awkward, stilted, and frustrating compared to face-to-face setting (Gutwin & Greenberg, 1998). Yet designers of CSCW systems lack tools that emulate the rich communication taken for granted in face-to-face settings. In particular the proximity issue and customisation to the new working mode are critical factors to address.

During the past ten years awareness is analysed as a yet more critical factor in design of collaborative systems (Gutwin & Greenberg, 1997) and has become a key factor to CSCW-design. Despite this, awareness is broadly interpreted within CSCW and is subject to considerable debate (Rodden, 1996). Awareness phrases such as 'awareness,' 'general awareness' and 'workspace awareness,' which emerged at the beginning of the awareness boom, have expanded into a multitude of different phrases of awareness. Each year, in journals and at conferences, workshops, and tutorials new awareness phases are proposed and debated.

On the background of reading 83 HCI articles dealing with awareness, and identifying 41 different awareness phrases[1], this paper focuses on the devolvement of the framework to get an overview of important areas in awareness in CSCW systems. The need for this is illustrated by the confusion that arise when one awareness phrase could have more than one definition - depending on the author(s), and two different phrases could have the same definition - again depending on the different author(s). Even when the phrases are defined, the definitions often overlap with other phrases and the differences between phrases are rarely clear. For example Dourish & Bellotti (1994) use the terms 'background awareness,' 'informal awareness' and 'peripheral awareness' interchangeably when referring to knowledge of ongoing activity necessary for maintaining informal communication. Further more, there are awareness phrases with no definition, so the reader has to decipher it from reading the text. We need to have the same understanding of each of the awareness phrases, before the concept losses it's importance in designing CSCW systems. Some have already pointed out that it seems like awareness hardly is a concept any longer (Schmidt, 2002).

## 2    Previous research on awareness

One of the early insights in *system awareness* proposed timing and focus of the collaborative situations as a key to understand the dynamics of awareness (Fuchs, Pankoke-Babatz & Prinz, 1995). They identify four different modes or states of awareness based on whether events are synchronous or asynchronous (i.e. events currently occurring or those that happened in the past) and coupled or uncoupled (i.e. events related to the users current work focus or not related). These modes of awareness help to 1) identify the types of knowledge users need to maintain and 2) to illustrate how completed activities and ongoing activities provide information to collaborating users. This event model is limited to activities in the virtual workspace, and fails to the surroundings in the interaction, which have an important impact on the understanding of the activity.

Hayashi, Hazama, Nomura, Yamada & Gudmundson (1999) discuss *activity awareness,* and divide it into three nodes: people (awareness of others), projects (organization awareness) and places (workspace awareness). Their approach delivers awareness information between activities, which are executed in individual workspace (p. 103). They define *activity awareness* as: "an awareness which gives workers indications of what is happening and what has happened recently in collaborative activities" (p. 99). *Awareness of others*, the people node, provides information about what tasks individuals are currently doing. *Organizational awareness*, centred in the project node, gives an indication of how the individual activities fit and contribute to the organization's goals – it is a mechanism for keeping track of the progress in an ongoing project. *Workspace awareness*, centred on the places node, gives an indication of what has or is happening in the virtual (i.e. chat rooms and message boards) or physical (i.e. conference rooms) workspace. Their approach delivers awareness information between activities, to extend the scope of awareness beyond that of a shared workspace. The key feature of activity awareness is based upon individual, rather than shared, activities. It can be seen as an uncoupled - asynchronous activity.

Lee, Gingerson & Danis (2000) suggest that awareness provides context and guidance for one's activities, in addition to enabling coordination and collaboration. They divide awareness into: 1) *people*, where questions as: "Is anyone around?" and "Who are they?" is asked. Key factors here are accessibility and availability information, and people and group information, 2) *activity*, where

---

[1] Empirical data on the 83 papers reviewed is available at URL: http://www.cbs.dk/staff/nina/paper.htm

the question "What are they doing?" are asked, and 3) *workspace*, where questions as "Who/what are they doing it with?" where the key factors explore context information including an up-to-the-moment understanding of the interactions within the shared workspace. The problem here is that they presented their framework in a tutorial at the CSCW conference in 2000, and therefore doesn't have a deeper description and argumentation for of each of the categories. But their categorisation seems reasonable, since it considers the broader spectra of interaction.

Gutwin & Greenberg have established a good foundation in workspace awareness (e.g. 1997, 1998). At the same time they have also used other terms/ phrases of awareness as informal and situational, to present a descriptive framework of awareness for small groups in shared-workspace groupware. The workspace awareness term is a specialisation of situation awareness, where the "situation" is well-defined--others' interaction with a shared workspace, whereas informal awareness is who is around, and available for collaboration. But here the limitation also is seen as a contribution to the workspace alone.

Awareness of people in the workspace was not the only concern we had in trying to develop a framework for awareness. We were not only dealing with the interaction in the workspace but would also consider all the surrounding factors, which are important for a successful interaction in the workspace. Following is a presentation of the categorisations we found was needed for a successful interaction in the virtual workspace.

## 3    The WAGC Framework

The work*space* facilitates interactions and acts as a bridge between distant people and objects. In order to work successfully in the work*space*, it is often necessary to be aware not only of the current activity in the workspace, but also the interactions between the people in the virtual group and the context in which it progresses. With this in mind, we attempted to determine the explicit awareness mechanisms needed in the new working situation – the virtual work*space*. The content of WAGC framework consist of the following argumentation.

**W**: Since people are interacting in the workspace they need the knowledge of the tasks that have been, are being and need to be done *within* the virtual environment – or *workspace awareness* as defined by Gutwin & Greenberg (eg. 1997; 1998) and repeatedly referenced throughout the literature. With few exceptions (i.e. Hayashi, 1999), most definitions of workspace awareness referred to their work and closely followed their concept, that workspace awareness is knowledge of who is working in the virtual space and what tasks they perform. This concept was very pervasive in the literature and was often referred to by different names. Our work*space* category covers several concepts since the awareness that are needed are depended on the specific work situation. Fuchs et al. (1995) suggest, these actives can be divided into four categories: synchronous, asynchronous, coupled and uncoupled, so each specific work situation can be supported by the awareness mechanisms needed to support the work situation. *Gaze awareness* (i.e. knowing where someone else is looking and an indication of attention focus) is an example of workspace awareness that is used to support a better interaction in synchronous and coupled activity.

**A**: When a user wants to join the virtual workspace they need to know if the people or objects they want to work with are available. This becomes a very important factor since interaction in the workspace may not be possible until people or objects are. *Availability awareness* is therefore necessary before interaction even can take place. This category of awareness spans both physical and virtual space as well as people and objects. Several researchers (i.e. Bellotti & Sellen, 1993) have examined issues surrounding availability awareness with a focus on how it relates to privacy, which also is worth considering.

**G**: Also there is a strong need to specify awareness mechanism which could give the team members a feeling of belonging to a group. We label this category *group awareness*. Non-verbal communications, spontaneously interactions, informal and physical presence are all elements of face-to-face interaction that can promote a sense of community. Root (1988) uses the term 'social browsing' to describe this type of interaction: "Moving around in the workplace presents frequent opportunities for both planned and unplanned interactions with local co-workers. We use the term 'social browsing' to describe this dynamic process of informal, in-person, mobility-based social interaction. We suggest that social browsing is a fundamental mechanism for developing and maintaining social relationships in the workplace" (p.27). When groups are geographically distributed, it is particular important not to neglect the need for informal interaction, spontaneous conversation and awareness of people and events at other sites (Bly et al., 1993). Therefore the *group awareness* category consists of actions taken to help group members get a feeling of belonging to a group. This category is action oriented toward establishing and maintaining group cohesion.

**C**: The context surrounding the interaction is an important factor, also when the work is done virtually. Group engagement in the workplace depends on a shared context (Gaver, 1991). We define context both as a physical, social and mental context. The physical context deals with the physical environment such as the room in which the individual works, the background noise and the physical resources they have at their disposal. The mental context is the individual's horizons[2], which they bring into the shared space. This horizon is very unique to each person but will overlap with others horizons in the sense of how well they know each other or how often they work together. The social context deals with the intangibles in the environment such as deadlines an individual might have on other projects, interpersonal relationships, and organizational support. These are all issues that can have either positive or negative effects on interactions in the work*space* but that are extremely hard to communicate across distances. For example, in a face-to-face environment, judging a co-worker's attitude, working style, or even competence may be as simple as observing them casually or even subconsciously, as they perform tasks. In a virtual environment this observation must be explicit. In any group, context is part of an ongoing history, an awareness of which can be important for a successful outcome of the interaction. All interactions depend, in part on the context of the situation in which they take place.

Although others have suggested conceptual frameworks for certain types of awareness (Fuchs et al., 1995; Hayashi et al., 1999; Gutwin et al., 1997; Lee et al., 2000), our proposed composition is *generic* encompassing both individual aspects *and* interactions in both physical and virtual workspaces.

# 4   Conclusion and perspectives

This paper categorizes the many awareness phrases that can be found in CSCW literature. In doing this we observed a need for the transport of information not only within virtual workspaces but also between the physical spaces that the virtual space bridges. The challenges this need suggests, calls for further research and insight in the coordination mechanisms of awareness. Our proposed

---

[2] The horizon is defined/seen as the individual persons accumulated knowledge

WAGC framework comprises four different awareness groups: workspace, availability, group, and context.

Our hypothesis is that awareness is a dynamic concept that has evolved from the concept of general awareness into the concept of workspace awareness, only to be separated by at least 41 different phrases and labels, which we categorise in four separate but closely related concepts. We propose the WAGC framework as a beginning for the dialog on the use of 'awareness,' in hopes that the chaos surrounding the concept might be stemmed.

# References

Bellotti, V. & Sellen, A. (1993). Design for privacy in ubiquitous computing environments. in *Proceedings of ECSCW'93, European Computer Supported Cooperative Work*, 61-76.

Bly, S.A., Harrison, S.R. & Iving, S. (1993). Media Spaces: Bringing people together in a video, audio, and computing environment. *Communications of the ACM, 36* (1). 28-46.

Dourish, P., & Bellotti, V. (1994). Networked multimedia support for informal collaboration. Retrieved January 13, 2002 from http://citeseer.nj.nec.com/dourish94networked.html

Fuchs, L., Pankoke-Babatz, U., & Prinz, W. (1995). Supporting cooperative awareness with local event mechanism: The groupdesk system. in *Proceedings of ECSCW'95, European Computer Supported Cooperative Work,* 247-262.

Gaver, W.W. (1991). Sound support for collaboration. in *Proceedings of ECSCW'91, European Computer Supported Cooperative Work*, 293-308.

Gutwin, C., & Greenberg, S. (1998). Effects of Awareness Support on Groupware Usability. in *Proceedings of CHI'98, Computer Human Interaction*, p. 511-518.

Gutwin, C., & Greenberg, S. (1997). Workspace Awareness. Position paper for the *ACM CHI'97 Workshop on Awareness in Collaborative Systems*, organized by Susan E. McDaniel and Tom Brinck, Atlanta, Georgia, March 22-27.

Hayashi, K., Hazama, T., Nomura, T., Yamada, T., & Gudmundson, S. (1999). Activity awareness: A framework for sharing knowledge of people, projects, and places. in *Proceedings of ECSCW'99, European Computer Supported Cooperative Work*, 99-118.

Lee, A., Gingerson, A., & Danis, C. (2000). Developing Web-based Collaborative Applications: Social and Technical Issues (Tutorial). in *CSCW'00, Computer Supported Cooperative Work.*

Rodden, T. (1996). Populating the application: A model of awareness for cooperative application. in *Proceedings of CSCW'96, Computer Supported Cooperative Work*, 87-96.

Root, W. R, (1988). Design of a Multi-Media Vehicle for Social Browsing. In Proceedings of *the 1988 ACM conference on Computer-supported cooperative work 1988*, Portland, Oregon, United States

Schmidt, K. (2002). The problem with "Awareness". *Computer Supported Cooperative Work*, vol. 11, 285-298.

# Automatic Behavioral Responses as a Measure of Immersion in Virtual Environments

*Joseph Cohn,*

US Naval Research
Laboratory
4555 Overlook Avenue,
Washington, DC 20375
cohn@ait.nrl.navy.mil

*Carey Balaban,*

Department of
Otolaryngology
Eye & Ear Institute Building
203 Lothrop Street
Pittsburgh, PA 15213

*Eric Muth,*

Department of
Psychology
Clemson University
410J Brackett Hall
Clemson, SC 29634
muth@clemson.edu

*Keith Brendley*

Artis, LLC
1117 N. 19th St.
Arlington, VA 22209
brendleyk@artisllc.com

*Roy Stripling,*

Strategic Analysis, Inc.
3601 Wilson Blvd.
Arlington, VA 22201
rstripling@sainc.com

## Abstract

We propose a theoretical framework involving monitoring automatic motor behaviors to develop quantitative or semi quantitative, unobtrusive metrics for cognitive engagement, or immersion, in Virtual Environments (VE). These behaviors include behaviors which are actively modulated in response to, or in anticipation of, changing sensory input. We provide initial support for this approach from the literature, developing this framework from results derived from our research efforts. Ultimately, it is expected that this framework will provide the basis for developing more effective VE training systems, as well as for developing unique applications for these devices.

## 1    Introduction

### 1.1    The Challenge

The use of simulators in training in the United States can be traced back to at least 1929 and Edwin Link's flight simulator. Technological innovations have brought increased fidelity to simulators, in particular for VE systems which claim such benefits as small footprint, rapid reconfiguration, and enhanced training delivery. While ultimately the purpose of a simulator is to train tasks in a cost-effective, efficient, practical, safe manner (Taylor, Lintern and Koonce, 1992), current doctrine often assumes that if the simulation accurately represents the real-world environment from a visual standpoint, then the training will transfer to the real world environment (Cohn, Helmick, Myers and Burns, 2000). Yet, there is currently little, if any, guidance, for determining when the VE scenario is close enough to the real world one being modeled to support these training goals. While some may suggest simply making any VE as hi fidelity as possible, in fact, it has been shown that low-visual fidelity, PC-based flight simulators, can be used as effective trainers provided that the necessary *realism* elements are afforded to the user (Gopher, Weil and Bareket, 1994).

## 1.2 Theoretical Framework

In order to resolve this problem, it is important to realize that VE represents a unique form of altered environment, in which the key essence of the information flow process is found, time and again, to be lacking. There are certainly many contributing factors, ranging from the over reliance of VE on the visual channel to convey information typically encoded through multiple sensory channels, to the simple fact that it is likely not possible to ever fully capture the rich tapestry of sensory information using modeling and simulation tools. Nevertheless, it is possible to identify key components that provide an indication of how closely the developed VE system actually parallels the real world upon which it is based.

This challenge is not novel. In fact, it has its antecedents in an operational concept termed *immersion*. Immersion denotes a collective measure of the degree to which a user is able to slide out of the physical world they inhabit sensorally and to slip into the one presented to them virtually, usually through a mix of sensory substitution and sensory packing. Current attempts at measuring immersion do not provide a satisfactory solution to this need. For the most part, the data they provide are subjective, derivative measures that can be highly biased by factors unrelated to VE quality. A potentially more satisfying approach is to realize that immersion in fact contains within it two components, a higher order intellectual one and a lower-level primal one. It is our contention that current measures of immersion address only the subjective, intellectual component without supporting the more objective, primal one. Importantly, this primal component can be measured in both the virtual and real world environments, to provide an unbiased indication of how well the information conveyed through the VE system to the user duplicates what would be conveyed in the real world setting.

## 2 Examples

### 2.1 Reaching During Real and Virtual Rotation

A very basic laboratory experiment illustrates this point well. When individuals are placed at the center of a fully enclosed, darkened room and rotated at constant –and hence, undetectable through vestibular end organs- velocity, and are subsequently asked to reach towards a point located in front of them, their movements are initially deflected away from the intended path. The reason for this reaching error is that the forward arm motion in a rotating environment generates velocity dependent forces, called Coriolis forces that act orthogonal to the movement. With repeated movements subjects learn to program corrections to their movements that compensate for the Coriolis forces and movements become accurate (Lackner and DiZio, 1994).

What is most intriguing about these results is that we encounter Coriolis forces every day. For example when we rotate our upper bodies while reaching for a cup of coffee, similar forces act upon the moving arm. Yet, we are unaware of these forces precisely because we have experienced them so often that our nervous system –cued through the self generating nature of the act- can preplan compensations for them, and our movements proceed flawlessly. In the rotating room environment, by contrast, the perceptual system is, by design, kept ignorant of the rotating reference frame and hence, cannot preplan these corrections. Only with repeated probing of the environment can the necessary corrections be incorporated into the motion.

This provides an ideal paradigm for evaluating a VE system from the perspective of the primal immersion component. Consider the initial reaching movement errors as an indication of the magnitude of the error that can arise when the perceptual system is not cued about body state. We would expect that, when the same experiment is run in a VE system that is by design meant to provide exactly the same motion cues, reaching errors of the same magnitude should actually occur in the opposite direction: the nervous system is now incorporating corrections for expected Coriolis forces, yet because the subject is actually stationary, these corrections are inappropriate. In fact, what is found is that initial reaching errors in this VE setting are only about one half the magnitudes of those in the actual rotating experiment, suggesting that at a fundamental level the VE system fails to fully provide the full suite of perceptual cues (Cohn, DiZio and Lackner, 2000; Cohn, DiZio & Lackner, 1995).

## 2.2   Low Level Entrainment

While these results suggest that certain primal responses might serve as indications of simulator fidelity, or alternatively, immersion, they are also derived from highly idealized settings. More commonly, a full scale simulation is developed, that will engage a range of sensory-perceptual systems. Consequently, it is useful to look at such a simulation, and examine what underlying responses one might expect to be evoked, both from the real and virtual system, and to then determine if such, is, in fact, the case. The power of this approach is illustrated through a second experiment that involves an evaluation of the degree to which more autonomic components of the central nervous system can bind themselves to the information being provided in the environment, for instance reflexive postural adjustments.

It is known that in the real environment, certain basic reflexes entrain to periodic real world stimuli. Recent work by (Duh, Lin et al 2002) suggests that there is a similar link between such reflexes as   postural stability and virtually   presented field of view. This work focused on ensemble changes in whole body center of pressure. Demonstrating an impact on more basic reflexes, which are relegated almost exclusively to subconscious control, would provide an even stronger indication of how well the information being presented by a VE can truly influence the perceptual system. We therefore chose to look at the degree to which a simple CAVE based VE system, projecting a scene oscillating at 0.2 Hz, would entrain respiration. Figure 1 provides

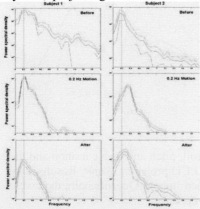

respiration data from 2 subjects, before during and after exposure to the CAVE. Before exposure, both subjects demonstrated a very broad frequency spectrum, with multiple peaks. During exposure, however, both subjects' respiration clearly was altered to show peaks at the both the frequency of the moving environment (0.2Hz) and twice the frequency of this motion (0.4Hz). Following exposure, this entrainment effect is still seen to influence overall respiration patterns.

**Figure 1:** Entrainment of respiration with Virtual scene motion.  CAVE scene oscillated front to back at 0.2Hz, while subject remained upright and   immobile.   Power spectrum of respiration for two subjects before (*top*), during (m*iddle)* and after  (*bottom)*exposure to VE.

## 2.3    Low Level Behavior Indicates Higher Order Engagement

While the examples provided thus far suggest that primal behavior responses might serve as an important indication of how well information is presented to the perceptual system by a VE it also seems likely that these lower level processes might reveal higher order ones such as cognitive engagement. That is, changes in automatic behaviors (e.g., postural adjustments and respiratory rate) are advantageous for assessing cognitive awareness because they are minimally intrusive and take advantage of the intricacy of these coordinated movements.   We know from personal experience that we can recognize changes in the movements of a driver (or a student/colleague in a lecture or seminar) that are associated with fatigue, drowsiness or lack of attention.   To illustrate this, we have constructed a chair to measure postural changes in personnel seated at instrument consoles, as may be found on a ship's bridge or inside an aircraft cockpit, while actively engaging in a virtually presented task. Preliminary results from this work suggest that changes in seated postural sway can indicate various levels of Cognitive engagement, such as: "Concentrating on task", "Normal awareness", or "Low awareness".

## 3    Extension: Modification of Behavior Through VE Intervention

The data in Figure 1 suggest that at a low level, VE systems might actually modify primal behaviors. While this has been demonstrated, in a small sample population (n=2) for a very basic behavior, an equally intriguing question is how might this general approach be exploited to modify more operationally relevant behaviors, such as motion sickness, which is in a sense a mismatch between externally provided perceptual information and the internal processing of this information. In terms of presenting  visual information to manipulate behavior, it is  known that visual stimuli can capture one's sense of body position (Nakamura & Shimojo, 2000; Cohn, Lackner & DiZio, 2000). Based on this, it was hypothesized that exposing subjects to an environment in which the virtually presented scene motion is coupled to the one experienced physically should reduce motion sickness. Contrastingly, exposing subjects to a VE in which the virtually presented scene motion differs from that experienced physically should lead to significant motion sickness.

 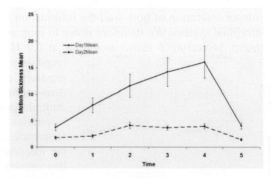

**Figure 2:** *Left:* The ship motion simulator (SMS), capable of roll, pitch and heave motion and the VE scene displayed on a wall screen inside the SMS using an LCD projector. *Right:* Results. Day 1, no motion coupling (solid) and with motion coupling (dashed lines).  Questionnaire data were collected every five minutes; thus the 'time' axis reflects these epochs.

To test this, data were collected on 11 subjects during exposure to a ship motion simulator, capable of roll, pitch and heave motion, with degree of motion sickness evaluated using the Motion Sickness Assessment Questionnaire (Gianoros, P.J et al, 2001). On the first day of exposure, subjects looked at a computer generated scene that was stationary relative to the motion of the ship motion simulator. On the second day of exposure, subjects looked at a computer generated scene that moved relative to the motion of the ship motion simulator. As Figure 2 suggests, subjects reported significantly stronger symptoms on Day 1 (no motion coupling) compared to Day 2 (motion coupling). While these data are confounded by time and possible adaptation effects, they provide preliminary evidence for the effectiveness of using Virtual Environments to modify automatic behavioral responses.

# 4    Conclusion

The current framework suggests a method for evaluating VE efficacy by examining the degree to which expected, basic, behaviors are elicited. While this technique has been explored for several different types of behavior, within a small range of simulation environments, it remains to be shown that metrics of this nature correlate to improved training performance on real world tasks. Yet, the power of this approach can not be ignored. Not only might it provide an objective measurement of VE design success, but as well, it could lead to new uses for VE systems above and beyond providing training.

## References

Cohn, J. V., DiZio, P., & Lackner, J. R. (2000). Reaching During Virtual Rotation: Context Specific Compensations for Expected Coriolis Forces. *Journal of Neurophysiology* 83(6), p 3230-3240.

Cohn JV, Helmick J, Meyers C & Burns J (2000). Training-Transfer Guidelines for Virtual Environments (VE). *Presented at 22nd Annual Interservice/Industry Training, Simulation and Education Conference, Orlando Fl.*

Cohn, J. V., DiZio, P., & Lackner, J. R. (1995). Reaching errors are made to stationary targets presented in full field moving virtual environments (VE). (From *1995 meeting of the Association for Research in Vision and Ophthalmology Abstracts,*, Abstract No. 1680.628).

Duh, H. B-L. Lin, J.J.W. Kenyon, R.V. Parker, D.E. & Furness, T.A. (2002). Effects of Characteristics of Image Quality in an Immersive Environment. *Presence* 11(3), 324-332.

Gianaros, Peter J.; Muth, Eric R.; Mordkoff, J. Toby; Levine, Max E.; Stern, & Robert M. (2001). A questionnaire for the assessment of the multiple dimensions of motion sickness. *Aviation, Space, & Environmental Medicine* 72(2) 115-119

Gopher D, Weil M, & Bareket T. (1994). Transfer of skill from a computer game trainer to flight. *Human Factors*,36:387-405.

Lackner, J.R., & DiZio, P.(1994). Rapid adaptation to Coriolis force perturbations of arm trajectory. *J. Neurophysiol.*, 72(1): 299-313.

Taylor HL, Lintern G & Koonce JM. (1994). Quasi-transfer as a predictor of transfer from simulator to airplane. *The Journal of General Psychology*, 120:257-276.

Nakamura, S & Shimojo, S. (2000). A slowly moving foreground can capture an observer's self-motion-- a report of a new motion illusion: Inverted vection. *Vision Research.* 40(21) 2915-2923.

# Interaction with Human Models in Virtual Environments

*Manfred Dangelmaier*

Fraunhofer IAO
Nobelstr. 12
70569 Stuttgart
manfred.dangelmaier@iao.fhg.de

*Oliver Stefani*

University of Stuttgart, IAT
Nobelstr. 12
70569 Stuttgart
oliver.stefani@iao.fhg.de

*Angelos Amditis*

Institute of Communication and Computer Systems
National Technical University of Athens
9,Iroon Polytechniou Str.
157 80 Zografos, Athens
angelos@esd.ece.ntua.gr

## Abstract

Three-dimensional human models are used in human factors and engineering to perform ergonomic evaluations of products and workplaces. Commonly they are presented to the user on a two-dimensional screen and are manipulated by keyboards and pointing devices. The evaluation procedures are not standardized. This way of working displays severe drawbacks concerning reliability, efficiency and expenses of the work process and its results. Within the European SAFEGUARD project a study was carried out to test a normative method of evaluating driver seats by human models in a stereoscopic immersive projection room by using spatial interaction devices. In order to provide a better guidance for the user, an electronic checklist and user support system was integrated into the virtual environment. It is concluded that an improvement was achieved, however there is still a potential for optimization, e. g. by introducing motion tracking of the user's body for controlling human models.

## 1    Introduction

Ergonomic quality has become a crucial issue for the success of many products in various markets or market segments, respectively. In parallel, competition leads to an increasing speed in product development cycles, while costs have to be kept to a minimum. These challenges are tackled by the Virtual Engineering approach. It aims at the creation and evaluation of virtual prototypes at early stages of the design process in order to eliminate design errors before they will cause high costs due to repeated work, and costly delays. Such approaches are also applied in the field of ergonomics. Ergonomic models - more precisely anthropometrical human models - such as JACK, SAFEWORK, RAMSIS and ANTHROPOS (e. g. Kraus et al., 1997; Krueger et al., 1997) are in use to perform three-dimensional anthropometrical evaluation and design tasks for products and mobile and stationary workstations. One of the main fields of application is the vehicle industry, using human models particularly to design the packaging of the vehicle or the cabin, respectively (Figure 1).

## 2    Problem

The state of the art for those applications of human models is characterized by the use of graphics workstations with one or more screens. The tools used for the anthropometrical analysis are implemented as supplements to various CAD systems ready to be used by the CAD designer. This way of implementation leads to a number of shortcomings and barriers to a wider spread application of such tools:

- The use of the tools requires both human factors knowledge and CAD skills. This combination is not frequently found on the labour market.
- A special training and a certain amount of experience is needed to be able to work with a particular human model. Human models cannot be operated intuitively.
- Performing a three-dimensional analyses by using two-dimensional displays is difficult and leads frequently to perspective errors.
- Furthermore positioning and animating the human model correctly in the virtual environment is a time-consuming task.

**Figure 1:** RAMSIS applied to evaluate car interior

Finally the procedures of ergonomic evaluation of a workplace - as applied in practice - are neither well defined nor well documented or standardized. This is the reason for evaluation results being poorly reproducible, incomparable and thus being not fully reliable.

## 3    Method

Firstly the use of Virtual Environments which provide a stereoscopic view of both the system and the human model seems a promising solution for the mentioned three-dimensional perception problem. Secondly direct manipulation of the human model's limbs by a three-dimensional "drag-and-drop" control can be expected to reduce training requirements and to lower the barriers to efficient use considerably. Thirdly a normative method for performing ergonomic analyses has to be used to solve the reliability problem.

All components for this approach have been provided during the recent few years. Virtual ANTHROPOS is the implementation of the ANTHOPOS human model for use in stereoscopic and 3D interactive virtual environments (Rössler et al., 1997, Deisinger et al., 2000). It is a general-purpose fully three-dimensional human model with an anthropometrical database allowing for modelling a variety of humans with respect to gender, nationality, body height, and proportions based on data of several international anthropometrical surveys. It provides analysis capabilities for anthropometrical fit, reach, vision, and postural workload. Virtual ANTHROPOS can be used with a stereoscopic projection wall or, even better, with a projection room. PC-based Virtual Reality systems (Bues et al. 2001) will more and more allow for affordable Virtual Environment solutions in Virtual Engineering.

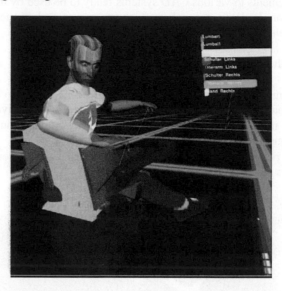

**Figure 2:** Virtual ANTHROPOS in the SAFEGUARD tractor seat

The interaction technology includes head-tacking which is required to create the correct stereoscopic view. Furthermore two tracked flying joysticks (Figure 3) are used to position the reference points and contact points of the human model in a natural and intuitive way by three-dimensional drag-and-drop related to the model geometry to be evaluated. Speech recognition is used as well to enhance the interaction capabilities and to reduce the requirement of complex model operations to a minimum, e. g. by allowing to select even hidden body parts for manipulation with the pointing devices.

A normative method for the evaluation with human models and with real persons is described by Dangelmaier et al. (1995) and Dangelmaier (2001). It defines how to use a human model in the context of an evaluation task within the iterative design process. It addresses all the required steps for a comprehensive analysis providing a checklist for the users. By means of that it both facilitates the usability for non-expert users and a well-defined procedure to improve reproducibility. The suggested procedures are not only suited for the evaluation of a merely virtual environment but can also be applied in (mixed) mock-ups and prototype vehicles.

For the SAFEGUARD (Action for Enhancement of Occupational Safety and Guarding of Health of Professional Drivers) project (QLRT-1999-30235) within the Fifth European Research Framework, the proprietary generic CHECKLIST software tool (Figure 4a) was used to develop and apply a normative checklist for the anthropometrical and cognitive evaluation of the driver seats designed within the project. This checklist tool, running on a PC, can be integrated by means of a remote virtual console into the virtual environment inside a stereo projection room

(Figure 4b). An anthropometrical seat evaluation was carried out by using VIRTUAL Anthropos (Figure 2).

# 4   Results

During the study it became clear that the standardization of the process by using a checklist helps clearly to provide better reproducible results. However, due to manual positioning and manual posture control of the human model, reproducibility is not perfect. Another source of errors is the limited precision of the magnetic tracking system used.

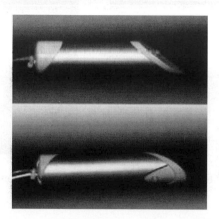

**Figure 3:** Pointing device for three-dimensional interaction

A further technical issue is the time lag in the control loop which includes the operator, the spatial tracking of the interaction devices, the position computation of the environmental and human model, and the rendering process of the images. This delay makes it difficult to use the system and requires some training. On the other hand the spatial drag-and-drop mechanism together with speech control is very easily comprehensible, easy to memorize, and easy to "automate" in terms of behavioural patterns, and also quick to perform. This may reduce the required initial training considerably compared to the customary method. There is also some indication (Dangelmaier, 2001, p. 123) that it might be possible to reduce the required time for an anthropometrical analysis of a driver's place by up to 66%. But this can be only achieved if the posture of the human model can be controlled directly by capturing the body posture of user. Spatial drag-and-drop with pointing devices turned out not to be quick enough.

The stereoscopic view proved to be helpful in better understanding the spatial situation. The possibility to control the view point and viewing direction in the virtual environment intuitively by head-tracking also facilitates usability compared to work with normal CAD workstations. This is in particular the case when the spatial situation is complex and cannot be shown in one single planar view.

# 5 Conclusions

It is concluded that the SAFEGUARD study encourages further research and development work related to the integration of human models in virtual environments and the use of checklist-based user support systems. It is useful to have the latter as an integrative part of the virtual environment. Technical deficiencies such as time lags in the control loop and imprecise tracking should be minimized, e. g. by using optical tracking systems and optimised hardware / software systems. Work with human models could be speeded up by using motion tracking of the user for posture control.

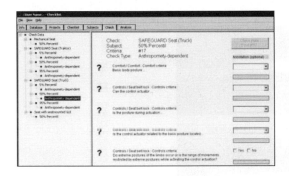 

**Figure 4:** Checklist tool for seat evaluation (a) and representation of a remote console in a stereoscopic projection room (b)

# References

Bues M., Blach R., Stegmaier S., Häfner H., Hoffmann H., & Haselberger F. (2001). Towards a Scalable High Performance Application Platform for Immersive Virtual Environments. In Fröhlich, Deisinger, Bullinger (Eds.), *Immersive Projection Technology and Virtual Environments 2001*, Springer: Wien, New York, 165-174.

Dangelmaier, M. (2001). Ein Verfahren zur ergonomischen Bewertung des Fahrerplatzes von Personenkraftwagen, Heimsheim: Jost-Jetter-Verlag.

Dangelmaier M., Bullinger, H.-J., & Kern, P. (1995). Evaluation of the Ergonomic Quality of Cars. In: *Advances in Industrial Ergonomics and Safety VII*, A.C. Bittner (ed.), P.C. Champney, Amsterdam, New York, Oxford: Taylor and Francis 1049-1056.

Deisinger J., Breining R. , & Rößler A. (2000). ERGONAUT: A tool for ergonomic analysis in virtual environments. In: Mulder J.D. , Liere van R., European Association for Computer Graphics, *EUROGRAPHICS*, Vienna University of Technology, Austrian Academy of Sciences, Virtual Environments 2000, Proceedings of the Eurographics Workshop, Wien: Springer 167-176.

Kraus W., Koos H.; & Lippmann R. (1997). Ergonomische Fahrerplatzanalyse bei MAN: eine Vergleichsstudie mit realen und virtuellen Menschen. In: *ATZ Automobiltechnische Zeitschrift* 99 (3), 156-161.

Krueger W., Seidl A., & Speyer, H. (1997). *Mathematische Nachbildung des Menschen: RAMSIS 3D Softdummy.* FAT-Schriftenreihe *135*, FAT, Frankfurt/M.

Rößler A., Lippmann R. and Bauer, W. (1997) Ergonomiestudien mit virtuellen Menschen. In: *Spektrum der Wissenschaft*, September 1997, Heidelberg.

# Developing a mixed reality co-visiting experience for local and remote museum companions

A. Galani, M. Chalmers
B. Brown, I. MacColl

Computing Science
University of Glasgow
Glasgow, UK
{areti, matthew, barry, ianm}
@dcs.gla.ac.uk

C. Randell

Computer Science
University of Bristol
Bristol, UK
cliff@cs.bris.ac.uk

A. Steed

Computer Science
University College London
London, UK
A.Steed@cs.ucl.ac.uk

## Abstract

This paper focuses on the first stage of the City project that concerns the design of a mixed reality system that may support co-visiting for local and remote museum visitors. We discuss the initial visitor studies, the prototype system and the user trials, with a focus on the role direct interaction and peripheral awareness have in the shaping of the visitor experience. The paper concludes with reflections on the use of the system and future plans.

## 1   Introduction

Social interaction among museum visitors shapes their engagement with the displays and influences their overall museum experience (McManus, 1987). Interaction can be intentional or unintentional, and may happen among members of the same group or among 'strangers'. Museological studies have emphasised the learning arising from social interaction during the visit (Falk & Dierking, 2000) and more recently on how visitors use the social activity around them in making sense of exhibits (vom Lehn, 2002), but social aspects of the visit remain unsupported for remote visitors visiting a museum's web site. Museum web sites often resemble digital brochures or databases, neglecting the need of the remote visitor for deeper engagement with collections and the museum experience (Galani, 2003).

Our research partially responds to the increasing number of remote visitors, which often outstrips the number of visitors to the traditional museum premises (Lord, 1999). It also adds a new dimension to museological research by bridging between local and remote visitors, and exploring the combination of traditional and digital media in the visitor experience. Our project, *City,* is concerned with how sociality may be achieved across different media, and explores new types of social interaction between on-site and off-site visitors. We develop digital spaces and artefacts that correspond to a traditional building or urban space, and the artefacts within it, and support spatial forms of interaction, such as positional awareness and gesture, as well as verbal forms such as talk over a shared audio channel. Our work is set within Equator, an interdisciplinary research collaboration (www.equator.ac.uk), and extends existing mixed reality (MR) (Koleva, Schnadelbach, Benford, & Greenhalgh, 2000) and augmented reality (AR) (Billinghurst, Karo, &

Poupyrev, 2001) research. Our work is also informed by museum–based research such as PARC's SottoVoce (Aoki et al., 2002).

In the next section, we relate our studies of social interaction in two cultural institutions in Glasgow to our system that supports a shared museum visit for a group of local and remote visitors. We then outline the user trials of the system, outline ongoing technological developments, and finally reflect on issues of social participation and engagement in museum MR experiences.

## 2 Elements of co-visiting

To initially investigate the activity of co-visiting in traditional museum settings, we used ethnographic methods such as unobtrusive observations and videotaping. Approximately 60 visitors in friend and family groups were observed in two cultural institutions in Glasgow, The Lighthouse, Centre of Architecture, Design and the City (www.thelighthouse.co.uk) and the House for an Art Lover (www.houseforanartlover.co.uk). The studies focused on visitors' social interaction with their companions and how this shaped their visiting behaviour.

The studies suggested that co-visiting involves the constant collaboration between co–visitors, in that they continually engage with each other and the exhibition, for example by highlighting elements of a display for each other and by mutually managing the pace of the visit. Visitors engage with the displays in the gallery by interacting with them individually and by seeing their friends doing so; their friends' activity informs their future engagement with displays. Similarly, they engage with each other in direct interactions that involve conversation and explicit collaboration, as well as subtle interactions that can be characterised as peripheral awareness. Co-visiting is constituted by continuous and harmonious interweaving of interactions of various degrees of engagement. In that respect, *flexibility* in the transition between and combination of such interactions is essential for a good experience (Cheverst et al. 2000).

Co-visitors both generate and use a range of resources when engaging with their friends and with displays. These resources are, on one hand, *visual* and *verbal cues*, such as orientation, gestural behaviour etc., and on the other hand, *shared content*. Visual and verbal cues may support both explicit communication and implicit personal engagement. The use of the resources does not follow a repetitive pattern but is contextually dependent on the situation or the activity in hand. Often the same resource, for example a pointing gesture, acquires different communicative roles in the course of a single visit. *Shared content* includes elements of the exhibition, e.g. labels, audio guides, displays, as well as people's previous experience. Shared content facilitates peoples' discussions and interpretations of the artefacts. In our studies, museum content was presented via a range of media, such as touch screens, audio guides, leaflets and also objects in collections. Co-visitors used all media inseparably and often made associations between content delivered in different media, e.g. audio narration and label text. They also used shared previous experience—built up during the visit, for example by jointly watching an introductory video, or which had been acquired before the visit in other museums, institutions and interactions.

## 3 The 'City' system

The design of the City system was informed by the visitor studies described above, as well as by technical and interactional goals. The prototype explored co-visiting by people who know each other and share an interest in museum visiting, but who may not always be able to visit together because of geographical separation. The City system was designed for visits to a specific exhibition: the Mackintosh Interpretation Centre (hereinafter 'Mack Room') in The Lighthouse. The Mack Room is devoted to the life and work of C.R. Mackintosh (1868-1928), Glasgow architect and artist. The exhibition combines textual and graphical displays with authentic artefacts, and over 20 screens presenting video and interactive material to visitors.

Figure 1                    Figure 2                                    Figure 3

The City co-visiting system combines virtual environments (VE), hypermedia technology, hand-held devices and ultrasound tracking technology, coordinated through Equator's shared tuple space infrastructure, EQUIP (MacColl, Millard, Randell, & Steed, 2002) to allow three visitors, one on-site and two remote, to visit the Mack Room simultaneously. An ultrasound positioning system and a wireless communications network is installed in the Mack Room and the Lighthouse respectively. The on-site visitor carries a PDA that is location–aware and displays the ongoing positions of all three visitors on a map of the gallery (Figure 1). The two off-site visitors use two different environments: a web-only environment and a VE. The web visitor uses a standard web browser with an applet that displays another gallery map (Figure 2). The VE visitor uses a first person, 3D display with avatars representing the other visitors (Figure 3). All visitors share an audio channel, and wear headphones and microphones. The system also supports multimedia information for the off-site visitors in the form of web pages that are dynamically presented upon movement in the map or VE. This automatic presentation schematically follows the spatial organisation of the exhibition, so that all visitors may 'look' at the same display in the same location. In that respect the system supports interaction around corresponding exhibits in the Mack Room and in digital form: 'hybrid exhibits' (Brown et al., 2003).

## 4    Evaluation and Discussion

A set of user trials of the prototype system was carried out in the Mack Room. The trials aimed to offer us an initial understanding of users' experience of the system in a museum setting, and how this might compare with traditional on-site co-visiting. Qualitative methods such as video and audio recordings and semi-structured interviews were used to analyse the users' experience. The experience was highly interactive and retained many of the elements of traditional museum co-visiting (Brown et al., 2003). The users of the system took advantage of the available shared resources, for example the audio channel and the map/VE, in order to effectively engage with their friends and the exhibition, but also contributed individual information to their interaction.

The trials also showed that direct interaction among the visitors was mainly achieved through verbal communication. Peripheral awareness was mainly facilitated by shared location and orientation, although sometimes one visitor would be on the periphery of talk between two others i.e. overhearing people discussing about an exhibit other than the one he or she was looking at. In that respect, the shared audio was used to compensate for the difficulty or unfamiliarity of gesture or deixis in this hybrid medium. The spatial arrangement of the information informed both the discussions between companions and individual assumptions of what others were viewing. Off-site visitors' direct engagement with the exhibits was achieved through both spatial (map– and VE–based) and informational (link-based) navigation, with a slight preference for the former.

Reflecting on the trials, as well as our design process, we aimed to balance a respect for the behavioural ecology of the setting with our awareness that our technologies cannot and need not replicate the traditional interactions between museum visitors. Our discussions about future design and development can be seen as involving three interlinked aspects of system design and user experience: the duration of use, the media used and the setting.

Firstly, we suggest that future trials run for a longer period of time in order to offer greater insight into the process of appropriation: the development of new forms of interaction to compensate for the lack of traditional ones or to take advantage of new affordances. Each set of trial participants only used the system for one hour, on average, and became relatively familiar with the system's resources for interaction but just began to appropriate them. Secondly, the use of media—especially audio—compensated for the coarse grain of spatial awareness and unfamiliar resources for gesture and deixis, but did not conform to traditional museum 'etiquette'. The PDA visitor spoke more loudly and often than is traditional in museums, and was relatively unaware of the reactions of visitors to the Mack Room who were not part of our trial. Even though the accepted behaviour in a setting often changes over time, we do not wish to promote a style of interaction akin to the (apparently increasing) use of mobile phones inside museums and classrooms. To better fit with this setting of use, we are considering ways to support communicative achievements such as deixis and peripheral awareness of activity through slightly different media. Even simple changes in the type of headphones used by the PDA visitor may make him or her more aware of his or her own audibility, and of other people's conversations and reactions. Lastly, we are aware that we could experiment with the same technology in a different setting. Slight modifications to our system now support its use outdoors in the city streets as well as inside the Mack Room. This lets us compare user experience in the current setting with other settings that may demand slightly different forms of peripheral awareness and engaged interaction.

From a museum studies perspective, MR systems such as the one presented in this paper may have immediate benefits in two interrelated areas: accessibility of collections and educational activities. In the system trials, the combination of location–specific information with participants' talk and motion proved to support visitors' engagement with the collections and each other. The remote visitors' interest in the exhibition was raised by this real–time feedback, and they expressed their desire to visit the traditional gallery space. On-site visitors also benefit from interaction with their on-line friends, e.g. they were often prompted to look at artifacts that they had missed at a first glance. In that respect, both local and remote visitors take advantage of the different perspectives imposed by their different media, and we suggest that maintaining a balance of diverse perspectives is essential. Moreover, the City system can support on-site or off-site guides in offering tours that address the needs and expectations of a range of local, remote and mixed audiences. More generally, MR systems may contribute to communication and collaboration between school and other educational groups. They may also support a rich contextualisation of collections, for example in the use of ethnographic material, by fostering direct communication between visitors and communities of origin.

# 5    Conclusion

This paper has presented the implementation and study of a mixed reality system that allows on-site and on-line visitors to share a museum visit. Informed by studies of visitors to traditional museum settings, we have aimed to support direct or engaged interaction between visitors, mutual and peripheral awareness of activity, and location–specific display of multimedia information. In particular, our work suggests that rich user experiences that span on–site and off–site visitors can be based on mixtures of ubiquitous computing, hypermedia, and map– and VE–based interaction. In our ongoing work, we continue to explore novel combinations of media that support traditional goals such as the mutual reinforcement between visitors' social interaction and their interpretation of cultural institutions, exhibitions and collections.

## Acknowledgements

We acknowledge the contribution made by our Equator colleagues C. Greenhalgh (U. Nottingham) and D. Millard (U. Southampton); the assistance of D. McKay (House for an Art Lover), L. Bennett and S. MacDonald (Lighthouse); the participation of the visitors who took part in our studies; the funding of the UK EPSRC; and Hewlett–Packard's 'Art & Science' donation.

## References

Aoki, P. M., Grinter, R. E., Hurst, A., Szymanski, M. H., Thornton, J. D., & Woodruff, A. (2002). *Sotto Voce: Exploring the Interplay of Conversation and Mobile Audio Spaces.* Proc. ACM CHI 2002.

Billinghurst, M., Karo, H., & Poupyrev, I. (2001). The MagicBook: A transitional AR interface. *Computer Graphics, 25,* 745-753.

Brown, B., MacColl, I., Chalmers, M., Galani, A., Randell, C., & Steed, A. (2003). *Lessons from the Lighthouse: Collaboration in a shared mixed reality system.* To appear in Proc. ACM CHI 2003.

Cheverst, K., Davies, N., Mitchell, K., Friday, A., & Efstratiou, C. (2000). *Developing a Context-aware Electronic Tourist Guide: Some Issues and Experiences.* Proc. ACM CHI 2000.

Falk, J. H., & Dierking, L. D. (2000). *Learning from Museums.* Walnut Creek: Altamira Press.

Galani, A. (2003). Mixed Reality Museum Visits: Using new technologies to support co-visiting for local and remote visitors. *Museological Review, 10,* (to appear).

Koleva, B., Schnadelbach, H., Benford, S., & Greenhalgh, C. (2000). *Traversable Interfaces Between Real and Virtual Worlds.* Proc. ACM CHI 2000.

Lord, M. (1999). Editorial. *Museum International, 51*(4), 3.

MacColl, I., Millard, D., Randell, C., & Steed, A. (2002). *Shared visiting in EQUATOR City.* Proc. ACM Collaborative Virtual Environments 2002, Bonn.

McManus, P. (1987). It's the Company You Keep...: The social determination of learning-related behaviour in a science museum. *Intl. J. Museum Management & Curatorship, 6,* 263-270.

vom Lehn, D. (2002). *Exhibiting Interaction: Conduct and Participation in Museums and Galleries.* Ph.D. thesis, King's College, University of London, UK.

# Development of a Mixed-Mock-Up-Simulator for Work Science Related Studies

*Lorenz Hagenmeyer, Martin Braun, Dieter Spath*

Fraunhofer Institute of Industrial Engineering (IAO)
Nobelstr. 12, 70569 Stuttgart, Germany
Lorenz.Hagenmeyer@iao.fhg.de

## Abstract

The goal of this research was to develop a Mixed-Mock-Up-(MMU)-Simulator on the basis of a common bridge crane as well as to prove that such a device is an appropriate means to integrate the human being with its individual perception, reactions, and behavior into computer based simulations of work systems. First, a brief introduction to work science related problems as well as an outline of virtual reality (VR) technologies and their capacities for work science is given. Then, the development of a MMU-simulator and the concept of an according empirical study are presented. Finally, the results of this study are presented and discussed in respect to current research. Results indicate that such a simulator is an appropriate means to integrate human factors in the computer based simulation of work systems.

## 1    Introduction

Increasing computing power at simultaneous price decline of the hardware promoted the application of computer based simulation techniques in many fields. The main advantage of such techniques is their low cost at high flexibility. Normally, simulated systems can be sufficiently well described by mathematical expressions. Whereas this characteristic represents no problem for most technical systems it does represent one for work systems. In this case a major part of the system is human. Human beings' individual perception, reactions, and behavior are hardly broadly describable by means of mathematic formulations. Consequently, they are not simulatable in most of their characteristics. One approach to make use of the advantages of computer based simulation techniques in work science related studies, is to integrate the human being in a such a simulation. In a so called "Operator-in-the-Loop"-configuration, it should be possible to analyze and optimize work systems as a whole with respect to their efficiency as well as with respect to their robustness against disturbing influences.

Thus, the question arises in which way the human being with its individual characteristics can be integrated in a computer based simulation. One main requirement to an according solution is a sufficient immersion, i.e. the technical capability of the system to supply the human perception channels with appropriate information. VR-technology is highly immersive and therefore seems to be the appropriate tool for such a task. The term *VR* as well as the term virtual environment (VE) constitute an interface for the special multisensorial and time-continuous integration of a user into a computer generated environment (Deisinger 2002). The critical related question in work sciences is to what extend work induced stress can be simulated in such an environment.

The aim of this research is to exemplarily develop a simulator that enables the integration of the human factors described above into a computer based simulation. Such a simulator has to include a highly immersive human-machine-interface and must allow for the assessment of work induced stress of the user by means of a standardized work task.

## 2 Work Science

The aim of work science is to improve the effectiveness of work systems by optimizing the inter-action of working humans, manufacturing resources, and work tasks. Therefore, careful analysis and reorganisation of work systems are conducted with consideration of the capacities of the working human as well as its individual and social needs. Also, work science aims to improve the efficiency and reliability of work systems by optimizing the human-machine-interface (HMI) as well as by influencing the behavior of working humans. Thus, work science aims for human and efficient goals at the same time.

Work science follows an interdisciplinary approach. The analysis, optimization, and evaluation of work systems uses methods of the scientific fields of engineering, occupational medicine, and psychology. Computer based simulations are increasingly often used in work sciences. For in-stance, new car concepts are analysed with respect to ergonomic aspects by means of virtual an-thropometrical human models before a physical mock-up is built. However, the restriction to the simulation of systems that can be sufficiently well described by mathematical algorithms remains.

## 3 Work system simulation on the example of a virtual bridge crane

An empiric study was conducted to determine to what extend work induced stress and strain can be simulated in a VE as well as to clarify in which way task and surrounding related stressors pre-sented in such an environment influence the stress and effectiveness of a working individual. Therefore a MMU-simulator was developed on the basis of a virtual bridge crane. Such a set-up is suitable as the functional range of a bridge crane is restricted and clearly structured.

### 3.1 Experimental Set-up

#### 3.1.1 Mixed-Mock-Up-Simulator

The VE used in this study[1] has been created in the fully immersive, cubic stereo-projection room *HyPi-6* (**Hy**brid **P**ersonal **I**mmersion System featuring **6** walls) of the Fraunhofer IAO. This system can be operated using either a PC-Cluster consisting of 12 PCs (passive projection) or using a graphic super computer of the type SGI Onyx 3400. Here, the active projection method was used where each wall is served by one projector. To provide both eyes with the appropriate part of the stereo picture shutter glasses were used. The computer generated pictures were projected at a solution of 1024x1024 dots on 5 walls of the room. The normally used

**Figure 1**: Screenshot of the virtual shop floor

head-tracking could be omitted in this study because the position of the participant's eyes was defined by the fixed position of the steering seat which will be described in further detail below.

A simulation application was programmed on the basis of the VR-software *Lightning* that had also been developed at IAO. In this application, a real shop floor has been depicted in the VE using a 3D-modelling-tool together with existing CAD-data and photographs for texturisation. A common bridge crane featuring a crab, cable, and load-magnet has been depicted within this shop floor (see

---

[1] based on research supported by the Deutsche Forschungsgemeinschaft (DFG), reference code BU 528-21

Fig. 1). The crane could be traversed in 3 axis. The oscillation behavior of the load on the cable was modeled close to reality by implementing a software-library for physics-simulation and collision detection.

The control of the crane was implemented by means of a common steering-seat of a crane which was fixed in position at the entrance to HyPI-6. The right lever grip controlled movement in x-y-direction (traversing) and the left lever grip controlled movement in z-direction (up-down). The load-magnet could be activated by an additional switch so that the participants could hook on and off the load by themselves (see Fig. 2). The haptic setting of this MMU-cockpit enhances the immersion for the participant.

In order to auralize the simulation, sounds of the surrounding of a real bridge crane were recorded and synchronized to the motor movements in the simulation. The sounds were played-back by two active studio speakers as well as by a subwoofer. Figure 3 illustrates the spatial configuration of the experimental set-up.

**Figure 2:** steering-seat of the MMU-simulator

### 3.1.2 Control task

A control task was developed on the basis of VDI-guideline 2194 (VDI 1995) for the professional training of crane drivers. The task featured 3 subtasks of increasing difficulty. It allowed a continuous performance measurement as well as the application of additional stressors such as impulses of white noise or reduction of visibility. The task included driving around pillars as well as 'hopping' over walls (see Fig. 4). Preliminary tests showed that the crane was very difficult to steer for a non-trained person. Thus, the oscillation of the load was decreased by shortening the cable.

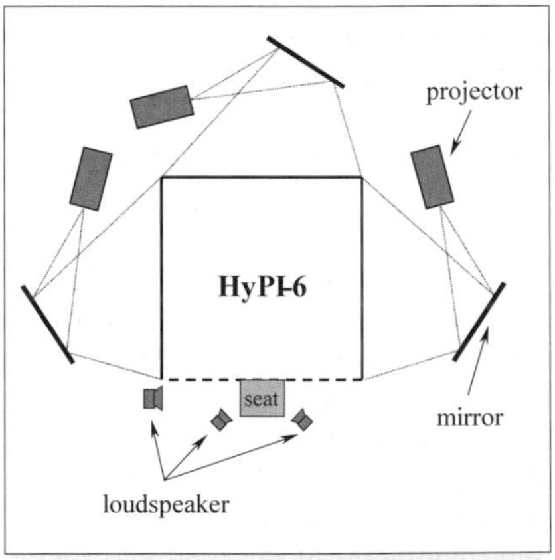

### 3.1.3 Parameters and measuring quantities

The set-up described above was designed to investigate performance and strain of a participant working in the simulated environment. These can be measured by objective and subjective means, where the first ones can not or just slightly be influenced by the participant.

To measure the performance of the participants their speed (defined as accomplished tasks per time) and their error rate were taken into account. The quotient of these 2

**Figure 3:** experimental set up

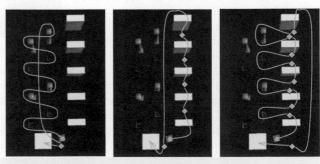

**Figure 4 :** Subtasks of the control task

quantities refers to the work efficiency and hence to the speed accuracy trade off (SATO).
In order to measure and assess strain, psycho-physiological, biochemical and subjective quantities were investigated. The first category included parameters such as the frequency of heart beat and the blood pressure. The second category included the level of different hormones that indicate strain such as Adrenalin, Noradrenalin and Cortisol. These quantities were investigated by collecting 12 blood samples in total which were analyzed in a medical laboratory. The state-scale of the State-Trait-Anxiety-Inventory (Laux, Glanzmann, Schaffner & Spielberger 1981) was used as a subjective quantity of strain. Also, an analog scale and a questionnaire regarding the level of presence (following Schubert et al. 2002) in the VE were employed for this purpose.

## 3.2    Execution of the experiments

At a total testing time per participant of about 3 hours only 2 participants could be tested per day. Therefore, only $N$=24 individuals (20 men, 4 women), aged 20 to 50 (mean age $M$=28), participated at the study. They were divided randomly into two groups. After a detailed introduction, a 5 minute test crane drive and a medical anamnesis a first set of data was collected. Having rested for 10 minutes the first group passed through the test course. After a resting period of 15 minutes they passed through the course again. This time, additional stressors such as strokes of white noise, reduction of visibility and irritating radio transmissions were applied. Fig. 5 gives a detailed depiction of the test procedure as well as the exact time at which quantitiers were measured. The second group was tested by the same procedure but the order of the 2 sessions was reversed, i.e. they first passed the session with additional stressors. A first analysis of the biochemical parameters of the first group showed a strong correlation between these parameters and all of the other ones. As the collection of blood samples is very elaborate it was omitted for the second group, so that the number of participants tested per day could be increased to 4.

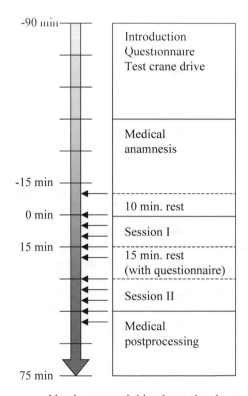

blood pressure & blood sample taken

**Figure 5:** Time table of test procedure

## 3.3    Results and Discussion

The simulator was built in a MMU-configuration with full functionality. Thus, immersion and resulting presence of the participants could be induced. A technical work environment could be represented close to reality. Some participants complained about bad resolution of depth; this is due to a systematic technical deficiency of VEs which is unable to serve all human perception channels. The resulting lack of information or resulting disinformation is known to cause perception problems in VR. Up to date there is no technical solution to that problem.
A method to measure strain was developed and applied under the conditions of the test setting with good success. The single measured values were combined to metavariables corresponding to resting period and test period with and without additional stressors by calculating arithmetic

means. These metavariables were compared by t-tests. The psycho-physiological as well as the biochemical and psychological quantities consistently showed a significant increase from resting periods to working periods thus indicating a significant increase of strain. Hence, work load related stress could simulated close to reality. The results of the presence questionnaire indicated that the participants felt being integrated in the VE well and on a constant level troughout all the tests.

Between the tests with and without additional stressors the parameters did not show a consistent and/or significant change. Thus, it cannot clearly be deducted that additional stress can be induced by applying additional stressors. However, the results of the group with reversed test order (with-without additional stressors) indicate that the effect of the additional stressors may be compensated by habituation and practice related effects. This finding is supported by the data related to the performance of the participants. Independently from the additional stressors, all participants showed a significant and strong increase in performance. Also, a playful trial-and-error approach to the assigned tasks could be observed.

This finding reveals the limitations as well as the possibilities of VR: the participants were always aware of being in a *virtual* environment, i.e. an environment without real consequences of their (technical) actions. Thus, strong emotional reactions such as panic or stress at a performance destroying level are not inducible to healthy persons in such an environment differently to real crane driving situations. Compromising the findings of Hacker (1997) the combination of a close to reality representation of work tasks and a trial-and-error approach to dangerous situations without the fear of real consequences may significantly enhance the learning effects with respect to professional formation and training of behavior in dangerous situations.

## 4  Summary and conclusions

A MMU-simulator was developed in order to investigate if human factors can be integrated into a computer based simulation. A study was conducted on this simulator. It was shown that such a device is suitable for a close to reality simulation of work tasks that include measurable performance and strain-reactions of the working user. Thus, it is proved that the integration of human factors in a computer based simulation is principally possible. However, the limitations as well as specific opportunities of such an integration have to be investigated in further research.

It has to be mentioned that strong emotional reactions can not be induced in VR due to the absence of real consequences of the actions of the user. However, this trait may include new and effective opportunities to professional formation and training for danger prevention.

## References

Deisinger, J (2002). Entwicklung eines hybriden Modelliersystems zur immersiven konzeptionellen Formgestaltung. Heimsheim: Jost-Jetter.

Hacker, W. (1997): Lernen. In: Luczak, H.; Volpert, W. (Eds.), *Handbuch der Arbeitswissenschaft* (pp. 439-443). Stuttgart: Schäffer-Pöschel.

Laux, L.; Glanzmann, P.; Schaffner, P.; Spielberger, C. D. (1981): State-Trait-Angstinventar. Weinheim: Beltz.

Schubert, T. et al. (2002): I-Group Presence Questionnaire (IPQ). I-Group, Internet: http://www.igroup.org. as of 1.5.2002

VDI (1995): Richtlinie VDI 2194 „Auswahl und Ausbildung von Kranführern". Verein Deutscher Ingenieure (VDI). Berlin: Beuth.

# Collaborative City-Planning System based on Augmented Reality

*Hirokazu Kato[1]*  *Keihachiro Tachibana[1]*  *Takeaki Nakajima[1]*
*Yumiko Fukuda[2]*  *Masaaki Tanabe[3]*  *Adrian D. Cheok[4]*

[1]Hiroshima City University
3-4-1 Ozuka-Higashi, Asaminami-ku,
Hiroshima 731-3194 Japan
kato@sys.im.hiroshima-cu.ac.jp
tatibana@sys.im.hiroshima-cu.ac.jp
nakajima@art.hiroshima-cu.ac.jp

[2]Hiroshima Institute of Technology
yfukuda@cc.it-hiroshima.ac.jp
[3]Knack Images Production Center
nachm@lime.ocn.ne.jp
[4]National University of Singapore
adriancheok@nus.edu.sg

## Abstract

In this paper, a collaborative augmented reality system for city planning is described. Miniature models, illustrations and graphical computer displays have been used for the comparison and consideration. Augmented reality technology enables users to consider of city plans more effectively and easily. One important issue of the augmented reality environment is how user can manipulate 3D structures that are displayed as virtual objects. It has to be intuitive and easy so that it may not disturb user's thought. We propose a new direct manipulation method based on the concept called tangible user interface. User holds a transparent cup upside down and can pick up, move or delete a virtual object by using it. From a brief user testing, it was reported that this interface is intuitive and easy to use and it does not prevent users from thinking or communicating with each other.

# 1 Introduction

Recently, many kinds of applications based on augmented reality have been studied (Azuma et al., 2001). Outdoor AR systems or Mobile AR systems are proposed and developed by some researchers in the situation that small but powerful computers and devices are available. However, desktop AR system is a still appropriate application target as a virtual object manipulation interface or face to face CSCW environment and the effectiveness are well known (Kiyokawa et al., 2000).

We have proposed some desktop AR systems so far (Kato et al., 2000). In the development of the systems, we figured out that intuitive interaction can be realized by designing interfaces based on the concept of tangible user interface that are proposed by prof. Ishii in MIT Media Lab (Ishii & Ullmer, 1997).

In this paper, we propose a collaborative city planning system as an application of desktop AR system. In a city planning, it is very important to arrange the layout of 3D objects such as buildings or trees and to see the relationship among them. Since 3D cue is required, consideration by using physical miniature models is a still important process as well as simulations in a 2D computer display.

However it is very hard to change the shape and size of miniature models and also environmental simulations such as sunshine simulation cannot be done with them while their manipulation is very easy. Therefore we developed an AR system that displays virtual objects and allows users to manipulate them easily as if there are real miniature models. Also it enables simulation of lighting conditions.

In this system, the method of virtual object manipulation is the most important issue. The system has to allow user to manipulate virtual objects as easily as the user manipulates real miniature models. We proposed virtual object manipulation interface using 'paddle' before (Kato et al., 2000). The interaction is very intuitive but it is not easy for the user to manipulate virtual objects as he/she wants. In this paper, we propose a new interface for virtual object manipulation using a physical 'cup'.

## 2 Overview

### 2.1 Hardware

This system uses a video see-through Head Mounted Display (HMD). The HMD has two small cameras at each eye position. Size of the color video images that are taken by the cameras is 640x480. These images are input into a PC with Pentium 4 2.0GHz CPU. Pose and position of the HMD is calculated through image processing and virtual objects are overlaid on the images. This 3D graphics processing is done by GeForce4 Ti 4600 GPU. The result is displayed on the left and right LCD panels in the HMD and the user can look at them. There are no other sensors so that the system can be used wherever it is carried. If the computer would be small enough, this system could be a mobile system.

Virtual miniature models for city planning are displayed on the 900x1200[mm] main table. 35 black squares are drawn on the table. These squares are used for the virtual object registration. The side table is used for the selection of virtual objects that can be moved on the main table. 12 black squares are also drawn on the side table for the registration. Figure 1 shows these tables setting. A transparent cup with a black square is used as an interface device for virtual object manipulation (figure. 2). User holds the cup upside down and can pick up, move or delete a virtual object by using it.

**Figure 1:** Tables setting

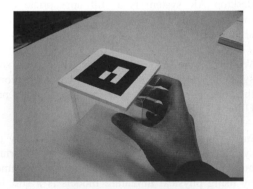

**Figure 2:** Cup interface

## 2.2 Software

In order for the registration using square markers, software library called ARToolKit that we have been developing is used (Kato & Billinghurst, 1999). This library detects square markers in an image and calculates pose and position of the camera from the appearance of the marker in real-time. If there is one marker at least in an image, this calculation can be done. In the task of city planning, user's hand sometimes overlaps a marker because many operations are done by user's hand on the table. Therefore many markers are drawn so that the camera can look at one marker at least. Unique coded pattern is drawn inside each marker. Since this pattern is trained to the computer, each marker can be identified.

The cup interface also has a marker. When there is the cup in a image, the relationship between the cup and the camera can be calculated. If there is a marker on a table in the same image, the relationship between the table and the camera is also calculated and then the relationship between the cup and the table can be calculated. Virtual object manipulations are executed based on either relationship between the cup and the camera or between the cup and the table.

Figure 3 shows an example of planning of a public square with some trees and facilities. The entire virtual public square is displayed on the main table. All virtual objects are represented in VRML format and they are properly rendered with shadow from user's viewpoint (figure 4 and 5). Since the light position can be changed like the sun movement, this system also enables shadow

**Figure 3:** A entire view of public square

**Figure 4:** A close view

**Figure 5:** Another close view from different view point

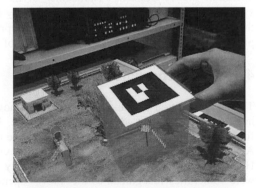

**Figure 6:** A chute picked up by the cup interface

simulation. In addition, user can arrange many kinds of facilities or plants freely by using the cup interface. Figure 6 shows that user is picking up a chute.

# 3 Manipulation Method

One important issue of the system is how user can manipulate virtual objects. It has to be intuitive and easy so that it may not disturb user's thought. We propose a new direct manipulation method based on the concept called tangible user interface. User can manipulate virtual objects directly and also correspondences between user's actions and system's reactions are very natural. The interface device is a simple transparent cup with a marker. User holds it upside down and manipulates virtual objects. The operations assigned to the cup are described below.

**1) Pick up**
When the cup covers a vurtual object on a table, the virtual object locks in the cup. The cup is picked up without slide on the table, and then the virtual object leaves the table and remains in the cup.

**2) Put**
The cup that has a virtual object inside is put on the table so that the virtual object can be arranged in appropriate position and direction. After the cup is picked up again, the virtual object is unlocked from the cup and remains on the table.

**3) Move**
After the cup covers a virtual object on a table, the cup is slid on the table. Then the virtual object moves with the cup and changes its position and direction. After the cup is picked up again, the virtual object remains on the table.

**4) Delete**
When the cup that has a virtual object is shaken, the virtual object disappears.

In addition, there are two modes on 'put' operation. They are 'once mode' and 'repeat mode'. Under the 'once mode', when a virtual object is put on a table, it disappears from the cup. Contrarily under the 'repeat mode', even if a virtual object is put on a table, it remains in the cup. This means that user can repeatedly put the same object on the table. Generally, the existence of operation modes decreases learnability of the interface and increases operation errors. In this system, the cup interface is semi-transparently rendered with different color that corresponds to the each operation mode (figure 6) or appropriate annotations appear on the cup. These ideas prevent mischief that result from the existence of operation mode.

# 4 User Experiences

The demonstration of our system has been shown at several informal meetings. Over 20 people have tried the system and given us feedback. Users had no difficulty with the interface. They found it natural and intuitive to hold and move the physical cup to manipulate the virtual objects. Also they could talk with other people easily even while they were manipulating virtual objects. However some people reported that when they tried to pick up a virtual object, their operations were sometimes recognized as the 'move' operation because the cup accidentally moved a little. Though we have to evaluate the system by strict user testing, the feedback suggested that proposed interaction method is useful and the system is suitable for the collaborative city planning.

# 5 Conclusion

In this paper, we propose a collaborative city planning system as an application of desktop AR system. In the city planning, it is very important to arrange the layout of 3D objects such as buildings and trees and to see the relationship among them. Since 3D cue is required, consideration by using physical miniature models is a still important process as well as simulations in a 2D computer display

However it is very hard to change the shape and size of miniature models and also environmental simulations such as sunshine simulation cannot be done with them while their manipulation is very easy. Therefore we developed an AR system that displays virtual objects and allows users to manipulate them easily as if there are real miniature models. Also it enables simulation of lighting conditions.

One important issue of the system is how users can manipulate 3D structures that are displayed as virtual objects. It have to be intuitive and easy so that it may not disturb thought of users. We proposed a direct manipulation method based on the concept called tangible user interface. User holds a transparent cup upside down and can pick up, move and erase virtual objects by using it. From a brief user testing, it was reported that this interface is intuitive and easy to use and it does not prevent users from thinking or communicating with each other.

In future issues, we will consider of more complex manipulation method than arranging virtual objects. The effectiveness of the system and the interaction method will also have to be evaluated by strict user testing.

## References

Azuma, R., Baillot, Y., Behringer, R., Feiner S., Julier S., & MacIntyre B. (2001). Recent Advances in Augmented Reality, IEEE Computer Graphics and Applications, 21, 6, 34-47.

Ishii, H., & Ullmer, B. (1997). Tangible Bits: Towards Seamless Interfaces between People, Bits and Atoms, Proc. of CHI 97, 234-241.

Kato, H., & Billinghurst, M. (1999). Marker Tracking and HMD Calibration for a Video-based Augmented Reality Conferencing System, Proc. of 2nd Int. Workshop on Augmented Reality, 85-94.

Kato, H., Billinghurst, M., Poupyrev, I., Imamoto, K., & Tachibana, K. (2000). Virtual Object Manipulation on a Table-Top AR Environment, Proc. of IEEE and ACM International Symposium on Augmented Reality 2000, 111-119.

Kiyokawa, K., Takemura, H., & Yokoya, N. (2000). SeamlessDesign for 3D Object Creation, IEEE MultiMedia, ICMCS '99 Special Issue, 7, 1, 22-33.

# No Silver Bullet but Basic Rules
## User Interface Design for Virtual Environments

*Christian Knöpfle*

Fraunhofer-IGD
Fraunhoferstraße 5
64283 Darmstadt / Germany
knoepfle@igd.fhg.de

## Abstract

In this paper we will present a classification scheme for applications, the 4W-model, which helps to derive requirements for the design of a VR user interface for a given application type. Furthermore we will present several basic rules for the design of a VR based interaction technique. Finally we outline how the 4W-model is used on the application area "Design Review" and how the software systems are designed.

## 1    Introduction

Each year, new ideas for the design of virtual reality user interfaces are presented and compared to existing approaches. This shows us, that this thematic area is still very vivid. But it gives us also the impression, that already very good approaches exist, which are ready for practical use.

But when we look at industry, especially the automotive, which was and still is the driving force of Virtual Reality, we can observe that only a very small amount of the so-called VR applications are interactive and make use of the developed ideas. VR is reduced to real-time rendering and stereo-viewing. Some do tracking, but it is not really essential for most applications. In fact, people from industry tried a number of interactive research prototypes but nearly all of them failed in convincing them, that this kind of interactivity helps them in their every-day work. The major reason was that for a long time research people (including ourselves) ignored the fact, that there is a very huge difference between "doing a demo" and "work with the system". The open question is now: What does that mean for interface design?

In this paper we will present a high-level approach to the design of a VR user interface and describe several basic rules, which interface designers should adhere to. We will also point out that the knowledge we gained from 2D-GUI design (WIMP interfaces in general) is not simply mapable on VR interfaces. The following guidelines are mainly derived from our 6 years VR experience with the automotive industry and supported by publications of other authors.

## 2    The 4W-Model

To design of a VR user interface mainly depends on several factors: functional requirements of the application, users, place, etc.. Thus the first step is to work out all requirements, which will influence the design. Based on these requirements, the interaction techniques can be determined.

For a structured approach, we developed the 4W-Model[1], which embraces four thematic areas, each characterized by a simple question. It is important to note that these areas are tightly linked together and heavily influence each other, as shown in the following image.

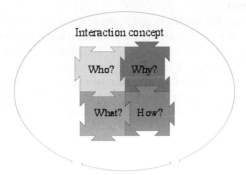

**Figure 1: The 4W-Model**

The 4W-Model should help VR developers to characterize the application on an abstract level and to work out the parameters, which are important for the design of the UI. Then he can evaluate his UI and check if all requirements will be met. It is important to note that the 4W model is not a guide, where everything is done automatically. It still needs a lot of brainwork, but it helps to approach the problem in a structured way.

In the following chapters the four "W" will be described in more detail and on the basis of a sample application we will show how one can use the 4W-model.

## 2.1   Who? – The user

The first step is to characterize the typical user of the system. Different aspects are important, some of them are outlined below:

- The *experience* of the user. For example the spacemouse is often used as input device in the CAD and VR domain. The big difference: CAD-users control the scene with the spacemouse, while VR users move the camera. Experience and mental model are different and for both user groups it is pretty hard to learn the other navigation model. Thus "experience" has to be understood in a more integral way and should not be limited to VR experience.

- If the *social position* of the user plays an important role (e.g. high level management), he will not use the system, if there is a chance, that he do not understand it and make a fool of himself. Thus the system as to be as simple as possible without complex mental models. An approach for a simple navigation interface was presented by [??], where a small model of a real car was used to move the virtual car around. The amount of time spend on the training of such a system is heavily linked to the maximum complexity of the system

- One should also think about the *physiognomy* as an important issue for the design of the user interface. Most users will be "normal humans" without special abilities, but there important differences, e.g. left and right handed people.

---

[1] for those who wonder, why the model is called "4W". In our native language german all questions start with a "W": Wer (Who), Wozu (Why), Was (How), Wie (What). Unfortunately this is not true for the English language.

The outcome of the characterization is now mapped on parameters of our system. Common factors influenced by it are system model (Norman, 1990), hardware, system control.

## 2.2 Why? – Intention

The second step is to find out his intention and motive to use the system, i.e. if he wants to use it for playing, exploring, investigating, learning or working. Of course, a mixture of this motives is possible (e.g. edutainment – playing and learning). Nonetheless a simple characterization is not sufficient. To go into more detail, it is a good approach to take the question "How?" into consideration. In the case a user wants to work with the system, it is important to know if he is just looking at the virtual objects and make decision on a *gut level*, or if he needs a huge number of functions to investigate the virtual model and then decide on *technical level*. In the latter case case the system needs an efficient menu system to offer all relevant functions to the user.

## 2.3 How? – Procedure

There are many work processes which have proofed themselves. Now, industrial companies want to use Virtual Reality in some of these work processes. It is obvious that the work procedures will change and that the people have to adopt to the new ones. It is important now (and worth to do it) to investigate the old procedure and figure out there advantages, disadvantages and how to *transfer* the best of it into the VR world.

Work procedures where VR is or should be introduced (i.e. digital prototypes for automotive industry), communication plays an important role, because several people have to work together o find solutions to problems. The old-style procedures supported communication fairly well, because people had pen, paper and a real prototype of a car they had to talk about. Thus it is important for a VR GUI designer to have a closer look on the verbal and non-verbal communication and the media, which is used to support the communication between the participants. Some interesting thoughts about media can be found in (Perry, 1998).

The time the system is operated by the users is also essential for the user interface. A UI designer should be aware of the fact that people can work for four to eight hours continuously with the system and that fatigue is an issue. Consequently heavy input devices, which have to be carried around by the user, or interaction technique, which require large movements of the body limbs should be avoided.

When investigating the work procedure, functional requirements can be derived too, e.g. precision of input devices, parts of the virtual scene , which should be selectable (faces, points on surface, points in free space, objects,).

## 2.4 What? - Content

For the selection of interaction techniques it is important to know the type of content presented in the VR application. Important parameters are the size of the scenery, number and sizes of single objects and the object type (surface, volume, simulation data)

## 3 Basic Rules for the Design of VR User Interfaces

When designing a single interaction technique or a complete interface for a VR application, several basic rules should be adhered to. Otherwise they might fail, when used in an industrial context In the following we will just list a few of them:

- **BIT:** It is commonly agreed, that VR user interfaces can be mapped onto the 4 basic interaction tasks (BIT) position/orientation, selection, quantification and text, introduced by (Foley et. al., 1990). Taking into account the action model of Norman (Norman, 1990), we can conclude that interface designers should try to identify a single interaction technique per BIT and build the whole interface based on this techniques.

This will help the users a lot in understanding the concept of the interface. Another advantage of the BIT concept is the simplified replacement of the interaction devices (e.g. from flystick to mouse and keyboard).

- **Make use of proprioception:** The knowledge on the position of one's extremities without explicitly looking at them is a powerful tool for the UI (M. Mine, 1997).
- **Minimal movement of users limbs**, especially when used continuously over a long time (fatigue).
- **Hardware and user jitter:** If people interact in free space, they will not be able to hold an interaction device at a specific position over a long time
- **Accuracy of interaction:** When interacting in free space, accuracy of limb movement is much lower than on a table top

# 4    Sample application: The VR Design Review

In this chapter we will describe how the 4W-model can be used. First we will shortly describe the application area itself. Then we show how the 4W-model is applied and finally we present the software system "VR-Design Review".

## 4.1    Application Area Design Review

There is a big trend in automotive industry towards digital prototyping. The goal is to avoid time and cost intensive physical prototypes and use virtual prototypes instead, because there are cheaper and faster to build. In theory this approach sounds simple and appropriate: A new car is modeled in a CAD system anyway, thus visualizing it should be straightforward. But looking at a prototype is not the only "function" people need. They have to measure distances, walk around the prototype, sketch on it, etc. On a real prototype, the procedures to do this were obvious and simple media could be used to carry them out (pen, tape measure). Since VR owns the attribute "natural and intuitive user interfaces", VR was introduced in this domain.

The prototypes are often used during design reviews, where various experts meet and evaluate the current design, try to find flaws and solutions. Since only virtual prototypes are available, a software system with a VR based user interface has to offer functions to the participants of the design review, which allows them to carry out their work properly. Besides, today Design Review is the major application in industry were virtual prototypes are used.

## 4.2    The 4W-model in action

Now we will classify the application "Design Review" with the 4W-Model. Due to space constraints we will do it only on a very rough level:

- **User:** The users do not have VR experience. Most of them are coming from the CAD modeling, thus they have experience with computers. CAD is their main work environment and they do not have time to get extensively trained in the use of the system.
- **Intention:** The participants have to "work" with the system. They will do their decision on extensive investigation. Therefore the system will offer numerous functions to support them in their work
- **Procedure:** The users will use the system for hours (fatigue), but not necessarily on an everyday basis (learning issue). Furthermore people have to work cooperatively at a single place.
- **Content:** In the case of automotive industry mainly model of cars, sometimes the whole car, sometimes only the engine bay.

Putting it all together, VR design review can be characterized as "Experts of different domains work together on virtual models to find solutions to problems".

### 4.3 The realisation: VR-Design Review

Some time ago we developed system for the VR Design Review with a focus on functionality (Knöpfle, 2000) and also a system with a focus on communication (Ehnes, 2001). The 4W-model and the basic rules were mainly derived and extracted from these two software systems. Due to space constraints we would like to point you to these papers.

## 5 Conclusion and Future Work

Today the 4W-model is an abstract but structured approach to user interface design for VR applications. If more and more application types are investigated using the 4W-model, we finally can come to a point where most application types will be classified and most classification parameters will be discovered . Then a style guide could be derived, maybe organized as a flow diagram, where interface designer has to answer questions and finally get the right set of interaction techniques.

Until then, the basic rules catalog should be extended.

### References

Ehnes, J., Knöpfle, C., Unbescheiden, M. (2001): „The Pen and Paper Paradigm - Supporting Multiple Users on the Virtual Table", in Proceedings of IEEE-VR 2001, Yokohama, Japan

Foley, J., van Dam, A., Feiner, S., Hughes, J. (1990): „Computer Graphics: Principles and Practice", Addison Welsey, Second Edition

Knöpfle, C., Voß, G.: "An intuitive VR user interface for Design Review", in Proceedings of Advanced Visual Interfaces 2000

Norman, D. (1990): „The design of everyday things", Currency Doubleday

Mine, M. (1997): „Moving objects in space: Exploiting proprioception in virtual environment interaction", in Proceedings of SIGGRAPH 1997

Perry, M., Sanderson, D. (1998): „Coordinating joint design work: the role of communication and artefacts", in „Design Studies", pp. 273-288, Elsevier Science Ltd.

# Comparison of Hand- and Wand Related Navigation in Virtual Environments

*Mikko Laakso*

Researcher in Telecommunications Software and Multimedia Laboratory
Helsinki University of Technology, P.O. Box 5400, FIN-02015 HUT, Finland
mlaakso@tml.hut.fi

## Abstract

In this paper, we describe the development of a new virtual environment navigation system and report the results of an experiment comparing two different input devices and navigation techniques. The input devices were our custom-made wand a data glove. The study consisted of seven different navigation tasks in virtual worlds and the evaluation was based on speed & accuracy and the ease of use. The experiment showed no significant differences in navigation performance between the two tools but proved that both of them are good input devices for navigation in a virtual environment.

## 1 Introduction

Virtual reality (VR) and virtual environments (VE) are becoming popular as media for example training, modeling and entertainment. The applications being developed for VEs run a wide spectrum. This paper concentrates on navigation in architectural environments; movement in worlds that are very similar to our own. Navigation in a virtual world should be practical, intuitive and simple. Unfortunately, it is very often far from that – for some reason the usability issues in VEs have been usually left with little attention. This situation must change, navigating through virtual worlds can no longer be considered a task reserved for experts only.

The purpose of this paper is to introduce a study in which two different navigation devices, a custom-made wand and data glove, were compared in VE navigation. The evaluation was based on speed & accuracy and the ease of use. The evaluation study consisted of seven different navigation tasks in virtual worlds. The purpose was to see how well they suit for VE navigation and find out if there emerges differences in navigation performance between the two tools.

## 2 Related Work

Many researchers have dealt with the open questions of navigation in immersive virtual worlds. For example, (Bowman, 2000) and (Bowman et al., 2001) have tested different virtual metaphors for their effect on a user's spatial orientation. (Kruijff, 2000) and (Darken & Goerger, 1999) have investigated and discussed wayfinding support and the use of VEs for navigation training. The results indicate generally that, if proper strategies are used, a technique giving the user more control produces better spatial orientation. However, one must not forget the cognitive load a technique induces – it must not be too complicated. (Bowman et al, 2001) states that the navigation methods should be evaluated according to specific quality factors rather than to their suitability for particular applications, as the requirements of applications may be very different. More research is needed to better identify the factors affecting to efficient VE navigation.

# 3 Methods

Our research group at Telecommunications Software and Multimedia Laboratory at Helsinki University of Technology (HUT) has since 1997 studied VR systems. We have built a full-scale virtual room called Experimental Virtual Environment (EVE) (Jalkanen, 2000), which is shown in Figure 1 (a) (see also *http://eve.hut.fi*). It is a model example of the typical immersive surround-screen VE system. The stereoscopic images are reflected into big screens with projectors and the users are watching these images with stereo shutter glasses. In addition, a 6 degree-of-freedom tracking device produces location information needed for computing the view perspectives etc.

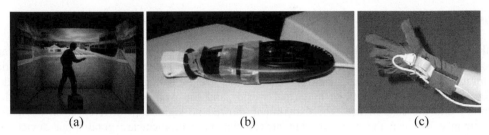

| (a) | (b) | (c) |

**Figure 1 (from left):** User exploring a virtual town in EVE , our custom wand and the data glove.

## 3.1 New Navigation System

In the spring 2000 a new research project called BS-VE (Gröhn et al., 2001) (Mantere, 2001) (Laakso, 2001) was started. It was aimed to develop methods and techniques for visualizing building services systems in 3D virtual reality environment. One objective of the project was to develop practical navigation methods that would allow users to navigate easily in virtual architectural environments. This required the adoption and testing of different interaction equipment and methods. The pilot study described in (Laakso & Repokari, 2002) showed that conventional input devices are not the best possible solution to be used in VE. But there is a number of specialized types of hardware devices that have been used in VR applications. For the second navigation study we chose to test out two different input devices: wand and data glove.

### 3.1.1 Wand

There were essentially two alternatives to get a wand: either to purchase it or build it ourselves. Commercial alternatives were naturally considered, but we decided to test what we can do with our existing equipment. The main requirements for a wand are some kind of a tracking device and some buttons. Both of these were already available in the EVE (MotionStar Tracker and wireless Logitech Mouse), so it was possible to construct a brand new wand variant just by combining these into one unit. Figure 1 (b) shows the current version of our custom-made wand.

### 3.1.2 Data Glove

The second input device studied was a data glove. Hand movement tracking combined with gesture recognition offers a natural and intuitive way for users to input information, for example navigational commands. We acquired two 5DT Data Gloves (left and right) from Fifth Dimension Technologies. To be able track the position of the hand, we attached one magnetic tracker receiver into the back of the glove with Velcro stickers. Figure 1 (c) shows the "updated" glove.

**Navigation techniques:** After checking several possible solutions for travel techniques we finally determined that the most natural way of navigating with the wand is probably the throttle/steering metaphor. In this technique the user pushes a button in the wand and moves the wand to the direction he wishes to move. Also the rotation can be handled simply by pressing another button and rotating the wand to the desired direction. The same metaphor can be used with the data glove also. Pressing the button can be substituted with some specific gesture, we used the closed hand (fist). For different movement requirements our navigation system has four different movement modes, namely WALK, DRIVE, FLY and HOVER. Some of them restrict some user movements.

## 3.2 Evaluation Study

To determine how practical these solutions were, an evaluation study was conducted to evaluate different users' navigation abilities with different tools. The purpose was to find out if there emerges differences in navigation performance between the two tools. The age of the twelve test subjects varied between 23 and 47 and they had immersive VE experience ranging from none to extensive. Before the actual test the subjects had the possibility to get familiar with the EVE. Each user completed seven trials. To be able to compare the results with our previous test the trials were the same as in our pilot test (Laakso & Repokari, 2002). The test trials included all the universal user tasks for navigation, namely exploration, search (naive/primed), maneuvering and specified trajectory movement (Bowman, 2000) (Kruijff, 2000). One of the main points was also that the trials had to be fun and feel quite realistic (some pictures shown in Figure 2). The test trials were:

- **Test 1 - The Tunnel:** The first test trial was navigation in a narrow and twisty tunnel. The task was to fly through the tunnel to the other end. This was not that easy, completing the test with little or no collisions required continuous attention.
- **Test 2 - Downtown Cruise:** The second task was a primed search in a small town. The user's starting point was at the edge of town and the task was to find a way (with DRIVE-mode) to a big statue (could be seen from far away) located in the downtown.
- **Test 3 - Search the House:** The third test trial was a general exploration task followed with a primed search. After examining a small detached house freely for 3 minutes the user was restored back to outside and was asked to find the bathroom.
- **Test 4 - Examine the Car:** The fourth test trial consisted of maneuvering. The user was standing next to a car and the task was to examine the car's (very small) license plate to see in which state it was registered on (it had a USA-style license plate).
- **Test 5 - Relocate the Car:** In the fifth trial, the user was again standing next to the car. This time the task consisted of specified trajectory movement followed with a primed search. The user was driven away (manually by the evaluator) from the starting point to a distant location and the task was to find back to the starting place.
- **Test 6 - Find the Visitor:** The sixth test trial took place in a town hall. Before the test, the user was shown a picture of a space alien, which was hiding somewhere in the building. The task was to find it; this time the user was allowed to travel through walls.
- **Test 7 - Catch the Plane:** In the last test trial the task was to catch a moving target (a plane). When the test started, the user could see the plane crossing the screen with high speed. The user was required to start to follow the plane and eventually catch it.

**Figure 2 (from left):** Screen shots from Test 1, Test 2, Test 3, Test 5, Test 6 and Test 7.

# 4 Results

In the study both qualitative and quantitative analysis were made. The time to complete each task was measured and the user routes or paths to complete the tasks were recorded, i.e. the user's location and orientation were continuously stored on disk. From this data it was possible to calculate several average values. Table 1 shows the completion times, traveled distances (in "pixels") and average speeds ("pixels" per second) for each test. Subject behavior was also recorded in written notes documenting the observations made by the evaluator and the comments made by the subjects during trials. All the subjects were interviewed with a semi-structured interview immediately after the tasks. This interview consisted of questions about the ease of use and how the subjects experienced the input devices as navigation tools in a virtual environment.

| | Wand | Glove |
|---|---|---|
| Test 1: The Tunnel | Completion time: 0.45,3<br>Distance traveled: 306,0 pixels<br>Average speed:　6,9 p/s | Completion time: 0.36,0<br>Distance traveled: 270,8 pixels<br>Average speed:　7,9 p/s |
| Test 2: Downtown Cruise | Completion time: 2.03,8<br>Distance traveled: 1683,5 pixels<br>Average speed:　14,2 p/s | Completion time: 1.32,6<br>Distance traveled: 1479,8 pixels<br>Average speed:　19,1 p/s |
| Test 3: Search the House | Completion time: 0.42,7<br>Distance traveled: 270,3 p. (in 3 min.)<br>Average speed:　1,5 p/s　(in 3 min.) | Completion time: 0.36,0<br>Distance traveled: 366,5 p. (in 3 min.)<br>Average speed:　2,0 p/s　(in 3 min.) |
| Test 4: Examine the Car | Completion time: 0.30,3<br>Distance traveled: 28,9 pixels<br>Average speed:　1,1 p/s | Completion time: 0,46,0<br>Distance traveled: 103,0 pixels<br>Average speed:　2,0 p/s |
| Test 5: Relocate the Car | Completion time: 3.09,2<br>Distance traveled: 1459,4 pixels<br>Average speed:　8,1 p/s | Completion time: 2.41,4<br>Distance traveled: 1580,9 pixels<br>Average speed:　10,3 p/s |
| Test 6: Find the Visitor | Completion time: 2.45,9<br>Distance traveled: 374,0 pixels<br>Average speed:　3,2 p/s | Completion time: 2.29,4<br>Distance traveled: 425,0 pixels<br>Average speed:　2,7 p/s |
| Test 7: Catch the Plane | Completion time: 0.16,8<br>Distance traveled: 1740,7 pixels<br>Average speed:　99,7 p/s | Completion time: 0.17,9<br>Distance traveled: 2594,6 pixels<br>Average speed:　119,2 p/s |

**Table 1:** Summary of measured average values by test and tool.

# 5 Conclusions and Discussion

The results indicate that both wand and glove are very good tools for general navigation. Glove would seem to be a little better in most cases, but statistical analysis on the data shows no significant differences. The comments and judgements made by the users support the results. Both wand and glove were popular tools. They were said to be easy to learn and use, since the wand had just two logical buttons and glove none. These two navigation tools feel natural because they are based on actual motion. It is an intuitive thought that movement results in movement. There is no unnecessary memory load, the user just moves the device to the direction he wishes to move. This is true especially with the glove: there is absolutely nothing extra to confuse the user. Not even any buttons, just fist and motion. Perhaps the only problem with the glove with the selected travel technique is that turning in place is not easy. In fact, it is almost impossible. This caused some problems to some users especially in trials that took place inside of buildings. They went accidentally through walls while just trying to turn. Also, some hardware restrictions emerged: the cable from the EVE's auxiliary PC to the glove was a bit too short. In some cases the users had to back up to be able to gather some speed again. A wireless system would obviously be better.

The test results show that the main objective was achieved: our navigation system is a flexible and practical tool for navigating in virtual architectural environments. Especially the kind of navigation that consists of continuous movement and rotation requires a very flexible travel technique and an input device that is easy to maneuver. Both wand and glove seem to fulfil these demands rather well. The most important goal was to make the navigation so intuitive that we can give the tool to any arbitrary visitor who comes to EVE and after 15 second explanation he or she can start navigating around the models without guidance. This goal was reached, especially the "fist-interface" in glove is almost self-explanatory. However, the wand is easier to circulate between multiple users than the glove, which requires a separate fitting/removing period. But the system is not perfect yet. For example, most of the users seemed to use the wand with both hands, although we had assumed that it would be quite easy to handle with just one hand. It is clear that the device needs some product development to make it more practical.

The conclusion is that the navigation in a virtual environment can be managed with both tools. Both wand and glove are good input devices and the travel techniques bound to them seem to be practical and easy to learn. However, they are different types of devices. Wand is an easy and cheap solution, since it is actually just an extended mouse. Glove is a more complicated device and has wider range of possibilities. An interesting question would be if it is possible to use wand for navigation and glove for other interaction in a virtual environment. More research is needed.

# 6   References

Bowman, D. (2000). Travel Techniques. *SIGGRAPH 2000 Course 36 – 3D User Interface Design: Fundamental Techniques, Theory and Practice,* SIGGRAPH 2000, New Orleans, USA.

Bowman, D., Johnson, D. and Hodges, L. (2001). Testbed Evaluation of Virtual Environments Interaction Techniques. *Presence: Teleoperators and Virtual Environment,* 10(1), 75-95.

Darken, R. and Goerger, S. (1999). The Transfer of Strategies from Virtual to Real Environments: An explanation for Performance Differences? *Proceedings of Virtual Worlds and Simulation '99,* 159-164.

Gröhn, M., Laakso, M., Mantere, M. and Takala T. (2001). 3D Visualization of Building Services in Virtual Environment. *Proceedings of IS&T/SPIE 13th International Symposium on Electronic Imaging (Conference 4297B),* San Jose, USA.

Jalkanen, J. (2000). Building a spatially immersive display - HUTCAVE. *Licenciate Thesis,* Helsinki University of Technology, Espoo, Finland.

Kruijff, E. (2000). Wayfinding, *SIGGRAPH 2000 Course 36 – 3D User Interface Design: Fundamental Techniques, Theory and Practice,* SIGGRAPH 2000, New Orleans, USA.

Laakso, M. (2001). Practical Navigation in Virtual Architectural Environments. *Master's Thesis,* Helsinki University of Technology, Espoo, Finland.

Laakso, M. and Repokari, L. (2002). Mouse Related Navigation in Virtual Environment - a Pilot Study. *SAICSIT 2002 Post-Graduate Research Symposium,* Port Elizabeth, South Africa.

Mantere, M. (2001). Visualization of Flow Data in Photo-realistic Virtual Environment. *Master's Thesis,* Helsinki University of Technology, Espoo, Finland.

# Phobia Treatment Using a Virtual Reality System

*Miguel Leitão*

ISEP / INESC Porto
R. Dr. Ant. Bernardino de Almeida, 431
4200-072 Porto, Portugal
jml@dee.isep.ipp.pt

*Vítor Cunha*

ISEP
R. Dr. Ant. Bernardino de Almeida, 431
4200-072 Porto, Portugal
vmc@dee.isep.ipp.pt

## Abstract

A phobia can be described as a persistent and recognized fear of a circumscribed stimulus and consequent avoidance of that stimulus. A typical phobia, known as acrophobia, occurs when the feared stimulus is heights.

The most common and most successful treatment for phobia illnesses is the graded exposure in-vivo. In this method, the phobic patient is exposed to a sequence of the feared stimuli. It is known that the correct intensity of the feared stimuli will lead to habituation and to the decrease of the fear. This method can present some important difficulties. A promising alternative is the graded exposure to virtual stimuli, using Virtual Reality environments. Virtual reality exposure has several advantages over exposure in vivo. The treatment can be conducted, privately, in the therapist office rather than in outside, where the real phobic situations are found. The treatment can then better assisted and more cost-effective. Also, virtual stimuli are easier to control, so the therapist can nicely grade the fear intensity.

This paper describes the Virtual Reality system (VRFobias) developed at ISEP and tested at the University of Minho. The main requisites of a virtual reality system, needed to the use in phobia therapy, are described. The results obtained so far and the new ideas aimed to the control of the fear intensity are also presented.

## 1    Introduction

Virtual Reality (VR) is a powerful interaction paradigm that uses computer representations to create people's sense of presence, so that they feel themselves in a virtual environment. The software nature of these representations enables virtual environments to be precisely manipulated in a way that is not possible in real world.

A VR interaction can only be achieved with dedicated hardware such as a Head Mounted Display (HMD) and other interaction devices like gloves or 3D mousse. Usually the HMD integrates a position sensor (tracker) that continuously reports the head position to the sensory generators. The visual sensation is obtained through the visualization of images generated in real time by the rendering process. This rendering process retrieves the state information from the database, specially the position of camera and other objects, and it generates according images. Usually, this process relies on dedicated hardware and specialised software libraries to produce images with fair quality in the available time slot. Several advanced and well-known techniques are used to speed up the image generation process. These techniques include hierarchical organisation of objects, distance clipping, view frustum culling, and level-of-detail selection. Due to the dominant role of vision in human affairs, this process has a determinant importance in the sensation of presence in the virtual environment (Slater, Usah & Steed, 1994).

## 1.1 Exposure Therapy

The most common and most successful treatment for phobia illnesses is the graded exposure in-vivo. Exposure therapy is based on the principle that people can get used to discomfort situations. In this method, the phobic patient is exposed to a sequence of the feared stimuli. It is known that the correct intensity of the feared stimuli will lead to habituation and to the decrease of the fear (Emmelkamp & Felten, 1985). This habituation process occurs naturally in over 95% of people.

Exposure consists of facing the feared situations gradually and progressively. It is important to stay facing the situation until the discomfort decreases. Usually, the therapist elaborates an individualized sequence of the feared situations and both the therapist and the patient are required to go to the local where the feared situation occurs.

This method can present some important difficulties. The feared stimuli can be hard to get (like animals), expensive (like plains). Also, the feared stimuli can be difficult to manage as required to obtain a precisely graded fear. In situations where the feared stimuli can only be present outside the therapist office, it can be very difficult to monitor the required psycho-physiologic variables like CRT or heart rate, to help the therapist to understand the fear level and to control of the exposure process.

Exposure may be done in vivo (in real life) or in imagination. In vivo exposure is usually more effective than imaginal exposure, but the imaginal exposure is sometimes suggested as an alternative, especially if the feared stimuli are difficult to obtain. This can be achieved using photos, images, records, books or any objects that recreate the fear.

## 1.2 VR exposure

VR exposure can be considered a new approach of imaginal exposure therapy, where the imagined environment is recreated form the real world using a Virtual Reality system.

In a VR exposure therapy, the patient is faced against virtual environments that are known to produce the undesired fear. The fears stimuli existent in the virtual environment are easier to control than the real world counter parts. This way, the therapist can nicely grade the fear intensity. Virtual reality exposure has also some other several advantages over exposure in vivo. The treatment can be conducted, privately, in the therapist office rather than in outside, where the real phobic situations are found. The treatment can then better assisted and more cost-effective.

Several researches, recently conducted, probed the functionality of this alternative. These researches address different phobic problems like the fear of spiders (Carlin, Hoffman, & Weghorst, 1997), fear of heights (Rothbaum et al., 1995), claustrophobia (Alcaniz el al., 1998) and fear of flying (Banos et al., 2002).

Most of these works usually try to reproduce the standard treatment method, but with a virtual exposure in a simulated environment. All the potential and versatility of the virtual reality environment is not explored at all. We expect that better results can be achieved if new methods can be defined for grading the feared stimuli.

# 2    Implementation

The VR system developed to test the virtual exposure therapy (VRFobias) is based on a SGI Octane MXE workstation that holds the scene database with descriptions of all objects within the virtual world. This workstation also runs the rendering process that generates the view presented to the phobic patient.

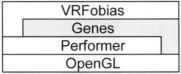

**Figure 1:** VRFobias development.

VRFobias was developed using Genes, as can be seen in figure 1. Genes is a house made Generic Environment Simulator. It is made on top of Performer, a well-known visual simulation software from Silicon Graphics aimed to visual simulation.

Genes allows real-time image and sound generation, together with some movement and behavior controllers. Genes includes powerful modeler and scene description language that integrates object geometry, light and sound properties, and physical characteristics. Genes is also being used in other visual simulation application like a realistic Driving Simulator (Leitão, Coelho & Ferreira, 1997).

The system requires a dedicated OCO (Octane Channel Option) as illustrated in figure 2, that allows to split the frame buffer into 4 independent channels.

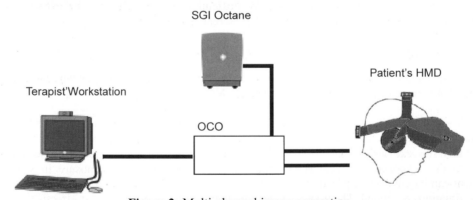

**Figure 2:** Multi-channel image generation.

With this module, three independent video channels are generated simultaneously. Two of these three video channels are used to generate a stereoscopic image presented to the patient by a Virtual Research VR6 Head-Mounted-Display. The third channel is available to the therapist that can choose his preferred viewpoint interactively. The therapist also receives, in real time, information about the state of the virtual environment.

In the experiments preformed for acrophobia treatment, the therapist controlled the virtual height of the patient interactively, using a traditional keyboard. So, the virtual camera used to generate the patient image is controlled in 4 degrees of freedom.

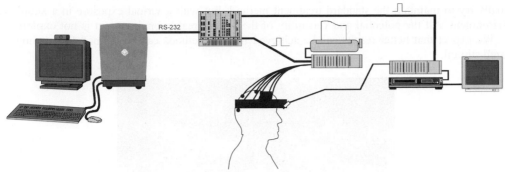

**Figure 3:** Synchronization with psycho-physiologic monitors

All the system is synchronized with several equipments for psycho-physiologic monitoring by a dedicate synchronizer module, as shown in figure 3. This module receives commands from the main workstation and it is also used in other VR applications. The acoustic environment is simulated with non-interactive playback of a real world sound recording.

It is known that the exposure therapy can be efficient with any environment or situation that reproduces a stimulus from the same type of the feared one. It is usually not necessary to use the exact feared stimuli. Even though, we decide to implement a computer model that reproduces the real world landscape that can be seen from an existent skyscraper. The computer model was developed using textures obtained from real world photographs. Figure 4 allows to compare the real world view with the virtual one.

**Figure 4**: Views or the real world (left) and the virtual environment (right).

This relation between the virtual and the real environment allowed the validation of the system by comparing the exposure in-vivo therapy to the VR exposure therapy.

Figure 5 shows a typical use during a VR exposure therapy of a acrophobic patient.

# 3    Conclusions

Several researches, recently conducted, probed the functionality of this alternative and some works are known that try to compare the virtual reality exposure to the exposure in-vivo. These works

1171

usually try to maintain the standard treatment method, but with a virtual exposure in a simulated environment. All the potential and versatility of the virtual reality environment is not explored at all. We expect that better results can be achieved if new methods can be defined for grading the feared stimuli.

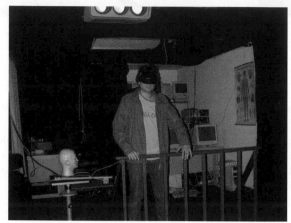

**Figure 5:** Acrophobia VR Expoure therapy.

Several experimental treatments, already performed, allowed the validation of the system as an efficient tool to the phobia treatment. Some innovative techniques, aimed to the control of the fear intensity, are now being implemented and tested. Using the data acquired from the psycho-physiologic monitors, we expect to automatically control, the virtual height of the phobic patient. This automatic control can be considered a first step towards the achievement of the phobia self treatment.

# References

Alcaniz, M., Banos, R.M., Botella, C., Perpina, C., Rey, A., & Villa, H. (1998). Virtual Reality Treatment of Claustropobia. *Behaviour Research & Therapy*, 36(2), 239-246.

Banos, R. M., Botella, C., Perpina, C., Alcaniz, M., Lozano, J.A., Osma, J., & Gallardo, M. (2002). Virtual reality treatment of flying phobia. *IEEE Transactions on Information Technology in Biomedicine* 6(3):206-212

Carlin, A. S., Hoffman, H. G. & Weghorst, S. (1997). Virtual reality and tactile augmentation in the treatment of spider phobia: A case study. *Behaviour Research and Therapy*, 35, 153-158.

Emmelkamp, P. M. G., & Felten, M. (1985). The process of exposure in vivo: cognitive and physiological changes during treatment of acrophobia. *Behaviour Research & Therapy*. 23, 219-223.

Leitão, J. M., Coelho, A., & Ferreira, F. N. (1997). DriS – A Virtual Driving Simulator. *Proceedings of the Second International Seminar on Human Factors in Road Traffic*, ISBN 972-8098-25-1

Rothbaum, B. O., Hodges, L. F., Kooper, R., Opdyke, D., Williford, J. & North, M. M. (1995). Virtual reality graded exposure in the treatment of acrophobia: a case report. *Behavior Therapy* 26(3), 547-554.

Slater, M., Usah, M., Steed, A. (1994). Depth of presence in virtual environments. *Presence: Teleoperators and Virtual Environments*, 3, 130-144.

# Being Confident – Development of a TV-based Tele-Assistance System

Joachim Machate

Ioannis Karaseitanidis

Maria F. Cabrera

User Interface Design GmbH
Teinacher Str. 38
D-71634 Ludwigsburg
Germany
joachim.machate@uidesign.de

I-Sense Group
ICCS
National Technical
University of Athens
Greece
gkara@esd.ece.ntua.gr

G.B.T - E.T.S.I. UPM
Ciudad
Universitaria, s/n
ES-28040 Madrid
Spain
chiqui@gbt.tfo.upm.es

## Abstract

People with special needs, e.g., disabled people or elderly people, often require assistance by other people in their daily activities. Approaches how they can be supported in managing their daily life by means of IT-based services are developed by the CONFIDENT project[1], which is sponsored by the European Commission. Besides the setting-up of a service network (Amditis, e.a., 2003) which suits the needs of the target group, the development of appropriate user interfaces in order to easily access the service network is of crucial importance for the success of the project. Multiple channels of communication are established so as to provide the users with a communication environment that helps them to live their life independently and according to their own choices. Bio-metric signal monitoring and the provision of commonly used ways of interaction, such as TV-based user interfaces, help them to organize their own way of life whilst being confident not being left alone in exceptional and emergency cases.

The paper is presenting the interaction concept of the CONFIDENT Daily Living Manager – the TV based access point to the tele-assistance system, its user interfaces and the user-centered design process(UCDP) which has been adopted by the project.

## 1   Introduction

The dramatic demographic changes of West-European cultures in the next decades are commonly well known. Their sociopolitical explosiveness is a matter of political and social discussions focusing primarily on the search for solutions for amendments of the social insurance systems and for new social care concepts for people in need of care. A self-organized European network of people concerned, who depend on the help of other people in their daily life, is discussing and developing concepts and guidelines for supporting a self-determined and self-organized life (ENIL, 1998). One precondition in order to help these people in taking their lives into their own hands and to life it according to their individual needs is that help and care is provided in the moment when they need it. One possible solution for achieving this is provided by the use o so-called tele-assistance systems. A prerequisite for a high quality in use of such a system is that it's

---

[1] Work presented in this paper has been funded by the European Commission's IST Programme as Project IST-2000-27600.

functionality is properly aligned to the needs of it's users, that it is available 24 hours per day, a high reliability is required in terms of the delivery of the social care services and an interaction concept which is understood by the users immediately and can be operated without any problems. Another criteria – especially when considering the specific situation of the potential users – is a visual design which is by no way stigmatizing, but which conveys to fun and enjoyment, and has the potential to become a prime candidate for being used by all family members. This would contribute to a real design-for-all philosophy.

## 2    Context of Use

The project started its development with a sound analysis of the context of use of the envisaged system (Bevan, 1995). As Bevan is stating, the quality of the context of use can be seen as the proper usability of a system. Usability should not only be seen as the quality of the interaction concept, but also as the degree of appropriateness of the functionality with regard to the user's needs, and furthermore, the quality how the system fits into the environment in which it is intended to be used. System development which takes into account the contextual boundaries in all of their aspects is also called contextual design, a term which is heavily assigned to the work of Beyer and Holtzblatt (Beyer & Holtzblatt, 1998).

Estimations are saying that 3.7 million adults in the U.S. and about 7 Million people in Europe are depending on daily assistance in managing their lives (La Plante, 1992). In order to understand how people in need of care organize their daily life and how their individual needs meet the possibilities of the social organizations, the CONFIDENT project set up three pilot sites being located in Greece, Northern Ireland, and Spain. Interviews and workshops were conducted with all relevant stakeholders (Kruchten, 2000) in all of the three pilot sites. In order to empower the users to actively participate in the development process, a project internal user form was set up. User forum discussions took place in all pilot sites, but also in common meetings with participants of different pilot sites and via emails. Besides the technical team, four major groups of user group stakeholders were identified:
- People in Need of Care
- Family Members
- Professional Assistants
- Service Provider

In compliance with the view of the European Network on Independent Living three basic areas of daily assistance activities were identified:
- Personal needs: body care, getting up, feeding, dressing up, going to bed, etc.
- House keeping needs: shopping, room cleaning, doing the washing, etc.
- Social needs: communication, work, leisure, entertainment, etc.

During the context analysis 190 people have been interviewed. In the interviews participated 79 severely disabled people, 75 family members, 30 professional assistants and 6 representatives from social service provider organizations. All the data and its analysis have been made publicly available as a project report (Cabrera, 2001). The context analysis brought about that the most significant needs of the interviewed people were support and organization of personal assistance and transportation. With regard to their preferences of system input modalities, most of the people in need of care preferred to have a remote control available and to use wearable sensors. The choice of their favorite means for system interaction can be well understood, if one is considering the fact that firstly the use of remote controls is rather common in assistive technology. Secondly, it is the primary means of interaction for the control of consumer electronic equipment. Also, their expressed wish to wear sensors for monitoring bio-signals can be based on the assumption that wearable sensors can enhance their feeling of safety. So, by wearing appropriate sensors people can be sure that in case of a health emergency situation, an alarm center will be automatically notified and help will be available as soon as possible. With this, they can live more autonomously whilst not diminishing safety aspects of their life.

## 3   Design Rationale

Having conducted all the interviews and discussions, it was clear for the project that the best way to provide access to the CONFIDENT services would be to put an ordinary TV-set at the disposal of the users. In order to be able to order a particular service, the TV-set needs to be connected to a set-top-box which has a 24 hours connection to the service center. A complete overview of the CONFIDENT system architecture and it's components can be taken from the paper "Towards an Informatics System Enabling Disabled People Universal Access to Information and Assistance Services" (Amditis, e.a., 2003). The TV-based component of the system which is forming the basis for the communication of the user with the care takers we call Dailiy Living Manager. The platform for using the Daily Living Manager is a LINUX-based set-top-box, a TV screen is used as the display. Using the system is accomplished by an ordinary remote control which is putting a set of arrow keys, an OK button and a menu button at the disposal of the user. The use of specifically adapted input devices, e.g., a joystick, is possible, as long as the signal transmission between set-top-box and remote input device is working, e.g. via a Bluetooth connection.

The development of the interaction concept of the Daily Living Manager was adopting an iterative user-centered approach. The start point was set by the results of the context analysis which led to the definition of a set of essential use cases. Based on this information, a first interactive prototype was developed. The prototype only showed the envisaged navigation and interaction concept without any underlying functionality in terms of access to services. The development of the prototype was based on ideas from the Inclusive Design Guidelines by Microsoft (Microsoft Corporation, 2001) which recommend to design a screen in such a way, that it's purpose and what to do with it should become apparent to the user by first sight. Also, recommendations to provide input assistants for inexperienced uses and users with special needs in order to manage complex tasks (Burmester, Machate & Klein, 1997) and guidelines for inclusive design (Nicolle & Abascal, 2001) were taken into account. Taking into consideration the fact that the system must be used via a remote control it was clear that free mouse pointer positioning in the way it is possible with ordinary graphical user interfaces (GUIs) would cause handling problems for any user, but particularly for the target user group. Results from comparative experiments undertaken by projects which investigated solutions for interactive TV showed clear benefits for a grid-based movement of the cursor by means of the remote control's arrow keys in contrast with solutions providing free pointer positioning on a TV screen by use of trackballs or comparable solutions. Another difficult to handle situation is the situation in which data browsing would be required in order to see all the data in a particular screen. Instead of using scroll bars like it is done in GUIs, control elements for the selection of data were adopting a fixed focus selection procedure. Using a fixed focus selection procedure means that the currently active input element has one area, called the input focus, which has a stable position on the screen. When an arrow key of the remote control is pressed by the user, it is not the cursor which is moving, but instead the displayed data are moved in accordance with the arrow key being used, so that a new item is drawn into the input focus and hence becoming selectable. Pressing the OK-button of the remote control would now lead to the activation of the item under focus.

The visual design of the Daily Living Manager screens was requiring a clear and distinct structure being far off the usual PC and Windows-like software application user interface design: A clearly emotional design being by no way stigmatizing but strengthening a positive and joyful attitude towards the interaction with the system (see Figure 1.).

The prototype was evaluated and discussed with the target users in all pilot sites, and results were made available to the user forum. Workshops with representatives from the user forum and the technical teams brought about new insights. Particularly, it became clear to the development team that the first prototype, although being based exactly on the requirements from the context analysis, was not exactly what our target users had in mind. A crucial point was that the degree of a service specification which a user was able to make with the prototype in order to call a particular service was not in the way the people in need of care usually make their requests. Furthermore, it's informational value was not sufficient for the service providers in order to be able to deliver the relevant service. So, the project team was facing the problem what went wrong. What apparently became clear was that during the context analysis the participants of the analysis only got a vague idea of what the envisaged system could do for them. When they got the chance to interact with a prototype, their ideas became more concrete and they got more substance. The re-design of the Daily Living Manager kept the basic ideas of the prototype, but extended the input possibilities for the service order assistants (see Figure 2).

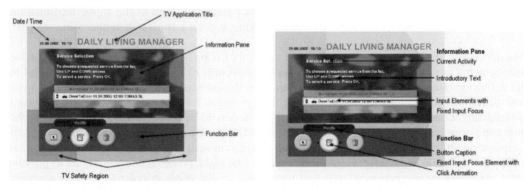

**Figure 1:** Screen Structure and Elements of the CONFIDENT Daily Living Manager

**Figure 2:** Service Order Assistant of the CONFIDENT Daily Living Manager

# 4 Conclusion

The development of the CONFIDENT Daily Living Manger clearly showed the advantages of a user-centered design approach and the importance of using early-stage prototypes which convey the product idea to their future users. Also, it gives an example that a sound and comprehensive context analysis could only form the starting point of the development, but that the success of the final version is depending on a participative design process in which the user is the central gear wheel. The visual design of the Daily Living Manager is a supreme example that the design of products for people with special needs can have the substance of a real design for all which can bring fun and enjoyment to all users no matter whether they are restricted in their abilities or not.

## Acknowledgements

The authors whish to thank the CONFIDENT Project Consortium for the possibility to present this work to a larger auditorium. Also, we want to especially express our thanks to Simone Keller for her beautiful visual design.

## References

Amditis, A., e.a., (2003). Towards an Informatics System Enabling Disabled People Universal Access to Information and Assistance Services. In *Proc. of HCI Int., 2003, Crete*, Mahwah, N.J.: Lawrence Erlbaum Associates.

Bevan, N. (1995). Usability is Quality of Use. In Y.Anzai, , H. Mori& K. Ogawa (Eds.), *Proc. of the 6th Int. Conf. on Human-Computer Interaction* (pp. 349-354). Amsterdam: Elsevier.

Beyer, H. & Holtzblatt, K. (1998). Contextual Design- Defining Customer-Centered Systems. San Francisco: Morgan Kaufmann Publishers, Inc.

Burmester, M., Machate, J., & Klein, J. (1997). Access for all. HEPHAISTOS - A Personal Home Assistant. In S. Pemberton (Ed.), *ACM-SIGCHI Human Factors in Computing Systems, CHI 97 Conf. Proc., Extended Abstracts* (pp. 36-37). Reading, Mass: Addison-Wesley.

Cabrera, M.F., e.a. (2001). D3: User Requirements. IST-2000-27600 CONFIDENT Public Deliverable. Madrid: Universitá Politéchnica de Madrid.

CONFIDENT (2003). Project Website. Retrieved Jan. 2, 2003, from www.confident-project.org.

Constantine, L.L. (2001). The Peopleware Papers: Notes on the Human Side of Software. Upper Saddle River, NJ: Prentice Hall.

ENIL (1998). "Shaping Our Futures". Conference on Independent Living sponsored by the European Network on Independent Living (ENIL), London, June 1998. Retrieved Jan. 2, 2003, from www.independentliving.org/docs2/enilfuture2.html

Kruchten, Ph. (2000). The Rational Unified Process: An Introduction. Second Edition. Boston: Addison-Wesley.

La Plante, M.P., & Miller, K.S. (1992). People with Disabilities in Basic Life Activities in the U.S. *ABSTRACT 3*, Disability Statistics Centre.

Microsoft Corporation (2001). Microsoft Inductive User Interface Guidelines. Retrieved Jan. 2, 2003, from msdn.microsoft.com/library/default.asp.

Nicolle, C. & Abascal, J. (Eds.) (2001). Inclusive Design Guidelines for HCI. London: Taylor & Francis.

# Cognitive Ergonomics in the Development of Virtual Reality: A Neurophysiological approach

*Ralph Mager, Robert Stoermer, Marcus F. Kuntze, Franz Mueller-Spahn, Angelos Amditis,, Evangelos Bekiaris, and Alex H. Bullinger*

Center of Applied Technologies in Neuroscience - Basel  (COAT-Basel),
University of Basel, Switzerland
Wilhelm-Klein-Strasse 27, 4025 Basel
ralph.mager@pukbasel.ch

## Abstract

Detection and prediction of attention and workload is a relevant issue in the field of human factors. In the development and definition of virtual environments there is a demand to identify relevant parameters which modulate performance and the underlying cognitive processing. Therefore the neurophysiological approach is introduced as  an important domain testing usability in VE applications. Neurophysiological parameters, i.e. event related brain potentials (ERP`s) are proposed to give objective information about cognitive processes in the course of the performance in VR. The neurophysiological measures can be employed to evaluate virtual environments and to shed light on side-effects like sensory mismatch and divided attention. Thus ERP`s are implemented in VR settings to monitor the trade-off of cognitive processing by sensory conflict. a practical example is given.

## 1    Background

The overall goal of cognitive ergonomics is to support and to enhance the relationship between cognitive sciences and technical information processing system developments. In VR development this issue is of special interest since the user is exposed to an artifical environment to a so far unknown extent with unknown implications for cognitive processes in human beings. There has been concern about the potential side effects and after-effects of the use of virtual reality technology.  It has been reported that some users suffer from long-term visual disturbances and "flashbacks" such as  other difficulties occur with re-entry into the real world (La Viola, 2000). The sensory conflict theory has been proposed as one candidate responsible for these problems (Reason & Brand, 1975; Regan, 1994;  Warwick-Evans & Beaumont, 1995).  Sensory conflict theory proposes that symptoms occur as a result of conflict between signals received by the three major spatial senses - the visual system, the vestibular system and non-vestibular proprioception.

In respect to these observations the question arises how VR and the inherent sensory stimulation of the user interferes with cognitive processes. This interference might have relevant implications for workload and work-efficiency in virtual environments. The hypothesis is advanced that established models of resource/performance relation are reshaped by sensory mismatch as incurred in VR.
To mirror this modulation of cognitive processes by VR most objectively we propose a neurophysiological monitoring available for applications in virtual environments. As parameters

electroencephalographic signals including event related potentials (ERP`s) are suggested to get access to the basic branches of information processing: automatic processing that is activated without active control or attention (like mismatch negativity) and controlled processing that depends on focussed attention (P300, see below). The P300 is observed as a large positive deflection following presentation of a target stimulus.

Given that there are a multitude of techniques available to address human factors problems and issues, there is a demand to explain the value of neurophysiology in human factors research. To the extent that information derived by neurophysiological measures is redundant with that obtained from other more established methods, neurophysiological measures will not gain wide acceptance. One has to realise the high costs to perform recordings, to install the equipment and to analyse the data later on. Therefore we will focus in this section on some relevant benefits of neurophysiological measures.

The temporal resolution of neurophysiological measures can not be achieved by any other method. In many human factors contexts, such as evaluation of virtual environments and their effects on performance and information processing, data can be collected on-line and analysed off-line allowing a reduced disturbance of the performance since there are no interventions in the course of the measurement. Principally one can segregate this offline approach from a number of human factors contexts that demand instantaneous data processing and interpretation to allow a kind of feedback-loop. We will focus on the former approach.

The application of neurophysiological techniques to issues of human factors is relatively new, but it can be already demonstrated that these measures provide useful insights into human performance and cognition in laboratory settings. The fact that applications in virtual environments are in any case close to criteria of a laboratory setting facilitates the utilisation of proper neurophysiological measures. A general interest in respect to human factors is the detection and prediction of relevant physiological parameters in the operator performance, which are crucial to efficiency of work or more generally to workability.

Neurophysiological studies have already been conducted to investigate predictions of resource models as mentioned above under non-VR conditions. And indeed basic predictions could be confirmed by neurophysiological methods. Kramer, Wickens and Donchin (Kramer et al. 1985) found that increasing the difficulty of a primary task resulted in a systematically decreased amplitude of the P300 component of the ERP elicited by a secondary task. Thus a model was developed so that the amplitude of the P300 component reflects the allocation of the resources. An increased difficulty or resource demand in the primary task would lead to decreased resources in the secondary task. This model could be confirmed by different groups (Hoffmann 1985; Kramer & Strayer 1988; McCallum et al. 1987; Strayer & Kramer 1990). Another important criterion is the reciprocity in P300 elicited by two concurrently performed tasks. This means that an increase in the amplitude of P300 elicited by events in one task should be accompanied by decreases in the amplitude of P300s elicited in another, concurrent task (Sirevaag et al. 1987; Wickens et al. 1983). The P300 reciprocity is an important issue since it provides converging support for the idea of resource trade-offs. A remarkable advantage is that P300 data can be recorded in the absence of overt responses, thus allowing a direct assessment of resource allocation. The limitation which can be derived is that two concurrently performed tasks will show trade-offs only when the tasks require the same types of processing resources or capacity.

Following this line of evidence ERP`s in VR are proposed to monitor the trade-off of cognitive processing by sensory conflict.

## 2    Method and an example

The measurement of brain electrical activity using the electroencephalography (EEG) provides a non-invasive and inexpensive method to directly measure brain function and make inferences about regional brain activity. It has many virtues as a direct measure of brain function that can be used in different applications in the bio-behavioural sciences, including studies of basic cognitive processes. Beside the spontaneous rhythmic oscillations of electrocortical activity, more recent research has concentrated on those aspects of the electrical potential that are specifically time locked to events, that is, on ERP`s. The ERP`s are regarded as manifestations of brain activities that occur in preparation for or in response to discrete events, be they internal or external to the subject.

To illustrate the evidence that an ERP (P300) provides an index of the allocation of resources is given below. ERP and the conventional EEG signals were recorded along the midsagittal line (Fz, Cz, Pz), from frontal (F3, F4), temporal (T3,T4) and temporoparietal (T5, T6,  P3, P4) leads and from both mastoids (A1, A2) using a 32 channel Neuroscan amplifier. The electrodes were referenced to the connected mastoid electrodes. One additional electrode was placed about the left outer canthus to detect eye movements (artifact-rejection). Non-polarizable Ag/AgCl electrodes were used. EEGs were digitized with a sampling rate of 512 Hz, parallel the stimulation markers were stored for further analysis.

During EEG recording a primary task was performed, the subjects had to identify  stimuli that are applied with a frequency of 0.5 Hz. Under condition I (black line) the visual stimulus ($V_1$, dotted line) is given without any distraction, under condition II (red line) the stimulus is given with a not task relevant distracting information ($V_2$). The diminished resources for the primary task are reflected by decreases in the P300 amplitude as indexed in the grey box . An auditive warning stimulus (A, solid line) precedes each visual stimulus separated by 300 ms. Every trace represents averaged recordings of  18 subjects each of them contributing several hundreds of  trials.

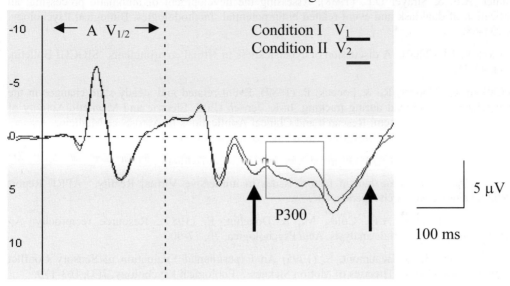

Figure 1: Event related potentials at electrode Cz in response to stimuli V1 and V2 reflecting the different resource allocation/ trade-off.

## 3     Conclusions

The proposed neurophysiological measures could be used in a wide variety of settings and across a number of different systems to provide a general indication of the mental workload and resource-allocation experienced by human operators in VE`s. Thus data might enable the VE designer to evaluate the overall magnitude of processing demands imposed upon the operator. This is suggested to be of relevance in VE development since sensory conflict will interfere with resource allocation and might evoke resource trade-offs as reflected in ERP`s. Without doubt neurophysiological measures will be limited to selected applications cause of high costs and analysis efforts. But we think that there is a need to develop objective parameters to evaluate VE`s and possible mismatch related sensory conflicts inherent to stay in VE.

## References

Hoffman, J., Houck, M., MacMillian, F., Simons, R., & Oatman, L. (1985). Event-related potentials elicited by automatic targets: a dual-task analysis. Journal of Experimental Psychology: Human Perception Performance, 11(1), 50-61.

Kramer A.F. (1991). Physiological metrics of mental workload: A review of recent progress. In Damos D. (Ed.), Multiple Task Performance, 279-328. London: Taylor & Francis

Kramer, A.F., Wickens, C.D., and Donchin, E. (1985) Processing of stimulus properties: evidence for dual-task integrality. J Exp Psychol Hum Percept Perform, 1985. 11(4): p. 393-408.

Kramer, A.F. & Strayer D.L. (1988). Assessing the development of automatic processing: an application of dual-task and event-related brain potential methodologies. Biological Psychology, 26, 231-68.

La Viola Jr, J.J. (2000). A discussion of cybersickness in virtual environments. SIGCHI Bulletin, 32(1), 47-56.

McCallum, C., Cooper, R., & Pocock, P. (1987). Event related and steady state changes in the brain related to workload during tracking. In K. Jensen (Ed.), Electric and Magnetic Activity of the Central Nervous System: Research and Clinical Applications in Aerospace Medicine. France: NATO AGARD. methodologies. Biological Psychology, 26(1-3), 231-67.

Reason, J.T. & Brand, J.J. (1975) Motion Sickness. London: Academic Press.

Regan, E.C. (1994) Some human factors issues in immersive Virtual Reality. APRE Report 94R027, Defence Research Agency, UK.

Sirevaag, E., Kramer, A.F., Coles, M., & Donchin, E. (1987). Resource reciprocity: An eventrelated brain potentials analysis. Acta Psychologica, 70, 77-90.

Warwick-Evans, L. & Beaumont, S. (1995) An Experimental Evaluation of Sensory Conflict Versus Postural Control Theories of Motion Sickness. Ecological Psychology, 7(3), 163-179.

Wickens, C.D., Kramer, A.F., Vanesse, L. & Donchin, E. (1983). The performance of concurrent tasks: A psychophysiological analysis of the reciprocity of information processing resources. Science, 221, 1080-2.

# A User Interface for Virtual Maintainability in Immersive Environments

*Luis Marcelino, Norman Murray, Terrence Fernando*

Centre for Virtual Environments
University of Salford
M5 4WT, UK

## Abstract

This paper presents a virtual maintenance simulation system from the user's perspective. Some of the identified requirements for such a system to be integrated in a product development process are discussed. Based on these requirements, a user interface was built that enable the user to perform assembly and disassembly operations. The interfaces presented include system control, navigation and selection/manipulation. These were implemented using a combination of 3D widgets and direct object interaction.

## 1    Introduction

Due to global competition, engineering companies are under increasing pressure to develop products in shorter periods of time at a reduced cost. Virtual Prototypes (VP) can be generated quickly from existing CAD data at a marginal cost. However, maintenance simulation on VP is still a challenging problem since it requires the simulation of physical realism on virtual products and advanced interfaces to assess the maintainability issues. This lack of simulation tools in current CAD/CAM systems, forces engineers to use physical prototypes to assess maintainability.

This paper presents the outcome of a research programme aimed at developing an advanced maintenance simulation environment. The simulation of the assembly and disassembly operations are supported by this environment using automatic constraint recognition and constraint management techniques. The underlying software architecture and the techniques used for simulating the assembly constraints were presented at HCII2001 (Fernando, Marcelino & Wimalaratne, 2001), and this paper reports the research challenges undertaken in implementing advanced immersive interfaces to support maintenance tasks directly on VPs using two handed direct interaction techniques.

The interfaces presented in this paper cover the three forms of user interaction: system control, navigation and selection/manipulation. The first two categories are implemented using virtual windows based on a 3D widgets approach, while manipulation of 3D objects is supported through direct and natural object interaction.

## 2    Virtual Maintainability

This section presents some of the research done in the area of VP with emphasis on virtual maintainability. After introducing some existing VP systems we enumerate the requirements for a virtual maintenance simulator, identified from our industrial partners needs. This section ends with a description of the case study used for assessing our VR system.

### 2.1    Background

The industry's interest in Virtual Prototyping has been a driving force in the research of virtual maintainability. This cooperation between the industry and the research community has resulted in different VP systems.

The Fraunhofer Institute for Computed Graphics (IGD) developed a system called Virtual Design II (Zachmann, 2000). That simplifies the creation of Virtual Environments. A major obstacle to the use of VR by design engineers is that VR need to be programmed. IGD developed a flexible script-language that VE authors can use to define the behaviour of virtual objects.

The Virtual Assembly Design Environment (VADE), developed at Washington State University, aims to support assembly planning and evaluation for design and manufacturing (Jayaram et. al., 1999). VADE supports parametric models and allow users to use both hands to assemble parts in an immersive environment.

## 2.2 Maintenance simulator requirements

To develop a system that meets the industry's expectation we performed various studies to gather the requirements for a virtual maintenance simulator. These requirements were then used to implement a system suitable for real industrial maintenance tasks.

An essential function of a Virtual Prototyping system is the ability to load any number of parts and assemblies from CAD files. Once the model is loaded into the system, engineers expect to be able to inspect the model at full scale. They want to look at the model from different perspectives just by walking around the model. Such contact with full-scale model provides engineers with a better understanding of the developing product and enables them to assess how easy it is to reach out for a part, or position it in an assembly.

A natural and intuitive style of interaction with the virtual objects would enable engineers to study and assess the assembly and disassembly procedures and determine different operations sequences while timing them. Support for two-hand interaction would enable engineers to assess the effect of using both hands in the assembly and disassembly procedures. The use of both hands does not always simplify the task as it can be cumbersome in confined spaces.

The interaction and inspection of virtual prototypes give an engineer the opportunity to study the effect of different values of clearance between mating parts and to demonstrate the existence of adequate space for spannering. Furthermore, a VR system can check for existence of interference or collisions between parts and reproduce restrictions imposed by adjacent hardware.

The implementation of a VR system that meets the requirements presented above is not straightforward. The models used in an industrial environment are usually complex with a large number of surfaces and the real-time nature of a VR application limits the available computation time. It is therefore necessary to compromise between accuracy and performance and to optimise the system according to the used hardware.

## 2.3 Industry's case study

The following case study description illustrates a concrete scenario for maintenance simulation. We chose this case study because it has a diverse set of tasks and is well documented. This case study corresponds to a real application, taken from an international project from the aerospace domain. The case study provided by SENER comprises the bottom area of an aircraft that hosts one engine electronic control box. The real working requirement on which the test case is based on was to demonstrate to the customer that the electronic box could be removed and replaced from the airframe in less than 60 minutes.

The instructions, as defined in the standard maintenance handbook, state that maintainer must start by disconnecting the cable looms pipes. The user must then remove the security bolts and slide the electronic control box over the guide brackets. To mount the electronic control box the reverse order must be followed.

## 3 System Overview

This section gives an overview of the developed system. It presents its architecture and summarises its functionality as a maintenance simulator.

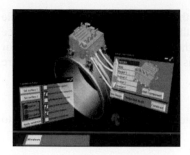

**Figure 1: An industry case study and the control windows**

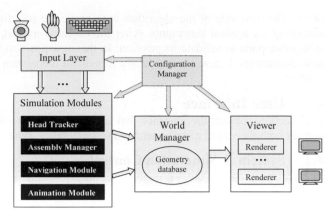

Figure 2: System Architecture

## 3.1 Software Architecture

The developed Virtual Prototyping software supports different hardware platforms. The modularity of the system means that performance can be maximised by distributing the computation load across multiple processors. Figure 2 shows a simplified description of the system architecture, which evolved from the one presented at HCII2001 (Fernando, et. al. 2001). The configuration of the VE is defined in configuration files that use simple primitives. These configuration files are loaded by the Configuration Manager, which then starts the requested processes.

The World Manager (WM) maintains two databases with all the geometry in the VE. One database is part of the geometric kernel, and the other is the scene graph. The incorporation of a geometric kernel creates the ability to load CAD files into the VE and save them back to CAD format without losing the assembly structure. The flexibility and power of the geometric kernel is complemented with a high performance scene graph, providing us with the best of both worlds. The Viewer creates the rendering processes and synchronises them. Whenever a change occurs in the scene graph, each process renders a frame and waits until the next change. The Input Layer is software API that allows multiple simulation modules to share the same input devices.

The simulation modules are shared object libraries that are loaded at run-time when selected. This modular approach allows us to develop many simple simulation modules that can then be combined to produce a complex environment. An example of a simple simulation module is the head movement where the tracker position is inserted into the user's head representation. The most complex simulation module is the Assembly Manager. This module uses geometric constraints and collision detection to simulate realistic interaction between objects.

## 3.2 System Functionality

The Assembly Manager is the simulation module used for maintenance simulation. It uses automatic geometric constraints to compute the inter-object collision response. The collision detection is provided by the PQP toolkit.

The user interacts with objects through manipulators (e.g. a virtual hand). A manipulator is a generic object to which a part can be attached. When the user moves the manipulator an attached part will try to move as well. The first step in the assembly simulation algorithm is to compute collisions on the new position that do not involve constrained surfaces. This intended position is used by the system to determine the user's intension: an assembly operation creates new contacts among mating parts while a disassembly operation reduces the number of collisions. The second step is to remove constraints that were broken by the user's movement, e.g. two against planes that move apart in directions opposite to their normals or two cylinders that are no longer inside each

1185

other. The next step of the algorithm is to move the part to the user's requested position while enforcing the applied constraints. After the object is moved, the part is checked for interference with other parts to validate its position. In this new position, the Assembly Manager searches for new constraints. If new constraints are found and the system is in fully automatic mode, they are applied.

# 4 User Interface

This section presents the interface that enables the user to interact with the environment and to control both the virtual environment and the simulation flow.

## 4.1 System & Simulation Control

The ability to control the VE enables the user to adjust the system to best approximate his/her needs. An example of this is the control over the lights in the VE, which creates the possibility to assess a task in different light conditions. Another advantage of virtual environments is their potential to enhance the user's acquaintance with the product. By scaling the world up it is possible to explore in detail parts of the model not possible otherwise. On the other hand scaling the world down can provide an overview of this same model. Light and scale control are available to the user through a control window implemented with 3D widgets. These widgets are 3D buttons and panels that can be combined in complex windows.

User can also specify the simulation level of detail. At present there are three simulation levels. The simplest mode allows manipulation with no further simulation. Objects can intersect each other and constraints are ignored. Another mode only has collision detection. Objects cannot intersect each others but constraints are ignored. The more advanced mode is simulations with collisions and constraints. Only in this mode it is possible to assemble mating parts.

## 4.2 Navigation

The ability for a user to walk naturally within an immersive environment is limited by the physical space inside or around the projection system. The user can not move beyond the walls of a CAVE or outside of the room where the VR system is. This limited space may not be sufficient for a large model and an alternative way of locomotion must be provided.

An important aspect in developing the maintenance environment is to ensure that the user has various navigational options to review the design easily. A navigation interface allows the user to move his/her physical space around in the virtual space. The user can lift the CAVE up and down or displace it, so it is in the best position to assess the maintenance operation. Other forms of navigation include target-based and free navigation.

## 4.3 Manipulation

Object manipulation is supported through direct interaction using manipulators. The assembly manager provides the user with two virtual hands and one virtual screwdriver. A ray casting pointer is another form of selection available.

The virtual hands and the virtual screwdriver are mapped to the user's real hands. To pick an object the user pinches two fingers while touching the object (Figure 5). The object is then attached to the hand and its movement is the result of the selected simulation. Even though the hand can pass through virtual objects, a picked object can not unless collisions are disabled. A difficulty arose when users wanted to pick up a bolt. Bolts can be smaller than the accuracy of the tracking system, making them hard to pick. An attempt to pick up a bolt usually resulted in picking an adjacent object or nothing at all. The screwdriver enables users to manipulate bolts without touching them. When a user moves the screwdriver, the assembly manager determines the closest bolt and highlights it (Figure 3). This enables a user to pick up a bolt just by moving the screwdriver to its vicinity. Once the user acknowledges which bolt to pick by pinching two fingers, the screwdriver snaps to the bolt's head (Figure 4).

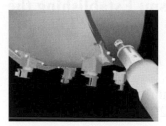
Figure 3: Screwdriver high-
lights the closest bolt

Figure 4: Removing a bolt

Figure 5: Removing a part
using the virtual hand

## 5    System Feedback

One of the key challenges in developing the system has been to provide the user with appropriate feedback about the collisions and recognition of various assembly relationships etc. This has been achieved through colour changes, use of transparency, sound and constraint visualisation to inform the user which constraints are being recognised. Such feedback proved to be very useful when several potential assembly constraints are being recognised within cluttered object spaces.

Recognized constraints are visualised with a descriptive icon, e.g. concentric constraints have an axis and two aligned cylinders, against constraint a plane with two opposing arrows. Besides these icons, surfaces of recognised constraints are highlighted blue.

Colliding surfaces are also highlighted and a sound is played. The identification of collisions enables engineers to identify parts liable to be damaged during maintenance operations.

Selected parts are completely highlighted, unlike constraints and collisions that only highlight relevant surfaces. As mentioned above, the selection of parts depends on the manipulator used, e.g. by touch or by proximity. Once the selected part is acknowledged, it is de-highlighted.

## 6    Evaluation and Future Work

A preliminary study involving the described case study revealed that the system is intuitive and has a small learning period. Novice users could lean how to remove the bolts and brackets from an airframe (figures 3-5) in less than 5 minutes, after which they could do it proficiently.

This experience showed us that users had some difficulty in associating the pinch to a picking action. Furthermore, the switch between virtual hand and virtual screwdriver was done by pinching the thumb and the pinkie finger. Some users found it hard to remember what finger did what in the first trial, but were comfortable with the interface after that.

A feature that could improve the user's performance is feedback to indicate what is stopping a manipulated part from moving. During the trials, several users tried to remove the left bracket (figure 3-5) before removing its four bolts. Because they were easy occluded by the bracket many users forgot them and tried unsuccessfully to remove the bracket.

The success of this experiment encouraged us to increase the complexity of the simulated maintenance task. We are currently planning a new experiment that will provide more details about the suitability of immersive VEs to simulate maintenance operations.

## References

Fernando, T., Marcelino, L., Wimalaratne, P., (2001), "Constraint-based Immersive Virtual Environment for Supporting Assembly and Maintenance Tasks", HCII2001, pp. 943- 946

Gabriel Zachmann, (2000), "Virtual Reality in Assembly Simulation". Ph.D. Dissertation, Technischen Universität Darmstadt

Jayaram, S; Jayaram, U; Wang, Y; Tirumali, H; Lyons, K; Hart, P; WADE: A Virtual Assembly Design Environment, IEEE Computer Graphics and Applications, Dec 1999, pp. 44-50.

# Interactivity, Control of Movement and Realism: Establishing the Factors Influencing Virtual Reality Training

*Eleanor Marshall [1], Sarah Nichols, John R Wilson*

VIRART, University of Nottingham
School of MMMEM, Nottingham, NG7 2RD, UK
[1] Epxem1@nottinham.ac.uk

## Abstract

Virtual environments have been used as a training tool since their development and it has always been considered a suitable application for this diverse medium. With this in mind it is surprising to consider how few have been developed into everyday training tools for not only training intensive technical or manual roles, but in particular less 'hands on' situations, such as health and safety training for new workers. This led to the consideration of a training environment that not only taught a particular task to a worker but also provided information that could be used in an unexpected situation within a workplace, such as fire drills, to ensure all workers have all the knowledge they require in such a situation. Three main factors that may be considered as major contributing factors to the presence, usability and resulting effectiveness of a virtual environment; Interactivity, Control of Movement and Realism were explored within this experiment to help establish what is required to create an effective and usable training virtual environment for such an application.

## 1    Introduction

Within virtual reality research a variety of factors have been explored as having a possible influence on the effectiveness of training using a virtual environment. Effectiveness can been defined as the extent to which a specified goal or task is achieved and may be quantified by factors such as presence, usability and task performance achieved by the user. (Kaur, Sutcliffe & Maiden 1998) explored interaction and provided guidelines for suitable interaction design as failure to understand and find existing interactions result in 'user frustration and poor usability'. Realism is considered an important factor in the formation of a sense of presence within a virtual environment although too much can result in a slowing down of the environment rendering to such an extent that it has a negative effect on presence (Nichols, Haldane & Wilson 2000). It has also been suggested that the type of task performed in the virtual environment will affect the presence experienced as a more engaging and interesting task may result in increased presence (Nichols et al., 2000). Increased realism and interaction may improve both the interest in the task and therefor the experience for the participant. The ability to navigate around a virtual environment as desired by the user is not only a defining aspect of a virtual environment but also an important influencing factor in both its usability and the presence experienced by the user. (Sheridan 1992) puts the ability to modify the physical environment (this includes both control of movement and interaction) and the control of relations between sensors and display as two of the three major aspects that determine the level of presence within a virtual environment. It has also been suggested that presence is an influencing factor on task performance (Barfield Zeltzer, Sheridan & Slater, 1995), real world knowledge transfer (Mania & Chalmers, 1999) and usability (Kalawsky, Bee & Nee 1999).

The aim of this experiment was to establish if a VR training tool is an effective alternative when real world training is not possible, too costly or difficult to practice. The experiment was designed to help establish if navigational, spatial information and effectiveness (i.e. fire exit location, safety equipment location) were sufficiently taught from a virtual environment and if so what features of that environment were the most important. With this in mind the experiment considered three major factors in environment design; the extent of interaction, control of movement and realism. This was done by testing participants in the different levels of each condition i.e. full interaction to no interaction, full control of movement to no control, virtual world to real world.

The virtual environment training effectiveness was tested by assessing real world knowledge transfer to establish if virtual reality training could be applied to the real world. The following four conditions were explored within the experiment.

1) Virtual Environment with full control of Interaction and Control of movement and knowledge transfer.
2) Virtual Environment with Control of movement NO interaction and knowledge transfer.
3) Virtual Environment with NO control over interaction or control of movement and knowledge transfer.
4) Video of the real world with NO control over interaction or control of movement and knowledge transfer.

Two other potential conditions that were not compared are firstly, complete reality with the training and testing all performed in the real world (with or without interaction and control). This was not done as this is the scenario that alternative training methods are being developed to avoid. Secondly, a virtual environment where interaction is possible but not control over movement. This was not possible to test as participants could not then choose where to go and therefore what to interact with, so results could not be directly compared with those of the other conditions.

Training effectiveness of each condition was measured by memory recall in the real world one-day and one-week after the training. Also assessed were usability and presence, to ascertain their influence on memory recall and how or if they vary between the four types of training.

## 2 Experimental Method

32 participants with a mean age of 27.9 were tested, eight within each condition (four of whom repeated the experiment in the real world one day after the training, and four one week later). None of them had been in the real world building or have used virtual environments before.
Instructions were provided to each participant and a short demonstration by the experimenter of the control and interaction devices, where applicable, to familiarize the participant with the interface. Participants completed:

- Pre-test demographics questionnaire.
- Post-test questionnaire to establish usability (for conditions using the virtual environment).
- Post-test questionnaire to establish presence including enjoyment (for all conditions).

The experimenter recorded:
- Time taken to complete the search task in conditions one and two (decided by the individual)
- Observation was done throughout to provide further information about their experience within the environment.

The environment was run on a PC with a joystick for navigation and a mouse for interaction (where applicable). The participant's actions were observed for analysis using a digital camcorder (also used to record the video of the real world for condition four), a quad mixer, scan converter and a TV/video combination (also used to demonstrate the video of the real world for condition four). A minidisk recorder was used for the real world testing to support written notes and also for structured interviews in the real world after the experiment had taken place.

The environment consisted of 14 enterable rooms on 2 floors. The virtual world was modeled on an existing manufacturing laboratory with the same layout, décor and position of major objects within the building. The task was for each participant to find 14 health and safety items on a list provided at the start (fire extinguishers, fire exits, first aid boxes and fire alarms) within the environment and note them down. They were not told that they would be required to later recall their location within the real world in the same manner. Figure one shows a view from within the virtual environment used for this experiment.

**Figure 1: View from within the virtual environment**

# 3 Results

## 3.1 Effectiveness of Navigational and Spatial Information Training

The number of search items found on average during viewing the environment for each condition decreased as control of movement and interaction for the participant was removed. This is possibly a result of the participants in conditions three and four not being able to control their viewpoint and therefore it was harder to see the search items. They were also unable to recheck areas that they have already been to, whereas in conditions one and two the participants could, and this often happened. Participant 32 commented on this when referring to condition four in retrospect during part two of the experiment.

> 'it [the video] might hover in a room and then disappear again and I'd go wait I haven't finished looking whereas if it was you, you would stay until you felt comfortable you had seen everything you wanted to.'

This consequently gave the participants who viewed these conditions a disadvantage for recalling the items in the real world as they had seen fewer in the virtual world. To compensate for the variance in the number of items spotted in each condition for each participant the values for the mean number of items found were considered using the items found in the real world as a percentage of items found in the virtual world.

The mean number of items found in the real world one day later showed a steady decline from the environment with the most participant movement control and interaction being the highest (condition one) to the environments with the least environment movement control being the lowest (conditions three and four) see figure 2. A surprisingly low level of recall one-week later was noted in condition one. This may be as nearly all the participants found all the items in the virtual

1190

world so the number of items recalled in the real world relative to this was lower than expected compared to the other conditions where fewer items were found in the virtual world, see figure 2.

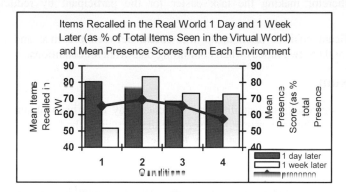

**Figure 2: Item recall and presence results**

This trend could be contributed to the level of interest and concentration involved with the environments where the participants maintained control. The higher presence level seen in these environments was considered as an indicator of this, see figure 2. This was explored using presence as a percentage of the maximum presence score possible from the questionnaire and using real world recall as a percentage of total items seen in the VW. No significant relationship indicating presence experienced is related to the number of items recalled in the real world was discovered despite similar patterns being indicated in the descriptive statistics see figure 2. (Rs = 0.0489, $N = 32$, p<0.05; two tailed)

Participants could spend as long as they wished in conditions one and two to search for the items whereas in conditions three and four participants each spent a fixed time viewing the display. The influence that this may have had on real world recall was explored. No significant relationship was found between the time spent in each environment and the number if items recalled in the real world. This indicates time is not a factor in the effectiveness of the virtual training and therefore may not be a factor when developing training environments and interaction methods.

The number of items recalled in the real world were considered in greater depth and none of the values found were significantly different between conditions though the number of items recalled in the real world one week later demonstrated greater variance between conditions, see figure 2.

## 3.2    The Relationship between Presence and Usability.

It was thought there would be a positive relationship between presence and usability as the easier and more intuitive the user finds the environment to interact with and react to the greater their sense of 'being there' or 'presence' (Sheridan, 1992). This theory was explored within all the experiment conditions.

Presence and usability values were calculated for each participant by assigning a value to each response on a Likert scale to each question, the higher the value indicating a higher level of presence or usability score. The values used are a percentage of the total presence and usability values obtainable from the questionnaire in each condition. Although the results found for each condition were not significantly different it was found that the highest presence score was from participants in condition two, see figure 2.

Condition three demonstrated the highest usability score, condition one the lowest. This may be a result of interaction and control being removed from the participant in condition three compared to condition one therefor making the task easier for the participant by reducing their active involvement.

A strong, significant positive relationship was found between presence and usability ($R_{obs}$ = 0.7580, $df$ = 20, p<0.05; two tailed) when combining data from all conditions.

## 4    Discussion

The most promising results in usability and presence may have been obtained from the VEC environment for a number of reasons. For example the participant had control over movement so can spend as long as they need searching for the items and retracing their steps, therefore consolidating the items' location to memory. It is also possible that interaction with the environment in condition one distracted from the search task and caused poorer results.

Results indicate that although the values are not significantly different effectiveness of navigational and spatial information training, usability, and presence were all higher in the conditions that allowed the participants control over their movement within the environment. These conditions also provided the most suitable environment for the participants to find the task items as they were able to retrace their steps in their own time. This indicates that control of movement is one of the most important factors in a training virtual environment and that realism and interaction seem to neither add to nor distract from this effectiveness. This could have many related consequences for the cost effectiveness of the development of training environments for such a purpose if realism and interaction within the environment can be reduced with no adverse effect on the training provided. Further research will be required to establish precise requirements and this research provides a useful start point and indicates that the right virtual environment could be an effective training tool.

## 5    References

Barfield, W., Zeltzer, D., Sheridan, T., & Slater, M. (1995). Presence and Performance within Virtual Environments. In W. Barfield & T. A. Furness (Ed.), *Virtual Environments and Advanced Interface Design*, (pp. 473-513). New York: Oxford University Press.

Kalawsky, R. S., Bee, S. T., & Nee, S. P. (1999). Human Factors Evaluation Techniques to Aid Understanding of Virtual Interfaces. *BT Technical Journal,* 17(1), 128-141.

Kaur, K., Sutcliffe, A. G., & Maiden, N. (1998). Applying Interaction Modeling to Inform Usability Guidance for Virtual Environments. *First International Workshop on Usability Evaluation for Virtual Environments, De Montford University, Leicester, UK.*

Mania, K. & Chalmers, A. (1999). Between real and unreal: investigating presence and task performance. *Proceedings of the Second International Workshop on Presence, Colchester, UK.*

Nichols, S., Haldane, C., & Wilson, J. R. (2000). Measurement of Presence and its Consequences in Virtual Environments. *International Journal of Human-Computer Studies,* 52, 471-491.

Sheridan, T. B. (1992). Musings on Telepresence and Virtual Presence. *Presence: Teleoperators and Virtual Environments,* 1(1), 120-126.

# Developing 3D UIs using the IDEAS Tool: A case of study

*J.P. Molina Massó, P. González López and M.D. Lozano Pérez*

LoUISE – Laboratory of User Interaction and Software Engineering
Escuela Politécnica Superior de Albacete, Universidad de Castilla-La Mancha
{ jpmolina | pgonzalez | mlozano }@info-ab.uclm.es

## Abstract

Success of three-dimensional games makes us imagine the interaction with computers as engaging as playing Doom. Indeed, PC hardware is ready to accommodate new uses of 3D graphics, such as new post-WIMP user interfaces. However, just as software engineering practice helps in the development of Windows applications and their interfaces, 3D user interfaces should take advantage of a systematic development process. In this paper, we present IDEAS, a UI development tool which allows the designer to create not only standard 2D UIs but also 3DUIs.

## 1    Introduction

First attempts in 3D user interfaces (3DUIs) were aimed to radically transform the current WIMP style and did not have the expected success. New proposals try to improve the existing windows environments with the addition of the third dimension. One example is the IBM RealPlaces guidelines (IBM), which describes a 3D user environment that consists of Places, each one containing groups of objects that support one or more tasks. Another example is "The Task Gallery" (Robertson et al., 2000), where the user can place windows on the walls, the ground or even the ceiling. There are also companies that are marketing 3D desktops, such as Win3D (http://www.clockwise3d.com) and 3DNA Desktop (http://www.3dna.net). Both transform the 2D desktop into a more intuitive 3D interface where the navigation is identical to a 3D video game.

However, it is widely known that the development of 3DUIs is significantly more difficult than WIMP UIs. The reason is that many designs are made by intuition, without reusing others' experience. A fundamental lesson learned in software engineering is that improvements in design require systematic work practices that involve well-founded methods.

In this paper, we present IDEAS (Lozano et al., 2000), a UI development methodology which not only supports the description of WIMP UIs but also the creation of 3DUIs. IDEAS structures the UI development process in four vertical levels: requirements, analysis, design and implementation. At each level, IDEAS offers a useful number of abstraction tools, and has been successfully applied to the development of WIMP applications, either for desktop or Web (Lozano et al., 2002). In order to describe 3DUIs, new concepts and notations have been added, which are the focus of the next section. Then, a case of study is used to demonstrate the IDEAS expressiveness, guiding the UI designer to model the system and then create both 2D and 3D UIs as the front-end of the system.

# 2    IDEAS as a 3DUI development tool

The idea of using the IDEAS methodology to develop 3DUIs in a systematic way was briefly discussed in a previous paper (Molina et al., 2002). We proposed to extend this methodology so the dialog model includes not only windows but also 3D concepts such as Rooms or Places, and the final GUI generation accommodates 3D world description languages too, such as VRML97 or X3D. The following paragraphs describe the changes brought to IDEAS at each abstraction level.

At requirements level, the UI developer can gather new information related to the third dimension of the space. As for the tasks, this information refers to the place where the tasks are carried out, data that can be added to the task template in the Task Model. On the other hand, users can express their preferences about the type of visualization of the tasks, which can also include a 3D graphical visualization, data that the developer compiles in the User Model.

At analysis level, no extensions have been added. On the opposite, the design level has been enriched with a *Map* that relates tasks to places in the 3D space. This map sketches the layout of the scene from a plan point of view or birds eye view. This is a 2D map because, as can be noticed in the 3D environments referenced in the introduction section, many of them constraint the user to navigate along the horizontal plane. This makes sense, as many of them simulate a real environment where gravity keeps the user on the ground, making navigation a more familiar task. In this map, a *Place* represent an area of the space where a task is carried out. A dialog from the Dialog Model or a set of them are attached to each place. A place has also a *Zone of Influence*, which is the region of the space where the user is able to manipulate the interaction objects of the dialog related with that place, and a *Preferred Point of View*, an optimal location of the observer to interact with that dialog. These concepts are similar to the ones introduced in the RealPlaces guidelines. Besides, the Dialog Model represents a hierarchy of tasks and subtasks which can be also reflected in the map by means of nesting places.

Finally, at the implementation level, the information compiled up to this point can now be used to generate a standard WIMP UI, or it can also be used to obtain a new 3DUI. In the last case, the Abstract Interaction Objects specified in the Dialog Model must be translated into concrete interaction widgets, which depends on which final platform will be used, such as VRML97 and JavaScript, Java3D or a virtual reality toolkit, such as Sense8 WorldUp. In any case, a 3D widget toolkit is needed, although the final presentation style relies on which guidelines are chosen.

# 3    A case of study: Multimedia zone

The Win3D desktop developed by ClockWise Corp. consists of several spaces, each one aimed to support a specific set of tasks. One of these spaces is the Multimedia room, where the user can play the music he or she likes and also choose a wallpaper to personalize his or her desktop. Based on this space, in this paper we propose a similar multimedia space as a case of study to show how the IDEAS can be used to develop both 2D and 3D UIs. In our Multimedia Zone, the user can play a music album, as in the previous example, selecting the song he or she wants to listen to and tuning the volume up and down. The wallpaper has been replaced by a slide projector, which allows the user to view a presentation, deciding when to go to the next slide or go back to the previous one. In the following sections, the development process will be detailed. Some of the diagrams shown here are image captures from the IDEAS CASE application, a tool developed to help the UI designer through the whole process.

## 3.1 Requirements

This level consists of three main models. The first one is the Use Case Model. In our case of study, the *Initial Use Case Diagram* should show that the "User" is the main actor and that he or she is able to "View a Presentation" or "Listen to Music". Each use case can be enriched by adding entities, rules or other use case that are included in them or are an extension of them. For instance, the "Listen to Music" use case has two entities, "Album" and "Track", and some rules, such as "No next song at the end of the album".

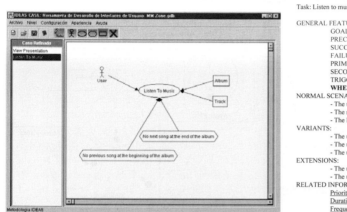

Task Model: Template (Example: Listen To Music)

Task: Listen to music.

GENERAL FEATURES:
    GOAL: Listen to a music album.
    PRECONDITION: The music album is available.
    SUCCESS CONDITIONS: The music album has been played.
    FAILURE CONDITIONS: The music album has not been played.
    PRIMARY ACTOR: User.
    SECONDARY ACTOR: None.
    TRIGGER ACTION: The user launchs the music player application.
    **WHERE: Multimedia zone.**
NORMAL SCENARIO:
    - The user makes the first song to be played.
    - The next song is played*.
    - The last song is played.
VARIANTS:
    - The user skips current song.
    - The user goes back to the previous song.
    - The user stops playing the music album.
EXTENSIONS:
    - The user turns the music up.
    - The user turns the music down.
RELATED INFORMATION:
    Priority: Normal.
    Duration: -
    Frequency: -

**Figure 1:** "Listen To Music" use case (left) and task template (right)

The next step is to identify user task from use cases, which represents the Task Model. In this case, only one task has been identified for each one of the two use cases. Thus, the "View a Presentation" use case corresponds to the task of the same name. In order to detail the actions involved, a *Task Template* is used. To support the development of 3D user interfaces, new information is added to that template, showing "where" the task is carried out. Following our example, the "Listen to Music" task is undertaken in the multimedia zone.

The last abstract model of this level is the User Model. With this model, the developer gathers information related with each user, such as the tasks that he or she is allowed to execute or the kind of visualization he or she prefers to use in each task. In our case of study, the user could choose to view the tasks not only as the classical 2D interface but also using 3D graphics.

## 3.2 Analysis

The first diagram generated at analysis level is the *Sequence Diagram*, which models the system behavior. Tasks and their related actions are the starting point to describe a sequence diagram, which also involves interface and entity classes. For instance, in the normal scenario of the "Listen to Music" task, the user can perform the action "Play album", which is received by the interface class and then to the system, which finds the first song to be played. The structure of classes are described in the *Role Model*, which uses UML class diagram to show the different classes and their relationships.

## 3.3 Design

The Dialog Model includes the generation of three different kinds of diagrams: the *Dialog Structure Diagram*, the *Component Specification Diagram* and the *Component Behavior Definition Table*. These diagrams allow the designer to specify the interaction spaces and the interaction objects corresponding to each task, as well as the links that the user follows to navigate from one space to another in order to complete his or her tasks. In our case of study, a main window is defined, "Multimedia Zone", where two options are given to the user, either "View a Presentation" or "Listen to Music". Selecting one takes the user to the window which corresponds to the desired task.

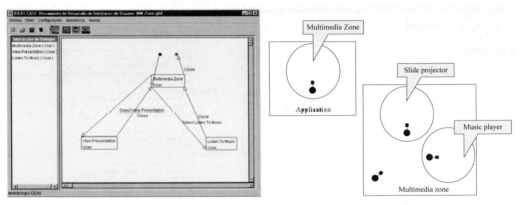

**Figure 2:** Dialog structure diagram (left) and map diagram (right)

These diagrams allows to generate a standard UI for a WIMP environment, but are not enough to develop new 3DUI. In order to accomplish this, a new diagram can be used: the *Map Diagram*. This Map relates parts of the previous diagrams with places in the 3D space. In our example, the "Multimedia Zone" is a *Place* which defines a *Zone of Influence* (represented with a circle in the map) and a *Preferred Point of View* (depicted as a Pinocchio camera). A place can also includes other places, which can be detailed in additional diagrams, as when we zoom in or out a picture. This creates a hierarchy of zones of influence and points of view, which can be used to help the user to navigate through the environment. Thus, the "Multimedia Zone" has also two more places which corresponds to the two main tasks: "View a Presentation" and "Listen to Music". This information will be used to generate the 3D user interface in the next level.

## 3.4 Implementation

At implementation level, the Presentation Model is built from the dialog model taking into account style guides. The interaction objects modeled in the dialog model were abstract ones (AIOs), representing *Controls*, *Presentors* and *Components*. These are now translated into concrete interaction objects (CIOs) depending on the final platform. Thus, a control can be translated into a button in a WIMP desktop interface, but also into a hyperlink in a Web interface. In any case, a classical 2D UI would be generated.

Thanks to the new abstraction models added to IDEAS Tool, the developer has now the chance to generate a new 3DUI. In our case of study, a 3DUI has been built taking into account the presentation style of the Win3D desktop and using VRML97 as the 3D world description language. A different color or texture is used for each place because differentiation helps the user to identify spaces, as can be learnt from studying cityscape environments (Pettifer et al., 1999). Dialogs are presented as *Interactive Walls*, where interaction objects can be found. In order to manipulate them, the user needs to be within the zone of influence of the place. The user can navigate to a place using arrow keys as in a 3D game, or just clicking on the target place to be positioned at the preferred point of view. The result is a new kind of interaction space that can be both productive and enjoyable.

**Figure 3:** 3D user interface

# 4    Conclusions

In this paper, we have presented IDEAS, an multi-paradigm development tool which helps in the design and creation of 2D and 3D UIs. The 3D concepts and notations are based on the latest proposals in 3D environments for PCs, 3D spaces where places have a strong relationship with tasks. The Map Diagram is one of the main changes brought to IDEAS, an abstract model which relates places in 3D space with the Dialog Model. Thus, following the development process given by IDEAS, the UI designer is not only able to generate classical WIMP UIs, but also new 3DUIs.

## References

IBM, RealPlaces Design Guide. http://www-3.ibm.com/ibm/easy/eou_ext.nsf/Publish/580

Lozano, M., González, P., Montero, F., Molina, J. P., & Ramos, I. (2002). A Graphical User Interface Development Tool. *Proceedingsof the 16th British HCI Conference*, 2, 62-65.

Lozano, M., Ramos, I., & González, P. (2000). User Interface Specification and Modeling in an Object Oriented Environment for Automatic Software Development. *IEEE 34th International Conference on Technology of Object-Oriented Languages and Systems, TOOLS-USA*, 373-381

Molina J.P., González, P., & Lozano, M. D. (2002). A Unified Envisioning of Future Interfaces. *Proceedings of the 2nd IASTED International Conference VIIP 2002*, 185-190.

Pettifer, S., West, A., Crabtree, A., & Murray, C. (1999). Designing shared virtual environments for social interaction. *User Centred Design and Implementation of Virtual Environments*. Available at http://www.cs.york.ac.uk/hci/kings_manor_workshops/UCDIVE/

Robertson, G., van Dantzich, M., Robbins, D., Czerwinski, M., Hinckley, K., Risden, K., Thiel, D., & Gorokhovsky, V. (2000). The Task Gallery: A 3D Window Manager. *Proceedings of CHI 2000*, ACM Press, 494-501.

# Visual Tracking for a Virtual Environment

*N. Murray, J. Y. Goulermas and T. Fernando*

Centre for Virtual Environments, Business House,
University of Salford, Salford, M5 4WT, U.K.
{n.murray, j.y.goulermas, t.fernando}@salford.ac.uk

## Abstract

The paper will discuss experiences of integrating a motion tracking system (Vicon Motion Systems 612) onto a workbench (Trimension V-Desk 6), and evaluate the advantages and disadvantages of such an approach including a comparison with the acoustic tracker originally fitted to the workbench (Intersense IS-900). The motion tracking system was used for tracking the head and hand held wand interaction device supplied with the original acoustic tracking system. Interaction capabilities with the system were extended by tracking the movements of the hand and recognising hand gestures as input into the virtual environment. The integrated system was then used within a geometric constraint based maintenance environment for the assembly and disassembly of engineering components.

## 1    Introduction

Tracking is the process of obtaining the location (position and orientation) of a moving object in real-time. After the display device, tracking is the probably the most important component within a large screen projection virtual environment system. Typically, users will need to have at least a tracked head so that their location with respect to the workbench or cave can be gauged and for interaction one or two hand-held devices are required so that the user can have a virtual representation of their hand and so be able to select and manipulate objects within the environment. With inaccurate tracking the process of interacting within the environment can prove cumbersome and detract from the experience of using the environment and lead to simulator sickness. This can be caused by proprioceptive conflicts, such as static limb location conflicts, dynamic visual delay (lag) and limb jitter or oscillation. The following are some of the criteria defining the requirements of a tracking system:

- accuracy – the error between the real location and the measured location. (Ribo et al., 2001) state that position should be within 1mm, and orientation errors should be less than 0.1 degrees.
- degrees of freedom – the capability of the system to capture up to 6 degrees of freedom (DOF)consisting of position and orientation (roll, pitch and yaw). VR systems typically need 6 DOF capabilities.
- range – the maximum working area within which the system can or needs to operate. This will be dependent on the size of the workbench or cave volume. These range from 1.5 metre to 3metre cubic volumes.
- update rate – maximum operating frequency of reporting positional values. VR systems need to maintain a frame rate of at least 25Hz. Therefore a tracking system should be able to track objects at a minimum of 25Hz.
- user requirements – it is preferable that the system should not restrict the mobility of the user so

it would be an advantageous for the tracking components to be wireless, furthermore they should be light and easy to hold/wear.

None of the current commercially available VR tracking systems fulfils all of the above criteria (see (Rolland et al., 2000) for a recent survey). The following section will describe how the Vicon system was initially integrated with the workbench as a substitute for the acoustic tracking. With this integration complete, the extra advantages offered by the system over the acoustic tracking were explored by the creation of alternate input devices. This involved the initial creation of a tracked hand and led to the development of a wireless glove based gesture recognition system. Finally, conclusions and future work are outlined.

## 2    System Integration

Current areas of use of the Vicon motion capture system have been life sciences, with applications such as clinical gait analysis, biomechanics research and sports science and within the area of visual arts for broadcast, post production and game development. This section will review the experience of integrating the Vicon system with a Trimension V-Desk 6 and comparisons with the Intersense acoustic tracking system.

The workbench was fitted with 6 cameras, four were placed in front of the user orientated towards the working area and two were placed behind the user looking over their shoulder, see Figure 1 (left) where four of the cameras have been highlighted. The cameras can either be fitted with visible light or infra red emitters for detection of the markers. As the workbench needs to be used in an unlit room the infra red emitters were attached the cameras. Components that are to be tracked are fitted with retro-reflective markers ranging in size from 4mm, 6.5mm and 12.5mm (see Figure 1 (right)).

Figure 1: Workbench with 4 of the 6 cameras marked (left) and Glasses & wrist prototypes (right).

For proper operation of the system, the cameras and the components within the real world need to be calibrated. Calibration of the cameras involves the tracking of a fixed size wand fitted with 2 retro-reflective markers at a known distance from one another within the operating volume of the cameras that takes about a minute to complete. The devices that are to be tracked also have the retro-reflective markers attached. The markers were placed on the required devices, in this case the shutter glasses (see Figure 1(right)) and the Intersense wand. Originally 5 12.5mm balls were placed on the devices. The locations of the balls were measured with respect to the origin of the device and using this information the system can calibrate the device for future real-time recognition. The 12.5mm balls can be seen attached to the wrist strap on the left of Figure 1(right). After successful testing of the setup the markers were replaced with 6 4mm balls. These are much smaller are much less noticeable when placed on the devices in question. These were calibrated, tested and compared with the larger balls and were found to produce results of the same accuracy.

With both the motion tracking system and the acoustic tracking systems in place, the accuracy of systems could be compared using the Intersense wand as it could also be tracked using the Vicon system. The wand was placed at various fixed positions within the operating space of the workbench and the static accuracy of wand was taken. For the Intersense tracker, values as those specified by the manufacturers were observed. That is the static accuracy or jitter observed at a fixed location fluctuated within a range of $\pm$ 2.0–3.0mm and with an angular error of 0.25 degrees. This is for the wired wand. For the wireless wand these values increase to $\pm$ 3.0–5.0mm and 0.5 degrees respectively. The optical tracking system satisfied the criteria outlined in the introduction. The system provided real-time simultaneous tracking with accuracy and jitter within the defined limits. The system is both light weight (several small retro-reflective markers) and requires no wires. The tracking is robust and only suffers from the line of sight problem - where the markers may be occluded from the cameras. This did not prove a problem as the user is always facing the workbench under normal operating conditions and so is facing 4 of the 6 cameras.

After the initial work to get the head and interaction devices tracked, further work was performed to track alternative input devices and to perform gesture recognition. The motion tracking system uses retro-reflective balls for tracking which can be attached to any device for tracking purposes. Tracked devices can be created by attaching the inexpensive markers to any object and calibrating the object. In this way the system can track a multitude of input devices. Work was thus carried out in creating new devices to be tracked by the system other than just a replacement for the input devices available for the acoustic tracking system (head and wand input device). Multiple glasses were marked so that as users took control at the workbench they would see the correct viewpoint within the workbench rather than having to change to the glasses with the tracking device fitted.

Further work looked at creating trackers for the wrist and fingers to provide a simple pinch hand for interaction within the virtual environment. This involved the creation of a wrist strap with markers placed on it to track the location of the wrist, and markers attached to the fingers via rings for detecting pinch gestures (similar to the operation of pinch gloves). The wrist strap was used to track the movement of the hand and was associated with a 3D pointer that was tracked at the workbench and used for ray casting within the environment for selecting objects. The pinch gesture operation was used to signify selection of the component. This has allowed for multiple tracked pointers within the system and has led to the extension of the system to support two handed interaction.

## 3    Gesture Recognition

To implement gesture recognition, we attach a number of 4mm markers on a fabric glove. As shown in Figure 2 (left), there is a small number of markers placed on the metacarpal and some on each individual finger. The latter markers are placed at the finger joints, so that the distance between immediately adjacent ones remains fixed; this allows the formation of rigid body segments as seen in Figure 2 (right). The metacarpal is considered a single almost rigid body, while on average there are three bodies per finger.

The VICON system is used in two stages. In the former, the user repeatedly performs a series of gesticulations which span the entire human kinematic range at all possible finger positions and angles. Subsequently, manual labelling of the markers takes place and the system calculates the kinematic information relating the trajectories of all markers of the entire motion trial to a predefined kinematic structure which correspond to the hand marker and body model. This is an off-line procedure and it only needs to take place once, unless the glove size and/or markers position change. The second stage is the real-time operation of the system which provides the spatio-temporal data, while the user is performing a series of gestures. During this phase, the captured marker trajectories are best-fit to the stored model information.

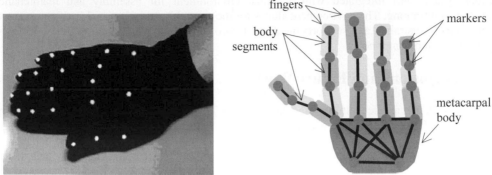

Figure 2: Marker attachments on a glove (left). Schematic placement of markers and formation of body segments in a typical glove marker set configuration (right).

We target recognition of two types of gestures: static gestures (postures) and dynamic (time-varying) ones. For both cases, we make use of Hidden Markov Models (HMM) (Rabiner, 1989), which have been previously used successfully in gesture recognition but utilising different data sensors; see for instance (Yoon, Soh, Bae & Yang, 2001), (Nam & Wohn, 1997). A HMM is a type of probabilistic state model whose states cannot be observed directly, but only through a sequence of observations. First-order HMMs follow the property that the current state only depends on the immediately preceding one. A HMM consists of the following elements: a set of hidden states $S=\{s_1,...s_N\}$, a set of observation symbols $V=\{v_1,...v_M\}$ a state transition matrix $A=(a_{ij})$, with $a_{ij}$ being the probability of moving from state $s_i$ to $s_j$, an observation probability matrix $B=(b_{ik})$, where $b_{ik}$ is the probability of emitting symbol $v_k$ at state $s_i$ and $\pi$ an initial probability distribution for each state $s_i$. In this way, a HMM is fully defined as $\lambda=(A,B,\pi)$.

There are three basic issues involved with a given model $\lambda$. The first, the *evaluation*, relates to the calculation of the probability of observing a given sequence of symbols $o=(o_1,...,o_T)$ for $T$ discrete time events, i.e. $P(o|\lambda)$. If $s$ defines some state sequence of length $T$, then we have:

$$P(o\mid\lambda)=\sum_s P(o\mid s,\lambda)\cdot P(s\mid\lambda) \tag{1}$$

In Equation 1, the probability of $o$ given a sequence $s$ is $\prod_{t=1}^{T} b_{s_t o_t}$, while the probability of having generated $s$ is $\pi_{s_1}\cdot\prod_{t=1}^{T-1} a_{s_t s_{t+1}}$. For the actual evaluation however, we use the Forward-Backward algorithm, which effectively alleviates the exponential complexity of Equation 1. The other two important issues related to HMMs are: *decoding*, that is how we can estimate the optimal state sequence $s$ given $o$ and $\lambda$, and, *learning*, that is how to estimate the three parameters of $\lambda=(\pi,A,B)$, given some sequence $o$, such that $P(o|\lambda)$ is maximised. These two questions are normally handled with the Viterbi and the Baum-Welch algorithms, respectively; see (Rabiner, 1989).

All data was preprocessed so that discrete observation sequences were obtained for training and testing the HMM. In order to achieve positional and rotational recognition invariance, we used normalised marker distances between selected pairs of markers for static gestures. For dynamic ones, a type of chain coding of the trajectory of the averaged marker positions (single point) was used. For a total of $G$ gestures, a set of $\lambda_1,...,\lambda_G$ models was stored, and each captured test frame was compared with all models to find the one yielding the highest value probability $P(o\mid\lambda_i)$. As mentioned in (Nam & Wohn, 1997), the recognised gesture can be used as symbolic commands for object description and/or action indication. Navigation, manipulation and environmental control are easily and more naturally achievable by using gesture recognition. The motion based tracking system was integrated with a virtual environment for assembly and maintenance simulation and training. The environment allows for the assembly and disassembly of components via geometric constraints and supports the simulation of the mechanisms (allowable movements) of the constructed components.

# 4    Conclusions and Future Work

The Vicon system was successfully integrated with the Trimension V-Desk 6 and was found to be more accurate than the Intersense acoustic tracker. The system offers further advantages in that it is possible to place the markers on any device for calibration and so allows the easy addition of extra input devices for minimal extra cost. These were used for tracking the wrist and then a simple pinch hand to a gesture recognition system utilising markers placed on a glove. For gesture recognition, the provided marker and body data were rich enough to allow off-line training and on-line recognition. Although accuracy depends on data pre-processing as well as the pattern recognition algorithm used, the utilisation of the provided 3D positional data was straightforward.

The system also satisfies the criteria that it should be minimally invasive as the markers are light weight and do not require wires. The disadvantages of the system are that it suffers from line of sight problems although this was not found to be a problem with the relatively small operating area around the workbench as the user is always facing the screen and so is facing 4 of the 6 cameras. Also, certain problems were caused in the real-time mode, where a number of frame sequences had missing marker information from the data stream. This problem could be improved by adding extra markers on the wrist for stability of recognition.

Further proposed work is to test the installation of the system within the Trimension Reactor cave. The cameras supplied by Vicon have decreased in size so their installation within the Reactor would be possible and would provide wireless interaction within the Reactor and so increase its usability. Within the Reactor the line of sight problem may become prominent but this issue would need to be investigated. Concerning the gesture recognition part, we aim to implement simultaneous recognition of static and dynamic gestures and integration with the environment.

# References

[1] Ribo, M., Pinz, A., and Fuhrmann, A. L. (2001). A new optical tracking system for virtual and augmented reality applications. In IEEE Instrumentation and Measurement Technology Conference.

[2] Rolland, J. P., Baillot, Y., and Goon, A. A. (2000). A survey of tracking technology for virtual environments. In Barfield, W. and Caudell, T., editors, Fundamentals of Wearable Computers and Augmented Reality.

[3] Rabiner, L. R. (1989). A tutorial on Hidden Markov Models and Selected Applications in Speech Recognition. *Proc. IEEE*, 77 (2), 257-286.

[4] Yoon, H. S., Soh, J., Bae, Y. J., and Yang H. S. (2001). Hand gesture recognition using combined features of location, angle and velocity. *Pattern Recognition*, 34, 1491-1501.

[5] Nam, Y. and Wohn K. (1997). Recognition of hand gestures with 3D, nonlinear arm movement. *Pattern Recognition Letters*, 18, 105-113.

# Mixed Systems: Combining Physical and Digital Worlds

*Laurence Nigay*
Université de Grenoble
CLIPS-IMAG BP 53
38041 Grenoble
cedex 9 France
laurence.nigay@imag.fr

*Emmanuel Dubois*[+]
3Université de Toulouse
IRIT, 118 route de Narbonne
31062 Toulouse
cedex 4 France
emmanuel.dubois@irit.fr

*Philippe Renevier*
Université de Grenoble
CLIPS-IMAG BP 53
38041 Grenoble
cedex 9 France
philippe.renevier@ imag.fr

*Laurence Pasqualetti*
FT R&D DIH/UCE
38-40 rue G. Leclerc
92794 Issy-les-Moulineaux France
laurence.pasqualetti@francetelecom.fr

*Jocelyne Troccaz*
Université de Grenoble
TIMC-IMAG, Faculty of Medicine
38706 La Tronche cedex France
jocelyne.troccaz@imag.fr

## Abstract

The growing interest of designers for mixed interactive systems is due to the dual need of users to both benefit from computers and stay in contact with the physical world. Based on two intrinsic characteristics of mixed interactive systems, target of the task and nature of augmentation, we identify four classes of systems. We then refine this taxonomy by studying the fluidity and the links between the physical and digital worlds.

## 1    Introduction

In recent years, mixed interactive systems have been the subject of growing interest. The growing interest of designers for this paradigm is due to the dual need of users to both benefit from computers and interact with the real world. A first attempt to satisfy this requirement consists of augmenting the real world with computerized information: This is the rationale for Augmented Reality (AR). Another approach consists of making the interaction with the computer more realistic. Interaction with the computer is augmented by objects and actions in the physical world. Examples involve input modalities (Nigay & Coutaz, 1995) based on real objects, such as Fitzmaurice et al's bricks (Fitzmaurice, Ishii & Buxton, 1995) or Ishii & Ulmer's phicons (Ishii & Ulmer, 1997). Ishii has described this interaction paradigm as the Tangible User Interface.

This paper presents our analysis of mixed interactive systems that combine the real and digital worlds. We clarify the notion of mixed interactive systems by defining four classes of systems. After presenting and illustrating our taxonomy, we then refine its classes of mixed systems by studying the fluidity and the links between the real and digital worlds. The discussion is illustrated using three systems developed at the University of Grenoble, whose main features are presented in the next section.

## 2    Illustrative examples

The first system, Mirror Pixel, enables a user to modify a digital drawing displayed on screen using a pen and a sheet of paper. To do so, as shown in Figure 1, a camera points to the sheet of paper used by the user to draw or write. On screen (Figure 1), the user can see the current drawing to be modified as well as her/hand and the pen. Displaying the hand and the tool superimposed on

---

[+] This work has been done while E. Dubois was a PhD student at the University of Grenoble in the laboratories CLIPS-IMAG and TIMC-IMAG.

the drawing enables the user to precisely modify the drawing. The user focuses on the screen and does not look at her/his hand. The system has been developed by C. Lachenal and is described in (Vernier, Lachenal, Nigay & Coutaz, 1999).

The second system, CASPER (Computer ASsisted PERicardial puncture) (Chavanon et al., 1997), is a system that we developed for computer assistance in pericardial punctures. The clinical problem is to remove a build up of fluid (water, blood) in the region around the heart (pericardium), the effect of which is to compress the heart. This procedure involves minimal access to the chest. CASPER allows pre-operative acquisition and modeling of a 3D stable region in the pericardial effusion from which a target is selected and a safe trajectory is planned. During the surgery, guidance is achieved through use of an optical localizer that tracks the needle position. Figure 2-a shows the application in use during the intervention (guidance step). The system transforms the signal from the needle localizer into a graphical representation of the position and the orientation of the needle. In the same window on screen, presented in Figure 2-b, the current position and orientation of the needle are represented by two mobile crosses, while one stationary cross represents the planned trajectory. When the three crosses are superimposed the executed trajectory corresponds to the planned one.

(a)        (b)

**Figure 1:** Mirror Pixel system       **Figure 2:** CASPER system

The third example, MAGIC (Mobile, Augmented reality, Group Interaction, in Context), is a generic hardware and software mobile platform for collaborative Augmented Reality. We first describe the hardware and then the software responsible for the fusion of the two worlds, the physical and digital worlds.

The hardware platform is an assembly of commercial pieces of hardware. We use a Fujitsu Stylistic 3400 pen computer. This pen computer is a PC, with a tactile screen having the size of a A4 sheet of paper. Its weight is 1,5 kg. Moreover, it has a video exit allowing the dual display, to which we connect a semi-transparent Head-Mounted Display (HMD). A camera is fixed between the two screens of the HMD (between the two eyes). The hardware platform also contains a magnetometer, which determines the orientation of the camera as well as a GPS which locates the mobile user. For sharing data amongst users and communication between users, a WaveLan network was added.

Based on the hardware platform described above, we designed and developed generic interaction techniques that enable the users to perform actions that involve both physical and digital objects. In order to smoothly combine the digital and the real, we create a gateway between the two worlds. This gateway (Figure 3) has a representation both in the digital world (displayed on the screen of the pen computer) and in the real environment (displayed on the HMD).

• Information from the physical environment is transferred to the digital world thanks to the camera carried by the user. The camera is positioned so that it corresponds to what the user is seeing, through the HMD. The real environment captured by the camera is displayed in the gateway window on the pen computer screen. Based on the gateway window, we allow the user to select or click on the real environment. The interaction technique is called "Clickable Reality".

• Information from the digital world is transferred to the real environment, via the gateway window, thanks to the HMD. For example the MAGIC user can drag a drawing or a picture stored in the database to the gateway window. The picture will automatically be displayed on the HMD on top of the physical environment. As shown in Figure 3, moving the picture using the stylus on the screen will move the picture on top of the physical environment. This is for example used by archaeologists in order to compare objects, for instance a physical object with a digital object from the database.

- Another technique, "Augmented Field", consists of superimposing an image of an object in its original real context (in the real world), thanks to the semi-transparent HMD. Because a picture is stored along with the location of the object, we can restore the picture in its original real context (2D location).

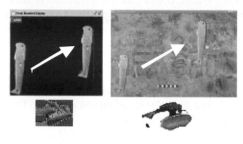 

**Figure 3:** Gateway between the two worlds: Comparing a physical object with a digital one

**Figure 4:** TROC: a mobile player is collecting a digital object.

Two systems have been developed using the MAGIC platform. The first one is dedicated to archaeological prospecting activities. It enables the archaeologists to perform ground analysis of the site and to communicate with other mobile archaeologists working in the site as well as with distant archaeologists. The complete software and its architecture are detailed in (Renevier & Nigay 2001). The second system based on the MAGIC platform is called TROC and is a collaborative game based on bartering. The mobile players must collect a set of digital objects such as a "blue cow" using physical cubes. Figure 4 presents a view through the HMD while a player is collecting a digital object using a physical cube.

## 3    Taxonomy of mixed interactive systems

We identify four classes of interactive mixed systems. We have defined these classes of systems on the basis of their intrinsic characteristics identified by first studying Computer Assisted Medical Interventions (Dubois et al.,1999). We first distinguish two classes of systems based on the two possible targets of the user's task, namely, the real world and the computer:

- In Augmented Reality systems (AR), interaction with the real world is augmented by the computer.
- In Augmented Virtuality systems (AV), interaction with the computer is augmented by objects and actions in the real world.

In Table 1, we locate these two classes along with graphical user interface (GUI) and purely physical tasks. Using CASPER and MAGIC, interaction with the real world is augmented by the computer. For example in both cases, the physical world is enriched by digital objects. They belong to the Augmented Reality class. On the other hand the mirror pixel system enables a user to modify a digital picture by drawing with a pen on a sheet of paper. The system is therefore an example of an Augmented Virtuality system.

**Table 1**: Four types of interaction

| | Interaction using physical objects | Interaction using digital object |
|---|---|---|
| Task in the physical world | Physical interaction | **Augmented Reality Mixed Interaction (1)** |
| Task in the digital world | **Augmented Virtuality Mixed Interaction (2)** | Graphical User Interface |

Then we consider the augmentation provided by the system. It can take on a number of different forms. If we refer to the Theory of Action (Norman 1986), augmentation can be dedicated to the execution phase (action) and/or to the evaluation phase (perception). For example the mirror pixel system augments the execution phase by defining new modalities involving real objects (a pen and a sheet of paper). On the other hand, augmented evaluation in the real world consists for example of superimposing digital objects on physical objects as in MAGIC.

To sum up, the *target of the task* (AR and AV) and *the nature of augmentation* constitute two orthogonal classification axes that further lead us to define four classes of mixed systems. The four classes are fully described and illustrated in (Dubois et al.,1999).

The above four classes of mixed interactive systems are not sufficient to fully capture the differences in terms of interaction between mixed systems. For example CASPER and MAGIC belong to the same class while the interaction techniques they offer are quite different. Due to the various sources of information from the computer as well as from the physical world, we need to additionally consider the perceptual and cognitive continuity that increases the fluidity between the two worlds:

> *Definition*: Perceptual continuity (Dubois, Nigay & Troccaz 2002) is verified if the user directly and smoothly perceives the different representations of a given concept. Cognitive continuity is verified if the cognitive processes that are involved in the interpretation of the different perceived representations lead to a unique interpretation of the concept resulting from the combination of the interpreted perceived representations.

When assessing continuity, the perceptual environments involved in the interaction must be identified as well as the different representations of involved objects. For example, in CASPER the surgeon has to look both at the real needle to avoid any distortion of the tool and at the screen to get the guiding information (represented as a stationary cross) according to the virtual needle (represented as two mobile crosses). Keeping the trajectory aligned and controlling the depth of the needle by referring to the visual display was difficult to implement. The required switch between the screen and the operating field was disturbing to the surgeon. Perceptual continuity is not verified here. Likewise, at the cognitive level, two cognitive processes are involved and are very different. Indeed, the representation of the needle on screen (the two mobile crosses) is bi-dimensional, while the position of the real needle is of course three-dimensional. The matching between the real object and its representation is far from direct. Cognitive continuity is transgressed. On the other hand, using MAGIC, the user is looking at the physical environment that is augmented by digital objects. There is only one perceptual environment. Nevertheless the digital objects are pictures, bi-dimensional objects, while the physical objects are three-dimensional. As a conclusion, the perceptual continuity is verified but at the cognitive level, the digital objects are not completely integrated in the physical environment. For example the TROC players told us that they have the feeling that the digital objects they are looking for (cats, cows, etc.) are floating in the air. Finally in the pixel mirror system, and as opposed to CASPER and MAGIC, perceptual and cognitive continuity is verified. The user focuses on the screen (one single perceptual environment) : the display of her/his hand using the tool on the top of the drawing enables precise modifications (cognitive continuity).

In addition to the perceptual and cognitive continuity, we also study the links between the two worlds. Two axes are relevant to characterizing these links: the owner of the link (i.e., he/she who is defining the link) and their static/dynamic character. For example, in the pixel mirror system, the designer conceived the link between the two worlds, by using a camera pointing to the user's hands. This link is static. Likewise, the designer of CASPER decided to combine the digital representation of the puncture trajectory with the representation of the current position of the needle thanks to a 3D localizer. Such a link is again static. On the other hand, using MAGIC, the users dynamically define new digital objects that are combined with physical objects. As pointed out in (Mackay 2000), instead of fixing the relationship between the two worlds during the design, "another strategy is to explicitly give the control to the users, allowing them to define and more importantly, continue to evolve, the relationship between the physical and virtual documents". A promising avenue to let the users specify such links is multimodal commands (McGee, Cohen & Wu, 2000). For example in our TROC system, the player could issue the voice command "this door is now a trap for others" while designating a door

## 4    Conclusion

We have presented a classification space that describes the properties of mixed interactive systems. This classification highlights four main characteristics of such systems: (1) the two possible targets of the user's task, namely, the real world / the computer, (2) the two possible types of augmentation, namely augmented execution / evaluation, (3) the perceptual and cognitive continuity between the physical and digital worlds,(4) the links between the physical and digital worlds, defined by the designer / the user and their static/dynamic character.

The contribution of our classification space is two-fold:

1.  Classes of existing mixed interactive systems are identified. We illustrated this point using three systems that we have developed.
2.  By identifying and organizing the various aspects of interaction, our framework should also help the designer to address the right design questions and to envision future mixed interactive systems.

## References

Chavanon, O., Barbe, C., Troccaz, J., Carrat, L., Ribuot, C., Blin, D. (1997). Computer Assisted Pericardial Punctures: animal feasibility study. In *Proceedings of CVRMed/MRCAS'97 LNCS 1205* (pp. 285-291), Springer-Verlag.

Dubois, E., Nigay, L., & Troccaz, J. (2002). Assessing Continuity and Compatibility in Augmented Reality Systems. *UAIS, International Journal on Universal Access in the Information Society*, Special Issue on Continuous Interaction in Future Computing Systems, 1(4), 263-273.

Dubois, E., Nigay, L., Troccaz, J., Chavanon, O., Carrat, L., (1999). Classification Space for Augmented Surgery. In *Proceedings of. Interact'99* (pp. 353-359), .IOS Press.

Fitzmaurice, G., Ishii, H., & Buxton, W. (1995). Bricks: Laying the Foundations for Graspable User Interfaces. In *Proceedings of CHI'95* (pp. 442-449), ACM Press.

Ishii, H., & Ulmer, B. (1997). Tangible Bits: Towards Seamless Interfaces between People, Bits and Atoms. In *Proceedings of CHI'97* (pp. 234-241), ACM Press.

MacKay, W. (2000). Augmented Reality: Dangerous Liaisons or the Best of Both Worlds? *In Proceedings of DARE'00* (pp. 170-171), ACM Press.

McGee, D., Cohen, P., & Wu, L. (2000). Something from nothing: Augmenting a paper-based work practice via multimodal interaction. *In Proceedings of DARE'00 Designing Augmented Reality Environments* (pp. 71-80), ACM Press.

Norman, D. (1986). Cognitive Engineering. In User Centered System Design, New Perspectives on Human-Computer Interaction (pp. 31-36), L. Erlbaum Associates.

Nigay, L., & Coutaz, J. (1995). A Generic Platform for Addressing the Multimodal Challenge. In *Proceedings of CHI'95* (pp. 98-105), ACM Press/ Addison-Wesley Publishing Co.

Renevier, P., & Nigay L. (2001). Mobile Collaborative Augmented Reality, the Augmente Stroll. In *Proceedings of EHCI'2001 LNCS 2254* (pp. 315-334), Springer-Verlag.

Vernier, F., Lachenal, C., Nigay, L., & Coutaz, J. (1999). Interface Augmentée par Effet Miroir. In *Actes de la Conférence IHM'99* (pp. 158-165), Cepadues Publ.

# Virtual Assembly Based on Stereo Vision and Haptic Force Feedback Virtual Reality[1]

George Nikolakis, George Fergadis and Dimitrios Tzovaras

Informatics and Telematics Institute
Centre for Research and Technology Hellas
1st Km Thermi-Panorama Road
57001 (PO Box 361)
Thermi-Thessaloniki, Greece
Dimitrios.Tzovaras@iti.gr

## Abstract

This paper presents the virtual reality application developed for the project called "A Virtual Assembly Environment for Industrial Planning" funded by General Secretariat for Research and Technology of Greece. The project aims at developing a highly interactive and extensible haptic VR system that allows the user to study and interact with various virtual objects and perform assembly planning in the virtual environment. An application has been developed based on the interface provided by the CyberTouch™ and the CyberGrasp ™ force feedback haptic devices. The main objective of this paper was to develop a complete assembly planning system for industrial applications based on force feedback haptic interaction techniques in simulated virtual reality environments. The challenging aspect of the proposed VR system is that of addressing visual and haptic feedback to the end user in order to emulate realistic assembly conditions.

## 1 Introduction.

In recent years there has been a growing interest in developing force feedback interfaces that allow people to access not only two-dimensional graphic information, but also information presented in 3D virtual reality environments (VEs). It is anticipated that the latter will be the most widely accepted, natural form of information interchange in the near future (Burdea 1994). The greatest potential benefits from VEs, built into current virtual reality (VR) systems, exist in such applications as design, planning, education, training, and communication of general ideas and concepts.

Other studies and products focusing on the same subject are the Virtual Assembly Design Environment (VADE), the PTC DIVISION™ MockUp & Reality and the Virtual Environment for General Assembly Simulation (VEGAS). VADE is a VR based engineering application that allows engineers to plan, evaluate, and verify the assembly of mechanical systems. This system focuses on utilizing an immersive virtual environment tightly coupled with commercial Computer Aided Design (CAD) systems (Sankar Jayaram et al. (1999)). PTC DIVISION™ MockUp & Reality is a commercial product for the graphical representation of mechanical products. It

---

[1] This work was supported by the GSRT projects PABET "Virtual Assembly Environment for Industrial Planning" and PRAXE VRSENSE " A Virtual Reality Tool for Advanced Information Services."

provides the user with limited interaction abilities. (http://www.ptc.com/). VEGAS provides engineers with a tool to investigate assembly feasibility. The application works with model data derived from popular CAD. Once in the virtual environment the engineers can interact with their models to perform assembly tasks as they would in the physical world. The assembly methods are being analyzed to inform the user if the parts fit together or if there are undesirable interferences taking place (Johnson T. (2000)).

In the present paper our intention is to combine visual and haptic information in innovative ways, in order to perform an assembly in the VE. The CyberTouch™ and CyberGrasp™ haptic devices (Virtex) were selected, based on its commercial availability and maturity of technology. In this paper we have developed an environment for performing a virtual assembly planning and specific software to support the communication of the applications with the peripheral devices. We have also integrated a new optimized collision detection algorithm (based on RAPID (Gottschalk, S., Lin, M. C., Manocha, D., (1996))) in the Virtual Hand Suite (VHS) software library ("Virtual Hand Suite 2000 user & programmers guide", 2000) in order to improve the performance of the whole system.

# 2 Assembly planning system

The main parts constructing the proposed assembly planning system presented in this paper are: (1) the scenario authoring tool, (2) the assembly simulation environment and (3) the assembly review part.

## 2.1 Authoring tool

The authoring tool is implemented to assist the user create the assembly scenarios. The scenario authoring tool provides: a) functionalities for the composition of 3D simulations, b) capability of composing, processing and storing scenarios, c) allows the user to import the 3D models of the objects constituting an assembly and d) has procedures for supporting assignment of special properties to the objects in the VE. The design data are stored in an XML formatted file in order to be reused by the same or other users.

The scenario authoring tool (figure1) supports:
- Data process (save, open, edit).
- Loading 3D models in VRML (Virtual Reality Modelling Language) format as parts of the assembly.
- Loading of parameters in the scene objects i.e. if we need a tool to assemble a part.
- Setting the assembled position and orientation of objects (final state after the object is assembled).
- Setting the initial position and orientation of objects in the scene (The construction starts from that state).
- Connection with the assembly simulation environment
- Possibility of storing and processing of the scenario parameters in XML format

The expected complexity of the scenario files, lead to the adoption of a specific standard as scenario format. The scenario file format selected is XML and its form is completely determined by an appropriate DTD (Document Type Definition) file. The authoring tool provides a user interface that allows the user create the XML scenario files.

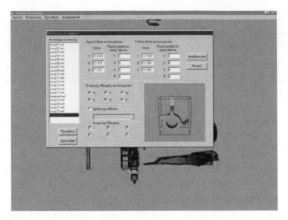

**Figure 1** Authoring tool environment

## 2.2 Assembly Simulation Environment

The proposed application allows the user to assemble parts of an object and record the assembly process for post-processing. The assembly procedure can be done using one or two hands (i.e. one or two haptic VR gloves CyberTouch or CyberGrasp). A position tracker (MotionStar Wireless Tracker of Ascesion Technologies Inc. (2000)) with one or two position sensors installed is used to detect the position and orientation of the user hands in the space.

Another element of virtual models is that of VE agents. VR agents are sophisticated software components with the capability to control a dynamic virtual environment and take actions based on a set of aims and rules. There are two kinds of agents implemented in the VR Assembly environment: a) the snap agent and b) the tool agents (a screwdriver and a wrench are implemented).

The snap agent is responsible to decide when two parts in the scene must connect. The aim of that agent is to place the components of the assembly in the correct position and to allow two or more components construct a new larger component that can act like any other component. The rule that snap agent uses to connect two objects is a distance threshold and a "first contact rule". The "first contact rule" detects which sides of the objects collide first (using bounding boxes). If the colliding sides are valid then the distance from the current position to the snapping position is calculated. When this distance is smaller than the distance threshold the objects snap to each other. The distance threshold used depends on the radius of the smaller bounding sphere of the objects.

The tools (figure 2) are components with the capability to control objects in the dynamic virtual environment. The tools aim to increase the immersion of the user in the VE. Unlike the snap agent, which is always active, the tools need to be activated by the user. Tools provide constraint to the object movement and allow the user do construction tasks. The virtual tools have the potential to increase user productivity by performing tasks on behalf of the user and increase the immersion of the user in the virtual environment. Furthermore use of virtual tools during assembly in the VE aids the user to detect possible construction difficulties related to the position and shape of the tools.

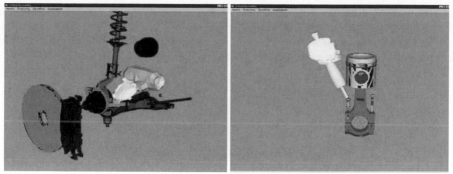
**Figure** 2 Assembly simulation environments.

The user receives haptic feedback when he/she touches the components (parts, tools) in the scene. The user also has the option to move and assemble the parts together in order to construct the final product. Haptic feedback from parts is provided to the user according to the fingers that touch the part. Tools also provide force feedback to the user. Force feedback from tools has a different form from other objects. Tool force feedback has to two states: low force state and high force state. The first, low force state is used when the user is holding the tool and the second, high force state, when the user is interacting with the objects in the scene using the tool. This way the user can distinguish when the tool is functioning. In order to provide the user with real time haptic feedback, as described, we need a fast and robust collisions detection algorithm. The algorithm must be fast so that the application can detect collision in real time and robust so that any geometry can be imported in the scene.

The 'RAPID' (Gottschalk S. et al. (1996)) collision detection algorithm is employed for this purpose. The procedure followed by rapid is the creation of appropriate structures (trees) before the initialization of the simulation. Based on these structures each object is split into smaller sub-objects and side information is saved for each sub-object in order for the collision detection algorithm to be able to calculate very fast the existence of collision or not. The use of these tree structures reduces the time needed for collision detection in two ways: a) it rejects pairs of objects that do not collide using simple controls, b) limits the controls between collided objects in small areas in their neighborhood. In order to minimize the time consumed for collisions between fingertips and objects a variation of the algorithm is implemented. Initially, all the hand is assumed as one object and the collision detection takes place between the hand and the objects. Fingertip collisions are examined only when hand-object collision is detected. The collision detection unit initializes with the geometries loaded from the data entry unit of the virtual reality platform. In the following, based on the position of the scene objects and the hand, returns, when called, the pairs of objects that collide.

In order to increase the immersion of the user a Head Mounted Display (HMD) is supported by the system providing 3D visual feedback during the assembly process. The HMD helps the user understand the exact position of the hand and the objects in the VE. Thus it helps the user avoid mistakes that may occur from illusions while working on a perspective 2d graphical environment.

## 2.3 Assembly review

The tool supports reviewing of the assembly plan, modification, post-processing and observation of statistical information about the planning. The user can also playback the assembly planning and view the assembly process from different 3D viewpoints (figure 3). The playback process can be stopped at any point, in order to further process and modify it. For each assembly plan performed a log file is created and stored. Data stored in the log file is then further processed to produce useful quantitative measurements (e.g. total path length for each part).

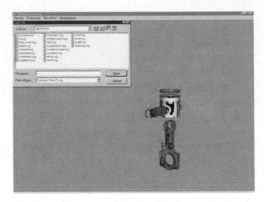

**Figure 3** The user selects a log file to review the assembly planning.

# 3    Conclusions

All aforementioned steps were deemed necessary in order to develop a realistic virtual assembly tool, which can offer adequate functionality for end-users to familiarize themselves with the technology and also to enable them to judge its potential and usefulness. The assembly tool has been installed in two Greek companies and evaluation of its performance by the users has shown that its performance is very satisfactory and also that it can lead to modernization of the assembly planning procedure.

## References

Ascension Technologies Corporation,(2000), MotionStar Wireless™ Insallation and operation guide.

Burdea, G. C. (1994). Force and touch feedback for virtual reality, *Wiley-Interscience Publication.*

Gottschalk, S., Lin, M. C., Manocha, D.,(1996), OBBTree: A Hierarchical Structure for Rapid Interface Detection, Proc. Of ACM Siggraph.

Johnson, T. (2000), "A General Virtual Environment for Part Assembly Method Evaluation", Master's Thesis, Department of Mechanical Engineering, Iowa State University (http://www.vrac.iastate.edu/~jmvance/ASSEMBLY/Assembly.html)

PTC DIVISION™ MockUp & Reality, http://www.ptc.com/products/sw_landing.htm

Sankar Jayaram, Uma Jayaram, Yong Wang, Hrishikesh Tirumali, Kevin Lyons, Peter Hart, (1999), VADE: A Virtual Assembly Design Environment, pp. 44-50, IEEE Computer Graphics and Applications

Immersion Technologies Inc., CyberGrasp Haptic Device, http://www.immersion.com/products/3d/interaction/cybergrasp.shtml

Immersion Technologies Inc. (2000), Virtual Hand Suite 2000 user & programmers guide, http://www.immersion.com/products/3d/interaction/software.shtml

# Virtual Environment Design for Gene Selection Using Gene Expression Data

*Kunihiro Nishimura[1], Shumpei Ishikawa[2], Koji Abe[2], Shuichi Tsutsumi[2], Hiroyuki Aburatani[2], Koichi Hirota[2], and Michitaka Hirose[2]*

{[1]Graduate School of Information Science and Technology, [2]Research Center for Advanced Science and Technology}, The University of Tokyo
4-6-1, Komaba, Meguro-ku, Tokyo, 153-8904, JAPAN
{kuni, abe, hirota, hirose}@cyber.rcast.u-tokyo.ac.jp,
{shumpei, shuichi, abura}@genome.rcast.u-tokyo.ac.jp

## Abstract

Virtual reality technology can help in the improvement of the effectivity of genome data analysis by enabling interactive visualization and providing a three-dimensional virtual environment. In the field of genome science, the use of gene expression data enables the selection of suitable genes for new drug development. In this paper, we discuss the virtual environment design for gene selection for drug discovery using gene expression data. We emphasize the importance of interactive visualization and propose two guidelines for virtual environment design: the use of a spatial cue and the utilization of the physical movement of the body. We implement the virtual environment according to the two guidelines proposed using immersive projection display and evaluate these guidelines.

## 1    Gene Selection in the Field of Genome Science

In genome science, microarray technology has enabled the simultaneous measurement of the gene expression level of more than 30,000 genes. Gene expression level is proportional to the number of mRNAs transcribed from genes, thus it indicates the activity of genes in a cell. It is expected that gene expression data will provide information on the mechanisms of disease progression and assist in drug discovery. For these purposes, the gene expression data of many normal and diseased tissues are being accumulated rapidly. The first step in the discovery of drugs for the treatment of diseases such as cancer is to identify genes that are involved in the disease, or to select target genes that are specifically active in the disease. The most simple gene selection process is the screening of genes by setting filtering thresholds of the gene expression level.

We are investigating means of helping genome scientists in the gene selection process by developing visualization and human-computer interaction technologies. Visualization is essential for providing a holistic view of the large amount of data, and interactivity is important in the gene selection process for determining many filtering parameters based on genome scientists' knowledge and judgments. A wide field of view is effective for the determination of thresholds by presenting much information to refer to. We consider that virtual reality technology can facilitate genome data analysis by providing a three-dimensional virtual environment that enables interactive visualization. In this paper, we propose a visualization methodology for drug discovery and discuss a virtual environment using immersive projection technology.

# 2 Virtual Environment Design

## 2.1 Virtual Environment for Genome Science

There are several approaches to genome science analysis using virtual reality technology. Ruthes et al. discussed about visualization and interaction with reference to a hierarchical tree such as phylogenetic and taxonomic trees in a virtual environment (Ruthes et al., 2000). Adams et al. introduced the visualization methodology of genome sequence annotations (Adams et al., 2002). However, the above two approaches do not fully use a three-dimensional virtual environment.

Stolk et al. proposed a methodology of visualizing hierarchical relationships within a gene family and networks of gene expression data (Stolk et al., 2002) using a CAVE display (Cruz-Neira et al., 1993). We have been studying the visualization technology applied to genome science using immersive projection technology, and have discussed about the visualization of gene expression data and the new cluster analysis of such data using a virtual environment (Kano et al., 2002). The above two approaches were implemented to the realization of on immersive projection display, because a virtual environment with a wide-view display has the advantage that it avoids overlapping of the large amount of information being presented. However, they were implemented ad-hoc and the virtual environment design was not fully discussed.

## 2.2 Gene Selection Process

The gene expression levels used are numeric, ranging from 1 to 10,000 and gene expression data is represented as a numerical matrix. Gene expression levels are log-normalized by each gene and are drawn in colour (red: high expression level, green: low expression level), which enables us to grasp the gene expression level intuitively.

The filtering process involves setting of the filtering thresholds for each tissue sample. For example, in order to select genes that are active specifically in a lung cancerous tissue sample compared with those in a normal sample, we set the thresholds such that the gene expression level is high in cancerous tissues and low in other normal tissues. There are two kinds of filtering threshold. One uses a raw gene expression level and another uses a relative gene expression level. The relative gene expression level defines the expression ratio of the normal tissue samples to the lung cancerous tissue samples, because the expression fold change is important in a biological data. Filtering process is selecting genes that satisfy gene expression level is higher than the minimum threshold and lower than the maximum threshold. Researchers select the filtering conditions that multiply "and" or "or" between two kinds of filtering process.

After the filtering process, an advanced analysis of the selected gene is required. Genes are categorized according to the localization of their encoded protein in the cell, because protein localization is essential for the development of effective drugs. For example, some drug targeted proteins are membrane proteins.

As described above, the process of gene selection is summarized in three operation steps. The first is setting the filtering thresholds of absolute gene expression level. The second is setting the filtering thresholds of relative gene expression level. The third is categorizing the selected genes based on the localization of their encoded proteins.

## 2.3 Guidelines for Virtual Environment Design

We propose two guidelines for virtual environment design with respect to gene selection.

### 2.3.1 To use a spatial cue

The first is to use a spatial cue to enable visualization of the data and the process of gene selection. The gene selection process requires setting of filtering thresholds of the gene expression level for all tissue samples. One operation of the selection process requires trial and error to set appropriate filtering thresholds. To support such process of trial and error, several types of reference information, that is, filtering threshold, the number of selected genes, and the selected gene expression level, should be presented and feedbacked in real time. Thus, one operation requires one workspace that satisfies interactivity and presentation of the reference information described above.

Each of the three gene selection processes requires a workspace as written above. In order to link each workspace sequentially, we propose to use a spatial cue. All workspaces for each operation step are placed spatially and each workspace is visually linked together. Each link is visualized as a metaphorical road. The road indicates the sequence of analysis and the branches of the road represent the decisions that have to be made during the course of the analytical process. Users walk along the road as they reach the workspace one by one. That is, the virtual environment for gene selection is composed of three workspaces and a metaphorical road linked each workspace.

A spatial cue solves the difficulty of understanding the analytical process. It enables intuitive navigation of the analysis process in a virtual environment by making good use of spatiality.

### 2.3.2 Utilization of the physical movement of the body

The second guideline is to utilize the physical movement of the body for setting filtering parameters. When visualized elements fit our body size, we can recognize the size interactively because we recognize the scale of objects in comparison with our body size. Thus, the system should enable users to manipulate filtering parameters using the physical movement of their body.

## 3    Virtual Environment for Gene Selection

We used the immersive projection display, CABIN, which has five screens and generates a 270-degree view, to develop a virtual environment (Hirose et al., 1999). The virtual environment was designed according to the guidelines discussed above. It has three workspaces.

The first and second workspaces are for setting filtering parameters (Figure 1). In order to set filtering parameters with the reference information, the visualized expression level of selected genes (described above), the number of selected genes, and filtering parameters are presented to users. The number of genes is represented by two bar graphs. One graph represents the number of selected genes versus all genes, and the other graph represents the number of selected genes versus 1,000 genes in order to increase the resolution. Filtering parameters are visualized as slide bars so that users can manipulate them interactively through the spatial interface by using their arms (Figure 2). Many of the parameters are visualized toward the same direction in the workspace, which enables simultaneous comparison of all the parameters. When filtering parameters are set or changed, genes are selected immediately and the presented information is updated. Users can use the menu when they need it.

The third workspace is for further analysis of the selected genes (Figure 3). To categorize these genes based on the localization of their encoded proteins in the cell, the expression level of selected genes, the number of selected genes, protein localization information are presented to users. The localization information of the protein that is encoded by the gene is used in the computer program for the prediction of protein localization sites in cells, called "PSORT" (Nakai and Horton, 1999). This information predicted 11 localization sites. We presented the results as a bar graph representing the probability of protein localization.

Three workspaces are linked together according to the analytical process that is visualized as a metaphorical road (Figure 4).

Spatial cues were used in visualizing the road and the arrangement of these three workspaces in the virtual environment. Physical movements were used for adjusting filtering parameters intuitively.

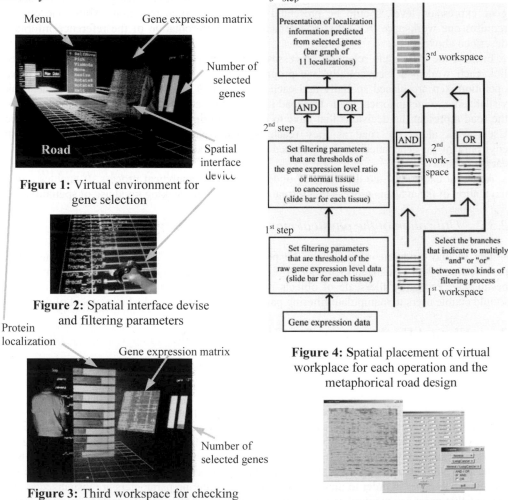

**Figure 1:** Virtual environment for gene selection

**Figure 2:** Spatial interface devise and filtering parameters

**Figure 3:** Third workspace for checking and categorizing selected gene

**Figure 4:** Spatial placement of virtual workplace for each operation and the metaphorical road design

**Figure 5:** GUI

## 4    Results and Evaluation

We used this system for gene selection in order to develop a therapeutic medicine for lung cancer. We use the gene expression data of 35 normal tissue samples and one lung cancerous tissue sample and selected 344 genes using two types of filtering method. Then we picked up 34 genes that are predicted to encode plasma membrane proteins for developing new drugs. We are testing whether these genes are appropriate for drug development by biological experiments.

We evaluated the usability of the virtual environment compared with the PC system and results are shown in Figure 6. We employed four male and one female genome scientists in their 20s and 30s as subjects. The subjects were required to evaluate the virtual environment compared to the PC system that has identical functions using GUI (Figure 5), and to answer 6 questions. The answer to Q1 and Q2 show that the understanding of the working process of gene selection in the virtual environment is easier than that in the PC system. We believe that that use of spatial cues as a metaphorical road is effective for the understanding. The answer to Q5 and Q6 show that the subjects committed fewer mistakes than those in the case of PC. We think that performing parameter setting using their own body movements causes this. On the other hand, the answer to Q3 and Q4 indicate that the subjects faced slight difficulty in adjusting filtering parameters in the virtual environment and did not improve their working efficiency. Throughout the experiment, we confirmed that our guidelines for virtual environment design can be effectively applied to analyses in the field of genome science from the viewpoint of understanding and interactivity. In future studies the interfaces should be improved in order to increase the working efficiency in a virtual environment.

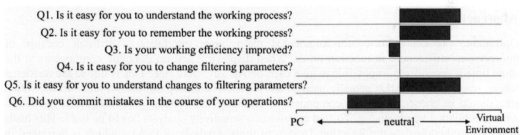

**Figure 6:** The results of subjective evaluation regarding the comparison between the PC and the virtual environment system

# References

Adams, R. M., Stancampiano, B., McKenna, M., and Small, D. (2002). Case Study: A Virtual Environment for Genomic Data Visualization, *13th IEEE Visualization 2002 Conference (VIS 2002)*, 513-516.

Cruz-Neira, C., Sandin, D.J., and DeFanti, T.A. (1993). Surround-Screen Proje ction-Based Virtual Reality: The Design and Implementation of the CAVE, *Computer Graphics (Proceedings of SIGGRAH '93), ACM SIGGRAPH*, 135-142.

Hirose, M., Ogi, T., Ishiwata, S., and Yamada, T. (1999). Development and Evaluation of the Immersive Multiscreen Display CABIN, *Systemsand Computers in Japan*, 30(1), 13-22.

Kano, M., Tsutsumi, S., Nishimura, K., Aburatani, H., Hirota, K., and Hirose, M. (2002). Visualization for Genome Function Analysis Using Immersive Projection Technology, IEEE Virtual Reality Conference 2002, 224-231.

Nakai, K. and Horton, P., (1999). PSORT: a program for detecting the sorting signals of proteins and predicting their subcellular localization, *Trends Biochem. Sci.*, 24(1) 34-35.

Ruths, D. A., Chen, E. S., and Ellis, L. (2000). Arbor 3D: an interactive environment for examining phylogenetic and taxonomic trees in multiple dimensions, *Bioinformatics*, 16(11), 1003-1009.

Stolk, B., Abdoelrahman, F., Koning, A., Wielinga, P., Neefs, J. M., Stubbs, A., Bondt, A. d., Leemans, P., von der Spek, P. (2002) Mining the human genome using virtual reality, *Proceedings of the Fourth Eurographics Workshop on Parallel Graphics and Visualization*, 17-21.

# RTSA – Reaction Time Sensitivity Analysis: A Methodology to Design an Augmented Reality User Interface for a Head Based Virtual Retinal Display

*Olaf Oehme[1], Britta Sommer[2], Holger Luczak[1]*

[1] Institute of Industrial Engineering and Ergonomics (IAW)
RWTH Aachen University
Bergdriesch 27, D - 52056 Aachen
GERMANY
o.oehme@iaw.rwth-aachen.de

[2] Institute for Pure and Applied Mathematics (IRAM)
RWTH Aachen University
Templergraben 55, D-52062 Aachen
GERMANY
britta@iram.rwth-aachen.de

## Abstract

Optimizing processing times for augmented reality user interfaces is difficult because of interpersonal differences in reaction times. When displaying several dialogue components at the same time it generally is impossible to give each one its advisable size. This is due to the restricted screen resolution of HMDs actually used in the industrial context. So, the problem arises, which size should be chosen for which component to minimize the average processing time needed to work through the whole dialog. To find an answer a sensitivity analysis has to be used. This leads to a new methodology – the Reaction Time Sensitivity Analysis (RTSA) – which is described in this article. It is applied to a case study for an augmented reality user interface. Optimal representation sizes insensitive to different user performances could be determined.

## 1    Introduction

Augmented Reality (AR) is a rather young technology. It allows an enrichment of the real world with additional virtual information (Luczak et al. 2002; 2003). For example, repair instructions for machine tools can be superimposed on the workers field of view (FOV) such that the use of manuals becomes superfluous (Oehme et al. 2002). So far, the design of user interfaces in this field has been dominated merely by technological aspects. First of all, the systems are supposed to work satisfyingly under the conditions imposed by the test design. There are only few projects that focus on the user's requirements (ARVIKA (2003) is an example for such a project).

So-called Head-Mounted Displays (HMD) are used to overlay the real world with virtual information. They can be worn just like glasses. Half-silvered mirrors allow seeing the real world as well as the virtual information. Most commonly integrated LC-Displays are employed, but usual cathode ray tubes can be found, too. Another possibility of realizing the visual overlay is the Virtual Retinal Display (VRD), which addresses the retina directly with a single laser stream of pixels. Furthermore, there are binocular and monocular HMDs with different resolutions and different numbers of available colors.

## 2    Problem

No matter which system is chosen, one problem remains: the results of research on standard computer displays cannot be transferred to see-through displays. One example for this reasons is the fact that the contrast between background and characters is rapidly changing with the user's direction of view. With a standard display this contrast always remains the same. The size of the displayed virtual information becomes crucial when the question of human perception is addressed. In Oehme et al. (2001) minimal and advisable representation sizes to provide successful information processing for different kinds of information can be found.

The problem is that the HMDs actually used in the industrial context have a rather low screen resolution. Thus, in general the information cannot be displayed in the advisable size. Nevertheless, the aim of any user-centered interface design is to guarantee a fast access to the given information and to keep the error rate as low as possible. Thus, always employing the minimal representation size is no suitable choice either.

The empirical studies mentioned above show the dependence of processing time and error rate of different kinds of information (e. g. characters, symbols etc.) on their representation sizes. Because of environmental, inter- and intrapersonal factors an uncertainty of this empirical data has to be taken into account. Hence, the value of the reaction time for any chosen representation size always ranges within some interval, whose maximum and minimum can be determined.

# 3 Methodology

Accordingly, facing the task to design the interface it is not possible to choose the optimal representation size for each kind of information immediately. A sensitivity analysis helps to overcome this obstruction. So, the Cost Tolerance Sensitivity Analysis (CTSA) methodology (e. g. Gerth et al. 2000), which is originally used in the field of Concurrent Engineering to determine the sensitivity of the costs with respect to certain independent variables in an early state of product design, has been adapted in a suitable way. Within the design of a user interface it can be employed, if the following conditions hold:

- *The dependence of the minimal and the maximal reaction time on the representation size can be approximated by differentiable functions for each dialog component.*
- *The frequency of occurrence of each dialog component is known.*
- *Dialog components appear one after the other, such that sequential processing of information can be assumed.*
- *Each dialog component has a well-defined position on the display. Dialog components never overlap.*

## 3.1 Reference design and resolution function

First of all, the basic design of the interface hast to be determined. That is, the relative positions of the dialog components with respect to each other have to be set. This basic design is designed here as reference design from now on. It might be a totally new design or the design of a prototype or of a preceding interface. The next step is to define the so called resolution function $P$. It describes the maximal number of pixels which is absorbed in horizontal direction when all dialog components are displayed at the same time. Let $k$ be the number of relevant dialog components, then $P$ is defined as

$$P = \beta_1 S_1 + \beta_2 S_2 + \beta_3 S_3 + \ldots + \beta_k S_k \qquad (1)$$

where $S_i$ is the representation size of the $i$-th dialog component - it is usually measured in vertical direction and given in arc min - and the coefficient $\beta_i$ is introduced to take the unit of arc min into pixel and to calculate the horizontal extension of the dialog component from the given (vertical) representation size.

## 3.2    Representation size boundaries and reaction time functions

The representation size of a dialog component cannot have arbitrary values. It is restricted by certain boundaries. Previous investigations have shown the existence of certain representation sizes that allow recognition, while the error rate is rather low (e.g. Oehme et al. 2001). These values shall be chosen as lower boundaries. The upper boundaries are determined by the finiteness of the field of view (FOV). Furthermore, it has to be taken into account that any enlargement of the virtually displayed information yields a more extensive occlusion of the real world. Thus, the recognizability of the real word is lessened.

The next step is to determine the average maximal and the minimal reaction time for each dialog component in dependence on the representation size. The necessary data can be obtained by suitable experiments. Now, a differentiable function has to be constructed that approximates or interpolates the minimal or maximal reaction times, respectively. It will be employed in the sensitivity analysis of the optimized representation sizes of the dialog components. The actual reaction time - representations size function (TS function) is supposed to be differentiable and to have values that lie between the graphs of these upper and lower time functions. It was already pointed out (cf. Montgomery 1997) that the choice of these confining functions merely depends on their suitability to describe the given data, while they should keep the optimization problem as simple as possible. Hyperbolic, inverse exponential or inverse power functions can be possible choices.

## 3.3    Optimization problem

The main goal of this investigation is to show, how to choose representation sizes for each component in such a way that the average time $T$ needed to process the whole reference dialog is minimized. Therefore, the objective function is defined by

$$T = \left[ \sum_{i=1}^{k} H_i \cdot T_i(S_i) \right] \qquad (2)$$

where $T_i$ is the reaction time function in dependence on the representation size $S_i$ and $H_i$ is the frequency of occurrence of the $i$-th dialog component. The objective function is to be minimized:

$$T_{min} = \min \left[ \sum_{i=1}^{k} H_i \cdot T_i(S_i) \right] . \qquad (3)$$

The domain of the objective function is given by the direct product of the domains of the TS functions $T_i$. Furthermore, the resolution of the display y is restricted, such that the co-domain of the resolution function (cf. Equation 1) is finite. One gets the constraints

$$S_{i,min} \leq S_i \leq S_{i,max}, \qquad P(S_i; \beta_i) \leq P_{max}, \qquad i = 1...k \qquad (4)/(5)$$

## 3.4    Sensitivity analysis

Now, the sensitivity of the optimized dialog components in dependence on the change of the time functions has to be determined. To this end, $2^k$ optimization problems have to be solved. The independent variables are the TS functions $T_i$ and their levels are the corresponding upper and lower time functions $T_i^{\pm}$. The latter two restrict the values the reaction time can have. Therefore, they are used to solve the optimization problem. The dependent variables are the sizes of the dialog components. Thus, for every experiment there are $k$ dependent variable measures, resulting in a $2^k \times k$ (statistical counterbalanced) experimental design matrix, respectively results matrix (see also Table 1).

The result of the sensitivity analysis are the optimized representation sizes for any combination of upper and lower time functions. For some dialog components the optimized representation sizes remain fixed, no matter which combination of time functions is chosen. In this case, this component is insensitive regarding the choice of the time functions and the optimized value of its representation can be fixed. If a dialog component is sensitive, then we cannot give any optimal value of its representation size by the means of this methodology.

# 4   Case Study

For this case study an AR User Interface for the maintenance area that employs a VRD (vertical FOV = 21.37 degrees) with a resolution of 640 × 480 pixels was designed. In addition to repair instructions, menu items and navigation arrows have to be displayed (Figure 1 left).

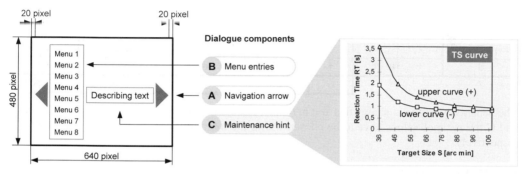

**Figure 1:** Reference Design for the AR User Interface; TS Curve for the Dialogue Component C

Thus, there are three relevant dialog components, A, B and C. Though they are supposed to appear in succession it is necessary that all three of them can be displayed simultaneously. It also has to be taken into account that the outer 20 pixels of the given VRD cannot be used because of a distortion in this area. This yields the constraint $P \leq 600$ pixel corresponding to Equation 5. The resolution function (in horizontal direction) corresponding to Equation 1 is given by

$$P_H = \beta_A S_A + \beta_B S_B + \beta_C S_C = 2.65^{-1}(\tfrac{1}{2} \cdot 2S_A + \tfrac{5}{7}(13S_B + 20S_C)). \tag{6}$$

The coefficients ½ and $^5/_7$ are due to the fact that the representation sizes are always given with respect to the vertical direction. The factor 2.65 transforms the representation sizes $S_{A,B,C}$, which are given in arc min, into numbers of pixels. The smallest representation size is chosen to be 36 arc min in accordance with Oehme et al. 2001. The biggest representation size is set to be 108 arc min. Hence, it is possible to display up to 13 letters for menu text or up to 20 letters for explanations. The experimental data obtained in this setting is used to construct the upper and lower time functions as the inverse of second order polynomials (compare Figure 1 right for an example of the upper and lower time curves corresponding to the dialog component C):

$$T_A^+ = (-1,11 \cdot 10^{-5}\ S_A^2 + 4,39 \cdot 10^{-3}\ S_A + 0,13)^{-1} \tag{7}$$

$$T_A^- = (-5,88 \cdot 10^{-5}\ S_A^2 + 2,56 \cdot 10^{-2}\ S_A + 5,34 \cdot 10^{-3})^{-1} \tag{8}$$

$$T_B^+ = (-5,35 \cdot 10^{-5}\ S_B^2 + 9,61 \cdot 10^{-3}\ S_B - 0,18)^{-1} \tag{9}$$

$$T_B^- = (-6,04 \cdot 10^{-5}\ S_B^2 + 1,21 \cdot 10^{-2}\ S_B - 0,21)^{-1} \tag{10}$$

$$T_C^+ = (-1,45 \cdot 10^{-4}\ S_C^2 + 3,15 \cdot 10^{-2}\ S_C - 0,67)^{-1} \tag{11}$$

$$T_C^- = (-2{,}64 \cdot 10^{-4} \, S_C^2 + 4{,}75 \cdot 10^{-2} \, S_C - 0{,}85)^{-1} \qquad (12)$$

The optimization problem is defined as in Equation 3, where the frequencies of occurrence are $H_A = 0.35$, $H_B = 0.22$, $H_C = 0.43$. In Table 1 the results of the $2^3$ optimizations are displayed.

**Table 1:** Factorial Design (+/-lower or upper TS function) and Optimization Results

| Optimization Run | Factorial Design | | | Optimized Representation Size $S_i$ [arc min] | | | $T_{min}$ [s] |
|---|---|---|---|---|---|---|---|
| | A | B | C | A | B | C | |
| 1 | - | - | - | 108,00 | 74,47 | 55,33 | 1,22 |
| 2 | + | - | - | 108,00 | 69,78 | 58,38 | 1,76 |
| 3 | - | + | - | 108,00 | 75,72 | 54,52 | 1,53 |
| 4 | + | + | - | 108,00 | 75,72 | 54,52 | 2,08 |
| 5 | - | - | + | 108,00 | 67,22 | 60,05 | 1,45 |
| 6 | + | - | + | 108,00 | 67,22 | 60,05 | 2,00 |
| 7 | - | + | + | 108,00 | 69,78 | 58,38 | 1,76 |
| 8 | + | + | + | 108,00 | 69,78 | 58,38 | 2,31 |
| | | | $\sigma$ | 0,00 | 3,57 | 2,32 | 0,36 |

The sensitivity analysis shows that the optimal value of the component A is 108 arc min. For B and C the optimal values have to be chosen within the interval [64.42, 74.89] and [51.34, 59.73], respectively.

# 5 References

ARVIKA (2003): Augmented Reality for Development, Production & Servicing. Retrieved February 15, 2003, from http://www.arvika.de

Gerth, R. ; Pfeifer, T. ; Oehme, O. (2000): Early Cost Tolerance Sensitivity Analysis for Inspection Planning. In: Human Factors and Ergonomics in Manufacturing, Vol. 10 (3). New York : John Wiley & Sons, Inc., 2000, pp. 309-329

Luczak, H. ; Oehme, O. (2002): Visual Displays - Developments of the Past, the Present and the Future. In: Luczak, H. ; Cakir, A.E. ; Cakir, G. (Eds): WWDU 2002 - Work With Display Units - World Wide Work. Proceedings of the 6th Int. Scientific Conference 2002, Berchtesgaden, May 22-25. Berlin : ERGONOMIC Inst. für Arbeits- und Sozialforschung, pp. 2-5

Luczak, H. ; Rötting, M. ; Oehme, O. (2003): Visual Displays. In: Jacko, Julie A. ; Sears, Andrew (Eds.): The Human-Computer Interaction Handbook – Fundamentals, Evolving Technologies and Emerging Applications. Mahawah, New Jersey : Lawrence Erlabaum Associates, 2003, pp. 187-205

Montgomery, D.C. (1997): Design and Analysis of Experiments. 4th Edition. New York : J. Wiley and Sons, 1997

Oehme, O ; Wiedenmaier, S. ; Schmidt, L. ; Luczak, H. (2001): Empirical Studies on an Augmented Reality User Interface for a Head Based Virtual Retinal Display. In: Smith, J. M. ; Salvendy, G. (eds.): Systems, Social and Internationalization Design Aspects of Human-Computer Interaction. Volume 1. New Yersey : Lawrence Erlbaum Associates, pp. 1026-1030

Oehme, O. ; Wiedenmaier, S. ; Schmidt, L. (2002): Evaluation eines Augmented Reality User Interfaces für ein binokulares Video See-Through Head Mounted Display. In: Proceedings of the Useware 2002 – Mensch-Maschine-Kommunikation/Design, Darmstadt, May 11-12. Düsseldorf : VDI Verlag (VDI-Berichte 1978), S. 35-40

# ThumbsUp: Integrated Command and Pointer Interactions for Mobile Outdoor Augmented Reality Systems

*Wayne Piekarski and Bruce H. Thomas*

Wearable Computer Laboratory
School of Computer and Information Science
University of South Australia
Mawson Lakes, SA, 5095. Australia
{wayne, thomas}@cs.unisa.edu.au

## Abstract

This paper presents a new user interface technology known as *ThumbsUp*, which we have designed and developed for use with mobile outdoor augmented reality systems. Using a simple pair of vision tracked pinch gloves and a new menuing system, a user is able to control an augmented reality system in outdoor environments under poor tracking conditions with a high level of accuracy. Highly interactive 3D augmented reality applications can now be operated outdoors with our new easy to use interface technology.

## 1 Introduction

*ThumbsUp* is our new user interface technology for use with mobile outdoor augmented reality (AR) systems. User interfaces to date for outdoor AR systems have been quite simple, but our investigations into modelling 3D geometry outdoors (Piekarski and Thomas 2001) have required complex user interfaces on par with what is currently available on desktop workstations. The ThumbsUp user interface technology utilises a tracked set of pinch gloves that combine command entry and 3D manipulation into one user interface device. ThumbsUp enables the user to enter commands via a hierarchical menu with mapped pinch gestures and perform 3D manipulations through the tracking of the user's thumbs relative to their head position, as shown in Figure 1 and Figure 2.

Operating user interfaces for mobile computers outdoors is inherently difficult due to the large and dynamic nature of outdoor environments. The computing technology must be mobile to allow the user to roam freely in this environment. Restrictions on size, performance, electric power consumption, weight,

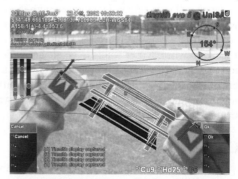

Figure 1 – Example showing a mobile AR user performing an interactive rotation operation on a virtual table at a distance outdoors, wearing the Tinmith-Endeavour backpack

Figure 2 – Immersive AR view of figure 1, showing the virtual table at a distance being rotated with the hands, implemented using new AR extensions to existing image plane techniques

and magnetic interference limit the options of devices for use outdoors. Although technology improves from year to year, we are designing user interfaces based on vision tracked gloves that takes advantage of technology available today. Other recent input devices for mobile user interfaces are implemented using a variety of hardware, such as with ultrasonics (Foxlin and Harrington 2000) or accelerometers (Cheok, Kuman and Prince 2002). Mobile computers can now perform realistic rendering for augmented reality; therefore good user interfaces are now required to support powerful new applications, with particular attention to the limitations of the technology outdoors.

One application domain we are currently investigating is outdoor augmented reality 3D modelling, where a user can capture the models of existing large structures (such as buildings), or prototype plans for new objects that may be constructed in the future. We see this form of application improving design and planning in areas such as landscape design, building construction, and surveying. In order to control a complex modelling application with many features, we developed the ThumbsUp user interface and evaluated it on a number of users to iteratively refine it. Other application areas are explored in systems such as the Touring machine (Feiner, MacIntyre and Hollerer 1997) and Studierstube system (Reitmayr and Schmalstieg 2001).

Our user interface is made up of three components: a 3D tracked pointer using gloves on the user's hands; a command entry system where the user's fingers interact with a menu for performing actions; and an augmented reality display that presents the results back to the user. These components are used to interact with a virtual environment, in this case implemented as outdoor augmented reality. The hands free nature of ThumbsUp does not require interaction props, allowing the user to freely move about the real world without restrictions. Our investigation into ThumbsUp has leveraged current research into different 3D interaction techniques, and complements rather than replaces existing techniques. Interaction techniques for outdoor augmented reality (OAR) are a subset of the augmented reality (AR) and virtual reality (VR) domains, which are a further subset of virtual environments (VE) and 3D interfaces.

Our interaction techniques use natural head and hand movements to specify operations during the construction of 3D graphical models outside. Using parts of the body such as the head and hands to perform gestures is a natural way of interacting with 3D environments, as humans are used to performing these actions when explaining operations to others and when dealing with the physical world. By using techniques such as the pointing and grabbing of objects in positions relative to the body, user interfaces can leverage the user's inbuilt knowledge (known as Proprioception) as to what their body is doing. Mine et al (Mine, Brooks and Sequin 1997) demonstrated that designing user interfaces to take advantage of these human proprioceptive capabilities produced improved results. We were also inspired by the elegant combination of commands and pointing by the *Put-That-There* system (Bolt 1980). The user interface we have developed uses similar techniques, with the focus being the user's region of interest framed by their current field of view. Commands and pointing both operate within this view, and building on this is the use of the user's physical presence (location and orientation) to aid with their interactions. For example, in a scenario where a user wants to create an outdoor scene of a garden, the order of operations would be as follows: first, specify the prefabricated object to create, such as a table; second, use AR image plane techniques to slide the table into position from different angles; third, scale and rotate the table (such as shown in Figure 1 and Figure 2); and finally, walk away from the table to preview at a distance its placement.

## 2 Current applications

As previously mentioned, one application domain we believe augmented reality will be used in the future for is modelling of 3D geometry, allowing the preview of non-existent objects, and the capture of existing geometry that can then be modified to view what the proposed changes will be. The Tinmith-Metro application (Piekarski and Thomas 2001) implemented simple building construction using the *infinite planes technique* (by placing down large planes and combining them with constructive solid geometry operations), and the placement and manipulation of *street furniture* objects, both with the user interface described in this paper. These techniques allow users to capture the geometry of outdoor objects without having to actually stand next to or on top of them. The user can model the objects from a distance, with partial occlusion of the real world as if the objects were physically present. This is an advantage over existing techniques, such as: 1) photo and laser based scanning, requiring a full view of the object; 2) using GPS waypoints, not working well near large buildings; and 3) standing on top of the building, possibly being not possible or too dangerous.

## 3 Interface overview

The user interface can be described as two separate components, a tracked 3D cursor for selection and manipulation, and a special menu for controlling the system and entering commands. The menu is fixed to the user's display and presents up to ten possible commands that are available at that moment. Eight of these commands are mapped to the fingers as shown in Figure 3, and the user activates a command by pressing the appropriate finger against the thumb. At this point, the menu refreshes to reflect the selection made, and the next series of commands are then made available to the user. *Ok* and *cancel* operations are indicated by pressing the fingers into the palm of the appropriate hand, depending on which is selected by the user as being their dominant hand, and these are indicated in the topmost boxes in the menu. The 3D cursor is implemented using vision tracking techniques (Kato and Billinghurst 1999) and fiducial markers placed on the tips of the thumbs. Using this tracking, and combining this with the previously described command system, the user interface can perform selection, manipulation, and creation operations by pointing into the virtual environment.

The design of the menu is based around users executing commands through direct finger mappings, without requiring them to lift up their hands to interact with the menu. This allows users to perform cursor operations with their hands without having to move them to execute commands. Traditional VR systems require the user to select a mode from a menu, and then interact with an object. With our design, operations may be performed without having to take the hands away from the task at hand.

Figure 3 – Each finger maps to a menu option, the user selects one by pressing the appropriate finger against the thumb, and does not rely on the position of the hands when navigating through the menu hierarchy

# 4 Cursor operations

Tinmith-Metro is our original AR outdoor 3D modelling application (Piekarski and Thomas 2001) that performed the placement of outdoor street furniture and the capture of simple building shapes. The Tinmith-Metro application extends previous image plane techniques (Pierce, Forsberg, Conway, Hong, Zeleznik and Mine 1997) to support object manipulation (translate, rotate, scale) and object selection in mobile augmented reality. Figure 1 and Figure 2 is an example showing a virtual table that has been placed down on the ground and is being manipulated into the correct position using these techniques.

This section discusses the features of the 3D cursor that forms an integrated part of the user interface, performing direct manipulation operations such as selection and object transformation. Interacting with 3D graphical objects in an outdoor environment is implemented using vision based hand tracking. To create new objects and then edit them (scale, rotate, translate, carving), we provide a number of interactions, using a combination of zero, one, and two handed input techniques, depending on what is most appropriate for the task. We implement each transformation technique as a separate command. This is on the assumption the user will wish to work with certain degrees of freedom without affecting the others. This constraining of degrees of freedom is useful to compensate for most users inability to maintain the exact position and orientation of their hands simultaneously (Hinckley, Pausch, Goble and Kassell 1994), and in environments with poor vision tracking.

Using a single hand, an object can be translated in the scene. To perform translation, it must be selected first, and the hand is brought into view so the cursor can be placed on top of the object. Using extended AR image plane techniques, a user is able to move the object against the view plane fixed to their head. By rotating the head and keeping the hands at the same point in the image plane, the object can be dragged around the user's body since our techniques maintain the same distance.

In order to provide more natural manipulation techniques for operations like scaling and rotation, it has been shown that using two hands for interaction can improve performance and accuracy. The two handed interaction ideas used for these transformations were initially pioneered in a study in 2D environments by (Buxton and Myers 1986). Although our work is different in that we are working at a distance in absolute coordinates (rather than directly on the object), the previous work is very useful in showing possible approaches, and how these tasks can be improved with two hands. We make use of the two hands by having the angle between the dominant and non-dominant hand control the rotation. This technique is also implemented using our AR extensions to image planes, and can be configured for either left or right hand dominance. Figure 1 and Figure 2 show a rotation operation being performed on a virtual object at a distance from the user. Since the tracking system used produces high quality position and low quality rotation values, the use of two hands allows rotations to be specified through only two position values, maximising the accuracy of the operation.

The user interface has very powerful command menus to perform a number of manipulation and creation operations without requiring the hands to be visible. The nudge commands allow the user to use precise manipulations based on fixed increments in order to accurately work with objects, and are most useful for altering the distance of an object (which is not possible using image planes since the distance is fixed) or for when very precise fixed movements are required. The eye cursor is used to create objects relative to the front of the user's body.

# 5 Conclusion

In this paper, we have presented a set of new user interface technologies we have developed for use in mobile outdoor augmented reality systems. Manipulation of 3D virtual artefacts in an outdoor setting requires different user interface technologies to traditional indoor AR and VR systems, due to the difference in tracking hardware and input devices. ThumbsUp is a new user interface technology that integrates a 3D cursor for selection and manipulation with a special menu for entering commands to support mobile outdoor augmented reality systems. The 3D cursor is controlled by a vision tracking system with fiducial markers placed on the tips of the thumb. The user interface can perform selection, manipulation, and creation operations by pointing into the virtual environment. A number of interaction modes (zero, one, or two handed input techniques) are provided to manipulate objects, such as translation, scaling, and rotation. The menu system is screen relative and presents up to ten possible commands that are available at that time. Each of these commands is mapped directly to the user's fingers, and the user activates a command by pressing the appropriate finger against the thumb.

# 6 Acknowledgements

The authors are very grateful for support provided by the following people: Rudi Vernik and Peter Evdokiou from the Defence Science Technology Organisation; the Division of ITEE and the School of CIS; Barrie Mulley, Benjamin Close, and Spishek and Arron Piekarski.

# 7 References

Bolt, R. A. (1980): "Put-That-There" : Voice and Gesture at the Graphics Interface. In *ACM SIGGRAPH 1980,* pp 262-270, Seattle, Wa, Jul 1980.

Buxton, W. and Myers, B. A. (1986): A Study In Two-Handed Input. In *CHI - Human Factors in Computing Systems,* pp 321-326, Boston, Ma, 1986.

Cheok, A. D., Kuman, K. G., and Prince, S. (2002): Micro-Accelerometer Based Hardware Interfaces for Wearable Computer Mixed Reality Applications. In *6th Int'l Symposium on Wearable Computers,* pp 223-230, Seattle, Wa, Oct 2002.

Feiner, S., MacIntyre, B., and Hollerer, T. (1997): A Touring Machine: Prototyping 3D Mobile Augmented Reality Systems for Exploring the Urban Environment. In *1st Int'l Symposium on Wearable Computers,* pp 74-81, Cambridge, Ma, Oct 1997.

Foxlin, E. and Harrington, M. (2000): WearTrack: A Self-Referenced Head and Hand Tracker for Wearable Computers and Portable VR. In *4th Int'l Symposium on Wearable Computers,* pp 155-162, Atlanta, Ga, Oct 2000.

Hinckley, K., Pausch, R., Goble, J. C., and Kassell, N. F. (1994): A Survey of Design Issues in Spatial Input. In *7th Int'l Symposium on User Interface Software Technology,* pp 213-222, Marina del Rey, Ca, Nov 1994.

Kato, H. and Billinghurst, M. (1999): Marker Tracking and HMD Calibration for a Video-based Augmented Reality Conferencing System. In *2nd Int'l Workshop on Augmented Reality,* pp 85-94, San Francisco, Ca, Oct 1999.

Mine, M., Brooks, F. P., and Sequin, C. H. (1997): Moving Objects In Space: Exploiting Proprioception In Virtual-Environment Interaction. In *ACM SIGGRAPH 1997,* pp 19-26, Los Angeles, Ca, Aug 1997.

Piekarski, W. and Thomas, B. H. (2001): Tinmith-Metro: New Outdoor Techniques for Creating City Models with an Augmented Reality Wearable Computer. In *5th Int'l Symposium on Wearable Computers,* pp 31-38, Zurich, Switzerland, Oct 2001.

Pierce, J. S., Forsberg, A., Conway, M. J., Hong, S., Zeleznik, R., and Mine, M. R. (1997): Image Plane Interaction Techniques in 3D Immersive Environments. In *1997 Symposium on Interactive 3D Graphics,* pp 39-43, Providence, RI, Apr 1997.

Reitmayr, G. and Schmalstieg, D. (2001): Mobile Collaborative Augmented Reality. In *Int'l Symposium on Augmented Reality,* pp 114-123, New York, NY, Oct 2001.

# When a House Controls its Master – Universal Design for Smart Living Environments

*Brigitte Ringbauer, Dr. Frank Heidmann, Jakob Biesterfeldt*

Fraunhofer Institute for Industrial Engineering
Nobelstr. 12, 70569 Stuttgart, Germany
{Brigitte.Ringbauer, Frank.Heidmann}@iao.fhg.de

## Abstract

The potential of smart living environments can only be tapped if the user is recognized as the key player in his networked environment. In LIVEfutura[1] the development of a fully integrated home environment is driven by a user-centred design approach to shift the focus in the development process on the user and task requirements. Networking, integration, automation, and profiling on the technical side are complemented by the development of an integrated user interface prototype to control all functions and services provided in the smart living environment. Empirical data for the acceptance of smart appliances is taken into account.

## 1 Introduction

Ubiquitous computing has begun to seep in our domestic world. From the mobile phone with organizer and payment functionality to the programmable coffee machine with internet access, humans are confronted more and more with technological innovations invented to make their life easier, more enjoyable and more secure.

One could think about a happy middle class family being supported by their networked home environment in terms of the easier organization and coordination of family tasks and appointments and increased comfort through personalized multimedia opportunities. But one could also think about the same family relying heavily on their house organizing functionality during a network breakdown or struggling with the setting of their multimedia system and having debates about conflicting personal profiles.

Therefore the question is, if an increase in smart technologies in the living environment automatically increases the quality of life of an inhabitant. Not necessarily. If not, why not?

The potential benefits of smart living environments are time-saving, increasing comfort, and security. But there are still some obstacles to overcome to tap the full potential of smart functions in the home environment:

- Technologies used and products and services provided are very heterogeneous and lack compatibility. This would be a precondition for integrated services and added value for the user.

---

[1] LIVEfutura (www.livefutura.de) is funded by the German Ministry for Education and Research, funding no. 01AK931.

- Users are confronted with a variety of interaction concepts and user interfaces for smart appliances which make intuitive usage nearly impossible.
- Especially in the home environment there is not only the rational component in the evaluation of the usefulness of services. Subjective factors like felt transparency or belief in the trustworthiness of a smart system have to be taken into account. Health issues play a major role as well.

These issues are dealt with in LIVEfutura. In this project an integrated overall home control concept covering all relevant application areas is specified and prototypically implemented: home automation, audio/video control, home appliances, home computer network, home telecommunication network, and even the private car. In LIVEfutura the control tasks are simplified and unified and the creation of complex control scenarios that can easily be configured, personalized, and activated is supported. LIVEfutura covers both, the interconnection of the different subnetworks via gateways and proxies and the realization of integrating user interfaces for overall home control, to overcome the problems described above. The goal is to provide an easy, secure and satisfying access to smart home services for all.

## 2 Usability challenges

Concerning smart home environments there are specific challenges as seen from a usability perspective:
- The user group for smart living environments is very heterogeneous: from children to the elderly, everyone should be supported by the functions and the user interface provided.
- The complexity of a fully networked home environment is quite high. The question is how to map different devices and services on a single interface without loosing consistency. How can the content be structured and named adequately for the user to keep transparency?
- There is little research about advantages and disadvantages of different input devices to monitor and control devices and functionalities at home. We presume, that several input devices will be used at the same time. Therefore consistency is an important issue here.
- Profiling is one possibility to manage the complex technical environment in a way adapted to users habits and needs. A crucial point from a usability perspective is how to generate a suitable user profile without forcing the user to undertake too much modifications and to keep him in control at any time.

To meet the challenges, a user-centred design approach as stated in the ISO 13407 (ISO, 1999) has to be taken so that the needs of target users in all stages of the development process are considered.

### 2.1 Methods

#### 2.1.1 Methods Overview
Methods used in the requirements analysis were extant systems analysis, competitive analysis and scenario technique. The output consisted of a full range of functions and services that make sense from a users perspective. With task analysis techniques, card sorting and wording tests these functions and services were categorized and an information architecture mapping the mental model of the target group was defined and built in a first mock-up GUI. Afterwards the layout, page partitioning, the display and control elements were redefinded according to the results of a quick layout review and user walkthrough. The next step was a usability test of the clickable GUI prototype. In addition the overall acceptance of smart house technology was investigated in user interviews.

## 2.1.2  Description of Methods

- *Scenario technique:* Following a scenario-based approach (Carroll, 2000; Kotonya & Sommerville, 1998), several scenarios have been worked out. A choice was made and the chosen scenarios were split into building blocks to identify the technical challenges to face within the development.
- *Extant Systems Analysis:* On the one hand, other smart home systems on the market or currently being developed were analysed and evaluated concerning degree of being user-centred. On the other hand, the real world system of home life was investigated with diary keeping and event logging over a one week period. The diaries were analyzed by the team and several reoccurring scenarios were singled out for further analysis.
- *Task analysis:* The data from the diary used in extant systems analysis was analysed with hierarchical task analysis and link analysis (Kirwan & Ainsworth, 1992). Both methods give the opportunity to see how often particular objects were used and actions were repeated.
- *Card Sorting and Wording tests:* Card sorting was done by six users to develop an information architecture mapping their mental structure and element names of potential users (Usability Net, 2003). From the various group headings a wording questionnaire was created to identify the most suitable terms for the information chunks in the interface.
- *Layout review with users:* Three different layouts for a GUI prototype have been worked out in paper and another six potential users had to walk through a task, chose one afterwards and give reasons for their decision.
- *Usability testing:* The clickable prototypical GUI developed was tested in a usability test to optimize information architecture and wording, the navigational concept, and the graphical layout. Ten potential users of the target group, aged between 25 and 57, six men and four women, tested the GUI in an 1-hour test, using a webpad as input device. The test was recorded by video, logfile and verbal protocols.
- *Acceptance of smart home technologies:* The acceptance of smart house technologies was investigated using thinking aloud protocols and data of an interview at the end of the usability test.

## 2.2  Results

### 2.2.1  User requirements

The usual definition of user groups by personal, social and demographic criteria is not sufficient in the smart home context. Assuming that different users with different backgrounds and different requirements live together under one roof (e.g. mother, father, son and daughter), the users have to be differentiated by types of household. Dual-career-households with children and single-parent-households were identified in the scenarios as a target group for LIVEfutura because they appear to have the greatest need for organization and time-saving technology (Meyer et al., 2001). They have a very limited time-budget and great demands on organizing and structuring their everyday life.

### 2.2.2  Tasks and functions

The task analysis showed which events in our everyday lives appear to occur automatically and happen every day at roughly the same time. These "domestic patterns" form the basis of so-called "settings". Settings combine a set of functionalities or services. For example, by activating the setting "absence", all electronic equipment and all lights will be switched off, the temperature will be reduced, all windows will close automatically and the burglar-alarm is being activated.

Domestic patterns clearly identified were the "weaking up" and the "coming home" scenario with nearly identical activities for several days.

The extant systems analysis also showed, that usual functions such as online-help, a back-button, a home-button, a calender, a notes-function, a display of time and date and display of the status of what the house "is doing" at that moment should be added to the desired elements of a smart home control.

### 2.2.3 Information architecture and navigational concept

In order to minimize the time needed to complete a certain task, the information architecture was designed to be rather broad than deep. That way, it is ensured that the user only needs a minimum amount of clicks to accomplish his goal. Screenshots of the prototype are shown in fig. 2 and fig. 3. The structuring of information and the wording of the household domains were the result of the card sorting and wording tasks. The navigational structure is built upon the task structure resulting from the task analysis,

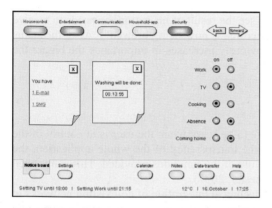

Fig.2: Notice board screen

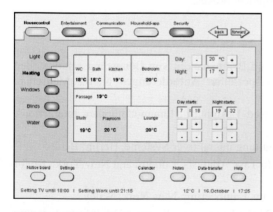

Fig.3: Housecontrol – Heating screen

### 2.2.4 Acceptance of smart home technologies

The overall acceptance turned out to be very high, 9 of 10 subjects said that they would like to live in a smart house. Most significant advantages named were: Saving time/money, relief of organizational stress, flexibility, mobility, security and help for elderly and disabled people. The "settings"-function, i.e. the combination of different functions, was most appreciated.

Disadvantages named ranged from high costs to fears of unauthorized access and fear of technical breakdown of the system. All test persons mentioned that they wished to be able to shut the system down and use everything conventionally at any time.

## 2.2.5   Usability Testing

All in all the usability test ran smoothly: all tasks were solved successfully and the prototype was judged as being easy to use. Probably this was partly due to the fact that the prototype was not very deep und the tasks not very complex.

There have been several problems with the wording ("Profile"="Settings", meaning of "Data transfer") and the concept of some functions (i.e. the difference between "notice board" and "notes"). Another result was, that the users expected an "ok" button located at the bottom right for every action. The most significant result the usability test revealed was that a feedback system is absolutely crucial. As the user often cannot directly see the result of his action, he needs to receive sufficient feedback on the input device (e.g. "Your coffee is now being brewed"). Also, the status of running applications has to be displayed, this obviously increases in importance the bigger the house and the bigger the family is.

## 2.3   Conclusion

Smart living begins in normal life and does not have to be scary when the user is in control of the supporting technology. Main factors are therefore the transparency of the whole application, the usability of the user interface and the supporting nature of the functions provided. This states even more for users with special needs and elderly (Burmester, 2001).

The LIVEfutura project shows that automation and technical integration of co-existing networks in smart living environments is accepted and appreciated by potential users if their requirements and needs are taken into account over the entire development process.

## References

Burmester, M. (2001). Optimierung der Erlern- und Benutzbarkeit von Benutzungsschnittstellen interaktiver Hausgeräte auf der Basis der speziellen Anforderungen älterer Menschen. Düsseldorf: VDI Verlag.

Carrol, J. M. (2000). Making Use. Scenario-Based Design of Human-Computer Interactions. Cambridge, MA: The MIT Press.

ISO (1999). ISO 13407: Benutzer-orientierte Gestaltung interaktiver Systeme. Berlin: Beuth.

Kirwan, B. and Ainsworth, L.K (1992). A Guide to Task Analysis. London and New York: Taylor and Francis.

Kotonya, G. & Sommerville, I. (1998). Requirements Engineering. New York: Wiley.

Meyer, S., Schulze, E., Helten, F. & Fischer B. (2001). Vernetztes Wohnen: Die Informatisierung des Alltagslebens. Berlin: Edition Sigma.

Usability Net. Card Sorting. Retrieved February 5, 2003, from www.usabilitynet.org/tools/cardsorting.htm

# Immersive 360-Degree Panoramic Video Environments: Research on Creating Useful and Usable Applications

*Albert A. Rizzo, Kambiz Ghahremani,*
Integrated Media Systems Center,
University of Southern California
3740 McClintock Ave, EEB 131, Los Angeles,
California, 90089-2561, USA
arizzo@usc.edu

*Larry Pryor and Susannah Gardner*
Annenberg School for Communication
School for Journalism
University of Southern California
3502 Watt Way, Los Angeles, California
90089-0281, USA

**Abstract:** Advanced methods for creating 360-degree Panoramic Video environments are producing new options for computer delivered information content. Panoramic Video (PV) camera systems could potentially serve as useful tools for creating virtual environment scenarios that would be difficult and/or labor intensive to produce using traditional computer graphic (CG) modeling methods or for applications where hi-fidelity capture of a real event is important. The thoughtful development of PV content in the future will require an understanding of how users can best observe, interact with, enjoy and benefit from dynamic PV scenarios. This paper presents observations from some of our research using a 360-degree PV camera. This system captures high-resolution panoramic video (>3Kx480) by employing an array of five video cameras that capture real world scenes over a combined 360-degrees of horizontal arc. During playback from a standard laptop, users wear a head-mounted display (HMD) and a 3DF head-tracking device that allows them to turn their heads freely to observe the desired portions of the panoramic scene. The current paper expands on our work in this area as presented in Rizzo et al. (2001). A brief technical description of this system is presented, followed by a detailing of the latest scenarios and applications that we have developed with the panoramic video system. The paper concludes with a description of our ongoing research testing human interaction and cognitive performance in a PV journalism application.

## 1. Introduction

Recent advances in Panoramic Video (PV) camera systems have produced new methods for the creation of virtual environments (James, 2001). With these systems, users can capture, playback and observe pictorially accurate 360-degree video scenes of "real world" environments. When delivered via an immersive head mounted display (HMD), an experience of presence within these captured scenarios can be supported in human users. This is in sharp contrast to the constrained delivery and passive viewing of television and video images that have been the primary mode for providing humans with a "virtual eye" into distant times and locations over the last fifty years. During this time, video technology has matured from gray-scale images to big-screen color and digitally processed imagery. Yet one aspect of both the content creation and delivery technology has remained unchanged— *the view is controlled at the source and is identical for all observers.* Along with traditional CG virtual environments (VEs), PV overcomes the passive and structured limitations of how imagery is presented and perceived. The recent convergence of camera, processing and display technologies make it possible for a user to have control and choice in their viewing direction. As opposed to mouse and keyboard methods for interacting with flatscreen panoramic content, users can more intuitively observe PV content via natural head movement within an HMD. Users of PV become virtual participants immersed in the observed scene, creating a new dimension in the way people perceive imagery within these types of VEs. However, when compared with CG-based VEs, PV has some limitations regarding functional interactivity. Whereas users operating within a CG-based VE scenario are usually capable of both 6DF navigation, and interaction with rendered objects, PV immersion allows mainly for observation of the scene from the fixed location of the camera with varying degrees of orientation control (i.e. pitch, roll and yaw). In spite of this limitation, the goals of certain application areas may be well matched to the assets available with this type of PV image capture and delivery system. Such application areas may include those that have high requirements for presenting real locations inhabited by real people (e.g. capture of an event of possible future historical significance). As well, alternative methods to support "pseudo-interaction" are possible by augmenting panoramic imagery with video overlays and CG objects. This paper will briefly present the technical details of our PV system, describe the scenarios we have captured thus far and highlight our research program for a PV journalism project called "User Directed News".

## 2. Brief system overview and technical description

Panoramic image acquisition is based on mosaic approaches developed in the context of still imagery. Mosaics are created from multiple overlapping sub-images pieced together to form a high resolution, panoramic, wide field-of-view image. Viewers often dynamically select subsets of the complete panorama for viewing. Several panoramic video systems use single camera images (Nayar, 1997), however, the resolution limits of a single image sensor reduce the quality of the imagery presented to a user. While still image mosaics and panoramas are common, we produce high-resolution panoramic video by employing an array of five video cameras viewing the scene over a combined 360-degrees of horizontal arc. The cameras are arrayed to look at a five-facet pyramid mirror. The images from neighboring cameras overlap slightly to facilitate their merger. The camera controllers are each accessible through a serial port so that a host computer can save and restore camera settings as needed. The complete camera system (Figure 1) is available from FullView, Inc. (FullView, 2001).

**Figure 1. FullView Panoramic Camera**

The five camera video streams feed into a digital recording and playback system that we designed and constructed for maintaining precise frame synchronization. All recording and playback is performed at full video (30Hz) frame rates. The five live or recorded video streams are digitized and processed in real time by a computer system. The camera lens distortions and colorimetric variations are corrected by the software application and a complete panoramic image is constructed in memory. With five cameras, this image has over 3000x480 pixels. From the complete image, one or more scaled sub-images are extracted for real-time display in one or more frame buffers and display channels. Figure 2 shows an example of the screen output with a full 360° still image extracted from the video.

The camera system was designed for viewing the images on a desktop monitor. With a software modification provided by FullView Inc. (FullView, 2001), we were able to create an immersive viewing interface with a SONY Glasstron head-mounted display (HMD). A single window with a resolution of 800x600 is output to the HMD worn by a user. A real-time (inertial-magnetic) orientation tracker (Intersense, 2001) is fixed to the HMD to sense the user's head orientation. The orientation is reported to the viewing application through an IP socket, and the output display window is positioned (to mimic pan and tilt) within the full panoramic image in response to the user's head orientation. View control by head motion is a major contributor to the sense of immersion experienced by the user. It provides the natural viewing control we are accustomed to without any intervening devices or translations.

**Figure 2. Still 360-degree PV image extracted from video footage taken at the Los Angeles Coliseum.**

## 3. Exploratory field testing and user testing

The capture, production and delivery of PV scenarios present unique challenges. Application development decisions require informed consideration of pragmatic issues involving the assets/limitations that exist with PV scenarios, the requirements of the application and how these two factors relate to user capabilities and preferences. Based on our initial field-testing experience, we outlined a series of guidelines and recommendations for the creation of PV scenarios that appeared in Rizzo et al., (2001). The areas covered in that paper dealt with pragmatic production issues, determination of suitable PV content, display and user interaction considerations, audio/computer graphic/PV integration issues and hardware options for maximizing accessibility. These recommendations were based on our experience in PV scenario production from a producer/developer standpoint and from user feedback provided by approximately 400-500 individuals at the time. Since then, we have continued to collect user feedback and have used this data to inform the design process in our evolving PV application research and development program.

Field-trials with the PV camera and user testing with acquired content has been conducted across a range of scenarios to explore feasibility issues for using this system with a variety of user applications. The following test scenarios were captured in order to assess the PV system across a range of lighting, external activity, camera movement and conceptual conditions. Informal evaluation of users' responses to these scenarios has been conducted with controlled experiments currently underway for some of the applications. Our PV scenarios have included:

1. An outdoor mall with the camera in a static position in daytime lighting with close background structures and moderate human foot traffic, both close-up and at a distance.

2. An outdoor ocean pier with the camera in a static position in extremely intense lighting conditions (direct intense late afternoon sunlight with high reflectance off of the ocean) with both long shots of activity on a beach and close-up activity of human foot traffic and amusement park structures on the pier

3. The interior of an outside facing glass elevator with the camera in a static position and the elevator smoothly rising 15 floors from a low light position (e.g., tree-shielded street level) to more intense lighting as the elevator ascended above the tree line.

4. Traveling on a canyon road with the camera mounted in the bed of a pickup truck for 30 minutes at speeds ranging from 0-40 mph under all daylight ranges of lighting (low shaded light to intense direct sun).

5. Same as #4, except at night on a busy well lit street (*Sunset Blvd. In Los Angeles),* and on a freeway traveling at speeds from 0-60 mph.

6. A USC Football game within the Los Angeles Coliseum from both static and moving positions in daytime lighting, with extreme close-ups of moving people and massive crowd scenes (40-60 thousand people).

7. An indoor rock concert in a theatre (*Duran Duran*) from a static position under a variety of extreme lighting conditions in the midst of an active crowd, slightly above average head level.

8. Two artistic projects were done in collaboration with the UCLA Digital Media Arts Department and the USC School of Fine Arts. The UCLA project involved the capture of dancers performing around the 360-degree field of view of the camera. Significant post-production work took place to display the panoramic capture within an immersive theatre that incorporated live dancers in a mixed reality installation. The USC project involved building a circular fish tank around the camera with live tropical fish swimming within and a coral reef photo serving as background on the outermost tank wall. The users wore an HMD that helped to create the illusion of being immersed within the swimming fish environment for one minute. Following this sequence, the coral reef photo background was manually removed to reveal the activity in the laboratory where the capture occurred creating a dramatic "breaking of the illusion" effect. This application also served as an early test for a future project in which the panoramic camera will be placed within a sealed plexiglass tube and lowered into a very large commercial aquarium exhibit.

9. Thirteen scenarios were created in an indoor office space for an "anger management in the workplace" application. In these scenarios, actors portrayed agitated and insulting co-workers who addressed the camera (and vis a vis, the clinical user wearing the HMD) with provocative and hostile verbal messages (Figure 3). The scenarios were designed to provide role playing environments for patients undergoing psychotherapy for issues relating to anger management in the workplace, or as it is commonly referred to as "Desk Rage" (Daw, 2001). The patients wearing the HMD in these scenarios have the opportunity to practice appropriate responses to the characters and employ therapeutic strategies for reducing rage responses. Traditional methods of therapy in this area have mainly relied on guided imagery or role-playing with the therapist. It was hypothesized that PV content could serve to create immersive simulations that patients will find more realistic and engaging, and research is currently underway to assess this with clinical users at The VRMH center in San Diego, CA (see: http://www.vrphobia.com/about.htm).

10. A Virtual "Mock-Party" with the camera in a static position in the center of an indoor home environment in the midst of an active party with approximately 30 participants (Figure 4). This "scripted" scenario was shot while systematically directing and controlling the gradual introduction of participants into the scene and orchestrating their proximity and "pseudo-interaction" with the camera. The scenario was created for a therapeutic application designed to conduct graded exposure therapy (Rothbaum & Hodges, 1999) with social phobics. We have also experimented with pasting "blue screen" capture of actors (using a single video camera in the lab) into the panoramic scenes. The actors address the camera with a spectrum of socially challenging questions that provide the clinical user with

opportunities to practice social verbal engagement in a psychologically safe environment. The separate capture and pasting of characters will allow the therapist to introduce a progressively more challenging level of social stress to the patient when deemed appropriate based on therapist monitoring of patient self-report and physiological responses. User testing on this project with clinical populations is anticipated to begin in June, 2003.

11. We have captured one PV test environment for a journalism project entitled "User Directed News". This project is being conducted in collaboration with USC's Annenberg School for Journalism and involves the PV capture of a news event with an in-the-field news reporter (Figure 5). The research aim of this project involves a comparison of groups of viewers who observe the news story either in the traditional single frame "talking head" newscaster format vs. full user immersion "in the scene" using a HMD. The project has thus far involved the capture of a news story on the situation of the Los Angeles' "homeless" population as part of a study on the usability, preference for use and information processing issues related to this potential journalism application. The details of this project will be presented below.

**Figure 3. "Desk Rage" – PV Therapy for Anger Management in the Workplace**

**Figure 4. PV "Mock-Party" for Social Phobia Exposure Therapy**

**Figure 5. PV/Journalism Project: Traditional Viewing (L) vs. Panoramic Viewing (R)**

## 4. User Directed News research program

The User Directed News project is based on the idea that as journalism moves into the 21st Century, new forms of information technology (IT) stand to revolutionize methods for acquiring, packaging, organizing and delivering newsworthy information content. With these advancements in IT will come both opportunities and challenges for creating systems that humans will find to be usable, useful and preferred options for interacting with newsworthy information content. While some of the basic issues that constrain journalistic methods over the centuries (word of mouth, print, radio, TV, etc.) remain relevant, new issues are emerging for how humans will effectively interact with the deluge of digital content that will continue to expand in both quantity and scope as we move further into the information age. One area where journalism could benefit from the emerging IT revolution is in the use of systems that are capable of capturing and delivering news events within more "immersive" viewing formats. As opposed to traditional "fixed-camera/talking head" capture and delivery of "on-the-scene" reporting of newsworthy events, the potential now exists to produce HMD-delivered 360 PV environments that allow the user to self-select what aspects of the event that they would like to observe and promote a sense of being immersed within the ongoing event. When immersed in this context, the user has the option to actively choose what aspects of the event that they are most interested in or are compelled to view. At the same time, *this would not eliminate the need for the news reporter*, but rather, would dynamically transform her role into that of a "news-guide", who could stroll freely around the camera and point out aspects of the scene that the user could either chose to view or not. Viewer choice is key here in that current single camera approaches can indeed follow a roving reporter around an event, but do not allow for the user-option to self-select what aspects of the event are most relevant to his interests. This approach may serve to transform the user from simply being a passive observer of "fixed" content, to an active participant in the news acquisition process. The combination of the "immersive" aspect of "being there" along with free choice of viewing may provide a new paradigm for how news is created and consumed. One can imagine many scenarios where use of such a system would be desirable (i.e., observing the activity at a Security Council meeting at the United Nations or presence at a post Academy Awards party in Hollywood).

However, a number of pragmatic and user-centered questions need to be addressed scientifically before a determination of the value of this system can be made. Some of these basic questions include: Will users generally prefer to have news delivered in this format? Does immersion and self-selection compel the user to prefer this method of being "involved" in the story? Will reporters be able to adapt to this more "free form" method of reporting and what challenges will this produce for reporters in delivering "stories" to users who may not chose to follow the information flow in a traditional fixed "linear" manner? Will choice of viewing interfere with the acquisition of the logical story line in a news report? Will users be able to recall key points of the reported event in a meaningful manner? What types of news events would this system be best suited for in terms of user preference and information processing issues and what are the key elements of newsworthy events that might predict successful outcomes for use of the system? Will users naturally explore the 360 environment and choose to use this option?

**Research Design Summary -** Research is currently underway to address some of these issues by way of a very basic analog "news event" user preference and information processing comparison study. The current study is based on the production of a panoramic news story on issues regarding the "homeless" in Los Angeles, which puts viewers in the midst of some of the harshest living conditions imaginable within a modern society. Within this environment, the camera was strategically positioned in the middle of a street that was lined with people, an array of tents and makeshift living quarters on the sidewalk. The reporter stood in a fixed position within the field of view of one of the five cameras and reported a 2-minute story, as is typical of on-the-scene reporting. With this content, two user groups are currently being compared:

**Group 1** simply views the feed from one of the cameras' field of view containing the reporter's delivery of the story, as is common practice in a standard reporting approach.

**Group 2** views the event within a HMD to create the sense of being within the event while having free choice to observe the PV scene from any perspective within the 360 arc. At the same time, users hear the *exact same verbal delivery* from the reporter as presented in Group 1.

Following exposure to the 2-minute story, users in both groups are tested on multiple measures of memory for the information presented in the story and on user preference for use of the system. Memory for the content of the news story is also being tested one week later. This allows us to compare groups on immediate acquisition/retention of content and on long-term recall/recognition retrieval. As well, head tracking data in Group 2 is being quantified to produce a metric of exploratory behavior within the 360 degree PV scene. We hypothesize that the sense of "being there" or "presence" will be enhanced in Group 2 by way of using an immersive HMD, and that this added engagement will increase recall by providing better contextual retrieval cues that leverage episodic memory processes. While the groups may not differ on measures of immediate memory, due to competing distraction effects nullifying immersion based gains in the HMD condition, we predict that when subjects are tested one week later, the contextual, episodic memory and presence effects will operate to produce much better recall/recognition retrieval. Early results from this study will be presented at the HCI conference.

## 5. References

Daw, J. (2001). Road Rage, Air Rage and Now "Desk Rage". *The Monitor of the Amer. Psych. Assoc.,* 32, (7).

FullView.com Inc. (2003). Retrieved February 15, 2003, from www.fullview.com

Intersense Inc. (2003). Retrieved February 15, 2003, from www.isense.com

James, M.S. (2001). 360-Degree Photography and Video Moving a Step Closer to Consumers. Retrieved March 23, 2001, from http://abcnews.go.com/sections/scitech/CuttingEdge/cuttingedge010323.html

Nayar, S. K. (1997). Catadioptric Omnidirectional Camera, Proc. of IEEE Computer Vision and Pattern Recognition (CVPR).

Rizzo, A.A., Neumann, U., Pintaric, T. and Norden, M. (2001). Issues for Application Development Using Immersive HMD 360 Degree Panoramic Video Environments. In M.J. Smith, G. Salvendy, D. Harris, & R.J. Koubek (Eds.), *Usability Evaluation and Interface Design* (pp. 792-796). New York:L.A. Erlbaum

Rothbaum, B.O. & Hodges, L.F. (1999). The use of Virtual Reality Exposure in the Treatment of Anxiety Disorders. *Behavior Modification*, 23 (4), 507-525.

Witmer, B.G. & Singer, M. J. (1998). Measuring presence in virtual environments: A Presence Questionnaire. *Presence: Teleoperators and Virtual Environments*, 7 (3), 225-240.

# Multimodal Interaction Techniques for Astronaut Training in Virtual Environments

*Jukka Rönkkö, Raimo Launonen*

Technical Research Centre
of Finland
02100 Espoo, Finland
jukka.ronkko@vtt.fi

*Seppo Laukkanen*

SenseTrix Oy
Kaivosmestarinkatu 8 C 29
02770 Espoo, Finland
seppo.laukkanen@sensetrix.com

*Enrico Gaia*

Alenia Spazio S.p.A.
Strada Antica di Collegno
253 10146 Torino Italy
egaia@to.alespazio.it

## Abstract

This paper describes multimodal interaction techniques used in a virtual environment training prototype for astronaut assembly sequence training. Astronaut training takes place on Earth, which makes the simulation of zero gravity conditions necessary. Performing assembly tasks on orbit differs dramatically from ground operation in terms of object handling as well as user movement. The use of virtual environment techniques in astronaut training offers potentially significant cost savings and quicker access to the training situation compared to methods like parabolic flights. The usefulness of such a training application depends largely on the high enough resemblance of the operation to the real situation. In this respect the selected combination of interaction techniques plays a vital role. This paper presents a possible interaction technique battery for astronaut training that utilises two-handed interaction, speech and gesture recognition in a physically simulated virtual environment. User feedback is provided via head tracked stereo graphics and audio cues.

## 1    Introduction

Astronaut training takes place on Earth, while their working environment is mostly on orbit, in this case at ISS (International Space Station). The work includes zero gravity conditions in restricted spaces. The assembly sequences of equipment should be trained on Earth, before the actual operation takes place on orbit. The training methods have traditionally included classroom training, multimedia presentations, physical mock-up and swimming pool based training as well as parabolic flights. However VEs (Virtual Environments) are assumed to bring a better sense of the surrounding working environment and user movement as well as object behavior compared to classroom and multimedia based training. On the other hand, compared to the parabolic flights, VE training simulation could offer a more economical and accessible training method.

The assembly tasks include procedures that require the users to perform actions such as handling freely floating objects, connecting and disconnecting parts, pushing buttons, turning switches on and off as well as opening and closing lids. While performing these operations, the user may have to move by grabbing handles placed along the walls of the corridors of the ISS modules and hook his or her feet on special loops to avoid floating away. The main focus in training is the correct sequence of these operations as well as becoming familiar with the general object behavior and the user movement in zero gravity. Fine-grained manipulation tasks requiring fingertip level accuracy

– like using a screwdriver – are not the center of interest within this context, because such tasks are assumed to be very well known to the trainees.

The target of this paper is to present a feasible set of interaction techniques for a training prototype application for the user movement and object handling in a physically simulated VE. This set of techniques should take into account multimodal interaction research, resemble real world behavior to a sufficient degree and avoid unnecessarily large cognitive load. Some initial user experiences are presented. The work is conducted within the VIEW (Virtual Interactive Workspaces of the Future) EU project (VIEW, 2003).

## 2 Related work

Foley et al. (Foley, Van Dam, Feiner & Hughes, 1996) suggest that computer graphics based interaction can be thought to be a composition of basic interaction tasks. This classification is refined by others (Bowman, 1999) so that the tasks include object indication, selection, orienting, moving, creation and deletion, text input as well as viewpoint and system control. The basic tasks can be implemented in a multimodal fashion, which means that multiple communication channels are used between the user and a computer, such as speech, hand and head tracking as well as gestures. The use of these channels should be complementary rather than redundant (Bowman, 1997). Billinghurst (Billinghurst, 1998) and Oviatt (Oviatt, 1999) point out other principles to be taken into account in designing multimodal input. For example speech should be used for non-graphical command and control, gestures for visuo/spatial input. VEs have been used in assembly simulation with multimodal input  (Zachmann, 2000) with views supporting Billinghurst's guidelines.

In terms of hand based manipulation two-handed interaction provides a means for more efficient interaction than using a single hand. However a person's hands do not operate totally independently of each other. In experiments run by Kelso, Southard and Goodman (Kelso, Southard, Goodman, 1979) the participants had to reach quickly for two targets with both hands. Even if the other object was more difficult to reach because of its size and distance, both hands reached their targets simultaneously. It has also been suggested (Schmidt, Zelaznik, Hawkins, Frank & Quinn, 1979) that when people make rapid simultaneous movements with two hands, some of the parameters regarding the movement have to be the same. In the field of interaction technique research it is suggested that object manipulation is efficient and does not posses an excessive cognitive load if the non-dominant hand (NDH) acts as a reference frame to the movements of the dominant hand (DH) (Hinckley, Pausch & Proffit 1997, Pierce et. al. 1997).

Virtual environment techniques have been used in astronaut and ground-based personnel training to prepare for the Hubble Space Telescope repair (Loftin & Kenney 1995). The conclusions were that the use of the VE simulation had a positive impact on acquiring the knowledge for the mission. However not very detailed description of interaction was presented. Montgomery, Bruyns and Wildermuth have studied VE training for biological microgravity experiments with and without haptic feedback (Montgomery, Buyns & Wildermuth, 2001).

## 3 Design of an interaction technique battery for astronaut training

The users of the assembly sequence training application prototype have technical background, but not necessarily very much experience in VEs. The main target of the application is to teach correct operational sequences, general object handling and user movement in zero gravity and in a small operational space. Fine-grained manipulation is not the main target of the application.

The interaction can be split up into three main categories including object manipulation, viewpoint and system control. The techniques have been designed bearing in mind the resemblance to the real world behaviour in training tasks. Therefore for example two-handed interaction is used. The behaviour of the objects is determined by rigid body dynamics including inertia properties and collision detection under user defined gravity conditions. For example it is not possible to pass objects through one another. This applies also to virtual hand manipulators described in the next chapter.

## 3.1 Object manipulation

Object manipulation is performed with virtual hand manipulators. The manipulators have their visual and physical presentations in the virtual world. The shape of the manipulators resembles real hands. Their position is determined as a combination of the position data from the trackers, navigational position of the user as well as rigid body physics calculations. This means that the difference of the current hand position and the desired position is fed into physics calculations as a force and torque that move the hand manipulator towards the desired direction and orientation. This ensures realistic virtual hand movement when they collide with other objects as well as clamping of the virtual hand movements if they collide with static objects.

The basic object manipulation tasks include object indication, selection, orienting and positioning. The inaccuracy of tracking systems as well as data glove finger flexure measurements have led us to discard at this point the real world way of touching, grabbing and moving of objects. Rather an object can be indicated by touching it with a sensor object. The sensor objects are boxes, which are bigger than the virtual hand objects. The sensor object moves as the virtual hand is moved, i.e. the sensor object surrounds the virtual hand object at all times. An indicated object can be selected and grabbed by clenching fingers into a fist gesture. After this, the object is attached to the virtual hand and can be moved and rotated. The object can be released by changing the gesture from the fist gesture to something else. When the user doesn't want to explicitly grab an object, the virtual hand and its fingers can still interact with other objects through rigid body dynamics. It is for example possible to push objects.

Objects can be connected together by grabbing objects with each hand, making them collide and issuing a speech command "connect" while they touch each other. In similar fashion two objects can be grabbed and disconnected with a speech command "disconnect". The user can also grab objects that are permanently attached to the surrounding structures, but they do not move. Connectable objects can be configured to connect only to certain other objects within the simulation.

The user is provided with additional feedback of his or her interaction by visual and audio cues. Indicating an object causes it to be highlighted. The highlight mechanism is implemented by creating a temporary, semitransparent replica of the original object, which is slightly larger than the original object and located at the same position. Collisions in general cause a sound being played. Also in the case of touching objects with the hand manipulators, a special sound is played. As objects are connected or disconnected different sounds are used as indicators of the completion of the operation.

## 3.2 Viewpoint control

The position of the user in the virtual space is a combination of the tracked head position and navigational position. The latter can be altered by a two-handed movement technique, in which DH (Dominant Hand) index and little finger gesture signifies forward movement, index and middle finger pointing gesture the willingness to turn left, little finger gesture is used to go up. Naturally there is no distinct up direction in zero gravity, up in this context means relative to the user's current reference frame. The NDH (Non-Dominant Hand) index and little finger gesture is used for backwards movement and index middle gesture to turn right. The NDH little finger gesture is used to for moving down relative to the current reference frame.

## 3.3 System control

System commands (such as exit) and controlling the simulation in general are handled via speech commands. For example commands that affect the gravity levels are issued via speech; gravity on/off, gravity 9.0. System control is a task that should not interfere to a large extent with the training itself. Using speech in system control allows other tasks (viewpoint control and handling of objects) to be carried out with little or no interruption. However if there are a lot of commands that should be memorised, it increases the learning curve of the system. Therefore a visual menu style mechanism is planned which can visually clarify commands that are available within a certain operational context.

## 4 Implementation

The training prototype and related interaction techniques are implemented on a PC cluster based rear projection VE system. The hardware system consists of two projection screens, two projectors per screen for passive stereo, two rendering PCs (2GHz, GeForce4 display cards, 512MB main memory) per screen, a dedicated PC for physics calculations and a central, control PC. In addition to this, a laptop is used for dataglove and speech input. These computers are connected via gigabit network cards and a gigabit switch. The user input is received via head and hand tracking, speech recognition and dataglove input for both hands. Optical ART tracker and Intersense IS600 ultrasonic tracker are used for the head and hand tracking. 5th Dimension datagloves are used for gesture input.

VEVIEW software platform is used to control the hardware and to implement the training prototype as well as the related interaction techniques. The VEVIEW software platform is being developed in the VIEW project (VIEW 2003). The prototype runs on Windows operating system.

## 5 Future work

Ongoing work includes the design and development of the playback of user action sequences and the interaction techniques related to it. More realistic user movement methods are also researched. The new method could use grabbing of handles attached to the corridor walls of the ISS. The prototype training application utilising the interaction technique battery will be user tested by comparing the use of the prototype to the use of a physical mock-up. This will give information both on the usefulness of the VE based training in astronaut assembly training in general as well as information on specific interaction techniques.

# 6 Conclusions

A multimodal technique battery for astronaut assembly training in a physically simulated VE was presented. The battery was designed bearing in mind related research in the field of multimodal interaction and VEs as well as the requirements of the application field of astronaut training including two-handed real world like interaction. The interaction techniques in the battery cover object indication, selection, position and orienting as well as viewpoint and system control. The presented techniques take advantage of hand and head position tracking, gesture and speech recognition. Position tracking and gestures are used in viewpoint as well as hand manipulator control. Speech input is utilised in system control.

# References

Billinghurst, M., Put That Where? (1998) Voice and Gesture at the Graphics Interface. *Computer Graphics*, 32 (4), 60-63.

Bowman D. A. & Hodges L. F. (1997) An evaluation of techniques for grabbing and manipulating remote objects in immersive virtual environments. *Proceedings of the 1997 symposium on Interactive 3D graphics* (pp. 35-38), New York: ACM Press.

Bowman D. A. (1999) Interaction techniques for common tasks in immersive virtual environments. Doctoral thesis. Georgia Institute of Technology.

Foley, James D., Van Dam A., Feiner Steven K. & Hughes John F. (1996) Computer Graphics: Principles and Practice in C. Addison-Wesley.

Hinckley, K., Pausch R. & Proffit D. (1997) Attention and visual feedback: the bimanual frame of reference. *Proceedings of the 1997 Symposium on Interactive 3D Techniques*, (pp. 121-126). New York: ACM Press.

Kelso, J. A. S., Southard, D. L., & Goodman, D. (1979). On the coordination of two handed movements. *Journal of Experimental Psychology: Human Perception and Performance*, 5, 229-238.

Loftin, R. B. & Kenney, P. (1995) Training the Hubble space telescope flight team. *Computer Graphics and Applications*, 15 (5), 31-37.

Montgomery, K., Bruyns C. & Wildermuth S. (2001). A virtual environment for simulated rat dissection: a case study of visualization for astronaut training. *Proceedings of the conference on Visualization 2001* (pp. 509-514). Piscataway, NJ, USA: IEEE Press.

Oviatt, S., Ten Myths of Multimodal Interaction. (1999) *Communications of the ACM*, 42 (11), 74-81.

Pierce J. S., Forsberg A. S., Conway M. J., Hong S., Zeleznik R. C. & Mine M. R. (1997) Image plane interaction techniques in 3D immersive environments. *Proceedings of the 1997 symposium on Interactive 3D graphics*. (pp. 39-43), New York: ACM Press.

Schmidt, R. A., Zelaznik, H.N., Hawkins, B., Frank, J.S. & Quinn, J.T.Jr. (1979) Motor output variability: A theory for the accuracy of rapid motor acts. *Psychological Review*, 86, 415-451.

VIEW (2003), Virtual and Interactive Environments for Workplaces of the Future. Retrieved February 10, 2003 from http://www.view.iao.fhg.de.

Zachmann, G. (2000) Virtual Reality in Assembly Simulation – Collision Detection, Simulation Algorithms and Interaction Techniques. Doctoral Dissertation. University of Darmstadt.

# The Improvement of the Perception of Space and Depth by the Help of Virtual Reality (Programmed by VRML and HTML)

*Cecília Sik Lányi, Ádám Tilinger,*
*Zsolt Kosztyán*

*Zsuzsanna Lányi*

University of Veszprém, Department of
Image Processing and Neurocomputing
Hungary
lanyi@almos.vein.hu, tilinger@vnet.hu
kzst@vision.vein.hu

Ophthalmologic Department of the
Ferenc Csolnoky Hospital
Hungary
lanyi.zs@freemail.hu

## Abstract

Most of the authors are of the opinion that the perception of space can be improved till the age of 15 or 16. In our experiment we have prepared virtual reality applications for this age group to improve their space perception. The students had to take part first in an orthoptic investigation (ophthalmologic screening, visual acuity measurement, space and depth investigation). Then they had to solve traditional paper tests. We have developed virtual reality (VRML) programs that help to learn how to solve the paper test, and a group of students practised using these programs. Then the paper test tasks had to be solved again, both by the group, who could practice using the VRML programs, and a control group, who used only the paper tests, and the results were compared with the results of the first test, taking also the results of the ophthalmologic test into consideration. Test results have shown that practising with the VRML tasks helped to improve the space perception of the students.
Our experiments have shown that VRML techniques are well suited to foster the ability of students to seeing in three dimensions. Even a student, who – according to the first ophthalmologic test – had absolutely no depth perception, had acquired some feeling of three dimensionality and space perception after practicing with our programs.

## 1   Introduction

Space perception is important not only in many professions but also in the everyday life (from becoming a pilot till navigating with a motorcar on a crowded road). Thus it is vital for the up growing generation to acquire good space and depth perception. Most of the authors are of the opinion that the perception of space can be improved till the age of 15 to 16 years old. On the other hand the Hungarian general education curriculum does not provide much opportunity for the students to develop their depth perception. Therefore we seeked for methods that could be used in schools to practice space and depth perception.
Virtual Reality pictures are not very much used in our schools, but we thought that by presenting tasks on the computer using these techniques, we could develop applications by the help of which student could increase their space perception skill.

In our experiment we have prepared virtual reality applications for a 10 to 15 years age group to improve its space perception. After evaluating a pilot program a VRML application has been developed intended for such students. The school children had to take part first in an orthoptic investigation (ophthalmologic screening, visual acuity measurement, space and depth investigation). Then they had to solve traditional paper tests. Virtual reality (VRML) programs, similar to the paper tests have been developed with animations of the proper solutions of the tasks. Then the test tasks had to be solved again, and the results were compared with the results of the first test. We seeked the answer on the question, whether the space perception has improved by virtual reality training or not.

Further on we studied what the relationship between the results of the ophthalmologic space-vision test and the amelioration of space vision by the use of practise was.

We have also tested the improvement of space vision using interactive, computer based VRML and Anaglyph programs built to supplement the ground school mathematics and geometry curriculum and investigated whether the use of solving tasks on the computer has an influence on the students mathematics and drawing accomplishments.

## 2    The Input Tests

The original paper test has been prepared by Andrea Kárpáti, László Séra and János Gulyás at the Eötvös Loránd University (Kárpáti A., Séra L. & Gulyás J., 2002). It contains 25 test sheets.

We have developed computerized versions of these tests, where our aim was to give to the students tasks that are not boring, keep their attention, are like games, but help to increase space and depth perception. It was important to do this is such a form that the programs can be run also on moderately up to day computers, as the hard-ware environment at the schools is mostly several years behind most modern requirements. Due to this some compromise solutions had to be elaborated.

First we searched for the types of tasks we should realize in computerized form. Based on these investigations the tests can be grouped into two capability factors, thus we suppose two partial capabilities. Taking the name from Eliot (Eliot 1987) these are recognition and manipulation. Eliot used these terms to group his tests, but according to our experience these terms describe not only the differences between the tests very clearly but describe also the partial capabilities needed to solve the tests.

## 2.1    Types of the Tests

Producing projections: changing the inner viewpoint, or the production of two-dimensional projections of three-dimensional figures (truncated bodies, shaded slings, wire frame sling figures) by mental rotation and drawing of these. The difficulty of the tasks changed from the concrete to the more abstract analytical tasks. Some of the tasks required an answer, while others were multiple-choice tasks or of mixed character. Reconstruction: Based on the projections the test person had to draw the axonometric picture of the body. In performing this task a complex synthetic process is done: a series of inner viewpoint changes are done while the test person compares his/her view with the original and corrects his/her drawing. The picture of the body is mentally reconstructed from the projections in the course of the process even that there is no visible picture. Thus one can say that this picture vibrates at the border of being a two- or three-dimensional picture. It is characterized by being both icon like and symbolic. Seeing the structure: In these tasks one had to match two-dimensional forms that were distorted due to a change in the perspective with pictures where the forms were shown in their proper proportions. The tasks could have been subdivided into two groups depending on whether one had to start from the picture showing the form in perspective or in frontal view. This task investigated whether the mental picture of the test person could reflect the real proportions of the body or not. Estimating the perception of the relations and ratios one could guess how accurate the inner mental picture was. This group of tasks permitted to evaluate also how accurate the mental corrections of those distortions were that were produced by the distortions due to the viewpoint and the representations. They point to the equilibrium of the perception and cognition, of the visual perception and thinking. Anaglyph pictures: The use of anaglyph pictures is a well-known technique to show bodies in three dimensions. We have prepared a simplified VRML version of this technique that enabled an easy preparation of anaglyph pictures, so that a large number of such graphs could be prepared to train the eyes of the students to perceive three-dimensionality. We looked for the answer on the question, whether the space perception could be improved by virtual reality or not. Further on we studied what the relationship between the results of the ophthalmologic space-vision test and the improvement of space vision by the use of practise is.

## 3 Description of the Softwares

The tests described in this chapter have been realized using two software programs. The main menu was written in HTML, in which VRML applications have been embedded. This has the advantage that no translation is needed to get an executable file; only a VRML browser has to be downloaded. Both programs start with the test tasks according to the main menu points, and show animated solutions.

This choice had the advantage that the programs could be executed even in schools with moderate computer hardware capacities. In our country it is general practice to have Windows operational systems running on the computers, and these have a built in browser (Internet Explorer). To build the three-dimensional objects we used the VRML language. This enables interactivity too, and can be used if a VRML browser (e.g. CosmoPlayer) has been downloaded.

The main menu starts with the selection of the task. The student can go through all the tasks one by one, or can select one to be studied at will.

All the parts enable the active participation of the student. This has been enabled by the possibility of direct manipulation. The student can not only be engaged in the solution of the task, but can also check whether he or she found the correct solution or not. The programs do not score the performance of the student, but help to get the correct solution by showing the best examples by the help of animations. The programs are user friendly, the student does not have to have computer knowledge, the use of the programs is easy.

The first program has eight task groups. Each task groups has several tasks too. The second program has 24 tasks.

Two tasks are shown in Figure 1. In the task shown on the left hand side the student had to find out which of the bodies depicted in blue could be put through the hole of the body depicted in red. Both the blue and the red body could be turned with the mouse. The task shown on the right hand side was the following: the student had to find out which of the basic forms shown in the lower part of the screen had to be used to build the bodies shown in the upper part of the screen. For some of these bodies two, for some three basic forms have to be used. The mouse could turn the bodies, and by clicking on the body the correct answer could be obtained.

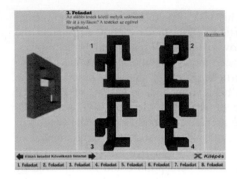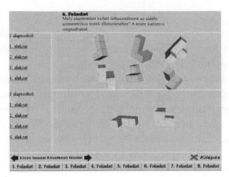

**Figure 1:** Two tasks from the group prepared for the children in the sixth and eight grade.

# 4    The Course of the Investigation

We have conducted several tests, using our pilot and final programs. In the first test 18 15 year-old high school students participated. In this test we wanted to clarify whether there is any difference if the students practice with a multimedia or a virtual reality program. First they performed a paper test, and then an ophthalmologic test was done. Half of the class used the multimedia program, the other half the virtual reality program. Then the paper test was repeated. We have investigated whether the multimedia or the virtual reality task helped more to get better scores in the second paper test, i.e. which method increased their space vision more. In our experiments we compare the above tests with the results of the ophthalmologic investigations and the school marks of the students in mathematics as well. Some studies show that space tests are useful to predict the achievements of students in subjects as geometry, mathematics and chemical studies (Ghiselli, 1973).

# 5    Evaluation

The two paper tests conducted before and after the training with the multimedia and virtual reality computer tests has been evaluated and compared. In the first test students had to solve 20 tests, and we evaluated their results by computing the number of good answers related both to the total number of tests and to the number of tests the students dealt with (due to the high number of tests and the relatively short time to perform the task, most students did not solve all the tests). In the final evaluation we used only the ratio: good answers/solved tests.

In the second paper test we have selected only seven test sheets, and evaluated these similarly to that of the first test. Figure 2 shows the paper test results of the high school students before and after practicing with the computer tests.

The average of the results for girls: first test: 61 %, second test 70 %; boys: first test: 74 %, second test: 82 %. Comparing the two groups, who practised using our programs and those who have not used it shows the following result: Those who could use the computer programs got 32 % better scores at the second test; those, who could use only the paper test got only 1 % better scores.

It is interesting to mention that for one student the ophthalmologic test showed absolutely no space vision, (on the right hand side in the third row from the top) despite of this his second test, after practicing with the computer tests, got slightly better.

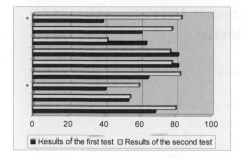 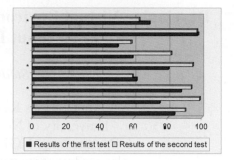

| | | | | |
|---|---|---|---|---|
| 0 | 20 | 40 | 60 | 80 | 100 |

■ Results of the first test □ Results of the second test

■ Results of the first test □ Results of the second test

**Figure 2:** Test results of the high school (15 years old) students: On the left hand side results of the girls, on the right hand side those of the boys can be seen. Test results of the first test are depicted in blue, of the second test in yellow. The students marked with a * practised using the computerized tests.

Based on the results of the first test a second program has been prepared. This was tested in two schools, in two classes (sixth grade, 12 year-old students: Group B and eighth grade, 14 year-old students). The class in eighth grade consisted of students with generally better scores as those in the sixth grade. Thus the total number of students, who participated in the tests, was 60. This is a relatively small number, but it permitted to draw some conclusions on the effectiveness of the programs. The results of the second test will reviewed at the conference.

## 6   Summary

The space-vision can be augmented in childhood. In our experiments virtual reality applications were used to test 15-year-old high school students and 12 - 14 year-old elementary school children. The high school students first solved traditional paper tests then they were screened by orthoptical tests (visual acuity, ophthalmic investigations, space- and depth-vision tests).
Two computer tests have been prepared: one for high school students, the other for elementary school students. Both tests were made using VRML virtual reality program and the main menu by HTML application. The students could practice using these tests, and after such practice sessions the paper tests were re-run. Our investigations have shown that the computer programs can help children to better space perception. Thus it can be recommended that such programs should be included in the curriculum of the 12 to 15 year-old children.

## References

Ghiselli E. E., 1973. The validity of aptitude tests in personnel selection, Personal Psychology, 26, 461-477.
Hodges L., 2001. Using the Virtual World to Improve Quality of Life in the Real World, Proceedings of the Virtual Reality 2001 Conference, Yokohama: IEEE Computer Society, 1-3.
Johnson A., Moher T., Ohlsson S., & Leigh J., 2001. Exploring Multiple Representation in Elementary School Science Education, Proceedings of the Virtual Reality 2001 Conference, Yokohama: IEEE Computer Society, 201-208.
Séra L.,Kárpáti A., & Gulyás J., 2002. A térszemlélet fejlődése és iskolai fejlesztése. Comenius Kiadó, Pécs, 2002

# Developing Virtual Environments Using Speech as an Input Device

*Alex W Stedmon*

VR Applications Research Team
University of Nottingham,
Nottingham, NG7 2RD, UK
alex.stedmon@nottingham.ac.uk

## Abstract

In order that early decisions can be made about the design, usability and evaluation of new Virtual Environments (VEs), a clear progression taking user needs forward into system development is required. To assist developers, stakeholders and researchers alike, the Virtual Environment Development Structure (VEDS) has been developed which is a holistic, user-centred, approach for specifying, developing and evaluating VE applications. This paper provides a summary of VEDS and considers the human factors issues associated with incorporating speech recognition into VR applications. Speech has been identified as a novel input device that can be used in isolation or in conjunction with other input devices such as a mouse, joystick, or data-gloves. Speech allows hands-free operation and maybe particularly useful for navigating through menus and short cut commands. Overall, the success of speech recognition depends on the user, system, task and environmental variables, all of which have significant effects on overall system performance.

## 1    VIEW of the Future

Virtual and Interactive Environments for Workplaces (VIEW) of the Future is a project funded by the European Union within the Information Society Technology (IST) program. The overall goal of VIEW is to develop best practice for the appropriate implementation and use of VEs in industrial settings for the purposes of product development, testing and training. Underpinning this is a clear emphasis on incorporating human factors throughout the conception and implementation of Virtual Reality (VR) applications by considering user- and applications-needs in developing initiatives that are fundamentally usable. In order that early decisions can be made about the design, usability and evaluation of new Virtual Environments (VEs), a clear progression taking user needs forward into system development is required. To assist developers, stakeholders and researchers alike, the Virtual Environment Development Structure (VEDS) has been developed which is a holistic, user-centred, approach for specifying, developing and evaluating VE applications (Wilson, Eastgate & D'Cruz, 2002). This paper provides a summary of VEDS and considers the human factors issues associated with incorporating Automatic Speech Recognition (ASR) into VR applications.

## 2    Speech as an input device

Of particular importance in the VIEW project is the need to examine which input devices are best suited to conducting specific tasks within VR applications. There is also a strong interest for investigating how participants navigate around VEs, as well as looking at how VE content can support usability.

Input devices can be defined as the medium through which users interact with a computer interface and, more specifically in the context of the present research, a VE (Stedmon, Patel, Nichols & Wilson, 2003). Currently, there is an increasing variety of input devices on the market that have been designed for VR use, which range from traditional mouse and joystick devices; to wands, data-gloves, speech, etc. With such a variety, this may lead to individuals selecting an inappropriate input device which could compromise the overall effectiveness of a VR application and the user's satisfaction in using the application.

For the purpose of the VIEW project, speech has been identified as a novel input device that can be used in isolation or in conjunction with other input devices such as a mouse, joystick, or data-gloves. Speech allows hands-free operation and maybe particularly useful for navigating through menus and short cut commands. Furthermore, speaker independent ASR systems support multi-user applications in collaborative VEs allowing many user to interact with each other and the VE at the same time.

## 3    Incorporating speech in the development of VR applications

In developing particular applications, it is important to recognise that VR may not be the only technology or the best potential solution. It is important, therefore, to consider from the outset what is already known about VR/VE attributes and consider how they match the application requirements and may compare to alternative technologies (D'Cruz, Stedmon, Wilson & Eastgate, 2002). Assuming that VR offers a sensible solution, the use of a formal development structure such as VEDS will assist the developers in building an application that satisfies application goals and user requirements; supports task and user analyses; and offers a balanced concept design. A summary of the main aspects of VEDS is summarised in D'Cruz, Stedmon, Wilson, Modern, & Sharples (2003). A more detailed explanation is presented in Wilson, et al, (2002).

### 3.1    Identifying application goals, priorities and constraints

At the beginning of any development process goals must be defined that provide the driving force of the VE building process. These goals may stem from a problem within the organisation, such as wastage, high costs or delays, or to increase competitiveness, support training, or reduce design life cycles (D'Cruz, et al, 2002). Through methods such as structured interviews, focus groups, and real world task analyses, the initial stages of VE development involve assessing which tasks and functions must be completed in the VE; determining user characteristics and needs; and allocating and dividing functions within the VE. As such, a VE can be specified in terms of its goals and expected user tasks, in relation to the complexity and balance between interactivity and exploration afforded. This specification should then be agreed by the VE development team and stakeholders (including end users). Further methods can be employed at this stage, such as storyboarding and virtual task analyses, specifying tasks to be performed by the participant within the VE. This is a critical component of the VE development process as there are trade-offs between the technical capabilities of VR applications and costs associated with VE complexity (related to sensory richness), update rate (related to sense of presence and any disturbing effects), and interactivity (numbers of objects that can be 'manipulated' in real-time, and how this is to be done).

At this stage of the development process the VR application goals, priorities and constraints can be identified before the building of a VE begins. As such, technical problems that may compromise the successful use of speech in VEs can be addressed. For example, the integration of

ASR software into the building of VEs can still be problematic. Using microphones that are too sensitive may pick up background noise and electrical interference that could affect the accuracy of processing speech commands. Hand-held microphones or head-mounted microphones can be heavy and obtrusive but blue-tooth technology can offer more versatility in terms of wireless mobility and integration of devices.

## 3.2    Detailing user requirements

Capturing user requirements can support the definition of expected (or unexpected) behaviours that underpin the successful use of a VE. Furthermore, understanding user requirements can also provide a basis for evaluation of a VE at a later stage of the design process. For all VE applications, evaluations should be made of both the environments themselves and also their use and usefulness (Wilson, et al, 2002). It is possible to divide such evaluations into examinations of validity, outcomes, user experience and process (D'Cruz, et al, 2003). This means that before VE building progresses too far, a more detailed examination can be made of how participants will respond to different elements of the VE; utilise its functionality; and comprehend all the interface elements to meet the application goals. As such, a basis can be set for encouraging participants to explore a VE and enable them to understand which elements may be interacted with, minimising dysfunctional participant behaviour and serious errors. Building on a sound understanding of user needs, it is important, that a VR application and the VE that is developed supports user interaction and overall application effectiveness. As such, Barfield, Baird & Bjorneseth (1998) state that a VE input device should account for the type of manipulations a user has to perform and that the device should be designed so that it adheres to natural mappings in the way in which the device is manipulated, as well as permit movements that coincide with a user's mental model of the type of movement required in a VE. Kalawsky (1996) argues that the design of an input device should match the perceptual needs of the user. As such, the integration of input devices should follow a user needs analysis to map their expectations onto the attributes of the overall VR system.

## 3.3    Task & user analyses

During the information gathering stages it is often useful to carry out task and user analyses. These involve a step-by-step 'walkthrough' of each scenario to note procedures of the task and what the user is doing and thinking, preferably by the real end users of the task to gather realistic data. This can be performed using techniques such as direct observation, photographic or video analysis, verbal protocol, interviews, and short questionnaires. This process can be problematic without a greater understanding of what type of information is required. Due to the limitations of current VR systems it is impossible and sometimes unnecessary to design a VE that reflects every interaction of the real world. Therefore only the information that affects the user's experience and addresses the goals of the application is necessary. Task and user analyses help to define the minimum type of interactivity and cues to interactivity that might be needed in the VE before it is built.

## 3.4    Concept design

To provide a better understanding of the goals of the VE, storyboarding methods can be applied. Focus groups can be held where HCI experts and developers assess the application as it may be presented in the VE. Several iterations of the storyboard might be used to highlight usability and user interface issues such as cues, feedback, levels of detail and interaction.

Within VIEW, speech has been primarily implemented to navigate and manipulate objects in VR training applications. Speech can be used to select menu commands, where, for example, specific commands can turn a gravity simulation on or off, and in terms of manipulation, objects can be selected, grabbed, or released (Stedmon, et al, 2003). If a speech metaphor is well designed then it can be intuitive, taking the burden of navigation or interaction control away from a physical input device, and increasing opportunities for multi-modal interaction. In addition, anecdotal evidence suggests that speech may not be best suited for specific actions such as navigation and so the best use of speech interfaces might be in combination with other input devices for a more integrated approach (Stedmon, et al, 2003),

## 4      Understanding the human factors of speech recognition

No strict guidelines exist for the design and development of ASR interfaces. It seems that various developers adopt their own best practise strategies and designs (Stedmon, et al, 2003). Within any system application human factors variables can be divided into four main areas:

- **users** - taking into account the variation between users according to variables such as age, gender, experience, & accent. Within VR applications, users may come from many backgrounds and with collaborative VEs users may interact using different languages. As such, these variables need to be understood if the overall VR application and integrated ASR system is to perform properly;
- **systems** - realistic recognition accuracies, response times, dialogue structures, & vocabularies. Within VR applications, ASR systems will generally be led by the real-world applications market. As such, the systems variables are generally defined by what the conventional market dictates and what technology can deliver;
- **tasks** - including any concurrent or competing tasks, and realistic levels of mental workload, stress, fatigue. Whether in the real world of a virtual world, ASR applications are very much application specific. As such, detailed analysis of the task needs to be carried out to address how and where speech input may support the user over more conventional input devices;
- **environments** - including physical attributes such as noise and vibration, and psycho-social aspects such as the presence of others. Within VR there are many technical environments that are used to present VEs. These can range from traditional headset or desk-top presentations; through to projection walls and VR CAVEs. In these situations variables such as acoustics and ambient temperature, as well as the presence of any other users, may affect the overall system performance and use acceptance of ASR in VR applications.

Overall, the success of speech recognition depends on the user, system, task and environmental variables, all of which have significant effects on overall system performance, as outlined above. These issues are explored further in Stedmon, et al (2003).

## 5      Conclusion

This paper describes the VEDS framework, which has been used to support development of VEs in industrial use. By careful, but flexible, use of such a framework, the consequences of some of the barriers to VR application can be minimised. Technical limitations can be addressed or their effects reduced by appropriate choice of technology and careful specification of the VE required to meet the user goals. Trade-offs, such as choices between visual fidelity and interactivity, can be resolved; and usability can be enhanced through selection of appropriate hardware and input devices. The research within the VIEW project will not only be an exploration of the key issues of using ASR technology in a VE but will hopefully lend itself to the development of an ASR protocol for VE applications, through the lifespan of the VIEW Project, and beyond.

# 6       Acknowledgement

The work presented in this project is supported by IST grant 2000-26089: VIEW of the Future.

# 7       References

Baber, C., Mellor, B., Graham, R., Noyes, J.M., & Tunley, C., (1996). Workload and the use of Automatic Speech Recognition: The effects of Time and Resource Demands. *Speech Communication*, 20, (1996), pp.37-53.

Barfield, W., Baird, K., & Bjorneseth, O. (1998). Presence in Virtual Environments as a Function of Type of Input Device and Display Update Rate. *Displays*, 19: 91-98.

D'Cruz, M., Stedmon, A.W., Wilson, J.R., & Eastgate, R. (2002). *From User Requirements to Functional and User Interface Specification: General Process.* University of Nottingham Report No.: IOE/VIRART/01/351.

D'Cruz, M., Stedmon, A.W., Wilson, J.R., Modern, P.J., & Sharples, G.J. (2003). Building Virtual Environments using the Virtual Environment Development Structure: A Case Study. In, *HCI International '03. Proceedings of the 10th International Conference on Human-Computer Interaction, Crete, June 22-27, 2003.* Lawrence Erlbaum Associates.

Linde, C., & Shively, R. (1988). Field Study of Communication and Workload in Police Helicopters: Implications for Cockpit Design". In, *Proceedings of the Human Factors Society 32nd Annual Meeting.* Human Factors Society, Santa Monica, CA. 237-241.

Kalawsky. R. (1996). Exploiting virtual reality techniques in education and training: Technological issues. *Prepared for AGOCG.* Advanced VR Research Centre, Loughborough University of Technology.

Stedmon, A.W., Patel, H., Nichols, S.C., & Wilson, J.R. (2003). A View of the Future? The Potential Use of Speech Recognition for Virtual Reality Applications. In, P. McCabe (ed). *Contemporary Ergonomics 2003. Proceedings of The Ergonomics Society Annual Conference, Cirencester. 2003.* Taylor & Francis Ltd. London.

Stedmon, A., D'Cruz, M., Tromp, J., & Wilson, J.R. (2003) Two Methods and a Case Study: Human Factors Evaluations for Virtual Environments. In, *HCI International '03. Proceedings of the 10th International Conference on Human-Computer Interaction, Crete, June 22-27, 2003.* Lawrence Erlbaum Associates.

Wilson, J.R., Eastgate, R.M. & D'Cruz, M. (2002) Structured Development of Virtual Environments. In K. Stanney (ed). *Handbook of Virtual Environments.* Lawrence Erlbaum Associates.

# Methodologies and Evidence in Support of a Human-Centred Approach to Virtual Environment Systems Design

Robert J. Stone
University of Birmingham
Edgbaston, B15 2TT, UK
r.j.stone@bham.ac.uk

## Abstract

Two case studies are described that emphasise the importance of developing a sound understanding of the needs and human performance characteristics of the end users of virtual environment/virtual reality training systems. Using design processes advocated in ISO 13407 (*Human-Centred Design Guidelines for Interactive Systems*, 1999), the impact of short, highly-focused human factors analyses of naval gunners and helicopter rear-door aircrew on the development of what have become the first *operational* VR trainers based on head-mounted display technologies in the UK is demonstrated.

## 1    Introduction

Today, the very fact that a commercial, off-the-shelf (COTS) personal computer, equipped with a low-cost graphics accelerator can out-perform some of its supercomputer "competitors" – at a fraction of the cost it takes to *maintain* those competitors – has rekindled interest in those commercial and industrial organisations who were once potential adopters of Virtual Reality (VR) for competitive advantage. Support has also been forthcoming as a result of recognition by many that VR should no longer be viewed as a *single* technology in its own right, or simply a "branch" of 3D graphics. Rather, VR has evolved to become an integral part of a rich range of "mixed interactive media" techniques, all of which strive to improve the interaction between the human and sophisticated computerised databases, or between the human and other spatially or geographically disparate human users. Human factors initiatives in the late 1990s, including those designed to promulgate international interactive media standards (eg. ISO 13407 - *Human-Centred Design Guidelines for Interactive Systems* - and associated post-1999 publications), are now at a stage where, in association with other, established civilian and defence documents, VR developers and system integrators can do much to avoid the historical pitfalls of "technology for technology's sake" and can deliver affordable technologies that *empower* human users in a wide range of applications. Central to this recognition is a process that elevates the status of the human user to becoming the principal driver of interactive systems projects and relegates "technology push" to a much lower place in the driving requirements league table.

This paper presents 2 recent VR case studies that have been the subject of a strong human-centred approach from the outset. Whilst the studies described herein are firmly rooted in the UK defence community, the cost-benefit and human performance results that are now emerging have enabled other, civilian market sectors (eg. aerospace, petrochemical, automotive and medical industries) to reassess the application of interactive synthetic or virtual media to their business processes.

# 2 Case Study 1 – Royal Navy Close-Range Weapons Training

Until April of 2001, HMS Cambridge, a coastal training base located in the south-west of England, had enjoyed a prestigious history of training in support of naval gunnery, exposing students to a wide variety of close- and medium-range weaponry, including the general-purpose machine gun, the 20mm and 30mm close-range weapons (the main foci of this case study) and the medium-calibre, 4.5" Mk.8 gun. However, the closing years of the $20^{th}$ Century brought with it a realisation that the capability to undertake coastal training using live ammunition was unaffordable, as a result of the cost of live ammunition (from \$50/46 Euro to \$1800/1760 Euro *per round*) and that of flying towed targets (\$7666/7100 Euro *per hour*).

The original Navy requirement was to develop synthetic training simulators for use with the inert weapons physically removed from HMS Cambridge to another UK land base at HMS Collingwood. It was suggested that a system based on an existing "bolt-on" aimer trainer (a PC-driven LCD module that attaches to the weapon sight, designed for use at sea), integrated with an additional, single large-screen projection display, would deliver the image quality and field of view required for close-range weapons training. However, a human factors appraisal of the tasks demanded of the students demonstrated that this solution could potentially lead to errors in the target acquisition and engagement process. As well as an issue of projection display resolution, with one pixel representing a span of only 7m at typical engagement ranges, a greater concern was the fact that the navy requirement demanded realistic interaction between the weapons aimers and the Weapons Director Visual (WDV). The WDV is located on a raised part of the deck (the Gunner Director's Platform, or GDP) and is responsible for relaying target type, bearing and range data from the ship's Operations Centre and supervising the final stages of the engagement procedure (using binoculars). It became obvious that, whilst both the student aimer and WDV would visually acquire the target at the same ("*x-y*") point on a simulator projection screen, their perception and cognition of range and bearing would differ, as a result of their different positions relative to the fixed screen. These problems suggested that the only realistic means of delivering effective close-range weapons training to multiple participants was to implement a semi-immersive VR set-up, networked to allow the participants to coexist within in the same synthetic environment without compromising their relative views of that environment (and, indeed, of each other's virtual bodies, or "avatars").

## 2.1 The Gunnery Simulator

One of the key issues here is the importance of using real equipment to augment the synthetic or VR experience. The head-mounted displays (HMDs) used – Kaiser *ProView* XL-50s – do not fully enclose the students' eyes, as would other headsets. Instead, their design affords students some peripheral vision in both azimuth and elevation. As well as the VR environment, students interact with the real weapons hardware. In the case of the 20mm weapon, the aimer is normally strapped into the shoulder rests, thereby helping to maintain a fixed relationship between the eyes, gun sight, locks and trigger. Similarly, in the 30mm weapon, where the aimer actually sits at a small control panel, the Kaiser HMD affords visual access to the real panel, as well as displaying it in the virtual reproduction of the weapon. In addition to the weapon aimers' positions, 2 trainee WDVs stand on a purpose-built GDP within the HMS Collingwood facility. The WDVs are also equipped with Kaiser HMDs, each having been modified by the addition of a small switch that serves to magnify the view available to the WDV, thereby simulating the use of binoculars. Aimers, WDVs and the weapons themselves are tracked in real time using an InterSense *IS-900* inertial-acoustic tracking system with ultrasonic *SoniStrips* mounted onto brackets arranged in a matrix above the gun and aimer in order to maximise tracking coverage.

Calling out and acting on instructions, the WDV and weapons aimers interact to engage incoming targets. Once a scenario has been run as a training exercise, it can be replayed by the instructors for debrief purposes. Students are able to undertake realistic firing exercises, engaging aircraft, surface vessel and missile targets as if located on an actual Royal Navy vessel. On firing the virtual 20/30mm weapon, realistic barrel motions and ammunition feed effects can be seen, together with smoke and tracer. Successfully destroyed aircraft explode and fall to the sea. Sound effects have also been provided, including weapons discharge and an ambient, background naval vessel noise. The Collingwood facility is based on a network of 5 Pentium 3, 1GHz, dual-processor PCs, equipped with nVIDIA GeForce2 graphics cards. The computers are distributed to support both the 20mm and 30mm student aimers and their respective WDVs. The fifth PC is allocated to the instructor and coordinates the distribution of engagement scenarios, once set up via a plan-view, menu-driven Scenario Control Interface The virtual images are rendered using VP Defence Limited's own *kRender* package, a scenegraph-based real-time renderer for Windows 2000.

## 2.2 Initial Results

It has been calculated by the Naval Recruitment & Training Agency that the close-range weapons simulator facility, delivered in 7 months and costing a total of $1.13 million (1 million Euro) has, in its first year of operation, saved the Royal Navy around $5.3 million (4.9 million Euro) – a figure based on the cost of the number of real rounds fired and target-towing aircraft hours at HMS Cambridge in a typical year prior to closure (Stone & Rees, 2002). Today, the simulator facilities at HMS Collingwood are booked up many weeks in advance and, even though the systems were developed to foster "safe firing of weapons during peacetime" (the aim being to train engagement procedures as opposed to firing accuracy), anecdotal reports being relayed from the fleet suggest that those students who have been exposed to the VR training are demonstrating superior marksmanship skills at sea. The Collingwood facility has recently been upgraded to include a separate GPMG training facility, based on a single-screen projection system and electro-mechanical tracking of the weapon cradle in azimuth and elevation.

## 3 Case Study 2 – RAF Helicopter Voice Marshalling Training

Helicopter voice marshalling (VM) aircrew play an integral role in search and rescue missions and in the delivery of military or humanitarian aid to remote, often hazardous environments, confined by natural features such as forests and mountains. In the UK, the majority of training for VM aircrew takes place at the air bases of RAF Shawbury and RAF Valley. Working out of the open rear doors of the RAF's *Griffin* HT1 (Bell 412) helicopters, VM aircrew verbally relay important flight commands to the pilot in order to guarantee the accuracy and safety of the aircraft's approach to a landing site or target object. The original goal behind the procurement by the RAF of a Virtual Reality (VR) VM simulator system was to use innovative technologies to improve the quality and efficiency of pre-flight ground-based training (typically based on briefings and very simplistic scale models of the operational environments), and to provide a more cost-effective mode of *remedial* training. Mastering the style and content of the vocal VM commands relayed to the helicopter pilot can take considerable time for certain individuals, as this style of dynamic spatial "working" may be unlike anything they have previously experienced. Making aircraft available for remedial training has become increasingly difficult due to flying restrictions, personnel resources and, of course, the actual cost of flying.

From the outset, the focus of the VM simulator project was to deliver a level of training fidelity that would generate confidence in the RAF instructors that the correct sensory cues were being delivered to the students (thereby going some way to guaranteeing positive skills transfer from the virtual to the real training environments). The need to undertake a basic human-centred analysis, focusing on current ground and in-flight training methodologies, was, then, considered to be of vital importance. This was conducted at RAF Valley and RAF Shawbury over a study period of just 3 days and culminated in the publication of a report, the contents of which set the standard for the training content and the definition of specific audio-visual cues (see Stone & McDonagh, 2002, for more details). One of the most important findings of this study was the observation of VM crewman head and torso movements made during marshalling procedures, indicating the fact that instructors and students were using physical features of the helicopter to obtain parallax cues, or to line up visually with distant objects in the environment. This behaviour, along with the requirement of the VM air crewman to make regular head-up motions to view the horizon, command a large visual surveillance volume around the helicopter and even look under the helicopter on occasions led to the decision to adopt a tracked, head-mounted display (HMD) solution (semi-immersive). In the case of over-sea operations, there is often a paucity of natural and man-made markers. Coastal features, breakwaters, coastal buildings, the painted bands of lighthouses, surface vessels and so on provide visual references, but cannot be guaranteed. The VM crewman makes regular head-up glances to the horizon, in order to avoid any descent drift that may be evident (fixating on the surface does not help in assessing changes in altitude). Also, in the situation where the helicopter is approaching a target, which can range in size from a small vessel to an even smaller pilot's survival raft, the VM crewman will use head motion in conjunction with occluding wave peaks and troughs to judge distances. Other cues include the foam or bubbles left by a peaking wave (these can last for up to 25 seconds and provide temporary range "gates" to the target).

## 3.1 The VM Simulator

The delivered simulators (3 in total) are based on a simple wooden framework representation of the rear door of the *Griffin*, such that the door height is the same as that in the real aircraft and that handholds are located reasonably accurately in parallel with the open doorframes. A wooden framework also provides an ideal distortion-free environment for the Polhemus Fastrak electromagnetic tracking system, mounted on the Kaiser *ProView XL-50* HMDs. Again, the Kaiser HMD was chosen on the basis that the peripheral vision of the real world provided by the headset would support users' interaction with the physical *safety* elements of the simulator's wooden framework (ie. harness and hand-holds) as well as providing visual access to the VE. Given the ranges over which VM procedures are active, the need for stereoscopic presentation is not necessary and the virtual imagery is presented *biocularly*. Each VM simulator is based on a single, stand-alone dual-processor Pentium IV 1.7 GHz PC, each equipped with 2 graphics cards – an nVIDIA GeForce 3 TI-500 for the handling of the 3D Virtual Environment rendering and a GeForce 2 PCI 32Mb for generating the instructor's Scenario Control Interface, based on plan, or "overview" displays of the virtual land and sea environments, with training targets selected by drop-down menus. As well as the overview display, a second monitor presents the instructor with a view of the VE as seen by the student. Once again, the virtual images are rendered using VP Defence Limited's *kRender* package, as described above. Topographical data (digital terrain elevation data) were procured from Ordnance Survey, formatted into a 3D surface mesh using 3DS

Max and endowed with a baseline level of visual fidelity (and accurate templates for man-made features) using aerial photographs purchased from *Getmapping.com*.

## 3.2 Initial Results

The VM training system is the UK's third *operational* defence VR trainer, the first being the Avionics Maintenance Facility for the RAF's *Tornado* F3 aircraft (Stone, 2001); the second has been described under Case Study 1 (and in Stone & Rees (2002)). Since the VR systems were installed in the Spring of 2002, the RAF has experienced a considerable reduction in the failure rate of VM students as a result of remedial training on the VR systems (particularly at RAF Valley, where most of the at-sea training is undertaken). Exposure to around 90 minutes of VR VM training has endowed remedial students with the spatial and verbal skills necessary to continue with – and pass – real training sorties. The simulators are attracting much attention from the international defence/Search-and-Rescue helicopter community (not to mention the offshore oil and gas community) and, at the time of writing, the system is being extended to incorporate a littoral (coastal) training environment, the collection, transit and depositing of under-slung loads, advanced confined area operations, night vision goggle effects, even basic pilot training and rear-door weapons simulation. The use of COTS equipment and re-use of software modules developed for previous projects enabled the simulators to be delivered only 6 months after contract award. The cost of all three simulator systems, including development, hardware and installation amounted to approximately $382,000 (350,000 Euro), equating to just 96 flying hours.

## 4 Conclusions

VR/VE technologies are now capable of delivering significant cost-benefits and improving the quality of educational delivery, for remedial as well as mainstream training régimes. However, it is now clear that, in order to deliver real-time synthetic training systems that are not only capable of fostering the appropriate skill sets of the end user population, but also guarantee effective transfer of skills from the virtual to real operational settings, the adoption of human-centred design processes is essential. Whilst there are, as yet, no VR-dedicated human factors standards published under a *single* cover, it is evident that the defence and civilian standardisation communities are producing generic guidelines of importance to the interactive media community as a whole.

## 5 References

Stone, R.J. (2001). Virtual Reality in the Real World: A Personal Reflection on 12 Years of Human-Centred Endeavour. *Proceedings of International Conference on Artificial Reality and Tele-Existence* (ICAT). Tokyo, pp.23-32.

Stone, R. J., & Rees, J. B. M. (2002). Application of Virtual Reality to the Development of Naval Weapons Simulators. *Proceedings of Interservice/Industry Training, Simulation & Education* (I/ITSEC) *Conference*. Orlando, 3-5 December, 2002 (published on CD-ROM).

Stone, R. J., & McDonagh, S. (2002). Human-Centred Development and Evaluation of a Helicopter Voice Marshalling Simulator. *Proceedings of Interservice/Industry Training, Simulation & Education* (I/ITSEC) *Conference*. Orlando, 3-5 December, 2002 (published on CD-ROM).

# Contribution to Task Representation in Model-Based User Interface Design: Application to New People-Organization Interactions

*Dimitri Tabary, Mourad Abed, Christophe Kolski*

LAMIH – UMR CNRS 8530
"Automated reasoning and Human-Machine Interaction" Research group
Le Mont Houy
F-59313 Valenciennes cedex 9 - FRANCE
{dimitri.tabary, mourad.abed, christophe.kolski}@univ-valenciennes.fr

## Abstract

In the current research approaches in Human-Computer Interaction, Model Based Design (MBD) is considered as very promising. With the aim to generate a prototype of the presentation and the human-computer dialogue of interactive systems, we propose a MBD Method named TOOD (Task Object Oriented Design). TOOD covers several stages of an interactive system lifecycle.

Using TOOD, it is possible to define a task model which allows to identify the objects the interactive system handles as well as the allocation of the functions between the users and the system. It is possible to formalize then the dynamic behaviour of the tasks and the objects by Object Petri Nets. Finally, the approach leads to the generation of an interface prototype derived from this task model. In this paper, we will also propose to apply it to current new interactions between people and organization.

## 1    Introduction

The current breakthrough of new technologies in communication and information places the emphasis on the domination of interactive systems at every level: interactive terminals for the general public, multi-site organization of companies systems, migration of complex system control rooms (air traffic control, nuclear power plant control...), development of interactive and personalized Internet sites... The increasing presence and democratisation of interactive systems brings about an even greater need for the automatic generation of human-machine interfaces which integrate the needs and habits of the users.

One of the solutions created through research is the generation of a Human-Machine Interface (HMI) prototype based on the Model-Based user interface Design (MBD) paradigm [Szekely, 1996]. Current research trends therefore increasingly suggest design methods using various models: user task models, models of the data handled by the system, user models, and presentation models [Szekely, 1996].

With the aim of generating a prototype for the presentation and dialogue of an interactive system, we suggest a user task object oriented design method named TOOD (Task Object Oriented Design). It defines a task model, which makes it possible to identify the objects handled as well as the distribution of tasks between the user and the machine. It then formalizes the dynamic behaviour of the tasks and the objects in order to lead to the generation of a HMI prototype derived directly from the task model. Due to lack of place, dynamic structural task model and the operational model of the method can't detail in this article. However, this is available in [Abed & al., 2003]

## 2   Related Work

In this section, we give a non-exhaustive review of existing representative approaches based on the MBD paradigm. Table 1 presents a certain number of methods, based on this paradigm, specifying the various models used to cover the development cycle. In the table, we mention the tools used by these methods and the part of the interactive system generated, along with the generation mode. We also specify the type of task model used in order to obtain the HMI generation.

Table. 1 Model Based Design Method

| MBD / Tool | Specification of need | Global Design | Detailed Design | Generation / type of generation | Type of Task Model |
|---|---|---|---|---|---|
| ADEPT /ADEPT [Markopoulos et al., 1996] | - Task model (TKS) - User model | Abstract interface model | Concrete interface model | Presentation and dynamics / Assisted | User centred |
| DIANE + [Tarby and Barthet, 1996] | - Task model - Domain model (OPAC) | Specification of goals, procedures and actions | Dialogue model | Presentation, dynamics and assistance / Automatic | User centred |
| TRIDENT /TRIDENT [Bodart et al., 1995] | - Task model - Entities / relations | Activity Chaining Graph | Definition of presentation units | Presentation and dynamics / Assisted | Computing tasks |
| MECANO/MO BI-D [Puerta, 1996] | - Task model - User model - Domain Model | Dialogue and interface model | | Presentation and dynamics/ Automatic | Computing tasks |
| TADEUS/TAD EUS [Schlungbaum, 1998] | - Task Model - User Model - Domain Model | Design of the dialogue presentation and the dynamics according to two types of inter-window and intra-window dialogue | | Presentation and dynamics / Assisted | Computing tasks |

The comparison shows that all of these methods are capable of generating an HMI presentation and dynamics more or less automatically. They all define a dialogue model, which defines the links between the objects of the domain and their representation in the interface. The major difference between these approaches is the effective integration or otherwise of a user model in order to define the dialogue. The TADEUS, MECANO and TRIDENT methods are concerned with a computing task model which precludes any integration of user knowledge in the model. However, these approaches define a user model which includes the major characteristics regarding the type of user and the level of experience required. On the other hand, DIANE+ and ADEPT methods base their generation on a user task model. They belong to the current trend of Task-based User Interface Design. They provide the generation of an interface based on user tasks.

This number of methods remains insufficient, from our point of view, and shows up the tendency by computer design specialists to favour formal models (e.g. entity-relation model, OOD), which allow them to automate the design process to the greatest degree possible, which unfortunately is not always the case with the user task models (declarative). Our work attempts, in this research trend, to contribute through a method based on the use of formal models which especially favour iterative design with assessments. The contribution of TOOD is to formalize the user task by jointly using the Object-Oriented approach (OO) and Object Petri Nets (OPN). The OO owes part of its success to the fact that it favours structuring because of the notion of class, objects, systems and sub-systems. The OPN combine user-friendly graphics with a mathematical formalism, which provides design possibilities for tools for the verification, validation, and assessment of performance and for prototyping.

The semantics of OO are therefore sufficiently complete and rich for the representation of the static part of the task. In TOOD, the dynamics part is taken over by the OPN.

In this chapter, we begin by giving TOOD development cycle. We continue by defining its task model. Finally, we give a succinct presentation of the operational model.

# 3    TOOD and the Cycle of HMI Development

The TOOD design process can be divided into four major stages [Abed & al., 2003]:

- The analysis of the existing system and user needs is based on the human activity; it forms the entry point and the basis for any new designs.
- The Structural Task Model (STM) is concerned with the description of the user tasks in the system to be designed. Two models are created at this level: the Static Structural Task Model (SSTM) and the Dynamic Structural Task Model (DSTM).
- The Operational Model (OM) makes it possible to specify the HCI objects in a Local Interface Model (LIM), as well as the user procedures in a User Model (UM) of the system to be designed. It uses the needs and characteristics of the structural task model in order to result in an Abstract Interface Model (AIM); the AIM describes the user's objectives and procedures.
- The creation of the HCI is concerned with the computerised implementation of the specifications resulting from the previous stage, supported by the software architecture defined in the Interface Implementation Model (IIM).

# 4    Case Study: Cinema Server

This example is an extract from the NIPO project (New People-Organization Interactions). NIPO concerns the use usually methods and tools for the new people-organization interactions in the new information and communication technologies. In this paper, we present a cinema server specification. The user is able to use the application cinema server to search a film, to reserve places, to search information.

## 4.1    Structural Task Model (STM)

### 4.1.1    Static Structural Task Model (SSTM)

The structural model enables the breakdown of the user's stipulated work with the interactive system into significant elements, called tasks. Each task is considered as being an *autonomous entity* corresponding to a goal or a sub-goal, which can be situated at various hierarchical levels. This goal remains the same in the various work situations. In order to perfect the definition, TOOD formalises the concept of tasks using an object representation model, in which the task can be seen as an **Object**, an instance of the **Task Class**. This representation consequently attempts to model the task class by a generic structure of coherent and robust data, making it possible to describe and organise the information necessary for the identification and performance of each task.

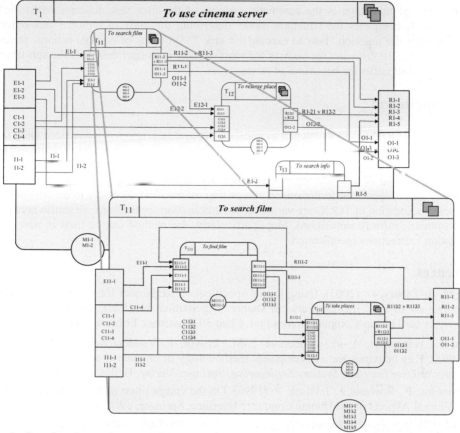

Fig.1: *Graphical specification of the task class "T1: to use cinema server" (structural model)*

The diagrams in figure 1 show a two-level breakdown for the example dealt with here. The diagram of the root task "T1: to use cinema server", gives rise to three sub-tasks T11, T12 and T13 which represent the three main functions of cinema server respectively: search film, reserving places and search information. It can be seen that task T11 can be activated by the event "E11–1: New search". This event calls upon one type of treatment corresponding to the interactive tasks: "T111 to find film".

### 4.1.2 Dynamic Structural Task Model (DSTM)

The dynamics of the task are defined by the dynamics of its body. The dynamic task model aims at integrating the temporal dimension (synchronization, concurrency, and interruption) by completing the static model. The dynamic behaviour of tasks is defined by a control structure, called TCS (Task Control Structure), based on a Object Petri Net (OPN) [Sibertin-Blanc, 2001]. It is merely a transformation of the static structure. The TCS describes the consumption of the input interface describer objects, the task activity, the release of describer objects from the output interface as well as the resource occupation, [Tabary, 2001].

## 4.2 Operational Model (OM)

The objective of this stage is the automatic passage of the user task descriptions to the HCI specification. It completes the structural model describing the body of terminal task-objects in order to answer the question "how to execute the task ?" (in terms of objects, actions, states and control structure). In this way, the body of the terminal task no longer describes the sub-tasks and their relations, but rather the behaviour of its resources.

## 5    Conclusion

The interest of TOOD lies in its attempt to translate the user task into a computerized task; i.e. the transformation of information resulting from task descriptions into a high level Human-machine Interface specification, with the aim of generating a concrete interface. In the TOOD method, this stage is divided overall into two major stages: the task model and the operational model. These models provide a support for formalization comprising information necessary to the design.

The cases of validation of TOOD are varied (office automation applications, air traffic control, fire control systems, railway simulation). More particularly, the method can be used in new people-organization interactions specification

## References

Abed, M. & Tabary, D. (2003). Using formal specification techniques for modelling of tasks and generation of HCI specifications. In Diaper, D., Stanton, N. (Eds), The handbook of task analysis for Human-Computer Interaction. Chap 30. Lawrence Erlbaum Ass., Mahwah, N.J.

Bodart, F., Hennebert, A.-M., Leheureux, J.-M., Provot, I., Vanderdonckt, J., Zucchinetti, G. (1995). Key activities for a development methodology of interactive applications. In *Critical Issues in User Interface Systems Engineering*, Springer-Verlag, Berlin, pp. 109-134.

Markopoulos, P., Rowson, J., Johnson, P. (1996). On the composition of interactor specifications. In Formal Aspects of the Human Computer Interface, Springer, eWiC series.

Puerta, A.R. (1996). The Mecano Project: Enabling User-Task Automation During Interface Development. In AAAI'96 Spring Symposium on Acquisition, Learning and Demonstration: Automating Tasks for Users. AAAI Press, Stanford, pp117-121.

Schlungbaum, E. (1998). Knowledge-based Support of Task-based User Interface Design in TADEUS, Proceedings CHI'98 ACM SIGCHI Conference on Human Factors in Computing Systems, Los Angeles, CA USA, ACM Press.

Sibertin-Blanc, C. (2001). Cooperative Objects: Principles, Use and Implementation. Concurrent Object-Oriented Programming and Petri Nets. In Agha, G. & al. (Eds), Advances in Petri Nets, pp. 16-46. Lecture Notes in Computer Science, no. 2001. Springer, Berlin.

Szekely, P. (1996). Retrospective and challenge for Model Based Interface Development. . In J. Vanderdonckt (ed.), Proceedings of CADUI'96, Presses Universitaires de Namur, Belgium, pp. xxi-xliv.

Tabary, D. (2001). Contribution à TOOD, une méthode à base de modèles pour la spécification et la conception des systèmes interactifs. Ph.D Thesis University of Valenciennes, France.

Tarby, J.-C., Barthet, M.-F. (1996). The Diane+ Method. In J. Vanderdonckt (ed.), Proceedings of CADUI'96, Presses Universitaires de Namur, Namur, Belgium, pp. 95–119.

## Acknowledgements
The authors thank the Region-Nord Pas-de-Calais and the FEDER (TACT NIPO Project).

# User-Centred Evaluation Criteria for a Mixed Reality Authoring Application

*Marjaana Träskbäck, Toni Koskinen, Marko Nieminen*

Helsinki University of Technology, Software Business and Engineering Institute
Information Ergonomics Research Group
P.O.Box 9600, FIN-02015 HUT, Finland
E-mail: Firstname.Lastname@hut.fi

## Abstract

Mixed Reality (MR) applications are beginning to emerge in various professional and leisure application domains. Currently the threshold for wide utilization of MR applications is that the authoring of MR applications consumes a lot of time and resources. Current MR authoring environments are designed for dedicated experts only.

In this article the objective is to specify the user-centred evaluation criteria for the MR authoring environment. This MR authoring environment is meant to be used by non-dedicated MR application developers (MR authors), for instance multimedia content providers. The study was conducted in the EU-funded project AMIRE: Authoring Mixed Reality.

The MR authoring tool for non-dedicated MR application developers has to be consistent with the commonly used GUI-tools and it has to be intuitive to use. Additionally the authoring tool should offer possibility to reuse MR components in order to enable cost effective and efficient development of MR applications.

## 1    Introduction

Authoring can be categorized by the amount of programming it requires from the authors and by the development style, which may or may not be primarily script-based. Many existing authoring systems assume that authors will posses a well-developed "programming mentality" and their applications consist of rigidly defined events in logical sequences and hierarchies. In some authoring systems applications are constructed by entering text into an editor and then compiling or interpreting the text. Other authoring systems have a graphical interface design editor, but they require extensive scripting to describe interactions. Some authoring systems attempt to dispense with a script entirely and use iconic constructs or menu commands for almost all operations. (Davies, P., Brailsford, T., 1994)

Currently authoring of MR requires a lot of time and resources. It has to be hard coded and the reusability of existing work is very low. Authoring in such way is not very cost effective and efficient. The criteria for authoring presented in this study consider the authoring in a component-based environment. Through reuse, adaptation, and combination of existing building blocks synergies of previous solutions can be utilized (Dörner, Geiger and Paelke, 2002). A term component in this study means a software component in the sense of the component theory

(Sametinger, 1997). It is a building block of a software system with an explicitly defined interface designed for reusability (Dörner et al., 2002).

Mixed reality is a particular subset of virtual reality related technologies that involve the merging of real and virtual worlds somewhere along the "virtuality continuum", which connects completely real environments to completely virtual ones (Milgram 1994). While mixed reality applications are increasingly appearing to various application fields (medical, military, entertainment, industry etc.)(Azuma, 1997, Azuma et al. 2001), more heterogeneous group of professionals takes part in the development process. In order to offer them the possibility to develop MR applications, we need to support them with appropriate tools. We need to find out the context in which they will develop the application and the level of knowledge they have in developing MR applications.

In order to make the tools usable for the diverse users of the authoring tool we need to consider usefulness of the tool. Usefulness can be divided into two categories: utility and usability. Utility offers information on what is needed for the functionality of the system to be sufficient and usability offers information how well users can use that functionality (Nielsen, 1993). In ISO 9241-11 usability is defined as being the extent to which a product can be used by specified users to achieve specified goals with effectiveness, efficiency and satisfaction in a specified context of use (ISO 9241-11). In this study the users are content providers of a multimedia company and the context is the work done in their office with desktop PC's or laptops and occasionally a site visit to the customers' environment.

The case study is conducted in the EU-funded project called AMIRE (Authoring Mixed Reality). The purpose of the project is to create an authoring tool that enables non-dedicated MR application developers efficiently create MR applications. This requires that the authoring is possible without detailed knowledge of the underlying technology. In the AMIRE, user-centred design approach is used to address this topic.

## 2    Objective of the Study

The objective of the study is to specify the user-centred evaluation criteria for the MR authoring environment for the non-dedicated MR application developers (MR authors). The evaluation criteria will be used to assess the forthcoming requirements, features and the implementation of the MR authoring environment. The main goal is to support the creation of such an MR authoring environment that enables widespread use of mixed reality features in future applications.

## 3    Method and Implementation of the Study

The research method applied in the study resembled a contextual inquiry (Beyer, H., Holtzblatt, K., 1998). Contextual inquiry is a combined ethnography-based on-site interview and observation method. The method was used to study the context-of-use of the MR author (the developers) in the MR provider organisation (a company called Talent Code Oy). On-site interviews and field observations were conducted in a real working environment of the MR provider. Two persons were interviewed and four persons were observed throughout their working day.

From the results of the field study, a context of use description of the MR author was created according to the model presented by ISO 9241-11 (1998). This model consists of the following components: a description of the users, tasks, equipment, and environment of the forthcoming MR authoring tool.

The user-centred evaluation criteria were created for the component based authoring tool that enables content providers to produce MR applications easily. It means that a typical developer in an MR provider organization can create the required MR applications using his/her existing skills and knowledge within a specified time.

# 4    Results

The contextual inquiry in the MR provider organisation revealed requirements that have to be met for the authoring tool to be usable for developers who are not familiar with a low level coding and implementation of the MR applications. The following issues contribute the evaluation criteria for the MR authoring tool.

## 4.1    User

Users of the forthcoming MR authoring environment are familiar with multimedia and web design. They are familiar with multimedia development applications like Macromedia Director. One user stated: "*The user interfaces that are currently in use are good and we are used to use them*". According to users the MR authoring tool should use the common user interface (UI) conventions.

## 4.2    Tasks and Tools

Working with customers is typically iterative prototyping in such a way that intermediate demonstration applications are presented. Therefore, the users suggested that the tool should support intermediate previews of the application. During the project the customer needs to see previews and snapshots of the future application. This is especially important in the project proposal phase when the customer commits the company for the project.

Usually a site visit is done at the beginning of the project to collect information. However, the content creation and editing itself should be possible to do off-site at the developer's office. If the work can be done in the authors own office environment, and if the customer can test the application on the site, the use efficiency of the MR application increases.

Content provision is handled on a high, domain specific level. In this context the domain means customer application domain, so that the author does not need to touch the source code. According to users the tool should be intuitive to use and it should resemble other commonly used tools (Macromedia Director and Flash in the AMIRE project). It should also be compatible with these tools. One user stated: "*We should be able to connect the MR-objects with other environments, like html-site or flash.*"

Furthermore the maintenance of the application must be easy and off-site maintenance should be possible. Application has to be robust and error resistant.

### 4.2.1  Use of Components

Components are the heart of the component-based mixed reality framework built in AMIRE. The key idea of the components is to be reusable. The new components should always be usable in several customer application domains.

MR application developers must be able to efficiently create and organise applications with the existing components. They need to be able find the specifications easily, to select components from the library and to include them in a new application, allowing them to view the system at a high level of abstraction. The components must be able to be customised to fit the particular requirements posed by other components and by the application itself. Authoring process must ensure that editing the properties of the MR components is readily available using easy-to-use standard API (Application Programming Interface) (i.e. property window in Macromedia Flash). The developer of a component-based application has to be able to connect the component also by using API. Insertion of a MR component into the framework and the building up connections between the components as well as changing the parameters needs to be easy. The following was stated by one user: *"I must be able change some easily available parameters without touching the source code"*

Applications may require some functionality that cannot be supported by the existing components. In order to add this functionality, application author must be able to ask for a new component from a component developer. The application developer must have a way to initialise and co-ordinate the gathering of the components needed for the application.

### 4.2.2  Programming

The authoring tool has to abstract the programming processes by defining the interfaces and basic behaviour of components. The level of programming required from the MR Provider should be equivalent to writing Flash Action Script in Macromedia Flash. An expert in the application field who is not familiar with programming has to be able to create an application with the authoring tool, i.e. no low level programming should be needed (i.e. C++). High-level programming is accepted (i.e. Java script, ECMA, Flash Action Script, Lingo). As a graphic designer stated: *"I do not want to touch the source code"*. The compiling after changing the parameters should be automatic like in Macromedia Flash.

## 4.3  Environment

From the organisational standpoint, the authoring tool should not change the MR author's current development, or production, process. The tool should also decrease the time required for the technical development phase of the MR applications. The production process cycles are short and there is a constant lack of time so the development of MR with current tools is not reasonable. Similarly, the tool should make cost-effective development possible; the authoring of MR should be handled with low extra costs.

## 5  Conclusions and Discussion

MR authors and MR content provider companies have their special characteristics that need to be taken into account in the development of the MR authoring environment. The user interface of the MR authoring application should be consistent with the widely used development environments.

Multimedia authoring is a closely related field to MR authoring, and it can offer intuitive, quickly learnable and easy-to-use metaphors, and interfaces for the developers. Intuitiveness can be increased with proper feedback, error messages and with a help function (context-sensitive help). The utilization of common user interfaces will help the introduction of the authoring tool in MR provider organisations. The tools must adapt to the surrounding organisational and production environment. Dramatic change in the production process can lead to the lack of use of the authoring tool. Cost efficiency and quick creation of iterative prototypes must be employed.

With the results presented in the article, it is possible to develop the authoring tool that is specially targeted for non-dedicated MR application developers in the content provider companies.

# 6    Future Work

At this point of the AMIRE project a prototype of the AMIRE authoring tool Is developed. This will be used in usability evaluations and it will be developed iteratively. The first version will be evaluated by the content provider that will design and author a MR demonstrator. The authoring tool will also be evaluated in usability tests.

# Acknowledgements

This work is conducted in the AMIRE project, which is funded by the European Commission in the IST programme (IST-2001-34024). The AMIRE website can be found at www.amire.net.

# References

Azuma, R. (1997). A survey of Augmented Reality. *In Presence: Teleoperators and Virtual Environments* 6, 4 (August 1997), 355-385.

Azuma, R., Baillot, Y., Behringer, R., Feiner, S., Julier, S., MacIntyre, B. (2001). Recent Advances in Augmented Reality. *IEEE Computer Graphics and Applications 21*, 6 (Nov/Dec 2001), 34-47.

Beyer, H., Holtzblatt, K. (1998). Contextual Design. San Fransisco: Morgan Kaufmann Publishers, Inc.

Davies P., Brailsford T. (1994) *New Frontiers of Learning – Guidelines for Multimedia Courseware Developers in Higher Education.* UCoSDA (ISBN 1-85889-062-4), 1994.

Dörner, R., Geiger, C. and Paelke, V. (2002). Authoring Mixed Reality – A Component and Framework-Based Approach. *First International Workshop on Entertainment Computing* (IWEC 2002) May 14 - 17, 2002 Makuhari, Chiba, JAPAN

ISO/IEC (1999) 9241-11 Ergonomic requirements for office work with visual display terminals (VTDs) – Part 11: Guidance on usability *ISO/TC* 159/SC 4

Milgram, P., Kishino, F. (1994). A taxonomy of mixed reality visual displays. *IEICE Transactions on Information Systems*, Vol. E77-D, No.12 December 1994.

Nielsen, J. (1993). Usability engineering. Boston: Academic Press Inc.

Sametinger, J. (1997). Software Engineering with Reusable Components. Springer Verlag.

# Continuity as a Usability Property

*Daniela Trevisan[1,2], Jean Vanderdonckt[2], Benoît Macq[1]*

Université Catholique de Louvain
[1] Communications and Remote Sensing Laboratory
Place du Levant, 2 – B-1348 Louvain-la-Neuve (Belgium)
{trevisan,macq}@tele.ucl.ac.be
[2] Unit of Information Systems, ISYS/BCHI
Place des Doyens, 1 – B-1348 Louvain-la-Neuve (Belgium)
{trevisan,vanderdonckt}@isys.ucl.ac.be

## Abstract

In this paper, we describe continuity as an important property to reach usability in many emerging systems such as multimodal and virtual reality systems. The approach is based on synchronization and integration characteristics existent between entities involved in the domain.

## 1    Introduction

In modern interaction techniques such as gesture recognition, animation and haptic feedback, the user is in constant interaction with the computing system. Thus the interaction is no longer based only on the discrete events but is based on a continuous process of information exchange. This is particularly the case for virtual environments where interaction remains continuous over time. As continuity of interaction may influence usability, it is important to consider it as an additional usability property, beyond existing usability guidelines that are applicable to virtual environments (Kaur, Sutcliffe, & Maiden, 1999) (Kaur, Maiden, & Sutcliffe, 1999).

In (Nigay, Dubois, & Troccaz, 2001), continuity is applied at the perceptual and cognitive levels. Perceptual continuity is verified if the user directly and smoothly perceives the different representations of a given entity. Cognitive continuity is verified if the cognitive processes that are involved in the interpretation of the different perceived representations are similar. Here we define the continuity as a capability of the system to promote a smooth interaction scheme with the user during task accomplishment considering perceptual, cognitive and functional aspects. The functional aspects correspond to those discontinuities that can occur between different functional workspaces, forcing the user to change and/or learn new modes of operation.

## 2    Continuity and usability properties

Gram & Cockton (1996) define a set of user-centred properties of interactive systems, which promote high usability quality from the user's perspective. The usability property establishes how well the users can interact with the system and meet their goal. Three external properties may to improve system usability: goal and task completeness, interaction flexibility, and interaction robustness. In this approach we are interested in addressing the robustness once that it refers to the user can avoid doing things you wish he/she had not done. Thus, interaction robustness covers all those properties that minimise the risk of task failure as observability, insistence, honesty, predictability, and access control, pace tolerance and deviation tolerance. To address the continuous interaction according the definition given here we should consider the observability and honesty properties.

The *observability* property evaluates if the system makes all relevant information potentially available to the user. Of course, not all information should be displayed all the time. In this case, *browsability* suggests that information which is not first-class information required to carry out the task may be accessible on-demand. Therefore, information that is not observable may become browsable. It corresponds to cognitive aspects. The honesty property measures if the dialog structure ensures that users correctly interpret perceived information. It corresponds to perceptual aspects. Thus, we should introduce the functional property to provide a complete analysis of continuous interaction contributing to the principle of interaction robustness. The functional property addresses the discontinuities between different functional workspaces. The next sub-section proposes an analysis based on synchronisation, integration. These are characteristics that result from interaction and relationships between all entities involved in the domain and contribute to design system in terms of continuity properties (See Figure 1). The honesty and observability properties involve respectively cognitive and perceptual aspects of continuity (see Table 1).

# 3    Temporal synchronisation of entities

Synchronisation is an event controlled by system that should be analysed between media, devices and tasks. Basically there are two types of temporal synchronisation: sequential (before relation) and simultaneous that can be equal, meets, overlaps, during, starts, or finishes relations according to description in (Allen, 1993). Regarding media synchronisation, it is possible to find all these kinds of temporal relationships and we can still consider the start- and end-points of events and distinguish the end of the event in natural (i.e., when the media object finishes its presentation) and forced (i.e., when an event explicitly stops the presentation of a media object). Devices synchronisation describes a way that devices will be available to the user interacts with them at a specific time. It raises the question of how the user interaction is with multiple devices. For example, if the system permits to select one object using a data glove and another with speech recognition at the same time, then there is simultaneous synchronisation. Tasks synchronisation can be simultaneous or sequential and performed by one user, by various users, by the system, by a third party, or by any combination.

# 4    Integration

We have considered three aspects about, which are: physical integration, spatial integration and insertion context of devices. The physical integration is controlled by the system and it describes how the user will receive feedback and how the media are distributed into output devices. It means that each media could be displayed in different displays or integrated within the same display. For example, overlapping real and virtual images in a head mounted display or showing sequences of images in a multiscreen device. Spatial integration concerns the spatial ordering and topological features of the participating visual objects. The spatial composition of objects can be performed by designer or by users or by the system for example in augmented reality systems it come from registration procedures by mixing correct way both information, real and digital. There is a spatial integration between media entities only when they are integrated into the same device. Insertion context of device can be peripheral or central according to the user's task. If the device is inserted in the central context of the user's task, she does not need to change her attention focus to perform the task. Otherwise if the user is changing the attention focus all time, then in this case the device is inserted in context peripheral of use.

# 5    Identifying Continuity

Norman's interaction model (Norman & Draper, 1986) show how the seven stage model can be used to ask design questions about the system regarding the human side of the interaction between human and system. In this model is possible to identify two main levels in the execution cycle of a task: execution and evaluation flows. The execution level consists of how the user will accomplish the task involving the temporal interaction synchronisation, the insertion context of input devices in the environment and the operation mode corresponding to the functional aspects. The evaluation level consists of three phases: user's perception, interpretation and evaluation. The perception corresponds to how the user perceives the system state involving the temporal interaction synchronisation; spatial and physical media-device integration and insertion context of output devices in the environment. The interpretation level consists of how much cognitive effort the user needs to understand the system state. It depends of what communication language or media type will be used by the system to provide the feedback to the user. The last phase corresponds to the evaluation of the system state by the user with respect to the goals.

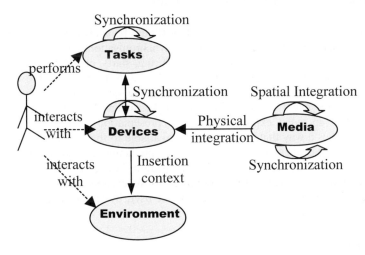

**Figure 1:** Continuity analysis based on synchronisation and integration characteristics

| Norman's Theory | Continuity properties | External properties [2] | Characteristics related |
|---|---|---|---|
| Interpretation and Evaluation level | Perception | Observability | Temporal synchronisation Integration Insertion context of output devices |
| | Cognitive | Honesty | Language and media used to represent the information |
| Execution level | Functional | | Temporal synchronisation Insertion context of input devices |

**Table 1:** Characteristics related to continuous interaction properties
according to the Norman's theory.

**Figure 2:** Example of potential discontinuity.

Figure 2 shows a potential source of discontinuity in Image-Guided Surgery (IGS): while a surgeon is operating on patient laying down on a table (the primary task and thus, the main focus of attention), additional information is displayed on TV screens and monitors. Those devices are not necessarily located closely to the patient's location, thus forcing the surgeon to switch attention from the patient to the various devices and then to come back on the main focus of attention. The more far the devices are from the main focus of attention, the more important the discontinuity may be induced.

# 6    Discussion

Regarding the model in Figure 1 there is a dependence between task and devices synchronisation that should be respected to improve continuity in the system. Insertion context of devices in the environment according to the user's task focus is a keep point to provide to the user a continuous functional interaction. The interaction model of Norman provides interaction analyses for systems in which the system state changes as a result of human actions. Therefore is interesting to consider another model such that suggest by (Massink & Faconti, 2002) supporting the analyses interaction from system side too

## Acknowledgments
We gratefully acknowledge the support from the Special Research Funds of "Université catholique de Louvain" under contract QANT01C6. The work described here is part of the VISME project (VIsual Scene composition with multi-resolution and modulation for a Multi-sources Environment dedicated to neuro-navigation). The web site of this project is accessible at http://www.isys.ucl.ac.be/bchi/research/visme.htm

# References

Allen, J.F. (1993). Maintaining knowledge about temporal intervals. *Communications of the ACM*, 26(11), 832–843.

Kaur.K., Sutcliffe, A., & Maiden, N. (1999). A design advice tool presenting usability guidance for virtual environments. In Proceedings of the Workshop on User Centred Design and Implementation of Virtual Environments (York, September 1999). Accessible at http://web.soi.city.ac.uk/homes/dj524/papers/kk99b.ps

Kaur, K., Maiden, N., & Sutcliffe, A. (1999). Interacting with virtual environments: an evaluation of a model of interaction. *Interacting with Computers*, 11, 403-426

Massink, M., & Faconti, G. (2002). A Reference Framework for Continuous Interaction. *Journal of Universal Access in the Information Society*, 1(4), 237–251.

Nigay; L., Dubois, E., & Troccaz, J. (2001). Compatibility and Continuity in Augmented Reality Systems. In Proceedings of Spring Days Workshop, Continuity in Future Computing Systems (Porto, April 23-24, 2001).

Norman, D.A., & Draper, S.W. (Eds.) (1986). User centered system design: New perspectives on human-computer interaction. Hillsdale: Lawrence Erlbaum Associates.

Gram, C., & Cockton, G. (Eds.) (1996). Design Principles for Interactive Software. London: Chapman & Hall.

# Model-Based Approach and Augmented Reality Systems

*Daniela Trevisan[1,2], Jean Vanderdonckt[2] and Benoît Macq[1]*

Université catholique de Louvain, 1348 Louvain-la-Neuve, Belgium
[1]Unité de Télécommunications et de Télédétection, Place du Levant, 2
[2]Unité de Systèmes d'Information, Place des Doyens, 2
{trevisan, vanderdonckt@isys.ucl.ac.be} {macq@tele.ucl.ac.be}

## Abstract

Methods to support guidance during design of conventional interfaces are not anymore valid for modeling, analyzing and designing virtual and augmented reality systems. Our work addresses this lack proposing an extension of the model-based approach to support design of interactive AR system where the continuous interaction can be assessed through the inter-model relations. An image guided surgery system is used as a case study to show how this methodology can be applied.

## 1    Introduction

An Augmented Reality (AR) system supplements the real world with virtual (computer-generated) objects that appear to coexist in the same space as the real world (Azuma, 1997). The application domains of AR reveal that the augmentation can take on a number of different forms by augmenting interaction, user's action and/or user's perception (Trevisan, Vanderdonckt & Macq, 2002). One of the central design aspects in human-computer interaction (HCI) concerns how combines real and virtual objects into a real environment where the user accomplishes her/his tasks. This limitation results in two different interfaces possibly inconsistent – one to deal with the physical space and another for the digital one. Consequently interaction discontinuities do break the natural workflow, forcing the user to switch between operation modes. For these reasons methods to support guidance during design of conventional interfaces are not anymore valid for modeling, analyzing and designing virtual and augmented reality systems.

Some classifications have been proposed in order to faster understanding of these spaces of interaction. The "Dimension space" proposes in (Graham et al., 2000) classifies entities in the system according of attention received, role, manifestation, I/O capacities and informational density. In (Dubois, Silva & Gray, 2002) the authors propose a notational support for the design of augmented reality systems based on ASUR and UMLi notations. The ASUR notation describes physical properties of components and their relationships. UMLi notation describes the structure and behavior of user interfaces.

This paper proposes the model-based approach (Vanderdonckt & Berquin, 1999, Paterno, 2000) to analyze interactions and relationships between the components of complex environments including multiplicity of interactions spaces, of systems, of inputs and outputs devices, and of users too. All synchronization and integration characteristics existent between the models should be evaluated. This is why we introduced inter-model relations to enable designers to assess the continuity in these terms and not in an isolated way.

This work is structured as follow. In the section 2 following the model-based approach we aim to identify the specification of the abstract users interface in terms of static structure and dynamic behavior in AR systems. Section 3 defines and analysis interaction in terms of continuity property using the Image Guided Systems as example. Section 4 gives some commentaries about the use of this methodology.

## 2    Models Description for AR Systems

Here we provide a description of the User, Task, Domain, Presentation, Dialog, Application and Platform models focusing on the AR requirements (Rekimoto & Nagao, 1995).

### 2.1    User Model

A user model represents the different characteristics of end users and the roles they are playing within the organization. Could hold habits, skills, experience, cognitive profile, preferences, customization parameters, etc. The user class has an ID, an user function containing information about the user (as experience level, profile, etc) and user position in the *SRU* – system reference of universe for those applications that require tracking the user. The user interacts with objects in the real world and/or in the virtual world through some kind of object (object tool) or some interactive device.

### 2.2    Task Model

The Task Model describes the tasks an end user performs and dictates what interaction capabilities must be designed. By analyzing the temporal relationships of a task model, it is possible to identify the set of tasks that are enabled over the same period of time. Thus, the interaction techniques supporting the tasks belonging to the same enable task set are logically candidate to be part of the same presentation unit though this criteria should not be interpreted too rigidly in order to avoid too modal users interfaces (Paterno, 2000).

### 2.3    Domain Model

The Domain Model defines the objects that a user can view, access and manipulate through a user interface (rendered by the presentation model) or directly in the physical world. Objects can be real (patients, paper document, a pen, a needle, etc.) or digital (images, sounds, etc.) and should support the task execution. Real or physical object is any object that has an actual objective existence and can be observed directly. Digital objects can be either real or virtual. Digital virtual objects are objects that exist in essence or effect, but not formally or actually it must be simulated like a rendered volume model of brain. However, a live video of a real scene, for example, is a digital real object. A typically screen-based computer system can only render a small amount of the total information on an output device as a head mounted display for example. For this reason the digital object visibility can be observable or browsable. Observability means that the object is always visible and browsability means that the object is visible only on demand (Gram, Christian & Cockton, 1996).

### 2.4    Presentation Model

Presentation Model is a representation of the visual, haptic and auditory elements that a user interface offers to its users. It is directly linked to how much cognitive effort the user needs to understand the system state. This model is composed by Presentation Unit (PU) which can be decomposed into one or many logic windows, which are supposed to be physically constrained by the user's screen and which may or may not be all displayed on the screen simultaneously (e.g.

spatial integration). Each PU is composed of at least one window called the basic window, from which it is possible to navigate to the other windows (Vanderdonckt & Bodart, 1993).

Regarding presentation model for AR systems you may also consider the *Physical* and *Spatial integration*. The physical integration describes how the visual media are distributed into outputs devices. It means that each media could be displayed into different displays or integrated within the same display. The spatial integration concerns the spatial ordering and topological features of the participating visual objects. The spatial composition aims at representing three aspects based on relationships between the objects: topological, directional and metric. The spatial integration in many AR applications comes from the registration procedures to mix real and digital information correctly (Azuma, 1997). There is a spatial integration between media objects only when there is some kind of simultaneous synchronization between media objects. Regarding insertion context of device it can be peripheral or central depending on the user's focus when a specific task is carried out. If the device is inserted in the central context of the user's task, s/he does not need to change her/his attention focus to perform the task. Otherwise if the user is changing the attention focus all time the device is inserted in context peripheral of use.

## 2.5 Dialog Model

Once the static arrangement of the abstract user interface is identified, the next step is to specify its dynamic behavior. Interaction Model defines how users can interact with the objects presentation (as push buttons, commands, etc), with interaction media (as voice input, touch screen, etc) and the reactions that the user interface communicates via these objects. The user's interaction with the environment is often continuous and the general opinion is that AR environments are hybrid systems and the behavior should be modeled as a combination of discrete and continuous components (Jacob, 1996).

## 2.6 Application and Platform Model

The application model here corresponds to the system class and can be composed of one or more computer-based system. Synchronization between different systems can be necessary to exchange information. The system class may synchronize events from devices according the task performed and also integrate different media sources into one or more presentation units mixing information from real and virtual world correctly[1].

A platform model brings abstraction for intended platform characteristics such as devices, monitor, CPU, speed, mobility, etc. A device model encapsulates attributes, events and primitives. For example: mouse (number of buttons, speed,...), glove (degrees of freedom,...), screen (resolution, colors, speed), etc. The device class has information about the media type that it can present according to his physical capabilities.

## 3 Defining and analyzing continuous interaction

The definition of continuity in interaction can be a thorny issue, since every interaction can be viewed as both continuous and discrete at some level of granularity. Here we define the continuity as a capability of the system to promote a smooth interaction scheme with the user during task accomplishment considering perceptual, cognitive and functional properties. This definition is expanded and revised from (Dubois, Nigay & Troccaz, 2002). The perceptual property evaluates if the system makes all relevant information potentially available to the user. The cognitive property

---

[1] It corresponds to the *Registration* procedures in AR systems discussed in (Azuma, 1997).

measures if the dialog structure ensures that users correctly interpret the perceived information. The functional property corresponds to the adaptability level of the user to change or learn new operation modes.

We take account an Image Guided Surgery (King, 2000) system as example to evaluate the interaction. In this kind of system the user interaction with the patient is augmented by the computer giving to the user extra information about the surgical planning (e.g. specifying details such as the locations of incisions, areas to be avoided and the diseased tissue). In the operation room a navigation workstation and a microscope display are used to guide the procedure based on the original and processed images but the patient should be the main focus of the user's task.

Analyzing the inter-model relations is possible to extract some assumptions regarding continuous interaction. For example, taking account the surgical navigation task which take place during the intra-operative phase simultaneously with the surgical procedures, we have:

- Task and User model: this task is accomplished by one surgeon;
- Task and Domain model: this task requires real (live video of the patient) and digital (pre-operative images, microscope position and warning messages) objects;
- Domain and Presentation Model: the Presentation unit presents the domain objects according to the device that will be used to display them. In this case we can get at least three presentation models. For presentation model 1 (see Figure 1) we have overlapping real image from patient and digital path line in a microscope display. In this case the warning message is displayed in context of information, that's mean not overlapping the region of interest (view of patient). For the presentation model 2 (see Figure 2) we have four views from pre-operative images displayed in a workstation display together with microscope positions. In this case the warning message is displayed in the top right of the screen, outside of visualization context. For the presentation model 3 (see Figure 3) we have all information displayed in a multiscreen device. Regarding cognitive property, the media language (see Figures 1, 2 and 3) used to transmit information to the surgeon is adequate and of easy interpretation providing cognitive continuity.

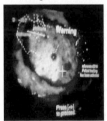 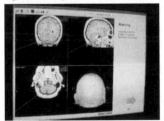 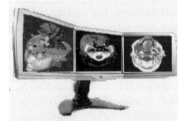

**Figure 1:** Presentation for the microscope display.  **Figure 2:** Presentation model for the workstation display.  **Figure 3:** Presentation model for the multiscreen display.

- Presentation and Platform Model: the presentations units are integrated into different devices according to their capabilities of render these presentations. For our example the microscope display (which is inserted in central context of user's task) shows one presentation model and the workstation display (which is inserted in peripheral context of user's task) shows another one with complementary information to guide the surgeon.
- User and Platform Model: the surgeon manipulates the microscope system through a manual remote control, the workstation system through a mouse and the patient through surgical tools. With this kind of interaction the user needs to stop one interaction to start another one causing a breaking down in the interaction flow and a functional

discontinuity. As both displays, microscope and workstation, are required to guide the surgeon during the surgery there is also a perceptive and functional discontinuity during the visualization once that the surgeon needs to change her/his attention focus to access all available information to perform the task.

# 4    Conclusion

In fact this approach intends help the designs envisage critical points during the design process of AR systems. The systematization of the process through the model-based approach enables designers to decompose the overall design interaction task into smaller and conceptually distinct tasks and then to assess the continuous interaction in these terms. In particular we intend to run a series of usability testing to identify the potential combinations of input and output modalities that users can accommodate maintaining continuity during the interaction.

# References

Azuma, R. T. (1997). A survey of Augmented Reality, Presence: Teleoperators and Virtual Environments, Vol 6, No. 4(August), pp. 355-385.

Dubois, E, Silva, P. P. and Gray, P. (2002). Notational Support for the Design of Augmented Reality Systems. Proceedings of DSV-IS'2002, Rostock, Germany, June 12-14, 2002.

Dubois, E. Nigay, L., Troccaz, J. (2002). Assessing continuity and compatibility in augmented reality systems. *Journal of Universal Access in the Information Society*, 1(4), 263-273, 2002.

Graham, T., C., N., Watts, L., A., Calvary, G., Coutaz, J., Dubois, E.,Nigay, L. (2000). A Dimension Space for the Design of Interactive Systems within their Physical Environments, DIS2000, 17-19 August 2000, ACMPubl. New York - USA, p. 406-416.

Gram, Christian & Cockton (1996), Design Principles for Interactive Software. Chapman & Hall, London.

Jacob, R. J. K. (1996). A visual language for non-WIMP user interfaces, in Proceedings IEEE Symposium on Visual Languages, pages 231-238. IEEE Computer Science Press, 1996.

King, P. Edwards, C. R., et al. (2000). Stereo Augmented Reality in the Surgical Microscope, Presence: Teleoperators and Virtual Environments, 9(4):360-368.

Paterno, F., (2000). Model-based Design and Evaluation of Interactive Applications, Springer-Verlag, Berlin.

Rekimoto, J. & Nagao, K. (1995). The World through the Computer: Computer Augmented Interaction with Real World Environments, User Interface Software and Technology (UIST '95).

Trevisan, D., Vanderdonckt, J., Macq, B. (2002). Analyzing Interaction in Augmented Reality Systems, Proceedings of ACM Multimedia'2002 International Workshop on Immersive Telepresence ITP'2002 (Juan Les Pins, 6 December 2002), G. Pingali, R. Jain (eds.), ACM Press, New York, 2002, pp. 56-59.

Vanderdonckt, J. and Bodart, F. (1993). Encapsulating Knowledge for Intelligent Interaction Objects Selection, Proc. of InterCHI'93, ACM Press, New York, 1993, pp. 424-429.

Vanderdonckt, J. and Berquin, P. (1999). Towards a Very Large Model-based Approach for User Interface Development, Proc. of 1st Int. Workshop on User Interfaces to Data Intensive Systems UIDIS'99, IEEE Computer Society Press, Los Alamitos, 1999, pp. 76-85.

# Flexible Force Grid Field for Three Dimensional Modeling

*Daisuke Tsubouchi[1], Tetsuro Ogi[2,3], Toshio Yamada[3], Hirohisa Noguchi[1]*

[1]Keio University, [2]University of Tokyo, [3]Gifu MVL Research Center
3-14-1, Hiyoshi, Kohoku-ku, Yokohama, 223-8522, Japan
tsubo@noguchi.sd.keio.ac.jp

## Abstract

In this study, we proposed a flexible force grid field method that can redefine an arbitrary force grid field around the user's fingertip in order to design objects in the three-dimensional virtual space. In this system, since the movement of the user's finger is restricted by the force that is applied from the grid points and the grid lines, he can easily define geometrically complex lines correctly and intuitively. This paper describes the basic principle, the implementation and the examples of the usage of the proposed flexible force grid field method.

## 1. Introduction

In order to design three-dimensional objects in the virtual space, it is required that the user can define points, lines and surfaces accurately as well as manipulate them intuitively. Although several interface devices such as a 3D mouse or a stylus pen are used in the virtual environment, it is difficult to point at an accurate position only by using the visual information in the three-dimensional space. In our previous work, we have proposed a force grid field method that supports the user to design three-dimensional objects in the virtual space using the force feedback [1].

In the first prototype system, the discrete grid points defined in the simple rectangular coordinates were used in the three-dimensional space. In this system, when the user's fingertip comes near a grid point, the attractive force toward the nearest grid point is applied to the user's finger. The force intensity is changed according to the distance between the fingertip position and the nearest grid point. This force is not so strong that the user can easily move his finger out of the influence of the force grid field. When the user moves his fingertip passing through several force grid points, he can recognize how his finger moved, and then he can define the lines or the shapes intuitively in the three-dimensional space. However, the simple rectangular force grid field is not always effective, when the directional characteristics of the grid field are not suitable for the user's task. This paper describes the concept and the implementation of the flexible force grid field method that can redefine the arbitrary force grid field freely around the user's finger to design complex objects in the virtual space.

## 2 Basic Principle of Force grid field

Figure 1 shows the system configuration of the force grid field system used in this study. In this system, the PHANToM[2] haptic interface device was used in order to display force feedback according to the user's action in the three-dimensional space.

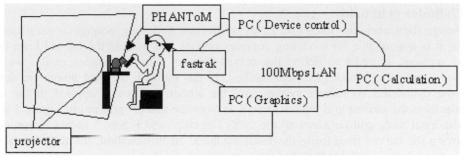

Figure 1. Force Grid Field Environment

## 2.1 Rectangular grid field

The basic principle of the force grid field using the simple rectangular coordinate system is shown in Figure 2. In this method, the attractive force is applied to the user's fingertip using the PHANToM around the grid points and the grid lines. When the user's finger is near the grid point, the attractive force is applied to it from the nearest grid point. In this case, the force intensity varies according to the distance between the fingertip position and the nearest grid point.

By utilizing the attractive force in the force grid field, the user can easily define lines. Figure 3 shows the example of the process of drawing lines in the rectangular force grid field. When the user is drawing lines along the grid lines in the rectangular force grid field, the attractive force that is perpendicular to the grid line is applied to the user's fingertip. Then, the user can easily and correctly define lines, because his finger is restricted along the grid lines. When the fingertip position is moved away from the grid line, the restriction force disappears so that the user can move his finger freely in the three-dimensional space to draw a next new line.

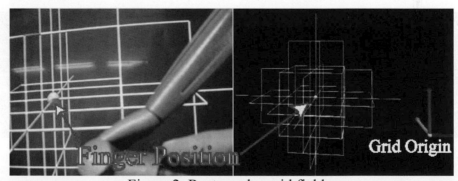

Figure 2. Rectangular grid field

Figure 3. Drawing lines on rectangular grid field

## 2.2. Cylinder grid field

Although the rectangular force grid fields is effective to design rectangular parallelepiped objects, it is not suitable for modeling complex objects that include the curved lines or the curved surfaces. In order to define the curved lines or curved surfaces easily, we have proposed a cylindrical force grid field method. In this method, the grid points are defined using the cylindrical coordinate system, and the attractive force is applied to the user's fingertip from the nearest grid circles as well as from the nearest grid point. Figure 4 shows the cylindrical force grid field around the user's fingertips and Figure 5 shows the processes of drawing the curved lines using the force feedback. In this method, the user was able to design cylindrical objects that included the curved lines and the curved surfaces easily.

Figure 4. Cylinder grid field

Figure 5. Drawing lines on cylinder grid field

## 3. Flexible force grid field

In the previous chapter, we discussed two kinds of force grid field methods. Although we can design specific models by these grid fields, they are not necessarily effective to design complex objects that contain arbitrary curved lines or curved surfaces because these grid fields have directional characteristics. The most suitable force grid field would vary according to the models being produced in the three-dimensional space. Therefore, in this study, a flexible force grid field method that can redefine arbitrary force grid field around the user's fingertip was developed. In this method, the local force grid field can be defined in the following processes. First, when the user double-clicks the button on the penholder of the PHANToM, an origin of the base grid field is defined at the fingertip position. The base grid field itself cannot be deformed but is used to define a new local grid field. As a base grid field, a basic rectangular force grid field is used as shown in figure 6. Next, when the user drags the cursor and releases it, z-axis of the new grid field is defined along the locus that extends from the origin of the base grid field to the release position. The shape of this axis is determined using the spline interpolation for the grid points through which the user's fingertip passes.

In this method, the other two axes (x-axis and y-axis) are defined using the rectangular coordinate system or the cylindrical coordinate system so that they cross mutually at right angles. The directions of these two axes are determined according to the posture of the PHANToM penholder when the user releases the button on it (Figure 7). Thus, the user can define an arbitrary new force grid field that contains the curved axes around the user's fingertip (figure 8, figure 9).

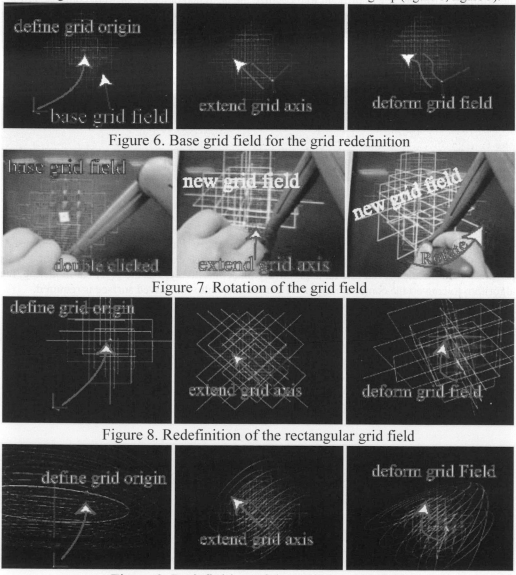

Figure 6. Base grid field for the grid redefinition

Figure 7. Rotation of the grid field

Figure 8. Redefinition of the rectangular grid field

Figure 9. Redefinition of the cylinder grid field

## 4.Examples

Figure 9 shows several examples of models designed by using the proposed force grid field method. Each model is composed of various types of lines that were defined in the various local force grid fields individually. In these examples, we used a basic rectangular grid for the base grid fields. But the rectangular base grid was not suitable to recognize the subtle distinction of the axis angles of the redefined new grid field. Therefore, it is desirable that the arbitrary types of base grid field can be used in order to redefine the new force grid field.

In addition, though the proposed force grid field was very useful to define the arbitrary lines at the arbitrary positions, it was difficult to recognize the designed objects that were composed of the lines in the three-dimensional space. Therefore, it is also desired to develop a method to design arbitrary surfaces directly in the virtual space.

Figure 10. Examples

## Conclusion

We proposed the method that can redefine an arbitrary force grid field around the user's fingertip in order to design objects in the three-dimensional virtual space. In this method, it is possible to define the curved line on the arbitrary place intuitively. For the practical modeling, the algorithm to convert the combinations of the curved lines into the curved surface is required.

## References

[1] Yamada T., Tsubouchi D, Ogi T, Hirose M, Desk-sized immersive Workplace Using force feedback grid interface: Proceedings of IEEE Virtual Reality 2002 Conference, pp.135-142, 2002

[2] Sensable Technologies inc., http://www.sensable.com

# Anticipation in a VR-based Anthropomorphic Construction Assistant

*Ian Voss*

University of Bielefeld
Artificial Intelligence Group
Faculty of Technology
D-33594 Bielefeld
ivoss@techfak.uni-bielefeld.de

*Ipke Wachsmuth*

University of Bielefeld
Artificial Intelligence Group
Faculty of Technology
D-33594 Bielefeld
ipke@techfak.uni-bielefeld.de

**Abstract**

We describe an implemented system that anticipates user instructions in a collaborative construction task in virtual reality. The functionality of the system is embodied by an anthropomorphic interface agent and enables the usage of functional names for pieces of an uncompleted target aggregate in natural language instructions. Based on an internal model of the construction state and salience ratings for aggregate parts, an anticipatory system behaviour is achieved that tolerates imprecise user instructions.

## 1   Introduction

In human-human communication one of the many facilitating features is that the interlocutor can anticipate what we want and thus enable efficient communication by bridging missing or underspecified information. We here describe an implemented system that anticipates user instructions in a virtual construction situation. It is a model-based anticipatory system according to the terms of [Ekdahl 2000]. The functionality of the system is embodied by the multimodal assembly expert MAX [Sowa, Kopp & Latoschik, 2000]. MAX acts as an anthropomorphic interface agent operating in large scale virtual reality. The agent is able to interact with the user in multimodal dialogue via speaking and gesturing to perform a collaborative construction task in a virtual environment. The anticipatory abilities together with MAX's abilities to handle discourse offer a system behaviour that is robust and allows imprecise user instructions to be compensated [Voss 2001]. The task performed is domain-specific and demonstrated using German instructions and the wooden toy kit "Baufix".

To give an idea Figure 1 on the left shows a partly-built aeroplane with the tail unit and the propeller still missing. There are three as yet unassembled parts in the construction scene, two three-hole bars and one bolt. An approach based on aggregate matching would allow recognition of the aggregate only when it is finished, here in Step 3 of Figure 1. Our anticipation model enables MAX to interpret the unassembled parts as parts of the target aggregate to be built, which allows the user to use the functional names of the parts from the beginning of a construction sequence.

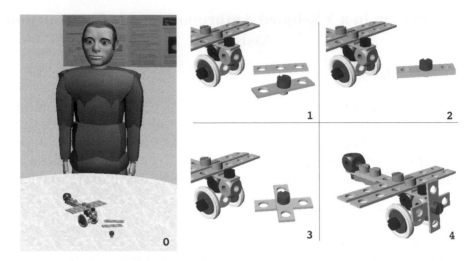

**Figure 1:** MAX in VR with partly-built aeroplane and unassembled parts on the left. On the right the progressive steps of the construction of a propeller are shown.

The interaction with MAX takes place in a fully immersive large scale virtual reality environment. The display technology used is a six channel three-sided high resolution stereo projection. Graphic rendering is done by a synchronized six-channel cluster based on distributed open-GL technology. The user interacts with the system by way of natural speech as well as deictic gesture. Data necessary for gesture interpretation are gained with the help of a marker-based real-time optical motion detection system. Wireless microphone and a speaker-independent incremental speech recognition system allow for natural speech input, auditory system reaction is done with an eight channel audio system to represent the location of an acoustic event in 3D space [Latoschik 2001].

## 2    Dynamic Conceptualization

The central idea for our solution is to use situated interpretations of knowledge bases to realize anticipation. The situated interpretation is done with respect to the state of construction as well as recent instructions to the system.

The knowledge bases used are of two kinds. The building parts are described in a "Baufix world" knowledge base (see Fig.2 on the left). Information about the different parts of the toy kit and all possible modes of interconnection are modelled in this knowledge base, such as the information that a possible connection between a bolt and a three-hole-bar would mean putting the shaft of the bolt through one of the holes with the head of the bolt not disappearing through the wooden bar. The possible models that can be build with these parts are described in an "Aeroplane world" knowledge base. It includes descriptions of functional aggregates, e.g., the tail unit or the propeller of an aeroplane (see Fig. 2 on the right).

Both these knowledge bases are implemented in the form of a hierarchically organized frame-structure [Jung 1998]. These knowledge bases are utilized in the process of virtual construction. For an adequate technical assistance of the user, a "join" action of two parts in the virtual construction scene is accompanied by the conceptualization of the newly formed aggregate. The conceptualization of an aggregate is done in the context of the construction situation of the toy aeroplane. Thus the actual interpretation of e.g. a wooden bar with three holes in it would be a propeller blade of an aeroplane when it is part of the completed and conceptualized assembly PROPELLER. (In another situation such a piece can be part of e.g. the tail unit.) A necessary prerequisite for a successful conceptualization is the completeness of an aggregate.

## 3   Anticipation

The frame-based knowledge representations mentioned above were originally designed to conceptualize, and thus to recognize, complete aggregates. This is necessary to technically follow the progress in construction. We now explain how a technical look-ahead in the construction situation is realized, by extending interpreting mechanisms to account for incomplete aggregates.

The "traditional" interpretation of the mentioned knowledge bases allows recognition of complete aggregates, i.e., only when each slot of a knowledge base concept is filled. Our new method enables us to continue using a frame-based representation even with some slots unfilled. When calculating anticipation, the role types are searched for the possible uses of a specific object type. Then this information is propagated back to the object type and where it is accumulated and standardised to generate salience ratings. This is done for all possible roles an object type can play.

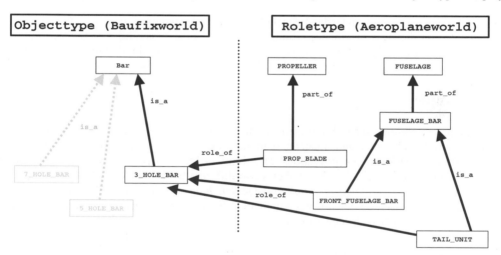

**Figure 2:** Knowledge bases of objects and roles

Anticipation is realized as a prospective interpretation of the possible places where to put a certain piece in the model. The interpretation of a single part is always done with respect to other as yet unassembled parts as well as to those parts already assembled to form an aggregate. Baufix parts receive salience ratings reflecting the likelihood of using this part in a certain role in one or another model. These ratings are dynamically calculated in the process of construction. Table 1 shows the ratings of the different parts and aggregates for the construction sequence of the propeller that is illustrated in Figure 1. Using this mechanism MAX is able to anticipate the possible roles of an object on the basis of the construction knowledge.

**Table 1:** Salience ratings for the anticipation of the parts and aggregates of the construction sequence in Figure 1

| Step 0 | Step 1 | Step 2 | Step 3 | Step 4 |
|---|---|---|---|---|
| PROP_BLADE 0,66 | PROP_BLADE 0,66 | PROP_BLADE 0,99 | PROP_BLADE 1,00 | PROP_BLADE 1,00 |
| PROP_BLADE 0,66 | PROP_BLADE 0,87 | PROP_BLADE 0,99 | PROP_BLADE 1,00 | PROP_BLADE 1,00 |
| PROP_BOLT 0,66 FUSELAGE_BOLT 0,66 TAILUNIT_BOLT 0,66 | PROP_BOLT 0,87 FUSELAGE_BOLT 0,66 TAILUNIT_BOLT 0,66 | PROP_BOLT 0,99 FUSELAGE_BOLT 0,66 TAILUNIT_BOLT 0,66 | PROP_BOLT 1,00 | PROP_BOLT 1,00 |
| AEROPLANE 0,68 | PROPELLER 0,66 FUSELAGE 0,20 AEROPLANE 0,68 | PROPELLER 1,00 AEROPLANE 0,68 | PROPELLER 1,00 AEROPLANE 0,68 | AEROPLANE 0,79 |

Anticipation is done to enable the user to use the functional names of the parts of an aggregate that is still under construction when giving instructions to MAX. In the case of a propeller as the target aggregate, the user might as well use the functional name "propeller-blade" and not just "three-hole-bar" in an instruction like "put the bolt in the middle hole of the propeller blade".

Anticipation is realized to understand functional names of aggregates or parts of them. This is implemented on the basis of salience ratings, which leaves the possibilities of two or more parts receiving the same ratings. In case of such ambiguities as well as simple underspecifications in user instructions, a dialogue system is used to clarify the user's intention.

## 4    Example

The anticipatory abilities of MAX represent an internal model of the state of construction. The full advantages of the robustness and situatedness in the construction situation can only be made use of when getting feedback for user instructions. The ability of MAX to ask context-sensitive questions is used to resolve ambiguities. Figure 3 shows two different behaviours of MAX in response to a user instruction like *"Put the tail-unit bolt in the middle hole of the five-hole bar"*. This instruction is even then underspecified when using anticipation because of identical salience ratings that do not enable MAX to choose the correct bolt by itself.

Question I: ***"The black or the white bolt?"***    Question II: ***"Which bolt do you want?"***

**Figure 3:** Context-sensitive question to the user

A user reaction to these questions shown in Figure 3 could be an underspecified instruction like *"The black one!"* for Question I or *"The left one!"* for Question II. The ellipsis in the answers is correctly interpreted with the use of a discourse memory.

## 5    Conclusion and Perspective

We here described a method to establish a mapping between an uncompleted aggregate, or even single parts of it, and the representation of the complete aggregate as described in the knowledge base. A corpus of human-human Baufix construction dialogues is available and exploited with respect to the usage of functional names or role types instead of object types. As anticipation is based on knowledge of role-type worlds, our approach is so far limited as to that no anticipatory assistance can be given for the construction of variants. The user is then still assisted in "join" actions when making use of the names of the object types. Discourse abilities that help disambiguating with or without the use of pointing gestures are not limited to the aeroplane world. The transfer to another domain is possible. Different role types have been modeled to show this such as a toy scooter. The transfer to another toy kit is limited to those toy kits making use of similar connection types between the building parts.

Follow-on work is underway to make construction episodes available for the next level of user instructions. It is intended to cope with instructions like "and now do the same thing on the other side" when just having built something that is needed in a similar way at another place on the aeroplane.

## 6    Acknowledgements

This work is part of the Collaborative Research Centre SFB 360 at the University of Bielefeld and partly supported by the Deutsche Forschungsgemeinschaft (DFG). Implementation assistance by Thies Pfeiffer is gratefully acknowledged.

## 7    References

Ekdahl, B. (2000), Anticipatory Systems and Reductionism. *Fifteen European Meeting on Cybernetic and Systems Research*, EMCSR 2000, Vienna, April 25-28.

Jung, B. (1998), Reasoning about Objects, Assemblies, and Roles in On-Going Assembly Tasks. *Distributed Autonomous Robotic Systems 3*, Springer-Verlag, 257-266.

Sowa, T., Kopp, S. & Latoschik, M. E. (2001), A Communicative Mediator in a Virtual Environment: Processing of Multimodal Input and Output. *Proc. of the International Workshop on Information Presentation and Natural Multimodal Dialogue*, pp. 71-74, Verona, Italy.

Latoschik, M.E. (2001), A gesture processing framework for multimodal interaction in virtual reality. In A. Chalmers and V. Lalioti, eds, *"1st Int. Conf. on Computer Graphics, Virtual Reality and Visualization in Africa, pp. 95-100"*. ACM_SIGGRAPH.

Voss, I. (2001), Anticipation in construction dialogues. In J. Vanderdonckt, A. Blandford and A. Derycke, eds., *"Interaction without frontiers"*, Proceedings of Joint AFIHM-BCS Conf. on Human-Computer Interaction. IHM_HCI 2001, Lille, France, volume 2 189-190.

# SR:DistoPointer
# Using a Tracked Laser-Range-Meter as an Augmented-Reality Ray-Pick Interaction Device

*Michael T. Wagner*
shared-reality.com
Heilmeyersteige 156/4
89075 Ulm / Germany
mtw@shared-reality.com

## Abstract

We present DistoPointer, a mixed reality ray pick interaction device which allows users of an augmented reality (AR) system to quickly register and align virtual objects in their real environment with a point and click interaction metaphor without requiring the AR system to have prior knowledge about the real world environment.

## 1   Introduction and Motivation

Augmented Reality (AR) attempts to enrich a user's real environment by adding spatially aligned virtual objects to it (see [1] for definitions ) .
The goal is to create the impression that the virtual objects are part of the user's real world environment. In our setting users of an AR system experience the augmented environment through a Head Mounted Display. An external tracking device tracks his or her head motion and the position of his or her interaction devices.

While the user perceives the virtual objects as integrative parts of his or her real world environment, the AR system itself has very little knowledge about its real world surroundings.
The information it gets from its tracking system is the position and orientation of the users head and single tracked items like interaction devices but it doesn't know about the surrounding walls or tables standing in the room.

A lot of tasks in augmented environments however involve registering virtual objects with real world items. In an interior design setting a user may want to place a virtual picture on the wall or put a virtual cupboard to a certain place on the floor. For placing virtual objects in the real environment a point and click interface is desirable,   a user can put a virtual picture to a certain position on a real wall by simply pointing his interaction device to that location and clicking a button.

Since the AR system has no knowledge about its real world surroundings it doesn't know about the wall and therefore cannot position the virtual picture correctly.
The tracking system can only tell the direction in which the user points with his or her interaction device -  the challenge is to find the picked point in the real world environment.

Our approach, the DistoPointer, finds this point by extending the virtual pick ray into the real world environment with a tracked laser-range-meter.

## 2  Related work

A common approach to represent real objects in AR systems is to use invisible representations of the environment (phantom objects), mainly to solve occlusion problems [2]. This approach would also allow to the find picked points in the real world environment because picked points on the phantom model correspond to real points in the environment, provided that the phantom model is correct and properly registered. Phantom models can either be modeled with a CAD tool by hand or generated in real-time by dedicated hardware [3].

Both approaches are not suitable for our setting since dedicated hardware is too expensive and building a phantom model by hand impedes ad-hoc sessions in an arbitrary environment.

## 3  System setup

### 3.1  sr:environment augmented reality system

The sr:environment augmented reality system [4] is one of the first commercially available component oriented AR systems. In this setting we are using the following configuration :

The user is wearing video-see-through head mounted display (HMD) [5] connected to a Laptop (P4 2Ghz, GeForce4, MS-Windows 2000) and is tracked outside-in by an high-end optical tracking system [6] which tracks also the DistoPointer.

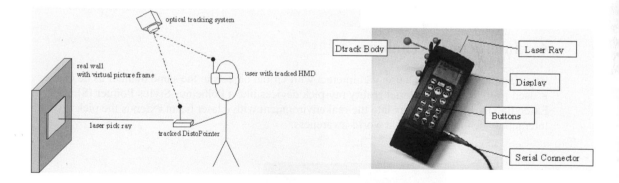

**Figure 1:** sr:environment AR system setup                    **Figure 2:** DistoPointer device

## 3.2  DistoPointer

The DistoPointer  is a modified Leica Disto Pro4a laser range meter [7] with an ART-Dtrack tracking body attached. The housing has a size of 188x70x47mm. A small display and several buttons are attached on the upper side.

The range meter is connected with a standard serial interface to the same laptop the user is wearing.

The Leica Disto pro4a laser range meter has a accuracy of +-1.5mm and an update rate of 1-6 Hz. It works in the range from 30cm to 100m.

## 3.3    Software components

We have implemented several software components for the sr:environment AR system which are used to

- control the DistoPointer and poll continuously measurement- and button data
- calculate the real world pick point
- estimate a translation plane with picked points over time (see Interaction Techniques)
- place and move virtual objects in the real environment

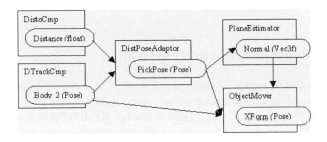

**Figure 3:** Softwarecomponents dealing with DistoPointer data

## 4    Interaction Techniques

By using the DistoPointer a user can interact with virtual objects in the same manner as he or she is used from using other virtual-reality ray-pick devices like a Polhemus Stylus Pointer [8]. Extending the virtual pick ray into the real environment with a laser beam extends the pick ray interaction techniques with real world-awareness.

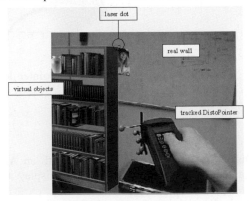

**Figure 4:** using DistoPointer to register virtual objects with real world pick point

## 4.1 Placing virtual objects by Pointing and Clicking

If a user wants to register a virtual object with a real world item he or she first selects a registration point on the virtual model and then simply points the DistoPointer on the desired spot and clicks a button on the device to translate the virtual model to the real world pick point.

If the user keeps the button pressed, DistoPointer translates the model to each new real world pick point it finds. The virtual object can be moved on the real world surface without reconstructing it in the AR system. Rotating the DistoPointer around the pick ray axis rotates the virtual model around its registration point.

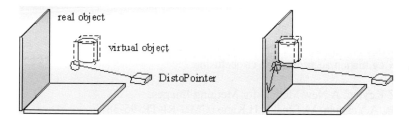

**Figure 5:** selecting a corner of the object's bounding box to register with real pick point

## 4.2 Placing virtual objects with alignment plane estimation

For some tasks it is desirable to align the virtual object with a real world surface. For example if the user wants to put a picture on the wall it is not sufficient to select a corner of the picture to register with the real pick point, furthermore the picture has to be aligned to be parallel with he wall. We have developed a plane estimation component which analyses the real world picked points over time to estimate an alignment plane for object placement.

## 4.3 Placing virtual objects on non-fixed real world objects

The interaction techniques listed in 4.1 and 4.2 work for fixed items in the real environment like walls or the floor but fail on movable objects like a table since the AR system doesn't know when the table moves. This problem can be solved by attaching a tracking body to the table and roughly defining its volume. When virtual objects are placed on picked points in such a movable volume the system moves them whenever the tracked object is moved.

## 5 Conclusion and outlook

In this paper we have presented DistoPointer, a mixed reality ray pick interaction device which allows users of an augmented reality (AR) system to quickly register and align virtual objects in their real environment. The laser beam of the distance meter produces a red dot on the surface it hits. Future work will include removing it in the video image before the user sees it in the HMD as well as implementing better support for movable objects. Another task is to use the estimated alignment planes to generate simple phantom models for occlusion.

# 6 Acknowledgements

We'd like to thank the people from Nagel Baumaschinen GmbH, Ulm for providing the Leica Disto Distance Meter and the DaimlerChrysler Virtual Reality Center for providing the ART Tracking system

# 7 References

[1] Milgram, P., Takemura, H., Utsumi, A., Kishino, F., Augmented Reality: A Class of Displays on the Reality-Virtuality Continuum. Proceedings of Telemanipulator and Telepresence Technologies. 1994. SPIE Vol. 2351, 282-292.

[2] Anton Fuhrmann    Gerd Hesian    François Faure    Michael Gervautz Computers and Graphics number 6 volume 23 pages 809-819 December 1999 or on http://www.cg.tuwien.ac.at/research/vr/occlusion

[3] Video-Rate Z Keying: A New Method for Merging Images, T.Kanade, K.Oda, A.Yoshida, M.Tanaka, H.Kano (CMU-RI-TR-95-38)

[4] Regenbrecht T., Wagner M., Baratoff G., A collaborative tangible augmented reality system, Virtual Reality 2002 6:151-166, Springer Press London

[5] Trivisio GmbH Homepage, retrieved February 11 2003, from http://www.trivisio.com/

[6] Advanced Realtime Technologies GmbH Homepage, retrieved February 11 2003, from http://www.ar-tracking.de/

[7] Leica Distance Meter Homepage, retrieved February 11 2003, from http://www.leica-geosystems.com/products/disto-laser-distance-meter/

[8] Polhemus Stylus Product description, retrieved February 11 2003, from http://www.polhemus.com/Products.htm

# Embedding Public Displays in Non-technical Artifacts: Critical Issues and Lessons Learned From Augmenting a Traditional Office Door Whiteboard With Ubiquitous Computing Technology

*Mikael Wiberg,*

Interaction Theory Lab, Department of Informatics,
Umeå University, 901 87 Umeå, Sweden.
mwiberg@informatik.umu.se

## Abstract

In this paper we present UbiqBoard, an augmented office door whiteboard that supports the situated nature of everyday activities, as an initial result from an ongoing project on embedding small public displays in traditionally non-technical artifacts. The UbiqBoard was informed by a three months naturalistic study of how people use traditional office door whiteboards to communicate. In particular, this paper reports some critical issues identified, and lessons learned, when augmenting public office door whiteboards with ubiquitous computing technology.

## 1    Introduction

Office door whiteboards, mounted directly on office doors or closely nearby, has been frequently used all over the world for the last couple of decades to communicate with potential visitors. Motivated by the recent trend within the HCI community towards embedded and ubiquitous computing our project focuses on the following research question: *What will the critical issues be when augmenting traditional office door whiteboards with ubiquitous computing technology, and what lessons can be learned from such an attempt?*

As background research we conducted a naturalistic pre-study (Solso, et al, 1998), with duration of three months, of both staff and visitors in a typical office environment. This part of the project was undertaken at an academic department with a staff of approximately 60 employees. During the study we gathered data on how they used small public whiteboards mounted on the office doors in their everyday activities. The data collected during the study consists of pictures taken of the office door whiteboards in use together with follow up interviews with both office workers and visitors to the department. Figure 1 illustrates four typical uses of the office door whiteboards. From left to the right, the four pictures illustrate how office door whiteboards were used to: 1) communicate *availability & current location*, (the text on the whiteboard says: "Thuesday 16/10 Working from home"), 2) communicate *current activities and present location* together with a *message left* by a visitor (notice that the visitor has used the small space left at the top of the whiteboard and has not erased the location message on the whiteboard although there was an eraser available next to the pen), (the text on the whiteboard says: "Usability study at Paregos the whole day Monday"), 3) support asynchronous *outeraction activities* (Nardi, et al, 2000), i.e. meta communication about when to establish interaction, (the text on the whiteboard says: "Meeting about D-essay? Would it be possible to meet me and Klaas this week?" together with signature and a phone number), and finally used as 4) *local discussion boards* for both work related, co-located collaboration (e.g. used as a shared sketch board by two colleagues standing in the hallway while talking about a project), as well as used asynchronously between the office worker and his/her visitors around leisure time related topics (the text on the forth whiteboard says: (in black text): -"What about Liverpool?" (followed by a scared smiley), and (in red text written by a visitor): –"...we didn't wanted to win anyway....").

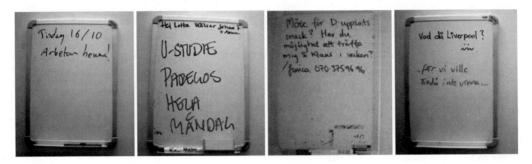

Fig 1. Four pictures illustrating how the office door whiteboards were used in very different ways[1]

Overall, we found that these whiteboards were most frequently used to communicate: 1) the Staffs *availability* (e.g. "I am busy, do not disturb"), 2) their *current activities* (e.g. "Meeting 10-12"), and 3) their *present location* (e.g. "At home", "At conference X", "Working from home today"). We also found that the displays were used for 4) *outeraction activities* (i.e. meta interaction for establishing interaction) (Nardi, et al, 2000), and finally used as 5) *local discussion boards* for local and informal mobile meetings, i.e. spontaneous, informal, and work related discussions (Wiberg, 2001), e.g. used by people passing by the office (sometimes, but not always together with the office worker) to draw models, and sketch on ideas, etc. The pre-study also revealed some critical issues, and typical problems with traditional office door whiteboards that inspired the design of our ubiquitous computing augmented office door whiteboard called UbiqBoard. The issues and problems observed were threefold: 1) People often put up their scheduled events, appointments, and when they will be back at the office again on their office door whiteboards but encounter difficulties with keeping to those times. 2) People frequently make remote phone calls to colleagues to ask them to make changes on their office door whiteboards according to unplanned changes of plans. This behavior illustrates a limitation to traditional office door whiteboards. Further, the necessary workarounds to handle this limitation might be disruptive to the colleagues since it relies on them to do the job and, it takes time out of their schedules. 3) The staff often leaves contact information (e.g. email addresses or phone numbers) on their whiteboards to enable other people to reach them when they are out of office. And, the other way around, i.e. visitors sometimes leaves small notes and contact information to the person sitting in that particular office. This is a potential problem since some information might not be public to anyone passing down the hallway (e.g. home telephone numbers or private messages). These three observations we interpreted as three more general problems with the use of traditional office door whiteboards. The first problem relates to the situated nature of everyday actions (Suchman, 1987), e.g. unplanned upcoming events, and opportunistic conversations. The second problem relates to the geographical boundedness of physical resources (i.e. the whiteboard), which is closely related

---

[1] The texts on the whiteboards are in Swedish, since the pre-study was conducted in Sweden, so for each picture we provide an English translation of the text.

to the first problem. Finally, the third problem can be interpreted as a matter of limitations to perform personal outeractions (Nardi, et al, 2000) via a public display (i.e. leave personal messages and contact information on a public whiteboard).

## 2    Requirement analysis and design implications

Based on the naturalistic pre-study of regular office door whiteboard use as outlined in the introduction we conducted a requirement analysis to formalize the critical issues identified into a few design implications. Based on the analysis we suggest that the three issues and problems identified can be met by: 1) Enable the user to change the whiteboards content dependent on upcoming situations and events *(Due to the situated nature of everyday activities).* 2) Enable these updates of the office door whiteboard from anywhere *(Due to the geographical boundedness of the traditional office door whiteboard).* 3) Enable two-way local and remote communication through the whiteboard between the office worker and their visitors *(to support private outeraction activities via a normally public display).*[2]

## 3    Technical implementation of UbiqBoard

To meet these requirement in the UbiqBoard prototype we augmented a traditional office door whiteboard with a small display embedded in an ordinary office door whiteboard[3]. We used an iPAQ PocketPC for the embedded display equipped with a WLAN pc-card (IEEE 802.11b) to enable the user to change the whiteboard content from anywhere. To access the Internet the office door UbiqBoard communicates over a wireless ad-hoc connection with a proxy server running on the stationary computer in the office. To feed the UbiqBoard with electricity a small hole was drilled in the office door for the AC-adapter. Further, to show new content on the display we used the Pocket Explorer to show an HTML-page with an automatic update HTML-tag (i.e. <HTTP-EQUIV="refresh"). To post new notes from the display the Pocket email client and a traditional "mailto:"-URL in the bottom of the display was used together with an SMTP-server running on the stationary office computer. Finally, to enable the user to update the UbiqBoard's content from "anywhere" we developed a PocketPC-based client that supports mobile updates. Figure 2 illustrates the current implementation. As the figure illustrates the UbiqBoard is designed as a single display groupware (SDG) (Stewart, et al, 1999). A visitor can easily see messages posted by the office worker and can also post messages to him/her by just clicking on the mailto-URL to invoke a new message to be sent via the Pocket email client. The display shows that the office worker is currently in a meeting ("Friday: Meeting 10-12") and that messages can be posted to

---

[2] It might seem odd to support two-way communication via the whiteboard. However, there are several reasons why this might be a more preferable solution then using e.g. mobile phones or a mobile email system. First, according to the first design implication, e.g. if someone is already waiting by the door for him/her to show up and the content on the whiteboard gets updated remotely due to unplanned changes of plans, two-way communication is almost necessary so that they can agree about a new time to meet. Second, two-way communication via the office door whiteboard might be a better solution then solving rescheduling over e.g. a mobile phone since they might not know about each others mobile phone numbers and, the numbers might be private (the whiteboard is a public medium *per se*). Third, the office door whiteboard is preferable as communication device due to the fact that it is already ubiquitous to any potential visitor, compared to a mobile solution that would require that every potential visitor carry such a device (e.g. a mobile phone).

[3] The reason for augmenting a traditional whiteboard instead of just using a large touch sensitive display was that we didn't wanted to ruin the natural "public device" affordance communicated by the traditional office door whiteboard, an affordance that might be hard to design into an ordinary full size display.

him/her (i.e. by clicking on the send_comment@msn.com mailto-URL below the scheduled event).

Figure 2. Two illustrations of UbiqBoard – the ubiq comp augmented office door whiteboard.

## 4    The UbiqBoard in use – some lessons learned

We have observed the UbiqBoard in use now over a period of two months. The observations was made by logging updates of the display, logging messages sent via the UbiqBoard, pictures taken of the whiteboard in use, and follow-up interviews with both office workers and some of the visitors. During these two months the UbiqBoard was frequently updated almost every day from different places (e.g. from the office, from home, from meeting rooms, etc) according to other unplanned situations (e.g. the message "Monday: Working from home" could easily be altered to "Thuesday: Working from home" if necessary without having to call somebody at work and ask them to change the content on the whiteboard). Thus, UbiqBoard meet requirement 1 and 2. Concerning the third requirement identified we noticed that a lot of visitors were interested in the augmented whiteboard and clicked a lot on the display to find out what could be done with it. However, most of them were reluctant to send any messages from the display. During some of the interviews with a few of the visitors we were told that the reason for this were two-folded. First, this is a new kind of communication media so they were unsure of whether they should get a direct reply to their messages or not. A phone is a very direct medium compared to email where a reply can come after several hours or days, but what about this medium? As one of the subjects said: -"What if the reply to my message gets posted on the public display after I have left?".

Some critical lessons learned from this evaluation includes: 1) The importance of building simple solutions to enable everyday use of the whiteboard, and avoid to much overhead work for the office worker. Further on, the system must be so simple to use that any potential visitor can use it without training. 2) The importance of communicating the underlying communication model (i.e. public postings possible to *post to* a public display (for the office worker) and private messages possible to *send from* the display (for the visitor). 3) A third lesson learned concerned the lack of content status indications. From our observations of the UbiqBoard in use we have seen that it is important to communicate to potential visitors whether the content is recently updated or not. A final issue relates to 4) Interaction response times and channel choices. This issue relates to distributed outeractions and availability issues (i.e. the person that is "Out of office" might be busy and thus unable to respond to notes sent to him/her from the display, and the visitor must be able to receive the feedback while standing in front of the display (not after he/she has left the office) or otherwise get reply's redirected via another channel (e.g. as an SMS to his/her mobile phone).

# 5    Related work

Related work near us includes research on shared, distributed, mobile, and virtual whiteboards (e.g. Pedersen et al, 1993; Stewart, et al, 1999), as well as recent empirical studies on the use of public artifacts and shared physical resources in informal workplace interaction (e.g. Bellotti & Bly, 1996). This paper contributes to their efforts by illustrating how a single display groupware (SDG) embedded in a traditional office door whiteboard can support lightweight interaction, outeraction, and remote display updates to support rescheduling of events due to unplanned upcoming events. This paper also contributes to empirical studies conducted of how people use meeting room whiteboards in collaboration by focusing on how small door mounted whiteboards are used to support informal communication, and spontaneous, but work related, collaboration and outeractions. This work might also add to current efforts on designing single (and public) display groupwares (SDGs) (e.g. Stewart et al, 1999) by illustrating how these concepts can be fruitfully applied to communicate around when to meet and collaborate instead of just supporting *the collaborative moment*, i.e. the times when people really sit down together to collaborate. In fact, these *collaborative moments* might be less frequent then the *outeraction moments* necessary to establish collaboration. Here, more studies are of course needed to validate this hypothesis.

# 6    Conclusion

In this paper we have reported from an ongoing project on embedding small public displays in traditionally non-technical artifacts. In particular, this paper has reported some critical issues and common problems identified with the use of traditional office door whiteboards and have described how these observations guided the design of UbiqBoard, and some lessons learned from that effort. Overall, the research reported in this paper contributes to a general trend within the HCI community towards ubiquitous and pervasive computing. More specifically, this work contributes to research on shared, distributed, mobile, and virtual whiteboards, as well as recent empirical studies on the use of public artifacts and shared physical resources in informal workplace interaction. Future work includes an iterative process of redesign of the UbiqBoard system according to these lessons learned followed by real use evaluations. One direction for the next redesign of UbiqBoard includes support for lightweight availability profiles (similar to those common in IM applications) to meet the forth lesson learned from this evaluation.

## References

1. Bellotti, V. and S. Bly (1996). Walking away from the desktop computer: Distributed collaboration and mobility in a product design team. In proceedings of ACM 1996 Conference on CSCW.

2. Nardi, B., Whittaker, S., Bradner, E. (2000). Interaction and Outeraction: Instant Messaging in Action. In Proceedings of Conference on Computer Supported Cooperative Work, 79-88. New York: ACM Press.

3. Pedersen, R., et al (1993). Tivoli: an electronic whiteboard for informal workgroup meetings; Proceedings of ACM 1993 conference on CHI.

4. Solso, R., Johnson, H. & Beal, K. (1998) Experimental psychology: A case approach, sixth edition, Addison Wesley Longman Inc.

5. Stewart, J., Bederson, B., Druin, A. (1999) Single display groupware: a model for co-present collaboration Proceeding of the ACM CHI conference.

6. Suchman, L. (1987). Plans and situated actions: The problem of human-machine communication. Cambridge, Cambridge University Press.

7. Wiberg, M. (2001) In between Mobile Meetings: Exploring seamless ongoing interaction support for mobile CSCW, PhD-thesis, Department of Informatics, Umeå University, Umeå, Sweden.

# Integration of 3D Sound Feedback into a Virtual Assembly Environment

*Ying Zhang, Norman Murray, Terrence Fernando*
Center for Virtual Environments, Information System Institute, University of
Salford, Greater Manchester, M5 4WT, UK
Y.zhang1@pgr.salford.ac.uk, {N.murray, T.Fernando}@salford.ac.uk

**Abstract:** This paper presents our approach for integrating 3D sound into virtual assembly environment (VAE) for the investigation of the effect of visual and 3D sound feedback on assembly task-performance in virtual environments (VE). This experimental platform brings together technologies such as a responsive workbench, constraint-based assembly environment, ultrasonic tracking, and a Huron 3D audio workstation under a common software platform. Several experiments are being conducted to explore and evaluate the effect of neutral, visual, auditory and integrated feedback mechanism on the sense of presence and task performance in the context of assembly simulation in VE.

## 1    Introduction

In the manufacturing area, Virtual Reality technology provides a useful method to interactively assess assembly-related engineering decisions [1,2]. Assembly is an interactive process between the operator and the handled objects, and hence the simulation environments must be able to react according to the users actions in real time. Furthermore, the action of the user and the reaction of the environments must be presented in a suitable manner, which should be in an intuitively comprehensible way. Therefore, it is of great importance to investigate the factors and information presentation modalities that affect the human performance in carrying out assembly task in VE. The multi-modal information integrated into VE can stimulate different senses, increase the users sense of immersion, and the amount of information that are accepted and processed by the operator. Consequently, the increase of useful feedback information may enhance the operator's efficiency and performance while interacting with VE. However, despite of recent efforts in modelling sound performance in VE [3,4,5,6,7], little research has been carried out to investigate the effect of visual, 3D sound and integrated feedback mechanisms on assembly task-performance in VE [8]. This paper presents the overall system architecture implemented for creating a multi-modal virtual assembly environment (VAE) and the approaches adopted to evaluate the user performance of the system.

## 2    Hardware configuration of the VAE

The hardware configuration of the VAE system is comprised of three major parts: visualization subsystem, auralization subsystem, and the real-time tracking systems. The visualization subsystem is centred on a L-shape responsive-workbench (Trimension's V-Desk 6) driven by SGI Onyx2 rack system. The auralization subsystem is based on a Huron PCI audio workstation. A set of TCP/IP protocol-based procedures was written to allow the VE host to transmit the attributes of the acoustic scene and the sound-triggering event to the auralization subsystem through a local area network. The VE host sends packets specifying the acoustic scene and events such as collisions and motions between the objects, including the position of the event, user and environmental attributes, achieved from the geometry of the assembly environment. From these packets, the auralization host generates a pair of auralization filters and sends them to the DSP boards. Based on an event-driven scheme for the presentation of objects interactions, the DSP board samples and processes audio sources with specified filters. Processed audio is then sent back to set of headphones within the VE area in analogue form through two coaxial cables.

## 3    Software architecture of the VAE

The software environment is a multi-thread system that runs on SGI IRIX platforms. It consists of an Interface/Configuration Manager, World Manager, Input Manager, Viewer, Sound Manager, Assembly Simulator and CAD Database (see Fig. 1). The Interface/Configuration Manager tracks all master processes to allow run time configuration of different modules.

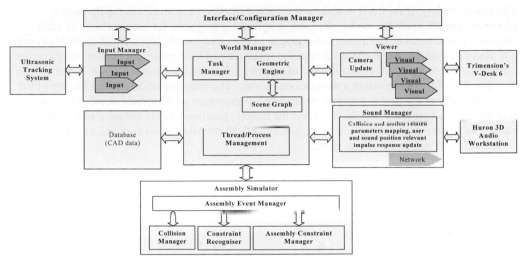

Fig. 1 Software Architecture

The World Manager is responsible for administrating the overall system. It coordinates the visualization, user inputs, databases, assembly simulation, and sound management via a network configured using TCP/IP protocol. The World Manager fetches the user input for manipulation, produces constrained motion using the assembly simulator, and passes the corresponding data (e.g. the position and orientation information of the objects and the user) to the Viewer and Sound Manager for visual and sound feedback generation. The new data is used to update the scene graph and control the sound server via the Sound Manager. The World Manager also has the responsibility to synchronize various threads such as rendering and collision detection. Extension to the OpenGL Optimizer has been made to view the scene in different display technology (e.g. CAVE, Reality Room and L-shape Workbench). The Viewer renders the scene to the selected display in the appropriate mode. Rendering is performed using parallel threads to provide real time response.

The Input Manager manages user-object interactions and establishes the data flow between the user input and the objects held by the World Manager. It supports devices such as pinch gloves and wands. This input describes the user actions/commands in the VE. Each device has its own thread to process data from it. These threads run in parallel with the rendering threads to achieve low latency. Once the assembly parts are loaded into the scene graph via the CAD interface, the Input Manager allows the user to select and manipulate objects in the environment. The Sound Manager gets the location data of the user (listener/viewer), the positions of the collisions and motions (sound sources), and the parameters relating to sound signal modulation from the World Manager and Assembly Simulator, and then uses the Huron API to manage the audio workstation via local network using the TCP/IP protocol.

The Assembly Simulator carries out the detection of collisions between the manipulated object and the surrounding objects, and supports interactive constraint-based assembly and disassembly operations. During object manipulation, the Assembly Simulator samples the position of the moving object to identify new constraints between the manipulated object and the surrounding objects. Once new constraints are recognized, new allowable motions are derived by the assembly simulator to simulate realistic motion of assembly parts. Parameters such as the accurate positions of the assembly parts are sent back to the World Manager, which defines their precise positions in the scene. When a constraint is recognized, the mating surface is highlighted to provide visual feedback, or 3D sound feedback would be rendered through the Sound Manager and the audio server.

## 4    Scene graph structure and rendering

The scene graph (virtual world) has been designed to maintain both geometric and polygonal data. Polygonal data is maintained to visualize the assembly component parts and scene using standard polygonal rendering techniques. It has been created by the integration of the Parasolid geometric kernel and the OpenGL Optimizer graphical toolkit. The scene graph structure is shown in Fig.2.

The initial loading of the CAD models to the virtual world is performed via the Parasolid geometric kernel. Once the models are loaded, extracting polygons for each surfaces of each assembly part creates an Optimizer scene graph. In the virtual world representation of this system, it is assumed that the user is always within the space termed *user space* and his/her body position is tracked with respect to the centre of the *user space*. The user may move the *user space* within the virtual world space by using a navigational device. The transformation node at the top of the *user space* allows the user to position the *user space* anywhere in the virtual world. The current *user space* structure maintains the head and the hand(s) positions within the *user space*. The tracker values are assigned directly to the hand and head transformation nodes to maintain their correct position within the virtual world.

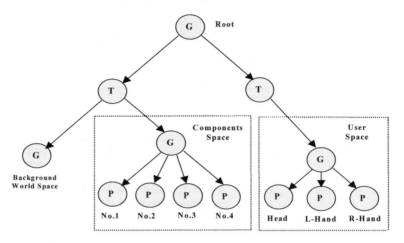

Fig. 2 Scene Graph Structure

The scene graph has two kinds of components, the *background world space* and *assembly components space*, the *background world space* is maintained to provide a visually realistic assembly scene. The *background world space* is rendered but not available for interaction. The scene acoustic models are associated with this space as well. The *assembly component space* maintains the assemblies, which are to be evaluated for assembly and disassembly.

## 5    Acoustic scene rendering and feedback

The acoustic presentation of the VAE requires producing a sound field, which can create the illusion of the sound reaching the user from any direction and distance, and carrying the signature of the environment in which the sound was propagated. The first aspect termed localization is important for navigation and orientation, and hence contributes mainly to the illusion of presence. The second aspect in terms of reverberation conveys important information about the environment shared by the sound source and the user, and therefore enhances essentially the degree of immersion. The third aspect concerns the sound synthesis from physically based motions. In the virtual assembly applications, all these aspects vary dynamically since collision sound may occur at any position and direction with respect to the user within the virtual space.

Since the interaction of the users with the VE and the assembly simulation require sufficient real time behaviour, computation time is limited to perform 3D sound simulation. Due to the limitation of the available processing time, detailed auralization has not been implemented in this work. In order to implement real time 3D sound within the limited computation power, some tradeoffs algorithms have to be adopted. In this research, B-format soundfield representation is used because it is a convenient method for creating and manipulating 3D sound in auralization subsystem. The B-format is essentially a four-channel audio format that can be recorded using a set of four coincident microphones arranged to provide one omni-directional channel (the W channel) and three figure-8 channels (the X, Y, Z channels). This set of X, Y, Z and W signals represent a first-order approximation to the soundfield at a point in the assembly scene. Binaural impulse responses are used to simulate the acoustic attributes of the assembly scene and headphone to display. The direct sound plus six first order reflections are calculated in real time and mixed to form the X, Y, Z, W soundfield·signals. The gain values used in this mixing are computed from the direction of arrival of each sound image at the user position and the parameters from the events such as the collision strength. A box that is used to calculate the six first-order reflections at run time approximates the volume of the geometry of the assembly scene. From the second order reflections to the reverberant tail of the impulse response are pre-computed depend on the environmental parameters such as geometry of the scene, materials of the scene boundary, locations/orientations of the sound sources/users etc. The binaural filters based on Head Related Transfer Functions (HRTFs) are static and do not change in real time in this set-up.

# 6 Unification of visual and acoustic presentation

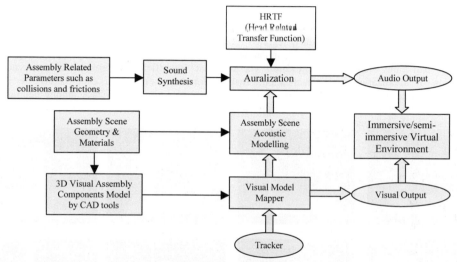

Fig. 3 System Information Flowgraph

The whole VAE system consists of two major parallel data streams: the visual stream and aural stream that process sound activation, sound synthesis and sound propagation in the scene to auralization in the user's ears. The two streams share common control and synchronization mechanisms and information sources. The outputs of the streams are what the user hears and sees in the immersive virtual environment. The visual aspects focus the geometry definition, motion description, and the physical properties of the assembly parts and visual feedback such as modification of colour, hue and saturation. The virtual world software extracts the spatial coordinates of the user and sound source positions that are transmitted via TCP/IP packets to an audio server (aural stream) that runs a separate world model with the required acoustic properties. The audio server then spatializes the audio according to the geometry information received and introduces the relevant scene acoustic attributes. The system components and the overall information flow are shown in Fig.3. The upper half of the figure shows the audio stream while the lower half shows the visual stream. Visual models are created using CAD tools, transformed and imported into this system with OpenGL Optimizer software. Acoustic models are generated using CATT-Acoustic software and then loaded into Huron audio workstation with the relevant API.

# 7 Task performance evaluation

In a typical VAE, the user is presented with an assembly scene where the virtual parts are initially located. The user can then perform assembly tasks, take decisions, make design changes, and perform a host of other engineering tasks in the VE. One of the aims of our integration of the 3D sound into the VAE is to investigate the effect of multi-modal user interface components, mainly visual and auditory, on task performance in VE. Initially, in this research, a peg-in-a-hole assembly and disassembly task (see Fig 4) is used to explore and evaluate the effectiveness of neutral, visual, auditory and integrated feedback mechanisms on the sense of presence and assembly task performance. The full assembly task has several phases (a) Placement of the peg to the upper surface of the plate (see Fig. 4a); (b) Collision between the bottom surface of the peg and the upper surface of the plate (see Fig. 4b); (c) Constraint recognition (see Fig. 4b); (d) Constraint motion on the plate (see Fig. 4c); (e) Alignment constraint between the peg cylinder and the hole cylinder (see Fig. 4d); (f) Constraint motion between two cylinders (see Fig. 4e); (g) Collision between the bottom surface of the peg ear and the upper surface of the plate (see Fig. 4f); and (h) Constraint recognition (see Fig. 4f). The different realistic 3D localized sounds or/and colour intensity of the colliding polygons are presented as the action cues for each of the aforementioned phases. The disassembly task is an inverse process (see Fig. 4g and Fig. 4h). Under each condition and for each subject, the assembly and disassembly Task Completion Time (TCT) is recorded and analysed. The initial experiment has investigated the effects of signal frequency on the localization acuity of 3D sound in VAE so as to select the best frequency signal for further presence and task performance experiments. The results indicated that a suitable frequency band is 800Hz to 1.5kHz.

## 8    Conclusion

A VAE system integrated with 3D sound feedback has been developed in order to explore and evaluate the effect of neutral, visual, auditory and integrated feedback mechanism on the sense of presence and task performance in the context of assembly simulation. Initial evaluation has been carried out using assembly cases for real time simplified auralization via B-format soundfield representation, and synchronous integration of auralization and visualization. Task performance evaluation experiments are currently conducted in order to address key research issues relating to the type of feedback among neutral, visual, auditory and integrated feedback mechanisms and the effect of using integrated visual and auditory mechanisms on the task performance in VE.

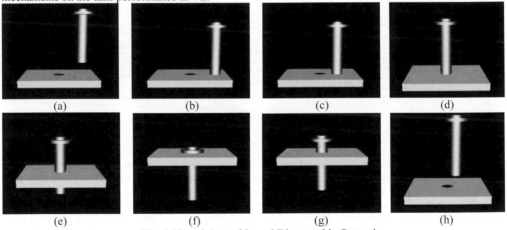

| (a) | (b) | (c) | (d) |
| (e) | (f) | (g) | (h) |

Fig. 4 Virtual Assembly and Disassembly Scenario

## References

[1]  F. Dai (Ed.), *Virtual Reality for Industrial Application,* Springer Verlag, Berlin, Heidelberg, 1998.
[2]  R. Steffan and T. Kuhlen, MAESTRO – A Tool for Interactive Assembly Simulation in Virtual Environments, *Proceedings of the joint IAO and EG workshop,* 16-18 May, 2001, Stuttgart, Germany.
[3]  D.R. Begault, *3D Sound for Virtual Reality and Multimedia,* Academic Press, Cambridge, Massachusetts, USA, 1994.
[4]  E.M. Wenzel, Localization in Virtual Acoustic Displays, *Presence*, Vol. 1, No. 1, 80-107, Winter 1992.
[5]  K. Doel, P.G. Kry and D.K. Pai, Physically-based Sound Effects for Interactive Simulation and Animation, *Proceedings of ACM SIGGRAPH'2001*, 12-17 August 2001, Los Angeles, CA, USA.
[6]  J.F.O'Brien, P.R.Cook and G.Essl, Synthesizing Sounds from Physically Based Motion, Proceedings of *ACM SiGGRAPH'2001*, Los Angeles, CA, USA.
[7]  J.K. Hahn, H.Fouad, L.Gritz and J.W. Lee, Integration Sounds and Motions in Virtual Environments, Presence, Vol.7 No.1, 67-77, February 1998.
[8]  Y.Kitamura, A. Yee and F. Kishino, A Sophisticated Manipulation Aid in a Virtual Environment using Dynamic Constraints among Object Faces, *Presence*, Vol.7, No.5, 460-477, October, 1998.

# A Bayesian Framework for Real-Time 3D Hand Tracking in High Clutter Background

*Hanning Zhou, Thomas S. Huang*

University of Illinois at Urbana-Champaign
405 N. Mathews, Urbana, Il 61801, U.S.A
hzhou,huang@ifp.uiuc.edu

## Abstract

Robust tracking of global hand motion in cluttered background is an important task in human-computer interaction and automatic interpretation of American Sign Language. It is still an open problem due to variant lighting, cluttered background and occlusion. In this paper, a Bayesian framework is proposed to incorporate the hand shape model, the skin color model and image observations to recover the position and orientation of hand in 3D space from monocular images. The efficiency and robustness of our approach has been verified with extensive experiments.

## 1 Introduction

Hand gestures can be a more natural way for humans to interact with computers. For instance, one can use his or her hand to manipulate virtual objects directly in a virtual environment. However, capturing human hand motion is inherently difficult due to its articulation and variability. One way to solve the problem is divide and conquer [Wu and Huang, 1999], i.e., decoupling hand motion into the global motion of a rigid model and finger movements and iteratively solve them until a convergence is reached. This approach demands (1) a robust algorithm to recover the 3D position and orientation of the hand; and (2) an efficient algorithm to recover the finger configuration.

The first problem has been extensively explored. [Black and Jepson, 1996] used an appearance-based model in the eigen space to recover the 2D position of a bounding box. The more general case of 3D curve matching has been addressed by [Zhang, 1994]. [Blake and Isard, 1998] uses the active contour to track global hand motion and recover global hand position. Their model is a deformable 2D planar curve, which cannot handle out-plane rotation. [O'Hagan et al., 2001] developed a real-time HCI system by tracking fingertips from stereo views, which requires clean background and the accuracy was only evaluated using a grid pattern.

In this paper, we propose a Bayesian framework to combine the *a priori* knowledge in the color histogram and the shape of hand with the observation from image sequences. Within this framework, an algorithm based on ICP (iterative closed point) [Zhang, 1994] is used to find the maximum likelihood solution of the position and orientation a rigid planar model of the hand in 3D space, given images captured with a single camera.

Section 2 establishes a Bayesian network to describe the generative model. Section 3 introduces a novel feature *likelihood edge* and the observation. In Section 4, the tracking problem is formulated as inference in the Bayesian network and solved with an ICP-based algorithm. Section 5 provides experimental results in both quantitative and visual forms. Section 6 concludes applicable situation and the limitation of this approach and future directions for extension and improvement.

## 2 Bayesian Hand Tracking

A monocular sequence of color images are given by perspective projection of human hand going through 3D rotation and translation.

*A priori* knowledge $\lambda$ includes: a rigid planer model for contour of the hand given as $\{s_i = (m_i, dm_i), i = 1, 2, \ldots n\}$, where $m_i = [u_i, v_i]$ is the 2D coordinates, $dm_i = [du_i, dv_i]$ is the normal direction (pointing from inner region to outer region) of the contour at $s_i$, and the initial pose (3D position and orientation) of the hand.

The tracking problem is formulated as finding the corresponding sequence of rotation-translation matrices $M = [R \mid T]$ with respect to the initial pose.

Figure 2 shows the Bayesian network describing the dependencies in the generative model, where $HE$ and $GE$ denote histgram edge and grayscale edge respectively, as defined in Section 3.

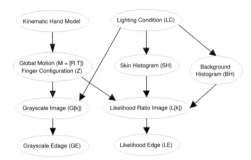

Figure 1: The Bayesian network for hand tracking.

The observed features LE and GE (edge points in likelihood ratio image and grayscale image) are generated from a distribution $p(edge|M, \lambda)$.

Assuming the *a priori* $p(M|\lambda)$ is uniformly distributed in the subspace of feasible transformations, by Bayesian rule $p(M|edge, \lambda) = \frac{p(edge|M, \lambda) \times p(M|\lambda)}{p(edge|\lambda)}$, we can maximize the *a posteriori* $p(M|edge, \lambda) = p(edge|M, \lambda) \times p(M|\lambda)$ by maximizing the likelihood $p(edge|M, \lambda)$ with respect to $M$, since $p(edge|\lambda)$ is independent of $M$.

Since $HE$ and $GE$ are independent given $M$ and $LC$, the likelihood function $p(edge|M, \lambda)$ can be decomposed as $p(edge|M, \lambda, LC) = p(HE|M, \lambda, LC) \times p(GE|M, \lambda, LC)$

## 3 Observation: Extract Matching Candidates

Each frame captured by the camera is an RGB image denoted by $I_k$, which is converted to a grayscale image $G_k$ and an HSI (hue, saturation and intensity) image $H_k$. The HSI image is further mapped to a *likelihood ratio image* $L_k$ by the function defined in Equation (1)

$$L_k(u, v) = \frac{p(H_k(u, v)|skin)}{p(H_k(u, v)|nonskin)} \tag{1}$$

Since $p(H_k(u, v)|skin) = \frac{p(skin|H_k(u,v))p(H_k(u,v))}{p(skin)}$ and

$p(H_k(u, v)|nonskin) = \frac{p(nonskin|H_k(u,v))p(H_k(u,v))}{p(nonskin)}$, Equation (1) can be evaluated as:

$L_k(u, v) = \frac{p(skin|H_k(u,v))p(nonskin)}{p(nonskin|H_k(u,v))p(skin)}$

[Jones and Rehg, 1999] used a standard likelihood ratio approach, but the quantitative information is lost during thresholding the likelihood ratio to decide pixel belongs to skin or nonskin region. To preserves all the sufficient statistics, we choose to use the likelihood ratio without thresholding. The candidate correspondences of the sample points $m_i$ are those edge points in the likelihood ratio image, called likelihood edge: $LE = \{sh_j = (le_j, dle_j), j = 1, 2, \ldots, n_{HE}\}$, where $le_j = [u_j, v_j]$ denotes 2D coordinates and $dle_j = [du_j, dv_j]$ denotes the gradient. Similarly, in the gray scale image $G_k$, grayscale edge: $GE = \{sg_j = (ye_j, dye_j), j = 1, 2, \ldots, n_{GE}\}$ are extracted.

## 4  Inference: Recover 3D Motion

Under moderate out-plane rotation and finger articulation, a human hand can be approximated as a planar object. Assuming the centroid of the hand model to be the origin of the world coordinate and the z-axis to be pointing out of the frontal side of the palm, the 2D coordinates for each sample point is $p_i \quad [x_i[1] = x[1], y_i[1] - \bar{y}[1], 0]^T$, where $(\bar{x}[1], \bar{y}[1])$ is the centroid of the model. The transformation from time instant 1 to $k$ can be expressed as

$$p_i[k] = Rp_i[1] + T, \quad i = 1 \ldots N \tag{2}$$

Given the correspondences of a planar object from two perspective views and the canonical planar model, an ICP-based algorithm [Zhang, 1994] can be used to iteratively search for correspondences and solve the Homography.

### 4.1  Matching With The Model

First use the homography matrix $H$ in the previous frame ($H$ is identity matrix in the very first frame) to warp the model with Equation (3) and (4). Then match the warped model points $m_i[k]$ with observed edge $g_j[k], l_j[k]$ can be formulated as the optimization problem:

$F_{GE}(i) = \arg\min_{j \in \wp} \{d(s_i[k], g_j[k])\}$

$F_{LE}(i) = \arg\min_{j \in \wp} \{d(s_i[k], l_j[k])\}$, where $\wp$ denotes the search region, and the distance measure is defined as follows:

$d(s_i, g_j) = w_1(m_i - g_j)^T \Sigma^{-1}(m_i - g_j) + w_2 (dm_i \cdot dg_j)$

$d(s_i, l_j) = w_1(m_i - l_j)^T \Sigma^{-1}(m_i - l_j) + w_2 (dm_i \cdot dl_j)$

In $d(s_i, g_j)$ The first term is the Mahalanobis distance between $m_i$ and $g_j$. In order to discriminate between edges belonging to two adjacent fingers, in the second term, we use the inner product of the normal directions. $w_1$ and $w_2$ are the weights for trade-off between position information and orientation information. This optimization problem is solved by a nearest-neighbor search.

### 4.2  Estimating 3D Homography Transformation

$H$ can be solved up to a scale with the four-point algorithm [Faugeras et al., 2001], given the correspondences. We warp the model $p_i'[1]$ with homography $H$

$$p_i'[k] = Hp_i'[1] \tag{3}$$

and project with intrinsic parameter matrix $\Pi_0$

$$m_i[k] = \frac{1}{z_i[k]} \Pi_0 p_i'[k] \tag{4}$$

and go back to the matching step to search for correspondences.

### 4.3 Final Pose: From 3D Homography to $R$ and $T$

When ICP algorithm converges, the 3D homography $H$ is solved from $m_i[k] = Hm_i[1]$, $i = 1, 2, \ldots N$ where $m_i[1]$ is all the sample points in the very first frame and $m_i[k]$ is those in the current frame, with the SVD-based algorithm [Faugeras et al., 2001]. From $H$, we can find unique solution for $R'$ and $T'$ (see Appendix I), which are between $p_i'[1]$ (the sample points in camera coordinate at time 1 ) and $p_i'[k]$. To find [R T] between $p_i[1]$ and $p_i[k]$ in the world coordinate as defined in Equation(2), we notice that

$$p_i[k] = Rp_i[1] + (R - I)[0, 0, d, 0]^T + T \tag{5}$$

thus $R = R'$ and $T = (R' - I)[0, 0, d, 0]^T + T'$.

## 5 Experimental Results

### 5.1 Quantitative Results for Global Motion

The mean square error (MSE) in tracking synthesized sequence is shown in Table 1, where $R(X)$, $R(Y)$ and $R(Z)$ denote rotation along $X, Y, Z$ axis respectively and $T(X), T(Y), T(Z)$ denotes translation along $X, Y, Z$ axis.

Table 1: MSE, the range of motion (ROM) and the percentage value of their ratio in each dimension.

|          | $R(X)$ | $R(Y)$ | $R(Z)$ | $T(X)$ | $T(Y)$ | $T(Z)$ |
|----------|--------|--------|--------|--------|--------|--------|
| MSE      | 1.22   | 1.13   | 0.40   | 1.78   | 1.11   | 1.06   |
| ROM      | 10.4   | 10.4   | 15.6   | 10     | 10     | 10     |
| Ratio(%) | 11.73  | 10.87  | 2.56   | 17.75  | 11.06  | 10.60  |

### 5.2 Demonstration using Real-world Data

We have implemented the system and it executes at 29 frames per second on a Pentium3 1.0GHz processor. Figure2 shows some snapshots from the various video clips.

Comparing with the hand tracker in [Blake and Isard, 1998], our tracker has the following advantages: (1) it can recover not only in-plane rotation but also out-plane rotation; (2) it is robust against clutter background; and (3) it is robust against lighting variance.

## 6 Conclusions

In this paper, we propose a Bayesian Framework to accommodate the a prior knowledge of the color and hand-shape within the observation from image sequences. Based on this framework, we introduce a new feature *likelihood edge* to combine color and edge information and use an ICP-based algorithm to find the maximum likelihood solution of the position and orientation of hand in 3D space. The ICP-based iteration alleviates the influence of noise, and the efficiency and robustness has been verified with extensive experiments.

Since we are using 2D model to approximate the hand, the matching of edge points will introduce considerable noise when extreme out-plane rotation of the hand occurs.

In order to drive a control using the recovered motion, it would be helpful to smooth the motion analysis result. We shall also extend our approach to articulated hand tracking by incorporating with the local finger tracker.

Figure 2: Snapshots of the tracking result. These and other AVI sequences are available at the author's web site http://www.ifp.uiuc.edu/~hzhou/histEdge.

## Acknowledgments

This work is supported by National Science Foundation Grant IIS 01-38965 and by National Science Foundation Alliance Program. The authors appreciate Ying Wu and John Lin for inspiring discussion and suggestions.

## References

[Black and Jepson, 1996] Black, M. and Jepson, A. (1996). Eigentracking: Robust matching and tracking of articulated object using a view-based representation. In *Proc. European Conference on Computer Vision*, volume 1, pages 343–356.

[Blake and Isard, 1998] Blake, A. and Isard, M. (1998). *Active Contours*. Springer-Verlag, London.

[Faugeras et al., 2001] Faugeras, O., Luong, Q.-T., and Papadopoulo, T. (2001). *Geometry of multiple images*. MIT press, Cambridge.

[Jones and Rehg, 1999] Jones, M. and Rehg, J. (1999). Statistical color models with application to skin detection. In *Proc. IEEE Conf. on Computer Vision and Pattern Recognition*, volume I, pages 274–280, Fort Collins.

[O'Hagan et al., 2001] O'Hagan, R., Zelinsky, A., and Rougeaux, S. (2001). Visual gesture interfaces to virtual environments. *Interacting with Computers*.

[Wu and Huang, 1999] Wu, Y. and Huang, T. S. (1999). Capturing articulated human hand motion: A divide-and-conquer approach. In *Proc. IEEE International Conference on Computer Vision*, pages 606–611, Corfu, Greece.

[Zhang, 1994] Zhang, Z. (1994). Iterative point matching for registration of free-form curves and surfaces. *International Journal of Computer Vision*, 13:119–152.

# Author Index

# Subject Index